AN

ILLUSTRATED FLORA

OF THE

NORTHERN UNITED STATES
AND CANADA

FROM NEWFOUNDLAND TO THE PARALLEL OF THE SOUTHERN BOUNDARY OF
VIRGINIA, AND FROM THE ATLANTIC OCEAN WESTWARD
TO THE 102D MERIDIAN

BY

NATHANIEL LORD BRITTON, Ph.D., Sc.D., LL.D.
DIRECTOR-IN-CHIEF OF THE NEW YORK BOTANICAL GARDEN; PROFESSOR IN COLUMBIA UNIVERSITY

AND

HON. ADDISON BROWN, A.B., LL.D.
PRESIDENT OF THE NEW YORK BOTANICAL GARDEN

THE DESCRIPTIVE TEXT
CHIEFLY PREPARED BY PROFESSOR BRITTON, WITH THE ASSISTANCE OF SPECIALISTS IN
SEVERAL GROUPS; THE FIGURES ALSO DRAWN UNDER HIS SUPERVISION

SECOND EDITION—REVISED AND ENLARGED

IN THREE VOLUMES

VOL. I.

OPHIOGLOSSACEAE TO POLYGONACEAE

FERNS TO BUCKWHEAT

DOVER PUBLICATIONS, INC., NEW YORK

Published in Canada by General Publishing Company, Ltd., 30 Lesmill Road, Don Mills, Toronto, Ontario.

Published in the United Kingdom by Constable and Company, Ltd., 10 Orange Street, London WC 2.

This Dover edition, first published in 1970, is an unabridged and unaltered republication of the second revised and enlarged edition as published by Charles Scribner's Sons in 1913 under the title *An Illustrated Flora of the Northern United States, Canada and the British Possessions.*

International Standard Book Number: 0-486-22642-5
Library of Congress Catalog Card Number: 76-116827

Manufactured in the United States of America
Dover Publications, Inc.
180 Varick Street
New York, N.Y. 10014

Contents of Volume I.

PTERIDOPHYTA   I

SPERMATOPHYTA   55

Gymnospermae   55

Angiospermae   68

Monocotyledones   68

Dicotyledones   577

Choripetalae   577

ENGLISH NAMES

FERNS AND FERN-ALLIES   I

SEED-BEARING PLANTS 55

Seeds naked 55

Seeds enclosed 68

Cotyledon one 68

Cotyledons two 577

Petals distinct, or none 577

SYMBOLS USED

° is used after figures to indicate feet.
′ is used after figures to indicate inches.
″ is used after figures to indicate lines, or twelfths of an inch.
´ over syllables indicates the accent, and the *short* English sound of the vowel.
ˋ over syllables indicates the accent, and the long, broad, open or close English sound of the vowel.

In the Metric System.

The metre = 39.37 inches, or 3 feet 3.37 inches. ⎫
The decimetre = 3.94 inches. |
The centimetre = $\frac{2}{5}$ of an inch, or $4\frac{3}{4}$ lines. ⎬very nearly.
The millimetre = $\frac{1}{25}$ of an inch, or $\frac{1}{2}$ a line. |
$2\frac{1}{2}$ millimeters = 1 line. ⎭

Introduction.

THE present work is the first complete Illustrated Flora published in this country. Its aim is to illustrate and describe every species, from the Ferns upward, recognized as distinct by botanists and growing wild within the area adopted, and to complete the work within such moderate limits of size and cost as shall make it accessible to the public generally, so that it may serve as an independent handbook of our Northern Flora and as a work of general reference, or as an adjunct and supplement to the manuals of systematic botany in current use.

The first edition (6000 copies) was exhausted during the period from 1896 to 1909. The continued public demand for the work has induced the authors to prepare and publish a second edition, which has been materially revised and enlarged. About 300 pages have been added to the text and the number of species illustrated has been increased from 4162 to 4666, besides many others redrawn for improvement. This increase of about one-eighth both in the text and in the number of plants figured is due in part to the more complete botanical exploration of the geographical area, in part to the more critical delimitation of species and in part to the introduction, in recent years, of additional alien species from the Old World and from the western and southern United States. Exploration and critical study have been greatly stimulated by the first edition, and much of the additional information now brought into the second edition was elicited by the use of the first, by students all over the country.

To all botanical students, a complete illustrated manual is of the greatest service; always useful, often indispensable. The doubts and difficulties that are apt to attend the best written descriptions will often be instantly solved by figures addressed to the eye. The greatest stimulus, moreover, to observation and study, is a clear and intelligible guide; and among the aids to botanical enquiry, a complete illustrated handbook is one of the chief. Thousands of the lovers of plants, on the other hand, who are not botanists and are not familiar with botanical terms or the methods of botanical analysis, will find in the illustrations of a complete work the readiest means of comparison and identification of the plants that grow around them; and through the accompanying descriptions they will at the same time acquire a familiarity with botanical language. By these facilities, not only is the study of our native plants stimulated and widened among all classes, but the enjoyment, the knowledge and the scientific progress derivable from these studies are proportionately increased.

Though most European countries have complete illustrations of the flora of their own territory, no similar work has hitherto been attempted here. Our illustrated works, some of them of great value, have been either sumptuous and costly monographs, accessible to comparatively few, or confined to special groups of plants, or have been works of a minor and miscellaneous character, embracing at most but a few hundred selected species, and from incompleteness, therefore, unsuited for general reference. Scarcely one-third of the species illustrated in the present work have ever been figured before. That no such general work has been previously attempted is to be ascribed partly, perhaps, to the imperfect exploration of our territory, and the insufficiency of the collections to enable such a work to be made approximately complete; partly to the great number of species required to be figured and the consequent difficulty and cost of the undertaking, and partly to the lack of any apparent demand for such a work sufficient to warrant the expense of the enterprise.

In the first edition, it was shown that many more species existed within the geographical area of the work than previous publications had recorded, and many collectors and students have, since its publication, been eager to detect and describe others. This enthusiasm for additional species had led, in some instances, to the descriptive publication as species, of a considerable number which appear to be not sufficiently different from plants already well known to warrant their recognition as distinct; some of these have been satisfactorily relegated to synonymy, while others have been recognized in this edition by brief notes in order to call attention to them and to indicate the necessity for their further study, in order to ascertain their true status. Similar notes have been entered relative to a few species of which the occurrence within the area has become known to the authors during the preparation and composition of the work, which has covered a period of nearly four years, a course which has been taken in order to supersede the need of an Appendix.

A few species illustrated in the first edition have been omitted, except by the entry of notes upon them, in the second, for reasons explained by such notes, mostly because they have been ascertained to be undistinguishable specifically from others.

The enterprise, projected by Judge Brown, and maintained and supervised by him throughout, has been prosecuted for the past twenty-two years. Its execution has been mainly the work of Dr. Britton. The text, founded upon a careful examination of living or herbarium specimens, has been chiefly prepared by him, with the assistance, however, of specialists in a few groups who have contributed the descriptions for certain families as stated in the footnotes. The figures also have been drawn by artists under his immediate supervision; except those of most of the grasses, drawn for the first edition by Mr. Holm, under the eye of Prof. Scribner, and those in the other families contributed by specialists who have supervised them; while the work in all its parts has been carefully revised by both authors. The keys to the genera and species, based upon a few distinctice characters, will, it is believed, greatly facilitate the determinations.

In preparing a new work of this character, the authors have felt that there should be no hesitation in adopting the matured results of recent botanical studies here and in Europe, so as to bring the work fully abreast of the knowledge and scientific conceptions of the time, and make it answer present needs. Although this involves changes in systematic order, in nomenclature, and in the division of families and genera, such as may seem to some to be too radical, no doubt is entertained that time will fully justify these changes in the judgment of all, and demonstrate that the permanent advantages to Botanical Science will far outweigh any temporary inconveniences, as has been already so fully shown in Ornithology and other zoological sciences.

The first edition was issued in three volumes, published consecutively in 1896, 1897 and 1898. The second edition is issued in three volumes simultaneously published.

Area.

The area of the work extends from the Atlantic Ocean westward, in general, to the 102d Meridian, a little beyond that of Gray's Manual, so as to include the whole of the State of Kansas; and northward from the parallel of the southern boundary of Virginia and Kentucky to the northern limits of Labrador and Manitoba. For convenience, the whole of Nebraska has been included, thus permitting the illustration of practically the entire Flora of the northern portion of the Great Plains. Western North and South Dakota are not included.

The Flora of Canada and the British possessions not being distinguishable by any well marked features from that of the adjacent parts of the United States, and not embracing more than about 400 additional species, it was deemed best to include this more northern territory, in order to present a manual of the whole Flora of the northeastern part of the continent, with the exception of that of Greenland and the Arctic Circle, which is much the same on both continents; nearly all the Arctic plants are, however, included, as but very few of them are strictly confined to the Arctic Zone.

Further botanical exploration will, doubtless, reveal additional species, especially along the southern and western boundaries, and in the north.

Figures.

Within the above area there are over 4600 recognized species, more than three times the number in Bentham's Illustrated Handbook of the British Flora. To illustrate all these in a work of moderate size and cost, only parts of each plant could usually be figured, and these mostly below life-size. To exhibit full-page illustrations would have added fourfold to the bulk of the work, and the consequent more limited sales would have necessarily increased the price in a much greater proportion, and thus have thwarted the primary objects, viz., to supply a work adapted to general circulation and use. On the other hand, it was found that any considerable further reduction of the figures in order to reduce the size of the work, would be at the sacrifice of the clearness and usefulness of the illustrations.

In the general plan adopted and in giving parts only of the larger plants, it has been the constant aim to make the reduction of each figure as little below life-size as possible, to select the most characteristic parts for illustration and to preserve the natural proportion. In these respects, it is believed, the present work will be found to be at least not inferior to that above named and often superior.

The cuts are all from original drawings for this work, ether from life or from herbarium specimens, though reference has constantly been made to published plates and figures. All have been first drawn life-size from medium-sized specimens, and afterwards reduced to the proportion indicated by the fraction near the bottom of each cut, most of them being from $\frac{1}{2}$ to $\frac{2}{3}$ of medium life-size. By this method the illustrations do not suffer from the use of a magnifier, but are improved by it and retain their full expression.

The large number of additional figures in the second edition and the incorporation into the main text of the appendix to the first edition, have necessitated the renumbering of the figures consecutively.

Enlargements of special parts are added in most of the illustrations in order to show more clearly the floral structure, or minute organs, or the smaller flowers. These are in various degrees of enlargement, not deemed necessary to be stated. The figures are uncolored, because coloring, except in costly work, obscures the fineness of linear definition and injures the cuts for descriptive and educational uses.

The Classification of Plants.

The Plant Kingdom is composed of four subkingdoms, divisions or primary groups:

1. Thallophyta, the Algae, Fungi and Lichens.
2. Bryophyta, the Mosses and Moss-allies.
3. Pteridophyta, the Ferns and Fern-allies.
4. Spermatophyta, the Seed-bearing plants.

Individuals are grouped, by similarity, into races; races into species; species into genera; genera into families; families into orders; orders into classes; classes into divisions or subkingdoms.

In addition to these main ranks, subordinate ones are sometimes employed, when closer grouping is desirable: thus a Class may be separated into Subclasses, as the Class Angiospermae into the Sublasses Monocotyledones and Dicotyledones; Families may be separated into Tribes, as in the treatment of Gramineae in the following pages; Genera are often separated into Subgenera; Species into Subspecies.

Critical field observations of plants in the wild state, supplemented by the cultivation side by side of species supposed to be distinct and by the lessons learned from experimental plant breeding, have developed the theory that many species, perhaps all, are composed of a greater or lesser number of races, differing from each other too little to cause them to be regarded as species, notwithstanding the fact that they may breed true from seed to such slight or trivial differentiations. It also seems to have been proved, by DeVries and others, that such differentiations may originate abruptly from seed, in a single generation, and remain constant for at least several generations thereafter if so isolated from their relatives as to prevent cross-pollination. These recently ascertained phenomena of mutation are most suggestive, and experimentation and observation concerning them are now occupying the attention of many students.

In the present edition of " Illustrated Flora," the view is taken that the races composing many species are often too numerous and too slightly characterized to be described so as to be recognized; many of them have been described as species and many more as varieties, and varieties of different degrees of differentation have been suggested. We here regard species alone as entitled to distinct botanical appelation; it has been suggested that races may be indicated numerically.

Other than the omission of descriptions of varieties, the general system of classification used in the first edition has been maintained in the second. A few new family groups and a number of genera have been separated or distinguished from their congeners.

The grouping of Races into Species, of Species into Genera, and of Genera into Families, though based upon natural characters and relationships, is not governed by any definite rule that can be drawn from nature for determining just what characters shall be sufficient to constitute a Species, a Genus or a Family. These groups are, therefore, necessarily more or less arbitrary and depend upon the judgment of scientific experts, in which natural characters and affinities, as the most important and fundamental factors, do not necessarily exclude considerations of scientific convenience. The practice among the most approved authors has accordingly been various. Some have made the number of genera and families as few as possible. This results in associating under one name species or genera that present marked differences among themselves. The present tendency of expert opinion is to separate more freely into convenient natural groups, according to similarity of structure, habit, form or appearance. While this somewhat increases the number of these divisions, it has the distinct advantage of decreasing the size of the groups, and thus materially facilitates their study. This view has been taken in the present work, following in most instances, but not in all, the arrangement adopted by Engler and Prantl in their great work, "Natürliche Pflanzen-familien," in which nearly all known genera are described.

Systematic Arrangement.

The Nineteenth Century closed with the almost unanimous scientific judgment that the order of nature is an order of evolution and development from the more simple to the more complex. In no department of Natural Science is this progressive development more marked or more demonstrable than in the vegetable life of the globe. Systematic Arrangement should logically follow the natural order; and by this method also, as now generally recognized, the best results of study and arrangement are obtained. The sequence of Families formerly adopted has become incongruous with our present knowledge; and it has for some time past been gradually superseded by truer scientific arrangements in the later works of many authors.

It now seems probable that continued investigation and consideration will again modify the sequence of various groups. Many suggestions in this regard have already appeared in botanical literature; notably, in our own country, those of Professor Charles E. Bessey.

The more simple forms are, in general, distinguished from the more complex, (1) by fewer organs or parts; (2) by the less perfect adaptation of the organs to the purposes they subserve; (3) by the relative degree of development of the more important organs; (4) by the lesser degree of differentiation of the plant-body or of its organs; (5) by considerations of antiquity, as indicated by the geological record; (6) by a consideration of the phenomena of embryogeny. Thus, the Pteridophyta, which do not produce seeds and which appeared on the earth in Silurian time, are simpler than the Spermatophyta; the Gymnospermae in which the ovules are borne on the face of a scale, and which are known from the Devonian period onward, are simpler than the Angiospermae, whose ovules are borne in a closed cavity, and which are unknown before the Jurassic.

In the Angiospermae the simpler types are those whose floral structure is nearest the structure of the branch or stem from which the flower has been metamorphosed, that is to say, in which the parts of the flower (modified leaves) are more nearly separate or distinct from each other, the leaves of any stem or branch being normally separated, while those are the most complex whose floral parts are most united. These principles are applied to the arrangement of the Subclasses Monocotyledones and Dicotyledones independently, the Mono-cotyledones being usually regarded as the simpler, as shown by the less degree of differentiation of their tissues, though their floral structure is not so very different nor their antiquity much greater, so far as present information goes. For these reasons it is considered that Typhaceae are the simplest of the Monocotyledones, and Orchidaceae the most complex; Saururaceae the simplest family of Dicotyledones in our area, and Compositae the most complex.

Inasmuch as evolution has not always been progressive, but some groups, on the contrary, have clearly been developed by degradation from more highly organized ones, and other groups have been produced by divergence along more than one line from the parent stock, no linear consecutive sequence can, at all points, truly represent the actual lines of descent.

Nomenclature.

The names of genera and species used in this work are in general accordance with the Code of Nomenclature recommended by the Nomenclature Commission of the Botanical Club of the American Association for the Advancement of Science, published in *Bulletin of the Torrey Botanical Club* **34**: 167–178, 1907, to which reference is made. The synonyms given under each species in this work include the recent current names, and thus avoid any difficulty in identification.

The necessity for rules of nomenclature arose from the great confusion that has existed through the many different botanical names for the same species or genera. Some species have had from 10 to 50 different names, and, worse still, different plants have often had the same name. For about 200,000 known species of plants there are not fewer than 700,000 recorded names. Such a chaotic condition of nomenclature is not only extremely unscientific, burdensome and confusing in itself, but the difficulty and uncertainty of identification which it causes in the comparative study of plants made it a serious and constant obstruction in the path of botanical inquiry.

The need of reform, and of finding some simple and fixed system of stable nomenclature, has long been recognized. This was clearly stated in 1813 by A. P. DeCandolle in his "Théorie Élémentaire de la Botanique" (pp. 228–250), where he declares priority to be the fundamental law of nomenclature. Most systematists have acknowledged the validity of this rule. Dr. Asa Gray, in his "Structural Botany," says (p. 348): "For each plant or group there can be only one valid name, and that always the most ancient, if it is tenable; consequently no new name should be given to an old plant or group, except for necessity."

This principle was applied to Zoology in the "Stricklandian Code," adopted in 1842 as Rules of the British Association, and revised in 1860 and 1865 by a committee embracing the most eminent English authorities, such as Darwin, Henslow, Wallace, Clayton, Balfour, Huxley, Bentham and Hooker. In American Zoology the same difficulties were met and satisfactorily overcome by a rigid system of rules analogous to those here followed and now generally accepted by zoologists and palaeontologists.

At an International Botanical Congress held at Paris in 1867, A. DeCandolle presented a system of rules which, with modifications, were adopted, and are the foundation of the present rules of the botanists of the American Association. These rules were in part adopted also by the International Botanical Congress held at Genoa in 1892, and by the Austro-German botanists at their meeting in September, 1904.

The Botanical Club of the American Association for the Advancement of Science adopted rules for Nomenclature at meetings held in 1892 and 1893, which were followed in our first edition. An International Botanical Congress assembled at Vienna in 1905, and materially modified the Paris rules of 1867, and another Congress was held at Brussels in 1910. In the present edition the Code of Nomenclature recommended by the American Commission in 1907, is closely followed, as above stated.

Types of Genera and Species.

The critical study of plants, resulting in the present knowledge by botanists of many more genera and species than formerly, has made necessary more exact definition and determination of both genera and species by basing them on types, a method previously reached in zoology. The following principles are contained in the Code of Nomenclature above referred to:

1. The nomenclatorial type of a species or subspecies is the specimen to which the describer originally applied the name in publication.

   (*a*) When more than one specimen was originally cited, the type or group of specimens in which the type is included may be indicated by the derivation of the name from that of the collector, locality or host.

   (*b*) Among specimens equally eligible, the type is that first figured with the original description, or in default of a figure the first mentioned.

   (*c*) In default of an original specimen, that represented by the identifiable figure or (in default of a figure) description first cited or subsequently published, serves as the type.

2. The nomenclatorial type of a genus or subgenus is the species originally named or designated by the author of the name. If no species was designated, the type is the first binomial species in order eligible under the following provisions:

(*a*) The type is to be selected from a subgenus, section or other list of species originally designated as typical. The publication of a new generic name as an avowed substitute for an earlier invalid one does not change the type of a genus.

(*b*) A figured species is to be selected rather than an unfigured species in the same work. In the absence of a figure, preference is to be given to the first species accompanied by the citation of a specimen in a regularly published series of exsiccatae. In the case of genera adopted from prebinomial authors (with or without change of name), a species figured by the author from whom the genus is adopted should be selected.

(*c*) The application to a genus of a former specific name of one of the included species, designates the type.

(*d*) Where economic or indigenous species are included in the same genus with foreign species, the type is to be selected from (1) the economic species or (2) those indigenous from the standpoint of the original author of the genus.

(*e*) The types of genera adopted through citations of nonbinomial literature (with or without change of name), are to be selected from those of the original species which receive names in the first binomial publication. The genera of Linnaeus' Species Plantarum (1753) are to be typified through the citations given in his Genera Plantarum (1754).

In the present edition, the type species of genera are cited or otherwise indicated.

English Names of Plants.

The general desire for some English name for the different plants described has been met so far as possible. All names in common use have been inserted, so far as they have come to the authors' knowledge, except such as were merely local, or where they were too numerous for insertion. An exception has also been made in a few instances where a common name, from its false suggestion, as in the name of Dog's-tooth Violet for Adder's-tongue, is calculated to mislead as to the nature of the plant. Where no previous names in common use could be found, the names given are founded on some characteristic circumstance of description, habitat, site or author.

In the first edition, many thousand popular names, compiled mostly by Judge Brown, were printed in the General Index only. In this edition, they are all carried into the body of the work in their appropriate places in connection with the descriptive text—a great convenience to those interested in plant-nomenclature. A few additional common names are given in this edition.

No similar compilation of American plant-names has been hitherto published in any other work. Many of them are not to be found in any general dictionaries. To the mass of the people they will afford, in connection with the illustrations, the readiest means of plant-identification.

The popular names are full of interest, from their origin, history and significance. Hundreds of them, brought to this country by the early English Colonists, are still in current use among us, though now obsolete in England. As observed in Britten and Holland's work cited below, " they are derived from a variety of languages, often carrying us back to the early days of our country's history, and to the various peoples who as conquerors or colonists have landed on our shores and left an impress on our language. Many of these old-world words are full of poetical associations, speaking to us of the thoughts and feelings of the people who invented them; others tell of the ancient mythology of our ancestors, of strange old medicinal usages, and of superstitions now almost forgotten."

Most of these names suggest their own explanation. The greater number are either descriptive or derived from the supposed uses, qualities or properties of the plants; many refer to their habitat, appearance or resemblance real or fancied to other things; others come from poetical suggestion, affection or association with saints or persons. Many are very graphic, as the western name, *Prairie Fire* (Castilleia coccinea) ; many are quaint or humorous, as *Cling-rascal* (Galium Aparine) or *Wait-a-bit* (Smilax rotundifolia) ; and in some the corruptions are amusing, as *Aunt Jerichos* (N. Eng.) from Angelica. The words *Horse, Ox, Dog, Bull, Snake, Toad* are often used as a prefix to denote size, coarseness, worthlessness or aversion. *Devil* or *Devil's* is used as a prefix for upwards of 40 of our plants, mostly expressive of dislike or of some traditional resemblance or association. A number of names have been contributed by the Indians, such as Chinquapin, Wicopy, Pipsissewa, Wankapin, etc. ; while the term Indian, evidently a favorite, is applied as a descriptive prefix to upwards of 80 different plants.

There should be no antagonism in the use of scientific and popular names, since their purposes are quite different. Science demands certainty and universality, and hence a single universal name for each plant. For this the Latin has been adopted, and the Latin name should be used, when only scientific objects are sought. But the vernacular names are a part of the growth and development of the language of each people. Though these names are sometimes indicative of specific characters and hence scientifically valuable, they are for the most part not at all scientific, but utilitarian, emotional or picturesque. As such, they are invaluable; not for science, but for the common intelligence, and the appreciation and enjoyment of the plant world. These names, in truth, reflect the mental attitude of each people, throughout its history, toward the plant kingdom; and the thoughts, suggestions, affections or emotions which it has aroused in them. If these are rich and multitudinous, as in the Anglo-Saxon race, so will the plant-names be also.

Usually the most common or the favorite plants have a variety of names; but this is noticeably otherwise with the Asters and the Golden-rods, of which there are about 125 species within our area, the common names of which, considering their abundance and variety, are comparatively few. The Golden-rods, without distinction, are also known as *Yellow-weed* or *Yellow-tops;* the Asters are called also *Frost-weed, Frost-flowers, Good-bye Summer* and by the Onandaga Indians, "It brings the Frost." A few like *Aster ericoides* have several interesting names, but most of the species in each genus resemble each other so much that not a quarter of the species have suggested to the popular apprehension any distinctive name; while other less showy plants, like the Pansy (*Viola tricolor*), the Marsh Marigold (*Caltha palustris*), the Spotted Touch-me-not (*Impatiens biflora*), Bluets (*Houstonia coerulea*) and others, have a score of different names.

In compiling these names, reference has been made to numerous general and special botanical works, to our state and local Floras, to Hobbs' Botanical Handbook (pharmaceutical), to Beal's, Scribner's and Pammel's works on Grasses, to Sudworth's Arborescent Flora, to Britten and Holland's Dictionary of English Plant Names (London, 1886), and to the valuable papers of Mrs. F. D. Bergen on Popular Plant Names in the Botanical Gazette for 1892, p. 365; for 1893, p. 420; for 1894, p. 429, and for 1896, p. 473. Prof. E. S. Burgess has also supplied about 100 popular names not before noted that are in use at Martha's Vineyard and in Washington, D. C.; and Mrs. Horner, of Georgetown, Mass., and Miss Bartlett, of Haverhill, Mass., have each contributed some.

Pronunciation.

In botanical names derived from Greek or Latin words, their compounds, or derivatives, the accent, according to the ordinary rule, is placed upon the penultimate syllable, if it is long in Latin quantity; otherwise, upon the antepenult. Many names, however, have been given to plants in honor of individuals, which, having nothing Latin about them except the terminal form, and the pronunciation given to them by botanical authors being diverse, are here accented like the names of the persons, so far as euphony will permit. This rule is followed because it is believed to agree with the prevailing usage among botanists in ordinary speech; because it is in accord with the commemorative object of such names, which ought not to be obscured by a forced and unnatural pronunciation; and because the test applied to words properly Latin, viz., the usage of the Latin poets, cannot be applied to words of this class. We therefore give Tórreyi, Vàseyi, Càreyi, Jàmesii, Alleni, rather than Torrèyi, Vasèyi, Carèyi, Jamèsii, Allèni.

The acute accent is used to denote the short English sound only; as in bát, bét, bíd, nót, nút; the grave accent, to denote either of the other English sounds, whether long, broad or open; as *a* in bàle, bàll, bàr, bàre, làud; *e* in ève, thère; *i* in pìne, pìque, machìne; *o* in nòte, mòve; *u* in pùre, rùde. The accent for the short or longer English sound is based upon current English usage, as given in the chief English dictionaries from Walker's to the most recent, and without reference to the supposed ancient pronunciation.

Much diversity has been found in botanical works in the accented syllable of many modern Latin adjectives ending in -inus, -ina, -inum, derived from Latin words. As these adjectives are derived from Latin roots and are regularly formed, their pronunciation should properly follow classical analogies. When signifying, or referring to, time, material, or inanimate substances, they should, therefore, according to Andrews & Stoddard's rule, have the penult usually short, and the accent on the antepenult; as in gossípina, cannábina, secálina, salícina, amygdálina, and other adjectives derived from plant names, like the classic nárdinus, cýprinus, fáginus. When these adjectives have other significations than those above referred to, the penult under the ordinary Latin rule is usually long and accented; as in lupulìna, leporìna, hystricìna, like the classic ursìna, canìna.

The Use of Capital Letters.

In accordance with the recommendations of the Nomenclature Commission of the Botanical Club of the American Association for the Advancement of Science, specific or varietal names derived from persons, or used as the genitive of generic names or as substantives, are printed with the initial capital letter. There is much difference of opinion as to the desirability of this practice, many botanists, and almost all zoologists, following the principle of writing all specific names with a small initial letter. Should this custom prevail, much information concerning the history and significance of the specific names would be lost. Thus in the Tulip-tree, *Liriodendron Tulipifera,* the specific name *Tulipifera* was the ancient generic name; and the same with *Lythrum Salicaria, L. Hyssopifolia, L. Vulneraria,* and many other species. In all other forms of writing, personal adjectives such as *Nuttallii, Engelmanni* or *Torreyi* are printed with capitals. We adhere to the ordinary literary usage.

Keys.

A general Key of the Orders and Families has been prepared by Dr. Britton according to the method followed in the Keys to the genera and species. This general Key has been elaborated on the natural method, dividing the two subkingdoms of plants described in the work into Classes, Subclasses, Orders and Families successively. The Orders are not described in the work itself, but their principal distinguishing characters are given in this key. The natural method adopted necessitates a considerable number of exceptions to statements, owing to the varying degree of development of floral organs in the derivation of plants from their ancestors; these exceptions are either noted under the headings or indicated by cross-references.

In using this key, or any of the keys to genera or to species, the student will often find, in the analysis of a plant that it does not provide all the information necessary for its determination; this is generally owing to the incomplete condition of the specimen collected; it may be in flower, while the characteristic differences between it and others are only to be found in the fruit, or *vice versa;* or the species may be dioecious, or polygamous, when its other organs, perchance the characteristic ones, must be sought on another individual, and there are various other causes for incompleteness. It is therefore earnestly recommended that collections be carefully made, seeking to reduce as far as possible this more or less necessary incompleteness. Where satisfactory material can not be obtained, it will usually be found possible to reach the desired analysis by following out two or more lines of the key, and by comparing the results reached with the descriptions to determine the family, genus or species. The illustrations provide an almost indispensable aid in such cases.

Assistance and Coöperation.

In the preparation of both the first edition and of the second we have had valued coöperation from many botanists, which is here gratefully acknowledged. The late Professor Thomas C. Porter contributed much to the first edition by suggestion, specimens, and the examination of proof sheets. Mr. Eugene P. Bicknell has contributed specimens studied for both editions and read the proof sheets of the first. Dr. John K. Small has assisted in the preparation of both editions, contributing the entire text of several families, and has read the proof sheets of the second. The Pteridophyte text was contributed to the first edition by the late Professor Lucien M. Underwood, and to the second edition by Mr. William R. Maxon. The text of the Grass Family has been written by Mr. George V. Nash for both editions; many of the drawings of grasses made by Mr. Theodore Holm for the first edition were supervised by Professor F. Lamson Scribner. The late Mr. Charles E. Smith critically examined the final proof sheets of the first edition. Mr. Frederick V. Coville has contributed the text of Juncaceae to both editions. The late Dr. Thomas Morong wrote the text of several families for the first edition. The text of the Carrot Family in both editions has been examined by Dr. J. N. Rose. Most of the drawings for the first edition were supervised by Dr. Arthur Hollick.

For the second edition Mr. Kenneth K. Mackenzie has contributed the text of *Carex,* and supplied many specimens for study; Mr. W. W. Eggleston has written the text of *Crataegus;* Dr. Ezra Brainerd has written the text of *Viola;* Dr. Per Axel Rydberg has aided in the determination of specimens; and many others have aided by specimens, notes and information.

Draughtsmen.

Most of the drawings for the first edition were executed by Mr. F. Emil; he made all the figures of the Pteridophyta, Gymnospermae, and nearly all of the Monocotyledones, with the exception of those of Gramineae, Melanthaceae, Liliaceae and Convallariaceae; also nearly all of the apetalous Choripetalae, and a considerable portion of the Sympetalae. Miss Millie Timmerman (now Mrs. Heinrich Ries) drew the bulk of the polypetalous Choripetalae, the enlarged parts being mostly inserted by Dr. Arthur Hollick; she also did some work on several of the sympetalous families. Mr. Joseph Bridgham drew the Melanthaceae, Liliaceae and Convallariaceae; also the Ericaceae, Primulaceae and several related families. Mr. Theodore Holm drew most of the Gramineae. Dr. Hollick has made some drawings and numerous enlargements of special parts throughout the work. Miss Mary Knight and Mr. Rudolph Weber have also contributed drawings.

The additional drawings needed for the second edition, and some corrections of the old ones, have been made by Mr. A. Mariolle, Miss Mary E. Eaton and Miss Rachel Robinson.

New York,
    April 15, 1913.

Abbreviations of the Names of Authors.

A. Benn. **Bennett,** Arthur.
A. Br. **Braun,** Alexander.
Adans. **Adanson.** Michel.
Ait. **Aiton,** William.
Ait. f. **Aiton,** William Townsend.
All. **Allioni,** Carlo.
Anders. **Andersson,** Nils Johan.
Andr. **Andrews,** Henry C.
Andrz. **Andrzejowski,** Anton Lukianowicz.
Angs. **Angström,** Johan.
Ard. **Arduino,** Luigi.
Arn. **Arnott,** George Arnold Walker.
Aschers. **Ascherson,** Paul Friedrich August.
Asch. & Graebn. **Ascherson,** P. F. A., and
  Graebner, Paul.
Aubl. **Aublet,** Jean Baptiste Christophore Fusée.
Aust. **Austin,** Coe Finch.
Bab. **Babington,** Charles Cardale.
Baldw. **Baldwin,** *William.*
Baill. **Baillon,** Henri.
Bartl. **Bartling,** Friedrich Gottlieb.
Bart. **Barton,** William P. C.
Bartr. **Bartram,** John.
Beauv. Palisot de **Beauvois,** A. M. F. J.
Benth. **Bentham,** George.
Benth. & Hook. **Bentham,** George, and **Hooker,**
  Joseph Dalton.
Berch. **Berchtold,** Friedrich von.
Bernh. **Bernhardi,** Johann Jacob.
Bess. **Besser,** Wilhelm S. J. G. von.
Bieb. **Bieberstein,** F. A. M. von.
Bigel. **Bigelow,** Jacob.
Bisch. **Bischoff,** Gottlieb Wilhelm.
Biv. **Bivona-Bernardi,** Antonio.
Boeckl. **Boeckeler,** Otto.
Boehm. **Boehmer,** George Rudolf.
Boiss. **Boissier,** Edmond.
Borck. **Borckhausen,** Moritz Balthazar.
Brack. **Brackinridge,** William D.
Brew. **Brewer,** William Henry.
Britt. **Britton,** N. L.
B. S. P. **Britton,** N. L.; **Sterns,** Emerson Alex-
  ander; **Poggenburg,** Justus.
Brongn. **Brongniart,** Adolphe Theodore.
Brot. **Brotero,** Felix de Avellar.
Buch. **Buchenau,** Franz.
Buckl. **Buckley,** Samuel Botsford.
Burgsd. **Burgsdorff,** Friedrich August Ludwig
  von.
Carr. **Carriere,** Élie Abel.
Casp. **Caspary,** Robert.
Cass. **Cassini,** Henri.
Cav. **Cavanilles,** Antonio José.
Celak. **Celakowsky,** Ladislav.
Cerv. **Cervantes,** Vicente.
Cham. **Chamisso,** Adalbert von.
C. & S., Cham. & Sch. **Chamisso and Schlech-**
  tendahl.
Chapm. **Chapman,** Alvan Wentworth.
Chois. **Choisy,** Jacques Denis.
C. Chr. **Christensen,** Carl.
Clairv. **Clairville,** Joseph Philippe de.
Clayt. **Clayton,** John.
Cogn. **Cogniaux,** Alfred.
Coult. (Dips.) **Coulter,** Thomas.
Coult. **Coulter,** John Merle.
C. & R. **Coulter,** J. M., and **Rose,** Joseph Nelson.
Darl. **Darlington,** William.

Davenp. **Davenport,** George Edward.
DC. **De Candolle,** Augustin Pyramus.
A. DC. **De Candolle,** Alphonse.
Decne. **Decaisne,** Joseph.
Desf. **Desfontaine,** René Louiche.
Desr. **Desroussoux,** Louis Auguste Joseph.
Desv. **Desvaux,** Nicaise Augustin.
Dicks. **Dickson,** James.
Dietr. **Dietrich,** David Nathaniel Friedrich.
Dill. **Dillen,** John Jacob.
Dougl. **Douglas,** David.
Drej. **Drejer,** Saloman Thomas Nicolai.
Dryand. **Dryander,** Jonas.
Dufr. **Dufresne,** Pierre.
Dumort. **Dumortier,** Barthélemy Charles.
Duperr. **Duperry,** Louis Isidore.
Durazz. **Durazzini.**
Eat. **Eaton,** Amos.
Eat. & Wr. **Eaton,** Amos, and **Wright,** John.
Eberm. **Ebermaier,** Karl Heinrich.
Eggl. **Eggleston,** Willard Webster.
Ehrh. **Ehrhart,** Friedrich.
Ell. **Elliott,** Stephen.
Endl. **Endlicher,** Stephen Ladislaus.
Engelm. **Engelmann,** George.
Esch. **Escholtz,** Johann Friedrich.
Fabr. **Fabricius,** Philipp Konrad.
Fisch. **Fischer,** Friedrich Ernst Ludwig von.
F. & M. **Fischer** and **Meyer,** C. A.
Foug. **Fougeroux,** Auguste Denis.
Forsk. **Forskal,** Pehr.
Forst. **Forster,** Johann Reinhold and George.
Fourr. **Fourreau,** Jules.
Fresen. **Fresenius,** Johann Baptist Georg Wolf-
  gang.
Froel. **Froelich,** Joseph Aloys.
Gaert. **Gaertner,** Joseph.
Gaertn. f. **Gaertner,** Carl Friedrich.
Gal. **Galeotti,** Henri.
Gaud. **Gaudichaud-Beaupré,** Charles.
Gey. **Geyer,** Carl Andreas.
Gill. **Gillies,** John.
Ging. **Gingins de Lassaraz,** Frédéric Charles
  Jean.
Glox. **Gloxin,** Benjamin Peter.
Gmel. **Gmelin,** Samuel Gottlieb.
Gmel. J. F. **Gmelin,** Johann Friedrich.
Gooden. **Goodenough,** Samuel.
Gren. & Godr. **Grenier,** Charles, and **Godron,**
  D. A.
Grev. **Greville,** Robert Kaye.
Griseb. **Grisebach,** Heinrich Rudolf August.
Gronov. **Gronovius,** Jan Frederik.
Guss. **Gussone,** Giovanni.
Hack. **Hackel,** Eduard.
Hall. **Haller,** Albert von.
Hamilt. **Hamilton,** William.
Hartm. **Hartman,** Carl Johann.
Hassk. **Hasskarl,** Justus Carl.
Hausskn. **Haussknecht,** Carl.
Haw. **Haworth,** Adrian Hardy.
HBK. **Humboldt,** Friedrich Alexander von;
  Bonpland, Aimé, and **Kunth,** Carl Sieges-
  mund.
Hegelm. **Hegelmaier,** Friedrich.
Hedw. f. **Hedwig,** Romanus Adolf.
Hell. **Hellenius,** Carl Niclas.
Heist. **Heister,** Lorenz.

xiv

Herb. **Herbert,** William.
Hitchc. **Hitchcock,** Albert Spear.
Hochst. **Hochstetter,** Christian Friedrich.
Hoffm. **Hoffman,** George Franz.
Hoffmg. **Hoffmansegg,** Johann Centurius.
Holz. **Holzinger,** Johann Michael.
Holl. **Hollick,** Arthur.
Hook. **Hooker,** William Jackson.
H. & A. **Hooker,** W. J., and **Arnott,** George A. Walker.
Hook. f. **Hooker,** Joseph Dalton.
Hornem. **Hornemann,** Jens Wilken.
Huds. **Hudson,** William.
Irm. **Irmisch,** Thilo.
Jacq. **Jacquin,** Nicholas Joseph.
Juss. **Jussieu,** Antoine Laurent.
A. Juss. **Jussieu,** Adrien de.
B. Juss. **Jussieu,** Bernard de.
Karst. **Karsten,** H.
Kit. **Kitaibel,** Paul.
Kl. **Klotsch,** Johann Friedrich.
Kuehl. **Kühlwein.**
L. **Linnaeus,** Carolus, or Carl von Linné.
L. f. **Linné,** Carl von (the son).
L'Her. **L'Heritier de Brutelle,** Charles Louis.
Laest. **Laestadius,** Lars Levi.
Lag. **Lagasca,** Mariano.
Lall. **Lallemant,** Charles.
Lam. **Lamarck,** Jean Baptiste Antoine Pierre Monnet.
Lamb. **Lambert,** Aylmer Bourke.
L. C. Rich. **Richard,** Louis Claude Marie.
Leavenw. **Leavenworth,** Melines C.
Ledeb. **Ledebour,** Carl Friedrich von.
Lehm. **Lehmann,** Johann Georg Christian.
Le Peyr. **Le Peyrouse,** Philippe.
Lepech. **Lepechin,** Iwan.
Lesp. & Thev. **Lespinasse,** G., and **Theveneau,** A.
Less. **Lessing,** Christian Friedrich.
Lestib. **Lestiboudois,** François Joseph.
Lightf. **Lightfoot,** John.
Lilj. **Liljeblad,** Samuel.
Lindl. **Lindley,** John.
Lodd. **Loddiges,** Conrad.
Loefl. **Loefling,** Pehr.
Lois. **Loiseleur-Deslongechamps,** Jean Louis Auguste.
Loud. **Loudon,** John Claudius.
Lour. **Loureiro,** Juan.
Mack. **Mackenzie,** Kenneth K.
MacM. **MacMillan,** Conway.
Marsh. **Marshall,** Humphrey.
Mars. **Marsson,** Theodor.
Mart. **Martens,** Martin.
Mart. & Gal. **Martens,** Martin, and **Galeotti,** Henri.
Max. **Maximilian,** Alexander Philipp, Prince of Wied.
Maxim. **Maximowicz,** Carl Johann.
Medic. **Medicus,** Friedrich Cassimir.
Meisn. **Meisner,** Carl Friedrich.
Mer. **Merat,** François Victor.
Mert. & Koch., M. & K. **Mertens,** Franz Karl, and **Koch,** Wilhelm Daniel Joseph.
Mett. **Mettenius,** George Heinrich.
Mey. **Meyer,** Ernest Heinrich Friedrich.
Michx. **Michaux,** André.
Michx. f. **Michaux,** François André.
Mill. **Miller,** Philip.
Millsp. **Millspaugh,** Charles Frederic.
Mitch. **Mitchell,** John.
Mont. **Montagne,** Jean François Camille.
Moric. **Moricand,** Moise Etienne.
Moq. **Moquin-Tandon,** Alfred.
Muell. Arg. **Müller,** Jean, of Aargau.
Muench. **Muenchhausen,** Otto von.
Muhl. **Mühlenberg,** Heinrich Ludwig.
Murr. **Murray,** Johann Andreas.
Neck. **Necker,** Noel Joseph de.
Newm. **Newman,** Edward.
Nestl. **Nestler,** Christian Gottfried.

Nutt. **Nuttall,** Thomas.
Ort. **Ortega,** Casimiro Gomez.
Pall. **Pallas,** Peter Simon.
Parl. **Parlatore,** Filippo.
P. Br. **Browne,** Patrick.
Pers. **Persoon,** Christian Hendrik.
Planch. **Planchon,** Jules Emile.
Plum. **Plumier,** Charles.
Poir. **Poiret,** Jean Louis Marie.
Poll. **Pollich,** Johann Adam.
Pourr. **Pourret,** Pierre André.
R. Br. **Brown,** Robert.
Raf. **Rafinesque-Schmaltz,** Constantino Samuel.
Redf. & Rand. **Redfield,** John H., and **Rand,** Edward S.
Reichb. **Reichenbach,** Heinrich Gottlieb Ludwig.
Retz. **Retzius,** Anders Johan.
Richards. **Richardson,** John.
Ridd. **Riddell,** John L.
Rivin. **Rivinus,** August Quirinus.
Robb. **Robbins,** J. W.
Roem. **Roemer,** Johann Jacoo.
R. & S. **Roemer,** J. J., and **Schultes,** Joseph August.
Roem. & Ust. **Roemer,** J. J., and **Usteri,** Paulus.
Rostk. **Rostkovius,** Friedrich Wilhelm Gottlieb.
Rottb. **Rottboell,** Christen Fries.
Roxb. **Roxburgh,** William.
R. & P. **Ruiz,** Lopez Hipolito, and **Pavon,** Josef.
Rupp. **Ruppius,** Heinrich Bernhard.
Rupr. **Ruprecht,** Franz J.
Rydb. **Rydberg,** Per Axel.
St. Hil. **St. Hilaire,** August de.
Salisb. **Salisbury,** Richard Anthony.
Sarg. **Sargent,** Charles Sprague.
Sartw. **Sartwell,** Henry P.
Sav. **Savi,** Gaetano.
Schk. **Schkuhr,** Christian.
Schlecht. **Schlechtendal,** Diedrich Franz Leonhard von.
Schleich. **Schleicher,** J. C.
Schleid. **Schleiden,** Matthias Jacob.
Schrad. **Schrader,** Heinrich Adolph.
Schreb. **Schreber,** Johann Christian Daniel von.
Schult. **Schultes,** Joseph August.
Sch. Bip. **Schultz Bipontinus,** Karl Heinrich.
Schum. **Schumacher,** Christian Friedrich.
Schwein. **Schweinitz,** Lewis David von.
Scop. **Scopoli,** Johann Anton.
Scribn. **Scribner,** Frank Lamson.
Scribn. & Ryd. **Scribner,** F. L., and **Rydberg,** P. A.
Ser. **Seringe,** Nicolas Charles.
Seub. **Seubert,** Moritz.
Sheld. **Sheldon,** Edmund P.
Shuttlw. **Shuttleworth,** Robert.
Sibth. **Sibthorp,** John.
Sieb. & Zucc. **Siebold,** Philipp Franz von, and **Zuccarini,** Joseph Gerhard.
Soland. **Solander,** Daniel.
Spreng. **Sprengel,** Kurt.
Steud. **Steudel,** Ernest Gottlieb.
Stev. **Steven,** Christian.
Sudw. **Sudworth,** George B.
Sw. **Swartz,** Olof.
S. Wats. **Watson,** Sereno.
Thuill. **Thuillier,** Jean Louis.
Thunb. **Thunberg,** Carl Peter.
Thurb. **Thurber,** George.
Torr. **Torrey,** John.
Torr. & Schw. **Torrey,** J., and **Schweinitz,** L. D.
Torr. & Hook. **Torrey,** John, and **Hooker,** William Jackson.
Tourn. **Tournefort,** Joseph Pitton de.
Tratt. **Trattinnick,** Leopold.
Trel. **Trelease,** William.
Traut. **Trautvetter,** Ernest Rudolph.
Trin. **Trinius,** Karl Bernhard.
Trin. & Rupr. **Trinius,** Karl B., and **Ruprecht,** F. J.

Tuckerm. **Tuckerman**, Edward.
Turcz. **Turczaninow**, Nicolaus.
Underw. **Underwood**, Lucien Marcus.
Vaill. **Vaillant**, Sébastien.
Vell. **Velloso**, José Marianno de Conceiçao.
Vent. **Ventenat**, Étienne Pierre.
Vill. **Villars**, Dominique.
Wahl. **Wahlenberg**, Georg.
Wahlb. **Wahlenberg**, Pehr Friedrich.
W. & K. **Waldstein**, Franz Adam von, and **Kitalbal**, Paul.
Wallr. **Wallroth**, Karl Friedrich Wilhelm.
Walp. **Walpers**, Wilhelm Gerhard.
Walt. **Walter**, Thomas.
Wang. **Wangenheim**, Friedrich Adam Julius von.
Wats. & Coult. **Watson**, Sereno, and **Coulter**, John Merle.

Web. **Weber**, Friedrich.
Wedd. **Weddell**, H. A.
Weinm. **Weinmann**, J. A.
Wender. **Wenderoth**, George Wilhelm Franz.
Wendl. **Wendland**, Johann Christoph.
Wettst. **Wettstein**, R. von.
Wigg. **Wiggers**, Friedrich Heinrich.
Willd. **Willdenow**, Carl Ludwig.
Wimm. **Wimmer**, Friedrich.
Wisliz. **Wislizenus**, A.
With. **Withering**, William.
Wolfg. **Wolfgang**, Jan.
Woodv. **Woodville**, William.
Wormsk. **Wormskiold**, M. von.
Wr. **Wright**, John.
Wulf. **Wulfen**, Franz Xavier.

GLOSSARY OF SPECIAL TERMS.

Acaulescent. With stem subterranean, or nearly so.

Accumbent. Cotyledons with margins folded against the hypocotyl.

Achene. A dry one-seeded indehiscent fruit with the pericarp tightly fitting around the seed.

Acicular. Needle-shaped.

Acuminate. Gradually tapering to the apex.

Acute. Sharp pointed

Adnate. An organ adhering to a contiguous differing one; an anther attached longitudinally to the end of the filament.

Adventive. Not indigenous, but apparently becoming naturalized.

Albumen. See *Endosperm*.

Alliaceous. Onion-like, in aspect or odor.

Alternate. Not opposite; with a single leaf at each node.

Alveolate. Like honeycomb; closely pitted.

Ament. A spike of imperfect flowers subtended by scarious bracts, as in the willows.

Amphibious. At times inhabiting the water.

Amphitropous. Term applied to the partly inverted ovule.

Amplexicaul. Clasping the stem, or other axis.

Anastomosing. Connecting so as to form a well-defined network.

Anatropous. Applied to an inverted ovule with the micropyle very near the hilum.

Androgynous. Flower clusters having staminate and pistillate flowers; in *Carex*, a spike with upper flowers staminate and lower pistillate.

Angiospermous. Pertaining to the Angiospermae; bearing seeds within a pericarp.

Anther. The part of the stamen which contains the pollen.

Antherid. The male organ of reproduction in Pteridophyta and Bryophyta.

Anthesis. Period of flowering.

Apetalous. Without a corolla.

Aphyllopodic. In *Carex*, with lower leaves bladeless or with rudimentary leaves only.

Apical. At the top, or referring to the top.

Apiculate. With a minute pointed tip.

Appressed. Lying against another organ.

Arborescent. Tree-like, in size or shape.

Archegone. The female reproductive organ in Pteridophyta and Bryophyta.

Areolate. Reticulated.

Areolation. The system of meshes in a network of veins.

Areole. A mesh in a network of veins.

Aril. A fleshy organ growing about the hilum.

Arillate. Provided with an aril.

Aristate. Tipped by an awn or bristle.

Aristulate. Diminutive of aristate.

Ascending. Growing obliquely upward, or upcurved.

Asexual. Without sex.

Assurgent. See *Ascending*.

Auricled. (*Auriculate*) with basal ear-like lobes.

Awn. A slender bristle-like organ.

Axil. The point on a stem immediately above the base of a leaf.

Axile. In the axis of an organ.

Axillary. Borne at, or pertaining to an axil.

Baccate. Berry-like.

Barbellate. Furnished with minute barbs.

Basifixed. Attached by the base.

Berry. A fruit with pericarp wholly pulpy.

Bilabiate. With two lips.

Bipinnate. Twice pinnate.

Bipinnatifid. Twice pinnatifid.

Blade. The flat expanded part of a leaf.

Bract. A leaf, usually small, subtending a flower or flower-cluster, or a sporange.

Bracteate. With bracts.

Bracteolate. Having bractlets.

Bractlet. A secondary bract, borne on a pedicel, or immediately beneath a flower; sometimes applied to minute bracts.

Bulb. A bud with fleshy scales, usually subterranean.

Bulblet. A small bulb, especially those borne on leaves, or in their axils.

Bulbous. Similar to a bulb; bearing bulbs.

Caducous. Falling away very soon after development.

Caespitose. Growing in tufts.

Callosity. A small, hard protuberance.

Callus. An extension of the inner scale of a grass spikelet; a protuberance.

Calyx. The outer of two series of floral leaves.

Campanulate. Bell-shaped.

Campylotropous. Term applied to the curved ovule.

Cancellate. Reticulated, with the meshes sunken.

Canescent. With gray or hoary fine pubescence.

Canaliculate. Channelled; longitudinally grooved.

Capitate. Arranged in a head; knob-like.

Capsular. Pertaining to or like a capsule.

Capsule. A dry fruit of two carpels or more, usually dehiscent by valves or teeth.

Carinate. Keeled; with a longitudinal ridge.

Carpel. The modified leaf forming the ovary, or a part of a compound ovary.

Caruncle. An appendage to a seed at the hilum.

Carunculate. With a caruncle.

Caryopsis. The grain; fruit of grasses, with a thin pericarp adherent to the seed.

Caudate. With a slender tail-like appendage.

Caudex. The persistent base of perennial herbs, usually only the part above ground.

Caudicle. Stalk of a pollen-mass in the Orchid and Milkweed Families.

Cauline. Pertaining to the stem.

Cell. A cavity of an anther or ovary.

Chaff. Thin dry scales.

Chalaza. The base of the ovule.

Chartaceous. Papery in texture.

Chlorophyll. Green coloring matter of plants.

Chlorophyllous. Containing chlorophyll.

Ciliate. Provided with marginal hairs.

Ciliolate. Minutely ciliate.

Cilium. A hair.

Cinereous. Ashy; ash-colored.

Circinnate. Coiled downward from the apex.

Circumscissile. Transversely dehiscent, the top falling away as a lid.

Clavate. Club-shaped.

Cleistogamous. Flowers which do not open, but are pollinated from their own anthers.

Cleft. Cut about halfway to the midvein.

Clinandrium. Cavity between the anther-sacs in orchids.
Cochleate. Like a snail shell.
Coma. Tuft of hairs at the ends of some seeds.
Commissure. The contiguous surfaces of two carpels.
Conduplicate. Folded lengthwise.
Confluent. Blended together.
Connate. Similar organs more or less united.
Connective. The end of the filament, between the anther-sacs.
Connivent. Converging.
Convolute. Rolled around or rolled up longitudinally.
Coralloid. Resembling coral.
Cordate. Heart-shaped.
Coriaceous. Leathery in texture.
Corm. A swollen fleshy base of a stem.
Corolla. The inner of two series of floral leaves.
Corona; Crown. An appendage of the corolla; a crown-like margin at the top of an organ.
Coroniform. Crown-like.
Corymb. A convex or flat-topped flower-cluster of the racemose type with pedicels or rays arising from different points on the axis.
Corymbose. Borne in corymbs; corymb-like.
Costate. Ribbed.
Cotyledon. A rudimentary leaf of the embryo.
Crenate. Scalloped; with rounded teeth.
Crenulate. Diminutive of crenate.
Crustaceous. Hard and brittle.
Cucullate. Hooded, or resembling a hood.
Culm. The stem of grasses and sedges.
Cuneate. Wedge-shaped.
Cusp. A sharp stiff point.
Cuspidate. Sharp-pointed; ending in a cusp.
Cyme. A convex or flat flower-cluster of the determinate type, the central flowers first unfolding.
Cymose. Arranged in cymes; cyme-like.
Deciduous. Falling away at the close of the growing period.
Decompound. More than once-divided.
Decumbent. Stems or branches in an inclined position, but the end ascending.
Decurrent. Applied to the prolongation of an organ, or part of an organ running along the sides of another.
Deflexed. Turned abruptly downward.
Dehiscence. The opening of an ovary, anther-sac or sporange to emit the contents.
Dehiscent. Opening to emit the contents.
Deltoid. Broadly triangular, like the Greek letter delta, Δ.
Dentate. Toothed, especially with outwardly projecting-teeth.
Denticulate. Diminutive of dentate.
Depauperate. Impoverished, small.
Depressed. Vertically flattened.
Dextrorse. Spirally ascending to the right.
Diadelphous. Stamens united into two sets.
Diandrous. Having two stamens.
Dichotomous. Forking regularly into two nearly equal branches or segments.
Dicotyledonous. With two cotyledons.
Didymous. Twin-like; of two nearly equal segments.
Diffuse. Loosely spreading.
Digitate. Diverging, like the fingers spread.
Dimorphous. Of two forms.
Dioecious. Bearing staminate flowers or antherids on one plant, and pistillate flowers or archegones on another of the same species.
Discoid. Heads of Compositae composed only of tubular flowers, rayless; like a disk.
Disk. An enlargement or prolongation of the receptacle of a flower around the base of the pistil; the head of tubular flowers in Compositae.
Dissected. Divided into many segments or lobes.
Dissepiment. A partition-wall of an ovary or fruit.

Distichous. Arranged in two rows.
Distinct. Separate from each other; evident.
Divaricate. Diverging at a wide angle.
Divided. Cleft to the base or to the mid-nerve.
Dorsal. On the back, or pertaining to the back.
Drupaceous. Drupe-like.
Drupe. A simple fruit, usually indehiscent with fleshy exocarp and bony endocarp.
Drupelet. Diminutive of drupe.
Echinate. Prickly.
Ellipsoid. A solid body, elliptic in section.
Elliptic. With the outline of an ellipse; oval.
Emarginate. Notched at the apex.
Embryo. A rudimentary plant in the seed.
Embryo-sac. The macrospore of the flowering plants, contained in the ovule.
Endocarp. The inner layer of the pericarp.
Endogenous. Forming new tissue within.
Endosperm. The substance surrounding the embryo of a seed; albumen.
Ensiform. Shaped like a broad sword.
Entire. Without divisions, lobes, or teeth.
Ephemeral. Continuing for only a day or less.
Epigynous. Adnate to or borne on the upper part of the ovary.
Epiphytic. Growing on other plants, but not parasitic.
Equitant. Folded around each other; straddling.
Erose. Irregularly margined, as if gnawed.
Evanescent. Early disappearing.
Evergreen. Bearing green leaves throughout the year.
Excurrent. With a tip projecting beyond the main part of the organ.
Exfoliating. Peeling off in layers.
Exocarp. The outer layer of the pericarp.
Exogenous. Forming new tissue outside the older.
Exserted. Prolonged past surrounding organs.
Exstipulate. Without stipules.
Extrorse. Facing outward.
Falcate. Scythe-shaped.
Farinaceous. Starchy, or containing starch.
Fascicle. A dense cluster.
Fascicled. Borne in dense clusters.
Fastigiate. Stems or branches which are nearly erect and close together.
Fenestrate. With window-like markings.
Fertile. Bearing spores, or bearing seed.
Fertilization. The mingling of the contents of a male and female cell.
Ferruginous. Color of iron-rust.
Fetid. Ill-smelling.
Fibrillose. With fibres or fibre-like organs.
Filament. The stalk of an anther; the two forming the stamen.
Filamentous. Composed of thread-like structures; thread-like.
Filiform. Thread-like.
Fimbriate. With fringed edges.
Fimbrillate. Minutely fringed.
Fistular. Hollow and cylindric.
Flabellate. Fan-shaped, or arranged like the sticks of a fan.
Flaccid. Lax; weak.
Flexuous. Alternately bent in different directions.
Floccose. With loose tufts of wool-like hairs.
Foliaceous. Similar to leaves.
Foliolate. With separate leaflets.
Follicle. A simple fruit dehiscent along one suture.
Follicular. Similar to a follicle.
Foveate. Foveolate. More or less pitted.
Free. Separate from other organs; not adnate.
Frond. The leaves of ferns.
Frutescent. Fruticose. More or less shrub-like.
Fugacious. Falling soon after development.
Fugitive. Plants not native, but occurring here and there, without direct evidence of becoming established.

Funiculus. The stalk of an ovule or seed.

Fusiform. Spindle-shaped.

Galea. A hood-like part of a perianth or corolla.

Galeate. With a galea.

Gametophyte. The sexual generation of plants.

Gamopetalous. With petals more or less united.

Gemma. A bud-like propagative organ.

Gibbous. Enlarged or swollen on one side.

Glabrate. Nearly without hairs.

Glabrous. Devoid of hairs.

Gladiate. Like a sword-blade.

Gland. A secreting cell, or group of cells.

Glandular. With glands, or gland-like.

Glaucous. Covered with a fine bluish or white bloom; bluish-hoary.

Globose. Spherical or nearly so.

Glomerate. In a compact cluster.

Glomerule. A dense capitate cyme.

Glumaceous. Resembling glumes.

Glume. The scaly bracts of the spikelets of grasses and sedges.

Granulose. Composed of grains.

Gregarious. Growing in groups or colonies.

Gynaecandrous. In *Carex*, a spike with upper flowers pistillate and lower staminate.

Gynobase. A prolongation or enlargement of the receptacle, supporting the ovary.

Habit. General aspect.

Habitat. A plant's natural place of growth.

Hastate. Halberd-shaped; like sagittate, but with the basal lobes diverging.

Haustoria. The specialized roots of parasites.

Head. A dense round cluster of sessile or nearly sessile flowers.

Herbaceous. Leaf-like in texture and color; pertaining to an herb.

Hilum. The scar or area of attachment of a seed or ovule.

Hirsute. With rather coarse stiff hairs.

Hispid. With bristly stiff hairs.

Hispidulous. Diminutive of hispid.

Hyaline. Thin and translucent.

Hypocotyl. The rudimentary stem of the embryo; also termed *radicle*.

Hypogynium. Organ supporting the ovary in some sedges.

Hypogynous. Borne at the base of the ovary, or below.

Hyponym. A generic or specific name untypified.

Imbricated. Overlapping.

Imperfect. Flowers with either stamens or pistils, not with both.

Incised. Cut into sharp lobes.

Included. Not projecting beyond surrounding parts.

Incumbent. With the back against the hypocotyl.

Indehiscent. Not opening.

Indusium. The membrane covering a sorus.

Inequilateral. Unequal sided.

Inferior. Relating to an organ which arises or is situated below another.

Inflexed. Abruptly bent inward.

Inflorescence. The flowering part of plants; its mode of arrangement.

Integument. A coat or protecting layer.

Internode. Portion of a stem or branch between two successive nodes.

Introrse. Facing inward.

Involucel. A secondary involucre.

Involucellate. With a secondary involucre.

Involucrate. With an involucre, or like one.

Involucre. A whorl of bracts subtending a flower or flower-cluster.

Involute. Rolled inwardly.

Irregular. A flower in which one or more of the organs of the same series are unlike.

Labiate. Provided with a lip-like organ.

Laciniate. Cut into narrow lobes or segments.

Lanceolate. Considerably longer than broad, tapering upward from the middle or below; lance-shaped.

Latex. The milky sap of certain plants.

Leaflet. One of the divisions of a compound leaf.

Legume. A simple dry fruit dehiscent along both sutures.

Lenticular. Lens-shaped.

Ligulate. Provided with or resembling a ligule.

Ligule. A strap-shaped organ, as the rays in Compositae.

Limb. The expanded part of a petal, sepal, or gamopetalous corolla.

Linear. Elongated and narrow with sides nearly parallel.

Lineolate. With fine or obscure lines.

Lobed. Divided to about the middle.

Loment. A jointed legume, usually constricted between the seeds.

Loculicidal. Applied to capsules which split longitudinally into their cavities.

Lodicules. Minute hyaline scales subtending the flower in grasses.

Lunate. Crescent-shaped.

Lyrate. Pinnatifid, with the terminal lobe or segment considerably larger than the others.

Macrosporange. Sporange containing macrospores.

Macrospore. The larger of two kinds of spores borne by a plant, usually giving rise to a female prothallium.

Marcescent. Withering but remaining attached.

Medullary. Pertaining to the pith or medulla.

Mericarp. One of the carpels of the Carrot Family.

Mesocarp. The middle layer of a pericarp.

Micropyle. Orifice of the ovule, and corresponding point on the seed.

Microsporange. Sporange containing microspores.

Microspore. The smaller of two kinds of spore borne by a plant, usually giving rise to a male prothallium; pollen-grain.

Midvein (Midrib). The central vein or rib of a leaf or other organ.

Monadelphous. Stamens united by their filaments.

Moniliform. Like a string of beads.

Monoecious. Bearing stamens and pistils on the same plant, but in different flowers.

Monstrous. Unusual or deformed.

Mucronate. With a short sharp abrupt tip.

Mucronulate. Diminutive of mucronate.

Muricate. Roughened with short hard processes.

Muticous. Pointless, or blunt.

Naked. Lacking organs or parts which are normally present in related species or genera.

Naturalized. Plants not indigenous to the region, but so well established as to have become part of the flora.

Nectary. A sugar-secreting organ.

Node. The junction of two internodes of a stem or branch, often hard or swollen, at which a leaf or leaves are usually borne.

Nodose. Similar to nodes or joints; knotty.

Nodulose. Diminutive of nodose.

Nut. An indehiscent one-seeded fruit with a hard or bony pericarp.

Nutlet. Diminutive of nut.

Obcordate. Inversely heart-shaped.

Oblanceolate. Inverse of lanceolate.

Oblong. Longer than broad with the sides nearly parallel, or somewhat curving.

Obovate. Inversely ovate.

Obovoid. Inversely ovoid.

Obsolete. Not evident; gone, rudimentary, or vestigial.

Obtuse. Blunt, or rounded.

Ochreae. The sheathing united stipules of Polygonaceae.

Ochreolae. The ochreae subtending flowers in the Polygonaceae.

Ochroleucous. Yellowish white.

Oösphere. The cell of the archegone which is fertilized by spermatozoids.

Operculate. With an operculum.

Operculum. A lid.

Orbicular. Approximately circular in outline.

Orthotropous. Term applied to the straight ovule, having the hilum at one end and the micropyle at the other.

Ovary. The ovule-bearing part of the pistil.

Ovate. In outline like a longitudinal section of a hen's egg.

Ovoid. Shaped like a hen's egg.

Ovule. The macrosporange of flowering plants, becoming the seed on maturing.

Palate. The projection from the lower lip of two-lipped personate corollas.

Palet. A bract-like organ enclosing or subtending the flower in grasses.

Palmate. Diverging radiately like the fingers.

Pandurate; Panduriform. Fiddle-shaped.

Panicle. A compound flower cluster of the racemose type, or cluster of sporanges.

Paniculate. Borne in panicles or resembling a panicle.

Papilionaceous. Term applied to the irregular flower of the Pea Family.

Papillose. With minute blunt projections.

Pappus. The bristles, awns, teeth, etc., surmounting the achene in the Chicory and Thistle Families.

Parasitic. Growing upon other plants and absorbing their juices.

Parietal. Borne along the wall of the ovary, or pertaining to it.

Parted. Deeply cleft.

Pectinate. Comb-like.

Pedicel. The stalk of a flower in a flower-cluster, or of a sporange.

Peduncle. Stalk of a flower, or a flower-cluster, or a sporocarp.

Pedunculate. With a peduncle.

Peltate. Shield-shaped; a flat organ with a stalk on its lower surface.

Penicillate. With a tuft of hairs or hair-like branches.

Perfect. Flowers with both stamens and pistils.

Perfoliate. Leaves so clasping the stem as to appear as if pierced by it.

Perianth. The modified floral leaves (sepals or petals), regarded collectively.

Pericarp. The wall of the fruit, or seed-vessel.

Perigynium. The utricle enclosing the ovary or achene in the genus *Carex.*

Perigynous. Borne on the perianth, around the ovary.

Peripheral. Pertaining to the periphery.

Persistent. Organs remaining attached to those bearing them after the growing period.

Petal. One of the leaves of the corolla.

Petaloid. Similar to petals; petal-like.

Petiolate. With a petiole.

Petiole. The stalk of the leaf.

Phyllode. A bladeless petiole or rachis.

Phyllopodic. In *Carex,* with lower leaves of the fertile culms normally blade-bearing.

Pilose. With long soft hairs.

Pinna. A primary division of a pinnately compound leaf.

Pinnate. Leaves divided into leaflets or segments along a common axis.

Pinnatifid. Pinnately cleft to the middle or beyond.

Pinnule. A division of a pinna.

Pistil. The central organ of a flower containing the macrosporanges (ovules).

Pistillate. With pistils; and usually employed in the sense of without stamens.

Placenta. An ovule-bearing surface.

Plicate. Folded into plaits, like a fan.

Plumose. Resembling a plume or feather.

Plumule. The rudimentary terminal bud of the embryo.

Pollen. Pollen-grain. Contents of the anther. See *Microspore.*

Pollinia. The pollen-masses of the Orchid and Milkweed Families.

Polygamous. Bearing both perfect and imperfect flowers.

Polypetalous. With separate petals.

Pome. The fleshy fruit of the Apple Family.

Procumbent. Trailing or lying on the ground.

Prophylla. Bractlets.

Prothallium. The sexual generation of Pteridophyta.

Puberulent. With very short hairs.

Pubescent. With hairs.

Punctate. With translucent dots or pits.

Pungent. With a sharp stiff tip.

Pyriform. Pear-shaped.

Raceme. An elongated determinate flower-cluster with each flower pedicelled.

Racemose. In racemes, or resembling a raceme.

Rachilla. The axis of the spikelet in grasses.

Rachis. The axis of a compound leaf, or of a spike or raceme.

Radiant. With the marginal flowers enlarged and ray-like.

Radiate. With ray-flowers; radiating.

Radicle. The rudimentary stem of the embryo; hypocotyl.

Radicular. Pertaining to the radicle or hypocotyl.

Raphe (Rhaphe). The ridge connecting the hilum and chalaza of an anatropous or amphitropous ovule; the ridge on the sporocarp of *Marsilea.*

Ray. One of the peduncles or branches of an umbel; the flat marginal flowers in Compositae.

Receptacle. The end of the flower stalk, bearing the floral organs, or, in Compositae, the flowers; also, in some ferns, an axis bearing sporanges.

Virgate. Wand-like.

Recurved. Curved backward.

Reflexed. Bent backward abruptly.

Regular. Having the members of each part alike in size and shape.

Reniform. Kidney-shaped.

Repand. With a somewhat wavy margin.

Reticulate. Arranged as a network.

Retrorse. Turned backward or downward.

Retuse. With a shallow notch at the end.

Revolute. Rolled backward.

Rhachis. See *Rachis.*

Rhizome. See *Rootstock.*

Ringent. The gaping mouth of a two-lipped corolla.

Rootstock. A subterranean stem, or part of one.

Rostellum. Beak of the style in Orchids.

Rostrate. With a beak.

Rosulate. Like a rosette.

Rotate. With a flat round corolla-limb.

Rugose. Wrinkled.

Runcinate. Sharply pinnatifid, or incised, the lobes or segments turned backward.

Sac. A pouch, especially the cavities of anthers.

Saccate. With a pouch or sac.

Sagittate. Like an arrow-head, with the lobes turned downward.

Samara. A simple indehiscent winged fruit.

Saprophyte. A plant which grows on dead organic matter.

Scabrous. Rough.

Scale. A minute, rudimentary or vestigial leaf.

Scape. A leafless or nearly leafless stem or peduncle, arising from a subterranean part of a plant, bearing a flower or flower-cluster.

Scapose. Having scapes, or resembling a scape.

Scarious. Thin, dry, and translucent, not green.

Scorpioid. Coiled up in the bud, unrolling in growth.

Secund. Borne along one side of an axis.

Segment. A division of a leaf or fruit.

Sepal. One of the leaves of a calyx.

Septate. Provided with partitions.
Septicidal. A capsule which splits longitudinally into and through its dissepiments.
Serrate. With teeth projecting forward.
Serrulate. Diminutive of serrate; serrate with small teeth.
Sessile. Without a stalk.
Setaceous. Bristle-like.
Setose. Bristly.
Silicle. A silique much longer than wide.
Silique. An elongated two-valved capsular fruit, with two parietal placentae, usually dehiscent.
Sinuate. With strongly wavy margins.
Sinuous. In form like the path of a snake.
Sinus. The space between the lobes of a leaf.
Sorus (Sori). A group or cluster of sporanges.
Spadiceous. Like or pertaining to a spadix.
Spadix. A fleshy spike of flowers.
Spathaceous. Resembling a spathe.
Spathe. A bract, usually more or less concave, subtending a spadix.
Spatulate. Shaped like a spatula; spoon-shaped.
Spermatozoids. Cells developed in the antherid, for the fertilization of the oösphere.
Spicate. Arranged in a spike; like a spike.
Spike. An elongated flower cluster or cluster of sporanges, with sessile or nearly sessile flowers or sporanges.
Spikelet. Diminutive of spike; especially applied to flower-clusters of grasses and sedges.
Spinose. With spines or similar to spines.
Spinule. A small sharp projection.
Spinulose. With small sharp processes or spines.
Sporange. A sac containing spores.
Spore. An asexual vegetative cell.
Sporocarp. Organ containing sporanges or sori.
Sporophyte. The asexual generation of plants.
Spreading. Diverging nearly at right angles; nearly prostrate.
Spur. A hollow projection from a floral organ.
Squarrose. With spreading or projecting parts.
Stamen. The organ of a flower which bears the microspores (pollen-grains).
Staminodium. A sterile stamen, or other organ in the position of a stamen.
Standard. The upper, usually broad, petal of a papilionaceous corolla.
Stellate. Star-like.
Sterigmata. The projections from twigs, bearing the leaves in some genera of Pinaceae.
Sterile. Without spores, or without seed.
Stigma. The summit or side of the pistil to which pollen-grains become attached.
Stipe. The stalk of an organ.
Stipitate. Provided with a stipe.
Stipules. Appendages to the base of a petiole, often adnate to it.
Stipulate. With stipules.
Stolon. A basal branch rooting at the nodes.
Stoloniferous. Producing or bearing stolons.
Stoma (Stomata). The transpiring orifices in the epidermis of plants.
Strict. Straight and erect.
Strigose. With appressed or ascending stiff hairs.
Strophiole. An appendage to a seed at the hilum.
Strophiolate. With a strophiole.
Style. The narrowed top of the ovary.
Stylopodium. The expanded base of a style.
Subacute. Somewhat acute.
Subcordate. Somewhat heart-shaped.
Subcoriaceous. Approaching leathery in texture.
Subfalcate. Somewhat scythe-shaped.
Subligneous. Somewhat woody in texture.
Subterete. Nearly terete.
Subulate. Awl-shaped.

Subversatile. Partly or imperfectly versatile.
Succulent. Soft and juicy.
Sulcate. Grooved longitudinally.
Superior. Applied to the ovary when free from the calyx; or to a calyx adnate to an ovary.
Suture. A line of splitting or opening.
Symmetrical. Applied to a flower with its parts of equal numbers.
Syncarp. A fleshy multiple or aggregate fruit.
Tendril. A slender coiling organ.
Terete. Circular in cross section.
Ternate. Divided into three segments, or arranged in threes.
Tetradynamous. With four long stamens and two shorter ones.
Thallus. A usually flat vegetative organ without differentiation into stem and leaves.
Thyrsoid. Like a thyrsus.
Thyrsus. A compact panicle.
Tomentose. Covered with tomentum.
Tomentulose. Diminutive of tomentose.
Tomentum. Dense matted wool-like hairs.
Torsion. Twisting of an organ.
Tortuous. Twisted or bent.
Tracheae. The canals or ducts in woody tissue.
Tracheids. Wood-cells.
Triandrous. With three stamens.
Tricarpous. Composed of three carpels.
Trimorphous. Flowers with stamens of three different lengths or kinds; in three forms.
Triquetrous. Three-sided, the sides channeled.
Truncate. Terminated by a nearly straight edge or surface.
Tuber. A thick short underground branch or part of a branch.
Tubercle. The persistent base of the style in some Cyperaceae; a small tuber.
Tuberculate. With rounded projections.
Turbinate. Top-shaped.
Uliginous. Inhabiting mud.
Umbel. A determinate, usually convex flower-cluster, with all the pedicels arising from the same point.
Umbellate. Borne in umbels; resembling an umbel.
Umbellet. A secondary umbel.
Umbelloid. Similar to an umbel.
Uncinate. Hooked, or in form like a hook.
Undulate. With wavy margins.
Urceolate. Urn-shaped.
Utricle. A bladder-like organ; a one-seeded fruit with a loose pericarp.
Valvate. Meeting by the margins in the bud, not overlapping; dehiscent by valves.
Vascular. Relating to ducts or vessels.
Vein. One of the branches of the woody portion of leaves or other organs.
Veinlet. A branch of a vein.
Velum. A fold of the inner side of the leaf-base in *Isoetes.*
Velutinous. Velvety; with dense fine pubescence.
Venation. The arrangement of veins.
Vernation. The arrangement of leaves in the bud.
Versatile. An anther attached at or near its middle to the filament.
Verticillate. With three or more leaves or branches at a node; whorled.
Vestigial. In the nature of a vestige or remnant.
Villous. With long soft hairs, not matted together.
Whorl. A group of three similar organs or more, radiating from a node. Verticil.
Whorled. See *Verticillate.*
Winged. With a thin expansion or expansions.

GENERAL KEY TO THE ORDERS AND FAMILIES

Subkingdom PTERIDOPHYTA.   I ∶ 1–54.

SPORES DEVELOPING INTO FLAT OR IRREGULAR PROTHALLIA, WHICH BEAR THE REPRODUCTIVE ORGANS (ANTHERIDIA AND ARCHEGONIA) ; FLOWERS AND SEEDS NONE.

1. Spores produced in sporanges, which are borne on the back of a leaf, in spikes or panicles, or in special conceptacles.   Order 1. FILICALES.

     * Spores all of one sort and size (isosporous families).

† Vernation erect or inclined ; sporanges in spikes, or panicles, opening by a transverse slit.
                Fam. 1. *Ophioglossaceae.*   1 : 1.

†† Vernation coiled ; sporanges reticulated, usually provided with a ring (annulus). Sporanges opening vertically.

   Sporanges panicled, with a rudimentary ring ; marsh ferns.   Fam. 2. *Osmundaceae.*   1 : 7.
   Sporanges sessile on a filiform receptacle ; leaves filmy, translucent.
                Fam. 3. *Hymenophyllaceae.*   1 : 8.
   Sporanges ovoid, in panicles, or spikes, provided with an apical ring. Fam. 4. *Schizaeaceae.*   1 : 9.
   Sporanges opening transversely, provided with a vertical ring ; borne in sori on the back or margin of a leaf.   Fam. 5. *Polypodiaceae.*   1 : 19.

     ** Spores of two sizes (microspores and macrospores).

Plants rooting in the mud ; leaves 4-foliolate, or filiform.   Fam. 6. *Marsileaceae.*   1 : 36.
Plants floating ; leaves entire, or 2-lobed.   Fam. 7. *Salviniaceae.*   1 : 37.

2. Spores produced in sporanges, which are clustered underneath the scales of a terminal cone-like spike ; stems jointed, rush-like.

                Order 2. EQUISETALES.

        One family.

                Fam. 8. *Equisetaceae.*   1 : 38.

3. Spores produced in sporanges, which are borne in the axils of scale-like or tubular leaves.

                Order 3. LYCOPODIALES.
Spores all of one sort and size.   Fam. 9. *Lycopodiaceae.*   1 : 42.
Spores of two sizes (microspores and macrospores).
   Leaves scale-like, 4–many-ranked, on branching stems.   Fam. 10. *Selaginellaceae.*   1 : 48.
   Leaves tubular, clustered on a corm-like trunk ; aquatic or mud plants.
                Fam. 11. *Isoetaceae.*   1 : 50.

Subkingdom SPERMATOPHYTA.   I : 55.

MICROSPORES (POLLEN-GRAINS) DEVELOPING INTO A TUBULAR PROTHALLIUM (POLLEN-TUBE) ; MACRO-SPORES (EMBRYO-SAC) DEVELOPING A MINUTE PROTHALLIUM, AND, TOGETHER WITH IT, REMAINING ENCLOSED IN THE MACROSPORANGE (OVULE) WHICH RIPENS INTO A SEED.

   Class 1. GYMNOSPERMAE.   Ovules not enclosed in an ovary.   I : 55–68.
Fruit a cone, with several or numerous scales, sometimes berry-like by their cohesion.
                Fam. 1. *Pinaceae.*   1 : 55.
Fruit (in our genus) a fleshy integument nearly enclosing the seed.   Fam. 2. *Taxaceae.*   1 : 67.

   Class 2. ANGIOSPERMAE.   Ovules enclosed in an ovary.   1 : 68.

Subclass 1.   MONOCOTYLEDONES.   I : 68.

EMBRYO WITH 1 COTYLEDON ; STEM WITH NO DISTINCTION INTO PITH, WOOD AND BARK ; LEAVES MOSTLY PARALLEL-VEINED.

1. Carpels 1, or more, distinct (united, at least partially, in Family 6, Scheuchzeriaceae, where they are mostly united until maturity, and Family 8, Vallisneriaceae, aquatic herbs, with monoecious or dioecious flowers) ; parts of the flowers mosty unequal in number.

     * Inflorescence various, not a true spadix.

† Flowers not in the axils of dry chaffy scales (glumes) ; our species aquatic or marsh plants.
‡ Endosperm mealy or fleshy ; perianth of bristles or chaffy scales ; flowers monoecious, spicate or capitate.

                Order 1. PANDANALES.
Flowers spicate, the spikes terminal.   Fam. 1. *Typhaceae.*   1 : 68.
Flowers capitate, the heads axillary to leaf-like bracts.   Fam. 2. *Sparganiaceae.*   1 : 69.

     ‡‡ Endosperm none, or very little ; perianth corolla-like, or herbaceous, or none.
Perianth wanting, or rudimentary.   Order 2. NAIADALES.
   Carpels distinct ; stigmas disk-like or cup-like.   Fam. 3. *Zannichelliaceae.*   1 : 74.
   Carpels united ; stigmas slender.
     Flowers axillary ; leaves spinose-dentate.   Fam 4. *Naiadaceae.*   1 : 89.
     Flowers on a spadix ; leaves grass-like.   Fam. 5. *Zosteraceae.*   1 : 90.
Perianth present, of 2 series of parts.

Carpels distinct.   Order 3. ALISMALES.
    Petals similar to the sepals; anthers mostly elongated.   Fam. 6. *Scheuchzeriaceae.* 1 : 91.
    Petals not similar to the sepals; anthers short.   Fam. 7. *Alismaceae.* 1 : 93.
Carpels united.   Order 4. HYDROCHARITALES.
    Ovary 1-celled with parietal placentae.   Fam. 8. *Vallisneriaceae.* 1 : 104.
    Ovary 6–9-celled.   Fam. 9. *Hydrocharitaceae.* 1 : 106.

  †† Flowers in the axils of dry chaffy scales (glumes), arranged in spikes or spikelets.
                      Order 5. GRAMINALES (*Glumiflorae*).
Fruit a caryopsis (grain); stems (culms) mostly hollow in our species. Fam. 10. *Gramineae.* 1 : 107.
Fruit an achene; stems (culms) solid.   Fam. 11. *Cyperaceae.* 1 : 295.
    (Order 6, PALMALES, including only the family *Palmaceae,* Palms, and Order 7, CYCLANTHALES,
including only the family *Cyclanthaceae,* are not represented in our territory.)
** Inflorescence a fleshy spadix, with or without a spathe; or plants minute, floating free, the flowers few or
        solitary on the margin or back of the thallus.
                      Order 8. ARALES (*Spathiflorae*).
Large herbs, with normal foliage and well-developed spadix.   Fam. 12. *Araceae.* 1 : 441.
Minute floating thalloid plants.   Fam. 13. *Lemnaceae.* 1 : 446.
2. Carpels united into a compound ovary; parts of the usually complete flowers mostly in 3's or 6's.
                  * Seeds with endosperm.

  † Flowers regular, or nearly so (corolla irregular in *Commelina* and *Pontederia*).
    ‡ Endosperm mealy; ovary superior.
                    Order 9. XYRIDALES (*Farinosae*).
          a. Ovary 1-celled.
Aquatic moss-like leafy herbs; flowers solitary.   Fam. 14. *Mayacaceae.* 1 : 450.
Erect rush-like herbs; flowers in terminal scaly heads or spikes. Fam. 15. *Xyridaceae.* 1 : 450.
Mud or aquatic herbs, the flowers subtended by spathes (*Heteranthera* in Pontederiaceae 1 : 462).
          b. Ovary 2–3-celled (except in some Pontederiaceae).
Flowers very small, densely capitate, monoecious or dioecious. Fam. 16. *Eriocaulaceae.* 1 : 453.
Flowers perfect.
    Epiphytes; leaves scurfy.   Fam. 17. *Bromeliaceae.* 1 : 456.
    Terrestrial or aquatic herbs; leaves not scurfy.
        Perianth of 2 series of parts, the outer (sepals) green, the inner (petals) colored.
                    Fam. 18. *Commelinaceae.* 1 : 457.
        Perianth 6-parted.   Fam. 19. *Pontederiaceae.* 1 : 462.
    ‡‡ Endosperm fleshy or horny; ovary superior or inferior.
                    Order 10. LILIALES.

  a. Ovary superior (except in *Aletris,* in the Liliaceae, and some species of *Zygadenus* in the
        Melanthaceae).
Perianth-segments distinct, green or brown, not petal-like; herbs with grass-like leaves and small
    flowers.   Fam. 20. *Juncaceae.* 1 : 465.
Perianth-segments distinct, or partly united, at least the inner petal-like.
    Fruit a capsule (except in *Yucca baccata,* where it is large, fleshy and indehiscent).
        Capsule mostly septicidal; plants rarely bulbous.   Fam. 21. *Melanthaceae.* 1 : 485.
        Capsule loculicidal (septicidal in *Calochortus*); plants mostly bulbous.
                    Fam. 22. *Liliaceae.* 1 : 495.
    Fruit a fleshy berry (except in *Uvularia* of the Convallariaceae).
        Erect herbs; tendrils none; flowers perfect.
            Leaves basal or alternate.   Fam. 23. *Convallariaceae.* 1 : 513.
            Leaves verticillate.   Fam. 24. *Trilliaceae.* 1 : 522.
        Vines, climbing by tendrils, or rarely erect; flowers dioecious, in axillary umbels.
                    Fam. 25. *Smilaceae.* 1 : 526.
          b. Ovary inferior, wholly or in part.
Stamens 3, opposite the inner corolla-segments.   Fam. 26. *Haemodoraceae.* 1 : 530.
Stamens 6 in our species.
    Erect perennial herbs; flowers perfect.   Fam. 27. *Amaryllidaceae.* 1 : 531.
    Twining vines; flowers dioecious.   Fam. 28. *Dioscoraceae.* 1 : 535.
Stamens 3, opposite the outer corolla-segments.   Fam. 29. *Iridaceae.* 1 : 536.
    †† Flowers very irregular; ovary inferior.
                    Order 11. SCITAMINALES.
One family represented in our territory.   Fam. 30. *Marantaceae.* 1 : 546.
    ** Seeds without endosperm, very numerous and minute; ovary inferior.
                  Order 12. ORCHIDALES (*Microspermae*).
Flowers regular; stem-leaves reduced to scales.   Fam. 31. *Burmanniaceae.* 1 : 546.
Flowers very irregular.   Fam. 32. *Orchidaceae.* 1 : 547.

Subclass 2. DICOTYLEDONES.   1 : 577.

EMBRYO NORMALLY WITH 2 COTYLEDONES; STEMS MOSTLY DIFFERENTIATED INTO PITH, WOOD AND BARK;
              LEAVES MOSTLY NET-VEINED.
      Series 1. CHORIPETALAE. 1 : 577 to 2 : 666.
    Petals distinct to the base, or wanting (exceptions noted Vol. 1 : 577).
A. Petals none, except in Portulacaceae and in most Caryophyllaceae, which are herbs with leaves
    nearly always opposite, the seeds with endosperm, and in the pistillate flowers of the walnuts
    (*Juglans*).
    1. Calyx none (except in the Juglandaceae, which are trees with odd-pinnate leaves).
Marsh herbs with perfect flowers in nodding spikes.   Order 1. PIPERALES.

One family only.

Fam. 1. *Saururaceae.* 1 : 577.

Trees or shrubs; staminate flowers, and sometimes also the pistillate, in aments.
Leaves odd-pinnate; fruit a nut enclosed in a husk.   Order 2. JUGLANDALES.
One family only.

Fam. 2. *Juglandaceae.* 1 : 578.

Leaves simple.
  Fruit 1-seeded.   Order 3. MYRICALES.
    Ovule erect, orthotropous.   Fam. 3. *Myricaceae.* 1 : 584.
    Ovule laterally attached, ascending, amphitropous. Fam. 4. *Leitneriaceae.* 1 : 586.
  Fruit many-seeded; seeds with a tuft of hairs at one end. Order 4. SALICALES.
One family only.

Fam. 5. *Salicaceae.* 1 : 587.

2. Calyx present.
* Flowers, at least the staminate ones, in aments.

Order 5. FAGALES.

Both staminate and pistillate flowers in aments.   Fam. 6. *Betulaceae.* 1 : 605.
Pistillate flowers subtended by an involucre, which becomes a bur or a cup in fruit.
Fam. 7. *Fagaceae.* 1 : 614.

** Flowers not in aments (in ament-like spikes in *Morus*), but variously clustered, rarely solitary.

  a. Flowers monoecious, dioecious or polygamous (sometimes perfect in *Ulmus*);
ovary superior, 1-celled.

Order 6. URTICALES.

Fruit not an achene; trees, shrubs or herbs; ovule pendulous.
  Trees with alternate leaves, the sap not milky.   Fam. 8. *Ulmaceae.* 1 : 625.
  Trees with alternate leaves and milky sap.   Fam. 9. *Moraceae.* 1 : 630.
Fruit an achene; herbs with small clustered greenish flowers.
  Ovule pendulous; styles or stigmas 2.   Fam. 10. *Cannabinaceae.* 1 : 633.
  Ovule erect or ascending; style or stigma 1.   Fam. 11. *Urticaceae.* 1 : 634.
  (Order 7, PROTEALES, extensively developed in the southern hemisphere, is not represented in
our area.)

  b. Flowers dioecious, or perfect; ovary inferior, at least in part.

Ovary 1-celled.   Order 8. SANTALALES.
  Tree-parasites, with opposite leaves or scales; fruit a berry. Fam. 12. *Loranthaceae.* 1 : 638.
  Root-parasites, or shrubs; leaves alternate in our genera; fruit a drupe, or nut.
Fam. 13. *Santalaceae.* 1 : 639.
Ovary several- (usually 6-) celled; flowers perfect.   Order 9. ARISTOLOCHIALES.
One family in our area.

Fam. 14. *Aristolochiaceae.* 1 : 641.

  c. Flowers mostly perfect in our genera (dioecious in some species of *Rumex* in Polygonaceae,
monoecious or dioecious in some Chenopodiaceae and Amaranthaceae); ovary superior.

    † Embryo straight, or nearly so; fruit an achene.

Order 10. POLYGONALES.

One family.

Fam. 15. *Polygonaceae.* 1 : 646.

    †† Embryo coiled, curved, or annular; fruit not an achene.

Order 11. CHENOPODIALES (*Centrospermae*).

Fruit a utricle; stipules none
  Bracts and sepals scarious.   Fam. 16. *Amaranthaceae.* 2 : 1.
  Bracts none, or not scarious.   Fam. 17. *Chenopodiaceae.* 2 : 8.
  Fruit a berry in our genus.   Fam. 18. *Phytolaccaceae.* 2 : 25.
Fruit a utricle; stipules present (except in *Scleranthus* which has subulate, opposite connate leaves).
Fam. 19. *Corrigiolaceae.* 2 : 26.
Fruit an anthocarp, the persistent base of the corolla-like calyx enclosing a utricle.
Fam. 20. *Nyctaginaceae.* 2 : 30.

Fruit a capsule, dehiscent by valves, or teeth.
  Capsule 2–several-celled; petals none.   Fam. 21. *Aizoaceae.* 2 : 34.
  Capsule 1-celled; petals mostly present.
    Sepals 2.   Fam. 22. *Portulacaceae.* 2 : 35.
    Sepals 5 or 4, distinct or united.
      Sepals distinct; ovary sessile.   Fam. 23. *Alsinaceae.* 2 : 41.
      Sepals united; ovary stipitate.   Fam. 24. *Caryophyllaceae.* 2 : 61.

B. Petals present (wanting in Ceratophyllaceae—aquatic herbs with whorled dissected leaves; in
many Ranunculaceae; in *Calycocarpum*—a dioecious vine of the Menispermaceae; in Lauraceae—alternate-leaved aromatic trees and shrubs; in Podostemaceae—aquatic herbs, the
simple flowers involucrate; in *Liquidambar*—a tree with palmately-lobed leaves and capitate
flowers of the Hamamelidaceae—in *Sanguisorba*—herbs with pinnate leaves of the Rosaceae;
in *Xanthoxylum*—trees with pinnate leaves of the Rutaceae; in Euphorbiaceae; in Callitrichaceae, Empetraceae and Buxaceae; in some of the Aceraceae and Rhamnaceae; in Thymeleaceae, Elaeagnaceae, and in some species of *Ludwigia* in Onagraceae and of *Nyssa* in Cor-
naceae).

  I. *Ovary superior, free from the calyx* (partly or wholly inferior in some Saxifragaceae,
in Grossulariaceae, Hamamelidaceae, Malaceae and Loasaceae).

1. Carpels solitary, or several or distinct (united in Nymphaeaceae); stamens mostly hypogynous
and more numerous than the sepals; sepals mostly distinct. Order 12. RANALES.

  * Aquatic herbs; floating leaves if present, peltate, or with a basal sinus.

Pistil 1; petals none; leaves whorled, dissected.   Fam. 25. *Ceratophyllaceae.* 2 : 75.

Carpels 3 or more; petals large; floating leaves not dissected.
  Carpels distinct.
    Carpels not in a receptacle.   Fam. 26. *Cabombaceae.* **2** : 75.
    Carpels in a fleshy receptacle.   Fam. 27. *Nelumbonaceae.* **2** : 76.
  Carpels united into a compound ovary.   Fam. 28. *Nymphaeaceae.* **2** : 77.

             ** Land or marsh plants (some Ranunculaceae aquatic).

Stamens numerous; sepals distinct; petals present (except in some Ranunculaceae and in *Calyco-*
  carpum of the Menispermaceae).
  Receptacle not hollow; leaves alternate (except in *Clematis*).
    Flowers perfect (except in some species of *Clematis* and *Thalictrum*).
      Fruit aggregate, cone-like; trees; sepals and petals in 3 series, or more, of 3.
                                   Fam. 29. *Magnoliaceae.* **2** : 80.
      Fruit not aggregate, the carpels separate, at least when mature.
        Anthers not opening by valves; pistils usually more than 1.
          Sepals 3; petals 6; shrubs or trees.   Fam. 30. *Annonaceae.* **2** : 83.
          Sepals 3–15; petals (when present) about as many; our species herbs or vines
          (*Xanthorrhiza* shrubby).   Fam. 31. *Ranunculaceae.* **2** : 84.
        Anthers opening by valves (except in *Podophyllum*); pistil 1.
                                     Fam. 32. *Berberidaceae.* **2** : 126.
    Dioecious climbing vines with simple leaves; fruit drupaceous.
                                   Fam. 33. *Menispermaceae.* **2** : 130.
  Receptacle hollow, enclosing the numerous pistils and achenes; opposite-leaved shrubs.
                                   Fam. 34. *Calycanthaceae.* **2** : 132.
Stamens 9 or 12, in 3 or 4 series of 3; anthers opening by valves; aromatic trees or shrubs with no
  petals, more or less united sepals, and 1 pistil.   Fam. 35. *Lauraceae.* **2** : 133.

2. Carpels 2 or more, united into a compound ovary; stamens hypogynous; sepals mostly distinct.
                    * Plants not insectivorous.
                    Order 13. Papaverales (*Rhoeadales*).
Sepals 2 (very rarely 3 or 4); endosperm fleshy.
  Flowers regular; stamens 8–many.   Fam. 36. *Papaveraceae.* **2** : 136.
  Flowers irregular; stamens 6.   Fam. 37. *Fumariaceae.* **2** : 141.
Sepals or calyx-segments 4–8; endosperm none.
  Capsule 2-celled by a longitudinal partition, usually 2-valved, rarely indehiscent; sepals and
    petals 4.   Fam. 38. *Cruciferae.* **2** : 146.
  Capsule 1-celled, of 2–6 carpels.
    Sepals and petals 4, regular, or petals irregular; capsule of 2 carpels, 2-valved.
                                     Fam. 39. *Capparidaceae.* **2** : 196.
    Sepals and petals 4–8, irregular; capsule of 3–6 carpels, 3–6-valved at the top; disk large.
                                     Fam. 40. *Resedaceae.* **2** : 199.

  ** Insectivorous plants, secreting a viscid liquid, with basal leaves and scapose flowers.
                    Order 14. Sarraceniales.
Ovary 3–5-celled; leaves hollow.   Fam. 41. *Sarraceniaceae.* **2** : 201.
Ovary 1-celled; leaves circinate in unfolding, the blade flat.   Fam. 42. *Droseraceae.* **2** : 202.

3. Carpels solitary, or several and distinct, or sometimes united; stamens mostly perigynous or
  epigynous; sepals mainly united or confluent with the concave receptacle. Order 15. Rosales.
  * Small aquatic fleshy herbs, with a spathe-like involucre, and a 2–3-celled capsule; perianth none.
                        Fam. 43. *Podostemaceae.* **2** : 205.
        ** Land or rarely swamp plants without an involucre.
        † *Endosperm present, usually copious and fleshy.*
Herbaceous plants.
  Carpels as many as the sepals.
    Carpels distinct, or united below, longitudinally dehiscent; succulent plants.
                                   Fam. 44. *Crassulaceae.* **2** : 205.
    Carpels united to the midlde, circumscissile; plants not succulent.
                                   Fam. 45. *Penthoraceae.* **2** : 211.
  Carpels fewer than the sepals.
    Carpels 3 or 4, united into a 1-celled ovary; staminodia present.
                                   Fam. 46. *Parnassiaceae.* **2** : 211.
    Carpels mostly 2, distinct, or only partly united; no staminodia.
                                   Fam. 47. *Saxifragaceae.* **2** : 214.
Shrubs or trees.
  Leaves opposite.   Fam. 48. *Hydrangeaceae.* **2** : 230.
  Leaves alternate.
    Fruit a 2–5-celled capsule.
      Capsule thin-walled, almost free from the calyx-tube (hypanthium).
                                Fam. 49. *Iteaceae.* **2** : 233.
      Capsule woody, or thick-walled, adnate to the calyx-tube.
        Ovule solitary, suspended; calyx-limb or calyx-limb and petals present.
                                Fam. 50. *Hamamelidaceae.* **2** : 234.
        Ovules several or numerous; no calyx-limb nor petals. Fam. 51. *Altingiaceae.* **2** : 235.
    Fruit a 1-celled berry.   Fam. 52. *Grossulariaceae.* **2** : 236.
    †† *Endosperm none, or very little* (copious in *Opulaster*, shrubs of the Rosaceae).
        ‡ Trees with broad leaves and small monoecious capitate flowers.
                                  Fam. 53. *Platanaceae.* **2** : 242.

‡‡ Flowers perfect (dioecious in *Aruncus* and in species of *Fragaria* of the Rosaceae; in *Gleditsia*
  and *Gymnocladus* of the Caesalpiniaceae, and rarely in some Fabaceae).

a. Flowers regular.

Pistils usually several or numerous (one only in *Cercocarpus* and some species of *Alchemilla* and *Aphanes;* in *Sanguisorba, Poteridium* and *Poterium*).

Carpels distinct, sometimes adnate to the calyx, ripening into follicles or achenes.
                 Fam. 54. *Rosaceae.* 2 : 242.
Carpels united, enclosed by the calyx-tube and adnate to it, the fruit a pome.
                 Fam. 55. *Malaceae.* 2 : 286.

Pistil only 1.
Ovary 2-ovuled; fruit a drupe; leaves simple.   Fam. 56. *Amygdalaceae* 2 : 322.
Ovary several-ovuled; fruit a legume; leaves 2–3-pinnate. Fam. 57. *Mimosaceae.* 2 : 330.

b. Flowers irregular (nearly or quite regular in *Gleditsia* and *Gymnocladus.*
       trees of the Caesalpiniaceae).

Fruit a legume; upper petal enclosed by the lateral ones in the bud; leaves compound, · mostly
 stipulate.   Fam. 58. *Caesalpiniaceae.* 2 : 334.
Fruit spiny, indehiscent; leaves simple, exstipulate. Fam. 59. *Krameriaceae.* 2 : 340.
Fruit a legume or loment; upper petal enclosing the lateral ones in the bud; leaves compound
 (sometimes 1-foliolate), stipulate.   Fam. 60. *Fabaceae.* 2 : 341.

4. Carpels united into a compound ovary; sepals mostly distinct.

* Stamens few, rarely more than twice as many as the sepals.

 † Stamens as many as the sepals or fewer, and opposite them, or more numerous.
 ‡ Ovules mostly pendulous, with the raphe toward the axis of the ovary.
            Order 16. GERANIALES.

Stamens more than one; land plants.
Filaments partially united (distinct in some Geraniaceae); herbs, the leaves not punctate.
Leaves not pinnately compound.
 Capsule at length splitting into its 5 carpels; leaves lobed or dissected.
             Fam. 61. *Geraniaceae.* 2 : 425.
 Capsule 2–5-celled, not splitting into its carpels.
  Stamens 2–3 times as many as the petals; leaves 3-foliolate in our species.
           Fam. 62. *Oxalidaceae.* 2 : 430.
  Stamens as many as the petals; leaves entire. Fam. 63. *Linaceae.* 2 : 435.
Leaves pinnately compound.   Fam. 66. *Zygophyllaceae.* 2 : 442.
Filaments distinct (united in some Balsaminaceae).
Flowers very irregular; calyx with a spurred or saccate sepal.
           Fam. 64. *Balsaminaceae.* 2 : 440.
Flowers regular;
Herbaceous plants with pinnately divided leaves. Fam. 65. LIMNANTHACEAE. 2 : 441.
 Our species trees or shrubs with compound leaves, often punctate; flowers dioecious ɔr
  polygamous.
  Leaves punctate.   Fam. 67. *Rutaceae.* 2 : 443.
  Leaves not punctate, but the bitter bark with oil-sacs.
           Fam. 68. *Simaroubaceae.* 2 : 445.
Flowers very irregular; petals 3; stamens usually 8; ours low herbs.
           Fam. 69. *Polygalaceae.* 2 : 446.
Flowers regular, often apetalous, small, monoecious or dioecious; carpels mostly 3; herbs or
 low shrubs, mostly with milky juice. Fam. 70. *Euphorbiaceae.* 2 : 452.
Stamen only 1; perianth none; styles 2; small aquatic or rarely terrestrial plants with opposite
 entire leaves.   Fam. 71. *Callitrichaceae.* 2 : 477.

 ‡‡ Ovules pendulous, with the raphe away from the axis of the ovary, or erect or ascending.
           Order 17. SAPINDALES.

Petals none (or 3 in *Empetrum*); flowers monoecious or dioecious; leaves evergreen.
Stamens mostly 3; low heath-like shrubs. Fam. 72. *Empetraceae.* 2 : 478.
Stamens 4–7; our species an herb with broad leaves and spiked flowers.
           Fam. 73. *Buxaceae.* 2 : 480.
Petals present; leaves deciduous, except in Cyrillaceae and some Ilicaceae.
Ovary 1-celled (in ours); fruit a small drupe. Fam. 74. *Anacardiaceae.* 2 : 480.
Ovary 2–several-celled.
Leaves simple, pinnately veined.
 Seeds not arilled.
  Fruit dry; flowers racemed, perfect. Fam. 75. *Cyrillaceae.* 2 : 485.
  Fruit a small drupe; flowers not racemed, mostly polygamo-dioecious; ovules pedulous.
           Fam. 76. *Ilicaceae.* 2 : 486.
 Seeds arilled; ovules erect; capsule fleshy. Fam. 77. *Celastraceae.* 2 : 490.
Leaves simple and palmately veined, or compound.
 Leaves opposite.
  Fruit a bladdery 3-lobed capsule. Fam. 78. *Staphyleaceae.* 2 : 493.
  Fruit of 2 winged samaras.   Fam. 79. *Aceraceae.* 2 : 494.
  Fruit a leathery capsule; flowers irregular; leaves digitately compound.
           Fam. 80. *Aesculaceae.* 2 : 498.
 Leaves alternate; fruit various.   Fam. 81. *Sapindaceae.* 2 : 500.

 †† Stamens as many as the sepals and alternate with them, opposite the petals when
    these are present; ovules erect.
           Order 18. RHAMNALES.

Shrubs, small trees, or vines; petals 4 or 5, or none; fruit a drupe or capsule.
           Fam. 82. *Rhamnaceae.* 2 : 501.
Vines, climbing by tendrils, rarely shrubs; petals caducous; fruit a berry.
           Fam. 83. *Vitaceae.* 2 : 505.

** Stamens usually very numerous (except in some Hypericaceae, in Elatinaceae, Violaceae and Passifloraceae); disk inconspicuous or none.

† Sepals valvate; placentae united in the axis.   Order 19. MALVALES.

Stamens in several sets; anthers 2-celled; embryo straight.   Fam. 84. *Tiliaceae.*   **2** : 511.
Stamens monadelphous; anthers 1-celled; embryo curved.   Fam. 85. *Malvaceae.*   **2** : 513.

†† Sepals or calyx-segments imbricated or convolute (except in Loasaceae, in which the calyx-tube is adnate to the ovary); placentae mainly parietal, sometimes united in the axis.
Order 20. HYPERICALES (*Parietales*).

Sepals distinct, mostly persistent.
  Endosperm little or none.
    Trees or shrubs with alternate leaves, and large solitary axillary flowers.
          Fam. 86. *Theaceae.*   **2** : 526.

    Herbs or low shrubs with opposite, rarely verticillate leaves.
      Leaves punctate or black-dotted, exstipulate.   Fam. 87. *Hypericaceae.*   **2** : 527.
      Leaves stipulate; minute or small marsh or aquatic herbs with axillary flowers.
          Fam. 88. *Elatinaceae.*   **2** : 537.

  Endosperm copious.
    Flowers regular, but the 2 outer sepals smaller; stamens numerous; ovules orthotropous.
          Fam. 89. *Cistaceae.*   **2** : 539.
    Flowers irregular, some often cleistogamous; stamens 5; ovules anatropous.
          Fam. 90. *Violaceae.*   **2** : 545.

Sepals more or less united into a gamosepalous calyx.
  A fringed crown in the throat of the calyx; our species vines; stamens 5; ovary free from the calyx.   Fam. 91. *Passifloraceae.*   **2** : 564.
  No crown; our species herbs; stamens numerous; ovary adnate to the calyx.
          Fam. 92. *Loasaceae.*   **2** : 565.

II. *Ovary inferior, adnate to the calyx, wholly, or in part* (except in Lythraceae and our Melastomaceae, where it is usually merely enclosed by it, and in Thymeleaceae and Elaeagnaceae, which are shrubs or trees, with no corolla).

1. Fleshy spiny plants, often with jointed stems, the leaves very small in our species, or none; calyx-segments and petals mostly numerous.   Order 21. OPUNTIALES.

One family.
          Fam. 93. *Cactaceae.*   **2** : 568.

2. Herbs, shrubs or trees, not fleshy nor spiny; calyx-segments and petals (when present) rarely more than 5.   Order 22. THYMELEALES.
  Petals none in our species; shrubs or trees; ovary 1-ovuled.
    Leaves green; seed pendulous.   Fam. 94. *Thymeleaceae.*   **2** : 574.
    Leaves silvery-scurfy; seed erect.   Fam. 95. *Elaeagnaceae.*   **2** : 575.
  Petals present (except in some Haloragidaceae, which are small aquatic herbs).
    Ovules several or numerous in each cavity of the ovary (except in Haloragidaceae and Trapaceae).   Order 23. MYRTALES (*Myrtiflorae*).
      Land or marsh plants, or, if aquatic, submerged leaves not dissected.
        Calyx-tube merely enclosing the ovary, but free from it (except at the base).
          Anthers longitudinally dehiscent.   Fam. 96. *Lythraceae.*   **2** : 577.
          Anthers opening by a terminal pore.   Fam. 97. *Melastomaceae.*   **2** : 582.
        Calyx-tube almost wholly adnate to the ovary.   Fam. 98. *Onagraceae.*   **2** : 584.
      Aquatic or amphibious herbs, the submerged leaves dissected (except in *Hippuris,* which has whorled narrow leaves and only 1 stamen.)
        Petioles of the broad floating leaves inflated; flowers rather large, white.
          Fam. 99. *Trapaceae.*   **2** : 611.
        Leaves most sessile; petioles, if present, not inflated; flowers small, greenish; seeds with 1 coat.   Fam. 100. *Haloragidaceae.*   **2** : 612.
    Ovules 1 in each cavity of the ovary.   Order 24. UMBELLALES (*Umbelliflorae*).
      Stamens 5; styles 2–5, rarely united; flowers umbellate or capitate.
        Fruit a fleshy berry or drupe.   Fam. 101. *Araliaceae.*   **2** : 616.
        Fruit dry when mature, splitting into two mericarps.   Fam. 102. *Ammiaceae.*   **2** : 619.
      Stamens 4; style 1; stigma 1; shrubs and trees; flowers not umbellate.
          Fam. 103. *Cornaceae.*   **2** : 660.

Series 2. GAMOPETALAE.
          2 : 666 to **3** : 560.

Petals more of less united.   (See exceptions noted on page 666, Vol. 2.)

A. **Ovary superior** (except in Vacciniaceae and Symplocaceae, in which it is partly or wholly inferior).

I. Stamens mostly free from the corolla, or adnate merely to its base (at the sinuses of the corolla in *Diapensia* and *Pyxidanthera* of the Diapensiaceae), as many as the lobes and alternate with them, or twice as many.   Order 1. ERICALES.

Stamens free from the corolla, or merely adnate to its base, not united into a tube.
  Ovary superior; fruit a capsule, or rarely drupaceous.
    Corolla essentially polypetalous.
      Ovary 3-celled; shrubs; leaves deciduous.   Fam. 1. *Clethraceae.*   **2** : 666.
      Ovary 4–5-celled; low, mostly evergreen perennials.   Fam. 2. *Pyrolaceae.*   **2** : 667.
    Corolla distinctly gamopetalous (except in *Monotropa* and *Hypopitys* of the Monotropaceae and *Ledum* of the Ericaceae).
      Herbaceous saprophytes without green leaves.   Fam. 3. *Monotropaceae.*   **2** : 673.
      Shrubs with normal, often evergreen leaves.   Fam. 4. *Ericaceae.*   **2** : 675.
  Ovary inferior, adnate to the calyx, forming a many-seeded berry in fruit.
          Fam. 5. *Vacciniaceae.*   **2** : 694.

Stamens borne at the sinuses of the corolla, or united in a 10-lobed tube.
Fam. 6. *Diapensiaceae.* **2** : 705.
II. Stamens borne on the corolla, as many as its lobes and opposite them, or twice as many, or more.
Herbs.   Order 2. PRIMULALES.
     Style 1 ; fruit a capsule.   Fam. 7. *Primulaceae.*   **2** : 707.
     Styles 5 ; fruit an achene or utricle.   Fam. 8. *Plumbaginaceae.*   **2** : 717.
Shrubs or trees.   Order 3. EBENALES.
     Stamens as many as the corolla-lobes.   Fam. 9. *Sapotaceae.*   **2** : 719.
     Stamens twice as many as the corolla-lobes, or more.
         Styles 2–8 ; flowers mostly monoecious or dioecious.   Fam. 10. *Ebenaceae.*   **2** : 720.
         Style 1, simple or lobed ; flowers mostly perfect.
             Stamens in several series.   Fam. 11. *Symplocaceae.*   **2** : 711.
             Stamens in 1 series.   Fam. 12. *Styracaceae.*   **2** : 711.
III. Stamens borne on the corolla, as many as its lobes or fewer, and alternate with them (in our
     species of *Fraxinus* and *Forestiera* of the Oleaceae there is no corolla).

                          * *Corolla not scarious, nerved.*
† Ovaries 2, distinct (except in some Loganiaceae, and in Gentianaceae and Menyanthaceae, in which
     the ovary is compound, with 2 cavities, or rarely more, or with 1 cavity and 2
     placentae) ; flowers regular ; stamens mostly adnate to only the
         lower part of the corolla ; leaves mostly opposite.
                         Order 4. GENTIANALES (*Contortae*).
     a. Stamens (usually 2), fewer than the corolla-lobes, or corolla none ; our species
                  trees or shrubs.
                         Fam. 13. *Oleaceae.*   **2** : 723.
     b. Stamens as many as the corolla-lobes ; mostly herbs.
Stigmas distinct ; juice not milky ; ovary 1, compound.
     Ovary 2-celled ; leaves stipulate, or their bases connected by a stipular line.
                       Fam. 14. *Loganiaceae.*   **2** : 729.
     Ovary 1-celled ; leaves not stipulate.
         Leaves opposite or rarely verticillate ; corolla-lobes convolute or imbricated in the bud.
                       Fam. 15. *Gentianaceae.*   **3** : 1.
         Leaves basal or alternate ; corolla-lobes induplicate-valvate in the bud ; marsh or aquatic
         herbs.   Fam. 16. *Menyanthaceae.*   **3** : 17.
Stigmas united ; juice milky ; ovaries 2 in our species.
     Styles united ; stamens distinct ; pollen of simple grains.   Fam. 17. *Apocynaceae.*   **3** : 19.
     Styles distinct ; stamens mostly monadelphous ; pollen-grains united into waxy masses.
                       Fam. 18. *Asclepiadaceae.*   **3** : 23.
†† Ovary 1, compound (2-divided in *Dichondra ;* in Boraginaceae and Labiatae mostly deeply
     4-lobed around the style) flowers regular or irregular ; stamens mostly adnate to
     the middle of the corolla-tube or beyond ; leaves opposite or alternate.
                       Order 5. POLEMONIALES (*Tubiflorae*).
     a. Corolla regular (irregular in *Echium* of the Boraginaceae).
Ovary not 4-lobed, the carpels not separating as separate nutlets at maturity.
     Ovary 2-divided.   Fam. 19. *Dichondraceae.*   **3** : 39.
     Ovary 2-celled, rarely 3-4-celled ; style 1, entire, 2-cleft, or 2-parted ; mostly twining vines.
         Leaves normal.   Fam. 20. *Colvolvulaceae.*   **3** : 40.
         White or yellowish parasitic vines, the leaves reduced to minute scales.
                       Fam. 21. *Cuscutaceae.*   **3** : 48.
     Ovary 3-celled ; stigmas 3, linear ; herbs, not twining.   Fam. 22 *Polemoniaceae.*   **3** : 52.
     Ovary 1-celled (2-celled in *Nama*) ; style 1, 2-lobed, or 2-parted ; herbs, not twining.
                       Fam. 23. *Hydrophyllaceae.*   **3** : 65.
Ovary deeply 4-lobed around the style, or not lobed (*Heliotropium*) ; carpels separating as nutlets.
                       Fam. 24. *Boraginaceae.*   **3** : 72.
     b. Corolla irregular, more or less 2-lipped (regular in Solanaceae, in *Mentha* and *Lycopus* of the
     Labiatae, and nearly or quite so in *Verbena* and *Callicarpa* of the Verbenaceae).
                   1. Carpels 1–2-seeded.
Ovary not lobed, 2-4-celled, the style apical ; carpels separating into 1-seeded nutlets.
                       Fam. 25. *Verbenaceae.*   **3** : 94.
Ovary 4-lobed around the style, the lobes ripening into 1-seeded nutlets.   Fam. 26. *Labiatae.*   **3** : 99.
                2. Carpels several–many-seeded (2-seeded in some Acanthaceae).
‡ Fruit a berry, or more commonly a capsule which is 1-2-celled, 2-valved, circumscissile, or irregularly
                  bursting, not elastically dehiscent.
Placentae axile.
     Ovary 2-celled, or rarely 3-5-celled.
         Flowers regular ; fertile stamens 5 (4 in *Petunia*) ; fruit a berry or capsule.
                       Fam. 27. *Solanaceae.*   **3** : 154.
         Flowers more or less irregular ; fertile stamens 2 or 4 (5 in *Verbascum*) ; fruit a capsule.
                       Fam. 28. *Scrophulariaceae.*   **3** : 172.
     Ovary 1-celled ; marsh or aquatic herbs with flowers on scapes.
                       Fam. 29. *Lentibulariaceae.*   **3** : 225.
Placentae parietal.
     Herbs, parasitic on the roots of other plants, the leaves reduced to scales, not green ; ovary
     1-celled.   Fam. 30. *Orobanchaceae.*   **3** : 233.
     Trees, vines, shrubs, or herbs, the foliage normal.
         Trees, shrubs, or woody vines ; capsule 2-celled ; seeds winged in our genera.
                       Fam. 31. *Bignoniaceae.*   **3** : 236.

Opposite-leaved herbs; capsule 1-celled in our genus; seeds wingless.

Fam. 32. *Martyniaceae.*   **3** : 239.

‡‡ **Capsule completely 2-celled, elastically loculicidally dehiscent; opposite-leaved herbs; placentae axile.**

Fam. 33.   *Acanthaceae.*   **3** : 239.

3. Ovary and fruit 1-celled with 1 erect orthotropous ovule and seed; herb with spicate flowers and reflexed fruits.

Fam. 34.   *Phrymaceae.*   **3** : 244.

** Corolla scarious, nerveless.*

Order 6. PLANTAGINALES.

Herbs with small spicate or capitate flowers; one family.   Fam. 35.   *Plantaginaceae.*   **3** : 245.

B. Ovary inferior.

I. *Anthers distinct.*

Stamens as many as the corolla-lobes and alternate with them (one fewer in *Linnaea* of the Caprifoliaceae), or twice as many; ovary compound, with 1 ovule or more in each cavity; leaves opposite, or verticillate.   Order 7. RUBIALES.

Stamens as many as the corolla-lobes.

Leaves always stipulate, usually blackening in drying.   Fam. 36. *Rubiaceae.*   **3** : 250.

Leaves usually exstipulate, not blackening in drying.   Fam. 37. *Caprifoliaceae.*   **3** : 267.

Stamens twice as many as the corolla-lobes; low herb with ternately divided leaves.

Fam. 38. *Adoxaceae.*   **3** : 283.

Stamens mostly fewer than the corolla-lobes; ovary 1-celled with 1 pendulous ovule, or 3-celled with 2 of the cavities without ovules.   Order 8. VALERIANALES (*Aggregatae*).

Ovary 3-celled, 2 of its cavities empty.   Fam. 39. *Valerianaceae.*   **3** : 284.

Ovary 1-celled; flowers densely capitate, involucrate.   Fam. 40. *Dipsacaceae.*   **3** : 288.

II. *Anthers united* (except in *Campanula* and *Specularia* of the Campanulaceae, in Ambrosiaceae, and in *Kuhnia* of the compositae).   Order 9. CAMPANULALES (*Campanulatae*).

Flowers not in involucrate heads; juice mostly milky.

Endosperm none; flowers monoecious or dioecious; our species vines.

Fam. 41. *Cucurbitaceae.*   **3** : 290.

Endosperm present, fleshy; flowers perfect.

Flowers regular   Fam. 42. *Campanulaceae.*   **3** : 293.

Flowers irregular.   Fam. 43. *Lobeliaceae.*   **3** : 299.

Flowers in involucrate heads.

Flowers all expanded into rays (ligulate); juice milky.   Fam. 44. *Cichoriaceae.*   **3** : 304.

Flowers all tubular, or the outer expanded into rays; juice very rarely milky.

Stamens distinct, or nearly so.   Fam. 45. *Ambrosiaceae.*   **3** : 338.

Stamens united by their anthers into a tube around the style (except in *Kuhnia*).

Fam. 46. *Compositae.*   **3** : 347.

ILLUSTRATED FLORA.

Subkingdom PTERIDOPHYTA.*

FERNS AND FERN-ALLIES.

Plants containing woody and vascular tissues in the stem and producing spores asexually, which, on germination, develop small mostly flat green structures called prothallia (gametophyte). On these are borne the sexual reproductive organs, the female known as archegones, the male as antherids. From the fertilization of the oösphere in the archegone by spermatozoids produced in the antherids, the asexual phase (sporophyte) of the plants is developed; this phase is represented by an ordinary fern, lycopod or horsetail.

This subkingdom comprises about 6,000 living species, of which more than three-fourths are confined to tropical regions. The number of extinct species known probably exceeds those living. They appeared on the earth in the early part of the Palaeozoic Era, reached great abundance in Carboniferous Time, but have since been mainly replaced by plants of higher organization, so that at present they form only a small proportion of the total flora. The time of year noted under the species indicates the season at which the spores are mature.

Family 1. OPHIOGLOSSÀCEAE Presl, Tent. Pterid. 6. 1836.

ADDER'S-TONGUE FAMILY.

Succulent plants consisting of a short fleshy rootstock bearing one or several leaves and numerous fibrous often fleshy roots. Leaves erect or pendent, consisting of a simple, palmately or dichotomously lobed, pinnately compound or decompound, sessile or stalked, sterile blade, and one or several separate stalked fertile spikes or panicles (sporophyls), borne on a common stalk. Sporanges formed from the interior tissues, naked, each opening by a transverse slit. Spores yellow, of one sort. Prothallia subterranean, usually devoid of chlorophyl and associated with an endophytic mycorhiza.

Five genera, the following well represented in both hemispheres; the others tropical.

Veins reticulate; sporanges cohering in a distichous spike.   1. *Ophioglossum.*
Veins free; sporanges distinct, borne in spikes or panicles.   2. *Botrychium.*

1. OPHIOGLÓSSUM [Tourn.] L. Sp. Pl. 1062. 1753.

Small terrestrial plants, with small, erect, fleshy, often tuberous, rootstocks bearing fibrous naked roots and 1–6 slender, erect leaves, these consisting usually of a short, cylindric common stalk, bearing at its summit a simple entire lanceolate to reniform sessile or short-stalked sterile blade with freely anastomosing veins and usually a single simple long-stalked spike, the sporophyl, formed of 2 rows of large coalescent sporanges; spores copious, sulphur yellow. Bud for the following season borne at the apex of the rootstock, exposed, distinct and free from the leaf of the present season. [Name from the Greek, signifying the tongue of a snake, in allusion to the form of the narrow spike.]

About 45 species of wide geographic distribution. Besides the following 4 others occur in the southern and western United States and Alaska. Type species: *Ophioglossum vulgatum* L.

Leaves usually solitary; sterile blade obtuse or acutish, never apiculate.   1. *O. vulgatum.*
Leaves often in pairs; sterile blade acutish or apiculate.
     Sterile blade elliptic, rarely ovate, apiculate; areoles broad.   2. *O. Engelmanni.*
     Sterile blade lanceolate, acutish, somewhat apiculate; areoles narrow.   3. *O. arenarium.*

* Text (except Equisetaceae and Isoetaceae) revised by William R. Maxon.

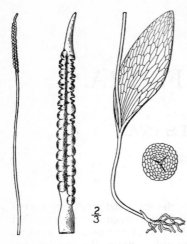

1. Ophioglossum vulgàtum L.
Adder's-tongue.   Fig. 1.

Ophioglossum vulgatum L. Sp. Pl. 1062. 1753.

Rootstock short, oblique or erect; leaves usually solitary, 3′–16′ long, the common stalk usually one-half or more above ground and constituting one-third to two-thirds the length of the plant; sterile blade lanceolate, oblanceolate or spatulate, elliptical, oblong or ovate, 1′–5′ long, ½′–2′ broad, sessile, obtuse or acutish, the middle areoles long and narrow, the outer ones shorter and hexagonal, with included veins; sporophyl ¾′–2′ long, borne on a stalk 4′–10′ long, solitary, apiculate from the prolongation of the axis.

In moist meadows and boggy thickets, Prince Edward Island to Ontario, south to Florida. Also in Europe and Asia. The genus is also called Adder's-fern or -spear. Snake's-tongue. Serpent's-tongue. May–Aug.

2. Ophioglossum Engelmánni Prantl.   Engelmann's
Adder's-tongue.   Fig. 2.

Ophioglossum Engelmanni **Prantl**, Ber. Deuts. Bot. Ges. **1**: 351. 1883.

Rootstock cylindric, with long brown roots; leaves commonly 2–5, mostly fertile, 3′–9′ long, the common stalk often mostly below the ground and usually sheathed by the more or less persistent bases of old leaves; sterile blade elliptic or rarely ovate, 1′–3½′ long, ½′–2′ broad, sessile, usually acute, apiculate, with wide oblique areoles containing numerous anastomosing or free veins; sporophyl 6″–12″ long borne on a stalk 1′–4′ long, apiculate; sporanges 12–27 pairs.

In damp, sterile places or on rocks in cedar woods, mainly in the Central States, from Indiana and Virginia to Louisiana, Texas and Arizona. April–Oct.

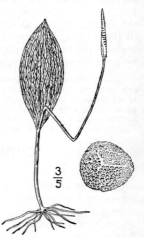

3. Ophioglossum arenàrium E. G. Britton.   Sand
Adder's-tongue.   Fig. 3.

O. arenarium E. G. Britton, Bull. Torr. Club, **24**: 555. *pl. 318.* 1897.

Rootstock slightly thickened, with numerous fleshy, spreading (sometimes proliferous?) roots; leaves slender but rigidly erect, single or often in pairs, mostly fertile, 2′–7′ long, the common stalk one-half its length below ground and constituting one-third or less the length of the plant; sterile blade lanceolate, with a long tapering base, or somewhat elliptic, 1′–2′ long, 3″–6″ broad, acutish or somewhat apiculate, with long, narrow areoles with a few indistinct included veinlets, the outer areoles shorter; sporophyl 6″–13″ long, borne on a slender stalk 2′–3½′ long, apiculate; sporanges 12–26 pairs.

Gregarious in a colony of many plants in sandy ground under trees at Holly Beach, New Jersey, the type locality. Also in New York and New Hampshire. July.

2. BOTRÝCHIUM Sw. Schrad. Journ. Bot. 1800[2] : 8. 1801.

Fleshy terrestrial plants, with stout erect rootstocks, bearing clustered, fleshy, often corrugated roots and 1 or sometimes 2 or 3 erect leaves, these consisting of a short cylindric wholly or partially hypogean common stalk, bearing at its summit a simple 1–3-pinnately compound or decompound free-veined sterile blade and a single long-stalked fertile spike or 1–4-pinnate panicle, the sporophyl, with numerous globular distinct sporanges in two rows, sessile or nearly so; spores copious, sulphur yellow. Bud for the following season at the apex of the rootstock, enclosed within the base of the common stalk, either wholly concealed or visible along one side. [Name in allusion to the grape-like arrangement of the sporanges.]

About 20 species, largely natives of the temperate regions of both hemispheres. Type species: *Botrychium Lunaria* (L.) Sw.

Buds of the following season wholly concealed within the base of the common stalk; sterile blade
    more or less fleshy; cells of the epidermis straight.
    Sporophyl and sterile blade both erect in the bud.   1. *B. simplex.*
    Sporophyl or sterile blade, or both, at least slightly bent over in bud.
        Buds glabrous; sterile blade pinnate (or, in no. 10, sometimes subternate); spores maturing
           in early summer.
           Sterile blade slightly bent over in bud, clasping the nearly erect sporophyl.
                Leaves usually stout, the sterile blade nearly sessile, oblong, with close (often
                  imbricate) segments.   2. *B. Lunaria.*
                Leaves slender, the sterile blade usually stalked, oblong to deltoid, with cuneate
                  mostly distant segments.
                    Sterile blade distinctly bent over at the tip in the bud, always pinnately divided;
                    segments 3–4 pairs.   3. *B. onondagense.*
                    Sterile blade with the tip slightly inclined in bud, entire, or with 1–3 pairs of
                    smaller segments.   4. *B. tenebrosum.*
        Sterile blade and sporophyl bent over in bud.
             Sterile blade distinctly stalked.   5. *B. neglectum.*
             Sterile blade closely sessile.   10. *B. lanceolatum.*
        Buds pilose; sterile blades subternately divided; spores maturing in late summer or autumn.
            Sterile blades membranous in drying; segments mostly acutish, serrulate to laciniate.
                Segments mostly acute or acutish, serrulate-dentate.   6. *B. obliquum.*
                Segments laciniate, often deeply so.   7. *B. dissectum.*
             Sterile blades thick, leathery in drying; segments obtuse, crenate to sinuate.
                Leaves 3′–7′ long; sterile blade at most 2′ broad; segments few. 8. *B. Matricariae.*
                Leaves 8′–18′ long; sterile blade 4′–8′ broad; segments numerous. 9. *B. silaifolium.*
Bud of the following season exposed along one side; sterile blade very thin; cells of the epi-
    dermis flexuous.   11. *B. virginianum.*

1. Botrychium símplex E. Hitchcock. Hitchcock's or Little Grape-fern. Fig. 4.

B. simplex E. Hitchcock, Amer. Journ. Sci. **6**: 310. 1823.

Leaves 2′–6′ long, slender and variable, the common stalk usually about half under ground; sterile blade and sporophyl straight in the bud. Sterile blade usually short-stalked, thickish, ovate, obovate or oblong, simple and roundish, or pinnately 3–7-lobed (rarely binate or ternate, the divisions pinnately lobed), the segments cuneate to somewhat lunulate, usually apart, the veins forking from the base; sporophyl long-stalked (often one-half or more the height of the plant), simple or 1–2-pinnate.

In meadows and pastures, Prince Edward Island to Mary-land, California and Oregon. Europe and Asia. May–June.

2. Botrychium Lunària (L.) Sw. Moonwort. Moon-fern. Fig. 5.

Osmunda Lunaria L. Sp. Pl. 1064. 1753.
B. Lunaria Sw. Schrad. Journ. Bot. 1800²: 110. 1801.

Leaves very fleshy, usually stout, 2′–12′ long, variable, the common stalk nearly all above ground and consti-tuting about one-half the length of the plant, the sterile blade bent over in the bud only at the apex, clasping the nearly erect sporophyl. Sterile blade nearly sessile, broadly oblong, once pinnately divided into 2 to 8 pairs of lunate subentire, crenate or somewhat incised, often close or imbricate segments, the radiating veins several times forked; sporophyl 2–3-pinnate, paniculate.

Newfoundland to Alaska, Connecticut, New York, Michi-gan, British Columbia and in the Rocky Mountains to Colorado. Europe and Asia. June–July.

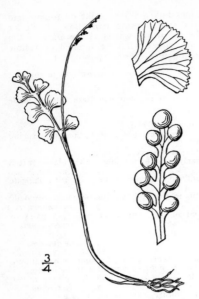

3. Botrychium onondagénse Underw.
Underwood's Moonwort.   Fig. 6.

Botrychium onondagense Underwood, Bull. Torr. Club, 30: 47. 1903.

Leaves 4′–6½′ long, slender, the common stalk slender, rather weak and spreading, 3′–4¾′ long, nearly all above ground, the sterile blade bent over in the bud only at the apex, clasping the nearly erect sporophyl. Sterile blade oblong, often narrowly so, ¾′–1½′ long, distinctly stalked (up to ⅜′), pinnately divided into 3 to 4 (casually 7) pairs of mostly distant broadly cuneate subentire to flabellately lobed segments; sporophyl ½′–1′ long, mostly 2-pinnate, borne upon a slender stalk 1′–1½′ long.

On shaded rocky slopes, near Syracuse, New York. Also in Montana and northern Michigan.

4. Botrychium tenebròsum A. A. Eaton.
Eaton's Grape-fern.   Fig. 7.

Botrychium tenebrosum A. A. Eaton, Fern Bull. 7: 8. 1899.

Leaves 1′–9′ long, slender, delicate and lax, shining, light or yellowish green, eventually decumbent and stramineous, the common stalk very long, usually more than half the length of the plant; buds rather small, the sporophyl erect, the tip of the sterile blade slightly inclined. Sterile blade short-stalked, simple, lobed, or usually with 1–3 pairs of distant, alternate, lunulate or cuneate, decurrent, usually entire segments, the apex emarginate; sporophyl short-stalked, simple or rarely a little branched, flattened, the large sporangia somewhat immersed in rows or groups on either side.

In rich moist woods and swamps, New York and New England.

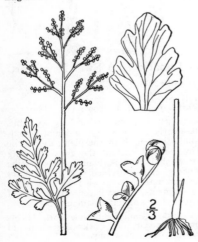

5. Botrychium negléctum Wood.
Wood's Grape-fern.   Fig. 8.

Botrychium neglectum Wood, Class Book Bot. ed. 2, 635. 1847
B. matricariaefolium of most American writers.

Leaves 2′–12′ long, often very fleshy, erect, bright green, the common stalk relatively stout, nearly all above ground, devoid of sheathing bases of previous years; buds stout, the sporophyl and sterile blade both bent over at the tip, the latter enfolding the former. Sterile blade short-stalked, ½′–2¾′ long, oblong, ovate or deltoid-ovate, acute, pinnate or deeply 2-pinnatifid, the segments oblong or ovate, obtuse, crenately lobed or divided, the divisions ovate-oblong; sporophyl 2–3-pinnate, with terete branches, the sporanges sessile or short-stalked.

In grassy woods and swamps, Nova Scotia to Maryland, west to South Dakota and Nebraska. Also in Washington and in Europe. May–June.

6. Botrychium oblìquum Muhl.
Ternate Grape-fern.   Fig. 9.

Botrychium obliquum Muhl.; Willd. Sp. Pl. 5 : 63. 1810.
Botrychium ternatum var. *obliquum* D. C. Eaton, Ferns N.
Am. 1 : 149. 1878.

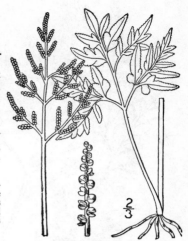

2/3

Leaves 6′–20′ long, usually robust, the common stalk
short and under ground; bud pilose, the sporophyl and
sterile blade bent down. Sterile blade usually long-
stalked, commonly 2′–5′ broad, subpentagonal, subternately 3-pinnatifid, or 3-pinnate below, the principal
divisions stalked; ultimate segments obliquely ovate or
oblong-lanceolate, acutish, the terminal ones elongate,
½′–1′ long; margins variously serrulate-dentate; sporophyl long-stalked, 3–4-pinnate, usually stout.

In moist woods and thickets, or open slopes, New Brunswick to Florida, Missouri and Minnesota. Very variable
especially in New York and New England. Several forms
have the divisions of the sterile blade longer-stalked and
lax, with fewer and rounded segments, or the segments
usually long and acute.

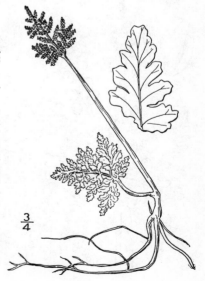

3/5

7. Botrychium disséctum Spreng. Cut-
leaved Grape-fern, or Moonwort. Fig. 10.

Botrychium dissectum Spreng. Anleit. 3 : 172. 1804.
Botrychium ternatum var. *dissectum* D. C. Eaton,
Ferns N. Am. 1 : 150. 1878.

Leaves 8′–16′ long, usually slender, the common stalk short, under ground; bud pilose, the
sporophyl and sterile blade both bent down.
Sterile blade long-stalked, subpentagonal, rarely
more than 6′ broad, subternately divided, the
basal divisions unequally and broadly deltoid,
decompound, the upper and secondary pinnae
deltoid-lanceolate, pinnate, with laciniate or deeply
cut pinnules, the ultimate divisions divergent,
narrow and incised; sporophyl 2–4-pinnate, usually
long-stalked.

In low woods and thickets or wooded slopes,
Maine to Virginia, Kentucky and Indiana. Congested forms closely resemble the preceding.

8. Botrychium matricàriae (Schrank)
Spreng. Grape-fern. Fig. 11.

Osmunda matricariae Schrank, Baier. Fl. 2 : 419. 1789.
Botrychium rutaceum Sw. Schrad. Journ. Bot. 1800[2]:
110. 1801.
Botrychium matricarioides Willd. Sp. Pl. 5 : 62. 1810.
Botrychium rutaefolium A. Br. in Doell, Rhein. Fl. 24.
1843.

Leaves single or in pairs, the fertile 3′–7′ long,
usually slender, fleshy, coriaceous in drying, somewhat glaucous, the common stalk ½′–2′ long, wholly
under ground; bud densely pilose, both sporophyl
and sterile blade bent over. Sterile blade stalked
(¾′–2′), triangular or subpentagonal, 1′–2′ broad,
nearly as long, subternately divided, 2–3-pinnate, the
basal pinnae nearly equalling the middle division,
the ultimate divisions few, oval or obliquely ovate,
rounded, the margins obscurely crenate or sinuate;
sporophyl long-stalked, large, 2–3-pinnate.

3/4

In old meadows and upon open hillsides, Labrador and
Newfoundland to New Brunswick, northern New Eng-
land and New York. Reported from northern Michigan.
Also in Europe. Aug.–Sept.

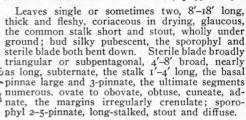

9. Botrychium silaifòlium Presl.
Leathery Grape-fern.   Fig. 12.

B. silaifolium Presl, Rel. Haenk. **1**: 76.   1825.
Botrychium ternatum subvar. *intermedium* D. C.
Eaton, Ferns N. Am. **1**: 149.   1878.
B. occidentale Underw. Bull. Torr. Club, **25**: 538.   1898.

Leaves single or sometimes two, 8'–18' long, thick and fleshy, coriaceous in drying, glaucous, the common stalk short and stout, wholly under ground; bud silky pubescent, the sporophyl and sterile blade both bent down.   Sterile blade broadly triangular or subpentagonal, 4'–8' broad, nearly as long, subternate, the stalk 1'–4' long, the basal pinnae large and 3-pinnate, the ultimate segments numerous, ovate to obovate, obtuse, cuneate, adnate, the margins irregularly crenulate; sporophyl 2–5-pinnate, long-stalked, stout and diffuse.

In moist meadows, sandy pastures and borders of low woods, northern New England to British Columbia, Oregon, Idaho and Minnesota. Aug.–Sept.

10. Botrychium lanceolàtum (S. G. Gmel.)
Angs.   Lance-leaved Grape-fern.   Fig. 13.

Osmunda lanceolata S. G. Gmel. Nov. Comment.
Acad. Petrop. **12**: 516.   1768.
B. lanceolatum Angs. Bot. Notiser, **1854**: 68.   1854.

Leaves 2'–12' long, fleshy, the common stalk nearly all above ground, long, usually three-fourths the length of the plant; sporophyl bent down in the bud, the sterile blade recurved upon it. Sterile blade sessile, ¾–2½' broad, nearly as long, either subternately parted with divisions acutely pinnatifid, or broadly deltoid, with 3–4 pairs of deeply pinnatifid pinnae, the segments ovate or ovate-oblong and lobed; sporophyl short-stalked, 2–3-pinnate, the branches usually stout and diffuse.

In meadows and moist woods, Nova Scotia to Alaska, New Jersey, Pennsylvania, Ohio, Colorado and Washington. Europe and Asia. June–July.

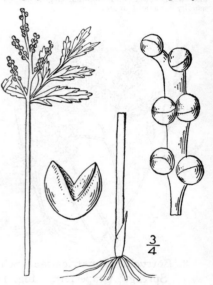

11. Botrychium virginiànum (L.) Sw.
Virginia Grape-fern.   Fig. 14.

Osmunda virginiana L. Sp. Pl. 1064.   1753.
Botrychium virginianum Sw. Schrad. Journ. Bot.
1800[2]: 111.   1801.
B. gracile Pursh, Fl. Am. Sept. 656.   1814.

Leaves 4'–2½° long, the common stalk slender, nearly all above ground, comprising one-half to two-thirds the length of the plant; bud pilose, both the sporophyl and sterile blade wholly bent down. Sterile blade nearly or quite sessile, spreading, membranous, deltoid, 2'–16' broad, nearly as long, ternate, the short-stalked primary divisions 1–2-pinnate, the numerous segments 1–2-pinnatifid, the ultimate segments oblong, toothed at the apex; sporophyl long-stalked, 2–3-pinnate.

In rich woods, Labrador to British Columbia, Washington, Arizona, and the Gulf states.   Mexico, Europe and Asia.   June–July.   Rattlesnake-fern, Hemlock-leaved-moonwort.

Family 2. **OSMUNDÀCEAE** R. Br. Prodr. Fl. Nov. Holl. **1**: 161. 1810.

ROYAL FERN FAMILY.

Large ferns with creeping or suberect rootstocks. Stipes winged at the base, the blades 1–2-pinnate or tripinnatifid, with free mostly forked veins extending to the margins. Sporanges naked, large, globose, mostly stalked, borne on modified contracted pinnae and nearly covering them or (in *Todea* and *Leptopteris,* Old World genera) in clusters (sori) on the lower surface of the pinnules or segments, opening in 2 valves by a longitudinal slit; ring wanting or mere traces of one near the apex.

Three living genera, *Osmunda* and the two mentioned.

1. **OSMÚNDA** [Tourn.] L. Sp. Pl. 1063. 1753.

Tall swamp or lowland ferns, the leaves in large crowns, long-stalked, the blades bipinnatifid or bipinnate, with regularly forked prominent veins, the fertile portions much contracted and devoid of chlorophyl, the short-stalked sporanges thin, reticulated, opening in halves, a few parallel thickened cells near the apex representing the rudimentary transverse ring. Spores copious, greenish. [From *Osmunder,* a Saxon name for the god Thor.]

Eight species, the following in North America. Type species: *Osmunda regalis* L.

Blades bipinnate, some of them fertile at the apex.   1. *O. regalis.*
Herbaceous blades bipinnatifid.

Pinnae of sterile blade with a tuft of tomentum at the base; blades normally dimorphous.
                                                                           2. *O. cinnamomea.*
Pinnae of sterile blade lacking a tuft of tomentum at the base; blades normally fertile only
  in the middle.   3. *O. Claytoniana.*

1. **Osmunda regàlis** L. Royal Fern. Fig. 15.

Osmunda regalis L. Sp. Pl. 1065. 1753.

Rootstock stout, bearing a cluster of several long-stalked leaves, 2°–6° high, the apical pinnae fertile, contracted, forming an upright terminal panicle, the pinnules linear-cylindric, greenish before maturity, dark brown and withering with age. Sterile pinnae 6′–12′ long, 2′–4′ wide, the pinnules oblong-ovate or lanceolate-oblong, sessile or slightly stalked, glabrous, finely serrulate, especially near the apex and occasionally crenate toward the truncate, oblique, or even cordate, base.

In low woods, swamps and marshes, Newfoundland to Florida, west to Mississippi, Nebraska and Saskatchewan. Also in Tropical America, Europe and Africa. May–July. Called also Royal Osmond. Bracken, Buckhorn-brake. King's-, flowering-, water-, tree-, snake- or ditch-fern. Bog-onion, Herb Christopher, Hartshorn-bush.

2. **Osmunda cinnamòmea** L. Cinnamon-fern. Fig. 16.

Osmunda cinnamomea L. Sp. Pl. 1066. 1753.

Rootstock very large, widely creeping, bearing a circular cluster of sterile leaves with one or more fertile ones within. Stipes 1° or more long, clothed with ferruginous tomentum when young, glabrous with age. Sterile blades 1°–5° long, oblong-lanceolate, deeply bipinnatifid, the pinnae linear-lanceolate, deeply pinnatifid into oblong obtuse segments, the margins usually entire. Fertile blade contracted, bipinnate, soon withering; sporanges cinnamon-colored.

In wet woods, swamps and low grounds, Newfoundland to Minnesota, the Gulf states and New Mexico. Also in Mexico, Brazil, the West Indies and eastern Asia. Forms occur with leaves variously intermediate between the fertile and sterile. May–June. Bread-root. Fiddle-heads. Swamp-brake.

3. Osmunda Claytoniàna L.
Clayton's Fern.   Fig. 17.

Osmunda Claytoniana L. Sp. Pl. 1066.   1753.
Osmunda interrupta Michx. Fl. Bor. Am. **2**: 273.
1803.

Rootstock stout, creeping; leaves 2°–6° long,
loosely tomentose when young, glabrous with age,
the outer ones usually sterile and spreading, the
inner erect and usually fertile in the middle.
Blades oblong-lanceolate, 1°–4° long; sterile pinnae
oblong-lanceolate, deeply cleft into ovate-oblong
close or slightly imbricate segments, the margins
usually entire; fertile pinnae 2–5 pairs, fully pinnate, the cylindric divisions very close, greenish
at first, dark brown, brittle and withering with
age.

In swamps and moist woods, Newfoundland to
Minnesota south to North Carolina, Kentucky and
Missouri. Ascends to 5000 ft. in Virginia. Also in
China and India. May–July. Interrupted- or Clay-
ton's-flowering-fern.

Family 3.   HYMENOPHYLLÀCEAE Gaud. in Freyc. Voy. 262.   1826.
FILMY-FERN FAMILY.

Membranaceous, mostly tropical small ferns, with slender often filiform creeping or rarely suberect rootstocks, the leaves usually much divided, the leaf-tissue
pellucid, usually of a single layer of cells. Sporanges sessile upon a filiform,
usually elongate receptacle, within an urn-shaped or tubular truncate or two-
lipped marginal indusium, terminal upon the veins; ring complete, transverse,
opening vertically.

Two genera, *Hymenophyllum* and the following, comprising some 450 or more species, abundant in the humid tropics and mainly epiphytic.

---

1. TRICHÓMANES L. Sp. Pl. 1097.   1753.

Blades entire, pinnatifid or lobed, or several times pinnately divided. Indusium tubular
or funnel-shaped, truncate or sometimes broadly two-lipped, the sporanges sessile, mostly
upon the lower portion of the slender often exserted receptacle. [Greek, in allusion to the
delicate hair-like ultimate segments of some of the species.]

About 210 species, mostly tropical. Besides the following, 3 species occur in the southern
United States. Type species: *Trichomanes crispum* L.

1. Trichomanes Boschiànum Sturm.
Filmy-fern. Bristle-fern. Fig. 18.

Trichomanes Boschianum Sturm; v. d. Bosch,
Ned. Kr. Arch. **5**²: 160. 1861.
Trichomanes radicans of American writers. Not
Sw.

Rootstocks filiform, wiry, tomentose, creeping. Stipes (petioles) ascending, 1′–3′ long,
naked or nearly so; blades 2′–8′ long, 8″–1½′
wide, membranaceous, lanceolate or ovate-
lanceolate, 2–3-pinnatifid; pinnae ovate, obtuse,
the upper side of the cuneate base parallel with
or appressed to the narrowly winged rachis;
segments toothed or cut into linear divisions;
indusia terminal on short lobes, 1–4 on a pinnule, the mouth slightly 2-lipped; receptacle
more or less exserted, bristle-like, bearing the
sessile sporanges mostly near the base.

On wet rocks, Kentucky to Florida and Alabama.

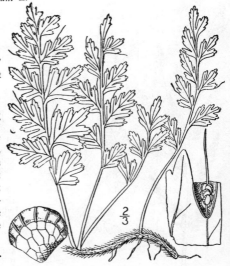

Family 4. **SCHIZAEÀCEAE** Reichenb. Consp. 39. 1828.

Climbing Fern Family.

Plants with erect, simple, pinnate or dichotomous, or vine-like, twining, elongate leaves, with stalked, alternate, paired and mostly palmately lobed or pinnate leafy divisions. Sporanges borne in double rows on narrow specialized lobes or segments, obovoid, pyriform or globose, sessile, provided with a transverse apical ring and opening vertically by a longitudinal slit.

Genera 4 or more; species about 125, mainly tropical.

Leaves short, tufted, rigid.   1. *Schizaea.*
Leaves elongate, climbing.   2. *Lygodium.*

1. **SCHIZAÈA** J. E. Smith, Mem. Acad. Turin **5**: 419. *pl. 19. f. 9.* 1793.

Mostly small plants, with erect or recurved slender filiform simple or dichotomously divided or cleft leaves. Sporanges in 2 rows along the close slender segments of small pinnate terminal spikes and partially protected by the narrowly reflexed indusiiform margin. [Greek, in allusion to the divided or deeply cleft leaf-blades of some species.]

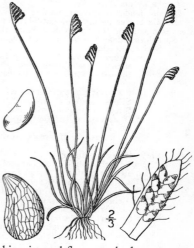

A genus of about 25 species, of wide geographic distribution, mostly in tropical regions. Type species: *Schizaea dichotoma* (L.) J. E. Smith.

1. **Schizaea pusílla** Pursh. Curly-grass. Fig. 19.

Schizaea pusilla Pursh, Fl. Am. Sept. 657. 1814.

Rootstock minute, horizontally creeping, the leaves tufted. Sterile leaves linear, very slender, flattened and tortuous. Fertile léaves longer, 3′–5′ high, the fertile portion terminal, consisting of about 5 pairs of crowded pinnate divisions, forming a distichous spike; sporanges ovoid or pyriform, sessile in two rows along the single vein of the narrow incurved linear divisions of the fertile spike, partially concealed by the incurved hairy margins.

In wet soil, pine barrens of central and eastern New Jersey, the historic region. Also in Newfoundland and Nova Scotia. Rare and local. Aug.–Sept.

2. **LYGÒDIUM** Sw. Schrad. Journ. Bot. 1800[2]: 106. 1801.

Twining vine-like ferns. Leaves elongate, the rachis wiry and flexuous; leafy parts consisting of the stalked palmately lobed or pinnate (or compound) secondary pinnae, borne in pairs upon short stalks arising alternately from the rachis. Sporanges borne on contracted divisions of the leaf, as short or elongate spikes, the lower surface bearing a double row of imbricate hood-like indusia fixed by their broad bases and concealing each 1 (rarely 2) sporanges. [Name Greek, in allusion to the flexible rachis.]

About 26 species, mostly of tropical distribution. Type species: *Lygodium scandens* (L.) Sw.

1. **Lygodium palmàtum** (Bernh.) Sw. Climbing-fern. Hartford-fern. Fig. 20.

Gisopteris palmata Bernh. Schrad. Journ. Bot. 1800[2]: 129. 1801.
Lygodium palmatum Sw. Syn. Fil. 154. 1806.

Rootstock slender, creeping. Stipes slender, flexible and twining; leaves 1°–3° long, their short alternate branches 2-forked, each fork bearing a nearly orbicular 4–7-lobed pinnule more or less cordate at the base with a narrow sinus; surfaces naked; fertile pinnules contracted, several times forked, forming a terminal panicle; sporanges solitary, borne on alternate veins springing from the flexuous midvein of the segments, each covered by a scale-like indusium.

In moist thickets and open woods, New Hampshire to Pennsylvania, south to Florida and Tennessee. Ascends to 2100 ft. in eastern Pennsylvania. Summer. Called also Creeping or Windsor-fern.

Family 5. **POLYPODIACEAE** R. Br. Prodr. Fl. Nov. Holl. 1: 145. 1810.

FERN FAMILY.

Leafy plants of various habit, the rootstocks horizontal and often elongate, or shorter and erect, the leaf-blades simple, once or several times pinnate or pinnatifid, or decompound, coiled in vernation. Sporanges borne on the under surface of the foliaceous leaf-blades, or upon slender or contracted, partially foliose or non-foliose leaves or parts of leaves, or, as in most of our species, in clusters (sori) upon the backs of the leaf-blades; distinctly stalked, provided with an incomplete vertical ring of thickened cells (the annulus), and opening transversely. Sori either with or without a membranous covering (indusium). Prothallia green.

About 145 genera and 4500 or more species of very wide geographic distribution. This family includes by far the greater number of living ferns.

Leaves strongly dimorphous, the fertile ones with divisions greatly contracted, brownish, berry-
    like or necklace-like.
      Sterile blades deeply pinnatifid; veins freely anastomosing.   1. *Onoclea.*
      Sterile blades deeply 2-pinnatifid; veins free.   2. *Matteuccia.*
Leaves mostly uniform; if dimorphous, the fertile blades flat, the divisions green, not as above.
    Sori dorsal upon the veins, not marginal.
      Sori roundish.
        Indusium wholly or partially inferior.
          Indusium wholly inferior, the divisions stellate or spreading.   3. *Woodsia.*
          Indusium attached by its base at one side of the sorus, hood-shaped, withering.
                                                      5. *Filix.*
        Indusium, if present, superior.
          Stipes jointed to the rootstock; indusia wanting.   20. *Polypodium.*
          Stipes continuous with the rootstock (not jointed); indusia present in most species.
            Indusium (present, in our species) orbicular-peltate, centrally attached.
                                            6. *Polystichum.*
            Indusium, if present, orbicular-reniform, attached at its sinus.
                                            7. *Dryopteris.*
      Sori oblong to linear.
        Sori in chain-like rows parallel to the midrib and rachises.
          Leaves uniform; veins free between the sori and margin.   8. *Anchistea.*
          Leaves dimorphous; veins of sterile blade freely anastomosing.   9. *Lorinseria.*
        Sori oblique to the midribs or irregularly disposed.
         Veins free; sori all oblique to the midribs.
          Sori confluent in pairs; indusia single, contiguous, appearing double. 10. *Phyllitis.*
          Sori single on the outer side of veinlet, or crossing it and recurved.
            Sori straight or slightly curved; leaves mostly evergreen. 12. *Asplenium.*
            Sori usually curved, often crossing the veinlet and recurved; leaves herbaceous.
                                        13. *Athyrium.*
         Veins freely anastomosing; sori variously disposed.   11. *Camptosorus.*
    Sori borne at or very near the margin.
      Sporanges borne within a special cup-shaped indusium.   4. *Dennstaedtia.*
      Sporanges not borne within a special cup-shaped indusium.
        Sori without indusia, somewhat protected by the revolute leaf-margin.
                                      19. *Notholaena.*
        Sori with indusia formed entirely or in part by the revolute or reflexed more or less
          modified leaf-margins.
          Sori distinct, borne on the under side of the reflexed lobes. 14. *Adiantum.*
          Sori wholly or partially confluent.
            Sori borne on a vein-like receptacle connecting the ends of the free veinlets;
            indusium double.   15. *Pteridium.*
           Sori borne at or near the ends of the free veinlets; indusia single.
            Leaves dimorphous.   16. *Cryptogramma.*
            Leaves uniform or nearly so.
              Sori confluent, forming a wide submarginal band; segments smooth or
              nearly so.   17. *Pellaea.*
             Sori distinct or contiguous; segments usually pubescent, tomentose or scaly.
                                        18. *Cheilanthes.*

1. ONOCLEA L Sp. Pl. 1062. 1753.

Coarse lowland ferns with leaves of two very dissimilar sorts borne separately upon a creeping rootstock, the sterile ones foliaceous and suberect, withering with frosts, the fertile ones rigidly erect, with pinnules greatly contracted into separate hard rounded berry-like divisions, these (until maturity) completely concealing the included sori, finally dehiscent and persistent throughout the winter. Sori roundish, on elevated receptacles, partially covered by delicate hood-shaped indusia fixed at the base of the receptacles. [Name ancient, not originally applied to this plant.]

A single species, *O. sensibilis* L.

1. Onoclea sensíbilis L. Sensitive Fern.
Fig. 21.

Onoclea sensibilis L. Sp. Pl. 1062. 1753.

Rootstock rather slender, copiously rooting. Fertile leaves 1°–2½° high, persistent over winter, the fertile portion bipinnate, much contracted, the short pinnules rolled up into closed berry-like bodies and forming a narrow close panicle. Sterile leaves 1°–4½° high, the blades broadly triangular, deeply pinnatifid, the rachis winged; pinnae lanceolate-oblong, entire, undulate, or the lower and sometimes the middle ones sinuate-pinnatifid; veins freely anastomosing, forming a somewhat regular series of narrow elongate areoles next the midvein and numerous smaller areoles between this series and the margin.

In moist soil, Newfoundland to Saskatchewan, south to Oklahoma and the Gulf states. Ascends to 3000 ft. in Virginia. Various intermediate forms between the sterile and fertile leaves occur. Sensitive to early frosts. Aug.–Nov.

2. MATTEÙCCIA Todaro, Giorn. Sci. Nat. Palermo 1: 235. 1866.

[STRUTHIOPTERIS Willd. 1809, not Weiss, 1770.]

Coarse lowland ferns with dissimilar leaves in a close crown upon a stout ascending rootstock. Sterile leaves tall, in a complete circle, the shorter fertile leaves appearing late in the season, borne within, rigidly erect, the pinnae closely contracted into necklace-like or pod-like divisions, these concealing the sori, finally dehiscent. Sori roundish, on elevated cylindrical receptacles, partly covered by delicate fugacious lacerate indusia attached below. [Named in honor of Carlo Matteucci, an Italian professor of physics.]

Species 3, the following, which is the generic type, and 2 Asiatic species.

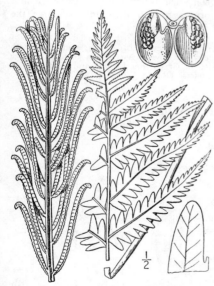

1. Matteuccia Struthiópteris (L.) Todaro.
Ostrich-fern. Fig. 22.

Osmunda Struthiopteris L. Sp. Pl. 1066. 1753.
Onoclea Struthiopteris Hoffm. Deutsch. Fl. 2: 11. 1795.
Struthiopteris germanica Willd. Enum. 1071. 1809.
Matteuccia Struthiopteris Todaro, Giorn. Sci. Nat. Palermo 1: 235. 1866.

Rootstock stout, ascending, with slender underground stolons. Fertile leaves 1°–1½° high, the pinnae dark brown, slightly crenate, contracted, with closely and widely revolute margins, the included sori crowded and confluent. Sterile leaves 2°–7° high, 6'–15' broad, broadly oblanceolate or spatulate, abruptly short-acuminate, gradually narrowed below the middle, the lower pinnae greatly reduced; pinnae narrow, deeply pinnatifid, glabrous, the segments oblong, obtuse, entire.

In moist thickets, especially along streams, Nova Scotia to Virginia, west to British Columbia and Iowa. Ascends to 2000 ft. in Vermont. Also in Europe and Asia. July–Oct.

3. WOÓDSIA R. Br. Prodr. Fl. Nov. Holl. 1: 158. 1810.

Small or medium-sized ferns, growing in rocky places, the rootstocks in dense tufts. Leaves numerous, the stipes often jointed above the base and separable, the blades 1–2-pinnate or deeply 3-pinnatifid. Sori roundish, borne on the simply-forked free veins. Indusia slight and often evanescent, inferior in attachment, either roundish and soon cleft into irregularly jagged lobes, or deeply stellate, the filiform divisions concealed beneath the sporanges or inflexed and partially covering them. [Named in honor of Joseph Woods, 1776–1864, an English architect and botanist.]

About 25 species, mainly of temperate or cold regions. Besides the following, another occurs in the southwestern United States. Type species: *Polypodium ilvense* L.

Indusium small or inconspicuous, the divisions narrow or filiform.
　Stipes jointed near the base; filiform divisions of the indusium more or less inflexed over the sporanges.
　　Blades with more or less rusty chaff underneath.　　　　　　1. *W. ilvensis.*
　　Blades glabrous or nearly so.
　　　Blades oblong-lanceolate; divisions of the indusium numerous.　2. *W. alpina.*
　　　Blades linear or linear-lanceolate; divisions of the indusium few.　3. *W. glabella.*
　Stipes not jointed; divisions of the indusium spreading, mostly concealed beneath the sporanges.
　　Puberulent, usually hispidulous; indusium deeply cleft into narrow flaccid segments.
　　　　　　　　　　　　　　　　　　　　　　　　　　4. *W. scopulina.*
　　Glabrous; indusium divided to the center into a few short whitish turgid beaded hair-like segments.　　　　　　　　　　　　　　　　　5. *W. oregana.*
Indusium ample; the divisions broad, early spreading.　　　　　6. *W. obtusa.*

1. **Woodsia ilvénsis** (L.) R. Br.　Rusty Woodsia.　Fig. 23.

Acrostichum ilvense L. Sp. Pl. 1071. 1753.

Woodsia ilvensis R. Br. Prodr. Fl. Nov. Holl. 1 : 158.　1810.

Rootstocks short, ascending, growing in masses, the leaves closely caespitose. Stipes short, stoutish, jointed near the base, rusty chaffy with narrow filiform scales; blades lanceolate, 4′–10′ long, pinnate, nearly glabrous above, more or less covered with rusty chaff beneath; pinnae crowded, sessile, pinnately parted, the crowded segments oblong, crenate; sori borne near the margins of the segments, somewhat confluent with age; indusium minute, concealed beneath the sorus, cleft into numerous filiform segments, these inflexed over the sporanges and inconspicuous.

On exposed rocks, Labrador to Alaska, south to North Carolina, Kentucky and Iowa. Ascends to 5000 ft. in New Hampshire. Also in Greenland, Europe and Asia. June–Aug. Ray's Woodsia, Oblong Woodsia.

2. **Woodsia alpìna** (Bolton) S. F. Gray.　Alpine Woodsia.　Fig. 24.

Acrostichum alpinum Bolton, Fil. Brit. 76.　1790.
Acrostichum hyperboreum Liljeb. Kgl. Vetensk. Akad. Nya Handl. **14**: 201.　1793.
Woodsia hyperborea R. Br. Prodr. Fl. Nov. Holl. **1**: 158.　1810.
W. alpina S. F. Gray, Nat. Arr. Brit. Pl. **2**: 17.　1821.

Rootstocks short, ascending, the leaves densely caespitose. Stipes slender, chestnut-colored, shining, somewhat chaffy below, jointed near the base; blades narrowly oblong-lanceolate, 2′–6′ long, 8″–12″ wide, scarcely narrower below the middle, deeply bipinnatifid; pinnae somewhat apart, cordate-ovate or triangular-ovate, pinnately 5–7-lobed, glabrous or very nearly so on both surfaces; sori near the margin, usually distinct; indusium as in the preceding species.

On moist rocks, Labrador to Alaska, Maine, northern New York and western Ontario. Also in Greenland. Ascends to 4200 ft. in Vermont. July–Aug. Called also Northern Woodsia, Flower-cup-fern.

3. Woodsia glabélla R. Br. Smooth Woodsia. Fig. 25.

Woodsia glabella R. Br. App. Franklin's Journ. 754. 1823.

Rootstocks small, ascending, densely clustered. Stipes very slender, usually stramineous, jointed above the base; blades delicate, linear or narrowly lanceolate, 2'–5' long, 4"–8" wide, once pinnate; pinnae deltoid to roundish-ovate, crenately lobed, glabrous, the lower pinnae remote, obtuse, often somewhat smaller than the middle ones; sori few, distinct or with age confluent; indusium minute, with 6–10 hair-like incurved or radiating segments.

On moist rocks, Labrador to Alaska, south to New Brunswick, northern New England, northern New York and British Columbia. Also in Greenland and arctic and alpine Europe and Asia. Summer.

4. Woodsia scopulìna D. C. Eaton. Rocky Mountain Woodsia. Fig. 26.

Woodsia scopulina D. C. Eaton, Can. Nat. **2**: 90. 1865.
Woodsia Cathcartiana Robinson, Rhodora **10**: 30. 1908.

Rootstock short, creeping, densely chaffy, the numerous leaves borne close together. Stipes 2'–6' long, not jointed, bright rusty or chestnut-colored at the base, paler above; blades lanceolate, 6'–12' long, finely glandular-puberulent and usually hispidulous with jointed whitish hairs; pinnae numerous, oblong-ovate, deeply pinnatifid into 10–16 oblong toothed segments, or fully pinnate, the larger pinnules nearly free and deeply incised; indusium concealed, cleft into narrow or slender spreading flaccid segments.

In crevices of rocks, Michigan and western Ontario to British Columbia, south in the Rocky Mountains to Arizona and in the Sierra Nevada to California. Also in Gaspé County, Quebec. Summer.

5. Woodsia oregàna D. C. Eaton. Oregon Woodsia. Fig. 27.

Woodsia oregana D. C. Eaton, Can. Nat. II. **2**: 90. 1865.

Rootstock short, creeping, chaffy, the numerous leaves very densely clustered. Stipes not jointed, brownish and chaffy below, paler or stramineous above, glabrous; blades 2'–10' long, elliptic-lanceolate, deeply bipinnatifid or partially bipinnate, the sterile shorter than the fertile; pinnae glabrous, deltoid-oblong, obtuse, deeply pinnatifid, the lower smaller and remote; segments oblong or ovate, obtuse, adnate or the largest nearly free, dentate or crenate, the teeth often revolute and covering the submarginal sori; indusia minute, concealed, consisting of a few short whitish turgid hair-like segments.

British Columbia and Athabasca to Manitoba, Wisconsin, northern Michigan, Nebraska, Oklahoma, Colorado, Arizona and California. Also in eastern Quebec. July–Aug.

6. Woodsia obtùsa (Spreng.) Torr. Blunt-
lobed Woodsia. Fig. 28.

Polypodium obtusum Spreng. Anleit. 3: 92. 1904.
Woodsia obtusa Torr. Cat. Pl. in Geol. Rep. N. Y. 195.
1840.

Rootstock short, creeping, with relatively few
leaves. Stipes not jointed, straw-colored, chaffy,
3'-6' long; blades broadly lanceolate, 6'-15' long,
minutely glandular-puberulent, nearly or quite 2-pin-
nate; pinnae rather remote, triangular-ovate or
oblong, pinnately parted into oblong obtuse crenate-
dentate segments, or usually pinnate, the lower pinnules free and parted nearly to the midveins; sori
nearer the margin than the midveins; indusia conspicuous, at first enclosing the sporanges, at length
splitting into several broad jagged spreading lobes.

On rocks, Nova Scotia and Maine to Wisconsin and
south to Georgia, Alabama, and Texas. Also in Alaska
and British Columbia. Variable. Ascends to 2200 ft.
in Virginia.

4. DENNSTAÈDTIA Bernh. Schrad. Journ. Bot. 1800²: 124.   1801.

[DICKSONIA in part of some authors, not L'Her. 1788.]

Mostly medium-sized ferns, with slender wide-creeping hairy rootstocks and scattered
2-3-pinnate erect leaves, 2°-6° high. Sori marginal, terminal upon the free veinlets, the
sporanges clustered upon a very small receptacle within a special cup-shaped indusium formed
in part of the more or less modified reflexed segment of the leaf-margin. [Name in honor
of August Wilhelm Dennstaedt.]

About 50 species mainly of tropical and subtropical regions. Type species: *D. flaccida*
(Forst.) Bernh.

1. Dennstaedtia punctilóbula (Michx.) Moore.   Hay-scented Fern.   Fig. 29.

Nephrodium punctilobulum Michx. Fl. Bor. Am. 2: 268.
1803.
Dicksonia pilosiuscula Willd. Enum. 1076. 1809.
Dicksonia punctilobula A. Gray, Man. 628. 1848.
Dennstaedtia punctilobula Moore, Ind. Fil. xcvii. 1857.

Rootstock slender, extensively creeping, not chaffy.
Stipes stout, chaffless, usually castaneous at the base;
blades 1°-3° long, 5'-9' wide, ovate-lanceolate to
deltoid-lanceolate, acute or acuminate, frequently
long-attenuate, usually 3-pinnatifid, thin and delicate,
the rachis and under surface minutely glandular and
pubescent; pinnae numerous, lanceolate, the segments ovate to oblong, close and deeply lobed, the
margins with oblique rounded teeth; sori minute,
each on a recurved tooth, usually one at the upper
margin of each lobe; sporanges few, borne within
the delicate cup-shaped indusium.

In various situations, most abundant on open hill-
sides, Nova Scotia and New Brunswick to Ontario and
Minnesota, south to Georgia, Alabama and Missouri.
Ascends to 5600 ft. in Virginia. Aug. Called also
Fine-haired-fern, Hairy dicksonia, Boulder-fern.

5. FÌLIX Adans. Fam. Pl. 2: 20, 558.   1763.

[CYSTÓPTERIS Bernh. Schrad. Neues Journ. Bot. 1²: 26. 1806.]

Delicate rock ferns with slender stipes, 2-4-pinnate blades, and roundish sori borne on
the backs of the veins. Indusium membranous, hood-like, attached by a broad base on its
inner side and partly under the sorus, early thrust back by the expanding sporanges and at
least partly concealed by them, withering, the sori thus appearing naked with age. Veins free.

About 10 species mainly natives of temperate regions; the following in North America. Type
species: *Polypodium bulbiferum* L.

Blades lanceolate, broadly lanceolate, or narrowly deltoid-lanceolate, 2-3-pinnate.
  Blades broadest at base, long-tapering, bearing bulblets beneath.   1. *F. bulbifera.*
  Blades scarcely broader at base, short-pointed; no bulblets.   2. *F. fragilis.*
Blades deltoid-ovate, 3-4-pinnate.   3. *F. montana.*

1. Filix bulbífera (L.) Underw. Bulblet
Cystopteris. Fig. 30.

Polypodium bulbiferum L. Sp. Pl. 1091. 1753.
Cystopteris bulbifera Bernh. Schrad. Neues Journ. Bot.
1²: 26. 1806.
Filix bulbifera Underw. Nat. Ferns, ed. 6, 119. 1900.

Rootstock short, somewhat chaffy at the apex.
Stipes clustered, 4′–6′ long, light-colored; blades
1°–2½° long, usually 3-pinnatifid, deltoid-lanceolate,
the gradually tapering narrow apex sometimes greatly
elongate; pinnae numerous, oblong-ovate to lanceolate-oblong, horizontal, pinnate; pinnules close or
somewhat apart, unequally oblong-ovate, obtuse, at
least the largest deeply pinnatifid and free, the others
more or less adnate and variously incised; rachis and
pinnae underneath bearing large fleshy bulblets,
these falling and giving rise to new plants; indusia
short, convex, truncate.

On wet rocks and in ravines, especially on limestone,
Newfoundland to Manitoba, Wisconsin and Iowa, south
to northern Georgia, Alabama and Arkansas. Ascends
to 3500 ft. in Virginia. July–Aug.

2. Filix frágilis (L.) Underw. Brittle
Fern. Fig. 31.

Polypodium fragile L. Sp. Pl. 1091. 1753.
Cystopteris fragilis Bernh. Schrad. Neues Journ.
Bot. 1²: 27. 1806.
Filix fragilis Underw. Nat. Ferns, ed. 6, 119. 1900.

Rootstock extensively creeping, chaffy, especially at the apex. Stipes 4′–10′ long, slender,
brittle; blades thin, broadly lanceolate, slightly
tapering below, 4′–10′ long, 2–3-pinnatifid or
pinnate; pinnae deltoid-lanceolate to deltoid-
ovate, acute, deeply pinnatifid or pinnate, the
segments ovate or oblong-ovate, pinnatifid or
incised, acutish, mostly decurrent upon the
usually winged rachis; indusia roundish or
nearly ovate, deeply convex, delicate.

On rocks and in moist grassy woods, Newfoundland and Labrador to Alaska, south to
Georgia, Alabama, Kansas, Arizona, and southern
California. Also in Greenland. Almost cosmopolitan in distribution and very variable. Ascends to 5000 ft. in New Hampshire. May–July.
Called also Bottle-, Brittle-, or Bladder-fern.

3. Filix montàna (Lam.) Underw.
Mountain Cystopteris. Fig. 32.

Polypodium montanum Lam. Fl. Franc. 1: 23.
1778.
Cystopteris montana Bernh.; Desv. Mém. Soc.
Linn. Paris 6: 264. 1827.
Filix montana Underw. Nat. Ferns, ed. 6, 119.
1900.

Rootstock slender, widely creeping, the
leaves few and distant. Stipes 6′–9′ long,
slender; blades broadly deltoid-ovate, 3–4-pin-
nate, about 4′–6′ long and broad, the basal
pinnae much the largest, unequally deltoid-
ovate, their inferior pinnules 1′–2′ long; pinnules deeply divided into oblong or ovate-
oblong lobes, these deeply toothed or again
pinnate; sori numerous; indusia ovate, deeply
convex, delicate, very early thrust back and
concealed or evanescent.

On rocks, Labrador and Quebec to British
Columbia and Alaska, south to the northern shore
of Lake Superior. Also in Colorado, and in
northern Europe and Asia. Aug. Called also
Wilson's-, Mountain-, or Bladder-fern.

6. POLÝSTICHUM Roth, Romer's Arch. Bot. 2¹: 106. 1799.

Coarse and usually rigid erect ferns of harsh texture, with pinnatifid to quadripinnatifid leaves borne typically in a crown upon a suberect or decumbent rootstock, the stipe not jointed to it. Sterile and fertile leaves similar, the vascular parts usually chaffy; divisions of the blade mainly auriculate and spinulose or mucronate, with free veins. Sori round; indusium superior, orbicular, attached at its middle. [Greek, signifying many rows, in allusion to the numerous regular rows of sori in *P. Lonchitis* (L.) Roth, the typical species.]

About 100 species, of wide distribution, mainly in temperate regions.

Leaves simply pinnate.
     Lower pinnae gradually much reduced; upper (soriferous) pinnae conform.
                                        1. *P. Lonchitis.*
     Lower pinnae scarcely reduced; upper (soriferous) pinnae of fertile fronds contracted.
                                        2. *P. acrostichoides.*
Leaves bipinnatifid or bipinnate.
     Leaves coriaceous, the pinnae deeply lobed at their base.   3. *P. scopulinum.*
     Leaves herbaceous, fully bipinnate.   4. *P. Braunii.*

1. Polystichum Lonchìtis (L.) Roth. Holly-fern. Fig. 33.

Polypodium Lonchitis L. Sp. Pl. 1088. 1753.
Aspidium Lonchitis Sw. Schrad. Journ. Bot. 1800²: 30. 1801.
Polystichum Lonchitis Roth, Röm. Arch. Bot. 2¹: 106. 1799.
Dryopteris Lonchitis Kuntze, Rev. Gen. Pl. 813. 1891.

Rootstock short, stout, densely chaffy. Stipes 1′–5′ long, bearing large ferruginous scales with smaller ones intermixed; blades rigid, coriaceous, evergreen, 6′–2° long, linear-lanceolate, once pinnate; pinnae numerous, close, broadly lanceolate-falcate, 1′–1½′ long, acute, strongly auricled on the upper side at the base, obliquely truncate below, notably spinulose-dentate, the lowest commonly triangular and shorter; sori large, borne usually in two rows, nearly equidistant between the margin and midrib, subconfluent with age; indusium entire.

On rocks, Labrador to Alaska, south to Nova Scotia, Ontario, Wisconsin, Montana and Washington, and in the mountains to Utah, Colorado and California. Also in Greenland, Europe and Asia. Called also Rough alpine fern. Aug.

2. Polystichum acrostichoìdes (Michx.) Schott.   Christmas-fern.   Fig. 34.

Nephrodium acrostichoides Michx. Fl. Bor. Am. 2: 267. 1803.
Aspidium acrostichoides Sw. Syn. Fil. 44. 1806.
Polystichum acrostichoides Schott, Gen. Fil. 1834.
Dryopteris acrostichoides Kuntze, Rev. Gen. Pl. 812. 1891.

Rootstock stout, creeping. Stipes 5′–7′ long, densely chaffy; blades lanceolate, 1°–2° long, 3′–5′ wide, rigid, evergreen, subcoriaceous, once pinnate; pinnae 1′–3′ long, narrowly oblong-lanceolate, somewhat falcate, acutish at the apex, half halberd-shaped at the base, with appressed, bristly teeth, the lower pinnae scarcely smaller, sometimes deflexed; fertile fronds contracted at the apex, the reduced pinnae soriferous, their under surface nearly covered with large contiguous sori in 2–4 rows, confluent with age; indusium entire, persistent.

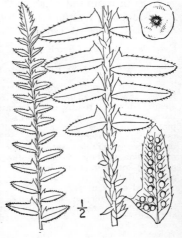

In woods and on hillsides, most abundant in rocky places, Nova Scotia to Ontario and Wisconsin, south to Texas and the Gulf states. Ascends to 2700 ft. in Maryland. July–Aug. Called also Christmas shield-fern.
Forms with cut-lobed or incised pinnae are known as var. *Schweinitzii;* occasional forms are 2-pinnatifid.

3. Polystichum scopulìnum (D. C. Eaton) Maxon. Eaton's Shield-fern. Fig. 35.

Aspidium aculeatum var. *scopulinum* D. C. **Eaton**, Ferns N. Am. **2**: 125. *pl. 62, f. 8.* 1880.

P. scopulinum Maxon, Fern Bull. **8**: 29. 1900.

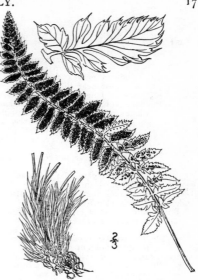

Rootstock stout, ascending, with numerous cord-like roots. Leaves 9'–17' long, the stipe 2'–5' long, densely chaffy at the base with both broad and narrow bright brown scales; blades 6'–12' long, linear to narrowly oblong-lanceolate, 1½'–2½' broad, coriaceous, the chaff largely deciduous from the rachis; pinnae numerous, 7"–15" long, 4"–8" broad at the base, ovate, obtuse, the basal portion pinnately lobed, the apical half serrate with pointed or aculeate teeth, the lower pinnae usually much reduced; sori near the midvein; indusium large, somewhat lobed, glabrous.

On rocky slopes, Washington to Idaho, Utah and Southern California. Gaspé county, Quebec.

4. Polystichum Braùnii (Spenner) Fée. Braun's Holly-fern. Prickly Shield-fern. Fig. 36.

Aspidium Braunii Spenner, Fl. Frib. **1**: 9. 1825.
A. aculeatum var. *Braunii* Doell, Rhein. Fl. 21. 1843.
Polystichum Braunii Fée, Gen. Fil. 278. 1850–52.
Dryopteris aculeata var. *Braunii* Underw. Native Ferns, ed. 4, 112. 1893.
Dryopteris Braunii (Spenner) Underw. in Br. & Brown, Ill. Fl. ed. 1, **1**: 15. 1896.

Rootstock stout, suberect. Stipes 4'–5' long, chaffy with both broad and narrow brown scales; blades lanceolate, 1°–2° long, herbaceous, 2-pinnate, the rachis chaffy; pinnae numerous, close, oblong-lanceolate, slightly broadest at the base, the middle ones 2½'–4' long, the lower gradually shorter; pinnules ovate to oblong, truncate and nearly rectangular at the base, mostly acute, sharply toothed, beset with long soft hair-like scales; sori small, mostly nearer the midvein than the margin; indusium small, entire.

In rocky woods, Nova Scotia to Alaska, to northern New England, the mountains of Pennsylvania, to Michigan and British Columbia. Ascends to 5000 ft. in Vermont. Aug.

7. DRYÓPTERIS Adans. Fam. Pl. 2: 20, 550. 1763.

[ASPIDIUM Sw. Schrad. Journ. Bot. **1800**[2]: 29, in part. 1801.]

Mainly woodland ferns, commonly of upright habit, the fertile and sterile leaves usually similar, not jointed to the rootstock. Blades 1–3-pinnate or dissected, with veins free in northern species, uniting occasionally or even freely in some of the southern. Sori round or rarely elliptical in outline, borne upon the veins, indusiate or non-indusiate, the indusium (if present) in northern species orbicular-reniform, fixed at its sinus; sporanges numerous.

A genus of several hundred species, widely distributed in the tropics, its limits variously understood. Besides the following, some 13 species occur in the southern and western United States. Type species: *Polypodium Filix-mas* L.

Indusia present (§ *Eudryópteris*).
  Texture membranous; veins simple or once forked.
    Lower pinnae gradually and conspicuously reduced.   1. *D. noveboracensis.*
    Lower pinnae scarcely reduced.
      Veins once or twice forked.   2. *D. Thelypteris.*
      Veins simple.   3. *D. simulata.*
  Texture firmer, sometimes subcoriaceous; veins freely forked.
    Blades 2-pinnatifid or 2-pinnate; segments not spinulose.
      Leaves small; rachis commonly chaffy throughout.   4. *D. fragrans.*
      Leaves larger, 1½°–5° high; rachis naked or deciduously chaffy.

Indusia flat, thin.
　　Blades narrow, linear-oblong to lanceolate ; sori nearly medial. 5. *D. cristata.*
　　Blades broader, narrowly oblong, ovate or triangular ovate ; sori near midvein.
　　　Apex attenuate ; pinnae broadest at base ; sori 3–7 pairs.   6. *D. Clintoniana.*
　　　Apex short-acuminate, often abruptly so ; pinnae broadest above the base ;
　　　　sori 6–10 pairs.   7. *D. Goldiana.*
Indusia convex, firm.
　　Sori near the margin.   8. *D. marginalis.*
　　Sori near the midvein.   9. *D. Filix-mas.*
Blades 2-pinnate to 3-pinnate ; segments spinulose or mucronate.
　Blades ovate-lanceolate, triangular, or broadly oblong, usually not narrowed below.
　　Indusia glabrous or nearly so ; pinnae usually somewhat oblique to the rachis, the
　　　lowest broadly and unequally ovate to triangular.
　　　Pinnules flat, decurrent ; sori terminal on the veinlets ; scales pale brownish.
　　　　　　　　　　　　　　　　　　　　　　　　　　　　　10. *D. spinulosa.*
　　　Pinnules concave, some not decurrent ; sori mostly subterminal ; scales dark brownish.
　　　　　　　　　　　　　　　　　　　　　　　　　　　　　11. *D. dilatata.*
　　Indusia glandular ; pinnae usually at right angles, the lowest unequally lanceolate to
　　　ovate-lanceolate.   12. *D. intermedia.*
　Blades elongate-lanceolate, usually narrowed below.   13. *D. Boottii.*
Indusia wanting (§ *Phegópteris*).
　Basal pinnae sessile or partially adnate : rachis more or less alate.
　　Blades usually longer than broad ; rachis and midveins freely chaffy ; under surfaces pilose.
　　　　　　　　　　　　　　　　　　　　　　　　　　　　　14. *D. Phegopteris.*
　　Blades usually broader than long ; rachis and midveins scarcely scaly ; under surfaces
　　　slightly pubescent.   15. *D. hexagonoptera.*
　Basal pinnae long-stalked ; rachis not alate.
　　Blades nearly horizontal, glabrous or nearly so, subternate, the basal pinnae approaching
　　　the terminal portion in size.   16. *D. Dryopteris.*
　　Blades suberect, copiously glandular, triangular-ovate, the basal pinnae considerably smaller
　　　than the terminal portion.   17. *D. Robertiana.*

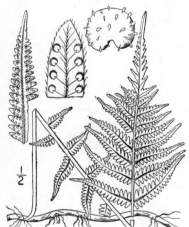

1. Dryopteris noveboracénsis (L.) A. Gray. New York Fern. Fig. 37.

Polypodium noveboracense L. Sp. Pl. 1091. 1753.
Aspidium noveboracense Sw. Schrad. Journ. Bot.
　1800[2]: 38. 1801.
Dryopteris noveboracensis A. Gray, Man. 632. 1848.

Rootstock slender, widely creeping. Stipes slender, short ; blades lanceolate, tapering both ways
from the middle, 1°–2° long, 4′–7′ wide, membranous, once pinnate, the apex long-acuminate ; pinnae
1½′–3½′ long, lanceolate, sessile, long-acuminate,
deeply pinnatifid, pilose along the midribs and veins,
especially beneath, ciliate, the lower (2–7) pairs
gradually shorter and deflexed, commonly distant,
the lowest auriculiform ; segments flat, oblong, obtuse, the basal ones often enlarged ; veins simple or
those of the basal lobes forked ; sori near the margin ; indusia small, delicate, glandular, withering.

　In moist woods and thickets, Newfoundland to Ontario and Minnesota, south to Georgia, Alabama and
Arkansas. Ascends to 5000 ft. in Virginia. Sometimes
sweet-scented in drying. July–Sept.

2. Dryopteris Thelýpteris (L.) A. Gray. Marsh Shield-fern. Fig. 38.

Acrostichum Thelypteris L. Sp. Pl. 1071. 1753.
Aspidium Thelypteris Sw. Schrad. Journ. Bot. 1800[2]: 40.
　1801.
Dryopteris Thelypteris A. Gray, Man. 630. 1848.

Rootstock slender, creeping, blackish. Leaves long-
stipitate, the blades lanceolate or oblong-lanceolate,
scarcely narrowed at base, 1°–2½° long, 4′–6′ wide,
short-acuminate, membranous, once pinnate ; pinnae
1½′–3′ long, linear-lanceolate, short-stalked or sessile,
horizontal or decurved, broadest at the base, short-
acuminate, pubescent or pilose beneath, deeply pinnatifid ; segments oblong, obtuse or appearing acute from
the strongly revolute margins ; veins regularly once or
twice forked ; sori nearly medial, crowded ; indusia
small, glabrous.

　In marshes and wet woods, rarely in dry soil, New
Brunswick to Manitoba, south to Florida, Louisiana and
Texas. Ascends to 2000 ft. in Vermont. Europe and
Asia. Summer. Wood-, Swamp-, Quill- or Marsh-Fern.

3. Dryopteris simulàta Davenp. Dodge's Shield-fern. Fig. 39.

Aspidium simulatum Davenp. Bot. Gaz. **19**: 495. 1894.
Dryopteris simulata Davenp. Bot. Gaz. **19**: 497. 1894. As synonym.

Rootstock wide-creeping, slender, brownish; stipes 6′–20′ long, straw-colored, dark brown at base, with deciduous scales; blades 8′–20′ long, 2′–7′ wide, oblong-lanceolate, membranous, once pinnate, little or not at all narrowed at the base, the apex abruptly acuminate, attenuate; pinnae 12–20 pairs, lanceolate, deeply pinnatifid, the segments oblique, oblong, obtuse, entire or lightly crenate, slightly revolute in the fertile leaf, ciliate, finely pubescent along the midribs; veins simple; sori rather large, somewhat apart, mostly nearer the margin than the midrib; indusia finely glandular, withering, persistent.

In woodland swamps, Maine to Maryland. Reported also from Missouri. Late summer.

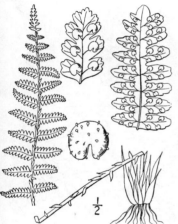

4. Dryopteris fràgrans (L.) Schott. Fragrant Shield-fern. Fig. 40.

Polypodium fragrans L. Sp. Pl. 1089. 1753.
Aspidium fragrans Sw. Schrad. Journ. Bot. 1800[2]:35. 1801.
Dryopteris fragrans Schott, Gen. Fil. 1834.

Rootstock stout, erect, densely chaffy with brown shining scales. Stipes 2′–4′ long, chaffy; blades lanceolate to narrowly oblanceolate, 3′–12′ long, firm, aromatic, nearly or quite 2-pinnate, the apex acute; pinnae numerous, ½′–1¼′ long, oblong-lanceolate to deltoid-lanceolate, usually subacute; segments oblong, obtuse, adnate, decurrent, deeply incised to subentire, nearly covered by the sori; indusium thin, very large, nearly orbicular, long-persistent, its margin ragged and sparingly glandular, the sinus narrow.

On rocks, Labrador to Alaska, south to Maine, New Hampshire, Vermont, New York, Wisconsin and Minnesota. Ascends to 4000 ft. in Vermont. Also in Greenland, Europe and Asia. Fragrant wood-fern.

5. Dryopteris cristàta (L.) A. Gray. Crested Shield-fern. Fig. 41.

Polypodium cristatum L. Sp. Pl. 1090. 1753.
Aspidium cristatum Sw. Schrad. Journ. Bot. 1800[2]: 37. 1801.
Dryopteris cristata A. Gray, Man. 631. 1848.

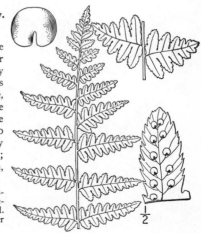

Rootstock stout, creeping, densely chaffy. Sterile leaves low, short-stipitate, spreading, much shorter than the fertile, evergreen. Fertile leaves rigidly erect, 1½°–3½° long, long-stipitate, withering; blades 1°–2½° long, 3′–6′ broad, linear-oblong to lanceolate, acuminate, deeply bipinnatifid, dark green; pinnae spaced, oblong-lanceolate to triangular-ovate or the lower ones subtriangular; deeply pinnatifid into 6–10 pairs of oblong to triangular-oblong, obtuse, finely serrate segments, the basal ones more deeply cut; sori nearly medial; indusia large, orbicular-reniform, glabrous.

In wet woods and swamps, Newfoundland to Saskatchewan, south to Virginia, Kentucky, Arkansas, Nebraska and Idaho. Ascends to 2700 ft. in Maryland. Also in Europe and Asia. July–Aug. Crested-fern or crested wood-fern.

6. Dryopteris Clintoniàna (D. C. Eaton) Dowell. Clinton's Fern. Fig. 42.

Aspidium cristatum var. *Clintonianum* D. C. Eaton in Gray, Man. ed. 5, 665. 1867.
Dryopteris cristata var. *Clintoniana* Underw. Native Ferns, ed. 4, 115. 1893.
Dryopteris Clintoniana Dowell, Proc. Staten Id. Assoc. Arts & Sc. 1 : 64. 1906.

Rootstocks stout, creeping, densely chaffy. Leaves $2\frac{1}{2}°-4\frac{1}{2}°$ high; stipes 1° or more long, straw-colored or brownish, with thin concolorous or often dark-centered scales; blades $1\frac{1}{2}°-3°$ long, 5'–10' broad, oblong to ovate-oblong, acute or acuminate, deeply bipinnatifid; pinnae apart, oblong-lanceolate, broadest at the base, or lower ones unequally elongate-triangular, deeply pinnatifid; segments oblong, usually obtuse, serrate, or the basal ones pinnately cut; sori 3–7 pairs, borne near the midvein; indusia orbicular-reniform, glabrous.

In swampy woods, Maine and Ontario to Wisconsin, and North Carolina. Often confused with the preceding and the following species.

7. Dryopteris Goldiàna (Hook.) A. Gray. Goldie's Fern. Fig. 43.

Aspidium Goldianum Hook. Edinb. Philos. Journ. **6**: 333. 1822.
Dryopteris Goldiana A. Gray, Man. 631. 1848.

Rootstock stout, ascending, chaffy. Leaves up to $5\frac{1}{2}°$ long, in a crown; stipes 10'–18' long, densely covered below with large lanceolate usually dark lustrous scales; lamina 2°–4° long, 10'–16' broad, ovate to oblong, short-acuminate, nearly glabrous, dark green above, nearly 2-pinnate; pinnae 6'–9' long, 1'–2' broad, broadly lanceolate to oblong-lanceolate, broadest above the base, acuminate, pinnatifid almost to the midrib; segments about 20 pairs, narrowly oblong, acute or subacute, subfalcate, serrate, the teeth appressed; sori 6–10 pairs, near the midrib, distinct; indusia glabrous, nearly orbicular, the sinus narrow.

In rich woods, New Brunswick to Minnesota, south to North Carolina, Tennessee and Iowa. Ascends to 5000 ft. in Virginia and to 2500 ft. in Vermont. July–Aug. Goldie's Wood-fern.

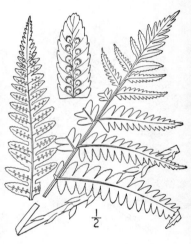

8. Dryopteris marginàlis (L.) A. Gray. Evergreen Wood-fern. Fig. 44.

Polypodium marginale L. Sp. Pl. 1091. 1753.
Aspidium marginale Sw. Syn. Fil. 50. 1806.
Dryopteris marginalis A. Gray, Man. 632. 1848.

Rootstock stout, woody, ascending, densely covered with bright brown shining scales, the leaves borne in a crown. Stipes 4'–10' long, chaffy below; blades ovate-oblong or ovate-lanceolate, chartaceo-coriaceous, $6'-2\frac{1}{2}°$ long, nearly or quite 2-pinnate, acuminate, usually a little narrowed at the base; pinnae numerous, sessile or nearly so, glabrous, 2'–5' long, the lowermost unequally deltoid-lanceolate, those above lanceolate to broadly oblong-lanceolate, acuminate; segments oblong or lanceolate, obtuse or subacute, subfalcate or falcate, subentire, crenate or pinnately lobed, partially adnate or the lowermost distinct; sori distant, close to the margin; indusia orbicular-reniform, glabrous.

In rocky woods and on banks Nova Scotia to British Columbia, south to Georgia, Alabama, Arkansas, Kansas and Oklahoma. Ascends to 5000 ft. in Virginia. **Leaves** evergreen. July–Aug. Marginal Shield-fern.

9. Dryopteris Fìlix-más (L.) Schott.　Male Fern.　Fig. 45.

Polypodium Filix-mas L. Sp. Pl. 1090.　1753.
Aspidium Filix-mas Sw. Schrad. Journ. Bot. 1800²: 38.
1801.
Dryopteris Filix-mas Schott, Gen. Fil.　1834.

Rootstock stout, woody, ascending or erect, chaffy. Leaves up to 4° high, in an erect crown; stipes 4′–10′ long, densely chaffy below; blades nearly evergreen, 1°–3° long, 6′–11′ broad, broadly oblong-lanceolate, acuminate, narrowed at the base, nearly or quite 2-pinnate; pinnae narrowly deltoid-lanceolate to oblong-lanceolate, acuminate; segments adnate, oblong, obtuse and biserrate, or partially adnate, ovate-oblong, acutish and deeply incised; sori numerous, large, nearer the midvein than the margin; indusia orbicular-reniform, glabrous.

In rocky woods, Newfoundland and Labrador to Alaska, south to Vermont, northern Michigan, South Dakota, Arizona and California. Aug. Also in Greenland. Numerous related forms of wide distribution are referred to this species; the type is European. The rootstock of this and the preceding species furnish the drug Filix-mas used as a vermifuge. Basket-fern. **Male shield-fern. Shield-roots. Bear's-paw-roots. Sweet or knotty brake.**

10. Dryopteris spinulòsa (Muell.) Kuntze.　Spinulose Shield-fern.　Fig. 46.

Polypodium spinulosum Muell. Fl. Fridr. 113. *f. 2.*　1767.
Aspidium spinulosum Sw. Schrad. Journ. Bot. 1800²: 38.
1801.
Dryopteris spinulosa Kuntze, Rev. Gen. Pl. **2**: 813.　1891.

Rootstock stout, creeping, chaffy. Leaves in an incomplete crown, the taller erect, the others spreading; stipes 4′–14′ long, with pale brownish scales; blades ½°–1½° long, 3½′–9′ broad, ovate-lanceolate to oblong, acuminate, deeply 2-pinnatifid; pinnae usually oblique, pinnately divided, the lower ones unequally deltoid, those above lanceolate from a broad base, acuminate; pinnules flat, oblong to lanceolate, acute, decurrent, pinnately cut, segments incised, teeth mucronate, falcate, appressed; sori submarginal, terminal on veinlets; indusia without glands.

In rich low woods, Labrador to Selkirk and Idaho, to Virginia and Kentucky. Also in Europe. Called also Narrow Prickly-toothed Fern.

11. Dryopteris dilatàta (Hoffm.) Gray.　Spreading Shield-fern.　Fig. 47.

Polypodium dilatatum Hoffm. Deutschl. Fl. **2**: 7.　1795.
Aspidium spinulosum var. *dilatatum* Hook. Brit. Fl. 444.
1830.
Dryopteris spinulosa var. *dilatata* Underw. Nat. Ferns, ed. 4, 116.　1893.

Rootstock creeping, or ascending. Leaves equal, spreading, in a complete crown; stipes ½°–1½° long, with dark brownish often darker-centered scales; blades ¾°–2¾° long, 4′–16′ broad, triangular to ovate or broadly oblong, acuminate, 3-pinnatifid; pinnae variable, the lower ones broadly and unequally ovate or triangular, those above lanceolate to oblong, acute or acuminate, the lowermost at least pinnately divided; pinnules convex, oblong to lanceolate, acute, the largest not decurrent, pinnately divided, segments pinnately lobed, teeth mucronate, straight or falcate, usually not appressed; sori mostly subterminal; indusia glabrous, or with a few glands.

A high mountain species of rocky woods, Newfoundland to Alaska, California, Idaho, Tennessee and North Carolina, Greenland. Also in Eurasia, Japan and the Madeira Islands. Broad Prickly-toothed Wood-fern.

12. Dryopteris intermèdia (Muhl.) Gray.   American Shield-fern.   Fig. 48.

$\frac{2}{5}$

Polypodium intermedium Muhl.; Willd. Sp. Pl. **5**: 262. 1810.
Aspidium americanum Davenp. Am. Nat. **12**: 714. 1878.
Dryopteris spinulosa var. *intermedia* Underw. Nat. Ferns, ed. 4, 116. 1893.

Rootstock creeping. Leaves equal, spreading in a complete crown; stipes 4′–14′ long, with light brownish or darker-centered scales; blades similar in size and shape to those of *D. spinulosa*, glandular-pubescent when young; pinnae usually at right angles to the rachis, the lower ones at least pinnate, unequally lanceolate to ovate-lanceolate; the upper ones lanceolate to oblong, acuminate; pinnules convex, oblong or lanceolate, acute, the largest not decurrent, pinnately divided, nearly at right angles; segments dentate, usually straight; sori submarginal, subterminal; indusia glandular.

In moist woods, Newfoundland to Wisconsin, south to North Carolina and Tennessee. Known only from eastern North America. Called also Common Wood-fern.

13. Dryopteris Boòttii (Tuckerm.) Underw.   Boott's Shield-fern.   Fig. 49.

Rootstock stout, ascending. Stipes 8′–12′ long, covered below with thin pale-brown scales; blade of fertile leaves elongate-oblong or lanceolate, acuminate, slightly narrowed toward the base, firm, bipinnate, or 3-pinnatifid, 1°–2½° long, 3′–5′ wide, the sterile ones commonly shorter and less divided; middle and upper pinnae lanceolate with a broad base, acuminate, those below unequally deltoid-lanceolate, the lowest elongate-triangular; pinnules oblong–ovate, constricted at the base, the lower ones nearly sessile, often pinnatifid, those above adnate and slightly decurrent upon the narrowly winged rachis, serrate, the margins spinulose throughout; sori numerous, distinct, medial or nearer the mid-vein; indusia orbicular-reniform, glandular.

In low woods and wet thickets, Nova Scotia to Minnesota, south to Virginia and West Virginia. July–Sept.

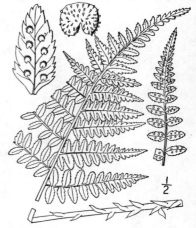

$\frac{1}{2}$

Several American writers regard *D. Boottii* as a natural hybrid between *D. cristata* and *D. intermedia*. Other supposed hybrids have recently been described, which have been confused either with *D. Boottii* or with species of which they were regarded as aberrant forms. The characters of these are such as to support strongly the hybridity hypothesis. They should be sought in localities exceptionally favorable to a mingling of the supposed parent forms. A list of these, including *D. Boottii*, follows:

Dryopteris Clintoniana × Goldiana Dowell, Bull. Torrey Club **35**: 137. 1908.
 Dryopteris Goldiana celsa Palmer, Proc. Biol. Soc. Wash. **13**: 65. 1899.
Dryopteris Clintoniana × intermedia Dowell, Bull. Torrey Club **35**: 136. 1908.
Dryopteris Clintoniana × marginalis Slosson, Bull. Torrey Club **37**: 202. 1910.
Dryopteris Clintoniana × spinulosa Benedict, Bull. Torrey Club **36**: 45. 1909.
Dryopteris cristata × Goldiana Benedict, Bull. Torrey Club **36**: 47. 1909.
Dryopteris cristata × intermedia Dowell, Bull. Torrey Club **35**: 136. 1908.
 Aspidium Boottii Tuckerm. Hovey's Mag. **9**: 145. 1843.
 Aspidium spinulosum var. *Boottii* D. C. Eaton in A. Gray, Man., ed. 5, 665. 1867.
 Dryopteris Boottii Underw. Nat. Ferns, ed. 4, 117. 1893.
 Dryopteris cristata intermedia Slosson, Fern Bull. **16**: 97. 1908.
Dryopteris cristata × marginalis Davenp. Bot. Gaz. **19**: 497. 1894, as syn.
 Aspidium cristatum × marginale Davenp. Bot. Gaz. **19**: 494. 1894.
Dryopteris cristata × spinulosa (Milde) C. Chr. Ind. Fil. 259. 1905.
 Aspidium cristatum × spinulosum Milde, Nov. Act. Acad. Leop.-Carol. **26**: 533. 1856.
Dryopteris Goldiana × intermedia Dowell, Bull. Torrey Club **35**: 138. 1908.
Dryopteris Goldiana × marginalis Dowell, Bull. Torrey Club **35**: 139. 1908.
Dryopteris Goldiana × spinulosa Benedict, Bull. Torrey Club **36**: 47. 1909.
Dryopteris intermedia × marginalis Benedict, Bull. Torrey Club **36**: 48. 1909.
Dryopteris marginalis × spinulosa Slosson, Fern Bull. **16**: 99. 1908.
 Dryopteris pittsfordensis Slosson, Rhodora **6**: 75. 1904.
 Nephrodium pittsfordense Davenp. Rhodora **6**: 76. 1904, as syn.
 Aspidium spinulosum × marginale Eggleston, Rhodora **6**: 138. 1904.

14. Dryopteris Phegópteris (L.) C. Chr.   Long Beech-fern.   Fig. 50.

Polypodium Phegopteris L. Sp. Pl. 1089. 1753.
Phegopteris polypodioides Fée, Gen. Fil. 243. 1850–52.
Phegopteris Phegopteris Underw.; Small, Bull. Torr.
   Club, **20**: 462. 1893.
Dryopteris Phegopteris C. Chr. Ind. Fil. 284. 1905.

Rootstock slender, creeping, somewhat chaffy.
Stipes stramineous, 6′–14′ long, blades triangular,
thin, mostly longer than wide, 4′–9′ long, 3′–8′ wide,
long-acuminate, pilose, especially on the veins beneath, the rachis and midribs with narrow rusty or
brownish scales; pinnae close, lanceolate or linear-
lanceolate, broadest above the base, acuminate, pinnately parted nearly to the rachis into oblong obtuse
entire or crenate close segments, the lowest pair
deflexed; basal segments, at least those of the upper
pinnae, adnate to the rachis and decurrent; sori small,
near the margin, non-indusiate.

Moist woods and hillsides, Newfoundland to Alaska,
the mountains of Virginia, Michigan to Washington.
Ascends to 4000 ft. in Vermont. Greenland, Europe
and Asia. Aug. Sun-fern. Common beech-fern.

15. Dryopteris hexagonóptera (Michx.) C. Chr.   Broad Beech-fern.   Fig. 51.

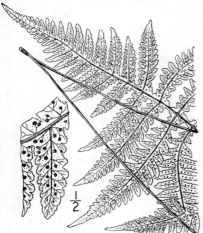

Polypodium hexagonopterum Michx. Fl. Bor. Am. **2**:
   271. 1803.
Phegopteris hexagonoptera Fée, Gen. Fil. 243. 1850–
   52.
D. hexagonoptera C. Chr. Ind. Fil. 270. 1905.

Rootstock slender, creeping, chaffy, somewhat
fleshy. Stipes 8′–18′ long, greenish or brownish
straw-colored; blades triangular, 7′–15′ broad, usually broader than long, acuminate, slightly pubescent, often glandular beneath; pinnae adnate to the
irregularly winged rachis, acuminate, the upper and
middle ones lanceolate, pinnatifid into numerous
obtuse oblong subentire or crenate segments, the
lowermost pinnae broader, unequally ovate to lanceolate-ovate with the middle pinnules elongate,
spaced, often deeply pinnatifid; sori mostly near the
margin, non-indusiate.

In dry woods and on hillsides, Quebec to Minnesota, Florida, Louisiana, Kansas and Oklahoma.
Aug. Called also Hexagon Beech-fern.

16. Dryopteris Dryópteris (L.) Britton.
Oak-fern.   Fig. 52.

Polypodium Dryopteris L. Sp. Pl. 1093. 1753.
Phegopteris Dryopteris Fée, Gen. Fil. 243. 1850–52.
Dryopteris Linneana C. Chr. Ind. Fil. 275. 1905.

Rootstock blackish, very slender, wide-creeping.
Stipes slender, straw-colored, 4′–12′ long, chaffy at
least below; blades thin, at right angles to the stipe,
nearly or quite glabrous, 4′–11′ broad, broadly triangular, subternate by the enlargement of the basal
pinnae, these triangular, very deeply 2-pinnatifid,
long-stalked; second pair of pinnae oblong or deltoid-
oblong, sessile and nearly pinnate, or (rarely) stalked
and 2-pinnatifid; upper pinnae gradually adnate, pinnatifid; segments oblong, blunt, entire to serrate-
crenate; sori near the margin, non-indusiate.

In moist woods, thickets and swamps, Newfoundland
and Labrador to Alaska, south to Virginia, Kansas,
Colorado and Oregon. Ascends to 2400 ft. in the
Catskills. Also in Greenland, Europe and Asia. Aug.
Pale-mountain, or tender three-branched-polypody.

2/3

17. Dryopteris Robertiàna (Hoffm.) C. Chr.
Scented Oak-fern. Fig. 53.

Polypodium Robertianum Hoffm. Deutschl. Fl. 2:
[add. 4]. 1795.
Phegopteris Robertiana A. Br.; Aschers. Fl. Brand. 2:
198. 1859.
Polypodium calcareum Sm. Fl. Brit. 1117. 1804.
Phegopteris calcarea Fée, Gen. Fil. 243. 1850–52.

Rootstock slender, creeping, branched. Stipes slender, straw-colored, 6′–13′ long; blades 6′–8′ long, 5′–7′ broad, copiously glandular, suberect, tri-angular-ovate; basal pinnae largest, 3′–4½′ long, unequally deltoid-ovate, long-stalked, 2-pinnatifid; second pair of pinnae distant, short-stalked or ses-sile, pinnate or 2-pinnatifid, deltoid-oblong; suc-ceeding pinnae sessile, narrower, mostly pinnatifid; segments close, oblong to elongate-oblong, the mar-gins subentire to crenate-dentate, reflexed; sori near the margin, non-indusiate, numerous.

On shaded limestone, Labrador to Alaska, New Brunswick and Iowa. Rare and local Also in Europe.

8. ANCHÍSTEA Presl, Epim. Bot. 71. 1851.

Coarse swamp ferns with wide-creeping prostrate or underground rootstocks, the leaves scattered and rigidly erect, the blades long-stalked and deeply bipinnatifid, the fertile ones similar in outline to the sterile. Veins united in a single series of elongate areoles next to the secondary rachis and midveins of the segments, the veinlets arising from these simple or once-forked, extending to the margin, almost invariably free. Sori superficial, borne on the inner side of the transverse vein forming the outer side of the areole, elongate-linear to oval, covered by convex indusia attached at the outer margin. [Name from the Greek, in allusion to the alliance with *Woodwardia*.]

A monotypic genus of eastern North America.

1. Anchistea virgínica (L.) Presl.
Virginia Chain-fern. Fig. 54.

Blechnum virginicum L. Mant. 2: 307. 1771.
Woodwardia virginica J. E. Smith, Mem. Acad.
Turin 5: 412. 1793.
Anchistea virginica Presl, Epim. Bot. 71. 1851.

Rootstock rather slender, creeping, spar-ingly branched, chaffy at the apex. Stipes stout, 1°–3° long, toward the base purplish brown and polished; blades 1°–2° long, 6′–9′ broad, oblong-lanceolate, acute, subcoriaceous, bipinnatifid; pinnae linear-lanceolate, usually alternate, oblique, glabrous, sessile, acuminate, 3′–6′ long, deeply pinnatifid into numerous and usually close ovate or oblong obtuse segments, their margins serrulate; sori along the sec-ondary rachis elongate-linear, those of the segments shorter, elliptical; indusia subentire or erose, extrorse, obscured at maturity.

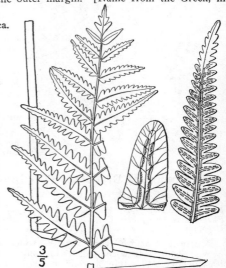

3/5

In swamps, often in deep water, Nova Scotia to Ontario and Michigan, south to Florida, Louisiana and Arkansas. Ascends to 1300 ft. in Pennsylvania. Also in Bermuda. June–July.

9. LORINSÈRIA Presl, Epim. Bot. 72. 1851.

Swamp ferns of medium size, with dimorphous leaves, the sterile ones spreading, with deeply pinnatifid blades, the veins copiously anastomosing; fertile leaves rigidly erect, the pinnae somewhat foliaceous, but greatly reduced in width, with a single series of elongate costal areoles and a few short excurrent veinlets. Sori in a single row, linear to elliptic, borne as in *Anchistea*, superficial, sometimes appearing immersed from the pustulate mem-branous leaf-tissue beneath. Indusium extrorse, firmly membranous, persistent and scarcely reflexed with age. [Name in honor of Gustav Lorinser, an Austrian physician and botanist.]

A monotypic genus of eastern North America.

1. Lorinseria areolàta (L.) Presl.
Net-veined Chain-fern. Fig. 55.

Acrostichum areolatum L. Sp. Pl. 1069. 1753.
Woodwardia angustifolia J. E. Smith, Mem.
    Acad. Turin 5: 411. 1793.
Lorinseria areolata Presl, Epim. Bot. 72. 1851.
Woodwardia areolata Moore, Ind. Fil. xlv. 1857.

Rootstock slender, widely creeping, chaffy,
with scattered leaves. Fertile leaves erect,
surpassing the sterile; stipes 1°–2° long, stout,
puplish-brown, lustrous; blades 6′–12′ long,
ovate-oblong; pinnae linear, distant, usually
connected by a slight wing. Sterile leaves
spreading; stipes 6′–14′ long, slender, greenish; blades 6′–15′ long, ovate-oblong to deltoid-
ovate, acuminate, membranous, deeply pinnatifid; pinnae linear-lanceolate to oblong-lanceolate, acute, lightly or sometimes deeply
sinuate, serrulate, usually connected by a variable wing, or the lower pairs free; veins
joined in numerous hexagonal areoles.

In swamps and moist soil, Maine to Florida,
Tennessee, Louisiana, and Arkansas; also in
Michigan. Aug.–Oct. Called also Netted chain-
fern. Various imperfectly fertile forms occur.

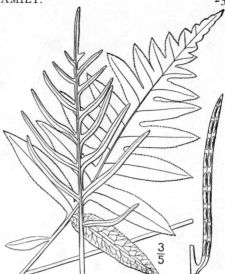

10. PHYLLÌTIS Ludwig, Inst. Hist. Phys. Reg.-Veg., ed. 2, 142. 1757.
[SCOLOPENDRIUM Adans. Fam. Pl. 2: 20. 1763.]

    Small or medium-sized ferns with deltoid, oblong or strap-shaped mostly entire leaves,
and linear elongate sori almost at right angles to the midrib and contiguous in pairs, one on
the upper side of a veinlet, the other on the lower side of the next contiguous veinlet of the
group above, the closely adjacent sori each with a narrow laterally attached indusium meeting
that of the other, the double sorus thus appearing to have a common indusium opening longitudinally along its middle. [Greek name of fern.]

    About 5 species, mainly of temperate regions, only the following known in North America.
Type species: *Asplenium Scolopendrium* L.

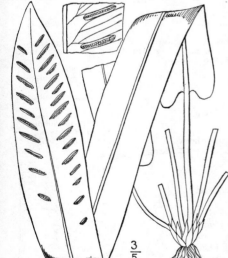

1. Phyllitis Scolopéndrium (L.) Newm.
Hart's-tongue. Fig. 56.

Asplenium Scolopendrium L. Sp. Pl. 1079. 1753.
Scolopendrium vulgare J. E. Smith, Mem. Acad.
    Turin 5: 421. 1793.
Phyllitis Scolopendrium Newm. Hist. Ferns, ed.
    2: 10. 1844.
Scolopendrium Scolopendrium Karst. Deutsch.
    Fl. 278. 1880–83.

Rootstock short, erect or ascending, chaffy
with light brown scales, the leaves in a spreading crown. Stipes 2′–6′ long, deciduously
fibrillose-chaffy; blades simple, linear-ligulate,
7′–18′ long, 1′–2½′ broad, bright green, firm,
cordate or auricled at the base, entire or lightly
sinuate, usually repand; veins once or twice
dichotomous near the midrib, free; pairs of
sori distinct, 2″–8″ long, the indusia whitish
at first, soon thrust back and wholly concealed
by the heavy lines of dark brown sporanges.

    Shaded limestone cliffs and depressions, in
central New York, near Woodstock, N. B., in
Bruce and Grey Counties, Ontario, and near
south Pittsburg, Tennessee. Very rare. Eurasia. Widely different forms are cultivated in
Europe. Snake-fern, Sea-weed fern.

11. CAMPTOSÒRUS Link, Hort. Berol. 2: 69. 1833.

    Slender ferns with narrow tapering simple entire or lightly sinuate leaves, bearing linear
or oblong sori several times longer than boad, irregularly scattered on either side of the
reticulate veins or sometimes crossing them, partly parallel to the midrib and partly oblique
to it, the outer ones more or less approximate in pairs. Indusium membranous. [Greek,
referring to the bent or curved sori.]

    Two species, the following, which is the generic type, the other of northern Asia.

1. Camptosorus rhizophýllus (L.) Link.
Walking-fern. Fig. 57.

Asplenium rhizophylla L. Sp. Pl. 1078, in part. 1753.
C. rhizophyllus Link, Hort. Berol. **2**: 69. 1833.

Rootstock short, usually creeping, somewhat chaffy. Stipes light green, 1′-6′ long, tufted, spreading; blades evergreen, 4′-9′ long, rather thin or somewhat chartaceous, simple, lanceolate, the bases usually cordate or auriculate, sometimes hastate, the basal auricles occasionally much elongate, the apex of the blade long-attenuate and usually filiform, rooting at the tip and giving rise to a new plant by the ultimate withering of the tissue, but 2-4 plants sometimes thus connected; sori usually numerous, irregularly placed.

In shaded situations, usually upon moist mossy rocks, preferring limestone, Quebec to Minnesota, Georgia, Alabama and Kansas. Ascends to 2500 ft. in Virginia. Aug.–Oct. Called also Walking-leaf.

12. ASPLÈNIUM L. Sp. Pl. 1078. 1753.

Large or small ferns of various habitat, with simple lobed or 1-3-pinnatifid or pinnate mostly uniform leaves, the veins free; scales of the rootstock firm, with thick-walled cells. Sori straight or sometimes slightly curved, oblong to linear, borne on the oblique veins, usually somewhat apart. Indusia invariably present, attached lengthwise along the veins, usually at the inner side. [Ancient Greek name, being a supposed remedy for the spleen.]

About 400 or more species of wide distribution. Besides the following, 9 species occur in Florida and 4 in the western United States. Type species: *Asplenium Trichomanes* L.

Blades pinnatifid, or pinnate only below, the apices long-attenuate.
  Stipe and rachis dark purplish brown throughout.   1. *A. ebenoides.*
  Stipe dark brownish below, green above; rachis green.   2. *A. pinnatifidum.*
Blades 1-3-pinnate, the apices not long-attenuate.
  Blades 1-pinnate only.
    Stipe and rachis blackish, reddish or purplish brown throughout.
      Sori short, nearer the margin than the midvein.   3. *A. resiliens.*
      Sori longer, medial or nearer the midvein.
        Fertile leaves rigidly erect; pinnae more or less auriculate.   4. *A. platyneuron.*
        Fertile leaves spreading like the sterile; pinnae not auriculate.   5. *A. Trichomanes.*
    Stipe dark only at the base, green above like the rachis.
      Blades small, 2′-8′ long, linear.   6. *A. viride.*
      Blades large, 1°-2½° long, lanceolate to lanceolate-ovate.   7. *A. pycnocarpon.*
  Blades 2-3-pinnatifid.
    Stipe and rachis green throughout.   8. *A. Ruta-muraria.*
    Stipe dark brownish, at least toward the base.
      Stipes dark at the base, greenish above; rachis green.
        Blades deltoid-ovate to deltoid-lanceolate.   9. *A. montanum.*
        Blades linear-lanceolate.   10. *A. fontanum.*
      Stipe and lower rachis (at least) dark chestnut-brown.   11. *A. Bradleyi.*

1. Asplenium ebenoìdes R. R. Scott.
Scott's Spleenwort. Fig. 58.

Asplenium ebenoides R. R. Scott, Journ. Roy Hort. Soc. 87. 1866.

Rootstock short, chaffy, with dark, shining scales. Stipes tufted, 1½′-7′ long, purplish brown; blades triangular-lanceolate, rarely almost linear, variable in outline and size, 3′-12′ long, 1′-3′ wide at the base, firm, tapering into a long narrow acuminate apex, pinnatifid, or commonly pinnate below, the segments or pinnae lanceolate from a broad base, acute or acuminate, variable in length, the lower sometimes shorter than those just above; sori straight or nearly so; indusium narrow, reflexed at maturity.

Rare. Vermont to Missouri, south to Virginia and Alabama, where it is self-perpetuating. Now proved by Miss Margaret Slosson to be a hybrid between *Camptosorus rhizophyllus* and *Asplenium platyneuron.* Ascends to 1400 ft. in Virginia.

2. Asplenium pinnatífidum Nutt
Pinnatifid Spleenwort. Fig. 59.

Asplenium pinnatifidum Nutt. Gen. **2**: 251. 1818.

Rootstock short-creeping, branched, conspicuously chaffy, with firm lanceolate dark brown iridescent scales. Stipes often densely clustered, polished, dark brown below, greenish above, 2′–5′ long; blades 3′–10′ long, rigidly herbaceous or coriaceous, narrowly deltoid-lanceolate, tapering upward to a long narrow or filiform sinuate apex, deeply pinnatifid or the lower parts pinnate, the basal pinnae or occasionally several pairs sometimes long attenuate like the apex; lobes or pinnae rounded or the lowest acuminate; sori commonly numerous, straight or slightly curved, copiously confluent with age.

On rocks, Connecticut and New York to Missouri, south to Georgia, Alabama, Arkansas and Missouri. Ascends to 3000 ft. in North Carolina. July–Oct.

$\frac{3}{5}$

3. Asplenium resíliens Kunze. Small Spleenwort.
Fig. 60.

Asplenium parvulum Mart. & Gal. Mém. Acad. Brux. 15[5]: 60. 1842, not Hook. 1840.
Asplenium resiliens Kunze, Linnaea **18**: 331. 1844.

Rootstock short, creeping, chaffy with black stiff scales. Stipes tufted, blackish and shining, 1′–2½′ long; blades firm, linear-oblong or linear-oblanceolate, 3′–10′ long, 5″–12″ wide, once pinnate; pinnae 2″–6″ long, mostly opposite, oblong, obtuse, entire or crenulate, auricled on the upper side or sometimes hastate-auriculate, nearly sessile, the middle ones the longest, the lower gradually shorter, distant and reflexed; rachis dark brown or black; sori oblong, short, nearly or quite straight, borne rather nearer the margin than the midrib, often strongly confluent with age and appearing as a broad submarginal band.

On limestone, Virginia to Florida, west to Kansas, Texas and New Mexico. Ascends to 2400 ft. in Virginia. Also in Mexico and Jamaica. June–Oct. Called also Little Ebony Spleenwort.

$\frac{1}{2}$

4. Asplenium platyneùron (L.) Oakes. Ebony Spleenwort. Fig. 61.

Acrostichum platyneuros L. Sp. Pl. 1069. 1753.
Asplenium ebeneum Ait. Hort. Kew. **3**: 462. 1789.
Asplenium platyneuron Oakes; D. C. Eaton, Ferns N. Am. **1**: 24. 1879.

Rootstock short-creeping, the fertile leaves upright, usually much surpassing the spreading sterile ones. Stipes densely tufted, purplish or reddish brown, shining, 1′–5′ long; blades linear-oblanceolate, 8′–15′ long, 1′–2½′ wide above the middle, firm, once pinnate, the rachis like the stipes; pinnae 20–40 pairs, lanceolate, subfalcate, alternate or partly so, sessile, crenate, serrate or incised, auricled on the upper side at the base and occasionally also on the lower; lower pinnae gradually smaller and oblong or triangular; sori numerous, oblique, linear-oblong, nearer the midvein than the margin, often confluent with age.

On rocks and banks, preferring limestone soil, Maine and Ontario to Colorado, southward to Texas and the Gulf states generally. Ascends to 4200 ft. in North Carolina. Erroneously ascribed to Jamaica. South African specimens, however, are identical. Several deeply incised or pinnatifid forms have been described from the United States. July–Sept.

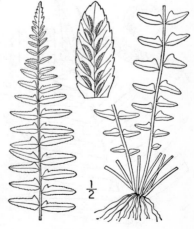

$\frac{1}{2}$

5. Asplenium Trichómanes L. Maiden hair Spleenwort. Fig. 62.

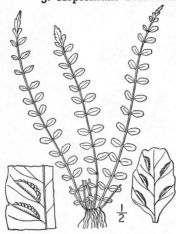

Asplenium Trichomanes L. Sp. Pl. 1080. 1753.

Rootstock short, nearly erect, chaffy with blackish scales. Stipes densely tufted, commonly numerous, 1′–2½′ long, purplish-brown and shining; blades linear, often somewhat reduced toward the base, 3′–8′ long, 4″–9″ wide, rather rigid, once pinnate, evergreen, the rachis dark brownish; pinnae mostly oval or roundish-oblong, inequilateral, partly opposite, partly alternate, or nearly all opposite, cuneate at the base, the margins slightly crenate; lower pinnae smaller and relatively broader, farther apart, often fan-shaped in outline; sori 3–6 pairs, short, commonly confluent at maturity; sporanges dark brown.

On rocks, preferring limestone, throughout nearly the whole of North America north of Mexico except the extreme north. Ascends to 2500 ft. in Vermont. Also in Europe and Asia. July–Sept. Called also Wall- or dwarf-spleenwort; water-wort, english maiden-hair.

6. Asplenium víride Huds. Green Spleenwort. Fig. 63.

Asplenium viride Huds. Fl. Angl. 385. 1762.

Rootstock stout, creeping, chaffy with brown nerve-less scales, the leaves usually borne in dense tufts. Stipes numerous, stout or sometimes very slender, brownish below, green above, 1′–3′ long; blades linear-lanceolate, 2′–8′ long, 4″–10″ wide, once pinnate, pale green, soft-herbaceous or almost membranous; rachis green; pinnae 12–20 pairs, roundish-ovate or rhombic, deeply crenate, obtuse, unequal-sided, broadly cuneate at the base, the lower side obliquely truncate; sori near the midvein, oblong, usually numerous and confluent, or sometimes fewer and somewhat apart.

On rocks, Quebec and New Brunswick to Alaska, south to Washington, Wyoming, and the Green Mountains of Vermont. Also in Europe and Asia. Summer.

7. Asplenium pycnocàrpon Spreng. Narrow-leaved Spleenwort. Fig. 64.

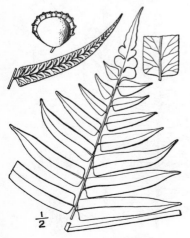

Asplenium angustifolium Michx. Fl. Bor. Am. **2**: 265. 1803. Not Jacq. 1786.
Asplenium pycnocarpon Spreng. Anleit. **3**: 112. 1804.
Athyrium pycnocarpon Tidestrom, Elys. Marianum 36. 1906.

Rootstock stout, creeping, rooting along its whole length. Stipes clustered, naked, dark brown at the base, green and somewhat fleshy above, 8′–15′ long; blades lanceolate to lanceolate-ovate, 1°–2½° long, once pinnate, glabrous, membranous; pinnae 20–30 pairs, 2′–5′ long, short-stalked, lightly crenulate, linear-oblong, attenuate, flaccid, obtuse or broadly cuneate at the base, those of the fertile blades usually smaller and considerably narrower than those of the fertile, often falcate; sori 20–30 pairs, close, linear, slightly curved, oblique; indusium firm, convex, concealed by the strongly confluent sori at maturity.

In moist woods and shaded ravines, Quebec to Wisconsin, south to Georgia, Alabama, Missouri and Kansas. Ascends to 1700 ft. in the Adirondacks and to 2300 in the Catskills. Aug. Swamp-spleenwort.

8. Asplenium Rùta-murària L.   Wall Rue Spleenwort.   Fig. 65.

Asplenium Ruta-muraria L. Sp. Pl. 1081. 1753.

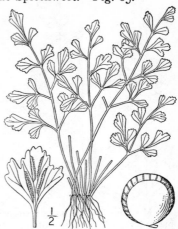

Rootstock short, creeping or ascending, the small evergreen leaves closely clustered. Stipes naked, 2′–3′ long, green throughout; blades ovate or deltoid-ovate, 2′–5′ long, glabrous, evergreen, 2–3-pinnate, at least below; pinnae and pinnules mostly alternate, stalked; pinnules very variable in shape, size and marginal cutting, commonly rhombic or obovate, obtuse, with the margins dentate or incised, but often cuneate-spatulate, the margins deeply fimbriate; veins flabellate; sori few, linear-oblong, confluent when mature and covering nearly the whole pinnule; membranous, delicate.

On limestone, Vermont to southern Ontario and Michigan, south to Alabama and Missouri. Ascends to 2100 ft. in Virginia. Also in Europe, Asia and northern Africa. July–Sept. Dwarf Spleenwort. Tentwort. Stone-rue. Stonefern. Rue-fern. White maiden-hair. A hybrid between this species and *A. Trichomanes*, described originally from European specimens, has been found also in Vermont.

9. Asplenium montànum Willd.   Mountain Spleenwort.   Fig. 66.

Asplenium montanum Willd. Sp. Pl. 5: 342. 1810.

Rootstock short, creeping, dark-chaffy at the apex. Stipes tufted, slender, naked, dark brown at the base, green above, 2′–4½′ long; blades deltoid-ovate to deltoid-lanceolate, acuminate, rather firm, evergreen, 1–2-pinnate; lower pinnae largest, deltoid, pinnate or pinnatifid, the lobes or segments ovate or rhombic-oblong, dentate, often narrowly cuneate; upper pinnae less divided, merely toothed or incised; rachis green, winged toward the apex; veins obscure; sori linear-oblong, short, the lower ones sometimes double, usually abundant, often confluent at maturity and concealing the narrow membranous indusia.

On dry and moist rocks, Connecticut and New York to Ohio, south to Georgia, Alabama and Arkansas. Ascends to 4500 ft. in North Carolina. June–Aug.

10. Asplenium fontànum (L.) Bernh.   Rock Spleenwort.   Fig. 67.

Polypodium fontanum L. Sp. Pl. 1089. 1753.
Asplenium fontanum Bernh. Schrad. Journ. Bot. **1799**[1]: 314. 1799.

Rootstock short, ascending, clothed with narrow dark scales at the apex. Stipes tufted, 1′–3′ long, somewhat blackish at the base, especially on the inner side, usually glabrous; blades linear-lanceolate, broadest above the middle, 2–3-pinnate, 3′–6′ long, 6″–1½′ wide, acuminate, gradually narrowed at the base, the lower pinnae often greatly reduced; rachis narrowly winged; pinnae 10–15 pairs, deltoid-lanceolate to ovate, or the lower ones fan-shaped and flabellately divided, the segments deeply dentate with spinulose teeth; sori short, only 1 to 4 on each segment, rarely confluent; indusia membranous, subentire.

On rocks, Lycoming Co., Pa., and Springfield, Ohio. One of the rarest ferns of the United States; common in Europe. Summer. Called Smooth Rock-spleenwort.

$\frac{1}{2}$

11. Asplenium Brádleyi D. C. Eaton. Bradley's Spleenwort. Fig. 68.

A. Bradleyi D. C. Eaton, Bull. Torr. Club **4**: 11. 1873.

Rootstock short, covered with dark narrow scales. Stipes tufted, slender, $2'-3\frac{1}{2}'$ long, dark chestnut-brown throughout, shining; blades oblong-lanceolate to oblong, acuminate or scarcely narrowed at the base, pinnate, with 8-12 pairs of short-stalked mostly oblong-ovate, obtuse pinnae, the lower pinnae often unequally deltoid, pinnatifid or pinnate with oblong obtuse lobes or pinnules, these toothed at the apex, the upper pinnatifid with dentate or nearly entire lobes; rachis brown or greenish above; sori short, borne near the midveins; indusia membranous, persistent.

On rocks, preferring limestone, New York to Georgia Alabama, Arkansas and Missouri. Local. July–Sept.

13. ATHÝRIUM Roth, Römer's Arch. Bot. 2^1: 105. 1799.

Medium-sized or large ferns with greenish succulent stipes and 1-3-pinnate or pinnatifid blades; veins free; scales of the rootstock delicate, of thin-walled cells. Sori usually curved, oblong to linear-oblong, or crossing the vein and recurved, sometimes unequally hippocrepiform, rarely roundish. Indusia shaped like the sorus, attached as in *Asplenium,* subentire to fimbriate, rarely vestigial and concealed. [Greek, shieldless, of doubtful application.]

A genus of about 85 species, mainly of tropical regions. *A. cyclosorum* occurs in western North America. Type species: *Athyrium Filix-foemina* (L.) Roth.

Blades bipinnatifid; segments lightly crenate-serrate.   1. *A. thelypteroides.*
Blades bipinnate; pinnules variously incised or deeply serrate.   2. *A. Filix-foemina.*

1. Athyrium thelypteròides (Michx.) Desv. Silvery Spleenwort. Fig. 69.

Asplenium acrostichoides Sw. Schrad. Journ. Bot. **1800**[2]: 54. 1801. Not *Athyrium acrostichoideum* Bory, 1836.
Asplenium thelypteroides Michx. Fl. Bor. Am. **2**: 265. 1803.
Athyrium thelypteroides Desv. Mém. Soc. Linn. Paris **6**: 266. 1827.

$\frac{1}{2}$

Rootstock slender, sinuous, creeping. Stipes $8'-16'$ long, straw-colored, somewhat chaffy below, at least when young; blades lanceolate, oblong-lanceolate, or ovate-oblong, $1°-3°$ long, $6'-12'$ wide, acute or acuminate, narrowed to the base, very deeply bipinnatifid; pinnae linear-lanceolate to oblong-lanceolate, sessile, acuminate, deeply pinnatifid into numerous oblong obtuse or subacute lightly serrate-crenate segments; sori crowded, curved or straight, the lower often double; indusium light-colored and shining when young.

In rich moist woods, Nova Scotia to Minnesota, Missouri and Georgia. Ascends to 5000 ft. in Virginia. Closely related forms occur in eastern Asia. Aug.–Oct.

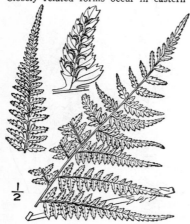

$\frac{1}{2}$

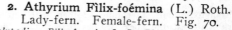

2. Athyrium Fìlix-foémina (L.) Roth. Lady-fern. Female-fern. Fig. 70.

Polypodium Filix-foemina L. Sp. Pl. 1090. 1753.
Asplenium Filix-foemina Bernh. Schrad. Neues Journ. Bot. 1^2: 26. 1806.
A. Filix-foemina Roth, Romer's Arch. 2^1: 106. 1799.

Rootstock creeping or ascending, slender for the size of the plant. Stipes tufted, $6'-12'$ long, straw-colored, brownish or reddish; blades broadly oblong-ovate to oblong-lanceolate, acuminate, $1°-3°$ long, 2-pinnate; pinnae lanceolate, acuminate, short-stalked or the upper ones sessile, $4'-8'$ long; pinnules oblong-lanceolate to broadly elliptical, incised or serrate, the lobes or teeth often again toothed, those toward the ends of the pinnae confluent; sori short; indusia straight or curved, sometimes horseshoe-shaped.

In woods and thickets, Newfoundland to British Columbia, the Gulf states, and California. Ascends to 6000 ft. in North Carolina, and to 2000 ft. in Vermont. Europe and Asia. June–Aug. Backache-brake.

14. **ADIANTUM** [Tourn.] L. Sp. Pl. 1094. 1753.

Graceful ferns of rocky hillsides, woods and ravines, with much divided leaves, the stipes and branches slender or filiform, rigid, polished, usually dark-colored and shining. Sori appearing marginal, borne at the ends of the free forking veins, on the under side of reflexed indusiiform marginal lobes of the pinnules or segments. [Name ancient.]

A genus of about 175 species, largely tropical American. Besides the following another occurs in Florida, one in Texas and one in California and Nevada. Type species: *A. Capillus-Veneris* L.
Blades ovate-lanceolate in outline, with a continuous main rachis.   1. *A. Capillus-Veneris*.
Blades reniform-orbicular, the two equal divisions with pinnate branches. 2. *A. pedatum*.

1. Adiantum Capíllus-Véneris L.
Venus-hair Fern.   Fig. 71.

Adiantum Capillus-Veneris L. Sp. Pl. 1096. 1753.

Rootstock creeping, rather slender, chaffy with light-brown scales. Stipes very slender, black, or nearly so and shining, 3'–12' long; blades ovate-lanceolate in outline, 2-pinnate below, simply pinnate above, membranous, 6'–2° long, 4'–12' wide at the base; pinnules and upper pinnae wedge-obovate or rhomboid, rather long-stalked, glabrous, the upper margin rounded and more or less deeply incised, the sterile lobes crenate or dentate-serrate, the fertile ones with lunate or transversely oblong indusia; main and secondary rachises and stalks of the pinnules black or dark brown.

In ravines, Virginia to Florida, west to Missouri, Utah and California. Also in South Dakota. Ascends to 1300 ft. in Kentucky. Also in tropical America, and in the warmer parts of the Old World. June–Aug. True or black maiden's-hair. Lady's-hair. Dudder-grass.

2. Adiantum pedàtum L.   Maiden-hair or Lock-hair Fern.   Fig. 72.

Adiantum pedatum L. Sp. Pl. 1095. 1753.

Rootstock slender, creeping, chaffy, rooting along its whole length. Stipes 9'–18' long, dark chestnut-brown, polished and shining, once forked at the summit; blades reniform-orbicular in outline, 8'–18' broad, membranous, the pinnae arising from the outer sides of the two equal branches, somewhat pedately arranged, the larger ones 6'–10' long, 1'–2' wide; pinnules oblong, triangular-oblong, or the terminal one fan-shaped, short-stalked, the lower margin entire and slightly curved, the upper margin cleft or lobed, the lobes bearing the linear-oblong, often short sori.

In woods, Nova Scotia and Quebec to Alaska, south to Georgia, Louisiana, Kansas; Rocky Mountains to Utah and California. Ascends to 5000 ft. in Virginia. Also in Asia. July-Sept. Most of the western and northwestern specimens and from the Gaspé region, Quebec, are referable to the var. *aleuticum* Rupr., characterized by its fewer and more strict pinnae and more deeply cleft pinnules and stouter suberect rootstock.

15. **PTERÍDIUM** Scop. Fl. Carn. 169. 1760.

Coarse ferns of open or partially shaded situations, the triangular or deltoid-ovate compound blades borne upon stout stipes, these scattered upon a slender freely branched woody rootstock creeping underground. Sori in a continuous marginal line, arising from a transverse vein-like receptacle connecting the ends of the forked free veins. Indusium double, the outer conspicuous, formed by the reflexed membranous margin of the blade; the inner obscure, delicate, borne upon the receptacle. [Greek name for ferns.]

Variously regarded as containing one or several species of the widest distribution, the several forms closely allied to the following, the generic type. *P. caudatum* occurs in Florida.

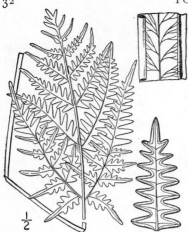

1. Pteridium aquilìnum (L.) Kuhn. Brake.
Bracken. Fig. 73.

Pteris aquilina L. Sp. Pl. 1075. 1753.
Pteridium aquilinum Kuhn, in Decken's Reisen III.
Bot. Ost.-Afrika 11. 1879.

Stipe 1°–3° long, straw-colored or brownish, rigid,
without chaff, swollen at the base. Blade 2°–4°
long, 1°–3° broad, triangular to deltoid-ovate, usually
subternate, the long-stalked basal pinnae and the
middle ones 2-pinnate, those above 2-pinnate to
lobed or simple; segments oblong to lanceolate, the
under surface glabrous or pubescent.

In thickets or open situations throughout most of
North America. Ascends to 5000 ft. in North Carolina.
Aug. Nearly cosmopolitan. July–Sept. Earnfern,
Eagle-fern, Lady-bracken, Adder-spit, Hog-brake.

The var. *pseudocaudatum* Clute, from Massachusetts
southward, has long linear pinnules, nearly simple.

16. CRYPTOGRÁMMA R. Br. App.
Franklin's Journ. 767. 1823.

Small mainly alpine or boreal ferns with dimorphous leaves, the stipes greenish or straw-
colored, the blades 2–3-pinnate, the fertile exceeding the sterile. Sori borne at or near the
ends of the free forking veins, at length confluent. Indusia formed of the altered reflexed
margin of the segment. [Greek, alluding to the sori hidden before maturity.]

Four species, the following and 2 of Europe and Asia. Type species: *C. acrostichoides* R. Br.

Rootstocks stout, clustered, ascending; fertile segments linear.   1. *C. acrostichoides.*
Rootstocks slender, creeping; fertile segments much broader.   2. *C. Stelleri.*

1. Cryptogramma acrostichoìdes R. Br.
American Rock-brake. Fig. 74.

Cryptogramma acrostichoides R. Br. App. Franklin's Journ. 767. 1823.

Rootstock stout, short, chaffy; leaves clustered, the fertile ones surpassing the sterile.
Stipes 2′–6′ long, chaffy below, those of the
sterile leaves slender, greenish and of the fertile stouter and stramineous; blades ovate or
ovate-lanceolate, thin, glabrous, 2–3-pinnate, the
sterile ones with the ultimate segments and
pinnules crowded, ovate, oblong or obovate,
obtuse, crenate or incised; fertile blades with
segments 3″–6″ long, 1″ or less wide, the thin
margins involute to the midrib at first, at
maturity expanded, exposing the sporanges.

Among rocks, Labrador to Alaska, south to
Lakes Huron and Superior, in the mountains to
Colorado and California. Summer.

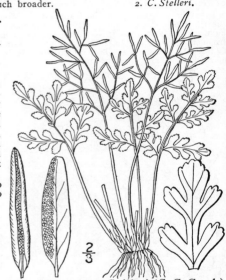

2. Cryptogramma Stélleri (S. G. Gmel.)
Prantl. Slender Cliff-brake. Fig. 75.

Pteris Stelleri S. G. Gmel. Nov. Com. Acad.
Petrop. 12: 519. *pl. 12. f. 1.* 1768.
Pellaea gracilis Hook. Sp. Fil. 2: 138. 1858.
Cryptogramma Stelleri Prantl, Engler's Bot.
Jahrb. 3: 413. 1882.

Rootstock slender, creeping, somewhat scaly.
Stipes scattered, 2′–5′ long, straw-colored or
pale brown, slightly chaffy below; blades thin-
membranous, ovate or oblong-ovate, 2′–5′ long,
1′–2′ wide; pinnae few, the lower nearly 2-pin-
nate, the middle pinnate, the upper simple;
segments of sterile blades ovate to obovate,
cuneate, crenately lobed, those of the fertile
linear-oblong or lanceolate; indusium broad.

On rocks, preferring limestone, Labrador to
Alaska, Pennsylvania, Iowa, Wisconsin and Colorado. Also in Asia. Aug.–Sept.

17. **PELLAÈA** Link, Fil. Hort. Berol. 59. 1841.

Rock-loving small or medium-sized ferns, with nearly uniform leaves, the blades 1–3-pinnate, smooth, the fertile divisions commonly narrower than the sterile. Sori roundish or elongate, on the free veins, usually confluent in a submarginal line. Indusium formed by the reflexed margins of the segments. [Greek, alluding to the dark-colored stipes.]

About 50 to 60 species of wide geographic distribution. Besides the following several occur in the western and southwestern United States. Type species: *Pellaea atropurpurea* (L.) Link.

Blades pinnate or 2-pinnate with large pinnules.   1. *P. atropurpurea.*
Blades small, 3-pinnate, the pinnules narrow.   2. *P. densa.*

1. Pellaea atropurpùrea (L.) Link.
Purple-stemmed Cliff-brake. Fig. 76.

Pteris atropurpurea L. Sp. Pl. 1076. 1753.
Pellaea atropurpurea Link, Fil. Hort. Berol. 59. 1841.
P. glabella Mett.; Kuhn, Linnaea **36**: 87. 1869,

Rootstock short, densely clothed with long-attenuate rusty scales. Stipes tufted, 2′–8′ long, dark purple, smooth, or, with the rachis, more or less pubescent with hair-like chaff; blades coriaceous, lanceolate, ovate-lanceolate or deltoid-ovate, 4′–12′ long, 2′–6′ wide, simply pinnate, or below 2-pinnate; pinnules and upper pinnae 1′–2′ long, glabrous, or sparsely fibrillose below, 3″ or less wide, short-stalked or sessile; veins obscure.

On rocks, preferring limestone, Ontario to British Columbia and Mackenzie, Georgia, Mississippi, Texas and California. Reported from northern Mexico. June–Sept. Clayton's Cliff-brake, Rock- or Winter-brake, Indian's Dream.

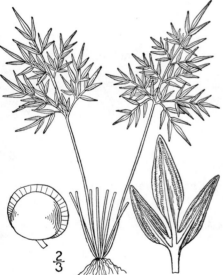

$\frac{2}{3}$

2. Pellea dénsa (Brack.) Hook. Oregon or Clayton's Cliff-brake. Fig. 77.

Onychium densum Brack. Fil. U. S. Expl. Exp. 120. 1854.
Pellaea densa Hook. Sp. Fil. **2**: 150. 1858.

Rootstocks slender, creeping, entangled, chaffy with narrow blackish scales. Stipes numerous, densely tufted, wiry, slender, light brown, 3′–9′ long; blades ovate or triangular-oblong, 1′–3′ long, densely 3-pinnate, the segments 3″–6″ long, linear, nearly sessile, acuminate or mucronate, those of the fertile blades tapering at each end, with narrowly recurved margins; indusium distinctly scarious; segments of the rarely sterile blades broader and serrate.

Mt. Albert, Gaspé, Quebec and Grey county, Ontario; British Columbia to Montana, Wyoming and California. Summer. Indian's Dream.

18. **CHEILÁNTHES** Sw. Syn. Fil.
126. 1806.

Small rock-loving ferns, mostly with pubescent, tomentose or scaly leaves, the blades uniform, 1–3-pinnate, the divisions often minute and bead-like. Sori terminal upon the veins, marginal, roundish and distinct, or somewhat confluent, often obscured by the hairy or scaly covering. Indusia formed of the revolute or reflexed usually modified margins of the segments. [Greek, in allusion to the marginal sori.]

About 100 or more species, of temperate and tropical regions. Besides the following numerous other species occur in the southwestern and western United States and in Mexico. Type species: *Cheilanthes micropteris* Sw.

Blades nearly glabrous.   1. *C. alabamensis.*
Blades hirsute or tomentose.
    Blades hirsute and glandular; indusia discontinuous.   2. *C. lanosa.*
    Blades tomentose; indusia mostly continuous.

Blades 2′–5′ long; stipes slender; indusia herbaceous.   3. *C. Feei.*
Blades 6′–15′ long; stipes stout, tomentose; indusia membranous.   4. *C. tomentosa.*

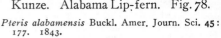

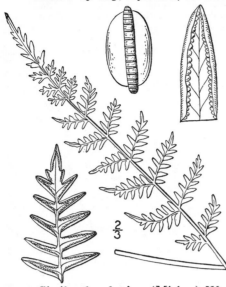

1. Cheilanthes alaménsis (Buckl.) Kunze. Alabama Lip-fern. Fig. 78.

Pteris alabamensis Buckl. Amer. Journ. Sci. **45**: 177. 1843.

C. alabamensis Kunze, Linnaea **20**: 4. 1847.

Rootstock creeping, rather stout and short, clothed with very slender hair-like dark ferruginous scales. Stipes black, 3′–7′ long, slender, wiry, villous at least towards the base with rusty hair-like scales; blades lanceolate, glabrous, 2′–10′ long, 2-pinnate; pinnae numerous, ovate-lanceolate, acuminate, very short-stalked, the lowest usually smaller than those above; pinnules oblong or triangular-oblong, mostly acute, often auriculate on the upper side at the base, or the larger ones on both sides and above more or less lobed; indusia pale, membranous, continuous or sometimes slightly interrupted by the incising of the pinnules.

On rocks, Virginia to Alabama, Illinois, Missouri, Arkansas and Arizona. Aug.–Oct.

2. Cheilanthes lanòsa (Michx.) Watt. Hairy Lip-fern. Fig. 79.

Nephrodium lanosum Michx. Fl. Bor. Am. **2**: 270. 1803.
Cheilanthes vestita Sw. Syn. Fil. 128. 1806.
C. lanosa Watt, Trimen's Journ. Bot. **12**: 48. 1874.

Rootstock short, creeping, with pale rusty-brown scales. Stipes tufted, wiry, chestnut-brown, 2′–4′ long, hirsute with rusty jointed hairs; blades herbaceous, oblong-lanceolate, 4′–9′ long, 1′–2′ wide, gradually attenuate to the apex, 2-pinnate; pinnae somewhat distant, especially the lower ones, deltoid-ovate to ovate-oblong, more or less densely hirsute like the stipe and rachis and usually somewhat glandular; pinnules in several pairs, close or somewhat apart, oblong, deeply pinnatifid into close roundish or oblong lobes, the margins of these forming separate herbaceous indusia.

On rocks, Connecticut and southern New York to Georgia, west to Kansas and Texas. Ascends to 1900 ft. in North Carolina. July–Sept. Clothed Lip-fern.

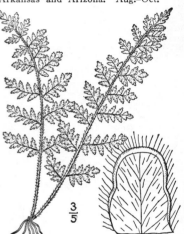

3. Cheilanthes Fèei Moore. Slender Lip-fern. Fig. 80.

Myriopteris gracilis Fée, Gen. Fil. 150. 1850–52.
C. lanuginosa Nutt.; Hook. Sp. Fil. **2**: 99. 1858.
C. gracilis Mett. Abh. Senck. Nat. Gesell. **3**: 80. 1859.
Cheilanthes Feei Moore, Ind. Fil. xxxviii. 1857.

Rootstock short, covered with narrow brown scales lined with black. Stipes densely tufted, slender, about as long as the leaves, at first covered with woolly hairs, at length nearly glabrous; blades ovate-lanceolate, 2′–5′ long, 1′–2′ wide, 2–3-pinnate, the upper surface slightly tomentose, the lower densely woolly with soft whitish-brown hairs; pinnae mostly oblong-ovate and contiguous, the lowermost deltoid-ovate and distant; pinnules pinnate or crenately pinnatifid into several pairs of crowded minute roundish segments, the reflexed margin forming an herbaceous indusium.

On rocks, Illinois and Minnesota to British Columbia, to Texas and New Mexico. July–Oct.

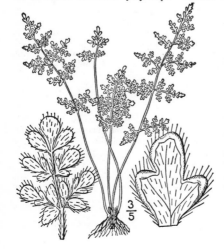

4. Cheilanthes tomentòsa Link. Woolly Lip-fern. Fig. 81.

Cheilanthes tomentosa Link, Hort. Berol. **2**: 42. 1833.

Rootstock stout, short, densely chaffy with rigid slender striped and concolorous bright brown scales. Stipes tufted, 4'–8' long, rather stout, densely brown-tomentose even when mature; blades oblong-lanceolate, 3-pinnate, 6'–18' long, densely tomentose, especially beneath, with brownish-white obscurely articulated hairs; pinnae and pinnules ovate-oblong or oblong-lanceolate, the ultimate pinnules distinct, usually obovate, about ½" long, the terminal ones sometimes twice as large as the others, the reflexed margin forming a narrow continuous indusium.

On rocks, Virginia to Georgia, Texas, Arizona and Mexico. July–Oct. Webby Lip-fern.

19. NOTHOLAÈNA R. Br. Prodr. Fl. Nov. Holl. **1**: 145. 1810.

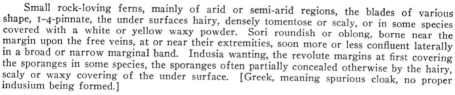

3/5

Small rock-loving ferns, mainly of arid or semi-arid regions, the blades of various shape, 1–4-pinnate, the under surfaces hairy, densely tomentose or scaly, or in some species covered with a white or yellow waxy powder. Sori roundish or oblong, borne near the margin upon the free veins, at or near their extremities, soon more or less confluent laterally in a broad or narrow marginal band. Indusia wanting, the revolute margins at first covering the sporanges in some species, the sporanges often partially concealed otherwise by the hairy, scaly or waxy covering of the under surface. [Greek, meaning spurious cloak, no proper indusium being formed.]

About 50 species of wide distribution. Besides the following numerous other species occur in the southwestern United States and Mexico. Type species: *Acrostichum Marantae* L.

3/4

1. Notholaena dealbàta (Pursh) Kunze. Powdery Notholaena. Fig. 82.

Cheilanthes dealbata Pursh, Fl. Am. Sept. 671. 1814.

Notholaena dealbata Kunze, Amer. Journ. Sci. (II.) **6**: 82. 1848.

Notholaena nivea var. *dealbata* Davenp. Cat. Davenp. Herb. Suppl. 44. 1883.

Rootstock short, chaffy with slender brown scales. Stipes closely tufted, wiry, very slender, shining, dark brown, 1'–4' long; leaves triangular-ovate, acute, broadest at the base, 1'–4' long, 3–4-pinnate, the rachis dark brown or blackish and wiry; pinnae ovate or deltoid-ovate, mostly with long slender stalks, the pinnules also mostly stalked; segments ovate-oblong, or somewhat elliptical by contraction, small, white and powdery on the lower surface.

On dry calcareous rocks, Missouri and Nebraska to Texas and Arizona. June–Sept.

20. POLYPÒDIUM [Tourn.] L. Sp. Pl. 1082. 1753.

Mainly shade-loving species of various habit, commonly epiphytic in the humid tropics, the leaves articulate to the creeping or ascending rhizome at the base of the stipe, the blades ranging from simple to bipinnate or several times pinnatifid, the veins free. Sori round or less commonly oval or elliptical, dorsal or sometimes terminal on the veins. Indusia wanting. [Greek, probably in allusion to the numerous knob-like prominences of the rootstock.]

As here limited to free-veined species, the genus comprises several hundred species, mainly of tropical and subtropical regions. Several additional species occur in the southern and western United States. Type species: *Polypodium vulgare* L.

Lower surface of the blade glabrous; plant green.   1. *P. vulgare.*
Lower surface of the blade densely scaly; plant grayish.   2. *P. polypodioides.*

1. Polypodium vulgàre L.   Common or
Golden Polypody. Fig. 83.

Polypodium vulgare L. Sp. Pl. 1085. 1753.

Rootstock slender, widely creeping, densely
covered with cinnamon-colored scales. Stipes
light colored, glabrous, 2′–6′ long; blades
ovate-oblong or narrowly oblong, subcoriaceous or chartaceous, evergreen, glabrous,
3′–10′ long, 1′–3′ wide, cut nearly to the rachis
into entire or slightly toothed, obtuse or subacute, linear or linear-oblong segments; sori
large, about midway between the midrib and
margins of the segments, upon the anterior
branch of the mostly 1–3-forked veins.

On rocks or rocky banks, occasionally on trees,
Labrador and Newfoundland to Manitoba and
Keewatin, south to Georgia, Alabama and Missouri. Ascends to 5600 ft. in Virginia. The
blade varies much in cutting, and numerous
forms have been described. One of these, the
var. *cambricum*, is notable for its broad pinnatifid segments. Male polypody, golden locks,
golden maiden's-hair. Adder's-, moss-, wood-,
male-, sweet-fern; Rock- or Stone-brake.

2. Polypodium polypodioìdes (L.) A.
S. Hitchcock. Gray Polypody. Fig. 84.

Acrostichum polypodioides L. Sp. Pl. 1068. 1753.
P. incanum Sw. Fl. Ind. Occ. 3: 1645. 1806.
Polypodium polypodioides A. S. Hitchcock, Rep.
Mo. Bot. Gard. 4: 156. 1893.

Rootstock widely creeping, woody, covered
with small brown scales. Stipes densely scaly,
1′–4′ long; blades oblong-lanceolate, acute, coriaceous, evergreen, 1′–7′ long, 1′–2′ wide, cut
very nearly or quite to the rachis into entire
oblong or linear-oblong obtuse segments, glabrous or nearly so on the upper surface, the
lower densely covered with gray peltate scales
with darker centers, as also the rachises; veins
indistinct, unconnected or casually joined.

On trees or less commonly on rocks, Pennsyl-
vania to Florida, west to Iowa, Kansas and Texas.
Widely distributed in tropical America. July–
Sept. Called also Hoary-, Scaly-, Tree-Polypody;
Rock-brake. Resurrection-fern.

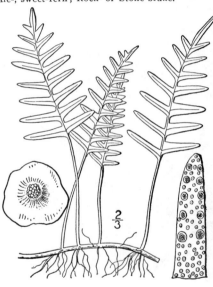

Family 6. **MARSILEÀCEAE** R. Br.
Prodr. Fl. Nov. Holl. **1**: 166.   1810.

Marsilea Family.

Perennial herbaceous plants rooting in mud, with slender creeping root-stocks
and 2- or 4-foliolate or filiform leaves. Asexual propagation consisting of sporocarps borne on peduncles which rise from the rootstock near the leaf-stalk or are
consolidated with it, containing both megaspores and microspores. The megaspores germinate into prothallia which bear mostly archegonia, while the microspores grow into prothallia bearing the antheridia.

Three genera and some 60 species of wide distribution known as *Pepperworts.*

1. MARSÍLEA L. Sp. Pl. 1099.   1753.

Marsh or aquatic plants, the leaves commonly floating on the surface of shallow water,
slender-petioled, 4-foliolate. Peduncles shorter than the petioles, arising from their bases or
more or less adnate to them. Sporocarps ovoid or bean-shaped, composed of two vertical
valves with several transverse compartments (sori) in each valve. [Name in honor of
Giovanni Marsigli, an Italian botanist, who died about 1804.]

About 53 species, widely distributed. Besides the following 2 or 3 others occur in Texas.
Sporocarps glabrous and purple when mature.   1. *M. quadrifolia.*
Sporocarps densely covered with hair-like scales.   2. *M. vestita.*

1. Marsilea quadrifòlia L   European
Marsilea or Pepperwort. Fig. 85.

Marsilea quadrifolia L. Sp. Pl. 1099. 1753.

Rootstock slender, buried in the muddy bottoms of shallow lakes or streams. Petioles usually slender, 2'–5' high, or when submerged sometimes elongated to 1° or 2°. Leaflets mostly triangular-obovate, variable in outline, 3"–8" long, 2"–6" wide, glabrous or rarely with scattered hairs when young, the margins entire; sporocarps 2 or rarely 3 on a branching peduncle which is attached to the petiole at its base, covered with short yellowish-brown hairs when young, becoming glabrous and dark purple when mature; sori 8 or 9 in each valve.

Bantam Lake, Litchfield Co., Conn.; thence introduced into other parts of the country, from Massachusetts to Maryland. Native of Europe and Asia.

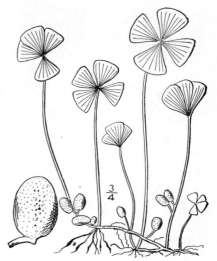

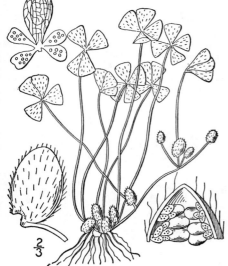

2. Marsilea vestìta Hook. & Grev.
Hairy Pepperwort. Fig. 86..

M. vestita Hook. & Grev. Ic. Fil. *pl. 159.* 1831.

Marsilea mucronata A. Br. Amer. Journ. Sci. (II.) **3**: 55. 1847.

Rootstocks slender, creeping. Petioles slender, 2'–5' high; leaflets similar to those of the preceding species, entire or toothed; sporocarps 2"–4" long, 2"–3" wide, with a short raphe, a short and blunt lower tooth and an acute and sometimes curved upper one, densely covered with soft spreading narrow hair-like scales or (in the forms known as *M. mucronata*) these short and appressed or almost wanting; sori 6–11 in each valve.

In wet sand or in shallow ditches, Florida to Kansas, Arizona and Mexico, California and British Columbia.

Family 7. SALVINIÀCEAE Reichenb. Consp. 30. 1828.
SALVINIA FAMILY.

Small floating plants with a more or less elongated and sometimes branching axis bearing apparently 2-ranked leaves. Sporocarps soft, thin-walled, borne 2 or more on a common stalk, 1-celled, with a central often branched receptacle, which bears megasporanges containing a single megaspore or microsporanges containing numerous microspores. The megaspores germinate into prothallia which bear archegones, the microspores into prothallia which bear antherids.

The family consists of two genera.
Leaves 6"–9" long, 2-ranked, on mostly simple stems.   1. *Salvinia.*
Leaves minute, closely imbricated on pinnately branching stems.   2. *Azolla.*

1. SALVÍNIA Adans. Fam. Pl. **2**: 15. 1763.

Floating annual plants with slender stems bearing rather broad 2-ranked leaves, these finely papillose on the upper surface. Sporocarps globose, depressed, 9–14-sulcate, membranous, arranged in clusters, 1 or 2 of each cluster containing 10 or more sessile megasporanges, each containing a single megaspore, the others containing numerous smaller globose pedicelled microsporanges with very numerous microspores. [Name in honor of Antonio Maria Salvini, 1633–1729, Italian scientist.]

About 13 species widely distributed. Only the following, the generic type, in the United States.

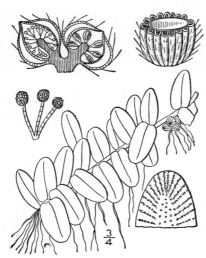

1. Salvinia nàtans (L.) Hoffm.   Salvinia.
Floating Moss.   Fig. 87.

Marsilea natans L. Sp. Pl. 1099.   1753.
Salvinia natans Hoffm. Deutschl. Fl. 2 : 1.   1795.

Leaves oblong, rather thick, obtuse or emarginate at the apex, rounded or cordate at the base, entire, spreading, 6′–12′ long, pinnately veined, bright green and papillose above, the lower surface densely matted with pellucid brown hairs; sporocarps 4–8 in a cluster, the upper ones containing about 10 megasporanges, each containing a single megaspore, the remainder containing numerous microsporanges each with numerous microspores; megaspores marked with 3 obtuse lobes, these meeting at the apex.

Bois Brulé Bottoms, Perry Co., Missouri, and near Minneapolis, Minn.   Introduced into ponds on Staten Island, N. Y.   Reported by Pursh in 1814 from central New York, but his exact station is unknown.   Widely distributed in Europe and Asia.

2. AZÓLLA Lam. Encycl. 1 : 343.   1783.

Minute moss-like reddish or green floating plants, with pinnately branched stems covered with minute imbricated 2-lobed leaves, and emitting rootlets beneath.   Sporocarps of two kinds borne in the axils of the leaves, the smaller ovoid or acorn-shaped, containing a single megaspore at the base and a few corpuscles above it whose character is not fully known, the larger globose, producing many pedicelled sporanges, each containing several masses of microspores which are often beset with a series of anchor-like processes of unknown function.   [Greek, signifying killed by drought.]

About 5 species of wide geographic distribution. Type species: *Azolla filiculoides* Lam.

1. Azolla coliniàna Willd.   Carolina
Azolla.   Fig. 88.

Azolla caroliniana Willd. Sp. Pl. 5 : 541.   1810.

Plants greenish or reddish, deltoid or triangular-ovate in outline, pinnately branching, sometimes covering large surfaces of water.   Leaves with ovate lobes, their color varying somewhat with the amount of direct sunlight, the lower usually reddish, the upper green with a reddish border. Megaspores minutely granulate, with three accessory corpuscles; masses of microspores armed with rigid septate processes.

Floating on still water, Ontario and Massachusetts to British Columbia, south to Florida, Arizona and Mexico.   Also in tropical America.   Naturalized in lakes on Staten Island, N. Y.

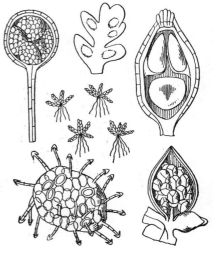

Family 8.   **EQUISETÀCEAE** Michx.
Fl. Bor. Am. 2 : 281.   1803.

HORSETAIL FAMILY.

Rush-like perennial plants, with mostly hollow jointed simple or often much-branched grooved stems, provided with a double series of cavities and usually with a large central one, the branches verticillate, the nodes provided with diaphragms.   Rootstocks subterranean.   Leaves reduced to sheaths at the joints, the sheaths toothed.   Sporanges 1-celled, clustered underneath the scales of terminal cone-like spikes.   Spores all of the same size and shape, furnished with 2 narrow strap-like appendages attached at the middle, coiling around the spore when moist and spreading when dry and mature, in the form of a cross (elaters).   Epidermis impregnated with silica, rough.   Prothallium on the surface of the ground, green, usually dioecious.

The family consists of the following genus:

1. EQUISÈTUM [Tourn.] L. Sp. Pl. 1061. 1753.

Characters of the family. [Name ancient, signifying horse-tail, in allusion to the copious branching of several species.] Called also Toad-pipe, Tad-pipe.

About 25 species, of very wide geographic distribution. Type species: *Equisetum fluviatile* L.

Stems annual; stomata scattered.
  Stems of two kinds, the fertile appearing in early spring before the sterile.
    Fertile stems simple, soon withering; sheaths of branches of sterile stems 4-toothed.
                                                         1. E. arvense.
    Fertile stems branched when old, only the apex withering.
      Branches of the stem simple, their sheaths 3-toothed.
                                               2. E. pratense.
      Branches compound.
                                               3. E. sylvaticum.
  Stems all alike; spores mature in summer; branches simple or none.
    Sheaths rather loose; branches usually long; stems bushy below, attenuate upwards.
      Central cavity very small; spike long.
                                          4. E. palustre.
      Central cavity about one-half the diameter of stem; spike short.
                                          5. E. littorale.
    Sheaths appressed; branches usually short.
                                          6. E. fluviatile.
Stems mostly perennial, evergreen; spikes tipped with a rigid point; stomata in regular rows.
  Stems tall, usually many-grooved.
    Stems rough and tuberculate, prominently ridged.
      Ridges with 1 line of tubercles; ridges of sheath tricarinate; stem stout.
                                        7. E. robustum.
      Ridges of the stem with 2 indistinct lines of tubercles; ridges of sheath obscurely 4-carinate; stem slender.
                                        8. E. hyemale.
    Stems not tuberculate; sheaths enlarged upward.
                                        9. E. laevigatum.
  Stems low, slender, tufted, usually 5–10-grooved.
    Central cavity small; sheaths 5–10-toothed.
                                      10. E. variegatum.
    Central cavity none; sheaths 3-toothed.
                                        11. E. scirpoides.

1. Equisetum arvénse L. Field Horsetail.
Fig. 89.

Equisetum arvense L. Sp. Pl. 1061. 1753.

Stems annual, provided with scattered stomata, the fertile appearing in early spring before the sterile. Fertile stems 4′–10′ high, not branched, soon withering, light brown, their loose scarious sheaths mostly distant, whitish, ending in about 12 brown acuminate teeth; sterile stems green, rather slender, 2′–2° high, 6–19-furrowed, with numerous long mostly simple verticillate 4-angled or rarely 3-angled solid branches, the sheaths of the branches 4-toothed, the stomata in 2 rows in the furrows.

In sandy soil, especially along roadsides and railways, Newfoundland and Greenland to Alaska, south to Virginia and California. Also in Europe and Asia. Ascends to at least 2500 ft. in Virginia. An occasional form in which the sterile stem bears a terminal spike is known as var. *serotinum*. Sterile stems sometimes very short and with long prostrate or ascending branches. Called also Cornfield Horsetail; Bottlebrush, Horse- or Snake-pipes; Cat's-tail. May.

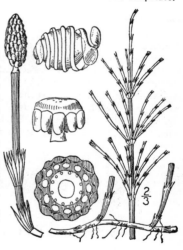

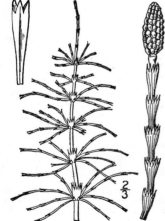

2. Equisetum praténse Ehrh. Thicket or Meadow Horsetail. Fig. 90.

Equisetum pratense Ehrh. Hanov. Mag. **9**: 138. 1784.

Stems annual, 8′–16′ high, with scattered stomata, the fertile appearing in spring before the sterile, branched when old, only its apex withering, the two becoming similar in age; stems rough, 8–20-ridged with narrow furrows and cylindric or cup-shaped sheaths; branches straight, rather short, simple, densely whorled, 3-angled or rarely 4–5-angled, solid; sheaths of the stem with about 11 short ovate-lanceolate teeth, those of the branches 3-toothed; rootstocks solid, acutely angled.

In sandy places, Nova Scotia and Rupert River to Minnesota, and Alaska, south to New Jersey, Iowa and Colorado. Also in Europe and Asia. July–Sept.

3. Equisetum sylváticum L.   Wood Horsetail.   Bottle-brush.   Fig. 91.

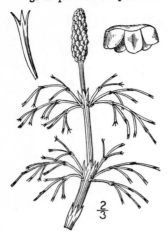

Equisetum sylvaticum L. Sp. Pl. 1061. 1753.

Stems annual, provided with scattered stomata, the fertile appearing in early spring before the sterile, at first simple, at length much branched and resembling the sterile, only its naked apex withering. Stems usually 12-furrowed, producing verticillate compound branches, the branchlets curved downward; sheaths loose, cylindric or campanulate, those of the stem with 8–14 bluntish teeth, those of the branches with 4 or 5 teeth, those of the branchlets with 3 divergent teeth; central cavity nearly one-half the diameter of the stem; branches and branchlets solid.

In moist sandy woods and thickets, Newfoundland and Greenland to Alaska, south to Virginia and Iowa. Also in Europe and Asia. May.

4. Equisetum palústre L.   Marsh Horsetail.   Fig. 92.

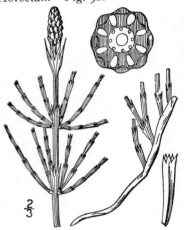

Equisetum palustre L. Sp. Pl. 1061. 1753.

Stems annual, slender, all alike, 10'–18' long, very deeply 5-9-grooved, the grooves separated by narrow roughish wing-like ridges, the central canal very small; sheaths rather loose, bearing about 8 subulate-lanceolate whitish-margined teeth; branches simple, few in the whorls, 4–7-angled, always hollow, barely sulcate, more abundant below than above, their sheaths mostly 5-toothed; spike rather long; stomata abundant in the furrows.

In wet places, Nova Scotia to Alaska, Connecticut, western New York, Illinois and Arizona. Also in Europe and Asia. July–Aug. Marsh-weed, Paddock-or Snake-pipes; Cat-whistles.

5. Equisetum littoràle Kuehl.   Shore Horsetail.   Fig. 93.

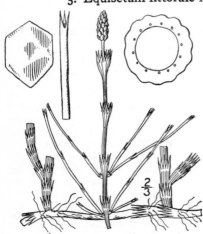

Equisetum littorale Kuehl. Beitr. Pflanz. Russ. Reichs, **4**: 91. 1845.

Stems annual, very slender, all alike, 8'–18' high, slightly roughened, 6–19-grooved, the ridges rounded, the central canal one-half to two-thirds the diameter; sheaths sensibly dilated above, the uppermost inversely campanulate, their teeth herbaceous, membranous at the margins, narrow, lanceolate; branches of two kinds, simple, some 4-angled and hollow, some 3-angled and solid, the first joint shorter or a trifle longer than the sheath of the stem; spike short with abortive spores, these commonly with no elaters.

On sandy river and lake shores, New Brunswick and Ontario to New Jersey and Pennsylvania, west to British Columbia. Also in Europe. Supposed to be a hybrid. Aug.–Sept.

6. Equisetum fluviátile L.   Swamp Horsetail.   Fig. 94.

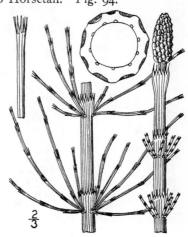

Equisetum limosum L. Sp. Pl. 1062. 1753.

Equisetum limosum L. Sp. Pl. 1062. 1753.

Stems annual, all alike, 2°–4° high, slightly 10–30-furrowed, very smooth, usually producing upright branches after the spores are formed, the stomata scattered. Sheaths appressed with about 18 dark brown short acute rigid teeth, air cavities wanting under the grooves, small under the ridges; central cavity very large; branches hollow, slender, smaller but otherwise much like the stems, short or elongated; rootstocks hollow.

In swamps and along the borders of ponds, Nova Scotia to Alaska, south to Virginia, Nebraska and Washington. Also in Europe and Asia. Water Horsetail, Paddock-pipes. May–June.

7. Equisetum robústum A. Br.   Stout Scouring-rush.   Fig. 95.

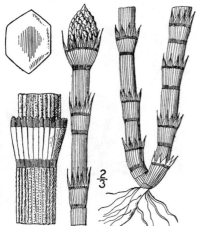

Equisetum robustum A. Br.; Engelm. Amer. Journ. Sci. **46**: 88. 1844.

Equisetum hyemale robustum A. A. Eaton, Fern Bull. **11**: 75. 1903.

Stems perennial, stout, tall, evergreen, 3°–11° high, sometimes nearly 1′ in diameter, 20–48-furrowed, simple or little branched. Ridges of the stem roughened with a single series of transversely oblong siliceous tubercles; sheaths short, nearly as broad as long, cylindric, appressed, marked with black girdles at the base, and at the bases of the dark caducous teeth; ridges of the sheath 3-carinate; branches when present occasionally fertile; spikes tipped with a rigid point.

In wet places, Ohio to Louisiana and Mexico, west to British Columbia and California. Also in Asia. May–June.

8. Equisetum hyemàle L.   Common Scouring-rush.   Fig. 96.

Equisetum hyemale L. Sp. Pl. 1062. 1753.

Stems slender, rather stiff, evergreen, 2°–4° high, with the stomata arranged in regular rows, rough, 8–34-furrowed, the ridges with two indistinct lines of tubercles, the central cavity large, from one-half to two-thirds the diameter; sheaths rather long, cylindric, marked with one or two black girdles, their ridges obscurely 4-carinate; teeth brown, membranous, soon deciduous; spikes pointed; stem rarely producing branches which are usually short and occasionally fertile; forms are sometimes found with longer sterile branches.

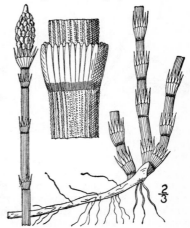

In wet places and on banks, especially along rivers and lakes, throughout nearly the whole of North America, Europe and Asia. The rough stems of this and related species are used for scouring floors. The species consists of numerous races. Called also Horsepipe, Mare's-tail, Shave-grass, Shave-weed, Pewterwort, Rough Horsetail, Dutch-rush, Gun-bright. May–June.

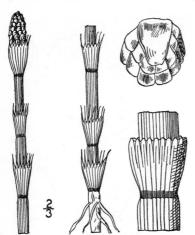

2/3

9. Equisetum laevigàtum A. Br. Smooth Scouring-rush. Fig. 97.

Equisetum laevigatum A. Br.; Engelm. Amer. Journ. Sci. 46: 87. 1844.

Stems 1°–5° high, simple or little branched, pale green, annual or persistent, 14–30-furrowed, the ridges almost smooth. Sheaths elongated and enlarged upward, marked with a black girdle at the base of the mostly deciduous, white-margined teeth and rarely also at their bases; ridges of the sheath with a faint central carina and sometimes with faint short lateral ones; stomata arranged in single series; central cavity very large, the wall of the stem very thin, spikes pointed.

Along streams and rivers, especially in clay soil, Ontario to New Jersey, North Carolina, Louisiana, British Columbia and the Mexican border. May–June.

10. Equisetum variegàtum Schleich. Variegated Equisetum. Fig. 98.

Equisetum variegatum Schleich. Cat. Pl. Helvet. 27. 1807.

Stems slender, perennial, evergreen, 6′–18′ long, rough, usually simple from a branched base, commonly tufted, 5–10-furrowed, the stomata borne in regular rows. Sheaths campanulate, distinctly 4-carinate, green, variegated with black above, the median furrow deep and excurrent to the teeth and downward to the ridges of the stem, the teeth 5–10, each tipped with a deciduous bristle; central cavity small, rarely wanting.

Labrador and Greenland to Alaska, south to Conneticut, western New York, Nebraska and Nevada. Also in Europe and Asia. Consists of several races. May–June.

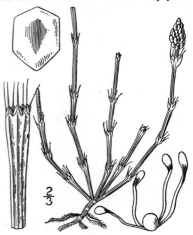

2/3

11. Equisetum scirpoìdes Michx. Sedge-like Equisetum. Fig. 99.

Equisetum scirpoides Michx. Fl. Bor. Am. 2: 281. 1803.

Stems perennial, evergreen, very slender or filiform, 3′–6′ long, somewhat rough, flexuous and curving, growing in slender tufts, mostly 6-furrowed with acute ridges, simple or branching from near the base. Sheaths 3-toothed, distinctly 4-carinate, the central furrow broad, the lateral narrow, the bristly teeth rather persistent; central cavity entirely wanting.

On moist or wet wooded banks, Labrador to Alaska, south to Pennsylvania, Illinois and British Columbia. Also in Europe and Asia. May–June.

2/3

Family 9. LYCOPODIÀCEAE Michx. Fl. Bor. Am. 2: 281. 1803.

Club-moss Family.

Somewhat moss-like, erect or trailing terrestrial herbs with numerous small lanceolate or subulate simple leaves, sometimes oblong or roundish, arranged in 2–many ranks, the stems often elongated, usually freely branching. Sporanges 1–3-celled, solitary in the axils of the leaves or on their upper surfaces. Spores uniform, minute. Prothallia (as far as known) mostly subterranean, with or without chlorophyll, monoecious.

Four genera and about 110 species. Besides the following, *Psilotum* occurs in Florida, the two other genera only in Australia.

1. LYCOPÒDIUM L. Sp. Pl. 1100. 1753.

Perennial plants with evergreen 1-nerved leaves arranged in 4–16 ranks. Sporanges coriaceous, flattened, reniform, 1-celled, situated in the axils of ordinary leaves or in those of the upper modified, bract-like ones, which are imbricated in sessile or peduncled spikes, opening transversely into 2 valves, usually by a line around the margin. Spores all of one kind, copious, sulphur-yellow, readily inflammable from the abundant oil they contain. [Greek, meaning wolf's-foot, perhaps in allusion to the branching roots of some species.]

About 100 species of wide geographic distribution, the largest occurring in the Andes of South America and in the Himalayas. Type species: *Lycopodium clavatum* L.

Sporophyls not closely associated in terminal spikes.
  Stems rigidly erect; leaves ascending, nearly uniform.   1. *L. Selago.*
  Stems ascending; leaves spreading or deflexed, longer or shorter in alternating zones.
    Leaves distinctly broadest above the middle, there usually erose-denticulate.
                                            2. *L. lucidulum.*
    Leaves linear or nearly so, entire or minutely denticulate.   3. *L. porophilum.*
Sporophyls closely associated in terminal spikes.
  Sporophyls similar to the foliar leaves in form and texture; sporanges subglobose.
    Sporophyls linear-deltoid, mostly entire; plants small.   4. *L. inundatum.*
    Sporophyls linear to lanceolate from a broader base; plants larger.
      Peduncles slender, the leaves incurved and mostly appressed; spikes slender, the sporophyls less than 3″ long, abruptly subulate, incurved.   5. *L. adpressum.*
      Peduncles very stout, the leaves more numerous and close, mostly ascending, not incurved; spikes stout, the sporophyls more than 4″ long, attenuate, ascending, spreading or reflexed.   6. *L. alopecuroides.*
  Sporophyls bract-like, very unlike the foliar leaves; sporanges reniform.
    Stems with numerous erect or assurgent leafy aerial branches, the spikes terminal upon some of these.
      Leaves of the ultimate aerial branches in 5 or more rows.
        Main stem creeping deep in the ground; aerial branches few, tree-like.
                                      7. *L. obscurum.*
        Main stem prostrate, or (in no. 10) a little below the surface; aerial branches numerous, not tree-like.
          Leaves of the ultimate aerial branches in 5 rows.   10. *L. sitchense.*
          Leaves of the ultimate aerial branches in more than 5 rows.
            Spikes solitary, sessile.   8. *L. annotinum.*
            Spikes one or several, on elongate peduncles.   12. *L. clavatum.*
      Leaves of the ultimate aerial branches in 4 rows.
        Spikes sessile upon leafy branches.   9. *L. alpinum.*
        Spikes borne upon bracteate peduncles, these terminal upon leafy branches.
          Leaves of the ultimate aerial branches adnate considerably more than half their length.
            Ultimate aerial branches conspicuously flattened; leaves of the under row greatly reduced, minute, deltoid-cuspidate.   14. *L. complanatum.*
            Ultimate aerial branches narrower and less flattened; leaves of the under row scarcely reduced, acicular.   15. *L. tristachyum.*
          Leaves of the ultimate aerial branches adnate about half their length or less.
                                    11. *L. sabinaefolium.*
    Stems without leafy aerial branches, the elongate peduncles arising directly from the prostrate stem.   13. *L. carolinianum.*

1. Lycopodium Selàgo L.   Fir Club-moss.
Fig. 100.

Lycopodium Selago L. Sp. Pl. 1102. 1753.

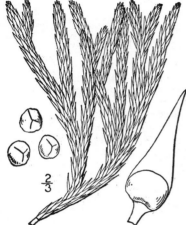

Stems rigidly erect from a short slender curved base, several times dichotomous, the densely foliaceous vertical branches forming mostly compact level-topped tufts 2′–6′ high; leaves nearly or quite uniform, very numerous, crowded, more or less appressed, or at least ascending, narrowly deltoid-lanceolate or somewhat acicular from a broader base, shining, pale green or yellowish, usually entire, acute, those bearing the sporanges (below the summit) a little shorter but not differing otherwise; plant frequently gemmiparous in the axils of the upper leaves.

On rocks, Labrador and Greenland to Alaska, south to the mountains of Maine, New Hampshire, Vermont and northern New York, on the summits of the higher Alleghenies to North Carolina, and to Michigan and Washington. Also in Europe and Asia. Autumn. Upright Club-moss, Fir-moss, Tree-moss, Fox-feet.

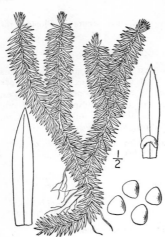

2. Lycopodium lucídulum Michx.   Shining Club-moss.   Fig. 101.

Lycopodium lucidulum Michx. Fl. Bor. Am. **2** : 284. 1803.

Stems rising 6'–10' from a curved or decumbent base, 1–3 times dichotomous, the branches forming a loose cluster of a few leafy vertical stems; leaves dark green, shining, wide-spreading or finally deflexed, acute, somewhat oblanceolate, broadest above the middle, there more or less erose-denticulate, tapering gradually to a narrower base, arranged in alternating zones of longer and shorter leaves, the latter more often bearing the sporanges, less denticulate, even entire; plant often gemmiparous, the gemmae early falling and giving rise to young plants.

In cold, damp woods, Newfoundland to British Columbia, south to South Carolina, Tennessee and Iowa. Ascends to nearly 5700 ft. in Virginia. Trailing evergreen, Moonfruit-pine. Aug.–Oct.

3. Lycopodium poróphilum Lloyd & Underw.   Lloyd's Club-moss.   Fig. 102.

Lycopodium porophilum Lloyd & Underw. Bull. Torrey Club **27** : 150. 1900.

Stems rising 2'–4' from a curved or decumbent base, 1–3 times dichotomous, the branches forming a rather close tuft of densely leafy vertical stems; leaves spreading or somewhat deflexed, entire or minutely denticulate, arranged in alternating series of longer and shorter, the former linear to linear-lanceolate, slightly broader above the middle, alternate, the latter distinctly broadest at the base, gradually tapering to an acuminate apex, and more often bearing the sporanges; plant often gemmiparous.

On partially shaded rocks, apparently preferring sandstone, Wisconsin to Indiana and Alabama; probably of wider distribution.

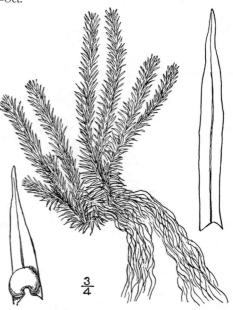

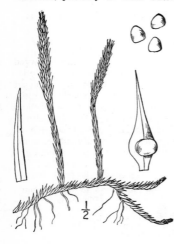

4. Lycopodium inundàtum L.   Bog or Marsh Club-moss.   Fig. 103.

Lycopodium inundatum L. Sp. Pl. 1102. 1753.

Plants small, with simple or 1–2-forked horizontal prostrate or slightly arched slender, often lax, leafy stems; peduncles 3"–2½' long, arising directly from the creeping stem, terminated by a slender spike ½'–1½' long, or the spike rarely subsessile; leaves of the stem linear-lanceolate, acute, mostly entire, curved upward, those of the peduncle more slender, spreading; sporophyls similar to the sterile leaves but wider at the base (linear-deltoid), spreading, entire or sometimes toothed just above the base.

In sandy bogs, Newfoundland to Alaska, south and west to New Jersey, Pennsylvania, Illinois, Michigan, Idaho and Washington. Also in Europe and Asia. Slender elongate forms, mainly from New England, are known as the var. *Bigelovii* Tuck.; they indicate a possible transition into the next species.

5. Lycopodium adpréssum (Chapm.)
Lloyd & Underw.   Chapman's
Club-moss.   Fig. 104.

Lycopodium inundatum var. *adpressum* Chapm.
   Fl. So. States, ed. 2. 671. 1883.
Lycopodium adpressum Lloyd & Underw. Bull.
   Torrey Club **27** : 153. 1900.
Lycopodium Chapmani Underw. Proc. U. S. Nat.
   Mus. **23** : 646. 1901.

Stems prostrate or slightly arching. 6'–16'
long, simple or rarely branched, leafy; peduncles 4'–12' long, slender, rigidly erect, arising
directly from the creeping stem, terminated by
a slender spike 9''–2¾' long; leaves of the
stem lanceolate-acuminate, curved upwards,
irregularly toothed, sometimes doubly so;
leaves of the peduncle more slender, incurved,
mostly appressed, yellowish green or stramineous, the lower ones sharply toothed, the
upper ones entire or nearly so; sporophyls
mostly incurved and subappressed, abruptly
subulate from an ovate more or less toothed
base.

   Moist banks and borders of swamps, New
York to the Gulf states, mainly near the coast.

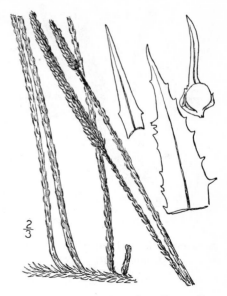

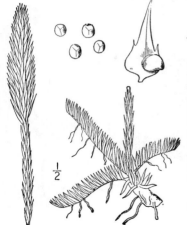

6. Lycopodium alopecuroìdes L.   Fox-tail
Club-moss.   Fig. 105.

Lycopodium alopecuroides L. Sp. Pl. 1102. 1753.

Stems stout, mostly recurved and more or less prostrate, elongate, 1°–2° long, densely leafy throughout;
peduncles very stout, 8'–13' long, erect, arising usually
from the arches of the sterile stems, terminating in
stout densely leafy spikes 9''–4' long, 4''–5'' thick;
leaves of the stem spreading, lanceolate-attenuate to
linear-subulate, conspicuously bristle-toothed, especially
below the middle, and hairy below near the base; leaves
of the peduncle similar, spreading or ascending; sporophyls similar but broader at the base, longer, with long
setaceous tips, ascending, spreading, or eventually reflexed, not hairy below.

   In pine-barren swamps, New York to Florida, near the
coast, west to Mississippi. Aug.–Oct. In tropical America.

7. Lycopodium obscùrum L.   Ground-pine.
Fig. 106.

Lycopodium obscurum L. Sp. Pl. 1102. 1753.
L. *dendroideum* Michx. Fl. Bor. Am. **2** : 282. 1803.

Main stem creeping horizontally, deep in the
ground, giving off a few distant upright aerial
branches, these 4'–10' high, tree-like, with numerous
bushy branches; leaves 8-ranked on the lower
branches, 6-ranked on the terminal, spreading, curved
upwards, linear-lanceolate, twisted, especially above,
the upper branches thus more or less dorsiventral,
sometimes conspicuously so; sporophyls broadly
ovate, acuminate, the margins scarious, erose.

   In moist woods, Newfoundland and Labrador to
Alaska, south to the mountains of North Carolina and
to Indiana. Ascends to 4000 ft. in Virginia. Also in
Asia. July–Sept. Spiral-pine, Tree-like-club-moss, Bunch-
evergreen, Crow-foot.

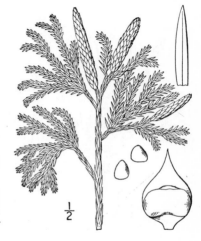

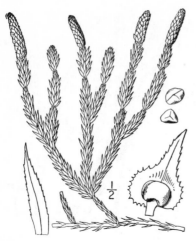

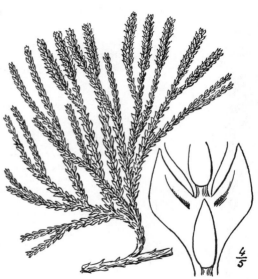

8. Lycopodium annótinum L.   Stiff Club-moss.   Fig. 107.

Lycopodium annotinum L. Sp. Pl. 1103. 1753.

Stems prostrate, creeping, 1°–3° or more long, stiff, rarely pinnately branching, leafy, with numerous aerial branches, these 5′–10′ high, simple or 1–3 times forked, the divisions mostly fertile; leaves uniform, 8-ranked, spreading horizontally or somewhat reflexed, with upward curved apices, lanceolate to linear-lanceolate, broadest at the middle or above, serrulate, pungent; spikes 1 or several, oblong-cylindric, ½′–1¼′ long, the sporophyls broadly ovate-subulate, with erose margins.

In woods and thickets, commonly in dry soil, Labrador to Alaska, south to Pennsylvania, Michigan, Colorado and Washington. Also in Europe and Asia. Mountain forms with more rigid pointed leaves have been separated as var. *pungens.* Autumn. Called also Interrupted club-moss.

9. Lycopodium alpìnum L.   Alpine Club-moss.   Fig. 108.

Lycopodium alpinum L. Sp. Pl. 1104. 1753.

Main stem prostrate, creeping (9′–1½° long) at or near the surface, with numerous ascending freely branched aerial stems 1½′–4′ high; branches crowded, glaucous, the fertile ones terete and longer than the others, with subulate leaves, the foliar ones strongly dorsiventral with leaves of 3 kinds in 4 rows, those of the upper row narrowly ovate, acute, those of the lateral rows thick, asymmetrical, falcate, the tips decurved, those of the under row trowel-shaped; spikes sessile, ⅜′–¾′ long; sporophyls ovate, erose, acute.

In woods, Quebec and Labrador to Washington and Alaska. Also in Europe and Asia. Cypress-moss. Heath-cypress, Savin-leaved club-moss.

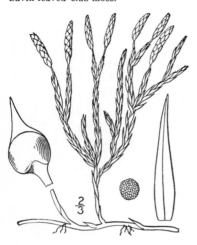

10. Lycopodium sitchénse Rupr.   Alaskan Club-moss.   Fig. 109.

Lycopodium sitchense Rupr. Beitr. Pfl. Russ. Reich. 3: 30. 1845.

Stems prostrate, 8′–15′ long, nearly superficial, sending up numerous aerial stems, these several times dichotomous, the branches terete, vertical, forming compact tufts 2′–3′ high, with few or numerous stronger projecting fertile branches; leaves of the branchlets 5-ranked, appressed or spreading and curved upwards, linear, thick, entire, acute; spikes sessile or upon short (up to ⅜′ long) minutely bracteate peduncles, solitary, cylindric; sporophyls ovate, acuminate or long-subulate, the margins erose.

In cold woods, Labrador and Quebec to Alaska, south to Washington, New York and northern New England.

11. Lycopodium sabìnaefòlium Willd.
Cedar-like Club-moss.   Fig. 110.

Lycopodium sabinaefolium Willd. Sp. Pl. **5**: 20. 1810.

Horizontal stems extensively creeping at or near the surface of the ground and occasionally branching, with numerous freely branched assurgent aerial stems, the branches of these 2'–4' long, loosely clustered, dorsiventrally flattened; leaves ascending, slender, subulate, nearly equal, in 4 rows upon the terminal and subterminal branchlets, those of the lateral rows slightly larger, thicker and more widely spreading than the usually subappressed leaves of the upper and lower rows; peduncles ¾'–2' long, slender, bracteate, terminal upon the main terete branches; spikes mostly solitary (casually 3), ½'–1¼' long, the sporophyls broadly ovate, acuminate, greenish, with scarious erose margins.

In cold mountain woods, Prince Edward Island, Quebec, northern New England and Ontario.

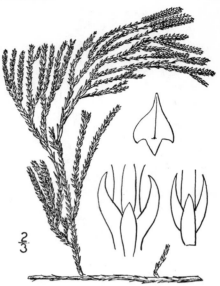

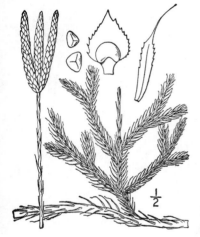

12. Lycopodium clavàtum L.   Running-pine.
Club-moss.   Fig. 111.

Lycopodium clavatum L. Sp. Pl. 1101. 1753.

Main stems prostrate, extensively creeping (3°–9°) along the ground, branching horizontally, with numerous very leafy ascending pinnately branched aerial stems; leaves crowded, many-ranked, linear, bristle-tipped, entire or denticulate, those of the main stems strongly denticulate; peduncles stout, 2½'–4½' long, channeled, with slender whorled or scattered denticulate bristle-tipped bracts, simple or several times forked near the summit; spikes linear-cylindric, the sporophyls deltoid-ovate, acuminate, or bristle-tipped, the margins scarious, erose.

In woods, Labrador to Alaska, south to North Carolina, Michigan and Washington. Also in Europe, Asia and tropical America. The spores of this species, and those of *L. complanatum*, furnish the inflammable power known as Lycopodium powder or vegetable sulphur, used in stage effects. Aug.–Oct. Called also running-moss. Foxtail. Buck's-horn. Buck's-grass. Staghorn-moss. Snakemoss. Wolf's-claws. Ground-pine. Toad's-tail. Lamb's-tails. Creeping-bur. Creeping Jennie. Coral-evergreen.

13. Lycopodium caroliniànum L.   Carolina
Club-moss.   Fig. 112.

Lycopodium carolinianum L. Sp. Pl. 1104. 1753.

Stems short, 1'–6' long, prostrate, pinnately branching, rooting below; leaves strongly dimorphic, those of the sides large, ovate-lanceolate, falcate, recurved, asymmetrical, acute, those of the upper side smaller, subulate from a broad base; peduncles 2'–8½' long, slender, with a few whorled or scattered subulate bracts; spikes ¾'–2' long, slender; sporophyls deltoid, acuminate, entire or somewhat erose.

In moist pine barrens, New Jersey to Florida and Louisiana near the coast.

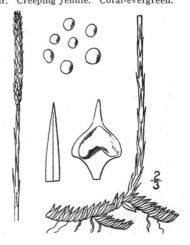

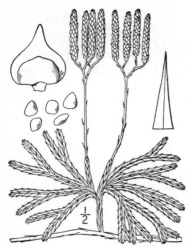

14. Lycopodium complanàtum L.   Trailing Christmas-green. Ground-pine. Fig. 113.

Lycopodium complanatum L. Sp. Pl. 1104. 1753.

Horizontal stems prostrate, wide-creeping, flattened above, sparingly branched, with numerous erect irregularly forked aerial stems, the branches of these broadly flattened, somewhat glaucous, 2–3-forked, the divisions few and somewhat apart or, more commonly, numerous, closer, and fan-like, leafy throughout, the leaves 4-ranked, minute and (excepting those of the under row) imbricate and strongly decurrent, those of the upper row narrow and incurved, of the lateral rows broad, with spreading tips, and of the under row minute, deltoid-cuspidate; peduncles slender, 1′–5′ long, bracteate, rarely simple, usually once or twice dichotomous, each branch terminating in a slender cylindric spike about 9″ long; sporophyls broadly ovate, acuminate.

In woods and thickets, Newfoundland to Alaska, south to North Carolina, Indiana, Minnesota and Idaho. Also in Europe and Asia. Ground-cedar, Festoon-pine, Crowfoot, Hogbed, Creeping Jennie.

15. Lycopodium tristàchyum Pursh. Ground-pine. Fig. 114.

L. tristachyum Pursh, Fl. Am. Sept. 653. 1814.
Lycopodium chamaecyparissus A. Br. in Döll, Rhein. Fl. 36. 1843.

Horizontal stems extensively creeping 1′–4′ below the surface of the ground, terete, sparingly branched, with numerous erect or assurgent repeatedly-forked aerial stems, the branches of these glaucous, narrow, somewhat flattened, with very numerous, crowded, erect divisions; ultimate divisions leafy throughout, the leaves 4-ranked, minute, imbricate, appressed, strongly decurrent, nearly equal and alike, those of the under row scarcely differing from the others, those of the lateral rows a little thicker, with the tips usually incurved downward; peduncles 3′–5′ long, bracteate, usually 2 (casually 3) times dichotomous at the summit; spike and sporophyls similar to those of the preceding.

In dryish open woods or clearings, usually in sandy soil, northern Maine to Minnesota and Georgia. Also in Europe. Early August.

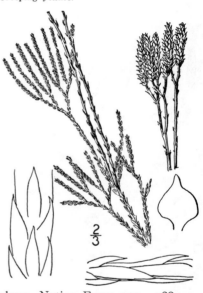

Family 10.   SELAGINELLÀCEAE Underw. Native Ferns 103. 1881.

Terrestrial, annual or perennial, moss-like plants with branching stems and scale-like leaves, which are many-ranked and uniform, or 4-ranked and of two types spreading in two planes. Sporanges 1-celled, solitary in the axils of leaves which are so arranged as to form more or less quadrangular spikes, some containing 4 megaspores (megasporanges), others containing numerous microspores (microsporanges), which develop into small prothallia, those from the megaspores bearing archegones, those from the microspores antherids.

The family consists of the following genus:

1.   SELAGINÉLLA Beauv. Prodr. Aetheog. 101. 1805.

Characters of family. [Name diminutive of Selago, an ancient name of some *Lycopodium*.]

About 340 species, widely distributed, most abundant in the tropics. Besides the following some five others occur in western North America. Type species: *Lycopodium selaginoides* L.
Stem-leaves all alike, many-ranked.
Stems compact with rigid leaves; spikes quadrangular.   1. *S. rupestris.*
Stems slender; leaves lax, spreading; spikes enlarged, scarcely quadrangular.   2. *S. selaginoides.*
Stem-leaves of 2 kinds, 4-ranked, spreading in 2 planes.   3. *S. apus.*

1. Selaginella rupéstris (L.) Spring.
Rock Selaginella.   Festoon-pine.
Fig. 115

Lycopodium rupestre L. Sp. Pl. 1101. 1753.
Selaginella rupestris Spring in Mart. Fl. Bras.
1²: 118.   1840.

Stems densely tufted, with occasional sterile runners and sub-pinnate branches, 1'-3' high, commonly curved when dry. Leaves rigid, appressed-imbricated, 1" or less long, linear or linear-lanceolate, convex on the back, more or less ciliate, many-ranked, tipped with a distinct transparent awn; spikes sessile at the ends of the stem or branches, strongly quadrangular, 6"-12" long, about 1" thick; bracts ovate-lanceolate, acute or acuminate, broader than the leaves of the stem; megasporanges and microsporanges borne in the same spikes, the former more abundant.

On dry rocks, New England and Ontario to Georgia and the middle West. Ascends to at least 2000 ft. in Virginia. Dwarf club-moss. Christmas-evergreen. Resurrection-plant. Aug.–Oct.

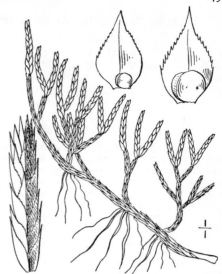

2. Selaginella selaginoìdes (L.) Link.
Low Selaginella.   Fig. 116.

Lycopodium selaginoides L. Sp. Pl. 1101.   1753.
S. spinosa Beauv. Prodr. Aetheog. 112.   1805.
Selaginella selaginoides Link, Fil. Hort. Berol.
158.   1841.

Sterile branches prostrate-creeping, slender, ½'-2' long, the fertile erect or ascending, thicker, 1'-3' high, simple; leaves lanceolate, acute, lax and spreading, sparsely spinulose-ciliate, 1"-2" long; spikes solitary at the ends of the fertile branches, enlarged, oblong-linear, subacute, 1' or less long, 2"-2½" thick; bracts of the spike lax, ascending, lanceolate or ovate-lanceolate, strongly ciliate.

On wet rocks, Labrador to Alaska, south to New Hampshire, Michigan and Colorado. Also in northern Europe, Greenland and Asia. Mountain-moss. Prickly club-moss. Summer.

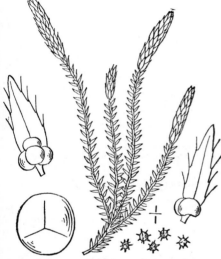

3. Selaginella àpus (L.) Spring.
Creeping Selaginella.   Fig. 117.

Lycopodium apodum L. Sp. Pl. 1105.   1753.
S. apus Spring in Mart. Fl. Bras. 1²: 119.   1840.

Annual, light green, stems prostrate-creeping, 1'-4' long, much branched, flaccid, angled on the face. Leaves minute, membranous, of 2 kinds, 4-ranked, spreading in 2 planes; upper leaves of the lower plane spreading, the lower reflexed, ovate, acute, serrulate, not distinctly ciliate; leaves of the upper plane ovate, short-cuspidate; spikes 3"-8" long, obscurely quadrangular; bracts ovate, acute, sometimes serrulate, acutely keeled in the upper half; megasporanges more abundant toward the base of the spike.

In moist shaded places, often among grass, Maine and Ontario to the Northwest Territory, south to Florida, Louisiana and Texas. Ascends to 2200 ft. in Virginia. July–Sept.

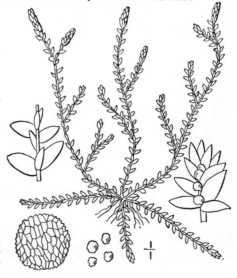

4

Family 11. **ISOETÀCEAE** Underw. Native Ferns, 104. 1881.

Aquatic or marsh plants rooting in the mud, with a short buried 2-lobed or 3-lobed trunk (stem) sending out abundant roots and sending up a compact tuft of rush-like leaves. Sporanges sessile in the axils of the leaves, some containing megaspores (megasporanges), others microspores (microsporanges); the former germinate into prothallia bearing only archegones, the latter into prothallia bearing usually only a single antherid.

1. **ISÒETES** L. Sp. Pl. 1100. 1753.

Submerged, amphibious or uliginous plants with a cluster of elongated awl-shaped leaves rising from a more or less 2–3-lobed fleshy short stem, the leaves with or without peripheral bast-bundles, with or without stomata, bearing a small membranous organ (ligule) above the base. Sporanges sessile in the excavated bases of the leaves, orbicular or ovoid, the sides more or less covered with a fold of the inner side of the leaf-base (velum). The sporanges of the outer leaves usually contain spherical, mostly sculptured macrospores, those of the inner ones contain minute powdery usually oblong microspores. [Name Greek, taken from Pliny, apparently referring to the persistent green leaves.]

The family consists of the following genus only.
About 60 species, widely distributed. Besides the following 2 are known from the southern United States, 7 from the Pacific Coast and 2 from Mexico. Owing to their aquatic habitat and apparently local distribution, these plants are popularly little known. The spores mature in summer and autumn. Type species: *Isoetes lacustris* L.

Leaves without peripheral bast bundles.
  Leaves without stomata; plants submerged.
    Leaves stiff and erect.   1. *I. macrospora.*
    Leaves slender and mostly recurved.
      Leaves about ½" in diameter.   2. *I. Tuckermanii.*
      Leaves at least 1½" in diameter.   3. *I. hieroglyphica.*
  Leaves with stomata; plants partially submerged, or emersed.
    Leaves green.
      Macrospores armed with spines.   4. *I. Braunii.*
      Macrospores without spines, merely crested or warted.
        Leaves 2′–3′ long; macrospores less than 550μ in diameter.   5. *I. saccharata.*
        Leaves 4′–8′ long; macrospores about 600μ in diameter.   6. *I. riparia.*
    Leaves reddish, or rarely olive green.   7. *I. foveolata.*
Leaves with stomata and bast bundles.
  Aquatic, but usually inhabiting the water's edge, sometimes completely emersed.
    Leaves from 12′–30′ long; macrospores with convolute labyrinthine ridges.
                                        8. *I. Eatoni.*
  Leaves shorter; macrospores otherwise marked.
    Bast bundles only 4, except in some forms of *I. Engelmanni.*
      Monoecious.
        Macrospores coarsely crested.   9. *I. Dodgei.*
        Macrospores merely reticulated.   10. *I. Engelmanni.*
      Polygamous; microspores rare.   11. *I. Gravesii.*
    Bast bundles 4, with extra ones near periphery.   12. *I. melanopoda.*
  Terrestrial, or rarely growing near water's edge.   13. *I. Butleri.*

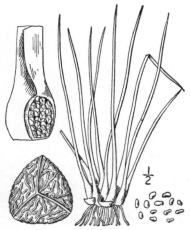

1. **Isoetes macróspora** Durieu. Lake Quillwort. Fig. 118.

Isoetes macrospora Durieu, Bull. Soc. Bot. France, 11: 101. 1864.
Isoetes heterospora A. A. Eaton, Fernwort Papers 8. 1900.

Submerged or rarely above water in dry seasons; leaves 10–30, rigid, rather thick, scarcely tapering, dark or olive green, obtusely quadrangular, 2′–6′ long; stomata none; peripheral bast-bundles wanting; sporange orbicular or broadly elliptic, unspotted; velum rather narrow; ligule triangular, short or somewhat elongated; macrospores 600–800 μ in diameter, marked all over with distinct or somewhat confluent crests, and bearing three converging ridges; microspores 30–46 μ long, smooth or papillose.

In 1°–5° of water, Labrador to the Northwest Territory, south to eastern Massachusetts and New Jersey. Formerly confused with *Isoetes lacustris* L., of the Old World.

2. Isoetes Tuckermáni A. Br. Tuckerman's Quillwort. Fig. 119.

Isoetes Tuckermani A. Br. in A. Gray, Man. Ed. 5, 676. 1867.

Isoetes Tuckermani borealis A. A. Eaton, Fernwort Papers 10. 1900.

Isoetes Harveyi A. A. Eaton, Fernwort Papers 11. 1900.

Submerged or rarely partly or wholly emersed during very dry seasons; leaves 10–40, very slender, tapering, olive-green, quadrangular, 2'–5' long, without peripheral bast-bundles, the outer recurved; sporange oblong, mostly white, its upper one-third covered by the velum; macroscpores 440–785 μ in diameter, with wavy somewhat parallel and branching ridges on the upper half, separated by the three converging ridges, the lower covered with an irregular network; microspores 26–42 μ long, nearly smooth, or with minute spines.

In ponds, Newfoundland to Massachusetts and Connecticut; clustered in shallow water.

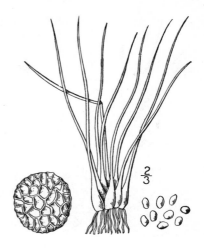

3. Isoetes hieroglýphica A. A. Eaton. Warty Quillwort. Fig. 120.

Isoetes hieroglyphica A. A. Eaton, Fernwort Papers 10. 1900.

Submerged with a bilobed trunk. Leaves 10–20, 2½'–3' long, blunt at the apex, without stomata; sporange spotted with dark cells, covered one-third to two-thirds by the velum; macrospores 486–720 μ in diameter, polished, covered with bold vermiform, subconfluent and somewhat reticulated ridges, becoming naked near the equator; microspores 31–44 μ in diameter, distinctly verrucose.

Nova Scotia to Quebec and Maine; usually growing in ponds and lakes.

4. Isoetes Braùnii Durieu. Braun's Quillwort. Fig. 121.

Isoetes Braunii Durieu, Bull. Soc. Bot. France, **11**: 101. 1864.

Isoetes echinospora var. *Braunii* Engelm. in A. Gray, Man. Ed. 5, 676. 1867.

Isoetes echinospora Boottii Engelm. in A. Gray, Man. Ed. 5, 676. 1867.

Isoetes echinospora robusta Engelm. Trans. St. Louis Acad. **4**: 380. 1882.

Isoetes echinospora muricata Engelm. in A. Gray, Man. Ed. 5, 676. 1867.

Submerged or in dry seasons emersed, leaves 10–70, tapering, soft, green, 3'–12' long, without peripheral bast-bundles, bearing stomata only toward the tip; sporange orbicular or broadly elliptic, spotted, one-half to three-fourths covered with the velum; macrospores 400–620 μ in diameter, covered with broad spinules which are often slightly confluent and incised at the tips; microspores 25–32 μ long, smooth.

Labrador and Greenland to Alaska, south to New Jersey, Pennsylvania and California.

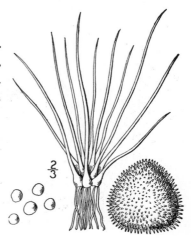

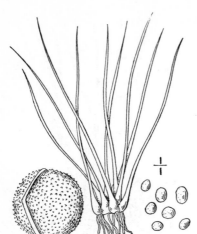

5. Isoetes saccharàta Engelm. Sugary Quill-wort. Fig. 122.

Isoetes saccharata Engelm. in A. Gray, Man. Ed. 5, 676. 1867.

Amphibious or uliginous with a flat depressed trunk. Leaves 10–30, green, pale at the base, spreading, 2′–3′ long, quadrangular, bearing numerous stomata; sporange oblong, unspotted, with a narrow velum covering only one-fourth or one-third of its surface; peripheral bast-bundles wanting; ligule triangular, rather short; macrospores 420–510 μ in diameter, with very minute distinct or rarely confluent warts as if sprinkled with grains of sugar; microspores sparingly papillose, 22–30 μ long.

In mud overflowed by the tides, eastern Maryland and District of Columbia.

6. Isoetes ripària Engelm. Riverbank Quillwort. Fig. 123.

Isoetes riparia Engelm.; A. Br. Flora, **29**: 178. 1846.

Amphibious or uliginous, usually emersed when mature; leaves 10–30, green, rather rigid, 4′–8′ long, quadrangular, bearing numerous stomata; peripheral bast-bundles wanting; ligule rather short, triangular; sporange mostly oblong, distinctly spotted with groups of brown cells, one-fourth to three-fourths covered with the velum; macrospores 450–756 μ in diameter, marked with distinct or anastomosing jagged crests or somewhat reticulate on the lower side; microspores 28–32 μ long, more or less tuberculate.

Borders of the lower Delaware River.

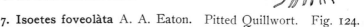

7. Isoetes foveolàta A. A. Eaton. Pitted Quillwort. Fig. 124.

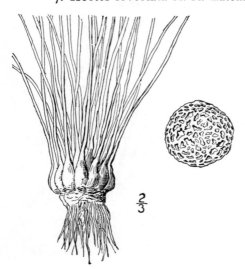

Isoetes foveolata A. A. Eaton; Dodge, Ferns and Fern Allies of New Eng. 38. 1896.
Isoetes foveolata pleurospora A. A. Eaton, Rhodora **5**: 280. 1903.

Amphibious from a bilobed or rarely trilobed base. Leaves 15–70, stout, 2′–6′ long, pinkish even when dry, or rarely dark green; stomata scattered, found only near the tips; no peripheral bast-bundles; monoicous or becoming dioicous; velum covering ¼ or ⅓ of the sporange; ligule round-ovate; sporanges thickly sprinkled with dark cells which are often collected in groups; macrospores 380–560 μ, covered beneath with very thick-walled reticulations, the openings appearing like little pits; reticulates elongate on the upper surface of the spore; microspores dark brown, 22–35 μ long, densely reticulate and usually slightly papillose.

In muddy banks, New Hampshire, Massachusetts and Connecticut

8. Isoetes Eatoni Dodge. Eaton's Quillwort. Fig. 125.

Isoetes Eatoni Dodge, Ferns and Fern Allies of New Eng. 39. 1896.

Amphibious from a large trunk $\frac{1}{2}'-1\frac{1}{4}'$ in diameter. Leaves of the submerged plant 30–200, varying in length up to 28', marked with an elevated ridge on the ventral side; leaves of the emersed plant shorter, 3–6; stomata abundant; peripheral bast-bundles irregular in occurrence or often wanting; velum covering $\frac{1}{4}$ of the sporange, polygamous; sporanges large, 0.4' by 0.15', pale spotted; macrospores small, 300–450 μ in diameter, marked with convolute labyrinthine ridges and cristate on the angles of the inner face; microspores 25–30 μ in diameter, minutely tuberculate.

In mud flats, New Hampshire to New Jersey.

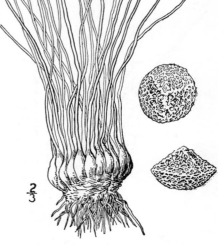

9. Isoetes Dódgei A. A. Eaton. Dodge's Quillwort. Fig. 126.

Isoetes Dodgei A. A. Eaton, Fern Bull. **6** : 6. 1898.
Isoetes canadensis A. A. Eaton, Proc. U. S. Nat. Mus. **23** : 650. 1901.

Plant amphibious from a bilobed trunk. Leaves 10–75, 8'–18' long when submersed, erect, or spirally ascending when scattered; emersed leaves 4'–6' long, tortuous and often interlaced, with numerous stomata and usually 4 bast-bundles; velum narrow, covering $\frac{1}{8}$ to $\frac{1}{4}$ of the sporange; sporanges thickly sprinkled with light brown cells; macrospores more numerous on submersed plants, globose, 500–675 μ in diameter, sparsely covered with irregular crests which at maturity separate into irregular groups leaving bare spaces, serrate or spinulose at the top; microspores more numerous on emersed plants, 22–44 μ, ashy, papillose or wrinkled.

Growing in mud flats, East Kingston, New Hampshire.

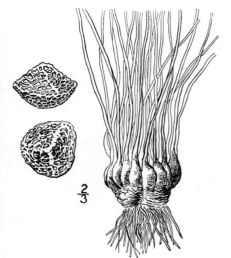

10. Isoetes Engelmánni A. Br. Engelmann's Quillwort. Fig. 127.

Isoetes Engelmanni A. Br. Flora, **29** : 178. 1846.
Isoetes Engelmanni fontana A. A. Eaton, Fern Bull. **13** : 52. 1905.
Isoetes Engelmanni gracilis Engelm. in A. Gray, Man. Ed. 5, 677. 1867.
Isoetes Engelmanni valida Engelm. in A. Gray, Man. Ed. 5, 677. 1867.

Amphibious, usually partly emersed when mature. Leaves 25–100, light green, quadrangular, tapering, 9'–20' long, bearing abundant stomata; peripheral bast-bundles present; monoecious; sporange oblong or linear-oblong, unspotted; velum narrow; macrospores 320–750 μ in diameter, covered with honeycomb-like reticulations; microspores 24–30 μ long, mostly smooth.

In ponds and ditches, rooting in mud, Maine to Virginia and Pennsylvania, Illinois and Missouri.

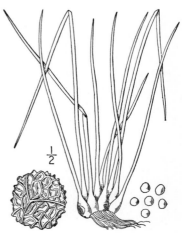

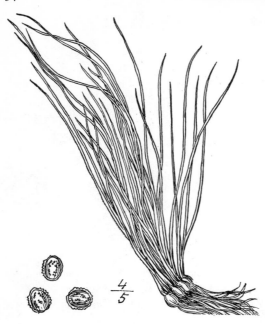

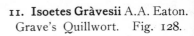

11. Isoetes Gràvesii A.A. Eaton. Grave's Quillwort. Fig. 128.

Isoetes Gravesii A. A. Eaton, **Fernwort**
Papers 14. 1900.

Polygamous; leaves 20-150, 5'-11' long and 1"-1½" in diameter, erect, reddish or dark green; sporanges with an abundance of light brown cells, ⅕-⅓ covered by the velum; macrospores 351-405 μ in diameter, the upper hemisphere depressed, covered with short truncate single columns; microspores 22-30 μ long, high-cristate or tuberculate.

Edges of ponds and streams; **Massachusetts and Connecticut.**

12. Isoetes melanópoda J. Gay. Black-based Quillwort. Fig. 129.

I. melanopoda J. Gay, Bull. Soc. Bot. France, **11** : 102. 1864.
Terrestrial with a subglobose deeply 2-lobed trunk. Leaves 15-60, slender, erect, bright green, with a blackish shining base, 5'-18' long, triangular, bearing stromata throughout, well developed peripheral bast-bundles, thick dissepiments and small air cavities within; ligule triangular, awl-shaped; sporange mostly oblong, spotted, with a narrow velum; polygamous; macrospores 250-400 μ in diameter, with low more or less confluent tubercles, often united into worm-like wrinkles, or almost smooth; microspores 23-30 μ long, spinulose.

In moist prairies and overflowed fields, Illinois to Iowa Oklahoma and California.

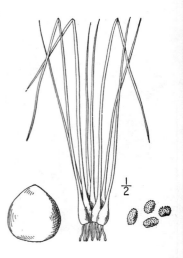

13. Isoetes Bútleri Engelm. Butler's Quillwort. Fig. 130.

Isoetes Butleri Engelm. Coult. Bot. Gaz. **3**: 1. 1878.

Terrestrial from a subglobose trunk. Leaves 8-60, bright green, paler at the base, triangular, 3'-7' long, bearing numerous stomata, and with well developed peripheral bast-bundles, thick dissepiments and small air cavities within; sporange usually oblong, spotted; velum very narrow or none; ligule small, triangular; dioecious; macrospores 400-630 μ in diameter, with distinct or confluent tubercles; "microspores 28-34 μ long, dark brown, papillose."

On rocky hillsides, Illinois and Kansas, **southward to** Tennessee and Oklahoma

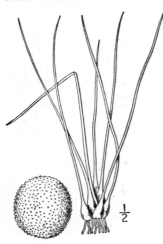

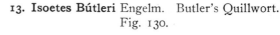

Subkingdom SPERMATÓPHYTA.

SEED-BEARING PLANTS.

Plants producing seeds which contain an embryo formed of one or more rudimentary leaves (cotyledons), a stem (hypocotyl, radicle), and a terminal bud (plumule), or these parts sometimes undifferentiated before germination. Microspores (pollen-grains) are borne in microsporanges (anther-sacs) on the apex or side of a modified leaf (filament). The macrosporanges (ovules) are borne on the face of a flat or inrolled much modified leaf (carpel) and contain one macrospore (embryo-sac); this develops the minute female prothallium, an archegone of which is fertilized by means of a tube (pollen-tube), a portion of the male prothallium sprouting from the pollen-grain.

The Seed-bearing plants form the most numerous group in existence, not less than 120,-000 species being known. The subkingdom was formerly known as Phanerogamia, or Phaenogamia and more recently as Anthophyta, this term signifying the presence of flowers, which characterizes most of the group. But the consideration that the spore-bearing organs of the Pine Family cannot well be regarded as flowers, and the fact that the production of seeds is the most characteristic difference between these plants and the Pteridophyta, are reasons which have led to the acceptance of the term here adopted.

There are two classes in the subkingdom, which differ from each other as follows:

Ovules and seeds borne on the face of a scale; stigmas none.   Class 1. GYMNOSPERMAE.
Ovules and seeds contained in a closed cavity (ovary).   Class 2. ANGIOSPERMAE.

Class 1. *GYMNOSPÉRMAE.*

Ovules (macrosporanges) naked, not enclosed in an ovary, this represented by a scale or apparently wanting. Pollen-grains (microspores) dividing at maturity into two or more cells, one of which gives rise to the pollen-tube (male prothallium), which directly fertilizes an archegone of the nutritive endosperm (female prothallium) in the ovule.

The Gymnosperms are an ancient group, first known in Silurian time. They became most numerous in the Triassic age. They are now represented by not more than 500 species of trees and shrubs.

There are three orders, Coniferales, Cycadales and Gnetales, the first of which is represented in our area by the Pine and Yew Families.

Family 1. **PINÀCEAE** Lindl. Nat. Syst. Ed. 2, 313. 1836.

PINE FAMILY. CONIFERS.

Resinous trees or shrubs, mostly with evergreen narrow entire or scale-like leaves, the wood uniform in texture, without tracheae, the tracheids marked by large depressed disks, the pollen-sacs and ovules borne in separate spikes (aments). Perianth none. Stamens several together, subtended by a scale; filaments more or less united; pollen-sacs (anthers) 2–several-celled, variously dehiscent; pollen-grains often provided with two lateral inflated sacs. Ovules with two integuments, orthotropous or amphitropous, borne solitary or several together on the surface of a scale, which is subtended by a bract in most genera. Fruit a cone with numerous, several or few, woody, papery or fleshy scales; sometimes berry-like. Seeds wingless or winged. Endosperm fleshy or starchy, copious. Embryo straight, slender. Cotyledons 2 or several.

About 25 genera and 240 species of wide distribution, most abundant in temperate regions.
Scales of the cone numerous (except in *Larix*); leaf-buds scaly.
  Cone-scales woody; leaves needle-shaped, 2–5 in a sheath.   1. *Pinus.*
  Cone-scales thin; leaves linear-filiform, scattered or fascicled, not in sheaths.
    Leaves fascicled on very short branchlets, deciduous.   2. *Larix.*
    Leaves scattered, persistent.
      Cones pendulous; leaves jointed to short persistent sterigmata.
        Leaves tetragonal, sessile.   3. *Picea.*
        Leaves flat, short-petioled.   4. *Tsuga.*
      Cones erect; sterigmata inconspicuous or none.   5. *Abies.*
Scales of the cone few (3–12); leaf-buds naked.
  Cone-scales spiral, thick; leaves deciduous.   6. *Taxodium.*
  Cone-scales opposite; leaves persistent.
    Cone oblong, its scales not peltate.   7. *Thuja.*
    Cone globose, its scales peltate.   8. *Chamaecyparis.*
    Fruit fleshy, berry-like, a modified cone.   9. *Juniperus.*

1. PÌNUS [Tourn.] L. Sp. Pl. 1000. 1753.

Evergreen trees with two kinds of leaves, the primary ones linear or scale-like, deciduous, the secondary ones forming the ordinary foliage, narrowly linear, arising from the axils
of the former in fascicles of 2–5 (rarely solitary in some western species), subtended by the
bud-scales, some of which are united to form a sheath. Staminate aments borne at the
bases of shoots of the season, the clusters of stamens spirally arranged, each in the axil of
a minute scale; filaments very short; anthers 2-celled, the sacs longitudinally dehiscent.
Ovule-bearing aments solitary or clustered, borne on the twigs of the preceding season, composed of numerous imbricated minute bracts, each with an ovule-bearing scale in its axil,
ripening into a large cone, which matures the following autumn, its scales elongating and
becoming woody. Seeds 2 on the base of each scale, winged above, the testa crustaceous.
[Name Celtic. The popular names of the species are much confused.]

About 100 species, natives of the northern hemisphere. In addition to the following, 25 others
occur in southern and western North America. Type species: *Pinus sylvestris* L., of Europe.
The group of which *Pinus Strobus* L. is the type is regarded by some authors as a distinct genus.

Leaves 5 in a sheath; cone-scales little thickened at the tip.   1. *P. Strobus.*
Leaves 2 or 3 in a sheath; cone-scales much thickened at the tip.
   Cones terminal or subterminal.
      Leaves 2 in a sheath; cones $1\frac{1}{2}'-2\frac{1}{2}'$ long, their scales pointless.   2. *P. resinosa.*
      Leaves 3 in a sheath; cones $4'-10'$ long, their scales prickle-tipped.
         Cones light, $6'-10'$ long; leaves $10'-16'$ long.   3. *P. palustris.*
         Cones very heavy and woody, $3'-4\frac{1}{2}'$ long; leaves $3'-6'$ long.   4. *P. scopulorum.*
   Cones lateral.
      Cone-scales with neither spine nor prickle; leaves in 2's.   5. *P. Banksiana.*
      Cone-scales tipped with a spine or prickle.
         Leaves some or all of them in 2's.
            Cones $1\frac{1}{2}'-2\frac{1}{2}'$ long, their scales tipped with prickles.
               Leaves stout, $1\frac{1}{2}'-2\frac{1}{2}'$ long.   6. *P. virginiana.*
               Leaves slender, $3'-5'$ long.   7. *P. echinata.*
            Cones $3\frac{1}{2}'-5'$ long, their scales tipped with very stout short spines.
                                            8. *P. pungens.*
     Leaves in 3's (very rarely some in 2's or 4's).
        Cones oblong-conic; leaves $6'-10'$ long; old sheaths $6''-10''$ long.   9. *P. Taeda.*
        Cones ovoid.
            Leaves $3'-5'$ long; cone-scales with stiff prickles.   10. *P. rigida.*
            Leaves $6'-10'$ long; cone-scales with small slender deciduous or obsolete prickles.
                                           11. *P. serotina.*

1. Pinus Stròbus L.   White Pine.   Weymouth Pine.   Fig. 131.

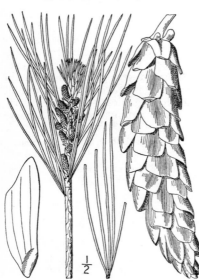

Pinus Strobus L. Sp. Pl. 1001.   1753.

A large forest tree, reaching a maximum height
of over $225°$ and a trunk diameter of $10\frac{1}{2}°$, the
bark nearly smooth except when old, the branches
horizontal, verticillate. Leaves 5 in a sheath, very
slender, pale green and glaucous, $3'-5'$ long, with
a single fibro-vascular bundle, the dorsal side
devoid of stomata; sheath loose, deciduous;
ovule-bearing aments terminal, peduncled; cones
subterminal, drooping, cylindric, often slightly
curved, $4'-6'$ long, about $1'$ thick when the scales
are closed, resinous; scales but slightly thickened
at the apex, obtuse and rounded or nearly truncate, without a terminal spine or prickle.

In woods, often forming dense forests, Newfoundland to Manitoba, south to Delaware, along the
Alleghanies to Georgia and to Illinois and Iowa.
Ascends to 4300 ft. in North Carolina and to 2500
ft. in the Adirondacks. Wood light brown or nearly
white, soft, compact, one of the most valuable of
timbers; weight per cubic foot, 24 lbs. June. Called
also Soft, Deal, Northern or Spruce-pine.

2. **Pinus resinòsa** Ait. Canadian Pine. Red Pine. Fig. 132.

Pinus resinosa Ait. Hort. Kew. **3**: 367. 1789.

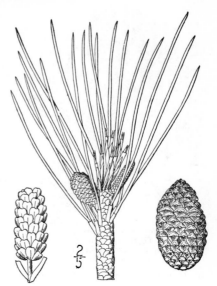

A tall forest tree, reaching a maximum height of about 150° and a trunk diameter of 5°, the bark reddish, rather smooth, flaky when old. Leaves 2 in each sheath, slender, dark green, 4'–6' long, with 2 fibro-vascular bundles; sheaths 6''–12'' long when young; staminate aments 6''–9'' long; cones subterminal spreading, oval-conic, 1½'–2½' long, usually less than 1' thick while the scales are closed; scales thickened at the apex, obtuse, rounded and devoid of spine or prickle.

In woods, Newfoundland to Manitoba, south to Massachusetts, Pennsylvania, Wisconsin and Minnesota. Wood compact, not strong, light red; weight per cubic foot 30 lbs. May–June. Called also hard- and norway-pine.

3. **Pinus palústris** Mill. Long-leaved Pine. Georgia Pine. Fig. 133.

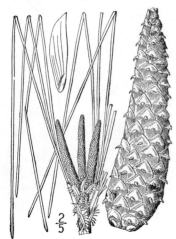

Pinus palustris Mill. Gard. Dict. Ed. 8, No. 14. 1768.
Pinus australis Michx. f. Hist. Arb. Am. **1**: 64. *pl. 6.* 1810.

A large tree, sometimes attaining a height of 120° and a trunk diameter of 5°, the bark nearly smooth. Leaves in 3's, slender, dark green, clustered at the ends of the branches, much elongated, 8'–16' long, with 2 fibro-vascular bundles; sheaths 1'–1¼' long; buds long; staminate aments rose-purple, 2'–3½' long, very conspicuous; cones terminal, spreading or erect, conic-cylindric, 6'–10' long, 2'–3' thick before the scales open; scales thickened at the apex, which is provided with a transverse ridge bearing a short central recurved prickle.

In sandy, mostly dry soil, often forming extensive forests, southern Virginia to Alabama, Florida and Texas, mostly near the coast. Wood hard, strong, compact, light red or orange; weight per cubic foot 44 lbs. This tree is the chief source of our turpentine, tar, rosin, and their derivatives. Also known as Southern, Yellow, Hard or Pitch Pine; Fat, Heart, Turpentine-pine; Virginia, Florida, Texas Yellow and Long-straw pine; Pine-broom and White Rosin-tree. March–April.

4. **Pinus scopulòrum** (Engelm.) Lemmon. Rock Pine. Fig. 134.

P. ponderosa scopulorum Engelm. in Brewer & Watson, Bot. Cal. **2**: 126. 1880.
P. scopulorum Lemmon Gard. & For. **10**: 183. 1897.

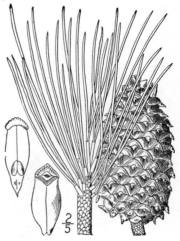

A large tree, attaining a maximum height of about 120° and a trunk diameter of 3½°. Branches widely spreading or somewhat drooping; bark nearly black, scaly; leaves in 3's (rarely some of them in 2's), rather stout, 3'–6' long; cones subterminal, very dense and heavy, ovoid-conic, 3'–4' long, 1½'–2½' thick; scales thickened at the apex, the transverse ridge prominent, with a short slender recurved prickle.

South Dakota to Nebraska, Texas, Utah and Arizona. Wood hard, strong, light brown; weight per cubic foot 29 lbs. United in first edition with *Pinus ponderosa* Dougl. April–May. Long-leaved, Red, Bull, Western pitch, and Gambier Parry's-pine.

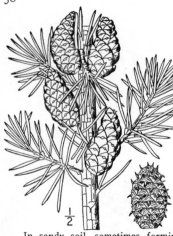

5. Pinus Banksiàna Lamb. Labrador Pine. Gray Pine. Fig. 135.

Pinus sylvestris var. *divaricata* Ait. Hort. Kew. **3**: 366. 1789.
Pinus Banksiana Lamb. Pinus, **1**: 7. *pl. 3.* 1803.
Pinus divaricata Gordon, Pinetum, 163. 1858.

A slender tree, usually 40–60° high, but sometimes reaching 100°, and a trunk diameter of 3°, the branches spreading, the bark becoming flaky. Leaves in 2's, stout, stiff, more or less curved, spreading or oblique, light green, crowded along the branches, seldom over 1' long; fibro-vascular bundles 2; cones commonly very numerous, lateral, oblong-conic, usually upwardly curved, 1'–2' long, 9"–15" thick when mature; scales thickened at the end, the transverse ridge a mere line with a minute central point in place of spine or prickle at maturity; young scales spiny-tipped.

In sandy soil, sometimes forming extensive forests, Nova Scotia to Hudson Bay and the Northwest Territory, south to Maine, northern New York, northern Illinois and Minnesota. Wood soft, weak, compact, light brown; weight per cubic foot 27 lbs. Also called Hudson Bay Pine, Northern scrub-pine; Black, Bank's-, Shore-, Jack- and Rock-pine; Unlucky-tree. May–June.

6. Pinus virginiàna Mill. Jersey Pine. Scrub Pine. Fig. 136.

Pinus virginiana Mill. Gard. Dict. Ed. 8, No. 9. 1768.
Pinus inops Ait. Hort. Kew. **3**: 367. 1789.

A slender tree, usually small, but sometimes attaining a height of 110° and a trunk diameter of 3°, the old bark dark colored, flaky, the branches spreading or drooping, the twigs glaucous. Leaves in 2's, dark green, rather stout and stiff, 1½'–2½' long, with 2 fibro-vascular bundles; young sheaths rarely more than 2½" long; cones commonly few, lateral, recurved when young, spreading when old, oblong-conic, 1½'–2½' long, their scales somewhat thickened at the apex, the low transverse ridge with a short more or less recurved prickle.

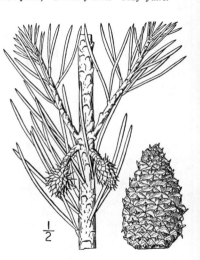

In sandy soil, Long Island, New York to Georgia, Alabama and southern Indiana and Tennessee, sometimes forming forests. Ascends to 3300 ft. in Virginia. Wood soft, weak, brittle, light orange; weight per cubic foot 33 lbs. April–May. Called also Short-shucks, Short-leaved or Short-shot Pine; Spruce, Cedar, Nigger and River-pine.

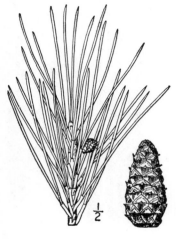

7. Pinus echinàta Mill. Yellow Pine. Spruce Pine. Fig. 137.

Pinus echinata Mill. Gard. Dict. Ed. 8, No. 12. 1768.
Pinus mitis Michx. Fl. Bor. Am. **2**: 204. 1803.

A forest tree, reaching a maximum height of about 120° and a trunk diameter of 4½°, the branches spreading, the old bark rough in plates. Leaves some in 2's, some in 3's, slender, not stiff, dark green, 3'–5' long, spreading when mature; fibro-vascular bundles 2; young sheaths 5"–8" long; cones lateral, oblong-conic, about 2' long, usually less than 1' thick when the scales are closed; scales thickened at the apex, marked with a prominent transverse ridge and armed with a slender small nearly straight early deciduous prickle.

In sandy soil, southern New York to Florida, west to Illinois, Kansas and Texas. Wood heavy, strong, orange; one of the most valuable timbers; weight per cubic foot 38 lbs. Also called Short-leaved or Short-shot Pine, and Bull, Carolina, Pitch, and Slash-pine. May–June.

8. Pinus púngens Lambert.   Table-Mountain Pine.   Hickory Pine.   Fig. 138.

Pinus pungens Lambert; Michx. f. Hist. Arb. Am. 1 : 61. pl. 5. 1810.

A tree with a maximum height of about 60° and trunk diameter of 3½°, the branches spreading, the old rough bark in flakes. Leaves mostly in 2's, some in 3's, stout and stiff, light green, 2½'-4' long, crowded on the twigs; fibro-vascular bundles 2; young sheaths 5''-8'' long; cones lateral, usually clustered, long-persistent on the branches, ovoid, 3½'-5' long, 2'-3' thick while the scales are closed, nearly globular when these are expanded; scales very thick and woody, their ends with a large elevated transverse ridge, centrally tipped by a stout reflexed or spreading spine 2''-2½'' long.

In woods, sometimes forming forests, western New Jersey and central Pennsylvania to Georgia and Tennessee. Ascends to 4000 ft. in North Carolina. Wood soft, weak, brittle, light brown; weight per cubic foot 31 lbs. May. Called also Prickly pine, Southern Mountain-pine.

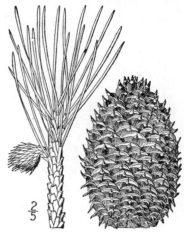

$\frac{2}{5}$

9. Pinus Taèda L.   Loblolly Pine.   Old-field Pine.   Fig. 139.

Pinus Taeda L. Sp. Pl. 1000. 1753.

A large forest tree, reaching, under favorable conditions, a height of 150° and a trunk diameter of 5°, the branches spreading, the bark thick and rugged, flaky in age. Leaves in 3's (rarely some of them in 2's), slender, not stiff, light green, ascending or at length spreading, 6'-10' long; fibro-vascular bundles 2; sheaths 8''-12'' long when young; cones lateral, spreading, oblong-conic, 3'-5' long, 1'-1½' thick before the scales open; scales thickened at the apex, the transverse ridge prominent, acute, tipped with a central short triangular reflexed-spreading spine.

Southern New Jersey to Florida and Texas, mostly near the coast, north through the Mississippi Valley to Arkansas. Wood not strong, brittle, coarse-grained, light brown; weight per cubic foot 34 lbs. Springs up in old fields or in clearings. Also called Frankincense, Sap, Torch, Slash, Swamp, Bastard, Long-straw or Indian-pine; Long-shucks; Foxtail, Shortleaf, and Rosemary pine. April–May.

$\frac{2}{5}$

10. Pinus rígida Mill.   Pitch Pine.   Torch Pine.   Fig. 140.

Pinus rigida Mill. Gard. Dict. Ed. 8, No. 10. 1768.

A forest tree reaching a maximum height of about 80° and a trunk diameter of 3°, the branches spreading, the old bark rough, furrowed, flaky in strips. Leaves in 3's (very rarely some in 4's), stout and stiff, rather dark green, 3'-5' long, spreading when mature; fibro-vascular bundles 2; sheaths 4''-6'' long when young; cones lateral, ovoid, 1½'-3' long, becoming nearly globular when the scales open, commonly numerous and clustered; scales thickened at the apex, the transverse ridge acute, provided with a stout central triangular recurved-spreading prickle.

In dry, sandy or rocky soil, New Brunswick to Georgia, west to southern Ontario, Ohio, West Virginia, Tennessee and Alabama. Ascends to 3000 ft. in Virginia. This forms most of the "pine barrens" of Long Island and New Jersey. Wood soft, brittle, coarse-grained, light reddish-brown; weight per cubic foot 32 lbs. Also called Sap, Hard, Yellow, and Black Norway or Candlewood-pine; produces shoots from cut stumps. April–May. Leaves sometimes only 1½' long on mountain trees.

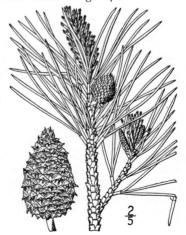

$\frac{2}{5}$

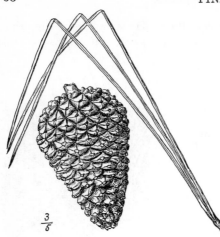

$\frac{3}{6}$

11. Pinus serótina Michx.　Pond Pine.
Fig. 141.

P. serotina Michx. Fl. Bor. Am. **2**: 105.　1803.

A tree of ponds and swamps, reaching a maximum height of about 75° and a trunk diameter of 3°, its trunk usually short, the bark fissured into small plates.　Leaves in 3's (rarely some in 4's), pale green, glaucous, 6′–10′ long, with 2 fibro-vascular bundles; sheaths about ½′ long; cones ovoid to globular-ovoid, about 2½′ long, the scales bearing a slender, incurved, usually deciduous prickle.

Atlantic coastal plain, southern New Jersey; Virginia to Florida.　Wood soft, brittle, coarse-grained; weight per cubic foot about 49 lbs.

Pinus sylvestris L., the Scotch Pine, of northern Europe, which resembles *P. resinosa* Ait. in having two needles to each sheath and unarmed cone-scales, is much planted for ornament and has become established on the coasts of Maine and Massachusetts.

2.　LÀRIX [Tourn.] Adans. Fam. Pl. **2**: 480.　1763.

Tall trees with horizontal or ascending branches and small narrowly linear deciduous leaves, without sheaths, in fascicles on short lateral scaly bud-like branchlets.　Aments short, lateral, monoecious, the staminate from leafless buds; the ovule-bearing buds commonly leafy at the base, and the aments red.　Anther-sacs 2-celled, the sacs transversely or obliquely dehiscent.　Pollen-grains simple.　Cones ovoid or cylindric, small, erect, their scales thin, spirally arranged, obtuse, persistent.　Ovules 2 on the base of each scale, ripening into 2 reflexed somewhat winged seeds.　[Name ancient, probably Celtic.]

About 9 species, natives of the north temperate and subarctic zones.　Besides the following, 2 others occur in western North America.　Type species: *Larix Larix* (L.) Karst., of Europe, much planted for ornament, and reported as established in Connecticut.

1. **Larix larícina** (Du Roi) Koch.　American Larch.　Tamarack.　Fig. 142.

Pinus laricina Du Roi, Obs. Bot. 49.　1771.
Pinus pendula Ait. Hort. Kew. **3**: 369.　1789.
Larix americana Michx. Fl. Bor. Am. **2**: 203.　1803.
Larix laricina Koch, Dendrol. **2**: Part 2, 263.　1873.

A slender tree, attaining a maximum height of about 100° and a trunk diameter of 3°, the branches spreading, the bark close or at length slightly scaly.　Leaves pale green, numerous in the fascicles, 5″–12″ long, about ¼″ wide, deciduous in late autumn; fascicles borne on short lateral branchlets about 2″ long; cones short-peduncled at the ends of similar branchlets, ovoid, obtuse, 6″–8″ long, composed of about 12 suborbicular thin scales, their margins entire or slightly lacerate.

In swampy woods and about margins of lakes, Newfoundland to the Northwest Territory, south to New Jersey, Pennsylvania, Indiana and Minnesota.　Wood hard, strong, very durable, resinous, light brown; weight per cubic ft. 39 lbs.　Called also Hackmatack, Hackmak, Black or Red Larch, Juniper Cypress.　March–April.

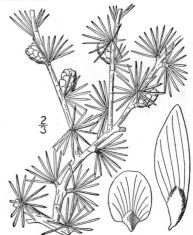

$\frac{2}{3}$

3.　PÌCEA Link, Abh. Akad. Wiss. Berlin, **1827**: 179.　1827–1830.

Evergreen conical trees, with linear short 4-sided leaves spreading in all directions, jointed at the base to short persistent sterigmata, on which they are sessile, falling away in drying, the bare twigs appearing covered with low truncate projections.　Leaf-buds scaly.　Staminate aments axillary, nearly sessile; anthers 2-celled, the sacs longitudinally dehiscent, the connective prolonged into an appendage; pollen-grains compound; ovule-bearing aments, terminal, ovoid or oblong; ovules 2 on the base of each scale, reflexed, ripening into 2 more or less winged seeds.　Cones ovoid to oblong, obtuse, pendulous, their scales numerous, spirally arranged, thin, obtuse, persistent.　[Name ancient.]

About 18 species, of the north temperate and subarctic zones.　Besides the following, 5 others occur in the northwestern parts of North America.　Type species: *Picea Abies* (L.) Karst., of Europe, which is much planted for ornament and is reported as spontaneous in Connecticut.

Twigs and sterigmata glabrous, glaucous; cones oblong-cylindric.　　　　　　1. *P. canadensis.*
Twigs pubescent, brown; cones ovoid or oval.
　　Leaves glaucous; cones persistent.　　　　　　　　　　　　　　　　2. *P. mariana.*
　　Leaves not glaucous; cones deciduous.　　　　　　　　　　　　　　　3. *P. rubens.*

1. Picea canadénsis (Mill.) B.S.P. White or Pine Spruce. Fig. 143.

Abies canadensis Mill. Gard. Dict. Ed. 8, No. 4. 1768.
Pinus alba Ait. Hort. Kew. **3**: 371. 1789.
Picea alba Link, Linnaea, **15**: 519. 1841.
Picea canadensis B.S.P. Prel. Cat. N. Y. 71. 1888.

A slender tree, attaining a maximum height of about 110° and a trunk diameter of 3°, but usually much smaller. Twigs and sterigmata glabrous, pale and glaucous; leaves light green, slender, 6″–8″ long, very acute; cones cylindric or oblong-cylindric, pale, 1½′–2′ long, 6″–8″ thick before the scales open; scales almost membranaceous, their margins usually quite entire; bracts incised.

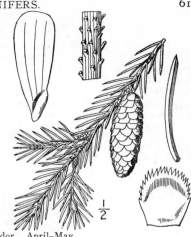

Newfoundland to Hudson Bay and Alaska, south to Maine, northern New York, Michigan and South Dakota. Wood soft, weak, light yellow; weight per cubic foot 25 lbs. Called also Cat Pine or Spruce; and Single, Black or Skunk-spruce. Sometimes with a skunk-like odor. April–May.

2. Picea mariàna (Mill.) B.S.P. Black Spruce. Fig. 144.

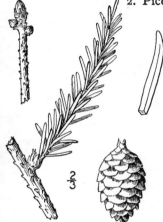

Abies mariana Mill. Gard. Dict. Ed. 8, No. 5. 1768.
Pinus nigra Ait. Hort Kew. **3**: 370. 1789.
Abies nigra Desf. Hist. Arb. **2**: 580. 1809.
Picea nigra Link, Linnaea, **15**: 520. 1841.
Picea mariana B.S.P. Prel. Cat. N. Y. 71. 1888.
Picea brevifolia Peck, Spruces of the Adirondacks 13. 1897.

A slender tree, sometimes 90° high, the trunk reaching a diameter of 2°–3°, the branches spreading, the bark only slightly roughened. Twigs pubescent; sterigmata pubescent; leaves thickly covering the twigs, deep green, glaucous stout, straight or curved, rarely more than ½′ long, obtuse or merely mucronate at the apex; cones oval or ovoid, 1′–1½′ long, persistent on the twigs for two or more seasons, their scales with entire or erose margins.

Newfoundland to Hudson Bay and the Northwest Territory, south to New Jersey, along the higher Alleghanies to North Carolina and to Michigan and Minnesota. Wood soft, weak, pale red or nearly white; weight per cubic foot 28 lbs. Called also Yew or Spruce Pine; He Balsam; Spruce Gum-tree; Juniper; and Blue, Double, White and Cat Spruce.

3. Picea rùbens Sargent. Red Spruce. Fig. 145.

Pinus rubra Lamb. Pinus, **1**: 43. *pl. 28.* 1803. Not Mill.
Picea rubra Dietr. Fl. Berl. **2**: 795. 1824.
Picea rubens Sargent, Silva N. A. **12**: 33. 1898.

A slender tree, sometimes reaching a height of 100° and a trunk diameter of 4°, the branches spreading, the bark reddish, nearly smooth. Twigs slender, sparingly pubescent; sterigmata glabrate; leaves light green, slender, straight or sometimes incurved, very acute at the apex, 5″–8″ long; cones ovoid or oval, seldom more than 1½′ long, deciduous at the end of the first season or during the winter, their scales undulate or lacerate.

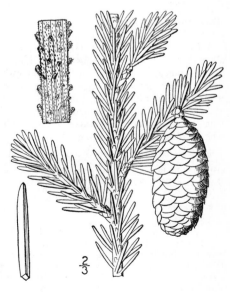

Newfoundland to northern New York, Minnesota and along the higher Alleghanies to Virginia and Georgia. Ascends to 4500 ft. in the Adirondacks. Wood similar to that of the preceding species. May–June.

Picea austràlis Small, of the high southern Alleghanies with very slender leaves, glabrous sterigmata and smaller cones, may be specifically distinct.

4. TSÙGA Carr. Trait. Conif. 185. 1855.

Evergreen trees with slender horizontal or drooping branches, flat narrowly linear scattered short-petioled leaves, spreading and appearing 2-ranked, jointed to very short sterigmata and falling away in drying. Leaf-buds scaly. Staminate aments axillary, short or subglobose; anthers 2-celled, the sacs transversely dehiscent, the connective slightly produced beyond them; pollen-grains simple. Ovule-bearing aments terminal, the scales about as long as the bracts, each bearing 2 reflexed ovules on its base. Cones small, ovoid or oblong, pendulous, their scales scarcely woody, obtuse, persistent. Seeds somewhat winged. [Name Japanese.]

About 7 species; the following in North America, 2 in northwestern North America, 3 or 4 Asiatic. Type species: *Tsuga Sieboldi* Carr. (*Abies Tsuga* Sieb. & Zucc.) of Japan.

Cones 6″–10″ long, their scales remaining appressed.   1. *T. canadensis.*
Cones 1′–1¼′ long, their scales widely spreading at maturity.   2. *T. caroliniana.*

1. Tsuga canadénsis (L.) Carr. Hemlock. Fig. 146.

Pinus canadensis L. Sp. Pl. Ed. 2, 1421. 1763.
Abies canadensis Michx. Fl. Bor. Am. 2: 206. 1803.
Tsuga canadensis Carr. Trait. Conif. 189. 1855.

A tall forest tree, sometimes 110° high, the trunk reaching 4° in diameter, the lower branches somewhat drooping, the old bark flaky in scales. Foliage dense; leaves obtuse, flat, 6″–9″ long, less than 1″ wide, dark green above, pale beneath, the petiole less than one-half as long as the width of the blade; cones oblong, obtuse, as long as or slightly longer than the leaves, their scales suborbicular, obtuse, minutely lacerate or entire, not widely spreading at maturity.

Nova Scotia to Minnesota, south to Delaware, along the Alleghanies to Alabama and to Michigan and Wisconsin. Ascends to 2000 ft. in the Adirondacks. One of the most ornamental of evergreens when young. Wood soft, weak, brittle, coarse-grained, light brown or nearly white; weight per cubic foot 26 lbs. Bark much used in tanning. April–May. Called also Spruce Pine, Hemlock Spruce.

2. Tsuga caroliniàna Engelm. Carolina Hemlock. Fig. 147.

Tsuga caroliniana Engelm. Coult. Bot. Gaz. 6: 223. 1881.
Abies caroliniana Chapm. Fl. S. States, Ed. 2, 650. 1883.

A forest tree attaining a maximum height of about 80° and a trunk diameter of 3½°, the lower branches drooping. Leaves narrowly linear, obtuse, rather light green above, nearly white beneath, 7″–10″ long, the petiole nearly as long as the width of the blade; cones 1′–1¾′ long, the scales firm but scarcely woody, oblong, obtuse, widely spreading at maturity.

Southwestern Virginia to South Carolina and Georgia in the Alleghanies. Wood soft, weak, brittle, light brown; weight per cubic foot about 27 lbs. A more graceful and beautiful tree than the preceding at maturity. Ascends to 4200 ft. in North Carolina. Called also Southern Hemlock. April.

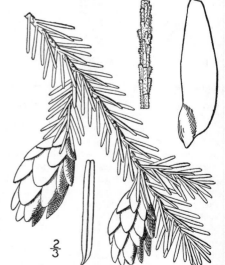

5. ÁBIES [Tourn.] Hill, Brit. Herb. 509. 1756.

Evergreen trees with linear flat scattered sessile leaves, spreading so as to appear 2-ranked, but in reality spirally arranged, not jointed to sterigmata, and commonly quite persistent in drying, the naked twigs marked by the flat scars of their bases. Staminate aments axillary; anthers 2-celled, the sacs transversely dehiscent, the connective prolonged into a short knob or point; pollen-grains compound. Ovule-bearing aments lateral, erect; ovules

2 on the base of each scale, reflexed, the scale shorter than or exceeding the thin or papery, mucronate or aristate bract.   Cones erect, subcylindric or ovoid, their scales deciduous from the persistent axis, orbicular or broader, obtuse.   [Ancient name of the firs.]

About 25 species, natives of the north temperate zone, chiefly in boreal and mountainous regions.   Besides the following, 8 others occur in the western parts of North America and 1 in Mexico.   Type species: *Pinus Picea* L., *Abies Picea* (L.) Lindley, of Europe.

Bracts surrulate, mucronate, shorter than the scales or but little longer.   1. *A. balsamea.*
Bracts aristate, reflexed, much longer than the scales.   2. *A. Fraseri.*

1. Abies balsàmea (L.) Mill.   Balsam Fir.   Fig. 148.

Pinus balsamea L. Sp. Pl. 1002.   1753.
Abies balsamea Mill. Gard. Dict. Ed. 8, No. 3.
1768.

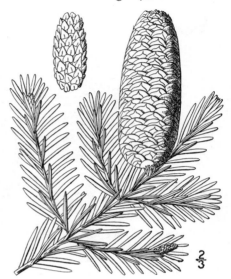

A slender forest tree attaining a maximum height of about 90° and a trunk diameter of 3°, usually much smaller and on mountain tops and in high arctic regions reduced to a low shrub.   Bark smooth, warty with resin "blisters."   Leaves fragrant in drying, less than 1″ wide, 6″–10″ long, obtuse, dark green above, paler beneath or the youngest conspicuously whitened on the lower surface; cones cylindric, 2′–4′ long, 9″–15″ thick, upright, arranged in rows on the upper side of the branches, violet or purplish when young; bracts obovate, serrulate, mucronate, shorter than the broad rounded scales.

Newfoundland and Labrador to Hudson Bay and Alberta, south to Massachusetts, Pennsylvania, along the Alleghanies to Virginia and to Iowa and Minnesota.   Ascends to 5000 ft. in the Adirondacks.   Wood soft and weak, light brown; weight per cubic foot 24 lbs.   *Canada balsam* is derived from the resinous exudations of the trunk.   Called also Fir-tree, Fir or Blister-pine, American Silver Fir, Single Spruce, Balm of Gilead.   May–June.

2. Abies Fràseri (Pursh) Poir.   Fraser's Balsam Fir.   Fig. 149.

Pinus Fraseri Pursh, Fl. Am. Sept. 639.   1814.
Abies Fraseri Poir. in Lam. Encycl. Suppl. 5: 35.
1817.

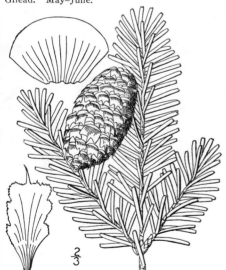

A forest tree, reaching a maximum size rather less than that of the preceding species, the smooth bark bearing similar resin "blisters."   Leaves, especially the younger, conspicuously whitened beneath, 5″–10″ long, nearly 1″ wide, emarginate or some of them obtuse at the apex; cones oblong-cylindric or ovoid-cylindric, 2′–3′ high, about 1′ thick, their scales rhomboid, much broader than high, rounded at the apex, much shorter than the papery bracts, which are reflexed, their summits emarginate, serrulate and aristate.

On the high Alleghanies of southwestern Virginia, West Virginia, North Carolina and Tennessee.   Wood similar to that of the northern species, but slightly lighter in weight.   Called also Double Spruce, She or Mountain Balsam. May.

6. TAXÒDIUM L. C. Rich. Ann. Mus. Paris, 16: 298.   1810.

Tall trees with horizontal or drooping branches, and alternate spirally arranged sessile linear or scale-like leaves, deciduous in our species, spreading so as to appear 2-ranked, some of the twigs commonly deciduous in autumn.   Leaf-buds naked.   Staminate aments very numerous, globose, in long terminal drooping panicled spikes, appearing before the leaves; anthers 2–5-celled, the sacs 2-valved.   Ovule-bearing aments ovoid, in small terminal clusters, their scales few, bractless, each bearing a pair of ovules on its base.   Cones globose or nearly so, the scales thick and woody, rhomboid, fitting closely together by their margins, each marked with a triangular scar at its base. Seeds large, sharply triangular-pyramidal. [Name Greek, referring to the yew-like leaves.]

Three known species, the following of southeastern North America, one Mexican. Type species: *Taxodium distichum* (L.) L. C. Rich.

Leaves linear, 2-ranked, spreading.   1. *T. distichum.*
Leaves awl-shaped, closely appressed to the twigs.   2. *T. ascendens.*

1. Taxodium dístichum (L.) L. C. Rich. Bald Cypress. Fig. 150.

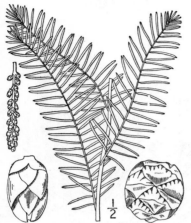

Cupressus disticha L. Sp. Pl. 1003. 1753.

T. distichum L. C. Rich. Ann. Mus. Paris, **16**: 298. 1810.

A large forest tree, attaining a maximum height of about 150° and a trunk diameter of 14°, the old bark flaky in thin strips. Leaves narrowly linear, flat, thin, 5″–10″ long, ½″ or less wide, rather light green, acute, those on some of the flowering branches smaller, scale-like; cones globose or slightly longer than thick, pendent at the ends of the branches, very compact, about 1′ in diameter; surfaces of the scales irregularly rugose above the inversely triangular scar; seeds 4″–5″ long.

In swamps and along rivers, southern New Jersey to Florida, west to Texas, north in the Mississippi Valley region to southern Indiana, Missouri and Arkansas. Wood soft, not strong, brown, very durable; weight per cubic foot 27 lbs. The roots develop upright conic "knees" sometimes 4° high and 1° thick. Called also White, Red, Black or Virginia Swamp-cypress; Sabino-tree. March–April.

2. Taxodium ascéndens Brongn. Pond Cypress. Fig. 151.

Cupressus disticha imbricaria Nutt. Gen. **2**: 224. 1818.

Taxodium ascendens Brongn. Ann. Sci. Nat. **30**: 182. 1833.

Taxodium imbricarium Harper, Bull. Torr. Club **29**: 383. 1902.

A tree with maximum height of about 80° and trunk diameter of about 3° above the greatly enlarged base, tapering upward, its thick fibrous bark deeply furrowed. Leaves awl-shaped, closely appressed to the slender twigs, 2″–5″ long, long-pointed, keeled above, concave beneath, the tips somewhat spreading; cones similar to those of *T. distichum.*

In ponds and swamps, southern Virginia to Florida and Alabama. Wood heavier and stronger than that of the Bald cypress.

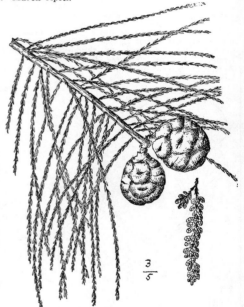

7. THÙJA L. Sp. Pl. 1002. 1753.

Evergreen trees or shrubs with frond-like foliage, the leaves small or minute, scale-like, appressed, imbricated, opposite, 4-ranked, those of the ultimate branchlets mostly obtuse, those of some of the larger twigs acute or subulate. Aments monoecious, both kinds terminal, the staminate globose; anthers opposite, 2–4-celled, the sacs globose, 2-valved. Ovule-bearing aments ovoid or oblong, small, their scales opposite, each bearing 2 (rarely 2–5) erect ovules. Cones ovoid or oblong, mostly spreading or recurved, their scales 6–10, coriaceous, opposite, not peltate, dry, spreading when mature. Seeds oblong, broadly or narrowly winged or wingless. [Name ancient.]

About 4 species, natives of North America and eastern Asia. Besides the following, another occurs from Montana, Idaho and Oregon to Alaska. Type species: *Thuja occidentalis* L.

1. Thuja occidentàlis L. White Cedar.
Arbor Vitae. Fig. 152.

Thuja occidentalis L. Sp. Pl. 1002. 1753.

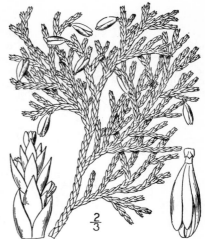

A conical tree, reaching a height of about 70° and a trunk diameter of 5°, the old bark deciduous in ragged strips. Scale-like leaves of the ultimate branchlets nearly orbicular, obtuse, $1''-1\frac{1}{2}''$ broad, the two lateral rows keeled, the two other rows flat, causing the twigs to appear much flattened; leaves of the older twigs narrower and longer, acute or acuminate; cones $4''-6''$ long, their scales obtuse; seeds broadly winged.

In wet soil and along the banks of streams, forming almost impenetrable forests northward, New Brunswick to James' Bay and Manitoba, south to New Jersey, along the Alleghanies to North Carolina, Tennessee and to Illinois and Minnesota. Ascends to 3500 ft. in the Adirondacks. Wood soft, brittle, weak, coarse-grained, light brown; weight per cubic foot 20 lbs. Called also False White and Feather-leaf Cedar. May–June.

8. CHAMAECÝPARIS Spach, Hist. Veg. 11: 329. 1842.

Evergreen trees, similar to the *Thujas,* with minute opposite appressed 4-ranked scale-like leaves, or those of older twigs subulate, and small monoecious terminal aments. Staminate aments as in *Thuja,* but the filaments broader and shield-shaped. Ovule-bearing aments globose, their scales opposite, peltate, each bearing 2–5 erect ovules. Cones globose, the scales thick, peltate, each bearing 2–5 erect seeds, closed until mature, each with a central point or knob. Seeds winged. [Greek, meaning a low cypress.]

About 6 species, the following of the eastern United States, 2 in western North America, the others Asiatic. Type species: *Chamaecyparis sphaeroidea* Spach.

1. Chamaecyparis thyoìdes (L.) B.S.P.
Southern White Cedar. Fig. 153.

Cupressus thyoides L. Sp. Pl. 1003. 1753.
Chamaecyparis sphaeroidea Spach, Hist. Veg. 11: 331. 1842.
Chamaecyparis thyoides B.S.P. Prel. Cat. N. Y. 71. 1888.

A forest tree, reaching a maximum height of about 90° and a trunk diameter of $4\frac{1}{2}$°. Leaves of the ultimate branchlets ovate, acute, scarcely $\frac{1}{2}''$ wide, those of the lateral rows keeled, those of the vertical rows slightly convex, each with a minute round discoid marking on the centre of the back, those of the older twigs narrower and longer, subulate; cones about $3''$ in diameter, blue, each of their closely fitting scales with a small central point; seeds narrowly winged.

In swamps, southern Maine and New Hampshire to northern New Jersey, south to Florida and Mississippi, mostly near the coast. Wood soft, weak, close-grained, light brown; weight per cubic foot 21 lbs. April–May. Called also Post or Swamp Cedar, Juniper.

9. JUNÍPERUS L. Sp. Pl. 1038. 1753.

Evergreen trees or shrubs with opposite or verticillate, subulate or scale-like, sessile leaves, commonly of 2 kinds, and dioecious or sometimes monoecious, small short axillary or terminal aments. Leaf-buds naked. Staminate aments oblong or ovoid; anthers 2–6-celled, each sac 2-valved. Ovule-bearing aments of a few opposite somewhat fleshy scales, or these rarely verticillate in 3's, each bearing a single erect ovule or rarely 2. Cones globose, berry-like by the coalescence of the fleshy scales, containing 1–6 wingless bony seeds. [Name Celtic.]

About 40 species, mostly natives of the northern hemisphere. Besides the following, 10 others occur in the western parts of North America. Type species: *Juniperus communis* L.

Leaves all subulate, prickly pointed, verticillate; aments axillary.
  Small erect tree or shrub; leaves slender, mostly straight.   **1.** *J. communis.*
  Low depressed shrub; leaves stouter, mostly curved.   **2.** *J. sibirica.*
Leaves of 2 kinds, scale-like and subulate, mostly opposite; aments terminal.
  Tree; fruit on short straight branches.   **3.** *J. virginiana.*
  Depressed shrub; fruit on short recurved branches.   **4.** *J. horizontalis.*

1. Juniperus commùnis L. Juniper. Fig. 154.

Juniperus communis L. Sp. Pl. 1040. 1753.

A low tree or erect shrub, sometimes attaining a height of 30° and a trunk diameter of 12′, usually smaller, the branches spreading or drooping, the bark shreddy. Leaves all subulate, rigid, spreading, or some of the lower reflexed, mostly straight, prickly pointed, verticillate in 3's, often with smaller ones fascicled in their axils, 5″–10″ long, less than 1″ wide, channeled and commonly whitened on the upper surface; aments axillary; berry-like cones sessile or very nearly so, dark blue, 3″–4″ in diameter.

On dry hills, Massachusetts to Alaska, south to New Jersey, North Carolina, Michigan, western Nebraska and in the Rocky Mountains to New Mexico. Ascends to 900 ft. in Pennsylvania. Also in Europe and Asia. The fruit, called Melmot berries, is used for flavoring gin. Called also Horse Savin, Hackmatack, Aiten. April–May. Fruit ripe Oct.

2. Juniperus sibírica Burgsd. Low Juniper. Fairy Circles. Fig. 155.

Juniperus sibirica Burgsd. Anleit. **2**: 124. 1787.
Juniperus nana Willd. Sp. Pl. **4**: 854. 1806.
J. communis depressa Pursh, Fl. Am. Sept. 646. 1814.
Juniperus communis var. *alpina* Gaud. Fl. Helv. **6**: 301. 1830.

A depressed or trailing rigid shrub, seldom over 18′ high, forming circular patches often 10° in diameter. Leaves similar to those of the preceding species, but stouter, similarly channeled and often whitened above, appressed-ascending, rather rigid, spiny tipped, 4″–6″ long, mostly incurved, densely clothing the twigs, verticillate in 3's; aments axillary; berry-like cones blue, 4″–5″ in diameter.

In dry, open places, Labrador to British Columbia, south to Massachusetts, New York, Michigan and in the Rocky Mountains to Colorado and Utah. Also in Europe and Asia. Although the characteristic growth in a depressed circular patch gives a very different aspect from the true Juniper, the plant may, perhaps, be better regarded as a race of *J. communis* L. April–May.

3. Juniperus virginiàna L. Red Cedar. Savin. Fig. 156.

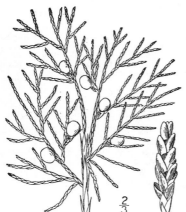

Juniperus virginiana L. Sp. Pl. 1039. 1753.

A tree, reaching a maximum height of about 100° and a trunk diameter of 5°, conic when young, but the branches spreading in age so that the outline becomes nearly cylindric. Leaves mostly opposite, all those of young plants and commonly some of those on the older twigs of older trees subulate, spiny-tipped, 2″–4″ long, those of the mature foliage scale-like, acute or subacute, closely appressed and imbricated, 4-ranked, causing the twigs to appear quadrangular; aments terminal; berry-like cones light blue, glaucous, about 3″ in diameter, borne on straight peduncle-like branchlets of less than their own length, 1–2-seeded, maturing the first season.

In dry soil, Nova Scotia to western Ontario and South Dakota, south to Florida and Texas. Wood soft, not strong, straight-grained, compact, odorous, red, the sapwood white; weight per cubic foot 31 lbs.; used in large quantities in the manufacture of lead pencils. April–May. Fruit ripe Sept.–Oct. Called also Red Savin or Juniper; Juniper-bush, Carolina Cedar, Pencil-wood.

Juniperus scopulòrum Sargent, the Rocky Mountain Red Cedar, which differs from *J. virginiana* mainly in maturing its fruit during the second season, has been reported from Nebraska.

4. Juniperus horizontàlis Moench.   Shrubby Red Cedar.   Creeping Juniper.   Fig. 157.

Juniperus horizontalis Moench, Meth. 699.   1794.
Juniperus prostrata Pers. Syn. 2: 632.   1807.
Juniperus Sabina var. *procumbens* Pursh, Fl. Am. Sept. 647.   1814.

A depressed, usually procumbent shrub, seldom more than 4° high.   Leaves similar to those of the preceding species, those of young plants and the older twigs of older plants subulate, spiny-tipped, those of the mature foliage scale-like, appressed, 4-ranked, acute or acuminate; aments terminal; berry-like cones light blue, somewhat glaucous, 4″–5″ in diameter, borne on recurved peduncle-like branchlets of less than their own length, 1–4-seeded.

On banks, Newfoundland to British Columbia, south to Massachusetts, northern New York, Minnesota and Montana.   Has been confused with *J. Sabina* of Europe. April–May.

$\frac{2}{3}$

Family 2.   **TAXÀCEAE** Lindl. Nat. Syst. Ed. 2, 316.   1836.

YEW FAMILY.

Trees or shrubs, resin-bearing except *Taxus*.   Leaves evergreen or deciduous, linear, or in several exotic genera broad or sometimes fan-shaped, the pollen-sacs and ovules borne in separate clusters or solitary.   Perianth wanting.   Stamens much as in the Pinaceae.   Ovules with either one or two integuments; when two, the outer one fleshy, when only one, its outer part fleshy.   Fruit drupe-like or rarely a cone.

About 10 genera and 75 species, of wide geographic distribution, most numerous in the southern hemisphere.   The Maiden-hair Tree, *Ginkgo biloba,* of China and Japan, with fan-shaped leaves, is an interesting relative of the group, now much planted for ornament.

1.   **TÀXUS** [Tourn.] L. Sp. Pl. 1040.   1753.

Evergreen trees or shrubs, with spirally arranged short-petioled linear flat mucronate leaves, spreading so as to appear 2-ranked, and axillary and solitary, sessile or subsessile very small aments; staminate aments consisting of a few scaly bracts and 5–8 stamens, their filaments united to the middle; anthers 4–6-celled.   Ovules solitary, axillary, erect, subtended by a fleshy, annular disk, which is bracted at the base.   Fruit consisting of the fleshy disk which becomes cup-shaped, red, and nearly encloses the bony seed.   [Name ancient.]

About 6 species, natives of the north temperate zone.   Besides the following, another occurs in Florida, one in Mexico and one on the Pacific Coast.   Type species: *Taxus baccata* L.

$\frac{2}{3}$

1. Taxus canadénsis Marsh.   American Yew.   Ground-hemlock.   Fig. 158.

Taxus baccata var. *minor* Michx. Fl. Bor. Am. 2: 245.   1803.
Taxus canadensis Marsh. Arb. Am. 151.   1785.
Taxus minor Britton, Mem. Torr. Club, 5: 19. 1893.

A low straggling shrub, seldom over 5° high.   Leaves dark green on both sides, narrowly linear, mucronate at the apex, narrowed at the base, 6″–10″ long, nearly 1″ wide, persistent on the twigs in drying; the staminate aments globose, 1″ long, usually numerous; ovules usually few; fruit red and pulpy, resinous, oblong, nearly 3″ high, the top of the seed not covered by the fleshy integument.

In woods, Newfoundland to Manitoba, south to New Jersey, in the Alleghanies to Virginia, and to Minnesota and Iowa.   Ascends to 2500 ft. in the Adirondacks.   April–May.   Called also Dwarf Yew, Shin-wood, Creeping Hemlock. Very different from the European Yew, *T. baccata,* in habit, the latter becoming a large forest tree, as does the Oregon Yew, *T. brevifolia.*

Class 2. *ANGIOSPÉRMAE.*

Ovules (macrosporanges) enclosed in a cavity (the ovary) formed by the infolding and uniting of the margins of a modified rudimentary leaf (carpel), or of several such leaves joined together, in which the seeds are ripened. The pollen-grains (microspores) on alighting upon the summit of the carpel (stigma) germinate, sending out a pollen-tube which penetrates its tissue and reaching an ovule enters the orifice of the latter (micropyle), and its tip coming in contact with a germ-cell in the embryo-sac, fertilization is effected. In a few cases the pollen-tube enters the ovule at the chalaza, not at the micropyle.

Cotyledon one; stem endogenous.   Sub-class 1. MONOCOTYLEDONES.
Cotyledons almost always two; stem (with rare exceptions) exogenous.
           Sub-class 2. DICOTYLEDONES.

Sub-class 1. *MONOCOTYLÉDONES.*

Embryo of the seed with but a single cotyledon and the first leaves of the germinating plantlet alternate. Stem composed of a ground-mass of soft tissue (parenchyma) in which bundles of wood-cells are irregularly imbedded; no distinction into wood, pith and bark. Leaves usually parallel-veined, mostly alternate and entire, commonly sheathing the stem at the base and often with no distinction of blade and petiole. Flowers mostly 3-merous or 6-merous.

Monocotyledonous plants are first definitely known in Triassic time. They constitute between one-fourth and one-third of the living angiospermous flora.

Family 1. **TYPHACEAE** J. St. Hil. Expos. Fam. 1: 60. 1805.

CAT-TAIL FAMILY.

Marsh or aquatic plants with creeping rootstocks, fibrous roots and glabrous erect, terete stems. Leaves linear, flat, ensiform, striate, sheathing at the base. Flowers monoecious, densely crowded in terminal spikes, which are subtended by spathaceous, usually fugacious bracts, and divided at intervals by smaller bracts, which are caducous, the staminate spikes uppermost. Perianth of bristles. Stamens 2-7, the filaments connate. Ovary 1, stipitate, 1-2-celled. Ovules anatropous. Styles as many as the cells of the ovary. Mingled among the stamens and pistils are bristly hairs, and among the pistillate flowers many sterile flowers with clavate tips. Fruit nutlike. Endosperm copious. Only the following genus:

1. TÝPHA [Tourn.] L. Sp. Pl. 971. 1753.

Characters of the family. [Name ancient.]

About 10 species, of temperate and tropical regions. Type species: *Typha latifolia* L.

Spikes with the pistillate and staminate usually contiguous, the former without bractlets; stigmas spatulate or rhomboid; pollen 4-grained.   1. *T. latifolia.*
Spikes with the pistillate and staminate usually distant, the former with bractlets; stigmas linear or oblong-linear; pollen in simple grains.   2. *T. angustifolia.*

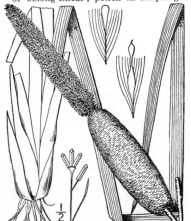

½

1. Typha latifòlia L. Broad-leaved Cat-tail. Fig. 159.

Typha latifolia L. Sp. Pl. 971. 1753.

Stems stout, 4°-8° high. Leaves 3''-12'' broad; spikes dark brown or black, the staminate and pistillate portions usually contiguous, each 3'-12' long and often 1' or more in diameter, the pistillate without bractlets; stigmas rhomboid or spatulate; pollen-grains in 4's; pedicels of the mature pistillate flowers 1''-1½'' long.

In marshes, throughout North America except the exterme north. Ascends to 1600 ft. in the Adirondacks and to 2200 ft. in Virginia. Also in Europe and Asia. June-July. Fruit, Aug.-Sept. Called also Great-Reed-mace, Cat-o'-nine-tail, Marsh Beetle, Marsh Pestle, Cat-tail Flag, Flax-tail, Blackamoor, Black-cap, Bull-segg, Bubrush, Water-torch, Candlewick.

2. Typha angustifòlia L.   Narrow-leaved Cat-
                        tail.   Fig. 160.

Typha angustifolia L. Sp. Pl. 971. 1753.

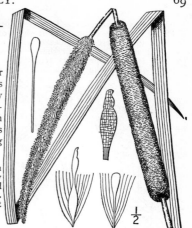

Stems slender, 5°–10° high. Leaves mostly narrower
than those of the preceding species, 2″–6″ wide; spikes
light brown, the staminate and pistillate portions usually
distant, the two together sometimes 15′ long, the pistillate, when mature 2″–8″ in diameter, and provided with
bractlets; stigmas linear or linear-oblong; pollen-grains
simple; pedicels of the mature pistillate flowers ½″ long
or less.

Abundant in marshes along the Atlantic Coast from
Nova Scotia to Florida, but also occurring rather rarely
inland, and in California. Also in tropical America and
South America. Also in Europe and Asia. June–July.
Fruit, Aug.–Sept. Called also Lesser Reed-mace and most
of the names of the preceding species.

Family 2. **SPARGANIÀCEAE** Agardh, Theor. Syst. Pl. 13.   1858.*

BUR-REED FAMILY.

Marsh or pond plants with creeping rootstocks and fibrous roots, erect or
floating simple or branched stems, and linear alternate leaves, sheathing at the
base. Flowers monoecious, densely crowded in globose heads at the upper part
of the stem and branches, the staminate heads uppermost, sessile or peduncled.
Spathes linear, immediately beneath or at a distance below the head. Perianth
of a few irregular chaffy scales. Stamens commonly 5, their filaments distinct;
anthers oblong or cuneate. Ovary sessile, mostly 1-celled. Ovules anatropous.
Fruit mostly 1-celled, nutlike. Embryo nearly straight in copious endosperm.

The family comprises only the following genus.

1. **SPARGÀNIUM** [Tourn.] L. Sp. Pl. 971.   1753.

Characters of the family. [Greek, referring to the ribbon-like leaves.]

About 22 species, of temperate and cold regions. Besides the following, 3 others occur in
western North America. Type species: *Sparganium erectum* L.

Achenes broadly obovoid or cuneate-obpyramidal, sessile, distinctly beaked; inflorescence com-
pound; fruiting heads 10″–15″ in diameter.   1. *S. eurycarpum.*
Achenes fusiform (in *S. minimum* somewhat obovoid, but short-beaked and short-stipitate).
Stipe and beak of the achene each 1″ long or more; fruiting heads 7½″ in diameter or more;
anthers 3–4 times as long as broad.
Beaks straight or slightly curved; stigmas linear.
Heads all axillary; beak shorter than the body of the achene; leaves keeled.
Achenes dull; stigmas 1″ long or less.
Inflorescence branched, the branches geniculate, bearing 3–7 staminate heads.
                                                                       2. *S. androcladum.*
Inflorescence simple, or, if branched, the branches strict and bearing 0–2 stami-
nate heads.   3. *S. americanum.*
Achenes glossy; stigmas 1¼″–1¾″ long.   4. *S. lucidum.*
Heads, at least some of them, supra-axillary.
Leaves, at least the middle ones, strongly triangular-keeled; stem usually erect, strict.
Fruiting heads over 10″ in diameter; leaves broad; bracts ascending-spreading;
beak fully as long as the body of the achene.
Leaves 3½″–7½″ wide, strongly veined; fruiting heads about 15″ in
diameter; achenes brown, shining, each gradually tapering into the beak.
                                                                       5. *S. simplex.*
Leaves 1½″–4″ wide, weakly veined; fruiting heads 10″–12½″ in diameter;
achenes green, dull, each abruptly contracted into the beak.
                                                                       6. *S. chlorocarpum.*
Fruiting heads rarely 10″ in diameter; leaves narrow; bracts almost erect.
Heads distant, nearly 10″ in diameter; achenes grayish-brown, distinctly
nerved.   7. *S. diversifolium.*
Heads approximate, about 7½″ in diameter; achenes dark olive-brown, not
nerved.   8. *S. acaule.*
Leaves not keeled, or only slightly so, narrow and slender; stem weak and often
floating; beak decidedly shorter than the body of the achene.
Leaves usually 2½″–5″ wide; leaves and bracts conspicuously scarious-margined;
fruiting heads 8½″–10″ in diameter; achenes gradually beaked.
                                                                 9. *S. multipedunculatum.*
Leaves 1½″–2″ wide; leaves and bracts not conspicuously scarious-margined;
fruiting heads about 7½″ in diameter; achenes abruptly beaked.
                                                                 10. *S. angustifolium.*

*Text revised by DR. JOHN KUNKEL SMALL.

Beaks gladiate-curved; stigma short oblong.   11. *S. fluctuans.*
Stipe and beak of the achene short or none, always less than ½″ long; fruiting heads about 5″
    in diameter; stigmas oblong; anthers 1½–2 times as long as broad.
Heads all sessile, or the lowest short-pedicelled, axillary; the staminate head distant from
    the pistillate ones; achenes short-beaked.   13. *S. minimum.*
Lower pistillate heads distinctly pedicelled and supra-axillary; the staminate head close to
    the upper pistillate one.   12. *S. hyperboreum.*

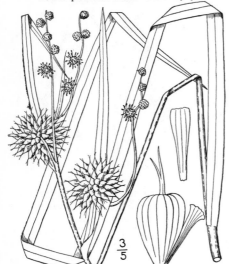

1. Sparganium eurycàrpum Engelm. Broad-fruited Bur-reed. Fig. 161.

Sparganium eurycarpum Engelm. in A. Gray, Man. Ed. 2, 430. 1856.

Stem stout, 3°–8° high, branching. Leaves linear, flat, slightly keeled beneath, the lowest 3°–5° long; staminate heads numerous; pistillate heads 2–4 on the stem or branches, sessile or more commonly peduncled, hard, compact and 10″–16″ in diameter; style 1; stigmas 1–2; nutlets sessile, 3″–5″ long, obtusely 4–5-angled, narrowed at the base, the top rounded, flattened or depressed, abruptly tipped with the style; scales as long or nearly as long as the fruit, often with 2 or 3 other exterior ones, somewhat spatulate, the apex rounded, denticulate or eroded.

In marshes and along streams, Newfoundland to British Columbia, Florida, Missouri, Utah and California. May–Aug.

2. Sparganium andrócladum (Engelm.) Morong. Branching Bur-reed. Fig. 162.

Sparganium simplex androcladum Engelm.; A. Gray, Man. Ed. 5, 481. 1867.
Sparganium androcladum Morong, Bull. Torrey Club **15**: 78. 1888.
Sparganium americanum androcladum Fernald & Eames, Rhodora **9**: 87. 1907.

Stem branching, 1°–3½° high; leaves 1½°–3½° long, triangular at the base; bracts similar to the leaves, bases slightly dilated, and but narrowly scarious-margined; inflorescence branched; branches and peduncles axillary, branches zigzag; fruiting heads 6″–12″ diameter; nutlets brown, dull, fusiform, 2½″–3″ long, terete or obtusely angled, often constricted at the middle; stigma linear, 1″ long.

In bogs or shallow water, Newfoundland to Minnesota, Florida and Louisiana. June–Aug.

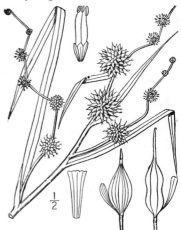

3. Sparganium americànum Nutt. Nuttall's Bur-reed. Fig. 163.

Sparganium americanum Nutt. Gen. **2**: 203. 1818.
Sparganium simplex Nuttallii Engelm.: A. Gray, Man. Ed. 5, 481. 1867.

Stem simple or nearly so, 1°–2½° high; leaves 1°–3½° long, keeled, somewhat scarious-margined near the base; bracts, at least the lower ones, similar to the leaves, but shorter, dilated and scarious-margined near the base; inflorescence usually simple, with the heads sessile, or the lower pistillate ones peduncled, the branches, when present, straight; fruiting heads 9″–12″ in diameter; nutlets brown, dull, fusiform, 2½″–3″ long; stigma oblong, seldom over ½″ long.

In low grounds or ponds, Nova Scotia and Ontario to Iowa, South Carolina and Oklahoma. June–Aug.

4. Sparganium lùcidum Fernald & Eames. Shining-fruited Bur-reed. Fig. 164.

Sparganium lucidum Fernald & Eames, Rhodora **9**: 87. 1907.

Stem stout, $2°–3\frac{1}{2}°$ high; leaves $1°–2°$ long, strongly keeled, $1\frac{1}{2}''–6''$ wide; bracts similar to the leaves but shorter; inflorescence simple or somewhat branched, with the branches, or heads, axillary, the main axis bearing 2–4 sessile pistillate heads and 6–10 staminate heads, the branches, when present, bearing 1 pistillate head and sometimes 1–4 staminate heads; fruiting heads about $15''$ in diameter; nutlets olive-brown, shining, the body fusiform, about $4''$ long; stigma linear, $1\frac{1}{4}''–1\frac{3}{4}''$ long.

In ponds and streams, Massachusetts to New York, Illinois and Missouri. July–Sept.

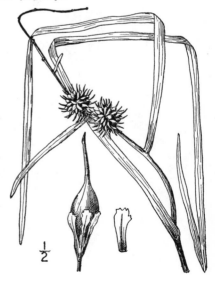

5. Sparganium símplex Huds. Simple-stemmed Bur-reed. Black-weed. Fig. 165.

Sparganium erectum β L. Sp. Pl. 971. 1753.
Sparganium simplex Huds. Fl. Aug. ed. 2. 401. 1788.

Stem rather stout, $1\frac{1}{2}°–2°$ high; leaves $1\frac{1}{2}°–3°$ long, triangular-keeled, $4''–8''$ wide; bracts flat or slightly keeled; inflorescence usually simple, the pistillate heads 2–5, at least some of them supra-axillary, the lower 1 or 2 peduncled, the staminate heads 4–8; fruiting heads about $15''$ in diameter; nutlets brown or sometimes greenish-brown, the body fusiform, $2\frac{1}{2}''–3''$ long, often constricted at the middle; stigma linear, about $1''$ long.

In lakes and streams, Quebec and Ontario, and in Washington and British Columbia. Also in Europe and Asia. July–Sept.

6. Sparganium chlorocàrpum Rydb. Green-fruited Bur-reed. Fig. 166.

Sparganium chlorocarpum Rydb. N. A. Fl. **17**[1]: 8. 1909.

Stem slender, $1°–2°$ tall, or sometimes floating; leaves $1°–2°$ long, at least the middle ones keeled, $1\frac{1}{2}''–3\frac{1}{2}''$ wide; bracts similar to the leaves, slightly, if at all, dilated or scarious at the base; inflorescence simple, the pistillate heads 2–4, sessile or the lowest one peduncled, the staminate heads 3–7; fruiting heads $10''–12''$ in diameter: nutlets green, rather dull, the body fusiform, $2''–2\frac{1}{2}''$ long; stigma less than $1''$ long.

In marshes and rivers, western New York to Iowa and Indiana. July–Sept.

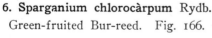

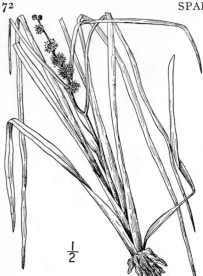

7. Sparganium diversifòlium Graebner.
Various-leaved Bur-reed.   Fig. 167.

Sparganium diversifolium Graebner, Schrift. Nat. Ges. Danzig. II. **9**: 335. 1895.

Stem rather stout, $\frac{3}{4}°$–$3°$ high; leaves narrow, $1\frac{1}{2}''$–$2\frac{1}{2}''$ wide, abruptly pointed, the lower ones flat, the upper ones convex on the back or sharply keeled near the base; bracts similar to the upper leaves; inflorescence simple, often nodding at the tip, the pistillate heads 1–3, distant, the staminate heads 1–6, distant; fruiting heads 10″ in diameter; nutlets grayish-brown, the body prominently-nerved, obovoid; stigma linear-lanceolate.

In bogs, Newfoundland, New Hampshire and Minnesota. Also in Europe.   July–Sept.

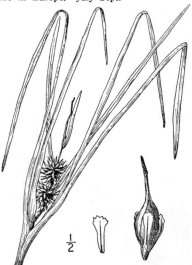

8. Sparganium acaùle (Beeby) Rydb. Stemless Bur-reed.   Fig. 168.

Sparganium simplex acaule Beeby; Macoun, Cat. Can. Pl. **5**: 367. 1890.
Sparganium diversifolium acaule Fernald & Eames, Rhodora **9**: 88. 1907
Sparganium acaule Rydberg, N. A. Fl. **17**¹: 8. 1909.

Stem rather slender, $\frac{1}{3}°$–$1°$ high; leaves very narrow, $1''$–$2''$ wide, triangular-keeled; bracts nearly similar to the leaves, but dilated and with broad scarious margins at the base; inflorescence simple, erect, the pistillate heads solitary or 2 or 3, and approximate, usually sessile, supra-axillary, the staminate 2–4, less crowded than the pistillate ones; fruiting heads 7″–8″ in diameter; nutlets olive-brown, the body not nerved, fusiform, $1\frac{1}{2}''$–$2''$ long.

In swamps and on muddy shores, Newfoundland to Iowa, South Dakota and Virginia.   July–Sept.

9. Sparganium multipedunculàtum (Morong) Rydb.   Many-stalked Bur-reed.   Fig. 169.

Sparganium simplex multipedunculata Morong, Bull. Torrey Club **15**: 79. 1888.
Sparganium multipedunculatum Rydb. Bull. Torrey Club **32**: 598. 1905.

Stem $1°$–$1\frac{3}{3}°$ high or less, or floating, rather slender; leaves narrow, $2\frac{1}{2}''$–$5''$ wide or rarely less, slightly keeled, dilated and scarious-margined at the base; bracts similar to the leaves, but relatively more dilated at the base; inflorescence usually simple, the pistillate heads 2–5, the lower 1 or 2 decidedly peduncled, some of them supra-axillary, the staminate heads 3–5, approximate to each other, but usually distant from the nearest pistillate one; fruiting heads 7″–10″ in diameter; nutlets brown, the body fusiform, about 2″ long; stigma linear, about $\frac{1}{2}''$ long.

In ponds and marshes, Mackenzie to western Ontario, Colorado, British Columbia and California. July–Sept.

10. Sparganium angustifòlium Michx. Narrow-leaved Bur-reed. Fig. 170.

Sparganium angustifolium Michx. Fl. Bor. Am. 2: 189. 1803.

Sparganium natans angustifolium Pursh, Fl. Am. Sept. 34. 1814.

Sparganium simplex angustifolium Torr. Fl. N. Y. 2: 249. 1843.

Stem floating and elongated, or occasionally $\frac{1}{2}°-1°$ high; leaves usually very narrow, $1\frac{1}{2}''-2''$ wide, not keeled; bracts various, the lower ones similar to the leaves, dilated and sometimes slightly scarious-margined at the base, the upper ones much shorter than the lower, lanceolate to ovate; inflorescence simple, the pistillate heads 2–4, the lower 1 or 2 usually on supra-axillary peduncles, the staminate heads usually approximate; fruiting heads $7''-8''$ in diameter; nutlets dirty-brown, except the reddish brown bases, the body constricted at the middle or above it; stigma $\frac{1}{2}''$ long or less.

In slow streams and ponds, Newfoundland to British Columbia, Connecticut, Pennsylvania, Colorado and California. Illustrated in first edition as *Sparganium simplex*. July–Sept.

11. Sparganium flúctuans (Morong) Robinson. Floating Bur-reed. Fig. 171.

Sparganium simplex fluitans Engelm.; A. Gray Man. Ed. 5, 481. 1867.

Sparganium androcladum fluctuans Morong, Bull. Torrey Club 15: 78. 1888.

Sparganium fluctuans Robinson, Rhodora 7: 60. 1905.

Stem floating, slender, usually elongated; leaves rather narrow, $2''-5\frac{1}{2}''$ wide, slightly, if at all, keeled, cellular-reticulate; bracts much shorter than the leaves, dilated and somewhat scarious-margined near the base; inflorescence usually branched, the main axis with 2–4 staminate heads, the branches usually bearing 1 pistillate and 2 staminate heads; fruiting heads about $10''$ in diameter; nutlets brown, the body fusiform, sometimes constricted at the middle; stigma obliquely oblong.

In ponds and cold lakes, Maine to Connecticut and Minnesota. July–Sept.

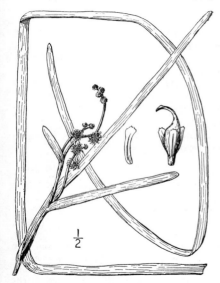

12. Sparganium hyperbòreum Laest. Northern Bur-reed. Fig. 172.

Sparganium natans submuticum Hartm. Handb. Skand. Fl. ed. 4: 312. 1843.

Sparganium hyperboreum Laest.; Beurl. Oefvers. Vet. Akad. Foerh. 9: 192. 1852.

Stem floating and elongated, or decumbent, or ascending and $4'-8'$ high; leaves light green, very narrow, $\frac{1}{2}''-2''$ wide, flat or slightly round-keeled near the base, in the case of floating plants sometimes greatly elongated; leaf-sheaths slightly dilated near the base, but not scarious-margined; pistillate heads 2–4, the lower 1 or 2 usually peduncled and supra-axillary; fruiting heads $4''-5''$ in diameter; nutlets dark-yellow, dull, the body ellipsoid; stigma oval.

In ponds and streams, Greenland to Newfoundland, Hudson Bay and Alaska Also in northern Europe and Asia. July–Sept.

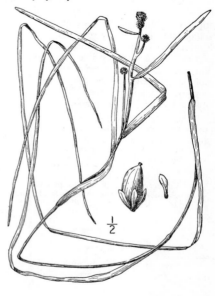

13. Sparganium mínimum Fries.　Small Bur-reed.
Fig. 173.

Sparganium natans Oeder. Fl. Dan. 2⁵: 5.　1764.　Not *S. natans*
　L.　1753.
Sparganium minimum Fries, Summa Veg. Scand. 2: 560.　1849.
Sparganium angustifolium A. Gray, Man. Ed. 5, 430.　1856.　Not
　S. angustifolium Michx.　1803.

Stem floating and sometimes elongated, or decumbent, ascending or erect, and relatively short; leaves dark-green, narrow, mostly 1″–3½″ wide, flat; upper leaf-sheaths somewhat
dilated, but not scarious-margined; pistillate heads solitary,
or 2 or 3 and placed about equally distant, axillary; fruiting
heads about 5″ in diameter; nutlets greenish-brown, dull, the
body broadly ellipsoid, usually constricted below the middle;
stigma obliquely oblong or oval.

In ponds and streams, Labrador to Alaska, New Jersey, Tennessee (?), Utah and Oregon.　Also in Europe and Asia.　June–Aug.

Family 3.　ZANNICHELLIÀCEAE Dumort. Anal. Fam. 61.　1829.*
PONDWEED FAMILY.

Perennial marine or fresh-water plants with floating or submerged leaves, or
both.　Leaf-blades petioled or sessile, capillary or expanded into a proper blade,
or rarely reduced to terete phyllodes.　Flowers perfect or monoecious, in sessile
or peduncled spikes, or in clusters in the axils of the leaves.　Perianth none, but
flowers sometimes enclosed in a hyaline sheath.　Androecium of 1–4 stamens.
Anthers extrorse, 1–2-celled, the connective sometimes becoming perianth-like.
Gynoecium of 1–4 distinct, 1-seeded carpels.　Fruits mostly nut-like or drupe-
like, sessile or stipitate.　Endosperm wanting.

About 4 genera and 70 species of wide geographic distribution, most abundant in temperate
regions.　The months noted in the descriptions indicate the fruting period.

Flowers perfect; stamens more than 1.
　Stamens 4; fruit sessile.　　　　　　　　　　　　　　　　1. *Potamogeton.*
　Stamens 2; fruit stalked.　　　　　　　　　　　　　　　2. *Ruppia.*
Flowers monoecious; stamen 1.　　　　　　　　　　　　　3. *Zannichellia.*

1.　POTAMOGÈTON L. Sp. Pl. 126.　1753.

Leaves alternate or the uppermost opposite, often of 2 kinds, submerged and floating,
the submerged mostly linear, the floating coriaceous, lanceolate, ovate or oval.　Spathes
stipular, often ligulate, free or connate with the base of the leaf or petiole, enclosing the
young buds and usually soon perishing after expanding.　Peduncles axillary, usually emersed.
Flowers small, spicate, green or red.　Perianth none.　Stamens 4.　Anthers sessile, the connective dilated, perianth-like (Fig. 186).　Ovaries 4, sessile, distinct, 1-celled, 1-ovuled, attenuated into a short erect or recurved style, or with a sessile stigma.　Fruit of 4 ovoid or subglobose drupelets, the pericarp usually thin and hard or spongy.　Seeds crustaceous, campylotropous, with an uncinate embryo thickened at the radicular end.　[Greek, in allusion to the
aquatic habitat.]　Water Spike.

About 65 well-defined species, natives of temperate regions.　Besides the following, about 3
others occur in the southern parts of North America.　Type species *Potamogeton natans* L.

Stipules axillary and free from the rest of the leaf.
　With floating and submerged leaves.
　　Submerged leaves bladeless.
　　　Nutlets more or less pitted.　　　　　　　　　　　1. *P. natans.*
　　　Nutlets not pitted.　　　　　　　　　　　　　　2. *P. Oakesianus.*
　　Submerged leaves with a proper blade.
　　　Submerged leaves of 2 kinds, lanceolate and oval or oblong
　　　　Uppermost broadly oval or elliptical, lowest lanceolate.　3. *P. amplifolius.*
　　　　Uppermost lanceolate and pellucid, lowest oblong and opaque. 4. *P. pulcher.*
　　　Submerged leaves all alike, capillary or linear-setaceous.
　　　　1-nerved or nerveless.　　　　　　　　　　　　25. *P. Vaseyi.*
　　　　3-nerved.　　　　　　　　　　　　　　　　　26. *P. lateralis.*
　　　Submerged leaves all alike, linear.

* Text of this family and of the two following ones contributed to the first edition by
the late Rev. Thomas Morong, revised for this edition by Mr. Norman Taylor.

Nearly the same breadth throughout, obtusely pointed, coarsely cellular-reticulated in
  the middle.   5. *P. epihydrus.*
  Broader at base, acute, without cellular-reticulation.   9. *P. heterophyllus.*
Submerged leaves all alike, lanceolate.
  Uppermost leaves petioled, lowest sessile.   6. *P. alpinus.*
  All the leaves petioled.
  Floating leaves large, broadly elliptic, rounded or subcordate at base.
                                                     11. *P. illinoensis.*
    Floating leaves narrowly elliptical, tapering at base. 7. *P. americanus.*
    Floating leaves mostly obovate or oblanceolate, tapering at base.
                                                     8. *P. Faxoni.*
  All the leaves sessile or subsessile.
    Fruit only 1 line long, obscurely 3-keeled.   10. *P. varians.*
    Fruit 1½ lines long, distinctly 3-keeled.   12. *P. angustifolius.*
With submerged leaves only.
  Without propagating buds and without glands.
  Leaves with broad blades, mostly lanceolate or ovate, many-nerved.
    Leaves subsessile or short-petioled, mostly acute or cuspidate. 13. *P. lucens.*
    Leaves semi-amplexicaul, obtuse and cucullate at the apex. 14. *P. praelongus.*
    Leaves meeting around the stem, very obtuse at the apex, not cucullate.
                                                     15. *P. perfoliatus.*
  Leaves with narrow blades, linear or oblong-linear, several-nerved.
    Leaves oblong-linear, 5–7-nerved, obtuse at the apex. 16. *P. mysticus.*
    Leaves narrowly linear, 3-nerved, acute at the apex. 21. *P. foliosus.*
  Leaves with narrow blades, capillary or setaceous, 1-nerved or nerveless.
                                                     17. *P. confervoides.*
With propagating buds or glands, or both.
  With buds, but without glands.
    Leaves serrulate, 3–7-nerved.   18. *P. crispus.*
    Leaves entire, with 3 principal and many fine nerves. 19. *P. compressus.*
  Commonly with glands, but no buds.
    Stems long-branching from the base; leaves lax, flat, 3-nerved, abruptly acute or
      cuspidate.   20. *P. Hillii.*
    Stems simple; leaves strict, revolute, 3–5-nerved, acuminate. 24. *P. rutilus.*
  With both buds and glands.
    Glands large and translucent; buds rare.   22. *P. obtusifolius.*
    Glands small, often dull; buds common.
      Leaves linear, 5–7-nerved.   23. *P. Friesii.*
      Leaves linear, 3-nerved.   27. *P. pusillus.*
      Leaves capillary, 1-nerved or nerveless.   28. *P. gemmiparus.*
Stipules adnate to the leaves or petioles.
  With both floating and submerged leaves.
  Submerged peduncles as long as the spikes, clavate, often recurved. 29. *P. diversifolius.*
  Submerged peduncles none, or at most hardly a line long. 30. *P. dimorphus.*
  With submerged leaves only.
  Stigma broad and sessile.
    Sheath of stipule less than 4″ long.   31. *P. filiformis.*
    Sheath of stipule more than 7″ long.   32. *P. interior.*
  Style apparent; stigma capitate.
    Fruit without keels or obscurely keeled.   33. *P. pectinatus.*
    Fruit strongly 3-keeled.
      Leaves entire, 3–5-nerved.   34. *P. interruptus.*
      Leaves minutely serrulate, finely many-nerved.   35. *P. Robbinsii.*

1. **Potamogeton nàtans** L. Common Floating Pondweed. Fig. 174.

Potamogeton natans L. Sp. Pl. 126. 1753.

Stems 2°–4° long, simple or sparingly branched.
Floating leaves thick, the blade ovate, oval or elliptic, 2′–4′ long, 1′–2′ wide, usually tipped with a short
abrupt point, rounded or subcordate at the base,
many-nerved; submerged leaves reduced to phyllodes
or bladeless petioles which commonly perish early
and are seldom seen at the fruiting period; stipules
sometimes 4′ long, acute, 2-keeled; peduncles as
thick as the stem, 2′–4′ long; spikes cylindric, very
dense, about 2′ long; fruit turgid, 2″–2¼″ long, about
1¼″ thick, scarcely keeled, narrowly obovoid, slightly
curved on the face; style broad and facial; nutlet
hard, more or less pitted or impressed on the sides,
2-grooved on the back; embryo forming an incomplete circle, the apex pointing toward the base.

In ponds and streams, Nova Scotia to British Columbia, New Jersey, Missouri and Nebraska. Also in
Europe and Asia. Called also Tench-weed, Batter-
dock, Deil's-spoons. July–Aug.

2. Potamogeton Oakesiànus Robbins. Oakes' Pondweed. Fig. 175.

Potamogeton Oakesianus Robbins in A. Gray, Man. Ed. 5, 485. 1867.

Stems very slender, often much branched from below. Floating leaves elliptic, mostly obtuse, rounded or slightly subcordate at the base, 1'–2' long, 5"–9" wide, 12–20-nerved; petioles 2'–6' long; submerged leaves mere capillary phyllodes, often persistent through the flowering season; peduncles 1'–3' long, commonly much thicker than the stem, mostly solitary; spikes cylindric, ½'–1' long; stipules acute, hardly keeled; fruit obovoid, about 1½" long, 1" thick, nearly straight on the face, 3-keeled, the middle keel sharp; style apical or subapical; sides of the nutlet not pitted, but sometimes slightly impressed; embryo circle incomplete, the apex pointing toward the base.

In still water, Anticosti to Wisconsin and New Jersey. Summer.

3. Potamogeton amplifòlius Tuckerm. Large-leaved Pondweed. Fig. 176.

Potamogeton amplifolius Tuckerm. Am. Journ. Sci. (II.) 6 : 225. 1848.
Potamogeton amplifolius ovalifolius Morong; A. Benn. Journ. Bot. 42 : 70. 1904.

Stems long, simple or occasionally branched. Floating leaves oval or ovate, abruptly pointed at the apex, rounded at the base, 2'–4' long, 1¼'–2' wide, many-nerved; petioles 3'–5' long; submerged leaves mostly petioled, the uppermost often elliptic or oval, 3'–6' long, 1'–2½' wide, the lowest lanceolate, often 8' long, with the sides closed and assuming a falcate shape; stipules tapering to a long sharp point, sometimes 4' long; peduncles thickened upward, 2'–8' long; spikes cylindric, 1'–2' long; fruit 2"–2½" long, 1¾" thick, turgid, the pericarp hard, obliquely obovoid, 3-keeled; face more or less angled; style subapical; embryo slightly incurved.

In lakes, Ontario to British Columbia, south to Georgia and Nebraska. July–Sept.

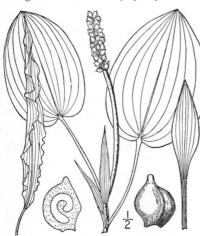

4. Potamogeton púlcher Tuckerm. Spotted Pondweed. Fig. 177.

Potamogeton pulcher Tuckerm. Am. Journ. Sci. 45 : 38. 1843.

Stems simple, terete, black-spotted, 1°–2° long. Floating leaves usually massed at the top on short lateral branches, alternate, ovate or round-ovate, subcordate, 2'–4½' long, 9"–3¼' wide, many-nerved; peduncles about as thick as the steam, 2'–4' long, spotted; submerged leaves of 2 kinds, the uppermost pellucid, lanceolate, long-acuminate, undulate, 3'–8' long, 6"–18" wide, tapering at the base into a short petiole, 10–20-nerved; the lowest much thicker, opaque, spatulate, oblong or ovate, on petioles ¼'–4' long; stipules obtuse or acuminate, 2-carinate; fruit 2"–2¼" long, 1½" thick, turgid, tapering into a stout apical style, the back sharply 3-kèeled; face angled near the middle, with a sinus below; embryo coiled.

In ponds and pools, Massachusetts to Georgia and Arkansas. July.

5. Potamogeton epihȳdrus Raf.   Nuttall's Pondweed.   Fig.

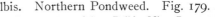

Potamogeton epihydrus Raf. Med. Repos. II. **5** : 354.
1808.
Potamogeton Nuttallii Cham. & Schl. Linnaea, **2** :
226. *pl. 6. f. 25.* 1827.
Potamogeton Claytonii Tuckerm. Am. Journ. Sci.
45 : 38. 1843.

Stems slender, compressed, 1°–6° long. Floating leaves opposite, elliptic to obovate, obtuse,
short-petioled, 1½′–3½′ long, 4″–12″ wide, many-
nerved; submerged leaves linear, 2-ranked, 2′–7′
long, 1″–3″ wide, 5-nerved, the 2 outer nerves
nearly marginal, the space between the 2 inner
and the midrib coarsely reticulated; stipules obtuse, hyaline, not keeled; peduncles 1′–5′ long;
spikes ½′–1′ long; fruit round-obovoid 1¼″–2″
long, 1″–1½″ thick, 3-keeled, the sides flat and
indistinctly impressed; style short, apical; embryo
coiled one and one-third times.

In ponds and streams, Newfoundland to British
Columbia, North Carolina and Iowa. Creek-grass.
June–Aug.

6. Potamogeton alpìnus Balbis.   Northern Pondweed.   Fig. 179.

Potamogeton alpinus Balbis, Misc. Bot. 13. 1804.
Potamogeton rufescens Schrad.; Cham. Adn. Fl.
Ber. **5.** 1815.

Plant of a ruddy tinge; stems simple or branched,
somewhat compressed. Floating leaves spatulate
or oblanceolate, obtuse, many-nerved, tapering
into petioles 1′–5′ long; submerged leaves semi-
pellucid, the lowest sessile, the uppermost petioled,
oblong-linear or linear-lanceolate, obtuse or rarely
acute, narrowed at the base, 3′–12′ long, 2″–9″
wide, 7-nerved; stipules broad, faintly 2-carinate,
obtuse or rarely acute; peduncles 2′–8′ long;
spikes 1′–1½′ long; fruit obovoid, lenticular, reddish, 1¾″ long, 1″ thick, 3-keeled, the middle keel
sharp, the face arched, beaked by the short recurved style; apex of the embryo pointing directly
to the basal end.

In ponds, Labrador to British Columbia, Florida
and California. Also in Europe. July–Aug.

7. Potamogeton americànus Cham. & Schl.   Long-leaved Pondweed.   Fig. 180.

Potamogeton fluitans Roth, Fl. Germ. **1** : 72. 1788?
Potamogeton americanus Cham. & Schl. Linnaea, **2** :
226. 1827.
Potamogeton lonchites Tuckerm. Am. Journ. Sci. (II.)
6 : 226. 1848.
Potamogeton lonchites noveboracensis Morong. Mem.
Torr. Club, **3** : Part 2, 20. 1893.

Stem terete, much branched, 3°–6° long. Floating
leaves rather thin, elliptic, pointed at both ends, 2′–6′
long, 6″–24″ wide, many-nerved, on petioles 2′–8′ in
length; submerged leaves pellucid, 4′–13′ long, 2″–12″
wide, rounded at the base or tapering into a petiole
1′–4′ long; stipules 1′–4′ long, acuminate, acute or
obtuse, strongly or faintly 2-carinate; peduncles
thickening upward, 2′–5′ long; spikes cylindric, 1′–3′
long; fruit about 2″ long, 1″–1½″ thick, obliquely
obovoid, the face nearly straight, the back 3-keeled,
the middle keel rounded or often with a projecting
wing under the style, not impressed on the sides;
embryo slightly incurved, apex pointing slightly inside of the base.

In ponds and slow streams, New Brunswick to Washington, Florida, West Indies, and California. July–Oct.

8. Potamogeton Fáxoni Morong. Faxon's Pondweed. Fig. 181.

Potamogeton Faxoni Morong, Mem. Torr. Club, **3**: Part 2, 22. 1893.

Floating leaves numerous, mostly obovate or oblanceolate, blunt-pointed or obtuse at the apex, narrowed at the base, often strikingly like those of *P. varians*, 2'-3½' long, 8"-12" wide, 13-17-nerved, on petioles 2'-6' long; submerged leaves oblong-lanceolate, acute or sometimes obtuse, 3'-5' long, 6"-12" wide, 5-13-nerved, often with an irregular areolation on each side of the midrib, borne on petioles ½'-2' in length; peduncles slightly thicker than the stem, 2'-5' long; spikes dense, 1'-2' long; fruit not collected.

Little Otter Creek and Lake Champlain, Ferrisburg, Vermont.

9. Potamogeton heterophýllus Schreb. Various-leaved Pondweed. Fig. 182.

Potamogeton heterophyllus Schreb. Spicil. Fl. Lips. 21. 1771.
Potamogeton heterophyllus var. *maximus* Morong, Mem. Torr. Club, **3**: Part 2, 25. 1893.
Potamogeton heterophyllus var. *longipedunculatus* Morong, Mem. Torr. Club, **3**: Part 2, 24. 1893.
Potamogeton heterophyllus graminifolius Morong. Mem. Torr. Club, **3**: Part 2, 24. 1893.
Potamogeton heterophyllus myriophyllus Morong, Mem. Torr. Club, **3**: Part 2, 24. 1893.
Potamogeton heterophyllus minimus Morong, Mem. Torr. Club, **3**: Part 2, 25. 1893.

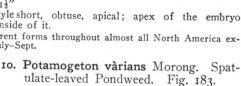

Stems slender, compressed, much branched, sometimes 12° long. Floating leaves pointed at the apex, rounded or subcordate at the base, 8"-4' long, 4"-14" wide, 10-18-nerved, on petioles 1'-4' long; submerged leaves pellucid, sessile, linear or linear-lanceolate, acuminate or cuspidate, rather stiff or flaccid, 1'-6½' long, 1"-8" wide, 3-7-nerved, the uppermost often petioled; peduncles often thickened upward, 1'-7' long, sometimes clustered; stipules spreading, obtuse, 8"-12" long; spikes 9"-1¼' long; fruit roundish or obliquely obovoid, 1"-1½" long, ¾"-1" thick, indistinctly 3-keeled; style short, obtuse, apical; apex of the embryo nearly touching the base, pointing slightly inside of it.

A very variable species, occurring in different forms throughout almost all North America except the extreme north. Also in Europe. July-Sept.

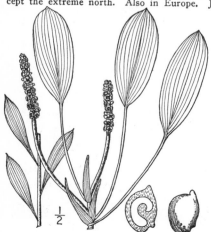

10. Potamogeton vàrians Morong. Spatulate-leaved Pondweed. Fig. 183.

Potamogeton gramineus var. (?) *spathulaeformis* Robbins in A. Gray, Man. Ed. 5, 487. 1867.
Potamogeton varians Morong; Fryer, Jour. Bot. **27**: 33. 1889.
Potamogeton spathulaeformis Morong, Mem. Torr. Club, **3**: Part 2, 26. 1893.

Stems many, branched, 2°-3° long. Floating leaves obovate or elliptic, abruptly acute at the apex, rather thin, 13-23-nerved, 1'-2½' long, 6"-13" wide, borne on slender petioles; submerged leaves pellucid, spatulate-oblong or linear-lanceolate, 2'-4' long, 3"-9" wide, 5-13-nerved, cuspidate or spinescent, sessile or subsessile, often reduced to phyllodes with a very narrow blade and a long acumination at the base and apex; peduncles often thickening upward, 1'-2' long; stipules obtuse, faintly keeled, the apex slightly hooded; spikes large; fruit about 1" long, roundish or obliquely ovoid, obscurely 3-keeled, with a curved or slightly angled face; embryo with the apex pointing slightly inside of the base.

In Mystic Pond, Medford, Mass. Also in Europe. Summer. Apparently a mere form of the preceding, or perhaps a hybrid between *P. angustifolius* and *P. heterophyllus.*

11. **Potamogeton illinoénsis** Morong. Illinois Pondweed. Fig. 184.

Potamogeton illinoensis Morong, Coult. Bot. Gaz. 5: 50. 1880.

Stems stout, much branched above. Floating leaves opposite, numerous, thick, 4′–5½′ long, 2′–3½′ wide, many-nerved, oval or broadly elliptic, short-pointed at the apex, rounded, subcordate or narrowed at the base; petioles 1′–4′ long; submerged leaves numerous, 4′–8′ long, 1′–2′ wide, 13–19-nerved, acuminate or the uppermost acute, mostly tapering at the base into a short broad flat petiole, rarely reduced to phyllodes; stipules 2′–3′ long, obtuse, strongly 2-carinate; peduncles 2′–4′ long; spikes 1′–2′ long; fruit roundish or obovoid, 1½″–2″ long, 1″–1½″ thick, dorsally 3-keeled; style short, blunt.

In ponds, Illinois to Iowa and Minnesota. Aug.

12. **Potamogeton angustifòlius** Berch. & Presl. Ziz's Pondweed. Fig. 185.

P. angustifolius Berch. & Presl, Rost. 19. 1821.
Potamogeton Zizii Roth, Enum. 1: 531. 1827.
Potamogeton lucens connecticutensis Robbins in A. Gray, Man. Ed. 5, 488. 1867.
Potamogeton angustifolius var. *Methyensis* A. Bennett, Journ. Bot. 29: 151. 1891.

Stems slender, branching. Floating leaves elliptic, 1½′–4′ long, 6″–12″ wide, many-nerved; petioles mostly short; submerged leaves mostly lanceolate or oblanceolate, thin, acute or cuspidate, 2′–6′ long, 3″–15″ wide, 7–17-nerved; stipules 6″–18″ long, obtuse, 2-keeled; peduncles thicker than the stem, 2½′–6′ long; spikes 1′–2′ long; fruit obliquely obovoid, 1¼″–2″ long, about 1″ thick, the face dorsally 3-keeled; style short, blunt, facial; apex of the embryo pointing directly to the base.

In lakes and streams, Quebec to California, Florida and Wyoming. Also in Europe. July–Aug.

13. **Potamogeton lùcens** L. Shining Pondweed. Cornstalk-weed. Fig. 186.

Potamogeton lucens L. Sp. Pl. 126. 1753.

Stems thick, branching below and often with masses of short leafy branches at the summit. Leaves all submerged, elliptic, lanceolate or the uppermost oval, shining, acute or acuminate and cuspidate, or rounded at both ends and merely mucronulate, sessile or short-petioled, 2½′–8′ long, 8″–20″ wide, the tips often serrulate; stipules 1′–3′ long, 2-carinate, sometimes very broad; peduncles 3′–6′ long; spikes 2′–2½′ long, cylindric, very thick; fruit about 1½″ long and 1¼″ thick, roundish, the face usually with a slight inward curve at the base; apex of the embryo pointing transversely inward.

In ponds, Nova Scotia to Florida, west to California and Mexico. Local. Also in Europe. Sept.–Oct.

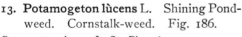

14. Potamogeton praelóngus Wulf. White-stemmed Pondweed. Fig. 187.

Potamogeton praelongus Wulf. in Roem. Arch. **3**: 331. 1805.

Stems white, flexuous, flattened, much branched, growing in deep water, sometimes 8° long. Leaves all submerged, oblong or oblong-lanceolate, semi-amplexicaul, bright green, 2′–12′ long, ½′–1¼′ wide, with 3–5 main nerves, stipules white, scarious, obtuse and commonly closely embracing the stem; peduncles 3′–20′ long, erect, straight, about as thick as the stem; spikes 1′–2′ long, thick, cylindric; fruit dark green, obliquely obovoid, 2″–2½″ long, 1½″–2″ thick, the back much rounded, often with the upper curve nearly as high as the style; the middle keel sharp; style short, obtuse, facial.

Nova Scotia to British Columbia, south to New Jersey, Minnesota and California. Also in Europe. Fruits in June and July, and usually withdraws its stems beneath the water as soon as the fruit is set.

15. Potamogeton perfoliàtus L.   Clasping-leaved Pondweed.   Fig. 188.

Potamogeton perfoliatus L. Sp. Pl. 126. 1753.
Potamogeton perfoliatus lanceolatus Robbins in A. Gray, Man. Ed. 5, 488. 1867. Not Blytt 1861.
Potamogeton perfoliatus Richardsonii A. Bennett. Journ. Bot. **27** : 25. 1889.
Potamogeton Richardsonii Rydb. Bull. Torrey Club **32** : 599. 1905.
Potamogeton bupleuroides Fernald, Rhodora **10** : 46. 1908.

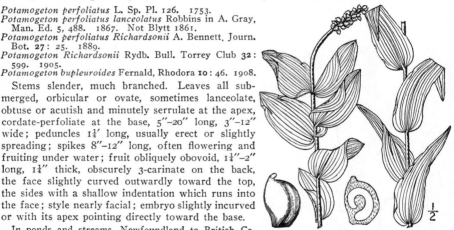

Stems slender, much branched. Leaves all submerged, orbicular or ovate, sometimes lanceolate, obtuse or acutish and minutely serrulate at the apex, cordate-perfoliate at the base, 5″–20″ long, 3″–12″ wide; peduncles 1¾′ long, usually erect or slightly spreading; spikes 8″–12″ long, often flowering and fruiting under water; fruit obliquely obovoid, 1¾″–2″ long, 1¼″ thick, obscurely 3-carinate on the back, the face slightly curved outwardly toward the top, the sides with a shallow indentation which runs into the face; style nearly facial; embryo slightly incurved or with its apex pointing directly toward the base.

In ponds and streams, Newfoundland to British Columbia, south to Florida and California. Also in Europe and Asia. July–Sept.

16. Potamogeton mýsticus Morong.   Mystic Pond Pondweed.   Fig. 189.

Potamogeton mysticus Morong, Coult. Bot. Gaz. **5** : 50. 1880.

Whole plant very slender and delicate, stems irregularly branching above, nearly filiform, terete, 1°–4° long. Leaves all submerged, scattered, oblong-linear, 1′–1½′ long, 1″–3″ wide, 5–7-nerved, obtuse and rarely with minute serrulations near the apex, abruptly narrowed at the base and sessile or partly clasping; stipules obtuse, about 6″ long, hyaline and with many fine nerves, mostly deciduous, but sometimes persistent and closely sheathing the stem; spikes few, capitate, 4–6-flowered, borne on erect peduncles 1′–2′ long; immature fruit obovoid, less than 1″ long, about ½″ wide, obscurely 3-keeled on the back, slightly beaked by the slender, recurved style.

Mystic Pond, Medford, and Miacount Pond, Nantucket, Mass. Aug.-Sept. Apparently a depauperate form of the preceding, and scarcely distinct from it. Perhaps a hybrid.

17. Potamogeton confervoìdes Reichb.   Alga-like Pondweed.   Fig. 190.

Potamogeton confervoides Reichb. Ic. Fl. Germ. & Helv. **7**:
13. 1845.
Potamogeton trichoides A. Gray, Man. 457. 1848. Not Cham.
Potamogeton Tuckermani Robbins; A. Gray, Man. Ed. *2*, 434.
1856.

Stems slender, terete, much branched, the upper branches
repeatedly forking, 6'–18' long. Leaves very delicate, flat,
setaceous, 1'–2½' long, the broadest scarcely ¼" wide, tapering to a long hair-like point, 1–3-nerved and often with a
few cross-veins; stipules bright green or yellowish; leaves delicate, obtuse, 2"–3" long; peduncles 2'–8' long, erect, somewhat thickened upward; spikes capitate, 3"–4" long; fruit
roundish-obovoid, 1"–1½" long and about as thick, the back
sometimes a little angular or sinuate, 3-keeled, the middle
keel sharp, the face notched near the base, the sides impressed with a shallow indentation which runs into the
notch of the face; apex of the embryo nearly touching the
base a little to one side.

In cold or mountain ponds, Maine and New Hampshire to
New Jersey and Pennsylvania. Aug.–Sept.

18. Potamogeton críspus L.   Curly Muck-weed.   Pondweed.   Fig. 191.

Potamogeton crispus L. Sp. Pl. 126. 1753.

Stems branching, compressed. Leaves 2-ranked,
linear-oblong or linear-oblanceolate, sessile or semi-
amplexicaul, obtuse at the apex, serrulate, crisped,
¾'–4' long, 3"–7" wide, 3–7-nerved, the midrib often
compound and the outer nerves very near the margin; stipules small, scarious, obtuse, early perishing;
peduncles 1'–2' long, frequently recurved in fruit,
sometimes very numerous; spikes about ½' long,
appearing very bristly with the long-beaked drupelets when in fruit; fruit ovoid, about 1½" long, 1"
or more wide, 3-keeled on the back, the middle keel
with a small projecting tooth near the base, the face
slightly curved, the style facial and nearly as long
as the drupelet; embryo small, its apex pointing
directly toward its base. The plant is mainly propagated by peculiar winter buds.

In fresh or salt water, about cities, Massachusetts to
Pennsylvania and Virginia. Also in Europe. Aug.

19. Potamogeton compréssus L.   Eel-grass Pondweed.   Fig. 192.

Potamogeton compressus L. Sp. Pl. 127. 1753.
Potamogeton zosteraefolius Schum. Enum. Pl. Saell. 50.
1801.

Stems much flattened, sometimes winged, widely
branching. Leaves linear, obtuse and mucronate or
short-pointed at the apex, 2'–12' long, 1"–2" wide, with
3 principal nerves and many fine ones; stipules scarious,
obtuse, finely nerved, soon perishing; peduncles 1½'–4'
long; spikes cylindric, about ½' long, 12–15-flowered;
fruit obovoid with a broad base, about 2" long, 1¼"–1½"
thick, 3-keeled on the back, the lateral keels rather
obscure; face arched, beaked with a short recurved
style; embryo slightly incurved. The plant is propagated by the terminal leaf-buds, which sink to the bottom, and rest during the winter.

In still or running water, New Brunswick to New York,
west to Oregon. Also in Europe. July–Aug. Grass-wrack.

6

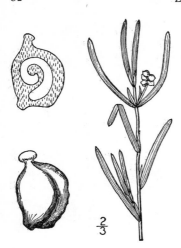

2/3

20. Potamogeton Híllii Morong.   Hill's Pondweed. Fig. 193.

Potamogeton Hillii Morong, Coult. Bot. Gaz. **6**: 290. 1881.

Stems slightly compressed, slender, widely branching,
1°–2° long. Leaves linear, acute or cuspidate, or often
almost aristate, 1'–2¼' long, ½''–1¼'' wide, 3-nerved, the
lateral nerves delicate and nearer the margins than the
midrib; stipules whitish, many-nerved, obtuse, 3''–5''
long; peduncles about ½' long, erect or slightly recurved,
more or less clavate; spikes capitate, 3–6-fruited; fruit
obliquely obovoid, obtuse at the base, about 2'' long,
1''–1¼'' thick, 3-carinate on the back, the middle keel
sharp and more or less undulate, flat on the sides, face
slightly arched; style nearly facial, short; embryo coiled.

In ponds, eastern New York to Michigan, south to
Pennsylvania and Missouri. There are two forms of the
species, the one 2-glandular at the base of the leaves, the
other glandless. July–Sept.

21. Potamogeton foliòsus Raf.   Leafy Pondweed.   Fig. 194.

Potamogeton foliosus Raf. Med. Rep. (II.) **5**:
  354. 1808.
Potamogeton pauciflorus Pursh, Fl. Am. Sept.
  121. 1814. Not Lam. 1789.
Potamogeton niagarensis Tuckerm. Am. Journ.
  Sci. (II.) **7**: 354. 1849.
Potamogeton foliosus niagarensis (Tuckerm.)
  Morong, Mem. Torr. Club, **3**: Part 2, 39.
  1893.

Stems flattened, much branched, 1°–3°
long. Leaves 1'–3' long, ½''–1'' wide, acute,
3–5-nerved, not glandular at the base;
stipules white, hyaline, obtuse or sometimes
acute, 6''–10'' long; peduncles more or less
clavate, erect, about ½' long; spikes 4–12-
flowered; fruit lenticular or nearly orbicular, about 1'' in diameter, 3-keeled on the
back, the middle keel winged, sinuate-
dentate, often with projecting shoulders or
teeth at each end, the face strongly angled
or arched, sharp, often with a projecting
tooth at the base; style apical.

In ponds and streams, New Brunswick to
British Columbia, south to Florida, New Mexico and California. July–Aug.

22. Potamogeton obtusifòlius Mert. & Koch.
Blunt-leaved Pondweed. Fig. 195.

Potamogeton compressus Wahl. Fl. Suec. **1**: 107. 1824.
  Not L. 1753.
Potamogeton obtusifolius Mert. & Koch, Deutsch. Fl.
  1: 855. 1823.

Stems usually slender, compressed, widely branching, especially above. Leaves linear, 2'–3' long, ½''–2''
wide, obtuse, often mucronate, usually 3-nerved with
a broad midrib, sometimes 5–7-nerved, 2-glandular
at the base, the glands large and translucent; stipules
white or scarious, many-nerved, obtuse, 6''–9'' long,
often as long as or longer than the internodes; peduncles numerous, 1'–1½' long, slender, erect; spikes
3''–4'' long, ovoid, 5–8-flowered; fruit obliquely obovoid, about 1½'' long and 1'' thick, 3-keeled; style
short, blunt, nearly facial.

In still water, Quebec to Minnesota, south to northern
New York and Kansas. Also in Europe. July–Aug.

½

23. Potamogeton Frièsii Ruprecht. Fries' Pondweed. Fig. 196.

Potamogeton compressus J. E. Smith, Engl. Bot. **3**: *pl. 418.* 1794. Not L. 1753.
Potamogeton pusillus var. *major* Fries, Novit. Ed. **2**, 48. 1828.
Potamogeton Friesii Ruprecht, Beitr. Pfl. Russ. Reichs, **4**: 43. 1845.
Potamogeton major Morong, Mem. Torr. Club, **3**: Part 2, 41. 1893. Not *P. pusillus* var. *major* M. & K. 1823.

Stems compressed, 2°–4° long, branching. Leaves 1½′–2½′ long, about 1″ wide, acute, obtuse or cuspidate at the apex, mostly 5-nerved, rarely 7-nerved, 2-glandular at the base, the glands small; stipules white, hyaline, finely nerved, obtuse or acute, 6″–12″ long; peduncles 1′–1½′ long, often thicker than the stem and sometimes thickening upward; spikes, when developed, interrupted; fruit quite similar to that of *P. pusillus*, but with a recurved style, usually with a shallow pit on the sides, and with the apex of the embryo pointing toward the basal end.

In still water, New Brunswick to New York, west to North Dakota and Iowa. Also in Europe. Propagating buds occasional. July–Aug.

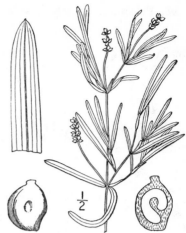

24. Potamogeton rùtilus Wolfg. Slender Pondweed. Fig. 197.

Potamogeton rutilus Wolfg.; R. & S. Mant. **3**: 362. 1827.

Stems very slender, 8′–24′ long, compressed, simple or nearly so. Leaves 1′–1½′ long, ¼″–½″ wide, acute or acuminate, strict, nearly erect, 3–5-nerved, revolute, the nerves prominent beneath, often 2-glandular at base and bright green; stipules acute, 6″–10″ long, often longer than the internodes and hiding the bases of the leaves above, persistent, becoming white and fibrous with age; peduncles 6″–18″ long; spikes 3″–5″ long, usually dense, but sometimes interrupted; fruit obliquely obovoid, about 1″ long and ⅓″ thick, obscurely keeled or the back showing only 2 small grooves; apex of the drupelet tapering into a short facial nearly straight recurved style; embryo circle not complete, the apex pointing towards or outside the base.

Anticosti and James Bay to Michigan and Minnesota. south to New York. Also in Europe. Propagating buds usually wanting.

25. Potamogeton Vàseyi Robbins. Vasey's Pondweed. Fig. 198.

Potamogeton Vaseyi Robbins in A. Gray, Man. Ed. **5**, 485. 1867.
Potamogeton Vaseyi var. *latifolius* Morong, Mem. Torr. Club, **3**: Part 2, 44. 1893.

Stems filiform, widely branching below, and with many short lateral branches above, 1°–1½° long, the emersed fertile forms in shallow water, and the more common sterile submerged forms in water from 6°–8° in depth. Floating leaves on the fertile stems only, coriaceous, in 1–4 opposite pairs, oval oblong or obovate, 4″–5″ long, 2″–3″ wide, with 5–9 nerves deeply impressed beneath, tapering at the base into petioles 3″–4″ long; submerged leaves capillary, 1′–1½′ long; stipules white, delicate, many-nerved, acute or obtuse, 2″–3″ long; peduncles 3″–6″ long, thickening in fruit; spikes 2″–3″ long, 2–6-fruited; fruit roundish-obovoid, about 1″ long and nearly as thick, 3-keeled, the middle keel rounded, tipped with a straight or recurved style.

Quebec to Wisconsin, south to southern New York. The plant is furnished with propagating buds. July–Aug.

26. Potamogeton lateràlis Morong.   Opposite-leaved Pondweed.   Fig. 199.

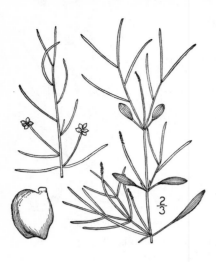

Potamogeton lateralis Morong, Coult. Bot. Gaz. **5**: 51. 1880.

Stems filiform, much branched. Floating leaves on sterile shoots only, coriaceous, elliptic, obtuse, 4″–5″ long, 1″–2″ wide, 5–7-nerved, the nerves deeply impressed beneath, usually in 1–3 opposite pairs which stand at right angles to the stem, on petioles 3″–10″ long; submerged leaves linear, acute, 1′–3′ long, ¼″–½″ wide, 1–3-nerved, 2-glandular at the base, but the glands small and often obsolete; stipules small, hyaline, many-nerved, obtuse, deciduous; peduncles and floating leaves lateral, with a peculiar appearance, widely spreading at maturity, sometimes recurved, thickening in fruit, 4″–15″ long; spikes capitate or often interrupted, 3–4-flowered; fruit obliquely obovoid, about 1″ long, lenticular, the back much curved and 2-grooved, the face arched and surmounted by the nearly sessile stigma; curve of the embryo oval, its apex nearly touching its base.

In lakes and slow streams, Massachusetts and Connecticut. Proliferous shoots at the summit of the stem and on the upper branches appear late in the season, as the plants are beginning to decay. July–Aug. A rare and local plant, which, in an incompletely developed state, when it lacks the broad floating leaves, has the aspect of *P. pusillus;* its affinities are probably with *P. Vaseyi* and *P. diversifolius.*

27. Potamogeton pusíllus L.   Small Pondweed.   Fig. 200.

Potamogeton pusillus L. Sp. Pl. 127. 1753.
Potamogeton panormitanus Biv. Sic. Pl. 1806–7.
Potamogeton pusillus polyphyllus Morong, Coult. Bot. Gaz. **5**: 51. 1880.
Potamogeton pusillus sturrockii A. Bennett in Hook. Stud. Fl. Ed. 3, 435. 1884.
Potamogeton pusillus panormitanus Morong, Mem. Torr. Club, **3**: Part 2, 46. 1893.

Stems filiform, branching, 6′–2° long. Leaves all submerged, linear, obtuse and mucronate or acute at the apex, 2-glandular at the base, 1′–3′ long, about ½″ wide, 1–3-nerved, the lateral nerves often obscure, or the leaf apparently nerveless; stipules short, hyaline, obtuse; peduncles usually 3″–9″, or rarely 3′ long; spikes 3–10-flowered; fruit obliquely ellipsoid, about 1″ long and ½″ thick curved and 2-grooved on the back or sometimes with 3 distinct keels, the face slightly arched, beaked by a straight or recurved style; apex of the embryo slightly incurved and pointing inside the base. Propagative buds occur in greater or less abundance.

In ponds and slow streams, New Brunswick to British Columbia, south to Virginia, Texas and California. Also in Europe. July–Aug. The forms listed in the above synonymy are all more or less distinctly, if inconstantly variable from the type. They are not sufficiently stable to merit specific recognition.

This is the commonest of the completely submerged Pondweeds. It may readily be distinguished from all other species of its group by its boat-shaped stipules which are usually twice as wide as the base of the leaf.

28. Potamogeton gemmíparus (Robbins) Morong. Thread-like Pondweed.
Fig. 201.

Potamogeton pusillus var. (?) *gemmíparus* Robbins in A. Gray, Man. Ed. 5, 489. 1867.

Potamogeton gemmíparus Morong, Coult. Bot. Gaz. **5**: 51. 1880.

Stems filiform, terete, branching, 5′–4° long. Leaves capillary, sometimes not as wide as the stem, often with no perceptible midrib, tapering to the finest point, 1′–3′ long, 2-glandular at the base; stipules ½′–1′ long, acute or obtuse, mostly deciduous; spikes interrupted, 3–6-flowered; peduncles filiform or sometimes slightly thickened, ½′–2′ long; fruit seldom formed, similar to that of *P. pusillus*.

In ponds, eastern Massachusetts and Rhode Island. It is commonly propagated by its abundant buds, the leaves and stems are often alike in thickness so that the plant seems to consist of threads. Aug.–Sept. A very slender form of the preceding and doubtfully distinct from it.

29. Potamogeton diversifòlius Raf. Rafinesque's Pondweed. Fig. 202.

Potamogeton hybridus Michx. Fl. Bor. Am. **1**: 101. 1803. Not Thuill. 1790.

Potamogeton diversifolius Raf. Med. Rep. (II.) **5**: 354. 1808.

Potamogeton diversifolius multidenticulatus Morong, Mem. Torr. Club, **3**: Part 2, 48. 1893.

Potamogeton diversifolius trichophyllus Morong, Mem. Torr. Club, **3**: part 2, 49. 1893.

Stems flattened or sometimes terete, much branched. Floating leaves coriaceous, the largest 1′ long by ½′ wide, oval or elliptic and obtuse, or lanceolate-oblong and acute; petioles generally shorter, but sometimes longer than the blades, filiform or dilated; submerged leaves setaceous, seldom over ¼″ wide, 1′–3′ long; stipules obtuse or truncate, 3″–5″ long, those of the floating leaves free, those of the submerged leaves sometimes adnate; emersed peduncles 3″–7″ long; submerged peduncles 2″–3″ long, clavate, as long as the spikes; emersed spikes 3″–5″ long, occasionally interrupted; fruit cochleate, rarely over ½″ long, 3-keeled, the middle keel narrowly winged and usually with 7–12 knob-like teeth on the margin, the lateral keels sharp or toothed; embryo coiled 1½ times.

In still water, Maine to Florida, west to California and Texas. June–Sept. A common and well-marked species which often covers large areas of water, practically to the exclusion of everything else. From *P. dimorphus*, its nearest relative, it may readily be distinguished by its distinctly stalked submerged spikes of flowers.

30. Potamogeton dimórphus Raf. Spiral Pondweed. Fig. 203.

Potamogeton dimorphus Raf. Month. Mag. Crit. Rev. 1: 358. 1817.
Potamogeton Spirillus Tuckerm. Am. Journ. Sci. (II.) 6: 228. 1848.
Potamogeton Spirillus curvifolius Peck, N. Y. State Mus. Rep. 49: 28. 1896.

Stems compressed, branched, 6′–20′ long, the branches often short and recurved. Floating leaves oval or elliptic, obtuse, the largest about 1′ long and ½′ wide, with 5–13 nerves deeply impressed beneath, their petioles often 1′ long; submerged leaves linear, 1½″–2″ long, about ½″ wide, mostly 5-nerved; stipules of the upper floating leaves free; those of the submerged leaves adnate to the blade or petiole; spikes above water 3″–5″ long, continuous, the lower mostly sessile, capitate and 1–10-fruited; fruit cochleate, roundish, less than 1″ long, flat and deeply impressed on the sides, 3-keeled on the back, the middle keel winged and rarely 4–5-toothed; style deciduous; embryo spiral, about 1½ turns.

In ponds and ditches, Nova Scotia and Ontario to Minnesota, south to Virginia, Missouri and Nebraska. June–Aug.

31. Potamogeton filifórmis Pers. Filiform Pondweed. Fig. 204.

Potamogeton filiformis Pers. Syn. 1: 152. 1805.
Potamogeton filiformis Macounii Morong; Macoun, Cat. Can. Pl. 4: 88. 1888.

Stems from a running rootstock, slender, 3°–20° long, filiform above, stout and thick towards the base. Leaves numerous, 2′–12′ long, ¼″–½″ wide, 1-nerved with a few cross veins; sheaths about 2″ long and the free part of the stipule shorter, scarious on the edges; flowers on long, often recurved peduncles, 2–12 in each whorl, the whorls ¾′–1′ apart; fruit 1″–1½″ long, slightly less than 1″ wide, the sides even, the back not keeled, the face nearly straight or obtusely angled near the top; stigma nearly or quite sessile, remaining on the fruit as a broad truncate projection. Embryo a complete spiral, the curved apex pointing inside the base.

In ponds and lakes, Anticosti to western New York and Montana. August.

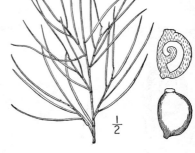

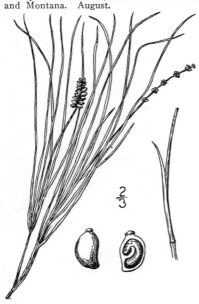

32. Potamogeton intèrior Rydb. Inland Pondweed. Fig. 205.

Potamogeton interior Rydb. Fl. Colorado 13. 1906.

Potamogeton marinus occidentalis Robb. Bot. King's Exp. 339. 1871.

Potamogeton filiformis occidentalis A. Benn. Ann. Conserv. Jard. Genev. 9: 102. 1905.

Stem slender, much branched and longer than in the preceding; leaves all submerged, capillary or narrowly linear, with an acute or more or less pungent apex, 2′–6′ long, ¼″–⅜″ wide, mostly 1-nerved; stipules adnate to the leaf-bases, the sheath at least 7″ long, the free part shorter; spikes few-flowered, often interrupted, ¼′–3½′ long; peduncles as thick as the stem, 1½′–7′ long; nutlets sometimes slightly pitted, without keels or inconspicuously 1-keeled; style almost invisible; embryo an incomplete spiral, the straight apex pointing directly towards the base.

Ontario to the Northwest Territory, south to Utah and Colorado.

33. Potamogeton pectinàtus L. Fennel-leaved Pondweed. Fig. 206.

Potamogeton pectinatus L. Sp. Pl. 127. 1753.

Stems slender, much branched, 1°–3° long, the branches repeatedly forking. Leaves setaceous, attenuate to the apex, 1-nerved, 1′–6′ long, aften capillary and nerveless; stipules half free, ½′–1′ long, their sheaths scarious on the margins; peduncles filiform, 2′–12′ long, the flowers in verticils; fruit obliquely obovoid, with a hard thick shell, 1½″–2″ long, 1″–1¼″ wide, without a middle keel, but with obscure lateral ridges on the back, plump on the sides and curved or occasionally a little angled on the face; style straight or recurved, facial; embryo apex pointing almost directly toward the basal end.

In fresh, brackish or salt water, Cape Breton to British Columbia, south to Florida, Texas and California. Also in Europe. Pondgrass. July–Aug.

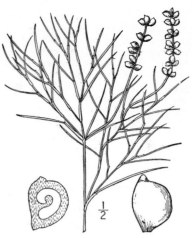

34. Potamogeton interrúptus Kitaibel. Interrupted Pondweed. Fig. 207.

Potamogeton interruptus Kitaibel in Schultes, OEst. Fl. Ed. 2, 328. 1814.
Potamogeton flabellatus Bab. Man. Bot. Ed. 3, 343. 1851.

Stems arising from a running rootstock which often springs from a small tuber, 2°–4° long, branched, the branches spreading like a fan. Leaves linear, obtuse or acute, 3′–5′ long, 1″–1¼″ wide, 3–5-nerved with many transverse veins; narrow, 1-nerved leaves on some plants, these acuminate, as *P. pectinatus;* stipules partially adnate to the leaf-blade, the adnate part ½′–1′ long, sometimes with narrowly scarious margins, the free part shorter, scarious, obtuse; peduncles 1′–2′ long; spikes interrupted; fruit broadly and obliquely obovoid, obtuse at the base, the largest 2″ long and nearly as broad, keeled and with rounded lateral ridges on the back, the face nearly straight; style facial, erect.

In ponds and streams, Michigan, Nebraska, Saskatchewan and Wisconsin. Also in Europe. August.

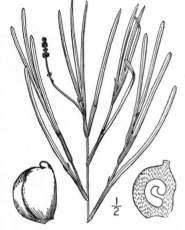

35. Potamogeton Robbínsii Oakes. Robbins' Pondweed. Fig. 208.

Potamogeton Robbinsii Oakes, Hovey's Mag. 7: 180. 1841.

Stems stout, wide-branching, 2°–4° long, from running rootstocks. Leaves linear, 3′–5′ long, 2″–3″ wide, acute, many-nerved, crowded in 2 ranks, minutely serrulate, auriculate at attachment with the stipule; stipules with the adnate portion and sheathing base of the leaf about ½′ long, the free part ½′–1′ long, acute, persistent, white, membranous, lacerate; peduncles 1′–3′ long, inflorescence frequently much branched, with 5–20 peduncles; spikes interrupted, ½′–1′ long, flowering under water; fruit obovoid, about 2″ broad and 1½″ wide, 3-keeled on the back, middle keel sharp, lateral ones rounded, face arched, sides with a shallow depression running into the face below the arch; style subapical, thick, slightly recurved; apex of the embryo pointing a little inside the basal end.

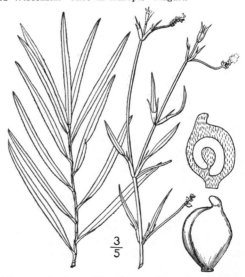

In ponds and lakes, New Brunswick to Oregon, south to New Jersey, Pennsylvania and Michigan. The plant is freely propagated by fragments of the stems which throw out rootlets from each joint, but this is the rarest of our species to form fruit. Aug.–Sept.

2. RÚPPIA L. Sp. Pl. 127. 1753.

Slender, widely branched aquatics with capillary stems, slender alternate 1-nerved leaves tapering to an acuminate apex, and with membranous sheaths. Flowers on a capillary spadix-like peduncle, naked, consisting of 2 sessile anthers, each with 2 large separate sacs attached by their backs to the peduncle, having between them several pistillate flowers in 2 sets on opposite sides of the rachis, the whole cluster at first enclosed in the sheathing base of the leaf. Stigmas sessile, peltate. Fruit a small, obliquely pointed drupe, several in each cluster and pedicelled; embryo oval, the cotyledonary end inflexed, and both that and the hypocotyl immersed. [Name in honor of Heinrich Bernhard Rupp, a German botanist.]

In the development, the staminate flowers drop off, the peduncle elongates, bearing the pistillate flowers in 2 clusters; after fertilization it coils up and the fruit is drawn below the water.

Three or four species, occurring in salt and brackish waters all over the world. Type species: *Ruppia maritima* L. The following are the only ones known to occur in North America.

Sheaths 3″–4″ long; drupes about 1″ long.   1. *R. maritima*.
Sheaths ½′–1½′ long; drupes 1½″–2″ long.   2. *R. occidentalis*.

1. Ruppia marítima L.   Sea- or Ditch-grass.   Tassel Pondweed.   Fig. 209.

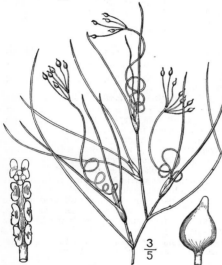

Ruppia maritima L. Sp. Pl. 127. 1753.
Ruppia curvicarpa A. Nelson, Bull. Torrey Club **26**: 122. 1899.

Stems often whitish, 2°–3° long, the internodes irregular, naked, 1′–3′ long. Leaves 1′–3′ long, ¼″ or less wide; sheaths 3″–4″ long, with a short free tip; peduncles in fruit sometimes 1° long; pedicels 4–6 in a cluster, ½′–1½′ long; drupes with a dark hard shell, ovoid, about 1″ long, often oblique or gibbous at the base, pointed with the long style, but varying much in shape; forms with very short peduncles and pedicels, and with broad, strongly marked sheaths occur.

Common along the coasts and in saline districts in the interior. Old World and South America. Tassel-grass. July–Aug.

2. Ruppia occidentàlis S. Wats.   Western Ruppia.   Fig. 210.

Ruppia occidentalis S. Wats. Proc. Am. Acad. **25**: 138. Sept. 1890.
Ruppia lacustris Macoun, Cat. Can. Pl. **5**: 372. Nov. 1890.

Stems stouter, 1°–2° long, the branching fan-like. Leaves 3′–8′ long, their large sheaths ½′–1½′ long; branches and leaves often thickly clustered at the nodes, the sheaths overlapping each other; drupes larger, 1½″–2″ long, ovoid or pyriform, borne on pedicels about 1′ long, the peduncles bright red when fresh and sometimes nearly 2° in length.

In saline ponds, Nebraska to British Columbia.

3. ZANNICHÉLLIA L. Sp. Pl. 969. 1753.

Stems, flowers and leaf-buds all at first enclosed in a hyaline envelope, corresponding to the stipule in *Potamogeton*. Staminate and pistillate flowers in the same axil; the staminate solitary, consisting of a single 2-celled anther, borne on a short pedicel-like filament; the pistillate 2–5. Ovary flask-shaped, tapering into a short style; stigma broad, hyaline, somewhat cup-shaped, its margins angled or dentate. Fruit a flattish falcate nutlet, ribbed or sometimes toothed on the back. Embryo bent and coiled at the cotyledonary end. [In honor of J. H. Zannichelli, 1662–1729, Italian physician and botanist.]

Two or three species of wide distribution in fresh water, the following typical.

1. Zannichellia palústris L. Horned Pondweed. Fig. 211.

Zannichellia palustris L. Sp. Pl. 969. 1753.
Z. intermedia Torrey; Beck Bot. 385. 1833.

Stems capillary, sparsely branched, the rhizome creeping, the roots fibrous. Leaves 1′–3′ long, ¼″ or less wide, acute, thin, 1-nerved with a few delicate cross-veins; spathe-like envelope separate from the leaves and fruits at maturity; fruits 2–6 in a cluster, 1″–2″ long, sometimes sessile, sometimes pedicelled, sometimes the whole cluster peduncled; style persistent, ½″–1″ long; plant flowering and ripening its fruit under water.

In fresh or brackish water, nearly throughout North America, except the extreme north. Widely distributed in the Old World. July–Sept.

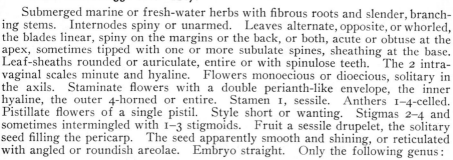

Family 4. **NAIADÀCEAE** (Lindl.) Asch. Linnaea **35** : 160. 1867.

Submerged marine or fresh-water herbs with fibrous roots and slender, branching stems. Internodes spiny or unarmed. Leaves alternate, opposite, or whorled, the blades linear, spiny on the margins or the back, or both, acute or obtuse at the apex, sometimes tipped with one or more subulate spines, sheathing at the base. Leaf-sheaths rounded or auriculate, entire or with spinulose teeth. The 2 intravaginal scales minute and hyaline. Flowers monoecious or dioecious, solitary in the axils. Staminate flowers with a double perianth-like envelope, the inner hyaline, the outer 4-horned or entire. Stamen 1, sessile. Anthers 1–4-celled. Pistillate flowers of a single pistil. Style short or wanting. Stigmas 2–4 and sometimes intermingled with 1–3 stigmoids. Fruit a sessile drupelet, the solitary seed filling the pericarp. The seed apparently smooth and shining, or reticulated with angled or roundish areolae. Embryo straight. Only the following genus :

1. **NÀIAS** L. Sp. Pl. 1015. 1753.

Characters of the family. Slender, branching, submerged aquatics. Flowers sessile or pedicelled. Sterile flowers with a double perianth, the exterior one entire or 4-horned at the apex, the interior one hyaline, adhering to the anther; stamen sessile or stalked, 1–4-celled, apiculate or 2-lobed at the summit. Fertile flowers of a single ovary which tapers into a short style; stigmas 2–4, subulate. Mature carpel sessile, ellipsoid, its pericarp crustaceous. Seed conformed to the pericarp, the raphe distinctly marked. [Greek, water-nymph.]

About 10 species in fresh water all over the world. There is one other American species known only from Florida and Cuba. Type species: *Naias marina* L.

Sheaths broadly rounded, their margins entire or with a few large teeth.   1. *N. marina.*
Sheaths narrowly and obliquely rounded, each margin with 5–10 minute teeth; leaves linear.
  Seeds shining, with 30–50 rows of faint reticulations.   2. *N. flexilis.*
  Seeds dull, with 16–20 rows of strongly marked reticulations.   3. *N. guadalupensis.*
Sheaths auriculate; leaves filiform.   4. *N. gracillima.*

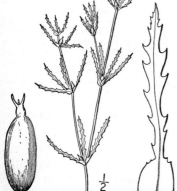

1. Naias marìna L. Large Naias. Fig. 212.

Naias marina L. Sp. Pl. 1015. 1753.
Naias major All. Fl. Ped. **2** : 221. 1785.
N. marina gracilis Morong, Coult. Bot. Gaz. **10** : 255. 1885.
Naias marina recurvata Dudley, Cayuga Fl. 104. 1886.
Naias gracilis Small, Fl. S. E. U. S. 40. 1903.

Dioecious, stem stout, compressed, commonly armed with teeth twice as long as their breadth. Leaves opposite or verticillate, 6″–12″ long, about 1″ wide, with spine-pointed teeth on each margin and frequently several along the back; sheaths with rounded lateral edges; fruit large, 2″–2½″ long, the pericarp as well as the seed rugosely reticulated, tipped with a long persistent style and 3 thread-like stigmas; seed not shining.

In lakes, Central New York to Florida, west to California. Summer. Also in Europe.

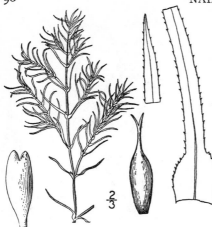

2. Naias fléxilis (Willd.) Rost. & Schmidt. Slender Naias. Fig. 213.

Caulinia flexilis Willd. Abh. Akad. Berlin, 95. 1803.
N. flexilis Rost. & Schmidt, Fl. Sed. 384. 1824.
Naias flexilis robusta Morong, Coult. Bot. Gaz. 10: 255. 1885.

Stem slender or stout, 3°–6° long, forking. Leaves linear, pellucid, acuminate or abruptly acute, ½′–1′ long, ½″–1″ wide, numerous and crowded on the upper parts of the branches, with 25–30 minute teeth on each edge; sheaths obliquely rounded with 5–10 teeth on each margin; fruit ellipsoid with very thin pericarp, 1″–2″ long, ¼″–½″ in diameter; style long, persistent; stigmas short; seed smooth, shining, straw-colored, sculptured, though sometimes quite faintly, with 30–40 rows of square or hexagonal reticulations scarcely seen through the dark pericarp.

In ponds and streams throughout nearly all North America. Also in Europe. Summer.

3. Naias guadalupénsis (Spreng.) Morong. Guadaloupe Naias. Fig. 214.

Caulinia guadalupensis Spreng. Syst. 1: 20. 1825.
Naias guadalupensis Morong, Mem. Torr. Club, 3: Part 2, 60. 1893.

Stem nearly capillary, 1°–2° long, widely branched from the base. Leaves numerous, 6″–9″ long, ¼″–½″ wide, acute, opposite or in fascicles of 2–5, frequently recurved, with sheaths and teeth like those of *N. flexilis* but generally with 40–50 teeth on each margin of the leaf; fruit about 1″ long; pericarp dark and strongly marked by 16–20 rows of hexagonal or rectangular reticulations which are transversely oblong; seed straw-colored.

In ponds and lakes, Pennsylvania to Oregon, Florida and Texas. Tropical America. July–Sept.

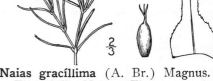

4. Naias gracíllima (A. Br.) Magnus. Thread-like Naias. Fig. 215.

Naias Indica var. *gracillima* A. Br.; Engelm. in A. Gray, Man. Ed. 5, 681. 1867.
Naias gracillima Magnus, Beitr. 23. 1870.

Monoecious, stem capillary, 6′–15′ long, much branched, the branches alternate. Leaves numerous, opposite or often fascicled in 3's–5's or more, setaceous, ½′–2′ long, usually with about 20 minute teeth on each margin; sheaths auricled, with 6 or 7 teeth on each auricle, the teeth standing upon setaceous divisions of the sheath; stigmas very short; fruit oblong-cylindric, ½″ long, ¼″ in diameter, slightly curved inwardly or straight, the pericarp straw-colored or purplish, marked by about 25 rows of irregularly oblong reticulations.

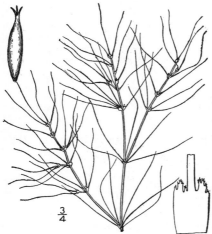

In pools and ponds, eastern Massachusetts to Delaware, Pennsylvania and Missouri. July–Sept.

Family 5. ZOSTERÀCEAE Demort. Anal. Fam. 65, 66. 1829.

EEL-GRASS FAMILY.

Perennial marine plants with creeping rootstocks and flattened, branching stems. Leaves all alternate, 2-ranked, linear, flat or complicate, acute or obtuse at the apex and sheathing at the base. Flowers monoecious or dioecious, arranged on a one-sided spadix and enclosed in a close fitting ultimately rupturing spathe. Perianth none, but some of the flowers covered by a hyaline envelope. Staminate flower of a single, sessile, 1-celled anther. Pistillate flower of two, united carpels, with a short or elongated style and 2 thread-like stigmas. Seeds ribbed or smooth.

Represented in North America by two genera, one Pacific and the following:

1. ZOSTÈRA L. Sp. Pl. 968. 1753.

Marine plants with slender rootstocks and branching compressed stems. Leaves 2-ranked, sheathing at the base, the sheaths with inflexed margins. Spadix linear, contained in a spathe. Flowers monoecious, arranged alternately in 2 rows on the spadix. Staminate flower merely an anther attached to the spadix near its apex, 1-celled, opening irregularly on the ventral side; pollen thread-like. Pistillate flower fixed on its back near the middle; ovary 1; style elongated; stigmas 2, capillary; mature carpels flask-shaped, membranous, rupturing irregularly beaked; seeds ribbed; embryo ellipsoid. [Greek, referring to the ribbon-like leaves.]

About 6 species of marine distribution, the following the type of the genus.

1. Zostera marìna L.   Eel-grass.   Grass-wrack or Sea-wrack.   Fig. 216.

Zostera marina L. Sp. Pl. 968. 1753.

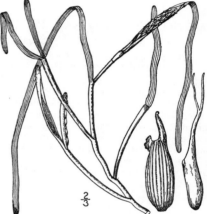

Leaves ribbon-like, obtuse at the apex, 1°–6° long, 1″–4″ wide, with 3–7 principal nerves. Spadix 1′–2½′ long; flowers about 3″ long, crowded, usually from 10–20 of each kind on the spadix; ovary somewhat vermiform; at anthesis the stigmas are thrust through the opening of the spathe and drop off before the anthers of the same spadix open; the anthers at anthesis work themselves out of the spathe and discharge the glutinous stringy pollen into the water; seeds cylindric, strongly about 20-ribbed, about 1½″ long and ½″ in diameter, truncate at both ends, the ribs showing very clearly on the pericarp.

In bays, streams and ditches along the Atlantic Coast from Greenland to Florida and on the Pacific from Alaska to California. Also on the coasts of Europe and Asia. Called also Wrack or Widgeon-grass; Sea, Sweet, Barnacle and Turtle-grass, Grass-weed, Tiresome-weed, Bell-ware, Drew. Summer.

$\frac{2}{3}$

Family 6.   SCHEUCHZERIÀCEAE Agardh, Theor. Syst. Pl. 44.   1858.

ARROW-GRASS FAMILY.

Marsh herbs with rush-like leaves and small spicate or racemose perfect flowers. Perianth 4–6-parted, its segments in two series, persistent or deciduous. Stamens 3–6. Filaments very short or elongated. Anthers mostly 2-celled and extrorse. Carpels 3–6, 1–2-ovuled, more or less united until maturity, dehiscent or indehiscent. Seeds anatropous. Embryo straight.

Four genera and about 10 species of wide geographic distribution.

Leaves all basal; flowers numerous on naked scapes, spicate or racemed.	1. *Triglochin.*
Stem leafy; flowers few in a loose raceme.	2. *Scheuchzeria.*

1. TRIGLÒCHIN L. Sp. Pl. 338. 1753.

Marsh herbs with basal half-rounded ligulate leaves with membranous sheaths. Flowers in terminal spikes or racemes on long naked scapes. Perianth-segments 3–6, concave, the 3 inner ones inserted higher up than the outer. Stamens 3–6; anthers 2-celled, sessile or nearly so, inserted at the base of the perianth-segments and attached by their backs. Ovaries 3–6, 1-celled, sometimes abortive; ovules solitary, basal, erect, anatropous. Style short or none. Stigmas as many as the ovaries, plumose. Fruit of 3–6 cylindraceous oblong or obovoid carpels, which are distinct or connate, coriaceous, costate, when ripe separating from the base upward from a persistent central axis, their tips straight or recurved, dehiscing by a ventral suture. Seeds erect, cylindraceous or ovoid-oblong, compressed or angular. [Greek, in allusion to the three-pointed fruit of some species.]

About 9 species, natives of the temperate and subarctic zones of both hemispheres. Type species: *Triglochin palustris* L. Only the following are known to occur in North America.

Carpels 3.	
Fruit linear or clavate, tapering to a subulate base.	1. *T. palustris.*
Fruit nearly globose.	2. *T. striata.*
Carpels 6; fruit oblong or ovoid, obtuse at the base.	3. *T. maritima.*

1. Triglochin palústris L.   Marsh Arrow-grass. Fig. 217.

Triglochin palustris L. Sp. Pl. 338. 1753.

Rootstock short, oblique, with slender fugacious stolons. Leaves linear, shorter than the scapes, 5′–12′ long, tapering to a sharp point; ligules very short; scapes 1 or 2, slender, striate, 8′–20′ high; racemes 5′–12′ long; pedicels capillary, in fruit erect-appressed and 2½″–3½″ long; perianth-segments 6, greenish-yellow; anthers 6, sessile; pistil of 3 united carpels, 3-celled, 3-ovuled; stigmas sessile; fruit 3″–3½″ long, linear or clavate; ripe carpels separating from the axis and hanging suspended from its apex, the axis 3-winged.

In bogs, Greenland to Alaska, south to New York, Indiana and Colorado. Also in Europe and Asia. July–Sept.

2. Triglochin striàta R. & P.   Three-ribbed Arrow-grass. Fig. 218.

Triglochin striata R. & P. Fl. Per. **3**: 72. 1802.
Triglochin triandra Michx. Fl. Bor. Am. **1**: 208. 1803.

Rootstocks upright or oblique. Scapes 1 or 2, more or less angular, usually not over 10′ high; leaves slender, slightly fleshy, nearly or quite as long as the scapes, ¼″–1″ wide; flowers very small, light yellow or greenish, in spicate racemes; pedicels ½″–1″ long, not elongating in fruit; perianth-segments 3, stamens 3; anthers oval, large; pistil of 3 united carpels; fruit subglobose or somewhat obovoid, about 1″ in diameter, appearing 3-winged when dry by the contracting of the carpels; carpels coriaceous, rounded and 3-ribbed on the back; axis broadly 3-winged.

In saline marshes, Maryland to Florida and Louisiana. Also in California and tropical America. June–Sept.

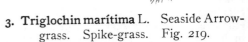

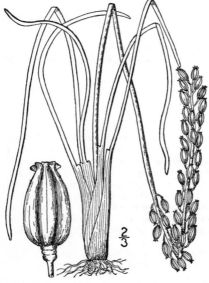

3. Triglochin marítima L.   Seaside Arrow-grass. Spike-grass. Fig. 219.

Triglochin maritima L. Sp. Pl. 339. 1753.
Triglochin elata Nutt. Gen. **1**: 237. 1818.
Triglochin maritima var. *elata* A. Gray, Man. Ed. 2, 437. 1852. In part.

Rootstock without stolons, often subligneous, the caudex thick, mostly covered with the sheaths of old leaves. Scape stout, nearly terete, 6′–2° high; leaves half-cylindric, usually about 1″ wide; raceme elongated, often 16′ or more long; pedicels decurrent, 1″–1½″ long, slightly longer in fruit; perianth segments 6, each subtending a large sessile anther; pistil of 6 united carpels; fruit oblong or ovoid, 2½″–3″ long, 1½″–2″ thick, obtuse at the base, with 6 recurved points at the summit; carpels 3-angled, flat or slightly grooved on the back, or the dorsal edges curving upward and winged, separating at maturity from the hexagonal axis.

In salt marshes, along the Atlantic seaboard from Labrador to New Jersey, and in fresh or saline marshes to Alaska, California and Mexico. Also in Europe and Asia. July–Sept.

2. SCHEUCHZÈRIA L. Sp. Pl. 338. 1753.

Rush-like bog perennials with creeping rootstocks, and erect leafy stems, the leaves elongated, half-rounded below and flat above, striate, furnished with a pore at the apex and a membranous ligulate sheath at the base. Flowers small, racemose. Perianth 6-parted, regularly 2-serial, persistent. Stamens 6, inserted at the base of the perianth-segments; filaments elongated; anthers linear, basifixed, extrorse. Ovaries 3 or rarely 4–6, distinct or connate at the base, 1-celled, each cell with 1 or 2 collateral ovules. Stigmas sessile, papillose or slightly fimbriate. Carpels divergent, inflated, coriaceous, 1–2-seeded, follicle-like, laterally dehiscent. Seeds straight or slightly curved, without endosperm. [Name in honor of Johann Jacob Scheuchzer, 1672–1733, Swiss scientist.]

A monotypic genus of the north temperate zone.

1. Scheuchzeria palústris L. Fig. 220.

Scheuchzeria palustris L. Sp. Pl. 338. 1753.

Leaves 4′–16′ long, the uppermost reduced to bracts; stems solitary or several, usually clothed at the base with the remains of old leaves, 4′–10′ tall; sheaths of the basal leaves often 4′ long with a ligule ½′ long; pedicels 3″–10″ long, spreading in fruit; flowers white, few, in a lax raceme; perianth-segments membranous, 1-nerved, 1½″ long, the inner ones the narrower; follicles 2″–4″ long, slightly if at all united at the base; seeds oval, brown, 2½″–3″ long with a very hard coat.

In bogs, Labrador to Hudson Bay and British Columbia, south to New Jersey, Pennsylvania, Wisconsin and California. Also in Europe and Asia. Summer.

Family 7. ALISMÀCEAE DC. Fl. Franc. 3: 181. 1805.*

WATER-PLANTAIN FAMILY.

Aquatic or marsh herbs, mostly glabrous, with fibrous roots, scapose stems and basal long-petioled sheathing leaves. Inflorescence racemose or paniculate. Flowers regular, perfect, monoecious or dioecious, pedicelled, the pedicels verticillate and subtended by bracts. Receptacle flat or convex. Sepals 3, persistent. Petals 3, larger, deciduous, imbricated in the bud. Stamens 6 or more; anthers 2-celled, extrorse or dehiscing by lateral slits. Ovaries numerous or rarely few, 1-celled, usually with a single ovule in each cell. Carpels becoming achenes in fruit in our species. Seeds uncinate-curved. Embryo horseshoe-shaped. Endosperm none. Latex-tubes are found in all the species, according to Micheli.

About 13 genera and 65 species, of wide distribution in fresh water swamps and streams.

Carpels borne in one series; achenes verticillate.	1. *Alisma.*
Carpels borne in several series; achenes capitate.	
Flowers perfect.	
Style not apical; fruit-heads not echinate; achene turgid, obscurely beaked.	2. *Helianthium.*
Style apical; fruit-heads echinate; achene flat, prominently beaked.	3. *Echinodorus.*
Flowers polygamous, monoecious or dioecious.	
Lower flowers of the inflorescence perfect.	4. *Lophotocarpus.*
Lower flowers of the inflorescence pistillate.	5. *Sagittaria.*

1. ALÍSMA L. Sp. Pl. 342. 1753.

Perennial or rarely annual herbs with erect or floating leaves, the blades several-ribbed, the ribs connected by transverse veinlets, or seemingly pinnately veined. Scapes short or elongated. Inflorescence paniculate or umbellate-paniculate. Flowers small, numerous on unequal 3-bracteolate pedicels, the petals white or rose-tinted. Stamens 6 or 9, subperigynous. Ovaries few or many, borne in one whorl on a small flat receptacle, ripening into flattened achenes which are 2–3-ribbed on the curved back and 1–2-ribbed on the sides. [Greek, said to be in reference to the occurrence of the typical species in saline situations.]

About 10 species, widely distributed in temperate and tropical regions. Only the following are known to occur in North America. Type species: *Alisma Plantago-aquatica* L.

Achenes longer than wide, grooved on the back, the inner edges not meeting in the whorl; peduncles and pedicels straight, ascending.	
Petals slightly longer than the sepals; corolla 1½″–2¼″ wide.	1. *A. subcordatum.*
Petals much longer than the sepals; corolla 5″–6½″ wide.	2. *A. brevipes.*
Achenes as wide as long, ridged on the back, the inner edges meeting in the whorl; peduncles and pedicels recurved in fruit.	3. *A. Geyeri.*

*Text revised by Dr. John Kunkel Small.

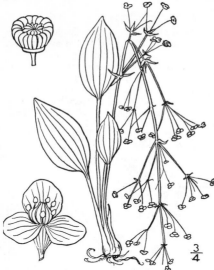

$\frac{3}{4}$

1. Alisma subcordàtum Raf. American Water-plantain. Fig. 221.

Alisma subcordatum Raf. Med. Repos. N. Y. **5**: 362. 1808.
Alisma Plantago Bigel. Fl. Bost. 87. 1814.
Alisma Plantago parviflorum Torr. Fl. N. U. S. 382. 1824.
Alisma Plantago americanum R. & S. Syst. **7**: 1598. 1830.

Plants erect; leaves oblong, elliptic, oval or ovate, or sometimes narrower, $1\frac{1}{4}'-6'$ long, usually abruptly pointed at the apex, cuneate to truncate, or cordate at the base, the petioles often longer than the blades; scapes $\frac{1}{3}°-3°$ tall, solitary or several together, the branches and pedicels in whorls of 3–10, variable in length, usually slender, sometimes filiform; bracts lanceolate or linear, often acuminate; sepals broadly ovate to suborbicular, obtuse; petals white or pinkish, $\frac{1}{2}''-1''$ long; achene-heads $1\frac{3}{4}''-2\frac{1}{4}''$ broad, the achenes obliquely obovate, $\frac{3}{4}''-1''$ long, the beak small, ascending.

In shallow water and mud, Massachusetts to Minnesota, Florida and Texas. Differs from the Old World *A. Plantago-aquatica* L., with which it has been united. Great Thrumwort, Mad-dog-weed, Deil's-spoons. June–Sept.

2. Alisma brévipes Greene. Western Water-plantain. Fig. 222.

Alisma brevipes Greene, Pittonia **4**: 158. 1900.
Alisma superbum Lunell, Bull. Leeds Herb, **2**: 5. 1908.

Plants similar to *A. subcordatum* in habit, but commonly larger; leaves oblong or oblong-lanceolate to ovate, $2'-7\frac{1}{2}'$ long, acute, sometimes abruptly pointed at the apex, rounded, truncate or subcordate at the base or sometimes gradually narrowed to the petiole which commonly exceeds the blade in length; scapes 3° tall or less, the branches and pedicels very numerous, except in small plants; bracts lanceolate or linear-lanceolate; sepals suborbicular or orbicular-ovate, mostly over $1\frac{1}{2}''$ long; petals white, $2\frac{1}{2}''-3''$ long; achene-heads $2\frac{1}{2}''-3\frac{1}{4}''$ broad, the achenes obovate, $1\frac{1}{4}''-1\frac{1}{2}''$ long.

In swamps and streams, Nova Scotia to Ontario, British Columbia, North Dakota and California. July–Sept.

$\frac{2}{5}$

$\frac{2}{3}$

3. Alisma Geyeri Torr. Geyer's Water-plantain. Fig. 223.

Alisma arcuatum Lunell, Bot. Gaz. **43**: 210. 1907. Not Michalet. 1854.
Alisma Geyeri Torr. in Nicollet, Rep. Hydrograph. Miss. Riv. 162. 1843.

Plants diffuse; leaves oblong, elliptic, oblong-lanceolate or ovate-lanceolate, or rarely linear, $2'-3\frac{1}{2}'$ long, acute or slightly acuminate at the apex, narrowed at the base, the petioles usually longer than the blades; scapes mostly $\frac{3}{4}°-1\frac{3}{4}°$ long, more or less diffusely spreading, the branches and pedicels relatively stout; bracts lanceolate; sepals orbicular-ovate, about $1\frac{1}{4}''$ long; petals pink, $1''-2''$ long; achene-heads $2\frac{1}{4}''-2\frac{3}{4}''$ broad, the achenes suborbicular, about $1''$ in diameter, the beak erect or nearly so.

In mud and shallow water, New York to North Dakota, Oregon and Nevada. July–Sept.

2. HELIÁNTHIUM Engelm.; Britton, Man. Ed. 2, 54. 1905.

Annual or perennial scapose marsh or aquatic herbs. Leaves erect or ascending, or floating, narrow and gradually narrowed into the petiole or broad and deeply cordate at the base, 3–several-ribbed. Scapes as long as the leaves or longer, terminating in a few-flowered whorl or a many-flowered panicle, the pedicels spreading or recurving in fruit. Flowers perfect. Sepals 3, broad, embracing the fruit-head or reflexed beneath it. Petals 3, mainly white or pink, about as long as the sepals. Stamens 6 or 9; filaments elongate; anthers very short, often broader than long. Carpels relatively few, borne in few series on an elevated receptacle. Style not apical, minute; stigma acute. Achenes forming a globular or depressed head, turgid, crested-ribbed, obscurely beaked or beakless. [Name from the Greek, meaning sunflower.]

Two known species, the following, and one in Cuba. Type species: *Echinodorus parvulus* Engelm.

Helianthium pàrvulum (Engelm.) Small. Dwarf Water-plantain. Fig. 224.

?Alisma tenellum Mart.; R. & S. Syst. Veg. **7**: 1600. 1830.
Echinodorus parvulus Engelm. in A. Gray, Man. Ed. 2, 438. 1856.
?Echinodorus tenellus Buchenau, Abh. Nat. Gesell. Bremen **2**: 18. 1868.
Helianthium tenellum Britton, Man. Ed. 2, 54. 1904.
Helianthium parvulum Small, N. A. Fl. **17**[1]: 45. 1909.

Plants 6′ tall or less; leaves linear to elliptic or oblong, 4″–15″ long, acute or acutish at the apex, 3-veined, gradually narrowed into the slender petioles which usually somewhat exceed the blade in length; scapes solitary or few together, mostly as long as the leaves or longer; pedicels mostly 2–8, recurved in fruit, 1¼″–2½″ long; sepals orbicular-ovate or deltoid-ovate, ¾″–2″ long; petals suborbicular, about as long as the sepals, emarginate at the apex; fruit-heads globular, 1½″–2″ in diameter, embraced by the persistent calyx; achenes ½″–¾″ long, the ribs obscurely crested.

In mud and shallow water, Massachusetts to Western Ontario, Minnesota, Florida, Texas and Mexico. Also in Cuba. April–Aug. This species was referred in the first edition of this work to *Alisma tenellum* Mart. a plant similar in habit, which appears to be confined to South America; it has been regarded by other authors as an *Echinodorus*.

3. ECHINÓDORUS Rich.; Engelm. in A. Gray, Man. 460. 1848.

Perennial or annual herbs with long-petioled, elliptic, ovate or lanceolate often cordate or sagittate leaves, 3–9-ribbed and mostly punctate with dots or lines. Scapes often longer than the leaves; inflorescence racemose or paniculate, the flowers verticillate, each verticil with 3 outer bracts and numerous inner bracteoles; flowers perfect; sepals 3, distinct, persistent; petals white, deciduous; receptacle large, convex or globose; stamens 12–30; ovaries numerous; style obliquely apical, persistent; stigma simple; fruit achenes, more or less compressed, coriaceous, ribbed and beaked, forming spinose heads. [Greek, in allusion to the spinose heads of fruit.]

About 14 species, mostly natives of America. Only the following are known in North America. Type species: *Alisma rostratum* Nutt.

Scapes reclining or prostrate; style shorter than the ovary; beak of achene short.   1. *E. radicans.*
Scapes erect; style longer than the ovary; beak of achene long.   2. *E. cordifolius.*

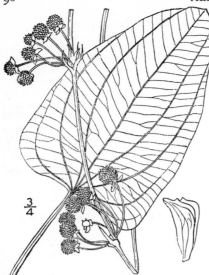

$\frac{3}{4}$

1. Echinodorus radìcans (Nutt.) Engelm. Creeping Bur-head. Fig. 225.

Sagittaria radicans Nutt. Trans. Am. Phil. Soc. (II.) **5**: 159. 1833–37.
Echinodorus radicans Engelm. in A. Gray, Man. Ed. 2, 438. 1856.

Leaves coarse, ovate, obtuse, cordate, 2′–8′ long, 1½′–7½′ wide, marked with short pellucid lines, nerves 5–9; cross–veins netted. Petioles sometimes 20′ long; scapes creeping, 2°–4° long, scabrous, often rooting at the nodes; verticils distant; bracts linear-lanceolate, acuminate, dilated at the base; pedicels 3–12, unequal, 1′–2½′ long, slender; sepals persistent, shorter than the heads; petals larger, obovate, about 3″ long; stamens about 20; style shorter than the ovary; achenes numerous, about 2″ long, 6–10-ribbed, with 2–several oval glands on each side and beaks about one-fourth their length; fruiting heads 4″ in diameter.

In swamps, Illinois to North Carolina and Florida, west to Missouri and Texas. June–July.

2. Echinodorus cordifòlius (L.) Griseb. Upright Bur-head. Fig. 226.

Alisma cordifolia L. Sp. Pl. 343. 1753.
Echinodorus rostratus Engelm. in A. Gray, Man. Ed. 2, 538. 1856.
Echinodorus cordifolius Griseb. Abh. Kon. Gesell. Wiss. Gott. **7**: 257. 1857.
Echinodorus cordifolius lanceolatus Mack. & Bush; Fl. Jackson Co. 10. 1902.

Leaves various, ovate, obtuse, cordate, 6′–8′ long and wide, in smaller plants sometimes lanceolate, acute at each end, 1′–2′ long; petioles angular, striate, 1′–10′ high; scapes 1 or more, erect, 5′–16′ tall; flowers 3–6 in the verticils; pedicels ¼′–½′ long, erect after flowering; bracts linear-lanceolate, acuminate, dilated at the base; sepals shorter than the heads; petals 2″–3″ long; stamens often 12; styles longer than the ovary; fruiting heads bur-like, 2″–3″ in diameter; achenes 1½″ long, narrowly obovate or falcate, 6–8-ribbed; beak apical, half the length of the achene.

In swamps and ditches, Illinois to Florida, west to Missouri and Texas. Also in tropical America. June–July.

$\frac{2}{3}$

4. LOPHOTOCARPUS T. Durand, Ind. Gen. Phan. X. 1888.

[LOPHIOCARPUS Miquel, Fl. Arch. Ind. **1**: Part 2, 50. 1870. Not Turcz. 1843.]

Perennial, bog or aquatic herbs with basal long-petioled sagittate or cordate leaves, simple erect scapes bearing flowers in several verticils or 2–3 at the summit, the lower perfect, the upper staminate. Sepals 3, distinct, persistent, erect after flowering and enclosing or enwrapping the fruit. Petals white, deciduous. Receptacle strongly convex. Stamens 9–15, hypogynous, inserted at the base of the receptacle. Filaments flattened. Pistils numerous; ovule solitary, erect, anatropous; style elongated, oblique, persistent. Achenes winged or crested. Embryo horseshoe-shaped. [Greek, signifying crested fruit.]

About 7 species, the following of eastern North America, the others in the Southern States, California and tropical America. Type species: *Sagittaria calycina* Engelm.

Leaves hastate or sagittate; plants of fresh-water ponds or marshes.
    Leaves with large basal lobes fully as long as the terminal one.   1. *L. calycinus.*
    Leaves with small basal lobes shorter than the terminal lobe.   2. *L. depauperatus.*
Leaves imperfect or obsolete, the phyllodia thick or partially flattened; plants of salt or brackish water.
    Phyllodia terete or nearly so, prominently nodose-septate.   3. *L. spongiosus.*
    Phyllodia flat, more or less spatulate, not prominently nodose.   4. *L. spathulatus.*

1. Lophotocarpus calycìnus (Engelm.) J. G. Smith. Large Lophotocarpus. Fig. 227.

Sagittaria calycina Engelm. in Torr. Bot. Mex. Bound.
Surv. 212. 1859.
Lophiocarpus calycinus Micheli, in DC. Monogr. Phan. **3**:
61. 1881.
Lophotocarpus calycinus (Engelm.) J. G. Smith, Mem.
Torrey Club **5**: 25. 1894.
Lophotocarpus calycinus maximus Robinson in A. Gray,
Man. Ed. 7, 84. 1908.

Plants mostly emersed; leaves $\frac{1}{2}°$–$2°$ tall, the petioles
stout, the blades sagittate, hastate or lunate, $2\frac{1}{2}'$–$12'$
long, the basal lobes usually longer than the broad
terminal lobe, usually caudate-acuminate; scapes shorter
than the leaves, the inflorescence usually simple, with
2–7 whorls; sepals suborbicular to orbicular-reniform,
becoming $4\frac{1}{2}''$–$7\frac{1}{2}''$ long; fruiting pedicels very thick,
usually elongate, mostly $1\frac{1}{4}''$–$2''$ long; fruit-heads $5\frac{1}{2}''$–$8''$
in diameter; achenes broadly cuneate, $1''$–$1\frac{1}{4}''$ long, the
beak stout, the dorsal wing thin.

In swamps, South Dakota to Delaware, Alabama, Texas
and New Mexico. July–Sept.

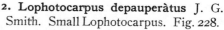

2. Lophotocarpus depauperàtus J. G. Smith. Small Lophotocarpus. Fig. 228.

Lophotocarpus depauperatus J. G. Smith, Rep. Mo.
Bot. Gard. **11**: 148. 1890.

Plants mostly emersed; leaves $\frac{1}{2}°$–$1°$ tall, the
petioles relatively slender, the blades oblong, ellip-
tic, sagittate or hastate, $\frac{3}{4}'$–$1\frac{1}{2}'$ long, including the
basal lobes which are usually more or less spread-
ing; scapes about one-half as long as the leaves,
mostly with 1 or 2 whorls; sepals suborbicular,
becoming $3''$–$3\frac{1}{2}''$ long; fruit-bearing pedicels
rather stout, $\frac{3}{4}'$–$1\frac{1}{4}'$ long; fruit-heads $3\frac{1}{2}''$–$4''$ in
diameter; achenes cuneate, fully $1''$ long, or rarely
shorter, the beak slender, the dorsal wing thin.

On margins of ponds, Wisconsin to Illinois, Missouri and Oklahoma. June–Sept.

3. Lophotocarpus spongiòsus (Engelm.) J. G. Smith. Spongy Lophotocarpus. Fig. 229.

Sagittaria calycina spongiosa Engelm. in A. Gray,
Man. Ed. 5, 493. 1867.
Lophotocarpus spongiosus (Engelm.) J. G. Smith,
Rep. Mo. Bot. Gard. **11**: 148. 1899.

Plants submerged; leaves $\frac{1}{3}°$–$\frac{2}{3}°$ tall, the
petioles stout and spongy, conspicuously no-
dose-septate, the blades spatulate, oblong, ellip-
tic, sagittate or hastate, $\frac{1}{2}'$–$1\frac{1}{4}'$ long, the basal
lobes, when present, more or less falcate;
scapes about one-half as long as the leaves or
less, the inflorescence simple, with one or two
whorls; sepals broadly ovate or orbicular-ovate,
becoming $5''$–$6''$ long; fruiting pedicels very
stout, $\frac{1}{3}'$–$\frac{3}{4}'$ long, or rarely longer; fruit-heads
$3\frac{1}{2}''$–$5''$ in diameter; achenes cuneate, $1''$–$1\frac{1}{4}''$
long, the beak short, at the top of the achene-
body, the dorsal wing thin.

On margins of brackish ponds and tide-water
marshes, New Brunswick to Virginia. July–Aug.

7

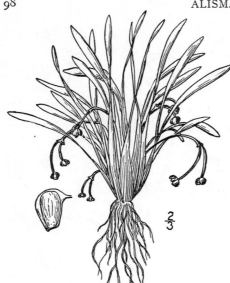

$\frac{2}{3}$

4. Lophotocarpus spathulàtus J. G. Smith. Spatulate Lophotocarpus. Fig. 230.

Lophotocarpus spathulatus J. G. Smith, Rep. Mo. Bot. Gard. 11: 149. 1899.

Plants aquatic; leaves less than 4' tall, the petioles stout, not conspicuously septate, the blades linear or spatulate dilations at the top of the petioles, or wanting; scape shorter than the leaves, stout but weak, the inflorescence with but one whorl; sepals ovate to orbicular-ovate, becoming 1½" long; fruiting pedicels stout, about 5" long or less; fruit-heads 2½"–3" in diameter; achens cuneate, ¾"–1" long, the beak much below the top of the achene-body, the narrow dorsal wing thin.

On sandy beaches above salt-water, Newburyport, Massachusetts. July–Sept.

5. SAGITTÀRIA L. Sp. Pl. 993. 1753.

Perennial aquatic or bog herbs, mostly with tuber-bearing or nodose rootstocks, fibrous roots, basal long petioled nerved leaves, the nerves connected by numerous veinlets, and erect, decumbent or floating scapes, or the leaves reduced to bladeless phyllodia (figs. 241, 242). Flowers monoecious or dioecious, borne near the summits of the scapes in verticils of 3's, pedicelled, the staminate usually uppermost. Verticils 3-bracted. Calyx of 3 persistent sepals, those of the pistillate flowers reflexed or spreading in our species. Petals 3, white, deciduous. Stamens usually numerous, inserted on the convex receptacle; anthers 2-celled, dehiscent by lateral slits; staminate flowers sometimes with imperfect ovaries. Pistillate flowers with numerous distinct ovaries, sometimes with imperfect stamens; ovule solitary; stigmas small, persistent. Achenes numerous, densely aggregated in globose or subglobose heads, compressed. Seed erect, curved; embryo horseshoe-shaped. [Latin, referring to the arrow-shaped leaves of some species, known generally as *Arrow-head* or *Arrow-leaf*.]

About 40 species, natives of temperate and tropical regions. Besides the following, some 18 others occur in southern and western North America. Type species: *Sagittaria sagittifolia* L.

Fertile pedicels slender, ascending, not reflexed in fruit.
　Leaf-blades sagittate or hastate.
　　Basal lobes one-fourth to one-half the length of the blade.
　　　Beak of the achene erect.
　　　　Achene long-beaked, the beak mostly ½ the length of the body or more.
　　　　　Achene obovate or orbicular-obovate, usually with 1 facial wing.　　1. *S. longirostra.*
　　　　　Achene cuneate, usually with 2 prominent facial wings.　　2. *S. Engelmanniana.*
　　　　Achene short-beaked, the beak mostly ¼ the length of the body or less.
　　　　　Achene with thick nearly equal wings, the blunt beak over the inner edge of the
　　　　　　ventral wing.　　　　　　　　　　　　　　　　　　　　3. *S. cuneata.*
　　　　　Achene with thin unequal wings, the sharp beak over the outer edge of the
　　　　　　ventral wing.　　　　　　　　　　　　　　　　　　　　4. *S. brevirostra.*
　　　Beak of the achene horizontal.
　　　　Achene with nearly even faces: bracts and pedicels glabrous.　　5. *S. latifolia.*
　　　　Achene faces prominently winged; bracts and pedicels pubescent.　6. *S. pubescens.*
　　Basal lobes two-thirds to three-fourths the length of the blade.　　7. *S. longiloba.*
　Leaves entire, or rarely hastate or cordate.
　　Filaments tapering upward; leaves seemingly pinnately veined.　　8. *S. ambigua.*
　　Filaments glabrous; bracts connate.
　　Filaments cobwebby-pubescent; bracts mostly distinct.　　　　　9. *S. falcata.*
　　Filaments abruptly dilated, pubescent; veins distinct to the base.　10. *S. rigida.*
　　　Fruiting heads sessile or very nearly so.
　　　Both staminate and pistillate flowers pedicelled.
　　　　Leaves with terete or 3-sided blades, often imperfectly developed.
　　　　　Achene with thick merely uneven facial wings or ridges, the beak erect.
　　　　　　　　　　　　　　　　　　　　　　　　　　　　　　11. *S. teres.*
　　　　　Achene with thin crested facial wings, the beak oblique.　12. *S. cristata.*
　　　　Leaves with flat blades.
　　　　　Filaments suborbicular; anthers longer than the filaments.　13. *S. Eatonii.*
　　　　　Filaments oblong; anthers about as long as the filaments.　14. *S. graminea.*
Fertile pedicels stout, reflexed in fruit; filaments dilated.
　Filaments pubescent; leaf-blades ovate or ovate-elliptic.　　　　15. *S. platyphylla.*
　Filaments glabrous; leaves linear-lanceolate or reduced to phyllodia.
　　Filaments about as long as the anthers; achene with 3 undulate or slightly toothed crests.
　　　　　　　　　　　　　　　　　　　　　　　　　　　　　　16. *S. subulata.*

Filaments much longer than the anthers; achene with 5-7 tuberculate or prominently toothed crests.   17. *S. lorata*.

1. Sagittaria longiróstra (Micheli) J. G. Smith. Long-beaked Arrow-head. Fig. 231.

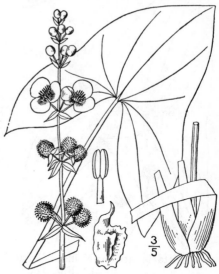

Sagittaria sagittaefolia var. *longirostra* Micheli in DC. Monog. Phan. 3: 69. 1881.
Sagittaria longirostra J. G. Smith, Mem. Torr. Club, 5: 26. 1894.

Monoecious, glabrous, scapes erect, rather stout, 1½°-3° tall. Leaves hastate or sagittate, 4'-12' long, abruptly acute at the apex, the basal lobes ovate or ovate-lanceolate or linear, acute, one-third to one-half the length of the blade; scape usually longer than the leaves, 6-angled below; bracts triangular-lanceolate, acuminate, 7''-15'' long, longer than the fertile pedicels; petals 8''-14'' long; filaments glabrous; styles curved, twice as long as the ovaries; achene obovate, about 2'' long, winged on both margins, the ventral margin entire or undulate, the dorsal eroded, its sides with a short crest, its beak stout, erect or recurved.

In swamps and along ponds, New Jersey and Pennsylvania to Florida and Alabama. July–Sept.

$\frac{3}{5}$

$\frac{1}{2}$

2. Sagittaria Engelmanniàna J. G. Smith. Engelmann's Arrow-head. Fig. 232.

Saggittaria variabilis var. (?) *gracilis* S. Wats. in A. Gray, Man. Ed. 6, 555. 1889. Not Engelm.
Sagittaria Engelmanniana J. G. Smith, Mem. Torr. Club, 5: 25. 1894.

Monoecious, glabrous, scape erect or ascending, slender, 8'-20' high. Leaves narrow, 1½'-8' long, 1''-4'' wide, acute or obtuse at the apex, the basal lobes narrowly linear, acuminate, one-third to one-half the length of the blade; scape striate, about as long as the leaves; bracts lanceolate, acute, shorter than the slender fertile pedicels, 4''-6'' long; flowers 7''-12'' broad; filaments glabrous; style about twice as long as the ovary; achene cuneate, 2'' long, winged on both margins and with 1-3 lateral wing-like crests on each face, the beak stout, erect, about ½'' long.

In shallow water, Massachusetts, Connecticut and Rhode Island. Arrow-leaf. Aug.–Sept.

3. Sagittaria cuneàta Sheldon. Arum-leaved Arrow-head. Fig. 233.

Sagittaria cuneata Sheldon, Bull. Torrey Club 20: 283. 1893.
Sagittaria arifolia Nutt.; J. G. Smith, Rep. Mo. Bot. Gard. 6: 32. 1894.
Sagittaria arifolia stricta J. G. Smith, Rep. Mo. Bot. Gard. 6: 34. 1894.

Glabrous or nearly so, terrestrial or submerged, scape weak, ascending or floating, ⅔°-2° long. Leaves sagittate, linear-lanceolate to ovate, 1¼'-6¾' long, acute or acuminate, long-petioled; phyllodia, when present, of two kinds, the one petiole-like and about as long as the leaves, the other lanceolate and clustered at the base of the plant; bracts lanceolate or linear-lanceolate, or rarely ovate-lanceolate; flowers 6''-12'' broad; achene cuneate-obovate, 1''-1¼'' long, the beak minute, erect over the ventral wing.

$\frac{1}{2}$

In mud or water, Nova Scotia and Maine to Quebec, British Columbia, Connecticut, Kansas, New Mexico and California. July–Sept. In the first edition both figures 196 and 197 and the descriptions apply to this species.

$\frac{1}{2}$

4. Sagittaria breviróstra Mack. & Bush. Short-beaked Arrow-leaf. Fig. 234.

Sagittaria variabilis diversifolia Engelm. in A. Gray, Man. Ed. 5, 493. 1867.
Sagittaria brevirosta Mack. & Bush, Rep. Mo. Bot. Gard. 16: 102. 1905.

Monoecious, glabrous, scape erect, $1\frac{1}{3}°-2\frac{2}{3}°$ tall, simple or branched. Leaves ovate to lanceolate, sagittate or hastate, 5′–17′ long, acute at the apex, the basal lobes lanceolate to linear-lanceolate, about $\frac{1}{2}$ the length of the blade; scape usually taller than the leaves, 4–6-angled; bracts lanceolate to linear-lanceolate, attenuate; flowers 1′ broad or less; filaments glabrous, slender; achene suborbicular to cuneate-obovate, $1\frac{1}{4}''-1\frac{1}{2}''$ long, broadly winged on both margins, and ridged on the sides, the short erect beak less than $\frac{1}{2}$ as long as the body.

In bogs and shallow streams, New Jersey and Pennsylvania to Florida and Tennessee. Aug.–Oct.

5. Sagittaria latifòlia Willd. Broad-leaved Arrowhead. Fig. 235.

Sagittaria latifolia Willd. Sp. Pl. 4: 409. 1806.
Sagittaria variabilis Engelm. in A. Gray, Man. 461. 1848.

Monoecious or sometimes dioecious, glabrous or nearly so, scape stout or slender, 4′–4° tall, simple or branched. Leaves exceeding variable in form and size, sometimes linear-lanceolate and acuminate at the apex, sometimes wider than long and obtuse; basal lobes from $\frac{1}{4}$ to $\frac{1}{2}$ as long as the blade; bracts acute, acuminate or obtuse, the upper ones sometimes united; flowers $1′-1\frac{1}{2}′$ wide; filaments slender, glabrous; achene 1″–2″ long, broadly winged on both margins, its sides even or 1-ribbed, the beak about one-third its length, horizontal or nearly so.

In shallow water, throughout North America, except the extreme north, extending to Mexico. Variable. July–Sept.

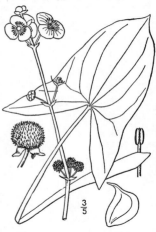

$\frac{3}{5}$

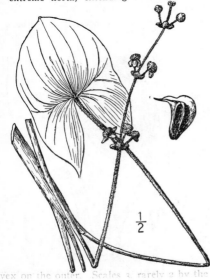

$\frac{1}{2}$

6. Sagittaria pubéscens Muhl. Hairy Arrow-leaf. Fig. 236.

Sagittaria pubescens Muhl. Cat. 86. 1813.
Sagittaria sagittifolia pubescens Torr. Comp. 356. 1826.
Sagittaria variabilis pubescens Engelm. in A. Gray, Man. Ed. 5, 493. 1867.
Sagittaria latifolia pubescens J. G. Smith, Mem. Torrey Club 5: 25. 1894.

Monoecious or rarely dioecious, pubescent, scape erect, 1°–2° tall, simple or rarely branched. Leaves ovate, deltoid or deltoid-lanceolate, mostly acute, rarely acuminate at the apex; basal lobes $\frac{1}{2}$ the length of the blades or nearly so; bracts obtuse, sometimes broadly rounded; flowers $\frac{1}{2}′-1′$ wide; filaments very slender, glabrous; achene $1\frac{1}{2}''-2''$ long, winged on both margins, its sides prominently winged, the beak about $\frac{1}{2}$ its length.

In bogs and shallow water, Nova Scotia to Wisconsin, Missouri and Kansas. July–Sept.

urface, convex on the outer. Scales 3, rarely 2 by the absence

7. Sagittaria longíloba Engelm.   Long-lobed Arrow-head.   Fig. 237.

Sagittaria longiloba Engelm. in Torr. Bot. Mex. Bound. Surv. 212. 1859.

Monoecious, glabrous, scape slender, simple or rarely branched, 1°–2° tall. Leaves long-petioled, the apex acute, the basal lobes linear-lanceolate, acuminate, about three-fourths the length of the blade; bracts lanceolate, acuminate, 3″–4″ long, much shorter than the very slender fertile pedicels which are longer than the sterile ones; stamens numerous, the filaments longer than the anthers; achene about 1″ long, quadrate-obovate, somewhat broader above than below, winged on both margins, its beak exceedingly short.

In shallow water, Nebraska to Colorado, south to Texas and Mexico.

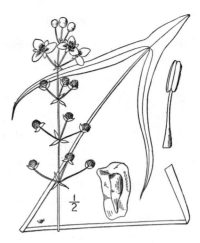

8. Sagittaria ambígua J. G. Smith.   Kansas Sagittaria.   Fig. 238.

Sagittaria ambigua J. G. Smith, Ann. Rep. Mo. Bot. Gard. **6**: 48. *pl. 17.* 1894.

Monoecious, glabrous, scape erect or ascending, simple or sparingly branched, 1°–2° high. Leaves lanceolate, entire, long-petioled, acute or acuminate at both ends, seemingly pinnately veined, really 5–7-nerved, 5′–8′ long, equalling or shorter than the scape; bracts lanceolate, acuminate, 5″–8″ long, much shorter than the slender fruiting pedicels, connate at the base, papillose; stamens 20–25; filaments glabrous, longer than the anthers; achene about 1″ long, oblong, curved, narrowly winged on both margins, its sides smooth and even, its beak short, oblique.

In ponds, Kansas and Oklahoma.

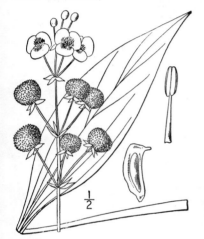

9. Sagittaria falcàta Pursh.   Scythe-fruited Sagittaria.   Fig. 239.

Sagittaria falcata Pursh, Fl. Am. Sept. 297. 1814.

Sagittaria lancifolia falcata J. G. Smith, Mem. Torrey Club **5**: 25. 1894.

Monoecious, glabrous to the inflorescence, scape erect, longer than the leaves, sometimes branched. Leaves erect or nearly so, almost linear to elliptic, the blades mostly 4′–16′ long, often slightly acuminate, much shorter than the petioles; whorls of the inflorescence few or many; flowers 9″–15″ broad; bracts ovate, less than 5″ long, obtuse or acutish, granular-papillose; filaments not dilated, pubescent; anthers shorter than the filaments; achene cuneate, about 1″ long, narrowly winged, the slender beak usually ascending.

In shallow water and swamps, Delaware to Florida, Texas and Mexico. *Sagittaria lancifolia* L., admitted as including this species in the first edition, is distinct from it, and inhabits Florida and tropical America.

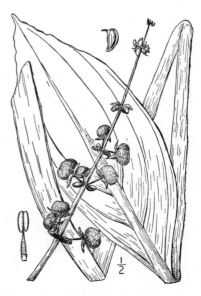

10. Sagittaria rígida Pursh.   Sessile-fruited Arrow-head.   Fig. 240.

Sagittaria rigida Pursh, Fl. Am. Sept. 397. 1814.
Sagittaria heterophylla Pursh, Fl. Am. Sept. 396. 1814. Not
Schreb. 1811.

Monoecious, glabrous, scape simple, weak, curving, ascending or decumbent, shorter than the leaves. Leaves
very variable, linear, lanceolate, elliptic or broadly ovate,
acute or obtuse at the apex, entire or with 1 or 2 short or
slender basal lobes; bracts ovate, obtuse, 2″–4″ long, united
at the base or sometimes distinct; heads of fruit sessile or
very nearly so; pedicels of the sterile flowers ½′–1′ long;
filaments dilated, mostly longer than the anthers, pubescent; achene narrowly obovate, 1½″–2″ long, winged on
both margins, crested above, tipped with a stout nearly
erect beak of about one-fourth its length.

In swamps and shallow water, Quebec to Minnesota, south
to New Jersey, Tennessee, Missouri and Nebraska. Petioles
rigid when growing in running water. July–Sept.

11. Sagittaria tères S. Wats.   Slender Sagittaria. Fig. 241.

Sagittaria teres S. Wats. in A. Gray, Man. Ed. 6, 555. 1890.

Monoecious, glabrous, scape slender, erect, simple, 6′–18′
long, bearing only 1–3 verticils of flowers. Leaves usually
reduced to elongated terete nodose phyllodia or some of
them short and bract-like, one or two of the longer ones
occasionally bearing a linear blade; bracts ovate, obtuse,
about 1½″ long, much shorter than the filiform fruiting
pedicels which are longer than the sterile ones; flowers
6″–8″ broad; stamens about 12, their dilated filaments
pubescent, shorter than the anthers; achene broadly obovate, 1″ long, the ventral margin winged, the dorsal
7–11-crested, the sides bearing several crenate crests, the
beak short, erect.

In ponds, Massachusetts to South Carolina. Aug.–Sept.

12. Sagittaria cristàta Engelm.   Crested Sagittaria.   Fig. 242.

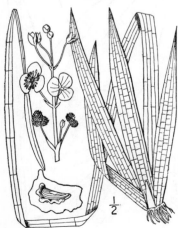

Sagittaria cristata Engelm.; Arthur, Proc. Davenport Acad.
4: 29. 1882.

Monoecious, scape slender, erect, 1°–2½° high, simple,
bearing 4 or 5 verticils of flowers at or above the surface of the water. Leaves long-petioled, spongy and
rigid, reduced to slender phyllodia or bearing linear-
lanceolate or elliptic blades 2′–4′ long and 3″–12″ wide;
bracts acute, 2″–4″ long, much shorter than the slender
fertile pedicels; flowers 8″–10″ broad; stamens about
24; filaments dilated, pubescent, at least at the middle,
longer than the anthers; achene obliquely obovate, the
dorsal margin with a broad crenate wing, the ventral
straight-winged, each side bearing 2 crenate crests, the
beak short, oblique.

In shallow water, Iowa and Minnesota. Phyllodia are
commonly developed from the nodes of the rootstock.
July–Aug.

13. Sagittaria Eatònii J. G. Smith.
Eaton's Sagittaria.　Fig. 243.

Sagittaria Eatonii J. G. Smith, Rep. Mo. Bot. Gard.
11: 150.　1899.

Monoecious, scape very slender, 4′–6′ tall,
Leaves represented by flat phyllodia which are
attenuate from broad bases and often also by
longer blade-tipped petioles, the blades linear
or narrowly linear-lanceolate, 10″–17½″ long,
acute or acuminate; bracts ovate, about 1″ long,
united at the base; pedicels of the pistillate
flowers filiform, mostly less than 5″ long, those
of the staminate flowers longer than the former;
sepals ovate to ovate-lanceolate, becoming 1½″–
2″ long; corolla 3½″–4½″ broad; filaments suborbicular, pubescent, much shorter than the
anthers; anthers suborbicular, conspicuously
larger than the filaments; fruits not seen.

On sandy shores, between low and high tide,
Massachusetts, Connecticut and Long Island, New
York.　July–Sept.

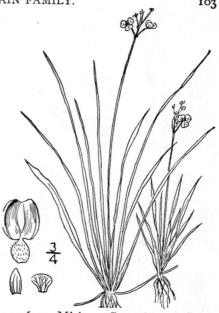

14. Sagittaria gramínea Michx.　Grass-leaved Sag-
ittaria.　Fig. 244.

Sagittaria graminea Michx. Fl. Bor. Am. 2: 190.　1803.

Monoecious or dioecious, glabrous, scape simple, erect, 4′–2°
tall.　Leaves long-petioled, the blades linear, lanceolate or
elliptic, acute at both ends, or rarely with spreading or recurved
basal lobes, 2′–6′ long, ½″–3″ wide, 3–5-nerved, the nerves
distinct to the base, some of them occasionally reduced to
flattened phyllodia; bracts ovate, acute, 1½″–3″ long, much
shorter than the slender or filiform fruiting pedicels, connate
to the middle or beyond; flowers 4″–6″ broad; stamens about
18; filaments dilated, pubescent, longer than or equalling the
anthers; achene obovate, ½″–1″ long, slightly wing-crested on
the margins and ribbed on the sides, the beak very short.

In mud or shallow water, Newfoundland to Ontario and South
Dakota, south to Florida and Texas.　Early leaves often purplish.
July–Sept.

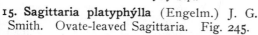

15. Sagittaria platyphýlla (Engelm.) J. G.
Smith.　Ovate-leaved Sagittaria.　Fig. 245.

Sagittaria graminea var. *platyphylla* Engelm. in A. Gray,
Man. Ed. 5, 494.　1867.
Sagittaria platyphylla J. G. Smith, Ann. Rep. Mo. Bot.
Gard. 6: 55. *pl. 26.*　1894.

Monoecious, glabrous, scape erect, simple, rather
weak, mostly shorter than the leaves.　Leaves rigid,
the blades ovate, ovate-lanceolate or ovate-elliptic,
short-acuminate or acute at the apex, rounded, gradually narrowed or rarely cordate or hastate at the
base, seemingly pinnately veined, 2′–6′ long; bracts
broadly ovate, acute, connate at the base, 2″–4″ long;
flowers 8″–14″ broad; fertile pedicels stout, divergent in flower, reflexed in fruit, ½′–2½′ long; filaments dilated, pubescent, rather longer than the
anthers; achene obliquely obovate, winged on both
margins, the dorsal margin somewhat crested, the
sides with a sharp wing-like ridge.

In swamps and shallow water, southern Missouri to
Mississippi and Texas.　Phyllodia, when present, oblong or oblanceolate.　July–Sept.

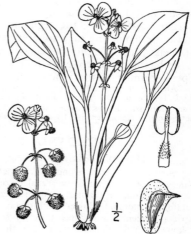

16. Sagittaria subulàta (L.) Buchenau.
Subulate Sagittaria. Fig. 246.

Alisma subulata L. Sp. Pl. 343.   1753.
Sagittaria pusilla Nutt. Gen. **2**: 213. 1818.
S. subulata Buchen. Abh. Nat. Ver. Bremen **2** : 490.   1871.

Monoecious or rarely dioecious, scape very slender,
2'–6' high, few-flowered, about equalling the leaves.
Leaves all reduced to rigid phyllodia or sometimes
bearing linear or linear-lanceolate blades, 1'–1½'
long; bracts united or partly separated; flowers 5"–8"
broad; fertile pedicels reflexed, longer than the
bracts in fruit; stamens about 8; filaments about
equalling the anthers, dilated, glabrous; achenes less
than 1" long, obovate, with narrow wings, 2 or 3
crests on each side, sometimes crenate; beak short.

In tide-water mud, New York to Florida and Alabama. July–Sept.

17. Sagittaria loràta (Chapm.) Small. Thong-
leaved Sagittaria. Fig. 247.

Sagittaria natans lorata Chapm. Fl. S. U. S. 449. 1860.
?Sagittaria natans (?) gracillima S. Wats. in A. Gray, Man.
   Ed. 6, 556. 1890.
?Sagittaria subulata gracillima J. G. Smith, in Mem. Torrey
   Club **5**: 26. 1894.
Sagittaria lorata Small, N. A. Fl. **17**¹: 52. 1909.

Monoecious, scape elongate, ⅓°–3½° long. Leaves with
bladeless petioles or blades when present floating, elliptic,
oblong or ovate-oblong, 7½"–25" long, rounded, subcordate
or hastate-truncate at the base; phyllodia flattened, strap-
like; whorls one or several; bracts thin, acuminate; sepals
becoming 1½"–2¼" long; corolla fully 10" wide; achenes
1"–1½" long, with 5–7 prominently dentate or crenate crests,
the beak erect or curved upward.

In ponds and streams, New Jersey to Florida. The form *S. natans gracillima* (the fruit of
which is unknown) in Massachusetts and Connecticut.

Butomus umbellàtus L., a plant of the related family Butomaceae, with many-ovuled ovaries,
rose-colored flowers and narrow ensiform leaves, native of Europe and Asia, has been found on
the shores of the St. Lawrence River, near Montreal.

Family 8. VALLISNERIÀCEAE Dumort. Anal. Fam. 54. 1829.
TAPE-GRASS FAMILY.

Submerged or floating aquatic herbs, the leaves various. Flowers regular,
mostly dioecious, appearing from an involucre or spathe of 1–3 bracts or leaves.
Perianth 3–6-parted, the segments either all petaloid or the 3 outer ones small
and herbaceous, the tube adherent to the ovary at its base in the pistillate flowers.
Stamens 3–12. Anthers 2-celled. Ovary 1-celled with 3 parietal placentae. Styles
3, with entire or 2-cleft stigmas. Ovules anatropous or orthotropous. Fruit
ripening under water, indehiscent. Seeds numerous, without endosperm.

About 6 genera and 25 species of wide distribution in warm and temperate regions. Besides
the following, another genus, *Halophila,* occurs on the coast of Florida.

Stem branched ; leaves whorled or opposite.   1. *Philotria.*
Acaulescent; stoloniferous; leaves grass-like, elongated.   2. *Vallisneria.*

1. PHILÒTRIA Raf. Am. Month. Mag. **2** : 175.   1818.

[ELODEA Michx. Fl. Bor. Am. **1** : 20. 1803. Not *Elodes* Adans. 1763.]

Stems submerged, elongated, branching, leafy. Leaves opposite or whorled, crowded,
1-nerved, pellucid, minutely serrulate or entire. Flowers dioecious or polygamous, arising
from an ovoid or tubular 2-cleft spathe. Perianth 6-parted, at least the 3 inner segments
petaloid. Staminate flowers with 9 stamens, the anthers oblong, erect. Ovary 1-celled with 3
parietal placentae. Stigmas 3, nearly sessile, 2-lobed. Fruit oblong, coriaceous, few-seeded.
[Name from the Greek, referring to the leaves, which are often whorled in three's.]

About 10 American species. Type species *Elodea canadensis* Michx.

Leaves oblong or ovate-oblong, mostly obtuse; staminate flowers unknown.   1. *P. canadensis.*
Leaves linear or oblong, acute; hermaphrodite flowers unknown.
    Leaves oblong or linear-oblong, 1"–1½" wide; spathe of the staminate flowers 2½"–3" long,
      anthers 1"–1¼" long.   3. *P. Nuttallii.*
    Leaves linear, rarely 1" wide; staminate spathe 1"–1½" long; anthers about ½" long.
      Leaves 5"–10" long; sepals and petals ¾"–1" long.   2. *P. angustifolia.*
      Leaves 2½"–4" long; sepals and petals ½"–¾" long.   4. *P. minor.*

1. Philotria canadénsis (Michx.) Britton. Water-weed.
Fig. 248.

Elodea canadensis Michx. Fl. Bor. Am. **1**: 20. 1803.
Elodea latifolia Casp. Jahrb. Wiss. Bot. **1**: 467. 1858.
?*Anacharis canadensis* Babingt, Ann. & Mag. Nat. Hist. II. **1**: 85.
    1848.
Philotria canadensis Britton, Science II. **2**: 5. 1895.

    Stem slender, 1°–3½° long, usually with short internodes. Leaves verticillate in 3's or 4's, or the lower ones in 2's, sessile, oblong or ovate-oblong, usually obtuse, 2½"–5" long, 1"–2" wide, minutely serrulate; staminate flowers unknown; flowers in the typical American form usually hermaphrodite, in the European (*Anacharis Alsinastrum*), pistillate; sheath tubular, 5"–7½" long; tube of the hypanthium 2'–4' long; sepals and petals elliptic, obtuse, about ¾" long; stamens usually 3, rarely 4–6, or reduced to mere filaments; anthers oblong, nearly sessile; stigmas 3, spreading, purplish, emarginate about equaling the petals and sepals.

    In ponds and slow streams, Quebec to Virginia and Minnesota. Naturalized in Europe. Called also Choke Pondweed.

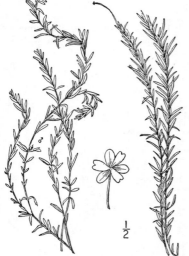

2. Philotria angustifólia (Muhl.) Britton.
Narrow-leaved Water-weed. Fig. 249.

Serpicula verticillata angustifolia Muhl. Cat. Pl. Am.
    Sept. 84. 1813.
?*Serpicula occidentalis* Pursh, Fl. Am. Sept. 33. 1814.
?*Apalanthe Schweinitzii* Planch. Ann. Sci. Nat. Bot. III.
    11: 76. 1849.
Philotria angustifolia Britton; Rydb. Bull. Agr. Exp.
    Sta. Colo. **100**: 15. 1906.

    Dioecious water plant; stem slender, flaccid, 1°–3½° long. Leaves in 3's or 2's, sessile, linear, 5"–10" long, about ½" wide, rarely ¾" wide, acute; spathe of the staminate flowers 1"–1½" long, sessile, ovoid; sepals and petals elliptic or oval, the former about 1" long, the latter smaller; anthers about ½" long; spathe of the pistillate flowers tubular, ½"–1" long; hypanthium 1¼'–4' long; sepals and petals elliptic, the former ¾"–1" long; stigmas 2-cleft.

    In streams New York and Pennsylvania to Florida.

3. Philotria Nuttàllii (Planch.) Rydb. Nuttall's
Water-weed. Fig. 250.

Serpicula verticillata Muhl. Cat. Pl. Am. Sept. 84. 1813.
    Not *S. verticillata* L. 1781.
Anacharis Nuttallii Planch. Ann. & Mag. Nat. Hist. II. **1**:
    85. 1848.
Philotria Nuttallii Rydb. Bull. Torrey Club **35**: 461. 1908.

    Dioecious water-plant; stem slender, 1°–3½° long; with the internodes often longer than the leaves. Leaves usually in 3's or the lower ones in 2's, sessile, oblong or lance-oblong, acute, 2½"–5" long, 1"–1½" wide, finely serrulate; spathe of the staminate flowers ovoid, sessile, 2½"–3" long; flower without a tube; sepals and petals oblong, the former scarcely exceeding the oblong anthers, which are 1"–1½" long; sheath of the pistillate flowers about 5" long, 2-cleft; tube of the hypanthium 2'–6' long; sepals and petals elliptic, ¾" long; filaments rudimentary; stigmas slightly exceeding the petals, 2-cleft at the apex.

    In slow streams and ponds, New York to Virginia. This figure was used for *P. canadensis* in our first edition, formerly confused with this species.

$\frac{3}{4}$

4. Philotria mìnor (Engelm.) Small.
Lesser Water-weed. Fig. 251.

Udora verticillata (?) *minor* Engelm.; Casp.
Jahrb. Wiss. Bot. 1: 465, as synonym under
Anacharis Nuttallii. 1858.
Philotria minor Small, Fl. SE. U. S. 47. 1903.

Dioecious water-plant; stems filiform, $1°-1\frac{3}{4}°$
long. Leaves in 3's or 2's, linear, $2\frac{1}{2}''-4''$ long,
$\frac{1}{2}''$ broad or less, acutish; staminate spathe
about $2\frac{1}{2}''$ long, ovoid, sessile; sepals and
petals oval, $\frac{1}{2}''-\frac{3}{4}''$ long; anthers about $\frac{1}{2}''$
long; pistillate spathe about $5''$ long; hypan-
thium-tube $\frac{3}{4}'-2'$ long (rarely longer); petals
and sepals elliptic, $\frac{1}{2}''-\frac{3}{4}''$ long; stigmas slender, longer than the sepals, deeply 2-cleft.

In ponds, lakes and slow streams. Wisconsin
to Ohio, Kentucky, Kansas and Arkansas.

2. VALLISNÈRIA L.
Sp. Pl. 1015. 1753.

Aquatic dioecious submerged perennials, with long grass-like floating leaves. Staminate
flowers with a 2–3-parted spathe on a short scape, numerous, nearly sessile on a conic receptacle; perianth 3-parted; stamens generally 2 (1–3). Pistillate flowers on a very long flexuous
or spiral scape, with a tubular, 2-cleft, 1-flowered spathe; perianth-tube adnate to the ovary,
3-lobed and with 3 small petals; ovary 1-celled with 3 parietal placentae; stigmas 3, nearly
sessile, short, broad, 2-toothed with a minute process just below each sinus; ovules numerous,
borne all over the ovary-wall, orthotropous. Fruit elongated, cylindric, crowned with the
perianth. [Named for Antonio Vallisneri, 1661–1730, Italian naturalist.]

A genus consisting of 2 species, the one of
wide distribution both in the Old World and the
New, the other confined to the Gulf States.
Type species: *Vallisneria spiralis* L.

1. Vallisneria spiràlis L. Tape-grass.
Eel-grass. Fig. 252.

Vallisneria spiralis L. Sp. Pl. 1015. 1753.

Plant rooting in the mud or sand, stoloniferous. Leaves thin, narrowly linear, 5-
nerved, obtuse, sometimes serrate near the
apex, $\frac{1}{2}°-6°$ long, $2''-9''$ wide, the 2 marginal
nerves faint; the staminate bud separates from
the scape at the time of flowering and ex-
pands upon the surface of the water; pistillate flower upon a long thread-like scape,
the spathe $\frac{1}{2}'-1'$ long, enclosing a single
white flower; ovary as long as the spathe,
after receiving the pollen from the staminate
flowers the scape of the pistillate contracts
spirally; ripe fruit $2'-7'$ long.

In quiet waters, New Brunswick and Nova
Scotia to North Carolina, west to South Dakota
and Indiana. The "Wild or Water Celery" of
Chesapeake Bay, and a favorite food of the can-
vas-back duck. Aug.–Sept.

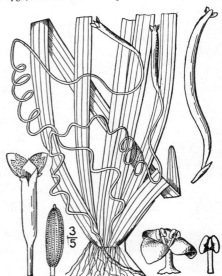

$\frac{3}{5}$

Family 9. HYDROCHARITÀCEAE Aschers.
Frog's-bit Family.

Aquatic or mud-inhabiting herbs, with broad or narrow leaves borne on a very
short stem. Flowers regular, monoecious or dioecious, arising from spathes of
distinct or united bracts. Perianth superior, 6-parted, the segments usually all
petaloid, the flower-tube adnate to the ovary in the pistillate flowers. Stamens
6–12, distinct or monodelphous. Ovary usually 6–9-celled. Styles 6–9. Ovules
numerous. Fruit somewhat fleshy, usually indehiscent. Seeds numerous.

About 8 genera and 20 species, in temperate and tropical regions. *Thalassia* in Florida.

1. **LIMNÒBIUM** L. C. Richard, Mem. Inst. Paris, **32**: 66. *pl. 8.* 1811.

Aquatic, stoloniferous herbs, the leaves fascicled at the nodes, petioled, broad, cordate. Flowers monoecious, white, arising from sessile or stipitate, 2-leaved, membranous spathes. Perianth 6-parted; segments petaloid, the 3 outer oblong to oval, the 3 inner oblong to linear. Staminate flowers 2–4 in a spathe, long-penduncled, the stamens united in a column bearing 6–12 anthers at different heights, sometimes producing only 9–12 staminodia, the filaments tipped with abortive anthers. Pistillate flowers sessile or short-peduncled with 3–6 vestigial stamens; ovary 6–9-celled with as many central placentae; stigmas as many as the cells, each 2-parted. Fruit a many-seeded berry. [Greek referring to the aquatic habitat.]

Species, 3 or 4, natives of America. Type species: *L. Bosci* L. C. Richard, the same as the following.

1. **Limnobium Spóngia** (Bosc.) L. C. Richard. Frog's-bit. Fig. 253.

Hydrocharis cordifolia Nutt. Gen. **2**: 241. 1818.
H. Spongia Bosc, Ann. Mus. Paris, **9**: 396, *pl. 30.* 1807.
Limnobium Spongia L. C. Richard, Mem. Inst. Paris, **32**: 66. *pl. 8.* 1811.
Limnocharis Spongia L. C. Richard; Steud. Nomencl. Ed. 2, Part 2, 45. 1841.

Blades of the leaves orbicular or broadly ovate, cordate or reniform, faintly 5–7-nerved and cross-veined, purplish and spongy beneath, 10″–2′ broad, on petioles 1′–10′ in length. Stolons rooting and sending up flowers and leaves at the nodes; peduncles of the staminate flowers 3′–4′ long, those of the pistillate flowers stouter, 1′–2′ long, nodding in fruit.

In shallow, stagnant water, Lake Ontario, to Florida, west to Illinois, Missouri and Texas. July–Aug.

Family 10. **GRAMÍNEAE** Juss. Gen. 28. 1789.*

GRASS FAMILY.

Annual or perennial herbs, of various habit, rarely shrubs or trees. Culms (stems) generally hollow, but occasionally solid, the nodes closed. Leaves sheathing, the sheaths usually split to the base on the side opposite the blade; a scarious or cartilaginous ring, naked or hairy, rarely wanting, called the ligule, is borne at the orifice of the sheath. Inflorescence spicate, racemose or paniculate, consisting of spikelets composed of two to many 2-ranked imbricated bracts, called scales (glumes), the two lowest in the complete spikelet always empty, one or both of these sometimes wanting. One or more of the upper scales, except sometimes the terminal ones, contains in the axil a flower, which is usually enclosed by a bract-like awnless organ called the palet, placed opposite the scale and with its back toward the axis (rachilla) of the spikelet, generally 2-keeled; sometimes the palet is present without the flower, and vice versa. Flowers perfect, pistillate, or staminate, sometimes monoecious or dioecious, subtended by 1–3 minute hyaline scales called the lodicules. Stamens 1–6, usually 3. Anthers 2-celled, versatile. Ovary 1-celled, 1-ovuled. Styles 1–3, commonly 2 and lateral. Stigmas hairy or plumose. Fruit a seed-like grain (caryopsis). Endosperm starchy.

About 4500 species, widely distributed throughout the world, growing in water and on all kinds of soil. Those yielding food-grains are called cereals. The species are more numerous in tropical countries, while the number of individuals is much greater in temperate regions, often forming extended areas of turf. The time of year noted is that of ripening seed.

A. Spikelets articulated below the empty scales or a subtending involucre, or attached to and
   deciduous with the internodes of a readily disarticulating rachis, 1-flowered, or if 2-flowered
   the lower imperfect, usually staminate; rachilla not extending beyond the uppermost scale.
   Spikelets round or dorsally compressed; hilum punctiform.
       Fruiting scale and palet hyaline, thin, much more delicate in structure than the thick-mem-
       branous to coriaceous empty scales.
           Spikelets unisexual, the pistillate borne in the lower, the staminate in the upper, part of
           the same spike.   I. MAYDEAE.
           Spikelets in pairs, one sessile, perfect, the other pedicellate, perfect, staminate or empty,
           sometimes reduced to a single scale or wanting.   II. ANDROPOGONEAE.
       Fruiting scale and palet never hyaline and thin, as firm as the empty scales, or firmer.
           Fruiting scale and palet membranous; spikelets naked, spiny (in ours).   III. ZOYSIEAE.

* Text contributed by MR. GEORGE V. NASH.

Fruiting scale and palet chartaceous or coriaceous, differing in color and appearance from
the remaining scales; spikelets sometimes enclosed in an involucre. IV. PANICEAE.
Spikelets laterally compressed; hilum linear. V. ORYZEAE.

B. Spikelets articulated above the empty scales (below them in nos. 38, 41, 49, 57 and 64) which
are persistent, 1–many-flowered; rachilla sometimes extending beyond the uppermost scale.
Culms herbaceous, hence annual; leaf-blades sessile, not articulated with the sheath.
Spikelets in panicles or racemes, usually upon distinct and often long pedicels.
Spikelets 1-flowered.
Empty scales 4; palets 1-nerved. VI. PHALARIDEAE.
Empty scales 2 (rarely 1); palet usually 2-nerved. VII. AGROSTIDEAE.
Spikelets 2–many-flowered.
Flowering scales usually shorter than empty ones, awn dorsal, bent. VIII. AVENEAE.
Flowering scales usually longer than the empty ones, awnless, or if awned the awn
terminal and straight, rarely dorsal. X. FESTUCEAE.
Spikelets borne in 2 rows:
On one side of a continuous axis, forming 1-sided spikes or racemes. IX. CHLORIDEAE.
On opposite sides of a continuous or sometimes articulated axis, forming equilateral
spikes (unilateral in *Nardus*). XI. HORDEAE.
Culms woody, perennial; leaf-blades petiolate, articulated to the sheath. XII. BAMBUSEAE.

Tribe I. MAYDEAE.

Pistillate spikelets imbedded in the internodes of the thick rachis. 1. *Tripsacum.*

Tribe II. ANDROPOGONEAE.

Internodes of the rachis of the racemes thickened, appressed to the pedicels of the primary spike-
lets, thus forming excavations for the reception of the secondary or sessile spikelets; fertile
flowering scales awnless. 2. *Coelorachis.*
Internodes not thickened, and without excavations for the reception of the spikelets.
Spikelets all perfect, awned.
Rachis of the racemes continuous; panicle axis short, racemes subflabellate. 3. *Miscanthus.*
Rachis articulated; panicle axis elongated. 4. *Erianthus.*
Sessile spikelets perfect, the pedicellate staminate or empty, awnless, sometimes wanting.
Inflorescence simple or compound, made up of 1 or more spike-like racemes which are
sessile or on very short peduncles.
Raceme single; pedicels and internodes of the rachis clavate, spongy, usually stout, with
a deep cup-shaped depression at the top. 5. *Schizachyrium.*
Racemes not single; pedicels and rachis-internodes filiform, or flat and linear, not
spongy, nor appendaged at the apex.
Racemes in pairs, or digitate in 5's or less, sessile, or only 1 pedunculate; pedicels
and internodes not sulcate nor with a median hyaline line. 6. *Andropogon.*
Racemes numerous, on an elongated axis, more or less pedunculate; pedicels and
internodes of the rachis with a median hyaline line. 7. *Amphilophis.*
Inflorescence decompound.
Pedicellate spikelet wanting. 8. *Sorghastrum.*
Pedicellate spikelet present. 9. *Holcus.*

Tribe III. ZOYSIEAE.

Spikelets in a terminal spike; second scale spiny. 10. *Nazia.*

Tribe IV. PANICEAE.

Spikelets without a subtending involucre of bristles or valves.
Spikelets all alike.
Palets not enlarged when mature.
Fruiting scale chartaceous the margins hyaline and flat.
Spikelets in slender racemes borne toward the summit of the stem. 11. *Syntherisma.*
Spikelets in an open panicle on long pedicels. 12. *Leptoloma.*
Fruiting scale indurated, rigid, the margins inrolled and not hyaline.
Opening in the fruiting scale turned toward the rachis.
Spikelets with a swollen ring-like callus at the base; fruiting scale mucronate or
awn-pointed. 13. *Eriochloa.*
Spikelets without a callus; fruiting scale not mucronate. 14. *Anastrophus.*
Opening in the fruiting scale turned away from the rachis.
Spikelets plano-convex, in secund racemes, usually of 3 scales. 15. *Paspalum.*
Spikelets unequally bi-convex, in panicles, or rarely in secund racemes; scales 4.
Scales or some of them awned; fruiting scale cuspidate. 16. *Echinochloa.*
Scales awnless.
Second scale like the third, few-nerved not broad and saccate. 17. *Panicum.*
Second scale unlike the third, 11–13-nerved, broad saccate. 18. *Sacciolepis.*
Palet in the axil of the third scale much enlarged and somewhat indurated when mature,
forcing the spikelet open. 19. *Steinchisma.*
Spikelets of 2 kinds, one in terminal panicles and not producing seed, the other subterranean
and perfecting seed. 20. *Amphicarpon.*
Spikelets with an involucre:
Of bristles, persistent; inflorescence a dense cylindric spike-like panicle. 21. *Chaetochloa.*
Of 2 spine-bearing valves enclosing the spikelets, deciduous with them. 22. *Cenchrus.*

Tribe V. ORYZEAE.

Spikelets unisexual; plants monoecious; tall aquatic grasses.
Pistillate spikelets ovate, at the base of each branch of the panicle. 23. *Zizaniopsis.*
Pistillate spikelets linear, on the upper branches of panicle. 24. *Zizania.*
Spikelets all perfect, broad, compressed; in swamps or wet grounds. 25. *Homalocenchrus.*

Tribe VI. Phalarideae.

Third and fourth scales
  Small and empty, or rudimentary, not awned; stamens 3.   26. *Phalaris.*
  Empty, awned upon the back; stamens 2.   27. *Anthoxanthum.*
  Subtending staminate flowers, stamens 3; fertile flowers, stamens 2.   28. *Savastana.*

Tribe VII. Agrostideae.

Flowering scale indurated at maturity, firmer than the empty scales.
  Spikelets with no basal callus; flowering scale awnless, margins inrolled. 29. *Milium.*
  Spikelets with a basal callus; flowering scale awned, the margins flat.
    Awn simple.
      Flowering scale broad, the awn deciduous; callus short, obtuse.
        Flowering scale glabrous, or pubescent with short hairs.   31. *Oryzopsis.*
        Flowering scale pubescent with copious long silky hairs.   30. *Eriocoma.*
      Flowering scale narrow, awn persistent; callus commonly acute.   32. *Stipa.*
    Awn 3-parted.   33. *Aristida.*
Flowering scale membranous, not firmer than the empty scales.
  Flowering scale with a terminal awn or awn-pointed, tightly enclosing the grain.
    Rachilla not prolonged beyond the base of the flowering scale; empty scales usually evident.
            34. *Muhlenbergia.*
    Rachilla extending beyond the base of the flowering scale as a bristle-like appendage; empty
      scales minute, the first sometimes wanting.   35. *Brachyelytrum.*
  Flowering scale awnless, or with a dorsal awn, loosely enclosing the grain.
    Spikelets readily deciduous entire at maturity.
      Empty scales awnless.   38. *Alopecurus.*
      Empty scales awned.   41. *Polypogon.*
    Spikelets not deciduous entire, the empty scales persistent, flowering scales usually deciduous.
      Empty scales awned.   37. *Phleum.*
      Empty scales awnless.
        Flowering scales 1-nerved.
          Panicle dense and spike-like, the spikelets markedly compressed laterally, ciliate
            on the keel.   36. *Heleochloa.*
          Panicle open or narrow, the spikelets not markedly laterally compressed, the
            keel glabrous.
            Grain loosely enclosed in the pericarp, from which it readily separates and
              falls at maturity; flowering scales with no hairs at the base.
              Empty scales minute; low arctic grass.   39. *Phippsia.*
              Empty scales evident.   40. *Sporobolus.*
            Grain adherent to the pericarp and not separating from it at maturity; flower-
              ing scale with a ring of long hairs at the base.   47. *Calamovilfa.*
        Flowering scales 3–5-nerved.
          Stamen 1; flowering scale stipitate; palet usually 1-nerved. 43. *Cinna.*
          Stamens 3; flowering scale sessile; palet 2-nerved.
            Rachilla not prolonged beyond the flowering scale.
              Empty scales shorter than flowering scale; spikelets large 42. *Arctagrostis.*
              Empty scales longer than the flowering scale; spikelets small.   44. *Agrostis.*
            Rachilla prolonged beyond the flowering scale.
              Prolongation of the rachilla glabrous; flowering scale glabrous at the base,
              and with a long awn just below the bifid apex.   48. *Apera.*
              Prolongation of the rachilla with long hairs; flowering scale awned at or
              below the middle.
                Flowering scale membranous; spikelets 4″ long or less.
                  45. *Calamagrostis.*
                Flowering scale chartaceous; spikelets 5″–6″ long. 46. *Ammophila.*

Tribe VIII. Aveneae.

Spikelets deciduous; lower flower perfect, upper staminate, awned.   49. *Nothoholcus.*
Spikelets not deciduous; empty scales persistent, flowering ones deciduous.
  Spikelets of 2 perfect flowers; rachilla not prolonged beyond the upper one. 50. *Aspris.*
  Spikelets 2–many-flowered; rachilla prolonged beyond the upper scale.
    Awn of flowering scale dorsal, inserted below the teeth.
      Flowers all perfect, or the upper ones staminate or wanting.
        Spikelets less than 6″ long; grain free, unfurrowed.
          Flowering scales convex; awn arising from or below the middle. 51. *Deschampsia.*
          Flowering scales keeled; awn arising from above the middle.   52. *Trisetum.*
        Spikelets over 6″ long; grain furrowed, usually adherent to the scales. 53. *Avena.*
      Upper flower perfect, lower staminate, its scale strongly awned.   54. *Arrhenatherum.*
    Awn from between the lobes or teeth of flowering scale, generally twisted. 55. *Danthonia.*

Tribe IX. Chlorideae.

Spikelets with perfect flowers, or sometimes some of them rudimentary or unisexual.
  Spikelets deciduous entire.
    Spikelets narrow, lanceolate; rachis produced beyond the upper spikelet. 57. *Spartina.*
    Spikelets broad, orbicular; rachis of the spike not produced.   64. *Beckmannia.*
  Spikelets not deciduous entire; empty scales persistent; flowering scales deciduous.
    One perfect flower in each spikelet (rarely 2 in no. 56).
      No empty scales above the flower.
        Spikes 2–6, digitate.   56. *Capriola.*
        Spikes many, scattered.   61. *Schedonnardus.*
      One to several empty scales above the flower.
        Lower empty scales 4; spike solitary, dense.   58. *Campulosus.*

Lower empty scales 2.
    Spikes in false whorls or closely approximate; scales long-awned.   59. *Chloris*.
    Spikes remote, or the lowest only approximate.
        Spikelets scattered or remote in long filiform spikes.   60. *Gymnopogon*.
        Spikelets crowded in short stout spikes.
            Spikes 4 or less; spikelets numerous, 25 or more.   62. *Bouteloua*.
            Spikes numerous, 12 or more; spikelets few, 12 or less.   63. *Atheropogon*.
2–several perfect flowers in each spikelet.
    Spikelets densely crowded; spikes digitate.
        Spikes with terminal spikelets.   65. *Eleusine*.
        Spikes with rachis extending beyond them in a naked point.   66. *Dactyloctenium*.
    Spikelets distinctly alternating; spikes remote.
        Branches of the inflorescence slender; spikelets less than 2″ long, numerous, the palets not gibbous.   67. *Leptochloa*.
        Branches of the inflorescence stout, rigid; spikelets 4″ long or more, few, the palets gibbous at the base.   68. *Acamptoclados*.
Spikelets with unisexual flowers, very unlike; plants dioecious.   69. *Bulbilis*.

Tribe X. Festuceae.

Rachilla with hairs longer than flowering scales enveloping them.   71. *Phragmites*.
Rachilla and flowering scales glabrous, or if hairy the hairs shorter than the scales.
    Stigmas barbellate; spikelets in clusters of 3–6 in axils of spinescent leaves.   70. *Munroa*.
    Stigmas plumose; spikelets not in the axils of leaves; inflorescence various.
        Spikelets of 2 forms, the fertile 1–3-flowered, surrounded by the sterile consisting of many empty pectinate scales.   91. *Cynosurus*.
        Spikelets all alike.
            Flowering scales 1–3-nerved, rarely with faint additional intermediate nerves.
                Flowering scales not coriaceous in fruit; seed beakless and not exserted.
                    Lateral nerves of the flowering scales pilose.
                        Internodes of rachilla long, the deeply 2-lobed flowering scale attached by a long pointed callus, which is pilose on the outer surface.   75. *Triplasis*.
                        Internodes of rachilla and callus of flowering scales short, the latter blunt.
                            Spikelets on pedicels of varying length, arranged in a contracted or open, simple or compound panicle.
                                Inflorescence a contracted or open panicle; leaf-blades not cartilaginous on margins.   73. *Tridens*.
                                Inflorescence a short congested raceme; leaf-blades with thick conspicuous cartilaginous margins.   74. *Erioneuron*.
                        Spikelets on short pedicels of approximately the same length, appressed to the long branches of the simple panicle.   77. *Diplachne*.
                  Lateral nerves of the flowering scales glabrous.
                      Callus of flowering scale conspicuously pubescent with long hairs.
                        Panicle contracted; flowering scales broadly oval, rounded at the apex.   79. *Rhombolytrum*.
                        Panicle open, diffuse; flowering scales lanceolate, acute.   76. *Redfieldia*.
                  Callus of the flowering scale glabrous.
                      Second empty scale very dissimilar from the first, broad and rounded at the summit.   81. *Sphenopholis*.
                      Second empty scale similar to the first.
                        Panicle narrow, branches appressed.
                          Panicle dull, interrupted; rachilla articulated.   78. *Molinia*.
                          Panicle shining, dense, spike-like; rachilla continuous.   82. *Koeleria*.
                      Panicle open, the branches more or less spreading.
                        Rachilla continuous; flowering scales deciduous in fruit.   80. *Eragrostis*.
                        Rachilla articulated; flowering scales and palets both deciduous with the rachilla internodes.   83. *Catabrosa*.
             Flowering scales coriaceous in fruit; seed beaked and exserted.   85. *Korycarpus*.
         Flowering scales 5–many-nerved.
            Flowering scales 3-toothed at the apex.   72. *Sieglingia*.
            Flowering scales not 3-toothed at the apex.
                Spikelets with 2 or more of the upper scales empty, broad and enfolding each other, forming a club-shaped mass.   84. *Melica*.
                Spikelets with the upper scales flower-bearing, or if empty similar in shape to the other scales.
                    Keels of the palet winged or with a linear appendage.   86. *Pleuropogon*.
                    Keels of the palet not winged or appendaged.
                      Stigmas placed at or near the apex of the ovary; flowering scales usually awnless, or awned in nos. 90 and 98.
                      Scales more or less strongly compressed and keeled.
                        Empty basal scales 3–6; spikelets flat, 2-edged.   87. *Uniola*.
                        Empty basal scales 2; spikelets somewhat flattened.
                          Spikelets unisexual; plant dioecious.   88. *Distichlis*.
                          Spikelets perfect.
                            Spikelets arranged in 1-sided dense capitate clusters at the end of the branches; flowering scales awned.   90. *Dactylis*.
                            Spikelets not arranged as above; flowering scales awnless.
                              Rachilla of the spikelets glabrous, or with webby hairs; flowering scales scarious-margined.
                                Spikelets cordate at the base, large.   89. *Briza*.
                                Spikelets not cordate, usually small.
                                    Empty scales projecting beyond the uppermost flowering ones; arctic grass.   93. *Dupontia*.

Empty scales shorter than the uppermost flowering
ones.   92. *Poa.*
Rachilla of the spikelets hirsute, extending into a hairy
appendage ; flowering scales membranous.
95. *Graphephorum.*
Scales rounded on the back, at least below.
Flowering scales with basal ring of hairs, apex toothed.
94. *Scolochloa.*
Flowering scales naked at the base.
Flowering scales obtuse or subacute and scarious at the apex,
usually toothed.
Manifestly 5–7-nerved ; styles present.   96. *Panicularia.*
Obscurely 5-nerved ; no styles.   97. *Puccinellia.*
Flowering scales acute, pointed, or apex awned. 98. *Festuca.*
Stigmas arising below apex of the ovary ; scales rarely awnless.
99. *Bromus.*

Tribe XI.   HORDEAE.

Stigma 1 ; spike unilateral ; spikelets 1-flowered, narrow.   100. *Nardus.*
Stigmas 2 ; spikes symmetrical.
Spikelets solitary at the notches of the rachis.
Flowering scales with their backs turned to the rachis.   101. *Lolium.*
Flowering scales with their sides turned to the rachis.
Spikelets 1- or 2-flowered, in slender articulate spikes.   102. *Lepturus.*
Spikelets 2–many-flowered, in stout inarticulate spikes.   103. *Agropyron.*
Spikelets 2–6 at each joint of the rachis ; scales mostly long-awned.
Spikelets 1-flowered, or with the rudiment of a second flower.   104. *Hordeum.*
Spikelets 2–many-flowered.
Rachis of the spike articulated, readily breaking up.   105. *Sitanion.*
Rachis of the spike continuous, not breaking up.
Empty scales a little smaller than the flowering ones.   106. *Elymus.*
Empty scales very small or none.   107. *Hystrix.*

Tribe XII.   BAMBUSEAE.

Tall canes with large flat spikelets in panicles or racemes.   108. *Arundinaria.*

1. TRÍPSACUM L. Syst. Nat. Ed. 10. 1261. 1759.

Tall perennial monoecious grasses with thick rootstocks, rather broad flat leaves and
spicate or racemose inflorescence. Spikelets 1- or 2-flowered, in terminal or axillary, solitary
or clustered, elongated spikes. Staminate spikelets in 2's at each node of the axis, 2-flowered,
consisting of four scales, the two outer coriaceous or membranous, the two inner thinner, the
palet hyaline ; stamens 3. Pistillate spikelets in excavations at the lower joints of the spike,
1-flowered ; stigmas exserted ; style slender. Grain partly enclosed in the excavations of the
spikes, covered in front by the horny exterior lower scale. [Name from the Greek, in allusion
to the polished outer scales.]

About 7 species, in tropical and temperate America. Type species : *Tripsacum dactyloides* L.

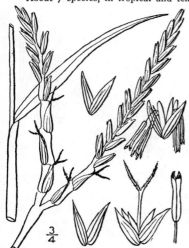

$\frac{3}{4}$

1. Tripsacum dactyloìdes L. Gama- Sesame- or Bull-grass. Fig. 254.

Coix dactyloides L. Sp. Pl. 972. 1753.
Tripsacum dactyloides L. Sp. Pl. Ed. 2, 1378. 1763.
Tripsacum dactyloides var. *monostachyum* Wood, Class-
book Ed. 2, 623. 1847.

Culms stout, erect, 4°–8° tall. Leaves smooth and
glabrous ; blades 1° or more long, ½′–1½′ wide, long-
acuminate, truncate or subcordate at the base ; spikes
terminal and in the upper axils, solitary or 2 or 3
together, 4′–9′ long, the lower spikelets pistillate, the
upper staminate and very numerous ; outer scales of
the staminate spikelets linear and obtuse, 3½″–5½″
long, faintly many-nerved ; exterior scale of the
pistillate spikelets horny, shining, closely appressed
in fruit.

In swamps or along streams, Rhode Island to Nebraska, south to Florida, Texas and Mexico, the southern
Bahamas, Haiti and South America. June–Sept.

2. COELÓRACHIS Brongn. in Duperr. Voy. Coq. Bot. Phan. 64. 1829.

Mostly tall perennials, with narrow flat leaves and cylindric jointed spikes, terminal and
from the upper axils. Spikelets in pairs at each node of the excavated rachis, one sessile

and perfect, the other with a pedicel and either staminate or empty. Scales of the perfect spikelet 4, the outermost thick and coriaceous, covering, together with the pedicel of the sterile spikelet, the excavation in the rachis; second scale chartaceous; third and fourth hyaline, the latter subtending the palet and perfect flower. Stamens 3. Styles distinct. Grain free. [Name Greek, meaning hollowed rachis.]

About 20 species, widely distributed in tropical and temperate countries.

Leaf-sheaths broad, compressed, keeled; plants without rootstocks.   1. *C. rugosa.*
Leaf-sheaths narrow, round, not keeled; plants with creeping rootstocks.   2. *C. cylindrica.*

3/4

1. Coelorachis ragòsa (Nutt.) Nash.
Wrinkled Joint-grass. Fig. 255.

Rottboellia rugosa Nutt. Gen. 1: 84. 1818.
R. corrugata Baldw. Am. Journ. Sci. 1: 355. 1819.
Manisuris rugosa Kuntze, Rev. Gen. 780. 1891.
Manisuris rugosa Chapmani Scribn. Mem. Torr. Club 5: 28. 1894.
C. rugosa Nash, N. A. Fl. 17: 86. 1909.

Smooth and glabrous. Culms 2°–5° tall, compressed, much-branched above, branches spreading; sheaths compressed; blades flat, acuminate, 6′–1½° long, 1″–5″ wide; racemes partially included in the sheath or more or less exserted, 1½′–2½′ long; outermost scale of the sessile spikelet oblong-ovate to ovate, 1¾″–2½″ long, strongly transversely rugose, the wrinkles continuous or interrupted.

In wet soil along the coast, southern New Jersey to Maryland, Florida and Texas. June–Sept.

2. Coelorachis cylíndrica (Michx.) Nash.
Pitted Joint-grass. Fig. 256.

Tripsacum cylindricum Michx. Fl. Bor. Am. 1: 60. 1803.
Rottboellia cylindrica Torr. Pac. R. R. Rep. 4: 159. 1857.
Manisuris cylindrica Kuntze, Rev. Gen. 779. 1891.
Coelorachis cylindrica Nash, N.A.Fl. 17: 85. 1909.

Culms from creeping rootstocks, round, 1°–3½° tall, slender; blades 1° long or less, ½″–1½″ wide; racemes 4′–8′ long, the rachis barely if at all contracted at the nodes; sessile spikelets 2¼″–2½″ long, about equalling the internodes, the first scale more or less pitted in longitudinal lines, or rarely unpitted, the pits often containing a subulate hair; pedicellate spikelets reduced to 1 or 2 short scales, the pedicel linear, shorter than the sessile spikelet and curved around its margin.

In sandy soil at low elevations, Georgia and Florida to Missouri, Oklahoma and Texas. Summer.

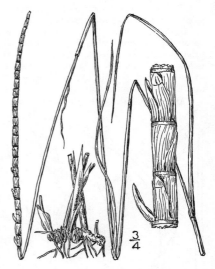

3/4

3. MISCÁNTHUS Anderss. Oefv. Sv. Vet.-Akad. Förh. 1855: 165. 1856.

Tall erect perennial grasses, with usually flat leaf-blades, and terminal ample commonly hairy panicles. Spikelets 1-flowered, unequally pedicellate, arranged in pairs along the continuous branches of the panicle, articulated with the pedicel. Scales 4; outer 2 larger, empty, membranous, muticous; third scale also empty but thinner; fourth scale thinly hyaline, subtending a perfect flower, 2-toothed at the apex, the awn arising from between the teeth, usually slender, often with a twisted column at the base and geniculate, sometimes straight, rarely very short or wanting; palet thin, hyaline. Stamens 3. Styles distinct. Stigmas plumose. [Greek, in allusion to the stalked spikelets.]

A genus of about 10 species, natives of the Old World. Type species: *Eulalia japonica* Trin.

1. Miscanthus sinénsis Anderss. Japanese Plume-grass. Fig. 257.

Saccharum polydactylon β Thunb. Fl. Jap. 43. 1784.
Saccharum japonicum Thunb. Trans. Linn. Soc. 2: 328, in part. 1794.
Erianthus japonicus Beauv.; R. & S. Syst. 2: 324. 1817.
Ripidium japonicum Trin. Fund. Agrost. 169. 1820.
Eulalia japonica Trin. Mem. Acad. St. Petersb. VI. 2: 333. 1832.
Miscanthus sinensis Anderss. Oefv. Sv. Vet.-Akad. Förh. 1855: 166. 1856.

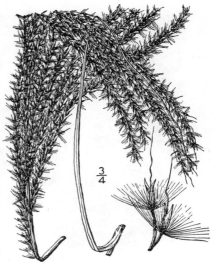

Stems 3°–6° tall; leaf-blades up to 3° long and 8″ wide; panicle 8′–16′ long, its branches erect or ascending; spikelets 2¼″–2½″ long, yellowish brown, shining, glabrous, encircled at the base with white or purplish hairs equaling or exceeding them, the awn 4″–5″ long, spirally twisted at the base.

Escaped from cultivation at Washington, D. C., and on Long Island; also in Florida. A native of China, Japan and the Celebes.

4. ERIÁNTHUS Michx. Fl. Bor. Am. 1: 54. 1803.

Tall generally robust perennial grasses, with long flat leaves, and perfect flowers in terminal panicles. Spikelets generally with a ring of hairs at the base, 2 at each node of the jointed rachis, one sessile, the other with a pedicel, generally 1-flowered. Scales 4, the two outer indurated, the inner hyaline, the fourth bearing a terminal straight or contorted awn; palet small, hyaline; stamens 3. Grain oblong, free, enclosed in the scales. [Greek, referring to the woolly spikelets.]

About 21 species, natives of the temperate and tropical regions of both hemispheres. Besides the following, three or four others occur in the Southern States. Type species: Anthoxanthum giganteum Walt.

Awns flat, closely spiral at the base, geniculate; apex of the fourth scale deeply 2-cleft.
  Basal hairs twice as long as the yellowish spikelets which are nearly concealed in the copious hairs of the cream-colored panicle.   1. E. divaricatus.
  Basal hairs sometimes equaling but not exceeding the brown spikelets which are plainly visible through the brown panicle.   2. E. contortus.
Awn terete, or flat only at the very base, not spiral at the base, straight; fourth scale usually entire, rarely shortly 2-toothed.
  Spikelets 2″–3″ long, exceeded by the basal hairs.   3. E. saccharoides.
  Spikelets 4″–4½″ long, much exceeding the basal hairs.   4. E. brevibarbis.

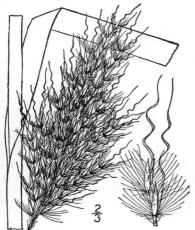

1. Erianthus divaricàtus (L.) Hitchc. Woolly Beard-grass. Fig. 258.

Andropogon divaricatus L. Sp. Pl. 1045. 1753.
Andropogon alopecuroides L. Sp. Pl. 1045. 1753.
Erianthus alopecuroides Ell. Bot. S. C. & Ga. 1: 38. 1816.
Erianthus divaricatus Hitchc. Contr. U. S. Nat. Herb. 12: 125. 1908.

Culms stout, erect, 6°–10° tall; nodes naked or barbed, the summit and the axis of the panicle densely pubescent with appressed long rigid silky hairs. Sheaths glabrous or hirsute; blades usually glabrous, 6′–2° long, ¼′–1′ wide, acuminate, narrowed and sometimes hairy on the upper surface near the base; panicle oblong, 7′–12′ long, 2′–3′ wide, branches 3′–5′ long, slender, loose; outer scales of the spikelet about 3″ long, exceeding the pedicel and about two-thirds as long as the basal hairs, lanceolate, acuminate; inner scales shorter, the awn 6″–8″ long, scabrous.

In damp soil, New Jersey to Oklahoma, south to Florida and Texas. Sept. Plume-grass (Tenn.).

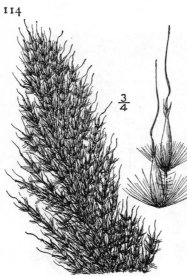

$\frac{3}{4}$

2. Erianthus contórtus Ell. Spiral-awned Beard-grass. Fig. 259.

Erianthus contortus Ell. Bot. S. C. & Ga. 1: 40. 1816.

Erianthus saccharoides contortus Hack. in DC. Monog. Phan. 6: 131. 1889.

Culms 3°–8° tall; leaf-sheaths smooth or rough, sometimes hirsute at the apex; blades 6′–32′ long, 2½″–10″ wide, smooth or rough; panicle 6′–16′ long, 1′–2½′ wide; spikelets crowded, equalling or exceeding the basal hairs, the outer 2 scales 3½″–4½″ long, pilose with long hairs, the fourth scale 2-cleft at the apex, the teeth long and subulate, the awn 7″–12″ long, the included portion spiral at the base, the remainder loosely spiral.

In moist soil, Delaware (?) and Maryland to Florida and Texas. Sept.–Oct.

3. Erianthus saccharoìdes Michx. Plume-grass. Fig. 260.

Erianthus saccharoides Michx. Fl. Bor. Am. 1: 55. 1803.

Erianthus compactus Nash, Bull. Torr. Club 22: 419. 1895.

Culms robust, erect, 3°–10° tall, barbed at the nodes, the summit and the axis of the panicle densely pubescent with appressed long rigid silky hairs. Sheaths glabrous or sparingly hairy below, densely pubescent at the throat with long more or less spreading silky hairs; blades glabrous or appressed-pubescent, 6′–2° long, ¼′–1′ wide, long-acuminate, somewhat narrowed towards the base; panicle lax, broadly oblong, 5′–15′ long, 2′–4′ wide, its branches 2′–4′ long, slender; outer scales of the spikelet about 2″–3″ long, a little exceeding the pedicel and about one-half as long as the basal hairs, lanceolate, acuminate; inner scales shorter, the awn 10″–12″ long, straight, scabrous.

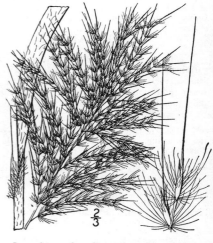

$\frac{2}{3}$

In moist sandy soil, southern New Jersey to Maryland, south to Florida and Texas. Also in Cuba. Aug.–Sept. Gama, or sesame grass.

4. Erianthus brevibàrbis Michx. Short-bearded Plume-grass. Fig. 261.

Erianthus brevibarbis Michx. Fl. Bor. Am. 1: 55. 1803.

Erianthus saccharoides sub-sp. *brevibarbis* Hack. in D. C. Monog. Phan. 6: 131. 1889.

Culms stout, erect, 3°–5° tall, the nodes naked or scantily barbed, the summit and axis of the panicle smooth or scabrous. Sheaths hirsute at the summit; blades rough, 6′–18′ long, 3″–5″ wide, acuminate; panicle linear-oblong, 8′–12′ in length, 1′–1½′ wide, the branches erect, 2′–5′ long; outer scales of the spikelet 4″–4½″ long, twice the length of the pedicel and equalling or twice as long as the basal hairs, lanceolate, long-acuminate; inner scales shorter, the awn 9″–12″ long, straight, scabrous.

In moist soil, Delaware to Florida, thence west to Louisiana. Autumn.

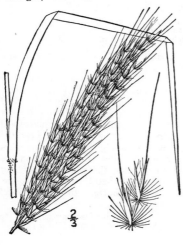

$\frac{2}{3}$

5. SCHIZACHÝRIUM Nees, Agrost. Bras. 331. 1829.

Annual or perennial grasses, tufted or from rootstocks, with flat or involute leaf-blades, and spikelike racemes, singly disposed, terminating the culm or its branches. Internodes of the articulated rachis cup-shaped or crowned at the apex with a toothed or bifid appendage. Spikelets in pairs at each node of the frequently hairy rachis, one sessile, the other pedicellate. Sessile spikelet dorsally compressed, of 4 scales; first scale 2-keeled, with the margins infolded; second scale 1-keeled; fourth scale usually 2-cleft at the apex, often almost to the base, bearing a perfect usually geniculate awn, the spiral column usually straight. Pedicellate spikelet flowerless, of 1 or 2 scales, rarely of 4 scales and bearing a staminate flower, or wanting. Stamens usually 3, very rarely 1 or 2. Styles distinct. Stigmas plumose. [Name Greek, referring to the deeply cleft flowering scale.]

About 35 species, mainly in tropical and warm temperate regions. Besides the following, others occur in the southern and southwestern parts of the United States. Type species: *Andropogon brevifolius* Sw.

Hairs at the apex of the rachis internodes short, ½″–1½″ long; plant usually green or purplish, rarely glaucous. 1. *S. scoparium*.
Hairs at the apex of the rachis internodes 2″–2½″ long; plant glaucous, the leaf-sheaths much compressed. 2. *S. littorale*.

1. Schizachyrium scopàrium (Michx.) Nash. Broom Beard-grass. Fig. 262.

Andropogon scoparium Michx. Fl. Bor. Am. 1 : 57. 1803.
Schizachyrium scoparium Nash, in Small, Fl. SE. U. S. 59. 1903.

Culms simple or much-branched, 1½°–4½° tall; sheaths smooth or scabrous, sometimes glaucous, glabrous or pubescent; blades 6′–1½° long, 1″–4″ wide, acuminate, glabrous or pubescent; racemes 1′–2½′ long, loose, on long-exserted slender peduncles; rachis slender, flexuous, the joints and pedicels ciliate with spreading hairs; outermost scale of sessile spikelet 2½″–3½″ long, acuminate, scabrous; awn spiral, more or less bent at point of exsertion, 4″–8″ long, scabrous; pedicellate spikelet reduced to a single awn-pointed scale.

In dry sandy fields. Maine to Saskatchewan and Montana, south to Florida, Texas and New Mexico. Broom-grass or -sedge; Bunch-grass; Red-stem or Blue-stem-grass; Big Blue-joint. Aug.–Oct.

2. Schizachyrium littoràle (Nash) Bicknell. Seacoast Beard-grass. Fig. 263.

Andropogon littoralis Nash, in Britton, Man. 69. 1901.
Schizachyrium littorale Bicknell, Bull. Torrey Club 35 : 182. 1908.

A densely tufted perennial, the innovations with glaucous leaves with much-compressed sheaths. Culms 2½°–3½° tall, compressed, branched; sheaths rough, keeled; blades up to 8′ long, 1½″–3½″ wide, rough, acute, strongly keeled; racemes usually 1′–1½′ long, the rachis commonly straight, the internodes long-ciliate on the margins, the hairs at the apex 2″–2½″ long, the pedicels, which are usually recurved, long-ciliate; sessile spikelets 4″–5″ long, linear-lanceolate, glabrous, the fourth scale shortly 2-toothed at the apex, ciliate, the awn 5″–7½″ long, the brown column tightly spiral, barely if at all exserted from the scales; pedicellate spikelet a single awned scale.

In sand along the coast, Nantucket to New York, south to Virginia. Summer and fall.

6. ANDROPÒGON L. Sp. Pl. 1045. 1753.

Perennial grasses with usually long narrow leaves, and terminal and axillary racemes. Spikelets in pairs at each node of the jointed hairy rachis, one sessile and perfect, the other with a pedicel, staminate, empty, or reduced to a single scale, or sometimes wanting. Perfect spikelet consisting of 4 scales, the outermost coriaceous, 2-keeled, the second keeled and acute, the two inner hyaline, the fourth more or less awned and subtending a palet and perfect flower. Stamens 1–3. Grain free. [Greek, in allusion to the bearded rachis.]

About 150 species, widely distributed in tropical and temperate regions. Besides the following, some 25 others in southern and western North America Type species: *Andropogon hirtum* L.

Pedicellate spikelets empty, of 1 or 2 scales, much smaller than the sessile spikelets, or wanting.
  Stamen 1; racemes, or some of them, included in the spathes; rachis internodes slender.
    Sheaths at the upper part of the culm not enlarged; racemes equally exserted.
      Inflorescence oblong; branches divided into corymbiform masses.   1. *A. glomeratus.*
      Inflorescence long, linear, little divided, not corymbiform.   2. *A. virginicus.*
    Sheaths at the summit or upper part of the stem much enlarged; racemes on one of the
      branches exserted much beyond the others.   3. *A. Elliottii.*
  Stamens 3; racemes usually exserted beyond the spathes; rachis internodes stouter.
    Intercarinal space of the first scale of the sessile spikelet nearly nerveless; terminal hairs
      of the internodes about twice their length.   4. *A. ternarius.*
    Intercarinal space with 2 or 3 nerves running the length of the scale; terminal hairs of the
      internodes about equaling them in length.   5. *A. Cabanisii.*
Pedicellate spikelets staminate, of 3 or 4 scales, equaling or exceeding the sessile spikelets.
  Rachis internodes copiously pubescent with long hairs.
    Awn perfect, with a defined column; culms tufted, or with short rootstocks.
      Sessile spikelets hispidulous; rachis hairs 1″ long or less.   6. *A. furcatus.*
      Sessile spikelets mostly glabrous; rachis hairs 1½″–2″ long, yellow.   7. *A. chrysocomus.*
    Awn imperfect, rarely spiral at the base; rootstocks long, horizontal.   8. *A. Hallii.*
  Rachis internodes glabrous, or with a few weak crimped hairs.   9. *A. paucipilus.*

1. Andropogon glomerátus (Walt.) B. S. P. Bushy Beard-grass. Fig. 264.

Cinna glomerata Walt. Fl. Car. 59. 1788.
Andropogon macrourum Michx. Fl. Bor. Am. **1**: 56. 1803.
Andropogon glomeratus B. S. P. Prel. Cat. N. Y. 67. 1888.
Andropogon corymbosus Nash, in Britton, Man. 70. 1901.

Culms erect, 1½°–3° tall, smooth, simple below, much branched above, upper nodes of branches barbed; sheaths compressed, scabrous, glabrous or pubescent; leaves 1″–2½″ wide, scabrous, long-acuminate, the basal two-thirds as long as to equalling the culm, those of the culm 6′–12′ long; branches elongated, forming a compact terminal inflorescence; racemes in pairs, 10″–15″ long, loose, protruding from the side or exserted from the apex of the scabrous spathes; rachis flexuous, the joints and pedicels pubescent with long spreading silky hairs; outermost scale of sessile spikelet about 2″–2½″ long; awn 6″–9″ long, scabrous; pedicelled spikelet reduced to a single scale or wanting.

Damp soil, Nantucket to southern New York, south to Florida and Mississippi. Sept.–Oct. Indian Beard-grass, Brook-grass, Bushy Blue-stem.

2. Andropogon virgínicus L. Virginia Beard-grass. Broom-sedge. Fig. 265.

Andropogon virginicus L. Sp. Pl. 1046. 1753.
Cinna lateralis Walt. Fl. Car. 59. 1788.
A. dissitiflorus Michx. Fl. Bor. Am. **1**: 57. 1803.
A. vaginatus Ell. Bot. S. C. & Ga. **1**: 148. 1817.
A. tetrastachyus Ell. Bot. S. C. & Ga. **1**: 150. 1817.

Culms erect, smooth, 1½°–4° tall, simple at base, branching above; sheaths, at least the lower, more or less hirsute; blades 6′–1½° long, 1″–3″ wide, long-acuminate, scabrous on the margins, usually hirsute above; branches of culm short, forming a loose and elongated inflorescence; racemes in pairs, occasionally 3 or 4, 10″–30″ long, loose, protruding from the side of the spathes, the rachis flexuous, slender, the joints and pedicels pubescent with long spreading silky hairs; outermost scale of sessile spikelet 1½″–2″ long; awn 4″–9″ long, straight, scabrous; pedicellate spikelet generally wanting, occasionally a rudimentary scale present.

In dry or moist fields, Massachusetts to Illinois, Florida and Texas; in the Bermudas, Bahamas and tropical America. Aug.–Sept.

3. Andropogon Ellióttii Chapm. Elliott's Beard-grass. Fig. 266.

Andropogon Elliottii Champ. Fl. S. States, 581.   1860.

Culms erect, 1°–3° tall, simple or sparingly branched above, the branches strongly bearded at the upper nodes; sheaths glabrous or loosely villous, the lower narrow, the upper elongated, inflated, imbricated; blades 2'–10' long, ½''–2½'' wide; racemes in pairs, rarely in 3's, 1'–2' long, loose, some of them long-exserted; rachis slender, flexuous, its joints and the pedicels pubescent with long spreading silky hairs; outermost scale of the sessile spikelet 1½''–2'' long, scabrous on the keel; awn 6''–9'' long, scabrous; pedicellate spikelet a minute scale or wanting.

In dry or moist places, southern New Jersey to Missouri, south to Florida and Texas. Aug.–Sept.

3/4

4. Andropogon ternàrius Michx.   Silvery Beard-grass.   Fig. 267.

2/3

Andropogon ternarius Michx. Fl. Bor. Am. **1**: 57.   1803.
Andropogon argenteus Ell. Bot. S. C. & Ga. **1**: 148.   1817.   Not DC. 1813.
Andropogon argyraeus Schultes, Mant. **2**: 450.   1824.
Andropogon Belvisii Desv. Opusc. 67.   1831.
Andropogon mississippiensis Scribn. & Ball, Bull. U. S. Dep. Agr. Agrost. **24**: 40.   1901.

Culms erect, 2°–4° tall, simple at base, generally much branched above; sheaths somewhat compressed, glabrous or pubescent; basal leaves 6'–1°; upper 2'–8' long, 1''–1½'' wide, acuminate, smooth to scabrous above, glabrous or pubescent beneath; racemes 1'–2' long, on more or less exserted slender peduncles; joints of the rachis and pedicels pubescent with long silky white spreading hairs; outermost scale of sessile spikelet 2½''–3½'' long, acuminate, scabrous; awn loosely spiral, 6''–12'' long, scabrous; pedicellate spikelet reduced to a minute lanceolate acuminate scabrous scale, which is early deciduous.

In dry sandy soil, Delaware to Missouri, south to Florida and Texas.   Sept.   Silver-beard.

5. Andropogon Cabanísii Hack.   Cabanis' Beard-grass.   Fig. 268.

Andropogon Cabanisii Hack. Flora **68**: 133.   1885.

Culms 2°–3½° tall, the branches in 1's or 2's; sheaths smooth or a little roughened; blades 10' long or less, 1''–2'' wide, smooth beneath, rough above; racemes in pairs on a peduncle which is densely barbed at the apex, 1½'–3' long, grayish, the hairs at the apex of the internodes about equalling them, rather scant; sessile spikelets 2½''–3½'' long, broadly lanceolate, tapering from the middle, the first scale strongly hispidulous and 2–5-nerved between the keels, the nerves running the entire length of the scale, the fourth scale bearing a very slender awn 1'–1¼' long, slightly spiral at the base; pedicellate spikelets of a single hispidulous scale.

In sandy places, Pennsylvania (according to Hackel) and Florida.

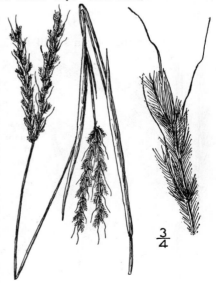

3/4

6. Andropogon furcàtus Muhl.   Forked Beard-grass.   Fig. 269.

Andropogon furcatus Muhl.; Willd. Sp. Pl. 4: 919.   1806.
Andropogon provincialis subvar. *furcatus* Hack. in DC. Mon. Phan. **5**: 442. 1889.

Culms erect, stout, smooth and glabrous, 3°–6° tall, simple at base, branched above.   Leaves glabrous or hirsute; blades smooth or rough, 6′–18′ long, 2″–7″ wide, acuminate; racemes 2–6, in pairs or approximate at the summit, 2′–5′ long; joints of rachis and pedicels ciliate with short hairs; outermost scale of sessile spikelet 3½″–5″ long, twice the length of the rachis-joints, scabrous; awn 3½″–7″ long, perfect; pedicelled spikelet consisting of 4 scales.

In dry or moist soil. Maine to Assiniboia, south to Florida and Texas and northern Mexico. Aug.–Sept. Broom-grass, Big Blue-stem, Blue-joint.

7. Andropogon chrysócomus Nash.   Yellow-haired Beard-grass.   Fig. 270.

Andropogon chrysocomus Nash, in Britton, Man. 70. 1901.

A tall usually stout grass, with extravaginal innovations.   Culms 2°–5° tall, the branches in 1's–3's; sheaths smooth and glabrous; blades up to 1° long, 3½″ wide or less, smooth beneath, a little roughened above; racemes in 2's–4's, 2′–3½′ long, stout, long-exserted, the hairs of the internodes and pedicels 1½″–2″ long, usually yellow; sessile spikelets 4″–5″ long, lanceolate, barbed at the base with hairs about ½″ long, the first scale hispid on the keels, the intercarinal space caniculate-depressed and hispidulous toward the apex, the fourth scale bearing a perfect geniculate awn 5″–6″ long; pedicellate spikelets equalling the sessile ones, awnless.

In dry usually sandy places, Nebraska and Colorado to Texas.   Summer and fall.

8. Andropogon Hàllii Hack.   Hall's Beard-grass.   Fig. 271.

Andropogon Hallii Hack. Sitz. Akad. Wiss. Wien, **89**: 127. 1884.
Andropogon geminatus Hack.; Beal, Grasses N. A. **2**: 55. 1896.

Stems from a creeping rootstock, 3°–6° tall, simple at base, branched above, smooth, more or less glaucous; leaves glaucous; blades 1° or less long, 3″–5″ wide, smooth; racemes 2 or 3, 1′–4′ long, the lateral ones often included in the spathes; joints of rachis and pedicels pubescent with spreading silky white or yellow hairs of about their own length; outermost scale of sessile spikelet 4″–6″ long, acuminate, glabrous at base, from sparingly to copiously silky-pubescent toward the apex, awnless or with a glabrous imperfect awn shorter than the scale; pedicellate spikelet consisting of 4 scales, the outermost generally larger than the corresponding scale of the sessile spikelet and subtending a palet and three stamens.

Dry sandy soil, North Dakota and Wyoming to Texas and Arizona. Aug.–Sept.   Turkey-foot Grass.

9. Andropogon paucípilus Nash. Few-haired Beard-grass. Fig. 272.

Andropogon paucipilus Nash, in Britton, Man. 70. 1901.

A glabrous perennial. Culms up to $3\frac{1}{2}°$ tall, sparingly branched above; blades erect, firm, usually somewhat glaucous, long-acuminate, the lower 8′–12′ long and $2\frac{1}{2}''$–$3\frac{1}{2}''$ wide; racemes in pairs, 2′–3′ long, the rachis internodes glabrous, or the margins with a few long weak crimped hairs; sessile spikelets lanceolate, about 5″ long, acuminate, the first scale sulcate on the back, the intercarinal space 2-nerved, the fourth scale with an imperfect awn less than $\frac{1}{2}$ its length; pedicellate spikelets staminate, a little smaller than the sessile ones, the first scale 9-nerved, not sulcate, the pedicels sparsely pilose with long weak crimped hairs.

In sand, Nebraska and Montana. Summer and fall.

$\frac{3}{4}$

7. AMPHÍLOPHIS Nash, in Britton, Man. 71. 1901.

Perennial grasses with usually flat leaf-blades and showy, often silvery white, panicles, the axis short or elongated. Racemes usually numerous, the internodes of the rachis and the pedicels with manifestly thickened margins, the median portion thin and translucent, the margins ciliate with long hairs. Spikelets dorsally compressed. Sessile spikelets of 4 scales, perfect, or rarely the lower pair or pairs staminate or empty; first scale 2-keeled, the margins narrowly inrolled; second scale 1-nerved; fourth scale stipe-like, the blade wanting, merging into a usually geniculate perfect, rarely imperfect, awn, or the awn rarely wanting. Pedicellate spikelets awnless, staminate and similar to the sessile ones, or empty and smaller than them. Stamens 3. Styles distinct. Stigmas plumose. [Greek, in reference to the hairs surrounding the spikelets.]

A genus of about 15 species. Besides the following several others occur in the United States. Type species: *Andropogon Torreyanus* Steud.

1. Amphilophis saccharòides (Sw.) Nash. Torrey's Beard-grass. Fig. 273.

Andropogon saccharoides Sw. Prodr. 26. 1788.
Andropogon glaucus Torr. Ann. Lyc. N. Y. 1 : 153. 1824. Not Muhl. 1817.
Andropogon Torreyanus Steud. Nomencl. Ed. 2, 93. 1841.
Andropogon saccharoides var. *Torreyanus* Hack. in DC. Monog. Phan. 6 : 495. 1889.

Culms erect, $1\frac{1}{2}°$–$3\frac{1}{2}°$ tall, simple or branched, glabrous, the nodes naked or barbed; sheaths glabrous, rarely pubescent, more or less glaucous; blades 3′–10′ long, 2″–4″ wide, long-acuminate, smooth and glabrous towards the base, scabrous on margins and at the apex, glaucous; racemes $1'$–$1\frac{1}{2}'$ long, in a terminal long-exserted panicle 2′–4′ long; joints of the rachis with a thin translucent median line; outermost scale of sessile spikelet $1\frac{1}{2}''$–2″ long, about equalling the terminal hairs of the rachis-joints, lanceolate, acute, pubescent at base with long silky hairs; awn 4″–8″ long, spiral, bent, scabrous; pedicellate spikelet reduced to a single narrow scale.

In dry soil, Missouri to Kansas and Colorado, northern South America, and in Jamaica. Feather Sedgegrass. Aug.–Sept.

$\frac{3}{4}$

8. SORGHÁSTRUM Nash, in Britton, Man. 71. 1901.

Generally tall perennial grasses, with long narrow flat leaves and terminal decompound panicles. Sessile spikelets consisting of 4 scales, the two outer indurated and shining, the inner hyaline, the fourth with a perfect, rarely imperfect, awn, and subtending a palet and perfect flower, or the palet sometimes wanting. Pedicellate spikelets wanting. Stamens 3. Styles distinct; stigmas plumose. Grain free. [Greek, resembling Sorghum.]

About 12 species, in temperate and tropical countries. Type species: *Sorgastrum avenaceum* (Michx.) Nash.

Awns 3 times as long as the spikelets or less; column straight, rarely geniculate. 1. *S. nutans.*
Awns 4–5 times as long as the spikelets, the column geniculate.   2. *S. Elliottii.*

1. Sorghastrum nùtans (L.) Nash. Indian-grass. Fig. 274.

Andropogon nutans L. Sp. Pl. 1045. 1753.

Andropogon avenaceum Michx. Fl. Bor. Am. **1**: 58. 1803.

Sorghum nutans A. Gray, Man. 617. 1848.

Sorghum avenaceum Chapm. Fl. S. States, 583. 1860.

Chrysopogon avenaceus Benth; Vasey, Grasses U. S. 20. 1883.

Sorghastrum nutans Nash, in Small, Fl. SE. U. S. 66. 1903.

Culms erect, 3°–8° tall, smooth, the nodes pubescent; sheaths glabrous, or the lower pubescent; blades 2° or less in length, 2″–8″ wide, long-acuminate, scabrous; panicle 4′–12′ long; branches 2′–4′ long, slender, erect-spreading; spikelets in pairs, or in 3's at the ends of the branches, erect or somewhat spreading; first scale of sessile spikelet 3″–4″ long, acute, pubescent with long hairs; second scale glabrous; awn 5″–10″ long, the column straight.

In dry fields, Maine to Manitoba, south to Florida and northern Mexico. Aug–Sept. Wood-grass. Bushy blue-stem. Wild oat-grass.

2. Sorghastrum Elliòttii (C. Mohr) Nash. Long-bristled Indian-grass. Fig. 275.

Sorghum nutans Linnaeanum Hack. in Mart. Fl. Bras. **2³**: 276. 1883.

Chrysopogon Elliottii C. Mohr, Bull. Torrey Club **24**: 21. 1897.

Sorghastrum Linnaeanum Nash, in Small, Fl. SE. U. S. 66. 1903.

Sorghastrum Elliottii Nash, N. Am. Fl. **17**: 130. 1912.

Culms 3°–4½° tall; sheaths smooth and glabrous; blades 1½° long or less, up to 5″ wide, very rough; panicle 6′–12′ long, the apex usually nodding, its branches erect or nearly so, at least the lower ones much exceeding the internodes of the axis, 2½′–3′ long, the ultimate divisions straight; spikelets 3″–4″ long, lanceolate, deep chestnut brown at maturity, hirsute, the awn 1′–1½′ long, the column· geniculate.

In dry soil, Virginia and Tennessee to Florida and Texas.

9. HÓLCUS L. Sp. Pl. 1047. 1753.

[SORGHUM Moench, Meth. 207. 1794.]

Annual or perennial grasses with long broad flat leaves and terminal ample panicles. Spikelets in pairs at the nodes, or in 3's at the ends of the branches, one sessile and perfect, the others pedicellate, and staminate or empty. Sessile spikelet consisting of 4 scales, the outer indurated and shining, obscurely nerved. the inner hyaline, the fourth awned and subtending a small palet and perfect flower, or palet sometimes wanting. Stamens 3. Styles distinct. Grain free. [Name Greek, taken from Pliny.]

About 10 species, of wide distribution in tropical and warm-temperate regions. Type species: *Holcus Sorghum* L.

1. Holcus halepénsis L. Johnson-grass. Evergreen Millet. Fig. 276.

Holcus halepensis L. Sp. Pl. 1047. 1753.
Andropogon halepensis Brot. Fl. Lusit. 1: 89. 1804.
Sorghum halepense Pers. Syn. 1: 101. 1805.

Culms erect, 3°–5° tall, simple or sometimes branched, smooth and glabrous; sheaths smooth; blades 2° or less long, ¼′–1′ wide, long-acuminate; panicle open, ½°–1½° long, the generally whorled branches spreading and naked towards the base; outer scales of sessile spikelet 2″–3″ long, ovate-lanceolate, usually purplish, pubescent with long appressed hairs; awn readily deciduous, 4″–8″ long, more or less bent; pedicellate spikelets of 4 scales, the outer two 2½″–3½″ long, membranous, 7–9-nerved, their inrolled margins ciliate, the inner two shorter and narrower, hyaline, sometimes with staminate flowers.

In fields and waste places, New Jersey and Pennsylvania to Kansas and Arizona, south to Florida and Texas. Widely distributed by cultivation in tropical America. Native of southern Europe and Asia. July–Sept. Maiden-cane, Egyptian Millet, Cuba, Syrian or St. Mary's-grass.

10. NÀZIA Adans. Fam. Pl. 2: 31. 1763.

[Tragus Hall. Hist. Stirp. Helv. 2: 203. 1768. Lappago Schreb. Gen. 55. 1789.]

Annual grasses, diffusely branched, with flat leaves and 1-flowered deciduous spikelets, either solitary or in clusters of 3–5 in a terminal spike. Scales of spikelet 2 or 3, the outermost small or wanting, the second rigid and covered with hooked prickles, the third membranous, subtending a palet and perfect flower.

Species 2 or 3, in tropical and temperate regions. Type species: *Cenchrus racemosus* L.

1. Nazia racemòsa (L.) Kuntze. Prickle-grass. Fig. 277.

Cenchrus racemosus L. Sp. Pl. 1049. 1753.
Lappago racemosa Willd, Sp. Pl. 1: 484. 1798.
Nazia racemosa Kuntze, Rev. Gen. Pl. 780. 1891.

Culms 2′–14′ tall, erect, simple to diffusely branched, smooth below, pubescent above. Sheaths smooth and glabrous; leaves 1′–3′ long, 1″–2″ wide, acuminate, rather strongly ciliate; spike 1′–4′ long, sometimes partially included in the somewhat inflated upper sheath; spikelets 1-flowered; first scale very small, almost hyaline; second scale coriaceous, 1½″ long, acute, 5-nerved, each nerve armed with a row of hooked prickles; third scale 1″ long, keeled, sharp-pointed 1-nerved, membranous, enclosing a palet of like texture and a perfect flower.

Occasional in ballast and waste places about the Atlantic seaports. Native of Europe and Asia. July–Sept. Burdock-grass.

11. SYNTHERÍSMA Walt. Fl. Car. 76. 1788.

[Digitaria Scop. Fl. Carn. Ed. 2, 1 · 52. 1772. Not Heist. 1763.]

Annual grasses with flat leaves, and spikelets borne in pairs or sometimes in 3's, in secund racemes which are digitate, in whorls, or approximate at the summit of the culm. Racemes with the rachis angled or winged. Scales of the spikelet 4, sometimes 3 by the suppression of the lowest one; the fourth or innermost scale chartaceous, the margins hyaline and not inrolled, subtending a palet of similar texture and a perfect flower. Stamens 3. Stigmas plumose. [Greek, crop-making, in allusion to its abundance.]

About 20 species, widely distributed in temperate and tropical regions. Type species: *Syntherisma praecox* Walt.

Rachis of the racemes wingless; first scale of spikelet wanting, or rudimentary.
 Racemes short, 1'–4' long; spikelets less than 1" long.   1. *S. filiforme.*
 Racemes exceeding 4' long, rarely shorter; spikelets over 1" long.   2. *S. villosum.*
Rachis of the racemes with the lateral angles broadly winged.
 Pedicels terete, glabrous or nearly so; first scale wanting.
  Leaves pubescent; second scale ½ as long as the spikelet or less.   3. *S. serotinum.*
  Leaves glabrous; second scale nearly as long as the spikelet.   4. *S. Ischaemum.*
 Pedicels sharply 3-angled, the angles strongly hispidulous; first scale minute.
  Spikelets about 1¼" long; third scale with nerves mostly hispid.   5. *S. sanguinale.*
  Spikelets about 1½" long; third scale with the nerves smooth.   6. *S. marginatum.*

1. Syntherisma filifórme (L.) Nash.   Slender Finger-grass.   Fig. 278.

Panicum filiforme L. Sp. Pl. 57. 1753.
Digitaria filiformis Koel. Descr. Gram. 26. 1802.
Syntherisma filiformis Nash, Bull. Torr. Club, **22** : 420. 1895.

Culms erect, 1°–4° tall, slender, smooth. Sheaths hirsute; blades 1'–8' long, ½"–2" wide, erect, hirsute or glabrous on the lower surface, rough on the upper; racemes 2–5, filiform, 1'–4' long, approximate at the summit of the culm, erect or nearly so; rachis 3-angled, very slender, not winged; spikelets less than 1" long, elliptic, pubescent, in pairs, occasionally in 3's; first scale rarely present; second three-fourths as long as or equalling the third, which equals the fourth.

Dry sandy soil, New Hampshire to Michigan, south to North Carolina and Oklahoma. July–Sept. Wire-grass.

2. Syntherisma villòsum Walt.   Southern Slender Finger-grass.   Fig. 279.

Syntherisma villosum Walt. Fl. Car. 77. 1788.

Culms densely tufted, 6'–4½° tall; sheaths, at least the lower ones, hirsute; blades 3'–10' long, 1½"–3" wide, hirsute on both surfaces; racemes 2–8, usually exceeding 4' long, rarely shorter, erect or ascending; spikelets over 1" long, elliptic, usually in 3's, the first scale wanting, the second 3-nerved, the fourth scale deep chestnut brown at maturity.

Sandy soil, Virginia to Missouri, south to Florida and Texas. June–Oct.

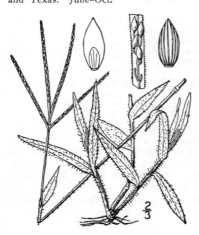

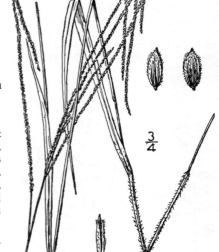

3. Syntherisma serótinum Walt.   Late-flowering Finger-grass.   Fig. 280.

Syntherisma serotinum Walt. Fl. Car. 76. 1788.
Panicum serotinum Trin. Gram. Panic. 166. 1826.

Culms slender, erect, often creeping and branching at the base, 8'–24' tall, smooth and glabrous; sheaths about one-half as long as the internodes, pilose with long spreading hairs; blades linear-lanceolate to lanceolate, 1'–4' long, 2"–4" wide, acuminate, pilose on both surfaces; inflorescence composed of 2–6 1-sided slender erect or ascending spike-like racemes 1'–4½' long, arranged singly, in pairs, or scattered and approximate; spikelets numerous, oval, about ¾" long and one-half as broad, acute, in pairs, in 2 rows on one side of a flat and winged rachis less than ½" wide; first scale wanting, the second about one-half as long as the spikelet, 3-nerved, the third scale 7-nerved, both appressed-pubescent on the margins.

Fields and roadsides, southern Pennsylvania and Delaware to Florida and Mississippi.

4. **Syntherisma Ischaèmum** (Schreb.) Nash. Small Crab-grass. Fig. 281.

Panicum lineare Krock. Fl. Sil. **1** : 95. 1787. Not L.
Panicum Ischaemum Schreb.; Schweigger, Spec. Fl.
 Erlang. 16. 1804.
Panicum glabrum Gaud. Agrost. **1** : 22. 1811.
Syntherisma linearis Nash, Bull. Torr. Club **22** : 420.
 1895.
Syntherisma humifusum Rydb. Mem. N. Y. Bot. Gard.
 1 : 469. 1900.
Syntherisma Ischaemum Nash, N. Am. Fl. **17** : 151. 1912.

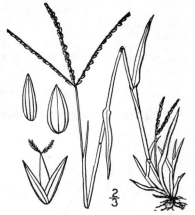

 Culms erect or decumbent, $\frac{1}{2}°$–$2°$ long, smooth and glabrous; leaves glabrous, the blades $1'$–$3'$ long, $1''$–$2''$ wide, acuminate; racemes 2–6, $2'$–$4'$ long, narrowly linear, digitate or approximate at the summit of the culm; rachis flat, winged; spikelets about $1''$ long, in pairs, sometimes in 3's; first scale rarely present, second and third as long as the fourth.

 In cultivated grounds and waste places, Nova Scotia to South Dakota, south to Florida and Kansas. Naturalized from Europe. July–Sept. Smooth Finger-grass.

5. **Syntherisma sanguinàle** (L.) Dulac. Large Crab-grass. Finger-grass. Fig. 282.

Panicum sanguinale L. Sp. Pl. 57. 1753.
Digitaria sanguinalis Scop. Fl. Carn. Ed. 2, **1** : 52. 1772.
Syntherisma praecox Walt. Fl. Car. 76. 1788.
Paspalum sanguinale Lam. Tabl. Encycl. **1** : 176. 1791.
Syntherisma sanguinale Nash, Bull. Torr. Club, **22** :
 420. 1895.

 Culms erect or decumbent, often rooting at the lower nodes, $1°$–$3°$ long, smooth. Sheaths, at least the lower, papillose-hirsute; blades $2'$–$6'$ long, $2''$–$5''$ wide, acuminate, more or less pubescent; racemes 3–10, narrowly linear, $2'$–$6'$ long, digitate or in approximate whorls at the summit of the culm; rachis flat, winged; spikelets $1\frac{1}{4}''$ long, in pairs, elliptic-lanceolate, acute, the first scale minute, rarely wanting, the second one-third to one-half as long as the spikelet, 3-nerved, the third 7-nerved.

 In cultivated or waste places, throughout North America, except the extreme north. Naturalized from Europe. Widely distributed as a weed in all cultivated regions. July–Aug. Hairy Finger-grass; Crowfoot or Pigeon-grass.

6. **Syntherisma marginàtum** (Link) Nash. Fringed Crab-grass. Fig. 283.

Digitaria marginata Link, Enum. Hort. Berol. **1** :
 226. 1827.
Panicum fimbriatum Link, Hort. Bot. Berol. **1** :
 226. 1827.
Syntherisma fimbriatum Nash, Bull. Torr. Club
 25 : 302. 1898.
Syntherisma marginatum Nash, N. Am. Fl. **17** :
 154. 1912.

 Culms $3°$ long or less, finally prostrate at the base and rooting at the lower nodes; sheaths, at least the lower, densely papillose-hirsute; blades $1\frac{1}{2}'$–$8'$ long, $2''$–$5''$ wide, more or less papillose-hirsute on both surfaces; racemes 3–10, $2'$–$7'$ long; spikelets $1\frac{1}{2}''$ long, elliptic-lanceolate, acute, in pairs, the first scale minute, the second scale 3-nerved, about $\frac{1}{2}$ as long as the spikelet, the third scale 7-nerved.

 Dry sandy soil, Maryland to Kansas, south to Florida and Texas. Also in tropical America. June–Sept.

12. LEPTOLÒMA Chase, Proc. Biol. Soc. Wash. 19: 191. 1906.

Perennial tufted grasses, with flat leaf-blades, and diffuse panicles, which break away when mature and act as tumble-weeds. Spikelets 1-flowered, solitary, or rarely in pairs. Scales 4, or sometimes 3 by the abortion of the first minute scale; second scale 3-nerved; third scale 5-7-nerved; fourth scale elliptic, acute, indurated in fruit, the delicate and hyaline margins flat, not inrolled, enclosing a palet of similar texture and a perfect flower. Grain free, enclosed in the scale and palet. [Greek, from the delicate hyaline margins of the fruiting scale.]

Species 4 or 5; besides the following typical one 3 or 4 others occur in Australia.

1. Leptoloma cognàtum (Schultes) Chase. Diffuse Crab-grass. Fig. 284.

Panicum nudum Walt. Fl. Car. 73. 1788?
Panicum divergens Muhl. Gram. 120. 1817. Not H.B.K. 1815.
Panicum cognatum Schultes, Mant. 2: 235. 1824.
Panicum autumnale Bosc; Spreng. Syst. 1: 320. 1825.
Leptoloma cognatum Chase, Proc. Biol. Soc. Wash. 19: 192. 1906.

Culms erect or decumbent, 1°-2° tall, generally much branched at the base, slender. Sheaths shorter than the internodes, the upper glabrous, the lower sometimes densely pubescent; leaves 1½'-4' long, 1"-3" wide, ascending, acuminate, glabrous; panicle 5'-12' long, bearded in the axils, the lower branches 4'-8' long, at first erect with the lower portion included in the upper sheath, finally exserted and widely spreading at maturity; spikelets lanceolate, about 1½" long, acuminate, glabrous or pubescent, on capillary pedicels of many times their length; first scale minute; second and third equal, acute, glabrous or sometimes villous, the fourth lanceolate, 1¼" long.

In dry soil, Illinois to Florida, Minnesota, Kansas and Arizona. Recorded from New Hampshire. July-Sept.

13. ERIÓCHLOA H.B.K. Nov. Gen. 1: 94. 1815.

[Helopus Trin. Fund. Agrost. 103. 1820.]

Perennial grasses with flat leaves, and short-pedicelled spikelets borne in secund spikes, which form a terminal panicle. Spikelets with an annular callus at the base and articulated to the pedicel. Scales 3, the two outer membranous, acute, the inner one shorter, indurated, and subtending a palet and a perfect flower. Stamens 3. Styles distinct. Stigmas plumose. Grain free. [Greek, signifying wool-grass.]

Species about 10, in tropical and temperate countries
Type species: *Eriochloa distachya* H.B.K.

1. Eriochloa punctàta (L.) W. Hamilton. Dotted Millet. Fig. 285.

Milium punctatum L. Amoen. Acad. 5: 392. 1759.
Eriochloa polystachya H.B.K. Nov. Gen. 1: 95. *pl. 31.* 1815.
Eriochloa punctata W. Hamilt. Prodr. Pl. Ind. Occ. 5. 1825.

Culms erect or ascending, 1°-3° tall, glabrous. Sheaths glabrous or sometimes pubescent; ligule a fringe of short white hairs; leaves 2'-10' long, 2"-3" wide, acuminate, glabrous or pubescent; spikes 4-25, 1'-2' long, sessile or nearly so; rachis pubescent; spikelets about 2" long, ovate-lanceolate, acuminate; outer scales pubescent with appressed silky hairs, the first a little exceeding the second, the third about 1" long, rounded at the apex and bearing a pubescent awn about ½" long.

Nebraska and Missouri to Mexico. Widely distributed in tropical America. Everlasting-grass.

14. ANÁSTROPHUS Schlecht. Bot. Zeit. 8: 681. 1850.

Perennial grasses, often with long creeping stolons which are thickly clothed with leaves bearing short blades, and erect stems. Spikes 1-sided, in pairs at the summit of the stem, or sometimes with an additional one a short distance below, or occasionally in scattered whorls, the rachis winged. Spikelets elliptic to lanceolate, obtuse or acute, glabrous or pubescent, singly disposed, articulated below the empty scales. Scales 3, the outer 2 membranous, 2–several-nerved, the third scale with its opening turned toward the rachis, chartaceous in flower, becoming indurated in fruit, enclosing a palet of similar texture and a perfect flower. Stamens 3. Styles distinct. Stigmas plumose. [Greek, in reference to the position of the spikelets.]

About 12 species, distributed in warm temperate and tropical regions. Type species: *Paspalum platyculmum* Thouar.

Spikelets not exceeding 1¼″ long, pubescent.   1. *A. compressus.*
Spikelets 2″–3″ long, glabrous.   2. *A. furcatus.*

1. Anastrophus compréssus (Sw.) Schlecht. Flat Joint-grass. Fig. 286.

Milium compressum Sw. Prod. 24. 1788.
Paspalum tristachyon Lam. Ill. 1: 176. 1791.
P. platycaule Poir. in Lam. Encycl. 5: 34. 1804.
Paspalum compressum Ness, in Mart. Fl. Bras. 2: 23. 1829.
Anastrophus compressus Schlecht.; Doell, in Mart. Fl. Bras. 2²: 102. 1877.

Stolons numerous, leafy, sometimes 2° long. Culms 4′–3° tall, slender, compressed, glabrous; sheaths loose; blades glabrous, sometimes ciliate, obtuse, those of the culm 2′–4′ long, 2″–4″ wide, those of the stolons about 1′ long, 1″–2″ wide; racemes in pairs, approximate at the summit of the long and slender stalk, or sometimes with an additional one below, 1′–4′ long; spikelets about 1″ long, obtuse or acute, the outer scales 3–5-nerved, or 2–4-nerved by the suppression of the midnerve.

Fields and roadsides, Virginia (?), Georgia and Florida to Louisiana. Widely distributed in tropical America. Carpet-grass. Louisiana-grass. Aug.– Sept.

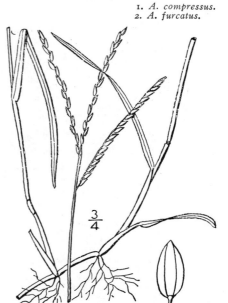

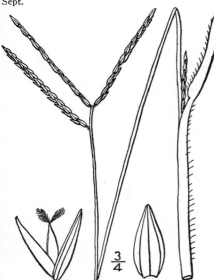

2. Anastrophus furcàtus (Fluegge) Nash. Flat Crab-grass. Fig. 287.

Paspalus furcatus Fluegge, Gram. Monog. 114. 1810.
Paspalum Elliottii S. Wats. in A. Gray, Man. Ed. 6, 629. 1890.
Paspalum paspaloides Scribn. Mem. Torrey Club 5: 29, in part. 1894. Not *Digitaria paspalodes* Michx. 1803.
A. paspaloides Nash, in Britton, Man. 75. 1901.
Axonopus furcatus Hitchc. Rhodora 8: 205. 1906.

Culms 1°–3½° tall; sheaths much compressed, keeled, glabrous or pubescent; blades 1° long or less, 3″–8″ wide, linear, glabrous, ciliate on the margins, or hirsute on both surfaces; spikes ascending, 1′–6′ long; spikelets 2″–3″ long and about ¾″ wide, acute, the first scale 5-nerved, the second usually 4-nerved by the suppression of the midnerve, the third scale ½–⅔ as long as the others.

In fields and woods, Maryland to Florida, thence west to Texas. July–Aug.

15. PÁSPALUM L. Syst. Ed. 10, 2: 855. 1759.

Perennial grasses of various habit, with generally flat leaves and 1-flowered spikelets borne in 2 rows in 1-sided spikes, which are single, in pairs, or panicled. Spikelets oblong to orbicular, flat on the inner surface, convex on the outer. Scales 3, rarely 2 by the absence of the outermost, the outer ones membranous, the inner one indurated and subtending a palet and perfect flower. Stamens 3. Styles separate; stigmas plumose. Grain ovoid or oblong, free. [Greek name for some grass, used by Hippocrates.]

About 160 species, widely distributed in tropical and temperate regions, most abundant in America. Type species: *Panicum dissectum* L.

Wings of the rachis broad, membranous, inrolled on the spikelets.
  Racemes 20 or more; spikelets elliptic, acute, about ¾" long, pubescent. 1. *P. mucronatum.*
  Racemes less than 10; spikelets oval, obtuse, 1" long or more, glabrous. 2. *P. dissectum.*
Wings of the rachis narrow, not membranous nor inrolled on the spikelets.
  Racemes 1–many, never conjugate at the summit of the culm, always distant one from the other.
    One to several raceme-bearing naked branches arising from the uppermost leaf-sheath.
      Leaf-blades glabrous on the lower surface, midnerve sometimes sparsely pubescent.
        Spikelets about ¾" long; leaves more numerous at the base of the stem.
                                                3. *P. longipedunculatum.*
        Spikelets 1" long or more; leaves scattered.
          Blades firm, appressed-pubescent on the upper surface with short hairs.
                                         4. *P. stramineum.*
          Blades thin, membranous, glabrous on the upper surface. 5. *P. ciliatifolium.*
      Leaf-blades densely pubescent on both surfaces.
        Pubescence of copious soft short appressed hairs.
          Stems prostrate; lower leaf-blades 4' long or less. 6. *P. psammophilum.*
          Stems erect; lower leaf-blades 6' long or more. 7. *P. Bushii.*
        Pubescence of long stiff spreading hairs.
          Culms long-hirsute below the racemes. 8. *P. pubescens.*
          Culms glabrous.
            Spikelets less than 1" long. 9. *P. setaceum.*
            Spikelets over 1" long.
               Racemes on the main culm 1, or sometimes 2; spikelets glabrous; leaf-blades
                 not thick. 10. *P. Muhlenbergii.*
               Racemes on the main culm 2 or 3; spikelets usually more or less pubescent; leaf-
                 blades thick. 11. *P. debile.*
  No branches arising from the uppermost leaf-sheath.
    Spikelets rounded or obtuse at the apex; not ciliate on the margins.
      Spikelets 1½" long or less.
        Fruiting scale white to yellowish.
          Spikelets singly disposed.
            Spikelets oval, ½ as thick as broad or more, the outer scales firm.
               Leaf-sheaths glabrous, or sometimes ciliate on the margins; blades glabrous, or
                 sparingly hirsute on the upper surface. 12. *P. laeve.*
               Leaf-sheaths as well as the blades hirsute. 13. *P. plenipilum.*
            Spikelets circular or nearly so, ¼–⅓ as thick as broa the outer scales thin and
               usually wrinkled. 14. *P. circulare.*
          Spikelets in pairs. 15. *P. laeviglume.*
        Fruiting scale deep seal brown. 16. *P. Boscianum.*
      Spikelets more than 1½" long.
        Leaf-blades short, the larger ones usually 6' long or less; racemes short.
                                     17. *P. difforme.*
        Leaf-blades long, exceeding 8'; racemes long. 18. *P. floridanum.*
    Spikelets acute, ciliate with very long hairs. 19. *P. dilatatum.*
Racemes conjugate at the summit of the culm, rarely in 3's or with an additional one a short distance below. 20. *P. distichum.*

1. Paspalum mucronàtum Muhl. Water Paspalum. Fig. 288.

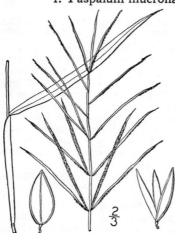

Paspalum paniculatum Walt Fl. Car. 75. 1788. Not L. 1759.

Paspalum mucronatum Muhl. Cat. 8. 1813.

Ceresia fluitans Ell. Bot. S. C. & Ga. 1: 109. 1817.

Paspalum fluitans Kunth, Rev. Gram. 24. 1829.

Culms ascending, 6'–3° long, from a floating or creeping base, branched. Sheaths very loose or inflated, smooth or scabrous, glabrous or pubescent; blades 3'–12' long, ¼'–1' wide, acuminate, scabrous; racemes 20–100, ½'–5' long, alternate or whorled, slender; rachis flat, thin, exceeding the spikelets, long-acuminate, scabrous, its margins nearly enclosing the spikelets; spikelets in two rows, ½"–¾" long, elliptic, pubescent; outer scales very thin, 2-nerved, the first one usually a little the longer.

In water or on mud, Virginia to southern Illinois and Missouri, south to Florida and Texas. Also in tropical America. Sept.

2. Paspalum disséctum L. Walter's Paspalum. Fig. 289.

Panicum dissectum L. Sp. Pl. 57. 1753.
Paspalum dissectum L. Sp. Pl. Ed. 2, 81. 1763.
Paspalum membranaceum Walt. Fl. Car. 75. 1788.
Not Lam. 1791.
Paspalum Walterianum Schultes, Mant. 2: 166. 1824.

Culms erect or ascending, much branched, smooth, creeping at the base. Sheaths a little inflated, smooth; blades 1½′–3½′ long, 2″–4″ wide, flat, smooth, acute; racemes 3–7, alternate, about 1′ long, the lower ones usually included in the upper sheath; rachis not exceeding the spikelets, flat, thin, 1″–1½″ wide, acute, smooth, many-nerved, its incurved margins partly enclosing the spikelets; spikelets about 1″ long, crowded in 2 rows, oval, obtuse, glabrous; outer scales 5-nerved; third scale lenticular, slightly shorter than the outer ones.

Moist or wet grounds, or sometimes in water, New Jersey to Florida and Texas. Sept.

3. Paspalum longipedunculàtum Le Conte. Long-stalked Paspalum. Fig. 290.

Paspalum longipedunculatum Le Conte, Journ. de Phys. **91**: 284. 1820.
Paspalum ciliatifolium brevifolium Vasey, Proc. Phila. Acad. Sci. **1886**: 285. 1886.

Stems 10′–2½° tall, leafy at the base. Sheaths glabrous, excepting on the ciliate margin; blades 1′–4′ long, 2″–4½″ wide, lanceolate to linear, glabrous, or shortly appressed-pubescent on the upper surface, ciliate along the margins and the mid-nerve; peduncles 1–2 from the upper sheath; racemes 1 or 2, 1′–3′ long, more or less curved, the rachis very narrow, more or less flexuous; spikelets in pairs, about ¾″ long, broadly obovate, the first scale 3-nerved, glabrous, the second 2-nerved by the suppression of the midnerve, glabrous.

Sandy soil, North Carolina and Kentucky to Florida. Aug.–Sept.

4. Paspalum stramíneum Nash. Straw-colored Paspalum. Fig. 291.

Paspalum stramineum Nash, in Britton, Man. 74. 1901.

A tufted branching perennial, with light yellowish green foliage, flat leaf-blades which are appressed-pubescent on the upper surface, and usually pubescent spikelets; culms 8′–3° tall; leaf-sheaths glabrous, excepting on the ciliate margins, the basal ones softly and densely pubescent; blades firm, erect, linear or lanceolate, with a few scattered long hairs in addition to the shorter pubescence on the upper surface, long-ciliate on the margins, 10′ long or less, 2½″–5″ wide; racemes 1½′–4′ long, on the main culm usually 2; spikelets in pairs, orbicular, about 1″ in diameter, the first scale 3-nerved, pubescent with short spreading glandular-tipped hairs, or sometimes glabrous, the second scale glabrous or nearly so, 2-nerved by the suppression of the midnerve.

In sandy places and fields, Nebraska and Missouri to Texas. Aug. and Sept.

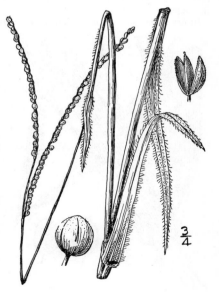

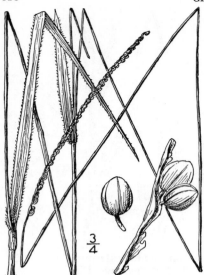

3/4

5. Paspalum ciliatifòlium Michx. Ciliate-leaved Paspalum. Fig. 292.

Paspalum ciliatifolium Michx. Fl. Bor. Am. **1** : 44. 1803.

A tufted branching perennial, with flat leaf-blades, which are glabrous excepting on the ciliate margins, and glabrous spikelets. Culms 1¼°–2½° tall; sheaths ciliate on the margin; blades 10′ long or less, 3″–8″ wide, linear to lanceolate; racemes single, or sometimes in pairs, 2′–4½′ long, the rachis ½″ wide or less; spikelets in pairs, about 1″ long and less than 1″ wide, oval to broadly obovate, the two outer scales 3-nerved, or the second one rarely 2-nerved by the suppression of the midnerve.

In rocky or sandy soil, Maryland to Florida and Mississippi. June–Aug.

6. Paspalum psammóphilum Nash. Prostrate Paspalum. Fig. 293.

Paspalum prostratum Nash, in Britton, Man. 74. 1901. Not Scribn. & Merr. 1901.
Paspalum psammophilum Nash; Hitchc. Rhodora **8** : 205. 1906.

A tufted branching softly pubescent perennial with prostrate culms forming dense mats, flat leaf-blades, and densely pubescent spikelets. Culms 1½°–3° long; sheaths softly and densely pubescent with short hairs; blades erect or nearly so, of medium texture, softly and densely pubescent on both surfaces, ciliate on the margins, lanceolate, up to 4′ long, 2½″–4″ wide; racemes on the main culm 2, rarely more or only 1, 2′–3′ long, the rachis less than ½″ wide; spikelets in pairs, 1″ long and a little less than 1″ wide, oval, the first scale 3-nerved, the second usually 2-nerved by the suppression of the midnerve.

In dry sandy soil, southern New York to Delaware.

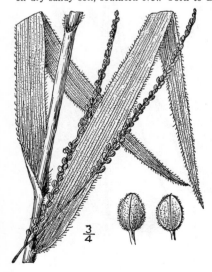

3/4

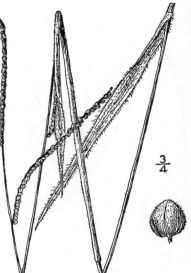

3/4

7. Paspalum Búshii Nash. Bush's Paspalum. Fig. 294.

Paspalum Bushii Nash, in Britton, Man. 74. 1901.

A tufted branching perennial, with flat ciliate leaf-blades, and pubescent spikelets. Culms erect or ascending, 1°–3° tall; sheaths, at least the lower ones, softly pubescent, ciliate on the margin; blades erect or ascending, rather firm in texture, softly and densely pubescent on both surfaces with short hairs and with a few long rather stiff hairs intermixed, lanceolate, 2′–8′ long, 2″–8″ wide; racemes usually in pairs on the main culm, 2½′–5′ long, the rachis less than ½″ wide; spikelets in pairs, about 1″ long, orbicular or nearly so, the empty scales densely pubescent, 3-nerved, or the second sometimes 2-nerved by the suppression of the midnerve.

In dry soil, Missouri and Nebraska to Texas.

8. Paspalum pubéscens Muhl.   Pubescent Paspalum.   Fig. 295.

Paspalum pubescens Muhl.; Willd. Enum. Hort. Berol. 89. 1809.

A tufted branching perennial, with flat pubescent leaf-blades, and glabrous spikelets. Culms 1¾°–2½° tall, densely pubescent below the racemes; sheaths glabrous, or sometimes pubescent on the margins or toward the apex, the basal ones sometimes pubescent all over; blades of medium texture, the pubescence long, spreading and rather stiff, linear, 10′ long or less, 1½″–3″ wide; racemes usually 1, rarely 2 on the main culm, 2½′–5′ long; spikelets in pairs, about 1″ long and ¾″ wide, broadly obovate, the 2 outer scales 3-nerved, or the second 2-nerved by the suppression of the midnerve.

In fields and dry woods, New York and New Jersey to Texas. Aug. and Sept.

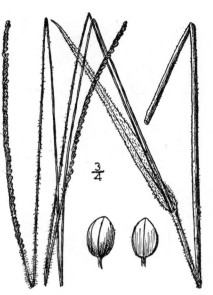

9. Paspalum setàceum Michx.   Slender Paspalum.   Fig. 296.

Paspalum setaceum Michx. Fl. Bor. Am. **1**: 43. 1803.

Stems 8′–2½° tall, slender. Sheaths and blades very pubescent, the latter 3′–8′ long, 1″–3″ wide, erect or ascending; racemes 1½′–3½′ long, more or less curved, generally solitary, occasionally 2, on a long-exserted slender peduncle, with 1 or 2 additional shorter peduncles from the same upper sheath; spikelets about ¾″ long, broadly obovate; the first scale 3-nerved, pubescent, with glandular hairs, the second scale 2-nerved, the midnerve rarely if ever present, glabrous or nearly so.

In dry fields, New Hampshire to Nebraska, Florida and Texas. Ascends to 2200 ft. in Virginia. Aug.–Sept. Beard-grass. Pitchfork-grass.

10. Paspalum Muhlenbérgii Nash.   Muhlenberg's Paspalum.   Fig. 297.

Paspalum Muhlenbergii Nash, in Britton, Man. 75. 1901.

A tufted branching perennial, with pubescent narrow leaf-blades, and glabrous spikelets. Culms at first erect, finally reclining, 1¾°–2½° long; sheaths commonly pubescent all over with long hairs, or sometimes only on the margins; blades flat, of medium texture, more or less pubescent on both surfaces with long hairs, linear to linear-lanceolate, 8′ long or less, 3½″–6″ wide; racemes single or in pairs, 2′–4′ long, the rachis less than ½″ wide; spikelets in pairs, about 1″ long and less than 1″ wide, oval or broadly obovate, the 2 outer scales 3-nerved, or the second one rarely 2-nerved by the suppression of the midnerve.

In fields or in sandy or stony ground, New Hampshire to Florida and Texas. Aug.–Oct.

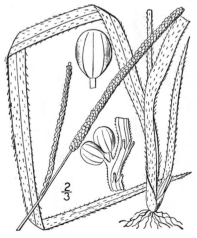

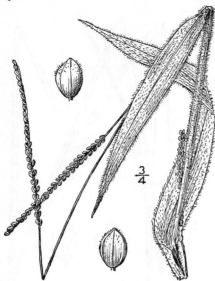

$\frac{3}{4}$

11. Paspalum débile Michx. Yellow-haired Paspalum. Fig. 298.

Paspalum debile Michx. Fl. Bor. Am. 1 : 44. 1803.
Paspalum supinum Bosc ; Poir. in Lam. Encycl. 5 : 29. 1804.
Paspalum dasyphyllum Ell. Bot. S. C. & Ga. 1 : 105. 1817.

A tufted branching yellowish green peren-nial, with flat leaf-blades which are pubescent with long yellowish hairs, and pubescent spike-lets. Culms 8′–2° tall, stout; leaves densely pubescent with long yellowish spreading hairs ; blades flat, thick, lanceolate, 8′ long or less, usually 5″–10″ broad ; racemes usually but little exserted, in 2′s or 3′s on the main culm, single on the branches, 1½′–4′ long, rather stout, the rachis about ½″ wide ; spikelets in pairs, about 1″ long and a little less than 1″ wide, broadly obovate, the first scale usually more or less pubescent with glandular-tipped hairs, 3-nerved, the second scale glabrous, commonly 2-nerved.

In dry sandy places, South Carolina to Mis-souri, south to Florida and Texas.   June–Sept.

12. Paspalum laève Michx. Field Paspalum. Fig. 299.

Paspalum laeve Michx. Fl. Bor. Am. 1 : 44. 1803.
Paspalum angustifolium Le Conte, Journ. de Phys. 91 : 285. 1820.
Paspalum australe Nash, in Britton, Man. 1039. 1901.

A nearly glabrous perennial, with flat leaf-blades and glabrous spikelets. Culms 1°–3° tall; leaf-sheaths glabrous, or hirsute on the margins, com-pressed; blades up to 16′ long, 2½″–4″ wide, erect or nearly so, often drooping at the apex, glabrous, or the upper surface more or less hairy; racemes 2–6, 1¾′–4′ long, the rachis less than ½″ wide; spikelets singly disposed, 1¼″–1½″ long and 1″–1¼″ wide, oval, the outer 2 scales firm, 5-nerved, the lateral nerves approximate.

In fields and sandy places, Maryland to Florida and Texas.

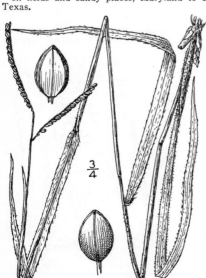

$\frac{3}{4}$

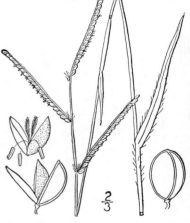

$\frac{2}{3}$

13. Paspalum plenípilum Nash. Many-haired Paspalum. Fig. 300.

Paspalum laeve pilosum Scribn. Bull. Univ. Tenn. Exp. Sta. 7 : 34. 1894. Not *P. pilosum* Lam. 1791.
Paspalum plenipilum Nash, in Britton, Man. 73. 1901.
Paspalum praelongum Nash, in Small, Fl. SE. U. S. 74. 1903.

A tufted pubescent perennial with flat leaf-blades and glabrous spikelets. Culms 1½°–3½° tall; leaf-sheaths tuberculate-hirsute with long spreading hairs, compressed; blades 1° long or less, 3″–5″ wide, erect, hirsute on both surfaces with long spreading hairs; racemes 2–4, spreading or ascending, 2′–4′ long, the rachis about ½″ wide; spikelets singly disposed, oval, about 1¼″ long and 1″ wide, the outer 2 scales 5-nerved, the lateral nerves near the margin, approximate.

In fields and along roadsides, New Jersey to Mis-souri, south to Florida and Alabama.   Aug.

14. **Paspalum circulàre** Nash. Round-flowered Paspalum. Fig. 301.

Paspalum circulare Nash, in Britton, Man. 73. 1901.

A tufted perennial with flat leaf-blades, and orbicular glabrous spikelets. Culms $1\frac{1}{2}°-3\frac{1}{2}°$ tall; leaf-sheaths tuberculate-hirsute with spreading or ascending hairs, compressed; blades erect, more or less hirsute on both surfaces, $1°$ long or less, $2\frac{1}{2}''-4''$ wide; racemes 2–4, erect or ascending, $2\frac{1}{2}''-4'$ long, the rachis about $\frac{1}{2}''$ wide; spikelets singly disposed, about $1\frac{1}{2}''$ in diameter, their thickness about one quarter their diameter, the outer 2 scales thin and usually wrinkled when dry, 5-nerved, the lateral nerves near the margin and approximate, quite distinct.

In fields, New York to Missouri, south to North Carolina and Texas. July–Sept.

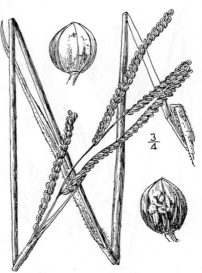

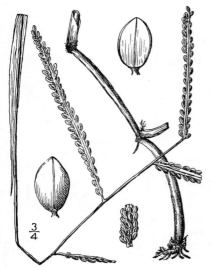

15. **Paspalum laeviglùme** Scribn. Smooth-scaled Paspalum. Fig. 302.

Paspalum remotum glabrum Vasey, Bull. Torrey Club **13** : 166. 1886. Not *P. glabrum* Poir. 1804.
Paspalum pubiflorum glabrum Vasey; Scribn. Bull. Tenn. Exp. Sta. **7** : 32. 1894.
Paspalum laeviglume Scribn.; Nash, in Small, Fl. SE. U. S. 75. 1903.

A stout glabrous perennial, usually rooting at the lower nodes, with flat leaf-blades, and glabrous spikelets. Culms $1\frac{1}{2}°-4\frac{1}{2}°$ tall, the nodes pubescent; leaf-sheaths glabrous, excepting the hirsute margins; blades $4'-16'$ long, $5''-10''$ wide, linear, glabrous on both surfaces; racemes 4–8, spreading or ascending, the lower ones commonly $2'-4'$ long; spikelets in pairs, $1\frac{1}{4}''-1\frac{1}{2}''$ long, about $1''$ broad, oval to broadly obovate, the first scale 3–5-nerved, the second 5–7-nerved.

In moist places, Maryland and Kentucky to North Carolina and Texas. June–Oct.

16. **Paspalum Bosciànum** Fluegge. Bosc's Paspalum. Fig. 303.

Paspalum virgatum Walt. Fl. Car. 75. 1788. Not L. 1753.
Paspalus Boscianus Fluegge, Gram. Monog. 170. 1810.
Paspalum purpurascens Ell. Bot. S. C. & Ga. **1** : 108. 1817.

A rather stout glabrous perennial with compressed culms, which often root at the lower nodes, flat leaf-blades, and glabrous spikelets. Culm $1\frac{1}{2}°-4°$ tall; leaf-sheaths compressed, glabrous, or the basal ones hirsute; blades of medium texture, hirsute above near the base, linear, $1°$ long or less, $1\frac{1}{2}''-5''$ wide; racemes 2–13, spreading or ascending, $1\frac{1}{2}''-4'$ long, the straight rachis $1''-1\frac{1}{4}''$ wide; spikelets in pairs, $1''-1\frac{1}{3}''$ long, $\frac{3}{4}''-1''$ wide, broadly obovate, the first scale 5-nerved, the second 3-nerved.

In meadows and moist places, Virginia and Tennessee to Florida, west to Texas. Aug. and Sept.

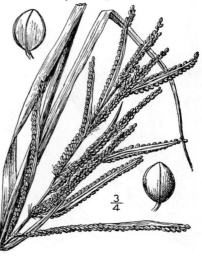

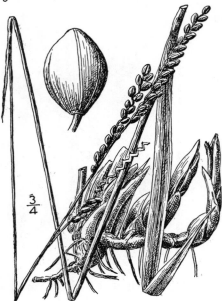

17. Paspalum difförme Le Conte. Le Conte's Paspalum. Fig. 304.

Paspalum difforme Le Conte, Journ. de Phys. **91** : 284. 1820.

A perennial, usually glaucous, grass, with short flat leaf-blades, and large glabrous spikelets. Culms $1\frac{1}{2}°-3°$ tall; leaf-sheaths glabrous, or the outer basal ones sometimes pubescent, the uppermost one usually blade-less; blades erect or ascending, thickish, glabrous, or the upper surface with long hairs, linear to linear-lanceolate, acuminate, commonly less than 6' long and 5" wide; racemes usually 2, or sometimes 1 or 3, rarely 4, erect or ascending, less than 4' long, the rachis often flexuous and about $\frac{1}{2}$" wide; spikelets singly disposed, sometimes in pairs, $1\frac{1}{2}"-2"$ long and $1\frac{1}{4}"-1\frac{1}{2}"$ wide, oval, the outer 2 scales 3-nerved, the third scale brownish when mature.

In sandy soil, New Jersey and Maryland to Florida and Texas. Aug.–Sept.

18. Paspalum floridànum Michx. Florida Paspalum. Fig. 305.

Paspalum floridanum Michx. Fl. Bor. Am. **1** : 44. 1803.
Paspalus macrospermus Fluegge, Gram. Monog. 172. 1810.
Paspalum arundinaceum Poir. in Lam. Encycl. Suppl. **4** : 310. 1816.
Paspalum floridanum glabratum Engelm.; Vasey, Contr. U. S. Nat. Herb. **3** : 20. 1892.

A tall perennial, sometimes glaucous, with long glabrous or hirsute leaves, and glabrous spikelets. Culms $3°-6\frac{1}{2}°$ tall; leaf-sheaths rather loosely embracing the culm; blades erect or nearly so, flat, rather firm, linear, $1°-2\frac{1}{2}°$ long, $3"-8"$ wide; racemes 3–6, rarely fewer, erect or neary so, $3'-6'$ long, the rachis about $\frac{3}{4}"$ wide; spikelets singly disposed or in pairs, $1\frac{3}{4}"-2\frac{1}{4}"$ long and $1\frac{1}{4}"-1\frac{3}{4}"$ wide, oval, the outer 2 scales 3-nerved.

In dry or moist soil, Delaware to Kansas, south to Florida and Texas. Sept.

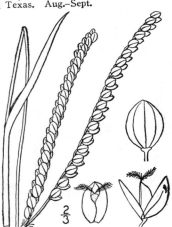

19. Paspalum dilatàtum Poir. Tall Paspalum. Fig. 306.

Paspalum dilatatum Poir. in Lam. Encycl. **5** : 35. 1804.
Paspalum ovatum Nees, Trin. Gram. Pan. 113. 1826.

Culms erect, $1\frac{1}{2}°-6°$ tall, smooth and glabrous. Sheaths compressed, smooth and glabrous; leaves 1° or less long, $1\frac{1}{2}"-6"$ wide, long-acuminate, rather scabrous on the margins, sometimes with a tuft of hairs at the base; racemes 3–8, $2'-5'$ long, erect or ascending, the rachis less than 1" wide, somewhat flexuous, scabrous; spikelets in pairs, about $1\frac{1}{2}"$ long, acute; outer scales 5–7-nerved, the first ciliate with long hairs on the margins, the second glabrous or sparsely ciliate, the third nearly orbicular, minutely punctate-striate.

In moist soil, Virginia and Tennessee to Florida and Texas. Aug.–Sept. Large Water-grass.

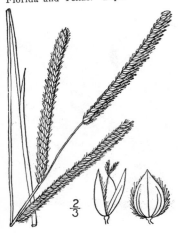

20. Paspalum dístichum L.   Joint-grass.   Fig. 307.

Paspalum distichum L. Syst. Nat. Ed. 10, 855. 1759.

Digitaria paspaloides Michx. Fl. Bor. Am. 1: 46. 1803.

Paspalum Michauxianum Kunth, Rev. Gram. 25. 1829.

Culms erect, 4′–2° tall, extensively creeping at the base. Sheats smooth, sometimes ciliate on the margins, or sparsely pubescent; blades flat, 1½′–5′ long, 1″–3″ wide, acuminate, smooth; racemes 1′– 2½′ long, in pairs, or occasionally with a third, the rachis flat, ½″–1″ wide, smooth; spikelets 1¼″–1½″ long, ovate, acute, nearly sessile in 2 rows, the outer scales 5-nerved, the first glabrous, the second appressed-pubescent, the acute third sparingly bearded at the apex.

On the seashore or along rivers, Virginia to Missouri, California and Washington, south to Florida, Texas and Mexico. Also in tropical America. Aug.– Sept. Knot-grass. Devil's-grass. Seaside-millet.

16. ECHINÓCHLOA Beauv. Agrost. 53. 1812.

Usually tall grasses, commonly annuals, with broad leaf-blades, and a terminal inflorescence consisting of 1-sided racemes. Spikelets 1-flowered, singly disposed, or in smaller racemes or clusters on the ultimate divisions of the inflorescence. Scales 4, the outer 3 membranous, hispid on the nerves, the third and usually also the second scale awned or awn-pointed, the awn often very long, the fourth scale indurated at maturity, shining, pointed, the margins thick and inrolled, enclosing a palet of similar texture, which is free at the tip, and a perfect flower. Stamens 3. Styles distinct. Stigmas plumose. [Greek, in reference to the hispid hairs of the spikelets.]

Species about 12, mostly in warm and tropical regions. Type species: *Panicum Crus-galli* L.

Sheaths glabrous.
     Spikelets 1½″ long, the second and third scales more or less awned.   1. *E. Crus-galli.*
     Spikelets 1″ long, the second and third scales merely awn-pointed.   2. *E. colona.*
Sheaths, at least the lower ones, densely papillose-hirsute.   3. *E. Walteri.*

1. Echinochloa Crús-gálli (L.) Beauv.   Barnyard-grass.   Cockspur-grass.   Barn-grass.   Water-grass.   Fig. 308.

Panicum Crus-galli L. Sp. Pl. 56. 1753.

Echinochloa Crus-galli Beauv. Agrost. 161. 1812.

Culms 2°–4° tall, often branching at base. Sheaths smooth and glabrous; blades 6′–2° long, ¼′–1′ wide, glabrous, smooth or scabrous; panicle composed of 5–15 sessile erect or ascending branches, or the lower branches spreading or reflexed; spikelets ovate, green or purple, densely crowded in 2–4 rows on one side of the rachis; second and third scales about 1½″ long, scabrous or hispid, the third scale more or less awned, empty, the fourth ovate, abruptly pointed.

In cultivated and waste places, throughout North America except the extreme north. Widely distributed as a weed in all cultivated regions. Naturalized from Europe. Aug.–Oct. Loose panic-grass.

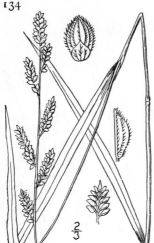

$\frac{2}{3}$

2. Echinochloa colòna (L.) Link. Jungle Rice.
Fig. 309.

Panicum colonum L. Syst. Ed. 10, 870. 1759.
Panicum Walteri Ell. Bot. S. C. & Ga. 1: 115. 1817. Not
Pursh, 1814.
Echinochloa colona Link, Hort. Berol. 2: 209. 1833.

Culms tufted, smooth and glabrous, 6′–2½° tall, often
decumbent and rooting at the lower nodes. Sheaths compressed, usually crowded; blades flat, 1′–7′ long, 1″–4″ wide;
inflorescence composed of 3–18 1-sided more or less spreading
dense racemes, ¼′–1¼′ long, disposed along a 3-angled rachis
and generally somewhat exceeding the length of the internodes; spikelets single, in pairs, or in 3's in 2 rows on one
side of the hispidulous triangular rachis, obovate, pointed,
the first scale about one-half as long as the spikelet, 3-nerved,
the second and third scales a little more than 1″ long, awnless, 5-nerved, hispid on the nerves, the fourth scale cuspidate.

Fields and roadsides, Virginia to Kansas, south to Florida
and Texas. Common in all tropical countries. March–Sept.

3. Echinochloa Wàlteri (Pursh) Nash. Salt-
marsh Cockspur-grass. Fig. 310.

Panicum hirtellum Walt. Fl. Car. 72. 1788. Not All.
1785.
Panicum Walteri Pursh, Fl. Am. Sept. 1: 66. 1814.
Panicum hispidum Muhl. Gram. 107. 1817.
Panicum Crus-galli var. *hispidum* Torr. Fl. N. Y. 2:
424. 1843.
Echinochloa Walteri Nash, in Britton, Man. 78. 1901.

Culms 3°–6° tall, robust, smooth. Sheaths, at least
the lower ones, papillose-hispid; blades 1° or more
long, ½′–1′ wide, generally smooth beneath, strongly
scabrous above; panicle 6′–18′ long, consisting of 10–40
ascending or spreading branches; spikelets ovate-
lanceolate, densely crowded in 2–4 rows on one side
of the scabrous and hispid rachis, brownish purple;
second and third scales about 1½″ long, scabrous
and hispid, tipped with upwardly barbed awns,
sometimes 10–20 times their length; fourth scale
ovate-lanceolate, acuminate.

In marshes and ditches along the coast, Ontario to
Rhode Island, Florida and Texas. Aug.–Oct.

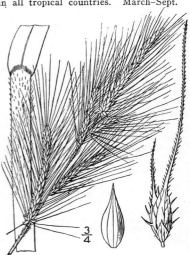

$\frac{3}{4}$

17. PÁNICUM L. Sp. Pl. 55. 1753.

Annuals or perennials of various habit, foliage and inflorescence. Spikelets 1–2-flow-
ered, when 2-flowered the lower one staminate only. Scales 4, awnless, the 3 lower membranous, empty, or the third with a staminate flower, varying in the same species; the inner
or fourth scale chartaceous, becoming indurated in fruit, the margins thick and inrolled,
shining, enclosing a palet of similar texture and a perfect flower. Stamens 3. Styles distinct.
Stigmas plumose. Grain free, enclosed in the hardened fruiting scale and palet. [Old Latin
name for some grass, probably millet, referring to its panicle, taken from Pliny.]

About 500 species, in temperate and tropical regions. The old English name, *Panic* or *Panic-
grass*, is often applied to any of the species. Type species: *Panicum miliaceum* L.

Spikelets on one side of the rachis on short pedicels forming 1-sided racemes.   I. PASPALOIDEA.
Spikelets arranged in panicles, the divisions sometimes strict and narrow but not 1-sided.
  Basal leaf-blades long and narrow, similar to those of the stem; no rosulate tufts of leaves
    in the fall; spikelets lanceolate to ovate, acute to acuminate, rarely obtuse.
    Spikelets manifestly tuberculate.   II. VERRUCOSA.
    Spikelets not tuberculate.
      Basal leaf-sheaths round or but little flattened, not keeled.
        Annuals.   III. CAPILLARIA.
        Perennial by long rootstocks or stolons.
          Rootstocks and stolons naked or with a few large scales.   IV. HALOPHILA.
          Rootstocks and stolons with numerous small broad scales.   V. VIRGATA.
      Basal leaf-sheaths much compressed, broad, keeled, often equitant. VI. AGROSTOIDEA.
  Basal leaf-blades unlike those of the culm, ovate to ovate-lanceolate; perennial by rosulate
    tufts which form in the fall at the base of the culms; spikelets elliptic to spheric, usually
    obtuse, rarely acute, never acuminate.   VII. DICHOTOMA.

I. PASPALOIDEA.

Spikelets ovate, acute, about 1¼″ long; an aquatic grass.   1. *P. hemitomon.*
Spikelets oval or obovate, obtuse, about 1½″ long.   2. *P. obtusum.*

II. Verrucosa.

A single species in our range; spikelets tuberculate.   3. *P. verrucosum.*

III. Capillaria.

Spikelets lanceolate or elliptic, 1¾″ long or less.
  Leaves glabrous.
  Leaves pubescent.   4. *P. dichotomiflorum.*
    Spikelets 1¼″ long or less, panicles broad.
      Panicle large and diffuse, occupying more than one half of the plant. 5. *P. capillare.*
      Panicle not occupying over one third of the plant, usually exserted from the sheath.
        Spikelets 1″ long or more; culms stout; blades 4″–5″ wide. 7. *P. Gattingeri.*
        Spikelets less than 1″ long; blades 1″–3″ wide.   8. *P. philadelphicum.*
    Spikelets 1½″–1¾″ long.
      Panicle narrow, its branches ascending or nearly erect.   9. *P. flexile.*
      Panicle open, diffuse, broader than long, its branches spreading or the lower ones reflexed.   6. *P. barbipulvinatum.*
Spikelets ovate, 2″–2½″ long.   10. *P. miliaceum.*

IV. Halophila.

Culms densely tufted; spikelets about 2¼″ long.   11. *P. amarulum.*
Culms scattered, from stout branching rootstocks; spikelets 2½″–3″ long.   12. *P. amarum.*

V. Virgata.

Stems tall, simple; panicle usually ample.   13. *P. virgatum.*

VI. Agrostoidea.

Rootstocks present.   14. *P. anceps.*
Rootstocks wanting.
  Fruiting scale sessile.
    Panicle broad, open, its branches spreading.
      Ligule naked; culms finally much-branched; spikelets numerous. 15. *P. agrostoides.*
      Ligule ciliate; culms simple or sparingly branched; spikelets few. 17. *P. longifolium.*
    Panicle oblong, narrow, dense, its branches erect.   16. *P. condensum.*
  Fruiting scale distinctly stalked.   18. *P. stipitatum.*

VII. Dichotoma.

Culms simple, or with basal branches and panicles only; not fasciculately branched later.
  Spikelets acute, usually over 1½″ long, the second and third scales extending beyond the fruiting scale.   19. *P. depauperatum.*
  Spikelets obtuse, rarely acutish, 1½″ long or less; outer scales not exceeding fruiting scale.
    Secondary panicles present.
      Panicle-branches erect; spikelets obtuse, about 1½″ long.   20. *P. perlongum.*
      Panicle branches spreading; spikelets acutish, 1″–1¼″ long. 21. *P. linearifolium.*
    Secondary panicles wanting.
      Leaf-sheaths glabrous or merely ciliate on the margins.
        Blades ciliate their entire length; plants yellowish green. 24. *P. ciliatum.*
        Blades not ciliate or sparingly so at the very base; plants green or grayish green.
          Spikelets less than 1″ long.
            Panicle much longer than broad; upper blades not smaller than the lower ones.
                32. *P. polyanthes.*
            Panicle nearly as long as broad, upper blades smaller than the lower ones.
                31. *P. sphaerocarpon.*
          Spikelets 1″ long or more.
            Blades linear, less than 2½″ wide.   22. *P. Werneri.*
            Blades linear-lanceolate, 3″–10″ wide.
              Panicle narrow, its branches appressed.   64. *P. xanthophysum.*
              Panicle broad, open, its branches spreading.   23. *P. Bicknellii.*
      Leaf-sheaths spreading-hirsute.
        Spikelets glabrous, ¾″ long; culms hirsute.   25. *P. strigosum.*
        Spikelets pubescent, about 1″ long; culms glabrous.   26. *P. laxiflorum.*
Culms simple only at first; later with fasciculate branches at the upper nodes.
  Middle blades of main culm less than 8″ wide, the base rounded to subcordate.
    Blades of the main culm usually elongated and narrowed at both ends.
      Plants large; culms over 3° long; leaf-blades up to 8′ long.
        Spikelets 1¼″ long, very acute, strongly nerved.   69. *P. scabriusculum.*
        Spikelets 1½″ long, barely acute, less strongly nerved.   70. *P. aculeatum.*
      Plants small; culms under 3° long; leaf-blades rarely exceeding 4′ long.
        Spikelets pubescent, narrowed at the base.
         Spikelets 1″ long; mature state of the blades involute.   28. *P. aciculare.*
         Spikelets 1¼″ long; mature blades flat, or involute only on the margins.
           Nodes not barbed; branches of the panicle spreading.   27. *P. angustifolium.*
           Nodes barbed; branches of the panicle ascending.   29. *P. consanguineum.*
        Spikelets glabrous, not narrow at the base.   30. *P. Bushii.*
    Blades of the culm not elongated nor conspicuously narrowed at the base.
      Spikelets less than 1½″ long.
        Spikelets glabrous.
          Spikelets strongly nerved, acute; blades up to 8′ long. 69. *P. scabriusculum.*
          Spikelets relatively obscurely nerved, obtuse; blades rarely exceeding 4′ long.
            Nodes densely barbed; spikelets about ¾″ long.   38. *P. microcarpon.*
            Nodes naked, or rarely the lowermost ones sparingly barbed.
              Spikelets usually less than ¾″ long.   42. *P. octonodum.*
              Spikelets 1″ long:
                Obovoid; culms puberulent.   60. *P. Nashianum.*
                Elliptic; culms glabrous.

Culms erect, the branches fasciculate near the middle.
35. *P. dichotomum.*
Culms at length long and trailing, branches fasciculate at all the nodes.
36. *P. lucidum.*
Spikelets 1¼″ long.   37. *P. yadkinense.*
Spikelets pubescent.
Sheaths glabrous, the margins ciliate, or the basal ones sometimes pubescent.
Blades velvety.   39. *P. annulum.*
Blades not velvety.
Ligule ½″ long or less.
Spikelets less than 1″ long.
Culms slender, the leaf-blades 1′ long or less; spikelets elliptic.
33. *P. ensifolium.*
Culms stouter, the leaf-blades larger; spikelets nearly globose.
Panicle longer than broad; upper leaf-blades not smaller than
the lower.   32. *P. polyanthes.*
Panicle nearly as long as broad; upper blades smaller than the
lower ones..
Blades less than 4″ wide, glabrous.   34. *P. tenue.*
Blades exceeding 4″ wide, ciliate on the margins toward
the base.   31. *P. sphaerocarpon.*
Spikelets over 1″ long.
Culms puberulent.   71. *P. Ashei.*
Culms glabrous.
Leaf-blades cordate, usually over 5″ wide. 72. *P. commutatum.*
Leaf-blades not cordate at the base, usually less than 5″ wide.
Blades erect, ciliate toward the base; fruiting scale not exceed-
ing the others.   40. *P. boreale.*
Blades spreading, glabrous; fruiting scale exceeding the other
scales.   41. *P. mattamuskeetense.*
Ligule 1″–2½″ long.
Spikelets over 1″ long.   48. *P. scoparioides.*
Spikelets less than 1″ long.
Panicle much longer than broad.
Spikelets less than ¾″ long, almost globose when mature,
sparsely pubescent with short hairs. 43. *P. paucipilum.*
Spikelets over ¾″ long, elliptic, densely pubescent with long hairs.
44. *P. spretum.*
Panicle as long as broad.   45. *P. Lindheimeri.*
Sheaths pubescent.
Sheaths merely puberulent.
Spikelets elliptic, over 1″ long.   71. *P. Ashei.*
Spikelets obovoid, 1″ long.   60. *P. Nashianum.*
Sheaths pubescent with longer hairs.
Plants velvety.
Spikelets over 1″ long.
Culms stout, a broad bare ring below each node; spikelets 1″–1¼″
long; primary panicles exceeding 3′ long. 67. *P. scoparium.*
Culms slender; no conspicuous bare ring; spikelets 1¼″–1½″ long;
panicle less than 3′ long.   **68.** *P. malacophyllum.*
Spikelets less than 1″ long.
Hairs on the sheaths long and shaggy; ligule over 1″ long.
55. *P. lanuginosum.*
Hairs on the sheaths short and inconspicuous; ligule less than ½″
long.   39. *P. annulum.*
Plants not velvety.
Spikelets ovate, pointed; blades and panicles usually 5′ long or more.
69. *P. scabriusculum.*
Spikelets not as above; blades and panicles shorter.
Spikelets less than 1″ long.
Pubescence spreading.
Blades glabrous above or nearly so. 54. *P. tennesseense.*
Blades pubescent on the upper surface.
Upper surface of blades with short appressed hairs.
47. *P. huachucae.*
Upper surface with long erect hairs 1½″ long or more.
Culms forming branches when primary panicles are
mature; spikelets ¾″ long.
Panicle 1′–1½′ long, its axis minutely pubescent.
52. *P. meridionale.*
Panicle 2′–3′ long; axis hirsute.   51. *P. implicatum.*
Culms forming branches before maturity of primary
panicles; spikelets 1″ long. 50. *P. praecocius.*
Pubescence not spreading.
Ligule obsolete; culms villous-puberulent, at least below.
Culms with 2 or 3 primary leaves, the blades 1½″ wide
or less, up to 3′–4′ long, branches crowded at the base.
53. *P. Owenae.*
Culms with 4–6 primary leaves, the blades 2″–4″ wide,
usually less than 3′ long, branches at all the nodes.
Culms erect, rigid; plants grey green.
58. *P. columbianum.*
Culms weak, fasciculately decumbent and forming
mats; plant blue-green. 59. *P. tsugetorum.*

Ligule ½" long or more ; culms, at least at the base, pubescent
    with long stiff appressed hairs.
    Blades glabrous on the upper surface ; spikelets about
        ½" long.   46. *P. leucothrix.*
    Blades pubescent on the upper surface with long hairs ;
        spikelets exceeding ½" long.   52. *P. meridionale.*
Spikelets 1" long or more.
    Pubescence spreading.
        Autumnal state erect ; pubescence papillose-hispid ; upper sur-
            face of blades glabrous or nearly so. 48. *P. scoparioides.*
        Autumnal state prostrate ; pubescence hirsute ; upper surface of
            blades long-hairy.   49. *P. villosissimum.*
    Pubescence not spreading.
        Panicle 1½'-2' long, oblong, dense ; spikelets 1" long.
                                                   56. *P. Addisoni.*
        Panicle 2½'-3½' long, broadly ovoid, open ; spikelets 1¼"
            long.   57. *P. Commonsianum.*
Spikelets 1½" long or more.
    Spikelets hirsute with long hairs, the first scale usually over ½ as long as the
        spikelet, often narrow and attenuate.   61. *P. Liebergii.*
    Spikelets glabrous, or if pubescent the hairs short.
        Leaf-blades glabrous, at least on one surafce.
            Panicle narrow, its branches usually appressed.   64. *P. xanthophysum.*
            Panicle broad, nearly as wide as long.
                Spikelets 1½" long ; blades commonly much elongated.
                    Blades rough, usually erect, not ciliate.   70. *P. aculeatum.*
                    Blades smooth, spreading, ciliate.   73. *P. mutabile.*
                Spikelets over 1½" long ; blades not elongated.
                    Blades softly and densely pubescent beneath with short hairs.
                        Culms papillose-hispid with appressed or ascending hairs ; blades
                            usually 6" wide or more.   66. *P. Ravenelii.*
                        Culms villous ; blades 5" wide or less. 63. *P. oligosanthes.*
                    Blades glabrous, or rarely puberulent beneath.
                                                                 62. *P. Scribnerianum.*
        Leaf-blades pubescent on both surfaces.
            Blades erect, less than 3" wide ; pubescence hispid.   65. *P. Wilcoxianum.*
            Blades spreading, exceeding 3" wide ; pubescence soft. 68. *P. malacophyllum.*
Middle blades of the main culm more than 8" wide, usually cordate and clasping at the base.
    Spikelets less than 1½" long.
        Blades glabrous on both surfaces.
            Spikelets less than 1" long ; culms simple.   32. *P. polyanthes.*
            Spikelets more than 1" long ; panicle nearly as broad as long ; culms branched.
                Sheaths papillose-hispid, especially the terminal ones. 76. *P. clandestinum.*
                Sheaths glabrous.   72. *P. commutatum.*
        Blades densely villous on both surfaces.   67. *P. scoparium.*
    Spikelets 1½" long or more.
        Panicle narrow, its branches appressed, rarely a little spreading. 64. *P. xanthophysum.*
        Panicle open, its branches spreading.
            Blades lanceolate, thick, glabrous above, densely pubescent on the lower surface with
                short spreading hairs.   66. *P. Ravenelii.*
            Blades thin, ovate-lanceolate, glabrous, or sometimes sparsely pubescent.
                Nodes barbed.   75. *P. Boscii.*
                Nodes naked.
                    Spikelets 1½" long ; blades rarely exceeding 8" wide. 73. *P. mutabile.*
                    Spikelets nearly 2" long ; blades exceeding 10" wide. 74. *P. latifolium.*

1. Panicum hemítomon Schult.   Maiden-cane. Simpson's-grass.   Fig. 311.

Panicum hemitomon Schult. Mant. **2**: 227. 1824.
Panicum carinatum Torr. Bost. Journ. Nat. Hist. **1**:
    137. 1835. Not Presl, 1830.
Panicum digitarioides Carpenter ; Steud. Syn. Pl. Gram.
    75. 1855.
Panicum Curtisii Chapm. Fl. S. States, 573. 1860.
    Not Steud. 1855.
Brachiaria digitarioides Nash, in Britt. Man. 77. 1901.

Glabrous, culms erect from a long and stout creeping rootstock, 3°–5° tall, simple, stout, smooth.
Sheaths smooth ; blades 4'–10' long, 4"–8" wide,
long-acuminate ; panicle linear, 6'–12' long, its
branches 1'–3' long, erect ; spikelets about 1¼" long,
ovate, acute ; first scale about one-half as long as
the spikelet, acute, 3-nerved ; second about 1" long,
5-nerved and a little exceeded by the 3-nerved third
one ; the fourth slightly shorter than the third.

In water, New Jersey to Florida and Texas. July–Aug.

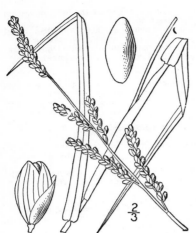

2. Panicum obtùsum H.B.K. Blunt Panic-grass. Wire- or Range-grass. Fig. 312.

Panicum obtusum H.B.K. Nov. Gen. 1: 98. 1816.

Brachiaria obtusa Nash, in Britt. Man. 77. 1901.

Glabrous, culms erect, 1°–2° tall, simple or branch-ing at base, smooth. Sheaths smooth; blades 2½′–9′ long, 1″–3″ wide, usually erect, long-acuminate; panicle linear, 2′–6′ long; branches ¾′–1½′ long, ap-pressed; spikelets about 1½″ long, crowded, oval or obovoid, obtuse, turgid; first scale shorter than the rest, obtuse, 5-nerved; second, third and fourth scales about equal, the second and third 5-nerved.

Usually in dry soil, Missouri to Arizona and Mexico. Vine Mesquite-, or Grape-vine-grass. July–Sept.

3. Panicum verrucòsum Muhl. Warty Panic-grass. Fig. 313.

Panicum verrucosum Muhl. Gram. 113. 1817.
Panicum debile Ell. Bot. S. C. & Ga. 1: 129. 1817. Not Desf. 1800.

Culms erect or decumbent, slender, generally much branched at base, 1°–6° long. Sheaths glabrous, much shorter than the internodes; ligule short, ciliate; blades 2′–7′ long, 1″–4″ wide, erect or ascending, glabrous, rough on the margins; panicle 3′–12′ long, its lower branches 2′–6′ long, naked below, strict and ascending, or lax and spreading, and smaller panicles sometimes produced at the lower part of the culm; spikelets about ¾″ long, elliptic, acutish; the first scale about one-quarter as long as the warty second and third, the fourth scale apiculate.

Moist soil, Massachusetts to Missouri, south to Florida and Texas, mostly near the coast. July–Sept.

4. Panicum dichotomiflòrum Michx. Spreading Witch-grass.   Fig. 314.

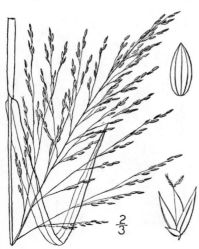

Panicum dichotomiflorum Michx. Fl. Bor. Am. 1: 48. 1803.
Panicum geniculatum Muhl. Gram. 123. 1817.

Culms at first erect, 1°–2° tall, simple, later de-cumbent and geniculate, 4°–6° long, branched at all the upper nodes. Sheaths loose, glabrous, some-what flattened; ligule ciliate; blades 6′–2° long, 2″–10″ wide, long-acuminate, scabrous on the mar-gins and occasionally on the nerves; panicle pyram-idal, 4′–16′ long, lower branches 3′–6′ long, at length widely spreading; spikelets 1″–1½″ long, crowded, lanceolate, acute, glabrous, sometimes purplish; first scale about one-fourth as long as the spikelet, enclosing its base; second and third scales about equal, acute, 5–7-nerved; fourth scale elliptic, shining, shorter than or equalling the third.

In wet soil, Maine to Nebraska, Florida, Texas and California. Also in the West Indies and continental tropical America. Formerly confused with *P. prolif-erum* Lam. Sprouting Crab-grass. July–Sept.

5. Panicum capillàre L.　Witch-grass. Tumble-weed.　Fig. 315.

Panicum capillare L. Sp. Pl. 58.　1753.

Culms erect or ascending, 1°–2° tall, simple or sometimes sparingly branched.　Sheaths papillose-hispid; blades 6′–1° long, 3″–8″ wide, more or less pubescent; terminal panicle generally 8′–14′ long, lower branches at first included in the upper sheath, finally exserted and spreading, 6′–10′ long; lateral panicles, when present, smaller; spikelets 1″–1¼″ long, acute; first scale one-fourth to one-half as long as the spikelet; second and third scales nearly equal, exceeding the fourth.

In dry soil, common as a weed in cultivated fields, Nova Scotia to North Dakota, south to Florida and Texas.　Also in Bermuda.　Old-witch-grass.　Tickle-grass.　Fool-hay.　July–Sept.

6. Panicum barbipulvinàtum Nash. Barbed Witch-grass.　Fig. 316.

Panicum barbipulvinatum Nash; Rydb. Mem. N. Y. Bot. Gard. 1 : 21.　1900.

Annual.　Culms 8′–18′ tall, smooth and glabrous; sheaths papillose-hispid with spreading hairs; blades up to 5′ long, 2″–5″ wide, lanceolate, hirsute; panicle occupying usually more than ½ of the plant, much-exserted, broader than long, its branches widely spreading or the lower ones reflexed, the pulvinus in the axils well-developed and strongly hirsute; spikelets 1½″–1¾″ long, acuminate, glabrous, the scales acuminate, the first ½ as long as the second which is longer than the third, the fruiting scale ⅔ as long as the spikelet.

In dry places, Wisconsin to British Columbia, Nebraska, Texas and California.　Aug. and Sept.

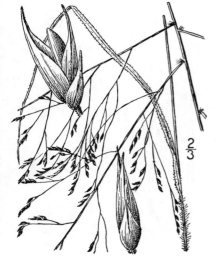

7. Panicum Gattingeri Nash.　Gattinger's Witch-grass.　Fig. 317.

Panicum capillare var. campestre Gattinger, Tenn. Fl. 94. 1887.　Not P. campestre Nees.　1826.
Panicum capillare Gattingeri Nash, in Britt. & Br. Ill. Fl. 1 : 123.　1896.
Panicum Gattingeri Nash, in Small, Fl. SE. U. S. 92.　1903.

Annual.　Culms slender, hispid, finally branched at all of the nodes and the branches again dividing, often prostrate at the base, 1°–2° long, or rarely depauperate and but a few inches high; sheaths papillose-hirsute; blades 6′ long or less, 2″–3″ wide, pubescent to nearly glabrous, erect; primary panicle 4′–6′ long, its branches ascending, the larger ones usually 2′–3′ long, the lateral panicles smaller; spikelets about 1″ long, elliptic, acute, glabrous.

In poor, often moist soil, Maine to North Carolina, Iowa and Missouri.　Aug.–Oct.　Illustrated for P. capillare L. in the first edition.

8. Panicum philadélphicum Bernh. Wood Witch-grass. Fig. 318.

Panicum capillare var. *minor* Muhl. Gram. 124. 1817.
Panicum capillare var. *sylvaticum* Torr. Fl. 149. 1824.
Not *P. sylvaticum* Lam. 1797.
Panicum diffusum Pursh, Fl. Am. Sept. **1**: 68. 1814.
Not Sw. 1788.
Panicum minus Nash, Bull. Torr. Club, **22**: 421. 1895.
Panicum *philadelphicum* Bernh.; Trin. Gram. Pan. 216.
1826.

Culms erect, or occasionally decumbent, 8′–2° long,
slender, often branched at base. Sheaths hirsute;
blades 2′–4′ long, 1″–3″ wide, erect, more or less
pubescent; panicle 4′–9′ long, its lower branches 3′–4′
long, spreading or ascending; spikelets about ¾″ long,
elliptic, acute, smooth, borne commonly in pairs at
the extremities of the ultimate divergent divisions of
the panicle; first scale about one-third as long as the
equal acute second and third ones, which barely
exceed the fourth.

In dry woods and thickets, New Brunswick to Wisconsin, Georgia, Texas and Oklahoma. Aug.–Sept.

9. Panicum fléxile (Gattinger) Scribn. Wiry Witch-grass. Fig. 319.

Panicum capillare var. *flexile* Gattinger, Tenn. Fl. 94.
1887.

Panicum flexile Scribn. Bull. Torr. Club, **20**: 476. 1893.

Culms erect, 6′–2° tall, rather stiff, slender, simple
or somewhat branched at base, bearded at the nodes.
Sheaths papillose-hispid; blades 4′–9′ long, 2″–3″
wide, erect, long-acuminate, pubescent or almost glabrous; panicle 4′–9′ long, narrowly ovoid to oblong
in outline, its branches ascending, the lower ones
2′–3½′ long; spikelets 1½″–1¾″ long, single on the
ultimate divisions of the panicle, acuminate; first
scale one-fourth to one-half as long as the spikelet;
second and third scales about equal, 5–7-nerved,
about one-third longer than the fourth scale.

In moist or dry soil, Ontario to South Dakota, south to
Florida and Texas. Aug.–Oct.

10. Panicum miliàceum L. Millet. Broom-corn Millet. Hirse-grass. Brown Millet. Fig. 320.

Panicum miliaceum L. Sp. Pl. 58. 1753.

Culms erect or decumbent, rather stout, 1° or more
tall, glabrous or hirsute. Sheaths papillose-hispid;
blades 5′–10′ long, ¼′–1′ wide, more or less pubescent;
panicle rather dense, 4′–10′ long; branches erect or
ascending; spikelets 2″–2½″ long, acuminate; first scale
about two-thirds as long as the spikelet, acuminate,
5–7-nerved; second scale 2″–2½″ long, acuminate, 13-
nerved, somewhat exceeding the 7–13-nerved acuminate
third one, which subtends an empty palet; fourth scale
shorter than the third, becoming indurated, obtuse.

In waste places, Maine to Florida, Michigan and California. Adventive from the Old World. July–Sept.

11. Panicum amàrulum Hitchc. & Chase. Southern Sea-beach Grass. Bitter Panic. Fig. 321.

Panicum amarulum Hitchc. & Chase, Contr. U. S. Nat. Herb. 15 : 96. 1910.

Smooth and glabrous, glaucous, the tufted culms $1\frac{1}{2}°-4\frac{1}{2}°$ tall; sheaths overlapping; blades $6'-1°$ long, $3''-6''$ wide, long-acuminate, thick and leathery, involute on the margins, at least toward the apex, the uppermost leaf exceeding the panicle; panicle contracted, $1°-2\frac{1}{2}°$ long, its branches erect; spikelets about $2\frac{1}{4}''$ long; the first scale one-half to two-thirds as long as the spikelet, the third somewhat longer than the second, usually with a palet and staminate flower, the fourth elliptic, about $1\frac{3}{4}''$ long.

On sea-beaches, Virginia to Florida and Mississippi; also in the Bahamas, Cuba and Jamaica. Confused in our first edition with the following species. Sept.–Nov.

12. Panicum amàrum Ell. Smaller Sea-beach Grass. Fig. 322.

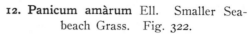

Panicum amarum Ell. Bot. S. C. & Ga. 1 : 121. 1817.
Panicum amarum var. *minor* Vasey & Scribn. Bull. U. S. Dep. Agr. Bot. 8 : 38. 1889.

Panicum amaroides Scribn. & Merr. Circ. U. S. Dep. Agr. Agrost. 29 : 5. 1901.

Glaucous and glabrous. Culms scattered, $1\frac{1}{2}°-3°$ tall, from a stout creeping rootstock; leaves thick and firm; sheaths overlapping; blades up to $1°$ long, $3''-5''$ wide, flat, or involute toward the apex; panicle $6'-1\frac{1}{2}°$ long, contracted, narrow, the short branches appressed; spikelets $2\frac{1}{2}''-3''$ long, commonly broad and stout, the first scale more than $\frac{1}{2}$ as long as the spikelet to nearly equalling it, the second and third scales about equal, all acute.

In sands along the coast, Connecticut and Long Island to Florida and Mississippi. Aug.–Oct.

13. Panicum virgàtum L. Switch-grass. Wild Red-top. Fig. 323.

Panicum virgatum L. Sp. Pl. 59. 1753.
P. virgatum var. *cubense* Griseb. Cat. Pl. Cub. 233. 1866.
Panicum virgatum var. *obtusum* Wood, Am. Bot. & Fl. 392. 1870.
Panicum virgatum var. *breviramosum* Nash, Bull. Torr. Club 23 : 150. 1896.

Culms erect from a creeping rootstock, $3°-6°$ tall, glabrous. Sheaths smooth and glabrous; blades elongated, $1°$ or more long, $3''-6''$ wide, flat, long-acuminate, narrowed toward the base, glabrous, rough on the margins; panicle $6'-20'$ long, the lower branches $4'-10'$ long, rarely shorter, spreading or ascending; spikelets ovate, acute to acuminate, $1\frac{1}{2}''-2\frac{1}{4}''$ long; first scale acuminate, about one-half as long as the spikelet, 3-5-nerved; second scale generally longer than the others, 5-7-nerved, the third similar and usually subtending a palet and staminate flower; fourth scale shining, shorter than the others.

In moist or dry soil, Maine to the Saskatchewan, south to Florida, Arizona and Costa Rica. Also in the West Indies. Thatch-grass, Wobsqua-grass, Black-bent. Aug.–Sept.

14. Panicum ánceps Michx. Beaked or Flat-stemmed Panic-grass. Fig. 324.

Panicum anceps Michx. Fl. Bor. Am. 1 : 48. 1803.
Panicum rostratum Muhl. Gram. 121. 1817.

Culms erect from a creeping scaly branched rootstock, 1½°–6° tall, much branched, compressed,
stout, smooth. Sheaths compressed, glabrous, or
the lower ones pubescent; blades 1° long or more,
2″–5″ wide, acuminate; ligule very short; panicles
pyramidal, 6′–16′ long; axis and ascending branches
scabrous; spikelets 1½″–1¼″ long, crowded, lanceolate,
acuminate, curved, longer than the scabrous pedicels;
first scale less than one-half as long as the spikelet;
second and third scales curved at the apex, much
exceeding the fourth scale which is minutely pubescent at the apex.

Moist soil, Rhode Island to Kansas, south to Florida
and Texas. July–Sept.
Panicum rhizōmatum Hitchc. & Chase, differing in
smaller spikelets and narrower panicle, occurs from Virginia to Florida and Texas.

15. Panicum agrostoìdes Spreng. Red-top Panic. Fig. 325.

Panicum agrostoides Spreng. Pugill. 2 : 4. 1815.

Culms erect, 1½°–3° tall, much branched, compressed, smooth. Sheaths compressed, glabrous, or
sometimes hairy at the throat; ligule very short,
naked; leaves 1° long or more, 2″–4″ wide, acuminate; panicles pyramidal, 4′–12′ long, terminating
the culm and branches; primary branches of the
panicle spreading, secondary appressed or divergent;
spikelets ¾″–1″ long, acute, straight, on usually
sparsely hairy pedicels; first scale 3-nerved, acute;
second and third scales 5-nerved, about twice as long
as the first and longer than the oval fourth scale,
which is sessile.

Wet ground, Maine to Minnesota, south to Florida
and Texas. July–Sept.

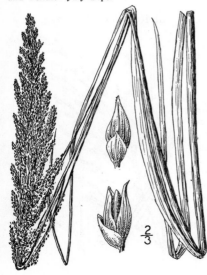

16. Panicum condénsum Nash. Dense Panic-grass. Fig. 326.

Panicum condensum Nash, in Small, Fl. SE. U. S.
93. 1903.

Culms 2½°–4½° tall; sheaths smooth and glabrous; ligule ½″ wide; blades 8′–20′ long, 4″–6″
wide, flat or folded; primary panicle up to 1°
long, narrowly oblong, the branches erect, the
spikelets densely arranged, the lower branches
naked at the base, the secondary panicles similar
but smaller and produced on long peduncles from
the upper sheaths; spikelets about 1″–1¼″ long,
glabrous, acute, on glabrous short pedicels.

Wet places and along streams, southern New Jersey
and Pennsylvania to Virginia and Florida; also in
the Bahamas, Cuba and Guadeloupe. Aug. and Sept.

17. Panicum longifòlium Torr. Long-leaved Panic-grass. Fig. 327.

Panicum longifolium Torr. Fl. U. S. 149. 1824.

Culms erect, 1°–2° tall, slender, simple, or occasionally with a single lateral panicle, flattened, smooth and glabrous. Sheaths smooth and glabrous; blades 8′–12′ long, 1″–2½″ wide, acuminate into a long, slender point, rough, glabrous; ligule short, pilose; panicles 5′–9′ long; primary branches long and slender, spreading, secondary very short, appressed, generally bearing 1–3 spikelets; spikelets 1″–1¼″ long, acute; first scale acute, about one-half as long as the acuminate second one; third scale equalling the second, acute, one-third longer than the elliptic obtuse fourth one, which is sometimes minutely pubescent at the apex.

Moist soil, Rhode Island to Maryland, Florida, Mississippi and Texas. Aug.–Sept.

18. Panicum stipitàtum Nash. Tall Flat Panic-grass. Fig. 328.

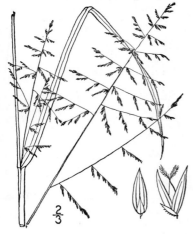

Panicum elongatum Pursh, Fl. Am. Sept. 69. 1814. Not Salisb. 1796.
Panicum stipitatum Nash, in Britt. Man. 83. 1901.

Culms erect, 3°–5° tall, much branched, stout, compressed, smooth. Sheaths smooth and glabrous, compressed; blades 1° long or more, 2″–4″ wide, acuminate, scabrous; panicles pyramidal, terminating the culm and branches, 4′–12′ long; primary branches spreading or ascending, the secondary appressed or divaricate; spikelets about 1¼″ long, crowded, acuminate; first scale acute or acuminate, one-third the length of the equal or nearly equal second and third; fourth scale narrowly elliptic, about one-half as long as the third and raised on a delicate stalk about ¼″ long.

Moist soil, southern New York and New Jersey to Kentucky, Missouri, Georgia and Louisiana. July–Sept.

19. Panicum depauperàtum Muhl. Starved Panic-grass. Fig. 329.

Panicum strictum Pursh, Fl. Am. Sept. 69. 1814. Not R. Br. 1812.
Panicum depauperatum Muhl. Gram. 112. 1817.
Panicum involutum Torr. Fl. U. S. 124. 1824.

Culms erect, 15′ tall or less, simple or branched at base. Sheaths glabrous or hirsute; blades erect, elongated, ½″–2″ wide, up to 8′ long, mostly crowded at base and equalling or one-half as long as the culm, the upper culm leaf often much exceeding the panicle; primary panicle generally much exserted from the upper sheath, 1′–3′ long, elliptic to linear, its branches ascending or erect; secondary panicles on very short basal branches and often concealed by the lower leaves; spikelets glabrous, 1¾″–2″ long.

In dry places, Maine to Minnesota, south to Georgia and Texas. June–Sept.

20. Panicum perlóngum Nash.   Long-stalked Panic-grass.   Fig. 330.

Panicum perlongum Nash, Bull. Torrey Club **26**: 575. 1899.

Culms 8′–16′ tall, simple; sheaths hirsute with long ascending hairs; blades elongated, linear, erect, papillose-hispid beneath, 1″–1½″ wide, the upper one commonly 3′–5½′ long; primary panicle long-stalked, much-exserted, generally extending beyond the apex of the upper leaf, 1½′–2½′ long, its branches erect or nearly so; spikelets about 1½″ long, about ½ as wide, glabrous or pubescent with a few scattered long hairs, strongly nerved.

On praries and in dry soil, Michigan to South Dakota, Manitoba and Texas. June–Aug.

21. Panicum linearifòlium Scribn.   Low White-haired Panic-grass.   Fig. 331.

Panicum Enslini Nash, in Britt. Man. 83. 1901. Not Trin. 1826.
Panicum linearifolium Scribn.; Nash, in Britt. & Br. Ill. Fl. 3: 500. 1898.

Culms tufted, slender, erect, smooth and glabrous, simple, 6′–16′ tall. Sheaths glabrous or pilose with long white hairs, longer than the internodes; ligule a ring of short hairs; blades elongated, smooth or rough, glabrous or more or less pilose, especially upon the lower surface, 3′–10′ long, 1″–2″ wide, the uppermost leaf the longest and often extending beyond the panicle; primary panicle loose and open, often long-exserted, 1½′–4′ long, its branches lax, ascending, secondary panicles small and contracted on very short culms and partly concealed by the bases of the long culms: spikelets 1″–1¼″ long, obtuse or acutish, pubescent with spreading hairs.

Dry soil, especially hillsides, Nova Scotia to Michigan, south to Georgia, Arkansas and Texas. May–July.

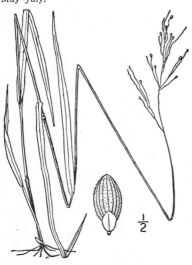

22. Panicum Wérneri Scribn.   Werner's Panic-grass.   Fig. 332.

Panicum Werneri Scribn.; Nash, in Britt. & Br. Ill. Fl. 3: 501. 1898.

Smooth and glabrous, light green. Culms tufted, erect, slender, simple or later sparingly branched, 10′–18′ tall; sheaths equalling or shorter than the internodes; ligule a ring of short hairs; blades erect, elongated, linear, acuminate, 2½′–4¼′ long, 1½″–2½″ wide, panicle finally long-exserted, loose and open, 2½′–3½′ long, its branches ascending; spikelets about 1″ long on longer hispidulous pedicels, oval, minutely and sparsely pubescent, the first scale orbicular, about one-quarter as long as the spikelet, 1-nerved, the second and third scales 7-nerved, the fourth scale oval, slightly apiculate.

Dry knolls in swamps. Maine to Ontario, Ohio, Missouri and Texas. June–July.

23. Panicum Bicknéllii Nash. Bicknell's Panic-grass. Fig. 333.

Panicum Bicknellii Nash. Bull. Torrey Club, **24**: 193. 1897.

Culms erect or decumbent at the base, slender, 8′–16′ tall, at length sparingly branched, the lower internodes puberulent, the nodes sparingly barbed. Sheaths generally longer than the internodes, ciliate on the margins, the lowermost pubescent; ligule a fringe of very short hairs; blades elongated, increasing in length toward the top of the culm, erect, linear-lanceolate, acuminate, narrowed toward the ciliate base, 7–9-nerved, primary leaves 3′–7′ long, 2½″–5″ wide; primary panicle 2½′–3′ long, its branches ascending, secondary panicles smaller, with appressed branches; spikelets obovate or oval, 1¼″–1½″ long, pubescent with short spreading hairs, the first scale 1-nerved, the second and third scales 9-nerved.

Dry wooded hills, Connecticut, New York and Pennsylvania to Georgia. July–Aug.

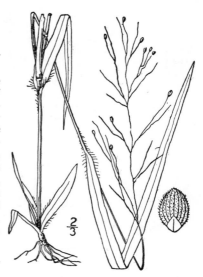

24. Panicum ciliàtum Ell. Dwarf Panic-grass. Fig. 334.

P. ciliatum Ell. Bot. S. C. & Ga. **1**: 126. 1817.

Plant yellowish green. Culms tufted, 4′–7′ tall, simple, glabrous; sheaths shorter than the internodes, ciliate on the margin, otherwise glabrous; blades up to 2½′ long, 2½″–5″ wide, glabrous on both surfaces, conspicuously ciliate, somewhat crowded at the base, narrowly elliptic, linear or lanceolate; panicle 1′–2′ long, broad, open, its axis and spreading branches hirsute; spikelets about 1″ long and ½ as wide, elliptic, pubescent with short spreading hairs, rarely nearly glabrous.

In sandy soil, southeastern Virginia to Florida and Mississippi. May to July.

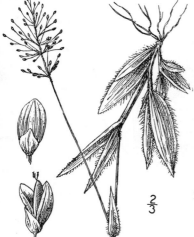

25. Panicum strigòsum Muhl. Rough-hairy Panic-grass. Fig. 335.

Panicum strigosum Muhl.; Ell. Bot. S. C. & Ga. **1**: 126. 1817.

Panicum longipedunculatum Scribn. Bull. Univ. Tenn. Exp. Sta. **7**: 53. 1894.

Culms 8′–18′ tall, simple, tufted, hirsute; leaves crowded at the base of the culm; sheaths much shorter than the internodes, hirsute; blades 1′–3′ long, 3″–4½″ wide, erect, lanceolate, papillose-ciliate on the margins, long-hirsute on the surface; panicle much-exserted, 1½′–3′ long, the axis long-hirsute, open, its branches spreading or ascending; spikelets ¾″ long or a little less, oval, glabrous.

Sandy woods, southeastern Virginia to Tennessee, Florida and Louisiana. Also in Cuba, Mexico and Guatemala. May–Aug.

26. Panicum laxiflòrum Lam. Lax-flowered Panic-grass. Fig. 336.

Panicum laxiflorum Lam. Encycl. **4**: 748. 1797.
Panicum xalapense H.B.K. Nov. Gen. & Sp. **1**: 103. 1816.

Culms erect, 8'-16' tall, simple, glabrate, the nodes barbed. Sheaths shorter than the internodes, hirsute with reflexed hairs; blades 2½'-5' long, 2''-5'' wide, erect, generally narrowed at base, long-acuminate, pubescent or glabrous, excepting the ciliate margin; panicle 2'-4' long, its axis and erect or spreading lax branches sometimes hirsute; spikelets about 1'' long, ellipsoid or narrowly obovoid, strongly pubescent; first scale minute, 1-nerved; second and third about equal, 9-nerved, very pubescent, as long as the shining obtuse minutely apiculate fourth one; third scale usually with an empty palet.

Moist soil, Maryland to Missouri, south to Florida and Mexico. Cuba and Haiti. June–Aug.

27. Panicum angustifòlium Ell. Narrow-leaved Panic-grass. Fig. 337.

Panicum angustifolium Ell. Bot. S. C. & Ga. **1**: 129. 1817.
Panicum consanguineum S. Wats. in A. Gray, Man. Ed. 6, 633, in part. 1890. Not Kunth, 1835.

Culms erect, 1°-2° tall, glabrous or pubescent toward the base, at first simple, later profusely branched above. Sheaths glabrous or the basal ones pubescent, those on the culm shorter than the internodes, those on the branches crowded; blades elongated, 3'-6' long, 1''-3'' wide, narrowed to the base, firm, glabrous, those of the culm distant, those of the branches shorter and crowded; primary panicle long-exserted, 1½'-3½' long, its branches spreading; lateral panicles smaller, shorter than the leaves; spikelets few, about 1¼'' long, elliptic to obovoid; first scale one-fourth to one-third as long as the spikelet; second and third oval, 9-nerved, pubescent; fourth oval, minutely pubescent at the apex.

Dry soil, Pennsylvania to Florida and Texas. June–Aug.

28. Panicum aciculàre Desv. Grisebach's Panic-grass. Fig. 338.

Panicum aciculare Desv.; Poir. in Lam. Encycl. **4**: 274. 1816.
?*Panicum neuranthum* Griseb. Cat. Pl. Cub. 232. 1866.

Culms tufted, slender, at length much branched, the primary simple, erect, glabrous or pubescent, 12'-30' tall. Sheaths glabrous, or the lower pubescent, the primary about one-half as long as the internodes, those on the branches overlapping; ligule a ring of hairs; blades smooth and glabrous, the primary erect, acuminate, 1'-4' long, 1''-2½'' wide, those on the branches shorter, erect or ascending, usually involute when dry, concealing the small secondary panicles; primary panicle 1'-4' long, its branches at first erect, at length widely spreading; spikelets numerous, broadly obovate, about 1'' long, densely pubescent with short spreading hairs, the second and third scales 7-nerved.

Dry or moist soil along the coast, New Jersey to Florida and Texas. Recorded from West Indies. June–Oct.

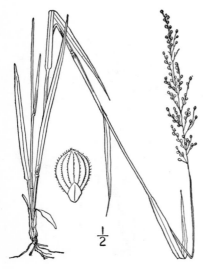

29. Panicum consanguíneum Kunth. Kunth's Panic-grass. Fig. 339.

Panicum villosum Ell. Bot. S. C. & Ga. 1 : 124. 1817.
Not Lam. 1791.
P. consanguineum Kunth, Rev. Gram. 1 : 36. 1829.

Culms 1°–2½° tall, finally much-branched, the nodes densely barbed with spreading hairs; sheaths shorter than the internodes, densely villous; blades erect, linear-lanceolate, the primary ones 1½″–3′ long, 1½″–5″ wide, those on the branches much smaller, flat, usually less than 2′ long; primary panicle 1½′–3′ long, oval, its branches ascending; spikelets about 1¼″ long and about ½ as wide, broadly obovoid, densely pubescent with spreading hairs.

In dry sandy soil, southeastern Virginia to Florida and Texas. June and July.

30. Panicum Búshii Nash. Bush's Panic-grass. Fig. 340.

Panicum Bushii Nash, Bull. Torr. Club, 26 : 568. 1899.

A tufted nearly glabrous perennial. Culms about 1° tall, finally much-branched; blades erect, linear, acuminate, very rough on the margins, ciliate at the base with a few long hairs, otherwise glabrous, the larger primary blades 3′–4′ long, 1½″–2″ wide; panicle much-exserted, 2½′–3′ long, its branches ascending; spikelets 1¼″ long and about 1″ wide, obovoid, glabrous.

In dry ground, Missouri. June and July.

31. Panicum sphaerocàrpon Ell. Round-fruited Panic-grass. Fig. 341.

Panicum sphaerocarpon Ell. Bot. S. C. & Ga. 1 : 125. 1817.

Culms generally erect, simple or somewhat branched at base, 10′–2° tall, smooth, or the nodes sometimes pubescent. Sheaths shorter than the internodes, or overlapping, glabrous, the margins ciliate; blades 2′–4′ long, 2″–7″ wide, acuminate, cordate-clasping at base, scabrous above, smooth beneath, the margins cartilaginous and minutely serrulate, ciliate towards the base; panicle ovoid, 2′–4′ long, about as broad as long; spikelets less than 1″ long, nearly spherical or somewhat longer than thick, obtuse, purple; first scale broadly ovate, obtuse; third and fourth scales three to four times as long as the first, suborbicular, 7-nerved; fourth scale oval, obtuse, ¾″ long.

Dry soil, Vermont to Kansas, south to Florida, Texas and Mexico, northern South America. July–Sept.

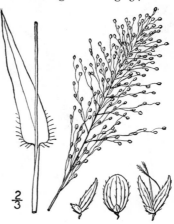

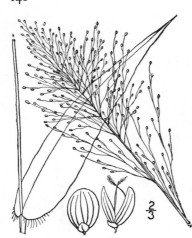

32. Panicum polyánthes Schultes. Small-fruited Panic-grass. Fig. 342.

Panicum polyanthes Schultes, Mant. 2 : 257. 1824.
Panicum microcarpon Muhl. Gram. 111, June, 1817. Not
Ell. Jan. 1817.

Culms generally erect, 2°–3° tall, simple, smooth.
Sheaths smooth, glabrous, longer than the internodes;
ligule none; blades 5′–8′ long, ½′–1′ wide, long-acuminate, smooth, cordate-clasping and sparingly ciliate at
the base; panicle 3′–8′ long, elliptic, longer than broad;
branches slender, ascending; spikelets ¾″ long, obovoid
to nearly spherical, numerous; first scale minute,
second and third about equal, 7-nerved, puberulent, the
fourth white and shining; palet of third scale usually
empty.

Woods and along thickets, southern New York to Oklahoma, south to Georgia and Texas. July–Sept.

33. Panicum ensifòlium Baldw. Small-leaved Panic-grass. Fig. 343.

Panicum ensifolium Baldw.; Ell. Bot. S. C. & Ga. 1 : 126. 1817.

Panicum Brittoni Nash, Bull. Torr. Club, 24 : 194. 1897.

Smooth and glabrous. Culms coarsely striate, finally
branched, tufted, slender, erect, rigid, 4′–8′ tall; sheaths
less than one-half as long as the internodes; ligule a ring
of short hairs; blades longer than the sheaths, those on
the culm up to 1¼′ long, the basal longer, ¾″–1½″ wide, erect,
acuminate, 5–7-nerved; panicle ¾′–1¼′ long, its branches
spreading or ascending; spikelets one-half as long as the
pedicels, or less, obovoid, obtuse, ¾″ long, the first scale
one-third as long as the spikelet, the second and third
scales 7-nerved, densely pubescent with spreading hairs.

Moist sand in the pine barrens, southern New Jersey to
Florida and Mississippi. May–July.

34. Panicum ténue Muhl. White-edged Panic-grass. Fig. 344.

Panicum tenue Muhl. Gram. 118. 1817.

P. albomarginatum Nash, Bull. Torr. Club, 24: 40. 1897.

Glabrous, excepting the spikelets. Culms densely
tufted, 8′–16′ tall, finally branched toward the base, the
upper part of the culm naked; leaves usually 2; sheaths
much shorter than the internodes; blades erect, thick,
stiff, lanceolate, with a prominent white thick margin,
usually 1½′ long or less, rarely longer, 1″–3½″ wide;
panicle ¾′–1½′ long, broadly ovate, open; spikelets elliptic, less than ¾″ long and about ½ as wide, pubescent
with short spreading hairs.

In pine lands, Dismal Swamp, Virginia, to Florida and
Louisiana; also in Cuba. June and July.

35. Panicum dichótomum L. Forked Panic-grass. Fig. 345.

Panicum dichotomum L. Sp. Pl. 58. 1753.
P. barbulatum Michx. Fl. Bor. Am. 1: 49. 1803.
P. gravius Hitchc. & Chase, Rhodora 8: 205. 1906.

Smooth and glabrous, or the lower nodes barbed. Culms erect, ½°–2° tall, at first simple, later profusely dichotomously branched at about the middle; blades light green, widely spreading, generally much narrowed toward the base, the primary ones distant, 2′–3′ long, 2″–3″ wide, those of the branches 1′ long or less, ½″–1″ wide, sometimes involute; primary panicle usually long-exserted, 1′–2′ long; branches lax, spreading, bearing few spikelets; secondary panicles smaller, not exceeding the leaves, their branches with very few spikelets; spikelets about 1″ long, ellipsoid, glabrous.

In woodlands and thickets, New Brunswick to Michigan, Florida and Texas. May–Aug.

36. Panicum lùcidum Ashe. Bog Panic-grass. Fig. 346.

P. lucidum Ashe, Journ. Mitch. Sci. Soc. 15: 47. 1898.

Culms slender, smooth and glabrous, 1½°–3° long, at length much elongated, dichotomously much branched and declining. Sheaths smooth and glabrous, or the lower ones pubescent, one-half the length of the internodes or less; ligule a short ring; blades erect, smooth and glabrous on both surfaces, lanceolate, principal nerves 5–7, the primary leaves 1′–2½′ long, 1″–5″ wide, those on the branches 1½′ or less long, concealing the small contracted panicles; primary panicle loose and open, 1½′–3′ long, its branches spreading or ascending, the lower ¾′–1½′ long; spikelets on elongated pedicels, scattered, 1″ long, oval to obovate, the scales glabrous, the first less than one-half as long as the spikelets.

Sphagnum bogs and wet woods, New York and southern New Jersey to Florida and Texas. June–Sept. Has been confused with *P. sphagnicola* Nash.

Panicum coeruléscens Hack., of the southern states, Bahamas and Cuba, differing by blue-green foliage and erect culms, is recorded from New Jersey and Virginia.

37. Panicum yadkinénse Ashe. Spotted-sheath Panic-grass. Fig. 347.

Panicum dichotomum var. *elatum* Vasey, Bull. U. S. Dep. Agr. Bot. 8: 31. 1889.
P. yadkinense Ashe, Journ. E. Mitch. Sci. Soc. 16: 85. 1900.

A glabrous perennial. Culms up to 3° tall, finally somewhat branched; sheaths much shorter than the internodes, usually white-spotted; ligule less than ½″ long; blades 3′–5′ long, 2″–6″ wide, glabrous; panicle 3′–5′ long, broadly ovate or oval, its branches long, ascending; spikelets about 1¼″ long and ½ as wide, acute, elliptic, glabrous, the second and third scales longer than the fruiting scale.

Moist woods and thickets, Pennsylvania to Georgia, Illinois and Louisiana. June–Aug.

Panicum roanokénse Ashe, of the southern states, differing by erect leaf-blades, and smaller turgid, strongly nerved spikelets, ranges north to Virginia.

38. Panicum microcàrpon Muhl. Barbed Panic-grass. Fig. 348.

P. *barbulatum* Nash, in Britt. & Br. Ill. Fl. **1**: 120. 1896. Not Michx. 1803.
P. *microcarpon* Muhl.; Ell. Bot. S. C. & Ga. **1**: 127. 1817.

Culms at first simple, erect, 2°–3° tall, later profusely branched for their whole length, 3°–4° long, prostrate or leaning, the nodes strongly barbed; blades smooth and glabrous, generally truncate or rounded at the base, the primary ones 3′–5′ long, about ¼′ wide, widely spreading, the lower ones usually reflexed, those of the branches ½′–2′ long, 1″–2″ wide; primary panicle 3′–5′ long, exserted, ovoid, its branches ascending, rigid; secondary panicles smaller, lax, not exceeding the leaves, the branches bearing few spikelets; spikelets about ¾″ long, ellipsoid, purple, glabrous; first scale about one-third as long as the spikelet, acute.

Moist soil, Massachusetts to Missouri south to Florida and Texas. Bearded Joint-grass. June–Aug.

Panicum nítidum Lam., differing by its larger pubescent spikelets, ranges from Virginia to Florida, the Bahamas and Texas.

39. Panicum ánnulum Ashe. Ringed Panic-grass. Fig. 349.

P. *annulum* Ashe, Journ. E. Mitch. Sci. Soc. **15**: 58. 1898.

Culms 16′–2½° tall, glabrous or sparingly pubescent, tufted, finally branched, the nodes densely barbed with spreading hairs, appearing like a ring; sheaths glabrous or the lower ones softly pubescent; ligule less than ½″ long; blades 2½′–5′ long, 3″–7″ wide, velvety pubescent on both surfaces; panicle 1½′–4′ long, open, its branches erect-ascending or ascending, rarely spreading; spikelets about 1″ long and nearly ½ as wide, elliptic, strongly pubescent with spreading hairs.

In dry rocky woods, New Jersey and Pennsylvania to Georgia, Missouri and Mississippi. June and July.

40. Panicum boreàle Nash. Northern Panic-grass. Fig. 350.

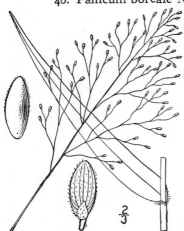

Panicum boreale Nash, Bull. Torr. Club, **22**: 421. 1895.

Culms at first erect and simple, 1°–2° tall, later decumbent and somewhat branched, smooth and glabrous. Sheaths shorter than the internodes, usually smooth, ciliate; ligule short, ciliate; blades 3′–5′ long, ¼′–½′ wide, erect, truncate or rounded at the sparsely ciliate base, acuminate; panicle 2′–4′ long, ovoid, its branches 1′–2′ long, spreading or ascending; spikelets 1″ long, about equalling the pedicels, ellipsoid, somewhat pubescent; first scale ovate, obtuse, about one-third as long as the spikelet; second and third scales oblong-ovate, 7-nerved, pubescent, equalling the fourth, which is oval, acute, and slightly more than ¾″ long; palet of third scale usually empty.

Moist soil, Newfoundland to Ontario south to New York, Indiana and Minnesota. June and July.

41. Panicum mattamuskeeténse Ashe. Clute's Panic-grass. Fig. 351.

P. mattamuskeetense Ashe, Journ. E. Mitch. Sci. Soc. **15**: 45. 1898.
P. Clutei Nash, Bull. Torrey Club **26**: 569. 1899.

Plant usually purplish. Culms tufted, $1\frac{1}{2}°-3\frac{1}{2}°$ tall, glabrous, the nodes sometimes puberulent or the lower ones barbed; sheaths loose, the upper ones glabrous excepting on the margins and occasionally toward the summit, the lower ones often softly pubescent; blades $2\frac{1}{2}'-5\frac{1}{2}'$ long, $3''-6''$ wide, firm, lanceolate, ascending or sometimes reflexed, glabrous; panicle $2\frac{1}{2}'-4'$ long, broad and open; spikelets about $1\frac{1}{4}''$ long, a little more than $\frac{1}{2}$ as wide, pubescent with short hairs.

Sandy borders of swamps and bogs, **Massachusetts to North Carolina.** July.

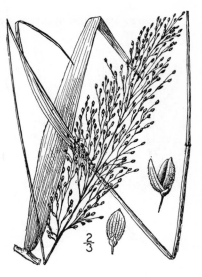

42. Panicum octonòdum J. G. Smith. Eight-jointed Panic-grass. Fig. 352.

P. octonodum J. G. Smith; Scribn. Bull. U. S. Dep. Agr. Agrost. **17**: 73. 1899.

Plant glabrous and usually purple. Culms erect, $2°-3\frac{1}{2}°$ tall, finally branched; sheaths much shorter than the internodes; ligule a narrow ring usually less than $\frac{1}{4}''$ wide; blades erect, firm, $1\frac{1}{2}'-4\frac{1}{2}'$ long, $2''-4''$ wide, lanceolate; panicle $3'-5'$ long, $\frac{3}{4}'-1\frac{1}{2}'$ .wide, dense, longer than broad. its branches erect or erect-ascending; spikelets less than $\frac{3}{4}''$ long and $\frac{1}{2}''$ wide, oval, glabrous.

In wet places, New Jersey to Florida and Texas. May–Aug.

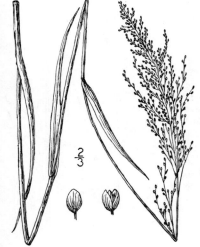

43. Panicum paucípilum Nash. Purple Panic-grass. Fig. 353.

P. paucipilum Nash, Bull. Torr. Club, **26**: 573. 1899.

Plant usually purple. Culms $2°-3\frac{1}{2}°$ tall, finally somewhat branched, smooth and glabrous; sheaths ciliate on the margin toward the summit, otherwise glabrous; ligule over $1''$ long; blades $2\frac{1}{2}'-3\frac{1}{2}'$ long, $2\frac{1}{2}''-3\frac{1}{2}''$ wide, erect or ascending, thickish, rather firm, sometimes minutely puberulent on the lower surface, usually with a few hair-bearing papillae at the base; panicle $2'-4'$ long, longer than broad, its branches erect or erect-ascending, rather dense; spikelets $\frac{3}{4}''$ long or a little less, a little over $\frac{1}{2}$ as wide, oval, pubescent with spreading hairs.

In wet soil, southern New Jersey to Florida and Mississippi. July and Aug.

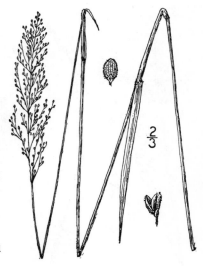

44. Panicum sprètum Schult. Eaton's Panic-grass. Fig. 354.

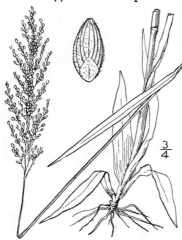

Panicum spretum Schult. Mant. **2**: 248. 1824.

Panicum Eatoni Nash, Bull. Torr. Club, **25**: 84. 1898.

Smooth and glabrous. Culms 1½°–3½° tall, erect, at length dichotomously branched and swollen at the nodes; sheaths much shorter than the internodes, usually more or less ciliate on the margins; ligule a ring of long hairs; blades erect, lanceolate, acuminate, 1½′–4′ long, 1½″–5″ wide; panicle finally long-exserted, dense and contracted, 3′–5′ long, 1¼′ or less broad, its branches erect-ascending; spikelets oval, from over ¾″ to nearly 1″ long, acutish, the first scale about one-third as long as the spikelet, pubescent, 1-nerved, the second and the third scales broadly oval when spread out, 7-nerved, densely pubescent with spreading hairs.

Along the coast, in damp or wet places, Maine to New Jersey and northern Indiana. May–Aug.

45. Panicum Lindheìmeri Nash. Lindheimer's Panic-grass. Fig. 355.

P. nitidum Nash, in Britt. & Br. Ill. Fl. **1**: 120. 1896. Not Lam. 1797.

P. Lindheimeri Nash, Bull. Torr. Club, **24**: 196. 1897.

Culms at first simple, 12′–18′ tall, later profusely dichotomously branched, 2°–3° long, glabrous or pubescent below. Sheaths less than half as long as the elongated internodes, glabrous, excepting the ciliate margin, or the lower sometimes pubescent; ligule over 2″ long; blades glabrous, or sparingly ciliate toward the base, sometimes puberulent below, the primary ones 1′–3′ long, 1½″–4″ wide, erect or ascending, sometimes reflexed, those of the branches ½′–1′ long, 1″ wide or less; primary panicle long-exserted, 1′–2½′ long, ovoid, as broad as long or nearly so, those of the branches smaller and exceeded by the leaves; spikelets about ¾″ long, obovoid, pubescent.

Common in dry sandy soil, Maine to Ontario and California, south to Florida and Texas. June–Aug.

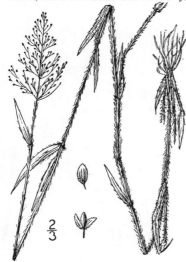

46. Panicum leùcothrix Nash. Roughish Panic-grass. Fig. 356.

P. leucothrix Nash, Bull. Torr. Club, **24**: 41. 1897.

Culms 1°–2° tall, densely tufted, erect, appressed papillose-hirsute, finally branched; sheaths similarly pubescent but the hairs more spreading; ligule 1½″ long; blades 1′–2′ long, 2″–3″ wide, lanceolate, erect or ascending, firm, softly pubescent on the lower surface, ciliate at the base, glabrous on the upper surface; primary panicle 1′–2′ long, rarely larger or smaller, broadly ovate, its branches ascending; spikelets a little over ½″ long and about ½ as wide, oval, pubescent with short-spreading hairs.

In usually dry sandy soil, southern New Jersey to Florida and Texas. Cuba. June and July.

Panicum Wrightiànum Scribn., of the southern states and Cuba, which differs in still smaller spikelets, is recorded from southern New Jersey.

47. Panicum huachùcae Ashe. Hairy Panic-grass. Fig. 357.

P. pubescens Nash, in Britt. & Br. Ill. Fl. **1**: 121. 1896. Not Lam. 1797.
P. huachucae Ashe, Journ. E. Mitch. Sci. Soc. **15**: 51. 1898.
P. huachucae silvicola Hitchc. & Chase, Rhodora, **10**: 64. 1908.

Culms at first erect and simple, later profusely branched and leaning or ascending, papillose-hirsute with ascending hairs, the nodes barbed; sheaths papillose-hirsute; ligule ¼″–2″ long; blades copiously pilose on the upper surface, densely pubescent on the lower, erect to spreading, firm or lax, those of the culm 2′–3′ long, those of the branches much shorter; primary panicle 1½′–4′ long, ovoid, the branches ascending or spreading; lateral panicles much smaller, not exceeding the leaves; spikelets about ¾″ long, pubescent.

In dry soil, Maine to South Dakota, Florida, Texas and California. June–Sept. Has been mistaken for *P. unciphyllum* Trin.

48. Panicum scoparioìdes Ashe.   Stiff Hairy Panic-grass. Fig. 358.

Panicum scoparioides Ashe, Journ. E. Mitch. Sci. Soc. **15**: 53. 1898.

Culms 1°–2½° tall, rather slender, pubescent with ascending hairs, finally branched; sheaths strongly papillose-hispid with ascending hairs; ligule 1″–1½″ long; blades 2′–4′ long, 2½″–4″ wide, lanceolate, ascending, glabrous on the upper surface, the lower surface more or less pubescent with scattered spreading hairs; panicle barely exserted, 2′–3′ long, its branches ascending; spikelets a little less than 1¼″ long and about ½ as wide, elliptic, pubescent.

In dry soil, Vermont to Pennsylvania, Delaware and Minnesota. July and August.

Panicum lánguidum Hitch. & Chase, of Maine, Massachusetts and New York, differs from this and related species by its pointed spikelets.

49. Panicum villosíssimum Nash.   White-haired Panic-grass. Fig. 359.

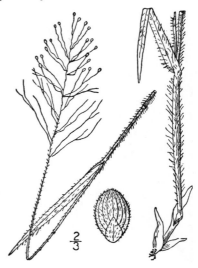

P. villosissimum Nash, Bull. Torr. Club, **23**: 149. 1896.
P. atlanticum Nash, Bull. Torr. Club, **24**: 346. 1897.

Papillose-pilose with long white spreading hairs. Culms tufted, at length branched, 12′–20′ tall, erect or ascending, a smooth ring below the nodes which are barbed with spreading hairs; sheaths shorter than the internodes; ligule a ring of hairs 1″–2½″ long; blades erect or ascending, rigid, thickish, lanceolate, 1¼′–4′ long, 2″–3½″ wide, acuminate, middle leaves the longest; panicle 1½′–3′ long, 1¼′–2¾′ wide, the branches and their divisions hispidulous; spikelets numerous, obovate to elliptic, about 1¼″ long, ¾″ wide, densely pubescent with short spreading hairs.

Dry soil, Massachusetts to Minnesota, Florida, Texas and Missouri. June–Aug.

Panicum pseudopubéscens Nash, differs in nearly appressed pubescence of the culms and glabrous upper leaf-surfaces. It ranges from Connecticut to Illinois, Florida and Mississippi.

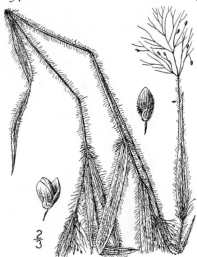

2/3

50. Panicum praecòcius Hitchc. & Chase. Early-branching Panic-grass. Fig. 360.

P. praecocius Hitch. & Chase, Rhodóra, **8** : 206. 1906.

Culms 6'–18' tall, slender, branching almost at once, the secondary panicles appearing before the primary ones are mature, strongly pubescent with long weak spreading hairs over 1½" long; sheaths similarly pubescent; ligule 1½"–2" long; blades 1½'–3' long, 2"–3½" wide, lanceolate, hirsute with long hairs on both surfaces, the hairs on the upper surface erect and over 2" long; primary panicle 1½'–2½' long and about as wide, its branches spreading or ascending; spikelets a little less than 1" long, obovoid, pubescent with long weak spreading hairs, the first scale ½ as long as the spikelet or a little less.

Dry places, Michigan to Texas. June–Aug.

51. Panicum implicàtum Scribn. Slender-stemmed Panic-grass. Fig. 361.

P. implicatum Scribn.; Nash, in Britt. & Br. Ill. Fl. **3** : 498. 1898.

Culms tufted, erect, 10'–22' tall, very slender, more or less pubescent, at length much branched. Sheaths shorter than the internodes, densely papillose-hirsute, at least the lower ones; ligule a ring of long hairs; blades erect, lanceolate, ¾'–2½' long, 1"–3" wide, at least the lower ones papillose-hirsute on both surfaces, especially beneath; panicle open, ovate, 1'–2¼' long, hirsute, its branches widely spreading; spikelets broadly obovate, obtuse, purplish, about ¾" long, the outer 3 scales pubescent with short spreading hairs, the first scale nearly one-half as long as the spikelet, broadly ovate, obtuse, 1-nerved, the second and third scales orbicular-oval, 7-nerved.

Dry soil, Nova Scotia to Minnesota, District of Columbia and Kentucky.

2/3

52. Panicum meridionàle Ashe. Matting Panic-grass. Fig. 362.

Panicum meridionale Ashe, Journ. E. Mitch. Sci. Soc. Soc. **15** : 59. 1898.
Panicum filiculme Ashe, Journ. E. Mitch. Sci. Soc. **15** : 59. 1898. Not Hack. 1895.
Panicum subvillosum Ashe, loc. cit. **16** : 86. 1900.
?Panicum albemarlense Ashe, loc. cit. 84. 1900.
Panicum oricola Hitchc. & Chase, Rhodora, **8** : 208. 1906.

Culms densely tufted, 4'–16' tall, later much-branched and often decumbent and forming mats, hirsute below with ascending or nearly erect hairs, the upper part of the culm puberulent; sheaths hirsute with ascending or somewhat spreading hairs; ligule commonly over 1" long; blades ¾'–3' long, 1"–3" wide, erect or nearly so, lanceolate, the upper surface with erect hairs over 1½" long, or sometimes nearly glabrous, the lower surface appressed-pubescent with shorter hairs; panicle up to 2' long, the axis puberulent or very shortly pilose, the branches spreading or ascending; spikelets from a little less than ¾" to nearly 1" long, pubescent.

Sandy places, Nova Scotia to Minnesota, Georgia and Missouri. June and July.

53. **Panicum Owenae** Bicknell. Mrs. Owen's Panic-grass. Fig. 363.

Panicum Owenae Bicknell, Bull. Torr. Club, **35**: 185. 1908.

Culms tufted, erect or ascending, 6'–1° tall, villous-puberulent below, later branched, the branches crowded at the base; leaves 2 or 3; sheaths puberulent and often pilose; ligule a ring of hairs ½" long or less; blades on the culm erect, up to 3'–4' long, 1½" wide or less, glabrous above, appressed-pubescent beneath, the basal blades shorter and broader; panicle up to 2½' long, the axis and the ascending or nearly erect branches puberulent; spikelets nearly 1" long, oval, densely pubescent with spreading hairs.

Sandy places, Nantucket, Mass. June–Sept.

54. **Panicum tennesseénse** Ashe. Tennessee Panic-grass. Fig. 364.

P. tennesseense Ashe, Journ. E. Mitch. Sci. Soc. **15**: 52. 1898.

Culms tufted, 10'–2° tall, slender, ascending, papillose-hirsute with long spreading hairs, finally much-branched and prostrate and forming broad mats; sheaths densely papillose-hirsute with long spreading hairs; ligule 2" long or more; blades 1½'–4' long, 2½"–5" wide, the upper surface glabrous or with a few long scattered hairs at the base, the lower surface densely and softly pubescent, the blades on the branches much shorter and spreading; panicle 1½"–4' long, ovate, its branches ascending; spikelets ¾" long or a little more, about ½ as wide, elliptic or obovoid, strongly pubescent with long spreading hairs.

In moist ground or in woods, Maine to Minnesota, south to Georgia and Texas. July and Aug.

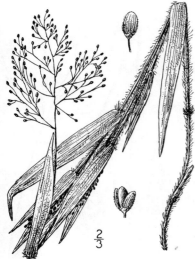

55. **Panicum lanuginòsum** Ell. Woolly Panic-grass. Fig. 365.

P. lanuginosum Ell. Bot. S. C. & Ga. **1** : 123. *1817.*
Panicum auburne Ashe, Bull N. C. Exp. Sta. **175**: 115. 1900.

Culms, sheaths and leaves villous with spreading hairs, those on the leaves and the upper part of the culm shorter. Culms leafy, tufted, 1°–2½° tall, erect, at length branched, a smooth ring below each barbed node; sheaths shorter than the internodes; ligule a ring of long hairs; blades erect, lanceolate, acuminate, 1½'–5' long, 2"–4½" broad; panicle ovate, 1½'–4' long, the axis pubescent, the branches ascending, the larger 1'–2' long; spikelets numerous, broadly obovate, from a little less than ¾" to nearly 1" long, the first scale orbicular, glabrous or pubescent, 1-nerved, the second and third scales nearly orbicular when spread out, 7–9-nerved, densely pubescent with spreading hairs.

Dry sandy soil, southern New Jersey to Florida, Louisiana and Texas. July and Aug.

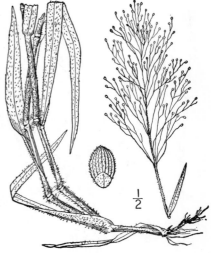

56. Panicum Addisònii Nash. Low Stiff Panic-grass. Fig. 366.

P. Addisonii Nash, Bull. Torr. Club, **25** : 83. 1898.

Culms 10′–15′ tall, rigid, tufted, erect or decumbent at the base, at length much branched, the branches erect, pubescent below with long nearly appressed hairs which decrease in length toward the summit where they are very short. Sheaths often longer than the internodes, appressed-pubescent, at least the lower ones; ligule a ring of hairs; blades erect, lanceolate, thickish, smooth and glabrous on both surfaces, rough on the margins, acuminate, 1′–3′ long, 1½″–3″ wide; panicle ovate to oblong, ¾′–2¼′ long, its branches spreading or ascending; spikelets obovate, 1″ long, the first scale acute or acutish, about one-half as long as the spikelet, 1-nerved, pubescent, second and third scales 9–11-nerved, densely pubescent with spreading hairs.

Sandy soil, Massachusetts to South Carolina. May–June.

57. Panicum Commonsiànum Ashe. Commons' Panic-grass. Fig. 367.

P. Commonsianum Ashe, Journ. E. Mitch. Sci. Soc. **15** : 55. 1898.

Culms tufted, 12′–20′ tall, pubescent below with long nearly appressed hairs, the upper portion glabrous or puberulent, finally branched, the nodes barbed with spreading hairs; sheaths, at least the lower ones, pubescent with long often appressed hairs; ligule ½″ long or more; blades 1½′–4′ long, 1½″–4″ wide, erect, lanceolate, appressed-pubescent on the lower surface with stiff hairs, the upper surface glabrous or with some long spreading hairs toward the base; panicle 1½′–3′ long, ovate, its branches spreading or ascending; spikelets 1¼″ to a little less than 3″ long and ½ as broad, obovoid, pubescent with spreading hairs, the first scale about ½ as long as the spikelet.

In dry sandy soil near the coast, Connecticut to North Carolina and Florida. June and July.

58. Panicum columbiànum Scribn. American Panic-grass. Fig. 368.

Panicum columbianum Scribn. Bull. U. S. Dept. Agric. Div. Agrost. **7** : 78. 1897.
P. psammophilum Nash, Bull. Torrey Club, **26** : 576. 1899.
P. columbianum var. *thinium* Hitchc. & Chase, Rhodora, **10** : 64. 1908.

Culms tufted, erect, softly pubescent, 8′–2° tall, at length dichotomously branched, the branches erect. Lower sheaths pubescent, the upper glabrous, the primary one-half as long as the internodes; ligule a ring of short hairs; blades lanceolate, erect, thickish and firm, glabrous above, the lower ones more or less pubescent beneath, the primary 1½′–2½′ long, 2″–3″ wide, those on the branches smaller; panicle small, ovate, 1′–1½′ long, its branches ascending; spikelets broadly obovate, a little more than ¾″ long, the outer 3 scales densely pubescent with spreading hairs, the first scale about one-half as long as the spikelet, 1-nerved, the second and third scales 7-nerved.

Fields and open woods, Maine to Virginia, Tennessee and Alabama. June–Sept.

59. Panicum tsugetòrum Nash. Hemlock Panic-grass. Fig. 369.

P. tsugetorum Nash, Bull. Torr. Club, **25**: 86. 1898.

Culms and sheaths pubescent with short appressed or ascending hairs intermixed toward the base with longer ones. Culms tufted, 1½° or less tall, somewhat slender, at length much branched and decumbent or prostrate; sheath shorter than the internodes; ligule a ring of hairs about ½″ long; blades erect or ascending, firm, lanceolate, 5-7-nerved, minutely appressed-pubescent beneath, smooth and glabrous above, or the upper primary leaves sometimes with a few long erect hairs, the primary leaves 1½′-3′ long, 2½″-4″ wide, those on the branches smaller and partly concealing the small panicles; primary panicles broadly ovate, 1½′-2½′ long, the branches spreading-ascending; spikelets broadly obovate, about ⅞″ long, the outer 3 scales pubescent, with short spreading hairs.

Dry soil in woods, Maine to Virginia, Illinois and Tennessee.

60. Panicum Nashiànum Scribn. Nash's Panic-grass. Fig. 370.

Panicum Nashianum Scribn. Bull. U. S. Dept. Agric. Div. Agrost. **7**: 79. 1897.
P. patulum Hitchc. Rhodora, **8**: 209. 1906.

Culms tufted, glabrous or puberulent, slender, 6′-15′ tall, at length much branched. Sheaths glabrous, or the lower pubescent, the primary about one-third as long as the internodes, those on the branches overlapping; ligule a short scarious ring; blades erect or ascending, lanceolate, acuminate, smooth and glabrous, ciliate, at least at the base, ¾″-2′ long, 1″-2½″ wide, the leaves of the branches smaller; primary panicle 1′-2′ long, the branches widely spreading; spikelets about 1″ long, obovate, the first scale 1-nerved, the second and third scales 7-nerved, densely pubescent with short spreading hairs.

Pine lands, Virginia to Florida and Mississippi; also in the West Indies. March–July.

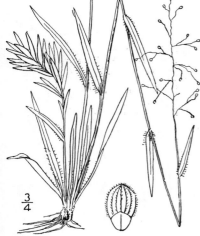

61. Panicum Liebérgii (Vasey) Scribn. Lieberg's Panic-grass. Fig. 371.

Panicum scoparium Lam. var. *Liebergii* Vasey, Bull. U. S. Dept. Agric. Div. Bot. **8**: 32. 1889.
P. Liebergii Scribn. Bull. U. S. Dept. Agr. Agrost. **8**: 6. 1889.

Culms erect, slender, glabrous, roughish, especially near the nodes, 1°-2° tall, at length branched. Sheaths papillose-hirsute with spreading hairs, usually longer than the internodes; ligule a short scarious ciliolate ring; blades erect or ascending, lanceolate, 2′-4′ long, 3″-6″ wide, acuminate at the apex, rounded at the partly clasping base, papillose-hispid beneath and sometimes sparingly so on the rough upper surface; panicle oblong, 2′-4′ long, its branches erect or ascending; spikelets 1½″-2″ long, oval, the outer three scales papillose-hirsute with long spreading hairs, the first scale about one-half as long as the spikelet, ovate, acute, 1-3-nerved, the second and third scales broadly oval when spread out, 7-9-nerved.

Dry soil, western New York to Manitoba and Kansas. June–July.

62. Panicum Scribneriànum Nash. Scribner's Panic-grass. Fig. 372.

Panicum scoparium S. Wats. in A. Gray, Man. Ed. 6, 632. 1890. Not Lam. 1797.
P. pauciflorum A. Gray, Man. 613. 1848. Not Ell. 1817.
P. Scribnerianum Nash, Bull. Torr. Club, 22: 421. 1895.

Culms erect, 6′–2° tall, simple or late in the season dichotomously branched above, sparingly pubescent. Sheaths strongly papillose-hispid, sometimes glabrate; blades 2′–4′ long, 3″–6″ wide, rounded or truncate at base, acuminate, more or less spreading, smooth above, scabrous beneath; panicles small, the primary one exserted, ovoid, 1½′–3′ long, the secondary ones much smaller and more or less included; branches of the primary panicle spreading, 8″–1½′ long, often flexuous; spikelets turgid, obovoid, a little over 1½″ long.

In dry or moist soil, Maine to British Columbia, south to Virginia, Texas and Arizona. June–Aug.

Panicum Hélleri Nash, of the south-central states, differs in being glabrous or nearly so, and with smaller spikelets. It is recorded from Missouri.

63. Panicum oligosánthes Schult. Few-flowered Panic-grass. Fig. 373.

P. pauciflorum Ell. Bot. S. C. & Ga. 1: 120. 1817. Not R. Br. 1810.
P. oligosanthes Schult. Mant. 2: 256. 1824.

Culms tufted, erect, 1°–2½° tall, villous, finally branched; sheaths, especially those on the branches, papillose-hispid, ciliate on the margin; blades erect or ascending, 2′–4′ long, 2½″–5″ wide, lanceolate to linear, softly and densely pubescent on the lower surface, the upper surface glabrous or with a few long hairs near the base; primary panicle 2′–4′ long, its branches ascending; spikelets 1¾″–2″ long and about ½ as wide, oval, pubescent.

In dry soil, New Jersey to Florida, Illinois and Texas. June–Sept.

⅔

¾

64. Panicum xanthóphysum A. Gray. Slender Panic-grass. Fig. 374.

P. xanthophysum A. Gray, Ann. Lyc. N. Y. 3: 233. 1835.
P. calliphyllum Ashe, Journ. E. Mitch. Sci. Soc. 15: 31. 1898.

Plant light green, becoming yellowish in drying. Culms erect, 1°–2° tall, simple. Sheaths sparingly papillose-pubescent; ligule very short; blades 3′–6′ long, ½′–¾′ wide, rounded at base, long-acuminate, erect, smooth and glabrous; panicle long-exserted, linear, 1½′–4′ long, its branches appressed, rarely somewhat ascending; spikelets few, 1½″–2″ long, obovoid, pubescent or rarely glabrous, first scale about one-half as long as the nearly equal obtuse second and third; fourth scale indurated and shining, elliptic or oval.

Dry soil, Quebec to Manitoba and Pennsylvania. June–Aug.

65. Panicum Wilcoxiànum Vasey. Wilcox's Panic-grass. Fig. 375.

Panicum Wilcoxianum Vasey, Bull. U. S. Dept. Agric. Bot. Div. 8: 32. 1889.

Culms erect, 6'–10' tall, sparingly pubescent. Sheaths papillose-hispid; ligule a ring of hairs; blades 1½'–3' long, less than 2" wide, long-acuminate, strongly pubescent with long hairs; panicle about 1½' long, about one-half as wide, oblong to ovoid, compact; branches less than 1' long, ascending, flexuous; spikelets 1¼"–1½" long, ellipsoid; first scale about one-quarter as long as the spikelet; second and third scales about equal, pubescent; fourth scale about as long as the third, obtuse.

In dry soil, North Dakota to Manitoba, Iowa and Kansas. July–Aug.

66. Panicum Ravenélii Scribn. & Merr. Ravenel's Panic-grass. Fig. 376.

P. scoparium Chapm. Fl. S. St. 575. 1860. Not Lam. 1797.
P. Ravenelii Scribn. & Merr. Bull. U. S. Dep. Agr. Agrost. 24: 36. 1900.

Culms tufted, erect, 16'–2° tall, finally branched, papillose-hispid below with spreading or ascending hairs, the pubescence above softer; sheaths densely papillose-hirsute with ascending hairs; blades 3'–5' long, 5"–10" wide, cordate at the clasping base, broadly lanceolate, erect or ascending, glabrous on the upper surface, densely and softly pubescent on the lower surface; panicle 3'–5' long, its branches ascending; spikelets about 2" long and nearly ½ as wide, obovoid, pubescent with rather weak hairs.

In woods, Maryland to Missouri, Florida and Texas. May–July.

67. Panicum scopàrium Lam. Velvety Panic-grass. Fig. 377.

Panicum scoparium Lam. Encycl. 4: 744. 1797.
P. pubescens Lam. Encycl. 4: 748. 1797.
P. viscidum Ell. Bot. S. C. & Ga. 1: 123. *pl. 7. f. 3.* 1817.

Culms erect, 2°–4° tall, simple or at length much branched above, villous. Sheaths shorter than the internodes, villous; blades generally narrowed, sometimes rounded or truncate at base, softly pubescent, those of the culm 4'–7' long, 5"–8" wide, distant, those of the branches 1'–2½' long, 2"–5" wide, crowded; primary panicle 3'–6' long, ovoid, branches ascending; secondary panicles much smaller, not exceeding the leaves; spikelets ovoid to oval, about 1¼" long, pubescent; first scale broadly ovate, about one-fourth as long as the spikelet; second and third scales nearly orbicular, 9-nerved, pubescent, the fourth oval, apiculate, 1" long.

Moist soil, Massachusetts to New Jersey, Pennsylvania, Florida, Oklahoma and Texas. Cuba. June–Aug.

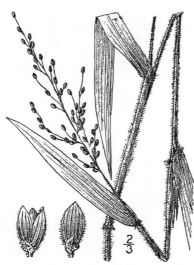

68. Panicum malacophýllum Nash. Soft-leaved Panic-grass. Fig. 378.

P. malacophyllum Nash, Bull. Torr. Club, **24**: 198. 1897.

Culms 1°–2° tall, slender, finally somewhat branched above, papillose-hirsute with long spreading hairs, the nodes densely barbed; sheaths papillose-hirsute with long spreading hairs; blades 2′–4′ long, 2½″–6″ wide, lanceolate, ascending, softly pubescent on both surfaces; panicle up to 2½′ long, its axis and spreading branches densely hirsute with spreading hairs; spikelets 1¼″–1½″ long, broadly oval or obovoid, densely hirsute with long spreading hairs.

In sandy woods, Missouri to Texas; also in Tennessee. May–July.

69. Panicum scabriúsculum Ell. Tall Swamp Panic-grass. Fig. 379.

P. scabriusculum Ell. Bot. S. C. & Ga. **1**: 121. 1817.

Culms up to 6° tall, roughened, glabrous or minutely pubescent, finally branched; sheaths nearly glabrous or papillose-hispid; ligule very short; blades 4′–8′ long, 4″–6″ wide, ascending or spreading, sometimes reflexed, linear-lanceolate, glabrous; panicle 4′–8′ long, its branches spreading or ascending; spikelets lanceolate, glabrous or minutely pubescent, strongly nerved, acute, about 1¼″ long and ½″ wide, ovate to ovate-lanceolate.

In swampy places and ponds, Maryland to West Virginia, Florida and Texas. May–Aug.

Panicum cryptánthum Ashe, differs by its smaller size, glabrous sheaths, and unbranched culms, and occurs from New Jersey to Florida and Texas.

70. Panicum aculeàtum Hitchc. & Chase. Tall Rough Panic-grass. Fig. 380.

P. aculeatum Hitchc. & Chase, Rhodora, **8**: 209. 1906.

Culms 3°–4° tall, tufted, rough, sometimes hispid below; sheaths papillose-hispid, or the upper ones glabrous; ligule a mere ring; blades 4″–8′ long, 4″–7″ wide, linear, elongated, stiff, ascending or erect, usually rough; panicle 3′–5′ long, nearly as wide, its branches spreading or ascending; spikelets 1½″ long, elliptic, glabrous or nearly so, the first scale ½–⅓ as long as the spikelet, the second and third scales neither so prominently nor so sharply nerved as in the above species.

In swampy woods, New York to District of Columbia and North Carolina. July.

71. Panicum Áshei Pearson.   Ashe's Panic-grass.   Fig. 381.

P. *Ashei* Pearson, Journ. E. Mitch. Sci. Soc. **15**: 35. 1898.

Culms tufted, 8′–16′ tall, erect, usually sparingly branched, rarely much-branched and prostrate, puberulent; sheaths puberulent, usually less than ½ as long as the internodes, ciliate on the margin; blades 2′–3′ long, rarely longer, 3″–5″ wide, occasionally broader, somewhat cordate at the base, erect or ascending, sometimes spreading, lanceolate, sparsely ciliate at the base with long hairs; panicle 2′–3′ long, its branches ascending; spikelets about 1¼″ long and ½″ wide, elliptic, obtuse, pubescent with rather long weak hairs.

In dry woods, Massachusetts to Michigan, south to Florida, Mississippi and Missouri. May–Aug.

72. Panicum commutàtum Schultes.   Variable Panic-grass.   Fig. 382.

P. *nervosum* Muhl. Gram. 116. 1817? Not Lam. 1797.
Panicum commutatum Schultes, Mant. **2**: 242. 1824.

Culms erect, 16′–2½° tall, rather slender, glabrous, puberulent at the nodes, simple, finally dichotomously branched above. Sheaths ciliate on the margin, and pubescent at the apex, otherwise glabrous; blades spreading or ascending, cordate and clasping at the base, 2′–5′ long, ¼′–1′ wide, ciliate at the base, glabrous or puberulent, those of the branches generally broader and more crowded than those of the main stem; panicle 2′–5′ long, lax, the branches spreading; spikelets 1¼″ to nearly 1½″ long, elliptic; second and third scales equal, 7-nerved, pubescent; fourth scale oval, obtuse, apiculate, about 1″ long.

In dry woods and thickets, Massachusetts to Missouri, Florida and Texas. June–Aug.

Panicum Joòrii Vasey, of the southeastern states and Mexico, differs by decumbent culms, leaves scarcely cordate and unsymmetrical, ranges north to Virginia.

73. Panicum mutábile Scribn. & Sm.   Tall Fringed Panic-grass.   Fig. 383.

P. *mutabile* Scribn. & Sm.; Nash, in Small, Fl. SE. U. S. 103. 1907.

Culms tufted, 16′–3° tall, glabrous or minutely puberulent below; sheaths glabrous, excepting the ciliate margin; blades 2½′–5′ long, 10″–20″ wide, horizontally spreading, conspicuously ciliate, especially the broader ones at the base, glabrous on the surfaces; panicle 3′–6′ long and nearly as wide; spikelets about 1½″ long, elliptic, pubescent, the first scale ⅓–½ as long as the spikelet.

In sandy soil, southeastern Virginia to Florida and Mississippi. June and July.

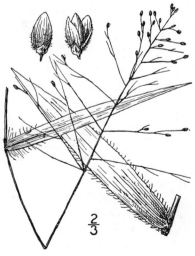

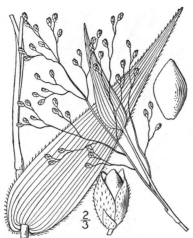

74. **Panicum latifòlium** L.   Broad-leaved
Panic-grass.   Fig. 384.

P. latifolium L. Sp. Pl. 58.   1753.
Panicum macrocarpon Le Conte; Torr. Cat. 91. 1819.

Culms 1°–3° tall, erect, simple, later somewhat
branched above, smooth; the nodes, at least the
upper ones, naked.   Sheaths smooth and glabrous,
excepting the pubescent ring at the apex and the
ciliate margin; blades 3′–7′ long, 9″–1½′ wide,
cordate-clasping at base, acuminate, smooth and glabrous or nearly so on both surfaces, ciliate; panicle
3′–6′ long, generally long-exserted, rarely included,
its branches with few spikelets and more or less
ascending; spikelets 1½″–2″ long, turgid, oval to
obovoid; second and third scales broadly oval, obtuse, 9-nerved, pubescent, the fourth oval, rather
acute, 1¾″ long.

In woods, Maine to Minnesota, south to North Carolina and Kansas.   July–Aug.

75. **Panicum Bóscii** Poir.   Bosc's Panic-grass.   Fig. 385.

P. Boscii Poir. Encycl. Suppl. **4**: 278.   1816.
P. latifolium Walt. Fl. Car. 73.   1788.   Not L. 1753.
Panicum Walteri Poir. in Lam. Encycl. Suppl. **4**: 282.   1816.
   Not Pursh, 1814.
P. Porterianum Nash, Bull. Torr. Club, **22**: 420.   1895.
P. pubifolium Nash, Bull. Torr. Club, **26**: 577.   1899.

Culms erect, 1°–2½° tall, simple, later somewhat dichotomously branched above, the nodes densely barbed.
Sheaths glabrous or softly pubescent; blades ovate to
broadly lanceolate, 2′–5′ long, ¾′–1¼′ wide, cordate-clasping
at base, acute, glabrous or softly pubescent; panicle included or somewhat exserted, 2½′–4′ long; branches spreading or ascending, bearing few elliptic short-pedicelled
appressed spikelets 2″–2½″ long; first scale one-third to
one-half as long as the pubescent and equal second and
third ones; fourth scale about as long as the third.

In woods, Massachusetts to Missouri, Oklahoma, Florida and
Texas.   June–Aug.

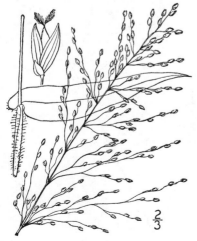

76. **Panicum clandestìnum** L.   Corn Grass.   Deer-tongue Grass.   Fig. 386.

Panicum clandestinum L. Sp. Pl. 58.   1753.
Panicum pedunculatum Torr. Fl. U. S. 141.   1824.
Panicum decoloratum Nash, Bull. Torr. Club, **26**: 570.
   1899.

Culms erect or ascending, 1½°–4° tall, rather stout,
simple at first, much branched later in the season.
Sheaths longer than the internodes, much crowded
on the branches, papillose-hispid, especially the
upper ones; blades 2′–8′ long, ½′–1¼′ wide, cordate-
clasping at base, acuminate, smooth and glabrous,
the margins ciliate at base; primary panicle sometimes long-exserted, 3′–5′ long, its branches ascending; panicles of the branches included in the sheaths,
rarely slightly exserted; spikelets 1″–1¼″ long, pubescent, elliptic; first scale about one-third as long
as the spikelet; second and third oval, acutish,
9-nerved, the fourth oval, obtuse, apiculate, whitish,
shining.

In thickets and moist places, Maine to Kansas, south
to Florida and Texas.   June–July.

18. SACCIÓLEPIS Nash, in Britt. Man. 89. 1901.

Perennial grasses with flat leaf-blades and terminal contracted panicles. Spikelets numerous, 1-flowered, articulated to the pedicels below the empty scales, readily deciduous when mature. Scales 4, the outer 3 membranous, the first scale small, the second one much larger than the rest, many-nerved, strongly saccate at the base, the fourth scale much shorter than the third, chartaceous, enclosing a palet of similar texture and a perfect flower. Stamens 3. Styles distinct. Stigmas plumose. Grain free. [Greek, in reference to the large saccate scale of the spikelet.]

Species 6 or 7, in both the Old World and the New. Type species: *Panicum gibbum* Ell.

1. Sacciolepis striàta (L.) Nash. Gibbous Panic-grass. Fig. 387.

Holcus striatus L. Sp. Pl. 1048. 1753.
Panicum striatum Lam. Ill. 1: 172. 1791.
Panicum gibbum Ell. Bot. S. C. & Ga. 1: 116. 1817.
Sacciolepis gibba Nash, in Britt. Man. 89. 1901.
S. striata Nash, Bull. Torr. Club, 30: 383. 1903.

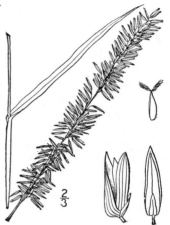

Culms erect from a creeping base, 2°–6° tall, dichotomously branched below. Lower sheaths densely hirsute, the upper generally glabrous; blades 3′–7′ long, 2″–10″ wide, usually spreading, more of less pubescent; panicle 3′–9′ long, dense and contracted; branches ½′–1′ long, erect; spikelets 1½″–2″ long, elliptic, somewhat acute; first scale about one-quarter as long as the spikelet; second scale gibbous at base, 11-nerved; third scale about equalling the second, 7-nerved, empty, the fourth one shorter than the second.

Swamps, New Jersey to Oklahoma, south to Florida and Texas. Also in the West Indies. July–Sept.

19. STEINCHÍSMA Raf. in Bull. Bot. Seringe 220. 1830.

Perennial tufted grasses, with flat leaf-blades, and loose open panicles. Spikelets 1-flowered, articulated to the pedicels below the empty scales, the outer 3 scales membranous, the first scale short, the second about as long as the spikelet, the third scale bearing in its axil a much enlarged and inflated papery palet which exceeds in length the fourth scale, the fourth scale indurated in fruit and enclosing a palet of similar texture and a perfect flower. Stamens 3. Styles long, united only at the base. Stigmas plumose. [Derivation unknown.]

Species 2. Type species: *Panicum hians* Ell.

1. Steinchisma hìans (Ell.) Nash. Gaping Panic-grass. Fig. 388.

S. hians Nash, in Small, Fl. SE. U. S. 105. 1903.

Panicum hians Ell. Bot. S. C. & Ga. 1: 118. 1817.

Glabrous, culms erect, 1°–2½° tall, generally simple, sometimes creeping at base, smooth. Blades 3′–5′ long, 1″–3″ wide, acuminate, generally erect; panicle 3′–8′ long; branches few, generally spreading, the longer ones often drooping, the lower naked below the middle; spikelets about 1″ long; fourth scale exceeded by the third and its usually empty palet which is much enlarged, generally forcing the spikelet wide open.

In moist ground, North Carolina to Missouri and Oklahoma, south to Florida and Texas. Aug.–Sept.

20. AMPHICÀRPON Raf. Am. Month. Mag. 2: 175. 1818.

Erect perennial grasses, with flat leaf-blades and spikelets of two kinds; one kind borne in terminal panicles, deciduous without perfecting fruit; the other solitary, terminating subterranean peduncles, and maturing seed. Scales 3, membranous, the innermost subtending a palet and a perfect flower; the scales of the subterranean spikelets become indurated and enclose the grain. Stamens 3. Stigmas plumose. [Greek, from the two kinds of spikelets.]

Species 2, one of them restricted to Florida. Type species: *Milium Amphicarpon* Pursh.

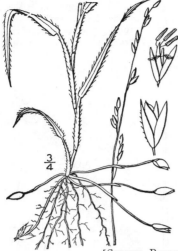

¾

1. Amphicarpon Amphicàrpon (Pursh) Nash. Pursh's Amphicarpon. Fig. 389.

M. Amphicarpon Pursh, Fl. Am. Sept. **1**: 62. *pl. 2*. 1814.
Milium ciliatum Muhl. Gram. 77. 1817.
Amphicarpum Purshii Kunth, Rev. Gram. 28. 1829.
A. Amphicarpon Nash, Mem. Torr. Club, **5**: 352. 1894.

Culms erect, 12′–18′ tall, slender, glabrous. Sheaths papillose-hirsute; ligule pilose; blades 1′–6′ long, 2″–6″ wide, erect, acuminate, hirsute and ciliate; panicle linear, 4′–6′ long, branches 3–4, erect, bearing few spikelets; spikelets about 2″ long, elliptic; outer scales 5-nerved, membranous, glabrous; subterranean spikelets ovoid in fruit, about 3″ long, acute, the scales all becoming much indurated.

In moist pine barrens, New Jersey; also in Florida (according to Chapman). Aug.–Sept.

21. CHAETÓCHLOA Scribn. Bull. U. S. Dep. Agr. Agrost. 4: 38. 1897.

[SETARIA Beauv. Agrost. 113. 1812. Not Ach. 1798.]

Mostly annual grasses with erect culms, flat leaf-blades, the inflorescence in spike-like panicles. Spikelets 1-flowered, or rarely with a second staminate flower, the basal bristles single or in clusters below the articulation of the rachilla, and therefore persistent. Scales of the spikelet 4, the three outer membranous, the third often subtending a palet and rarely a staminate flower; the inner or fourth scale chartaceous, often becoming indurated in fruit, subtending a palet of similar texture and a perfect flower. Stamens 3. Styles distinct, elongated. Stigmas plumose. Grain free, enclosed in the scales. [Greek, bristly-grass.]

Species about 35, in temperate and tropical regions. Type species: *Setaria longiseta* Beauv.

Bristles downwardly barbed.   1. *C. verticillata*.
Bristles upwardly barbed.
   Inflorescence racemose; second scale shorter than the spikelet; bristles 5–16, involucrate.
      Annual; spikelets exceeding 1½″ long; bristles yellowish brown.   2. *C. glauca*.
      Perennial; spikelets 1½″ long or less; bristles green, yellowish, or purple. 3. *C. imberbis*.
   Inflorescence paniculate; second scale as long as the spikelet; bristles 1–3, not involucrate.
      Fruiting scales dull, faintly rugose, obtuse, rather thin.
        Inflorescence 1′–3½′ long, ½′ thick or less; spikelets about 1″ long; bristles green.
                                                     4. *C. viridis*.
       Inflorescence 4′–9′ long, ½′–2′ thick; spikelets about 1½″ long; bristles usually purple.
                                                    5. *C. italica*.
      Fruiting scales shining, perfectly smooth, very acute, hard.   6. *C. magna*.

1. Chaetochloa verticillàta (L.) Scribn. Foxtail-grass. Fig. 390.

Panicum verticillatum L. Sp. Pl. Ed. 2, 82. 1762.
Setaria verticillata Beauv. Agrost. 51. 1812.
Chamaeraphis verticillata Porter, Bull. Torr. Club, **20**: 196. 1893.
Ixophorus verticillatus Nash, Bull. Torr. Club, **22**: 422. 1895.
C. verticillata Scribn. Bull. U. S. Dep. Agr. Agrost. **4**: 39. 1897.

Culms erect or decumbent, 1°–2° tall, more or less branched. Sheaths glabrous; blades 2′–8′ long, ¼′–½′ wide, scabrous above; spikes 2′–3′ long; spikelets about 1″ long, equalled or exceeded by the downwardly barbed bristles; first scale less than one-half as long as. the spikelet, 1-nerved; second and third scales 5–7-nerved, equalling the oval fourth one; palet of third scale empty.

About dwellings and in waste places, Nova Scotia and Ontario to New Jersey, Missouri and Nebraska. Naturalized from Europe. Bristly or Brown Foxtail; Rough Bristle-grass. July–Sept.

⅔

2. Chaetochloa glaùca (L.) Scribn.   Yellow Foxtail.   Pigeon-grass.   Fig. 391.

Panicum glaucum L. Sp. Pl. 56.   1753.
Setaria glauca Beauv. Agrost. 51.   1812.
Chamaeraphis glauca Kuntze, Rev. Gen. Pl. 767.   1891.
Ixophorus glaucus Nash, Bull. Torr. Club, **22** : 423.   1895.
Chaetochloa glauca Scribn. Bull. U. S. Dep. Agr. Agrost.
   4 : 39.   1897.

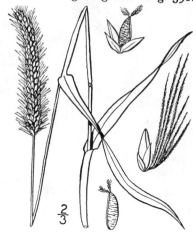

Culms erect or sometimes decumbent, 1°–4° tall,
more or less branched.   Sheaths glabrous; blades
2′–6′ long, 2″–4″ wide; spikes 1′–4′ long; spikelets
1¼″–1½″ long, oval, much shorter than the upwardly barbed yellowish brown bristles; first scale
1–3-nerved, somewhat shorter than the 5-nerved
second one; third scale 5-nerved, equalling the
fourth which is coarsely transversely rugose, very
convex, V-shaped in cross-section, about twice as
long as the second; palet of third scale usually
empty.

In waste places and cultivated grounds, throughout
North America except the extreme north.   Often a
troublesome weed.   Naturalized from Europe.   Yellow
or Glaucous Bristly Foxtail.   July–Sept.

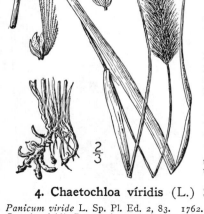

3. Chaetochloa imbérbis (Poir.) Scribn.   Perennial Foxtail-grass.   Fig. 392.

Panicum imberbe Poir. in Lam. Encycl. Suppl. **4** : 272.
   1816.
C. imberbis Scribn. Bull. U. S. Dep. Agr. Agrost. **4** : 39.
   1897.
C. versicolor Bicknell, Bull. Torr. Club, **25** : 105.   1898.
C. occidentalis Nash, in Britt. Man. 90.   1901.

Culms single or somewhat tufted, from a branching rootstock, 1°–3° tall; leaf-sheaths glabrous, compressed, keeled; blades up to 1° long and 4″ wide,
glabrous, or nearly so; inflorescence 1′–3′ long,
7″–10″ wide, the bristles 4″–6″ long; spikelets
1¼″–1½″ long, the first scale about ½ as long as the
spikelet, 3-nerved, the second scale 3–5-nerved, the
third scale 5-nerved, the fourth scale strongly transverse-rugose, elliptic, often purple-tipped.

In moist or saline soil, Massachusetts to Kansas, south
to Florida and Texas.   Also in tropical America, and in
the Bahamas.

4. Chaetochloa víridis (L.) Scribn.   Green Foxtail-grass.   Fig. 393.

Panicum viride L. Sp. Pl. Ed. 2, 83.   1762.
Setaria viridis Beauv. Agrost. 51.   1812.
Chamaeraphis viridis Porter, Bull. Torr. Club, **20** : 196.
   1893.
Ixophorus viridis Nash, Bull. Torr. Club, **22** : 423.   1895.
Chaetochloa viridis Scribn. Bull. U. S. Dep. Agr. Agrost.
   4 : 39.   1897.

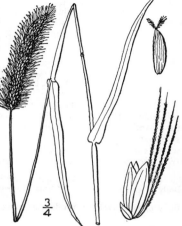

Culms erect or ascending, 1°–3° tall, simple or
branched.   Sheaths glabrous; blades 3′–10′ long, 2″–6″
wide, usually scabrous above; spikes 1′–4′ long; spikelets about 1″ long, elliptic, much shorter than the green
or sometimes yellowish, upwardly barbed bristles; first
scale less than one-half as long as the spikelet,
1–3-nerved; second and third scales 5-nerved; fourth
scale finely and faintly transversely rugose, or pitted,
striate, only moderately convex, equalling or slightly
exceeding the second; palet of third scale usually empty.

In waste places and cultivated grounds, throughout North
America except the extreme north, and often a trouble-
some weed.   Naturalized from Europe.   Green Bottle-grass,
Wild Millet, Pigeon-grass.   July–Sept.

5. Chaetochloa itálica (L.) Scribn. Italian Millet. Hungarian Grass. Fig. 394.

Panicum italicum L. Sp. Pl. 56. 1753.
Setaria italica R. & S. Syst. **2**: 493. 1817.
Chamaeraphis italica Kuntze, Rev. Gen. Pl. 768. 1891.
Ixophorus italicus Nash, Bull. Torr. Club, **22**: 423. 1895.
C. italica Scribn. Bull. U. S. Dep. Agr. Agrost. **4**: 39. 1897.

Culms erect, 2°–5° tall. Sheaths smooth or scabrous; blades 6′–1° or more in length, ¼′–1½′ wide, generally scabrous; spikes 4′–9′ long, ½′–2′ thick, usually very compound; spikelets about 1½″ long, elliptic, equalled or exceeded by the upwardly barbed generally purplish bristles; first scale less than one-half as long as the spikelet, 1–3-nerved; second and third 5–7-nerved; fourth scale equalling or somewhat exceeding the second, finely and faintly transverse-rugose, or pitted, striate, only moderately convex; palet of third scale minute or wanting.

In waste places, escaped from cultivation, Quebec to Minnesota, south to Florida and Texas. Native of the Old World. German or hungarian millet. Golden or cat-tail millet. July–Sept.

6. Chaetochloa mágna (Griseb.) Scribn. Giant Foxtail-grass. Fig. 395.

Setaria magna Griseb. Fl. Brit. W. I. 554. 1864.

C. magna Scribn. Bull. U. S. Dep. Agr. Agrost. **4**: 39. 1897.

Culms 4°–16° tall, stout; sheaths densely hirsute on the margins, otherwise glabrous; blades up to 3° long, 1′–2′ wide, very rough on both surfaces; panicles 8′–2° long, 1′–2½′ thick, nodding above; spikelets 1″ long, the first scale about ½ as long as the spikelet, 3-nerved, the second scale as long as the spikelet, 7-nerved, the fourth scale oval, very acute, smooth and shining.

In swamps, Delaware and Virginia to Florida and Texas; also in Cuba. July and Aug.

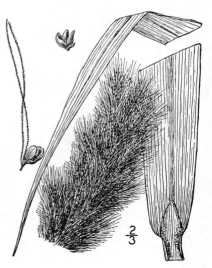

22. CÉNCHRUS L. Sp. Pl. 1049. 1753.

Annual or perennial grasses with usually flat leaves. Inflorescence in spikes. Spikelets subtended by a spiny involucre which is deciduous with them at maturity. Scales 4; the first hyaline; the second and third membranous, the latter sometimes having a palet and staminate flower in its axil; the fourth chartaceous, subtending a palet of similar structure which encloses a perfect flower. Stamens 3. Styles united below. Stigmas plumose. Grain free, enclosed in the scales. [Ancient Greek name for some grass, probably millet.]

About 20 species, in tropical and temperate regions. Type species: *Cenchrus echinatus* L.
Body of the involucre 3″–4″ broad, pubescent with very long hairs, the spines commonly 2½″–4″ long.
                                                                          1. *C. tribuloides.*
Body of the involucre rarely exceeding 2½″ broad, pubescent, the spines 1½″–2″ long.
                                                                          2. *C. carolinianus.*

1. Cenchrus tribuloìdes L. Bur-grass. Hedgehog-grass. Fig. 396.

Cenchrus tribuloides L. Sp. Pl. 1050. 1753.
C. macrocephalus Scribn. Bull. U. S. Dep. Agr. Agrost.
17: 110. 1899.

Culms at first erect, up to 1° tall, later elongated,
trailing and much-branched; sheaths glabrous, excepting the ciliate margin, compressed, very loose,
the upper one often partly enclosing the inflorescence; blades flat or complanate, smooth and glabrous, 4′ long or less, 2″–4″ wide; spikes stout, 1′–2′
long; involucres 5–12, the body 3″–4″ broad, pubescent with very long hairs, the spines usually 2½″–4″
long; spikelets 3″–3½″ long, not exserted beyond
the involucre.

In sands along the coast, Long Island and New Jersey
to Florida and Mississippi. Bear-grass, Sand-spur,
Sand-bur. Aug. and Sept.

$\frac{2}{3}$

2. Cenchrus caroliniànus Walt. Small Bur-grass. Fig. 397.

Cenchrus carolinianus Walt. Fl. Car. 79. 1788.

Culms at first erect, later prostrate and forming mats,
8′–2° long or more, branched; sheaths glabrous, excepting the ciliate margin, compressed; blades 2½′–5′ long,
2″–4″ wide, smooth or rough, usually flat; spikes 1′–3′
long; involucres 6–20, the body rarely exceeding 2½″
broad, pubescent with relatively short hairs, the spines
1½″–2″ long; spikelets 3″–3½″ long, usually not exserted beyond the involucre.

In dry sandy places, Maine to Wisconsin and California,
south to Florida and Mexico; also in the Bahamas and
tropical America. Figured for *C. tribuloides* L. in our first
edition. June–Sept.

$\frac{2}{3}$

23. ZIZANIÓPSIS Doell & Aschers. in Mart. Fl. Bras. 2: Part 2, 12. 1871.

Tall aquatic monoecious grasses, with long flat leaf-blades and paniculate inflorescence.
Spikelets 1-flowered, the staminate borne at the top of the branches, the pistillate at the
base. Scales 2, nearly equal, membranous, the outer one in the pistillate spikelets broad,
acute and bearing an awn. Stamens 6. Styles united.
Grain nearly globose, the pericarp readily separable.
[Name in allusion to the resemblance of this grass
to *Zizania*.]

A monotypic genus, of temperate and tropical America.
Type species: *Zizania miliacea* Michx.

1. Zizaniopsis miliàcea (Michx.) Doell & Aschers. Zizaniopsis. Fig. 398.

Zizania miliacea Michx. Fl. Bor. Am. 1: 74. 1803.
Z. miliacea Doell & Aschers.; Baill. Hist. Pl. 12: 293.
1893.

Culms 4°–15° tall from a long and creeping rootstock, robust, glabrous. Sheaths loose, glabrous;
ligule 4″–7″ long, thin-membranous; blades 1° long
or more, ½′–1′ wide, smooth, glabrous; panicle dense,
1°–1½° long, narrow; branches erect; staminate
spikelets 3″–4″ long, the outer scale 5-nerved, the
inner 3-nerved, both acute; pistillate spikelets about
3″ long, the outer scale about equalling the inner,
bearing an awn 1″–3″ long, scabrous, 5-nerved;
inner scale 3-nerved, acute.

Swamps, Virginia to Ohio (according to Riddell),
south to Florida and Texas. June–July.

$\frac{3}{4}$

24. ZIZÀNIA L. Sp. Pl. 991. 1753.

A tall aquatic monoecious grass with long flat leaf-blades and an ample panicle. Spikelets 1-flowered, the pistillate borne on the upper branches of the panicle, the staminate on the lower. Scales 2, membranous, the outer somewhat longer, acute in the staminate, long-awned in the pistillate spikelets. Stamens 6. Styles nearly distinct. Grain linear. [From an ancient Greek name for Darnel.]

A monotypic genus of North America and Asia. Type species: *Zizania aquatica* L.

1. Zizania aquática L. Wild Rice. Indian Rice. Water Oats. Fig. 399.

Zizania aquatica L. Sp. Pl. 991. 1753.

Zizania palustris L. Mant. 295. 1771.

Culms erect from an annual root, 3°–10° tall, smooth and glabrous. Sheaths loose, glabrous; ligule about ¼′ long, thin-membranous; blades 1° or more long, ¼′–1½′ wide, more or less roughened, especially above, glabrous; panicle 1°–2° long, the upper branches erect, the lower ascending or spreading; staminate spikelets 3″–6″ long, scales acute or awn-pointed, outer 5-nerved, the inner 3-nerved; scales of the linear pistillate spikelets 4″–12″ long, the outer one 5-nerved, with an awn 1′–2′ long, the inner narrower, 3-nerved, awn-pointed.

In swamps, New Brunswick to Manitoba, south to Florida and Texas. Canada-rice. Water-rice. June–Oct.

25. HOMALOCÉNCHRUS Mieg, in Soc. Phys.-med. Basil, Act. Helv. 4: 307. 1760.

[LEERSIA Soland.; Sw. Prod. 21. 1788. Not Hedw. 1782.]

Marsh grasses with flat narrow generally rough leaf-blades, and paniculate inflorescence. Spikelets 1-flowered, perfect, strongly flattened laterally, and usually more or less imbricated. Scales 2, chartaceous, the outer one broad and strongly conduplicate, the inner much narrower. Stamens 1–6. Styles short, distinct. Stigmas plumose. Grain ovoid, free. [Greek, in reference to the supposed resemblance of these grasses to Millet.]

About 5 species, natives of temperate and tropical countries. Besides the following, 2 others occur in the southern United States. Type species: *Phalaris oryzoides* L.

Spikelets oblong, their width less than one-half their length, somewhat imbricated.
    Spikelets 1¼″–1½″ long; panicle-branches usually rigid.   1. *H. virginicus.*
    Spikelets 2″–2½″ long; panicle-branches generally lax.   2. *H. oryzoides.*
Spikelets oval, their width more than one-half their length, much imbricated.   3. *H. lenticularis.*

1. Homalocenchrus virgínicus (Willd.) Britton. White Grass. Fig. 400.

Leersia virginica Willd. Sp. Pl. 1: 325. 1797.
Asprella virginica R. & S. Syst. 2: 266. 1817.
Homalocenchrus virginicus Britton, Trans. N. Y. Acad. Sci. 9: 14. 1889.

Culms glabrous, decumbent, 1°–3° long, much branched, slender, smooth. Sheaths usually shorter than the internodes; ligule short; blades 2′–6′ long, 1″–8″ wide, acute, usually narrowed toward the base, scabrous; terminal panicle finally long-exserted, 3′–8′ long, its branches generally spreading, usually naked below the middle; lateral panicles smaller and usually included; spikelets 1¼″–1½″ long, about ½″ wide, oblong, appressed; outer scale hispid on the keel and margins; inner scale hispid on the keel; stamens 1 or 2.

Swamps or wet woods, Maine to Ontario, Florida and Texas. White or False rice. White grama. Aug.–Sept.

2. Homalocenchrus oryzoìdes (L.) Poll.   Rice
Cut-grass. Fig. 401.

Phalaris oryzoides L. Sp. Pl. 55. 1753.
Homalocenchrus oryzoides Poll. Hist. Pl. Palat. **1** : 52.   1776.
Leersia oryzoides Sw. Prodr. 21.   1788.

Culms glabrous, decumbent, 1°–4° long, much branched, rather stout, smooth. Sheaths shorter than the internodes, very rough; ligule very short; blades 3′–10′ long, 2″–5″ wide, acute, narrowed toward the base, scabrous; terminal panicle 5′–9′ long, finally long-exserted, its branches lax, naked at the base, at first erect, later more or less widely spreading; lateral panicles generally included; spikelets 2″–2½″ long, about ¾″ wide, elliptic; scales pubescent, the outer one hispid on the keel and on the margins; inner scale much narrower, hispid on the keel; stamens 3; anthers yellow.

In swamps and along streams, often forming dense tangled masses, Newfoundland to Oregon, south to Florida and Texas. Also in the temperate parts of Europe and Asia. False grass. Rice's-cousin. Aug.–Sept.

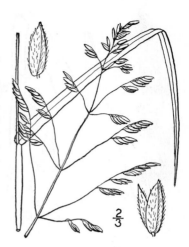

3. Homalocenchrus lenticulàris (Michx.) Scribn.
Catch-fly Grass. Fig. 402.

Leersia lenticularis Michx. Fl. Bor. Am. **1** : 39.   1803.

H. lenticularis Scribn. Mem. Torr. Club, **5** : 33.   1894.

Culms glabrous, erect, 2°–4° tall, usually simple, smooth. Sheaths shorter than the internodes, scabrous; ligule very short; blades 4′–12′ long, 4″–10″ wide, acute, more or less narrowed at the base, scabrous; panicle 4½′–9′ long, finally exserted, its branches lax, naked below, at first erect, later spreading; spikelets much imbricated, 2″–2½″ long, 1″–1¾″ wide, broadly oval; scales smooth or sparingly hispid-scabrous, the outer one strongly 3-nerved, hispid on the keel and margins, the inner much narrower, strongly 1-nerved, hispid on the keel; stamens 2.

Wet grounds. Virginia to Minnesota, south to Florida and Texas.   July–Sept.

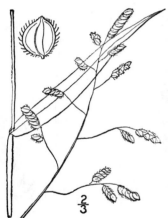

26. PHÁLARIS L. Sp. Pl. 54.   1753.

Annual or perennial grasses with flat leaf-blades, the inflorescence spike-like, capitate or a narrow panicle. Spikelets crowded, 1-flowered. Scales 5, the first and second about equal in length, strongly compressed laterally, usually wing-keeled; third and fourth scales much smaller or reduced to mere rudiments; fifth scale subtending a palet similar to itself and a perfect flower. Stamens 3. Styles distinct. Stigmas plumose. Grain oblong, free, smooth, enclosed in the scales. [Greek, alluding to the shining grain.]

About 10 species, mostly natives of southern Europe. Besides the following, 3 others occur in the United States. Type species : *Phalaris arundinacea* L.

Outer scales not winged; inflorescence a narrow panicle.   1. *P. arundinacea.*
Outer scales broadly winged; inflorescence a spike or spike-like panicle.
   Spikelets narrow; third and fourth scales much reduced, rigid, subulate, hairy.
                                                                       2. *P. caroliniana.*
   Spikelets broad; third and fourth scales thin-membranous, broadly lanceolate, glabrous or
     sparingly hairy.   3. *P. canariensis.*

3/4

1. Phalaris arundinàcea L.   Reed Canary-grass.
Fig. 403.

Phalaris arundinacea L. Sp. Pl. 55. 1753.

Glabrous, culms erect, 2°–5° tall, simple, smooth. Sheaths shorter than the internodes; ligule 1″–3″ long, obtuse, membranous; blades 3½′–10′ long, 3″–8″ wide, acuminate, smooth or scabrous; panicle 3′–8′ long, dense, its branches ½′–1½′ long, erect or sometimes slightly spreading; spikelets 2½″–3″ long; outer scales scabrous, 3-nerved; third and fourth scales less than one-half as long as the fifth, subulate, rigid, hairy; fifth scale about three-fourths as long as the spikelet, chartaceous, pubescent with long appressed silky hairs, subtending a palet of similar texture and a perfect flower.

In moist or wet soil, Nova Scotia to British Columbia, south to New Jersey and Colorado. Also in Europe and Asia. Lady-grass, Spires, Doggers, Sword-grass, Ladies' or Bride's-laces, London-lace. July–Aug. The Ribbon-grass or Painted-grass of cultivation, the so-called variety *picta*, has leaves variegated with green and white stripes, is a derivative of this species, and sometimes escapes from gardens.

2. Phalaris caroliniàna Walt.   Carolina Canary-grass.   Fig. 404.

Phalaris caroliniana Walt. Fl. Car. 74. 1788.
Phalaris intermedia Bosc.; Poir. in Lam. Encycl. Suppl.
  1: 300. 1810.
Phalaris americana Ell. Bot. S. C. & Ga. 1: 101. 1817.

Culms 1°–3½° tall, erect or sometimes decumbent at base, simple or somewhat branched, smooth or roughish, glabrous. Sheaths usually shorter than the internodes; ligule 1″–3″ long, rounded, thin-membranous; blades 2′–6′ long, 2″–5″ wide, smooth or slightly scabrous; spike-like panicle 1′–4′ long, dense, its branches about ½′ long, erect; spikelets 2½″ long, the outer scales more or less scabrous, 3-nerved, wing-keeled; third and fourth scales less than one-half as long as the fifth, subulate, hairy; fifth scale about two-thirds as long as the spikelet, acuminate, pubescent with long appressed silky hairs.

In moist soil, South Carolina to Missouri and California, south to Florida, Texas and Arizona. Southern or Wild Canary-grass. Ribbon-grass, California Timothy, Southern Reed-grass. Fox-tail-grass. June–Aug.

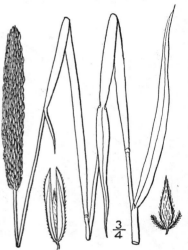

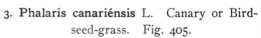

3/4

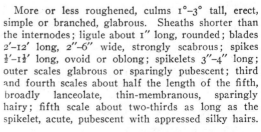

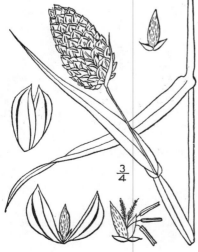

3/4

3. Phalaris canariénsis L.   Canary or Bird-seed-grass.   Fig. 405.

Phalaris canariensis L. Sp. Pl. 54. 1753.

More or less roughened, culms 1°–3° tall, erect, simple or branched, glabrous. Sheaths shorter than the internodes; ligule about 1″ long, rounded; blades 2′–12′ long, 2″–6″ wide, strongly scabrous; spikes ½′–1½′ long, ovoid or oblong; spikelets 3″–4″ long; outer scales glabrous or sparingly pubescent; third and fourth scales about half the length of the fifth, broadly lanceolate, thin-membranous, sparingly hairy; fifth scale about two-thirds as long as the spikelet, acute, pubescent with appressed silky hairs.

In waste places, Nova Scotia to Ontario, Virginia, Missouri and Colorado. Naturalized from Europe. The grain is the common food of canary birds. July–Aug.

27. ANTHOXÁNTHUM L. Sp. Pl. 28. 1753.

Fragrant annual or perennial grasses, with flat leaf-blades and spike-like panicles. Spikelets 1-flowered, narrow, somewhat compressed. Scales 5; the two outer acute or produced into a short awn, the first shorter than the second; third and fourth scales much shorter, 2-lobed, awned on the back; the fifth scale shorter than the others, obtuse. Stamens 2. Styles distinct. Stigmas elongated, plumose. Grain free, enclosed in the scales. [Greek, referring to the yellow hue of the spikelets in some species.]

A genus of 4 or 5 species, natives of Europe. Type species: *Anthoxanthum odoratum* L.

Perennial; third and fourth scales pubescent nearly to the apex, the awn of the latter arising about one-fifth above the base. 1. *A. odoratum.*

Annual; third and fourth scales pubescent only below the middle, the awn of the latter arising about one-third above the base. 2. *A. Puelii.*

1. Anthoxanthum odoràtum L. Sweet Vernal-grass. Fig. 406.

Anthoxanthum odoratum L. Sp. Pl. 28. 1753.

Culms 1°–2° tall, erect, simple or branched, smooth and glabrous. Sheaths shorter than the internodes; ligule 1″–2″ long, acute, membranous; blades ½′–6′ long, 1″–3″ wide, glabrous or nearly so; spike-like panicles 1′–2½′ long, branches short, erect or ascending; spikelets 4″ long, crowded; outer scales acute, glabrous or pubescent, the first 1-nerved, half as long as the second which is 3-nerved; the third and fourth very hairy, the former with an awn longer than itself inserted about the middle, the fourth scale bearing near the base an awn more than twice its length; fifth scale about two-thirds as long as the fourth, obtuse or rounded at the apex, and bearing a fertile flower.

In fields and meadows throughout nearly the whole of North America. Very fragrant in drying. Naturalized from Europe. Spring- or Prim-grass. June–July.

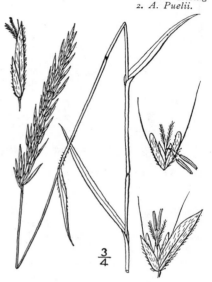

¾

⅔

2. Anthoxanthum Puélii Lecoq & Lamotte. Long-awned Vernal-grass. Fig. 407.

A. Puelii Lecoq & Lamotte, Cat. Pl. Auver. 385. 1848.

Culms up to 1° tall, slender, often branching above the base, leaves smooth and glabrous; sheaths shining; ligule scarious, obtuse, about 1″ long; blades flat, up to 4′ long and 2″ wide; panicle ½′–1½′ long, spike-like; spikelets, exclusive of the awns, about 3″ long, the first scale about one-half as long as the second, the third and fourth scales about 1½″ long, pubescent below the middle, the former with an awn twice as long as itself, the fourth bearing an awn 2½–4 times as long as itself.

Sparingly introduced, or escaped from cultivation in waste places, New England to Ontario and Pennsylvania. A native of Europe. May to August.

28. SAVASTÀNA Schrank, Baier. Fl. 1: 337. 1789.

[HIEROCHLOË J. G. Gmel. Fl. Sib. 1: 101. 1747.]

Aromatic perennial grasses, with flat leaf-blades and contracted or open panicles. Spikelets 3-flowered, the terminal flower perfect, the others staminate. Scales 5; the first and second nearly equal, acute, glabrous; the third and fourth somewhat shorter, obtuse, entire, emarginate, 2-toothed or 2-lobed, with or without an awn, enclosing a palet and stamens; fifth scale often produced into a short awn, enclosing a palet and perfect flower. Stamens in the staminate flowers 3, in the perfect 2. Styles distinct. Stigmas plumose. Grain free, enclosed in the scales. [Name in honor of Francesco Eulalio Savastano.]

About 8 species, natives of temperate and cold regions. Type species: *Savastana hirta* Schrank.

Third and fourth scales not awned,
Entire, culms 1°–3° tall.
Panicle 4′ long or less, its branches 2′ long or less; blades short, broad. 1. *S. odorata.*

Panicle 6' long or more, its branches 4'–8' long; blades long, narrow.   2. *S. Nashii.*
Erose-truncate, culms 6' tall or less.   3. *S. pauciflora.*
Third and fourth scales awned.   4. *S. alpina.*

1. Savastana odoràta (L.) Scribn. Holy Grass. Seneca-grass. Fig. 408.

Holcus odoratus L. Sp. Pl. 1048. 1753.
Hierochloa borealis R. & S. Syst. 2 : 513. 1817.
Hierochloa odorata Wahl. Fl. Ups. 32. 1820.
Savastana odorata Scribn. Mem. Torr. Club, 5 : 34. 1894.

Glabrous, culms 1°–2° tall, erect, simple, smooth. Sheaths smooth; ligule 1″–2″ long; lower blades elongated, 4'–8' long, 1″–3″ wide, scabrous, the upper ones ½'–2' long; panicle 2'–4' long, its branches 1'–2' long, usually spreading, naked below; spikelets yellowish-brown and purple, 2″–3″ long; first and second scales about equal, glabrous; third and fourth villous and strongly ciliate, entire, awn-pointed, the fifth smaller than the others, villous at the apex.

Labrador and Newfoundland to Alaska, south to New Jersey, Iowa and Colorado. Also in northern Europe and Asia. June–July. This and other sweet-scented grasses are strewn before the churches in northern Europe, whence the name Holy-grass. Also known as Vanilla-grass.

2. Savastana Náshii Bicknell. Nodding Vanilla-grass. Fig. 409.

Savastana Nashii Bicknell, Bull. Torr. Club, 25 : 104. *pl. 328.* 1898.

Plant smooth, glabrous and shining. Culms erect, slender, simple, 2°–3° tall. Sheaths overlapping, striate; ligule scarious, 2″–3″ long; blades erect or ascending, elongated, a little roughened above, the culm leaves 5 or 6, 2'–8' long, 2″–3″ wide, acuminate; panicle long-exserted, loose and open, 7'–17' long, its apex nodding, the capillary branches drooping, the larger 3'–7' long, in pairs, the divisions more or less flexuous; spikelets 2½″–4″ long, on capillary pedicels; scales 5, the outer 2 empty, abruptly long-acuminate, the first 1-nerved, the second 3-nerved, the third and fourth scales about 2½″ long, rough, ciliate on the margins with ascending hairs, 5-nerved, acute, usually awn-pointed, the fifth scale smaller, smooth, hispidulous at the apex, sometimes awn-pointed.

Along brackish marshes, New York City. July–Aug.

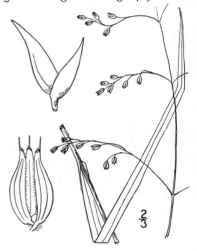

3. Savastana pauciflòra (R. Br.) Scribn. Arctic Holy Grass. Fig. 410.

Hierochloa pauciflora R. Br. App. Parry's Voy. 293. 1824.

S. pauciflora Scribn. Mem. Torr. Bot. Club, 5 : 353. 1894.

Glabrous, culms 6' high or less, erect, simple, smooth. Sheaths mostly at the base of the culm, overlapping; ligule about ½″ long; blades smooth, the basal ones 1'–2' long, ½″ wide, involute at least when dry; culm-blades ½' long or less, 1″ wide, flat; panicle less than 1' long, contracted; spikelets few, 1½″–2″ long; first and second scales 1½″–2″ long, smooth and glabrous; third and fourth shorter, scabrous, erose-truncate, the fifth shorter than the others, obtuse, villous at the apex.

Arctic America. Summer.

4. Savastana alpìna (Sw.) Scribn. Alpine Holy Grass. Fig. 411.

Holcus alpinus Sw.; Willd. Sp. Pl. **4**: 937. 1806.
Hierochloa alpina R. & S. Syst. **2**: 515. 1817.
Savastana alpina Scribn. Mem. Torr. Club, **5**: 34. 1894.

Glabrous and smooth, culms 6'-18' tall, erect, simple. Sheaths shorter than the internodes; ligule less than 1" long; lower blades elongated, 3'-6' long, about 1" wide, the upper much shorter, ½'-2' long, 1"-2" wide; panicle ¾-1½' long, contracted, branches short, erect or ascending; occasionally the panicle is larger with longer and spreading branches; spikelets 2½"-3½" long, crowded; first and second scales glabrous, 2½"-3½" long; third and fourth shorter, scabrous, ciliate on the margins, the former bearing an awn about 1" long, the latter with a more or less bent awn about 3" long; fifth scale shorter than the others, acute, usually awn-pointed, villous at the apex.

Greenland to Alaska, south to the high mountains of New England and New York. Also in northern Europe and Asia. July–Aug.

29. MÍLIUM L. Sp. Pl. 61. 1753.

Annual or perennial grasses, with flat leaf-blades and terminal lax panicles. Spikelets 1-flowered. Scales 3, obtuse, not awned; the outer about equal; the third thin-membranous, at length rigid, glabrous or pubescent, awnless, and with the margins inrolled; palet scarcely shorter. Stamens 3. Styles short, distinct. Stigmas plumose. Grain ovoid or oblong, free, tightly enclosed in the rigid and shining scale and palet. [Latin name for Millet.]

Species 5 or 6, chiefly in Europe and Asia. Type species: *Milium effusum* L.

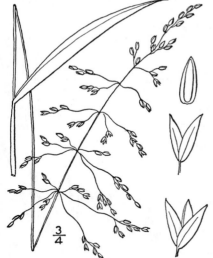

1. Milium effùsum L. Tall Millet-grass.
Fig. 412.

Milium effusum L. Sp. Pl. 61. 1753.

Glabrous throughout, culms 2°-6° tall, erect, simple, smooth. Sheaths shorter than the internodes; ligule 1½"-3" long, truncate, erose-dentate; blades 3'-9' long, 3"-8" wide, narrowed toward the base, acuminate, smooth or scabrous; panicle 3'-10' in length, lax, its branches 2'-3' long, slender, somewhat flexuous, naked at base and dividing above the middle, at length widely spreading; spikelets 1¼"-1½" long; outer scales equal, smooth or scabrous, the third scale shorter, smooth, white.

In woods, Cape Breton Island to Ontario, south to Massachusetts, Pennsylvania and Illinois. Also in northern Europe and Asia. June–July.

30. ERIÓCOMA Nutt. Gen. 1: 40. 1818.

Perennial tufted grasses, with usually involute leaves and a contracted or open panicle. Spikelets 1-flowered. Scales 3; outer 2 membranous, glabrous; third scale firmer, becoming hard in fruit, densely pubescent with long silky hairs, and bearing a terminal readily deciduous awn, the callus at the base of the scale short and obtuse. Stamens 3. Stigmas plumose. Grain free, enclosed in the scale. [Greek, referring to the copious silky hairs of the flowering scale.]

A small genus of 2 or possibly more species, natives of western North America. Type species: *Stipa membranacea* Pursh.

1. Eriocoma cuspidàta Nutt.   Wild or Indian Millet.   Silky Grass.   Fig. 413.

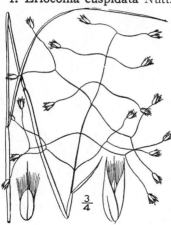

Stipa membranacea Pursh, Fl. Am. Sept. 728. 1814. Not L. 1753.
Eriocoma cuspidata Nutt. Gen. 1: 40. 1818.
Oryzopsis cuspidata Benth.; Vasey, Special Rep. U. S. Dept. Agric. 63: 23. 1883.
O. membranacea Vasey, Grasses S. W. Part 2, *pl. 10.* 1891.

Culms glabrous, 1°–2° tall, erect, rigid, simple. smooth. Sheaths usually shorter than the internodes, smooth or somewhat rough; ligule 1″–2″ long, acute; leaves 6′–12′ long, less than 1″ wide, involute, stiff, smooth or somewhat scabrous; panicle 6′–1° long, diffuse, generally partially included in the upper sheath, its branches widely spreading and many times forked, the ultimate divisions flexuous; outer scales of the spikelet 3″–4″ in length, long-acuminate, glabrous; third scale about one-half as long, acute, densely pubescent with long silky erect hairs nearly twice its own length, the awn 2″–3″ long.

On prairies, Saskatchewan to Washington, Iowa, Texas and Mexico.   Bunch-grass.   May–July.

31. ORYZÓPSIS Michx. Fl. Bor. Am. 1: 51.   1803.

[URACHNE Trin. Fund. Agrost. 109.   1818.]

Usually tufted grasses, with flat or convolute leaves and paniculate inflorescence. Spikelets 1-flowered, broad. Scales 3; the two lower about equal, obtuse or acuminate; the third scale shorter or a little longer, broad, bearing a terminal awn which is early deciduous, the callus at the base of the scale short and obtuse, or a mere scar. Stamens 3. Styles distinct. Stigmas plumose. Grain oblong, free, tightly enclosed in the convolute scale. [Greek, in allusion to the supposed resemblance of these grasses to rice.]

About 24 species, distributed through temperate and subtropical regions, rarely extending into the tropics. Besides the following, some 7 others occur in the western parts of North America. Type species: *Oryzopsis asperifolia* Michx.

Spikelet, exclusive of awn, 1¼″–2″ long.
   Awn not 1″ long, shorter than the scale; outer scales 1½″–2″.   1. *O. pungens.*
   Awn 3″–4″ long, more than twice as long as the scale; outer scales about 1¼″ in length.
                                                                                        2. *O. micrantha.*
Spikelet, exclusive of awn, 3″–4″ long.
   Culms nearly naked, leaves all crowded at the base; panicle 2′–3′ long, its branches 1′ in
      length or less, erect.   3. *O. asperifolia.*
   Culms leafy to the top; panicle 6′–12′, branches 2′–4′ long, spreading.   4. *O. racemosa.*

1. Oryzopsis púngens (Torr.) Hitchc.   Slender
Mountain-rice.   Fig. 414.

Milium pungens Torr.; Spreng. Neue Entd. **2**: 102. 1821.
Oryzopsis canadensis Torr. Fl. N. Y. **2**: 433. 1843.
Oryzopsis juncea B.S.P. Prel. Cat. N. Y. 67. 1888.
Oryzopsis pungens Hitchc. Contr. U. S. Nat. Herb. **12**: 151. 1908.

Culms glabrous, 6′–2° tall, erect, slender, simple, smooth. Sheaths shorter than the internodes, usually crowded at the base of the culm; ligule about 1″ long, decurrent; blades smooth or scabrous, erect, involute, the basal about one-half the length of the culm, occasionally equalling it, filiform, those of the culm 1′–4′ long, the uppermost often very small or reduced to the sheath only; panicle 1′–2½′ long, the branches ½′–1′ in length, erect or ascending, the lower half naked; spikelets 1½″–2″ long, the outer scales about equal, glabrous, whitish; third scale about the same length or a little longer, pubescent with short appressed silky hairs, the awn less than 1″ long.

In dry rocky places, Pennsylvania to Labrador and British Columbia.   May–June.

2. Oryzopsis micrántha (Trin. & Rupr.) Thurb.　Small-flowered Mountain-rice. Small Indian Millet.　Fig. 415.

Urachne micrantha Trin. & Rupr. Mem. Acad. St. Petersb. (VI.) **5**: 16.　1842.
O. micrantha Thurb. Proc. Phila. Acad. **1863**: 78.　1863.

Culms glabrous, $1°-2\frac{1}{2}°$ tall, erect, slender, simple, smooth. Sheaths shorter than the internodes; ligule about $\frac{1}{2}''$ long, truncate; blades erect, scabrous, the basal one-half the length of the culm, less than $\frac{1}{2}''$ wide, usually more or less involute, the culm leaves $2'-8'$ long, $\frac{1}{2}''-1''$ broad, the larger attenuate into a long slender point; panicle $3'-6'$ long, the branches finally spreading, the lower ones $1'-2'$ long, naked for about two-thirds their length; spikelets $1''-1\frac{1}{4}''$ long, the outer scales about equal, acute, glabrous; third scale shorter, glabrous, bearing an awn $3''-4''$ long.

In cañons and on dry hills, Saskatchewan to Nebraska, New Mexico and Arizona.　June–July.

3. Oryzopsis asperifòlia Michx.　White-grained Mountain-rice.　Fig. 416.

Oryzopsis asperifolia Michx. Fl. Bor. Am. **1**: 51.　1803.
Urachne asperifolia Trin. Unifl. **1**: 174.　1824.

Culms glabrous, $10'-20'$ tall, erect, simple, smooth or scabrous. Sheaths $1'-2'$ long, crowded at base; ligule very short, truncate; blades erect, scabrous, especially above, the basal ones elongated, often equalling or exceeding the culm, $2''-4''$ wide, attenuate into a long point, the 1 or 2 culm-blades much reduced, less than $\frac{1}{2}'$ long; panicle $2'-3'$ long, contracted, the branches $1'$ in length or less, erect; spikelet, exclusive of awns, $3''-4''$ long; outer scales glabrous, usually apiculate, the first somewhat shorter; third scale whitish, equalling the second or a little shorter, sparingly pubescent, the awn $3\frac{1}{2}''-5''$ long.

In woods, Newfoundland to British Columbia, south to New Jersey, Pennsylvania, Minnesota and in the Rocky Mountains to New Mexico.　May–June.

4. Oryzopsis racemòsa (J. E. Smith) Ricker.　Black-fruited Mountain-rice.　Fig. 417.

Milium racemosum J. E. Smith, in Rees, Cyclop. **23**: no. **15**. 1813.
Oryzopsis melanocarpa Muhl. Gram. 79.　1817.
Urachne racemosa Trin. Unifl. **1**: 174.　1824.
Oryzopsis racemosa Ricker; Hitchc. Rhodora **8**: 210.　1906.

Glabrous, culms $1\frac{1}{2}°-3°$ tall, erect, simple, roughish. Sheaths smooth or scabrous, the lower ones usually longer, the upper slightly shorter than the internodes; ligule very short; blades $5'-12'$ long, $2''-7''$ wide, narrowed toward the base, acuminate at apex into a long slender point, scabrous especially above; panicle branched or nearly simple, $3'-12'$ long, its branches $2'-4'$ long, spreading or ascending, the lower half naked; outer scales of the spikelet about equal, $3''-4''$ in length, acute; third scale shorter, acute, dark colored, sparingly pubescent, the awn $8''-12''$ long.

Rocky woods, Maine to Ontario, south to Maryland and Kentucky.　July–Aug.

32. STÌPA L. Sp. Pl. 78. 1753.

Generally tall grasses, the leaf-blades usually convolute, rarely flat, the inflorescence panic-ulate. Spikelets 1-flowered, narrow. Scales 3; the two outer narrow, acute or rarely bearing an awn, the third rigid, convolute, with a hairy callus at the base, and bearing a more or less bent persistent awn, which is spiral at the base. Palet 2-nerved. Stamens 3, rarely fewer. Styles short, distinct. Stigmas plumose. Grain narrow, free, tightly enclosed in the scale. [Greek, in allusion to the tow-like plumes of some species.]

A genus of about 120 species, distributed throughout the temperate and tropical zones. Besides the following, some 20 others occur in the southern and western parts of North America. Type species: *Stipa pennata* L.

Outer scales of the spikelet 2″–6″ long:
  Obtuse or blunt-pointed, 2″ in length.   1. *S. canadensis.*
  Acute, 4″–6″ in length.
      Awn less than five times the length of the scale.   2. *S. viridula.*
      Awn more than seven times the length of the scale.   3. *S. avenacea.*
Outer scales of the spikelet 10″ long or more.
  Base of panicle usually included in the upper sheath; third scale 4″–6″ long; awn slender, curled.
                                                                                      4. *S. comata.*
  Panicle exserted from the upper sheath; third scale 7″–12″ long, bent.   5. *S. spartea.*

1. Stipa canadénsis Poir.   Macoun's or Richardson's Feather-grass.   Fig. 418.

3/4

Stipa juncea Michx. Fl. Bor. Am. 1: 54. 1803. Not L. 1753.
Stipa canadensis Poir. Encycl. 7: 452. 1806.
Stipa Richardsonii A. Gray, Man. Ed. 2, 249. 1856. Not Link, 1833.
Stipa Macounii Scribn.; Macoun, Cat. Can. Pl. 5: 390. 1890.

Culms glabrous, 1°–2° tall, erect, simple, slen-der, smooth or somewhat scabrous. Sheaths shorter than the internodes; ligule about 1″ long, obtuse or truncate; blades 2′–5′ long, ½″–1″ wide, flat, becoming involute-setaceous in drying, sca-brous; panicle 2′–5′ long, contracted, the branches 1′–2′ long, erect, naked below; spikelets borne at the ends of the branches; outer scales about 2″ long, obtuse or blunt-pointed, glabrous; third scale somewhat shorter, pubescent with long ap-pressed silky hairs, callus obtuse; awn 4″–5″ long, contorted.

Shaded places, New Brunswick to northern New York, Ontario, Saskatchewan and northward. July.

2. Stipa virídula Trin.   Feather Bunch-grass.
Fig. 419.

Stipa viridula Trin. Mem. Acad. St. Petersb. (VI.) 2: 39. 1836.
Stipa spartea Hook. Fl. Bor. Am. 2: 237. 1840. Not Trin. 1831.

Glabrous, culms 1½°–3° tall, erect, simple, smooth. Sheaths shorter than the internodes; ligule 1″–2″ long; blades smooth or scabrous, the basal ones involute-filiform, one-third to one-half as long as the culm, those of the culm 3′–9′ long, broader; panicle spike-like, strict and erect, branches appressed; outer scales of spike-let 3″–4″ long, long-acuminate, glabrous; third scale shorter, more or less pubescent with long appressed silky hairs, callus acute; awn ¾′–1¼′ long, bent, loosely spiral at base.

Meadows and prairies, Athabasca to Minnesota and Colorado. Wild Oat-grass. July–Aug.

2/3

3. Stipa avenàcea L.   Black Oat-grass.   Fig. 420.

Stipa avenacea L. Sp. Pl. 78. 1753.
Stipa barbata Michx. Fl. Bor. Am. 1: 53. 1803.
Stipa virginica Pers. Syn. 1: 99. 1805.
Stipa bicolor Pursh, Fl. Am. Sept. 73. 1814.

Culms glabrous, 1°–2½° tall, erect or leaning, simple, smooth. Sheaths shorter than the internodes; ligule about 1″ long, obtuse; blades involute-filiform, smooth beneath, scabrous above, the basal one-third to one-half the length of the culm, those of the culm 3′–5′ long; panicle 5′–8′ long, loose, the branches lax, erect or finally spreading, naked below; outer scales of the spikelet 4″–5″ long, acute, glabrous; third scale a little shorter, scabrous near the summit, black, pilose at base and with a ring of short hairs at the top, otherwise smooth and glabrous; callus hard, acute; awn 1½′–2½′ long, bent, loosely spiral below.

In dry woods, Massachusetts to Wisconsin, Florida and Texas. Feather-grass. May–June.

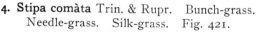

4. Stipa comàta Trin. & Rupr.   Bunch-grass. Needle-grass.   Silk-grass.   Fig. 421.

Stipa comata Trin. & Rupr. Mem. Acad. St. Petersb. (VI.) 5: 75. 1842.

Glabrous, culms 1°–2° tall, erect, simple, smooth. Sheaths usually longer than the internodes, smooth or scabrous, the uppermost very long and inflated, enclosing the base of the panicle; ligule 1″–2″ long, obtuse; blades smooth or somewhat scabrous, the basal involute-filiform, one-quarter to one-half as long as the culm, the culm blades 3′–6′ long, a little broader than the basal ones, involute; panicle 6′–9′ long, loose, the branches 3′–5′ in length, erect-ascending, naked at base; outer scales of the spikelet 9″–12″ long, glabrous, acuminate into an awn 2″–4″ in length; third scale 4″–6″ long, callus acute; awn 4′–8′ in length, slender, curled, spiral and pubescent below.

Dry places, Yukon to Iowa, Texas and California. Porcupine- or Blow-out-grass.   Needle and Thread. June–July.

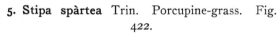

5. Stipa spàrtea Trin.   Porcupine-grass.   Fig. 422.

S. spartea Trin. Mem. Acad. St. Petersb. (VI.) 1: 82. 1831.

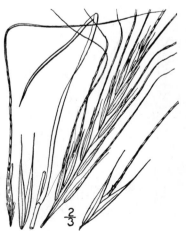

Culms glabrous, 2°–4° tall, erect, simple, smooth. Sheaths longer than the internodes, smooth or somewhat scabrous; ligule 1″–2″ long, obtuse; blades smooth beneath, scabrous above, the basal one-third to one-half as long as the culm, 1″ wide or less, usually involute, those of the culm 6′–12′ long, about 2″ wide, generally flat, attenuate into a long slender point; panicle finally long-exserted, 4′–10′ in length, its branches 3′–6′ long, erect, naked below; outer scales of spikelet 12″–18″ long, acuminate into a long slender point, glabrous; third scale 7″–12″ long, callus acute; awn 4′–8′ long, stout, usually twice bent, tightly spiral and pubescent below, doubly spiral about the middle.

On prairies, British Columbia to Michigan and Kansas. June–July.

33. ARÍSTIDA L. Sp. Pl. 82. 1753.

Grasses varying greatly in habit and inflorescence. Leaf-blades narrow, often involute-setaceous. Spikelets narrow, 1-flowered. Scales 3, narrow, the two outer carinate; the third rigid and convolute, bearing three awns occasionally united at the base, the lateral awns rarely wanting or reduced to rudiments. Palet 2-nerved. Stamens 3. Styles distinct. Stigmas plumose. Grain free, tightly enclosed in the scale. [Latin, from *arista,* an awn.]

About 120 species, in the warmer regions of both hemispheres. The English name *Three-awned Grass* is applied to all the species. Type species: *Aristida adscensionis* L.

Awns not articulated to the scale.
  Central awn coiled at the base.
    First scale usually equalling or sometimes slightly shorter than the second scale, which is
      commonly 3½"–4½" long.   1. *A. dichotoma.*
    First scale much shorter than the second (often but little more than ½ as long), which
      is 5"–8" long.
      Lateral awns short, straight and erect, ½"–1" long, the central awn usually more than
        five times their length, its straight portion 2½"–4" long.   2. *A. Curtissii.*
      Lateral awns more or less spreading, usually a little spiral at the base, the central awn
        from ½ again to twice their lenth, its straight portion 5"–8" long. 3. *A. basiramea.*
  Central awn not coiled at the base.
    Panicle narrow, linear to oblong, branches short (or long in no. 13), erect or ascending.
      Central awns and sometimes the lateral ones also stronlgy reflexed, the bend semicircular.
                                          4. *A. ramosissima.*
      Central awn from erect to spreading with no such bend at the base.
        First scale much shorter than the second, usually about ½ as long.
          Spikelets crowded, 4–6 on the short branches, which are spikelet-bearing to the
            base or nearly so.   5. *A. fasciculata.*
          Spikelets not crowded, usually 1–3 on branches naked at the base.
            Second scale of spikelet 8" long or less, equalling or exceeding flowering scale.
              Panicle simple or nearly so, its branches bearing 1 spikelet; culms com-
                monly naked above.   6. *A. Fendleriana.*
              Panicle compound, its branches bearing 2 or more spikelets; culms usually
                leafy.   7. *A. Wrightii.*
          Second scale of the spikelet 10" long or more, 1½–2 times as long as the
            flowering scale.   8. *A. longiseta.*
      First scale from a little shorter than to exceeding the second.
        Spikelets exceeding 10" long; first scale 5–7-nerved.   9. *A. oligantha.*
        Spikelets less than 8" long; first scale 1–3-nerved.
          Leaf-sheaths glabrous or sparsely pubescent.
            First scale generally shorter than or equalling the second.
              Flowering scale 2½"–3" long, its central awn usually 3"–5" long.
                                      10. *A. gracilis.*
              Flowering scale 3½"–4½" long, its central awn exceeding 7" in length.
                                    11. *A. intermedia.*
            First scale exceeding the second.   12. *A. purpurascens.*
          Leaf-sheaths, at least the lower ones, densely wooly.   13. *A. lanosa.*
    Panicle diffuse and open, the branches very long and widely spreading. 14. *A. divaricata.*
Awns articulated to the scale, united at the base into a spiral column.
  Column conspicuous, 3" long or more.   15. *A. tuberculosa.*
  Column inconspicuous, 1" long or less.   16. *A. desmantha.*

$\frac{2}{3}$

1. Aristida dichótoma Michx. Poverty-grass. Fig. 423.

Aristida dichotoma Michx. Fl. Bor. Am. 1: 41. 1803.

Culms 6'–2° tall, erect, slender, dichotomously branched, smooth or roughened. Sheaths much shorter than the internodes, loose, smooth and glabrous; ligule very short, ciliate; blades 1'–3' long, less than 1" wide, involute, acuminate, usually scabrous; spike-like racemes or panicles 2'–5' long, slender; spikelets about 3" long; outer scales nearly equal or the lower somewhat shorter, usually awn-pointed; third scale shorter than the second, the middle awn horizontal, coiled at base, the terminal straight portion 2"–3" long, the lateral awns 1" long or less, erect.

Dry sandy soil, Maine to Nebraska, south to Georgia and Texas. Aug.–Sept.

2. Aristida Curtíssii (A. Gray) Nash. Curtiss's Triple-awned Grass. Fig. 424.

A. dichotoma var. *Curtissii* A. Gray, Man. Ed. 6, 640.
1890.
Aristida Curtissii Nash, in Britt. Man. 94. 1901.

Culms tufted, 8'–20' tall, branched; blades 1½'–6' long, ½''–1'' wide, sometimes sparsely pilose above near the base; panicle 2'–4' long, the branches erect; spikelets commonly 5''–6'' long, rarely longer, the first scale much shorter than the second which usually about equals the body of the flowering scale, rarely somewhat exceeding it, the flowering scale 3½''–5½'' long, the lateral awns very short, ½''–1'' long, straight and erect, usually less than ⅓ as long as the central awn which has the straight portion 2½''–4'' long.

In dry soil, Missouri and Kansas to Oklahoma; also in Virginia. Sept.–Oct.

3. Aristida basiràmea Engelm. Forked Triple-awned Grass. Beard-grass. Fig. 425.

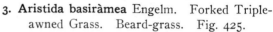

Aristida basiramea Engelm.; Vasey, Coult. Bot. Gaz. **9**: 76. 1884.

Glabrous and smooth, culms 6'–18' tall, erect, slender, much branched. Sheaths shorter than the internodes, loose; ligule very short, ciliate; blades 2'–6' long, about 1'' wide, involute-setaceous; spike-like panicle 3'–5' long; first scale of spikelet half to three-quarters as long as the second, which is 5''–7'' in length, both awn-pointed; third scale shorter than the second; middle awn 6''–9'' long, coiled at base, horizontal, lateral awns one-quarter to one-half as long, erect or divergent, somewhat spiral at the base.

In dry fields, Illinois to Minnesota and Nebraska. July–Sept.

4. Aristida ramosíssima Engelm. Branched Aristida. Fig. 426.

A. ramosissima Engelm.; A. Gray, Man. Ed. 5, 618. 1867.

Aristida ramosissima var. *uniaristata* A. Gray, Man. Ed. 5, 618. 1867.

Glabrous, culms 6'–2° tall, erect, slender, branched, smooth. Sheaths much shorter than the internodes, loose; ligule very short; blades 1½'–3' long, 1'' wide or less, flat, attenuate into a long point, smooth beneath, scabrous above; spikelets few, borne in loose spikes from 2'–4' in length; first scale awn-pointed; second scale 8''–10'' in length, exceeding the first, terminated with an awn 1''–3'' long; third scale as long as the second; middle awn about 1' long, horizontal or reflexed and forming a hook, the lateral awns erect, 1''–2'' long, rarely wanting.

In dry soil, Indiana to Missouri and Tennessee. July–Sept.

$\frac{2}{3}$

5. Aristida fasciculàta Torr. Triple-awned Beard-grass. Needle-grass. Fig. 427.

Aristida fasciculata Torr. Ann. Lyc. N. Y. 1 : 154. 1824.
Aristida dispersa Trin. & Rupr. Mem. Acad. St. Petersb. (VI.) **5** : 129. 1842.

Glabrous, culms 1°–2° tall, erect, slender, branched, smooth. Sheaths shorter than the internodes, ligule short, ciliate; blades 2′–6′ long, 1″ wide or less, flat, attenuate into a long point, smooth or scabrous; panicle 3′–7′ long, at first strict, the branches finally more or less spreading; first scale of spikelet 1-nerved, or occasionally with an obscure additional nerve on each side, shorter than the second scale; third scale equalling or longer than the second; awns divergent, the middle one 4″–8″ long, the lateral ones shorter.

Dry soil, Kansas to California and Mexico. Dog-town Grass. Purple beard-grass. Aug.–Sept.

6. Aristida Fendleriàna Steud. Fendler's Triple-awned Grass. Fig. 428.

A. Fendleriana Steud. Syn. Gram. 420. 1855.

Culms densely tufted, 6′–10′ tall, erect, rigid, simple; leaves confined to the base of the culm; sheaths with a tuft of hairs on each side at the apex; blades involute, often curved, ¼″ in diameter, those on the culm usually 2, up to 2′ long, the basal ones longer; panicle 3′–4′ long, strict, its branches short and appressed and usually bearing but a single spikelet; spikelets 6″–7½″ long, the 2 outer scales 1-nerved, the first scale about ½ as long as the second, the flowering scale 4½″–6″ long, equalling or a little shorter than the second scale, the awns ascending, the central one 1′–2′ long, the lateral awns a little shorter.

In dry sandy soil, South Dakota to Utah, Texas and New Mexico. Figured in our first edition for *A. purpurea* Nutt., which has not as yet been detected within our limits.

$\frac{2}{3}$

$\frac{2}{3}$

7. Aristida Wrìghtii Nash. Wright's Triple-awned Grass. Fig. 429.

A. Wrightii Nash, in Small, Fl. SE. U. S. 116. 1903.

Culms tufted, 1°–2° tall, simple, leafy; blades involute, those on the culm usually 3 or 4, 1′–8′ long, often curved, as are the commonly longer basal ones; panicle 4′–8′ long, its branches more or less spreading, the longer usually bearing 2–4 spikelets; spikelets 6″–7½″ long, the 2 outer scales 1-nerved, the first scale about ½ as long as the second, the flowering scale 5″–6″ long, usually a little shorter than the second scale, the awns ascending, the central one 1′–1¼′ long, the lateral awns a little shorter.

In dry sandy soil, Kansas to Texas and New Mexico. July, Aug.

8. Aristida longisèta Steud. Long-awned Aristida. Fig. 430.

A. longiseta Steud. Syn. Gram. 420. 1855.

Culms tufted, 8′–16′ tall, simple; blades 1′–4′ long, involute; panicle 4′–8′ long, its branches usually ascending, bearing generally 1 spikelet, or in the longer branches sometimes 2 spikelets; spikelets 10″–12½″ long, the 2 outer scales 1-nerved, the first one about ½ as long as the second, the flowering scale 6″–8″ long, occasionally a little shorter, from a little over ½ to ⅔ as long as the second scale, the awns ascending, the central one 2½′–4½′ long, the lateral ones equalling it or a little shorter.

In dry sandy soil, Montana and Washington to Nebraska, Texas and Mexico. July, Aug.

9. Aristida oligántha Michx. Few-flowered Aristida. Fig. 431.

Aristida oligantha Michx. Fl. Bor. Am. 1: 41. 1803.

Glabrous, culms 1°–2° tall, erect, slender, dichotomously branched, smooth or roughish. Sheaths exceeding the internodes, loose; ligule very short, minutely ciliate; blades 1′–6′ long, ½″–1″ wide, smooth, the larger ones attenuate into a long slender point; spikelets few, borne in a lax spike-like raceme or panicle; first scale 5-nerved, occasionally 7-nerved at base, acuminate or short-awned, equalling or somewhat shorter than the second, which bears an awn 2″–4″ long; third scale shorter than the first, awns divergent or spreading, the middle one 1½′–2½′ long, the lateral somewhat shorter.

Dry soil, New Jersey to Nebraska, and Texas. Aug.–Sept. Ant-rice.

10. Aristida grácilis Ell. Slender Triple-awned Grass. Fig. 432.

Aristida gracilis Ell. Bot. S. C. & Ga. 1: 142. 1817.

Aristida gracilis var. *depauperata* A. Gray, Man. Ed. 5, 618. 1867.

Glabrous and smooth throughout, culms 6′–2° tall, erect, simple or branched. Sheaths shorter than the internodes; ligule very short; blades 1′–4′ long, 1″ wide or less, flat, or involute when dry; panicle spike-like, 3′–7′ long, slender; spikelets about 3″ long; outer scales equal, or the lower somewhat shorter, awn-pointed; third scale about equalling the second, generally mottled, middle awn horizontal, the terminal straight portion 3″–7″ in length, the lateral awns 1″–3″ long, erect.

Dry soil, New Hampshire to Missouri, south to Florida and Texas. Aug.–Sept.

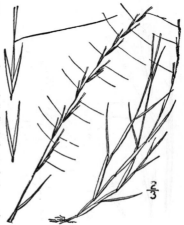

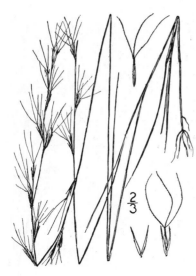

11. Aristida intermèdia Scribn. & Ball.   Plains Aristida.   Fig. 433.

A. *intermedia* Scribn. & Ball, Bull. U. S. Dep. Agr. Agrost. 24: 44. 1901.

Culms slender, finally branching, 1°–2½° tall; sheaths glabrous or sparsely hirsute; blades 2′–6′ long, 1″ wide or less, erect, involute; panicle 8′–16′ long, slender, its branches appressed; spikelets 4″–5″ long, the empty scales manifestly awned, about equal, the flowering scale strongly hispidulous above the middle, equalling or exceeding the empty scales, the awns spreading, the middle one 7″–13″ long, the lateral ones shorter.

In sandy soil, Iowa and Kansas to Mississippi and Texas. Aug.–Oct.

12. Aristida purpuráscens Poir.   Arrow-grass. Broom-sedge.   Fig. 434.

Aristida purpurascens Poir. in Lam. Encycl. Suppl. 1: 452. 1810.

Glabrous and smooth, culms 1°–2½° tall, erect, simple or sparingly branched at the base.   Sheaths longer than the internodes, crowded at the base of the culm; ligule very short; blades 4′–8′ long, about 1″ wide, flat, or becoming involute in drying, attenuate into a long point; spike-like panicles 5′–18′ long, strict, or sometimes nodding, its branches appressed; outer scales of spikelet awn-pointed, the first longer than the second; the third scale from two-thirds to three-quarters as long as the first, middle awn 9″–12″ long, horizontal, the lateral awns somewhat shorter, erect or divergent.

In dry soil, Massachusetts to Minnesota, south to Florida and Texas.   Sept.–Oct.

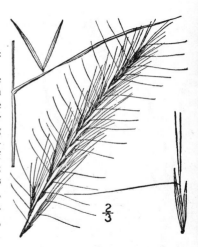

13. Aristida lanòsa Muhl.   Woolly Triple-awned Grass.   Fig. 435.

Aristida lanata Poir. in Lam. Encycl. Suppl. 1: 453.   1810. Not Forsk. 1775.

Aristida lanosa Muhl. Gram. 174.   1817.

Culms 2°–4° tall, erect, simple, smooth and glabrous. Sheaths longer than the internodes, crowded at the base of the culm, woolly; ligule very short, minutely ciliate; blades 1° long or more, about 2″ wide, attenuate into a long slender point, smooth beneath, scabrous above; panicle 1°–2° long, strict, branches erect or occasionally somewhat spreading; outer scales of the spikelet awn-pointed, the first 5″–7″ long, exceeding the second; third scale slightly shorter than the second, middle awn 8″–12″ long, usually horizontal, the lateral awns about two-thirds as long, erect or divergent.

Dry sandy soil, Delaware to Florida, Oklahoma and Texas.   Aug.–Sept.

14. Aristida divaricàta H. & B. Spreading Triple-awned Grass. Fig. 436.

Aristida divaricata H. & B.; Willd. Enum. Hort. Berol. 99. 1809.

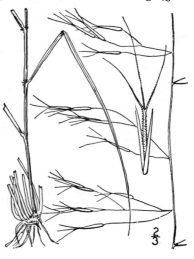

Culms 1½°–3° tall, tufted, erect. Sheaths overlapping, rough, usually with a tuft of long hairs on each side at the apex; ligule a short ciliate ring; blades smooth beneath, rough above, those of the culm 6′–12′ long, 1″–2″ wide, erect or ascending; the sterile shoots from one-third to one-half as long as the culm, the leaves narrower; panicle comprising one-half of the plant, or more, often included at the base, its branches rigid, at length widely spreading; spikelets, exclusive of the awns, about ½′ long, numerous; empty scales acuminate, usually awn-pointed; flowering scale commonly slightly shorter than the empty ones, firm, sometimes spotted with purple, hispidulous above; awns not articulated to the scale, the lateral ones shorter than the central, which is 6″–10″ long; callus pilose.

Dry sandy soil, Kansas to Arizona and Mexico.

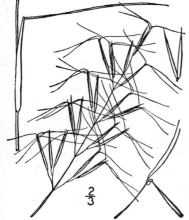

15. Aristida tuberculòsa Nutt. Sea-beach Triple-awned Grass. Fig. 437.

Aristida tuberculosa Nutt. Gen. 1 : 57. 1818.

Glabrous, culms 6′–2° tall, erect, dichotomously branched, smooth. Sheaths shorter than the internodes; ligule short, ciliate; blades 5′–9′ long, about 1″ wide, attenuate into a long slender point, smooth beneath, scabrous above; panicle 5′–8′ long, branches slender, ascending; outer scales of the spikelet about equal, awned, the third scale shorter; awns divergent or reflexed, more or less coiled, united at the base into a column 3″–6″ long which is articulated to the scale.

Sandy soil, especially on sea-beaches, Massachusetts to Georgia. Also about the Great Lakes. Long-awned Poverty-grass. Aug.–Sept.

16. Aristida desmántha Trin. & Rupr. Western Triple-awned Grass. Fig. 438.

Aristida desmantha Trin. & Rupr. Mem. Acad. St. Petersb. (VI.) 5 : 109. 1842.

Culms 1°–2° tall, erect, branched, smooth and glabrous. Sheaths shorter than the internodes, smooth, glabrous or the lower sometimes pubescent; ligule short; blades 6′–12′ long, less than 1″ wide, attenuate into a slender point, smooth beneath, scabrous above; panicle about 6′ long, the branches slender, ascending; outer scales of the spikelet about equal, the third one shorter; awns spreading or reflexed, somewhat coiled, united at base into a column less than 1″ long, which is articulated to the scale.

In dry soil, Nebraska to Texas. Aug.–Sept.

34. MUHLENBÉRGIA Schreb.; Gmel. Syst. Nat. 2: 171. 1791.

[VASEYA Thurb. Proc. Acad. Phila. 1863: 79. 1863.]

Mostly perennial grasses, with flat or convolute leaves and paniculate inflorescence. Rootstocks often scaly. Spikelets 1-flowered, very rarely 2-flowered. Scales 3, very rarely 4; the outer ones empty, membranous or hyaline, acute and sometimes awned; third scale 3–5-nerved, subtending a palet and perfect flower, obtuse, acute, or very often produced into a capillary awn; palet 2-keeled. Stamens often 3. Styles distinct. Stigmas plumose. Callus minute. Grain narrow, free, tightly enclosed in the scale. [In honor of Henry Muhlenberg, 1756–1817, North American botanist.]

About 60 species, chiefly natives of America, a few Asiatic. Type species: *Muhlenbergia Schreberi* Gmel.

Panicle contracted, narrow, often slender, its branches erect or appressed.
  Outer scales ¼ as long as the flowering scale or less.
    First scale minute, often wanting; flowering scale with an awn twice its length or less.
                                                            1. *M. Schreberi.*
    First scale about ⅔ as long as the second; flowering scale with an awn 3–4 times its length.
                                                            2. *M. palustris.*
  Outer scales more than ¼ as long as the flowering scale.
    Plants with numerous and conspicuous rootstocks covered with short appressed scales.
      Flowering scale awnless, or sometimes awn-pointed.
        Outer scales ovate to broadly lanceolate, cuspidate, about ½ as long as the flowering scale.   3. *M. sobolifera.*
        Outer scales subulate, equalling or exceeding the flowering scale, awn-pointed or awned.
          Outer scales about equal in length to the flowering scale, about 1½″ long, sharp-pointed.   4. *M. mexicana.*
          Outer scales exceeding the flowering scale, generally twice its length, about 2½″ long, awned.   5. *M. racemosa.*
      Flowering scales long-awned, the awn usually twice as long as the scale, sometimes shorter.
        Outer scales ½–⅔ as long as the flowering scale, ovate to broadly lanceolate, cuspidate.   6. *M. tenuiflora.*
        Outer scales equalling the flowering scale, subulate.
          Basal hairs not more than ½ as long as the flowering scale.
            Spikelets consisting of 3 scales and 1 perfect flower.   7. *M. umbrosa.*
            Spikelets consisting of 4 scales, the third with a perfect flower, the fourth empty and awned.   8. *M. ambigua.*
          Basal hairs as long as the flowering scale.   9. *M. comata.*
    Plants without rootstocks, or these rarely present and with few scattered long scales.
      Annual; outer scales less than ½ as long as the spikelet, rounded or truncate at the erose apex.   10. *M. simplex.*
      Perennial; outer scales ½ as long as the spikelet or more, acute or acuminate.
        Culms simple at the base, finally much-branched above; ligule about 1″ long, acutish; outer scales about ½ as long as the spikelet, acute.   11. *M. brevifolia.*
        Culms branched at the very base only, simple above; ligule ¼″ long or less, erose-truncate; outer scales more than ½ as long as the flowering scale, attenuate.   12. *M. cuspidata.*
Panicle open, its branches long and spreading, slender.
  Culms 16′ tall or less, from much-branched rootstocks; blades 2′ long or less.
    Secondary branches of the panicle single; basal leaves short, numerous, strongly recurved.   13. *M. gracillima.*
    Secondary branches of the panicle fasciculate; basal leaves few, not recurved.   14. *M. pungens.*
  Culms 20′ tall or more; rootstocks none; blades elongated.   15. *M. capillaris.*

2/3

1. Muhlenbergia Schréberi Gmel. Nimble Will. Dropseed- or Wire-grass. Satin-grass. Fig. 439.

M. Schreberi Gmel. Syst. Nat. 2: 171. 1791.

M. diffusa Willd. Sp. Pl. 1: 320. 1798.

Glabrous, culms 1°–3° long, decumbent, or often prostrate or creeping and ascending, very slender, diffusely branched. Sheaths shorter than the internodes, loose; ligule short, fringed; blades 1½′–3½′ long, ½″–2″ wide, scabrous; panicle 2′–8′ long, slender, somewhat lax, its branches 1′–2′ long, erect; outer scales of the spikelet minute, the lower one often wanting; the third scale, exclusive of the awn, about 1″ long, strongly scabrous, particularly upon the nerves; the awn ½″–2″ in length.

On dry hills and in woods and waste places, Maine to Minnesota, south to Florida and Texas. Aug.–Sept.

2. Muhlenbergia palústris Scribn. Swamp Drop-seed. Fig. 440.

M. palustris Scribn. Bull. U. S. Dep. Agr. Agrost. **11**: 47. 1898.

M. Schreberi palustris Scribn. Rhodora, **9**: 17. 1907.

Culms slender, weak, 2°–3° long; sheaths smooth and glabrous; blades erect, 1′–2′ long, about 1″ wide, smooth beneath and rough above; panicle slender, contracted, 4′–6′ long, its branches appressed; spikelets, exclusive of the awn, 1¼″–1½″ long, the first scale about ⅔ as long as the second which is ¼ as long as the spikelet, the flowering scale about 1¼″ long, shortly 2-toothed at the apex, and bearing an awn between the teeth 3–4 times its length, the callus hairy.

In swampy ground, District of Columbia and Illinois. Sept.

3. Muhlenbergia sobolífera (Muhl.) Trin. Rock Dropseed. Fig. 441.

Agrostis sobolifera Muhl.; Willd. Enum. 95. 1809.
Muhlenbergia sobolifera Trin. Unifl. 189. 1824.

Glabrous, culms 2°–3° tall, erect, slender, simple, or sparingly branched above, smooth. Sheaths smooth, those of the culm shorter than the inter-nodes, those of the branches overlapping and crowded; ligule very short, truncate; blades rough, those of the culm 4′–6′ long, 1½″–3″ wide, those of the branches 1′–3′ long, about 1″ wide; panicle 3′–6′ in length, slender, its branches ¾′–1′ long; outer scales about ½″ long, half to two-thirds the length of the spikelet, equal, or the lower somewhat shorter, acute, scabrous, especially on the keel; third scale scabrous, obtuse, 3-nerved, the middle nerve usually excurrent as a short point.

Rocky woods, New Hampshire to Minnesota, south to Virginia, Tennessee and the Indian Territory. Sept.–Oct.

4. Muhlenbergia mexicàna (L.) Trin. Satin-grass. Wood-grass. Fig. 442.

Agrostis mexicana L. Mant. **1**: 31. 1767.
Agrostis filiformis Willd. Enum. 95. 1809.
Muhlenbergia mexicana Trin. Unifl. 189. 1824.
M. foliosa Trin. Gram. Unifl. 190. 1824.

Glabrous, culms 2°–4° long, erect, or often prostrate, much branched, smooth. Sheaths shorter than the internodes, excepting at the extremities of the branches, where they are crowded and overlapping, smooth or scabrous; blades scabrous, those of the culm 4′–6′ long, 1″–3″ wide, the branch leaves smaller; panicle 2′–6′ long, contracted, its branches spike-like, 1′–2′ long, erect or appressed; spikelets 1¼″–1½″ long; outer scales some-what unequal, exceeding the flowering one, or slightly shorter, acuminate or short-awned, scabrous especially on the keel; third scale acuminate, scabrous, particu-larly toward the apex.

In swamps and borders of fields, New Brunswick to Wyoming, south to North Carolina and Texas. Knot-root grass. Aug–Sept.

5. Muhlenbergia racemòsa (Michx.) B.S.P.   Wild Timothy.   Satin-grass.
Fig. 443.

Agrostis racemosa Michx. Fl. Bor. Am. **1**: 53.   1803.
Muhlenbergia glomerata Trin. Unifl. 191.   1824.
Muhlenbergia racemosa B.S.P. Prel. Cat. N. Y. 67.   1888.

Culms 1°–3° tall, erect, usually much branched, smooth and glabrous. Sheath smooth, those of the culm shorter than the internodes, those of the branches overlapping and often crowded; ligule about ½″ long, erose-truncate; blades 2′–5′ long, 1″–3″ wide, scabrous; panicle 2′–4½′ in length, usually dense and interrupted, the branches ½′–1′ long, erect or appressed, the spikelets much crowded; outer scales of the spikelet acuminate, 2″–3″ long, including the awn, smooth or scabrous, especially on the keel; third scale one-half to two-thirds as long, acuminate, the strongly scabrous midrib excurrent in a short point.

In wet places, Newfoundland to British Columbia, south to New Jersey, Maryland and New Mexico. Aug.–Sept.

6. Muhlenbergia tenuiflòra (Willd.) B.S.P.   Slender Satin-grass.   Fig. 444.

Agrostis tenuiflora Willd. Sp. Pl. **1**: 364.   1798.

Agrostis pauciflora Pursh, Fl. Am. Sept. **1**: 63.   1814.

Muhlenbergia Willdenovii Trin. Unifl. 188.   1824.

M. tenuiflora B.S.P. Prel. Cat. N. Y. 67.   1888.

Glabrous, culms 2°–3° tall, erect, slender, simple or sparingly branched, smooth. Sheaths usually shorter than the internodes; ligule short and truncate; blades 2½′–7′ long, 1″–4″ wide, narrowed toward the base, acuminate, scabrous; panicle 5′–9′ long, slender, its branches 1′–3½′ long, appressed; outer scales of the spikelet unequal, half to two-thirds the length of the third one, awn-pointed, scabrous; third scale 1¼″–1½″ long, scabrous, bearing an awn 2–4 times its length.

In rocky woods, Massachusetts to Minnesota, Alabama and Texas. Slender-flowered Dropseed. Aug.–Sept.

7. Muhlenbergia umbròsa Scribn.   Wood or Woodland Dropseed.   Fig. 445.

Agrostis diffusa Muhl. Gram. 64.   1817.   Not Host, 1809.
Agrostis sylvatica Torr. Fl. U. S. **1**: 87.   1824.   Not L. 1763.
Muhlenbergia sylvatica Torr. Cat. Pl. N. Y. State, 188. 1840.
M. umbrosa Scribn. Rhodora, **9**: 20.   1907.

Culms 1°–3° tall, erect, branched, smooth or somewhat scabrous. Sheaths smooth or slightly scabrous, those of the culm shorter than the internodes, those of the branches overlapping and often crowded; ligule about ½″ long, erose-truncate; blades 2′–7′ long, 1″–3″ wide, rough; panicle 3′–7′ in length, somewhat lax, the branches 1′–3′ long, erect or ascending; outer scales of the spikelet 1¼″–1½″ long, awn-pointed, scabrous; third scale equalling or somewhat exceeding the outer ones, strongly scabrous, attenuate into a slender awn 2–4 times its length.

In moist woods and along streams, New Brunswick to South Dakota, south to North Carolina and Oklahoma. Aug.–Sept.

8. Muhlenbergia ambígua Torr. Minnesota Dropseed. Fig. 446.

Muhlenbergia ambigua Torr. Nicollet's Rep. 164. 1843.

Glabrous, culms 1° tall or lower, erect, branched, smooth. Sheaths shorter than the internodes; ligule about ½″ long erose-truncate; blades 1′–3′ long, 1″–2″ wide, scabrous; panicle 1′–3′ long, rigid, its branches ½′–1′ long, dense, appressed; outer scales of the spikelet awn-pointed, unequal, the longer about 2″ in length and exceeding the body of the third scale which is scabrous, villous, and attenuate into an awn 2–3 times its length; a fourth narrow awned scale is nearly always present.

Along a lake shore in Minnesota.

9. Muhlenbergia comàta (Thurb.) Benth. Hairy Dropseed. Fig. 447.

Vaseya comata Thurb. Proc. Phila. Acad. 1863: 79. 1863.

M. comata Benth.; Vasey, Cat. Grasses U. S. 39. 1885.

Culms 1°–2½° tall, erect, slender, smooth and glabrous. Sheaths shorter than the internodes, smooth or slightly scabrous; ligule about ½″ long, truncate, naked or minutely ciliate; blades 2½′–5′ long, 1″–2″ wide, erect, flat, rough; panicle often tinged with purple, 2′–4′ in length, dense, branches ½′–1½′ long, erect; outer scales of the spikelet equal, or the second a little the longer, smooth, scabrous on the keel; third scale shorter, smooth and glabrous, bearing an awn 2–3 times its length, the basal hairs silky, erect, fully as long as the scale.

On prairies, Montana to Washington, south to Kansas (?) and Colorado. Aug.–Sept.

10. Muhlenbergia símplex (Scribn.) Rydb. Slender Dropseed. Fig. 448.

Sporobolus simplex Scribn. Bull. U. S. Dep. Agr. Agrost. 11: 48. 1898.

M. simplex Rydb. Bull. Torr. Club, 32: 600. 1905.

A smooth and glabrous annual. Culms slender, up to 1° tall, but usually ½ that height; ligule about 1″ long, acute; blades erect, up to 2′ long, ½″ wide, flat, involute; panicle slender, sometimes interrupted below, 1′–2½′ long, the slender branches appressed; spikelets, exclusive of the short awn when present, a little over 1″ long, the outer scales less than ½ as long as the spikelet, rounded or truncate at the erose apex, the flowering scale very acute and often with a short point or awn.

In meadows and along brooks, Montana to Nebraska and Colorado. Aug. and Sept.

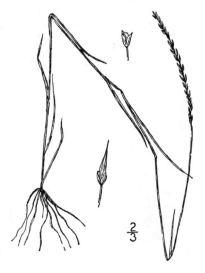

11. Muhlenbergia brevifòlia (Nutt.) Nash. Short-leaved Rush-grass. Fig. 449.

Agrostis brevifolia Nutt. Gen. 1 : 44. 1818.

Sporobolus Richardsonis Merrill, Rhodora, **4** : 46. 1902.

Sporobolus brevifolius Scribn. Mem. Torr. Club, **5** : 39. 1895.

Smooth and glabrous, culms 6′-18′ tall, arising from a horizontal rootstock, erect, slender, decumbent and branching above. Sheaths much shorter than the internodes; ligule ¾″-1″ long, acutish; blades ½′-2′ long, involute-setaceous; panicle ½′-3′ in length, usually about 1½′, linear, its branches ¼-½′ long, erect or appressed; spikelets 1¼″-1½″ long, the outer scales unequal, about one-half as long as the third, scabrous on the keel and at the apex; third scale long-acuminate, sometimes cuspidate, scabrous toward the apex.

In meadows and along rivers, Anticosti Island and Maine to British Columbia, south in the mountains to New Mexico and California. Summer.

12. Muhlenbergia cuspidàta (Torr.) Nash. Prairie Rush-grass. Fig. 450.

Vilfa cuspidata Torr.; Hook. Fl. Bor. Am. **2** : 238. 1840.

Sporobolus cuspidatus Wood, Bot. & Fl. 385. 1870.

Sporobolus brevifolius Scribn. Mem. Torr. Club, **5** : 39. In part. 1894.

Smooth and glabrous, culms 1°-2° tall, erect, simple above, branched at the base. Sheaths shorter than the internodes; ligule a mere ring, ¼″ long or less, erose-truncate; blades 1′-4′ long, less than 1″ wide at the base, erect, involute-setaceous, at least when dry; panicle 1½′-5′ in length, slender, its branches ¼-1′ long, appressed; spikelets 1¼″-1½″ long, the outer scales half to three-quarters as long, acuminate or cuspidate, scabrous on the keel; third scale long-acuminate and cuspidate, sparingly scabrous.

In dry soil, Manitoba to Alberta, south to Missouri and Kansas. Aug.-Sept.

13. Muhlenbergia gracíllima Torr. Filiform Dropseed. Fig. 451.

M. gracillima Torr. Pac. R. R. Rept. **4** : 155. 1875.

Glabrous, culms 4′-14′ tall, from a slender creeping rootstock, erect, slender, simple, rigid. Sheaths smooth; ligule 1″-2″ long, entire and acuminate, or variously cleft, with acuminate teeth; blades 1′-2′ long, involute-setaceous, smooth or somewhat scabrous, rigid, the basal numerous, usually strongly recurved, the 1-3 culm blades erect or ascending; panicle 2′-9′ in length, open, the branches finally widely spreading, 1′-3′ long, filiform; spikelets about as long as the filiform pedicels which are clavate-thickened at the apex; outer scales unequal, usually awn-pointed or short-awned, slightly scabrous; third scale 1¼″-1½″ long, longer than the outer ones, sometimes twice as long, scabrous; awn 1″-2″ long.

On prairies, Kansas to Colorado, south to Texas and Arizona. Sept.-Oct.

14. Muhlenbergia púngens Thurb.
Prairie Dropseed. Fig. 452.

Muhlenbergia pungens Thurb. Proc. Acad. Phila.
1863: 78. 1863.

Culms 6'–15' tall from a creeping rootstock,
erect from a decumbent branching base, rigid,
minutely pubescent. Sheaths overlapping,
crowded at the base of the culm, scabrous;
ligule a ring of soft silky hairs; blades 1'–2'
long, involute-setaceous, rigid, scabrous; panicle 3'–6' in length, open, the branches 2'–2½'
long, single, distant, much divided from near
the base, the divisions apparently fascicled;
spikelets on long pedicels, which are clavate-
thickened at the apex; outer scales, when mature, equalling or often shorter than the body
of the third one, scabrous, especially on the
keel; third scale, when mature, ¾"–1" long, scabrous, the awn shorter than its body.

On prairies, Nebraska to Utah, south to Texas
and Arizona. Blow-out-grass. Aug.–Sept.

15. Muhlenbergia capillàris (Lam.) Trin. Long-awned Hair-grass. Fig. 453.

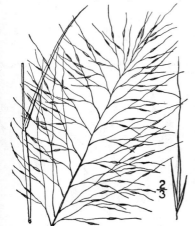

Stipa capillaris Lam. Tabl. Encycl. 1: 158. 1791.

Muhlenbergia capillaris Trin. Unifl. 191. 1824.

Glabrous, culms 1½°–4° tall, erect, simple, smooth or
nearly so. Sheaths smooth, the lower short and overlapping, the upper ones much longer; ligule about 2"
in length; blades 6'–1° long, 1"–2" wide, scabrous;
panicle 7'–1° in length or more, diffuse, the capillary
branches 4'–8' long, at length widely spreading; spikelets on long hair-like pedicels which are clavate-
thickened at the apex; outer scales unequal, acute or
short-awned, slightly scabrous; third scale, exclusive
of the awn, 2" long, about twice as long as the first
one, scabrous, the awn 3"–9" in length.

In dry sandy or rocky soil, Massachusetts to Kansas,
Florida and Texas. Bahamas and Cuba. Panicle usually
light purple. Sept.–Oct.

Muhlenbergia glabriflorus Scribn., an imperfectly known species, is reported from Illinois. It
is said to resemble *M. mexicana,* and to differ from that species in the glabrous softer scales.

35. BRACHYÉLYTRUM Beauv. Agrost. 39. 1812.

A tall grass with flat leaves and a narrow panicle. Spikelets 1-flowered, narrow, the
rachilla produced beyond the flower and sometimes bearing a minute scale at the summit.
Scales 3; the outer small and inconspicuous, the lower often wanting; the third much longer,
rigid, 5-nerved, acuminate into a long awn; palet scarcely shorter, rigid, sulcate on the back,
2-nerved. Stamens 2. Styles short, distinct. Stigmas plumose, elongated. Grain oblong,
free, enclosed in the scale and palet. [Greek, in allusion to the minute outer scales.]

A monotypic genus of eastern North America.

1. Brachyelytrum eréctum (Schreb.) Beauv.
Bearded Short-husk.   Fig. 454.

Muhlenbergia erecta Schreb. Besch. Gras. **2**: 139. *pl.
50.* 1772–9.
Brachyelytrum erectum Beauv. Agrost. 155. 1812.
Brachyelytrum aristatum R. & S. Syst. **2**: 413. 1817.
Brachyelytrum aristatum var. *Engelmanni* A. Gray, Man.
Ed. 5, 614. 1867.

Culms 1°–3° tall, erect, slender, simple, smooth or
rough, pubescent at and near the nodes. Sheaths
shorter than the internodes, scabrous toward the
apex, more or less villous especially at the throat;
ligule about ¾″ long, irregularly truncate; leaves 2′–5′
long, 3″–9″ wide, acuminate at both ends, scabrous;
panicle 2′–6′ in length, slender, branches 1′–3′ long,
erect or appressed; outer scales of the spikelet unequal, the upper less than one-third as long as the
flowering scale, the lower minute or wanting; third
scale, exclusive of the awn, 4½″–6″ long, 5-nerved,
scabrous, especially on the midnerve, the awn erect,
9″–12″ long; rachilla produced beyond the flower
about half the length of the third scale and lying in
the groove of the palet.

Moist places, Newfoundland to Minnesota, south to Georgia and Kansas. Ascends to 5000 ft.
in North Carolina. July–Aug.

36. HELEÓCHLOA Host, Gram. 1: 23. *pl. 29, 30.* 1801.
[CRYPSIS Lam. Tabl. Encycl. 1: 166. 1791. Not Ait. 1789.]

Perennial tufted grasses with flat leaves and spicate or paniculate inflorescence. Spikelets 1-flowered. Scales 3; the 2 outer empty, somewhat unequal, membranous, acute, ciliate-
keeled; the third scale similar, a little longer; palet shorter, hyaline, 2-nerved. Stamens 3.
Styles distinct. Stigmas plumose. Grain oblong, free, loosely enclosed in the scale. [Greek,
signifying meadow-grass.]

About 8 species, chiefly natives of the Mediterranean region, one or two also widely distributed
through middle Europe and Asia. Type species: *Heleochloa alopecuroides* Host.

1. Heleochloa schoenoìdes (L.) Host. Rush-
like Timothy; Rush Cat's-tail Grass. Fig. 455.

Phleum schoenoides L. Sp. Pl. 60. 1753.
Crypsis schoenoides Lam. Tabl. Encyl. 1: 166. *pl. 42.* 1791.
Heleochloa schoenoides Host, Gram. 1: 23. *pl. 30.* 1801.

Glabrous, culms 4′–18′ tall, erect or sometimes decumbent at the base, branched, smooth. Sheaths about half
the length of the internodes, the upper loose, the one
immediately below the spike inflated and usually partially enclosing it; ligule a ring of short hairs; leaves
1′–3′ long, 1″–2″ wide, flat, acuminate, smooth beneath,
scabrous above; spikelets 1¼″ long, the empty scales
acute, compressed, ciliate-keeled, 1-nerved, the lower
shorter than the upper; third scale equalling or longer
than the second, acute, compressed, ciliate-keeled,
otherwise glabrous, 1-nerved; palet shorter, obtuse.

In waste places, southern New York to Delaware and
Pennsylvania. Naturalized from Europe. July–Aug.

37. PHLÈUM L. Sp. Pl. 59. 1753.

Annual or perennial grasses with flat leaf-blades and spicate inflorescence. Spikelets
1-flowered. Scales 3; the 2 outer empty, membranous, compressed, keeled, the apex obliquely
truncate, the midnerve produced into an awn; the third scale much shorter, broader, hyaline,
truncate, denticulate at the summit; palet narrow, hyaline. Stamens 3. Styles distinct,
somewhat elongated. Stigmas plumose. Grain ovoid, free, enclosed in the scale and palet.
[Name Greek, taken from Pliny; originally applied to some very different plant.]

About 10 species, inhabiting the temperate zones of both hemispheres. The following only
are natives of North America. The English name *Cat's-tail Grass* is applied to all the species.
Type species: *Phleum pratense* L.

Spikes usually elongated, cylindric; awns less than one-half the length of the outer scales; upper sheath not inflated.   1. *P. pratense.*
Spikes not elongated, ovoid to oblong and cylindric; awns about one-half the length of the outer scales; upper sheath inflated.   2. *P. alpinum.*

1. **Phleum praténse** L.   Timothy.   Herd's-grass.

Fig. 456.

Phleum pratense L. Sp. Pl. 59. 1753.

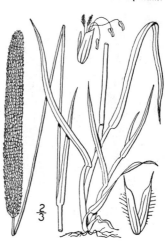

Glabrous and smooth or very nearly so throughout, the culms 1°–4° tall erect, simple. Sheaths usually exceeding the internodes, sometimes shorter, the upper one long and not inflated, or very slightly so; ligule 1″–2″ long, rounded; blades 3′–9′ long, 2″–3″ wide, smooth or scabrous; spike usually elongated, cylindric, 1½′–7′ in length, 2½″–4″ in diameter; outer scales of the spikelet, exclusive of the awn, 1¼″ long, ciliate on the keel, the awn less than half their length.

In fields and meadows nearly throughout North America. Also in Europe and Asia. Widely cultivated for hay. The scales are sometimes modified into small leaves. Meadow Cat's-tail. Rat-tail. Soldier's-feather. July–Aug.

2. **Phleum alpìnum** L.   Mountain Timothy or Foxtail.   Fig. 457.

Phleum alpinum L. Sp. Pl. 59. 1753.

Glabrous, culms 6′–18′ tall, erect or sometimes decumbent at the base, simple, smooth. Sheaths often much shorter than the internodes, sometimes longer, the upper one usually much inflated; ligule about 1″ long, truncate; blades smooth beneath, scabrous above, the lower 2′–3′ long, 1″–4″ wide; upper leaf generally very short, less than 1′ long; spike short, ovoid to oblong and cylindric, ½′–2′ in length, 3″–6″ in diameter; outer scales of the spikelet, exclusive of the awn, 1½″ long, strongly ciliate on the keel, the awn about one-half their length.

Labrador to Alaska, south to the mountains of New Hampshire, Vermont, Arizona and California. Also in northern Europe, Asia, and in Patagonia. Summer. Alpine cat's-tail.

38.   **ALOPECÙRUS** L. Sp. Pl. 60. 1753.

Annual or perennial grasses with erect or decumbent culms, usually flat leaf-blades, and spicate inflorescence. Spikelets 1-flowered, flattened; scales 3, the 2 lower empty, acute, sometimes short-awned, more or less united below, compressed-keeled; keel ciliate or somewhat winged; third scale truncate or obtuse, hyaline, 3-nerved, awned on the back, subtending a perfect flower and usually a palet; palet hyaline, acute, sometimes wanting. Stamens 3. Styles distinct or rarely united at the base. Stigmas elongated, hairy. [Greek, signifying Fox-tail Grass, in allusion to the spikes.]

About 30 species, principally natives of the north temperate zone. Besides the following, some 4 others occur in western North America. Type species: *Alopecurus pratensis* L.

Outer scales of spikelet united for half their length, keel smooth to hispid.   1. *A. myosuroides.*
Outer scales of the spikelet united for one-quarter their length or less, long-ciliate on the keel.
  Scales 1″–1¼″ in length.
    Awn inserted at ¼ above the base of flowering scale, exserted from the spikelet about 1″.
        2. *A. geniculatus.*
    Awn inserted at or about middle of scale, barely exserted from spikelet. 3. *A. aristulatus.*
  Scales 2″–3″ in length.
    Spike 1½′–2½′ long; outer scales glabrous or sparingly pubescent on the lateral nerves.
        4. *A. pratensis.*
    Spike 1½′ long or less; outer scales villous.   5. *A. alpinus.*

1. Alopecurus myosuroìdes Huds. Slender Fox-tail. Fig. 458.

Alopecurus myosuroides Huds. Fl. Angl. 23. 1762.
Alopecurus agrestis L. Sp. Pl. Ed. 2, 89. 1762.

Smooth or slightly scabrous, culms 1°–2° tall, erect, simple. Sheaths shorter than the internodes; ligule 1″ long, truncate; blades 1½′–7′ long, 1″–3″ wide, scabrous, especially above; spike 1½′–4′ long, 2″–4″ thick; outer scales of the spikelet united at the base for about half their length, narrowly wing-keeled, 2″–2½″ long, the nerves smooth or scabrous, sometimes hispid below, especially on the keel; third scale equalling or slightly exceeding the outer ones, smooth and glabrous, the awn inserted near the base, about twice its length, bent.

In waste places and ballast, southern Massachusetts, New York, New Jersey and Pennsylvania. Adventive from Europe. Native also of Asia. July–Aug. Mousetail; Bennet-weed. Black bent. Black couch-grass. Hunger-grass.

2. Alopecurus geniculàtus L. Marsh Foxtail. Fig. 459.

Alopecurus geniculatus L. Sp. Pl. 60. 1753.

Glabrous or very nearly so, culms 6′–18′ tall, usually decumbent at the base, simple or sparingly branched, smooth. Sheaths usually shorter than the internodes, loose or somewhat inflated; ligule 1½″–3″ long; blades rarely exceeding 3′ long, ½″–2″ wide, rough, especially above; spikes 1′–3′ in length, 2″–4″ thick; outer scales of the spikelet slightly united at the base, 1¼″–1½″ long, obtuse or subacute, smooth, glabrous except on the pubescent lateral nerves and strongly ciliate keel; third scale somewhat shorter, obtuse, smooth and glabrous, the awn inserted about ¼ above the base of the scale, and extending for about 1″ beyond the spikelet.

In wet soil, Newfoundland to Kansas, south to Florida and Texas. Also in Europe and Asia. Introduced. Sometimes found on ballast. July–Sept. Water or floating foxtail. Flote-grass.

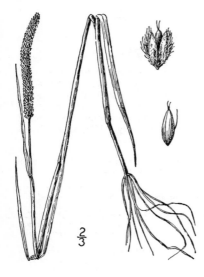

3. Alopecurus aristulàtus Michx. Short-awned Foxtail. Fig. 460.

A. aristulatus Michx. Fl. Bor. Am. 1: 43. 1803.

A. geniculatus var. *aristulatus* Torr. Fl. U. S. 1: 97. 1824.

Culms 6′–20′ tall, erect or somewhat decumbent at the base; sheaths shorter than the internodes; blades up to 6′ long, 1″–2″ wide, rough; spikes 1′–3′ long, 2″–3″ in diameter; spikelets 1″–1¼″ long, the outer scales slightly united at the base, pubescent, acutish, the keel ciliate, the flowering scale a little shorter, obtuse, glabrous, the awn inserted at or near the middle of the scale, barely exserted beyond the spikelet.

Wet meadows, Maine to Alaska, Pennsylvania and California. June–Aug.

4. Alopecurus praténsis L.   Meadow Foxtail.
Fig. 461.

Alopecurus pratensis L. Sp. Pl. 60. 1753.

Nearly or quite glabrous, slender, culms $1°-2\frac{1}{2}°$ tall, erect, simple. Sheaths usually much shorter than the internodes, loose or somewhat inflated; ligule about $\frac{1}{2}''$ long, erose-truncate; blades $1\frac{1}{2}-3\frac{1}{2}'$ long, $1''-3''$ wide, scabrous, at least above; spikes $1\frac{1}{2}-2\frac{1}{2}'$ in length, $4''-6''$ thick; outer scales of the spikelet united at the base for about one-quarter their length, $2''-3''$ long, acute, glabrous except the sparingly pubescent lateral nerves and the strongly ciliate keel; third scale slightly shorter, obtuse, smooth and glabrous, the awn inserted about quarter way up the scale and exceeding it.

In meadows, Newfoundland to southern New York, New Jersey and Ohio. Naturalized from Europe. June–July.

5. Alopecurus alpìnus J. E. Smith.   Alpine Foxtail.
Fig. 462.

Alopecurus alpinus J. E. Smith, Engl. Bot. *pl. 1126.* 1803.

Culms glabrous and smooth or nearly so, $5'-2°$ tall, erect, sometimes decumbent at the base, simple. Sheaths generally shorter than the internodes, loose, often inflated; ligule $1''-2''$ long, rounded at the apex; blades $1'-7'$ long, $1''-3''$ wide, smooth beneath, slightly scabrous above; spike $1\frac{1}{2}'$ in length or less, $3''-6''$ thick; outer scales of the spikelet united only at the base, $2''$ long, obtuse, villous and ciliate; third scale about equalling the outer ones, obtuse, glabrous except at the villous apex, the awn inserted about one-third the way up, a little exceeding the scale.

Greenland and Labrador to Alaska. Also in arctic and alpine Europe and Asia. Summer.

39. PHÍPPSIA R. Br. Suppl. App. Parry's Voy. 285. 1824.

A low annual tufted grass, with flat leaf-blades and spike-like panicles. Spikelets 1-flowered; scales 3; the 2 outer empty, minute, the first often wanting; the third scale thin-membranous, keeled. Palet somewhat shorter, 2-keeled. Stamen 1, rarely 2 or 3. Styles short, distinct. Stigmas plumose. Grain oblong, enclosed in the scale and palet, which readily split and allow it to drop out. [In honor of John Constantine Phipps, 1744–1792, Arctic navigator.]

A monotypic genus of the arctic regions.

1. Phippsia álgida (Soland.) R. Br.   Phippsia.   Fig. 463.

Agrostis algida Solander, in Phipps' Voy. 200. 1810.
Phippsia algida R. Br. Suppl. App. Parry's Voy. 285.
1824.

Smooth and glabrous throughout, culms $1'-5'$ tall, erect, simple; ligule $\frac{1}{2}''$ long; blades $1'$ in length or less, $\frac{1}{4}''-1''$ wide, obtuse; panicle $\frac{1}{4}''-1\frac{1}{2}'$ in length, contracted; branches $\frac{1}{4}'-\frac{3}{4}'$ long, erect or appressed; spikelets $\frac{1}{2}''-\frac{3}{4}''$ long; outer scales minute, unequal, acutish, the first often wanting; third scale broad, 1-nerved, obtuse, or sub-truncate and somewhat erose, the palet about two-thirds as long, broad, 2-keeled, erose-truncate.

Arctic and alpine regions of both the Old World and the New. Summer.

40. SPORÓBOLUS R. Br. Prodr. Fl. Nov. Holl. 1: 169. 1810.

Perennial or rarely annual grasses, with flat or convolute leaf-blades and open or contracted panicles. Spikelets generally small, 1-flowered, occasionally 2–3-flowered. Scales in the 1-flowered spikelets 3, membranous; the 2 outer empty, the first somewhat shorter; the third scale equalling or longer than the empty ones; palet 2-nerved. Stamens 2–3. Styles very short, distinct. Stigmas plumose. Grain free, and often early deciduous. [Greek, referring to the deciduous grain.]

About 100 species, in tropical and temperate regions, very numerous in America. Besides the following, several others occur in the southern and western United States. Type species: *Agrostis indica* L.

Panicle contracted.
  Annuals.
    Spikelets 2″ long; flowering scale pubescent.   1. *S. vaginaeflorus.*
    Spikelets 1¼″–1½″ long; flowering scale glabrous.   2. *S. neglectus.*
  Perennials.
    Plants tufted; no rootstocks.
      Panicle occupying but a small part of the plant.
        Leaves glabrous or nearly so.
          Flowering scale pubescent.
            Palet long-acuminate, much longer than the flowering scale. 3. *S. clandestinus.*
            Palet simply acute, about as long as the flowering scale.   4. *S. canovirens.*
          Flowering scale glabrous.
            Spikelets 2″ long; inflorescence slender.   5. *S. Drummondii.*
            Spikelets 2½″–3″ long; inflorescence stout.   6. *S. asper.*
        Leaves, at least the lower ones, papillose-hirsute.   7. *S. pilosus.*
      Panicle occupying ⅓–½ of the plant.   8. *S. angustus.*
    Plants with long creeping rootstocks.   9. *S. virginicus.*
Panicle open, its branches spreading, at least at maturity (sometimes contracted in no. 15).
  Annuals.
    Empty scales but little shorter than the flowering scale, usually pubescent.
                                         10. *S. confusus.*
    Empty scales ½ as long as the flowering scale, glabrous.   11. *S. uniflorus.*
  Perennials.
    Culms tufted; no rootstocks.
      Spikelets 1″–1½″ long; empty scales ovate to lanceolate.
        Panicle branches verticillate.
          Spikelets ¾″ long, green.   12. *S. argutus.*
          Spikelets 1¼″–1½″ long, purple.   13. *S. gracilis.*
        Panicle branches alternate.
          Leaf-sheaths naked, or sparingly ciliate at the throat; panicle usually exserted.
                                         14. *S. airoides.*
         Leaf-sheaths densely pilose at the throat; base of the panicle generally included.
                                       15. *S. cryptandrus.*
      Spikelets 2″–3″ long; first scale subulate, much narrower than the second.
                                     16. *S. heterolepis.*
    Culms from long running rootstocks.
      First scale of the spikelet ½ as long as the second or less.   17. *S. texanus.*
      First scale about equalling the second.
        Culms erect, simple; leaf-blades elongated.   18. *S. torreyanus.*
        Culms decumbent and branched below; leaf-blades short. 19. *S. asperifolius.*

⅔

1. Sporobolus vaginaeflòrus Torr. Sheathed Rush-grass. Fig. 464.

Vilfa vaginaeflora Torr.; A. Gray, Gram. and Cyp. No. 3. 1834.
Sporobolus vaginaeflorus Torr.; Wood, Classbook, 775. 1861.
Sporobolus minor Vasey; A. Gray, Man. Ed. 6, 646. 1890.

Culms 8′–18′ tall, erect, slender, smooth or scabrous. Sheaths usually inflated, about half as long as the internodes; ligule very short; blades 1″ wide or less, smooth and glabrous beneath, scabrous and hairy near the base above, attenuate into a slender involute point, the lower elongated, the upper 1′–3′ long, setaceous; panicles ¾′–2′ in length, the terminal one exserted or sometimes partially included, strict, the branches ½′ long or less, erect, the lateral ones enclosed in the sheaths; spikelets 1¾″–2¼″ long, the outer scales unequal, acuminate, smooth, the lower one shorter; third scale scabrous, especially toward the apex, about as long as the second and equalling or slightly exceeded by the very acute palet.

In dry soil, southern Maine to South Dakota, south to Georgia and Texas. Southern poverty-grass. Aug.–Sept.

2. **Sporobolus necléctus** Nash. Small Rush-grass. Fig. 465.

Sporobolus neglectus Nash, Bull. Torr. Club, **22** : 464. 1895.

Culms 6′–12′ tall, erect from a usually decumbent base, slender, often much branched, smooth and glabrous. Sheaths about half as long as the internodes, inflated; ligule very short; blades 1″ wide or less at the base, smooth and glabrous beneath, scabrous and hairy near the base above, attenuate into a slender point, the lower elongated, the upper 1′–3′ long, setaceous; terminal panicle 1′–2½′ in length, usually more or less included in the upper sheath, strict; lateral panicles enclosed in the sheaths; spikelets about 1½″ long, the outer scales acute, the lower one slightly shorter; third scale acute, glabrous, a little longer than the second and about equalling the acute palet.

In dry soil, New Brunswick to North Dakota, Virginia and Missouri. Aug.–Sept.

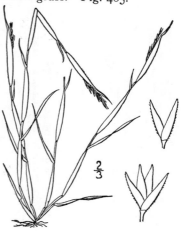

3. **Sporobolus clandestìnus** (Spreng.) Hitchc. Rough Rush-grass. Fig. 466.

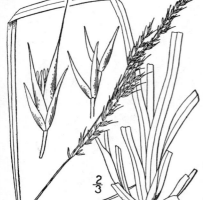

Agrostis clandestina Spreng. Fl. Hal. Mant. **32**. 1807.
S. clandestinus Hitchc. Contr. U. S. Nat. Herb. **12** : 150. 1908.

Culms 2°–5° tall, erect, simple, smooth and glabrous. Sheaths shorter than the internodes; ligule a mere ring, less than ¼″ long, naked; blades 3′–15′ long, 1″–2″ wide at the base, attenuate into a long slender involute tip, smooth and glabrous beneath, scabrous above, or somewhat hairy at the base; panicle 2′–5′ in length, linear, strict, its branches 1′–2′ long, appressed; spikelets 3″–4″ long, the outer scales unequal, acute; third scale pubescent at the base, much longer than the second and greatly exceeded by the long-acuminate almost awned palet.

In dry soil, Connecticut to Missouri, south to Florida and Texas. Described and figured as *S. asper* in our first edition. Prairie-grass. Aug.–Sept.

4. **Sporobolus canóvirens** Nash. Grey-green Rush-grass. Fig. 467.

Sporobolus canovirens Nash, in Britt. Man. 1042. 1901.

Culms 1°–3° tall, erect; leaf-blades 10′ long or less, ½″–1½″ wide, attenuate and filiform above; panicle 2′–5′ long; spikelets 2½″–3″ long, the scales acuminate, the empty ones unequal, the flowering scale appressed-pubescent below with long hairs, about equalling or a little exceeded by the acute palet.

Sandy soil, Tennessee to Missouri and Mississippi. Sept.

5. Sporobolus Drummóndii (Trin.) Vasey.
Drummond's Rush-grass. Fig. 468.

Vilfa Drummondii Trin. Mem. Acad. St. Petersb.
VI. 5²: 106. 1840.

S. Drummondii Vasey, Cat. Grasses U. S. 44. 1885.

Culms 1½°–3° tall, erect, slender; leaf-blades 1°
long or less, ½″–1½″ wide, attenuate and filiform at
the apex; panicle 4′–6′ long, slender; spikelets
about 2″ long, the empty scales acute, the first
shorter than the second, the flowering scale glabrous, acute or obtusish, longer than the second
one and about equalling the acutish palet.

In dry soil, Missouri to Louisiana and Texas.
Sept.–Oct.

Sporobolus attenuàtus Nash has been reported as
introduced along railroads in Jackson Co., Mo. It is
related to the above and may be distinguished by its
smaller spikelets about 1½″ long.

6. Sporobolus ásper (Michx.) Kunth. Long-leaved Rush-grass. Fig. 469.

Agrostis aspera Michx. Fl. Bor. Am. **1**: 52. 1803.
Agrostis longifolia Torr. Fl. U. S. **1**: 90. 1824.
S. asper Kunth. Rev. Gram. **1**: 68. 1829.
S. longifolius Wood, Class-book, 775. 1861.

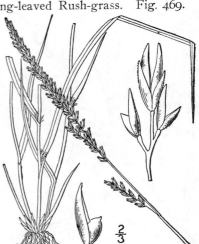

Culms 1½°–3½° tall, erect, simple or occasionally
branched, smooth and glabrous. Sheaths shorter
than the internodes; ligule very short, minutely
ciliate; leaves 4′–18′ long, 1″–2″ wide at the base,
attenuate into a long slender involute tip, smooth
and glabrous beneath, scabrous and hairy at the
base above; panicle more or less included in the
upper sheath, 3′–10′ in length, linear, strict, the
branches 1′–2′ long, erect; spikelets 2½″–3″ long;
outer scales unequal, acutish, glabrous, the lower
shorter; third scale glabrous, acutish or obtuse, exceeding the second and equalling or a little shorter
than the obtuse palet.

In dry soil, Maine to South Dakota and Texas.
Prairie-grass. Aug.–Sept.

7. Sporobolus pilòsus Vasey. Hairy Rush-grass. Fig. 470.

Sporobolus pilosus Vasey, Coult. Bot. Gaz. **16**: 26. 1891.

Culms 1°–1½° tall, erect, rigid, stout, smooth and glabrous. Sheaths shorter than the internodes, crowded
and overlapping at the base of the culm; ligule very
short, minutely ciliate; blades 3′–6′ long, 1″–2″ wide at
base, erect, rigid, attenuate into a slender involute tip,
the lower papillose-hirsute on both sides, the upper
usually glabrous beneath, scabrous above and somewhat hairy near the base; panicle 2′–3′ in length, included at the base, erect, strict, its branches ½′–1′ long,
erect; spikelets 2½″ long, the outer scales unequal, glabrous, obtuse, the lower shorter; third scale obtuse,
glabrous, somewhat exceeding the second and equalling
or a little longer than the obtuse palet.

In dry soil, Kansas and Missouri. Aug.–Sept.

8. Sporobolus angústus Buckley. Dense Rush-grass. Smut-grass. Fig. 471.

S. indicus Nash, in Ill. Fl. Ed. 1, 1 : 154. 1896. Not R. Br. 1810.

S. angustus Buckley, Proc. Phila. Acad. 1862 : 88. 1863.

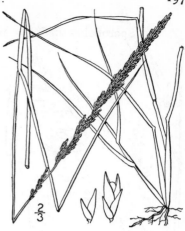

Glabrous and smooth throughout, culms 1°–4° tall, erect, tufted, simple or rarely sparingly branched. Sheaths few, long, shorter than the internodes; ligule a ring of very short hairs; blades 1″–3″ wide, attenuate into a long slender point, the lower 8′–1° long, the upper shorter; panicle 4′–15′ in length, usually elongated, narrow, spike-like; spikelets ¾″–1″ long, the outer scales unequal, about half as long as the third, obtuse, smooth and glabrous, the lower one shorter and often erose-truncate; third scale acute, somewhat exceeding the obtuse or acutish palet.

In meadows and waste places, Virginia to Florida, west to Missouri and California. Also in Bermuda, the Bahamas and the West Indies. Black-seed, Swamp-grass, Sweet-grass. July–Sept.

9. Sporobolus virgínicus (L.) Kunth. Sea-shore Rush-grass. Fig. 472.

Agrostis virginica L. Sp. Pl. 63. 1753.
Sporobolus virginicus Kunth, Rev. Gram. 1 : 67. 1829.

Culms 6′–2° tall, erect or sometimes decumbent, simple or branched at the base, smooth and glabrous. Sheaths numerous, short, overlapping and crowded at the lower part of the culm, smooth, glabrous or sometimes pilose on the margins and at the throat; ligule a ring of short hairs; blades 1′–8′ long, 2″ wide or less at the base, distichous, acuminate into a long point, involute on the margins and at the apex, smooth beneath, scabrous above or sometimes sparingly hairy; panicle 1′–3′ long, 2″–5″ thick, dense and spike-like, usually exserted; spikelets 1″–1¼″ long, the outer scales about equal, acute, smooth and glabrous; third scale smooth and glabrous, acute, slightly shorter than the second and about equalling the obtuse palet.

On sandy shores, Virginia to Florida, Texas, Mexico and South America. West Indies. Aug.–Sept.

10. Sporobolus confùsus (Fourn.) Vasey. Vasey's Dropseed. Fig. 473.

Vilfa confusa Fourn. Mex. Pl. Gram. 101. 1881.

S. confusus Vasey, Bull. Torr. Club, 15 : 293. 1888.

Culms tufted, 4′–12′ tall, slender; blades 2′ long or less, not over ¾″ wide; panicle open, 1′–8′ long, its slender branches spreading or ascending; spikelets about 1¼″ long, on capillary pedicels which are abruptly thickened at the apex, the empty scales shorter than the flowering scale, glabrous or pubescent, the flowering scale usually pubescent.

Usually in wet places, Montana and Nebraska to Mexico. June–Sept.

11. Sporobolus uniflòrus Muhl. Late-flowering Dropseed. Fig. 474.

Poa uniflora Muhl. Descr. Gram. 151. 1817.
Agrostis serotina Torr. Fl. U. S. 1: 88. 1824.
Sporobolus serotinus A. Gray, Man. 577. 1848.
S. uniflorus Muhl.; Scribn. & Merr. Circ. U. S. Dep. Agr. Agrost. 27: 5. 1900.

Glabrous and smooth or very nearly so, culms 6′–18′ tall, from an annual root, erect, slender, simple. Sheaths short, confined to the lower part of the culm; ligule less than ½″ in length, irregularly truncate; blades ⅓″ wide or less, slightly scabrous above, flat, the basal one-third to half the length of the culm, those of the culm 2′–4′ long; panicle 3′–9′ in length, the branches capillary, erect or ascending, the lower 1′–2½′ long; spikelets about ⅝″ long, the outer scales subequal, obtuse, smooth or sometimes sparingly scabrous; third scale twice the length of the outer ones, acuminate.

In wet sandy soil, Maine to Ontario and Michigan, south to New Jersey. Sept.–Oct.

12. Sporobolus argùtus (Nees) Kunth. Pointed Dropseed-grass. Fig. 475.

Vilfa arguta Nees, Agrost. Bras. 2: 395. 1829.
Sporobolus argutus Kunth, Enum. 1: 215. 1833.

Culms 1° tall or less, erect, or somewhat decumbent at the base, simple or sometimes branched, smooth and glabrous. Sheaths shorter than the internodes, their margins sometimes hirsute at the top; ligule a ring of short hairs; blades 1′–2′ long, 1″–2″ wide at the base, acuminate, smooth and glabrous beneath, scabrous and often sparingly hairy at the base above; panicle 1½′–3′ in length, the branches ½′–1′ long, verticillate, at first appressed, finally widely spreading; spikelets ¾″ long; outer scales smooth and glabrous, the first rounded or obtuse, one-quarter the length of the acute second one; third scale about equalling the second, acute.

In sandy and rocky places, Kansas and Colorado, south to Texas and Mexico. Also in the West Indies. July–Sept.

13. Sporobolus grácilis (Trin.) Merrill. Purple Dropseed-grass. Wire-grass. Fig. 476.

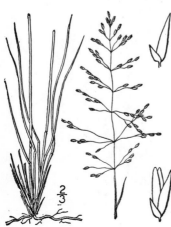

Agrostis juncea Michx. Fl. Bor. Am. 1: 52. 1803. Not Lam. 1783.
Vilfa gracilis Trin. Mem. Acad. St. Petersb. VI. 5²: 74. 1840.
Sporobolus junceus Kunth, Rev. Gram. 1: 68. 1829.
S. ejuncidus Nash, in Britt. Man. 106. 1901.
S. gracilis Merrill, Rhodora, 4: 48. 1902.

Glabrous and smooth throughout, culms 1°–2° tall, tufted, erect, slender, simple. Sheaths shorter than the internodes; ligule very short; blades filiform or setaceous, the basal 6′–1° long, numerous, those of the culm few, 1′–3′ long; panicle 3′–7′ in length, open, the branches verticillate, the lower 1′–2′ long, widely spreading; spikelets 1¼″–1½″, purple, the outer scales very unequal, the first obtuse or acutish, one-fourth to one-third the length of the acute second one; third scale subacute or blunt, equalling the second and the obtuse palet.

Dry sandy soil, Virginia to Florida, west to Texas. Rush-grass. Aug.–Sept.

14. Sporobolus airoìdes Torr.   Hair-grass Dropseed.   Fig. 477.

Agrostis airoides Torr. Ann. Lyc. N. Y. 1 : 151. 1824.
S. airoides Torr. Pac. R. R. Rept. 7 : Part 3, 21. 1856.

Culms 1½°–3° tall, erect, simple, smooth and glabrous.
Sheaths generally shorter than the internodes, sometimes sparsely ciliate at the throat; ligule very short;
blades smooth beneath, scabrous above and sometimes
sparingly hairy near the base, ½″–1½″ wide at the base,
attenuate into a long slender involute point, the basal
about one-half as long as the culm, the upper culm
leaves 2′–5′ in length; panicle 5′–15′ long, usually exserted, the branches alternate or the upper verticillate,
at length widely spreading, the lower 3′–7′ long; spikelets ¾″–1″ long, the scales acute, glabrous, the outer
unequal, the lower one about half as long as the upper;
third scale equalling the second and the palet.

Prairies, Nebraska to Montana, California and Texas.
Rush-grass. Salt-grass. Fine-top salt-grass. Aug.–Sept.

15. Sporobolus cryptándrus (Torr.) A Gray.   Sand Dropseed.   Fig. 478.

Agrostis cryptandra Torr. Ann. Lyc. N. Y. 1 : 151. 1824.
Sporobolus cryptandrus A. Gray, Man. 576. 1848.

Culms 1½°–3½° tall, erect, simple or sometimes branched
at the base, smooth and glabrous. Sheaths smooth, with
a dense pilose ring at the summit, the lower short, crowded
and overlapping, the upper much longer, generally enclosing the base of the panicle; ligule a ring of short hairs;
blades 3′–6′ long, 1″–2″ wide, flat, glabrous beneath, scabrous above, long-acuminate; panicle 6′–10′ in length, the
base generally included in the upper sheath, rarely entirely
exserted, the branches spreading or ascending, alternate,
the lower 1½′–3′ long; spikelets 1″–1¼″ long, the scales
acute, glabrous, the outer scabrous on the keel, the lower
one-third as long as the upper; third scale somewhat longer
or shorter than the second.

In sandy soil, Massachusetts to Montana, Pennsylvania and
Mexico. Prairie-grass. Aug.–Oct.

16. Sporobolus heterólepis A. Gray.   Northern Dropseed.   Fig. 479.

Vilfa heterolepis A. Gray, Ann. Lyc. N. Y. 3 : 233. 1835.
Sporobolus heterolepis A. Gray, Man. 576. 1848.

Culms 1°–3° tall, erect, simple, smooth and glabrous.
Sheaths sometimes sparingly pilose at the summit, the
lower short, loose, and overlapping, the upper much
elongated and tight to the culm; ligule a ring of short
hairs; blades involute-setaceous, glabrous, the margins
and upper part of the midrib very rough, the basal about
three-fourths the length of the culm, occasionally equalling it, those of the culm shorter; panicle 3′–10′ in length,
its branches erect or ascending, alternate or sub-verticil-
late, the lower 1½′–3½′ long; spikelets 2″–2¾″ long, the
scales smooth and glabrous, the outer unequal, acuminate,
the lower subulate, about half the length of the broad
second one, often awn-pointed; third scale obtuse or
acute, shorter than the second or occasionally equalling it.

In dry soil, Quebec to Saskatchewan, south to Connecticut, Pennsylvania, Missouri and Texas. Bunch-grass. Aug.–
Sept.

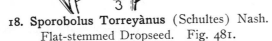

17. **Sporobolus texànus** Vasey. Texas Drop-seed. Fig. 480.

S. texanus Vasey, Contr. U. S. Nat. Herb. **1** : 57. 1890.

Culms 1°–2° tall, branching below; sheaths crowded, the lower ones papillose-hirsute; blades erect, firm, 1½′–6′ long, 1½″–2½″ wide, smooth beneath, very rough above; panicle included at the base, the upper branches finally widely spreading, 2′–4′ long; spikelets a little exceeding 1″ long, on long slender pedicels; scales smooth and glabrous, the first scale narrow, acuminate, less than ½ as long as the second which equals the third scale.

In dry places, Kansas to Mexico.   July–Sept.

18. **Sporobolus Torreyànus** (Schultes) Nash. Flat-stemmed Dropseed. Fig. 481.

Agrostis compressa Torr. Cat. Pl. N. Y. 91. 1819. Not Willd. *1790.*
A. Torreyana Schult. Mant. **2** : 203. 1824.
Sporobolus compressus Kunth, Enum. **1** : 217. 1830.
Sporobolus Torreyanus Nash, in Britt. Man. 107. 1901.

Culms 1°–2° tall, from a horizontal rootstock, stout, simple, much compressed, smooth and glabrous. Sheaths compressed, overlapping, sometimes scabrous at the summit; ligule very short; blades 5′–10′ long, 1″ wide or less, folded, slightly rough; panicle 4′–10′ in length, the branches erect or ascending, the lower 2′–3′ long; spikelets about ⅞″ long; outer scales subequal, obtuse or somewhat acute, smooth and glabrous; third scale obtuse and apiculate, strongly scabrous, slightly exceeding the outer ones.

In bogs, Long Island and in the pine barrens of New Jersey.   Sept.–Oct.

19. **Sporobolus asperifòlius** (Nees & Meyen) Thurber. Rough-leaved Dropseed. Fig. 482.

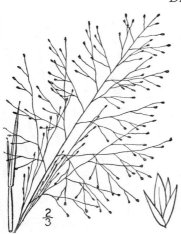

Vilfa asperifolia Nees & Meyen; Trin. Mem. Acad. St. Petersb. (VI.) **6** : 95. 1840.
S. asperifolius Thurber; S. Wats. Bot. Cal. **2** : 269. 1880.

Culms 6′–18′ tall, erect from a decumbent and branched base, smooth and glabrous. Sheaths short, crowded and overlapping, the upper usually enclosing the base of the panicle; ligule ¼″ long, erose-truncate; blades numerous, 1′–3½′ long, 1″–1½″ wide at the base, acuminate, strict, often erect, flat, glabrous, smooth beneath, very rough above; panicle 3′–8′ in length, included at the base, rarely entirely exserted, the capillary branches spreading or ascending, the lower 2′–4′ long; spikelets occasionally 2–3-flowered, ¾″ long; outer scales subequal, acute, glabrous, sparingly scabrous; third scale obtuse or acute, glabrous, somewhat exceeding the second.

Dry soil, Saskatchewan to British Columbia, south to Missouri and Mexico.   Aug.–Sept.

41. POLYPÒGON Desf. Fl. Atl. 1: 66. 1798.

Mostly annual grasses, with decumbent or rarely erect culms, flat leaf-blades and spike-like panicles. Spikelets 1-flowered; scales 3; the 2 outer empty, each extended into an awn; third scale smaller, generally hyaline, short-awned from below the apex, subtending a palet and perfect flower; palet shorter than the scale. Stamens 1–3. Styles short, distinct. Stigmas plumose. Grain free, enclosed in the scale and palet. [Greek, in allusion to the many long awns which resemble a beard.]

About 10 species, widely distributed in temperate and warm regions, rare in the tropics. Type species: *Alopecurus Monspeliensis* L.

1. Polypogon Monspeliénsis (L.) Desf.
Annual Beard-grass. Fig. 483.

Alopecurus Monspeliensis L. Sp. Pl. 89. 1753.
P. Monspeliensis Desf. Fl. Atl. 1: 67. 1798.

Culms 2° tall or less, erect from a usually decumbent base, smooth and glabrous. Sheaths generally shorter than the internodes, loose, sometimes slightly scabrous; ligule 1½″–4″ long; blades 1½′–6′ long, 1½″–3″ wide, scabrous, especially above; panicle 1′–4′ in length, dense and spike-like, the branches ½′ in length, ascending; spikelets crowded; outer scales about 1″ long, obtuse, slightly bifid, scabrous, bearing a more or less bent awn 2″–3″ long; third scale much shorter, erose-truncate, hyaline, bearing a delicate awn about ¼″ long, inserted below the apex.

In waste places, Maine to Georgia, and in Texas, mostly near the coast. Very abundant in western North America, from British Columbia to Mexico. Naturalized from Europe. Native also of Asia. July–Sept.

42. ARCTAGRÓSTIS Griseb. in Ledeb. Fl. Ross. 4: 434. 1853.

Perennial grasses with flat leaves and contracted panicle. Spikelets 1-flowered. Scales 3; the 2 outer empty, unequal, somewhat acute, membranous; the third scale exceeding the second, subtending a palet and perfect flower, obtuse; palet obtuse, 2-nerved. Stamens 2 or 3. Styles distinct, short. Stigmas plumose. Grain oblong, free, enclosed in the scale and palet. Seed adherent to the pericarp. [Latin, signifying an arctic *Agrostis*-like grass.]

A genus of 5 or 6 species, inhabiting arctic and sub-arctic regions. Type species: *Colpodium latifolium* R. Br.

1. Arctagrostis latifòlia (R. Br.) Griseb.
Arctagrostis. Fig. 484.

Colpodium latifolium R. Br. Suppl. App. Parry's Voy. 286. 1824.
A. latifolia Griseb. in Ledeb. Fl. Ross. 4: 434. 1853.

Culms 6′–2° tall, erect, or sometimes decumbent at the base, simple, smooth and glabrous. Sheaths shorter than the internodes; ligule 2″ long, truncate; blades 1′–7′ long, 1″–4″ wide, usually erect, scabrous; panicle 1½′–8′ long, narrow, its branches ½′–2′ in length, ascending or erect; spikelets 1½″–2″ long; outer scales unequal, acutish, the lower about two-thirds to three-fourths the length of the upper; third scale obtuse, exceeding the second, hispid on the keel.

Greenland to Hudson Bay and Alaska. Also in arctic Europe and Asia. Summer.

43. CÍNNA L. Sp. Pl. 5. 1753.

Tall grasses with flat leaf-blades and panicled spikelets. Spikelets 1-flowered. Scales 3; the 2 outer empty, keeled, acute; the third scale similar, but usually short-awned on the back, subtending a palet and a stalked perfect flower; palet a little shorter, 1- or 2-nerved. Stamen 1. Styles short, distinct. Stigmas plumose. Grain narrow, free, enclosed in the scale and palet. Seed adherent to the pericarp. [Greek, from Dioscorides.]

Four known species, inhabiting temperate regions of Europe and North America. Besides the following, another occurs in the western United States. Type species: *Cinna arundinacea* L.

Panicle narrow at maturity, its filiform branches erect or drooping; spikelets 2½″–3″ long; first scale much shorter than the second.   1. *C. arundinacea.*
Panicle open, its capillary branches flexuous and drooping; spikelets 1½″–2″ long; first scale about equalling the second.   2. *C. latifolia.*

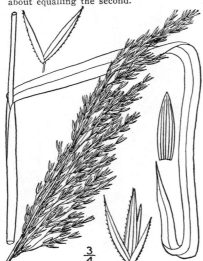

1. Cinna arundinàcea L. Wood or Sweet Reed-grass. Fig. 485.

Cinna arundinacea L. Sp. Pl. 5. 1753.

Culms 2°–5° tall, erect, simple, smooth and glabrous. Sheaths usually shorter than the internodes, overlapping at the base of the culm, smooth or roughish; ligule 1″–2″ long, truncate; blades 6′–1° long, 2″–7″ wide, scabrous; panicle 6′–12′ in length, usually contracted, sometimes purple, the filiform branches erect or drooping, the lower 1½′–4½′ long; spikelets 2½″–3″ in length, the scales acute, scabrous, especially on the keel, the first one shorter than the second; third scale slightly exceeded or equalled by the second, usually bearing an awn about ¼″ long from the 2-toothed apex.

In moist woods and swamps, Nova Scotia to Ontario, Georgia and Texas. Ascends to 1700 ft. in North Carolina. Indian Reed-grass. Aug.–Sept.

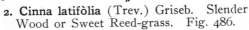

2. Cinna latifòlia (Trev.) Griseb. Slender Wood or Sweet Reed-grass. Fig. 486.

Agrostis latifolia Treviran, in Goeppert, Beschr. d. Bot. Gart. Breslau, 82. 1830.
Cinna pendula Trin. Mem. Acad. St. Petersb. (VI.) 6: 280. 1841.
C. latifolia Griseb. in Ledeb. Fl. Ross. 4: 435. 1853.

Culms 2°–4° tall, erect, usually slender, simple, smooth and glabrous. Sheaths shorter than the internodes, sometimes slightly scabrous; ligule 1″–2″ long; blades 4′–10′ long, 2″–6″ wide, scabrous; panicle 5′–10′ in length, open, the capillary branches generally spreading, flexuous and often drooping, the lower 1½′–5′ in length; spikelets 1½″–2″ long; scales scabrous, the outer acute, strongly hispid on the keel, the first about equalling the second; third scale usually exceeded by the second and bearing a rough awn ½″–1″ long from the 2-toothed apex.

In damp woods, Newfoundland to British Columbia, New Jersey and Washington and in the Alleghanies to North Carolina, and in the Rocky Mountains to Colorado and Utah. Also in northern Europe. Ascends to 5000 ft. in the Adirondacks. Aug.–Sept.

44. AGRÓSTIS L. Sp. Pl. 6. 1753.

Annual or perennial tufted grasses with flat or bristle-like leaves and paniculate inflorescence. Spikelets 1-flowered. Scales 3; the 2 outer empty, membranous, keeled, acute; the third shorter, obtuse, hyaline, sometimes bearing a dorsal awn, subtending a perfect flower; palet shorter than the scale, sometimes minute or wanting. Stamens usually 3. Styles distinct, short. Stigmas plumose. Grain free, enclosed in the scale. Seed adherent to the pericarp. [Name Greek, referring to the field habitat of many species.]

A genus of about 100 species, distributed throughout the world, numerous in temperate regions; 15 others are found in western North America. Type species: *Agrostis alba* L.

Palet conspicuous, at least one-half as long as the scale.
Panicle open in flower, branches long and spreading; an upland grass. 1. *A. alba.*

Panicle dense and contracted, spike-like, its branches short and appressed ; a grass of brackish
    marshes and wet sands.   2. *A. maritima.*
Palet inconpicuous, minute, or wanting.
    Branches of the contracted panicle short, spikelet-bearing to the base.   3. *A. asperifolia.*
    Branches of the panicle slender, naked below.
        Flowering scale awned.
            Awn flexuous, barbellate, twice the length of the ¾″ spikelet.   4. *A. Elliottiana.*
            Awn stouter, glabrous, rigid, usually bent, not twice the length of the spikelet.
                Branches of the panicle usually ascending ; spikelets 1″ long.
                    Culms 6′ tall or less ; panicle 1′ long or less ; alpine grass.   5. *A. rupestris.*
                    Culms 1°–2° tall ; panicle exceeding 2′ long.   6. *A. canina.*
                Branches of the mature panicle spreading ; spikelets 1¼″–1½″ long.
                    Awn exserted ; panicle branches glabrous or hispidulous.   7. *A. borealis.*
                    Awn short ; panicle branches hispid.   8. *A. geminata.*
        Flowering scale awnless, or very rarely with a short awn.
            Culms weak, decumbent or prostrate at the base ; blades lax.   9. *A. Schweinitzii.*
            Culms and blades erect.
                Branches of the panicle capillary, elongated, usually dividing above the middle, the
                    spikelets often crowded at the extremities.
                    Spikelets ¾″–1″ long ; blades short.   12. *A. hyemalis.*
                    Spikelets 1¼″–1½″ long ; blades elongated.   11. *A. altissima.*
                Branches of the panicle not elongated, usually dividing at or below the middle.
                    Spikelets about 1″ long ; a grass of low elevations.   10. *A. perennans.*
                    Spikelets 1¼″–1½″ long ; a high mountain grass.   13. *A. oreophila.*

1. **Agrostis álba** L. Red-top. Fiorin. Herd's-grass. Fig. 487.

Agrostis alba L. Sp. Pl. 63. 1753.
Agrostis vulgaris With. Bot. Arr. Brit. Pl. Ed.
    3, 132. 1796.
A. alba var. *aristata* A. Gray, Man. 578. 1848.
Agrostis alba var. *vulgaris* Thurber in A. Gray,
    Man. Ed. 6, 647. 1890.

Culms 8′–2½° tall, erect or decumbent at the
base, often stoloniferous, simple, smooth and
glabrous. Sheaths usually shorter than the
internodes, often crowded at the base of the
culm ; ligule 4″ long or less ; blades 2′–8′ long,
1″–3″ wide, scabrous ; panicle 2′–9′ in length,
contracted or open, green or purplish, the
branches ascending or erect, the lower 1′–3′
long ; spikelets 1″–1¼″ long ; outer scales
about equal, acute, smooth and glabrous, except on the hispid or scabrous keel ; third
scale shorter, rarely awned near the base, the
palet at least one-third its length.

Fields and meadows nearly throughout North
America, extensively cultivated for fodder. Naturalized from Europe, and perhaps also native
northward. White-top ; White, Marsh or Creeping Bent ; Black Quitch, Tussocks, Water Twitch,
Fine John ; Monkey's, Burden's or Summer
Dew-grass ; Conch or Bonnet-grass. July–Sept.

3/4

3/4

2. **Agrostis marítima** Lam. Dense-flowered Bent-grass. Fig. 488.

A. maritima Lam. Encycl. 1 : 61. 1783.
Agrostis coarctata Ehrh. ; Hoffm. Deutsch. Fl. Ed. 2,
    1 : 37. 1800.
A. alba maritima Meyer, Chloris Hanov. 656. 1836.

Glabrous. Culms tufted, erect, or decumbent at
the base and often rooting at the lower nodes,
smooth, 12′–20′ tall, at length branching ; sheaths
shorter than the internodes ; ligule scarious, ½″–1″
long ; blades erect, rough on both surfaces, 1½′–3½′
long, 1½″ or less wide ; panicle dense and contracted, 1½′–4′ long, ¼′–½′ thick, its branches erect,
the longer 1¼′ long or less ; spikelets numerous,
crowded, acute at both ends and lanceolate when
closed, 1″–1¼″ long, on shorter hispidulous pedicels
which are much thickened at the apex ; empty scales
acute, hispidulous on the upper part of the keel,
especially in the first scale ; flowering scale hyaline,
about three-quarters as long as the spikelet, den-
ticulate at the truncate or rounded apex ; palet about one-half as long as the scale.

Wet sands or brackish marshes along the coast, Maine and Quebec to Delaware. Also in Europe. July–Sept.

3. **Agrostis asperifòlia** Trin. Rough-leaved Bent-grass. Fig. 489.

Agrostis asperifolia Trin. Mem. Acad. St. Petersb. (VI.) 6: Part 2, 317. 1845.

Culms 1°–3° tall, erect, or sometimes decumbent at the base, simple, smooth and glabrous. Sheaths usually shorter than the internodes, smooth or roughish; ligule 1″–3½″ long, more or less decurrent; blades 1′–8′ long, 1″–4″ wide, generally erect, flat or involute, scabrous; panicle contracted, 2½′–10′ in length, often interrupted or glomerate, the branches 1½′–3′ in length, erect, spike-let-bearing to the base; spikelets crowded, 1″–1¼″ long, the outer scales subequal, scabrous, especially on the keel; third scale about three-fourths the length of the second, obtuse or subacute; palet minute.

Manitoba to Washington, south to western Texas and California. Northern Red-top. Aug.–Sept.

4. **Agrostis Elliottiàna** Schultes. Elliott's Bent-grass. Fig. 490.

Agrostis arachnoides Ell. Bot. S. C. & Ga. 1: 134. 1817. Not Poir. 1810.
Agrostis Elliottiana Schultes, Mant. 2: 202. 1824.

Culms 5′–14′ tall, erect, slender, simple, smooth and glabrous. Sheaths shorter than the internodes, smooth or slightly scabrous, strongly striate; ligule 1″ long; blades rough, ½′–2′ long, 1″ wide or less; panicle 2′–5′ in length, usually narrow, sometimes open, the branches slender, naked below, erect or ascending, the lower 1′–1½′ long; spikelets ¾″ long; outer scales subequal, scabrous on the keel, acute; third scale about three-quarters as long as the first, erose-truncate, acute or 2-toothed, bearing a very finely filiform flexuous bar-bellate awn, 2–4 times its length, inserted just below the apex; palet short.

In dry soil, South Carolina to Kentucky and Kansas, Florida and Texas. Spider Bent-grass. May–July.

5. **Agrostis rupéstris** Allioni. Rock Bent-grass. Fig. 491.

Agrostis rupestris Allioni, Fl. Pedem. 2: 237. 1785.

Culms tufted, 6′ or less tall, slender, erect, or decumbent at the base, smooth and glabrous. Sheaths longer than the internodes; ligule about ½″ long; blades smooth and glabrous, those on the culm 1′ or less long, the basal leaves from one-third to one-half as long as the culms; panicle contracted, 1′ or less long, its axis and branches smooth, the latter erect or nearly so, spikelet-bearing above the middle; spikelets about 1″ long; empty scales about equal, 1-nerved, acute, usually purple, hispidulous on the keel; flowering scale shorter, hyaline, den-ticulate at the obtuse or truncate apex, bearing about the middle a dorsal scabrous awn a little over 1″ long; palet wanting.

Labrador. Also in Europe. Summer.

6. Agrostis canìna L. Brown Bent-grass. Fig. 492.

Agrostis canina L. Sp. Pl. 62. 1753.

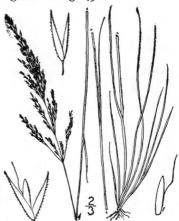

Culms 1°–2° tall, erect, slender, simple, smooth and glabrous. Sheaths shorter than the internodes; ligule ½″–1½″ long; blades 1′–3′ in length, 1″ wide or less, scabrous; panicle 2′–7′ in length, contracted in fruit, the branches slender, naked below, ascending or spreading in flower, the lower 1′–2½′ long; spikelets 1″ long, on appressed pedicels, the outer scales subequal, acute, strongly scabrous on the keel; third scale about two-thirds the length of the first, obtuse, smooth and glabrous, bearing a straight or somewhat bent dorsal awn 1″–2″ long, inserted just above the middle; palet minute or none.

In meadows, Newfoundland to Alaska, south to Pennsylvania and Tennessee. Native northward; naturalized from Europe southward. Rhode Island or Dog Bent-grass; Fine-top, Furze-top. Much used for lawns. July–Sept.

7. Agrostis boreàlis Hartm. Red Bent-grass. Fig. 493.

Agrostis rubra L. Sp. Pl. 62. 1753.
Agrostis borealis Hartm. Scand. Fl. Ed. 3, 17. 1838.
A. rupestris Chapm. Fl. S. States, 551. 1860. Not All. 1785.
Agrostis rubra var. *americana* Scribn.; Macoun, Cat. Can. Pl. **5** : 391. 1890.

Smooth or very nearly so, glabrous, culms 6′–2° tall, erect or sometimes decumbent at the base, simple. Sheaths usually shorter than the internodes; ligule 1″ long; blades 2′–4′ long, ½″–1½″ wide; panicle 2½′–5′ in length, open, the branches generally widely spreading and more or less flexuous, rarely erect, the lower 1′–2¼′ long; spikelets 1¼″–1½″ long, the outer scales acute, scabrous on the keel; third scale shorter than the first, obtuse, bearing a usually bent dorsal awn 2″–2½″ long, inserted below the middle.

Summits of the highest mountains of New England, New York and North Carolina. Europe. Summer.

Agrostis paludòsa Scribn., of Labrador, differs by an awnless flowering scale.

8. Agrostis geminàta Trin. Twin Bent-grass. Fig. 494.

Agrostis geminata Trin. Gram. Unifl. 207. 1824.

A. hiemalis geminata Hitchc. Bull. U. S. Dep. Agr. Pl. Ind. **68** : 44. 1905.

Culms 2′–4′ tall, tufted; leaf-sheaths smooth and glabrous, overlapping; ligule 1″–1½″ long; blades smooth and glabrous, erect, usually complanate; panicle 2½′–5′ long, usually included at the base, its branches very rough, ascending; spikelets about 1½″ long, the first scale longer and broader than the second, the third scale about ⅔ as long as the first.

Labrador to Alaska. This may differ specifically from the true *A. geminata* Trin. Summer.

9. Agrostis Schweinítzii Trin. Thin-grass. Fig. 495.

Agrostis Schweinitzii Trin. Mem. Acad. St. Petersb. **6²**: 311. 1841.

Culms 1°–2½° long from a decumbent or prostrate base, weak, slender, simple or sparingly branched above, smooth and glabrous; ligule ½″ long; blades 2′–6′ long, 1″–2″ wide, lax, scabrous; panicle 4′–8′ in length, open, the branches 1′–2′ long, widely spreading, the branchlets and pedicels divergent; spikelets ¾″–1″ long, the outer scales acute, scabrous on the keel; third scale about three-quarters the length of the first, smooth and glabrous, not awned; palet small or wanting.

In shaded damp places, Quebec to Wisconsin, south to South Carolina and Kansas. Ascends to 6600 ft. in North Carolina. Panicle usually light green, sometimes purplish. Twin-grass. This species was described and figured as *Agrostis perennans* in our first edition. July–Sept.

10. Agrostis perénnans (Walt.) Tuckerm. Upland Bent-grass. Fig. 496.

Cornucopiae perennans Walt. Fl. Car. 74. 1788.
A. perennans Tuckerm. Am. Journ. Sci. **45**: 44. 1843.
Agrostis intermedia Scribn. Bull. Torr. Club, **20**: 476. 1893. Not. Balb. 1801.
A. pseudo-intermedia Farwell, Ann. Rep. Com. Parks & Boul. Detroit **11**: 46. 1900.
A. Scribneriana Nash, in Small, Fl. SE. U. S. 126. 1903.

Culms 1°–3° tall, erect, simple, smooth and glabrous. Sheaths smooth, those at the base of the culm often crowded and overlapping; ligule 1″–2″ long; blades 4′–9′ long, 1″–3″ wide, scabrous; panicle 4′–9′ in length, the branches 1½′–3′ long, ascending, dividing at or below the middle, the divisions divergent, the pedicels appressed; spikelets about 1″ long, the outer scales acute or acuminate, scabrous on the keel; third scale about three-fourths the length of the first, smooth; palet small or wanting.

In dry soil, Massachusetts and New York to New Jersey, Tennessee and Missouri. Aug.–Oct.

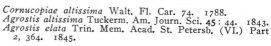

11. Agrostis altíssima (Walt.) Tuckerm. Tall Bent-grass. Fig. 497.

Cornucopiae altissima Walt. Fl. Car. 74. 1788.
Agrostis altissima Tuckerm. Am. Journ. Sci. **45**: 44. 1843.
Agrostis elata Trin. Mem. Acad. St. Petersb. (VI.) Part 2, 364. 1845.

Culms 2°–4° tall, erect, simple, smooth, usually stiff. Sheaths overlapping, scabrous, the upper one elongated; ligule 1″–2″ long; blades elongated, 6′–1° in length, 1″–1½″ wide, scabrous; panicle 7′–9′ long, the branches ascending or erect, somewhat scabrous, the lower 2′–4′ in length, spikelet-bearing at the extremities; spikelets 1¼″–1½″ long, the outer scales acute, scabrous on the keel; third scale shorter, obtuse, scabrous, occasionally bearing a short awn; palet small or wanting.

In sandy swamps, Long Island and New Jersey to Florida and Mississippi. Panicle usually purplish. Tall Thin-grass. Aug.–Oct.

12. Agrostis hyemàlis (Walt.) B.S.P. Rough Hair-grass. Fool-hay. Silk-grass. Fig. 498.

Cornucopiae hyemalis Walt. Fl. Car. 73. 1788.
Agrostis scabra Willd. Sp. Pl. 1: 370. 1798.
Agrostis hyemalis B.S.P. Prel. Cat. N. Y. 68. 1888.

Culms 1°–2° tall, erect, slender, simple, smooth and glabrous. Sheaths generally shorter than the internodes; ligule 1″–2″ long; blades 2′–5′ long, ¼″–1½″ wide, usually erect, roughish; panicle 6′–2° long, usually purplish, the capillary scabrous branches ascending, sometimes widely spreading, or often drooping, the lower 3′–6′ long, dividing above the middle, the divisions spikelet-bearing at the extremities; spikelets ¾″–1″ long, the outer scales acute, scabrous toward the apex and on the keel; third scale two-thirds the length of the first or equalling it, obtuse, rarely bearing a short awn; palet usually very small.

In dry or moist soil, nearly throughout North America except the extreme north. Tickle-grass. Fly-away, Rough or Rough-leaved Bent-grass. July–Aug.

Agrostis antecèdens Bicknell, of eastern Massachusetts, differs in having the spikelets clustered at the ends of the branches.

13. Agrostis oreóphila Trin. New England Bent-grass. Fig. 499.

?Agrostis novae-angliae Tuckerm. Hovey's Mag. 9: 143. April, 1843.
?Agrostis altissima var. *laxa* Tuckerm. Am. Journ. Sci. 45: 44. October, 1843.
A. oreophila Trin. Mem. Acad. St. Petersb. VI. 6²: 323. 1845.

Culms 8′–15′ tall, erect, simple, smooth and glabrous. Sheaths longer than the internodes, generally overlapping; ligule 1″ long; blades 1′–3½′ long, 1″ wide or less, erect, usually involute, scabrous; panicle 3½′–7′ in length, open, the branches spreading or ascending, dividing at or below the middle, the divisions divergent, the pedicels often appressed; spikelets 1¼″–1½″ long, the outer scales acute, strongly scabrous on the keel; third scale somewhat shorter, obtuse.

Newfoundland, south to the high mountains of New England, New York and North Carolina.

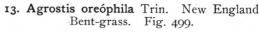

45. CALAMAGRÓSTIS Adans. Fam. Pl. 2: 31. 1763.

[DEYEUXIA Clarion; Beauv. Agrost. 43. *pl. 9. f. 9, 10.* 1812.]

Generally perennial grasses, of various habit, with flat leaf-blades and paniculate inflorescence. Spikelets 1-flowered, the rachilla usually prolonged beyond the flower and pubescent. Scales 3; the 2 outer empty, carinate, membranous; the third scale hyaline, shorter than the outer, obtuse, usually copiously long-hairy at the base, or rarely the hairs scanty or short, and bearing a straight, bent or twisted dorsal awn; palet shorter, 2-nerved. Stamens 3. Styles short, distinct. Stigmas plumose. Grain free, enclosed in the scale. Seed adherent to the pericarp. [Greek, signifying Reed-grass.]

A genus of about 150 species, widely distributed throughout temperate and mountainous regions, and particularly numerous in the Andes. Besides the following, some 25 others occur in the western parts of North America. The English name *Small-reed* is applied to any of the species. Type species: *Arundo Calamagrostis* L.

Prolongation of the rachilla hairy its whole length.
    Awn strongly bent, exserted, hairs of the callus usually much shorter than the scale.
        Leaf-sheaths naked at the summit, rarely bearded; panicle tinged with purple; empty scales rather thick.
            Basal hairs ¼ as long as the flowering scale or less.   1. *C. Pickeringii.*
            Basal hairs about ½–⅔ as long as the flowering scale.   2. *C. lacustris.*
        Leaf-sheaths bearded at the summit; panicle pale; empty scales thin.
            Spikelets 2″–3″ long; callus hairs sparse; palet about equalling the scale. 3. *C. Porteri.*
            Spikelets 1¾″–2″ long; callus hairs copious; palet shorter than the scale. 4. *C. perplexa.*
    Awn straight, included, hairs of the callus little if any shorter than the scale.
        Panicle open, the lower rays widely spreading.
            Spikelets 2″–3″ long, very acuminate.   5. *C. Langsdorfii.*
            Spikelets 1½″–2″ long; panicle usually loosely flowered. 6. *C. canadensis.*
            Spikelets 1″–1¼″ long; panicle rather densely flowered. 7. *C. Macouniana.*

Panicle more or less contracted.
    Culms and almost filiform leaf-blades soft, not rigid.   **8.** *C. neglecta.*
    Culms and wide leaf-blades hard, rigid.
        Panicle elongated, loosely flowered; culms not tufted, or little so. 9. *C. inexpansa.*
        Panicle short, dense and spike-like; culms strongly tufted.
            Panicle narrow, much interrupted below; awn much shorter than the scale.
                                10. *C. labradorica.*
            Panicle thick, continuous, or little interrupted; awn about equalling the scale.
                                11. *C. hyperborea.*
Prolongation of the rachilla hairy only at the summit.   12. *C. cinnoides.*

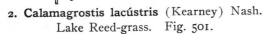

1. Calamagrostis Pickeríngii Gray. Pickering's Reed-grass. Fig. 500.

C. sylvatica var. *breviseta* A. Gray, Man. 582. 1848.
C. Pickeringii A. Gray, Man. Ed. 2, 547. 1856.
C. breviseta Scribn. Mem. Torr. Club, **5**: 41. 1894.
C. breviseta debilis Kearney, Bull. U. S. Dep. Agr. Agrost. **11**: 25. 1898.

Culms 12′–18′ tall, erect, rigid, simple, scabrous below the panicle. Sheaths smooth and glabrous, the lower overlapping, the upper one elongated; ligule 1″–3″ long; blades 1½′–4′ long, 2″ wide, erect, smooth beneath, rough above; panicle 3′–4½′ in length, the branches ascending or erect, the lower 1′–1½′ long; spikelets 1½″–2″ long, purple tinged, the outer scales acute, scabrous on the keel; third scale shorter than the second, obtuse, scabrous, the basal hairs very short; awn bent, not twisted, equalling or slightly exceeding the scale.

In wet places, Newfoundland to the mountains of New England and northern New York. Occurs in the alpine region of the White Mountains. Aug.–Sept.

2. Calamagrostis lacústris (Kearney) Nash. Lake Reed-grass. Fig. 501.

C. breviseta lacustris Kearney, Bull. U. S. Dep. Agr. Agrost. **11**: 25. 1898.
C. Pickeringii lacustris Hitchc. in Gray, Man. Ed. 7, 134. 1908.

Culms 1½°–3° tall, from rather stout rootstocks; leaf-sheaths sometimes bearded at the summit; blades 4′–8′ long, 1″–2″ wide, sometimes involute; panicle up to 6′ long, its branches short and erect; spikelets about 1½″ long, the empty scales acute, strongly hispidulous on the keel, the flowering scale rather thin, the awn attached ¼–⅓ way above the base, the basal hairs one-half to two-thirds as long as the scale, the palet markedly shorter than the flowering scale.

Mountains of New England and along the Great Lakes to Minnesota.

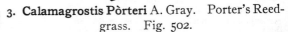

3. Calamagrostis Pòrteri A. Gray. Porter's Reed-grass. Fig. 502.

C. Porteri A. Gray, Proc. Am. Acad. **6**: 79. 1862.

Culms 2°–4° tall, erect, simple, smooth and glabrous. Sheaths shorter than the internodes, slightly scabrous, with a villous ring at the summit; ligule 2″–2½″ long; blades 6′–12′ long, 2″–4″ wide, rough; panicle 4′–8′ in length, the branches erect, the lower 1′–2′ long; spikelets 2″–2½″ long, the outer scales strongly scabrous, acute; third scale shorter than or equalling the second, obtuse, scabrous, the lateral basal hairs about one-third the length of the scale, those at the back short or wanting; awn bent, about equalling the scale, the lower part twisted.

In dry woods, southern New York to Virginia. Aug.–Sept.

4. Calamagrostis perpléxa Scribn. Wood Reed-grass. Fig. 503.

C. nemoralis Kearney, Bull. U. S. Dept. Agric. Agrost. **11**: 36. 1898. Not Philippi, 1896.
Calamagrostis perplexa Scribn. Circ. U. S. Dept. Agric. Agrost. **30**: 7. 1901.

Culms 3°–5° tall, erect; leaf-sheaths glabrous, excepting the usually pubescent summit; blades flat, rather thin, rough, sometimes glabrous on the upper surface, lax, up to 1° long, 1½″–3″ wide; panicle 3′–5′ long, ½–1′ wide, contracted, acute, its slender, somewhat flexuous hispidulous branches erect or nearly so; spikelets 1½″–2″ long, the empty scales lanceolate to oblong-lanceolate, acuminate, sometimes keeled, the flowering scale ovate-oblong, about as long as the second empty scale, obscurely toothed at the apex or entire, rather firm, the awn attached near the base and extending somewhat beyond the scale, stout, bent near the middle, somewhat twisted at the base, the callus hairs white, about three-fourths as long as the scale.

In dry rocky woods, Maine and western New York.

5. Calamagrostis Langsdórfii (Link) Trin. Langsdorf's Reed Bent-grass. Fig. 504.

Arundo Langsdorfii Link. Enum. **1**: 74. 1821.
C. Langsdorfii Trin. Unifl. 225. *pl. 4. f. 10.* 1824.

Culms 2°–4° tall, erect, simple, smooth or roughish. Sheaths shorter than the internodes; ligule 1″–3″ long; blades 4′–12′ long, 2″–4″ wide, scabrous; panicle 2′–6′ in length, the branches ascending or sometimes erect, the lower 1′–2′ long, naked at the base; spikelets 2″–3″ long, the outer scales acuminate, strongly scabrous; third scale equalling or shorter than the second, scabrous, the stout awn as long as or a little exceeding the copious basal hairs which are usually somewhat shorter than the scale.

In meadows and on rocks, Greenland to Alaska, south in the mountains to North Carolina, Michigan, New Mexico and California. Also in northern Europe and Asia. Purple-top. Northern blue-joint. Summer.

6. Calamagrostis canadénsis (Michx.) Beauv.　Blue-joint Grass.　Fig. 505.

Arundo canadensis Michx. Fl. Bor. Am. **1**: 73. 1803.
Calamagrostis canadensis Beauv. Agrost. 15. 1812.
C. canadensis acuminata Vasey, Bull. U. S. Dep. Agr. Agrost. **5**: 26. 1897.

Culms 2°–5° tall, erect, simple, smooth or somewhat scabrous. Sheaths shorter than the internodes; ligule 1″–3″ long; blades 6′–1° long or more, 1″–4″ wide, rough; panicle 4′–7′ in length, open, usually purplish, the branches spreading or ascending, the lower 1½′–3′ long, naked at the base; spikelets 1½″–2″ long, the outer scales equal or subequal, acute, strongly scabrous; third scale equalling or slightly shorter than the second, scabrous, the awn delicate and equalling the copious basal hairs which are about as long as the scale or some of them shorter.

In swamps and wet soil, Newfoundland to British Columbia, south to North Carolina, New Mexico and California. Ascends to 5000 ft. in the Adirondacks. Bluestem. July–Sept.

7. Calamagrostis Macouniàna Vasey.   Macoun's Reed-grass.   Fig. 506.

Deyeuxia Macouniana Vasey, Coult. Bot. Gaz. **10**: 297. 1885.

Calamagrostis Macouniana Vasey, Contr. U. S. Nat. Herb. **3**: 81. 1892.

Culms 2°–3° tall, erect, simple, smooth and glabrous. Sheaths shorter than the internodes; ligule 1″ long; blades 3′–7′ long, 1″–2½″ wide, erect, acuminate, scabrous; panicle open, 3′–4½′ in length, the branches ascending, or sometimes erect, the lower 1′–1½′ long, naked at the base; spikelets 1″ long, the outer scales acute, scabrous, the first shorter than the second; third scale equalling the second, the awn a little exceeding it; basal hairs about as long as the scale.

Manitoba to Missouri, west to Washington.   Summer.

8. Calamagrostis neglécta (Ehrh.) Gaertn.   Narrow Reed-grass.   Fig. 507.

Arundo neglecta Ehrh. Beitr. **6**: 137. 1791.
Calamagrostis neglecta Gaertn. Fl. Wett. **1**: 94. 1799.
Calamagrostis stricta Beauv. Agrost. 15. 1812.
C. neglecta borealis Kearney, Bull. U. S. Dept. Agr. Agrost. **11**: 35. 1898.

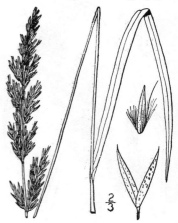

Culms 1°–2½° tall, erect, simple, slender, from a slender rootstock. Sheaths shorter than the internodes; ligule ½″ long or less, truncate; blades narrow, soft, smooth, sometimes involute, the basal one-third as long as the culm, those of the culm 2′–5′ long, erect; panicle contracted, 2½′–4′ in length, the branches 1′ long or less, erect; spikelets 2″ long, the outer scales acute; third scale obtuse, about three-fourths as long as the second and a little longer than the basal hairs; awn attached at or below the middle.

Shores and mountains, Labrador to Alaska, south to northern Maine, Wisconsin, Colorado and Oregon.   Also in Europe and Asia.   Yellow-top, Pony-grass.   Summer.

9. Calamagrostis inexpánsa A. Gray.   Bog Reed-grass.   Fig. 508.

C. confinis A. Gray, Man. Ed. 2, 547. 1856.   Not Nutt. 1818.
Calamagrostis inexpansa A. Gray; Torr. Fl. U. S. **2**: 445. 1843.

Culms 1½°–3° tall, erect, simple, smooth or rough. Sheaths shorter than the internodes; ligule about 1″ long; blades 2″ wide or less, rough, flat, or involute at the apex, the basal often one-half to two-thirds as long as the culm, the stem leaves 2′–10′ long; panicle contracted, 2½′–9′ in length, the branches 1′–2′ long, erect; spikelets about 2″ long, the scales somewhat scabrous, the outer acute; third scale obtuse, the basal hairs equalling it or three-fourths as long; awn more or less bent, from a little shorter to slightly longer than the scale.

In bogs, New York and New Jersey to South Dakota and Colorado.   Aug.–Sept.

10. Calamagrostis labradórica Kearney. Labrador Reed-grass. Fig. 509.

C. *labradorica* Kearney, Bull. U. S. Dep. Agr. Agrost.
11: 38. 1898.

Culms 1°–2° tall, rather stout; leaf-sheaths glabrous; ligule ¾″–1½″ long; blades up to 8′ long, 1½″ wide or less, very involute, filiform toward the apex, erect, glabrous on the lower surface; panicle 2′–4′ long, less than ½′ wide, linear to oblong-lanceolate, much interrupted below, strict, its stout branches short and appressed; spikelets about 2″ long, the empty scales ovate to ovate-lanceolate, acute, firm, purple or purplish, the flowering scale broad, rough on the back, the awn attached at or below the middle, slender, erect, straight.

Rocks on the seashore, Labrador. July.

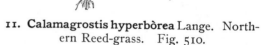

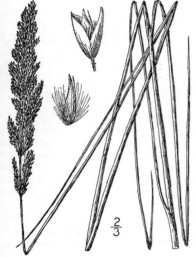

11. Calamagrostis hyperbòrea Lange. Northern Reed-grass. Fig. 510.

C. *hyperborea* Lang, Fl. Dan. 50: *pl. 3.* 1880.
C. *robusta* Vasey, Contr. U. S. Nat. Herb. 3: 82. 1892.
C. *hyperborea elongata* Kearney, Bull. U. S. Dep. Agr.
     Agrost. 11: 40. 1898.
C. *hyperborea americana* Kearney, Bull. U. S. Dep. Agr.
     Agrost. 11: 41. 1898.

Culms 1¾°–3° tall, rigid, densely tufted; leaf-sheaths smooth and glabrous; blades rough on both surfaces, flat, or often involute toward the apex, stiff, 4′–12′ long, 2½″ or less wide; panicle contracted, 3′–6′ long, its short branches erect or ascending; spikelets 1½″–2″ long, the empty scales scabrous, acute, the flowering scale with the callus-hairs from a little shorter than to nearly equalling it, the awn about equalling the scale.

Meadows and swamps, Greenland to Alaska, south to Pennsylvania, Colorado and California. Very variable. June–Aug.

12. Calamagrostis cinnoìdes (Muhl.) Scribn. Nuttall's Reed-grass. Fig. 511.

Arundo cinnoides Muhl. Gram. 187. 1817.
C Nuttalliana Steud. Syn. Pl. Gram. 190. 1855.
C. cinnoides Scribn. Mem. Torr. Club, 5: 42. 1895.

Culms 3°–5° tall, erect, simple, smooth and glabrous. Sheaths shorter than the internodes, smooth or rough, the lower sometimes sparingly hirsute, and rarely with a villous ring at the summit; ligule 1″–2″ long; blades 4′–1° long or more, 2″–5″ wide, attenuate into a long point, scabrous, occasionally sparingly hirsute; panicle 3′–7′ in length, contracted, the branches erect, the lower 1′–2′ long; spikelets 3″–4″ long; scales strongly scabrous, the outer about equal, acuminate and awn-pointed; third scale shorter, obtuse, the basal hairs one-half to two-thirds its length; awn stout, exceeding or equalling the scale; prolongation of the rachilla bearing a terminal tuft of hairs.

In moist soil, Maine to Ohio, south to Georgia and Alabama. Ascends to 2000 ft. in Pennsylvania. Reed Bent-grass. Wild Oats. July–Aug.

46. AMMÓPHILA Host. Gram. Austr. 4: 24. *pl. 41.* 1809.

Tall perennial grasses with flat leaf-blades, convolute above, and dense spike-like panicles. Spikelets 1-flowered, the rachilla prolonged beyond the flower and hairy. Scales 3, rigid, chartaceous, acute, keeled; the 2 outer empty, the lower 1-nerved, the upper 3-nerved; third scale 5-nerved, with a ring of short hairs at the base, subtending a chartaceous 2-nerved palet and a perfect flower. Stamens 3. Styles distinct. Stigmas plumose. Grain free, loosely enclosed in the scale and palet. [Greek, sand-loving, from the habitat of these grasses.]

Two species, the following widely distributed along the fresh and salt-water shores of the northern hemisphere, the other European. Type species: *Arundo arenaria* L.

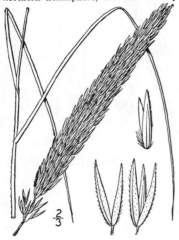

2/3

1. Ammophila arenària (L.) Link. Sea Sand-reed. Sea Mat-weed. Marram. Fig. 512.

Arundo arenaria L. Sp. Pl. 82. 1753.
Calamagrostis arenaria Roth, Fl. Germ. 1: 34. 1788.
Ammophila arundinacea Host. Gram. Austr. 4: 24. 1809.
Ammophila arenaria Link Hort. Berol. 1: 105. 1827.

Glabrous, culms 2°-4° tall, erect, rigid, stout, smooth, arising from a long horizontal branching rootstock. Sheaths smooth, the lower short, crowded and overlapping, the upper longer; ligule a mere ring; blades 6'-1° long or more, rigid, attenuate into a long slender involute point, smooth beneath, scabrous above; spike-like panicle dense, 4'-12' in length, 6"-8" thick, its branches 1½' long or less, appressed; spikelets 5"-6" long, the scales scabrous, about equal in length, the third usually with the rudiment of an awn just below the apex; basal hairs 1"-2" long.

In sands of the sea coast from Newfoundland to North Carolina, and inland along the shores of the Great Lakes. Also on the coasts of northern Europe. Reed- or Sea-shore-bent. Beach-grass. Spires. Sea Sand-grass, Sea-reed. Aug.–Sept.

47. CALAMOVÍLFA Hack. True Grasses 113. 1890.

Tall grasses with stout horizontal rootstocks, elongated leaf-blades, which are involute at the apex, and paniculate inflorescence. Spikelets 1-flowered; rachilla not prolonged beyond the flower. Scales 3, 1-nerved, acute, the 2 outer unequal, empty; third scale longer or shorter than the second, a ring of hairs at its base; palet strongly 2-keeled. Stamens 3. Styles distinct. Stigmas plumose. Grain free. Seed adherent to pericarp. [Greek, a reed-like grass.]

Species 4 or 5 in the temperate and subtropical regions of North America. Type species: *Arundo brevipilis* Torr.

Flowering scale and palet glabrous.　　　　　　　　　　　　　　　1. *C. longifolia.*
Flowering scale and palet pubescent.
　Spikelets 2"-2½" long; a plant of southern New Jersey.　　　2. *C. brevipilis.*
　Spikelets 3½"-4" long; a plant of the western United States.　3. *C. gigantea.*

1. Calamovilfa longifòlia (Hook.) Hack. Long-leaved Reed-grass. Fig. 513.

Calamagrostis longifolia Hook. Fl. Bor. Am. 2: 241. 1840.
C. longifolia Scribn. in Hack. True Grasses 113. 1890.
Calamovilfa longifolia magna Scribn. & Merr. Circ. U. S. Dep. Agr. Agrost. 35: 3. 1901.

Culms 2°-5° tall, erect, simple, stout, smooth and glabrous. Sheaths crowded and overlapping, glabrous or sometimes hairy; ligule a ring of hairs about 1" long; blades 8'-1° long or more; panicle generally narrow, often 1° long or more, commonly pale, the branches erect, or occasionally open with the branches somewhat spreading; spikelets 2½"-3" long; scales acute, the first shorter than the second, the third glabrous, a little longer or shorter than the second, the copious basal hairs from ⅔ as long as to nearly equalling the scale; palet slightly shorter than the third scale.

In sandy places, western Ontario to Mackenzie, south to northern Indiana, Kansas and Colorado. Big Sand-grass. Carizzo. July–Sept.

3/4

2. Calamovilfa brevípilis (Torr.) Hack. Short-haired Reed-grass. Purple Bent-grass. Fig. 514.

Arundo brevipilis Torr. Fl. U. S. 1 : 95. 1824.
Calamagrostis brevipilis Beck, Bot. North. & Mid. St. 401. 1833.
Calamovilfa brevipilis Hack. True Grasses 113. 1890.

Glabrous and smooth or very nearly so, culms 2°–4° tall, erect, simple. Sheaths shorter than the internodes; ligule a ring of very short hairs; blades 6′–12′ long, 1½″ wide or less, attenuate into a long slender involute tip, smooth beneath, slightly scabrous above; panicle open, 5′–10′ in length, the branches ascending, the lower 2′–4′ long; spikelets 2″–2½″ long; scales acute, scabrous toward the apex, the outer unequal, the first one-half as long as the second; third scale exceeding the second, pubescent on the lower half of the keel; basal hairs one-third the length of the scale; palet nearly equalling the scale, pubescent on the lower half of the keel.

In swamps, pine barrens of New Jersey. Aug.–Sept.

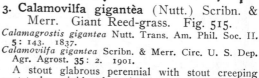

3. Calamovilfa gigantèa (Nutt.) Scribn. & Merr. Giant Reed-grass. Fig. 515.

Calamagrostis gigantea Nutt. Trans. Am. Phil. Soc. II. 5 : 143. 1837.
Calamovilfa gigantea Scribn. & Merr. Circ. U. S. Dep. Agr. Agrost. 35 : 2. 1901.

A stout glabrous perennial with stout creeping rootstocks, rigid often involute leaf-blades, and large open panicle. Culms 3°–6° tall; leaf-sheaths exceeding the internodes; blades up to 1° long or more; panicle 1°–2½° long, the spreading or ascending branches naked at the base, the longer up to 1° long; spikelets 3½″–4″ long, the empty scales acute, the first shorter than the second, the third scale a little longer or shorter than the second scale, long-hairy on the back and keel, the basal hairs copious.

In sandy places, Kansas to Arizona.

48. APÈRA Adans. Fam. Pl. 2 : 495. 1763.

Annual grasses with narrow flat leaf-blades, and ample open or contracted panicles. Spikelets 1-flowered, small, the rachilla prolonged beyond the flower into a bristle. Scales 3; the 2 outer empty, unequal, thin, membranous, keeled, acute; the third scale a little shorter, membranous, bearing a long slender awn inserted just below the shortly 2-toothed apex; palet a little shorter than the scale, 2-keeled, 2-toothed. Stamens 3. Styles distinct, short. Stigmas plumose. Grain narrow, free, included in the scale. Seed adherent to the pericarp. [Greek, signifying not mutilated, whole or entire; application uncertain.]

Two species, natives of Europe and western Asia. Type species: *Agrostis Spica-venti* L.

1. Apera Spìca-vénti (L.) Beauv. Silky Bent-grass. Windlestraw. Fig. 516.

Agrostis Spica-venti L. Sp. Pl. 61. 1753.
Apera Spica-venti Beauv. Agrost. 151. 1812.

Culms 1°–2° tall, erect, simple, slender, smooth and glabrous. Sheaths usually longer than the internodes, the upper one generally including the base of the panicle; ligule 1″–3″ long; blades 1′–7′ long, ½″–2″ wide, scabrous; panicle 3′–9′ in length, the branches erect or ascending, capillary, 1½′–3′ long; outer scales of the spikelet 1″–1¼″ long, acute, smooth and shining; third scale hairy or nearly smooth, bearing a dorsal scabrous awn 3″–4″ long; rudiment at the end of the rachilla less than ¼″ long.

In waste places and on ballast, Maine to southern New York and Pennsylvania. Adventive from Europe. Wind-bent, Wind-grass, Corn-grass. June–July.

49. NOTHOHÓLCUS Nash.

[HOLCUS L. Sp. Pl. 1047, in part. 1753.]

Annual or perennial grasses with flat leaf-blades and spike-like or open panicles. Spike-lets deciduous, 2-flowered; lower flower perfect, upper staminate. Scales 4; the 2 lower empty, membranous, keeled, the first 1-nerved, the second 3-nerved and often short-awned; flowering scales chartaceous, that of the upper flower bearing a bent awn. Palet narrow, 2-keeled. Stamens 3. Styles distinct. Stigmas plumose. Grain oblong, free, enclosed in the scale. [Greek, alluding to the reference of these grasses to the genus *Holcus*.]

About 8 species, natives of the Old World. Type species: *Holcus lanatus* L.

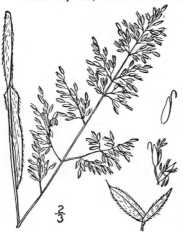

1. Nothoholcus lanàtus (L.) Nash. Velvet-grass. Meadow or Woolly Soft-grass. Fig. 517.

Holcus lanatus L. Sp. Pl. 1048. 1753.

Softly and densely pubescent, light green, culms 1½°–3° tall, erect, often decumbent at the base, simple. Sheaths shorter than the internodes; ligule ½″–1″ long; blades 1′–6′ long, 2″–6″ wide; spikelets 2″ long, the empty scales white-villous, the upper awn-pointed; flowering scales 1″ long, smooth, glabrous and shining, the lower sparsely ciliate on the keel, somewhat obtuse, the upper 2-toothed and bearing a hooked awn just below the apex.

In fields, meadows and waste places, Nova Scotia to Ontario and Illinois, North Carolina and Tennessee. Also on the Pacific Coast. Naturalized from Europe. Velvet Mesquite. Old White-top. Salem-, Rot-, Dart- or Feather-grass. Whites. Yorkshire-fog. White-timothy, Calf-kill. June–Aug.

50. ÁSPRIS Adans. Fam. 2 : 496. 1763.

[AIRA L. Sp. Pl. 63, in part. 1753.]

Mostly annual grasses with narrow leaf-blades and contracted or open panicles. Spikelets small, 2-flowered, both flowers perfect. Scales 4; the 2 lower empty, thin-membranous, acute, subequal, persistent; the flowering scales usually contiguous, hyaline, mucronate or 2-toothed, deciduous, bearing a delicate dorsal awn inserted below the middle; palet a little shorter than the scale, hyaline, 2-nerved. Stamens 3. Stigmas plumose. Grain enclosed in the scale and palet, and often adhering to them. [Greek, from Theophrastus.]

Six or seven species, natives of Europe. Type species: *Aira praecox* L.

Panicle open; flowering scales about 1″ long; plants 5′–10′ tall.   1. *A. caryophyllea*.
Panicle contracted; flowering scales about 1½″ long; plants 2′–4′ tall.   2. *A. praecox*.

1. Aspris caryophýllea (L.) Nash. Silvery Hair-grass. Fig. 518.

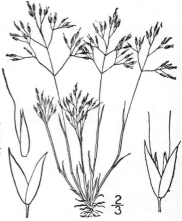

Aira caryophyllea L. Sp. Pl. 66. 1753.

Smooth and glabrous throughout, culms 5′–10′ tall, erect from an annual root, simple, slender. Sheaths mostly basal; ligule 1½″ long; blades ½′–2′ long, involute-setaceous; panicle 1′–4′ in length, silvery, shining, open, the branches spreading or ascending, the lower 1′ long or less; spikelets 1″–1¼″ long, the empty scales acute; flowering scales very acute, 2-toothed, 1″ long, bearing an awn 1½″–2″ long.

In fields and waste places, eastern Massachusetts to Ohio and Virginia. Also on the Pacific Coast. Local. Naturalized from Europe. Mouse-grass. May–July.

2. Aspris praècox (L.) Nash.   Early Hair-grass. Fig. 519.

Aira praecox L. Sp. Pl. 65. 1753.

Glabrous and smooth throughout, culms 2′–4′ tall, erect, from an annual root, simple, rigid. Sheaths clothing the whole culm, the upper one often enclosing the base of the panicle; ligule about 1½″ long; blades 1′ long or less, involute-setaceous; panicle contracted, strict, ½′–1′ in length; spikelets about 1½″ long, the empty scales acute; the flowering scales acuminate, 2-toothed, about 1½″ long, bearing an awn 1½″–2″ long.

In dry fields, southern New Jersey and Pennsylvania to Virginia. Naturalized from Europe. May–July.

51. DESCHÁMPSIA Beauv. Agrost. 91. *pl. 18. f. 3.* 1812.

Perennial grasses with flat or involute leaf-blades, and contracted or open panicles. Spikelets 2-flowered, both flowers perfect, the hairy rachilla extended beyond the flowers or rarely terminated by a staminate one. Scales 4 (rarely more), the 2 lower empty, keeled, acute, membranous, shining, persistent; the flowering scales of about the same texture, deciduous, bearing a dorsal awn, the apex toothed. Palet narrow, 2-nerved. Stamens 3. Styles distinct. Stigmas plumose. Grain oblong, free, enclosed in the scale. [In honor of J. C. A. Loiseleur-Deslongchamps, 1774–1849, French physician and botanist.]

About 20 species, inhabiting cold and temperate regions, a few occurring in the high mountains of the tropics. Besides the following, some 6 others occur in the western parts of Nort America. Type species: *Aira caespitosa* L.

Upper flowering scale reaching or extending beyond the apex of the empty scales.
Flowering scales about 1¼″ long, erose-truncate; leaves flat.   1. *D. caespitosa.*
Flowering scales about 2″ long, acute or obtuse; leaves involute.   2. *D. flexuosa.*
Empty scales extending much beyond the upper flowering scale.   3. *D. atropurpurea.*

1. Deschampsia caespitòsa (L.) Beauv. Tufted Hair-grass.   Fig. 520.

Aira caespitosa L. Sp. Pl. 64. 1753.
D. caespitosa Beauv. Agrost. 160. *pl. 18. f. 3.* 1812.

Culms 2′–4° tall, erect, simple, smooth and glabrous. Sheaths much shorter than the internodes; ligule 1″–3″ long; blades flat, 1″–1½″ wide, smooth beneath, strongly scabrous above, the basal ones numerous, one-quarter to one-half as long as the culm, those of the culm 2′–6′ long; panicle open, 3′–9′ in length, the branches widely spreading or ascending, often somewhat flexuous, naked at the base, the lower 2′–5′ long; spikelets 1¾″–2″ long; flowering scales about 1¼″ long, erose-truncate at the apex, the awns somewhat shorter or a little longer, the upper scale reaching to or extending beyond the apices of the empty ones.

Newfoundland to Alaska, south to New Jersey, Illinois, Minnesota and in the Rocky Mountains and Sierra Nevada to New Mexico and California, mostly in wet soil. Also in Europe and Asia. Hassock-grass, Bullpoll, Bullpates, Windlestraw. July–Aug.

2. Deschampsia flexuòsa (L.) Trin. Wavy Hair-grass. Fig. 521.

3/4

Aira flexuosa L. Sp. Pl. 65. 1753.
Deschampsia flexuosa Trin. Bull. Acad. Sci. St.
    Petersb. 1 : 66. 1836.

Glabrous throughout, culms 1°–2½° tall, erect, slender, simple, smooth. Sheaths much shorter than the internodes; ligule 1″ long or less; blades involute-setaceous, smooth beneath, scabrous above, the basal very numerous, one-fifth the length of the culm or less, those of the culm 1′–3′ long; panicle open, 2′–8′ in length, the branches ascending or erect, sometimes widely spreading, naked at the base, flexuous, the lower 1½′–5′ long; spikelets 2¼″–2½″ long; flowering scales about 2″ long, acutely toothed at the apex; awns bent and twisted, much exceeding the scale; upper scale reaching to or extending beyond the apices of the empty ones.

In dry soil, Greenland and Newfoundland to Ontario, south to North Carolina and Tennessee. Ascends to 5100 ft. in the Adirondacks. Also in Europe. Wood Hair-grass. July–Aug.

3. Deschampsia atropurpùrea (Wahl.) Scheele. Mountain Hair-grass. Fig. 522.

Aira atropurpurea Wahl. Fl. Lapp. 37. 1812.

D. atropurpurea Scheele, Flora 27 : 56. 1844.

Glabrous and smooth or very nearly so, culms 6′–18′ tall, erect, simple, rigid. Sheaths shorter than the internodes; ligule 1″ long or less, truncate; blades 1″–2″ wide, erect, sometimes slightly scabrous above, the basal 2½′–5′ long, those of the culm shorter; panicle contracted, usually purple or purplish, 1′–2′ in length, the branches erect, or sometimes ascending, the lower ½′–1½′ long; spikelets 2½″ long; flowering scales about 1¼″ long, erose-truncate at the apex; awns bent and much longer than the scales; upper scale much exceeded by the very acute outer ones.

On alpine summits, from New England to Colorado and Oregon, north to Labrador and Alaska. Also in Europe. July–Aug.

3/4

52. TRISÈTUM Pers. Syn. 1 : 97. 1805.

Mostly perennial tufted grasses, with flat leaf-blades and spike-like or open panicles. Spikelets 2–4-flowered, the flowers all perfect, or the uppermost staminate; rachilla glabrous or pilose, extended beyond the flowers. Scales 4–6, membranous, the 2 lower empty, unequal, acute, persistent; flowering scales usually shorter than the empty ones, deciduous, 2-toothed, bearing a dorsal awn below the apex, or the lower one sometimes awnless. Palet narrow, hyaline, 2-toothed. Stamens 3. Styles distinct. Stigmas plumose. Grain free, enclosed in the scale. [Latin, referring to the three bristles (one awn and two sharp teeth) of the flowering scales in some species.]

About 60 species, widely distributed in temperate or mountainous regions. Besides the following, about 8 others occur in the western parts of North America. Type species : Avena striata Lam.

Flowering scales all bearing long dorsal awns.
    Panicle contracted, dense; flowering scales 2½″ long or less.   1. T. spicatum.
    Panicle open, loose; flowering scales 2½″ long or more.   2. T. flavescens.
Lower flowering scale not bearing a long dorsal awn, a rudiment sometimes present.
                                                         3. T. pennsylvanicum.

1. **Trisetum spicàtum** (L.) Richter. Narrow False Oat. Fig. 523.

Aira spicata L. Sp. Pl. 64. 1753.
Aira subspicata L. Syst. Veg. Ed. 10, 673. 1759.
Avena mollis Michx. Fl. Bor. Am. 1: 72. 1803.
Trisetum subspicatum Beauv. Agrost. 180. 1812.
T. spicatum Richter, Pl. Europ. 1: 59. 1890.

Softly pubescent or glabrous, culms 6'–2° tall, erect, simple. Sheaths usually shorter than the internodes; ligule ½''–1' long; blades 1'–4' long, ½''–2'' wide; panicle spike-like, 1'–5' in length, often interrupted below, its branches 1¼' or less long, erect; spikelets 2–3-flowered, the empty scales hispid on the keel, shining, the second about 2½'' long, the first shorter; flowering scales 2''–2½'' long, acuminate, scabrous, each bearing a long bent and somewhat twisted awn.

In rocky places, Labrador to Alaska, south on the mountains to North Carolina, New Mexico and California. Also in Europe and Asia. Downy Oat-grass. Aug.–Sept.

2. **Trisetum flavéscens** (L.) Beauv. Yellow False Oat. Fig. 524.

Avena flavescens L. Sp. Pl. 809. 1753.
Trisetum pratense Pers. Syn. 1: 97. 1805.
T. flavescens Beauv. Agrost. 88. 1812.

Culms 1½°–2½° tall, erect, simple, smooth and glabrous. Sheaths shorter than the internodes, more or less pubescent; ligule ½'' long; blades 1½'–5' long, 1''–3'' wide, scabrous, sometimes sparingly hairy; panicle open, 2'–5' in length, the branches ascending or erect, somewhat flexuous, naked below, the lower 1'–2' long; spikelets 3–4-flowered; empty scales smooth and glabrous, the second acute, 2½'' long, the first about half as long, narrower, acuminate; flowering scales 2½''–3'' long, scabrous, bearing a long bent and twisted awn.

Introduced into Missouri and Kansas. Native of Europe and Asia. Panicle yellow, turning dull brown. Golden, Tall or Yellow Oat-grass. July–Aug.

3. **Trisetum pennsylvánicum** (L.) Beauv. Marsh False Oat or Oat-grass. Fig. 525.

Avena pennsylvanica L. Sp. Pl. 79. 1753.
Avena palustris Michx. Fl. Bor. Am. 1: 72. 1803.
Trisetum pennsylvanicum Beauv.; R. & S. Syst. 2: 658. 1817.
Trisetum palustre Torr. Fl. U. S. 1: 126. 1824.

Culms 1°–3° tall, erect, simple, slender and often weak, smooth and glabrous. Sheaths shorter than the internodes, sometimes scabrous; ligule ½'' long; blades 1'–6' long, 1''–3'' wide, rough; panicle 2'–8' in length, yellowish, narrow, the branches ascending, the lower 1'–2' long; spikelets 2-flowered; outer scales smooth, shining, subequal, the second 2''–2½'' long; flowering scales 2''–2½'' long, scabrous, the lower not long-awned, but a rudimentary awn sometimes present, the upper with a long bent and twisted awn.

In swamps and wet meadows, Massachusetts to Illinois, south to Florida and Louisiana. Ascends to 3500 ft. in Virginia. Panicle sometimes loose and nodding. June–July.

53. AVÈNA L. Sp. Pl. 79. 1753.

Annual or perennial grasses, with usually flat leaf-blades and panicled spikelets. Spikelets 2–many-flowered, or rarely 1-flowered; lower flowers perfect, the upper often staminate or imperfect. Scales 4–many (rarely 3); the 2 lower empty, somewhat unequal, membranous, persistent; flowering scales deciduous, rounded on the back, acute, generally bearing a dorsal awn, the apex often 2-toothed. Palet narrow, 2-toothed. Stamens 3. Styles short, distinct. Stigmas plumose. Grain oblong, deeply furrowed, enclosed in the scale and palet, free or sometimes adherent to the latter. [Old Latin name for the Oat.]

About 50 species, widely distributed in temperate regions, chiefly in the Old World. Type species: *Avena sativa* L.

Spikelets, exclusive of the awns, 8″ long or more; annuals.
    Flowering scales more or less hispid, the awn with a pronounced spiral column; rachilla hispid.
                                                        1. *A. fatua.*
    Flowering scales glabrous, awnless or with a straight awn slightly spiral at the base.
                                                        2. *A. sativa.*
Spikelets, exclusive of the awns, less than 8″ long; perennials.
    Empty basal scales much shorter than the spikelet; flowering scales herbaceous, with the awn inserted near the apex.
    Flowering scales with a ring of hairs at the base; awn equalling or exceeding the scale.
                                                      3. *A. Torreyi.*
    Flowering scales naked at the base; awn not more than ½ as long as the scale.
                                                      4. *A. Smithii.*
    Empty scales, at least the second one, as long as the spikelet or nearly so; flowering scales scarious and hyaline above, the awn inserted about the middle.
                                                     5. *A. Hookeri.*

1. Avena fátua L. Wild Oat. Fig. 526.

Avena fatua L. Sp. Pl. 80. 1753.

Culm 1°–4° tall, erect, simple, stout, smooth and glabrous. Sheaths smooth, or scabrous at the summit, sometimes sparingly hirsute, the lower often overlapping; ligule 1″–2″ long; blades 3′–8′ long, 1″–4″ wide; panicle open, 4′–12′ in length, the branches ascending; spikelets 2–4-flowered, drooping; outer scales ¾–1′ in length, smooth, enclosing the flowering scales; flowering scales 6″–9″ long, with a ring of stiff brown hairs at the base, pubescent with long rigid brown hairs, bearing a long bent and twisted awn.

In fields and waste places, Ontario and Ohio (according to Hitchcock) and westward to Missouri; abundant on the Pacific Coast. Naturalized from Europe or Asia. Havercorn. Poor Oat. Hever. Drake. July–Sept.

⅔

Avena stérilis L., a native of Europe, is reported as occurring sparingly as an adventive plant in New Jersey and near Philadelphia, Penn. It can be distinguished from the above by its larger spikelets and longer awns.

2. Avena sativa L. Oats. Fig. 527.

Avena sativa L. Sp. Pl. 79. 1753.

A glabrous annual. Culms up to 3° tall; blades flat, up to 1° long and ½′ wide, acuminate; panicle 4′–9′ long, its branches ascending; spikelets, exclusive of the awns, 8″–12″ long, the empty scales broad, acute, the flowering scales glabrous, awnless, or with an imperfect awn which is rarely a little spiral at the base.

Persisting in old fields and as a weed along roadsides and waste places. A native of Europe and Asia.

⅔

3. Avena Tórreyi Nash.   Purple Oat.   Fig. 528.

Avena striata Michx. Fl. Bor. Am. 1: 73. 1803. Not
  Lam. 1783.
Trisetum purpurascens Torr. Fl. U. S. 127. 1824.
  Not *Avena purpurascens* DC. 1813.

Culms 1°–2° tall, erect, simple, slender, smooth
and glabrous. Sheaths shorter than the internodes,
smooth or slightly scabrous; ligule ½″ long or less;
blades erect, 1′–6′ long, 1″–3″ wide, smooth beneath,
usually scabrous above; panicle 2½–5′ in length, lax,
the branches erect or ascending, naked below, the
lower 1′–2½′ long; spikelets 3–6-flowered, the
empty scales smooth, the second 3″–3½″ in length,
3-nerved, the first two-thirds to three-quarters as
long, 1-nerved; flowering scales 3″–4″ long, with a
ring of short hairs at the base, strongly nerved,
scabrous; awns as long as the scales or longer.

In woods, New Brunswick to British Columbia, Pennsylvania, Minnesota and Colorado. July–Aug.

4. Avena Smíthii Porter.   Smith's Oat.   Fig. 529.

Avena Smithii Porter; A. Gray, Man. Ed. 3, 640. 1867.
Melica Smithii Vasey, Bull. Torr. Club, 15: 294. 1888.

Culms 2½°–5° tall, erect, simple, scabrous. Sheaths
shorter than the internodes, very rough; ligule 2″ long;
blades 4′–8′ long, 3″–6″ wide, scabrous; panicle 6′–12′ in
length, the branches finally spreading; spikelets 3–6-flow-
ered; empty scales smooth, the second 3″–4″ in length,
5-nerved, the first shorter, obscurely 3-nerved; flowering
scales 5″ long, naked at the base, strongly nerved, scabrous, bearing an awn one-fourth to one-half their
length.

Northern Michigan and Isle Royale.   Summer.

5. Avena Hoókeri Scribn.   Hooker's Oat.   Fig. 530.

A. pratensis americana Scribn. in Macoun, Cat. Can. Pl.
  2: 243. 1888.
A. Hookeri Scribn. True Grasses 123. 1890.
A. americana Scribn. Bull. U. S. Dep. Agr. Agrost.
  7: 183. 1897.

A glabrous perennial. Culms tufted, 6′–18′ tall,
erect; sheaths keeled; blades erect, up to 4′ long,
flat, thick, the midnerve thickened as are the rough
margins, linear, acute, those on the culm up to 2″
wide, those of the innovations much narrower;
panicle contracted, 2′–4′ long, its branches erect;
spikelets, exclusive of the awns, 6″–7″ long, the
empty scales acute, scarious above, the second
equalling the spikelet or nearly so, the flowering
scale 4½″–6″ long, brown and firm at the base, scarious above the middle, acute, the awn inserted
about the middle, about ½′ long and bent near the
middle, spiral at the base.

On ridges and hillsides, Saskatchewan to South
Dakota, west to Alberta and Colorado. June–Aug.

Avena pubéscens Huds., of Europe, reported as adventive in Vermont and New Jersey, can be
distinguished by its pubescent foliage and the very long hairs on the rachilla.

54. ARRHENATHÈRUM Beauv. Agrost. 55. *pl. 11. f. 5.* 1812.

Tall perennial grasses, with flat leaf-blades and contracted or open panicles. Spikelets 2-flowered; lower flower staminate, upper perfect; rachilla extended beyond the flowers. Scales 4, the 2 lower empty, thin-membranous, keeled, very acute or awn-pointed, unequal, persistent, flowering scales rigid, 5–7-nerved, deciduous, the first bearing a long bent and twisted dorsal awn, inserted below the middle, the second unawned; palet hyaline, 2-keeled. Stamens 3. Styles short, distinct. Stigmas plumose. Grain ovoid, free. [Greek, referring to the awn of the staminate scale.]

Six species, natives of the Old World.   Type species: *Avena elatior* L.

1. Arrhenatherum elàtius (L.) Beauv. Oat-grass. Fig. 531.

Avena elatior L. Sp. Pl. 79.   1753.
A. avenaceum Beauv. Agrost. 152.   Name only.   1812.
A. elatius Beauv.; M. & K. Deutsch. Fl. 1 : 546.   1823.

Glabrous, culms 2°–4° tall, erect, simple. Lower sheaths longer than the internodes; ligule 1″ long; blades 2½′–12′ long, 1″–4″ wide, scabrous; panicle 4′–12′ in length, contracted, the branches erect, the lower 1′–2′ long; empty scales finely roughened, the second 4″ long, the first shorter; flowering scales about 4″ long.

In fields and waste places, Newfoundland to Ontario and Minnesota, south to Georgia and Tennessee and Nebraska. Also on the Pacific Coast. Naturalized from Europe. Tall, or False Oat-grass; Pearl-, Hever-, Evergreen-, Button- or Onion-grass; Button-, Butter- or Onion-twitch; Grass of the Andes. June-Aug.

55. DANTHÒNIA DC. Fl. Fran. 3 : 32. 1805.

Mostly perennial grasses, with flat or convolute leaf-blades and contracted or open panicles. Spikelets 3–many-flowered, the flowers all perfect, or the upper staminate; rachilla pubescent, extending beyond the flowers. Scales 5–many, the 2 lower empty, keeled, acute, subequal, persistent, generally extending beyond the uppermost flowering one; flowering scales rounded on the back, 2-toothed, deciduous, the awn arising from between the acute or awned teeth, flat and twisted at base, bent; palet hyaline, 2-keeled near the margins, obtuse or 2-toothed. Stamens 3. Styles distinct. Stigmas plumose. Grain free, enclosed in the scale. [Name in honor of Etienne Danthoine, a Marseilles botanist of the last century.]

A genus of about 100 species, widely distributed in warm and temperate regions, chiefly in South Africa. Type species: *Avena spicata* L.

Spikelets, exclusive of the awns, less than ½′ long.
  Teeth of flowering scales merely acute, not awned; panicle contracted, spike-like.   1. *D. spicata.*
  Teeth of the flowering scales long-awned; panicle commonly open, its branches spreading,
    and usually reflexed at flowering time.   2. *D. compressa.*
Spikelets, exclusive of the awns, exceeding ½′ long.
  Panicle loose and open, green; awns 5″ long or more.
    Foliage and flowering scales pubescent.   3. *D. sericea.*
    Foliage glabrous; flowering scales pubescent on margins and base only.   4. *D. epilis.*
  Panicle contracted, spike-like, dense, purple-variegated; awns 3″–4″ long.   5. *D. intermedia.*

1. Danthonia spicàta (L.) Beauv. Common Wild Oat-grass. Fig. 532.

Avena spicata L. Sp. Pl. 80.   1753.
Danthonia spicata Beauv.; R. & S. Syst. 2 : 690.   1817.

Culms 1°–2½° tall, erect, simple, smooth and glabrous, nearly terete. Sheaths shorter than the internodes, glabrous or often sparingly pubescent below; ligule very short; blades rough, 1″ wide or less, usually involute, the lower 4′–6′ long, the upper 1′–2′ long; inflorescence racemose or paniculate, 1′–2′ in length, the pedicels and branches erect or ascending; spikelets 5–8-flowered; empty scales 4″–5″ long, glabrous; flowering scales broadly oblong, sparingly pubescent with appressed silky hairs, the teeth about ½″ long, acute or short-pointed, the bent spreading awn closely twisted at the base, loosely so above.

In dry soil, Newfoundland to South Dakota, south to North Carolina and Texas. Ascends to 3000 ft. in Virginia. June-grass. July–Sept.

2. Danthonia compréssa Austin. Flattened Wild Oat-grass. Fig. 533.

Danthonia compressa Austin; Peck, Rept. Reg. N. Y. State
　Univ. 22: 54. 1869.
Danthonia Alleni Austin, Bull. Torr. Club 3: 21. 1872.

Culms 1½°–3° tall, erect, slender, simple, flattened,
smooth and glabrous. Sheaths shorter than the internodes; ligule pilose; blades 1″ wide or less, rough, lax,
the basal from one-third to one-half the length of the
culm; lower culm leaves 6′–8′ long, the upper 3′–6′;
panicle open, 2½′–4′ in length, the lower branches generally spreading; spikelets 5–10-flowered; empty scales
5″–6″ long, glabrous; flowering scales oblong, with a
ring of short hairs at base, pubescent with appressed
silky hairs, the awn erect or somewhat bent, strongly
twisted below, slightly so above, the teeth 1″–1½″ long,
acuminate, awned.

In woods, Maine to New York, south to North Carolina
and Tennessee. Ascends to 6000 ft. in North Carolina.
Tennessee Oat-grass. July–Sept.

3. Danthonia serícea Nutt. Silky Wild Oat-grass. Fig. 534.

Danthonia sericea Nutt. Gen. 1: 71. 1818.

Culms 1½°–3° tall, simple, glabrous. Sheaths shorter
than the internodes, usually villous; ligule pilose;
blades rough and more or less villous, 1″–1½″ wide,
the basal one-quarter to one-half the length of the
culm, usually flexuous, those of the culm 1′–4′ long,
erect; panicle 2½′–4½′ in length, contracted, the branches
erect or ascending; spikelets 4–10-flowered; empty
scales 7″–8″ long, glabrous; flowering scales oblong,
strongly pubescent with long silky hairs, the awn erect
or somewhat bent, closely twisted below, loosely so
above, the teeth 1″–1½″ long, acuminate, awned.

In dry sandy soil, Massachusetts to Pennsylvania, south
to Florida and Mississippi. May–July.

4. Danthonia épilis Scribn. Smooth Wild Oat-grass. Fig. 535.

Danthonia glabra Nash, Bull. Torr. Club 24: 42. 1897.
　Not Philippi, 1896.
Danthonia epilis Scribn. Circ. U. S. Dep. Agr. Agrost.
30: 7. 1901.

Glabrous. Culms erect, tufted, 16′–28′ tall, slightly
roughened just below the panicle and puberulent below
the brown nodes; sheaths usually shorter than the internodes; ligule densely ciliate with long silky hairs;
blades smooth excepting at the apex, 1″–2″ wide, erect,
those on the sterile shoots 6′ or more long, the culm
leaves 2′–4′ long; panicle 2′–3′ long, contracted; spikelets, including awns, 9″–10″ long, 5–10-flowered, on
hispidulous appressed pedicels; empty scales acuminate; flowering scales 2½″–3″ long to the base of the
teeth, pilose on the margins below and sometimes sparingly so on the midnerve at the base, the remainder of
the scale glabrous, teeth including the awns, 1″–1½″
long, the central awn 4½″–6″ long, more or less
spreading.

In swamps, southern New Jersey to Georgia. May–July.

5. Danthonia intermèdia Vasey. Vasey's Wild Oat-grass. Fig. 536.

D. intermedia Vasey, Bull. Torr. Club 10 : 52. 1883. .

A glabrous tufted perennial. Culms 4′–18′ tall; blades up to 6′ long and 2″ wide, often involute; panicle 1½″–2″ long, contracted, dense, spike-like, variegated with purple, its branches short and appressed; spikelets 7″–8″ long, exclusive of the awns, the empty scales broad, acuminate, variegated with purple, the flowering scales 3½″–4″ long, pubescent only on the margins below the middle and at the base, the teeth acute and usually awned, the central awn 3″–4″ long.

Hillsides and meadows, Quebec ; northern Michigan ; Saskatchewan to British Columbia, Washington and Oregon, and southward in the mountains to Colorado. July and Aug.

56. CAPRÌOLA Adans. Fam. Pl. 2 : 31. 1763.

[CYNODON Rich.; Pers. Syn. 1 : 85. 1805.]

Perennial grasses with short flat leaf-blades and spicate inflorescence, the spikes digitate. Spikelets 1-flowered, secund. Scales 3; the 2 lower empty, keeled; flowering scale broader, membranous, compressed; palet a little shorter than the scale, hyaline, 2-keeled. Stamens 3. Styles distinct. Stigmas short, plumose. Grain free. [Name mediaeval Latin for the wild goat that feeds on this grass in waste rocky places.]

Four known species, of which three are Australian, the following widely distributed. Type species : *Panicum Dactylon* L.

1. Capriola Dáctylon (L.) Kuntze. Bermuda-grass. Scutch-grass. Dog's-tooth Grass. Fig. 537.

Panicum Dactylon L. Sp. Pl. 58. 1753.
Cynodon Dactylon Pers. Syn. 1 : 85. 1805.
Capriola Dactylon Kuntze, Rev. Gen. Pl. 764. 1891.

Culms 4′–12′ tall, erect, from long creeping and branching stolons, smooth and glabrous. Sheaths glabrous or somewhat hairy, crowded at the bases of the culms and along the stolons; ligule pilose; blades 1′–2′ long, 1″–2″ wide, flat, rigid, smooth beneath, scabrous above; spikes 4–5, ½′–2′ in length, digitate; rachis flat; spikelets 1″ long; outer scales hispid on the keel, narrow, the first shorter than the second, about two-thirds as long as the broad and strongly compressed third one.

In fields and waste places, Massachusetts and southern New York to Missouri, Florida and Mexico. West Indies and South America. Cultivated for pasture. Naturalized from Europe. Wire-grass, Cane-grass, Bahamagrass, Indian Doob. July–Sept.

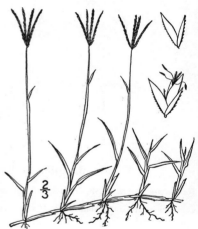

57. SPARTÌNA Schreb.; Gmel. Syst. 123. 1791.

Perennial glabrous grasses, with long horizontal rootstocks, flat or involute leaves, and an inflorescence of one-sided spreading or erect alternate spikes. Spikelets 1-flowered, narrow, deciduous, borne in two rows on the rachis, articulated with the very short pedicels below the scales. Scales 3; the 2 outer empty, keeled, very unequal; the third subtending a perfect flower, keeled, equalling or shorter than the second; palet often longer than its scale, 2-nerved. Stamens 3. Styles filiform, elongated. Stigmas filiform, papillose or shortly plumose. Grain free. [Greek, referring to the cord-like leaves of some species.]

About 7 species, widely distributed in saline soil, a few in fresh-water marshes. Type species : *Spartina Schreberi* Gmel.

First scale awn-pointed, equalling the third; second long-awned.   1. *S. Michauxiana.*
First scale acute, shorter than the third, usually one-half as long.
  First scale strongly scabrous-hispid on the keel.
    Leaves ½′ wide or more, flat.   2. *S. cynosuroides.*
    Leaves ¼′ wide or less.
      Spikes ascending or erect; leaves narrow, involute; coast plant.   3. *S. patens.*
      Spikes appressed; leaves usually flat at the base; western species.   4. *S. gracilis.*
  First scale smooth on the keel or occasionally lightly scabrous.   5. *S. stricta.*

1. Spartina Michauxiàna Hitchc. Tall Marsh-grass. Fig. 538.

S. cynosuroides Gray, Man. 585. 1848. Not Roth, 1806.
Spartina Michauxiana Hitchc. Contr. U. S. Nat. Herb. **12**: 153. 1908.

Culms 2°–6° tall, erect, simple, smooth. Sheaths long, overlapping, those at the base of the culm crowded; ligule a ring of hairs; blades 1° long or more, 3″–7″ wide, scabrous on the margins, becoming involute in drying, attenuate into a long slender tip; spikes 5–30, 2′–5′ long, often on peduncles ½′–1′ in length, ascending or erect; rachis rough on the margins; spikelets much imbricated, 6″–7″ long; outer scales awn-pointed or awned, strongly hispid-scabrous on the keel; third scale as long as the first, the scabrous midrib terminating just below the emarginate or 2-toothed apex; palet sometimes exceeding the scale.

In swamps and streams of fresh or brackish water, Nova Scotia to Saskatchewan, south to New Jersey, Texas and Colorado. Sometimes glaucous. Called also Fresh-water Cord-grass, Bull-grass, Upland Creek-stuff. Aug.–Oct.

2. Spartina cynosuroìdes (L.) Roth. Salt Reed-grass. Fig. 539.

Dactylis cynosuroides L. Sp. Pl. 71. 1753.
Trachynotia polystachya Michx. Fl. Bor. Am. **1**: 64. 1803.
T. cynosuroides Michx. Fl. Bor. Am. **1**: 64. 1803.
S. cynosuroides Roth, Catalect. **3**: 10. 1806.
Spartina polystachya Ell. Bot. S. C. & Ga. **1**: 95. 1817.

Culms 4°–9° tall, erect, stout, simple, smooth. Sheaths overlapping, those at the base of the culm crowded; ligule a ring of hairs; blades 1° long or more, ½′–1′ wide, flat, scabrous at least on the margins, attenuate into a long slender tip; spikes 20–50, ascending, often long-peduncled, 2′–4′ in length, the rachis rough on the margins; spikelets much imbricated, 4″–5″ long, the outer scales acute, strongly scabrous-hispid on the keel, the first half the length of the second; third scale scabrous on the upper part of the keel, obtuse, longer than the first and exceeded by the palet.

In salt and brackish marshes, Connecticut to Florida and Mississippi. Creek-thatch, Creek-stuff. Aug.–Oct.

3. Spartina pàtens (Ait.) Muhl. Salt-meadow Grass. Fig. 540.

Dactylis patens Ait. Hort. Kew. **1**: 104. 1789.
Spartina patens Muhl. Gram. 55. 1817.
Spartina juncea Ell. Bot. S. C. & Ga. **1**: 94. 1817.
S. caespitosa A. A. Eaton, Bull. Torr. Club **25**: 338. 1898.

Culms 1°–3° tall, smooth. Lower sheaths overlapping and crowded; ligule a ring of short hairs; blades ½°–1° long, 1″–2″ broad, involute, attenuate into a long tip, smooth and glabrous beneath; spikes 2–10, 1′–2′ long, usually ascending, more or less peduncled, the rachis slightly scabrous; spikelets 3″–4″ long; outer scales acute, scabrous-hispid on the keel, the first usually rather less than one-half as long as the second; third scale somewhat scabrous on the upper part of the keel, emarginate or 2-toothed at the apex, longer than the first and exceeded by the palet.

On salt meadows and sandy beaches, Newfoundland to Quebec, Florida and Texas. This and *Juncus Gerardi*, the "Black Grass," furnish most of the salt meadow hay of the Atlantic Coast. Fox-grass, Rush Salt-grass, Three-fork-grass, White-rush, Salt-marsh-grass. Aug.–Oct.

4. Spartina grácilis Trin.   Inland Cord-grass.   Fig. 541.

S. *gracilis* Trin. Mem. Acad. St. Petersb. (VI.) **6**: 110. 1840.

Culms 1°–3° tall, erect, simple, smooth. Sheaths overlapping, those at the base of the culm short and crowded; ligule a ring of short hairs; blades 1° long or less, 1″–3″ wide, flat or involute, attenuate into a long tip; spikes 4–8, 1′–2′ long, appressed, more or less peduncled; spikelets 3″–4″ long; outer scales acute, scabrous-hispid on the keel, the first half the length of the second; third scale obtuse, slightly shorter than the second and about equalling the obtuse palet.

In saline soil, Saskatchewan to British Columbia, south to Kansas and California. Slender Cord-grass. Aug.–Sept.

5. Spartina strícta (Ait.) Roth.   Smooth or Salt Marsh-grass.   Fig. 542.

Dactylis maritima Walt. Fl. Car. 77. 1788. Not Curt. 1785.
Dactylis stricta Ait. Hort. Kew. **1**: 104. 1789.
Spartina stricta Roth, Neue Beitr. 101. 1802.
Spartina alterniflora Lois. Fl. Gall. **2**: 719. 1807.
Spartina glabra Muhl. Gram. 54. 1817.
Spartina stricta alterniflora A. Gray, Man. Ed. 2, 552. 1856.
Spartina stricta maritima Scribn. Mem. Torr. Club **5**: 45. 1894.

Culms 1°–9° tall, erect, simple, smooth. Sheaths overlapping, those at the base shorter and looser, much crowded; ligule a ring of short hairs; blades up to 2° long, 2″–7″ wide at the base, involute, at least when dry; spikes 3–5, erect or nearly so, 1′–2′ long, or slender and 3′–5′ long; spikelets 6″–8″ long, loosely to rather densely imbricated; empty scales acute or acutish, 1-nerved, the first shorter than the second, which exceeds or equals the third which is glabrous or pubescent; palet longer than the third scale.

Very variable. Common, in some one of its forms, along the coast from Maine to Florida and Texas. Also on the coast of Europe. Spart-grass, Twin Spike-grass, Low Creek-stuff. Creek-sedge or -thatch. Aug.–Oct.

58. CAMPULÒSUS Desv. Bull. Soc. Philom. **2**: 189.   1810.

[Ctenium Panzer, Deutsch. Akad. Muench. **1813**: 288. *pl. 13*.   1814.]

Tall pungent-tasted grasses, with flat or convolute narrow leaves and a curved spicate inflorescence. Spikelets borne pectinately in two rows on one side of the flat curved rachis, 1-flowered. Lower 4 scales empty, the first very short, hyaline; the second, third, fourth and fifth awned on the back, the latter subtending a perfect flower and palet, the uppermost scales empty. Stamens 3. Styles distinct. Stigmas plumose. Grain oblong, free, loosely enclosed in the scale. [Greek, in allusion to the curved spike.]

Seven known species, four of them American, the others in the eastern hemisphere. Type species: *Chloris monostachya* Michx.

1. Campulosus aromáticus (Walt.) Scribn.
Toothache-grass. Fig. 543.

Aegilops aromatica Walt. Fl. Car. 249. 1788.
Ctenium americanum Spreng. Syst. 1: 274. 1825.
Campulosus aromaticus Scribn. Mem. Torr. Club 5:
45. 1894.

Culms 3°–4° tall, erect, simple, smooth or somewhat scabrous. Sheaths shorter than the internodes, rough; ligule 1″ long, truncate; blades 1′–6′
long, 1″–2″ wide, flat or involute, smooth; spike
terminal, solitary, curved, 2′–4′ long, the rachis
extended into a point; spikelets about 3″ long;
second scale thick and rigid, awn-pointed, bearing
just above the middle a stout horizontal or recurved
awn; third, fourth and fifth scales membranous,
scabrous, awned from below the 2-toothed apex, the
fifth subtending a perfect flower, the others empty.

In wet soil, especially in pine barrens, Virginia to
Florida and Mississippi. Lemon-grass, Wild Ginger.
July–Sept.

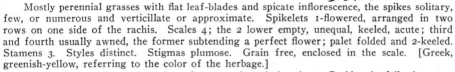

$\frac{3}{4}$

59. CHLÒRIS Sw. Prodr. 25. 1788.

Mostly perennial grasses with flat leaf-blades and spicate inflorescence, the spikes solitary,
few, or numerous and verticillate or approximate. Spikelets 1-flowered, arranged in two
rows on one side of the rachis. Scales 4; the 2 lower empty, unequal, keeled, acute; third
and fourth usually awned, the former subtending a perfect flower; palet folded and 2-keeled.
Stamens 3. Styles distinct. Stigmas plumose. Grain free, enclosed in the scale. [Greek,
greenish-yellow, referring to the color of the herbage.]

About forty species, mostly natives of warm and tropical regions. Besides the following some
10 others occur in the southern United States. Type species: *Agrostis cruciata* L.

$\frac{2}{3}$

1. Chloris verticillàta Nutt. Windmill-grass.
Prairie Chloris. Branching Foxtail. Fig. 544.

Chloris verticillata Nutt. Trans. Am. Phil. Soc. (II.)
5: 150. 1833–37.

Culms 6′–18′ tall, erect, or decumbent and rooting
at the lower nodes, smooth, glabrous. Sheaths
shorter than the internodes, smooth, or roughish at
the summit; ligule a ring of short hairs; blades 1′–3′
long, 1″–2″ wide, obtuse, often apiculate, scabrous;
spikes slender, usually spreading, 2′–4½′ long, in one
or two whorls, or the upper ones approximate;
spikelets, exclusive of the awns, about 1½″ long, the
first scale about one-half the length of the second;
the third 1″ long, obtuse, ciliate on the nerves, especially on the lateral ones, bearing just below the apex
a scabrous awn about 2½″ long; fourth scale as long
as or shorter than the third, awned near the usually
truncate apex.

On prairies, Missouri to Colorado and Texas. May–
July.

Chloris élegans H.B.K., common from New Mexico to
California and southward, has been reported from Kansas. Distinguished from the above by its short stout
spikes and the tuft of long hairs at the summit of the
lateral nerves of the flowering scales.

60. GYMNOPÒGON Beauv. Agrost. 41. *pl. 9. f. 3.* 1812.

Perennial grasses with flat and usually short rigid leaf-blades and numerous slender alternate spikes. Spikelets 1-flowered, almost sessile, the rachilla extended and bearing a small
scale which is usually awned. Scales 3 or 4; the 2 lower empty, unequal, narrow, acute;
third broader, fertile, 3-nerved, slightly 2-toothed at the apex, bearing an erect awn; the
fourth empty, small, awned; palet 2-keeled. Stamens 3. Styles distinct. Stigmas plumose.
Grain linear, free, enclosed in the rigid scale. [Greek, naked-beard, referring to the prolongation of the rachilla.]

Six species, all but one American. Type species: *Andropogon ambiguus* Michx.

Spikes bearing spikelets their whole length; awn longer than flowering scale.   1. *G. ambiguus.*
Spikes bearing spikelets above the middle; awn shorter than flowering scale.   2. *G. brevifolius.*

1. Gymnopogon ambíguus (Michx.) B.S.P. Broad-leaved or Naked Beard-grass. Fig. 545.

Andropogon ambiguus Michx. Fl. Bor. Am. 1: 58. 1803.
Gymnopogon racemosus Beauv. Agrost. 164. 1812.
G. ambiguus B.S.P. Prel. Cat. N. Y. 69. 1888.

Culms 12′–18′ tall, erect, or decumbent at the base, simple or sometimes sparingly branched, smooth and glabrous. Sheaths short, glabrous, excepting a villous ring at the summit, crowded at the base of the culm; ligule very short; blades 1′–4′ long, 2″–6″ wide, lanceolate, acute, cordate at the base, spreading, smooth or a little scabrous above; spikes slender, spikelet-bearing throughout their entire length, at first erect, the lower 4′–8′ long, at length widely spreading; spikelets, exclusive of awns, 2″–2½″ long; first scale shorter than the second; third scale exceeded by the second, the callus at the base hairy, the awn 2″–3″ long.

In dry sandy soil, southern New Jersey to Kansas, south to Florida and Texas. Aug.–Oct.

2. Gymnopogon brevifòlius Trin. Short-leaved Beard-grass. Fig. 546.

Gymnopogon brevifolius Trin. Unifl. 238. 1824.

Culms 1°–2° long, from a decumbent base, simple, slender, smooth and glabrous. Sheaths shorter than the internodes, sometimes crowded near the middle of the culm; ligule very short; blades 1′–2′ long, 1″–4″ wide, usually spreading, lanceolate, acute, cordate at the base; spikes very slender, spikelet-bearing above the middle, the lower 4′–6′ long, at first erect, finally widely spreading; spikelets, exclusive of the awns, 1½″ long; first scale shorter than the second; third scale equalling or exceeded by the second, short-awned, sparingly villous or glabrous, the callus hairy.

In dry soil, New Jersey to Florida, west to Mississippi. Aug.–Oct.

61. SCHEDONNÀRDUS Steud. Syn. Pl. Gram. 146. 1855.

An annual grass with branching culms, narrow leaf-blades and slender spikes arranged along a common axis. Spikelets 1-flowered, sessile and alternate on the rachis. Scales 3; the 2 lower empty, narrow, membranous, acuminate; the flowering scale longer, of similar texture; palet narrow, shorter. Stamens 3. Styles distinct. Stigmas plumose. Grain linear, free, enclosed in the rigid scale. [Greek, resembling the genus *Nardus*.]

A monotypic genus of central North America. Type species: *Schedonnardus texanus* Steud.

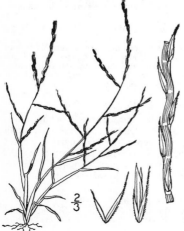

1. Schedonnardus paniculàtus (Nutt.) Trelease. Schedonnardus. Fig. 547.

Lepturus paniculatus Nutt. Gen. 1: 81. 1818.
Schedonnardus texanus Steud. Syn. Pl. Gram. 146. 1855.
Schedonnardus paniculatus Trelease, Branner & Coville, Rep. Geol. Surv. Ark. 1888: Part 4, 236. 1891.

Culms 8′–18′ tall, erect, slender, rigid, branching at the base, scabrous. Sheaths crowded at the base of the culm, compressed, smooth and glabrous; ligule 1″ long, truncate; blades 1′–2′ long, 1″ wide or less, flat, usually erect; spikes numerous, rigid, widely spreading, alternate, the lower 2′–4′ long, the axis and branches triangular; spikelets 1¼″–1½″ long, sessile and appressed, alternate; scales hispid on the keel, the second longer than the first and exceeded by the acute third one.

Open ground, North Dakota and Montana to Illinois, Texas and New Mexico. Texas Crab-grass, Wire-grass. July–Sept.

62. BOUTELOÙA Lag. Var. Cienc. y Litter. 2: Part 4, 134. 1805.

Annual or perennial grasses with flat or convolute leaf-blades and numerous spikelets in one-sided spikes. Spikelets 1–2-flowered, arranged in two rows on one side of a flat rachis, the rachilla extended beyond the base of the flowers, bearing 1–3 awns and 1–3 rudimentary scales. Two lower scales empty, acute, keeled; flowering scale broader, 3-toothed, the teeth awn-pointed or awned; palet hyaline, entire or 2-toothed. Stamens 3. Styles distinct. Stigmas plumose. Grain oblong, free. [In honor of Claudius Boutelou, a Spanish botanist.]

About 30 species, particularly numerous in Mexico and in the southwestern United States.

Rachilla bearing the rudimentary scales and awns glabrous; second scale strongly papillose-hispid on the keel. 1. *B. hirsuta.*
Rachilla bearing the rudimentary scales and awns with a tuft of long hairs at the apex; second scale scabrous and sparingly long-ciliate on the keel. 2. *B. oligostachya.*

1. Bouteloua hirsùta Lag. Hairy Mesquite-grass. Fig. 548.

B. hirsuta Lag. Var. Cienc. y Litter. 2: Part 4, 141. 1805.

Culms 6'–20' tall, erect, simple or sometimes sparingly branched at the base, smooth and glabrous. Sheaths mostly at the base of the culm, the lower short and crowded, the upper longer; ligule a ring of short hairs; blades 1'–5' long, 1" wide or less, erect or ascending, flat, scabrous, sparingly papillose-hirsute near the base, especially on the margins; spikes 1–4, ½'–2' long, usually erect or ascending, the rachis extending beyond the spikelets into a conspicuous point; spikelets numerous, 2½"–3" long, pectinately arranged; first scale hyaline, shorter than the membranous second one, which is strongly papillose-hirsute on the keel; third scale pubescent, 3-cleft to the middle, the nerves terminating in awns; rachilla without a tuft of hairs under the rudimentary scales and awns.

In dry soil, especially on prairies, Illinois to South Dakota and Mexico; also in Florida. Bristly Mesquite, Black Grama; Buffalo-grass. July–Sept.

2. Bouteloua oligostàchya (Nutt.) Torr. Grama-grass. Mesquite-grass. Fig. 549.

Atheropogon oligostachyus Nutt. Gen. 1: 78. 1818.
B. oligostachya Torr.; A. Gray, Man. Ed. 2, 553. 1856.

Culms 6'–18' tall, erect, simple, smooth and glabrous. Sheaths shorter than the internodes; ligule a ring of short hairs; blades 1'–4' long, 1" wide or less, involute, at least at the long slender tip, smooth or scabrous; spikes 1–3, 1'–2' long, often strongly curved, the rachis terminating in a short inconspicuous point; spikelets numerous, pectinately arranged, about 3" long; first scale hyaline, shorter than the membranous second one, which is scabrous and sometimes long-ciliate on the keel, and sometimes bears a few papillae; third scale pubescent, 3-cleft, the nerves terminating in awns; rachilla with a tuft of long hairs under the rudimentary scales and awns.

On prairies, Wisconsin to North Dakota, south to Texas and Mexico. Blue or Common Grama, Buffalo-grass. July–Sept.

63. ATHEROPÒGON Muhl.; Willd. Sp. Pl. 4: 937. 1806.

Perennial grasses with narrow flat leaf-blades and an inflorescence composed of numerous short scattered 1-sided spreading or reflexed spikes. Spikelets 1-flowered, crowded in 2 rows, sessile, imbricated, the rachilla articulated above the empty scales and extending beyond the flower, its summit bearing scales or awns. Scales 3 or more, the lower 2 empty, unequal, acute, narrow, keeled, the third scale thicker and broader, enclosing a narrow 2-toothed hyaline palet and a perfect flower, 3-toothed at the apex, the teeth more or less awned, the small upper scales minute, awned. Stamens 3. Styles distinct. Stigmas plumose. Grain free, enclosed in the scale. [Greek, in reference to the awns of the flowering scales.]

Species about 15, natives of temperate and tropical regions. Type species: *Atheropogon apludoides* Muhl.

1. Atheropògon curtipéndulus (Michx.) Fourn. Tall Grama-grass. Fig. 550.

Chloris curtipendula Michx. Fl. Bor. Am. 1: 59. 1803.
Bouteloua racemosa Lag. Var. Cienc. y Litter. 2: Part 4, 141. 1805.
Bouteloua curtipendula Torr. Emory's Rep. 153. 1848.
Atheropogon curtipendulus Fourn. Mex. Pl. Gram. 138. 1881.

Culms 1°–3° tall, erect, simple, smooth and glabrous. Sheaths shorter than the internodes; ligule a ring of short hairs; leaves 2′–12′ long, 2″ wide or less, flat or involute, rough, especially above; spikes numerous, 3″–8″ long, widely spreading or reflexed; spikelets 4–12, divergent from the rachis, 3½″–5″ long, scales scabrous, especially on the keel, the first shorter than or equalling the second; the third 3-toothed, the nerves extended into short awns; rachilla bearing at the summit a small awned scale, or sometimes a larger 3-nerved scale, the nerves extended into awns; anthers vermilion or cinnabar-red.

In dry soil, Connecticut to North Dakota and Wyoming, south to New Jersey, Tennessee, Mississippi and Mexico. Side-oats Grama, Mesquite-grass. July–Sept.

64. BECKMÁNNIA Host, Gram. Austr. 3: 5. *pl. 6.* 1805.

A tall erect grass with flat leaf-blades and erect spikes borne in a terminal panicle. Spikelets 1–2-flowered, globose, compressed. Scales 3 or 4; the 2 lower empty, membranous, saccate, obtuse or abruptly acute; the flowering scales narrow, thin-membranous; palet hyaline, 2-keeled. Stamens 3. Styles distinct. Stigmas plumose. Grain oblong, free, enclosed in the scale and palet. [In honor of Johann Beckmann, 1739–1811, teacher of Natural History at St. Petersburg.]

A monotypic genus of the north temperate zone. Type species: *Phalaris erucaeformis* L.

1. Beckmannia erucaefórmis (L.) Host. Beckmann's Grass. Slough-grass. Fig. 551.

Phalaris erucaeformis L. Sp. Pl. 55. 1753.
B. erucaeformis Host, Gram. Austr. 3: 5. 1805.
Beckmannia erucaeformis var. *uniflora* Scribn.; Wats. & Coult. in A. Gray, Man. Ed. 6, 628. 1890.

Glabrous, culms 1½°–3° tall, erect, simple, smooth. Sheaths longer than the internodes, loose; ligule 2″–4″ long; blades 3′–9′ long, 2″–4″ wide, rough; panicle 4′–10′ in length, simple or compound, the spikes about ½′ long; spikelets 1″–1½″ long, 1–2-flowered, closely imbricated in two rows on one side of the rachis; scales smooth, the outer saccate, obtuse or abruptly acute; flowering scales acute, the lower generally awn-pointed, the upper rarely present.

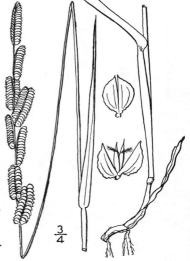

In wet places, western Ontario to Alaska, south to Iowa, Colorado and California. July–Sept.

65. ELEUSÌNE Gaertn. Fruct. & Sem. 1: 7. *pl. 1.* 1788.

Tufted annual or perennial grasses, with flat leaf-blades and spicate inflorescence, the spikes digitate or close together at the summit of the culm. Spikelets several-flowered, sessile, closely imbricated in two rows on one side of the rachis, which is not extended beyond them; flowers perfect or the upper staminate. Scales compressed, keeled; the 2 lower empty; the others subtending flowers, or the upper empty. Stamens 3. Styles distinct. Stigmas plumose. Grain loosely enclosed in the scale and palet. [From the Greek name of the town where Ceres was worshipped.]

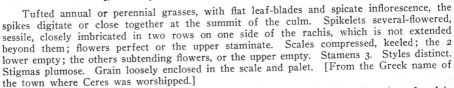

Species 6, natives of the Old World. Besides the following, two others have been found in ballast fillings about the eastern seaports. Type species: *Cynosurus coracanus* L.

1. Eleusine índica (L.) Gaertn. Wire-grass. Crab-grass. Yard-grass. Fig. 552.

Cynosurus indicus L. Sp. Pl. 72. 1753.

Eleusine indica Gaertn. Fruct. & Sem. 1: 8. 1788.

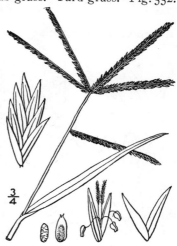

Culms 6′–2° tall, tufted, erect, or decumbent at the base, smooth and glabrous. Sheaths loose, overlapping and often short and crowded at the base of the culm, glabrous or sometimes sparingly villous; ligule very short; blades 3′–12′ long, 1″–3″ wide, smooth or scabrous; spikes 2–10, 1′–3′ long, whorled or approximate at the summit of the culm or one or two sometimes distant; spikelets 3–6-flowered, 1½″–2″ long; scales acute, minutely scabrous on the keel, the first 1-nerved, the second 3–7-nerved, the others 3–5-nerved.

In fields, dooryards and waste places all over North America except the extreme north. Naturalized from the warmer regions of the Old World. Dog's-tail or Goose-grass. Crop-grass. June–Sept.

66. DACTYLOCTÈNIUM Willd. Enum. 1029. 1809.

Annual grasses with flat leaf-blades and spicate inflorescence, the spikes in pairs or digitate. Spikelets several-flowered, sessile, closely imbricated in two rows on one side of the rachis which is extended beyond them into a sharp point. Scales compressed, keeled, the 2 lower and the uppermost ones empty, the others subtending flowers. Stamens 3. Styles distinct, short. Stigmas plumose. Grain free, rugose, loosely enclosed in the scale. [Greek, referring to the digitately spreading spikes.]

A genus of a few species, natives of the warmer parts of the Old World. Type species: *Cynosurus aegyptius* L.

1. Dactyloctenium aegýptium (L.) Willd. Crowfoot or Yard-grass. Egyptian Grass. Fig. 553.

Cynosurus aegyptius L. Sp. Pl. 72. 1753.
Eleusine aegyptia Pers. Syn. 1: 87. 1805.
Dactyloctenium aegyptiacum Willd. Enum. 1029. 1809.

Culms 6′–2° long, usually decumbent and extensively creeping at the base. Sheaths loose, overlapping and often crowded, smooth and glabrous; ligule very short; blades 6′ in length or less, 1″–3″ wide, smooth or rough, sometimes pubescent, ciliate toward the base; spikes in pairs, or 3–5 and digitate, ½′–2′ long; spikelets 3–5-flowered; scales compressed, scabrous on the keel, the second awned, the flowering ones broader and pointed.

In waste places and cultivated ground, southern New York, Pennsylvania and Virginia to Illinois and California, south to Florida and Mexico. Widely distributed in tropical America. Naturalized from Asia or Africa. Crab-grass. Finger-comb-grass. July–Oct.

67. LEPTÓCHLOA Beauv. Agrost. 71. *pl. 15. f. 1.* 1812.

Usually tall annual grasses, with flat leaf-blades and numerous spikes forming a simple panicle. Spikelets usually 2–many-flowered, flattened, alternating in two rows on one side of the rachis. Scales 4–many; the 2 lower empty, keeled, shorter than the spikelet; the flowering scales keeled, 3-nerved. Palet 2-nerved. Stamens 3. Styles distinct. Stigmas plumose. Grain free, enclosed in the scale and palet. [Greek, in allusion to the slender spikes.]

About 12 species, natives of the warmer regions of both hemispheres. Besides the following, 3 others occur in the southern United States. Type species: *Cynosurus virgatus* L.

Flowering scales ¾" long, the hairs on the nerves long and copious. 1. *L. filiformis.*
Flowering scales less than ½" long, the hairs on the nerves short and scant. 2. *L. attenuata.*

1. Leptochloa filifórmis (Lam.) Beauv. Slender Grass. Feather- or Salt-grass. Fig. 554.

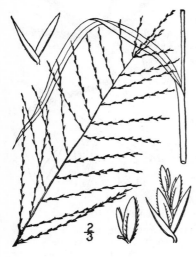

Festuca filiformis Lam. Ill. 1: 191. 1791.

Eleusine mucronata Michx. Fl. Bor. Am. 1: 65. 1803.

Leptochloa filiformis Beauv. Agrost. 166. 1812.

Leptochloa mucronata Kunth, Rev. Gram. 1: 91. 1829.

Culms 1°–3° tall, erect, branched, smooth and gla-brous. Sheaths shorter than the internodes, smooth and glabrous; ligule short, lacerate-toothed; blades 2'–8' long, 1"–3" wide, scabrous; spikes numerous, slender, rigid, spreading or ascending, the lower 2'–6' long; spikelets usually 3-flowered, about 1" long, the empty scales shorter than the spikelet, acute, 1-nerved, slightly scabrous on the keel; flowering scales 2-toothed at the apex, ciliate on the nerves.

In dry or moist soil, Virginia to Illinois, and Cali-fornia, south to Florida and Mexico. Also in tropical America. July–Sept.

2. Leptochloa attenuàta Nutt. Sharp-scaled Leptochloa. Fig. 555.

L. attenuata Nutt.; Steud. Syn. Pl. Gram. 209. 1855.

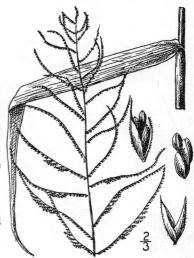

Culms tufted, branching at the base, 8'–16' tall, rarely taller; lower sheaths sparingly pilose; blades 1½'–4' long; spikes numerous, finally spreading, up to 2' long, rarely longer; spikelets 1¼"–1½" long, the scales 5 or 6, the outer 2 subulate, acuminate, awn-pointed, usually about equal, or the first shorter than the second, which reaches or extends beyond the apex of the second flowering scale.

In sandy river bottoms, Illinois to Oklahoma and Louisiana. Sept. and Oct.

68. ACAMPTÓCLADOS Nash, in Small, Fl. SE. U. S. 139. 1903.

Perennial tufted grasses with stiff culms and a panicle composed of scattered distant widely spreading rigid branches. Spikelets scattered and distinct, singly disposed in two rows, sessile, 4–6-flowered. Scales 6–8, firm; two lower scales empty, acuminate, about equal, the first 1-nerved, the second usually 3-nerved; flowering scales very acute, becoming harder in fruit, 3-nerved, the nerves glabrous, the lateral nerves vanishing at the margin below the apex; palet compressed, its two nerves ciliolate, gibbous at the base, obtuse at the apex. Stamens 3. Styles distinct. Stigmas plumose. [Greek, in reference to its rigid branches.]

A monotypic genus of south central United States.

1. Acamptoclados sessilispìcus (Buckley) Nash. Stiff Prairie-grass. Fig. 556.

Eragrostis sessilispica Buckley, Proc. Acad. Phila. **1862**: 97. 1862.
Diplachne rigida Vasey, Grasses S. W., Part 2. *pl. 41.* 1891.
A. sessilispicus Nash, in Small, Fl. SE. U. S. 140. 1903.

Culms 8'–3½° tall, erect, simple, smooth and glabrous. Sheaths short, crowded at the base of the culm, smooth, pilose at the summit; ligule a ring of short hairs; blades 2'–6' long, ¾"–1½" wide, rough above, glabrous or sparingly pilose beneath; panicle 8'–16' in length, the branches stout, rigid, widely diverging; spikelets scattered, closely sessile, appressed, 5–12-flowered, 4"–7" long, empty scales about equal in length, acute; flowering scales very acute, about 2" long, the lateral nerves very prominent.

Prairies, Kansas to Texas.   Aug.–Sept.

69. BÚLBILIS Raf. Am. Month. Mag. 4: 190.   1819.

[BUCHLOË Engelm. Trans. St. Louis Acad. 1: 432. *pl. 14. figs. 1–17.* 1859.]

A perennial stoloniferous monoecious or apparently dioecious grass with flat leaf-blades and spicate inflorescence. Staminate spikelets borne in two rows on one side of the rachis, the spikes at the summit of the long and exserted culms. Pistillate spikelets in spike-like clusters of 2 or 3, on very short culms, scarcely exserted from the sheath. Stamens 3. Styles distinct, long. Stigmas elongated, short-plumose. Grain ovate, free, enclosed in the scale. [Name apparently from the supposed bulb-like base of old plants.]
A monotypic genus of central North America.

1. Bulbilis dactyloìdes (Nutt.) Raf. Buffalo-grass. Early Mesquite. Fig. 557.

Sesleria dactyloides Nutt. Gen. 1: 65. 1818.
Buchloë dactyloides Engelm. Trans. St. Louis Acad. 1: 432. 1859.
Bulbilis dactyloides Raf.; Kuntze, Rev. Gen. Pl. 763. 1891.

Culms bearing staminate flowers 4'–12' tall, erect, slender, naked above, smooth and glabrous; those bearing pistillate flowers ½'–3' long, much exceeded by the leaves; ligule a ring of short hairs; blades 1" wide or less, more or less papillose-hirsute, those of the staminate culms 1'–4' long, erect, those of the stolons and pistillate culms 1' long or less, spreading; staminate spikes 2 or 3, approximate; spikelets 2"–2½" long, flattened, 2–3-flowered, the empty scales 1-nerved, the flowering 3-nerved; pistillate spikelets ovoid, the outer scales indurated.

On plains and prairies, Minnesota to Saskatchewan, south to Arkansas, Texas and northeastern Mexico. A valuable fodder grass. June–July.

70. MUNRÒA Torr. Pac. R. R. Rept. 4: 158.   1856.

Low diffusely branched grasses, with flat pungently pointed leaf-blades crowded at the nodes and the ends of the branches. Spikelets in clusters of 3–6, nearly sessile in the axils of the floral leaves, 2–5-flowered, the flowers perfect. Two lower scales empty, lanceolate, acute, 1-nerved, hyaline; flowering scales larger, 3-nerved; 1 or 2 empty scales sometimes present above the flowering ones; palet hyaline. Stamens 3.   Styles distinct, elongated. Stigmas barbellate or short-plumose. Grain free, enclosed in the scale and palet. [In honor of Gen. William Munro, English agrostologist.]

Three known species, the following typical one of the plains of North America, the others South American.

3/4

1. **Munroa squarròsa** (Nutt.) Torr. Munro's Grass. False Buffalo-grass. Fig. 558.

Crypsis squarrosa Nutt. Gen. 1 : 49. 1818.
M. squarrosa Torr. Pac. R. R. Rept. 4: 158. 1856.

Culms 2′–8′ long, tufted, erect, decumbent or prostrate, much branched, smooth or rough. Sheaths short, crowded at the nodes and ends of the branches, smooth, pilose at the base and throat, sometimes ciliate on the margins; ligule a ring of hairs; blades 1′ long or less, ½″–1″ wide, rigid, spreading, scabrous, pungently-pointed; spikelets 2–5-flowered, the flowers perfect; empty scales 1-nerved, shorter than the flowering scales which are about 2½″ long, 3-toothed, the nerves excurrent as short points or awns, tufts of hairs near the middle; palets obtuse.

On dry plains, Saskatchewan to Nebraska and northern Mexico. Aug.–Oct.

71. PHRAGMÌTES Trin. Fund. Agrost. 134. 1820.

Tall perennial reed-like grasses, with broad flat leaf-blades and ample panicles. Spikelets 3–several-flowered, the first flower often staminate, the others perfect; rachilla articulated between the flowering scales, long-pilose. Two lower scales empty, unequal, membranous, lanceolate, acute, shorter than the spikelet; the third scale empty or subtending a staminate flower; flowering scales glabrous, narrow, long-acuminate, much exceeding the short palets. Stamens 3. Styles distinct, short. Stigmas plumose. Grain free, loosely enclosed in the scale and palet. [Greek, referring to its hedge-like growth along ditches.]

Three known species, the following of the north temperate zone, one in Asia, the third in South America. Type species: *Arundo Phragmites* L.

1. **Phragmites Phragmìtes** (L.) Karst. Common Reed-grass. Fig. 559.

Arundo Phragmites L. Sp. Pl. 81. 1753.
Phragmites communis Trin. Fund. Agrost. 134. 1820.
Phragmites Phragmites Karst. Deutsch. Fl. 379. 1880–83.

Culms 5°–15° tall, erect, stout, from long horizontal rootstocks, smooth and glabrous. Sheaths overlapping, loose; ligule a ring of very short hairs; blades 6′–1° long or more, ½–2′ wide, flat, smooth, glabrous; panicle 6′–1° long or more, ample; spikelets crowded on the ascending branches; first scale 1-nerved, half to two-thirds as long as the 3-nerved second one; flowering scales 5″–6″ long, 3-nerved, long-acuminate, equalling the hairs of the rachilla

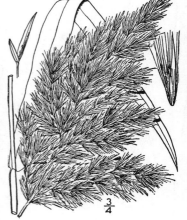

3/4

In swamps and wet places nearly throughout the United States, extending north to Nova Scotia, Manitoba and British Columbia. Also in Europe and Asia. Rarely ripening seed. Pole-, Bog- or Dutch-reed. Spires. Bennels. Wild Broom-corn. Aug.–Oct.

72. SIEGLÍNGIA Bernh. Syst. Verz. Pfl. Erf. 40. 1800.

Perennial tufted grasses with flat leaf-blades and a narrow simple panicle. Spikelets few, 3–5-flowered, the rachilla internodes short. Scales 5–7, the lower 2 empty, equalling or nearly as long as the spikelet, the flowering scales very firm, rounded on the back, obscurely 9-nerved, the nerves more prominent at the shortly 3-toothed apex, pilose near the margins, the callus short and obtuse, pilose; palet 2-keeled, ciliate on the keels. Stamens 3. Styles short, distinct. Stigmas plumose. Grain free, enclosed in the scale. [Named in honor of Prof. Siegling, German botanist.]

A monotypic genus of the Old World. Type species: *Sieglingia decumbens* Bernh.

1. Sieglingia decúmbens (L.) Kuntze. Heath- or Heather-grass. Fig. 560.

Festuca decumbens L. Sp. Pl. 75. 1753.
Triodia decumbens Beauv. Agrost. 76. 1812.
S. decumbens Bernh. Syst. Verg. Erf. 1 : 20, 44. 1800.

Culms 6'–18' tall, erect, often decumbent at the base, simple, smooth and glabrous. Sheaths shorter than the internodes, villous at the summit; ligule a ring of very short hairs; blades smooth beneath, usually scabrous above, ½"–1½" wide, the basal 3'–6' long, those of the culm 1'–3' long; panicle 1'–2' long, contracted, the branches 1' long or less, erect; spikelets 3–5-flowered, 3"–5" long, the joints of the rachilla very short; lower scales equalling the spikelet, acute; flowering scales broadly oval, ciliate on the margins below, obtusely 3-toothed, with two tufts of hair on the callus.

Introduced into Newfoundland. Native of Europe and Asia. Moor-grass. Summer.

73. TRÌDENS R. & S. Syst. 2 : 34. 1817.

[Tricuspis Beauv. Agrost. 77. 1812. Not Pers. 1807.]

Usually perennial grasses, with flat or involute leaf-blades, and the inflorescence composed of open or contracted and sometimes spike-like panicles. Spikelets 3–many-flowered, the flowers perfect or the upper ones staminate. Scales 5–many, membranous, sometimes firmer, the 2 lower empty, keeled, obtuse to acuminate, usually shorter than the rest, sometimes longer; flowering scales 3-nerved, the midnerve or all the nerves excurrent, the midnerve and the lateral nerves or the margins pilose, the apex entire or shortly 2-toothed, the teeth obtuse to acute, the callus short and obtuse; palet shorter than the scale, compressed, 2-keeled. Stamens 3. Styles short, distinct. Stigmas plumose. [Latin, in reference to the teeth of the flowering scales.]

Species about 30, natives chiefly of temperate regions. Type species : *Poa coerulescens* Michx.

Panicle open, the branches spreading and often drooping.   1. *T. flava.*
Panicle contracted, spike-like.
　Second empty scale 1-nerved; flowering scales about 2" long.   2. *T. stricta.*
　Second empty scale 3–5-nerved; flowering scales 2½"–3" long.   3. *T. elongata.*

1. Tridens flàva (L.) Hitchc. Tall Red-top. Fig. 561.

Poa flava L. Sp. Pl. 68. 1753.
Poa seslerioides Michx. Fl. Bor. Am. 1 : 68. 1803.
Tricuspis seslerioides Torr. Fl. N. Y. 2 : 463. 1843.
Sieglingia seslerioides Scribn. Mem. Torr. Club 5 : 48. 1894.
Tridens flava Hitchc. Rhodora, 8 : 210. 1906.

Culms 2°–5° tall, erect, often viscid above. Sheaths sometimes villous at the summit, equalling or shorter than the internodes; ligule a ring of very short hairs; blades 4'–1° long or more, 3"–6" wide, flat, attenuate into a long tip, smooth beneath, scabrous above; panicle 6'–18' long, the branches finally ascending or spreading, the lower 4'–10' long, usually dividing above the middle; spikelets 4–8-flowered, 3"–4" long, purple; joints of the rachilla short; empty scales glabrous, obtuse, generally slightly 2-toothed; flowering scales oval, the nerves pilose, excurrent as short points.

In fields, Massachusetts and New York to Kansas, south to Florida and Texas. July–Sept.

2. **Tridens strícta** (Nutt.) Nash.   Narrow Three-toothed Grass.   Fig. 562.

Windsoria stricta Nutt. Trans. Am. Phil. Soc. (II.) **5**:
147. 1833–37.
Tricuspis stricta A. Gray, Proc. Phila. Acad. **1862**: 335.
1863.
Sieglingia stricta Kuntze, Rev. Gen. Pl. 789. 1891.
Tridens stricta Nash, in Small, Fl. SE. U. S. 143. 1903.

Culms 1½°–4° tall, erect. Sheaths shorter than the
internodes; ligule a ring of short hairs; blades 6′–1°
long or more, flat, long-acuminate, smooth beneath,
scabrous above; spike-like panicle 5′–12′ in length,
the branches appressed, the lower 1′–2′ long; spikelets 4–10-flowered, 2″–3″ long, the joints of the
rachilla very short; lower scales usually about two-
thirds as long as the spikelet, rarely extending beyond the flowering scales, acute, glabrous; flowering
scales ovate, the nerves pilose for more than half
their length, the middle and often the lateral excurrent as short points.

Moist soil, Missouri and Kansas to Mississippi and Texas.   July–Oct.

3. **Tridens elongàta** (Buckley) Nash.   Long-panicled Three-toothed Grass.
Fig. 563.

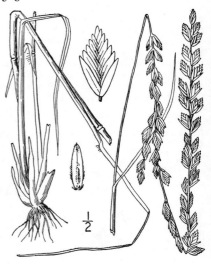

Uralepis elongata Buckley, Proc. Phila. Acad. Sci.
1862: 89. 1863.
Sieglingia elongata Nash, in Britt. & Br. Ill. Fl. **3**:
504. 1898.
Tricuspis elongata Nash, Britt. Man. 127. 1901.
Tridens elongata Nash, in Small, Pl. SE. U. S. 143.
1903.

Culms 1°–3° tall, tufted, erect, rough. Sheaths
rough, longer than the internodes, a ring of hairs
at the apex, the lower sheaths usually sparingly
papillose-pilose; blades rough, usually involute
when dry, 3′–10′ long, 1″–2″ wide; panicle narrow, 5′–10′ long, ½′ wide, its branches erect, 1½′
or less long; spikelets 10–12-flowered, 4½″–6″
long, the empty scales scabrous, hispidulous on
the midnerve, the first 1-nerved, the second
3-nerved; flowering scales about 3″ long, obtuse
at the scabrous apex, 3-nerved, the lateral nerves
vanishing at or below the apex, the midnerve
usually excurrent in a short point, all the nerves
pilose below the middle.

Prairies, Missouri to Colorado, Arizona and Texas.   June–Aug.

74. **ERIONEÙRON** Nash, in Small, Fl. SE. U. S. 143.   1903.

Perennial tufted grasses, with thick linear leaf-blades having thickened white margins,
and dense contracted almost capitate panicles. Spikelets several–many-flowered; empty
basal scales 2, narrow, acuminate; flowering scales broad, 3-nerved, pubescent on the nerves
below, and sometimes also on the body of the scale at the base, with long silky white hairs,
the apex acuminate, entire or slightly 2-toothed, the awn terminal or arising between the
minute teeth. Stamens 3. Styles short, distinct. [Greek, in reference to the hairy nerves
of the flowering scale.]

A monotypic genus of the southwestern United States and Mexico. Type species: *Uralepis
pilosa* Buckley.

1. Erioneuron pilòsum (Buckley) Nash.   Sharp-scaled Erioneuron.   Fig. 564.

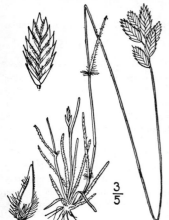

Uralepis pilosa Buckley, Proc. Phila. Acad. Nat. Sci. **1862** : 94. 1863.
Sieglingia pilosa Nash, in Britt. & Br. Ill. Fl. **3** : 504. 1898.
Erioneuron pilosum Nash, in Small, Fl. SE. U. S. 144. 1903.

Culms tufted, 2½'–12' tall, smooth and glabrous, the sterile shoots 4' tall or less. Sheaths smooth, a tuft of hairs on each side at the apex, much shorter than the internodes; ligule a ring of short hairs; blades strict or curved, thick, linear, obtuse, 1-nerved, the margins white, serrulate, 1½' long or less, less than 1" wide, folded, at least when dry, pubescent with long hairs, especially beneath; panicle almost racemose, long-exserted, ½'–1½' long; spikelets 3–10, crowded, 8–12-flowered; empty scales acuminate, 1-nerved; flowering scales 3"–3¼" long, acuminate, 3-nerved, the midnerve generally excurrent in a short point, all the nerves (the lateral at the top and bottom, the midnerve below the middle), the callus, and the base pilose.

Dry soil, Kansas to Nevada and Mexico. April–Sept.

75. TRÍPLASIS Beauv. Agrost. 81. 1812.

Grasses with narrow, flat or involute leaf-blades and contracted or open panicles. Spikelets shortly pedicelled, 2–6-flowered, the glabrous rachilla articulated between the flowers, the internodes very long. Scales 4–8, membranous, the lower 2 empty, keeled, the flowering scales dorsally rounded at the base, 3-nerved, the lateral nerves pilose, deeply 2-lobed at the apex, long-awned between the lobes, the callus long and subulate, pubescent on the outer surface; palet 2-keeled, the keels long-ciliate. Stamens 3. Styles short, distinct. Stigmas plumose. Grain free, enclosed in the scale. [Greek, referring to the 3 divisions of the flowering scales.]

Species 3, natives of the eastern and southern parts of North America.   Type species: *Triplasis americana* Beauv.

1. Triplasis purpùrea (Walt.) Chapm.   Sand-grass.   Fig. 565.

Aira purpurea Walt. Fl. Car. 78. 1788.
Tricuspis purpurea A. Gray, Man. 589. 1848.
Triplasis purpurea Chapm. Fl. S. U. S. 560. 1860.
Sieglingia purpurea Kuntze, Rev. Gen. Pl. 789. 1891.

Culms 1°–3° tall, erect, prostrate or decumbent, smooth and glabrous or the nodes pubescent. Sheaths shorter than the internodes, rough; ligule a ring of short hairs; blades ½'–2½' long, 1" wide or less, rigid, scabrous, sometimes sparsely ciliate; panicle 1'–3' in length, the branches rigid, finally widely spreading, the lower ¾'–1½' long; spikelets 2–5-flowered, 2½"–4" long, the joints of the rachilla half as long as the flowering scale; lower scales glabrous; flowering scales oblong, 2-lobed at the apex, the lobes erose-truncate, the nerves strongly ciliate, the middle one excurrent as a short point; palets long-ciliate on the upper part of the keel.

In sand, especially on sea beaches, Maine to Texas, and along the Great Lakes. Also from Illinois and Nebraska to Texas. Plant acid. Aug.–Sept.

76. REDFIÈLDIA Vasey, Bull. Torr. Club **14** : 133. 1887.

A tall perennial grass, with long narrow leaf-blades and an ample panicle. Spikelets 1–3-flowered, the flowers all perfect. Empty scales 2, about equal, shorter than the spikelet, 1-nerved; flowering scales membranous, 3-nerved, with a ring of hairs at the base. Palet 2-nerved, shorter than the scale. Stamens 3. Styles long, distinct. Stigmas short, plumose. Grain oblong, free. [In honor of John H. Redfield, 1815–1895, American naturalist.]

A monotypic genus of the western United States.   Type species: *Graphephorum flexuosum* Thurb.

3/4

1. Redfieldia flexuòsa (Thurb.) Vasey. Redfield's-grass. Fig. 566.

Graphephorum (?) *flexuosum* Thurb. Proc. Acad. Phila. 1863 : 78. 1863.
R. flexuosa Vasey, Bull. Torr. Club **14**: 133. 1887.

Culms 1½°–4° tall, erect from a long horizontal rootstock, simple, smooth and glabrous. Sheaths smooth, the lower short and overlapping, often crowded, the upper much longer; ligule a ring of short hairs; blades 1°–2° long, 1″–2″ wide, involute; panicle ample and diffuse, 8′–22′ in length, the branches finally widely spreading, flexuous, the lower 3′–8′ long; spikelets about 3″ long, 1–3-flowered, the empty scales acute, glabrous; flowering scales with a ring of hairs at the base, minutely scabrous, twice the length of the empty ones, acute, the middle nerve usually excurrent as a short point.

On prairies, South Dakota to Colorado and Oklahoma. Blow-out-grass. Aug.–Sept.

77. DIPLÁCHNE Beauv. Agrost. 80. *pl. 16. f. 9.* 1812.

Tufted grasses, with narrow flat leaf-blades and long slender spikes arranged in an open panicle, or rarely only one terminal spike. Spikelets several-flowered, narrow, sessile or shortly pedicelled, erect. Two lower scales empty, membranous, keeled, acute, unequal; flowering scales 1–3-nerved, 2-toothed and mucronate or short-awned between the teeth. Palet hyaline, 2-nerved. Stamens 3. Styles distinct. Stigmas plumose. Grain free, loosely enclosed in the scale and palet. [Greek, referring to the 2-toothed flowering scales.]

About 15 species, natives of the warmer regions of both hemispheres. Besides the following species, about 6 others occur in the southern and western parts of North America. Type species: *Festuca fascicularis* Lam.

Awn less than ⅓ as long as the flowering scale.
  Spikelets 2″–4″ long, the flowering scales acute or obtuse at the 2-toothed apex, lateral nerves
    often excurrent.   1. *D. fascicularis.*
  Spikelets 5″–6″ long, the flowering scales acuminate at the usually entire apex, the lateral nerves
    rarely excurrent.   2. *D. acuminata.*
Awn ½ as long as the flowering scale or more.   3. *D. maritima.*

1. Diplachne fasciculàris (Lam.) Beauv. Salt-meadow Diplachne. Clustered Salt-grass. Spike-grass. Fig. 567.

Festuca fascicularis Lam. Tabl. Encycl. **1**: 189. 1791.
Diplachne fascicularis Beauv. Agrost. 160. 1812.

Culms 1°–2½° tall, erect, ascending, or rooting at the lower nodes, finally branched, smooth and glabrous. Sheaths shorter than the internodes, loose, smooth or rough, the upper one longer and enclosing the base of the panicle; ligule 1″–2″ long; blades 3′–8′ long, 1″–1½″ wide, scabrous; panicle 4′–12′ in length, often exceeded by the upper leaf, the branches erect or ascending, the lower 2′–5′ long; spikelets 8–10-flowered, 2″–4″ long; lower scales glabrous, rough on the keel; flowering scales, exclusive of the awn, 1½″–2″ long, the midnerve extending into an awn ½″ long or less.

In brackish marshes, Florida to Texas, and up the Mississippi to Illinois and Missouri. Also in the West Indies. Aug.–Oct.

3/4

2. Diplachne acumina̍ta Nash. Sharp-scaled Diplachne. Fig. 568.

D. acuminata Nash, in Britt. Man. 128. 1901.

Culms tufted, 1°–2° tall, finally branching; blades erect, 4′–1° long, 2½″ wide or less, usually involute when dry, very rough; racemes numerous, erect or ascending, the larger 3′–6′ long; spikelets 5″–6″ long, the scales 8–11, the flowering scales 3″–3½″ long, acuminate at the entire or occasionally slightly 2-toothed apex, the lateral nerves rarely slightly excurrent, the midnerve extending into an awn ¾″ long or less.

Wet or moist soil, Arkansas and Missouri to Nebraska and Colorado. June–Aug.

$\frac{2}{3}$

$\frac{2}{3}$

3. Diplachne marítima Bicknell. Long-awned Diplachne. Fig. 569.

Festuca procumbens Muhl. Gram. 160. 1817.
Diplachne procumbens Nash, in Small, Fl. SE. U. S. 145. 1903. Not Arech. 1896.
D. maritima Bicknell, Bull. Torrey Club 35 : 195. 1908.

Culms tufted, finally branching, 8′–16′ tall; blades erect, 3′–8′ long, 2″ wide or less, involute when dry; racemes numerous, erect, the larger 2′–3′ long; spikelets about 5″ long; scales 8–10, the empty ones usually awned or awn-pointed, the flowering scales, exclusive of the awn, 2¼″–4½″ long, acuminate at the slightly 2-toothed apex, the midnerve extending into an awn ½ or more as long as the scale.

Brackish marshes and shores, Massachusetts to South Carolina ; also on the shore of Onondaga Lake, N. Y. Aug.–Oct.

78. AÍRA L. Sp. Pl. 63. 1753.

[Molinia Schrank, Baier. Fl. 1 : 100. 1789.]

Perennial tufted grasses, with narrow flat leaf-blades and paniculate inflorescence. Spikelets 2–4-flowered. Two lower scales empty, somewhat obtuse or acute, unequal, shorter than the spikelet ; flowering scales membranous, rounded on the back, 3-nerved ; palets scarcely shorter than the scales, obtuse, 2-keeled. Stamens 3. Styles short. Stigmas short, plumose. Grain oblong, free, enclosed in the scale and palet. [Greek name for *Lolium temulentum.*]

A genus of a few species, natives of Europe and Asia. Type species : *Aira coerulea* L.

1. Aira coerùlea L. Purple Melic- or Moor Grass. Lavender-grass. Indian-grass. Fig. 570.

Aira coerulea L. Sp. Pl. 63. 1753.
Molinia coerulea Moench, Meth. 183. 1794.

Culms 1°–3½° tall, erect, simple, smooth and glabrous. Sheaths overlapping and confined to the lower part of the culm, smooth and glabrous ; ligule a ring of very short hairs ; blades 4′–1° long or more, 1″–3″ wide, erect, acuminate, smooth beneath, slightly scabrous above ; panicle 3′–10′ in length, green or purple, the branches usually erect, 1′–4′ long ; spikelets 2–4-flowered, 2½″–4″ long ; empty scales acute, unequal ; flowering scales about 2″ long, 3-nerved, obtuse.

Sparingly introduced on ballast and in waste places, Maine to New York. Adventive from Europe. Aug.–Sept.

$\frac{2}{3}$

79. **RHOMBÓLYTRUM** Link, Hort. Berol. **2**: 296. 1823.

Perennial grasses, with usually flat leaf-blades, and a narrow contracted spike-like panicle. Spikelets numerous. Scales several, the outer 2 empty ones 1-nerved, the flowering scales broad, rounded at the apex, 3-nerved, the nerves glabrous, the lateral ones vanishing below the margin, the midnerve at the margin or sometimes excurrent as a short tip, the callus pilose; palet 2-keeled. Stamens 3. Styles short, distinct. Stigmas plumose. Grain free, enclosed in the scale. [Greek, in reference to the round flowering scales.]

Species 4 or 5, natives of warm countries. Type species: *Rhombolytrum rhomboideum* Link.

1. **Rhombolytrum albéscens** (Vasey) Nash.
White Prairie-grass. Fig. 571.

Triodia albescens Vasey, Bull. U. S. Dept. Agric. Div. Bot. **12**: Part 2, 33. 1891.
Sieglingia albescens Kuntze; L. H. Dewey, Contr. U. S. Nat. Herb. **2**: 538. 1894.
R. albescens Nash, in Britt. Man. 129. 1901.

Culms tufted, erect, smooth and glabrous, 12′–20′ tall, the sterile shoots one-half as long as the culm or more. Sheaths shorter than the internodes, smooth; ligule a ring of short hairs; blades smooth beneath, roughish above, acuminate, 2½–11′ long, 1″–2″ wide; panicle dense and contracted, white, 2½–5′ long, ¼–¾′ broad, its branches erect or ascending, 1′ or less long; spikelets about 7–11-flowered, 2″–2½″ long, the empty scales white, 1-nerved, about equal; flowering scales about 1½″ long, 3-nerved, the lateral nerves vanishing below the apex, all the nerves glabrous, the midnerve excurrent in a short scabrous point, denticulate and irregularly and obscurely lobed at the truncate apex.

Prairies, Kansas to New Mexico and Texas. Aug.–Sept.

80. **ERAGRÓSTIS** Beauv. Agrost. 70. *pl. 14. f. 11.* 1812.

Annual or perennial grasses, rarely dioecious, from a few inches to several feet in height, the spikelets in contracted or open panicles. Spikelets 2–many-flowered, more or less flattened. Two lower scales empty, unequal, shorter than the flowering ones, keeled, 1-nerved, or the second 3-nerved; flowering scales membranous, keeled, 3-nerved; palets shorter than the scales, prominently 2-nerved or 2-keeled, usually persisting on the rachilla after the fruiting scale has fallen. Stamens 2 or 3. Styles distinct, short. Stigmas plumose. Grain free, loosely enclosed in the scale and palet. [Greek, signifying probably a *Love-grass.*]

A genus of about 120 species, widely distributed throughout all warm and temperate countries. Besides the following, some 15 others occur in the southern and western parts of North America. Type species: *Briza Eragrostis* L.

Culms not creeping; plants with perfect flowers.
  Annuals.
    Spikelets 2–5-flowered, 1″–1½″ long.
      Culms branched only at base; pedicels and branches of panicle long and capillary.
        Flowering scales 1¼″–1½″ long; culms rarely over 20′, slender.   1. *E. capillaris.*
        Flowering scales 2″–2½″ long; culms 2° tall or more, usually stout.   2. *E. hirsuta.*
      Culms branched above the base; pedicels and branches of the panicle short. 3. *E. Frankii.*
    Spikelets 5–many-flowered, 1½″–8″ long.
      Spikelets ¾″ wide or less.
        Flowering scales thin, usually bright purplish, the lateral nerves faint or wanting;
          spikelets about ½″ wide.   4. *E. pilosa.*
        Flowering scales firm, usually dull purple or green, the lateral nerves very prominent;
          spikelets about ¾″ wide.   5. *E. Purshii.*
      Spikelets 1″ wide or more.
        Lower flowering scales about ¾″ long; spikelets 1″ wide.   6. *E. Eragrostis.*
        Lower flowering scales 1″–1¼″ long; spikelets 1¼″–1½″ wide. 7. *E. major.*
  Perennials.
    Spikelets not clustered.
      Branches of the open panicle stiff, widely spreading, at least when old.
        Pedicels long, commonly at least the length of the spikelets.   8. *E. pectinacea.*
        Pedicels commonly much shorter than the spikelets.
          Blades elongated; branches of the panicle long and ˙slender; spikelets scattered,
            6–25-flowered.   9. *E. refracta.*
          Blades not elongated; branches of the panicle short and stout, rigid, the spikelets
            crowded, 5–12-flowered.   10. *E. curtipedicellata.*
      Branches of the elongated panicle erect or ascending, capillary.   11. *E. trichodes.*
    Spikelets clustered on the very short erect or ascending branches.   12. *E. secundiflora.*
Culms extensively creeping; plants dioecious.
  Flowering scales less than 1″ long, glabrous.   13. *E. hypnoides.*
  Flowering scales 1½″–2″ long, pubescent.   14. *E. Weigeltiana.*

1. Eragrostis capilláris (L.) Nees. Lace-grass. Tiny Love-grass. Fig. 572.

Poa capillaris L. Sp. Pl. 68. 1753.
Poa tenuis Ell. Bot. S. C. & Ga. 1 : 156. 1817.
Eragrostis capillaris Nees, Agrost. Bras. 505. 1829.
E. tenuis Ell.; Steud. Syn. Gram. 273. 1855.

Culms 8'–18' tall, erect, slender, sparingly branched
at the base, smooth and glabrous. Sheaths short,
overlapping and crowded at the base of the culm,
glabrous or sparingly hairy, the upper enclosing the
base of the panicle; ligule a ring of very short hairs;
blades 3'–10' long, 1"–2" wide, long-acuminate,
smooth beneath, scabrous above and sparingly hirsute near the base; panicle diffuse, 4'–15' in length,
the branches capillary, spreading or ascending, 1½'–5'
long; spikelets ovate, 2–4-flowered, little flattened,
1"–1½" long; empty scales about equal, acute; flowering scales acute, the lower ¾" long, the lateral
nerves obscure.

In dry places, New Hampshire to Kansas, south to
Georgia and Texas. Aug.–Sept.

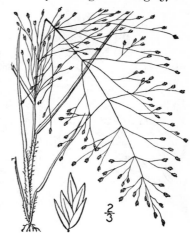

2. Eragrostis hirsùta (Michx.) Nees. Stout Love-grass. Fig. 573.

Poa hirsuta Michx. Fl. Bor. Am. 1 : 68. 1803.

Eragrostis hirsuta Nees, Agrost. Bras. 508. 1829.

Culms densely tufted, rather stout, 2°–4½° tall;
sheaths, at least the lower ones, strongly papillose-
hispid, each with a tuft of hairs at the apex;
blades of the lower leaves 1°–2° long, less than
5" wide, long-acuminate, flat; panicle 20'–3° long,
diffuse, its branches finally widely spreading;
spikelets 3–5-flowered, 1½"–2" long, the flowering
scales 1"–1¼" long.

In dry fields, thickets and woodlands, Virginia (according to Kearney) to Florida and Texas. July–
Sept.

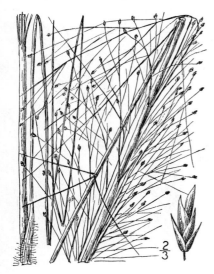

3. Eragrostis Fránkii Steud. Frank's Love-grass. Fig. 574.

Eragrostis Frankii Steud. Syn. Pl. Gram. 273. 1855.
E. erythrogona Nees; Steud. Syn. Pl. Gram. 273. 1855.

Glabrous, culms 6'–15' tall, tufted, erect, or often
decumbent at the base, branched, smooth. Sheaths
loose, shorter than the internodes; ligule a ring of
hairs; blades 2'–5' long, 1"–2" wide, smooth beneath,
scabrous above; panicle 2'–6' in length, open, the
branches ascending, the lower 1'–1½' long; spikelets
ovate, 3–5-flowered, 1"–1½" long; empty scales acute,
the first shorter than the second; flowering scales acute,
the lower ¾" long, the lateral nerves obscure.

In moist places, Massachusetts to Minnesota, Mississippi,
Louisiana and Kansas. Short-stalked Meadow-grass. Sept.–
Oct.

4. Eragrostis pilòsa (L.) Beauv.   Small Tufted Love-grass.   Fig. 575.

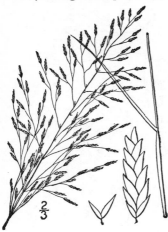

Poa pilosa L. Sp. Pl. 68. 1753.
Eragrostis pilosa Beauv. Agrost. 162. 1812.

Culms 6′–18′ tall, tufted, erect, slender, branched,
smooth and glabrous. Sheaths shorter than the internodes, smooth, sometimes pilose at the throat; ligule
a ring of short hairs; blades 1′–5′ long, 1″ wide or less,
smooth beneath, scabrous above; panicle 2′–6′ in length,
the branches at first erect, finally widely spreading,
1′–1½′ long, often hairy in the axils; spikelets 5–12-flow-
ered, 1½″–3″ long, about ½″ wide; lower scales acute,
the first one-half as long as the second; flowering scales
acute, the lower ¾″ long, thin, usually purplish, the
lateral nerves faint or wanting.

Waste places or cultivated ground, Massachusetts to Michigan and Kansas, south to Florida and Texas. Naturalized
from Europe. Also in the West Indies. Aug.–Sept.

5. Eragrostis Púrshii Schrad.   Pursh's Love-grass.   Fig. 576.

?Poa caroliniana Spreng. Mant. Fl. Hal. 33. 1807.
Eragrostis Purshii Schrad. Linnaea, 12 : 451. 1838.
E. caroliniana Scribn. Mem. Torr. Club 5 : 49. 1895.

Culms 6′–18′ tall, tufted, usually decumbent at the
base and much branched, smooth and glabrous. Sheaths
loose, shorter than the internodes, smooth and glabrous; ligule a ring of short hairs; blades 1½′–3½′ long,
1″ wide or less, smooth beneath, rough above; panicle
open, 3′–8′ long, the branches spreading, 1′–2½′ long,
naked in the axils; spikelets 5–15-flowered, dull purple
or green, 1½″–4″ long, about ¾″ wide; empty scales
acute, the lower about two-thirds as long as the upper,
scabrous on the keel; flowering scales acute, firm, the
lower ones ¾″ long, the lateral nerves prominent.

In dry places, Maine to Ontario and North Dakota,
south to Florida and Texas. Southern Spear-grass. Aug.–
Sept.

6. Eragrostis Eagróstis (L.) Karst.   Low Love-grass.   Fig. 577.

Poa Eragrostis L. Sp. Pl. 68. 1753.
Eragrostis poaeoides Beauv. Agrost. 162. 1812.
Eragrostis minor Host, Fl. Austr. 1 : 135. 1827.
Eragrostis Eragrostis Karst. Deutsch. Fl. 389. 1880–83.

Culms seldom over 15′ tall, tufted, usually decumbent
and much branched, smooth and glabrous. Sheaths
loose, shorter than the internodes, smooth, sometimes
a little pubescent, sparingly pilose at the throat; ligule
a ring of short hairs; blades 1′–2½′ long, ½″–2″ wide,
smooth beneath, rough above and somewhat pilose near
the base; panicle 2′–4½′ in length, the branches spreading or ascending, ¾′–1½′ long; spikelets 8–18-flowered,
3″–5″ long, about 1″ wide; empty scales acute, the first
two-thirds as long as the second; flowering scales obtuse, ¾″ long, the lateral nerves prominent.

In waste places or cultivated ground, Massachusetts, New
York and Pennsylvania. Locally naturalized from Europe.
July–Sept.

7. Eragrostis màjor Host.   Strong-scented Love-grass.   Fig. 578.

E. major Host, Gram. Austr. **4**: 14. *pl. 24.* 1809.

Eragrostis poaeoides var. *megastachya* A. Gray, Man. Ed. 5, 631. 1867.

Culms 6′–2° tall, erect, or decumbent at the base, usually branched, smooth and glabrous. Sheaths shorter than the internodes, smooth, sparingly pilose at the throat; ligule a ring of short hairs; blades 2′–7′ long, 1″–3″ wide, flat, smooth beneath, scabrous above; panicle 2′–6′ in length, the branches spreading or ascending, 1′–2′ long; spikelets 8–35-flowered, 2½″–8″ long, about 1½″ wide, very flat; empty scales acute, the first slightly shorter than the second; flowering scales obtuse, 1″–1¼″ long, the lateral nerves prominent.

In waste and cultivated places nearly throughout the United States, and in Ontario.   Naturalized from Europe. Unpleasantly scented, handsome.   Candy-grass.   Aug.–Sept.

⅔

8. Eragrostis pectinàcea (Michx.) Steud.   Purple Love-grass.   Fig. 579.

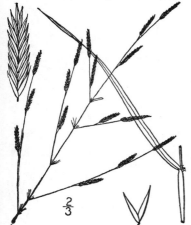

Poa pectinacea Michx. Fl. Bor. Am. **1**: 69. 1803.
Eragrostis pectinacea Steud. Syn. Pl. Gram. 272. 1855.
Eragrostis pectinacea var. *spectabilis* A. Gray, Man. Ed. 5, 632. 1867.

Culms 1°–2½° tall, erect or ascending, rigid, simple, smooth and glabrous. Sheaths overlapping, smooth, glabrous or villous, the upper one often enclosing the base of the panicle; ligule a ring of hairs; blades 5′–12′ long, 2″–4″ wide, smooth beneath, scabrous above and sparingly villous at the base; panicle 6′–24′ in length, purple or purplish, the branches 3′–10′ long, strongly bearded in the axils, widely spreading or the lower often reflexed; spikelets 5–15-flowered, 1½″–4″ long, on pedicels of at least their own length; scales acute, the empty ones about equal, the flowering ones about ⅞″ long, their lateral nerves very prominent.

In dry soil, Maine to South Dakota, south to Florida and Texas.   Pink-grass.   False Red-top.   Aug.–Sept.

⅔

9. Eragrostis refrácta (Muhl.) Scribn.   Meadow Love-grass.   Fig. 580.

Poa refracta Muhl. Gram. 146. 1817.

Eragrostis campestris Trin. Bull. Acad. Sci. St. Petersb. **1**: 70. 1836.

E. refracta Scribn. Mem. Torr. Club **5**: 49. 1894.

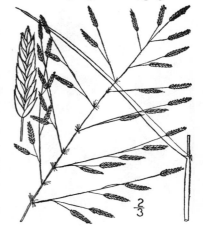

Culms 1°–3° tall, erect, slender, simple, smooth and glabrous. Sheaths overlapping, smooth and glabrous; ligule a ring of short hairs; blades 5′–12′ long, 1″–2″ wide, smooth beneath, rough above, and villous toward the base; panicle 8′–20′ long; branches slender, 4′–10′ long, at length widely spreading, the axils often bearded; spikelets 6–25-flowered, 2½″–6″ long, on pedicels shorter than themselves; empty scales acute, the first somewhat shorter than the second; flowering scales very acute, ¾″–1″ long, the lateral nerves prominent.

In sandy soil, Delaware and Maryland to Florida, west to Texas.   Aug.–Sept.

⅔

10. Eragrostis curtipedicellàta Buckley. Short-stalked Love-grass. Fig. 581.

Eragrostis curtipedicellata Buckley, Proc. Acad. Phila. 1862: 97. 1862.

Culms 6'-3° tall, erect, rigid, simple, smooth and glabrous. Sheaths overlapping, smooth, pilose at the summit; ligule a ring of short hairs; blades 2'-8' long, 1"-2" wide, smooth beneath, scabrous above; panicle 4'-12' in length, the branches widely spreading, 1½'-4½' long; spikelets 5-12-flowered, 1½"-3" long, on pedicels of less than their own length; scales acute, the empty ones somewhat unequal, the flowering ones about ⅞" long, scabrous on the midnerve, their lateral nerves prominent.

Prairies, Kansas to Texas. Aug.–Sept.

11. Eragrostis trichòdes (Nutt.) Nash. Hair-like Love-grass. Fig. 582.

Poa trichodes Nutt. Trans. Am. Phil. Soc. (II.) **5**: 146. 1833–37.
Eragrostis tenuis A. Gray, Man. Ed. 2, 564, in part. 1856.
E. trichodes Nash, Bull. Torr. Club **22**: 465. 1895.

Culms 2°-4° tall, erect, simple, smooth and glabrous. Sheaths overlapping, smooth, pilose at the throat; ligule a ring of very short hairs; blades 6'-28' long, 1"-2" wide, smooth beneath, slightly scabrous above, attenuate into a long slender tip; panicle 9'-26' in length, narrow and elongated, the branches erect or ascending, capillary, subdividing, somewhat flexuous, 3'-7' long; lower axils sometimes bearded; spikelets usually pale, 3-10-flowered, 2½"-4½" long; lower scales very acute, about equal; flowering scales acute, the lower ones 1¼"-1½" long, their lateral nerves manifest.

In dry sandy soil, Illinois to Nebraska, south to Texas. Blow-out-grass. Aug.–Sept.

12. Eragrostis secundiflòra Presl. Clustered Love-grass. Fig. 583.

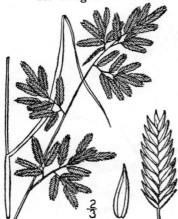

Poa interrupta Nutt. Trans. Am. Phil. Soc. (II.) **5**: 146. 1833–37. Not Lam. 1791.

Eragrostis secundiflora Presl, Rel. Haenk. **1**: 276. 1830.

Eragrostis oxylepis Torr. Marcy's Report, 269. 1854.

Smooth and glabrous, culms 6'-3° tall, erect, simple. Sheaths shorter than the internodes; ligule a ring of hairs; blades 2'-12' long, 1"-2" wide; panicle 1½'-6' in length, the branches ½'-1½' long, erect or ascending; spikelets crowded or clustered, sessile or nearly so, strongly flattened, 8-40-flowered, 3"-10" long, 1"-2½" wide; lower scales acute, about equal; flowering scales 1½"-1¾" long, acute, usually purple-bordered, the lateral nerves prominent.

In dry soil, Kansas and Colorado, south to Texas and Florida. Aug.–Sept.

13. Eragrostis hypnoìdes (Lam.) B.S.P.   Smooth Creeping Love-grass.   Fig. 584.

Poa hypnoides Lam. Tabl. Encycl. 1: 185. 1791.
Eragrostis reptans Nees, Agrost. Bras. 514. 1829.
Eragrostis hypoides B.S.P. Prel. Cat. N. Y. 69. 1888.
Ne-eragrostis hypnoides Bush, Trans. St. Louis Acad. 13: 180. 1903.

Culms 1′–18′ long, extensively creeping, branched, smooth and glabrous, the branches erect or ascending, 1′–6′ high. Sheaths shorter than the internodes, villous at the summit; ligule a ring of short hairs; blades 2′ long or less, ½″–1″ wide, flat, smooth beneath, rough above; spikelets dioecious, 10–35-flowered, 2″–8″ long; lower scales unequal, the first one-half to two-thirds as long as the second; flowering scales about 1¼″ long, the lateral nerves prominent; scales of the pistillate flowers more acute than those of the staminate.

On sandy or gravelly shores, Vermont and Ontario to Washington, south to Florida and Mexico. Also in tropical America. Aug.–Sept.

14. Eragrostis Weigeltiàna (Reichenb.) Bush.   Hairy Creeping Love-grass.   Fig. 585.

Poa Weigeltiana Reichnb. Mem. Acad. St. Petersb. VI. 1: 40. 1831.
Poa capitata Nutt. Trans. Am. Phil. Soc. II. 5: 146. 1837.
Eragrostis capitata Nash, in Britt. Man. 1042. 1901.
Ne-eragrostis Weigeltiana Bush, Trans. St. Louis Acad. 13: 178. 1903.
E. Weigeltiana Bush, Trans. St. Louis Acad. 13: 180. 1903.

Plants dioecious.   Culms branching and creeping, rooting at the nodes which send up branches 2½′–4′ tall; sheaths, at least those on the branches, pubescent; blades spreading or ascending, ½–1½′ long, 1½″ wide or less, flat, lanceolate, pubescent; panicle ¾–1½′ long, nearly or quite as broad, oval; spikelets crowded, clustered, pubescent, 12–30-flowered, 3″–7″ long, the flowering scales 1½″–2″ long.

In sandy, usually wet, soil, Nebraska to Louisiana and Texas; also in Mexico and northern South America. July–Oct.

81. SPHENÓPHOLIS Scribn. Rhodora 8: 142. 1906.

[Eatonia Endlich. Gen. 99. 1837. Not Raf. 1819.]

Tufted perennial grasses, with flat or involute leaf-blades and usually contracted panicles. Spikelets 2–3-flowered, the rachilla extended beyond the flowers.   Two lower scales empty, shorter than the spikelet, the first linear, acute, 1-nerved, the second much broader, 3-nerved, obtuse or rounded at the apex, or sometimes acute, the margins scarious; flowering scales narrower, generally obtuse.   Palet narrow, 2-nerved.   Stamens 3.   Styles distinct, short. Stigmas plumose.   Grain free, loosely enclosed in the scale and palet.   [Greek, referring to the wedge-shaped second scale of the spikelet.]

A genus of 7 or 8 species, confined to North America.   Type species: *Aira obtusata* Michx.

Empty scales unequal, the first shorter and about one-sixth as wide as the second.
    Second scale obovate, often almost truncate.   1. *E. obtusata.*
    Second scale oblanceolate, obtuse or abruptly acute.   2. *E. pallens.*
Empty scales equal, the first not less than one-third as wide as the second.   3. *E. nitida.*

$\frac{3}{4}$

1. Sphenopholis obtusàta (Michx.) Scribn. Early Bunch-grass. Fig. 586.

Aira obtusata Michx. Fl. Bor. Am. **1** : 62. 1803.
Eatonia obtusata A. Gray, Man. Ed. 2, 558. 1856.
Eatonia pubescens Scribn. & Merr. Circ. U. S. Dep. Agr. Agrost. **27** : 6. 1900.
E. robusta Rydb. Bull. Torr. Club **32** : 602. 1905.
S. obtusata Scribn. Rhodora **8** : 144. 1906.

Culms 1°–2½° tall, erect, simple, often stout, smooth and glabrous. Sheaths shorter than the internodes, usually more or less rough, sometimes pubescent; ligule ½″–1″ long; blades 1′–9′ long, 1″–4″ wide, scabrous; panicle 2′–6′ in length, dense and generally spike-like, strict, the branches 1½′ long or less, erect; spikelets crowded, 1¼″–1½″ long; empty scales unequal, often purplish, the first narrow, shorter than and about one-sixth as wide as the obtuse or almost truncate second one; flowering scales narrow, obtuse, ¾″–1″ long.

In dry soil, Maine to Saskatchewan, Florida and Arizona. Prairie-grass. June–Aug.

2. Sphenopholis pallens (Spreng.) Scribn. Tall Eaton's Grass. Fig. 587.

Aira pallens Spreng. Fl. Hal. Mant. 33. 1807.
Eatonia pennsylvanica A. Gray, Man. Ed. 2, 558, in part. 1856.
S. pallens Scribn. Rhodora **8** : 145. 1906.

Usually glabrous, culms 1°–3° tall, erect, simple, slender, smooth. Sheaths shorter than the internodes; ligule ¾″ long; blades 2½″–7′ long, 1″–3″ wide, rough; panicle 3′–7′ in length, contracted, often nodding, lax, its branches 1′–2½′ long; spikelets 1½″–1¾″ long, usually numerous, somewhat crowded and appressed to the branches; empty scales unequal, the first narrow, shorter than and about one-sixth as broad as the obtuse or abruptly acute second one, which is smooth, or somewhat rough on the keel; flowering scales narrow, acute, 1¼″ long, rarely awned.

In hilly woods or moist soil, Newfoundland to British Columbia, Georgia and Texas. June–July.

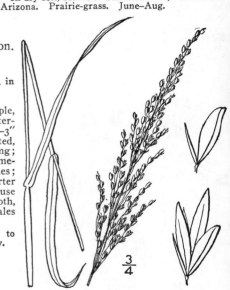

$\frac{3}{4}$

$\frac{3}{4}$

3. Sphenopholis nítida (Spreng.) Scribn. Slender Eaton's-grass. Fig. 588.

Aira nitida Spreng. Fl. Hal. Mant. 32. 1807.
Eatonia Dudleyi Vasey, Coult. Bot. Gaz. **11** : 116. 1886.
Eatonia nitida Nash, Bull. Torr. Club **22** : 511. 1895.
E. glabra Nash, in Britt. Man. 1043. 1901.
S. nitida Scribn. Rhodora **8** : 144. 1906.

Glabrous, culms 1°–2° tall, erect, very slender, smooth. Sheaths shorter than the internodes, generally pubescent; ligule ¼″ long; blades ½″–3′ long, 1″ wide or less, often pubescent, the uppermost very short; panicle 2′–6′ in length, lax, the branches spreading at flowering time, afterwards erect, 1′–2½′ long; spikelets not crowded, 1½″ long; empty scales smooth, the first about one-third as wide as and equalling the second, which is obtuse or almost truncate, often apiculate; flowering scales narrow, 1″–1¼″ long, obtuse or acutish, smooth.

In dry woods, Vermont to Michigan, Georgia and Mississippi. May–June.

82. KOELÈRIA Pers. Syn. 1 : 97. 1805.

Tufted annual or perennial grasses, with flat or setaceous leaf-blades and mostly spike-like panicles. Spikelets 2–5-flowered. Two lower scales empty, narrow, acute, unequal, keeled, scarious on the margins; the flowering scales 3–5-nerved. Palet hyaline, acute, 2-keeled. Stamens 3. Styles very short. Stigmas plumose. Grain free, enclosed in the scale and palet. [In honor of Georg Ludwig Koeler, German botanist.]

About 15 species of wide geographic distribution. The following, which may contain two forms, occurs in North America. Type species: *Poa nitida* Lam.

1. Koeleria cristàta (L.) Pers. Koeler's-grass. Crested Hair-grass. Fig. 589.

Aira cristata L. Sp. Pl. 63. 1753.
Koeleria cristata Pers. Syn. 1 : 97. 1805.
Koeleria nitida Nutt. Gen. 1 : 74. 1818.
Koeleria cristata var. *gracilis* A. Gray, Man. 591. 1848.

Culms 1°–2½° tall, erect, simple, rigid, smooth, often pubescent just below the panicle. Sheaths often shorter than the internodes, smooth or scabrous, sometimes hirsute; ligule ½″ long; blades 1′–12′ long, ½″–1½″ wide, erect, flat or involute, smooth or rough, often more or less hirsute; panicle 1′–7′ in length, pale green, usually contracted or spike-like, the branches erect or rarely ascending, 1′ long or less; spikelets 2–5-flowered, 2″–3″ long, the scales rough, acute, the empty ones unequal; flowering scales 1½″–2″ long, shining.

In dry sandy soil, especially on prairies, Ontario to British Columbia, south to Pennsylvania, Texas and California. Also in Europe and Asia. Very variable. Prairie June-grass. July–Sept.

83. CATABRÒSA Beauv. Agrost. 97. *pl. 19. f. 8.* 1812.

A perennial grass, with soft flat leaf-blades and an open panicle. Spikelets usually 2-flowered. Two lower scales empty, thin-membranous, much shorter than the flowering ones, unequal, rounded or obtuse at the apex; flowering scales membranous, erose-truncate. Palet barely shorter than the scale. Stamens 3. Styles distinct. Stigmas plumose. [Greek, in allusion to the erose top of the flowering scales.]

A monotypic genus of arctic and mountainous regions of the northern hemisphere. Type species: *Aira aquatica* L.

1. Catabrosa aquática (L.) Beauv. Water Whorl-grass. Fig. 590.

Aira aquatica L. Sp. Pl. 64. 1753.
Catabrosa aquatica Beauv. Agrost. 157. 1812.

Smooth and glabrous, culms 4′–2° tall, erect, from a creeping base, bright green, flaccid. Sheaths usually overlapping, loose; ligule 1½″–2½″ long; blades 1½″–5′ long, 1″–3″ wide, flat, obtuse; panicle 1′–8′ in length, open, the branches whorled, spreading or ascending, very slender, ½′–2′ long; spikelets 1¼″–1¾″ long, the empty scales rounded or obtuse, the first about half as long as the second, which is crenulate on the margins; flowering scales 1″–1¼″ long, 3-nerved, erose-truncate at the apex.

In water or wet soil, Labrador and Newfoundland to Alaska, south to Nova Scotia, Nebraska and Colorado. Also in Europe and Asia. Water-grass, Water Hair-grass. Summer.

84. MÉLICA L. Sp. Pl. 66. 1753.

Perennial grasses, with usually soft flat leaf-blades and contracted or open panicles. Spikelets 1–several-flowered, often secund, the rachilla extended beyond the flowers and usually bearing 2–3 empty club-shaped or hooded scales, convolute around each other. Two lower scales empty, membranous, 3–5-nerved; flowering scales larger, rounded on the back, 7–13-nerved, sometimes bearing an awn, the margins more or less scarious; palets broad, shorter than the scales, two-keeled. Stamens three. Styles distinct. Stigmas plumose. Grain free, enclosed in the scale and palet. [Name used by Theophrastus for Sorghum; said to be in allusion to the sweet culms of some species.]

About 30 species, inhabiting temperate regions. Besides the following, some 15 others occur in the Rocky Mountains and on the Pacific Coast. Type species: *Melica ciliata* L.

Terminal scales of the spikelet differing in shape from those below, forming a hood-shaped mass
    which is much shorter than the other scales.
    Spikelets 2-flowered, the second empty scale nearly as long as the spikelet, the flowering scales
        terminating on the same plane.   1. *M. mutica.*
    Spikelets usually 3-flowered, the second empty scale much shorter than the spikelet, the second
        flowering scale terminating beyond the apex of the first.   2. *M. nitens.*
Terminal scales like the others in shape, forming a convolute but not hood-shaped mass, which equals
    or extends beyond the apex of the other scales.   3. *M. Porteri.*

1. Melica mùtica Walt. Narrow Melic-grass. Fig. 591.

Melica mutica Walt. Fl. Car. 78. 1788.
Melica mutica var. *glabra* A. Gray, Man. Ed. 5, 626.
    1867.
M. altissima Walt. Fl. Car. 78. 1788. Not L. 1753.
Melica diffusa Pursh, Fl. Am. Sept. 77. 1814.
Melica mutica var. *diffusa* A. Gray, Man. Ed. 5, 626.
    1867.

Culms 1°–3° tall, erect, usually slender, simple, smooth and glabrous. Sheaths often overlapping, rough; ligule 1″–2″ long; blades rough, 4′–9′ long, 1″–5″ wide; panicle 3½′–10½′ in length, narrow, the branches spreading or ascending, 1′–2′ long; spikelets about 2-flowered, 3½″–4½″ long, nodding, on more or less flexuous pubescent pedicels; empty scales very broad, acutish to obtuse, the first shorter than the second, which is nearly as long as the spikelet or sometimes equals it; flowering scales 3″–4″ long, generally very obtuse, scabrous.

In rich soil, Maryland to Wisconsin, south to Florida and Texas. June–July.

2. Melica nìtens Nutt. Tall Melic-grass. Fig. 592.

Melica nitens Nutt.; Piper, Bull. Torr. Club **32** : 387. 1905.

Culms 1½°–4° tall, erect, simple, smooth and glabrous. Sheaths shorter than the internodes, the lower often overlapping; ligule 1″–2″ long; blades 4′–7′ long, 2″–4″ wide, rough; panicle 6½′–8½′ in length, open, the branches spreading or ascending, the lower 1½′–3′ long; spikelets usually numerous, about 3-flowered, 4½″–5½″ long, nodding, on slender, more or less flexuous pubescent pedicels; empty basal scales very broad, obtuse or acutish, the first shorter than the second, which is generally much exceeded by the spikelet; flowering scales 3½″–4½″ long, acute or obtuse, scabrous.

Woods and cliffs, Pennsylvania to Nebraska and Texas. Erroneously called *Melica diffusa* Pursh, in our first edition, that name proving to be a synonym of the preceding species. May–June.

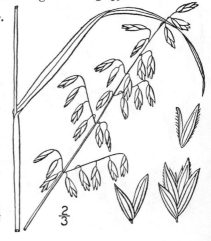

3. **Melica Pòrteri** Scribn. Small Melic-grass. Fig. 593.

Melica mutica var. *parviflora* Porter; Porter & Coulter, Fl. Colo. 149. 1874.
Melica Porteri Scribn. Proc. Acad. Phila. **1885**: 44. *pl. 1. f. 17, 18.* 1885.
M. parviflora Scribn. Mem. Torr. Club **5**: 50. 1894.

Culms 1½°–2½° tall, erect, simple, smooth and glabrous. Sheaths short, overlapping, more or less rough; ligule 1″ long; blades 5′–9′ long, 1″–2″ wide, rough; panicle 5′–7′ in length, contracted, the branches erect, the lower 1′–2′ long; spikelets few, 4–5-flowered, 5″–6½″ long, nodding, on somewhat flexuous strongly pubescent pedicels; lower scales obtuse or acutish, the first shorter than the second, which is much exceeded by the spikelet; flowering scales 3½″–4″ long, acutish, scabrous.

Cliffs and hillsides, Iowa to Missouri, Colorado, Arizona and Texas.

85. **KORYCÀRPUS** Zea, Act. Matrit. 1806.

[Diarina Raf. Journ. Bot. **2**: 169. 1809.]
[Diarrhena Beauv. Agrost. 142. 1812.]

Erect grasses, with long flat leaf-blades and narrow paniculate or racemose inflorescence. Spikelets 3–5-flowered, the rachilla readily disarticulating between the flowers. Upper scales empty, convolute. Two lower scales empty, the first narrow, 3-nerved, acute, the second broader, 5-nerved; flowering scales broader than the lower ones, acuminate or mucronate, rounded on the back, finally coriaceous and shining, 3-nerved. Palet 2-keeled. Stamens 2, rarely 1. Styles short, distinct. Stigmas plumose. Grain beaked, free. [Greek, in allusion to the beaked grain.]

Two known species, the following North American, the other Japanese. Type species: *Korycarpus arundinaceus* Zea.

1. **Korycarpus arundinàceus** Zea. American Korycarpus. Fig. 594.

Festuca diandra Michx. Fl. Bor. Am. **1**: 67. *pl. 10.* 1803. Not Moench, 1794.
Korycarpus arundinaceus Zea, Act. Matrit. 1806.
Diarrhena americana Beauv. Agrost. 142. *pl. 25. f. 11.* 1812.
Korycarpus diandrus Kuntze, Rev. Gen. Pl. 772. 1891.

Culms 1½°–4° tall, erect, simple, very rough below the panicle. Sheaths overlapping, confined to the lower part of the culm, smooth or a little rough at the summit, sometimes pubescent; ligule very short; blades 8′–24′ long, 5″–9″ wide, long-acuminate at the apex, usually scabrous; panicle often reduced to a raceme, 2′–7½′ in length, the branches erect, 1′–2′ long; spikelets 3–5-flowered, 6″–8″ long, the lower scales unequal, the first shorter than the second, which is much exceeded by the spikelet; flowering scales somewhat abruptly acuminate; palets shorter than the scales and exceeded by the beaked grain.

In rich woods, Ohio to South Dakota, south to Georgia and Texas. Aug.–Sept.

86. **PLEUROPÒGON** R. Br. App. Parry's Voy. 289. 1824.

Erect grasses with flat leaf-blades and racemose inflorescence. Spikelets 5–14-flowered; flowers perfect, or the upper staminate. Two lower scales empty, unequal, thin-membranous, 1-nerved, or the second imperfectly 3-nerved; flowering scales longer, membranous, 7-nerved, the middle nerve excurrent as a short point or awn. Palet scarcely shorter than the scale, 2-keeled, the keels winged or appendaged. Stamens 3. Styles short. Stigmas plumose. Grain free, enclosed in the scale and palet. [Greek, side-beard, from the appendages to the palets.]

Three known species, the following arctic, the others Californian. Type species: *Pleuropogon Sabinii* R. Br.

1. Pleuropogon Sabínii R. Br.   Sabine's Pleuropogon.   Fig. 595.

P. Sabinii R. Br. App. Parry's Voy. 289.   1824.

Smooth, culms 6' or less tall, erect, simple, glabrous. Sheaths one or two; ligule 1″ long; blades ¼′-1′ long, erect, glabrous; raceme 1′-2′ in length; spikelets 3-6, 5-8-flowered, about 5″ long, on spreading or reflexed pedicels 1″ or less in length; lower scales smooth, the first acute, shorter than the obtuse second; flowering scales oblong, 2″-2½″ long, erose-truncate at the scarious summit, scabrous, the midnerve sometimes excurrent as a short point; palet slightly shorter than the scale, truncate and somewhat 2-toothed at the apex, bearing an awn-like appendage on each keel near the middle.

Arctic regions of both the Old World and the New. Summer.

87.   UNÍOLA L. Sp. Pl. 71.   1753.

Erect and often tall grasses with flat or convolute leaf-blades and paniculate inflorescence. Spikelets 3–many-flowered, flat, 2-edged, the flowers perfect, or the upper staminate. Scales flattened, keeled, sometimes winged, rigid, usually acute; the lower 3-6 empty, unequal; the flowering scales many-nerved, the uppermost scales often smaller and empty; palets rigid, 2-keeled. Stamens 1-3. Styles distinct. Stigmas plumose. Grain compressed, free, loosely enclosed in the scale and palet. [Name diminutive of *unus*, one, of no obvious application.]

About 10 species, natives of America. Besides the following, 2 others occur in the southeastern United States. Type species: *Uniola paniculata* L.

Spikelets about ¼′ in length; panicle spike-like.   1. *U. laxa.*
Spikelets exceeding ½′ in length; panicle open.
   Panicle lax, the branches pendulous; spikelets on long capillary pedicels.   2. *U. latifolia.*
   Panicle strict, the branches erect, rigid; spikelets on short stout pedicels.   3. *U. paniculata.*

1. Uniola láxa (L.) B.S.P.   Slender Spike-grass.   Fig. 596.

Holcus laxus L. Sp. Pl. 1048.   1753.

Uniola gracilis Michx. Fl. Bor. Am. 1 : 71.   1803.

Uniola laxa B.S.P. Prel. Cat. N. Y. 69.   1888.

Smooth and glabrous, culms 1½°-4° tall, erect, simple, slender. Sheaths shorter than the internodes; ligule very short; blades 5′-15′ long, 1″-3″ wide, usually erect, flat, attenuate into a long tip, smooth or slightly rough; panicle spike-like, 4′-12′ in length, erect, strict, or nodding at the summit, the branches erect, 1′-2′ long; spikelets short-stalked or nearly sessile, 3-6-flowered, about 3″ long; lower scales much shorter than the flowering ones, which are 1½″-2″ long, acuminate, spreading in fruit; palet arched, about two-thirds as long as the scale; stamen 1.

Sandy soil, Long Island to Kentucky, south to Florida and Texas, mostly near the coast. Ascends to 900 ft. in North Carolina. Union-grass. Aug.–Sept.

2. Uniola latifòlia Michx.   Broad-leaved
        Spike-grass.   Fig. 597.

U. latifolia Michx. Fl. Bor. Am. 1: 70.   1803.

Culms 2°–5° tall, erect, simple, smooth and
glabrous. Sheaths shorter than the internodes;
ligule ¼″ long, lacerate-toothed; blades 4′–9′
long, ¼′–1′ wide, flat, narrowed into a somewhat
rounded, often ciliate base, acuminate at the
apex, smooth, excepting on the margins; panicle
lax, 5¾′–10′ in length, its branches filiform and
pendulous, the lower 2′–5′ long; spikelets many-
flowered, oblong to ovate, ¾′–1¼′ long, on long
capillary pendulous pedicels; lower scales much
smaller than the flowering ones, which are 4½″–6″
long, ciliate-hispid on the winged keel; stamen 1.

In moist places, Pennsylvania and Delaware to
Kansas, south to Florida and Texas. Ascends to
2000 ft. in North Carolina. Wild Oats. Aug.–Sept.

3. Uniola paniculàta L.   Sea Oats.   Spike-grass.   Beach-grass.   Fig. 598.

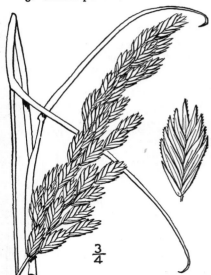

Uniola paniculata L. Sp. Pl. 71.   1753.

Glabrous throughout, culms 3°–8° tall, erect,
simple, smooth. Sheaths often longer than the
internodes; ligule a ring of hairs about ½″ long;
blades 1° long or more, about ¼′ wide, involute
when dry, attenuate into a long slender tip; panicle 9′–1° in length or more, the branches erect
or ascending, strict, rigid, the lower 2½′–5′ long;
spikelets many-flowered, short-pedicelled, ovate to
oval when mature, ½′–1′ long; lower scales much
shorter than the flowering ones, which are 4″–5″
long and scabrous on the keels; stamens 3.

In sands of the seacoast, Virginia to Florida and
west to Texas. Also in the Bahamas and other West
Indies and South America. Spikelets persistent into
the winter. Seaside Oats. Oct.–Nov.

88.   DÍSTICHLIS Raf. Journ. Phys. **89**: 104.   1819.

Dioecious grasses, with rigid culms creeping or decumbent at the base, flat or convolute
leaf-blades and spike-like paniculate inflorescence. Spikelets flattened, more numerous on
the staminate plants than on the pistillate, 6–16-flowered; rachilla continuous in the staminate
spikelets, articulated in the pistillate. Two lower scales empty, narrow, keeled, acute, shorter
than the flowering ones; flowering scales broader, many-nerved, acute, rigid; palets 2-keeled.
Stamens 3. Styles thickened at the base, rather long, distinct. Stigmas long-plumose. Grain
free, enclosed in the scale and palet. [Greek, signifying two-ranked, probably in reference
to the spikelets.]

Four known species, natives of America, inhabiting the seacoast or alkaline soil; one of them is
also found in Australia. Type species: *Distichlis maritima* Raf.

1. **Distichlis spicàta** (L.) Greene.   Marsh Spike-grass.   Alkali-grass.   Salt-grass.
Fig. 599.

Uniola spicata L. Sp. Pl. 71.   1753.
Distichlis maritima Raf. Journ. Phys. **89**: 104.   1819.
Uniola stricta Torr. Ann. Lyc. N. Y. **1**: 155.   1824.
D. spicata Greene, Bull. Cal. Acad. **2**: 415.   1887.
Distichlis spicata var. *stricta* Scribn. Mem. Torr. Club
   5: 51.   1894.

Glabrous throughout, culms 3′-2° tall, erect from
a horizontal rootstock, or often decumbent at the
base. Sheaths overlapping and often crowded; ligule
a ring of very short hairs; blades ½′-6′ long, 1″-2″
wide, flat or involute; panicle dense and spike-like,
¾′-2½′ in length, the branches 1′ long or less, erect;
spikelets 6–16-flowered, 4″-9″ long, pale green; empty
scales acute, the first 1–3-nerved, two-thirds as long
as the 3–5-nerved second one; flowering scales 1½″-
2½″ long, acute or acuminate.

On salt meadows along the coast from Nova Scotia
to Texas, in saline soil throughout the interior, and
on the Pacific Coast north to British Columbia. Also
in the Bahamas and other West Indies. The main figure
is that of the staminate plant. June–Sept.

89. BRÌZA L. Sp. Pl. 70.   1753.

Annual or perennial grasses, with flat or convolute leaf-blades and open or rarely contracted panicles. Spikelets large, flattened, tumid, many-flowered, nodding, the flowers perfect.
Scales thin-membranous, strongly concave, the 2 lower empty, 3–5-nerved, somewhat unequal;
flowering scales imbricated, broader than the empty ones, 5–many-nerved; uppermost scales
often empty; palets much shorter than the scales, hyaline, 2-keeled or 2-nerved. Stamens 3.
Styles distinct.   Stigmas plumose.   Grain usually free, enclosed in the scale and palet.   [Greek
name for some grain, perhaps rye.]

About 12 species, natives of the Old World and temperate South America.   Type species:
Briza minor L.

Perennial; ligule ½″ long or less, truncate; spikelets 5–12-flowered, 2″-2½″ long.   1. *B. media.*
Annual; ligule 1″ long or more, acute; spikelets 3–6 flowered, 1″-1½″ long.   2. *B. minor.*

1. **Briza mèdia** L.   Quake-grass.   Quaking-
grass.   Fig. 600.

Briza media L. Sp. Pl. 70.   1753.

Smooth and glabrous, culms 6′-2° tall, erect,
from a perennial root, simple. Sheaths shorter
than the internodes; ligule ½″ long or less, truncate; blades 1′-3′ long, 1″-2½″ wide; panicle 1½″-5′
in length, the capillary branches spreading or as-
sending, 1′-2½′ long; spikelets 2″-2½″ long, orbicular to deltoid-ovate, 5–12-flowered; scales scarious-
margined, the lower ones about 1″ long; flowering
scales 1″-1½″ long, broader than the lower ones,
widely spreading.

In fields and waste places, Ontario to Massachu-
setts and Rhode Island.   Naturalized from Europe.
Native also of Asia. Maidenhair, Shakers, Cow-quake,
Lady's-hair, Wag-wanton.   Pearl-, Fairy-, Dodder-
Dithering- or Jockey-grass.   June–July.

2. Briza mìnor L. Lesser or Smaller
Quaking-grass. Fig. 601.

Briza minor L. Sp. Pl. 70. 1753.

Smooth and glabrous, culms 4′–15′ tall, erect
from an annual root, simple. Sheaths shorter
than the internodes; ligule 1″–3″ long, acute;
blades 1′–5′ long, 1″–4″ wide, sometimes scabrous;
panicle 2′–5′ in length, open, the capillary branches
spreading or ascending, 1′–2½′ long; spikelets
3–6-flowered, 1″–1½″ long, about 2″ broad, truncate at the base; scales scarious-margined, the
lower ones about 1″ long; flowering scales much
broader and deeply saccate, about ¾″ long.

In ballast and waste places, New Jersey to Virginia; common in California, and widely distributed
in tropical America. Adventive or naturalized from
Europe. June–July.

90. DÁCTYLIS L. Sp. Pl. 71. 1753.

Tall perennial grasses, with flat leaf-blades and paniculate inflorescence. Spikelets
3–5-flowered, short-pedicelled, in dense capitate clusters, the flowers perfect or the upper
staminate. Two lower scales empty, thin-membranous, keeled, unequal, mucronate; flowering
scales larger than the empty ones, rigid, 5-nerved, keeled, the midnerve extended into a point
or short awn; palets shorter than the scales, 2-keeled. Stamens 3. Styles distinct. Stigmas
plumose. Grain free, enclosed in the scale and palet. [Name used by Pliny for some grass
with finger-like spikes.]

A genus of several species, natives of Europe and Asia. Type species: *Dactylis glomerata* L.

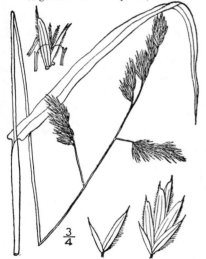

1. Dactylis glomeràta L. Orchard-grass.
Cock's-foot. Fig. 602.

Dactylis glomerata L. Sp. Pl. 71. 1753.

Culms 2°–4° tall, tufted, erect, simple, smooth
and glabrous. Sheaths shorter than the internodes,
smooth or rough; ligule 1″–2″ long; blades 3′–9′
long, 1″–3″ wide, flat, scabrous; panicle 3′–8″ in
length, the branches spreading or ascending in
flower, erect in fruit, the lower 1′–2½′ long, spikelet-
bearing from above or below the middle; spikelets
in dense capitate clusters, 3–5-flowered; lower
scales 1–3-nerved, the first shorter than the second;
flowering scales 2″–3″ long, rough, pointed or short-
awned, ciliate on the keel.

In fields and waste places, New Brunswick to
British Columbia, south to Florida and California.
Naturalized from Europe and cultivated for fodder.
Dew- or Hard-grass. June–July.

91. CYNOSÙRUS L. Sp. Pl. 72. 1753.

Annual or perennial tufted grasses, with flat leaf-blades and dense spike-like inflorescence.
Spikelets of two kinds, in small clusters; lower spikelets of the clusters consisting of narrow
empty scales, with a continuous rachilla, the terminal spikelets of 2–4 broader scales, with an
articulated rachilla and subtending perfect flowers. Two lower scales in the fertile spikelets
empty, 1-nerved, the flowering scales broader, 1–3-nerved, pointed or short-awned; upper
scales narrower, usually empty. Scales of the sterile spikelets pectinate, spreading, all empty,
linear-subulate, 1-nerved. Stamens 3. Styles distinct, short. Stigmas loosely plumose. Grain
finally adherent to the palet. [Greek, signifying dog's tail, referring to the spike.]

About 5 species, natives of the Old World. Type species: *Cynosurus cristatus* L.

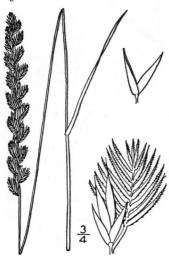

$\dfrac{3}{4}$

1. Cynosurus cristàtus L.   Dog's-tail-grass.
Fig. 603.

Cynosurus cristatus L. Sp. Pl. 72. 1753.

Culms 1°–2½° tall, erect, slender, simple, smooth and glabrous. Sheaths shorter than the internodes; ligule ½″ long, truncate; blades 1½′–5′ long, ½″–2″ wide, smooth, glabrous; spike-like panicle 2′–4′ in length, 2½″–6″ wide, long-exserted; spikelets arranged in clusters, the terminal fertile, the lower larger and sterile; scales of the former about 1½″ long, pointed or short-awned, the scales of the sterile spikelets very narrow, pointed, strongly scabrous on the keel.

In fields and waste places, Newfoundland to Ontario, southern New York and New Jersey. Adventive from Europe. Hendon Bent, or Crested Dog's-tail-grass. Leghorn-straw-grass. June–Aug.

92. PÒA L. Sp. Pl. 67. 1753.

Annual or perennial grasses with flat or convolute leaves and contracted or open panicles. Spikelets 2–6-flowered, compressed, the rachilla usually glabrous; flowers perfect, or rarely dioecious. Scales membranous, keeled; the 2 lower empty, 1–3-nerved; the flowering scales longer than the empty ones, generally with a tuft of cobwebby hairs at the base, 5-nerved, the marginal nerves usually pubescent, often also the dorsal one; palets a little shorter than the scales, 2-nerved or 2-keeled. Stamens 3. Styles short, distinct. Stigmas plumose. Grain free, or sometimes adherent to the palet. [Name Greek, for grass or herbage.]

A genus of about 150 species, widely distributed in all temperate and cold regions. The English name *Meadow-grass* is often applied to most of the species. Besides the following some 50 others occur in the western parts of North America. Type species: *Poa pratensis* L.

Annuals.
    Flowering scales distinctly 5-nerved, not webby at the base.   1. *P. annua.*
    Flowering scales 3-nerved, or obscurely 5-nerved, webby at base.   2. *P. Chapmaniana.*
Perennials.
    Innovations extravaginal, plants often bearing rootstocks or stolons.
        Culms tufted, usually densely so.
            Rootstock slender; plants often stoloniferous.
            Culms less than 8′ tall; low arctic or alpine grasses.
                Flowering scales pubescent all over.   3. *P. abbreviata.*
                Flowering scales glabrous, or slightly pubescent.   4. *P. laxa.*
        Culms exceeding 8′ tall.
            Flowering scales not webby at the base.
                Flowering scales glabrous below between the nerves.   6. *P. glauca.*
                Flowering scales pubescent below between the nerves.
                      Panicle narrow, contracted, its branches short, erect.   22. *P. arida.*
                      Panicle open, branches long, flexuous, spreading.   13. *P. autumnalis.*
            Flowering scales webby at the base.
                Flowering scales glabrous; culms manifestly compressed.   9. *P. debilis.*
                Flowering scales somewhat pubescent; culms round or little compressed.
                    Lateral nerves of the flowering scale glabrous.
                      Plant yellowish green; flowering scales 1¼″ long 8. *P. trivialis.*
                      Plant green; flowering scales 1¾″ long.   15. *P. alsodes.*
                  Lateral nerves of the flowering scale pubescent.
                    Lower half of the flowering scales densely villous between the nerves;
                      arctic grass.   18. *P. cenisia.*
                    Lower half of the flowering scales glabrous between the nerves (some-
                    times somewhat pubescent in No. 14).
                      Spikelets 2″ long or less; panicle-branches dividing and spikelet-
                      bearing at or below the middle.
                      Intermediate nerves of the flowering scale obscure.
                        Panicle erect, rarely exceeding 5′, branches ascending.
                          10. *P. crocata.*
                        Panicle drooping, up to 1°, or more, branches spreading.
                          11. *P. triflora.*
                    Intermediate nerves prominent.
                      Midnerve of the flowering scales pubescent only below; spike-
                      lets crowded on the branches.   12. *P. pratensis.*
                    Midnerve pubescent its whole length; spikelets scattered on
                    the spreading often reflexed branches.   14. *P. sylvestris.*
                Spikelets 2½″ long or more; panicle-branches usually dividing and
                spikelet-bearing only at the end.
                  Flowering scales very webby at the base, nerves strongly pilose;
                    innovation leaves much shorter than culms.   16. *P. Wolfii.*
                  Flowering scales little webby; nerves sparsely pilose; innovation
                  leaves equalling or exceeding culms.   17. *P. brachyphylla.*
    Rootstocks short and stout; no stolons.

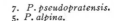

Ligule acute, 2″ long ; flowering scales lanceolate, 1½″ long.   7. *P. pseudopratensis.*
Ligule truncate, 1″ long ; flowering scales ovate, 2″ long.   5. *P. alpina.*
Culms not tufted ; rootstocks long, creeping.
    Flowering scales naked or with short hairs at the base.
        Culms compressed, slender ; plant bluish green ; spikelets 1½″–3″.   19. *P. compressa.*
        Culms not compressed, stout ; spikelets above 5″ long.   21. *P. eminens.*
    Flowering scales with webby hairs at the base longer than scale.   20. *P. arachnifera.*
Innovations intravaginal, hence plants without rootstocks or stolons.
    Flowering scales strigose below, hispidulous above.   23. *P. Buckleyana.*
    Flowering scales hispidulous all over.
        Ligules 1½″–2″ long, glabrous.   24. *P. laevigata.*
        Ligules ½″–1″ long, hispidulous on the outside.   25. *P. confusa.*

1. Poa ánnua L.   Annual or Dwarf Meadow-grass.   Low Spear-grass.   Fig. 604.

Poa annua L. Sp. Pl. 68. 1753.

Culms 2′–1° tall, from an annual root, erect or decumbent at the base, somewhat flattened, smooth. Sheaths loose, usually overlapping ; ligule about 1″ long ; blades ½′–4′ long, ¾″–1½″ wide, smooth ; panicle ½′–4′ in length, open, branches spreading, ¼′–1½′ long, naked at the base ; spikelets 3–5-flowered, 1½″–2½″ long ; lower scales smooth, the first narrow, acute, 1-nerved, about two-thirds as long as the broad and obtuse 3-nerved second one ; flowering scales 1¼″–2½″ long, distinctly 5-nerved, the nerves pilose below.

In waste and cultivated places nearly throughout North America. Naturalized from Europe. Native also of Asia. May-, Six-weeks- or Causeway-grass. May–Oct.

2. Poa Chapmaniàna Scribn.   Chapman's Spear-grass.   Fig. 605.

P. cristata Chap. Fl. S. States, 562. 1860. Not Walt. 1788.
P. Chapmaniana Scribn. Bull. Torr. Club **21** : 38. 1894.

Culms 3′–6′ tall, erect from an annual root, simple, rigid, smooth and glabrous. Sheaths tight, mostly at the base of the culm ; ligule ½″ long, truncate ; blades ½′–1′ long, ½″ wide or less, smooth ; panicle 1′–2′ in length, the branches usually erect, sometimes spreading or ascending, ¾′ long or less, naked at the base ; spikelets 3–7-flowered, 1¼″–1½″ long ; lower scales about equal, 3-nerved, acute ; flowering scales webbed at the base, obtuse, 3-nerved, sometimes with two additional obscure nerves, the prominent ones sometimes pilose for three-fourths their length.

In dry soil, Virginia to Iowa, south to Florida and Mississippi. April–May.

3. Poa abbreviàta R. Br.   Low Spear-grass.   Fig. 606.

Poa abbreviata R. Br. Bot. App. Parry's Voy. 287. 1824.

Culms 6′ tall or less, erect, simple, smooth and glabrous. Sheaths and leaves crowded at the base of the culm ; ligule ½″ long ; blades ½′–1′ long, ½″ wide ; panicle contracted, ½′–1′ long, branches very short and erect ; spikelets 3–5-flowered, 2½″ long ; lower scales acute, smooth and glabrous ; flowering scales about 1½″ long, obtuse, strongly pubescent all over, the intermediate nerves very obscure.

Arctic America from Greenland and Labrador to the Pacific. Summer.

4. Poa láxa Haenke. Wavy Meadow-grass. Mountain Spear-grass. Fig. 607.

P. laxa Haenke, in Jirasek, Beob. Riesengeb. 118. 1791.

Smooth and glabrous, culms 1° tall or less, erect, simple. Sheaths often overlapping; ligule about 1″ long; blades 1′–3′ long, ½″–1″ wide, acuminate; panicle 1′–3′ in length, the branches usually erect, sometimes ascending, 1′ long or less; spikelets 3–5-flowered, 2″–2½″ long; lower scales usually 3-nerved, acute, glabrous, rough on the keel at its apex; flowering scales 1½″–1¾″ long, obtuse, 3-nerved, or sometimes with an additional pair of obscure nerves, the midnerve pilose on the lower half, rough above, the lateral ones pilose for one-third their length.

Greenland to Alaska, south to the high mountains of New England and New York. Also in Europe and Asia. Summer.

5. Poa alpìna L. Alpine or Mountain Spear-grass. Fig. 608.

Poa alpina L. Sp. Pl. 67. 1753.

Smooth and glabrous, culms 4′–18′ tall, erect, simple. Sheaths shorter than the internodes; ligule 1″ long, truncate; blades 1′–3′ long, 1″–2″ wide, abruptly acute; panicle 1′–3′ in length, the branches generally widely spreading, 1′ long or less; spikelets 3–5-flowered, 2½″–3″ long; lower scales broad, glabrous, rough on the keel, acute; flowering scales about 2″ long, obtuse, pilose for half their length, pubescent between the nerves toward the base.

Newfoundland and Labrador to Alaska, south to Nova Scotia, Quebec, Lake Superior and Washington, and in the Rocky Mountains to Colorado. Also in Europe and Asia. Summer.

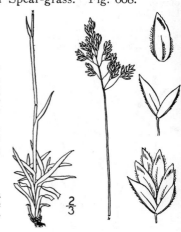

6. Poa glaùca Vahl. Glaucous Spear-grass. Fig. 609.

Poa glauca Vahl, Fl. Dan. *pl. 964.* 1790.
Poa caesia J. E. Smith, Eng. Bot. *pl. 1719.* 1807.

Culms 6′–2° tall, erect, rigid, glabrous, somewhat glaucous. Sheaths overlapping, confined to the lower half of the culm; ligule 1″ long; blades 1′–2′ long, 1″ wide or less, smooth beneath, scabrous above; panicle 1′–3′ in length, open, the branches erect or ascending, ½′–1½′ long; spikelets 2–4-flowered, 2½″–3″ long; empty basal scales acute, 3-nerved, glabrous, rough on the upper part of the keel; flowering scales 1½″–1¾″ long, obtuse or acutish, rough, not webbed at the base, the lower half of the midnerve and marginal nerves silky-pubescent, the intermediate nerves obscure and occasionally sparingly pubescent at the base.

Greenland and Labrador to Maine and White Mountains of New Hampshire and Minnesota. Also in Europe. July and August.

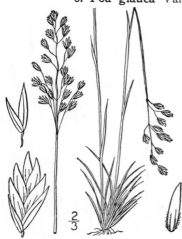

7. **Poa pseudopraténsis** Scribn. & Ryd. Prairie Meadow-grass. Fig. 610.

Poa pseudopratensis Scribn. & Rydb. Contr. Nat. Herb. 3: 531. 1896.

Culms 1°–2½° tall, erect, simple, smooth and glabrous. Sheaths shorter than the internodes, smooth or slightly rough; ligule 2″ long, acute, decurrent; blades 1″–3″ wide, smooth beneath, a little rough above and on the margins, those of the culm 1′–3½′ long, the basal 6′–10′ in length; panicle 2′–5′ long, open, the branches spreading or ascending, 1′–2′ long; spikelets 3–5-flowered, 3″–4″ long, exceeding their pedicels; lower scales nearly equal, acute, 3-nerved; flowering scales acutish, about 1½″ long, rough above, 5-nerved, pubescent between the nerves below, the marginal nerves and midnerve silky-pubescent about half their length.

Manitoba and Assiniboia to Nebraska and Colorado.

8. **Poa triviàlis** L.　Rough-stalked Meadow-grass. Fig. 611.

Poa trivialis L. Sp. Pl. 67. 1753.

Culms 1°–3° tall, usually more or less decumbent at the base, simple, smooth or slightly scabrous. Sheaths usually shorter than the internodes, rough; ligule 2″–3″ long, acutish; blades 2′–7′ in length, 1″–2″ wide, generally very rough; panicle 4′–6′ long, open, the branches usually spreading or ascending, 1′–2′ long; spikelets 2- or sometimes 3-flowered, 1½″ long, exceeding their pedicels; scales acute, the empty basal ones rough on the keel, the lower 1-nerved, shorter than the 3-nerved upper; flowering scales 1″–1½″ long, webbed at the base, 5-nerved, the midnerve silky-pubescent below, the lateral nerves naked, the intermediate ones prominent.

In meadows and waste places, Newfoundland to Ontario, South Carolina and Louisiana. Naturalized from Europe. Fowl or Round-stalked Meadow-grass, Natural grass, Bird-grass. June–Aug.

9. **Poa débilis** Torr.　Weak Spear-grass. Fig. 612.

Poa debilis Torr. Fl. N. Y. 2: 459. 1843.

Culms 1°–2½° tall, erect, slender, simple, somewhat flattened, smooth and glabrous. Sheaths compressed, much shorter than the internodes; ligule ½″–1″ long; blades 1′–4½′ long, 1″ wide or less, erect, smooth beneath, rough above; panicle 2′–6′ in length, open, often nodding at the top, the branches erect or ascending, sometimes spreading, 1½′–3′ long; spikelets 2–4-flowered, 1½″–2″ long, their pedicels longer; empty scales unequal, acute, the first 1-nerved, shorter than the 3-nerved second one; flowering scales 1½″ long, obtuse, sparingly webbed at the base, 5-nerved, the nerves naked.

In woods, Quebec and Ontario to Rhode Island, Pennsylvania, Illinois and Iowa. June–Aug.

$\frac{2}{3}$

10. Poa crocàta Michx. Wood Meadow-grass. Northern Spear-grass. Fig. 613.

Poa caesia var. *strictior* A. Gray, Man. Ed. 5, 629. 1867.
P. crocata Michx. Fl. Bor. Am. 1: 68. 1803.

Culms 6′-2° tall, erect, simple, slender, sometimes rigid, smooth and glabrous. Sheaths usually shorter than the internodes; ligule ½″-1″ long, truncate; blades 1′-4′ long, 1″ wide or less, erect, smooth or rough; panicle 2′-5′ in length, open, the branches erect or ascending, rarely spreading, 1′-2′ long; spikelets 2-5-flowered, 1½″-2½″ long; lower scales acute or acuminate, 1-3-nerved; flowering scales obtuse or acute, 1″-1¼″ long, faintly 5-nerved, somewhat webby at base, the midnerve and the marginal nerves silky-pubescent on the lower half.

Labrador to Yukon, Vermont, Minnesota and Alberta, and in the mountains to Colorado and Arizona. June–Aug.

Poa nemoràlis L., a grass of Europe and Asia, may be found as an occasional introduction. It may be distinguished from the above by its much narrower empty scales.

11. Poa triflòra Gilib. False Red-top. Fowl Meadow-grass. Fig. 614.

Poa serotina Ehrh. Beitr. 6: 83. Name only. 1791.
P. triflora Gilib. Exercit. 531. 1792.

Culms 1½°-5° tall, erect, simple or rarely branched, smooth, glabrous. Sheaths usually shorter than the internodes, smooth and glabrous; ligule 1″-2″ long; blades 2′-6′ long, 1″-2″ wide, smooth or rough; panicle 6′-13′ in length, open, the branches spreading or ascending, 2′-5′ long, divided and spikelet-bearing above the middle; spikelets 3-5-flowered, 1½″-2″ long, exceeding their pedicels; lower scales acute, glabrous, rough above on the keel, the lower usually 1-nerved, the upper 3-nerved; flowering scales obtuse, somewhat webby at the base, 1″-1½″ long, silky-pubescent on the lower half of the marginal nerves and the midnerve, the intermediate nerves obscure or wanting.

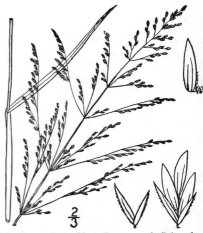

$\frac{2}{3}$

In swampy places, Newfoundland to Vancouver Island, south to New Jersey and Colorado. Also in Europe and Asia. Duck-grass. July–Aug. Formerly confused with *Poa flava* L.

$\frac{2}{3}$

12. Poa praténsis L. Kentucky Blue-grass. June-grass. Fig. 615.

Poa pratensis L. Sp. Pl. 67. 1753.
P. pratensis var. *angustifolia* Kunth, Enum. 1: 353. 1833.

Glabrous, culms 1°-4° tall, from long running rootstocks, erect, simple, smooth. Sheaths often longer than the internodes; ligule ¾″ long or less, truncate; blades smooth or rough, ½″-3″ wide, those of the culm 2′-6′ in length, the basal much longer; panicle 2½′-8′ in length, usually pyramidal, the branches spreading or ascending, sometimes flexuous, 1′-3′ long, divided and spikelet-bearing above the middle; spikelets 3-5-flowered, 2″-2½″ long, exceeding their pedicels; scales acute, the lower unequal, glabrous, rough on the keel, the lower 1-nerved, the upper 3-nerved; flowering scales 1½″ long, webbed at the base, 5-nerved, the marginal nerves and midnerve silky-pubescent below, the intermediate ones naked.

In meadows, fields and woods, almost throughout North America. Widely cultivated for hay and pasture. Also in Europe and Asia. In North America probably indigenous only in the northern and mountainous regions. Variable. Natural or green-grass. Common meadow-grass. June–Aug.

13. **Poa autumnàlis** Muhl.   Flexuous Spear-grass.   Fig. 616.

Poa autumnalis Muhl.; Ell. Bot. S. C. & Ga. 1: 159. 1817.

P. flexuosa Muhl. Gram. 148. 1817. Not J. E. Smith, 1803.

Culms 1°–3° tall, erect, slender, simple, smooth and glabrous. Sheaths usually much shorter than the internodes; ligule ½″ long; blades 1″ wide or less, smooth beneath, rough above, those of the culm 1½′–6′ long, the basal much longer; panicle 3′–9′ in length, the branches long and slender, spikelet-bearing at the extremities, 2′–5′ long; spikelets 3–5-flowered, 2½″–3″ long; empty basal scales acute, the first 1-nerved, narrow, shorter than the broad 3-nerved second; flowering scales rounded or retuse at the apex, 1½″–2″ long, not webbed at the base, pubescent on the lower part, 5-nerved, the midnerve silky-pubescent for three-fourths its length.

In woods, New Jersey to Missouri, south to Florida and Texas. March–May.

14. **Poa sylvéstris** A. Gray.   Sylvan Spear-grass.   Fig. 617.

Poa sylvestris A. Gray, Man. 596. 1848.

Culms 1°–3° tall, erect, slender, simple, slightly flattened, smooth, glabrous. Sheaths shorter than the internodes; ligule ½″ long or less; blades smooth beneath, rough above, 1″–3″ wide, those of the culm 1½′–6′ in length, the basal much longer; panicle 3′–7′ in length, the branches spreading or ascending, often reflexed in age, 1½′–3′ long, spikelet-bearing at the extremities; spikelets 2–4-flowered, 1″–2″ long; empty basal scales acute, the lower 1-nerved, the upper longer and 3-nerved; flowering scales about 1¼″ long, webbed at the base, obtuse, often pubescent below, 5-nerved, the midnerve pubescent nearly its entire length and the marginal nerves below the middle.

In thickets and meadows, New York to Wisconsin, Nebraska, south to Florida and Texas. Branches of the panicle sometimes reflexed when old. June–July.

15. **Poa alsòdes** A. Gray.   Grove Meadow-grass. Fig. 618.

Poa alsodes A. Gray, Man. Ed. 2, 562. 1856.

Culms 8′–2½° tall, erect, slender, simple, smooth and glabrous. Sheaths usually longer than the internodes; ligule ½″ long; blades usually rough, 1″–2″ wide, those of the culm 2′–8′ in length, the basal longer; panicle 3½′–8′ in length, the branches spreading or ascending, 1½″–3′ long, spikelet-bearing at the ends; spikelets 2–3-flowered, about 2½″ long; scales very acute, the empty basal ones unequal, the lower 1-nerved, the upper 3-nerved; flowering scales about 2″ long, webbed at the base, the midnerve pubescent near the base, the marginal nerves naked, the intermediate ones very faint.

In woods and thickets, Quebec to Minnesota, south to North Carolina and Tennessee. May–June.

16. Poa Wólfii Scribn.   Wolf's Spear-grass.
Fig. 619.

Poa Wolfii Scribn. Bull. Torr. Club **21**: 228.   1894.

Culms 2°–3° tall, erect, slender, simple, smooth and glabrous. Sheaths shorter than the internodes; ligule ½″ long; blades 1″ wide or less, smooth beneath, rough above, those of the culm 2′–4′ in length, the basal much longer; panicle 3′–6′ in length, lax, its branches erect or ascending, flexuous, 1½′–2½′ long; spikelets 2–4-flowered, 2½″–3″ long; scales acute, the lower unequal, 3-nerved, glabrous, rough on the keel, the first shorter than the second; flowering scales about 2″ long, copiously webbed at the base, 5-nerved, the marginal and midnerves silky-pubescent for more than half their length, the intermediate nerves prominent, naked.

Illinois to Minnesota and Tennessee.

17. Poa brachyphýlla Schult.   Short-leaved Spear-grass.   Fig. 620.

Poa brevifolia Muhl. Gram. 138.   1817.   Not DC. 1806.
Poa brachyphylla Schult. Mant. **2**: 304.   1824.

Culms 1°–3° tall. Sheaths often shorter than the internodes; ligule ½″–1½″ long; leaves smooth beneath, rough above, 1″–2″ wide, abruptly acute, those of the culm ½′–4′ long, the uppermost sometimes almost wanting; basal leaves usually equalling or nearly as long as the culm; panicle 2½′–5′ in length, open, the branches ascending, widely spreading or often reflexed, 1½′–3′ long, spikelet-bearing at the ends; spikelets 3–6-flowered, 2½″–3½″ long; empty basal scales unequal, acute, glabrous, the lower 1-nerved, the upper 3-nerved; flowering scales slightly webbed at the base, 2″–2½″ long, obtuse, 5-nerved, the keel and marginal nerves sparingly pubescent, the intermediate nerves prominent, naked.

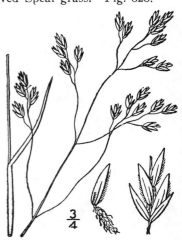

In rocky woods, southern New York to Illinois, Georgia and Tennessee.   April–June.

18. Poa cenísia All.   Arctic Spear-grass.   Fig. 621.

Poa cenisia All. Auct. Fl. Ped. 40.   1789.

Smooth and glabrous, culms 4′–15′ tall, erect, slender, simple. Sheaths shorter than the internodes; ligule 1″ long or less, truncate; blades 1′–4′ long; ½″–1″ wide; panicle 1′–4′ in length, open, the branches generally widely spreading and more or less flexuous, 1′–2½′ long; spikelets 3–5-flowered, 2½″–3½″ long; lower scales acute or acuminate, 1–3-nerved; flowering scales about 2″ long, faintly 5-nerved, the nerves short-pilose on the lower half, minutely pubescent between the nerves, somewhat webbed at the base.

Greenland and Labrador to Alaska, south in the higher peaks of the Rocky Mountains to Colorado.   Also in Europe. Summer.

19. Poa compréssa L.   Wire-grass.   Flat-stemmed Meadow-grass.   English Blue-grass.   Fig. 622.

Poa compressa L. Sp. Pl. 69.   1753.

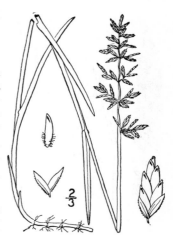

Pale bluish green, glabrous, culms 6'–2° tall, decumbent at the base, from long horizontal rootstocks, smooth, much flattened. Sheaths loose, flattened, shorter than the internodes; ligule ½″ long; blades 1'–4' long, about 1″ wide, smooth beneath, rough above; panicle usually contracted, the branches erect or ascending, 1' long or less, spikelet-bearing nearly to the base; spikelets 3–9-flowered, 1½″–3″ long; lower scales acute, 3-nerved; flowering scales 1″–1¼″ long, obscurely 3-nerved, the nerves sparingly pubescent toward the base.

Waste places and cultivated grounds and woods almost throughout North America. Ascends to 2100 ft. in Virginia. Naturalized from Europe. Native also of Asia. Varies from weak and slender to quite stiff. Squitch-grass. June–Aug.

20. Poa arachnífera Torr.   Texas Blue-grass.   Fig. 623.

Poa arachnifera Torr. Marcy's Exped. 301.   1853.

Culms tufted, 1°–3° tall, smooth and glabrous, from running rootstocks; sterile shoots from one-half as long as the culms to equalling them. Sheaths longer than the internodes, smooth or roughish, hyaline on the margins; ligule a short membranous ring; blades linear, erect, usually folded when dry, smooth beneath, rough above, 1½'–9' long, 1½″–3″ broad, abruptly acute; panicle dense and contracted, sometimes interrupted below, 3'–6½' long, ½'–1½' broad, its branches ascending or erect; spikelets numerous, 4–7-flowered, the scales acuminate, the empty ones hispidulous on the midnerve; flowering scales 2″–2½″ long, often pointed, pubescent at the base with copious long cobwebby hairs, 5-nerved, the midnerve and lateral nerves pilose below the middle.

Prairies, Kansas to New Mexico, south to Louisiana and Texas. Also introduced in Florida. April–May.

21. Poa éminens Presl.   Large-flowered Spear-grass.   Fig. 624.

P. eminens Presl, Rel. Haenk. 1 : 273.   1830.
P. glumaris Trin. Mem. Acad. St. Petersb. (VI.) 1 : 379. 1831.

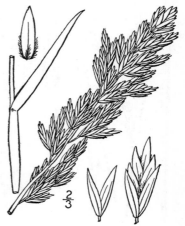

Smooth and glabrous, culms 6'–3° tall, erect or assurgent, simple. Sheaths loose, usually shorter than the internodes; ligule ½″ long, truncate; blades 4'–10' long, 1″–4″ wide; panicle 4'–10' in length, the branches erect or ascending, 1'–2' long; spikelets 3–5-flowered, 4″–6″ long; lower scales about equal, acute, slightly scabrous on the keel, the first 1–3-nerved, the second 3-nerved, rarely 5-nerved; flowering scales 3″–4″ long, usually acutish, scabrous, 5–7-nerved, pubescent at base and on the lower part of the midnerve and lateral nerves, not webbed.

Beaches and shores, Labrador to Quebec, Washington and Alaska.   Summer.

22. Poa árida Vasey. Prairie or Bunch Spear-grass. Fig. 625.

Poa andina Nutt.; S. Wats. Bot. King's Exp. 388. 1871. Not Trin. 1836.
Poa arida Vasey, Contr. U. S. Nat. Herb. **1**: 270. 1893.
Poa pratericola Rydb. & Nash; Rydb. Mem. N. Y. Bot. Gard. **1**: 51. 1900.

Culms 1°–2° tall, erect, rigid, simple, smooth and gla-brous. Sheaths usually overlapping, smooth or somewhat roughish; ligule 1″–2″ long, acute; blades smooth beneath, rough above, ½″–1″ wide, flat or folded, pungently pointed, those of the culm ½′–1′ long, erect, the basal leaves 3′–6′ long; panicle contracted, 2′–5′ in length, the branches erect, spikelet-bearing nearly to the base, 1½′ long or less; spike-lets 4–7-flowered, 2½″–3½″ long; lower scales nearly equal, acute, 3-nerved; flowering scales 1½″–2″ long, erose-trun-cate at apex, strongly silky-pubescent on the nerves for half their length, the lower part very pubescent between the nerves; intermediate nerves very obscure.

On prairies, Kansas to Utah, Wyoming, North Dakota and Manitoba. July–Sept.

23. Poa Buckleyàna Nash. Buckley's Spear-grass. Fig. 626.

Poa tenuifolia Buckley, Proc. Acad. Phila. **1862**: 96. 1862. Not A. Rich. 1851.
Poa Buckleyana Nash, Bull. Torr. Club **22**: 465. 1895.

Culms 6′–2° tall, erect, rigid, simple, smooth and glabrous. Sheaths shorter than the internodes; ligule 2″–3″ long, acute; blades 1′–4′ long, about 1″ wide, erect, flat, or becoming involute, smooth or rough; panicle 1′–4′ in length, contracted, the branches erect, 1½′ long or less, spikelet-bearing nearly to the base; spikelets 2–5-flowered, 2″–3″ long; scales acute, the lower nearly equal, scabrous on the keel; flowering scales about 2″ long, obtuse or acutish, sparingly pubes-cent on the nerves below, sometimes slightly hispid toward the base between the nerves.

South Dakota to Kansas, California and British Columbia. Bunch Red-top. Oregon Blue-grass. July–Aug.

24. Poa laevigàta Scribn. Smooth Spear-grass. Fig. 627.

Poa laevis Vasey, Contr. U. S. Nat. Herb. **1**: 273. 1893. Not Barb. 1877.

Poa laevigata Scribn. Bull. U. S. Dep. Agr. Agrost. **5**: 31. 1897.

Culms densely tufted, 1½°–2½° tall, erect, slender, the innovations 4′–8′ long; sheaths smooth and gla-brous; ligule 1½″–2″ long, glabrous; blades narrow and involute, 4′ long or less; panicle slender, 3′–6′ long, its larger branches 1′–2′ long; spikelets 2½″–3½″ long, 3–4-flowered, the flowering scales 1½″–2″ long, hispidulous all over, obtuse to acutish.

On dry hillsides and in meadows, Quebec to Wash-ington, south in the mountains to Colorado. June–Aug.

25. Poa confùsa Rydb. Tufted Spear-
grass. Fig. 628.

P. confusa Rydb. Bull. Torr. Club **32** : 607. 1906.

Culms densely tufted, 1½°–3° tall, erect, slender;
sheaths smooth and glabrous; ligule ½″–1″ long,
obtuse or acutish, hispidulous on the outside;
blades up to 8′ long, 1″–1½″ wide, flat or involute,
puberulent; panicle narrow, 4′–6′ long, its branches
short and appressed; spikelets about 4″ long, commonly 4-flowered, the empty scales shining, strigu-
lose above, the flowering scales narrow, a little
less than 2″ long, obtuse or rounded at the apex,
rounded on the back below, hispidulous.

In open places and on hills, Nebraska and Montana
to Colorado. June–Aug.

93. DUPÓNTIA R. Br. Parry's Voy. App. 290. 1824.

Low grasses, with flat leaf-blades and generally narrow panicles. Spikelets 2–4-flowered,
the flowers all perfect. Two lower scales empty, extending beyond the flowering scales,
membranous; flowering scales entire, membranous, with a tuft of hairs at the base.
Stamens 3. Styles distinct. Stigmas plumose. [Name in honor of J. D. Dupont, French
botanist.]

Two arctic species, both circumboreal. Type species: *Dupontia Fisheri* R. Br.

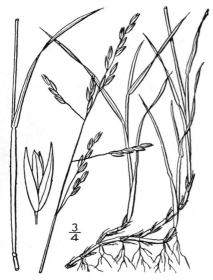

1. Dupontia Físheri R. Br. Fisher's Dupontia. Fig. 629.

D. Fisheri R. Br. Parry's Voy. App. 291. 1824.

Smooth and glabrous, culms 5′–12′ tall, erect,
simple. Sheaths overlapping; ligule 1″ long or
less; blades 1′–6′ long, 1″–2″ wide, flat; panicle
usually contracted, 1½′–3½′ long, the branches
less than 1½′ long, erect, or sometimes ascending; spikelets few, about 2-flowered, 3″–4″ long;
empty basal scales thin, generally acute, the first
1-nerved, somewhat shorter than the second,
which is usually 3-nerved, the lateral nerves often
vanishing at about the middle; flowering scales
2½″–3″ long, 1-nerved or obscurely 3-nerved;
basal hairs about ½″ long.

Arctic regions of North America. Also in arctic
Europe and Asia. Summer.

94. SCOLÓCHLOA Link, Hort. Berol. 1 : 136. 1827.

Tall aquatic or marsh grasses, with flat leaf-blades and ample panicles. Spikelets
2–4-flowered, the flowers perfect. Two lower scales empty, thin-membranous, 3–5-nerved;
flowering scales rigid, with a tuft of hairs at the base, rounded on the back, 5–7-nerved, some
of the nerves usually excurrent as short points; palets about equalling the scales, 2-nerved.
Stamens 3. Styles very short. Stigmas plumose. Grain hairy at the apex. [Greek, referring
to the prickle-like projecting nerves of the flowering scales.]

Species 2, in the north temperate zones of both continents. Type species: *Arundo festucacea*
Willd.

3/4

1. Scolochloa festucàcea (Willd.) Link.
Prickle Fescue.   Fig. 630.

Arundo festucacea Willd. Enum. 1 : 126.   1809.

S. festucacea Link, Hort. Berol. 1 : 137.   1827.

Graphephorum festucaceum A. Gray, Ann. Bot. Soc.
Can. 1 : 57.   1861.

Culms 3°–5° tall, erect, smooth and glabrous.
Sheaths often overlapping; ligule 1″–2″ long;
blades 7′–1° long or more, 2″–4″ wide, flat, scabrous on the margins; panicle 8′–12′ in length,
usually open, the branches ascending, naked at
the base, the lower 3′–4′ long; spikelets 3″–4″
long; empty basal scales acute, the first shorter
than the second; flowering scales scabrous,
7-nerved.

Wet places, Iowa and Nebraska, north to Manitoba
and Saskatchewan.   July–Aug.

95. GRAPHÉPHORUM Desv. Bull. Soc. Philom. 2 : 189.   1810.

Slender erect grasses, with flat leaf-blades and a usually contracted nodding panicle.
Spikelets 2–4-flowered, flattened, the rachilla hirsute and extending beyond the flowers. Two
lower scales empty, somewhat shorter than the flowering scales, thin-membranous, acute,
keeled; flowering scales membranous, obscurely nerved, entire, sometimes short-awned just
below the apex.   Stamens 3.   Styles distinct.   Stigmas plumose.   Grain glabrous.   [Greek,
pencil-bearing, referring to the tuft of hairs at the end of the rachilla.]

Three or four species, natives of northern North America.   Type species : *Aira melicoides* Michx.

1. Graphephorum melicoìdeum (Michx.)
Beauv.   Graphephorum.   Fig. 631.

Aira melicoides Michx. Fl. Bor. Am. 1 : 62.   1803.
Graphephorum melicoideum Beauv. Agrost. 164. *pl. 15.
f. 8.*   1812.
Dupontia Cooleyi A. Gray, Man. Ed. 2, 556.   1852.
Graphephorum melicoides var. *major* A. Gray, Ann.
Bot. Soc. Can. 1 : 57.   1861.

Culms 1°–2½° tall, erect, simple, rough just below
the panicle.   Sheaths usually shorter than the internodes, smooth, or the lower often villous; ligule
1″ long or less, truncate; blades 1½′–9′ long, 1″–2″
wide, long-acuminate, rough; panicle 2′–6′ in length,
the top usually nodding, the branches erect, 1′–2′
long; spikelets 2–4-flowered, 2½″–3″ long; scales
scabrous on the keel, the empty ones unequal, the
first 1-nerved or obscurely 3-nerved, shorter than
the 3-nerved second; flowering scales 3–5-nerved,
acute.

In wet soil, Anticosti Island to Ontario, south to
Maine, Vermont and Michigan.   Aug.–Sept.

3/4

96. PANICULÀRIA Fabr. Enum.
Hort. Helmst. 373.   1763.

[GLYCÈRIA R. Br. Prodr. Fl. Nov. Holl. 1 : 179.   1810.]

Mostly perennial grasses, often tall, with flat leaf-blades and paniculate inflorescence.
Spikelets few–many-flowered, terete or somewhat flattened.   Two lower scales empty, obtuse
or acute, 1–3-nerved; flowering scales membranous, rounded on the back, 5–9-nerved, the
nerves disappearing in the hyaline apex.   Palets scarcely shorter than the scales, rarely
longer, 2-keeled.   Stamens 2 or 3.   Styles distinct.   Stigmas plumose.   Grain smooth, enclosed in the scale and palet, free, or when dry slightly adhering to the latter.   [Latin,
referring to the panicled spikelets.]

About 20 species, widely distributed in North America, a few in Europe and Asia.   Type species :
Poa aquatica L.

Spikelets ovate or oblong, 4″ long or less.
  Flowering scales very broad, obscurely or at least not sharply nerved.
    Panicle open, the branches ascending or spreading, often drooping.
      Spikelets 3–5-flowered ; lowest flowering scale about 1″ long.   1. *P. laxa.*
      Spikelets 5–12-flowered ; lowest flowering scale about 1½″ long.   2. *P. canadensis.*
    Panicle contracted, the branches erect.   3. *P. obtusa.*
  Flowering scales narrow, sharply and distinctly 7-nerved.
    Panicle elongated, its branches erect or appressed.   4. *P. Torreyana.*
    Panicle not elongated, open, its branches spreading or drooping, rarely erect.
      Scales about 1″ long, obtuse or rounded at the apex.
        Spikelets 1½″ long or less ; branches of the panicle often drooping.
            5. *P. nervata.*
        Spikelets 2″–3″ long ; branches of the panicle ascending or spreading.
            6. *P. grandis.*
      Scales 1¼″–1½″ long, truncate and denticulate at the apex.   7. *P. pallida.*
Spikelets linear, 6″ long or more.
  Flowering scales 1½″–2½″ long, obtuse, equalling or exceeding the obtuse palet.
    Flowering scales firm, hispidulous all over, truncate at the apex. 8. *P. septentrionalis.*
    Flowering scales thin, hispidulous on the nerves only, obtuse at the apex.
            9. *P. borealis.*
  Flowering scales 3″–4″ long, usually shorter than the acuminate palet.
    Flowering scales obtuse, about 3″ long, a little exceeded by the palet. 10. *P. fluitans.*
    Flowering scales acute, about 4″ long, much exceeded by the palet. 11. *P. acutiflora.*

1. **Panicularia láxa** Scribn.   Northern Manna-grass.   Fig. 632.

Panicularia laxa Scribn. Bull. Torr. Club **21** : 37.   1894.

Glyceria laxa Scribn. ; Redf. & Rand, Fl. Mt. Desert, 180.
1894.

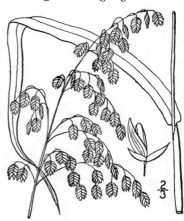

Culms 2°–4° tall, erect, simple, smooth or slightly
scabrous. Sheaths overlapping, rough; ligule ½″–1″
long; blades 8′–15′ long, 2″–4″ wide, very rough;
panicle 7′–9′ in length, the branches spreading or ascending, the lower 3′–6′ long; spikelets 3–5-flowered,
about 2″ long; empty scales unequal, scarious, acute,
1-nerved, the first one-half to two-thirds the length of
the second; flowering scales broad, about 1″ long,
twice the length of the second scale, obtuse, obscurely 7-nerved.

In water or wet soil, Nova Scotia to Maine, New Jersey
and Pennsylvania.   Aug.

2. **Panicularia canadénsis** (Michx.) Kuntze.   Rattlesnake-grass.   Fig. 633.

Briza canadensis Michx. Fl. Bor. Am. **1** : 71.   1803.
Glyceria canadensis Trin. Mem. Acad. St. Petersb. (VI.)
   1 : 366.   1831.
Panicularia canadensis Kuntze, Rev. Gen. Pl. 783.   1891.

Culms 2°–3° tall, erect, simple, smooth or slightly
scabrous. Sheaths shorter than the internodes, those
at the base of the culm overlapping; ligule 1″ long,
truncate; blades 6′–1° long or more, 2″–4″ wide,
rough; panicle 5½′–10′ in length, the branches spreading, ascending or often drooping, 2½′–5′ long; spikelets
5–12-flowered, 2½″–4″ long, flattened, turgid; empty
scales unequal, acute, 1-nerved; flowering scales, broad,
1½″–2″ long, obtuse or acutish, obscurely 7-nerved.

In swamps and marshes, Newfoundland to Minnesota,
south to New Jersey and Kansas.   The handsomest species
of the genus.   Ascends to 5000 ft. in the Adirondacks.
Tall Quaking-grass.   July-Aug.

3. Panicularia obtùsa (Muhl.) Kuntze. Blunt Manna-grass. Fig. 634.

Poa obtusa Muhl. Gram. 147. 1817.
Glyceria obtusa Trin. Mem. Acad. St. Petersb. (VI.) 1: 366. 1831.
Panicularia obtusa Kuntze, Rev. Gen. Pl. 783. 1891.

Culms 1°–3° tall, erect, simple, smooth and glabrous. Sheaths sometimes rough, strongly striate, the lower overlapping; ligule very short; blades 6′–15′ long, 2″–4″ wide, usually stiff, erect or ascending, smooth beneath, more or less scabrous above; panicle 3′–8′ in length, contracted, dense, the branches erect; spikelets 3–7-flowered, 2″–3″ long; empty scales acute, scarious, 1-nerved; flowering scales about 1½″ long, broad, obtuse, obscurely 7-nerved.

In swamps, Nova Scotia and New Brunswick to New York and central Pennsylvania, south to Maryland and North Carolina (according to Kearney). Ascends to 2300 ft. in the Catskill Mountains. July–Aug.

4. Panicularia Torreyàna (Spreng.) Merrill. Long Manna-grass. Fig. 635.

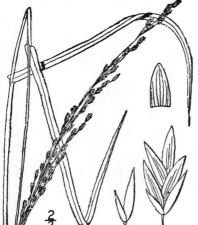

Poa elongata Torr. Fl. U. S. 1: 112. 1824. Not Willd. 1809.

P. Torreyana Spreng. Neue Entdeck. 2: 104. 1821.

Glyceria elongata Trin. Bull. Acad. Sci. St. Petersb. 1: 68. 1836.

Panicularia elongata Kuntze, Rev. Gen. Pl. 783. 1891.

P. Torreyana Merrill, Rhodora 4: 146. 1902.

Glyceria Torreyana Hitchc. Rhodora 8: 211. 1906.

Culms 2°–3° tall, erect. Sheaths often shorter than the internodes; blades lax, 6′–12′ long, 1½″–3″ wide, long-acuminate, smooth beneath, rough above; panicle elongated, contracted, narrow, usually nodding at the summit, 6′–12′ in length, the branches erect or appressed, 1′–2½′ long; spikelets 3–4-flowered, 1½″–2″ long; empty scales unequal, acute, 1-nerved; flowering scales narrow, about 1″ long, obtuse or acutish, 7-nerved.

In wet woods, Maine and Quebec to Minnesota, south to North Carolina and Kentucky. Ascends to 4000 ft. in the Adirondacks. Aug.–Sept.

5. Panicularia nervàta (Willd.) Kuntze. Meadow-grass. Nerved Manna-grass. Fig. 636.

Poa nervata Willd. Sp. Pl. 1: 389. 1798.
Glyceria nervata Trin. Mem. Acad. St. Petersb. (VI.) 1: 365. 1831.
Panicularia nervata Kuntze, Rev. Gen. Pl. 783. 1891.

Culms 1°–3° tall, erect, slender, simple, smooth and glabrous. Sheaths often shorter than the internodes, usually more or less rough; ligule ½″ long, truncate; blades 6′–12′ long, 2″–5″ wide, acute, smooth beneath, rough above; panicle 3′–8′ in length, open, the branches filiform, spreading, ascending or often drooping, rarely erect, 2′–5′ long; spikelets 3–7-flowered, 1″–1½″ long; empty scales obtuse, 1-nerved; flowering scales about ¾″ long, obtuse or rounded, with 7 sharp distinct nerves and evident furrows between.

In wet places, Newfoundland to British Columbia, south to Florida and Mexico. Ascends to 4000 ft. in Virginia. Panicle often purple. Meadow Spear-grass, Fowl-grass. June–Sept.

6. Panicularia grándis (S. Wats.) Nash. Reed Meadow-grass. Tall Manna-grass. Fig. 637.

Poa aquatica var. americana Torr. Fl. U. S. 1 : 108. 1824.
Glyceria grandis S. Wats. in A. Gray, Man. Ed. 6, 667. 1890.
P. americana MacMillan, Met. Minn. 81. 1892.

Culms 3°–5° tall, erect, stout, simple, smooth and glabrous. Sheaths loose, smooth, or sometimes rough; ligule 1″–2″ long, truncate; blades 7′–1° long or more, 3″–8″ wide, usually smooth beneath, rough above; panicle 8′–15′ in length, its branches spreading, ascending or rarely erect, 4′–8′ long; spikelets 4–7-flowered, 2″–3″ long; empty scales acute, 1-nerved; flowering scales about 1″ long, obtuse or rounded at the apex, sharply and distinctly 7-nerved, the furrows between the nerves evident.

In wet soil, Nova Scotia to Alaska, south to Pennsylvania, Colorado and Nevada. Ascends to 2100 ft. in Pennsylvania. White Spear-grass, Water Meadow-grass. June–Aug.

7. Panicularia pállida (Torr.) Kuntze. Pale Manna-grass. Fig. 638.

Windsoria pallida Torr. Cat. N. Y. 91. 1819.
Glyceria pallida Trin. Bull. Acad. Sci. St. Petersb. 1 : 68. 1836.
Panicularia pallida Kuntze, Rev. Gen. Pl. 783. 1891.
G. pallida Fernaldii Hitchc. Rhodora 8 : 211. 1906.

Pale green, culms 1°–3° long, assurgent, simple, smooth and glabrous. Sheaths loose, shorter than the internodes; ligule 2″–3″ long, acute; blades 2′–6′ long, 1″–2″ wide, smooth beneath, rough above; panicle 1½′–7′ in length, the branches spreading, ascending or rarely erect, often flexuous, 1′–2′ long; spikelets 4–8-flowered, 2½″–3½″ long; empty scales unequal, the first 1-nerved, obtuse, shorter than the 3-nerved and truncate second; flowering scales 1¼″– 1½″ long, truncate and denticulate at the apex, sharply and distinctly 7-nerved, with plain furrows between the nerves.

In shallow water, Nova Scotia and New Brunswick to Minnesota, south to North Carolina and Tennessee. Ascends to 2000 ft. in Pennsylvania. July–Aug.

8. Panicularia septentrionàlis (Hitchc.) Bicknell. American Flote-grass or Floating Manna-grass. Fig. 639.

Glyceria septentrionalis Hitchc. Rhodora 8 : 211. 1906.
Panicularia septentrionalis Bicknell, Bull. Torrey Club 35 : 196. 1908.

Culms 2°–5° long, flattened, erect or decumbent, usually stout, simple, smooth and glabrous, often rooting from the lower nodes. Sheaths loose, generally overlapping, smooth or rough; ligule 2″–3″ long; blades 5′–1° long or more, 2″–6″ wide, scabrous, often floating; panicle 9′–1½° long, the branches, at least the lower ones, at first appressed, later ascending, and 3′–6′ long; spikelets linear, 7–13-flowered, 4″–12″ long; empty scales unequal, 1-nerved, the lower acute or obtuse, the upper obtuse or truncate; flowering scales 1½″–2¼″ long, oblong, rounded or truncate at the erose apex, more or less scabrous, sharply 7-nerved.

In wet places or in water, Vermont and Quebec to British Columbia, south to North Carolina, Louisiana and Texas. Previously confused with P. fluitans. July–Sept.

3/4

9. Panicularia boreàlis Nash.   Northern Manna-grass.   Fig. 640.

Glyceria fluitans var. *angustata* Vasey, Proc. Port. Soc. Nat. Hist. **2**: 91.   1895.   Not *G. angustata* T. Fries, 1869.
P. borealis Nash, Bull. Torr. Club, **24**: 348.   1897.
Glyceria borealis Batch. Proc. Manch. Inst. **1**: 74. 1900.

Glabrous.   Culms erect from a creeping base, 1½°–5° tall; sheaths overlapping, smooth or roughish, the uppermost one enclosing the base of the panicle; ligule 2½″–7½″ long, membranous; blades linear, abruptly acuminate, 3½′–21′ long, 1″–5″ wide; panicle slender, narrow, the exserted portion 6′–20′ long, its branches appressed or nearly so, the lower in 2's of 3's, the longer of which bear 5–12 spikelets; spikelets 5″–9″ long, 7–13-flowered, appressed; outer two scales empty, 1-nerved, smooth and shining, unequal; flowering scales thin, 2″–2½″ long, 7-nerved, the nerves hispidulous, a broad scarious margin at the obtuse and erose apex; palet hyaline, shortly 2-toothed at the obtuse apex.

In shallow water, Newfoundland to Alaska, south to New York, Minnesota, Iowa and Oregon, and in the mountains to Colorado.   June–Aug.

10. Panicularia flùitans (L.) Kuntze.   Floating Manna-grass.   Sweet-grass.   Fig. 641.

Festuca fluitans L. Sp. Pl. 75.   1753.
Glyceria fluitans R. Br. Prod. **1**: 179.   1810.
P. fluitans Kuntze, Rev. Gen. 782.   1891.
P. brachyphylla Nash, Bull. Torr. Club **24**: 349.   1897.

Culms erect from a creeping base, 2°–3° tall; sheaths generally longer than the internodes, almost closed, the uppermost one enclosing the base of the panicle; blades linear, acuminate, 2½′–5′ long, 2″–2½″ wide; panicle slender, its branches appressed or nearly so, the lower in 2's or 3's, the longer of which bear 2 or 3 spikelets; spikelets compressed-cylindric, 10″–14″ long, 7–12-flowered; flowering scales hispidulous all over, 7-nerved, about 2½″ long, the obtuse apex obscurely and irregularly few-toothed; palet acuminate, a little exceeding the scale.

In shallow water, Gulf of St. Lawrence and near New York City.   Perhaps introduced.   Common in Europe. Flote-grass, Russia-grass, Manna Croup-grass, Poland Manna.   June–July.

2/3

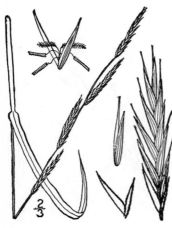

2/3

11. Panicularia acutiflòra (Torr.) Kuntze.   Sharp-scaled Manna-grass.   Fig. 642.

Glyceria acutiflora Torr. Fl. U. S. **1**: 104.   1824.

Panicularia acutiflora Kuntze, Rev. Gen. Pl. 782.   1891.

Culms 1°–2° tall, flattened, erect from a decumbent base, simple, smooth and glabrous.   Sheaths loose, generally a little exceeding the internodes, smooth and glabrous; ligule 2″ long, truncate; blades 3′–6′ long, 2″–3″ wide, smooth beneath, rough above; panicle 6′–12′ in length, the branches erect or appressed, 2′–4′ long; spikelets linear, 5–12-flowered, 1′–1¾″ long; empty scales acute, smooth; flowering scales about 4″ long, lanceolate, acute, scabrous, exceeded by the long-acuminate palets.

In wet places, Maine to Delaware and Ohio.   June–Aug.

97. PUCCINÉLLIA Parl. Fl. Ital. 1: 366. 1848.

Perennial grasses, with flat or involute leaf-blades and contracted or open panicles. Spikelets 3–several-flowered. Lower scales empty, obtuse or acute, unequal; flowering scales obtuse or acute, rounded on the back, 5-nerved, the nerves very obscure or almost wanting. Palet about equalling the scale. Stamens 3. Styles wanting. Stigmas sessile, simply plumose. Grain compressed, usually adhering to the palet. [Name in honor of Benedetto Puccinelli, Italian botanist.]

About 14 species, in all temperate regions. Type species: *Poa distans* L.

Flowering scales 1¼″ long or more; plants stoloniferous.
   Lower flowering scales 1½″–2″ long; spikelets commonly 4–many-flowered.   1. *P. maritima.*
   Lower flowering scales not exceeding 1½″ long, usually less; spikelets generally 2–4-flowered.
                                            2. *P. angustata.*
Flowering scales less than 1¼″ long; plants without stolons.
   Second empty scale less than one-half as long as the first flowering scale, broad, usually obtuse
    or truncate.
      Panicle-branches naked below, spikelet-bearing toward the apex; flowering scales 1″ long or
      less, truncate at the apex.   3. *P. distans.*
      Panicle-branches spikelet-bearing to the base; flowering scales 1″–1¼″ long, acutish or
      obtuse at the apex.   4. *P. fasciculata.*
   Second empty scale more than one-half as long as the first flowering scale, usually narrow, obtuse
    or acute.   5. *P. airoides.*

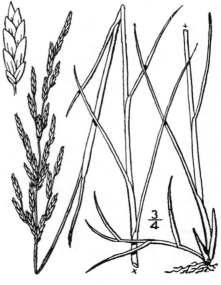

1. Puccinellia marítima (Huds.) Parl.
Goose-grass. Sea Spear-grass.
Fig. 643.

Poa maritima Huds. Fl. Angl. 35. 1762.
Glyceria maritima M. & K. Deutsch. Fl. 1: 588. 1823.
Puccinellia maritima Parl. Fl. Ital. 1: 370. 1848.

Stoloniferous, smooth, glabrous, culms 6′–2° tall, erect, or decumbent at the base, simple. Sheaths usually exceeding the internodes; ligule ½″–1″ long; blades ½–5′ long, 1″ wide or less, flat to involute; panicle 2′–6′ in length, open, the branches ascending, or rarely erect, 1′–2′ long; spikelets 3–10-flowered, 3″–6″ long; empty scales unequal, the first usually 1-nerved, the second 3-nerved; flowering scales 1½″–2″ long, broad, obtuse or truncate.

In salt marshes and on sea beaches, Nova Scotia to Rhode Island. Also on the Pacific coast, and on the coasts of Europe and Asia. Sea Meadow-grass. July–Aug.

2. Puccinellia angustàta (R. Br.) Nash.
Arctic Meadow-grass. Fig. 644.

Poa angustata R. Br. App. Parry's Voy. 287. 1824.
Panicularia angustata Scribn. Mem. Torr. Club 5: 54. 1894.
Puccinellia maritima var. *minor* S. Wats. in A. Gray, Man. Ed. 6, 668. 1890.
P. angustata Nash, Bull. Torr. Club 22: 512. 1895.

Smooth and glabrous, culms 4′–12′ tall, erect, simple. Sheaths usually overlapping; ligule 1″ long; blades ½–2½′ long, 1″ wide or less; panicle 1′–2′ in length, contracted, the branches short and erect or appressed; spikelets 2–7-flowered, 3″–4″ long; empty scales obtuse or rounded at the apex, the first 1-nerved, the second 3-nerved; flowering scales 1¼″–1½″ long, usually purplish, rounded at the apex.

Greenland and Hudson Bay to Alaska, south to Connecticut. Also in Europe and Asia. Summer.

$\frac{3}{4}$

3. Puccinellia dístans (L.) Parl. Spreading Meadow-grass. Sweet-grass. Fig. 645.

Poa distans L. Mant. 32. 1767.
Glyceria distans Wahl. Fl. Ups. 36. 1820.
Puccinellia distans Parl. Fl. Ital. 1 : 367. 1848.

Culms 1°–2° tall, decumbent at the base, tufted, smooth and glabrous. Sheaths often shorter than the internodes, smooth and glabrous; ligule ½″–1″ long; blades up to 4′ long, 1″–2″ wide, flat or folded, usually stiff and erect, smooth beneath; panicle 3′–7′ in length, open, the branches finally spreading, whorled, the lower up to 4½′ long, sometimes reflexed, naked below; spikelets crowded, 3–6-flowered, 1½″–2½″ long; empty scales obtuse or acute, 1-nerved, the second exceeding the first and less than half the length of the first flowering scale, the truncate flowering scales ¾″–1″ long.

On salt meadows, sea beaches and in waste places, Nova Scotia to Delaware. Probably naturalized from Europe. Sea-meadow-grass. July–Aug.

4. Puccinellia fasciculàta (Torr.) Bicknell. Torrey's Meadow-grass. Fig. 646.

Poa fasciculata Torr. Fl. U. S. 1 : 107. 1824.
P. fasciculata Bicknell, Bull. Torr. Club 35 : 197. 1908.

Culms 1°–2° tall; sheaths smooth and glabrous; ligule about ½″ long, truncate; blades erect, up to 5′ long, 1½″–3″ wide, smooth beneath, rough above; panicle 3′–5′ long, its branches spikelet-bearing to the base, usually ascending; spikelets about 2″ long, the empty scales obtuse or acute, the second one less than one-half as long as the first flowering scale, the flowering scales 1″–1¼″ long, obtuse or acutish, glabrous or nearly so.

Salt marshes, Nantucket to New Jersey. May and June.

$\frac{2}{3}$

5. Puccinellia airoìdes (Nutt.) Wats. & Coult. Slender Meadow-grass. Fig. 647.

Poa airoides Nutt. Gen. 1 : 68. 1818.
Panicularia distans airoides Scribn. Mem. Torr. Club 5 : 54. 1894.
Puccinellia airoides Wats. & Coult. in A. Gray, Man. Ed. 6, 668. 1890.

Culms 1°–4° tall, erect, simple, smooth and glabrous. Sheaths usually longer than the internodes; ligule 1″ long; blades 2′–6′ long, 1½″ wide or less, flat or involute, usually erect, smooth beneath, rough above; panicle open, its branches slender, spreading or ascending, rarely erect, the lower 2′–3½′ long and often reflexed; spikelets scattered, 1–7-flowered, 1½″–3″ long; empty scales unequal, the first acute, 1-nerved, the second obtuse or acute, 3-nerved, more than half the length of the obtuse flowering scales, which are 1″–1¼″ long.

In saline soil, southwestern Ontario to the Northwest Territory, south to Kansas and Nevada. July–Aug.

Puccinellia Bôrreri (Bab.) Hitchc. is reported as growing on ballast and in waste places from Delaware to Nova Scotia. It is related to *P. fasciculata* (Torr.) Bicknell.

98. FESTÙCA L. Sp. Pl. 73. 1753.

Mostly tufted perennial grasses, with flat or convolute leaf-blades and paniculate inflorescence. Spikelets 2–several-flowered. Two lower scales empty, more or less unequal, acute, keeled; flowering scales membranous, narrow, rounded on the back, 5-nerved, usually acute, and generally awned at the apex. Palet scarcely shorter than the scale. Stamens 1-3. Styles very short, distinct. Stigmas plumose. Grain glabrous, elongated, often adherent to the scale or palet. [Latin, stalk or straw.]

A genus of about 100 species, widely distributed, particularly numerous in temperate regions. Type species: *Festuca ovina* L.

Spikelets perfect; stigma-branches toothed, bilateral.
　Empty scales membranous, green, narrow, the second one 3-5-nerved.
　　Leaf-blades involute or folded, 1″ wide or less.
　　　Annuals; stamens 1 or 2.
　　　　Awn not longer than flowering scale; spikelets 5-many-flowered.　　1. *F. octoflora.*
　　　　Awn more than twice as long as flowering scale; spikelets 2-5-flowered.
　　　　　First empty scale half as long as the second or less.　　　2. *F. Myuros.*
　　　　　First empty scale more than half as long as the second.　　　3. *F. sciurea.*
　　　Perennials; stamens 3.
　　　　Innovations extravaginal; plants with rootstocks or stolons.　　　4. *F. rubra.*
　　　　Innovations intravaginal; plants densely tufted, no rootstocks or stolons.
　　　　　Awns more than half as long as membranous flowering scales.　　5. *F. occidentalis.*
　　　　　Awns less than one-half as long as the coriaceous flowering scales.
　　　　　　Flowering scales short-awned; leaf-blades setaceous.
　　　　　　　Culms 8′ tall or more; culm blades long.　　　6. *F. ovina.*
　　　　　　　Culms 6′ long or less; culm blades short.　　　7. *F. brachyphylla.*
　　　　　　Flowering scales awnless; leaf-blades capillary.　　　8. *F. capillata.*
　　Leaf-blades flat, 2″ wide or more.
　　　Flowering scales awnless or short-awned.
　　　　Flowering scales 2½″-3½″ long, spikelets 5-10-flowered.　　　9. *F. elatior.*
　　　　Flowering scales 2″ long or less, spikelets 3-6-flowered.
　　　　　Spikelets very broad; panicle branches spikelet-bearing from middle or below.
　　　　　　　　　　　　　　　　　　　　　　　　　　　　　　10. *F. Shortii.*
　　　　　Spikelets lanceolate; branches elongated, spikelets at the end. 11. *F. nutans.*
　　　Flowering scales with an awn twice their length or more.　　　12. *F. gigantea.*
　Empty scales broad, scarious, with broad hyaline margins, thin, 1-nerved; base of the culm
　　clothed with dry leafless sheaths.　　　13. *F. altaica.*
Spikelets unisexual; stigma-branches arising from all sides; dioecious.　　14. *F. confinis.*

1. Festuca octoflòra Walt. Slender Fescue-grass. Fig. 648.

Festuca octoflora Walt. Fl. Car. 81. 1788.
Festuca tenella Willd. Enum. 1: 113. 1809.

Culms 4′-18′ tall, erect, from an annual root, slender, rigid, simple, smooth and glabrous. Sheaths usually shorter than the internodes; ligule very short; blades 1½′-3′ long, involute; raceme or simple panicle often one-sided, 1′-6′ in length, contracted, its branches erect, or rarely ascending; spikelets 6-13-flowered, 3″-5″ long; empty scales acute, smooth, the first 1-nerved, more than half the length of the 3-nerved second one; flowering scales, exclusive of awns, 1½″-2½″ long, usually very scabrous, acuminate into an awn nearly as long as the body, or sometimes awnless; stamens 2.

Dry sandy soil, Quebec to British Columbia, south to Florida, Texas and California. June-Aug.

2. Festuca Myùros L. Rat's-tail Fescue-grass. Fig. 649.

Festuca Myuros L. Sp. Pl. 74. 1753.

Smooth, glabrous, culms 1°-2° tall, erect from an annual root, slender, simple. Sheaths often shorter than the internodes, the upper sometimes enclosing the base of the panicle; ligule ½″ long, truncate; blades 2′-5′ long, subulate, involute, erect; panicle usually one-sided, 4′-12′ in length, contracted, sometimes curved, its branches appressed; spikelets 3-6-flowered; empty scales very unequal, acute, smooth, the first 1-nerved, less than half as long as the 3-nerved second one; flowering scales, exclusive of the awns, 2″-3″ long, narrow, scabrous, acuminate into an awn much longer than the body; stamen 1.

In waste places and fields, New Hampshire to New Jersey and Ohio. Also on the Pacific coast. Naturalized from Europe. Mouse-tail, Capon's-tail grass. June-July.

3. Festuca sciùrea Nutt.   Southern Fescue-grass.   Fig. 650.

Festuca sciurea Nutt. Trans. Am. Phil. Soc. **5** : 147.   1837.

Culms 4′–20′ tall, slender; blades 2′ long or less, less than $\frac{1}{2}''$ wide; panicle slender, $1\frac{1}{2}$′–6′ long, its branches erect or appressed; spikelets 3–5-flowered, the first scale more than one-half as long as the second, the flowering scales appressed-pubescent, about $1\frac{1}{2}''$ long, exclusive of the awn which is 2–3 times as long as the scale.

In dry soil, Virginia to Florida, Oklahoma and Texas. Squirrel Fescue.   June–Aug.

4. Festuca rùbra L.   Red Fescue-grass.   Fig. 651.

Festuca rubra L. Sp. Pl. 74.   1753.

Culms $1\frac{1}{2}°$–$2\frac{1}{2}°$ tall, from running rootstocks, erect, simple, smooth and glabrous.   Sheaths usually shorter than the internodes; ligule very short, truncate; basal blades involute-filiform, 3′–6′ long; culm blades shorter, erect, flat or involute in drying, minutely pubescent above; panicle 2′–5′ in length, sometimes red, open at flowering time, contracted in fruit; spikelets 3–10-flowered, 4″–6″ long; lower scales acute, unequal, the first 1-nerved, shorter than the 3-nerved second; flowering scales about 3″ long, obscurely 5-nerved, sometimes scabrous, bearing awns of less than their own length.

Labrador to Alaska and Virginia, south, especially on the mountains, to Tennessee and Colorado.   Also in Europe and Asia.   Summer.

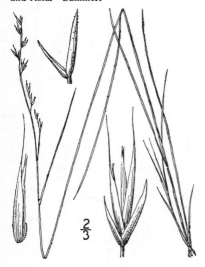

5. Festuca occidentàlis Hook.   Western Fescue-grass.   Fig. 652.

F. occidentalis Hook. Fl. Bor. Am. **2** : 249.   1840.

Culms densely tufted, $1\frac{1}{2}°$–3° tall, erect, slender, smooth and glabrous; blades filiform, soft, up to 4′ long, the basal ones numerous; panicle 3′–8′ long, loose; spikelets 3–5-flowered, the empty scales unequal, variable, the flowering scales membranous, glabrous, $2\frac{1}{2}''$–3″ long, bearing an awn more than half their length.

In woods, Michigan to British Columbia and California.   May–July.

6. Festuca ovìna L.   Sheep's Fescue-grass.
Fig. 653.

Festuca ovina L. Sp. Pl. *73.* 1753.
Festuca ovina duriuscula Hack. Monog. Fest. Europe 89. 1882.

Smooth, glabrous, culms 6'-2° tall, erect, tufted, slender, rigid, simple; no rootstocks. Sheaths usually crowded at the base of the culm; ligule auriculate, short; blades filiform or setaceous, those of the culm few, 1'-3' long, erect, the basal ones numerous; panicle 1½'-6' long, often one-sided, narrow, its branches short, usually erect or appressed; spikelets 3–5-flowered; empty scales unequal, acute, the first 1-nerved, the second 3-nerved; flowering scales 1½"-3" long, smooth, acute, short-awned.

In fields and waste places, New Hampshire to North Dakota, New Jersey; Kentucky and Iowa. Variable. Probably indigenous northward, but mostly naturalized from Europe. Native also of Asia. Black-twitch-grass. Hard Fescue. June–July.

The so-called var. **vivípara,** a state of this grass with the scales wholly or partly transformed into small leaves, is found on the mountains of New England and in arctic America.

7. Festuca brachyphýlla Schultes.   Short-leaved Fescue-grass.   Fig. 654.

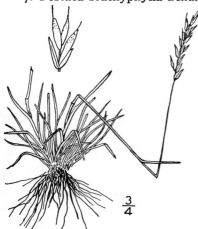

Festuca brevifolia R. Br. Append. Parry's Voy. Suppl. 289. 1824. Not Muhl. 1817.
Festuca brachyphylla Schultes, Mant. **3:** Addit. 1, 646. 1827.
Festuca ovina L. var. *brevifolia* S. Wats. in King's Rep. U. S. Geol. Expl. 40th Paral. **5:** 389. 1871.

Smooth and glabrous. Culms densely tufted, 6' or less tall, slender, erect, much exceeding the short basal leaves; sheaths coarsely striate; ligule a short scarious ring; blades very narrow, involute, at least when dry; those on the culm ½' or less long, erect or ascending; panicle 1' or less long, nearly simple, its branches appressed; spikelets 2–4-flowered, the empty scales acuminate, the first 1-nerved, the second 3-nerved; flowering scales acute or acuminate, rough toward the apex, 2"-2½" long, exclusive of the scabrous awn which is ½"-1¼" long.

Newfoundland to British Columbia, the higher mountains of Vermont, and the Rocky Mountains to Colorado. Summer.

8. Festuca capillàta Lam.   Filiform Fescue-grass.   Fig. 655.

Festuca capillata Lam. Fl. Franc. **3:** 598. 1778.

Festuca ovina capillata Hack. Bot. Centrb. **8:** 405. 1881.

Densely tufted. Culms erect with a decumbent base, 6'-15' tall, slender, smooth and glaucous, shining; sheaths smooth, longer than the internodes, confined to the base of the culm; ligule a short membranous ring; blades filiform, smooth or rough, the basal ones from one-third to one-half as long as the culm, the culm leaves 1'-1½' long; panicle contracted, ½'-2' long, its branches erect, ½' or less long; spikelets 2"-2½" long, 4-5-flowered; outer scales empty, unequal, the first acuminate, the second acute; flowering scales about 1¼" long, unawned, acute.

Fields and roadsides, Newfoundland to New Jersey and Michigan. Introduced from Europe. June–July.

9. Festuca elàtior L.   Tall or Meadow Fescue-grass.   Fig. 656.

Festuca elatior L. Sp. Pl. 75. 1753.

Festuca pratensis Huds. Fl. Angl. 37. 1762.

F. elatior var. *pratensis* A. Gray, Man. Ed. 5, 634. 1867.

Culms 2°–5° tall, erect, simple, smooth and glabrous. Sheaths shorter than the internodes; ligule very short; blades 4′–15′ long, 2″–4″ wide, flat, smooth beneath, more or less rough above; panicle 4′–14′ in length, often nodding at the top, simple to very compound, the branches ascending or erect, 2′–8′ long; spikelets 5–9-flowered, 4½″–6″ long; empty scales acute, the first 1–3-nerved, the second 3–5-nerved; flowering scales acute or short-pointed, smooth and glabrous, 2½″–3″ long, indistinctly 5-nerved.

In fields and waste places throughout the United States and southern Canada. Naturalized from Europe and cultivated for hay. Variable. Dover-grass, Randall or Evergreen-grass. Frisky (Meadow)-grass. July–Aug.

10. Festuca Shórtii Kunth.   Short's Fescue-grass.   Fig. 657.

Festuca Shortii Kunth; Wood, Class-book 794. 1861.

Festuca nutans var. *palustris* Wood, Bot. & Fl. 399. 1873.

Culms 2°–4° tall, erect, simple, smooth and glabrous. Sheaths much shorter than the internodes; ligule very short; blades 5′–10′ long, 1″–3″ wide, flat, smooth beneath, rough above; panicle 3′–7′ in length, open, the branches spreading or ascending, rarely erect, spikelet-bearing from the middle or below, the lower 1½′–3½′ long; spikelets broadly obovate, when mature, 3–6-flowered, 2½″–3″ long; empty scales acute, unequal, scabrous on the nerves, the first 1–3-nerved, the second 3-nerved; flowering scales about 2″ long, smooth, obtuse or acutish, faintly nerved.

In woods and thickets, Pennsylvania to Iowa, south to Georgia and Texas. July–Aug.

11. Festuca nùtans Willd.   Nodding Fescue-grass.   Fig. 658.

Festuca nutans Willd. Enum. 1 : 116. 1809.

Culms 2°–3° tall, erect, simple, slender, glabrous or sometimes pubescent. Sheaths much shorter than the internodes, glabrous or pubescent; ligule very short; nodes black; blades 4′–12′ long, 2″–3″ wide, rather dark green, flat, smooth beneath, rough above; panicle 4′–9′ in length, its branches at first erect, the lower 2½′–5′ long, finally spreading and nodding, spikelet-bearing only at the ends; spikelets lanceolate, 3–5-flowered, 2½″–3″ long; empty scales acute, scabrous on the keel, the first 1-nerved, shorter than the 3-nerved second; flowering scales about 2″ long, smooth, acute, very faintly nerved.

In rocky woods, Nova Scotia to Minnesota, south to Florida and Texas. Ascends to 2300 ft. in Virginia. June–Aug.

12. Festuca gigantèa (L.) Vill.   Great Fescue-grass.   Fig. 659.

Bromus giganteus L. Sp. Pl. 77. 1753.

Festuca gigantea Vill. Hist. Pl. Dauph. 2: 110. 1787.

Culms 2°–4° tall, erect, simple, smooth and glabrous. Sheaths usually overlapping; ligule 1″ long; blades 5′–1° long or more, bright green, 2″–6″ wide, flat, rough; panicle 7′–12′ in length, loose, narrow, the branches erect or ascending, the lower 2′–4′ long; spikelets 3–7-flowered; empty scales acuminate, smooth and glabrous, the first 1–3-nerved, shorter than the 3–5-nerved second; flowering scales, exclusive of awns, about 3″ long, faintly 5-nerved, slightly scabrous, minutely 2-toothed at the apex, bearing an awn 6″–8″ long.

In waste places, Maine to southern New York. Adventive from Europe. July–Aug.

13. Festuca altàica Trin.   Rough Fescue-grass.   Fig. 660.

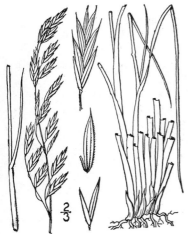

F. altaica Trin. in Ledeb. Fl. Alt. 1: 109. 1829.

Culms 1°–3° tall, erect, simple, usually rough below the panicle. Sheaths overlapping, smooth; ligule a ring of very short hairs; blades rough, 1″ wide or less, those of the culm 1′–3′ long, erect, the basal flat, much longer and readily deciduous from the sheaths, involute in drying; panicle 3′–4′ in length, open, its branches ascending or the lower widely spreading; spikelets 3–5-flowered, about 4″ long; empty scales scarious, unequal, smooth, the first 1-nerved, the second longer, 3-nerved; flowering scales about 3″ long, scabrous, often bearing a short awn 1″ long or less.

Labrador to Alaska, south to Quebec, North Dakota and British Columbia. Summer. Mistaken for *Festuca scabrélla* Torr. in our first edition.

14. Festuca confìnis Vasey.   Watson's Fescue-grass.   Fig. 661.

Poa Kingii S. Wats. Bot. King's Exp. 387. 1871.
Festuca Kingii Scribn. Bull. U. S. Dep. Agr. Agrost. **5**: 36. 1897. Not *F. Kingiana* Endlich. 1855.
F. confinis Vasey, Bull. Torr. Club **11**: 126. 1884.
F. Watsoni Nash, in Britt. Man. 148. 1901.

Culms tufted, erect, rigid, the base clothed with dry leafless sheaths; sheaths smooth and glabrous; leaves erect, stiff, smooth beneath, rough above, 10′ long or less, 1″–2″ wide, those on the culm much shorter than those of the innovations; panicle strict, narrow, 4′–5′ long, its branches erect; spikelets usually 3-flowered, 3½″–4″ long; the scales acute, the flowering scales strongly hispidulous.

Meadows, Montana to Nebraska, Colorado and California. June and July.

99. BRÒMUS L. Sp. Pl. 76. 1753.

Annual or perennial grasses, with flat leaf-blades and terminal panicles, the pedicels thick-ened at the summit. Sheaths sometimes not split. Spikelets few–many-flowered. Two lower scales empty, unequal, acute; flowering scales rounded on the back, or sometimes compressed-keeled, 5–9-nerved, the apex usually 2-toothed, generally bearing an awn just below the summit; palet shorter than the scale, 2-keeled. Stamens usually 3. Stigmas ses-sile, plumose, inserted below a hairy cushion-like appendage at the top of the ovary. Grain adherent to the palet. [Greek name for a kind of oats.]

About 60 species, most numerous in the north temperate zone. Besides the following, some 14 others occur in the western parts of North America. Type species: *Bromus secalinus* L.

Lower empty scales 1-nerved, the upper 3-nerved.
　Awns longer than the flowering scales; low annuals, 1½° tall or less.
　　Flowering scales strigose, 4″–6″ long. 　　　　　　　　　　　　　　　1. *B. tectorum.*
　　Flowering scales sparsely hispidulous, 6″ long or more.
　　　Spikelets usually single on the long naked spreading branches. 　　2. *B. sterilis.*
　　　Spikelets several on the branches which are divided and spikelet-bearing above the middle.
　　　　　　　　　　　　　　　　　　　　　　　　　　　　　　　　　3. *B. madritensis.*
　Awns shorter than the flowering scales, or wanting; perennials 1½° tall or more.
　　Flowering scales awned.
　　　Leaf-sheaths strongly retrorse-hirsute. 　　　　　　　　　　　　　4. *B. asper.*
　　　Leaf-sheaths glabrous or softly pubescent.
　　　　Blades 2″–6″ wide; panicle branches more or less spreading or drooping.
　　　　　Flowering scales pubescent on the margins only. 　　　　　　5. *B. ciliatus.*
　　　　　Flowering scales pubescent all over the back. 　　　　　　　6. *B. purgans.*
　　　　Blades less than 2″ wide; panicle branches erect. 　　　　　　7. *B. erectus.*
　　　Flowering scales awnless or merely awn-pointed. 　　　　　　　　8. *B. inermis.*
Lower empty scale 3-nerved, the second one 5–9-nerved (3-nerved in no. 9).
　Flowering scales rounded on the back, at least below.
　　Perennials; flowering scales densely pubescent with long silky hairs.
　　　Second empty scale 3-nerved; flowering scales 5″–6″ long. 　　　9. *B. Porteri.*
　　　Second empty scale 5–7-nerved; flowering scales about 4″ long. 10. *B. Kalmii.*
　　Annuals.
　　　Flowering scales awned.
　　　　Flowering scales pubescent with soft appressed hairs, not dense. 11. *B. hordeaceus.*
　　　　Flowering scales glabrous, or minutely roughened.
　　　　　Awns straight.
　　　　　　Fruiting scales with strongly inrolled margins, the nerves obscure; leaf-sheaths
　　　　　　　glabrous. 　　　　　　　　　　　　　　　　　12. *B. secalinus.*
　　　　　　Fruiting scales with the margins not inrolled, the nerves prominent; leaf-
　　　　　　　sheaths softly and densely pubescent.
　　　　　　　Spikelets broadly lanceolate, usually over 3″ wide, 1 or sometimes 2 on
　　　　　　　　the longer branches. 　　　　　　　　　　　13. *B. racemosus.*
　　　　　　　Spikelets lanceolate, usually less than 3″ wide, several on the longer
　　　　　　　　branches. 　　　　　　　　　　　　　　14. *B. arvensis.*
　　　　　Awns bent near the base, divergent.
　　　　　　Spikelets less than 2½″ broad in flower. 　　　　　　　15. *B. patulus.*
　　　　　　Spikelets exceeding 2½″ broad in flower. 　　　　　　　16. *B. squarrosus.*
　　　　Flowering scales awnless or awn-pointed, nearly as broad as long. 17. *B. brizaeformis.*
　Flowering scales compressed, keeled.
　　Flowering scales pubescent; awn 2″–3″ long. 　　　　　　　　　18. *B. breviaristatus.*
　　Flowering scales minutely roughened; awn less than 1″ long or wanting. 19. *B. unioloides.*

1. Bromus tectòrum L.　Downy Brome-grass.　Fig. 662.

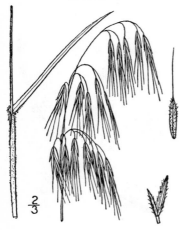

Bromus tectorum L. Sp. Pl. 77. 1753.

Culms 6′–2° tall, erect from an annual root, simple, smooth and glabrous. Sheaths usually longer than the internodes, at least the lower ones softly pubescent; ligule 1″–2″ long; blades 1′–4′ long, 1″–2″ wide, softly pubescent; panicle 2′–6′ in length, open, the branches slender and drooping, somewhat one-sided; spikelets numerous, 5–8-flowered, on capillary recurved slender pedicels; empty scales acuminate, usually rough or hir-sute, the first 1-nerved, the second longer, 3-nerved; flowering scales 4″–6″ long, acuminate, 7-nerved, usually rough or hirsute; awn 6″–8″ long.

In fields and waste places, Maine to Ontario, Maryland, Ohio and Missouri. Naturalized from Europe. Sometimes a troublesome weed. May–July.

$\frac{2}{3}$

2. Bromus stérilis L. Barren Brome-grass. Haver-grass. Black-grass. Fig. 663.

Bromus sterilis L. Sp. Pl. 77. 1753.

Culms 1°–2° tall, erect, simple, smooth and glabrous. Sheaths usually shorter than the internodes, smooth or rough, the lower sometimes pubescent; ligule 1″ long; blades 3′–9′ long, 1″–3″ wide, usually more or less pubescent; panicle 5′–10′ in length, the branches ascending or often widely spreading, not one-sided, stiff; spikelets few, 5–10-flowered, spreading or pendulous; empty scales acuminate, glabrous, the first 1-nerved, the second longer, 3-nerved; flowering scales 6″–8″ long, acuminate, 7-nerved, scabrous on the nerves, the awn 7″–12″ long.

In waste places and ballast, eastern Massachusetts to District of Columbia, Ohio, Arkansas and Colorado. Also on the Pacific Coast, and in Jamaica. Naturalized or adventive from Europe. Native also of Asia. June–July.

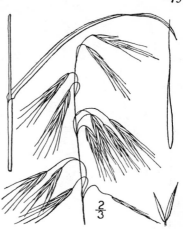

3. Bromus madriténsis L. Compact Chess. Fig. 664.

Bromus madritensis L. Amoen. Acad. 4: 265. 1755.

Culms 1°–2° tall, tufted, smooth and glabrous. Sheaths smooth, glabrous, or the lower ones pubescent, usually shorter than the internodes; ligule scarious, lacerated, 1″–2″ long; blades 2½′–8′ long, 1½″–3″ wide, rough above, often pubescent on both surfaces; panicle dense, 3′–6′ long, its rough branches erect or ascending, the longer 2′–3′ long; spikelets numerous, 1½″–2′ long, including the awns, the scales acuminate, scarious on the margins, the first scale 1-nerved, the second 3-nerved; flowering scales, exclusive of the awn, 6″–7″ long, sparsely and minutely appressed-pubescent, the apex acuminately 2-toothed, 5-nerved, bearing an erect or divergent awn 6″–9″ long.

Waste places, Michigan and Virginia; also in California. Locally adventive from Europe. Summer.

Bromus rùbens L. and **Bromus máximus** L., European species, are both reported as sparingly introduced in one or two localities.

4. Bromus ásper Murr. Hairy Brome-grass. Fig. 665.

Bromus asper Murr. Prodr. Stirp. Goett. 42. 1770.

Culms 2°–6° tall, erect, simple, rough. Sheaths shorter than the internodes, strongly retrorse-hirsute, especially the lower; ligule 1½″ long; blades 8′–1° long or more, 3″–6″ wide, rough or often hirsute; panicle 6′–12′ in length, open, the branches usually drooping; spikelets 5–10-flowered, 1′–1½′ long; empty scales acute, scabrous on the nerves, the first 1-nerved, the second longer, 3-nerved; flowering scales about 6″ long, acute, hispid near the margins and on the lower part of the keel; awn 3″–4″ long.

In waste places, New Brunswick to Michigan and Kentucky. Naturalized from Europe. July–Aug.

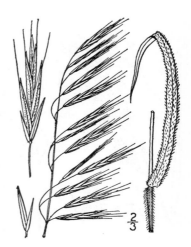

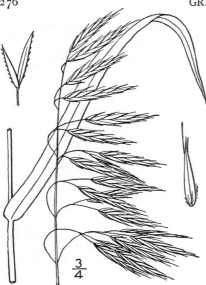

5. Bromus ciliàtus L.   Fringed Brome-grass. Wood Chess or Cheat. Fig. 666.

Bromus ciliatus L. Sp. Pl. 76. 1753.

Culms 2°–4° tall, erect, simple, glabrous or pubescent. Sheaths often shorter than the internodes, smooth or rough, often softly pubescent, or the lower sometimes sparingly hirsute; ligule very short; blades 4′–12′ long, 2″–6″ wide, smooth beneath, scabrous and often pubescent above; panicle open, 4′–10′ in length, its branches lax, widely spreading or often drooping; spikelets 5–10-flowered, 1′ long or less; empty scales very acute, glabrous, rough on the keel, the first 1-nerved, the second longer, 3-nerved; flowering scales 4″–6″ long, obtuse or acute, 5–7-nerved, appressed-pubescent on the margins; awn 2″–4″ long.

In woods and thickets, Newfoundland to Manitoba, New York, New Jersey, Minnesota and Texas. Variable. Hairy Brome-grass, Swamp-chess. July–Aug.

6. Bromus púrgans L.   Hairy Wood Chess. Wild Chess. Fig. 667.

Bromus purgans L. Sp. Pl. 76. 1753.
B. purgans latiglumis Shear, Bull. U. S. Dep. Agr. Agrost. **23**: 40. 1900.
B. incanus Hitchc. Rhodora **8**: 212. 1906.

Culms 2°–5° tall, erect, glabrous or pubescent at the nodes. Leaf-sheaths longer or shorter than the internodes, more or less pubescent, often furnished with a conspicuous pilose ring at the summit; blades 6′–12′ long, 2″–8″ broad, glabrous or pubescent on the upper surface, smooth or rough beneath; panicle 6′–1° long, loose, often nodding; spikelets 7–12-flowered, 10″–12″ long, the empty scales narrow, acuminate, sparsely pubescent, the lower 1-nerved, the upper 3-nerved, the flowering scales lanceolate, acute, usually 5-nerved, 5″–6″ long, appressed-pubescent all over on the back, the straight awn 2″–3″ long.

Woods and banks, Vermont to Montana, south to Florida and Texas. June–Aug.

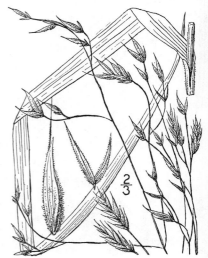

7. Bromus eréctus Huds.   Upright Brome-grass. Fig. 668.

Bromus erectus Huds. Fl. Angl. 39. 1762.

Culms 2°–3° tall, erect, simple, slender, smooth and glabrous. Sheaths shorter than the internodes, smooth and glabrous, or slightly pubescent; ligule ½″ long, erose-truncate; blades sparingly pubescent, 1″–2″ wide, those of the culm 4′–8′ long, the basal about 1° long, very narrow; panicle 3′–7′ in length, the branches erect or ascending, the lower 1′–3′ long; spikelets ½′–1½′ long, sometimes purplish, 5–10-flowered; empty scales acuminate, the first 1-nerved, the second longer, 3-nerved; flowering scales 5″–6″ long, acuminate, very rough-pubescent, 5-nerved, the intermediate nerves faint; awn 2″–3″ long.

In waste places, Maine to Ontario and New York. Adventive from Europe. July–Aug.

8. Bromus inérmis Leyss. Hungarian or Awnless Brome-grass. Fig. 669.

Bromus inermis Leyss. Fl. Hal. 16. 1761.

Culms tufted, $2\frac{1}{2}°-3\frac{1}{2}°$ tall. Sheaths smooth and glabrous; blades 6'–10' long, 2''–3'' wide, smooth and glabrous; panicle 6'–10' long, oblong, the axis hispidulous, the hispidulous branches ascending in clusters; spikelets 10''–$12\frac{1}{2}$'' long, about $2\frac{1}{2}$'' wide, oblong, erect, the first scale 1-nerved, the second 3-nerved, the flowering scales 5''–6'' long, 5–7-nerved, awnless or rarely awn-pointed.

In fields and waste places, South Dakota, Ohio and Colorado. Introduced from Europe. June and July.

9. Bromus Pòrteri (Coulter) Nash. Porter's Chess. Fig. 670.

Bromus Kalmii var. *Porteri* Coulter, Man. Bot. Rocky Mt. Region 425. 1885.
Bromus Porteri Nash, Bull. Torr. Club **22**: 512. 1895.

Culms $1\frac{1}{2}°-3°$ tall, erect, simple, pubescent below the nodes. Sheaths shorter than the internodes, glabrous or sometimes softly pubescent; ligule $\frac{1}{2}$'' long, truncate; blades 1''–3'' wide, rough, those of the culm 4'–9' long, the basal narrow and about one-half of the length of the culm; panicle 3'–6' in length, its branches drooping and flexuous, at least when old, the nodes of the axis pubescent; spikelets 5–10-flowered, 9''–15'' long, on slender flexuous pedicels; empty scales pubescent, the first narrower than the second, both 3-nerved; flowering scales 5''–6'' long, obtuse, 5–7-nerved, densely pubescent with long silky hairs; awn 1''–2'' long.

In dry soil, South Dakota to Montana, south to western Nebraska, New Mexico and Arizona. July–Aug.

10. Bromus Kàlmii A. Gray. Kalm's Chess. Wild Chess. Fig. 671.

Bromus Kalmii A. Gray, Man. 600. 1848.

Culms $1\frac{1}{2}°-3°$ tall, erect, simple, smooth and glabrous. Sheaths shorter than the internodes, more or less pubescent; ligule very short; blades $2\frac{1}{2}$'–7' long, 1''–4'' wide, sparingly pubescent; panicle 2'–6' in length, open, its branches usually flexuous; spikelets 6–10-flowered, 6''–12'' long, on slender flexuous pedicels; empty scales pubescent, the first narrow, acute, 3-nerved, the second longer, broad, obtuse or mucronate, 5–7-nerved; flowering scales about 4'' long, 7–9-nerved, densely silky pubescent, the awn 1''–$1\frac{1}{2}$'' in length.

In woods and thickets, Quebec to Manitoba, New Jersey, Pennsylvania and Missouri. July–Aug.

11. Bromus hordeàceus L. Soft Chess. Fig. 672.

Bromus hordeaceus L. Sp. Pl. 77. 1753.

Bromus mollis L. Sp. Pl. Ed. 2, 112. 1762.

Culms 8'–3° tall, erect, often slender, usually pubescent below the panicle. Sheaths shorter than the internodes, mostly pubescent; ligule ½" long; blades 1'–7' long, 1"–3" wide, pubescent; panicle generally contracted, its branches erect or ascending, 1'–2' long; spikelets appressed-pubescent, on short pedicels; empty scales acute, the first 3-nerved, the second longer, 5–7-nerved; flowering scales broad, obtuse, 3½"–4½" long, 7–9-nerved, bearing an awn 3"–4" in length between the obtuse or acute teeth.

In fields and waste places, Nova Scotia to British Columbia, California and North Carolina. Locally adventive from Europe. Soft Brome. Haver-grass. Blubber-, Hooded-, Bull-, Lob- or Lop-grass. July–Aug.

12. Bromus secálinus L. Cheat. Chess. Fig. 673.

Bromus secalinus L. Sp. Pl. 76. 1753.

Culms 1°–3° tall, erect, simple, smooth and glabrous. Sheaths usually shorter than the internodes, generally glabrous; ligule ½" long, erose; blades 2'–9' long, 1"–3" wide, smooth or rough, sometimes hairy; panicle 2'–8' in length, open, its branches ascending or drooping; spikelets turgid, glabrous, erect or somewhat pendulous, 6–10-flowered; empty scales scabrous toward the apex, the first 3-nerved, acute, the second longer and broader, 7-nerved, obtuse; flowering scales 3"–4" long, broad, turgid, obtuse, rough toward the apex, the nerves obscure, awnless, or bearing a straight awn 4" long or less between the obtuse short teeth; palet about equalling the scale.

In fields and waste places almost throughout temperate North America, often a pernicious weed in grain fields. Naturalized from Europe. Native also of Asia. Smooth Rye-brome. Cock-grass. June–Aug.

13. Bromus racemòsus L. Upright Chess. Smooth Brome-grass. Fig. 674.

Bromus racemosus L. Sp. Pl. Ed. 2, 114. 1762.

B. commutatus Schrad. Fl. Germ. 1 : 353. 1806.

Culms 1°–3° tall, erect, simple, smooth and glabrous, or sparingly pubescent below the panicle. Sheaths shorter than the internodes, glabrous or pubescent; ligule 1" long; blades 1'–9' long, ½"–4" wide, pubescent; panicle 1'–10' in length, the branches erect or ascending, the lower sometimes 2½' long; spikelets erect, 5–11-flowered; empty scales acute, the first 3-nerved, the second longer and broader, 5–9-nerved; flowering scales broad, 3½"–4½" long, obtuse, smooth and shining, the nerves prominent; awn straight, 3"–4" in length; palet considerably shorter than the scale.

In fields and waste places all over the United States and British America. Naturalized from Europe. Native also of Asia. June–Aug.

14. Bromus arvénsis L. Field Chess or Brome. Fig. 675.

Bromus arvensis L. Sp. Pl. 77. 1753.

Culms erect, 1°–3° tall, smooth and shining, glabrous except at or near the brown nodes. Sheaths shorter than the internodes, softly and densely pubescent with short reflexed hairs; ligule scarious, ½″–1½″ long; blades erect or ascending, more or less hirsute on both surfaces, 3′–6′ long, 2″–3″ wide; panicle ample, 5′–9′ long, its rough branches erect or ascending, rarely spreading, branching and spikelet-bearing above the middle, the longer 3′–6′ long; spikelets, including the awns, 9″–12″ long, lanceolate, somewhat shining, the scales membranous, scarious on the margins, minutely and sparsely appressed-pubescent toward the acute apex, papillose along the nerves, the first scale 3-nerved, the second 5-nerved; flowering scales broadest at the middle, 5-nerved, 3½″–4″ long, bearing an erect awn of about the same length.

Fields and waste places, New York to Michigan, Missouri and Florida. Locally adventive from Europe. Summer.

15. Bromus pátulus M. & K. Spreading Brome-grass. Fig. 676.

Bromus patulus M. & K. in Roehl. Deutsch. Fl. 1 : 684. 1823.

Culms 1°–1½° tall. Sheaths softly pubescent; blades up to 6′ long and about 2″ broad, pubescent; panicle 5′–8′ long, diffuse, somewhat drooping; spikelets drooping, on slender pedicels, lanceolate, 10″–12″ long, about 2½″ broad, glabrous, the first scale 3-nerved, the second one 5-nerved, the flowering scales 9-nerved, 3½″–4½″ long, emarginate at the apex, the awn 4″–5″ long, usually twisted and divaricate at maturity, inserted below the apex of the scale.

Sparingly introduced into Massachusetts, South Dakota and Colorado. July and Aug.

16. Bromus squarròsus L. Corn Brome. Fig. 677.

Bromus squarrosus L. Sp. Pl. 76. 1753.

Culms 8′–18′ tall, erect, simple, smooth and glabrous. Sheaths shorter than the internodes, softly pubescent; ligule ½″ long; blades 1′–5′ long, 1″–2″ wide, softly pubescent; panicle 2′–6′ in length, open, the branches ascending or drooping, often flexuous; spikelets nodding, 6–12-flowered, on slender pedicels; empty scales obtuse or acutish, the first 5-nerved, the second longer, 7–9-nerved; flowering scales 4½″–5½″ long, obtuse, shining, minutely scabrous; awn inserted below the apex, about as long as the scale, bent at the base and divergent.

In ballast and waste places about the eastern seaports. Fugitive or adventive from Europe. July–Aug.

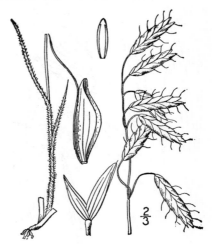

17. Bromus brizaefórmis Fisch. & Mey.
Quakegrass or Awnless Brome.   Fig. 678.

Bromus brizaeformis Fisch. & Mey. Ind. Sem. Hort.
Petrop. **3**: 30.   1836.

Culms 8′–2° tall, erect, simple, often slender,
smooth and glabrous.   Sheaths shorter than the
internodes, the lower pubescent with soft villous
hairs; ligule 1″ long, erose-truncate; blades 1′–7′
long, 1″–3″ wide, pubescent; panicle 1½′–8′ in length,
open, the branches ascending or often drooping,
flexuous; spikelets few, ½′–1′ long, laterally much
compressed; empty scales very obtuse, often purplish, glabrous or minutely pubescent, the first 3–5-
nerved, the second larger, 5–9-nerved; flowering
scales 3″–4″ long, very broad, obtuse, 9-nerved, shining, glabrous or sometimes minutely pubescent, unawned.

Sparingly introduced from Massachusetts to Michigan,
Delaware and Indiana; also from Montana to British
Columbia, California and Colorado.   Native of northern Europe and Asia.   Briza-like brome.
July–Aug.

18. Bromus breviaristàtus (Hook.) Buckl.
Short-awned Chess, or Brome.   Fig. 679.

Ceratochloa breviaristata Hook. Fl. Bor. Am. **2**: 253.   1840.
B. breviaristatus Buckl. Proc. Acad. Phila. **1862**: 98.   1862.

Culms 1°–4° tall, erect, simple, smooth or rough,
sometimes pubescent below the panicle.   Sheaths pubescent, at least the lower ones, which are often overlapping; ligule 1″ long, truncate; blades 6′–1° long or more,
2″–6″ wide, rough and often pubescent; panicle 4′–15′
in length, its branches erect or ascending, the lower
2′–6′ long; spikelets 5–10-flowered; empty scales acute,
pubescent, the first 3–5-nerved, the second longer, 5–9-
nerved; flowering scales compressed, keeled, 6″–7″ long,
acute, 7–9-nerved, appressed-pubescent; awn 2″–3″ long.

In dry soil, Manitoba to British Columbia, south to Iowa,
Arizona and California, and as an occasional escape from
cultivation eastward.   July–Aug.

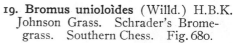

19. Bromus unioloìdes (Willd.) H.B.K.
Johnson Grass.   Schrader's Brome-
grass.   Southern Chess.   Fig. 680.

Festuca unioloides Willd. Hort. Berol. **1**: 3. *pl. 3.*
1806.
B. unioloides H.B.K. Nov. Gen. **1**: 151.   1815.
Bromus Schraderi Kunth, Enum. **1**: 416.   1833.

Culms 6′–3° tall, erect, simple, smooth and glabrous.   Sheaths usually shorter than the internodes, the lower often overlapping, smooth or
rough, and glabrous or frequently pubescent;
ligule 1″–2″ long; blades 3′–13′ long, 1″–4″ wide,
usually rough, at least above; panicle 2′–10′ in
length, the branches erect or ascending, or the
lower branches of the larger panicles widely
spreading; spikelets much compressed, 6–10-flow-
ered; empty scales acute, the first 3–5-nerved, the
second longer, 5–9-nerved; flowering scales 6″–8″
long, very acute, minutely scabrous, bearing an
awn less than 1″ long or awnless.

Missouri to the Indian Territory, Texas, Georgia,
Florida and Mexico.   Widely distributed in tropical
America.   Rescue- or Wild Brome-grass.   May–July.

100. NARDUS L. Sp. Pl. 53. 1753.

A low perennial tufted grass, with setaceous rigid leaf-blades and a terminal one-sided slender spike. Spikelets 1-flowered, narrow, sessile and single in each notch of the rachis. Scales 2, the lower empty, adnate to the rachis, or almost wanting, the upper flower-bearing, narrow, with involute and hyaline margins; palet narrow, 2-nerved. Stamens 3. Style elongated, undivided. Stigma elongated, short-papillose. Grain linear, glabrous, enclosed in the scale, usually free. [Greek name of spikenard, of uncertain application.]

A monotypic genus of the Old World. Type species: *Nardus stricta* L.

1. Nardus stricta L. Wire-bent. Mat-grass. Nard. Fig. 681.

Nardus stricta L. Sp. Pl. 53. 1753.

Culms 5'-15' tall, erect, simple, rigid, roughish. Sheaths usually at the base of the culm; ligule ½" long, rounded; blades setaceous, stiff, rough, the 1 or 2 culm leaves about 1' long, erect, the basal ones numerous, 2'-5' long; spike 1'-3' in length, strict; spikelets 1-flowered, 3"-4" long, arranged alternately in 2 rows on one side of the erect slender rachis, often purplish; lower scale empty, very short, adnate to the rachis, sometimes almost wanting; flowering scale 3"-4" long, scabrous, long-acuminate or short-awned.

Introduced into Newfoundland and at Amherst, Mass. Adventive from Europe. Black Bent. Mat. Reed-grass. July–Aug.

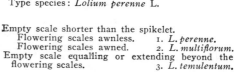

101. LOLIUM L. Sp. Pl. 83. 1753.

Annual or perennial grasses, with flat leaf-blades and terminal spikes. Spikelets several-flowered, solitary, sessile and alternate in the notches of the usually continuous rachis, compressed, the edge of the spikelet (backs of the scales) turned toward the rachis. Scales rigid; lower scale empty in the lateral spikelets, and the 2 lower empty in the terminal; flowering scales rounded on the back, 5-7-nerved; palets 2-keeled. Stamens 3. Styles distinct, very short. Stigmas 2, plumose. Grain adherent to the palet. [Latin name for Darnel.]

About 6 species, natives of the Old World. Type species: *Lolium perenne* L.

Empty scale shorter than the spikelet.
    Flowering scales awnless.   1. *L. perenne.*
    Flowering scales awned.   2. *L. multiflorum.*
Empty scale equalling or extending beyond the
    flowering scales.   3. *L. temulentum.*

1. Lolium perénne L. Ray-grass. Red Ray. Rye-grass. Ever-grass. Fig. 682.

Lolium perenne L. Sp. Pl. 83. 1753.

Smooth and glabrous, culms 6'-2½° tall, erect, simple. Sheaths shorter than the internodes; ligule very short; blades 2'-5' long, 1"-2" wide; spike 3'-8' in length; spikelets 5-10-flowered, 4"-6" long, the empty scale shorter than the spikelet, strongly nerved; flowering scales 2"-3" long, obscurely nerved, acuminate.

In waste places and cultivated grounds almost throughout the northern United States and southern British America. Naturalized from Europe. Native also of Asia. Erroneously called Darnel, this name belonging to *Lolium temulentum*. Crap. Perennial Rye. Red Darnel. Red Dare. White Nonesuch. English Blue-grass. July–Aug.

2/3

2. Lolium multiflòrum Lam. Awned or Italian Rye-grass. Fig. 683.

Lolium multiflorum Lam. Fl. Franc. **3**: 621. 1778.
Lolium italicum A. Br. Flora **17**: 259. 1834.

Culms tufted, 2°–3° tall. Sheaths usually shorter than the internodes, smooth and glabrous; blades 4'–8' long, 1½''–4'' wide, smooth and glabrous; spikes 8'–12' long; spikelets 20–30, the empty scale shorter than the spikelet, 7''–10'' long, strongly nerved, the flowering scales bearing an awn equalling or shorter than itself.

In fields and waste places, New York, New Jersey, Missouri and Iowa. June–Aug.

3/4

3. Lolium temuléntum L. Darnel. Poison Darnel. Ivray. Fig. 684.

Lolium temulentum L. Sp. Pl. 83. 1753.

Glabrous. Culms 2°–4° tall, erect, simple, smooth. Sheaths overlapping or shorter than the internodes; ligule 1'' long or less; blades 4'–10' in length, 1''–3'' wide, smooth beneath, rough above; spike 4'–12' in length; spikelets 4–8-flowered, 5''–9'' long, the strongly nerved empty scale equalling or extending beyond the obscurely nerved flowering scales, which are awned or awnless.

In waste places and cultivated grounds, locally naturalized or adventive from Europe, New Brunswick to Michigan, Georgia and Kansas. Abundant on the Pacific Coast. Locally a troublesome weed. Bearded Darnel. Sturdy Ryle. Tare. Drunk. Drawke. Dragge. Neale. Cheat. June–Aug.

102. LEPTÙRUS R. Br. Prodr. Fl. Nov. Holl. **1**: 207. 1810.

Usually low annual grasses, with narrow leaf-blades and strict or curved elongated slender spikes. Spikelets 1–2-flowered, sessile and single in alternate notches of the jointed rachis. Empty scales 2, rarely 1, narrow, rigid, acute, 5-nerved; flowering scales much shorter, hyaline, keeled, one side turned to the rachis. Palets hyaline, 2-nerved. Stamens 3, or fewer. Styles short, distinct. Stigmas 2, plumose. Grain narrow, glabrous, free, enclosed in the scale. [Greek, referring to the narrow spikes.]

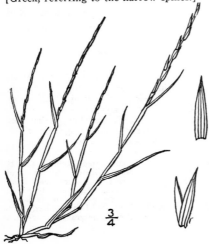

3/4

Species 5 or 6, natives of the Old World. Type species: *Lepturus repens* R. Br.

1. Lepturus filifórmis (Roth) Trin. Slender Hard-grass. Fig. 685.

Rottboellia filiformis Roth, Catal. **1**: 21. 1797.
L. filiformis Trin. Fund. Agrost. 123. 1820.

Culms 3'–12' long, decumbent, much branched, smooth and glabrous. Sheaths loose, shorter than the internodes; ligule ½'' long, auriculate; blades ½'–2' long, 1'' wide or less, usually involute, smooth beneath, rough above; spikes 1'–6' in length, slender, strict or curved; spikelets 2''–2½'' long; empty scales acute; flowering scales about 1¾'' long, 1-nerved.

In waste places and brackish marshes, southern Pennsylvania to Virginia, near or along the coast. Adventive from Europe. Summer.

103. AGROPYRON J. Gaertn. Nov. Comm. Petrop. **14**: Part 1, 539. 1770.

Annual or perennial grasses, with flat or involute leaf-blades and terminal spikes. Spikelets 3–many-flowered, sessile, single and alternate at each notch of the usually continuous rachis, the side of the spikelet turned toward the rachis. Two lower scales empty; flowering scales rigid, rounded on the back, 5–7-nerved, usually acute or awned at the apex; palets 2-keeled, the keels often ciliate. Stamens 3. Styles very short, distinct. Stigmas plumose. Grain pubescent at the apex, usually adherent to the palet. [Greek, referring to the growth of these grasses in wheat fields.]

About 50 species, in all temperate regions. Type species: *Agropyron cristatum* J. Gaertn.

Culms not densely tufted; plants with creeping rootstocks or stolons.
  Spikelets glabrous or hispidulous.
    Empty scales strongly 5–11-nerved.
      Empty scales attenuate into an awn or awn-point; plant green.
        Under surface of the leaf-blades smooth, the upper surface often pubescent; an
        introduced weed.   1. *A. repens.*
        Under surface of leaf-blades very rough, the upper surface glabrous; a western grass.
                                         2. *A. pseudorepens.*
      Empty scales rather abruptly narrowed to a blunt point; plant glaucous.
                                          3. *A. pungens.*
      Empty scales usually faintly 1–3-nerved, sometimes 5-nerved.   4. *A. Smithii.*
  Spikelets densely pubescent.   5. *A. dasystachyum.*
Culms densely tufted; plants with no rootstocks or stolons.
  Awn shorter than the flowering scale.
    Empty scales broad above the middle.   6. *A. biflorum.*
    Empty scales narrowed from below the middle.   7. *A. tenerum.*
  Awn much longer than the flowering scale.   8. *A. caninum.*

1. Agropyron rèpens (L.) Beauv. Couch-grass. Quitch-grass. Fig. 686.

Triticum repens L. Sp. Pl. 86. 1753.
Agropyron repens Beauv. Agrost. 146. 1812.

Culms 1°–4° tall, from a long jointed running rootstock. Sheaths usually shorter than the internodes, smooth and glabrous; ligule very short; blades 3′–12′ long, 1″–5″ wide, smooth beneath, rough above; spike 2′–8′ in length, strict; spikelets 3–7-flowered; empty scales strongly 5–7-nerved, usually acute or awn-pointed, sometimes obtuse; flowering scales smooth and glabrous, acute or short-awned at the apex.

In fields and waste places, almost throughout North America except the extreme north. Naturalized from Europe and often a troublesome weed. Very variable. Native also of Asia. Quitch-, Twitch-, or Witch-grass. Stroil. Quichens. Squitch. Wickens. Shelly-, Knot-, Dog-, Shear- or Quack-grass. Blue-joint. Slough- or Pond-grass. False Wheat. Colorado blue-grass. July–Sept.

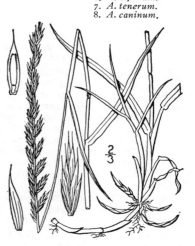

2. Agropyron pseùdo-rèpens Scribn. & Sm. False Couch-grass. Fig. 687.

Agropyron pseudorepens Scribn. & Sm. Bull. U. S. Dept. Agric. Div. Agrost. **4**: 34. 1897.

Light green. Culms 1°–3° tall, erect, smooth and glabrous, from a running rootstock; sheaths shorter than the internodes, smooth; ligule a short membranous ring; blades erect, prominently nerved, rough on both surfaces, acuminate, the culm leaves 3′–8′ long, 2″–3″ wide, the basal leaves about one-half as long as the culms; spikes 3′–8′ long, strict; spikelets 5″–8″ long, 3–7-flowered, a little compressed, appressed to the rachis which is hispidulous on the margins; empty scales lanceolate, equalling or somewhat shorter than the spikelet, acuminate and often awn-pointed, 5–7-nerved, the nerves hispidulous; flowering scales 5-nerved, roughish toward the apex, usually awn-pointed.

Rich river bottoms, British Columbia and Athabasca, to Arizona, Texas, Missouri and Iowa. July–Aug.

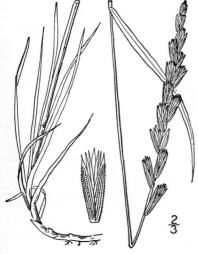

$\frac{2}{3}$

3. **Agropyron púngens** (Pers.) R. & S. Coast Wheat-grass. Fig. 688.

Triticum pungens Pers. Syn. 1 : 109. 1805.
Agropyron pungens R. & S. Syst. 2 : 753. 1817.
Agropyron tetrastachys Scribn. & Sm. Bull. U. S. Dept.
     Agric. Div. Agrost. 4 : 32. 1897.

Glaucous. Culms rigid, slender, erect, $1\frac{1}{2}°-3°$ tall,
from a running rootstock; sheaths shorter than the internodes, smooth and glabrous; ligule wanting; blades
erect, acuminate, $5'-8'$ long, $2''$ or less wide, smooth
beneath, glaucous above, scabrous on the margins;
spikes long-exserted, $3'-5'$ long, 4-sided; spikelets
crowded, $6''-10''$ long, 6-11-flowered, appressed to the
4-angled articulated rachis, the angles hispidulous;
empty scales lanceolate, $5''$ long, rough on the keel,
5-7-nerved; flowering scales lanceolate, keeled, rough
toward the apex, acute, awn-pointed or short-awned.

Sandy beaches, coast of Maine. Introduced from
Europe. July–Aug.

4. **Agropyron Smíthii** Rydb. Western Wheat-grass. Fig. 689.

A. repens glaucum Scribn. Mem. Torr. Club, **5** : 57,
     in part. 1894.
A. spicatum Scribn. & Sm. Bull. U. S. Dept. Agric.
     Div. Agrost. 4 : 33, in part. 1897.

Pale green, glaucous. Culms $1\frac{1}{2}°-4°$ tall, erect,
from a slender creeping rootstock, smooth and
glabrous; sheaths shorter than the internodes,
smooth; ligule a short membranous ring; blades
erect, $2'-8'$ long, $2''-4''$ wide, acuminate, very
scabrous above, smooth beneath, becoming involute when dry; spike long-exserted, strict, $4'-8'$
long; spikelets crowded, divergent from the
rachis, compressed, lanceolate when closed, $\frac{1}{2}'-1'$
long, 6-12-flowered; empty scales acuminate,
awn-pointed, shorter than the spikelet, hispidulous on the keel; flowering scales $5''-6''$ long,
acute or awn-pointed, glabrous or sparsely pubescent.

Moist land, Manitoba and Minnesota to British
Columbia, south to Missouri and Texas.

$\frac{2}{3}$

$\frac{2}{3}$

5. **Agropyron dasystàchyum** (Hook.) Vasey. Northern Wheat-grass. Fig. 690.

Triticum repens var. *dasystachyum* Hook. Fl. Bor. Am. **2** :
     254. 1840.
Agropyrum dasystachyum Vasey, Spec. Rept. U. S. Dept.
     Agric. **63** : 45. 1883.
Agropyron subvillosum E. Nels. Bot. Gaz. **38** : 378. 1904.

Glaucous, culms $1°-3°$ tall, erect, from long running
rootstocks, simple, smooth and glabrous; sheaths shorter
than the internodes; ligule very short; blades $2'-9'$ long,
$1''-3''$ wide, flat, or becoming involute in drying, smooth
beneath, rough above; spike $2\frac{1}{2}'-7'$ in length; spikelets
4-8-flowered; empty scales 3-5-nerved, lanceolate, acuminate or short-awned, $3''-4\frac{1}{2}''$ long; flowering scales broadly
lanceolate, 5-nerved, $4\frac{1}{2}''-6''$ long, acute or short-awned,
densely villous.

Hudson Bay to the Yukon, south to the Great Lakes,
Nebraska and Colorado. Summer.

6. Agropyron biflòrum (Brignoli) R. & S. Purplish Wheat-grass. Fig. 691.

Triticum biflorum Brignoli, Fasc. Pl. Foroj. 18.　1810.
Agropyron biflorum R. & S. Syst. **2**: 760.　1817.
Agropyrum violaceum Vasey, Spec. Rept. U. S. Dept. Agric.
63: 45.　1883.

Culms 6′–2° tall, erect, simple, smooth and glabrous.
Sheaths usually shorter than the internodes; ligule very
short; blades 2′–6′ long, 1″–3″ wide, flat or involute, rough
or sometimes smooth beneath; spike 1′–4′ in length, occasionally
longer, 2″–3″ broad; spikelets 3–6-flowered; empty
scales broad, usually purplish, scarious on the margins,
5–7-nerved, 4″–6″ long, acute or acuminate, sometimes awn-
pointed, rarely long-awned; flowering scales often purplish,
5–7-nerved, scarious on the margins, 4″–6″ long, acuminate
or short-awned, the awn rarely as long as the body.

Nova Scotia to British Columbia, south to the mountains of
New England, New York and Pennsylvania, and in the Rocky
Mountains to Colorado.　Ascends to 5500 ft. in the White
Mountains.　Also in northern Europe and Asia.　Summer.

7. Agropyron ténerum Vasey. Slender Wheat-grass. Fig. 692.

A. tenerum Vasey, Coult. Bot. Gaz. **10**: 258.　1885.

Agropyron novae-angliae Scribn. Contr. Bot. Vt. **8**: 103.
1900.

Glabrous, culms 2°–3° tall, erect, simple, often
slender, smooth.　Sheaths usually shorter than the
internodes, glabrous; ligule very short; blades 3′–10′
long, 1″–2″ wide, flat or involute, rough; spike 3′–7′
in length, usually narrow and slender; spikelets 3–5-
flowered; empty scales 4″–6″ long, acuminate or
short-awned, 3–5-nerved, scarious on the margins;
flowering scales 5″–6″ long, 5-nerved, awn-pointed
or short-awned, scarious on the margins, often rough
toward the apex.

In dry soil, Newfoundland to British Columbia, south
to Kansas, Colorado and California.　July–Aug.

8. Agropyron canìnum (L.) R. & S. Bearded or Awned Wheat-grass. Fibrous-rooted Wheat-grass. Fig. 693.

Triticum caninum L. Sp. Pl. 86.　1753.
Agropyrum caninum R. & S. Syst. **2**: 756.　1817.
Agropyrum unilaterale Cassidy, Bull. Colo. Agric. Exp.
Sta. **12**: 63.　1890.
A. Richardsoni Schrad. Linnaea **12**: 467.　1838.

Culms 1°–3° tall, erect, simple, smooth and glabrous.　Sheaths usually shorter than the internodes,
smooth, the lower sometimes pubescent; ligule short;
blades 3′–9′ long, 1″–3″ wide, smooth beneath, rough
above; spike 3′–8′ in length, sometimes one-sided,
often nodding at the top; spikelets 3–6-flowered;
empty scales 4½″–6″ long, 3–5-nerved, acuminate,
awn-pointed or bearing an awn 1″–3″ long; flowering
scales 4″–5″ long, usually scabrous toward the apex,
acuminate into an awn sometimes twice their own
length.

New Brunswick to the Yukon, south to North Carolina, Tennessee, Iowa and Colorado.　Also in Europe
and Asia.　Native northward; southward locally naturalized from Europe.　Dog's-tooth grass.　July–Aug.

104. HÓRDEUM (Tourn.) L. Sp. Pl. 84. 1753.

Annual or perennial grasses, with flat leaf-blades and terminal cylindric spikes. Spikelets 1-flowered, rarely 2-flowered, usually in 3's at each joint of the rachis, the lateral short-stalked and imperfect; rachilla produced beyond the flower, the lower empty scales often reduced to awns and forming an apparent involucre around the spikelets. Empty scales rigid; flowering scales rounded on the back, 5-nerved at the apex, awned; palet scarcely shorter than the scale, 2-keeled. Stamens 3. Styles very short, distinct. Grain usually adherent to the scale, hairy at the summit. [Latin name for Barley.]

About 20 species, widely distributed in both hemispheres. Type species: *Hordeum vulgare* L.

Lateral spikelets abortive.
  Flowering scales, exclusive of awns, 3″–4″ long.
    Awn of the flowering scale ½′ long or less.
      All the empty scales of each cluster bristle-like.   1. *H. nodosum.*
      Four of the empty scales of each cluster dilated above the base.   2. *H. pusillum.*
    Awn of the flowering scale 1′ long or more.   3. *H. jubatum.*
  Flowering scales, exclusive of awns, about 6″ long.   4. *H. murinum.*
Lateral spikelets with perfect flowers.   5. *H. Pammelii.*

1. Hordeum nodòsum L.   Meadow Barley.
Fig. 694.

Hordeum nodosum L. Sp. Pl. Ed. 2, 126. 1762.
Hordeum pratense Huds. Fl. Angl. Ed. 2, 56. 1762.

Culms 6′–2° tall, erect, or sometimes decumbent, simple, smooth and glabrous. Sheaths shorter than the internodes; ligule ¼″ long, truncate; blades 1½′–5′ long, 1″–3′ wide, flat, rough; spike 1′–3½′ in length; spikelets usually in 3's, the central one containing a palet and perfect flower, the lateral enclosing a staminate or rudimentary flower, or a palet only; empty scales of each cluster awn-like; flowering scale of the central spikelets 3″–4″ long exclusive of the awn, which is 3″–6″ long, the corresponding scale in the lateral spikelets much smaller and short-stalked.

In meadows and waste places, Indiana to Minnesota, Alaska, Texas and California. Also in Europe and Asia. June–July.

2. Hordeum pusíllum Nutt.   Little Barley.
Fig. 695.

Hordeum pusillum Nutt. Gen. 1: 87. 1818.

Culms 4′–15′ tall, erect, or decumbent at the base, smooth and glabrous. Sheaths loose, usually shorter than the internodes, smooth and glabrous, the upper often enclosing the base of the spike; ligule very short; blades ½′–3′ long, ½″–2″ wide, erect, smooth beneath, rough above; spike 1′–3′ in length; spikelets usually in 3's, the central one containing a palet and perfect flower, the lateral imperfect; scales awned, the empty ones scabrous, those of the central spikelet and the lower ones of the lateral spikelets dilated above the base; flowering scale smooth, that of the central spikelet 3″–4″ long, short-awned, the corresponding scale in the lateral spikelets smaller and very short-stalked.

In dry soil, Indiana to Tennessee, Wyoming, California, Texas and Louisiana; also sparingly introduced along the coast from Virginia to Florida. June–July.

3. Hordeum jubàtum L. Squirrel-tail Grass. Fig. 696.

Hordeum jubatum L. Sp. Pl. 85. 1753.

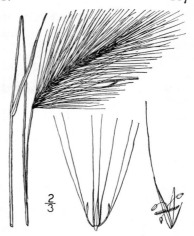

Culms 10′-2½° tall, erect, simple, usually slender, smooth and glabrous. Sheaths usually shorter than the internodes, generally loose, smooth and glabrous; ligule ½″ long or less; blades 1′-5′ long, 1″-2″ wide, erect, rough; spike 2′-4′ in length; spikelets usually in 3's, the central one containing a palet and perfect flower, the lateral imperfect; empty scales consisting of slender rough awns 1′-2½′ long; flowering scale of the central spikelet 3″-4″ long, scabrous at the apex, bearing a slender rough awn 1′-2½′ long; the corresponding scale in the lateral spikelets short-awned, about 3″ long including its pedicel, sometimes reduced to a rudiment.

In dry soil, Ontario to Alaska, south to Illinois, Texas and California. Naturalized in the east from Labrador and Quebec to New Jersey and Pennsylvania. July–Aug.

4. Hordeum murìnum L. Wall or Way Barley. Way Bent. Fig. 697.

Hordeum murinum L. Sp. Pl. 85. 1753.

Culms 6′-2° tall, erect, or decumbent at the base, smooth and glabrous. Sheaths loose, shorter than the internodes on the long culms, overlapping on the short ones, the uppermost often inflated and enclosing the base of the spike; ligule very short; blades 1′-6′ long, 1″-3″ wide, rough; spikes 2′-4′ in length; spikelets usually in 3's; scales awned, the empty ones awn-like, scabrous, those of the central spikelet broader and ciliate on the margins, bearing awns 9″-12″ long, those of the lateral spikelets similar, with the exception of the second scale, which is not ciliate; flowering scales scabrous at the apex, bearing an awn about 1′ long, those of the lateral spikelets about 6″ long, the corresponding scale in the central spikelet somewhat smaller.

On ballast and sparingly in waste places, Massachusetts to District of Columbia. Also from Arizona to Utah, California and British Columbia. Adventive or naturalized from Europe. Wild- or Mouse-barley. Squirrel-tail. June–July.

5. Hordeum Pammélii Scribn. & Ball. Pammel's Barley. Fig. 698.

H. Pammelii Scribn. & Ball, Ia. Geol. Surv. Suppl. Rep. 1903: 335. *f. 237.* 1904.

Perennial. Culms 2°-3° tall; leaves glabrous; sheaths smooth, shorter than the internodes; blades 4′-8′ long, 2″-4″ wide, rough; spikes 3′-6′ long, 10″-15″ in diameter; lateral spikelets nearly sessile, each with a perfect flower, the central spikelet with 2 perfect flowers, the empty scales subulate, long-awned, the flowering scales lanceo-late, about 3″ long, exclusive of the awn, which is two to three times their length.

On prairies, Illinois to South Dakota and Wyoming. July and Aug.

105. SITANION Raf. Journ. de Phys. 89: 103. 1819.

Tufted grasses, with flat or involute leaf-blades, and a terminal dense spike with the rachis articulated and readily breaking up. Spikelets numerous, in 2's or 3's at each node, 2–5-flowered; empty scales entire or divided, the divisions extending often to the base, the scales or their divisions bearing long slender awns; palet 2-keeled. Stamens 3. Styles distinct, short. Stigmas plumose. Grain adherent to the palet. [Greek, a kind of food.]

Species 12, or perhaps more, mainly natives of the western United States. Type species: *Sitanion elymoides* Raf.

1. Sitanion elymoïdes Raf.   Long-bristled Wild Rye.   Fig. 699.

Sitanion elymoides Raf. Journ. Phys. **89**: 103.   1819.
Elymus Sitanion Schultes, Mant. **2**: 426.   1824.
Elymus elymoides Swezey, Neb. Fl. Pl. 15.   1891.
S. brevifolium J. G. Smith, Bull. U. S. Dep. Agr. Agrost. **18**: 17.   1899.
S. longifolium J. G. Smith, Bull. U. S. Dep. Agr. Agrost. **18**: 18.   1899.

Culms 1°–2° tall, erect. Sheaths smooth or rough, sometimes hirsute, usually overlapping, the upper one often inflated and enclosing the base of the spike; blades 2'–7' long, ½"–2" wide, often stiff and erect, usually rough, sometimes hirsute, flat or involute; spike 2'–6' in length; spikelets 1–5-flowered; empty scales entire, awl-shaped; flowering scales 4"–5" long, 5-nerved, scabrous, bearing a long slender divergent awn 1½'–3' in length, the apex of the scale sometimes 2-toothed.

In dry soil, Wyoming to western Missouri, Texas, Arizona and Nevada. July–Aug.

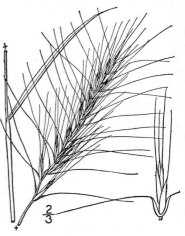

106. ÉLYMUS L. Sp. Pl. 83.   1753.

Tall grasses, with usually flat leaf-blades and dense terminal spikes. Spikelets 2–several-flowered (rarely 1-flowered), sessile, usually in pairs, occasionally in 3's or more, in alternate notches of the continuous or jointed rachis, the empty scales forming an apparent involucre to the cluster. Two lower scales empty, attached obliquely, narrow, acute or awned, entire; flowering scales shorter, rounded on the back, 5-nerved, usually bearing an awn. Palet a little shorter than the scale, 2-keeled. Stamens 3. Styles very short, distinct. Stigmas plumose. Grain sparsely hairy at the summit, adherent to the palet. [Greek, to roll up, referring to the involute palet.]

About 40 species, natives of temperate regions.   Type species: *Elymus arenarius* L.

Empty scales of the same length, equalling or longer than the flowering scales.
  Spikelets appressed; spike narrow, slender.
    Flowering scales glabrous.
      Blades less than 3" wide, rarely exceeding 2"; spikelets 1–3-flowered, the flowering scales
        4"–5" long, with an awn as long or a little longer.   1. *E. Macounii.*
      Blades 3"–7" wide; spikelets 3–6-flowered, the flowering scales 5"–6" long, with an awn
        one and a half to twice as long.   2. *E. glaucus.*
    Flowering scales appressed-hispid.   3. *E. vulpinus.*
  Spikelets spreading; spike broad and stout.
    Flowering scales muticous, or with awn rarely over a quarter as long as the scale.
      Spikelets villous.   4. *E. arenarius.*
      Spikelets glabrous.
        Empty scales subulate.   5. *E. condensatus.*
        Empty scales broad and flat, indurated at the base.   6. *E. curvatus.*
    Flowering scale with an awn as long as itself or longer.
      Empty scales linear-lanceolate to linear.
        Empty scales manifestly indurated, usually curved or bowed at the white base.
          Awn rarely exceeding one and a half times the length of the flowering scale; awn
            of the empty scales usually short.
            Flowering scales glabrous or hispidulous.
              Spike long-exserted, its own length or more, from the narrow upper sheath.
                Plant green; flowering scales glabrous or hispidulous; leaf-blades lax,
                  commonly exceeding 2" wide; a plant of the interior. 7. *E. jejunus.*
                Plant grey green, glaucous; flowering scales papillose; leaf-blades stiff,
                  2" wide or less; a plant of the brackish marshes. 8. *E. halophilus.*
              Spike included in the broad, inflated upper sheath. 9. *E. virginicus.*
             Flowering scales hirsute. 10. *E. hirsutiglumis.*
          Awn exceeding twice the length of flowering scale; awn of empty scales very long.
            Spikelets hirsute. 11. *E. australis.*
            Spikelets glabrous or hispidulous. 12. *E. glabriflorus.*
      Empty scales not indurated, not white at the base, straight. 13. *E. canadensis.*
        Flowering scales hirsute.
        Flowering scales glabrous or hispidulous. 14. *E. brachystachys.*

Empty scales narrowly subulate.
  Spikelets hirsute.   15. *E. striatus.*
  Spikelets glabrous or hispidulous.   16. *E. arkansanus.*
Empty scales variable in length, from a short point to longer than the spikelet, even in the same spike.
                                                                        17. *E. diversiglumis.*

1. Elymus Macoùnii Vasey.   Macoun's Wild Rye. Fig. 700.

Elymus Macounii Vasey, Bull. Torr. Club **13**: 119. 1886.

Culms 1°–3° tall, erect, simple, smooth and glabrous.
Sheaths shorter than the internodes; ligule very short,
truncate; blades 2′–6′ long, 1″–2½″ wide, rough, especially
above; spike 2′–5′ in length, narrow, slender, often somewhat flexuous; spikelets appressed to the rachis, single
at each node, or the lower sometimes in pairs, 1–3-flow-
ered; empty scales (occasionally 3) awl-shaped, 3-nerved,
rough, 3″–4″ long, bearing a slender straight rough awn,
3″–5″ in length; flowering scales 3½″–5″ long, rough
toward the apex, bearing a slender straight awn 3″–5″ long.

Prairies, Manitoba to Athabasca, Minnesota, Missouri and
New Mexico.   July–Aug.

2. Elymus glaùcus Buckl.   Smooth Wild Rye. Fig. 701.

Elymus glaucus Buckl. Proc. Acad. Phila. **1862**: 99. 1862.
Elymus americanus V. & S.; Macoun, Cat. Can. Pl. **4**: 245.
    1888.
Elymus sibiricus var. *americanus* Wats. & Coult. in A. Gray,
    Man. Ed. 6, 673. 1890.

Culms 2°–5° tall, erect, simple, smooth and glabrous.
Sheaths often shorter than the internodes, usually
glabrous, rarely pubescent; ligule 1″ long or less; blades
4′–12′ long, 2″–8″ wide, smooth beneath, sometimes
rough above; spike 3′–8′ in length, narrow, slender;
spikelets appressed to the rachis, 3–6-flowered; empty
scales narrowly lanceolate, 4″–6″ long, acuminate or
awn-pointed, rigid, 3–5-nerved; flowering scales smooth
or slightly rough, 5″–6″ long, bearing a slender straight
rough awn 6″–9″ in length.

In moist soil, Ontario and Michigan to British Columbia,
south to Colorado and California.   June–Aug.

3. Elymus vulpìnus Rydb.   Rydberg's Wild Rye. Fig. 702.

E. vulpinus Rydb. Bull. Torr. Club **36**: 540. 1909.

Culms 1½°–2½° tall, erect, slender, tufted,
smooth and glabrous. Sheaths smooth, usually
shorter than the internodes; ligule a short membranous ring; blades erect, 4′–6′ long, 1″–3″ wide,
acuminate, smooth beneath, scabrous on the margins and sometimes also sparsely so above; spike
slender, long-exserted, 4′–6′ long; spikelets usually
in pairs, rarely single, somewhat crowded, appressed to the rachis, more or less compressed,
4–6-flowered; empty scales 4″–5″ long, shorter
than the spikelet, acuminate into a shorter awn;
flowering scales 4″–5″ long, acuminate, bearing
a slender scabrous awn, 4″–5″ long.

Bottom lands, western Nebraska. Admitted into
first edition of this work as *Agropyron Gmelini*
Scribn. & Sm.

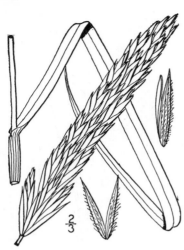

4. Elymus arenàrius L. Downy Lyme-grass. Sea Lyme-grass. Fig. 703.

Elymus arenarius L. Sp. Pl. 83. 1753.

Culms 1½°–8° tall, erect, simple, usually softly pubescent at the summit. Sheaths smooth and glabrous, often glaucous, those at the base overlapping, the upper shorter than the internodes; ligule very short; blades 3′–1° long or more, 1½″–5″ wide, flat, or becoming involute, smooth beneath, rough above; spike 3′–10′ in length, usually strict; spikelets 3–6-flowered, frequently glaucous; empty scales 8″–14″ long, 3–5-nerved, acuminate, more or less villous; flowering scales 8″–10″ long, acute or awn-pointed, 5–7-nerved, usually very villous.

On shores, Greenland and Labrador to the Northwest Territory and Alaska, Maine, New Hampshire, Lake Superior and Washington. Also in Europe and Asia. Narrow Bent, Rancheria-grass. Marram sea-grass. Summer.

5. Elymus condensàtus Presl. Smooth Lyme-grass. Fig. 704.

Elymus condensatus Presl, Reliq. Haenk. 1: 265. 1830.

Culms 2°–10° tall, erect, simple, smooth and glabrous. Sheaths smooth and glabrous, the upper ones shorter than the internodes; ligule 2″–3″ long, truncate; blades 6′–1° long or more, 3″–12″ wide, scabrous, at least above; spike 4′–15′ in length, usually stout, strict, often interrupted below, sometimes compound at the base; spikelets 3–6-flowered, 2–several at each node of the rachis; empty scales awl-shaped, 4½″–6″ long, 1-nerved, usually rough; flowering scales 4″–5″ long, generally awn-pointed, usually rough, sometimes smooth.

In wet saline situations, Alberta to British Columbia, south to northwestern Nebraska, Arizona and California. Western or Giant Rye-grass. Bunch-grass. July–Aug.

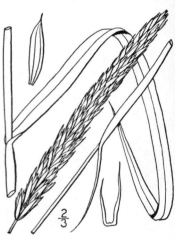

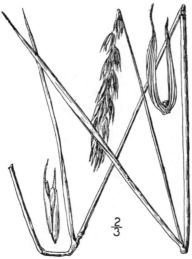

6. Elymus curvàtus Piper. Short-awned Wild Rye. Fig. 705.

E. virginicus submuticus Hook, Fl. Bor. Am. 2: 255. 1840.

Elymus curvatus Piper, Bull. Torr. Club 30: 233. 1903.

Culms 2°–3° tall, smooth and glabrous; leaves glabrous; blades up to 10′ long and 4″ wide, flat and lax or sometimes stiff and involute; spike 2′–5′ long, 3″–5″ in diameter, usually long-exserted; spikelets 5″–6″ long, glabrous, the empty scales broad, strongly nerved, manifestly indurated at the thickened curved base, muticous or more commonly short-awned, the awn less than quarter as long as the scale, the flowering scales muticous or short-awned as in the empty scales.

Low grounds, Saskatchewan to Iowa, Missouri and Kansas. July–Sept.

7. Elymus jejùnus (Ramaley) Rydb. Western Wild Rye. Fig. 706.

Elymus virginicus jejunus Ramaley, Minn. Bot. Stud.
  1: 114. 1894.
Elymus jejunus Rydb. Bull. Torr. Club **36**: 539. 1909.

Culms 2°–4° tall, slender, smooth and glabrous;
sheaths smooth and glabrous; blades rough, flat,
up to 8′ long and 5″ wide; spikes 1½′–4′ long,
5″–6″ in diameter, on long slender peduncles;
spikelets, exclusive of the awns, about 6″ long,
the empty scales indurated and somewhat curved
at the base, glabrous, linear, strongly nerved,
short-awned, the flowering scales hispidulous or
almost glabrous, bearing an awn equalling or
longer than themselves.

Along creeks and rivers in poor soil, Minnesota and
North Dakota to Nebraska. July and Aug.

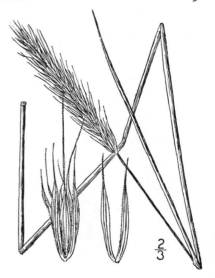

8. Elymus halóphilus Bicknell. Salt Marsh Wild Rye. Fig. 707.

Elymus halophilus Bicknell, Bull. Torr. Club **35**: 201. 1908.

Culms tufted, erect, rigid, 1½°–2½° tall; sheaths glabrous; blades up to 6′ long and 2″ wide, stiff, erect,
rough, becoming involute when dry; spike erect, long-
exserted, 1½′–4′ long; empty scales 3½″–5″ long, attenuate into a slender hispidulous awn, hispidulous
on the prominent nerves, the flowering scales 3″–3½″
long, papillose, attenuate into a slender hispidulous awn.

Salt marshes, Massachusetts to Staten Island. Plant
grey-green, glaucous. July–Sept.

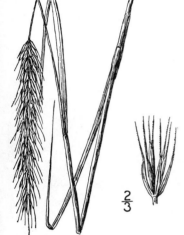

9. Elymus virgínicus L. Terrell-grass. Virginia Wild Rye. Fig. 708.

Elymus virginicus L. Sp. Pl. 84. 1753.

Culms 2°–3° tall, erect, simple, smooth and glabrous.
Sheaths usually shorter than the internodes, often
overlapping on the lower part of the culm, smooth,
sometimes pubescent, the uppermost often inflated and
enclosing the peduncle and the base of the spike;
ligule very short; blades 5′–14′ long, 2″–8″ wide,
rough; spike 2′–7′ in length, broad, stout, upright;
spikelets divergent from the rachis, 2–3-flowered; empty
scales thick and rigid, lanceolate, 8″–12″ long, including the short awn, 5–7-nerved; flowering scales
3″–4″ long, glabrous, bearing a rough awn 2″–6″ in
length.

In moist soil, especially along streams, Nova Scotia to
Manitoba, south to Florida and Texas. Ascends to 2000 ft.
in North Carolina. Virginia Lyme-grass. July–Aug.

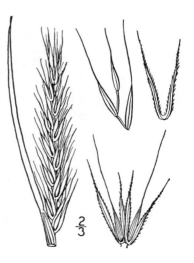

10. Elymus hirsutiglùmis Scribn.   Strict Wild Rye.   Fig. 709.

Elymus canadensis var. *intermedius* Vasey; Wats. & Coult. in A. Gray, Man. Ed. 6, 673.   1890.
Elymus hirsutiglumis Scribn. U. S. Dep. Agr. Agrost. **11**: 58.   1898.

Culms erect from a perennial root, 2°–3° tall, smooth and glabrous. Sheaths longer than the internodes, smooth, the uppermost often inflated and enclosing the base of the spike; ligule a short membranous ring; blades 7′–12′ long, 4″–9″ wide, acuminate, very rough on both surfaces; spikes 2½′–6′ long, stout, the rachis pubescent; spikelets crowded, in pairs, 2–5-flowered; empty scales linear, 5″–6″ long, thick, 3–5-nerved, the nerves hirsute, acuminate into a scabrous awn as long as or shorter than the scales; flowering scales lanceolate, 5-nerved, appressed-hirsute, 4″–5″ long, acuminate into a rough awn 6″–8″ long.

River banks, Maine to Virginia, Tennessee, Missouri and Nebraska. July–Aug.

11. Elymus austràlis Scribn. & Ball.   Southern Wild Rye.   Fig. 710.

Elymus australis Scribn. & Ball, Bull. U. S. Dep. Agr. Agrost. **24**: 46. *f. 20.* 1901.

Culms 3°–4° tall, erect; sheaths glabrous or hirsute; blades up to 1° long, 5″–8″ wide, rough, sometimes hirsute on the upper surface; spike 4′–6′ long, 1′–1½′ in diameter over all, exserted; empty scales thick, indurated and curved at the base, usually hirsute, long-attenuate into a long awn, the flowering scales 4″–5″ long, hirsute, bearing a hispidulous awn 1′–1¼′ long.

Moist woods and thickets, Connecticut to Missouri, south to Florida and Arkansas. June–Aug.

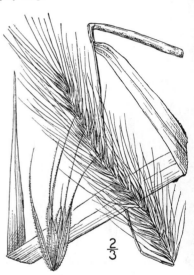

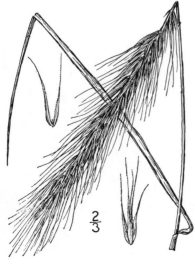

12. Elymus glabriflòrus (Vasey) Scribn. & Ball.   Smooth Southern Wild Rye.   Fig. 711.

Elymus canadensis glabriflorus Vasey; Dewey, Contr. U. S. Nat. Herb. **2**: 550.   1894.
Elymus glabriflorus Scribn. & Ball, Bull. U. S. Dep. Agr. Agrost. **24**: 49. *f. 23.* 1901.

Culms 2°–3° tall, erect, stout; sheaths glabrous or hirsute; blades up to 1° long, 3″–5″ wide, flat or nearly so, rough, sometimes sparsely hirsute on the upper surface; spike 4′–6′ long, stout, sometimes nodding; spikelets 2–3 at each node, the empty scales thick, indurated and somewhat curved at the base, strongly nerved, sometimes ciliate on the margins, attenuate into a long hispidulous awn, the flowering scales glabrous or hispidulous, bearing a long hispidulous awn.

Low woods or thickets, Pennsylvania to Iowa, south to Florida, Texas and New Mexico. June–Aug.

13. Elymus canadénsis L.   Nodding Wild Rye.   Canada Lyme-grass.   Fig. 712.

Elymus canadensis L. Sp. Pl. 83.   1753.
Elymus glaucifolius Willd. Enum. 1 : 131.   1809.
Elymus canadensis var. glaucifolius Torr. Fl. U. S. 1 : 137.   1824.
Elymus robustus Scribn. & Sm. Bull. U. S. Dep. Agr. Agric. 4 : 37.   1897.

Culms 2½°–5° tall, erect, simple, smooth and glabrous. Sheaths usually overlapping; ligule very short; blades 4'–1° long or more, 2''–10'' wide, rough, sometimes glaucous; spike 4'–12' in length, broad, stout, often nodding, its peduncle much exserted; spikelets divergent from the rachis, 3–5-flowered; empty scales narrowly lanceolate, rigid, 3–5-nerved, 8''–16'' long, including the long slender rough awns; flowering scales 4''–7'' long, hirsute, bearing a slender scabrous straight or divergent awn 10''–25'' in length.

On river banks, Nova Scotia to Alberta and Washington, south to New Jersey, West Virginia, Missouri and Arizona. Ascends to 2100 ft. in Virginia. July–Aug.

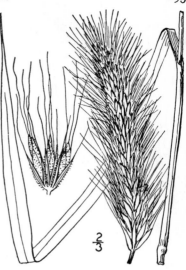

14. Elymus brachýstachys Scribn. & Ball.   Short-spiked Wild Rye.   Fig. 713.

Elymus brachystachys Scribn. & Ball, Bull. U. S. Dep. Agr. Agrost. 24 : 47. f. 21.   1901.

Culms 1°–3° tall, erect; sheaths glabrous; blades up to 8' long, 3''–6'' wide, rough, or sometimes smooth below; spike 3'–6' long, 1'–1½' in diameter, long-exserted; spikelets 3–5-flowered, in pairs, the empty scales hispidulous, flat, narrowly lanceolate, 3–5-nerved, attenuate into a hispidulous awn about three times their length, the flowering scales glabrous or hispidulous, bearing an awn 10''–20'' long.

Moist grounds, Maryland to Michigan, South Dakota, Texas and Mexico. July and Aug. Figured in our first edition as E. canadensis L. from which it is now distinguished.

15. Elymus striàtus Willd.   Slender Wild Rye.   Dennett-grass.   Fig. 714.

Elymus striatus Willd. Sp. Pl. 1 : 470.   1797.
Elymus striatus var. villosus A. Gray, Man. 603.   1848.
Elymus striatus Ballii Pammel, Ia. Geol. Surv. Suppl. Rep. 1903 : 347.   1904.

Culms 2°–3° tall, erect, slender, simple, smooth, glabrous. Sheaths usually shorter than the internodes, glabrous or hirsute; ligule very short; leaves 5'–9' long, 2''–5'' wide, smooth or slightly rough beneath, pubescent above; spike 2½'–4½' in length, broad, slender, dense; spikelets divergent from the rachis, 1–3-flowered; empty scales awl-shaped, 9''–12'' long, including the slender rough awn, 1–3-nerved, hirsute; flowering scales about 3'' long, hirsute, bearing a slender rough awn 8''–15'' in length.

In woods and on banks, Maine to North Dakota, North Carolina and Texas. Spike often nodding. June–July.

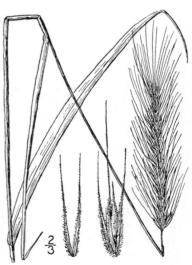

$\frac{2}{3}$

17. Elymus diversiglùmis Scribn. & Ball. Various-glumed Wild Rye. Fig. 716.

Elymus diversiglumis Scribn. & Ball, Bull. U. S. Dep. Agr. Agrost. **24**: 48. *f. 22*. 1901.

Culms $2\frac{1}{2}°$–4° tall, stout, erect; sheaths glabrous; blades up to 10′ long, 3″–6″ wide, rough; spike 4′–6′ long, flexuous; spikelets 2-flowered, in pairs, the empty scales subulate, variable in length, from a short awn-like body up to $\frac{1}{2}$′ long or more, the flowering scales 4″–5″ long, hispidulous and sparingly hirsute, the hispidulous awn 10″–15″ long.

Wisconsin and Minnesota to North Dakota and Wyoming. July and Aug.

107. HÝSTRIX Moench, Meth. 294. 1794.

Usually tall grasses, with flat leaf-blades and terminal spikes. Spikelets 2-several-flowered, in pairs, rarely in 3's, at each node of the rachis. Empty scales wanting, or sometimes appearing as mere rudiments; flowering scales narrow, convolute, rigid, rounded on the back, 5-nerved above, terminating in an awn; palet scarcely shorter than the scale, 2-keeled. Stamens 3. Styles very short, distinct. Stigmas plumose. Grain oblong, adhering to the palet when dry. [Greek name of the Porcupine, referring to the long awns.]

16. Elymus arkansànus Scribn. & Ball.   Smooth Slender Wild Rye. Fig. 715.

Elymus arkansanus Scribn. & Ball, Bull. U. S. Dep. Agr. Agrost. **24**: 45. *f. 19*. 1901.

Culms 2°–3° tall, slender, erect; sheaths glabrous, rarely the lower pubescent; blades up to 8′ long, 2″–4″ wide, rough on the lower surface, appressed-pubescent on the upper surface; spike $2\frac{1}{2}$′–4′ long, nodding, long-exserted; spikelets 2-flowered, in pairs, the empty scales subulate, indurated at the base, hispidulous, attenuate into a long awn; flowering scales glabrous or hispidulous, bearing a straight hispidulous awn 10″–20″ long.

In woods and on banks, New Jersey and Staten Island to Iowa and Arkansas; also at Hot springs, South Dakota. July and Aug. Figured in our first edition as *E. striatus* Willd., from which it is now distinguished.

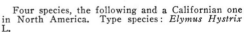

$\frac{2}{3}$

Four species, the following and a Californian one in North America. Type species: *Elymus Hystrix* L.

1. Hystrix Hýstrix (L.) Millsp. Bottlebrush Grass. Fig. 717.

Elymus Hystrix L. Sp. Pl. Ed. 2, 124. 1762.
Asprella Hystrix Willd. Enum. 132. 1809.
Gymnostichum Hystrix Schreb. Beschr. Gras. **2**: 127. *pl. 47*. 1810.
Hystrix Hystrix Millsp. Fl. W. Va. 474. 1892.

Culms 2°–4° tall, erect, simple, smooth and glabrous. Sheaths usually shorter than the internodes; ligule very short; blades $4\frac{1}{2}$′–9′ long, 3″–6″ wide, smooth beneath, rough above; spike 3′–7′ in length; spikelets at length widely spreading, 4″–6″ long, exclusive of the awns; empty scales awn-like, usually present in the lowest spikelet; flowering scales 4″–6″ long, acuminate into an awn about 1′ in length.

In rocky woods, New Brunswick to Ontario, south to Georgia, Illinois and Nebraska. Spikelets easily detached, even when young. Bottle-rush. June–July.

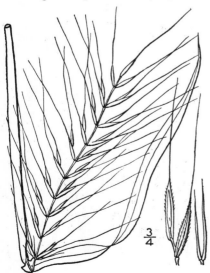

$\frac{3}{4}$

108. **ARUNDINÀRIA** Michx. Fl. Bor. Am. **1** : 73. 1803.

Arborescent or shrubby grasses, with simple or branched culms and flat short-petioled leaf-blades which are articulated with the sheath. Spikelets borne in panicles or racemes, 2–many-flowered, large, compressed. Empty scales 1 or 2, the first sometimes wanting; flowering scales longer, not keeled, many nerved; palets scarcely shorter than the scales, prominently 2-keeled. Lodicules 3. Stamens 3. Styles 2 or 3. Stigmas plumose. Grain furrowed, free, enclosed in the scale and palet. [From *Arundo,* the Latin name of the Reed.]

About 24 species, of Asia and America. Type species: *Arundinaria macrosperma* Michx.

Spikelets borne on radical shoots of the year ; culms 14° tall or less.	1. *A. tecta.*
Spikelets borne on the old culms, which are 16° tall or more.	2. *A. macrosperma.*

1. **Arundinaria técta** (Walt.) Muhl. Scutch Cane. Small Cane. Fig. 718.

Arundo tecta Walt. Fl. Car. 81. 1788.
Arundinaria tecta Muhl. Gram. 191. 1817.

Culms 3°–14° tall, erect, shrubby, branching at the summit, smooth and glabrous. Sheaths longer than the internodes, smooth or rough, ciliate on the margins; ligule bristly; blades lanceolate, 3½′–8′ long, 4″–12″ wide, flat, more or less pubescent beneath, glabrous above; racemes terminal, or on short leafless culms; spikelets 7–10-flowered, 1′–1½′ long, on pedicels 1′ in length or less, which are sometimes pubescent; empty scales unequal, the first usually very small, sometimes wanting; flowering scales 6″–10″ long, acute or acuminate.

In swamps and moist soil, Maryland to Indiana, Missouri, Florida and Texas. Switch-cane. Reed. Cane-brake. May–July.

$\frac{3}{4}$

2. **Arundinaria macrospérma** Michx. Giant Cane. Fig. 719.

A. macrosperma Michx. Fl. Bor. Am. **1** : 74. 1803.

Culms woody, 16°–30° tall, finally branched above; sheaths ciliate on the margins, otherwise glabrous; blades lanceolate, smooth or roughish, 1° long or less, the larger 10″–15″ wide, those on the ultimate divisions smaller and crowded at the summit of the branches; inflorescence on the old wood, the spikelets 1½′–2½′ long, on slender more or less leafy branches, the flowering scales glabrous or hirsute, acuminate.

Forming "cane brakes" along rivers and swamps, Virginia to Florida, west to Louisiana, and along the Mississippi River and its tributaries as far north as Kentucky, Tennessee and Missouri. May–July.

$\frac{2}{3}$

Family 11. **CYPERÀCEAE** J. St. Hil. Expos. Fam. **1** : 62. 1805.

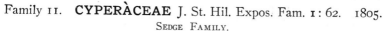

SEDGE FAMILY.

Grass-like or rush-like herbs. Stems (culms) slender, solid (rarely hollow), triangular, quadrangular, terete or flattened. Roots fibrous (many species perennial by long rootstocks). Leaves usually with closed sheaths. Flowers perfect or imperfect, arranged in spikelets, one (rarely 2) in the axil of each scale (glume, bract), the spikelets solitary or clustered, 1–many-flowered. Scales 2-ranked or spirally imbricated, persistent or deciduous. Perianth hypogynous, composed of bristles, or interior scales, rarely calyx-like, or entirely wanting. Stamens 1–3, rarely more. Filaments slender or filiform. Anthers 2-celled. Ovary 1-celled, sessile or stipitate. Ovule 1, anatropous, erect. Style 2–3-cleft or rarely simple or minutely 2-toothed. Fruit an achene. Endosperm mealy. Embryo minute.

About 75 genera and 3200 species, widely distributed. The dates give time of perfecting fruit.

* Fertile flowers perfect.

† Basal empty scales of the spikelets none, or not more than 2 (except in *Eriophorum*).
Scales of the spikelets 2-ranked; bristles none.
    Spikelets with only 1 perfect flower.   1. *Kyllinga.*
    Spikelets with 2 to many perfect flowers.   2. *Cyperus.*
Scales of the spikelets spirally imbricated.
  Base of the style persistent as a tubercle on the achene.
    Spikelet 1; culm leafless; bristles usually present.   3. *Eleocharis.*
    Spikelets several or numerous; culms leaf-bearing; bristles none.   4. *Stenophyllus.*
  Base of the style not persistent as a tubercle.
    Flowers without any inner scales.
      Base of the style swollen; bristles none.   5. *Fimbristylis.*
      Base of the style not swollen; bristles usually present.
        Bristles 6 to many, silky, much elongated.   6. *Eriophorum.*
        Bristles short, or little elongated, smooth or barbed.   7. *Scirpus.*
    Flowers with 1 or more inner scales.
      Flowers with 3 broad, stalked scales alternating with barbed bristles.   8. *Fuirena.*
      Flowers with 1 or 2 hyaline scales; bristles none.
        Flowers with 2 convolute inner scales.   9. *Lipocarpha.*
        Flowers with a single minute inner scale.   10. *Hemicarpha.*
          †† Basal empty scales of the spikelets 3 or more.
Style 2-cleft.
  Spikelets breaking up into 1-fruited joints; bristles present; scales 2-ranked. 11. *Dulichium.*
  Rachis of the spikelets not jointed, persistent; scales spirally imbricated.
    Spikelets flattened, clustered in a single involucrate head; bristles none.   12. *Dichromena.*
    Spikelets not flattened, variously clustered.
      Spikelets few-flowered; bristles usually present.   13. *Rynchospora.*
      Spikelets many-flowered; bristle none.   14. *Psilocarya.*
Style 3-cleft; bristles none.   15. *Mariscus.*

** All the flowers imperfect.

Pistillate flower subtended by a flat scale; achene bony.   16. *Scleria.*
Pistillate flower enclosed in a perigynium or enwrapped by a concave or convolute scale.
  Pistillate flower partly enwrapped by a scale.   17. *Kobresia.*
  Pistillate flower wholly enclosed by a perigynium.
    Leaves more than one, with sheath, ligule and midvein.   18. *Carex.*
    Leaf one, without sheath, ligule or midvein.   19. *Cymophyllus.*

1. KÝLLINGA Rottb. Descr. & Ic. 12. *pl. 4. f. 3, 4.* 1773.

Annual or perennial sedges, with slender triangular culms, leafy below, and with 2 or more leaves at the summit forming an involucre to the strictly sessile, simple or compound dense head of spikelets. Spikelets numerous, compressed, falling away from the axis of the head at maturity, consisting of only 3 or 4 scales, the 1 or 2 lower ones small and empty, the middle one fertile, the upper empty or staminate. Joints of the rachis wingless or narrowly winged. Scales 2-ranked, keeled. Perianth none. Stamens 1–3. Style 2-cleft, deciduous from the summit of the achene. Achene lenticular. [In honor of Peter Kylling, a Danish botanist of the seventeenth century.]

About 45 species, natives of tropical and temperate regions. Besides the following, 2 others occur in the southern United States. Type species: *Kyllinga monocéphala* Rottb.

$\frac{3}{4}$

1. Kyllinga pùmila Michx. Low Kyllinga. Fig. 720.

Kyllingia pumila Michx. Fl. Bor. Am. 1 : 28. 1803.

Annual, culms densely tufted, filiform, erect or reclined, 2'–15' long, mostly longer than the leaves. Leaves light green, roughish on the margins, usually less than 1″ wide, those of the involucre 3–5, elongated, spreading or reflexed; head oblong or ovoid-oblong, 3″–4″ long, simple or commonly with 1 or 2 smaller ones at the base; spikelets about 1½″ long, flat, 1-flowered, the 2 empty lower scales more or less persistent on the rachis after the fall of the rest of the spikelets; scales ovate, acuminate or acute, thin, about 7-nerved, the fertile one with a rough keel; stamens 2; style 2-cleft; achene lenticular, obtuse.

In moist or wet soil, Delaware to Florida, Illinois, Kansas, Texas and Mexico; West Indies and tropical continental America. Aug.–Sept.

2. CYPÈRUS [Tourn.] L. Sp. Pl. 44. 1753.

Annual or perennial sedges. Culms in our species simple, triangular, leafy near the base, and with 1 or more leaves at the summit, forming an involucre to the simple or compound, umbellate or capitate inflorescence. Rays of the umbel sheathed at the base, usually very unequal, one or more of the heads or spikes commonly sessile. Spikelets flat or subterete, composed of few or many scales, the scales falling away from the wingless or winged rachis as they mature (nos. 1–23), or persistent and the spikelets falling away from the axis of the head or spike with the scales attached (nos. 24–37). Scales concave, conduplicate or keeled, 2-ranked, all flower-bearing or the lower 1 or 2 empty. Flowers perfect. Perianth none. Stamens 1–3. Style 2–3-cleft, deciduous from the summit of the lenticular or 3-angled achene. [Ancient Greek name for these sedges.]

About 600 species, widely distributed in tropical and temperate regions. Besides the following, some 50 others occur in the southern United States. The English names *Galingale* or *Galangal* and *Sweet Rush* are sometimes applied to all the species. Type species: *Cyperus esculentus* L.

Style 2-cleft; achene lenticular, not 3-angled; scales falling from the rachis; spikelets flat.
Achene one-half as long as the scale; umbel nearly or quite simple.
  Spikelets yellow; superficial cells of the achene oblong.   1. *C. flavescens.*
  Spikelets green or brown; superficial cells of the achene quadrate.
    Scales obtuse or obtusish, appressed.
      Scales membranous, dull; style much exserted.   2. *C. diandrus.*
      Scales subcoriaceous, shining; style scarcely exserted.   3. *C. rivularis.*
    Scales acute, somewhat spreading at maturity.
      Achene narrowly obovate; spikelets ½'–1½' long.   4. *C. filicinus.*
      Achene linear-oblong; spikelets 3"–9" long.
        Scales ovate, brownish; umbel usually subcapitate.   5. *C. microdontus.*
        Scales oblong-lanceolate, greenish; umbel usually loose. 6. *C. paniculatus.*
Achene nearly as long as the scale; umbel sometimes much compound. 7. *C. sabulosus.*
** *Style 3-cleft; achene 3-angled.*
Scales falling away from the persistent rachis of the flattened spikelets.
  Wings of the rachis, if present, permanently adnate to it.
    Scales tipped with recurved awns; low annual, 1'–6' tall. 8. *C. inflexus.*
    Scales acute or obtuse, not awned.
      Wings of the rachis none or very narrow.
        Stamens 2 or 3; spikelets linear-oblong, 2½"–12" long.
          Annual; culms smooth, 2'–20' long.
            Scales sharply acuminate.   9. *C. compressus.*
            Scales blunt, mucronulate.   10. *C. Iria.*
          Perennial; culms 1°–2½° tall.
            Heads oblong; spikelets erect or ascending; culms rough. 11. *C. Schweinitzii.*
            Heads short; spikelets more or less spreading; culms smooth.
              Scales broadly ovate; achene 1" long. 12. *C. Houghtoni.*
              Scales oblong-ovate; achene 1¼" long. 13. *C. Bushii.*
        Stamen 1; spikelets ovate, 2"–4" long.
          Tall perennial; achene linear; scales acutish. 14. *C. pseudovegetus.*
          Low annual; achene oblong; scale-tips recurved. 15. *C. acuminatus.*
      Wings of the rachis distinct.
        Low annual, adventive from Europe; scales brown. 16. *C. fuscus.*
        Tall indigenous perennials (no. 17 sometimes annual?).
          Lower leaves reduced to pointed sheaths. 17. *C. Haspan.*
          Leaves all elongated-linear.
            Scales mucronate, reddish brown or green. 18. *C. dentatus.*
            Scales acute or obtuse, not mucronate.
              Scales wholly or partly purple-brown; achene linear.
                Scales tightly appressed. 19. *C. rotundus.*
                Tips of the scales free. 20. *C. Hallii.*
              Scales straw-colored; achene obovoid. 21. *C. esculentus.*
  Wings of the rachis separating from it as interior scales; annuals.
    Spikes loose; spikelets 3"–10" long. 22. *C. erythrorhizos.*
    Spikes dense, cylindric; spikelets 1½"–2½" long. 23. *C. Halei.*
Spikelets falling away from the axis of the spikes, the lower pair of scales commonly persistent.
  Annuals; spikelets elongated, nearly terete.
    Scales imbricated or but slightly distant; achene obovoid.
      Scales thin, dull brown; spikelets slender. 24. *C. speciosus.*
      Scales rigid, yellow-brown; spikelets stout. 25. *C. ferox.*
    Scales very distant; achene linear-oblong; spikelets very slender. 26. *C. Engelmanni.*
  Perennial by hard, tuber-like basal corms, spikelets more or less flattened.
    Achene narrowly linear-oblong, 3–4 times as long as thick.
      Spikelets flat, several–many-flowered. 27. *C. strigosus.*
      Spikelets subterete, few-flowered.
        Spikelets 6"–12" long, loosely spicate, the lower reflexed. 28. *C. refractus.*
        Spikelets 1½"–6" long, densely capitate or spicate.
          Spikelets all reflexed; culms rough. 29. *C. retrofractus.*
          Spikelets spreading or obovoid, the lower reflexed; culms smooth.
            Heads oblong or cylindric.
              Spikelets 2"–5" long, at least the lower reflexed.
                Head oblong or short-cylindric; lower spikelets reflexed.
                      30. *C. lancastriensis.*
              Head obovoid; all but the upper spikelets reflexed.
                    31. *C. hystricinus.*

Spikelets 1½″–2″ long, the lower spreading.   32. *C. Torreyi.*
    Heads globose.   33. *C. ovularis.*
Achene oblong or obovoid, about twice as long as thick.
    Rachis wingless or very narrowly winged.
      Heads globose.   34. *C. filiculmis.*
      Heads oblong.   35. *C. cayennensis.*
    Rachis-wings membranous, broad.
      Scales firm, not appressed; spikelets loosely capitate.   36. *C. Grayi.*
      Scales thin, closely appressed; spikelets densely capitate.   37. *C. globulosus.*

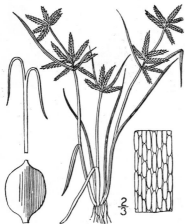

1. Cyperus flavéscens L. Yellow Cyperus. Galingale. Fig. 721.

Cyperus flavescens L. Sp. Pl. 46. 1753.

Annual, culms very slender, tufted, leafy below, 3′–12′ tall, mostly longer than the leaves. Leaves 1″–1½″ wide, smooth, the longer usually exceeding the inflorescence; clusters terminal and sessile or on 1–4 short rays; spikelets in 3's–6's, linear, subacute, yellow, many-flowered, flat, 4″–9″ long, 1½″–2″ broad; scales ovate, obtuse, 1-nerved, appressed, twice as long as the orbicular-ovate black obtuse lenticular shining achene; stamens 3; style deeply 2-cleft, its branches slightly exserted; superficial cells of the achene oblong.

In marshy ground, New York to Michigan, Florida, Mexico and Costa Rica. Also in the Old World, in Bermuda, Cuba and Dominica. Reported from Maine. Aug.–Oct.

2. Cyperus diándrus Torr. Low Cyperus. Fig. 722.

Cyperus diandrus Torr. Cat. Pl. N. Y. 90. 1819.
Cyperus diandrus elongatus Britton, Bull. Torr. Club 19: 226. 1892.

Annual, culms tufted, slender, 2′–15′ tall. Leaves about 1″ wide, those of the involucre usually 3, the longer much exceeding the spikelets; clusters sessile and terminal, or at the ends of 1–3 rays; spikelets 4″–9″ long, linear-oblong, acute, flat, many-flowered; scales ovate, green, brown, or with brown margins, obtuse, 1-nerved, appressed, membranous, dull; stamens 2 or 3; style 2-cleft, its branches much exserted; achene lenticular, oblong, subacute, gray, not shining, one-half as long as the scale, its superficial cells quadrate, about as long as wide.

In marshy places, New Brunswick to Minnesota, south to South Carolina and Kansas. Cypress-grass. Galingale or Galangal. Aug.–Oct.

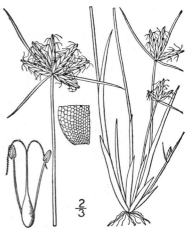

3. Cyperus rivulàris Kunth. Shining Cyperus. Fig. 723.

Cyperus rivularis Kunth, Enum. 2: 6. 1837.
Cyperus diandrus var. (?) *castaneus* Torr. Ann. Lyc. N. Y. 3: 252. 1836. Not *C. castaneus* Willd. 1798.

Similar to the preceding species, culms slender, tufted, 4′–15′ tall. Umbel usually simple; spikelets linear or linear-oblong, acutish, 4″–10″ long; scales green or dark brown or with brown margins, appressed, firm, subcoriaceous, shining, obtuse; stamens mostly 3; style 2-cleft, scarcely exserted; achene oblong or oblong-obovate, lenticular, somewhat pointed, dull, its superficial cells quadrate.

In wet soil, especially along streams and ponds, Maine to southern Ontario and Michigan, south to Missouri, North Carolina and Kansas. Aug.–Oct.

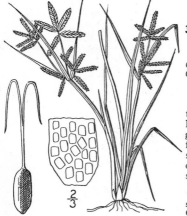

4. Cyperus filicìnus Vahl. Nuttall's Cyperus.
Fig. 724.

Cyperus filicinus Vahl, Enum. **2**: 332. 1806.
C. Nuttallii Eddy ; Spreng. Neue Entd. 1 : 240. 1820.

Annual, culms slender, tufted, 4'–18' tall, equalling or often longer than the leaves. Leaves of the involucre 3–5, spreading, the larger often 5' long; umbel simple or slightly compound, 3–7-rayed; spikelets rather loosely clustered, linear, very acute, flat, spreading, ½'–1½' long, 1"–1½" wide; scales yellowish-brown with a green keel, oblong, acute, rather loosely spreading at maturity; stamens 2; style 2-cleft, its branches somewhat exserted; achene lenticular, narrowly obovate, obtuse or truncate, dull, light brown, one-third to one-half as long as the scale, its superficial cells quadrate.

Salt marshes, Maine to Mississippi. Also in Bermuda. Aug.–Oct.

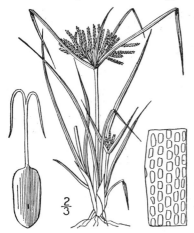

5. Cyperus microdóntus Torr. Coast Cyperus.
Fig. 725.

Cyperus microdontus Torr. Ann. Lyc. N. Y. **3** : 255. 1836.

Annual, similar to the preceding species, culms very slender, tufted, sometimes 20' high, usually lower. Leaves about 1" wide, those of the involucre much elongated; umbel commonly simple, sessile, capitate, or 1–6-rayed; spikelets linear, acute, 3"–9" long, less than 1" wide, yellowish-brown; scales ovate, acute, brownish, thin, appressed when young, spreading at maturity; stamens 2; style 2-cleft, its branches much exserted; achene lenticular, linear-oblong, short-pointed, light brown, one-half as long as the scale, its superficial cells quadrate.

In wet soil, on or near the coast, New Jersey to Florida and Texas. Aug.–Oct.

6. Cyperus paniculàtus Rottb. Panicled Cyperus. Fig. 726.

Cyperus paniculatus Rottb. Descr. & Icon. 40. 1773.

Cyperus Gatesii Torr. Ann. Lyc. N. Y. **3** : 255. 1836.

Annual; bright green; culms tufted, slender, 5 dm. high or less, bluntly 3-angled. Leaves mostly shorter than the culm, 2" wide or less, sometimes very narrow, those of the involucre 3–5, usually much elongated; umbel 2–8-rayed, the rays often 2' long; spikelets linear or linear-lanceolate, 3"–7½" long, about ½" wide, greenish-yellow; scales acute, oblong to oblong-lanceolate; achene oblong, ½" long, about one-half as long as the scale, its superficial cells quadrate.

Moist soil, Virginia to Florida, Arkansas and Texas. Also in tropical America. Aug.–Oct.

7. Cyperus sabulòsus Mart. & Schrad.   Elegant Cyperus.   Fig. 727.

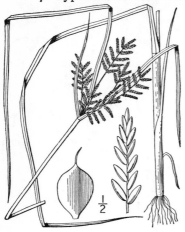

Cyperus flavicomus Vahl, Enum. **2** : 360. 1806. Not Michx.
Pycraeus sabulosus Mart. & Schrad.; Mart. Fl. Bras. **2**¹ :
    10. 1842.
Cyperus sabulosus Mart. & Schrad.; Boeckl. Linnaea **35** :
    468. 1867-68.

Annual, culms stout or slender, 1°–3° tall, leafy below. Leaves smooth, or rough-margined, 2″–3″ wide,
those of the involucre 3–8, the longer ones much exceeding the inflorescence; umbels few–several-rayed,
often compound; primary rays ½′–2½′ long; spikelets
numerous, usually densely clustered, linear, acute,
4″–10″ long, 1″–1½″ wide, flat, many-flowered, spreading; scales oblong, obtuse, thin, dull, yellowish-brown,
scarious-margined, faintly 3-nerved; stamens 3; style
2-cleft, little exserted; achenes obovate, lenticular,
black, mucronate, not shining, nearly as long as the
scales and often persistent on the rachis after these
have fallen away.

In wet or moist sandy soil, Virginia to Florida and
Louisiana. Also in Brazil. Aug.–Oct.

8. Cyperus infléxus Muhl.   Awned Cyperus.
Fig. 728.

Cyperus inflexus Muhl. Gram. 16. 1817.

Cyperus aristatus Boeckl. Linnaea, **35** : 500, in part. 1868.
Not Rottb. 1773.

Annual, culms slender or almost filiform, tufted, 1′–6′
tall, about equalled by the leaves. Leaves 1″ wide or less,
those of the involucre 2–3, exceeding the umbel; umbel
sessile, capitate, or 1–3-rayed; spikelets linear-oblong,
6–10-flowered, 2″–3″ long; scales light brown, lanceolate,
rather firm, strongly several-nerved, tapering into a long,
recurved awn, falling from the rachis at maturity; stamen
1; style 3-cleft; rachis narrowly winged, the wings persistent; achene 3-angled, brown, dull, narrowly obovoid
or oblong, obtuse, mucronulate.

In wet, sandy soil, New Brunswick to the Northwest Territory and British Columbia, south to Florida, Texas, California
and Mexico. Fragrant in drying. July–Sept.

9. Cyperus compréssus L.   Flat Cyperus.   Fig. 729.

Cyperus compressus L. Sp. Pl. 46. 1753.

Annual, tufted, culms slender, erect or reclining,
smooth, 3′–10′ long. Leaves light green, about 1″ wide,
those of the involucre 2–3, the longer exceeding the
spikelets; umbel capitate or with 2–3 short rays; spikelets narrowly lanceolate, acute, 4″–10″ long, 1½″–2″
wide, very flat, many-flowered; scales light green with
a yellow band on each side, ovate, acuminate, firm,
keeled, several-nerved, falling away from the narrowly-
winged rachis at maturity; stamens 3; style 3-cleft;
achene sharply 3-angled, obovoid, obtuse, dull, brown,
about one-third as long as the scale.

In fields and waste places, southern New York to Florida,
west to Missouri and Texas. Also in tropical America and
in the warmer parts of Asia and Africa. Aug.–Oct.

10. Cyperus Ìria L. Yellow Cyperus. Fig. 730.

Cyperus Iria L. Sp. Pl. 45. 1753.

Annual with fibrous roots; culms tufted, 4'–20' tall, 3-angled. Basal leaves 1"–3½" wide, shorter than the culm or equalling it, those of the involucre similar, the longer ones surpassing the inflorescence; umbel several-rayed, usually compound; spikelets numerous, spicate, narrowly oblong, 2"–6" long, about 1" wide, flattened, several–many-flowered; rachis nearly wingless; scales obovate, yellow to yellow-brown, blunt, mucronate, 3–5-nerved on the back; stamens 2 or 3; achene oblong-obovoid, trigonous, about ⅓" long.

Cultivated ground, Hempsted, Long Island; swamps, North Carolina to Florida and Texas. Naturalized from Asia.

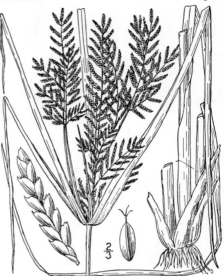

11. Cyperus Schweinítzii Torr. Schweinitz's Cyperus. Fig. 731.

C. Schweinitzii Torr. Ann. Lyc. N. Y. 3 : 276. 1836.

Perennial by the thickened corm-like bases of the culms, tufted, culms rather slender, rough, at least above, 1°–2½° tall, about equalled by the light green leaves. Leaves 1"–2½" wide, rough-margined, those of the involucre 3–7, erect, the longer exceeding the inflorescence; umbel simple, 3–9-rayed, the rays erect, sometimes 4' long; spikelets flat, in rather loose ovoid spikes, which are sessile and at the ends of the rays, linear-oblong, 6–12-flowered, 4"–8" long; scales convex, light green, ovate, acute or acuminate, 9–13-nerved, falling away from the rachis at maturity; stamens 3; style 3-cleft; achene sharply 3-angled, oblong, brown, acute at each end, nearly as long as the scale, its superficial cells quadrate.

In sandy soil, especially along lakes and streams, western New York and southern Ontario to the Northwest Territory, Iowa, Minnesota and Missouri. Aug.–Oct.

12. Cyperus Hoùghtoni Torr. Houghton's Cyperus. Fig. 732.

C. Houghtoni Torr. Ann. Lyc. N. Y. 3 : 277. 1836.

Perennial by tuber-like corms, culms very slender, smooth, erect, 1°–2° tall. Leaves shorter than the culm, 1" wide or less, smooth, those of involucre 3–5, the longer much exceeding the umbel; umbel simple, 1–5-rayed, the rays mostly short, their sheaths 2-toothed; spikelets loosely capitate, linear, compressed, acute, 4"–8" long, about 1" wide, 11–15-flowered, falling away from the axis when mature; scales chestnut brown, firm, somewhat spreading, shining, oblong, obtuse, truncate or apiculate, strongly about 11-nerved; rachis very narrowly winged; stamens 3; style 3-cleft; achene broadly oblong, less than twice as long as thick, 3-angled, brown, apiculate, nearly as long as the scale.

In sandy soil, Massachusetts to Manitoba, Virginia, Michigan and Wisconsin. July–Aug.

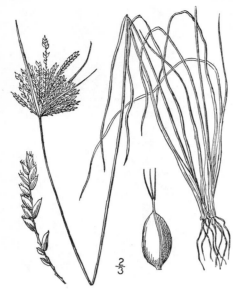

13. Cyperus Búshii Britton. Bush's Cyperus. Fig. 733.

Cyperus Bushii Britton, Man. 1044. 1901.

Perennial by tuber-like corms. Leaves 1½″–2″ wide, smooth; culms smooth, 1°–2° high, longer than the leaves; longer involucral bracts much exceeding the umbel; umbel capitate, or with 1–5 rays; spikelets loosely capitate, flat, linear, acute, 4″–8″ long; scales firm, shining, oblong, mucronate, strongly about 11-nerved falling away from the persistent axis of the spikelet at maturity; achenes oblong, 3-angled, nearly twice as long as thick, apiculate, two-thirds as long as the scale, sometimes persistent after the scales fall.

Sandy soil, Minnesota to Idaho, Missouri, Texas and Colorado. Resembling *C. filiculmis*, but the axis of the spikelet is persistent after the scales fall away. July–Sept.

14. Cyperus pseudovégetus Steud. Marsh Cyperus. Fig. 734.

Cyperus pseudovegetus Steud. Syn. Pl. Cyp. 24. 1855.
Cyperus calcaratus Nees; S. Wats. in A. Gray, Man. Ed. 6, 570. 1890.

Perennial by thickened tuber-like joints of the rootstocks, culm rather stout, 1°–4° high, often equalled by the leaves. Leaves 1½″–2″ wide, smooth, nodulose, the midvein prominent; leaves of the involucre 4–6, spreading, the longer much exceeding the inflorescence; umbel several-rayed, compound, the primary rays often 4′ long; spikelets ovate, flat, many-flowered, light green, densely capitate, 2″–3″ long; scales keeled, conduplicate, 1-nerved, curved, acute, longer than the linear 3-angled slightly stalked achene; stamen 1; style 3-cleft.

In marshes, New Jersey to Florida, Kentucky, Missouri, Kansas and Texas. Also in the Bahamas. July–Sept.

15. Cyperus acuminàtus Torr. & Hook. Short-pointed Cyperus. Fig. 735.

Cyperus acuminatus Torr. & Hook. Ann. Lyc. N. Y. 3: 435. 1836.

Annual, culms very slender, tufted, 3′–15′ tall, longer than or equalling the leaves. Leaves light green, usually less than 1″ wide, those of the involucre much elongated; umbel 1–4-rayed, simple; rays short; spikelets flat, ovate-oblong, obtuse, 2″–4″ long, many-flowered, densely capitate; scales oblong, pale green, 3-nerved, coarsely cellular, conduplicate, with a short sharp more or less recurved tip; stamen 1; style 3-cleft; achene sharply 3-angled, gray, oblong, narrowed at each end, about one-half as long as the scale.

In moist soil, Illinois to South Dakota, Louisiana, Iowa. Kansas, Texas, Oregon and California. July–Oct.

16. Cyperus fúscus L.   Brown Cyperus.
Fig. 736.

Cyperus fuscus L. Sp. Pl. 46. 1753.

Annual, culms slender, tufted, 6′–15′ high, longer than or equalled by the leaves. Leaves rather dark green, about 1″ wide, those of the involucre 4–6, the longer much exceeding the inflorescence; umbel several-rayed, somewhat compound, the rays short; spikelets linear, 2″–7″ long, less than 1″ wide, many-flowered, acute; scales ovate, subacute, becoming dark brown or remaining greenish on the keel, faintly about 3-nerved on the back, separating from the narrowly winged rachis as they mature; stamens 2 or 3; style 3-cleft; achene sharply 3-angled, oblong, pointed at each end, nearly as long as the scale.

Waste grounds and in ballast, eastern Massachusetts to New Jersey and Maryland. Adventive from Europe. July–Sept.

17. Cyperus Háspan L.   Sheathed Cyperus.   Fig. 737.

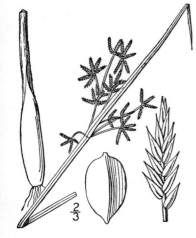

Cyperus Haspan L. Sp. Pl. 45. 1753.

Perennial by short rootstocks (sometimes annual?), roots fibrous, culms slender, weak, tufted, 1°–3° high. Lower leaves reduced to membranous acuminate sheaths, those of the involucre about 2, usually less than 1″ wide, commonly little exceeding or shorter than the inflorescence; umbel several-rayed, simple or compound, the longer rays 1′–2′ long; spikelets few, capitate, linear, acute, many-flowered, 3″–6″ long, about ½″ wide; scales oblong or oblong-lanceolate, reddish-brown, acute, mucronulate, keeled, 3-nerved; rachis narrowly winged; stamens 3; style 3-cleft, scarcely exserted; achene 3-angled, broadly obovoid, obtuse, nearly white, very much shorter than the scale.

In swamps, Virginia to Florida and Texas, mostly near the coast. Also in tropical America and in the warmer parts of Europe, Asia and Australia. A long basal leaf is rarely developed. July–Sept.

18. Cyperus dentàtus Torr.   Toothed Cyperus.   Fig. 738.

Cyperus dentatus Torr. Fl. U. S. 1 : 61. 1824.
C. dentatus ctenostachys Fernald, Rhodora **8** : 126. 1906.

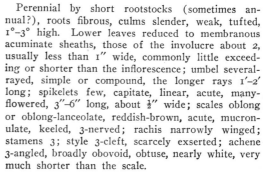

Perennial by scaly rootstocks which sometimes bear small tubers, culms rather stiff, 8′–20′ tall, longer than or equalled by the leaves. Leaves keeled, 1″–2″ wide, those of the involucre 3–4, one or two of them usually exceeding the inflorescence; umbel several-rayed, somewhat compound; longer rays 1′–3′ long; spikelets linear, very flat, many-flowered, mostly blunt, 5″–10″ long, nearly 2″ wide; scales light reddish-brown, ovate-lanceolate, thin, keeled, 5–7-nerved, mucronate, separating from the rachis when mature, their tips spreading, causing the spikelet to appear toothed; stamens 3; style 3-cleft, the branches exserted; achene 3-angled, obtuse, mucronate, light brown, much shorter than the scale.

In sandy swamps and on shores, Maine to northern New York, West Virginia and South Carolina. Scales often modified into tufts of small leaves. Aug.–Oct.

19. Cyperus rotúndus L. Nut-grass. Fig. 739.

Cyperus rotundus L. Sp. Pl. 45. 1753.
Cyperus Hydra Michx. Fl. Bor. Am. 1: 27. 1803.

Perennial by scaly tuber-bearing rootstocks, culm rather stout, 6′–20′ high, usually longer than the leaves. Leaves 1½″–3″ wide, those of the involucre 3–5, the longer equalling or exceeding the inflorescence; umbel compound or nearly simple, 3–8-rayed, the longer rays 2′–4½′ long; spikelets linear, closely clustered, few in each cluster, acute, 4″–10″ long, 1″–1½″ wide; scales dark purple-brown or with green margins and center, ovate, acute, closely appressed when mature, about 3-nerved on the keel; stamens 3; style 3-cleft, its branches exserted; achene 3-angled, about one-half as long as the scale.

In fields, Virginia to Florida, Missouri, Kansas and Texas. Adventive in southern New York and in ballast deposits at eastern seaports. Tropical America, and the Old World. A troublesome weed in the South. Coco-grass, Round-root. July–Sept.

20. Cyperus Hàllii Britton. Hall's Cyperus. Fig. 740.

C. Hallii Britton, Bull. Torr. Club 13: 211. 1886.

Perennial by scaly rootstocks, culm rather stout, 2°–3° tall, about equalled by the leaves. Basal leaves 2″–3″ wide; involucral leaves 3–6, the longer very much exceeding the inflorescence; umbel compound, its longer rays 3′–4′ long, the raylets sometimes 1′ long; spikelets numerous, loosely clustered, linear, 7–15-flowered, 5″–8″ long, 1″–1½″ wide; involucels setaceous; scales ovate, acute, strongly 7–9-nerved, dark reddish-brown or with lighter margins, their tips not appressed; stamens 3; style 3-cleft, its branches much exserted; achene linear-oblong, 3-angled, about one-half as long as the scale.

Kansas and Oklahoma to Texas. July–Sept.

21. Cyperus esculéntus L. Yellow Nut-grass. Fig. 741.

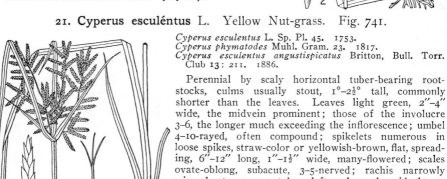

Cyperus esculentus L. Sp. Pl. 45. 1753.
Cyperus phymatodes Muhl. Gram. 23. 1817.
Cyperus esculentus angustispicatus Britton, Bull. Torr. Club 13: 211. 1886.

Perennial by scaly horizontal tuber-bearing rootstocks, culms usually stout, 1°–2½° tall, commonly shorter than the leaves. Leaves light green, 2″–4″ wide, the midvein prominent; those of the involucre 3–6, the longer much exceeding the inflorescence; umbel 4–10-rayed, often compound; spikelets numerous in loose spikes, straw-color or yellowish-brown, flat, spreading, 6″–12″ long, 1″–1½″ wide, many-flowered; scales ovate-oblong, subacute, 3–5-nerved; rachis narrowly winged; stamens 3; style 3-cleft; achene obovoid, obtuse, 3-angled.

In moist fields, New Brunswick to Minnesota, Nebraska, Florida and Texas. Also on the Pacific Coast from California to Alaska, in tropical America, and widely distributed in the Old World. Sometimes a troublesome weed. The species consists of numerous races, differing in length and width of spikelets. Edible galingale. Earth-almond. Rush-nut. Aug.–Oct.

22. Cyperus erythrorhìzos Muhl.   Red-rooted Cyperus.   Fig. 742.

Cyperus erythrorhizos Muhl. Gram. 20. 1817.

Annual, culms tufted, stout or slender, 3′–2° tall. Leaves 1½″–4″ wide, rough-margined, the lower longer than or equalling the culm, those of the involucre 3–7, some of them 3–5 times as long as the inflorescence; umbel mostly compound, several-rayed; spikelets linear, subacute, 3″–10″ long, less than 1″ wide, compressed, many-flowered, clustered in oblong, nearly or quite sessile spikes; scales bright chestnut brown, oblong-lanceolate, mucronulate, appressed, separating from the rachis at maturity, the membranous wings of the rachis separating as a pair of hyaline interior scales; stamens 3; style 3-cleft; achene sharply 3-angled, oblong, pointed at both ends, pale, one-half as long as the scale.

In wet soil, especially along streams, southern Ontario to Massachusetts, Florida, Minnesota, Kansas, Texas and California. Aug.–Oct.

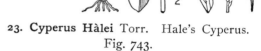

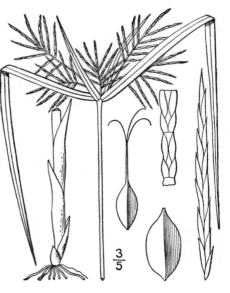

23. Cyperus Hàlei Torr.   Hale's Cyperus. Fig. 743.

C. Halei Torr.; Britton, Bull. Torr. Club **13**: 213. 1886.

Annual, culm stout, 2°–3° tall, about equalled by the leaves. Leaves 3″–4″ wide, very rough-margined, those of the involucre 5–8, much elongated; umbel compound, several-rayed; spikes cylindric, sessile or very nearly so, exceedingly dense, ½″–1′ long; spikelets very numerous, linear, 1½″–2½″ long, ½″ wide, spreading; scales brown, keeled, indistinctly 5-nerved, oblong, mucronulate, separating from the rachis at maturity, the wings of the rachis separating as a pair of hyaline scales, as in the preceding species; stamens 3; style 3-cleft; achene 3-angled, minute.

In swamps, southern Missouri to Tennessee, Louisiana and Florida. July–Sept.

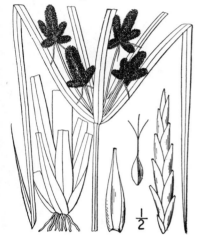

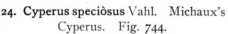

24. Cyperus speciòsus Vahl.   Michaux's Cyperus.   Fig. 744.

Cyperus speciosus Vahl, Enum. **2**: 364. 1806.
C. Michauxianus Schult. Mant. **2**: 123. 1824.

Annual, culms usually tufted, 1′–2° tall, reddish toward the base. Leaves rough-margined, 1½″–2½″ wide, shorter than or equalling the culm, the midvein prominent; leaves of the involucre much exceeding the umbel; umbel compound or nearly simple, 3–7-rayed, the primary rays ¼′–5½′ long; involucels narrow; spikelets subterete, very narrowly linear, loosely or densely clustered, 4″–12″ long, less than 1″ thick, 10–30-flowered, falling away from the axis at maturity; scales dull brown or reddish, thin, densely imbricated, ovate, obtuse or acute, faintly 3–5-nerved on the back; rachis-wings broad, clasping the achene, persistent; stamens 3; style 3-cleft, slightly exserted; achene pale, 3-angled, about one-half as long as the scale.

In marshes, Massachusetts to Ohio and South Dakota, south to Florida, Kansas, Texas and California. July–Sept. Sometimes flowering when 1′ high; variable in the overlapping of the scales.

3/5

25. Cyperus fèrax L. C. Richard. Coarse Cyperus. Fig. 745.

Cyperus ferax L. C. Rich, Act. Soc. Hist. Nat. Paris 1: 106. 1792.

Annual, closely related to the preceding species, but with smooth-margined, shorter and broader leaves, those of the involucre sometimes but little exceeding the inflorescence, the scales of the spikelets less imbricated. Umbel simple or somewhat compound, often compact, the rays mostly short; spikelets linear, subterete, 10–20-flowered, 8″–12″ long, about 1″ thick, falling away from the axis at maturity; scales ovate-oblong, appressed, slightly or scarcely imbricated, obtusish or acute, rather firm, green and 7–9-nerved on the back, yellowish on the sides; stamens 3; style 3-cleft; rachis broadly winged; achene 3-angled, narrowly obovoid, obtuse.

In wet soil, Massachusetts to Florida. California, and widely distributed in tropical America. Aug.–Oct.

26. Cyperus Engelmánni Steud. Engelmann's Cyperus. Fig. 746.

Cyperus Engelmanni Steud. Syn. Pl. Cyp. 47. 1855.

Annual, culms slender, 6′–2½° tall. Leaves elongated, 2″–3″ wide, flaccid, roughish on the margins, those of the involucre 4–6, the longer exceeding the umbel; umbel often compound, the raylets very short; spikelets often densely crowded, very narrowly linear, subterete, 6″–12″ long, 5–15-flowered; rachis narrowly winged; scales greenish-brown, oblong, obtuse, thin, faintly 3–5-nerved on the back, distant, the successive ones on each side of the spikelet separated by a space of about one-half their length; stamens 3; style 3-cleft; achene linear-oblong, 3-angled, two-thirds as long as the scale.

In wet soil, Massachusetts to southern Ontario and Wisconsin, south to New Jersey and Missouri. Aug.–Oct.

3/4

27. Cyperus strigòsus L. Straw-colored Cyperus. Fig. 747.

Cyperus strigosus L. Sp. Pl. 47. 1753.
C. strigosus capitatus Boeckl. Linnaea **36**: 347. 1869–70.
C. strigosus compositus Britton, Bull. Torr. Club **13**: 212. 1886.
C. strigosus robustior Kunth, Enum. **2**: 88. 1837.

Perennial by basal tuber-like corms, culm rather stout, 1°–3° tall. Leaves somewhat rough-margined, 2″–3″ wide, the longer ones of the involucre much exceeding the umbel; umbel several-rayed, compound or nearly simple, some of the primary rays often 4′–6′ long, their sheaths terminating in 2 bristles; involucels setaceous; heads oblong to subglobose; spikelets flat, linear, 3″–12″ long, 1″ wide or less, 7–25-flowered, separating from the axis at maturity; scales straw-colored, oblong-lanceolate, subacute, strongly severalnerved, appressed, or at length somewhat spreading; stamens 3; style 3-cleft; achene linear-oblong, 3-angled, acute, about one-third as long as the scale.

In moist meadows, swamps or along streams,

3/5

Maine and Ontario to Minnesota, Florida and Texas. Lank galingale. Nut-grass. Ground-moss. Contains several races, differing in size, length of rays, length and width of spikelets. Aug.–Oct.

28. Cyperus refráctus Engelm.   Reflexed Cyperus.   Fig. 748.

C. refractus Engelm.; Boeckl. Linnaea **36**: 369. 1869–70.

Perennial by tuber-like corms, culm stout, smooth, 1°–3° tall. Leaves 2½″–4″ wide, rough-margined, elongated; umbel 6–13-rayed, usually compound, the longer rays sometimes 8′ long, their sheaths terminating in 1 or 2 short teeth; involucels setaceous; raylets filiform; spikelets very narrowly linear, loosely spicate, acute, flattish, 5″–12″ long, ½″ thick, 3–6-flowered, the upper spreading, the lower reflexed; scales yellowish-green, oblong-lanceolate, obtuse, closely appressed, 9–11-nerved, thin; stamens 3; style 3-cleft, its branches much exserted; achene narrowly linear, obtuse, apiculate, about 5 times as long as thick, and one-half as long as the scale.

In dry fields, New Jersey to Georgia, Missouri and Texas. July–Sept.

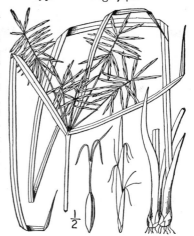

29. Cyperus retrofráctus (L.) Torr.   Rough Cyperus.   Fig. 749.

Scirpus retrofractus L. Sp. Pl. 50. 1753.
Cyperus retrofractus Torr.; A. Gray, Man. 519. 1848.
Cyperus dipsaciformis Fernald, Rhodora **8**: 127. 1906.

Perennial by tuber-like corms, culm slender, rough-puberulent, at least above, mostly longer than the puberulent leaves, 1°–3° tall. Leaves 1½″–2½″ wide, those of the involucre 4–7, the longer not greatly exceeding the umbel; umbel simple; rays very slender, nearly erect, or spreading, 2′–6′ long, their sheaths 2-toothed; heads oblong or obovoid; spikelets linear-subulate, 3″–6″ long, about ½″ thick, 1–3-flowered, all soon strongly reflexed; flowering scales lanceolate, acute, the upper one subulate, all strongly several-nerved; stamens 3; style 3-cleft; achene linear, 3-angled, obtuse, apiculate, two-thirds as long as the scale.

In dry, sandy soil, southern New Jersey to Florida, west to Kentucky, Missouri and Texas. July–Sept.

30. Cyperus lancastriénsis Porter.   Lancaster Cyperus.   Fig. 750.

C. lancastriensis Porter; A. Gray. Man. Ed. 5, 555. 1867.

Perennial by ovoid or oblong corms, culm slender, smooth, mostly longer than the leaves, 1°–2½° tall. Leaves 2″–3″ wide, those of the involucre 4–7, the longer much exceeding the inflorescence; umbel simple, 5–9-rayed, the longer rays 2′–4′ long, their sheaths nearly truncate; heads oval, obtuse, ½′–1′ long; spikelets densely clustered, 4″–5″ long, linear, subterete, 2–4-flowered, the lower reflexed, the middle ones spreading, all separating from the axis at maturity; scales green, strongly several-nerved, the flowering ones lanceolate, subacute; stamens 3; style 3-cleft; achene linear, obtuse, apiculate, 2–3 times as long as thick, two-thirds as long as the scale.

In dry fields, New Jersey and Pennsylvania to Georgia, Missouri and Alabama. July–Sept.

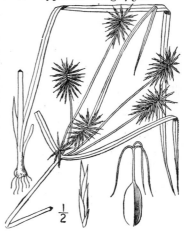

$\frac{2}{3}$

31. Cyperus hystricìnus Fernald. Bristly Cyperus. Fig. 751.

Cyperus hystricinus Fernald, Rhodora **8**: 127. 1906.

Perennial by corms and rootstocks; culms rather stout, smooth throughout, 3° tall or less. Leaves smooth, 1″–3″ wide, the basal ones shorter than the culm, those of the involucre about as long as the umbel; rays 14 or fewer, the longer 7′ long or less; heads obovoid, or obovoid-cylindric, $\frac{1}{2}′–1\frac{1}{2}′$ long; spikelets yellowish-brown, subulate, 2″–4″ long, the uppermost spreading, all the others strongly reflexed, bearing 1 or 2 achenes; fertile scales strongly nerved; stamens 3; achene linear, trigonous, about $1\frac{1}{4}″$ long, 3–4 times as long as the scale.

Dry sandy soil, New Jersey and eastern Pennsylvania to Georgia and Texas. Except for the smooth culm and leaves closely resembling *C. retrofractus*. July–Sept.

32. Cyperus Tórreyi Britton. Pine-barren Cyperus. Fig. 752.

Mariscus cylindricus Ell. Bot. S. C. & Ga. **1**: 74. 1816.
Cyperus cylindricus Britton, Bull. Torr. Club **6**: 339. 1879. Not Boeckl. 1859.
C. Torreyi Britton, Bull. Torr. Club **13**: 215. 1886.

Perennial by small hard corms, culms slender, smooth, usually tufted, 4′–18′ tall, longer than the leaves. Leaves smooth, 1″–1½″ wide, the longer ones of the involucre much exceeding the umbel; umbel simple, several-rayed, the rays short, or the longer 1′–2½′ long, the sheaths 2-toothed; heads very dense, cylindric, $\frac{1}{4}′–\frac{1}{2}′$ long, 2″–4″ in diameter; spikelets 1½″–2″ long, flattish, 1–2-flowered, spreading or the lower reflexed; scales green, oblong; rachis winged; stamens 3; style 3-cleft; achene linear-oblong, 3-angled, apiculate, slightly more than one-half as long as the scale.

In sandy pine barrens and on the sea shore, southern New York to Florida, west to Missouri and Texas. July–Sept.

$\frac{2}{3}$

33. Cyperus ovulàris (Michx.) Torr. Globose Cyperus. Fig. 753.

Kyllingia ovularis Michx. Fl. Bor. Am. **1**: 29. 1803.
Cyperus ovularis Torr. Ann. Lyc. N. Y. **3**: 278. 1836.

Perennial by hard tuber-like corms, culms usually strict, smooth, 8′–2½° tall, longer than the leaves. Leaves smooth, 2″–3″ wide, the longer ones of the involucre exceeding the umbel; umbel mostly simple, few-rayed, the rays rarely more than 2½′ long; sheath of the rays truncate or slightly toothed; heads globose or sometimes a little longer than thick, 4″–7″ in diameter, very dense, the spikelets radiating in all directions; spikelets 2″–3½″ long, usually 3-flowered, separating from the axis and leaving a scar at maturity; rachis winged; scales ovate or ovate-lanceolate, obtuse or subacute, green, strongly several-nerved; stamens 3; style 3-cleft; achene linear-oblong, 3-angled, 2–3 times as long as thick.

In dry fields and on hills, southern New York to Florida, west to Illinois, Kansas and Texas. Hedge-hog Club-rush. July–Sept.

$\frac{2}{3}$

34. Cyperus filicúlmis Vahl.   Slender Cyperus.   Fig. 754.

Cyperus filiculmis Vahl, Enum. **2** : 328. 1806.
C. filiculmis macilentus Fernald, Rhodora **8** : 128. 1906.
C. macilentus Bicknell, Bull. Torr. Club **35** : 478. 1908.

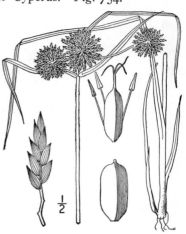

Perennial by hard oblong corms, culm smooth, slender or almost filiform, ascending or reclined, 6'–18' long, usually longer than the rough-margined leaves. Leaves 1"–2" wide, keeled, those of the involucre, or some of them, much exceeding the inflorescence; spikelets densely clustered in 1–7 globose heads, linear, acute, 4–11-flowered, subterete or compressed, 1½"–6" long, 1" wide or less; rachis wingless; scales ovate, acute or obtuse, pale green, strongly 7–11-nerved, appressed; stamens 3; style 3-cleft; achene oblong or obovoid, 3-angled, obtuse, apiculate, dull gray, two-thirds as long as the scale, about twice as long as thick.

In dry fields and on hills, Maine to Ontario, Minnesota, Florida, Kansas, Texas and Mexico. June–Aug. In its northern range the spikelets are fewer-flowered than in the south.

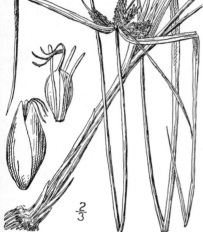

35. Cyperus cayennénsis (Lam.) Britton.   Cayenne Cyperus.   Fig. 755.

Kyllingia cayennensis Lam. Ill. **1** : 149. 1791.

Mariscus flavus Vahl, Enum. **2** : 374. 1806.

C. flavus Boeckl. Linnaea **26** : 384. 1869–70.

C. cayennensis Britton, Bull. Dept. Agric. Jam. **5** : Suppl. 1, 8. 1907.

Perennial by short rootstocks; culms trigonous, smooth, 2° high or less. Leaves flat, 2"–3½" wide, the basal ones often as long as the culm, the largest one of the involucre sometimes half as long; spikelets 2½" long or less, turgid, spicate, the spikes oblong, mostly sessile in a terminal cluster, 5"–8" long, obtuse; scales ovate, striate, the lowest persistent; style 3-cleft; achene trigonous, oblong-obovoid, about half as long as the scale.

Waste grounds, Camden, N. J., Philadelphia, Penna., and in the southern states. Adventive from tropical America.

36. Cyperus Gràyi Torr.   Gray's Cyperus.   Fig. 756.

Cyperus Grayi Torr. Ann. Lyc. N. Y. **3** : 268. 1836.

Perennial by thick hard oblong or ovoid corms, culms tufted, ascending or reclined, stiff, smooth, very slender, 6'–20' long. Leaves shorter than the culm, bright green, 1" wide or less, those of the involucre 4–8, the longer somewhat exceeding the umbel; umbel 4–10-rayed, simple, the longer rays 3'–4' long; sheaths of the rays truncate or nearly so; spikelets 2½"–5" long, loosely capitate, compressed, linear, rigid, spreading; scales green, ovate, obtuse or subacute, strongly 13–15-nerved, rather widely spreading when old; joints of the rachis broadly winged; stamens 3; style 3-cleft; achene oblong or oblong-obovoid, obtuse, apiculate, about two-thirds as long as the scale.

In sands of the sea shore and in pine barrens, New Hampshire to Florida. July–Sept.

$\frac{1}{2}$

37. **Cyperus globulòsus** Aubl.   Baldwin's Cyperus.   Fig. 757.

C. globulosus Aubl. Pl. Guian. 1 : 47.   1775.
Mariscus echinatus Ell. Bot. S. C. & Ga. 1 : 75.   1816.
Cyperus Baldwinii Torr. Ann. Lyc. N. Y. 3 : 270.   1836.
Cyperus echinatus Wood, Class-book 734.   1863.

Perennial by tuber-like corms, culm slender, smooth, erect, mostly longer than the leaves.   Leaves pale green, 1½″-2″ wide, those of the involucre 5-10, the longer usually much exceeding the umbel; umbel simple, 6-13-rayed, the rays filiform, their sheaths short, mucronate; spikelets 2″-3″ long, linear, flat, densely or loosely capitate in globose heads; scales thin, pale green, appressed, ovate-lanceolate, acute, 9-13-nerved, with narrow scarious margins; joints of the rachis broadly winged; stamens 3; style 3-cleft; achene oblong-obovoid, obtuse, one-half as long as the scale, about twice as long as thick.

In dry soil, sometimes a weed in cultivated fields, Virginia to Florida, west to Missouri and Texas.   Also in Bermuda and in tropical America.   July–Aug.

3. **ELEÓCHARIS** R. Br. Prodr. Fl. Nov. Holl. 1 : 224.   1810.

Annual or perennial sedges.   Culms simple, triangular, quadrangular, terete, flattened or grooved, the leaves reduced to sheaths or the lowest very rarely blade-bearing.   Spikelets solitary, terminal, erect, several–many-flowered, not subtended by an involucre.   Scales concave, spirally imbricated all around.   Perianth of 1-12 bristles, usually retrorsely barbed, wanting in some species.   Stamens 2-3.   Style 2-cleft and achene lenticular or biconvex, or 3-cleft and achene 3-angled, but sometimes with very obtuse angles and appearing turgid.   Base of the style persistent on the summit of the achene, forming a terminal tubercle. [Greek, referring to the growth of most of the species in marshy ground.]

About 140 species, widely distributed.   Besides the following, some 20 others occur in the southern and western parts of North America.   Type species: *Scirpus palustris* L.

   1. Spikelet scarcely or not at all thicker than the culm ; scales coriaceous.
Culm stout ; spikelet many-flowered.
    Culm terete, nodose.   1. *E. interstincta.*
    Culm quadrangular, continuous.   2. *E. mutata.*
Culm slender, triangular, continuous ; spikelet few-flowered, subulate.   3. *E. Robbinsii.*
   2. Spikelet manifestly thicker than the culm (except in No. 13) ; scales membranous.
*Style mostly 2-cleft ; achene lenticular or biconvex.
Upper sheath scarious, hyaline ; plants perennial by slender rootstocks.
    Scales pale green or nearly white ; achene ¼″ long.   4. *E. flaccida.*
    Scales dark reddish-brown ; achene ½″ long.   5. *E. olivacea.*
Upper sheath truncate, oblique or toothed, not scarious.
    Annual, with fibrous roots.
      Achene jet black.
        Culms 1′-3′ tall ; achene ¼″ long ; bristles 2-4.   6. *E. atropurpurea.*
        Culms 3′-10′ tall ; achene ½″ long ; bristles 5-8.   7. *E. capitata.*
      Achene pale brown.
        Spikelet ovoid or oblong.
          Tubercle narrower than the top of the achene.   8. *E. ovata.*
          Tubercle about as broad as the top of the achene.   9. *E. obtusa.*
        Spikelet oblong-cylindric ; tubercle broad, low.   10. *E. Engelmanni.*
    Perennial by horizontal rootstocks.
      Scales pale green to straw-color.   11. *E. macrostachya.*
      Scales brown to purple-brown.
        Tubercle flattened-conic ; spikelet thicker than the culm.   12. *E. palustris.*
        Tubercle swollen, bulb-like ; spikelet not thicker than the culm. 13. *E. Smallii.*
**Style 3-cleft ; achene 3-angled or turgid.
    Achene reticulated or cancellate.
      Spikelet compressed ; culm filiform.   14. *E. acicularis.*
      Spikelet terete ; culm slender.
        Achene transversely cancellate ; bristles none.   15. *E. Wolfii.*
        Achene reticulated ; bristles present, stout.
          Tubercle conic, smaller than the achene.   16. *E. simplex.*
          Tubercle cap-like, as large as or larger than the achene. 17. *E. tuberculosa.*
    Achene smooth or papillose.
      Achene smooth, white ; culms capillary.   18. *E. Torreyana.*
      Achene papillose or smooth, brown, black or yellow.
        Tubercle depressed or short-conic.
          Achene smooth.
            Tubercle flat, covering the top of the black achene.   19. *E. melanocarpa.*
            Tubercle ovoid-conic, acute, contracted at the base.   20. *E. albida.*
          Achene papillose.
            Achene 3-ribbed on the angles.   21. *E. tricostata.*
            Achene obtuse-angled, not ribbed.

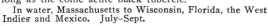

Culms filiform; scales obtuse.	**22.** *E. tenuis.*
Culms flat; scales acute.	**23.** *E. acuminata.*
Tubercle subulate or narrowly pyramidal.	
Culms filiform, wiry, densely tufted, 4′–10′ long.	**24.** *E. intermedia.*
Culms flattened, slender, 1°–2° long.	**25.** *E. rostellata.*

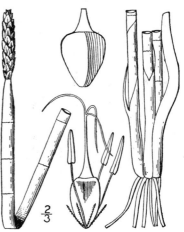

1. Eleocharis interstíncta (Vahl) R. & S. Knotted Spike-rush. Fig. 758.

Scirpus interstinctus Vahl, Enum. **2**: 251. 1806.
Scirpus equisetoides Ell. Bot. S. C. & Ga. **1**: 79. 1816.
Eleocharis interstincta R. & S. Syst. **2**: 148. 1817.
Eleocharis equisetoides Torr. Am. Lyc. **3**: 296. 1836.

Perennial by stout rootstocks, sometimes tuberiferous; culms terete, hollow, nodose, papillose, 1½°–3° tall, the sterile ones sharp-pointed. Sheaths oblique, membranous, brown or green, the lower sometimes bearing short blades; spikelet terete, cylindric, many-flowered, subacute, 1′–1½′ long, 2″ in diameter, not thicker than the culm; scales ovate, orbicular or obovate, obtuse or the upper acute, narrowly scarious-margined, faintly many-nerved, persistent; bristles about 6, rigid, retrorsely barbed, as long as the body of the achene or shorter; stamens 3; style 3-cleft, exserted; achene obovoid, brown, shining, with minute transverse ridges, obtusely trigonous, 2 or 3 times as long as the conic acute black tubercle.

In water, Massachusetts to Wisconsin, Florida, the West Indies and Mexico. July–Sept.

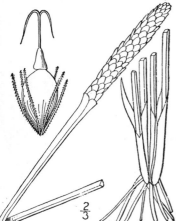

2. Eleocharis mutàta (L.) R. & S. Angled Spike-rush. Fig. 759.

Scirpus mutatus L. Am. Acad. **5**: 391. 1760.
Scirpus quadrangulatus Michx. Fl. Bor. Am. **1**: 30. 1803.
Eleocharis quadrangulata R. & S. Syst. **2**: 155. 1817.
Eleocharis mutata R. & S. Syst. **2**: 155. 1817.

Perennial by stout rootstocks, sometimes tuberiferous; culms sharply 3-4-angled, stout, not nodose, papillose, 2°–4° tall. Sheaths purplish-brown or green, membranous, sometimes bearing short blades; spikelet terete, acute, cylindric, 1′–2′ long, 2″ in diameter, many-flowered, about as thick as the culm; scales coriaceous, broadly ovate or obovate, obtuse or the upper subacute, scarious-margined and sometimes with a narrow brown band within the margins, faintly many-nerved, persistent; bristles about 6, rigid, retrorsely barbed, about as long as the achene; stamens 3; style 3-cleft; achene obovoid, biconvex or slightly angled on the back, minutely cancellate, about twice as long as the conic acute tubercle.

In ponds, streams and swamps, Massachusetts to New Jersey, Ontario, Michigan, Alabama, Missouri, Texas and Guatemala. West Indies and South America. July–Sept.

3. Eleocharis Robbínsii Oakes. Robbins' Spike-rush. Fig. 760.

Eleocharis Robbinsii Oakes, Hovey's Mag. **7**: 178. 1841.

Perennial by slender rootstocks, culms slender, 3-angled, continuous, 6′–2° long, sometimes producing numerous filiform flaccid sterile branches from the base. Sheaths appressed, obliquely truncate; spikelet subulate, few-flowered, not thicker than the culm, 6″–10″ long, 1″ in diameter; scales lanceolate or oblong-lanceolate, obtuse or subacute, strongly concave, faintly several-nerved, persistently clasping the rachis, narrowly scarious-margined; style 3-cleft; bristles 6, equalling the achene and tubercle, retrorsely barbed; achene obovoid, light brown, biconvex or very obtusely angled on the back, somewhat longer than the conic-subulate flattened tubercle, which has a raised ring around its base.

In shallow water, Nova Scotia to Michigan, south to Florida. Aug.–Sept.

4. Eleocharis fláccida (Rchb.) Urban. Pale Spike-rush. Fig. 761.

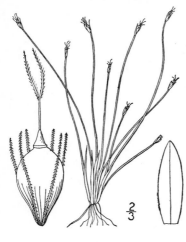

Scirpus flaccidus Rchb.; Spreng. Tent. Suppl. **3**: 1828.
1828.
Eleogenus ochreatus Nees, in Mart. Fl. Bras. **2**: Part 1, 102.
1842.
Eleocharis ochreata Steud. Syn. Pl. Cyp. 79. 1855.
Eleocharis flaccida Urban, Symb. Ant. **2**: 165. 1900.

Perennial by very slender rootstocks, culms very slender, or filiform, erect, pale green, 3-angled, 2′-10′ tall. Upper sheath with a white, hyaline, scarious limb; spikelet oblong or ovoid, subacute, 2-3 times as thick as the culm, about 2″ long, 1¼″ in diameter, several-flowered; scales pale green, oblong-lanceolate, obtuse or the upper acute, thin, hyaline with a faint midvein; style 2-cleft; bristles about 6, slender, retrorsely barbed, as long as or somewhat longer than the achene, or wanting; achene ¼″ long, lenticular, obovate, smooth, brown, 2-4 times as long as the conic acute tubercle, which is often constricted at the base.

In wet soil, New Jersey and Delaware to Florida and Mississippi. Also in tropical America. Aug.–Sept.

5. Eleocharis olivàcea Torr. Bright green Spike-rush. Fig. 762.

Eleocharis olivacea Torr. Ann. Lyc. N. Y. **3**: 300. 1836.

Perennial by running rootstocks, often tufted and matted, culms very slender, bright green, erect or reclining, flattened, 1′-4′ long. Upper sheath with a white hyaline limb; spikelet ovoid, acute or obtuse, much thicker than the culm, several–many-flowered, about 2″ long, 1″ in diameter; scales ovate, thin, acute, reddish-brown, with a green midvein and narrow, scarious margins; stamens 3; style 2-cleft; bristles 6-8, slender, retrorsely barbed, longer than the achene and tubercle; achene obovoid, similar to that of the preceding species but twice as large, 3-4 times the length of the conic acute tubercle.

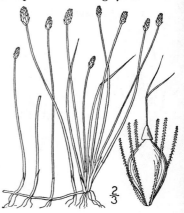

In wet soil, Maine to southern Ontario, Michigan, Pennsylvania, South Carolina and Kansas. Aug.–Sept.

6. Eleocharis atropurpùrea (Retz) Kunth. Purple Spike-rush. Fig. 763.

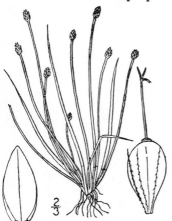

Scirpus atropurpureus Retz, Obs. **5**: 14. 1789.

Eleocharis atropurpurea Kunth, Enum. **2**: 151. 1837.

Annual, roots fibrous, culms tufted, very slender, 1′-3½′ high. Upper sheath 1-toothed; spikelet ovoid, many-flowered, subacute, 1½″-2″ long, 1″ in diameter or less; scales minute, ovate-oblong, obtuse or the upper acute, persistent, purple-brown with green midvein and very narrow scarious margins; stamens 2 or 3; style 2-3-cleft; bristles 2-4, fragile, white, minutely downwardly hispid, about as long as the achene; achene jet black, shining, ¼″ long, smooth, lenticular; tubercle conic, minute, depressed but rather acute, constricted at the base.

In moist soil, Nebraska and eastern Colorado to Central America, east to Iowa and Florida; widely distributed in tropical America, Europe and Asia. July–Sept.

7. Eleocharis capitàta (L.) R. Br.   Capitate Spike-rush.   Fig. 764.

Scirpus capitatus L. Sp. Pl. 48. 1753.
Eleocharis capitata R.Br. Prodr. Fl. Nov. Holl. 1 : 225. 1810.
Eleocharis dispar E. J. Hill, Bot. Gaz. 7 : 3. 1882.
E. capitata dispar Fernald, Rhodora 8 : 129. 1906.

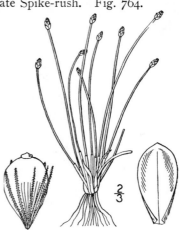

Annual, roots fibrous, culms densely tufted, nearly
terete, almost filiform, rather stiff, 2'–10' tall. Upper
sheath 1-toothed; spikelet ovoid, obtuse, much thicker
than the culm, 1½''–2½'' long, 1''–1½'' thick, many-
flowered; scales broadly ovate, obtuse, firm, pale or
dark brown with a greenish midvein, narrowly scarious-margined, persistent; stamens mostly 2; style
2-cleft; bristles 5–8, slender, downwardly hispid, as
long as the achene; achene obovate, jet black, smooth,
shining, nearly ½'' long; tubercle depressed, apiculate,
constricted at the base, very much shorter than the
achene.

In moist soil, Maryland to Florida, west to Indiana and
Texas. Widely distributed in tropical regions. July–Sept.

8. Eleocharis ovàta (Roth) R. & S.   Ovoid Spike-rush.   Fig. 765.

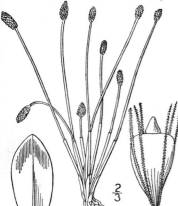

Scirpus ovatus Roth, Catal. Bot. 1 : 5. 1797.
Eleocharis ovata R. & S. Syst. 2 : 152. 1817.
Eleocharis diandra Wright, Bull. Torr. Club 10 : 101. 1883.

Annual, roots fibrous, culms tufted, slender or filiform, rather deep green, nearly terete, mostly erect,
2'–16' tall. Upper sheath 1-toothed; spikelet ovoid or
oblong, obtuse, many-flowered, 2''–5'' long, 1''–1½'' in
diameter; scales thin, oblong-orbicular, very obtuse,
brown with a green midvein and scarious margins;
bristles 6–8, sometimes fewer or wanting, deciduous,
usually longer than the achene; stamens 2 or 3; style
2-3-cleft; achene pale brown, shining, lenticular, obovate-oblong, smooth, ½'' long or more; tubercle deltoid or depressed, acute, about one-forth as long as
the achene and narrower.

In wet soil, New Brunswick to Ontario, Michigan and
Connecticut. Also in Europe. July–Sept.

9. Eleocharis obtùsa (Willd.) Schultes.   Blunt Spike-rush.   Fig. 766.

Scirpus obtusus Willd. Enum. 76. 1809.

Eleocharis obtusa Schultes, Mant. 2 : 89. 1824.

E. obtusa jejuna Fernald, Proc. Am. Acad. 34 : 492.
1899.

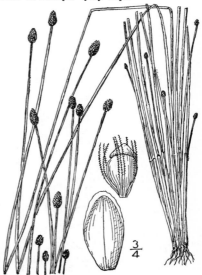

Annual, with fibrous roots; culms tufted,
slender to filiform, 1'–18' long. Upper sheath
1-toothed; spikelet ovoid to ovoid-oblong, 1''–6½''
long, 1''–2½'' thick, densely many-flowered; scales
broadly obovate, obtuse, brown with pale margins; achene pale brown, obovate, smooth, lenticular, about ½'' long; tubercle deltoid or depressed, nearly or quite as wide as the top of
the achene; bristles mostly longer than the achene.

In wet soil, Cape Breton Island to Minnesota,
British Columbia, Florida, Texas and Oregon. July–
Sept.

10. Eleocharis Engelmánni Steud.   Engelmann's Spike-rush.   Fig. 767.

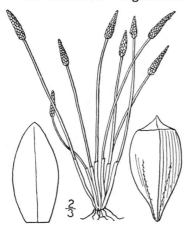

Eleocharis Engelmanni Steud. Syn. Pl. Cyp. 79. 1855.
Eleocharis ovata var. *Engelmanni* Britton, Journ. N. Y. Micros. Soc. **5**: 103. 1889.
E. monticola Fernald, Proc. Am. Acad. **34**: 496. 1899.

Annual, similar to the preceding species, but culms commonly taller, sometimes 18′ high. Upper sheath obliquely truncate or 1-toothed; spikelet oblong-cylindric or ovoid-cylindric, obtuse or subacute, 2″–8″ long, 1″–1½″ in diameter, many-flowered; scales pale brown with a green midvein and narrow scarious margin, ovate, obtuse, deciduous; style 2-cleft; bristles about 6, not longer than the achene, or wanting; achene broadly obovate, brown, smooth, lenticular; tubercle broad, low, covering the top of the achene.

In wet soil, Massachusetts to Indiana, South Dakota, Washington, New Jersey, Texas and California. July–Sept.

11. Eleocharis macrostàchya Britton.   Pale Spike-rush.   Fig. 768.

E. macrostachya Britton; Small, Fl. SE. U. S. 184. 1903.

Perennial by rootstocks, pale green. Culms tufted, rather stout, sometimes twisted, 4° high or less; spikelet lanceolate-cylindric, about 1″ long or less, acute, many-flowered; scales oblong-ovate to oblong-lanceolate, acute, light green to straw-color, with a somewhat darker midvein; bristles as long as the achene and tubercle, or shorter, sometimes very short; style 2-cleft; achene obovate, lenticular, 1.5 mm. long, brown, the cap-like tubercle small, yellow.

In wet soil, Missouri to Louisiana, Nevada, California and Jalisco. Aug.–Sept.

12. Eleocharis palústris (L.) R. & S.   Creeping Spike-rush.   Fig. 769.

Scirpus palustris L. Sp. Pl. 47. 1753.
Eleocharis palustris R. & S. Syst. **2**: 151. 1817.
Eleocharis palustris var. *vigens* Bailey; Britton, Journ. N. Y. Micros. Soc. **5**: 104. 1889.
E. glaucescens Willd. Enum. 76. 1809.

Perennial by horizontal rootstocks, culms stout or slender, terete or somewhat compressed, striate, 1½°–5° tall. Basal sheaths brown, rarely bearing a short blade, the upper one obliquely truncate; spikelet oblong to ovoid-cylindric, 3″–12″ long, 1½″–2″ in diameter, many-flowered, thicker than the culm; scales ovate-oblong or ovate-lanceolate, purplish-brown with scarious margin and a green midvein, or pale green all over; bristles usually 4, slender, retrorsely barbed, longer than the achene and tubercle, sometimes wanting; stamens 2–3; style 2-3-cleft; achene lenticular, smooth, yellow, over ½″ long; tubercle conic-triangular, constricted at the base, flattened, one-fourth to one-half as long as the achene.

In ponds, swamps and marshes, Labrador to British Columbia, south to Florida, Texas and California. The species consists of many races, the culms slender to stout, the tubercle narrow or quite broad. Also in Europe and Asia. Aglet-headed rush. Aug.–Sept.

13. Eleocharis Smàllii Britton.   Small's Spike-rush.   Fig. 770.

E. Smallii Britton, Torreya 3: 23. 1903.

Perennial by rootstocks; culms rather stout, about 2° high, and 1″–1½″ thick; top of the basal sheath oblique; spikelet cylindric to conic-cylindric, acute, about 8″ long, about as thick as the culm; scales lanceolate-oblong, acuminate; bristles very slender, equalling the achene and tubercle or a little longer; achene dark brown, obovate, turgid-lenticular, somewhat shining, nearly 1″ long, rounded at the top, the tubercle bulb-like, constricted at the base, one-fourth as long as the achene and about one-half as wide, rather abruptly tipped.

Valley of the Susquehanna River, Pennsylvania. Aug.–Sept.

14. Eleocharis aciculàris (L.) R. & S. Needle, or Least, Spike-rush.   Fig. 771.

Scirpus acicularis L. Sp. Pl. 48. 1753.
Eleocharis acitularis R. & S. Syst. 2: 154. 1817.

Perennial by filiform stolons or rootstocks, culms tufted, finely filiform or setaceous, obscurely 4-angled and grooved, weak, erect or reclining, 2′–8′ long. Sheaths truncate; spikelet compressed, narrowly ovate or linear-oblong, acute, broader than the culm, 3–10-flowered, 1½″–3″ long, ½″ wide; scales oblong, obtuse or the upper subacute, thin, pale green, usually with a narrow brown band on each side of the midvein, deciduous, many of them commonly sterile; bristles 3–4, fragile, fugacious, shorter than the achene; stamens 3; style 3-cleft; achene obovoid-oblong, pale, obscurely 3-angled with a rib on each angle and 6–9 lower intermediate ribs connected by fine ridges; tubercle conic, acute, one-fourth as long as the achene.

In wet soil, Newfoundland to British Columbia, New Jersey, Missouri, Mexico and California. Also in Europe and Asia. Sometimes entirely sterile. July–Sept.

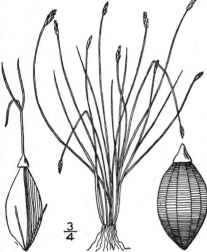

15. Eleocharis Wòlfii A. Gray.   Wolf's Spike-rush.   Fig. 772.

Scirpus Wolfii A. Gray, Proc. Am. Acad. 10: 77. 1874.
Eleocharis Wolfii A. Gray; Britton, Journ. N. Y. Micros. Soc. 5: 105. 1889.

Perennial by short rootstocks, culms very slender, erect, flattened and 2-edged, 8′–18′ tall. Upper sheath oblique, scarious, hyaline-tipped; spikelet oblong or ovoid-oblong, terete, acute, thicker than the culm, 2″–3″ long, nearly 1″ in diameter; scales ovate, obtuse or the upper acute, thin, pale green with purplish-brown bands, tardily deciduous; bristles none (or perhaps early deciduous); style 3-cleft; achene obovoid, obscurely 3-angled, longitudinally 9-ribbed, the ribs transversely connected by minute ridges; tubercle depressed-conic, much shorter than the achene.

In wet meadows, Illinois, Minnesota and Iowa. June–Aug.

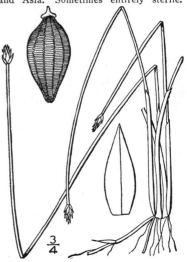

16. Eleocharis símplex (Ell.) A. Dietr.   Twisted Spike-rush.   Fig. 773.

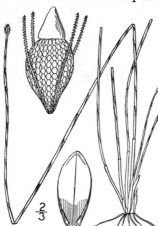

Scirpus simplex Ell. Bot. S. C. & Ga. **1** : 76.   1816.

Scirpus tortilis Link, Jahrb. **3** : 78.   1820.

Eleocharis tortilis Schultes, Mant. **2** : 92.   1824.

Eleocharis simplex A. Dietr. Sp. Pl. **2** : 78.   1833.

Annual, roots fibrous, culms tufted, filiform, sharply 3-angled, pale green, erect or reclining, twisting when old, 1°–1½° long. Sheaths obliquely truncate, 1-toothed; spikelet ovoid or oblong, subacute, several-flowered, 2″–3″ long, about 1″ thick, much thicker than the culm; scales firm, pale, ovate, mostly obtuse; bristles 4–6, rigid, retrorsely barbed, about equalling the achene and tubercle; stamens 3; style 3-cleft; achene obovoid, obscurely 3-angled, strongly reticulated, longitudinally about 18-ribbed; tubercle cap-like or conic, truncate at the base, one-fourth to one-half as long as the achene.

In wet soil, New Jersey to Florida and Texas, near the coast. July–Sept.

17. Eleocharis tuberculòsa (Michx.) R. & S.   Large-tubercled Spike-rush. Fig. 774.

Scirpus tuberculosus Michx. Fl. Bor. Am. **1** : 30.   1803.

Eleocharis tuberculosa R. & S. Syst. **2** : 152.   1817.

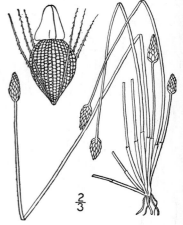

Annual, culms tufted, slightly compressed, very slender, rather stiff, striate, bright green, 8′–2° tall. Upper sheath obliquely truncate or 1-toothed; spikelet ovoid, obtuse or subacute, many-flowered, 3″–6″ long, nearly 2″ in diameter; scales broadly ovate, obtuse, pale greenish-brown with a darker midvein, broadly scarious-margined, firm, tardily deciduous; bristles 6, rigid, downwardly or rarely upwardly barbed, about as long as the achene and tubercle; stamens 3; style 3-cleft; achene obovoid, pale, trigonous, strongly reticulated, longitudinally about 18-ribbed; tubercle cap-like or conic, nearly or quite as large as the achene.

In wet soil, Massachusetts to Pennsylvania, Florida and Texas, near the coast.   July–Sept.

18. Eleocharis Torreyàna Boeckl.   Torrey's Spike-rush.   Fig. 775.

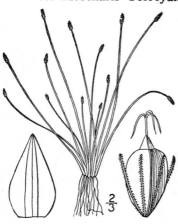

Eleocharis Torreyana Boeckl. Linnaea **36** : 440.   1870.

Annual, culms finely filiform, densely tufted, somewhat 4-sided, erect or reclining, often proliferous by developing secondary culms in the axils of the spikelet, sometimes rooting at the summit, 2′–8′ long. Upper sheath obliquely truncate; spikelet oblong, subacute, terete or nearly so, much thicker than the culm, many-flowered, 1½″–2½″ long; scales ovate, acute, brownish-red with a green midvein and lighter margins, early deciduous except the lowest which is commonly larger than the others, persistent and bract-like; bristles 3–6, slender, shorter than or equalling the achene; stamens 3; style 3-cleft; achene white, 3-angled, obovoid, smooth, minute; tubercle conic-pyramidal, much shorter than the achene.

In wet sandy soil, Connecticut to Florida and Texas, mostly near the coast. Also in Cuba. Confused in the first edition of this work with *E. microcàrpa* Torr. of the southern states.   June–Aug.

19. Eleocharis melanocàrpa Torr. Black-fruited Spike-rush. Fig. 776.

Eleocharis melanocarpa Torr. Ann. Lyc. N. Y. **3**: 311. 1836.

Perennial by short rootstocks, culms flattened, striate, tufted, slender, erect, wiry, 10′–20′ tall, sometimes proliferous. Upper sheath truncate, 1-toothed; spikelet oblong or cylindric-oblong, obtuse, 3″–6″ long, 1½″–2″ in diameter, many-flowered, thicker than the culm; scales ovate, obtuse, brown, with a lighter midvein and scarious margins; bristles 3–4, fragile, downwardly hispid, equalling or longer than the achene, fugacious or perhaps sometimes wanting; stamens 3; style 3-cleft; achene 3-angled, obpyramidal, black, smooth, its superficial cells nearly quadrate; tubercle depressed, covering the summit of the achene, light brown, pointed in the middle.

In wet sandy soil, eastern Massachusetts and Rhode Island to Florida, near the coast. Also in northern Indiana. July–Sept.

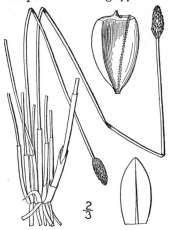

20. Eleocharis álbida Torr. White Spike-rush. Fig. 777.

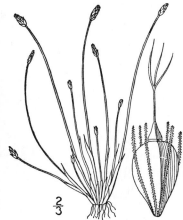

Eleocharis albida Torr. Ann. Lyc. N. Y. **3**: 304. 1836.

Annual, roots fibrous, culms very slender, tufted, nearly terete, striate, erect, 4′–8′ tall. Upper sheath very oblique and toothed on one side; spikelet ovoid-globose or oblong, obtuse, 2″–4″ long, 1½″–2″ in diameter, many-flowered, thicker than the culm; scales pale green or nearly white, rather firm, ovate, obtuse, deciduous; bristles about 6, downwardly barbed, persistent, as long as the achene; stamens 3; style 3-cleft; achene broadly obovoid, nearly black when ripe, 3-angled, smooth; tubercle ovoid-conic, contracted or truncate at the base, about one-fourth as long as the achene.

In wet soil, Maryland to Florida, Texas and eastern Mexico, near the coast. Recorded from Jamaica. June–Aug.

21. Eleocharis tricostàta Torr. Three-ribbed Spike-rush. Fig. 778.

Eleocharis tricostata Torr. Ann. Lyc. N. Y. **3**: 310. 1836.

Perennial by short rootstocks, culms very slender, erect, compressed, striate, 1°–2° tall. Upper sheath obliquely truncate, toothed on one side; spikelet oblong, becoming oblong-cylindric, obtuse, many-flowered, 5″–9″ long, 1″–1¼″ in diameter; scales ovate, thin, deciduous, obtuse, brown with a green midvein and scarious margins; bristles none; stamens 3; style 3-cleft; achene obovoid, 3-angled, brown, dull, papillose, strongly ribbed on each of its angles; tubercle conic, acute, light brown, constricted at the base, minute, very much shorter than the achene.

In wet soil, eastern Massachusetts to southern New York and Florida. July–Sept.

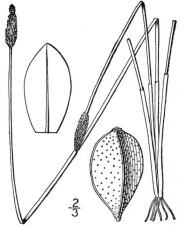

22. Eleocharis ténuis (Willd.) Schultes.   Slender Spike-rush.   Fig. 779.

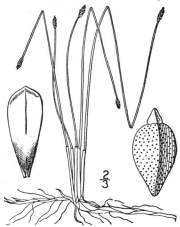

Scirpus tenuis Willd. Enum. 1 : 76. 1809.

Eleocharis tenuis Schultes, Mant. 2 : 92. 1824.

Eleocharis nitida Fernald, Rhodora 1 : 76. 1906.

Perennial by rootstocks, culms tufted, filiform, mostly erect, 4-angled with concave sides, 8'–16' tall. Upper sheath obliquely truncate, toothed on one side; spikelet narrowly oblong, mostly acute, many-flowered, thicker than the culm, 3"–5" long, about 1" in diameter; scales thin, obovate or ovate-oblong, obtuse, the midvein greenish, the margins scarious; bristles 2–4, shorter than the achene, fugacious or wanting; achene obovoid, obtusely 3-angled, yellowish-brown, papillose; stamens 3; style 3-cleft; tubercle conic, short, acute.

In wet soil, Cape Breton Island to Ontario and Manitoba, south to Florida and Texas. The achenes are more or less persistent on the rachis of the spikelet after the fall of the scales. Poverty-grass. Kill-cow. May–July.

23. Eleocharis acumináta (Muhl.) Nees.   Flat-stemmed Spike-rush.   Fig. 780.

Scirpus acuminatus Muhl. Gram. 27. 1817.
Eleocharis compressa Sulliv. Am. Journ. Sci. 42 : 50. 1842.
Eleocharis acuminata Nees, Linnaea 9 : 294. 1835.

Perennial by stout rootstocks, similar to the preceding species but stouter, culms flattened, striate, slender but rather stiff, tufted, 8'–2° tall. Upper sheath truncate, sometimes slightly 1-toothed; spikelet ovoid or oblong, obtuse, thicker than the culm, many-flowered, 3"–6" long; scales oblong or ovate-lanceolate, acute or the lower obtusish, purple-brown with a greenish midvein and hyaline white margins, deciduous; bristles 1–5, shorter than or equalling the achene, fugacious, or wanting; stamens 3; style 3-cleft, exserted; achene obovoid, very obtusely 3-angled, light yellowish brown, papillose, much longer than the depressed-conic acute tubercle.

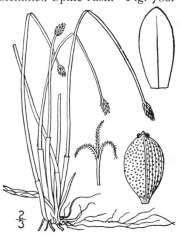

In wet soil, Anticosti to Manitoba, Washington, Georgia, Louisiana, Missouri and Nebraska. Achenes persistent on the rachis as in *E. tenuis.* June–Aug.

24. Eleocharis intermèdia (Muhl.) Schultes.   Matted Spike-rush.   Fig. 781.

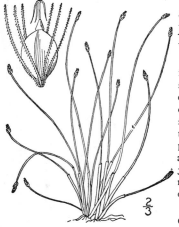

Scirpus intermedius Muhl. Gram. 31. 1817.
Eleocharis intermedia Schultes, Mant. 2 : 91. 1824.
E. intermedia Habereri Fernald, Rhodora 8 : 130. 1906.
E. Macounii Fernald, Proc. Am. Acad. 34 : 497. 1899.

Annual, roots fibrous, culms filiform, densely tufted, reclining or ascending, grooved, 4'–12' long. Upper sheath obliquely truncate, toothed on one side; spikelet ovoid-oblong, acute, 8–20-flowered, thicker than the culm; scales oblong-lanceolate, obtuse or the upper subacute, light purple-brown with a green midvein, tardily deciduous or the lower one persistent; bristles persistent, downwardly barbed, longer than the achene and tubercle, sometimes wanting; stamens 3; style 3-cleft; achene 3-angled, obovoid, light brown, finely reticulated; tubercle conic to conic-subulate, very acute, one-fourth to one-half as long as the achene.

In marshes, Quebec to Minnesota, south to New Jersey, Ohio, Illinois and Iowa. July–Sept.

25. Eleocharis rostellàta Torr.   Beaked
    Spike-rush.   Fig. 782.

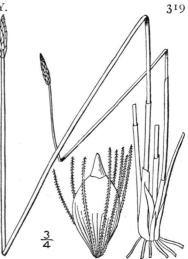

Scirpus rostellatus Torr. Ann. Lyc. N. Y. 3 : 318.   1836.
Eleocharis rostellata Torr. Fl. N. Y. 2 : 347.   1843.

Perennial by a short caudex, culms slender,
flattened, wiry, the fertile erect or ascending, the
sterile reclining and rooting at the summit, grooved,
1°–5° long.   Upper sheath truncate; spikelet oblong,
narrowed at both ends, thicker than the culm,
10–20-flowered, 3″–6″ long, about 1″ in diameter;
scales ovate, obtuse or the upper acute, green with a
somewhat darker midvein, their margins slightly
scarious; bristles 4–8, retrorsely barbed, longer than
the achene and tubercle; stamens 3; style 3-cleft;
achene oblong-obovoid, obtusely 3-angled, its surface
finely reticulated; tubercle conic-subulate, about one-
half as long as the achene or shorter, capping its
summit, partly or entirely falling away at maturity.

In marshes and wet meadows, New Hampshire, Vermont and New York to British Columbia, Florida, Texas,
Mexico and California.   Also in Cuba.   Aug.–Sept.

4. STENOPHÝLLUS Raf. Neog. 4.   1825.

Mostly annual sedges, with slender erect culms, leafy below, the leaves narrowly linear
or filiform, with ciliate or pubescent sheaths.   Spikelets umbellate, capitate or solitary, subtended by a 1–several-leaved involucre, their scales spirally imbricated all around, mostly
deciduous.   Flowers perfect.   Perianth none.   Stamens 2 or 3.   Style 2–3-cleft, glabrous,
its base much swollen and persistent as a tubercle on the achene as in *Eleocharis*.   Achene
3-angled, turgid or lenticular.   [Greek, referring to the narrow leaves.]

A genus of some 90 species, natives of temperate and warm regions.   Besides the following, 6
others occur in the southern United States.   Type species: *Scirpus stenophyllus* Ell.

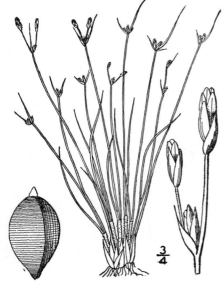

1. Stenophyllus capillàris (L.) Britton.
    Hair-like Stenophyllus.   Fig. 783.

Scirpus capillaris L. Sp. Pl. 49.   1753.
Fimbristylis capillaris A. Gray, Man. 530.   1848.
Stenophyllus capillaris Britton, Bull. Torr. Club
21 : 30.   1894.

Annual, roots fibrous, culms filiform, densely
tufted, erect, grooved, smooth, 2′–10′ tall.
Leaves filiform, roughish, much shorter than
the culm, their sheaths more or less pubescent
with long hairs; involucral leaves 1–3, setaceous, shorter than, or one of them exceeding
the inflorescence; spikelets narrowly oblong,
somewhat 4-sided, 2½″–4″ long, less than 1″
thick, several in a terminal simple or compound,
sometimes capitate, umbel, or in depauperate
forms solitary; scales oblong, obtuse or emarginate, puberulent, dark brown with a green
keel; stamens 2; style 3-cleft; achene yellow-
brown, narrowed at the base, very obtuse or
truncate at the summit, ¼″ long, 3-angled, transversely wrinkled; tubercle minute, depressed.

In dry or moist soil, Maine to southern Ontario,
Minnesota, Florida, Texas, California and tropical
America.   July–Sept.

5. FIMBRÍSTYLIS Vahl, Enum. 2 : 285.   1806.

Annual or perennial sedges.   Culms leafy below.   Spikelets umbellate or capitate,
terete, several to many-flowered, subtended by a 1–many-leaved involucre, their scales spirally
imbricated all around, mostly deciduous, all fertile.   Perianth none.   Stamens 1–3.   Style
2–3-cleft, pubescent or glabrous, its base much enlarged, falling away from the summit of
the achene at maturity.   Achene lenticular, biconvex, or 3-angled, reticulated, cancellate, or
longitudinally ribbed or striate in our species.   [Greek, in allusion to the fringed style of
some species.]

A large genus, the species about 125, widely distributed. Besides the following, some 5 others occur in southern and western North America. Type species: *Fimbristylis acuminata* Vahl.

Style 2-cleft; achene lenticular or biconvex.
  Culms 8′–3° tall; spikelets umbellate; style mostly pubescent.
    Perennial; leaves involute.
      Scales glabrous.
        Scales chestnut-brown, shining, coriaceous.   1. *F. castanea.*
        Scales yellow-brown, membranous, dull.   2. *F. interior.*
      Scales, at least the lower, pubescent or puberulent.   3. *F. puberula.*
    Annual; roots fibrous; leaves flat.   4. *F. Baldwiniana.*
  Culms 1′–4′ tall, very slender; spikelets capitate; style glabrous below.   5. *F. Vahlii.*
Style 3-cleft; achene 3-angled.
  Umbel mostly simple; spikelets ovoid to oval; achene reticulated.   6. *F. geminata.*
  Umbel mostly compound; spikelets linear; achene smooth or nearly so.   7. *F. autumnalis.*

1. **Fimbristylis castànea** (Michx.) Vahl. Marsh Fimbristylis. Fig. 784.

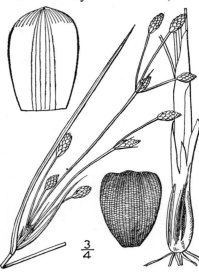

Scirpus castaneus Michx. Fl. Bor. Am. **1**: 31. 1803.
Fimbristylis castanea Vahl, Enum. **2**: 292. 1806.
F. spadicea castanea A. Gray, Man. Ed. 5, 566. 1867.

Perennial by a thickened base, glabrous, culms stiff, slender, 3-angled, wiry, 1°–3° tall, usually longer than the strongly involute rigid leaves. Leaves about 1″ wide when unrolled, their sheaths often brown; leaves of the involucre 3–6, erect, the longer sometimes exceeding the usually compound umbel; umbel several-rayed, the rays nearly erect, 2′–6′ long; central spikelets of the umbels and umbellets sessile, the others pedicelled; spikelets ovoid or ovoid-cylindric, acute, 2½″–6″ long, about 1″ in diameter; scales oval, obovate, or orbicular, obtuse or orbicular, obtuse or subacute, coriaceous, glabrous, dark brown with a green midvein; stamens 2; style 2-cleft; achene lenticular, obovate, brown, reticulated.

$\frac{3}{4}$

In marshes and shallow water, New York to Florida, along the coast. Bermuda. Erroneously referred in first edition to the tropical American *F. spadicea* (L.) Vahl, which has longer spikelets. July–Sept.

2. **Fimbristylis intèrior** Britton, n. sp. Plains Fimbristylis. Fig. 785.

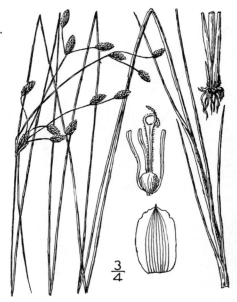

Perennial, with short stolons, the base of the culm slightly thickened. Culms loosely tufted to solitary, very slender, smooth, striate, somewhat compressed, 1°–2° high; leaves rough-margined, involute, at least toward the tip, 1″ wide or less, shorter than the culms; bracts of the involucre mostly shorter than the umbel, ciliate; umbel a little compound, its rays filiform; spikelets few to several, ovoid to ovoid-oblong, acutish, 5″ long or less, many-flowered; scales yellow-brown, ovate, striate, mucronate or the lower ones awned, glabrous, dull; stamens 3; achene broadly obovate, blunt, cancellate in many rows, chestnut-brown, nearly ½″ long.

$\frac{3}{4}$

Colorado and Nebraska to Texas. Type collected by Geo. E. Osterhout at Sterling, Logan County, Colorado, Aug. 13, 1896.

3. Fimbristylis pubérula (Michx.) Vahl.
Hairy Fimbristylis.   Fig. 786.

Scirpus puberulus Michx. Fl. Bor. Am. **1**: 31. 1803.

F. puberula Vahl. Enum. **2**: 289. 1806.

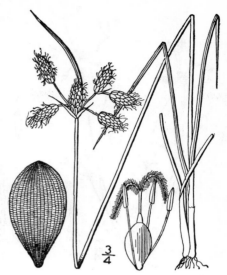

Perennial by stout rootstocks, culms slender, 3-angled, 8′–2° tall, usually exceeding the leaves. Leaves involute, less than 1″ wide when unrolled, often more or less pubescent, their sheaths green; leaves of the involucre 2–4, short; umbel simple or compound, the rays ½′–2′ long; central spikelets sessile; spikelets oblong, obtuse or subacute, 3″–5″ long, 1″–1½″ in diameter; scales thin, brown with a lighter midvein, broadly oblong or nearly orbicular, dull, puberulent, obtuse or mucronate; stamens 2–3; style 2-cleft; achene obovate or oblong, biconvex, pale brown, longitudinally striate and reticulated.

Fields and meadows, southern New York to Florida and Louisiana. Also from Ontario, Michigan and Illinois to Kansas and Texas. Mistaken in our first edition for *F. castànea.* July–Sept.

4. Fimbristylis Baldwiniàna Torr   Weak Fimbristylis.   Fig. 787.

F. Baldwiniana Torr. Ann. Lyc. N. Y. **3**: 344. 1836.

Annual, roots fibrous, culms slender, flattened, striate, densely tufted, erect or ascending, 2′–15′ long, usually longer than the leaves. Leaves flat, about ½″ wide, glabrous or sparingly ciliate, pale green and appearing glaucous, those of the involucre 3–5, one of them often exceeding the umbel; umbel simple or slightly compound, the central spikelet sessile; spikelets ovoid or ovoid-oblong, 3″–6″ long, about 1″ in diameter; scales ovate, thin, pale greenish-brown, subacute or mucronulate; stamen 1; style 2-cleft, pubescent; achene biconvex, obovoid, light brown, longitudinally ribbed, the ribs tubercled and connected by very fine cross-lines.

In moist soil, southern Pennsylvania to Florida, west to Illinois, Missouri and Texas. Included in our first edition in the southern and tropical American *F. laxa* Vahl. July–Sept.

5. Fimbristylis Vàhlii (Lam.) Link.   Vahl's Fimbristylis.   Fig. 788.

Scirpus Vahlii Lam. Tabl. Encycl. **1**: 139. 1791.

F. Vahlii Link, Hort. Berol. **1**: 287. 1827.

F. congesta Torr. Ann. Lyc. N. Y. **3**: 345. 1836.

Annual, culms very slender, densely tufted, compressed, striate, erect or ascending, 1′–4′ high, longer than or equalling the leaves. Leaves setaceous or almost filiform, rough, those of the involucre 3–5, erect, much exceeding the simple capitate cluster of 3–8 spikelets; spikelets oblong-cylindric, obtuse, 2″–4″ long, about ½″ thick, many-flowered; scales lanceolate, pale greenish-brown, acuminate; stamen 1; style 2-cleft, glabrous below; achene minute biconvex, yellowish-white, cancellate by longitudina and transverse ridges.

In moist soil, Missouri to Texas, east to North Carolina and Florida. California, Central and South America. Also in ballast about the eastern seaports. July–Oct.

$\frac{3}{4}$

6. Fimbristylis geminàta (Nees) Kunth.
Low Fimbristylis. Fig. 789.

Trichelostylis geminata Nees, in Mart. Fl. Bras. 2¹: 80. 1842.
F. Frankii Steud. Syn. Pl. Cyp. 111. 1855.
F. Frankii brachyactis Fernald, Rhodora 11: 180. 1909.

Annual, tufted, glabrous, low, 6'–8' high or less. Culms very slender, compressed; basal leaves about ½" wide, usually shorter than the culms, sometimes equalling them; involucral bracts 2 or 3, not longer than the inflorescence; umbel simple or somewhat compound; spikelets, or most of them, sessile, capitate, or some short-peduncled, or in capitate clusters at the base of the culms, ovoid or oval, 3" long or less; scales dull, green-brown, ovate, mucronulate; style-branches 3; style smooth; achenes rather larger than in *F. autumnalis,* distinctly reticulated, sometimes granular-tuberculate.

In moist soil, Maine to Ontario, Tennessee and Louisiana. July–Oct.

7. Fimbristylis autumnàlis (L.) R. & S.
Slender Fimbristylis. Fig. 790.

Scirpus autumnalis L. Mant. 2: 180. 1771.
Fimbristylis autumnalis R. & S. Syst. 2: 97. 1817.

Annual, roots fibrous, culms very slender, densely tufted, flat, roughish on the edges or smooth, erect, ascending or spreading, 1'–15' long, usually much exceeding the leaves. Leaves narrowly linear, flat, ½"–1" wide, long-acuminate, glabrous, those of the involucre 2–3, usually all shorter than the umbel; umbel compound or decompound (in dwarf forms sometimes reduced to a solitary spikelet), the primary rays ¼'–1½' long, the secondary filiform; spikelets linear-oblong, acute, 2"–5" long, ½" thick or less, several–many-flowered; scales ovate-lanceolate, appressed, subacute, strongly mucronate, greenish-brown, the midvein prominent; stamens 1–3; styles 3-cleft; achene obovoid, nearly white, 3-angled with a ridge on each angle, very finely reticulated and sometimes roughened.

$\frac{2}{3}$

In moist soil, Connecticut to Illinois, Florida and Texas. Also in tropical America. June–Sept.

6. ERIÓPHORUM L. Sp. Pl. 52. 1753.

Bog sedges, perennial by rootstocks, the culms erect, triangular or nearly terete, the leaves linear, or 1 or 2 of the upper ones reduced to bladeless sheaths. Spikelets terminal, solitary, capitate or umbelled, subtended by a 1–several-leaved involucre, or naked. Scales spirally imbricated. Flowers perfect. Perianth of 6 or apparently numerous, smooth soft bristles, which are white or brown, straight or crisped, and exserted much beyond the scales at maturity. Stamens 1–3. Style 3-cleft. Achene 3-angled, oblong, ellipsoid or obovoid. [Greek, signifying wool-bearing, referring to the soft bristles.]

About 15 species, in the northern hemisphere. Besides the following, 1 occurs in Alaska. Type species: *Eriophorum vaginatum* L. The species are called *Cotton-grass* or *Cotton-rush.*

Spikelet solitary; involucral leaf short or none.
  Bristles 6, simple, white, crisped.   1. *E. alpinum.*
  Bristles 6, each 4–6-cleft, thus appearing numerous.
    Plants stoloniferous.
      Scales with very narrow pale margins.   2. *E. Scheuchzeri.*
      Scales with broad pale margins.   3. *E. Chamissonis.*
    Plants tufted, not stoloniferous.
      Upper sheath inflated; culm rough at the top.   4. *E. callithrix.*
      Upper sheath not inflated; culm smooth.   5. *E. opacum.*
Spikelets several, involucrate by 1 or several leaves.
  Leaves triangular-channeled throughout.
    Blade of the upper stem-leaf not longer than the sheath.   6. *E. gracile.*

Blade of the upper stem-leaf much longer than the sheath.   7. *E. tenellum.*
Leaves flat, at least below the middle.
    Scales with a prominent midvein ; stamens 3.
        Midvein not prominent at the tip of the scale.   8. *E. angustifolium.*
        Midvein prominent to the tip of the scale.   9. *E. viridicarinatum.*
    Scales striate-nerved ; stamen 1.   10. *E. virginicum.*

1. Eriophorum alpìnum L. Alpine Cotton-grass. Fig. 791.

Eriophorum alpinum L. Sp. Pl. 53. 1753.
E. hudsonianum Michx. Fl. Br. Am. 1 : 34. 1803.
Scirpus hudsonianus Fernald, Rhodora 8 : 161. 1906.

Perennial by short rootstocks, sending up numerous filiform triangular roughish culms, 6′–10′ high. Leaves subulate, 3″–10″ long, triangular, channeled, borne very near the base of the culm, the lower sheaths often scarious and bladeless; spikelet solitary, terminal, small, erect; involucral bract subulate, mostly shorter than the spikelet, sometimes wanting; young spikelet ovoid-oblong, subacute; scales oblong-lanceolate, yellowish-brown, firm, obtuse or subacute, the midvein slender; bristles 6, white, crisped, 4–7 times as long as the scale; achene narrowly obovoid-oblong, brown, apiculate, dull.

In bogs and on high mountains, Newfoundland to Hudson Bay and British Columbia, south to Connecticut, northern New York and Michigan. Also in Europe and Asia. Summer.

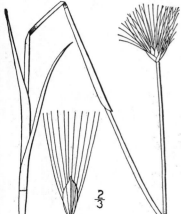

2. Eriophorum Scheùchzeri Hoppe. Scheuchzer's Cotton-grass. Fig. 792.

E. Scheuchzeri Hoppe, Taschenb. 1800 : 104. 1800.
E. capitatum Host, Gram. Aust. 1 : 30. *pl. 38.* 1801.

Stoloniferous; sheaths all blade-bearing or only the upper one bladeless; culms slender, smooth, nearly terete, 10′–16′ tall. Leaves filiform, channeled, usually much shorter than the culm; spikelet solitary, terminal, erect; involucre none; scales ovate-lanceolate or the inner ones linear-lanceolate, long-acuminate, purple-brown, membranous, with narrow, pale margins; bristles white, weak, nearly straight, 4–5 times as long as the scales; achene obovoid-oblong, acute, brown, dull, nearly ½″ long, subulate-beaked.

In bogs, Newfoundland and Labrador to Alberta, Alaska and British Columbia. Also in Europe and Asia. Summer.

3. Eriophorum Chamissònis C. A. Meyer. Russet Cotton-grass. Fig. 793.

E. Chamissonis C. A. Meyer; Ledeb. Fl. Alt. 1 : 79. 1829.
E. Chamissonis albidum Fernald, Rhodora 7 : 84. 1905.
Eriophorum russeolum Fries, Novit. Mant. 3 : 67. 1842.

Stoloniferous; culms solitary or little tufted, terete or somewhat triangular, erect, smooth, 4′–2½° tall, mostly longer than the leaves. Upper sheath inflated, bladeless, mucronate, rarely with a short subulate blade, usually borne below the middle of the culm; leaves filiform, triangular-channeled, mucronate, 1′–4′ long, or those of sterile shoots much longer; spikelet solitary, erect; involucre none; scales ovate-lanceolate, acuminate, thin, purplish-brown with broad white margins; bristles bright reddish-brown or white, 3–5 times as long as the scale; achene oblong, narrowed at each end, apiculate.

In bogs, Newfoundland to Quebec, New Brunswick, Ontario, Montana, Washington and British Columbia. Also in Europe and Asia. June–Aug.

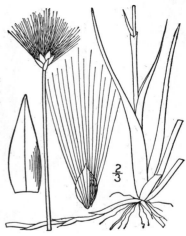

4. Eriophorum cállithrix Cham. Sheathed Cotton-grass. Fig. 794.

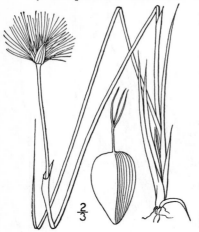

E. *vaginatum* Torr. Fl. 65. 1824. Not L.
E. *callithrix* Cham.; C. A. Meyer, Mem. Sav. Etrang.
1 : 203. 1831.

Plants not stoloniferous; culms tufted, stiff, obtusely triangular, forming tussocks, slender, 8′–20′ tall, leafless, except at the base, rough at the top, bearing 2 or 3 distant inflated sheaths, the upper one usually above the middle. Leaves stiff, filiform, triangular, channeled, slightly rough, shorter than or sometimes overtopping the culm; involucral leaf wanting; śpikelet solitary, erect; scales ovate-lanceolate or the lowest lanceolate, acuminate, purple-brown to nearly black, thin; bristles white or red-brown, straight, glossy, 4–5 times as long as the scale; anthers linear; achene obovoid, obtuse, brown, minutely apiculate.

In bogs, Newfoundland to Alaska, south to Massachusetts, Pennsylvania, Wisconsin and Manitoba. Also in Asia. Hare's-tail. Cotton-grass or -rush. Canna-down. Catlocks. Moss-crops. June–Aug.

5. Eriophorum opàcum (Björnst.) Fernald. Close-sheathed Cotton-grass. Fig. 795.

E. *vaginatum opacum* Björnst. Grunddr. af. Pit. Lappm. Växt. 35. 1856.

E. *opacum* Fernald, Rhodora 7 : 85. 1905.

Loosely tufted, not stoloniferous. Culms slender, terete or nearly so, 1°–2° high, smooth; basal leaves elongated, filiform-channeled; stem-leaves reduced to 2 or 3 close sheaths, the lower one sometimes with a short blade; spikelet solitary, erect; scales thin, ovate-lanceolate or the inner ones linear-lanceolate, acuminate; bristles white or brownish; achene obovate-oblong, apiculate.

Bogs, Maine and Massachusetts to Ontario, Alberta, the Yukon Territory and British Columbia. Also in Europe and Asia. Summer.

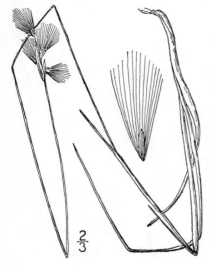

6. Eriophorum grácile Koch. Slender Cotton-grass. Fig. 796.

E. *gracile* Koch; Roth, Catal. Bot. 2 : 259. 1800.
E. *triquetrum* Hoppe, Taschenb. 1800 : 106. 1800.

Culms slender, smooth, nearly terete, spreading or reclining, 2° long or less. Leaves triangular-channeled, the basal ones mostly wanting at flowering time, those of the culm 2 or 3, the upper one with a blade shorter than its sheath, 1¾′ long or less; involucral leaf about ½′ long; spikelets 2–4, rarely 6, the slender peduncles pubescent, mostly less than 1′ long; scales ovate, grey to nearly black, acutish, the midvein prominent; achenes obovate-oblong, about 1″ long; bristles bright white, ½′–¾′ long.

In bogs, Quebec to British Columbia, New York, Pennsylvania, Iowa, Nebraska, Colorado and California. Also in Europe and Asia. June–Aug.

7. Eriophorum tenéllum Nutt. Rough Cotton-grass. Fig. 797.

Eriophorum tenellum Nutt. Gen. Add. 1818.
E. paucinervium A. A. Eaton, Bull. Torr. Club **25** : 341. 1898.

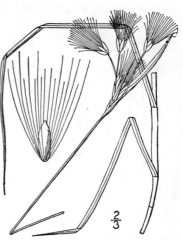

Culm slender, stiff, obtusely triangular, rough on the angles above, 1°–2½° tall, the sheaths all blade-bearing. Leaves narrowly linear, 1″ wide or less, triangular-channeled, rough-margined, the upper longer than its sheath; involucral leaf commonly only 1, stiff, erect; spikelets 3–8, capitate or subumbellate, the longer-peduncled ones drooping; scales ovate or oblong, obtuse or subacute, pale yellow to brown, the midvein rather strong, often with a weaker nerve on each side; bristles numerous, bright white, 8″–12″ long, 4–6 times as long as the scale; achene linear-oblong, acute, pointed, 1½″ long.

In bogs, Newfoundland to Hudson Bay, New Jersey, Pennsylvania and Illinois. Formerly confused with the next preceding species. June–Sept.

8. Eriophorum angustifòlium Roth. Tall Cotton-grass. Fig. 798.

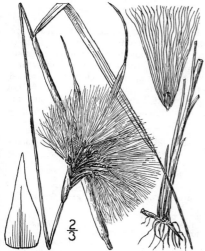

Eriophorum polystachyon L. Sp. Pl. 52, in part. 1753.
E. angustifolium Roth, Tent. 1 : 24. 1788.

Culm stiff, smooth, obtusely triangular above, nearly terete below, 1°–2° tall or less, all the sheaths blade-bearing. Leaves flat, at least below the middle, roughish-margined, 1½″–4″ wide, tapering to a triangular channeled rigid tip, the upper shorter than or rarely overtopping the culm, those of the involucre 2–4, often black at the base, the longer commonly equalling or exceeding the inflorescence; spikelets 2–12, ovoid, or oblong, clustered in a terminal umbel; rays filiform, smooth; scales ovate-lanceolate, acute or acuminate, purple-green or brown, the midvein not extending to the tip; bristles numerous, bright white, about 1′ long, 4–5 times as long as the scale; achene obovoid, obtuse, light brown.

In bogs, Newfoundland to Alaska, Maine, Illinois, Colorado and Oregon. Also in Europe and Asia. *E. polystàchyon* L. is confined to the Old World. June–Aug.

9. Eriophorum viridicarinàtum (Engelm.) Fernald. Thin-leaved Cotton-grass. Fig. 799.

E. latifolium viridicarinatum Engelm. Am. Journ. Sci. 46 : 103. 1844.
Eriophorum polystachyon latifolium A. Gray, Man. 529. 1848. Not *E. latifolium* Hoppe.

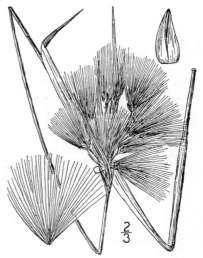

Similar to the preceding species but with thinner green, grass-like leaves, which are flat, except at the tip, 1″–3″ wide, the upper one 6′ long or less, those of the involucre not black at the base; spikelets usually more numerous, sometimes as many as 30, the rays finely hairy, elongated or sometimes very short; scales ovate-lanceolate, the midvein extending to the tip, sometimes slightly excurrent; achene oblong-obovoid; bristles white or yellowish-white.

In wet meadows and bogs, Newfoundland to British Columbia, New York, Georgia, Ohio and Michigan.

10. Eriophorum virgínicum L.   Virginia Cotton-grass.   Moss Crop.   Fig. 800.

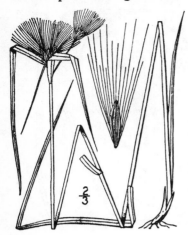

Eriophorum virginicum L. Sp. Pl. 53.   1753.

E. virginicum album A. Gray, Man. Ed. 5, 566.   1867.

Culm stiff, rather slender, obtusely triangular above, terete below, smooth, $1\frac{1}{2}°-4°$ tall, rather leafy. Leaves narrowly linear, flat, $1''-2\frac{1}{2}''$ wide, rough-margined, somewhat channeled toward the apex, the upper often overtopping the culm, those of the involucre 2–4, spreading or deflected, $2'-6'$ long, 1 or 2 of them much longer than the spikelets; spikelets several or numerous in a dense terminal capitate cluster usually broader than high, erect or the outer ones spreading; scales ovate, acute, brown with a green center, about 5-nerved; bristles numerous, dingy brown, rarely white, about 3 times as long as the scale; achene linear-oblong, acute, apiculate, light brown.

In bogs, Newfoundland to Manitoba, south to Florida and Nebraska. June–Sept.

7. SCÍRPUS L. Sp. Pl. 47.   1753.

Annual or perennial very small or very large sedges, with leafy culms or the leaves reduced to basal sheaths. Spikelets terete or somewhat flattened, solitary, capitate, spicate or umbellate, subtended by a 1–several-leaved involucre or the involucre wanting in some species. Scales spirally imbricated all around, usually all fertile, the 1 or 2 lower sometimes empty. Flowers perfect. Perianth of 1–6, slender or rigid, short or elongated, barbed, pubescent or smooth bristles, or none in some species. Stamens 2 or 3. Style 2–3-cleft, not swollen at the base, wholly deciduous from the achene, or its base persistent as a subulate tip. Achene triangular, lenticular or plano-convex. [Latin name of the Bulrush, said to be from *sirs,* the Celtic word for rushes.]

About 150 species of wide geographic distribution. Besides the following, some 10 others occur in the southern and western United States. Type species: *Scirpus lacustris* L.

1. Spikelet solitary, terminal, bractless or subtended by a single bract or short leaf.
  No involucral bract.
    Culms $1'-2'$ high; achene smooth; plant of saline soil.   1. *S. nanus.*
    Culms $3'-10'$ high; achene reticulated; plant of fresh-water marshes. 2. *S. pauciflorus.*
  Involucral bract present, erect.
    Bract shorter than or but little exceeding the spikelet; plants not aquatic.
      Culm terete; leaf of upper sheath subulate; bristles smooth.   3. *S. caespitosus.*
      Culms triangular; leaf of upper sheath linear; bristles upwardly barbed.
        Leaves shorter than the culm; scales acute.   4. *S. Clintoni.*
        Leaves about as long as culm; scales cuspidate or awned.   5. *S. planifolius.*
    Bract at least twice as long as the spikelet; plant aquatic.   6. *S. subterminalis.*
2. Spikelets normally more than 1, usually several or numerous, often appearing lateral; involucral bract only 1.
  Spikelets few, 1–12, appearing lateral.
    Culms not sharply 3-angled; achene plano-convex; annuals.
      Achene strongly transversely rugose.   7. *S. Hallii.*
      Achene smooth or very slightly roughened.   8. *S. debilis.*
    Culms sharply 3-angled; plants perennial by rootstocks.
      Achene plano-convex; bristles shorter than or equalling the achene.
        Spikelets acute, much overtopped by the slender involucral leaf; scales awned.
          9. *S. americanus.*
        Spikelets obtuse; involucral leaf short, stout; scales mucronulate.
          10. *S. Olneyi.*
      Achene 3-angled, ridged on the back.
        Bristles longer than the achene; involucral leaf erect.   11. *S. Torreyi.*
        Bristles as long as the achene; involucral leaf abruptly bent.   12. *S. mucronatus.*
  Spikelets several or numerous, umbelled; tall sedges.
    Culm sharply triangular, equalled by the long leaves.   13. *S. etuberculatus.*
    Culm terete; leaves reduced to sheaths.
      Styles 2-cleft; achene lenticular.
        Achenes $1''$ long, nearly as long as the scales; spikelets ovoid. 14. *S. validus.*
        Achenes about $1\frac{1}{2}''$ long, distinctly shorter than the scales; spikelets oblong-cylindric.
          15. *S. occidentalis.*
      Style 3-cleft; achene trigonous.   16. *S. heterochaetus.*
  Spikelets several, spicate.   17. *S. rufus.*

3. Spikelets several, capitate or umbellate, large; involucral leaves 2 or more.
   Achene lenticular or plano-convex; spikelets sessile or some stalked.
      Scales short-awned; achene lenticular.   18. *S. paludosus.*
      Scales long-awned; achene plano-convex.   19. *S. robustus.*
   Achene trigonous; spikelets mostly long-stalked.
      Achene sharply and nearly exactly trigonous.   20. *S. fluviatilis.*
      Achene with one face broader than the other two.
         Achene obovoid-orbicular; leaves 1″–2″ wide; spikelets ovoid.   21. *S. Fernaldi.*
         Achene obovoid; leaves 4″–8″ wide; spikelets narrowly cylindric. 22. *S. novae-angliae.*
4. Spikelets very numerous in compound umbels or umbelled heads, small; involucral leaves several;
   tall sedges.
   Bristles downwardly barbed; spikelets in umbelled heads.
      Spikelets ovoid or oblong, 1½″–2½″ long.
         Bristles equalling or slightly exceeding the achene; leaves 3″–8″ wide.
            Style 3-cleft; achene 3-angled; bristles 6.
               Spikelets 3–8 in each head; bristles barbed throughout; scales obtuse.
                               23. *S. sylvaticus.*
               Spikelets 8–20 in each head; bristles not barbed below; scales acute or awned.
                  Plant dark green; scales acute.   24. *S. atrovirens.*
                  Plant pale; scales rough-awned.   25. *S. pallidus.*
             Style 2-cleft; achene plano-convex; bristles 4.   26. *S. microcarpus.*
         Bristles flexuous, twice as long as the achene; leaves 2″–3″ wide. 27. *S. polyphyllus.*
      Spikelets cylindric, 3″–5″ long; style 3-cleft.   28. *S. Peckii.*
   Bristles smooth or slightly pubescent; umbel mostly decompound.
      Bristles shorter than or scarcely exceeding the scales.
         Bristles about as long as the achene; scales subacute.   29. *S. divaricatus.*
         Bristles much longer than the achene; scales mucronate.   30. *S. lineatus.*
      Bristles much exserted beyond the scales when mature.   31. *S. cyperinus.*

1. **Scirpus nànus** Spreng. Dwarf Club-rush. Fig. 801.

Scirpus nanus Spreng. Pug. 1: 4. 1815.

Eleocharis pygmaea Torr. Ann. Lyc. N. Y. **3**: 313. 1836.

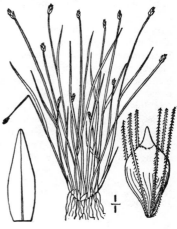

Annual, roots fibrous, culms filiform, flattened, grooved, tufted, erect or ascending, 1′–2′ high, bearing a scarious bladeless sheath near the base. Spikelet solitary, terminal, ovoid-oblong, rather acute, 3–8-flowered, 1″–1½″ long, not subtended by a bract; scales ovate or lanceolate, pale green, the lower obtuse, the upper subacute; bristles about 6, downwardly barbed, mostly longer than the achene; stamens 3; style 3-cleft; achene oblong, 3-angled, pale, pointed at each end, smooth.

Muddy places in salt marshes, Cape Breton Island to Florida and Texas, and about salt springs in New York, Michigan and Minnesota. Also on the Pacific Coast of North America and on the coasts of Europe, northern Africa, Cuba and Mexico. July–Sept.

2. **Scirpus pauciflòrus** Lightf. Few-flowered Club-rush. Fig. 802.

Scirpus pauciflorus Lightf. Fl. Scot. 1078. 1777.

Eleocharis pauciflorus Link, Hort. Berol. **1**: 284. 1827.

Perennial by filiform rootstocks, culms very slender, little tufted, 3-angled, grooved, leafless, 3′–10′ tall, the upper sheath truncate. Spikelet terminal, solitary, not subtended by an involucral bract, oblong, compressed, 4–10-flowered, 2″–3″ long, nearly 1″ wide; scales brown with lighter margins and midvein, lanceolate, acuminate; bristles 2–6, hispid, as long as the achene or longer; stamens 3; style 3-cleft; achene obovoid-oblong, gray, rather abruptly beaked, its surface finely reticulated.

In wet soil, Anticosti to Maine, Ontario, western New York, Illinois, Minnesota, British Columbia and California, south in the Rocky Mountains to Colorado. Also in northern Europe and Asia. July–Oct.

3. Scirpus caespitòsus L.   Tufted Club-rush.   Deer-hair.   Fig. 803.

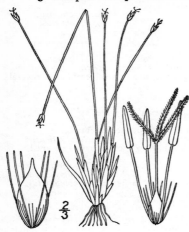

Scirpus caespitosus L. Sp. Pl. 48. 1753.

Perennial, culms smooth, terete, densely tufted, light green, erect or ascending, almost filiform, wiry, 4′–15′ long. Basal sheaths numerous, membranous, imbricated, acuminate, the upper one bearing a short very narrow blade; spikelet solitary, terminal, few-flowered, ovoid-oblong, about 2″ long, subtended by an involucral leaf or outer scale of about its own length; scales yellowish-brown, ovate, obtuse or sub-acute, deciduous; bristles 6, smooth, longer than the achene; stamens 3; style 3-cleft; achene oblong, smooth, 3-angled, brown, acute.

In bogs and on moist rocks, Greenland to Alaska, south to the mountains of New England, the Adirondacks, western New York, Illinois, Minnesota and British Columbia, in the Rocky Mountains to Colorado, and on the higher summits of the southern Alleghanies. Also in Europe and Asia. June–Aug.

4. Scirpus Clíntoni A. Gray.   Clinton's Club-rush.   Fig. 804.

S. Clintoni A. Gray, Am. Journ. Sci. (II.) **38** : 290. 1864.

Perennial, culms tufted, triangular, very slender, erect, 4′–15′ tall, roughish on the angles. Lower sheaths imbricated, one or more of them bearing short subulate blades, the upper one bearing a flat, narrowly linear blade shorter than the culm; spikelet solitary, terminal, ovoid, few-flowered, 1½″–2″ long, subtended by a subulate involucral bract of less than its own length or somewhat longer; scales ovate, pale brown, acute or the outer one awned; bristles 3–6, filiform, upwardly barbed, as long as the achene or longer; style 3-cleft; achene oblong, brown, sharply 3-angled, smooth, obtuse.

In dry fields and thickets, New Brunswick to western New York and Michigan, and in North Carolina. Local. June–Aug.

5. Scirpus planifòlius Muhl.   Wood Club-rush.   Fig. 805.

Scirpus planifolius Muhl. Gram. 32. 1817.

Perennial, culms triangular, slender, tufted, rather weak, roughish on the angles, 6′–15′ tall. Lower sheaths bearing short subulate blades, the upper with a flat narrowly linear rough-margined leaf about as long as the culm; spikelet solitary, terminal, ovoid-oblong, acute, several-flowered, subtended by a short involucral bract; scales ovate-lanceolate, yellowish-brown with a green midvein, which is extended beyond the acute apex into a sharp cusp; bristles 4–6, upwardly barbed, about equalling the achene; stamens 3; style 3-cleft, pubescent; achene oblong, 3-angled; smooth, light brown, rather obtuse.

In woods and thickets, Vermont and Massachusetts to Delaware, the District of Columbia, western New York and Missouri. May–July.

6. Scirpus subterminàlis Torr. Water Club-rush. Fig. 806.

Scirpus subterminalis Torr. Fl. U. S. 1 : 47. 1824.

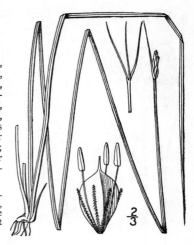

Perennial, aquatic, culms slender, terete, nodulose, 1°–3½° long. Leaves very slender, channeled, 6′–2° long, ¼″–¾″ wide; spikelet solitary, terminal, oblong-cylindric, narrowed at each end, several-flowered, 3″–7″ long, subtended by a subulate erect involucral leaf, ½′–2½′ long, thus appearing lateral; scales ovate-lanceolate, acute, membranous, light brown with a green midvein; bristles about 6, downwardly barbed, as long as the achene or shorter; stamens 3; style 3-cleft to about the middle; achene obovoid, 3-angled, dark brown, smooth, rather more than 1″ long, obtuse, abruptly beaked by the slender base of the style.

In ponds and streams or sometimes on their borders, Newfoundland to the Northwest Territory and British Columbia, South Carolina, Pennsylvania, Michigan and Idaho. The so-called variety *terrestris* is an emersed form with erect culms and shorter spikelets. July–Aug.

7. Scirpus Hàllii A. Gray. Hall's Club-rush. Fig. 807.

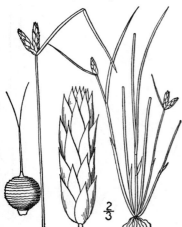

Scirpus Hallii A. Gray, Man. Ed. 2, Add. 1863.

S. supinus var. *Hallii* A. Gray, Man. Ed. 5, 563. 1867.

Annual, culms very slender, smooth, tufted, obtusely triangular, erect, striate, 5′–12′ tall. Lower sheaths oblique, and acuminate or mucronate on one side, the upper one commonly bearing a filiform blade, ¾′–2½′ long; spikelets capitate in clusters of 1–7, oblong-cylindric, obtuse, many-flowered, 3″–6″ long, about 1″ thick, appearing lateral by the extension of the solitary involucral leaf which is 1′–4′ long; scales ovate-lanceolate, light greenish brown, acuminate, keeled, cuspidate by the excurrent tip of the midvein; bristles wanting; stamens mostly 2; achene obovate-orbicular or slightly broader than high, black, planoconvex, mucronulate, strongly wrinkled transversely, about ¼″ in diameter.

In wet soil, Massachusetts to Florida, west to Illinois, Colorado, Texas and Mexico. Also in eastern Asia. The lowest sheaths occasionally subtend a flower with very long styles. July–Sept.

8. Scirpus débilis Pursh. Weak-stalked Club-rush. Fig. 808.

Scirpus debilis Pursh, Fl. Am. Sept. 55. 1814.
Scirpus Smithii A. Gray, Man. Ed. 5, 503. 1867.

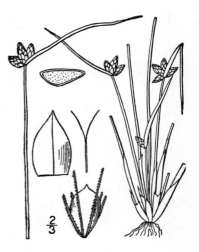

Annual, smooth, culms slender, obtusely triangular or nearly terete, tufted, erect or ascending, 6′–2° high. Sheaths obliquely truncate, the upper one rarely bearing a short subulate blade; spikelets capitate in clusters of 1–12, ovoid-oblong, subacute, many-flowered, appearing lateral, the solitary involucral leaf narrowly linear, 1½′–4′ long, erect or divergent; scales light yellowish-brown with a green midvein, broadly ovate, obtuse or acute; bristles 4–6, downwardly barbed, somewhat unequal and about as long as the achene or short or wanting; stamens 2–3; style 2-cleft or rarely 3-cleft; achene plano-convex, broadly obovate or orbicular, smooth or slightly roughened, dark brown, shining, obtuse, mucronulate.

In wet soil, Maine to Ontario, Minnesota, Georgia, Alabama and Nebraska. July–Sept.

9. Scirpus americànus Pers. Three-square. Chair-maker's Rush. Sword-grass. Fig. 809.

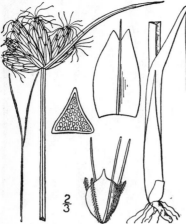

Scirpus americanus Pers. Syn. 1: 68. 1805.
Scirpus pungens Vahl, Enum. 2: 255. 1806.
Scirpus americanus longispicatus Britton, Trans. N. Y. Acad. Sci. 11: 78. 1892.
S. Olneyi contortus Eames, Rhodora 2: 220. 1907.

Perennial by long rootstocks, culms sharply triangular with concave sides or one of the sides nearly flat, erect, stiff, 1°–4° tall. Leaves 1–3, narrowly linear, keeled, shorter than the culm; spikelets oblong-ovoid, acute, 4''–12'' long, capitate in clusters of 1–7, appearing as if lateral; involucral leaf solitary, slender, 1½'–4' long; scales broadly ovate, brown, often emarginate or sharply 2-cleft at the apex, the midvein extended into a subulate awn sometimes 1'' long, the margins scarious, ciliolate or glabrous; bristles 2–6, downwardly barbed, shorter than or equalling the achene; stamens 3; style usually 2-cleft; achene obovate, plano-convex, smooth, dark brown, mucronate.

In fresh water and brackish swamps, temperate North America, north to Newfoundland. Also in South America and Europe. June–Sept.

10. Scirpus Ólneyi A. Gray. Olney's Bulrush. Fig. 810.

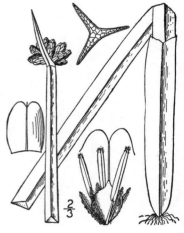

Scirpus Olneyi A. Gray, Bost. Journ. Nat. Hist. 5: 238. 1845.

Similar to the preceding species, perennial by long rootstocks, culms stout, sharply 3-angled with concave sides, 2°–7° tall. Leaves 1–3, 1'–5' long, or sheaths sometimes leafless; spikelets capitate in dense clusters of 5–12, oblong or ovoid-oblong, obtuse, 2½''–4'' long, the solitary involucral leaf short, stout, erect, ½'–1¾' long; scales oval or orbicular, dark brown with a green midvein, emarginate or mucronulate, glabrous; bristles usually 6, slightly shorter than or equalling the achene, downwardly barbed; stamens 2–3; style 2-cleft; achene obovate, plano-convex, brown, mucronate.

In salt marshes, New Hampshire to Florida, Texas, Mexico and California, extending north along the Pacific Coast to Oregon. Also in Michigan and Arkansas and in the West Indies. June–Sept.

11. Scirpus Tórreyi Olney. Torrey's Bulrush. Fig. 811.

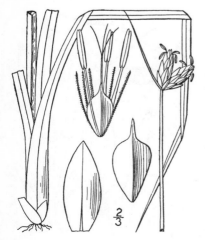

Scirpus Torreyi Olney, Proc. Providence Frank. Soc. 1: 32. 1847.

Perennial by short or slender rootstocks, culms sharply 3-angled, rather slender, nodulose, 2°–4° tall. Leaves narrowly linear, elongated, nodulose, light green, sometimes overtopping the culm; spikelets 1–4, in an apparently lateral capitate cluster, oblong, narrowed at each end, 5''–8'' long; involucral leaf 2'–6' long, erect; scales ovate or lanceolate, shining, chestnut-brown, glabrous, obtuse or the upper acute, mucronulate; bristles about 6, downwardly barbed, longer than the achene; stamens 3; style 3-cleft; achene obovoid, smooth, shining, light brown, 3-angled, one of its sides broader and flatter than the others.

In swamps, Maine to Rhode Island and Pennsylvania, west to Minnesota and Manitoba. July–Sept.

12. Scirpus mucronàtus L.   Bog Bulrush.
Fig. 812.

Scirpus mucronatus L. Sp. Pl. 50. 1753.

Perennial, culms stout, somewhat tufted, sharply 3-angled, smooth, 1°–3° tall. Spikelets 5–12 in a capitate cluster, oblong, obtuse, many-flowered, 4″–9″ long, rather more than 1″ in diameter, subtended by the solitary linear abruptly spreading involucral leaf; scales broadly ovate, obtuse, light brown with a narrow green midvein, mucronate; bristles 6, stout, rigid, downwardly barbed, as long as the achene; stamens 3; style 3-cleft; achene obovoid, smooth, shining, dark brown, 3-angled, two of the sides narrower and more convex than the third.

In a swamp in Delaware county, Pennsylvania. Probably adventive or fugitive from Europe. Widely distributed in the Old World. July–Sept.

13. Scirpus etuberculàtus (Steud.) Kuntze.   Canby's Bulrush.   Fig. 813.

Scirpus maritimus var. *cylindricus* Torr. Ann. Lyc. N. Y. 3: 325. 1836.
Rhynchospora etuberculata Steud. Syn. Pl. Cyp. 143. 1855.
S. Canbyi A. Gray, Am. Journ. Sci. (II.) 38: 289. 1864.
S. etuberculatus Kuntze, Rev. Gen. Pl. 758. 1891.
S. cylindricus Britton, Trans. N. Y. Acad. Sci. 11: 79. 1892.

Perennial by stout rootstocks, culm stout, sharply 3-angled above, 3°–6° high, the linear nodulose keeled and channeled dark green leaves nearly or quite as long. Involucral leaf solitary, 4′–10′ long, erect; spikelets in an apparently lateral simple or compound umbel, drooping, oblong-cylindric, acutish, 6″–10″ long; primary rays of the umbel 1′–4′ long, bracted by 1 or more subulate-linear leaves; scales ovate or ovate-lanceolate, pale brown with scarious margins, acute, mucronulate; bristles 6, stout, rigid, about as long as the achene, serrate; stamens 3; style 3-cleft; achene obovoid, 3-angled, light brown, smooth, abruptly subulate-pointed.

In ponds and swamps, Maryland to Florida and Louisiana, mostly near the coast. Pole or Pool-rush. July–Sept.

14. Scirpus válidus Vahl.   American Great
Bulrush.   Mat-rush.   Fig. 814.

Scirpus validus Vahl, Enum. 2: 268. 1806.

Perennial by stout rootstocks, culm stout, terete, smooth, erect, 3°–9° tall, sometimes nearly 1′ in diameter, sheathed below, the upper sheath occasionally extended into a short leaf. Involucral leaf solitary, erect, shorter than the umbel, appearing as if continuing the culm; umbel compound, appearing lateral, its primary rays slender, spreading, ½′–2½′ long, bracts linear-lanceolate, pubescent; spikelets oblong-conic, sessile or some of them peduncled, in capitate clusters of 1–5, obtuse or acute, 2½″–6″ long, 1½″–2″ in diameter; scales ovate to suborbicular, slightly pubescent, with a rather strong midvein which is sometimes excurrent into a short tip; bristles 4–6, downwardly barbed, equalling or longer than the achene; stamens 3; style 2-cleft; achene plano-convex, obovate, nearly as long as the scale, gray to brown, abruptly mucronate, a little more than 1½′ wide.

In ponds and swamps, throughout North America, except the extreme north, and in the West Indies. The Old World *S. lacustris* L., with which our plant has been confused, has a 3-cleft style. Black-rush. Bolder or Boulder Bast. Tule. June–Sept.

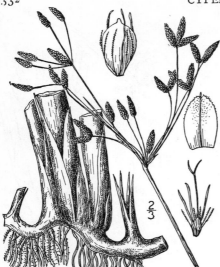

15. Scirpus occidentàlis (S. Wats.) Chase. Viscid Great Bulrush. Fig. 815.

S. lacustris occidentalis S. Wats. Bot. Cal. **2** : 218. 1880.
S. occidentalis Chase, Rhodora **6** : 68. 1904.

Similar to *S. validus,* tall, the culms firmer in texture, the margins of the basal sheaths becoming fibrillose. Involucral leaf shorter than the compound umbel; primary rays rather stiff; bracts viscid at the tip; spikelets clustered in 2's to 7's, or solitary, oblong-cylindric, 10″ long or less, about 2″ thick, acute or bluntish; scales ovate, short-awned, viscid above; style 2-cleft; achene biconvex, obovate, dull, nearly 1″ wide, much shorter than the scale.

Borders of lakes and streams. Newfoundland to British Columbia, New York, Missouri, Utah and California. July–Sept.

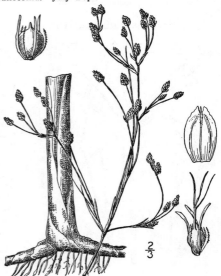

16. Scirpus heterochaètus Chase. Pale Great Bulrush. Fig. 816.

S. heterochaetus Chase, Rhodora **6** : 70. 1904.

Perennial by rather stout rootstocks; culms slender, sheathed below, 6° high or less. Involucral leaf much shorter than the compound umbel; primary rays slender, 4′ long or less; bracts acuminate, glabrous; spikelets solitary, ovoid to ellipsoid, acutish, 4″–7″ long, about 2½″ thick; scales ovate, glabrous, often erose-margined; style 3-cleft; bristles 2–4, unequal, as long as or shorter than the achene; achene about 1″ wide, obovate, yellowish, shorter than the scale.

Borders of lakes and in marshes, Vermont and Massachusetts to Oregon and Nebraska. July–Sept.

Scirpus califórnicus (C. A. Meyer) Britton, of the western and southern states and tropical continental America, admitted in the first edition of this work, has not been definitely established as growing within our limits. It has plumose bristles.

17. Scirpus rùfus (Huds.) Schrad. Red Clubrush. Fig. 817.

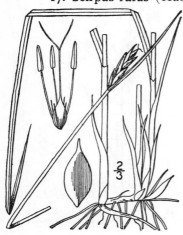

Schoenus rufus Huds. Fl. Angl. Ed. 2, 15. 1778.
Scirpus rufus Schrad. Fl. Germ. **1** : 133. 1806.
Blysmus rufus Link, Hort. Berol. **1** : 278. 1827.

Perennial by slender rootstocks, culms tufted, smooth, slender, erect, somewhat compressed, 3′–15′ tall. Leaves half-terete, smooth, shorter than the culm, channeled, ½′–3′ long, less than 1′ wide, the lowest reduced to bladeless sheaths; spikelets red-brown, few-flowered, narrowly ovoid-oblong, subacute, about 3″ long, erect in a terminal 2-ranked spike ½′–1′ long; involucral leaf solitary, erect, narrowly linear, equalling or longer than the spike; scales lanceolate, acute, 1-nerved; bristles 3–6, upwardly barbed, shorter than the achene, deciduous; stamens 3; style 2-cleft; achene oblong, pointed at both ends, light brown, plano-convex or slightly angled in front, 1½″–2″ long.

In marshes, Newfoundland, New Brunswick, Nova Scotia and Quebec to James' Bay; Northwest Territory. Also in northern Europe. Summer.

18. Scirpus paludòsus A. Nelson. Prairie Bulrush. Fig. 818.

Scirpus campestris Britton, in Britton and Brown, Ill. Fl. Ed. 1, 1: 267. 1896. Not Roth, 1795.
S. paludòsus A. Nelson, Bull. Torr. Club **26**: 5. 1899.
Scirpus interior Britton, Man. Ed. 2, 178. 1905.

Perennial by slender rootstocks, culm slender, smooth, sharply triangular, 1°–2° tall. Leaves usually pale green, smooth, shorter than or overtopping the culm, 1″–2″ wide, those of the involucre 2 or 3, the longer much exceeding the inflorescence; spikelets 3–10 in a dense terminal simple head, oblong-cylindric, mostly acute, 8″–12″ long, 2½″–4″ in diameter; scales ovate, membranous, puberulent or glabrous, pale to brown, 2-toothed at the apex, the midvein excurrent into an ascending or spreading awn about 1″ long; bristles 1–3, much shorter than the achene or none; style 2-cleft; achene lenticular, obovate or oblong-ovate, mucronulate, yellow-brown.

Salt marshes, Quebec to New Jersey, about salt springs inland and on wet prairies and plains, Manitoba and Minnesota to Oregon, Nebraska, Kansas, Nevada and Mexico. May–Aug.

19. Scirpus robústus Pursh. Salt Marsh Bulrush. Fig. 819.

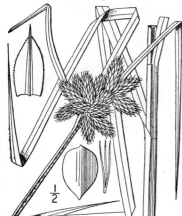

Scirpus robustus Pursh, Fl. Am. Sept. 56. 1814.
Scirpus maritimus var. *macrostachyus* Michx. Fl. Bor. Am. 1: 32. 1803. Not *S. macrostachyus* Lam.

Perennial by large rootstocks, culm stout, stiff, sharply 3-angled with flat sides, smooth, 2°–5° tall. Leaves equalling or overtopping the culm, dark green, smooth, 2½″–5″ wide, the midvein prominent; involucral leaves 2–4, elongated, erect, similar to those of the culm, often 1′ long; spikelets ovoid-oblong, obtuse or subacute, stout, 8″–12″ long, 4″–5″ in diameter, 6–20 together in a dense often compound terminal cluster; scales ovate, brown, puberulent, thin, lacerate or 2-toothed at the apex, the midvein excurrent into an, at length, reflexed awn, 1½″–2½″ long; bristles 1–6, fragile, shorter than the achene or none; stamens 3; style 3-cleft; achene compressed, very flat on the face, convex or with low ridge on the back, obovate-orbicular, dark brown, shining, 1½″ long.

In salt marshes, Nova Scotia to Texas. Spurt-grass. Sea Club-rush. July–Oct.

20. Scirpus fluviátilis (Torr.) A. Gray. River Bulrush. Fig. 820.

Scirpus maritimus var. *fluviatilis* Torr. Ann. Lyc. N. Y. **3**: 324. 1836.
Scirpus fluviatilis A. Gray, Man. 527. 1848.

Perennial by large rootstocks, culm stout, smooth, sharply triangular with nearly flat sides, 3°–6° tall. Leaves 4″–8″ wide, smooth, equalling or overtopping the culm, attenuate to a very long tip, the midvein prominent; those of the involucre 3–5, erect or spreading, some of them 5′–10′ long; spikelets in a terminal umbel, solitary, or 2–3 together at the ends of its long spreading or drooping rays, or the central spikelets sessile, oblong-cylindric, acute, 8″–12″ long, about 3½″ in diameter; scales ovate, scarious, puberulent, the midvein excurrent into a curved awn 1½″–2″ long; bristles 6; rigid, downwardly barbed, about as long as the achene; style 3-cleft; achene sharply 3-angled, obovoid, rather dull, short-pointed, 2″ long.

In shallow water along lakes and streams, Quebec to Minnesota, New Jersey, Nebraska and Kansas. River Club-rush. June–Sept.

21. Scirpus Férnaldi Bickwell. Fernald's Bulrush. Fig. 821.

S. Fernaldi Bicknell, Torreya 1: 96. 1901.

Perennial; culms rather pale green, slender, sharply 3-angled, 2½° tall or less. Leaves 1″–3″ wide, the upper equalling or surpassing the inflorescence, those of the involucre 3 or 4, the longest one 5′ long or less; spikelets ovoid, 5″–8″ long, sessile in a terminal cluster and solitary at the ends of the slender umbel-rays; scales finely puberulent, acuminate, entire or lacerate, the recurved awn 1½″–6″ long; bristles as long as the achene or shorter; style 3-cleft; achene obovoid-cuneate, about 1½″ long and thick, trigonous, with rounded angles, yellow-brown and shining.

Shore of Somes Sound, Mt. Desert, Maine. July–Aug.

22. Scirpus nòvae-ángliae Britton. New England Bulrush. Fig. 822.

S. novae-angliae Britton, in Britton and Brown, Ill. Fl. Ed. 1, 3: 509. 1898.

Perennial by rootstocks; culm stout, erect, 4°–7° tall, sharply 3-angled, the sides flat or nearly so. Leaves long, 4″–6″ wide, somewhat roughish on the margins when dry, the lowest reduced to pointed sheaths, those of the involucre 2–5, the longer of them much exceeding the inflorescence; spikelets narrowly cylindric, acute, ¾′–2′ long, less than ¼′ thick, solitary or 2–5 together at the ends of the rays of the umbel, the rays 1′–4′ long; scales awned; bristles 2–4, shorter than the grayish-white dull obovate achene, which is distinctly 3-angled; stamens 3; style 3-cleft.

In fresh water and brackish marshes, Massachusetts to New York.

23. Scirpus sylváticus L. Wood Bulrush or Clubrush. Fig. 823.

Scirpus sylvaticus L. Sp. Pl. 51. 1753.

Perennial by long rootstocks; culm triangular, stout, smooth, 4°–6° tall, often overtopped by the upper leaves. Leaves flat, 5″–8″ wide, rough on the margins, more or less rugulose, the midvein prominent, those of the involucre 5–8, the larger similar to those of the culm, often 1° long or more; umbel terminal, very large, sometimes 8′ broad, about 3 times compound, the spikelets ovoid or ovoid-oblong, mostly acute, 1½″–2½″ long, borne in capitate clusters of 2–8 at the ends of the raylets; bractlets of the involucels small, scarious, linear or lanceolate; scales ovate-oblong, obtuse, brown with a green centre; bristles 6, downwardly barbed, slightly exceeding the achene; stamens 3; style 3-cleft; achene oblong, 3-angled, obtuse, nearly white, mucronulate, not shining.

In swamps, Maine to Georgia and Michigan. Also in Europe and Asia. June–Aug.

24. Scirpus atróvirens Muhl. Dark-green Bulrush. Fig. 824.

Scirpus atrovirens Muhl. Gram. 43. 1817.

S. georgianus Harper, Bull. Torr. Club **27** : 331. 1900.

Perennial by slender rootstocks; culms triangular, rather slender, leafy, 2°-4½° high. Leaves elongated, more or less nodulose, rough on the margins, dark green, 3"-6" wide, one or two of them usually exceeding the inflorescence; umbel 1-2-compound or simple; spikelets ovoid-oblong, acute, 2"-5" long, densely capitate in 6's-20's at the ends of the rays or raylets; involucels short; scales greenish-brown, ovate-oblong, acute, the midvein excurrent; bristles usually 6, downwardly barbed above, naked below, about as long as the achene, or shorter or wanting; stamens 3; style 3-cleft; achene oblong-obovoid, 3-angled, pale brown, dull.

In swamps, Nova Scotia to Saskatchewan, south to Georgia and Louisiana. June-Aug.

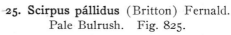

25. Scirpus pállidus (Britton) Fernald. Pale Bulrush. Fig. 825.

S. atrovirens pallidus Britton, Trans. N. Y. Acad. Sci. **9** : 14. 1889.

S. pallidus Fernald, Rhodora **8** : 162. 1906.

Perennial, the rootstocks short, stout; culms stout, triangular, 3°-4° high. Leaves elongated, pale, 3"-7" wide, somewhat nodulose; umbel mostly compound; spikelets oblong to oblong-cylindric, numerous in very dense capitate clusters; scales pale, ovate, acute, tipped with an awn half as long as the body; bristles 6, downwardly barbed, about as long as the oblong, trigonous achene.

Wet grounds, especially along streams, Manitoba to Nebraska, Kansas, Texas, Wyoming and New Mexico. Reported from Minnesota. Summer.

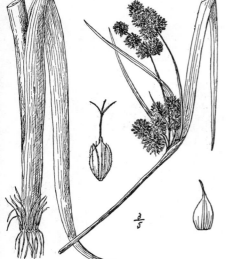

26. Scirpus microcàrpus Presl. Small-fruited Bulrush. Fig. 826.

Scirpus microcarpus Presl, Rel. Haenk. **1** : 195. 1828.
Scirpus sylvaticus var. *digynus* Boeckl. Linnaea **36** : 727. 1870.
Scirpus rubrotinctus Fernald, Rhodora **2** : 20. 1900.

Perennial, the culms 3°-5° tall, often stout, overtopped by the rough-margined leaves, the sheaths often tinged with red. Longer leaves of the involucre usually exceeding the inflorescence; spikelets ovoid-oblong, acute, 1½"-2" long, 3-25 together in capitate clusters at the ends of the usually spreading raylets; scales brown with a green midvein, blunt or subacute; bristles 4, barbed downwardly nearly or quite to the base, somewhat longer than the achene; stamens 2; style 2-cleft; achene oblong-obovate, nearly white, plano-convex or with a low ridge on the back, pointed.

In swamps and wet woods, Newfoundland to Alaska, south to Connecticut, northern New York, Minnesota, Nevada and California. July-Sept.

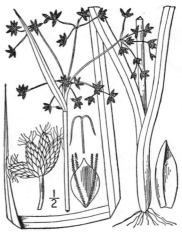

27. Scirpus polyphýllus Vahl. Leafy Bulrush. Fig. 827.

Scirpus polyphyllus Vahl, Enum. **2** : 274. 1806.

Perennial by slender rootstocks; culms slender, sharply triangular, 1½°–4° tall, very leafy, the leaves 2″–3″ wide, exactly 3-ranked, inconspicuously nodulose, rough-margined, the upper rarely overtopping the culm; leaves of the involucre 3–6, the longer commonly somewhat exceeding the inflorescence; umbel more or less compound; spikelets ovoid, 1½″–3″ long, rarely oblong and 4″ long, capitate in 3's–10's at the ends of the raylets; scales ovate, bright brown, mostly obtuse, mucronulate; bristles 6, mostly flexuous or twice bent, downwardly barbed above the middle, twice as long as the achene; stamens 3; style 3-cleft; achene obovoid, 3-angled with a broad face and narrower sides, short-pointed, dull.

In swamps, wet woods and meadows, Massachusetts to Minnesota, south to Georgia, Tennessee and Arkansas. Some of the scales of the spikelets occasionally develop into linear leaves. July–Sept.

28. Scirpus Péckii Britton.   Peck's Bulrush. Fig. 828.

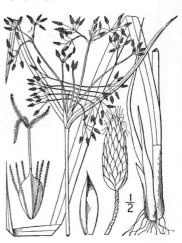

S. Peckii Britton, Trans. N. Y. Acad. Sci. **11** : 82. 1892.

Perennial by rootstocks; culms slender, triangular, 1½°–4½° tall, leafy. Leaves elongated, 2″–5″ wide, rough-margined, the upper overtopping the culm, those of the involucre 3–5, dark-colored at the base, the longer of them exceeding the inflorescence; umbel 1–2-compound, large; spikelets cylindric, obtusish, 3″–5″ long, in capitate clusters of 2–10 at the ends of the raylets or some of them distinctly peduncled; scales dark brown, keeled, mucronate, falling early; bristles 4–6, downwardly barbed from below the middle to the summit, longer than the achene; style 3-cleft; achene 3-angled, ¼″ long, oblong, narrowed at each end, slender-beaked.

In swamps, New Hampshire to Connecticut and New York. July–Sept.

29. Scirpus divaricàtus Ell.   Spreading Bulrush. Fig. 829.

S. divaricatus Ell. Bot. S. C. & Ga. **1** : 88. *pl. 2. f. 4.* 1816.

Perennial. Roots fibrous, culms obtusely triangular, smooth, rather slender, 2½°–4° tall. Leaves 2″–4″ wide, rough-margined, the upper and those of the involucre not exceeding the inflorescence; umbel decompound, the primary rays very slender, sometimes 6′ long, widely spreading or drooping; raylets filiform; involucels setaceous; spikelets mostly solitary at the ends of the raylets, sessile or peduncled, linear-oblong, obtuse, 3″–6″ long, ½″ thick; scales ovate, greenish-brown, subacute or obtuse, with a prominent midvein and scarious margins; bristles 6, flexuous, longer than the achene, somewhat pubescent, not barbed, shorter than the scales; stamens 3; style 3-cleft; achene sharply 3-angled, oblong, narrowed at both ends, apiculate, nearly white, not shining.

In swamps, Virginia to Kentucky, Missouri, Florida and Louisiana. The spikelets sometimes partially develop into tufts of leaves. June–Aug.

30. Scirpus lineàtus Michx.   Reddish Bulrush.   Fig. 830.

Scirpus lineatus Michx. Fl. Bor. Am. **1** : 32.   1803.

Perennial by stout rootstocks; culms rather slender, triangular, erect, $1°-4\frac{1}{2}°$ high, leafy, the upper leaves and those of the involucre not exceeding the inflorescence. Leaves $2''-4''$ wide, light green, flat, rough-margined; umbels terminal and commonly also axillary, decompound, the rays very slender, becoming pendulous; spikelets mostly solitary at the ends of the slender raylets, oblong, obtuse, $3''-5''$ long, about $1''$ in diameter; scales ovate or oblong, reddish-brown with a green midvein, their tips slightly spreading; bristles 6, weak, smooth, entangled, much longer than the achene, equalling the scales or slightly protruded beyond them at maturity; stamens 3; style 3-cleft; achene oblong or oblong-obovoid, pale brown, narrowed at both ends, 3-angled, short-beaked.

In swamps and wet meadows, Ontario to New Hampshire, Georgia, Oregon, Kansas and Texas.   June–Sept.

31. Scirpus cypérinus (L.) Kunth.   Wool-grass.   Fig. 831.

Eriophorum cyperinum L. Sp. Pl. Ed. 2, 77.   1762.
Scirpus cyperinus Kunth, Enum. **2** : 170.   1837.
S. Eriophorum Michx. Fl. Bor. Am. **1** : 33.   1803.
S. pedicellatus Fernald, Rhodora **2** : 16.   1900.

Perennial by stout rootstocks; culms stout or slender, smooth, obtusely triangular or nearly terete, stiff, leafy, $2°-6°$ tall. Leaves elongated, $2''-3''$ wide, rough-margined, the upper often overtopping the culm, those of the involucre 3–6, their bases often brown or black, the longer much exceeding the terminal, compound umbel; spikelets ovoid-oblong, obtuse, $1\frac{1}{2}''-5''$ long, in capitate clusters of 3–15 at the ends of the raylets, or some or all of them stalked; scales ovate or lanceolate, acute or subacute; bristles 6, entangled, smooth, much longer than the achene, much exserted beyond the scales and brown or reddish at maturity; stamens 3; style 3-cleft; achene 3-angled, oblong, slender-beaked, nearly white.

In swamps, Newfoundland to Ontario, Saskatchewan, Florida and Louisiana.   Clump-head grass.   Aug.–Sept.   Consists of many races with spikelets stalked or sessile.

Scirpus atrocínctus Fernald, characterized by black bases of the involucral leaves, is of northern range and may be specifically distinct.

Scirpus Lòngii Fernald, recently published as a distinct species of the New Jersey pine-barrens, appears to be the same as *S. atrocinctus*.

8. FUIRÈNA Rottb. Descr. & Ic. 70. *pl. 19. f. 3.*   1773.

Perennial sedges, with leafy triangular culms (in a southern species the leaves reduced to inflated sheaths) and many-flowered terete spikelets in terminal and axillary clusters, or rarely solitary. Scales spirally imbricated all around, awned, the 1 or 2 lower, commonly empty. Flowers perfect. Perianth of 3 ovate-oblong or cordate-ovate, stalked, often awned scales, usually alternating with as many downwardly barbed bristles. Stamens 3. Style 3-cleft, not swollen at the base, deciduous. Achene stalked or nearly sessile, sharply 3-angled, acute or mucronate, smooth. [In honor of George Fuiren, 1581–1628, Danish physician.]

About 30 species, natives of warm-temperate and tropical regions. Besides the following, 4 others occur in the southern United States.   Type species: *Fuirena umbellata* Rottb.

Perianth-scales awned from the apex or awnless.
    Annual; perianth-scales long-awned.   1. *F. squarrosa.*
      Perennial; perianth-scales short-awned or awnless.   2. *F. hispida.*
Perianth-scales awned on the back below the apex.   3. *F. simplex.*

3/4

2. Fuirena híspida Ell. Hairy Fuirena. Fig. 833.

Fuirena hispida Ell. Bot. S. C. & Ga. **1**: 579. 1821.
F. squarrosa hispida Chapm. Fl. S. States 514. 1860.

Perennial by short rootstocks which often bear tubers; culms glabrous or pubescent, 8′–2½° high. Leaves flat, both the blades and the sheaths more or less densely hirsute; spikelets 2–8 together in capitate terminal and usually also axillary clusters, similar to those of the preceding species, the scales with spreading or recurved awns; perianth-scales deltoid-ovate, cordate to rounded at the base, stalked, tipped with a short smooth awn or merely mucronate; bristles mostly downwardly barbed, shorter or longer than the achene.

Wet grounds, New York(?), New Jersey to Florida, Kentucky, Indian Territory and Texas. June–Oct.

3/4

1. Fuirena squarròsa Michx. Umbrella-grass. Fig. 832.

F. squarrosa Michx. Fl. Bor. Am. **1**: 37. 1803.

F. squarrosa var. *pumila* Torr. Fl. U. S. **1**: 68. 1824.

Annual, with fibrous roots, the rootstocks very short or none; culms tufted, glabrous or nearly so, 2′–1° tall. Leaves flat, nearly or quite glabrous or the lower sheaths pubescent; spikelets sessile and 1–10 together in terminal and often also lateral capitate clusters, ovoid or ovoid-oblong, acute or obtuse, 3″–6″ long, about 2½″ in diameter; scales ovate or oblong, brown, pubescent, mostly obtuse, 3-nerved, tipped with a stout spreading or recurved awn of nearly their own length; perianth-scales oblong to ovate, long-stalked, usually narrowed at both ends, tapering into a slender terminal downwardly barbed awn; bristles mostly longer than the achene.

In wet meadows and marshes, Massachusetts to Florida and Louisiana. Also in Michigan and Indiana. July–Sept.

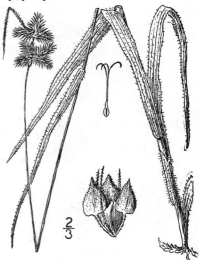

2/3

3. Fuirena símplex Vahl. Western Umbrella-grass. Fig. 834.

Fuirena simplex Vahl, Enum. **2**: 384. 1806.

Fuirena squarrosa var. *aristulata* Torr. Ann. Lyc. N. Y. **3**: 291. 1836.

Similar to the preceding species, rootstock short, thick; culms slender, 5′–2° tall, glabrous. Leaves flat, glabrous or ciliate; scales tipped with a spreading or reflexed awn; sepals ovate-oblong, obtuse and usually notched at the apex, obtuse, truncate or subcordate at the base, longer or shorter than their stalks, awned on the back from below the apex, the awn varying in length, smooth or downwardly barbed; bristles retrorsely hispid, equalling or exceeding the sessile or short-stalked achene.

In moist soil, Nebraska, Missouri, Kansas, Texas and Mexico. Also in Cuba. June–Sept.

9. LIPOCÀRPHA R. Br. App. Tuckey Exp. Congo, 459. 1818.

Low annual sedges, with slender tufted culms leafy at the base, and terete many-flowered spikelets in a terminal head, subtended by a 1–several-leaved involucre. Scales firm, spirally imbricated all around, all fertile or several of the lower ones empty, at length deciduous. Flowers perfect, with a small hyaline scale on each side; bristles none. Stamens 1–2; anthers 4-celled. Style 2–3-cleft, deciduous, its base not swollen. Achene plano-convex or 3-angled. [Greek, alluding to the thick sepals in some species.]

About 15 species, widely distributed in warm and tropical regions. Type species: *Hypaelytrum argenteum* Vahl.

1. Lipocarpha maculàta (Michx.) Torr. American Lipocarpha. Fig. 835.

Kyllingia maculata Michx. Fl. Bor. Am. 1: 29. 1803.
L. maculata Torr. Ann. Lyc. N. Y. 3: 288. 1836.

Annual, glabrous, roots fibrous, culms tufted, grooved, compressed, smooth, longer than the narrowly linear somewhat channeled leaves, 3'–10' tall. Leaves of the involucre 2–4, the larger 1'–5' long; spikelets ovoid-oblong, obtuse, 2½"–3" long, 1" in diameter, 2–6 together in a terminal capitate cluster; scales rhombic or lanceolate, acute at the apex, curved, the sides nearly white, or flecked with reddish-brown spots, the midvein green; exterior sepal convolute around the achene, nerved, hyaline; stamen 1; achene oblong, yellowish, contracted at the base.

In wet or moist soil, Virginia to Florida. Near Philadelphia probably adventive. Cuba, Panama. July–Sept.

10. HEMICÀRPHA Nees & Arn. Edinb. New Phil. Journ. 17: 263. 1834.

Low tufted mostly annual sedges, with erect or spreading, almost filiform culms and leaves, and terete small terminal capitate or solitary spikelets subtended by a 1–3-leaved involucre. Scales spirally imbricated all around, deciduous, all subtending perfect flowers, a single hyaline inner scale between the flower and the rachis of the spikelet; bristles none. Stamen 1. Style 2-cleft, deciduous, not swollen at the base. Achene oblong, turgid or lenticular. [Greek, in allusion to the single inner scale.]

About 5 species, natives of temperate and tropical regions. Besides the following, another occurs in the western United States. Type species: *Hemicarpha Isólepis* Nees.

Scales with a short tip or mucronate.   1. *H. micrantha.*
Scales abruptly narrowed into an awn about as long as the body.   2. *H. aristulata.*

1. Hemicarpha micrántha (Vahl) Pax. Common Hemicarpha. Fig. 836.

Scirpus micranthus Vahl, Enum. 2: 254. 1806.

Hemicarpha subsquarrosa Nees, in Mart. Fl. Bras. 2: Part 1, 61. 1842.

H. Drummondii Nees, in Mart. Fl. Bras. 2¹: 61. 1842.

H. micrantha Pax in E. & P. Nat. Pflf. 2²: 105. 1887.

Annual, glabrous, culms densely tufted, compressed, grooved, diffuse or ascending, 1'–5' long, mostly longer than the setaceous smooth leaves. Spikelets ovoid, many-flowered, obtuse, about 1" long, capitate in 2's–4's or solitary; involucral leaves, or one of them, usually much exceeding the spikelets; scales brown, obovate, with a short blunt tip; achene obovate to oblong, obtuse, mucronulate, little compressed, light brown, its surface minutely cellular-reticulated.

In moist, sandy soil, New Hampshire to Ontario, Washington, Florida, Texas, Mexico and South America. July–Sept.

¾

2. Hemicarpha aristulàta (Coville) Smyth.
Awned Hemicarpha. Fig. 837.

H. micrantha aristulata Coville, Bull. Torr. Club **21**: 36. 1894.
H. aristulata Smyth, Trans. Kans. Acad. Sci. **16**: 163. 1899.
H. intermedia Piper; Piper & Beattie, Fl. Pal. Reg. **36**: 1901.

Similar to the preceding species; culms 8' high or less, longer than the setaceous leaves; involucral leaves 1–3, sometimes nearly 1' long. Spikelets ovoid, 2″–4″ long; scales rhombic-obovate, brown, rather abruptly contracted into a subulate spreading or somewhat recurved awn about as long as the body; inner scale larger than that of *H. micrantha;* style short; achene narrowly obovate, black.

Wet, sandy soil, Kansas to Wyoming, Colorado and Texas; California and Washington.

H. occidentàlis A. Gray, a species of California and Oregon, with larger subglobose heads and lanceolate scales, is erroneously recorded from western Ontario.

11. DULÍCHIUM L. C. Richard; Pers. Syn. 1: 65. 1805.

A tall perennial sedge, with terete hollow jointed culms, leafy to the top, the lower leaves reduced to sheaths. Spikes axillary, peduncled, simple or compound. Spikelets 2-ranked, linear, many-flowered, breaking up into 1-fruited joints at maturity. Scales 2-ranked, carinate, conduplicate, decurrent on the joint below. Flowers perfect. Perianth of 6–9 retrorsely barbed bristles. Stamens 3. Style 2-cleft at the summit, persistent as a linear-oblong beak on the summit of the achene. [Name said to be from *Dulcichimum,* a Latin name for some sedge.] A monotypic genus.

1. Dulichium arundinàceum (L.) Britton.
Dulichium. Fig. 838.

Cyperus arundinaceus L. Sp. Pl. 44. 1753.
Cyperus spathaceus L. Syst. Ed. 12, **2**: 735. 1767.
Dulichium spathaceum Pers. Syn. 1: 65. 1805.
Dulichium arundinaceum Britton, Bull. Torr. Club **21**: 29. 1894.

Culm stout, 1°–3° tall, erect. Leaves numerous, 3-ranked, flat, 1′–3′ long, 2″–4″ wide, spreading or ascending, the lower sheaths bladeless, brown toward their summits. Spikes shorter than or the uppermost exceeding the leaves; peduncles 2″–12″ long; spikelets narrowly linear, spreading, 6″–12″ long, about 1″ wide, 6–12-flowered; scales lanceolate, acuminate, strongly several-nerved, appressed, brownish; bristles of the perianth rigid, longer than the achene; style long-exserted, persistent.

In wet places, Newfoundland to Ontario, Minnesota, Washington, Florida and Texas. Also in Costa Rica. Aug.–Oct.

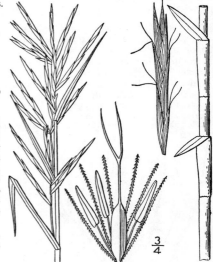

¾

12. DICHRÓMENA Michx. Fl. Bor. Am. 1: 37. 1803.

Leafy-stemmed sedges, perennial by rootstocks, the spikelets crowded in a terminal head involucrate by the upper leaves, which are often white at the base. Spikelets compressed, several–many-flowered. Scales spirally imbricated all around, several of them with imperfect flowers, or empty. Perianth none. Stamens 3. Style 2-cleft, its branches subulate. Achene lenticular, transversely rugose, crowned with the broad persistent base of the style (tubercle). [Greek, alluding to the two-colored involucral leaves.]

About 20 species, natives of America. Besides the following, 2 others occur in the southwestern United States. Type species: *Dichromena leucocephala* Michx.

Leaves of the involucre linear; tubercle truncate at the base.   1. *D. colorata.*
Leaves of the involucre lanceolate, long-acuminate; tubercle decurrent on the edges of the achene.   2. *D. latifolia.*

1. Dichromena coloràta (L.) Hitchcock. Narrow-leaved Dichromena. Fig. 839.

Schoenus coloratus L. Sp. Pl. 43. 1753.
D. leucocephala Michx. Fl. Bor. Am. 1 : 37. 1803.
Dichromena colorata A. S. Hitchc. Ann. Rep. Mo. Bot. Gard. 4 : 141. 1893.

Glabrous, culm slender, erect, rather sharply triangular, 1°–2° tall. Leaves distant, narrowly linear, about 1″ wide, much shorter than the culm, those of the involucre 4–6, reflexed when mature, yellowish-white at the base; head globose, 6″–10″ in diameter; spikelets narrowly oblong, acute; scales membranous, lanceolate, nearly white, 1-nerved, subacute at the apex; achene obovate, brown, papillose or wrinkled transversely, nearly truncate at the summit, compressed, covered by the truncate-based tubercle.

In moist sandy soil, New Jersey to Florida and Texas. Bermuda; tropical America. June–Sept.

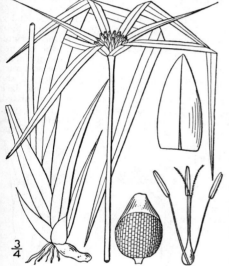

2. Dichromena latifòlia Baldw. Broad-leaved Dichromena. Fig. 840.

Dichromena latifolia Baldw.; Ell. Bot. S. C. & Ga. 1 : 90. 1816.

Culm stout, obtusely triangular or nearly terete, the leaves lanceolate or linear-lanceolate, tapering gradually to a long-acuminate apex from a broad base, 1½″–4″ wide, sometimes overtopping the culm, but the lowest much shorter, those of the involucre 7–10, strongly reflexed when old. Head globose, 6″–9″ in diameter; spikelets oblong, subacute; scales ovate-lanceolate, nearly white, rather obtuse; achene nearly orbicular in outline, pale brown, faintly wrinkled transversely and longitudinally, so as to appear reticulated; tubercle decurrent on the margins of the achene.

In wet pine barrens, Virginia to Florida and Texas. June–Aug.

13. RYNCHÓSPORA Vahl, Enum. 2: 229. 1806.

Leafy sedges, mostly perennial by rootstocks, with erect 3-angled or terete culms, narrow flat or involute leaves, and ovoid oblong or fusiform, variously clustered spikelets. Scales thin, 1-nerved, imbricated all around, usually mucronate by the excurrent midvein, the lower empty. Upper flowers imperfect, the lower perfect. Perianth of 1–20 (mostly 6) upwardly or downwardly barbed or scabrous bristles, wanting in some species (no. 2). Stamens commonly 3. Style 2-cleft, 2-toothed or rarely entire. Achene lenticular or swollen, not 3-angled, smooth or transversely wrinkled, capped by the persistent base of the style (tubercle), or in some species by the whole style. [Greek, referring to the beak-like tubercle.]

About 200 species, widely distributed, most abundant in warm regions. Besides the following, some 35 occur in the southern United States. Type species: *Rynchospora aurea* Vahl. BEAK-SEDGE.

Style entire or 2-toothed, persistent as a long-exserted subulate beak. 1. *R. corniculata.*
Style deeply 2-cleft, only its base persistent as a tubercle.
  Bristles minute or wanting. 2. *R. pallida.*
  Bristles plumose. 3. *R. oligantha.*
  Bristles downwardly barbed or rarely smooth.
    Scales white or nearly so; bristles 9–15. 4. *R. alba.*
    Scales brown; bristles 6.
      Leaves filiform; achene oblong. 5. *R. capillacea.*
      Leaves narrowly linear, flat; achene obovate.
        Bristles equalling the achene; tubercle one-half as long or less. 6. *R. Knieskernii.*
        Bristles reaching or exceeding the end of the tubercle, which is as long as the achene.
          Spikelets few–several in numerous rather loose clusters. 7. *R. glomerata.*
          Spikelets very numerous in 2–6 very dense globose heads. 8. *R. axillaris.*
  Bristles upwardly barbed.
    Spikelets numerous in 2–6 very dense globose heads. 8. *R. axillaris.*

Spikelets few–several in rather loose clusters.
  Achene smooth.
    Leaves setaceous ; achene obovate, shining.   9. *R. fusca.*
    Leaves narrowly linear.
      Achene broadly oval.   10. *R. gracilenta.*
      Achene narrowly obovate.   11. *R. Smallii.*
  Achene transversely wrinkled.
    Spikelets ovoid, in erect cymose clusters ; achene longer than the bristles.
      Leaves flat ; spikelets nearly or quite sessile.   12. *R. cymosa.*
      Leaves involute ; spikelets distinctly pedicelled.   13. *R. Torreyana.*
    Spikelets spindle-shaped, in drooping panicles ; bristles long.   14. *R. inexpansa.*

1. Rynchospora corniculàta (Lam.) A. Gray. Horned Rush. Fig. 841.

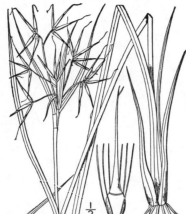

Schoenus corniculatus Lam. Tabl. Encycl. **1** : 137. 1791.
R. corniculata A. Gray, Ann. Lyc. N. Y. **3** : 205. 1835.
Rhynchospora corniculata macrostachya Britton, Trans. N.
    Y. Acad. Sci. **11** : 84. 1892.
R. macrostachya Torr. Ann. Lyc. N. Y. **3** : 206. 1835.

Culm obtusely triangular, stout or slender, smooth, 3°–7° tall. Leaves flat, broadly linear, 6′–18′ long, 3″–8″ wide, rough-margined; umbels terminal and axillary, sometimes 1° broad, usually compound; spikelets spindle-shaped, 4″–6″ long in flower, capitate at the ends of the rays and raylets; primary rays slender, sometimes 6′ long; scales lanceolate, thin, acute, light brown; bristles about 6, subulate or filiform, rigid, upwardly scabrous, shorter or longer than the achene; style subulate, entire or minutely 2-toothed at the apex, 2–4 times longer than the achene, upwardly scabrous, ½′–1′ long, persistent and much exserted beyond the scales when mature; achene obovate, flat, 2″ long, dark brown, smooth, its surface minutely cellular-reticulated.

In swamps, Massachusetts to Florida, west to Ohio, Missouri, Kansas and Texas. Consists of numerous races, differing in length of bristles and inflorescence. July–Sept.

2. Rynchospora pállida M. A. Curtis. Pale Beaked-rush. Fig. 842.

R. pallida M. A. Curtis, Am. Journ. Sci. (II.) **7** : 409. 1849.

Rootstocks slender, culms sharply triangular, 1½°–2½° tall. Leaves ½″–1″ wide, flattish, nearly smooth, the lowest reduced to many-nerved lanceolate acuminate scales; spikelets numerous, spindle-shaped, narrow, 2″–3″ long, aggregated in a compound convex terminal head, or occasionally also in a filiform-stalked cluster from the upper axil; uppermost leaves subulate, little exceeding the spikelets; scales pale greenish-brown, lanceolate, acuminate; bristles minute and early deciduous, or wanting; style 2-cleft; achene lenticular, obovate-oblong, smooth, brown, somewhat shining, ½″ long, tipped by a short tubercle.

In bogs, New Jersey to North Carolina. Aug.–Sept.

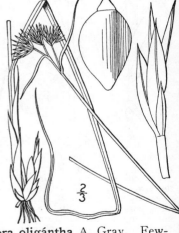

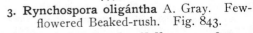

3. Rynchospora oligántha A. Gray. Few-flowered Beaked-rush. Fig. 843.

R. oligantha A. Gray, Ann. Lyc. N. Y. **3** : 212. 1835.

Rootstocks short, culms tufted, almost thread-like leafy only toward the base, 6′–16′ tall. Leaves filiform resembling and shorter than the culm or sometimes equalling it; spikelets 1–4, terminal, narrowly oblong acute, 3″–4″ long, sessile or peduncled, subtended by 1 or 2 filiform bracts; scales ovate, pale brown, acute cuspidate; bristles usually 6, densely plumose below the middle, upwardly scabrous above, equalling or shorter than the achene; style 2-cleft; achene obovoid-oblong obtuse, turgid-lenticular, pale brown, dull, transversely wrinkled; tubercle with a flat depressed border and a flattened conic acute central projection about one-fifth as long as achene.

In wet soil, New Jersey to Florida and Texas. June–Aug

4. Rynchospora álba (L.) Vahl. White Beaked-rush. Fig. 844.

Schoenus albus L. Sp. Pl. 44. 1753.
Rynchospora alba Vahl, Enum. **2** : 236. 1806.
Rynchospora alba macra Clarke; Britton, Trans. N. Y.
    Acad. Sci. **11** : 88. 1892.

Pale green, rootstocks short, culms slender or almost filiform, glabrous, 6′–20′ tall. Leaves bristle-like, $\frac{1}{4}''$–$\frac{1}{2}''$ wide, shorter than the culm, the lower very short; spikelets several or numerous, in 1–4 dense corymbose terminal and axillary clusters, narrowly oblong, acute at both ends, 2″–3″ long; scales ovate or ovate-lanceolate, white, acute; bristles 9–15, down-wardly barbed, slender, about as long as the achene and tubercle; style 2-cleft; achene obovate-oblong, smooth, pale brown, lenticular; tubercle triangular-subulate, flat, one-half as long as the achene.

In bogs, Newfoundland to Alaska, south to Florida, Kentucky, Minnesota, Idaho and California. Also in northern Europe and Asia. June–Aug.

5. Rynchospora capillàcea Torr. Capillary Beaked-rush. Fig. 845.

Rhynchospora capillacea Torr. Comp. 41. 1826.
Rynchospora capillacea laeviseta E. J. Hill, Am. Nat. **10** : 370. 1876.

Culms filiform, tufted, glabrous, 6′–20′ tall. Leaves filiform, less than $\frac{1}{4}''$ wide, much shorter than the culm, the lower very short; spikelets few, in 1–3 terminal and axillary loose clusters, oblong, acute at both ends, 2″–3″ long; scales ovate-oblong, chestnut-brown, keeled, mucronate; bristles 6, or sometimes 12, slender, downwardly barbed, or sometimes smooth, about equalling or becoming longer than the achene and tubercle; achene narrowly oblong, short-stalked, light brown, minutely wrinkled, lenticular; style 2-cleft; tubercle compressed, triangular-subulate, dark brown, about one-half as long as the achene.

In bogs, New Brunswick to Ontario, Minnesota, New Jersey, Pennsylvania, Indiana and Missouri. July–Aug.

6. Rynchospora Knieskérnii Carey. Knieskern's Beaked-rush. Fig. 846.

R. Knieskernii Carey, Am. Journ. Sci. (II.) **4** : 25. 1847.

Culms slender, tufted, smooth, 8′–18′ tall. Leaves narrowly linear, flat, about $\frac{1}{2}''$ wide, much shorter than the culm; spikelets numerous, in several distant compact clusters, oblong, acute, about 1″ long; scales chestnut-brown, ovate; bristles 6, downwardly barbed, equalling the achene; achene obovate, lenticular, brown, minutely wrinkled; style 2-cleft; tubercle triangular-subulate, pale, one-half as long as the achene or less and slightly decurrent on its edges.

Pine barrens, New Jersey to Virginia. July–Aug.

3/4

7. Rynchospora glomeràta (L.) Vahl.
Clustered Beaked-rush.　Fig. 847.

Schoenus glomeratus L. Sp. Pl. 44.　1753.
Rynchospora glomerata Vahl, Enum. **2**: 234.　1806.

Rootstocks slender, culms smooth, triangular, slender or rather stout, $\frac{1}{2}°-4\frac{1}{2}°$ high. Leaves flat, $\frac{1}{2}''-2''$ wide, rough-margined, shorter than the culm; spikelets several or numerous, in 2–7 corymbose-capitate axillary rather loose clusters, oblong, narrowed at both ends, $1\frac{1}{2}''-2''$ long; scales lanceolate, rich dark brown; bristles 6, downwardly barbed, rarely smooth, longer than or equalling the achene and tubercle; achene obovate, lenticular, smooth, dark brown; tubercle subulate, about as long as the achene.

In moist soil, New Brunswick to Ontario, Michigan, Arkansas, Florida and Texas. Consists of numerous races, differing in size, in width of leaves and in development of the inflorescence. July–Sept. False Bog-rush.

8. Rynchospora axillàris (Lam.) Britton.
Capitate Beaked-rush.　Fig. 848.

Schoenus axillaris Lam. Tabl. Encycl. **1**: 137.　1791.
Rhynchospora cephalantha A. Gray, Ann. Lyc. N. Y. **3**: 218.　1835.
R. axillaris Britton, Bull. Torr. Club **15**: 104.　1888.
Rynchospora axillaris microcephala Britton, Trans. N. Y. Acad. Sci. **11**: 89.　1892.

Culms stout, 3-angled, 2°–4° tall. Leaves flat, keeled, $1''-1\frac{1}{2}''$ wide; spikelets spindle-shaped, $2\frac{1}{2}''-3''$ long, exceedingly numerous, in several short-peduncled axillary and terminal very dense globose heads $4''-12''$ in diameter; scales dark brown, ovate-oblong, acute; bristles usually 6, longer than or equalling the achene and tubercle, downwardly or rarely upwardly barbed; achene obovate, brown, smooth, lenticular; tubercle subulate, about as long as the achene, somewhat decurrent on its edges.

In swamps, Long Island to Florida and Louisiana, near the coast. Cuba. Southern races have much smaller heads and smaller achenes than northern ones. July–Sept.

3/4

9. Rynchospora fúsca (L.) Ait.　Brown Beaked-rush.　Fig. 849.

Schoenus fuscus L. Sp. Pl. Ed. 2, 1664.　1763.
R. fusca Ait. Hort. Kew, Ed. 2, **1**: 127.　1810.

Rootstocks short, culms slender, 3-angled, smooth, tufted, 6′–18′ tall. Leaves setaceous, channeled, scarcely $\frac{1}{2}''$ wide, much shorter than the culm; spikelets spindle-shaped, acute, about $2\frac{1}{2}''$ long, several, or rather numerous, in 1–4 loose clusters; scales oblong-lanceolate, brown, shining, concave; bristles 6, upwardly barbed, often unequal, the longer ones usually exceeding the achene and tubercle; achene narrowly obovate, turgid-lenticular, smooth, shining; tubercle triangular-subulate, nearly as long as the achene, its margins serrulate or nearly smooth.

In bogs, Newfoundland to Delaware and Florida, west along the St. Lawrence and Great Lakes to Michigan. Also in Europe. July–Aug.

2/3

Rynchospora filifòlia Torr., with long filiform leaves and much smaller achenes, of the Southeastern States and Cuba, has recently been found to range northward into southern New Jersey.

10. Rynchospora gracilénta A. Gray. Slender Beaked-rush. Fig. 850.

R. gracilenta A. Gray, Ann. Lyc. N. Y. **3**: 216. 1835.

Culms very slender or filiform, smooth, obtusely triangular, 1°–2° tall. Leaves flat or becoming involute in drying, rather less than 1″ wide, elongated but shorter than the culm; spikelets narrowly ovoid, acute, 2″ long, few, in 1–4 loose clusters, the lower clusters borne on filiform stalks; scales ovate, brown, mucronate; bristles 6, upwardly barbed, equalling the achene and tubercle; achene broadly oval or nearly orbicular, dark brown, lenticular, dull, smooth; tubercle narrowly subulate, flat, widened at the base, pale, about as long as the achene.

In pine barren swamps, southern New York to Florida and Texas, near the coast. June–Aug.

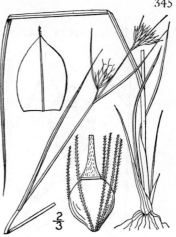

11. Rynchospora Smàllii Britton. Small's Beaked-rush. Fig. 851.

R. Smallii Britton; Small, Fl. SE. U. S. 1321. 1903.

Culms rather stout, 4° high or less, tufted. Leaves flat, 1″–2½″ wide; spikelets several or numerous in several rather loose axillary clusters, 1½″–2″ long, fusiform, their scales brown; bristles upwardly barbed, as long as the achene and tubercle; achene narrowly obovate, brown, smooth, shining, about ¾″ long, the smooth tubercle about one-half as long.

Bogs and damp hillsides, New Jersey and Pennsylvania to North Carolina. July–Aug.

12. Rynchospora cymòsa Ell. Grass-like Beaked-rush. Fig. 852.

Rynchospora cymosa Ell. Bot. S. C. & Ga. **1**: 58. 1816.
Schoenus cymosus Muhl. Gram. 8. 1817.

Light green, culms tufted, sharply 3-angled, smooth, 1°–2° tall. Leaves flat, narrowly linear, grass-like, 1½″–2″ wide or the basal ones broader, the uppermost sometimes overtopping the culm; spikelets ovoid-oblong, acute, 1¾″ long, sessile or nearly so, capitate in 2's–7's on the ultimate branches of the axillary and terminal clusters; bracts setaceous; scales dark brown, broadly ovate or suborbicular; bristles 6, upwardly barbed, shorter than the achene; achene broadly obovate to oblong, lenticular, transversely wrinkled; style 2-cleft; tubercle conic, one-fourth to one-third as long as the achene.

Moist soil, New Jersey to Illinois, Arkansas, Florida and Texas. Also in the West Indies and South America. June–Aug.

Rynchospora compréssa Chapm., of the southeastern states and reported from Missouri, differs by a broader lid-like tubercle.

Rynchospora rariflòra (Michx.) Ell., of the Southeastern States, Cuba and Jamaica, recently found at Cape May, New Jersey, has filiform culms and leaves and filiform-peduncled spikelets.

13. **Rynchospora Torreyàna** A. Gray.　Torrey's Beaked-rush.　Fig. 853.

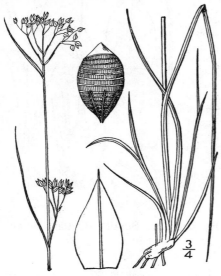

Rhynchospora Torreyana A. Gray, Ann. Lyc. N. Y. **3**: 197.　1835.

Culms terete or obscurely 3-angled, smooth, slender, 1½°–3° tall.　Leaves involute, the lower 1½″–2″ wide at the base and elongated, the upper bristle-like, distant; spikelets ovoid, 1½″ long, peduncled, numerous, in 1–4 loose distant clusters; scales brown, ovate, mucronate; bristles 6, upwardly barbed, shorter than the oblong-obovate transversely wrinkled lenticular achene; style 2-cleft; tubercle flat, conic, one-fourth to one-third as long as the achene.

In wet pine barrens, New Hampshire and Massachusetts to South Carolina and Georgia.　July–Aug.

14. **Rynchospora inexpánsa** (Michx.) Vahl.　Nodding Beaked-rush.　Fig. 854.

Schoenus inexpansus Michx. Fl. Bor. Am. **1**: 35. 1803.

R. inexpansa Vahl, Enum. **2**: 232.　1806.

Rootstocks slender, culms tufted, smooth, slender, 3-angled, 2°–3° tall.　Leaves smooth, 1½″ wide or less, flat, the lower elongated, the upper bristle-like, remote; spikelets spindle-shaped, acute at both ends, about 3″ long, numerous, in 1–4 narrow finally drooping panicles; scales brown, lanceolate, acuminate; bristles 6, upwardly hispid, very slender, about twice as long as the achene; achene narrowly oblong, transversely wrinkled; style 2-cleft; tubercle flat, triangular-subulate, one-half as long as the achene

Moist soil, Virginia to Louisiana.　June–Aug.

14. **PSILOCÀRYA** Torr. Ann. Lyc. N. Y. **3**: 359.　1836.

Annual sedges, with fibrous roots, slender leafy stems and ovoid or oblong, many-flowered terete spikelets in terminal and axillary, mostly compound umbels, the rays and raylets bracted at the base.　Scales of the spikelets spirally imbricated all around, membranous, deciduous. Flowers perfect.　Perianth none.　Stamens 1 or 2.　Style 2-cleft, enlarged at the base.　Achene lenticular or biconvex, smooth or transversely wrinkled, capped by the persistent base of the style (tubercle), or nearly the whole style persistent as a beak.　[Greek, referring to the absence of perianth-bristles.]

About 10 species, natives of temperate and tropical America.　Besides the following, another occurs in the southeastern United States.　Type species: *Psilocarya scirpoides* Torr.

Achene strongly wrinkled, much longer than the subacute tubercle.	1. *P. nitens.*
Achene smooth or but little wrinkled; tubercle subulate.	2. *P. scirpoides.*

1. **Psilocarya nìtens** (Vahl) Wood.   Short-beaked Bald-rush.   Fig. 855.

Scirpus nitens Vahl. Enum. **2** : 272.   1806.

P. rhynchosporoides Torr. Ann. Lyc. N. Y. **3** : 361. 1836.

Rhynchospora nitens A. Gray, Man. Ed. 5, 568.   1867.

Psilocarya nitens Wood, Bot. & Fl. 364.   1870.

Glabrous, culms tufted, slightly angled, 3′–2° tall.   Leaves narrowly linear, about 1″ wide, smooth, sometimes overtopping the culm, sheathing at the base, the midvein prominent; umbels mostly loose; spikelets ovoid, 2″–3″ long, rather less than 1″ in diameter; scales brown, broadly ovate, thin, 1-nerved, obtuse, acute or apiculate; achene lenticular, nearly orbicular, light brown, strongly wrinkled transversely; tubercle shorter than the achene, subacute, 2-lobed at the base.

In wet soil, Long Island, N. Y., Cape May, N. J., and Delaware to Florida and Texas, near the coast, and in Indiana.   July–Oct.

2. **Psilocarya scirpoìdes** Torr.   Long-beaked Bald-rush.   Fig. 856.

P. scirpoides Torr. Ann. Lyc. N. Y. **3** : 360.   1836.

Rhynchospora scirpoides A. Gray, Man. Ed. 5, 568. 1867.

Similar to the preceding species but smaller, usually less than 1° high.   Umbels commonly more numerous; spikelets oblong or ovoid-oblong; achene nearly orbicular in outline, biconvex, not as flat as that of *P. nitens,* dark brown, faintly transversely wrinkled or smooth, sometimes longitudinally striate, slightly contracted at the base into a short stipe; tubercle subulate, as long as or sometimes longer than the achene, its base decurrent on the edges.

In wet soil, eastern Massachusetts, Rhode Island and northern Indiana.   Perhaps a race of the preceding species.   July–Sept.

15.   **MARÍSCUS** (Hall.) Zinn, Cat. Hort. Goett. 79.   1757.

[Cladium P. Br. Civ. & Nat. Hist. Jam. 114.   Hyponym.   1756.]

Perennial leafy sedges, similar to the *Rynchosporas,* the spikelets oblong or fusiform, few-flowered, variously clustered.   Scales imbricated all around, the lower empty, the middle ones mostly subtending imperfect flowers, the upper usually fertile.   Perianth none.   Stamens 2 or sometimes 3.   Style 2–3-cleft, deciduous from the summit of the achene, its branches sometimes 2–3-parted.   Achene ovoid or globose, smooth or longitudinally striate.   Tubercle none.   [Greek, referring to the branched inflorescence of some species.]

About 40 species, natives of tropical and temperate regions.   Type species : *Schoenus Mariscus* L.

Leaves smooth, about 1″ wide.   1. *M. mariscoides.*
Leaves serrulate, 3″–10″ wide.   2. *M. jamaicensis.*

1. Mariscus mariscoìdes (Muhl.) Kuntze. Twig-rush. Water Bog-rush. Fig. 857.

Schoenus mariscoides Muhl. Gram. 4. 1817.
Cladium mariscoides Torr. Ann. Lyc. N. Y. 3: 372. 1836.
M. mariscoides Kuntze, Rev. Gen. Pl. 755. 1891.

Culm slender, erect, rather stiff, obscurely 3-angled, smooth, 1½°–3° tall. Leaves about 1″ wide, concave, with a long compressed tip, nearly smooth; umbels 2 or 3, compound, the 1 or 2 axillary, slender stalked; spikelets oblong, narrowed at both ends, acute, 2½″ long, capitate in 3's–10's on the raylets; scales chestnut-brown, ovate or ovate-lanceolate, acute, the midvein slightly excurrent; upper scale subtending a perfect flower with 2 stamens and a filiform 3-cleft style, the next lower one with 2 stamens and an abortive ovary; achene ovoid, acute, finely longitudinally striate, about 1″ long.

In marshes, Nova Scotia to Ontario and Minnesota, south to Florida, Kentucky and Iowa. July–Sept.

2. Mariscus jamaicénsis (Crantz) Britton. Saw-grass. Fig. 858.

Cladium jamaicense Crantz, Inst. 1: 362. 1766.
Schoenus effusus Sw. Prodr. 19. 1788.
Cladium effusum Torr. Ann. Lyc. N. Y. 3: 374. 1836.

Culm stout, 3°–9° high, bluntly 3-angled. Leaves very long, 3″–10″ wide, minutely serrulate on the margins; umbels several or numerous, decompound, forming large panicles; spikelets mostly 2–5 together at the ends of the raylets, narrowly ovoid, acute, 2″–2½″ long; uppermost scale subtending a perfect flower; stamens 2; achene ovoid, abruptly sharp-pointed, wrinkled, narrowed to the base, 2 mm. long.

In swamps, Virginia to Florida and Texas and in the West Indies. Aug–Sept.

16. SCLÈRIA Berg, Kongl. Acad. Sv. Handl. 26: 142. pl. 4, 5. 1765.

Leafy sedges, mostly perennial by rootstocks, the spikelets small, clustered in terminal, or terminal and axillary fascicles, or sometimes interruptedly spicate. Flowers monoecious, the staminate and pistillate spikelets separated or borne in the same clusters. Fertile spikelets 1-flowered. Staminate spikelets many-flowered. Scales imbricated all around, the 1–3 lower and sometimes also the upper ones of the fertile spikelets empty. Perianth none. Style 3-cleft, slender or sometimes swollen at the base, deciduous. Ovary supported on a disk (hypogynium), or this wanting. Stamens 1–3. Achene globose or ovoid, obtuse, crustaceous or bony, white in our species. [Greek, in allusion to the hard fruit.]

About 200 species, natives of tropical and temperate regions. Besides the following, some 8 others occur in the southern United States. Type species: Scleria flagellum-nigròrum Berg.

Spikelets in terminal, or terminal and lateral clusters; achene supported on a hypogynium.
  Achene smooth.
    Hypogynium supporting 8 or 9 small tubercles under the achene.   1. S. oligantha.
    Hypogynium covered with a rough white crust.   2. S. triglomerata.
  Achene reticulated or irregularly rugose.
    Culms erect or ascending; achene not hairy; peduncles short.   3. S. reticularis.
    Culms spreading; achene hairy; peduncles filiform.   4. S. setacea.
  Achene papillose.
    Hypogynium supporting 6 distinct tubercles.   5. S. pauciflora.
    Hypogynium supporting 3 entire, notched or 2-lobed tubercles.   6. S. ciliata.
Spikelets interruptedly glomerate-spicate; no hypogynium.   7. S. verticillata.

1. Scleria oligántha Michx.   Few-flowered Nut-rush.   Fig. 859.

Scleria oligantha Michx. Fl. Bor. Am. 2 : 167.   1803.

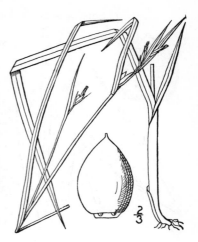

Rootstocks thick, hard, clustered; culms slender, erect, sharply 3-angled, nearly smooth, 1½°–2½° tall, the angles somewhat winged. Leaves smooth or slightly rough at the apex, 2″–3″ wide, the lower short, acute, the upper elongated; clusters terminal, usually also 1 or 2 axillary, and filiform-stalked; bracts slightly ciliate or glabrous; achene ovoid, obtuse but sometimes pointed, bright white, smooth, shining; hypogynium a narrow obtusely triangular border supporting 8 or 9 small tubercles under the achene.

In moist soil, District of Columbia and Virginia to Florida, Arkansas and Texas.   June–Aug.

2. Scleria triglomeràta Michx.   Tall Nut-rush.   Whip-grass.   Fig. 860.

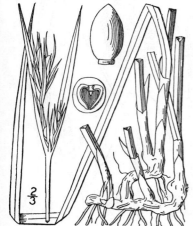

Scleria triglomerata Michx. Fl. Bor. Am. 2 : 168.   1803.
Scleria triglomerata var. *gracilis* Britton, Ann. N. Y. Acad. Sci. 3 : 230.   1885.   Not *S. gracilis* Ell. 1824.

Rootstocks hard, stout, clustered; culms 3-angled, slender or rather stout, erect or ascending, rough or nearly smooth on the angles, 1½°–3° tall. Leaves flat, smooth or slightly rough-margined, glabrous or nearly so, 1½″–2½″ wide, the lower short, acute, the upper tapering to a long tip, rarely exceeding the culm; flower-clusters terminal, and usually also 1 or 2 smaller ones from the axils; bracts glabrous or slightly ciliate; achene ovoid or ovoid-globose, obtuse but somewhat pointed, bony, obscurely 3-angled, smooth, bright white, shining, ¾″–1½″ high, supported on a low obtusely triangular, papillose-crustaceous hypogynium.

In meadows and thickets, Vermont to Ontario and Wisconsin, south to Florida, Arkansas and Texas.   July–Sept.

3. Scleria reticulàris Michx.   Reticulated Nut-rush.   Fig. 861.

Scleria reticularis Michx. Fl. Bor. Am. 2 : 167.   1803.
Scleria reticularis obscura Britton, Ann. N. Y. Acad. Sci. 3 : 232.   1885.

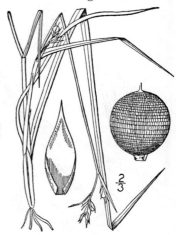

Rootstocks small; culms very slender, erect, 3-angled, 1°–2½° tall. Leaves narrowly linear, smooth, glabrous or nearly so, 1″–1½″ wide, not overtopping the culm; spikelets in a terminal cluster and 1–3 remote short-stalked axillary rather loose ones; bracts glabrous; achene globose, crustaceous, dull white when mature, reticulated by longitudinal and transverse ridges, ½″ in diameter, glabrous, the reticulations sometimes very obscure; hypogynium 3-lobed, its lobes appressed to the base of the achene.

In moist meadows, eastern Massachusetts to Florida and in northern Indiana.   July–Sept.

4. Scleria setàcea Poir.　Torrey's Nut-rush.　Fig. 862.

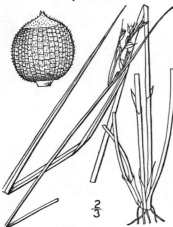

S. setacea Poir. in Lam. Encycl. **7**: 4.　1806.
Scleria Torreyana Walp. Ann. **3**: 696.　1852–53.
Scleria laxa Torr. Ann. Lyc. N. Y. **3**: 376.　1836.　Not. R. Br. 1810.
Scleria reticularis pubescens Britton, Ann. N. Y. Acad. Sci. **3**: 232.　1885.

Culms weak but rather thick, spreading or diffuse, 3-angled, nearly or quite smooth, 1°–2½° long. Leaves linear, nearly flat, smooth, glabrous, 1½''–4'' wide, not exceeding the culm; spikelets in a loose terminal cluster, and 1–3 filiform-stalked smaller axillary ones; bracts glabrous; achene globose, somewhat pointed, nearly 1'' in diameter, irregularly rugose with low ridges sometimes spirally arranged or reticulated, pubescent; hypogynium 3-lobed, the lobes appressed to the base of the achene.

In moist soil, Connecticut (?); Long Island to Florida, Indiana, Missouri, Texas and Mexico. Also in Cuba and Porto Rico.　June–Aug.

5. Scleria pauciflòra Muhl.　Papillose Nut-rush.　Fig. 863.

Scleria pauciflora Muhl.; Willd. Sp. Pl. **4**: 318.　1805.

Rootstocks thick, hard, clustered; culms slender, rather stiff, erect, usually tufted, glabrous or sparingly pubescent, 3-angled, 9'–2° tall. Leaves very narrowly linear, erect, less than 1'' wide, the lower short, the upper elongated and often overtopping the culm, their sheaths often densely puberulent; spikelets in a small terminal cluster and sometimes also in 1 or 2 axillary short-stalked ones; bracts ciliate or glabrous; achene oblong or globular, ½'' in diameter or rather more, crustaceous, papillose, the lower papillae elongated and reflexed; hypogynium a narrow obtusely triangular border supporting 6 small, distinct tubercles somewhat approximate in pairs, sometimes with 3 additional smaller intermediate ones.

In dry soil, New Hampshire to Ohio, Missouri, Kansas, Florida and Texas. Consists of several races, differing mainly in pubescence.　June–Sept.

6. Scleria ciliàta Michx.　Hairy Nut-rush.　Fig. 864.

S. ciliata Michx. Fl. Bor. Am. **2**: 167.　1803.

S. Elliottii Chapm. Fl. S. States, **531**.　1860.

S pauciflora Elliottii Wood, Bot. & Fl. 368.　1871.

Rootstocks rather stout, clustered; culms stout to slender, 3° long or less, 3-angled, glabrous or pubescent. Leaves ½''–3'' wide, glabrous or pubescent; spikelets in a terminal cluster and often in 2 or 3 stalked lateral ones; bracts ciliate; scales ciliate or glabrous; achene subglobose or globose-ovoid, about 1½'' in diameter, roughened by unequal papillae or short ridges; hypogynium an obtusely 3-angled border supporting 3 entire, 2-notched or 2-lobed tubercles.

Pine barrens and meadows, Virginia to Missouri, Florida and Texas. Also in Cuba and Santo Domingo. Consists of several races, differing in pubescence and in the form of the tubercles.

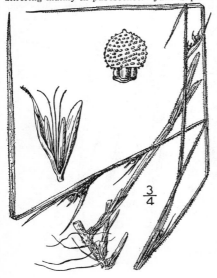

7. Scleria verticillàta Muhl. Low Nutrush. Fig. 865.

S. verticillata Muhl.; Willd. Sp. Pl. 4: 317. 1805.
Hypoporum verticillatum Nees, Linnaea 9: 303. 1835.

Annual, roots fibrous; culms very slender or filiform, 3-angled, smooth or nearly so, erect, 4'–2° tall. Leaves very narrowly linear, ¼''–½'' wide, erect, shorter than the culm, the lower very short; sheaths sometimes pubescent; spikelets in several separated clusters, the inflorescence simple or sparingly branched; bracts bristle-like; scales glabrous; achene globose, ½'' in diameter, crustaceous, usually tipped with the base of the style, marked by sharp distinct transverse ridges, or somewhat reticulated by additional longitudinal ridges; hypogynium none.

In moist meadows, eastern Massachusetts to Ontario, Minnesota, Missouri, Florida, Texas, Mexico. Bahamas and Cuba. Plant, especially the roots, fragrant in drying. July–Sept.

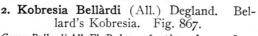

17. KOBRÈSIA Willd. Sp. Pl. 4: 205. 1805.

Slender arctic and mountain sedges, with erect culms, and 1–2-flowered spikelets, variously clustered. Stamens 3. Perianth-bristles or perigynium wanting. Ovary oblong, usually spicate, narrowed into a short style; stigmas 3, linear. Achene obtusely 3-angled, sessile. [Name in honor of Von Kobres, a naturalist of Augsburg.]

About 30 species, widely distributed in arctic and mountainous regions. Type species: *Kobresia scirpina* Willd. The generic name is sometimes written *Cobresia*.

| Spikes several, clustered. | 1. *K. bipartita.* |
| Spike solitary. | 2. *K. Bellardi.* |

1. Kobresia bipartita (All.) Della Torre. Arctic Kobresia. Fig. 866.

Carex bipartita All. Fl. Ped. 2: 265. *pl. 89. f. 5.* 1785.
Kobresia caricina Willd. Sp. Pl. 4: 206. 1805.
K. bipartita Della Torre, Anl. Alpenbl. 330. 1882.

Culms solitary or tufted, smooth or very nearly so, 4'–12' tall. Leaves about ½'' wide, infolded at least in drying, usually shorter than the culm, the old sheaths becoming fibrillose; spike 1' long or less, composed of several or numerous linear appressed or ascending spikelets; scales somewhat serrulate on the keel, rather more than ½'' long; mature achenes slightly longer than the scales.

Greenland to Alberta and the Canadian Rocky Mountains. Also in Europe and Asia. The name *Carex bipartita* All. is doubtfully associated with this plant. Summer.

2. Kobresia Bellàrdi (All.) Degland. Bellard's Kobresia. Fig. 867.

Carex Bellardi All. Fl. Ped. 2: 264. *pl. 92. f. 2.* 1785.
Kobresia scirpina Willd. Sp. Pl. 4: 205. 1805.
Elyna spicata Schrad. Fl. Germ. 1: 155. 1806.
K. Bellardi Degland, in Loisel, Fl. Gall. 2: 626. 1807.
Elyna Bellardi C. Koch, Linnaea 21: 616. 1848.

Densely tufted, culms very slender, 4'–18' tall, longer than the very narrow leaves. Old sheaths fibrillose, brown; margins of the leaves more or less revolute; spike subtended by a short bract or bractless, densely flowered or sometimes interrupted below, 8''–15'' long, 1½''–2'' in diameter; achenes rather less than 1'' long, ½'' thick, appressed.

In arctic America from Greenland to Bering Sea and Alberta, south in the Rocky Mountains to Colorado. Also in Europe and Asia. Summer.

18. CÁREX L. Sp. Pl. 972. 1753.*

Grass-like sedges, perennial by rootstocks. Culms mostly 3-angled, often strongly phyllo-podic, or aphyllopodic. Leaves 3-ranked, the upper elongated or very short (bracts) and subtending the spikes of flowers, or wanting. Flowers monoecious or dioecious, solitary in the axils of bracts (scales). Spikes either wholly pistillate, wholly staminate, androgynous or gynaecandrous. Perianth none. Staminate flowers of 3 stamens, the filaments filiform. Pistillate flowers of a single pistil with a style and 2 or 3 stigmas, surrounded by the peri-gynium, which completely encloses the achene or is rarely ruptured by it in ripening. Achene 3-angled, lenticular or plano-convex. Racheola occasionally developed.

Species over 1000, widely distributed, most abundant in the temperate zones. Besides the following about an equal number occur in western and southern North America. Specimens can only be satisfactorily determined when nearly or quite mature. Type species: *Carex pulicaris* L.

A. Spike one, androgynous; perigynia glabrous, not margined; leaves acicular, their sheaths striate, conspicuously clothing base of stem; stigmas 2 or 3.   1. NARDINAE.

B. Spikes one to very numerous; if one, plant not as above.
    1. ACHENES LENTICULAR AND STIGMAS 2; LATERAL SPIKES SESSILE; TERMINAL SPIKE PARTLY PIS-TILLATE, OR IF STAMINATE, THE LATERAL SPIKES SHORT OR HEADS DIOECIOUS. VIGNEA.
        a. Spike one, androgynous, orbicular to short ovoid.   2. CAPITATAE.
        b. Spikes one to very numerous; if one, not as above.
        † Rootstocks long-creeping, the culms arising 1–few together.
Spikes several, densely aggregated into a globular ovoid head, appearing like one spike; perigynia membranaceous.   3. INCURVAE.
Spikes not as above; perigynia not membranaceous.
    Spike one, staminate, pistillate or androgynous.   4. DIOICAE.
    Spikes more than one.
        Perigynia not thin or wing-margined, the beak obliquely cut, in age often bidentate.
            Culms becoming decumbent and branching.   5. CHORDORHIZEAE.
            Culms not branching.   6. DIVISAE.
        Perigynia thin or wing-margined, the beak bidentate.   7. ARENARIAE.
        †† Culms caespitose, but plants sometimes stoloniferous, or with slender rootstocks.
        * Spikes always androgynous.
Perigynia strongly compressed, not whitish-green.
    Perigynia 1″–2½″ long, the beak not exceeding the body.
        Spikes usually ten or less, green or reddish-brown tinged; sheaths loose, or if tight not red-dotted or transversely rugulose.   8. MUHLENBERGIANAE.
        Spikes numerous, yellow or brown; sheaths tight, the opaque part either red-dotted or trans-versely rugulose.
            Perigynia plano-convex, yellowish; opaque part of leaf-sheath transversely rugulose, often not red-dotted.   9. MULTIFLORAE.
            Perigynia thick, much rounded on outer, somewhat on inner surface, brownish; opaque part of leaf-sheath red-dotted, not transversely rugulose.   10. PANICULATAE.
    Perigynia 2″–4½″ long, spongy at base; beak much longer than body. 11. STENORHYNCHAE.
Perigynia scarcely compressed, nearly terete, whitish-green.   12. TENELLAE.
    * Spikes gynaecandrous, rarely staminate or pistillate.
Perigynia ascending or appressed, the body not margined.
    Perigynia 2″ long or less, puncticulate.   13. CANESCENTES.
    Perigynia longer, not puncticulate.   14. DEWEYANAE.
Perigynia body with thin or winged margins.
    Perigynia spongy at base, usually spreading at maturity, thin-margined. 15. STELLULATAE.
    Perigynia not spongy at base, not widely spreading at maturity, wing-margined. 16. OVALES.
    2. ACHENES TRIANGULAR OR LENTICULAR; IF LENTICULAR, THE LOWER LATERAL SPIKES CONSPICU-OUSLY PEDUNCLED OR WITH STAMINATE TERMINAL SPIKE AND ELONGATED LATERAL SPIKES. EU-CAREX.
        a. Scales bract-like; achenes strongly constricted at the base. 17. PHYLLOSTACHYAE.
        b. Scales not bract-like; achenes not strongly constricted at the base
        † *Spike normally one, the perigynia reflexed, or rounded and beakless at the apex.*
Perigynia rounded at apex, beakless, glabrous; scales persistent.   18. POLYTRICHOIDEAE.
Perigynia beaked, strongly reflexed; scales soon deciduous.   19. PAUCIFLORAE.
        †† *Spikes one to many; when one, the perigynia neither reflexed nor rounded.*
        * Perigynia both coriaceous and shining, the beak obliquely cut.   20. NITIDAE.
        ** Perigynia not both coriaceous and shining.
            ! Spike one; perigynia triangular, glabrous, not flattened.   21. RUPESTRES.
            !! Spikes one to many; when one, perigynia not as above.
            °Perigynia triangular, membranous, closely enveloping the achene, essentially nerveless, or 2-ribbed, pubescent or puberulent at least at base of beak, stipitate; bracts sheathless or nearly so.
Spikes normally one.
    Spikes androgynous; leaf-blades filiform.   22. FILIFOLIAE.
    Spikes dioecious; leaf-blades not filiform.   23. SCIRPINAE.
Spikes normally two or more.
    Perigynia obtusely triangular; foliage not pubescent.
        Young achenes mitrate at apex; lowest scales rough-awned; perigynia not slender-beaked.
                                                          24. MITRATAE.
        Young achenes, lowest scales, and perigynia not as above.   25. MONTANAE.
    Perigynia acutely triangular; foliage usually pubescent.   26. TRIQUETRAE.
            °°Perigynia not as above; or if so, bracts strongly sheathing.
        § Lowest bract strongly green-sheathing; perigynia beakless to beaked, entire, oblique or emarginate at apex; or long-beaked and apex hyaline, becoming bidentate, teeth weak; achenes triangular, or if rarely lenticular, the perigynia dull and subterete.

* Text contributed by Mr. KENNETH K. MACKENZIE.

Spike normally solitary.   27. PICTAE.
Spikes two or more.
  (1) Neither perigynia nor scales blackish.
    Bracts with obsolete or rudimentary blades.
      Lower spikes nearly radical; scales abruptly cuspidate.   28. PEDUNCULATAE.
      Lower spikes not radical; scales not abruptly cuspidate.
        Leaf-blades flat; perigynia puberulent to pubescent.   29. DIGITATAE.
        Leaf-blades filiform; perigynia glabrous.   30. ALBAE.
    Bracts with well-developed blades.
      Pistillate spikes short-oblong to linear, erect, or if drooping the spikes short and the
        perigynia acutely triangular; terminal spike staminate.
        Achenes lenticular; styles two.   31. BICOLORES.
        Achenes triangular; styles three.
          Perigynia with few to many strong nerves or nerveless.
            Perigynia tapering at base, triangular, closely enveloping the achene.
              Rootstock long-creeping.   32. PANICEAE.
              Rootstock not long-creeping.   33. LAXIFLORAE.
            Perigynia rounded at base, suborbicular in cross-section, loosely enveloping the
              achene.   34. GRANULARES.
          Perigynia finely many-striate.
            Perigynia tapering at base, constricted at apex, obtusely triangular, closely
              enveloping the achene.   35. OLIGOCARPAE.
            Perigynia rounded at both ends, cross-section suborbicular.   36. GRISEAE.
      Pistillate spikes elongated, linear to cylindric, slender-peduncled, the lower drooping.
        Perigynia beakless or short-beaked; terminal spike gynaecandrous. 37. GRACILLIMAE.
        Perigynia conspicuously or strongly beaked.
          Culms strongly reddish-tinged at base, aphyllopodic.
            Leaves glabrous; spikes very slender.   38. DEBILES.
            Leaves pubescent; spikes dense.   39. FLEXILES.
          Culms not strongly reddish-tinged at base, phyllopodic.
            Spikes slender, few-flowered; perigynia 2″ long or less, not inflated, the beak
              not becoming bidentate.   40. CAPILLARES.
            Spikes dense, many-flowered; perigynia longer, more or less inflated; the beak
              becoming bidentate.   41. LONGIROSTRES
  (2) Perigynia and scales strongly blackish.   42. FRIGIDAE.
§§ Lowest bract sheathless to strongly green-sheathing; if green-sheathing, achenes lenticular and
perigynia not dull and subterete, or perigynia with strongly bidentate non-hyaline apex and stiff teeth.
= Perigynia or foliage (at least the lower sheaths) pubescent; perigynia beakless, or the beak not
  strongly bidentate; achenes triangular.
  Terminal spike gynaecandrous.   43. VIRESCENTES.
  Terminal spike staminate.   44. PALLESCENTES.
= = Perigynia and foliage glabrous, or if pubescent, the perigynia strongly bidentate; achenes trian-
  gular or lenticular.
Perigynia rough-papillose or granular, beakless; or if beaked, the orifice not bidentate; achenes
  triangular.
  Perigynia beakless or very short-beaked.   45. TRACHYCHLAENAE.
  Perigynia conspicuously beaked.   46. ANOMALAE.
Perigynia glabrous or pubescent, neither papillose nor granular.
  Perigynia beakless or very short-beaked; achenes triangular.
    Terminal spike, if staminate, without rough-awned scales.
      Terminal spike staminate, lateral ones drooping on slender peduncles, at least at maturity.
        Perigynia glaucous, flattened; spikes not linear-cylindric.   47. LIMOSAE.
        Perigynia not glaucous, not flattened in our species; spikes narrow. 48. SCITAE.
      Terminal spike gynaecandrous, or if staminate, the lateral ones strictly erect.
        Scales dark-tinged.   49. ATRATAE.
        Scales not dark-tinged.   50. SHORTIANAE.
      Terminal spike staminate; scales rough-awned.   51. PALUDOSAE.
  Perigynia with strongly bidentate beak, or if not, the achenes lenticular.
    Achenes lenticular; perigynia dull.
      Scales obtuse to acuminate, not long-aristate; achenes not constricted. 52. RIGIDAE.
      Scales broad, long-aristate or in some arctic species acute or obtuse; achenes strongly
      constricted at the middle.   53. CRYPTOCARPAE.
    Achenes triangular, or if (rarely) lenticular, the perigynia not dull.
      Perigynia coriaceous, little if any inflated, often pubescent; bracts sheathless. 54. HIRTAE.
      Perigynia membranous or papery, from little to much inflated, never pubescent (rarely
        hispidulous); or if slightly coriaceous the lower bract long-sheathing.
        Perigynia little inflated, abruptly beaked, pistillate scales reddish or chestnut brown;
          lower bract strongly sheathing.   55. FLAVAE.
        Perigynia little to much inflated, pistillate scales not reddish brown or chestnut, or if
          somewhat so, lower bract not strongly sheathing.
          Perigynia lanceolate or lance-subulate, tapering into the beak, many-nerved.
            Perigynia-teeth reflexed; perigynia green, early deciduous. 56. COLLINSIAE.
            Perigynia-teeth not reflexed; perigynia yellowish-green.   57. FOLLICULATAE.
          Perigynia broader, abruptly contracted into beak, usually strongly ribbed.
            Perigynia less than 5″ long.
              Perigynium-body ovoid or globose, not truncately contracted.
                Perigynia coarsely ribbed or nerveless.   58. PHYSOCARPAE.
                Perigynia finely and closely ribbed.   59. PSEUDOCYPEREAE.
              Perigynia with an obovoid or obconic body truncately contracted into the
                prominent beak.   60. SQUARROSAE.
            Perigynia 5″ long or longer.   61. LUPULINAE.

1. NARDINAE. Represented by 1 species.   1. *C. nardina.*
2. CAPITATAE. Represented by 1 species.   2. *C. capitata.*
3. INCURVAE. Represented by 1 species.   3. *C. incurva.*
4. DIOICAE. Represented by 1 species.   4. *C. gynocrates.*
5. CHORDORRHIZEAE. Represented by 1 species.   5. *C. chordorrhiza.*

6. DIVISAE.

Heads not dioecious; styles short; perigynia short-beaked.
  Leaves narrowly involute.   6. *C. stenophylla.*
  Leaves 1″–2″ wide, flat above.   7. *C. camporum.*
Heads normally dioecious; styles long; perigynia long-beaked.   8. *C. Douglasii.*

7. ARENARIAE.

Perigynia thin-margined.   9. *C. Sartwellii.*
Perigynia wing-margined
  Spikes numerous; head heavy.   10. *C. arenaria.*
  Spikes few (3–8); head slender.   11. *C. siccata.*

8. MUHLENBERGIANAE.

1. Sheaths tight, often thickened at mouth; inconspicuously if at all septate-nodulose.
  Perigynia corky-thickened at base, usually widely radiating or reflexed at maturity.
    Perigynia beak smooth; scales acuminate, deciduous; spikes mostly approximate.
      Perigynia body broadly ovate, bi-convex.   12. *C. retroflexa.*
      Perigynia body lanceolate or ovate-lanceolate, plano-convex.   13. *C. texensis.*
    Perigynia beak minutely roughened; scales obtuse or acutish, persistent. 14. *C. rosea.*
  Perigynia not corky-thickened at base, spreading or ascending.
    Scales tinged with reddish-purple; perigynia more than 2″ long.   15. *C. muricata.*
    Scales not tinged with reddish-purple; perigynia 2″ or less long.
      Head 7½″–18″ long, the lower spikes distinct.
        Perigynia spreading, 1½″ long; bracts not broadly dilated at base; scales about length
        of and narrower than perigynia, short-awned.   16. *C. Muhlenbergii.*
        Perigynia ascending, 2″ long; bracts broadly dilated at base; scales (especially lower)
        exceeding and as wide as perigynia, strongly awned.   17. *C. austrina.*
      Head 4″–10″ long, the spikes densely capitate.
        Scale body about length of broadly ovate perigynia.   18. *C. mesochorea.*
        Scale body much exceeded by perigynia.
          Perigynia elliptic-ovate or narrower; leaves 1¼″–2″ wide. 19. *C. cephalophora.*
          Perigynia orbicular-ovate; leaves ½″–1″ wide.   20. *C. Leavenworthii.*
2. Sheaths loose and membranous, easily breaking, conspicuously septate-nodulose.
  Culms sharply triangular, not flattened or winged.
    Perigynia dull, its beak ¼–⅓ length of orbicular body.   21. *C. gravida.*
    Perigynia deep green, its beak ½ length of ovate body or more.
      Bracts not developed; scales one half length of body of perigynia. 22. *C. cephaloidea.*
      Some bracts developed; scales equalling body of perigynia or longer.
        Leaves 1½″–2″ wide; spikes aggregated or approximate.   23. *C. aggregata.*
        Leaves 2¼″–5″ wide; lower spikes strongly separate.   24. *C. sparganioides.*
  Culms narrowly winged, more or less flattened.
    Perigynia faintly nerved on outer face.   25. *C. alopecoidea.*
    Perigynia strongly nerved on outer face.   26. *C. conjuncta.*

9. MULTIFLORAE.

Leaves exceeding culms; perigynia beak equalling body.   27. *C. vulpinoidea.*
Culms exceeding leaves; perigynia beak shorter than body.
  Perigynia ovate to suborbicular.   28. *C. annectens.*
  Perigynia lanceolate to ovate-lanceolate.   29. *C. setacea.*

10. PANICULATAE.

Leaves ½″–1½″ wide; perigynia tapering into beak, rounded or truncate at base.
  Spikes approximate or little separate, the lower simple or nearly so; perigynia dark-brown,
  rounded on inner face, 1″–1¼″ long.   30. *C. diandra.*
  Spikes strongly separate, the lower compound; perigynia light-brown, nearly flat on inner face,
  1¼″–1½″ long.   31. *C. prairea.*
Leaves 2″–4″ wide; perigynia very abruptly short-beaked, tapering at base. 32. *C. decomposita.*

11. STENORHYNCHAE.

Perigynia 2″–2½″ long; beak 1–2 times length of body.   33. *C. stipata.*
Perigynia 3″–3½″ long; beak 3–4 times length of body.   34. *C. crus-corvi.*

12. TENELLAE. Represented by 1 species.   35. *C. disperma.*

13. CANESCENTES.

Lowest bract bristle-form, much prolonged, many times exceeding its 1–5-flowered spike; spikes
  widely separate.   36. *C. trisperma.*
Lowest bract much shorter or none; spikes several–many-flowered, the upper approximate.
  Spikes 2–4, subglobose, closely approximate, forming an ovate or suborbicular head; perigynia
  scarcely beaked; scales white-hyaline.   37. *C. tenuiflora.*
  Spikes 1–many, the lower more or less strongly separate, the head elongate; perigynia short to
  strongly beaked; scales darker.
    Perigynia broadest near middle; beak short, smooth or moderately serrulate.
      Perigynia beak smooth or very nearly so; scales very obtuse to acutish, strongly reddish-
      brown tinged; spikes closely approximate or in *C. norvegica* the lower remote.
        Spikes 1 or more, approximate, the lower less than 2″ wide.
          Spike 1, rarely with a smaller one at base.   38. *C. ursina.*
          Spikes 2–4.
            Terminal spike strongly clavate at base; culms rough at apex only.

Perigynia ovate, abruptly beaked.
Leaves flat, $\frac{1}{2}''-1\frac{1}{2}''$ wide; culms strict, erect.   39. *C. Lachenalii.*
Leaves involute, $\frac{1}{4}''-\frac{3}{4}''$ wide; culms weak, slender.   40. *C. amphigena.*
Perigynia lanceolate, long-tapering.   41. *C. glareosa.*
Terminal spike little clavate; culms usually very rough.   42. *C. Heleonastes.*
Lower spikes widely separate, $2''-2\frac{1}{2}''$ wide.   43. *C. norvegica.*
Perigynia beak serrulate, or if smooth the scales acutish to cuspidate and scarcely if at
all reddish-brown tinged; lower spikes remote.
Glaucous; leaves $1''-2''$ wide; spikes many-flowered, the scarcely beaked appressed-
ascending perigynia with emarginate or entire orifice.   44. *C. canescens.*
Not glaucous; leaves $\frac{1}{2}''-1\frac{1}{4}''$ wide; spikes fewer-flowered, the distinctly beaked,
loosely spreading perigynia with minutely bidentate orifice.   45. *C. brunnescens.*
Perigynia ovate, broadest near base, the beak conspicuous, strongly serrulate.   46. *C. arcta.*

14. DEWEYANAE.

Spikes oblong-ovoid; perigynia nerveless or nearly so, sharply margined above; achenes $1''$ long;
scales greenish-white.   47. *C. Deweyana.*
Spikes linear; perigynia noticeably or strongly nerved, little margined above; achenes $\frac{7}{8}''$ long or
less; scales brownish-tinged.   48. *C. bromoides.*

15. STELLULATAE.

Spike one (rarely with a small additional one).   49. *C. exilis.*
Spikes more than one.
Perigynia soon ruptured by elongating achene and appearing glume-like.   50. *C. elachycarpa.*
Perigynia not soon ruptured, remaining normal.
Perigynia broadest near base, the beak serrulate.
Perigynia beak $\frac{1}{4}-\frac{1}{3}$ length of body, the teeth very short, the suture on inner side incon-
spicuous; scales very obtuse to acutish.
Leaves flat, $\frac{1}{2}''-1''$ wide, usually shorter than culm; perigynia little nerved.
51. *C. interior.*
Leaves usually involute, $\frac{1}{4}''$ wide, usually exceeding culm; perigynia strongly nerved.
52. *C. Howei.*
Perigynia beak longer, strongly bidentate, the suture conspicuous; scales sharper.
Perigynia body lanceolate to broadly ovate, narrowed into beak more than half its
length; perigynia usually inconspicuously nerved on inner face.   53. *C. Leersii.*
Perigynia body suborbicular, abruptly contracted into beak less than half its length;
perigynia conspicuously nerved on inner face.
Perigynia lightly nerved on both faces; scales acute to short acuminate; leaves
usually less than $1''$ wide; culms slender, sharply triangular.   54. *C. incomperta.*
Perigynia strongly nerved on both faces; scales acutish to acute; leaves $1''-2''$
wide, the lower very short; culms stout, obtusely triangular.   55. *C. atlantica.*
Perigynia broadest near middle, the beak smooth.   56. *C. rosaeoides.*

16. OVALES.

1. Perigynia subulate, at least three times as long as wide, less than $\frac{3}{4}''$ wide, wing near base almost
obsolete.
Bracts much elongated, many times exceeding head, leaf-like; perigynia $2\frac{1}{2}''$ long, the beak 2-3
times length of body.   57. *C. sychnocephala.*
Bracts much shorter and reduced; perigynia $2\frac{1}{4}''$ long or less.
Tips of perigynia equalling or scarcely exceeded by the obtuse or acute dark brown scales;
leaves $1\frac{1}{4}''-2''$ wide; culms 20'-40' high.   58. *C. oronensis.*
Tips of perigynia conspicuously exceeding the acute or acuminate light brown scales; leaves
$\frac{1}{2}''-1\frac{1}{2}''$ wide; culms 5'-24' high.   59. *C. Crawfordii.*
2. Perigynia lanceolate to reniform; wing near base narrow to very wide, but never almost obsolete.
A. Perigynia narrowly to broadly lanceolate, at least $2\frac{1}{2}$ times as long as broad; tips of peri-
gynia prominently exceeding scales.
Spikes $2''-7\frac{1}{2}''$ long; perigynia $3\frac{1}{4}''$ long or less; achene short-oblong.
Leaves at most $1\frac{1}{2}''$ wide, those of sterile shoots few, ascending.   60. *C. scoparia.*
Leaves broader, those of sterile shoots very numerous, widely spreading.
Tips of perigynia appressed or ascending; spikes $3\frac{1}{2}''-6''$ long.   61. *C. tribuloides.*
Tips of perigynia widely spreading or recurved; spikes $2''-4''$ long.
Inflorescence dense, oblong; culm stiff, stoutish.   62. *C. cristatella.*
Inflorescence loose, elongate; culm weak, slender.   63. *C. projecta.*
Spikes $7\frac{1}{2}''-13''$ long; perigynia $3\frac{1}{2}''-5''$ long; achenes linear-oblong. 64. *C. muskingumensis.*
B. Perigynia ovate-lanceolate or broader, at most twice as long as broad, or if narrower tips of
perigynia equalled by scales.
a. Perigynia strongly exceeding scales or if nearly equalled by them much wider.
† Perigynia narrowly to broadly ovate, $1\frac{1}{2}''-2''$ long, nerveless or few-nerved on inner face,
the tips not appressed.
Perigynia brownish; spikes closely aggregated, rounded at base.   65. *C. Bebbii.*
Perigynia green; spikes contiguous to widely separate, usually clavate at base.
Leaves $1''$ wide or less.   66. *C. straminea.*
Leaves $1\frac{1}{4}''-3''$ (averaging $2''$) wide.   67. *C. normalis.*
†† Perigynia ovate to reniform, $2''$ or more long, or if shorter with closely appressed tips.
Spikes closely aggregated, the scales dark brown or blackish.   68. *C. macloviana.*
Spikes approximate to widely separate, the scales lighter.
Perigynia spreading-ascending; spikes green or brownish.
Spikes approximate or scattered, the head stiff; scales obtuse or acutish.
Perigynia $2''-2\frac{3}{4}''$ long, thickish, nerveless or obscurely nerved on inner face.
69. *C. festucacea.*
Perigynia $2\frac{3}{4}''-3\frac{3}{4}''$ long, very thin, prominently about 10-nerved on inner
face.   70. *C. Bicknellii.*
Spikes in a moniliform flexuous head; scales long-pointed.
71. *C. hormathodes.*

Perigynia closely appressed, or if somewhat spreading-ascending, the spikes whitish
   or silvery green.
   Spikes approximate, the head stiff ; lateral spikes rounded or little clavate at base.
      Perigynia ovate, tapering into beak half length of body or more.
                                                                72. *C. suberecta.*
      Perigynia broader, abruptly contracted into beak about third length of body.
         Scales long acuminate or aristate ; perigynia more than 2″ long and nearly
            1½″ wide or more ; achenes stipitate.   73. *C. alata.*
         Scales obtuse or acutish : perigynia 1½″–2″ long (rarely slightly more)
            and 1½″ or less broad ; achenes nearly sessile. 74. *C. albolutescens.*
   Spikes in a moniliform flexuous head ; lateral spikes long clavate. 75. *C. silicea.*
b. Scales very slightly shorter to slightly longer than perigynia and concealing them.
   Scales strongly dark-tinged, narrowly hyaline-margined ; head stiff, the spikes approximate,
      many-flowered.   76. *C. leporina.*
   Scales usually not dark-tinged, strongly hyaline margined ; if dark-tinged not as above.
      Inflorescence stiff, the spikes closely approximate.
         Spikes elliptic, tapering at base ; perigynia closely appressed ; bracts scale-like.
                                                                77. *C. xerantica.*
         Spikes suborbicular, rounded at base ; perigynia loosely ascending ; lower 1–2 bracts
            prominent, stiff.   78. *C. adusta.*
      Inflorescence not stiff, often flexuous and moniliform.
         Perigynia nerveless on inner face or faintly nerved.
            Perigynia lance-ovate, 2¼″–3¼″ long, 1″ wide or less ; leaves about 1″ wide.
                                                                79. *C. praticola.*
            Perigynia ovate, 2″–2½″ long, 1″–1½″ wide ; leaves 1½″–2″ wide.
                                                                80. *C. aenea.*
         Perigynia strongly nerved on inner face.   81. *C. foenea.*

17. PHYLLOSTACHYAE.

Scales (except lowest) not leaf-like, not enveloping perigynia, green with hyaline margins.
   Body of perigynia oblong ; pistillate flowers usually 3–10.   82. *C. Willdenovii.*
   Body of perigynia globose ; pistillate flowers usually 2–3.   83. *C. Jamesii.*
Scales leaf-like, half-enveloping perigynia, without conspicuous hyaline margins. 84. *C. durifolia.*
         18. POLYTRICHOIDEAE.   Represented by 1 species.   85. *C. leptalea.*

19. PAUCIFLORAE.

Racheola absent or rudimentary, not conspicuously exserted.   86. *C. pauciflora.*
Racheola present, conspicuously exserted.   87. *C. microglochin.*
      20. NITIDAE.   Represented by 1 species.   88. *C. supina.*
      21. RUPESTRES. Represented by 1 species.   89. *C. rupestris.*
      22. FILIFOLIAE. Represented by 1 species.   90. *C. filifolia.*
      23. SCIRPINAE. Represented by 1 species.   91. *C. scirpoidea.*
      24. MITRATAE.   Represented by 1 species.   92. *C. caryophyllea.*

25. MONTANAE.

1. None of the culms short and hidden among the bases of the leaves.
   Aphyllopodic and not stoloniferous ; lower sheaths but little fibrillose. 93. *C. communis.*
   Phyllopodic and often long-stoloniferous.
      Perigynia about equalling scales, 1¼″ long or less.
         Staminate spike stout ; lower sheaths usually strongly fibrillose.
            Long stoloniferous ; staminate spike 6″–12″ long.   94. *C. pennsylvanica.*
            Short stoloniferous ; staminate spike 2″–6″ long.   95. *C. varia.*
         Staminate spike not over ½″ thick ; sheaths little fibrillose. 96. *C. novae-angliae.*
      Perigynia twice length of scales, 1½″–2″ long.   97. *C. albicans.*
2. Many of the culms short and hidden among the bases of the leaves.
   Lower bract exceeding the culm, leaf-like.
      Staminate spike 1″–3″ long ; perigynia 1½″ long, or less. 98. *C. deflexa.*
      Staminate spike 3″–6″ long ; perigynia 2″ long, or more.   99. *C. Rossii.*
   Lower bract exceeded by the culm, scale-like.
      Pistillate and staminate spikes contiguous ; culms aphyllopodic. 100. *C. nigromarginata.*
      Lower pistillate spikes widely separate ; culms phyllopodic.
         Perigynia 2″ long or less, puberulent ; leaves slender, light green, ascending, 1½″ wide
            or less.   101. *C. umbellata.*
         Perigynia longer, glabrous except the long beak ; leaves stiff, deep green, spreading, wider.
                                                                102. *C. tonsa.*
                                                                103. *C. hirtifolia.*
      26. TRIQUETRAE.   Represented by 1 species.   104. *C. picta.*
      27. PICTAE.   Represented by 1 species.   105. *C. pedunculata.*
      28. PEDUNCULATAE. Represented by 1 species.

29. DIGITATAE.

Staminate spike 1½″–3″ long ; scales obtuse, one-half length of perigynia.   106. *C. concinna.*
Staminate spike 6″–10″ long ; scales acute to acuminate, exceeding perigynia. 107. *C. Richardsonii.*
      30. ALBAE.   Represented by 1 species.   108. *C. eburnea.*

31. BICOLORES.

Mature perigynia whitish, ellipsoid, not fleshy or translucent.   109. *C. Hassei.*
Mature perigynia orange or brownish, broader, fleshy, translucent. 110. *C. aurea.*

32. PANICEAE.

Perigynia beak none or very short, often bent.
   Leaves 1″ wide or less, involute or folded ; leaves and perigynia very glaucous ; bract sheaths
      short ; spikes approximate.   111. *C. livida.*
   Leaves 1″–3″ wide, flat ; leaves green or bluish-green ; bract sheaths long ; spikes distant.
      Perigynia turgid ; peduncle of staminate spike smooth ; leaves bluish-green.   112. *C. panicea.*
      Perigynia not turgid ; peduncle of staminate spike rough ; leaves green.

Culms phyllopodic, not strongly purplish tinged at base; rootstocks very slender, deep-seated.
  Fertile culm blades usually 6–10, $1\frac{1}{2}''$–$3\frac{1}{2}''$ wide; perigynia more than $1\frac{1}{2}''$ long; spikes oblong or linear-oblong.   113. *C. Meadii.*
  Fertile culm blades usually 3–5, $1''$–$1\frac{3}{4}''$ wide; perigynia less than $1\frac{1}{2}''$ long; spikes linear.   114. *C. tetanica.*
  Culms aphyllopodic, purplish at base; plants loosely stoloniferous. 115. *C. colorata.*
Perigynia beak straight, prominent, $\frac{1}{4}$–$\frac{1}{2}$ length of body.
  Culms phyllopodic; lower sheaths not purplish; pistillate spikes 2–3, spreading. 116. *C. vaginata.*
  Culms aphyllopodic; lower sheaths purplish; pistillate spikes 1–2, erect. 117. *C. polymorpha.*

33. Laxiflorae.

1. Sheaths and base of culm strongly purplish; staminate scales purplish.
  Leaf-blades of fertile culms rudimentary; perigynia $2\frac{1}{4}''$ long or less.   118. *C. plantaginea.*
  Leaf-blades of fertile culms developed; perigynia $2\frac{3}{4}''$ long or more.   119. *C. Careyana.*
2. Sheaths not purplish tinged, the base of culms but rarely so; staminate scales never purplish.
  A. Perigynia acutely triangular, short tapering at base.
    Leaf-blades very smooth (except edges), the larger $6''$ wide or more, those of fertile culm much smaller than those of sterile; perigynia smooth.   120. *C. platyphylla.*
    Leaf-blades hispidulous on veins, $5''$ wide or less; those of fertile culm moderately smaller than those of sterile; perigynia minutely roughened.
      Staminate spike sessile or nearly so; peduncles short, erect.   121. *C. abscondita.*
      Staminate spike usually strongly pedunceld; lower peduncles capillary.
        Perigynia short-beaked; second bract and leaves usually exceeding culm; blades $1\frac{1}{4}''$–$2\frac{1}{2}''$ wide, erect.   122. *C. digitalis.*
        Perigynia beakless or nearly so; second bract and leaves usually exceeded by culm; blades $2''$–$4''$ wide, spreading.   123. *C. laxiculmis.*
  B. Perigynia obtusely triangular, long tapering at base, smooth.
    Pistillate scales very truncate; blades $7\frac{1}{2}''$–$20''$ wide; culms very strongly flattened and wing-margined.   124. *C. albursina.*
    Pistillate scales acuminate to strongly cuspidate; blades narrower.
      Spikes elongated, 8–20-flowered, the lower normally on erect-stiff peduncles.
        Perigynia obovoid; beak abruptly bent, minute; sterile shoots developing conspicuous culms.
          Culms stout; sheaths loose, the margins crisped; staminate spike short-stalked or sessile.   125. *C. blanda.*
          Culms slender; sheaths rather tight, the margins little if at all crisped; staminate spike long or rarely short-stalked.   126. *C. laxiflora.*
        Perigynia obovoid or fusiform with straight or oblique conspicuous beak; sterile shoots reduced to tufts of leaves.
          Culms slightly wing-angled, stout, densely cespitose; basal leaves generally less than $1°$ long, their sides not parallel; perigynia appressed-ascending.
             127. *C. anceps.*
          Culms scarcely wing-angled, slender, often loosely cespitose; basal leaves generally more than $1°$ long, their sides parallel; perigynia spreading-ascending.
             128. *C. striatula.*
      Spikes short, 5–10-flowered, the lower on capillary spreading or drooping peduncles.
             129. *C. styloflexa.*

34. Granulares.

Culms tufted; bracts elongated, overtopping spikes; staminate spike short-stalked.
  Basal leaves $2\frac{1}{2}''$–$8''$ wide; lower bract not extending beyond upper spikes; perigynia narrowly obovoid.   130. *C. Shriveri.*
  Basal leaves $1\frac{1}{2}''$–$4\frac{1}{2}''$ wide; lower bract extending beyond upper spikes; perigynia elliptic obovoid.   131. *C. granularis.*
Rootstocks long-creeping; bracts short, rarely overtopping spikes; staminate spikes long-stalked.
             132. *C. Crawei.*

35. Oligocarpae.

Sheaths smooth; perigynia $1\frac{3}{4}''$–$2''$ long.   133. *C. oligocarpa.*
Sheaths rough-pubescent; perigynia $2\frac{1}{4}''$–$2\frac{1}{2}''$ long.   134. *C. Hitchcockiana.*

36. Griseae.

Perigynia elliptic, $\frac{3}{4}''$ wide; leaves $2''$ wide or less; bract sheaths rough and peduncles very rough.
  Culms $2\frac{1}{2}'$ high or less, much exceeded by leaves; staminate spike nearly sessile.
             135. *C. katahdinensis.*
  Culms $6'$–$36'$ high, exceeding leaves or shorter; staminate spike long-stalked. 136. *C. conoidea.*
Perigynia oblong, $1''$ wide; leaves $1\frac{1}{2}''$–$3\frac{1}{2}''$ wide; bract sheaths smooth and peduncles nearly so.
  Leaves not glaucous; larger spikes less than 12-flowered; lower scales usually equalling perigynia.
    Leaves $1''$–$2''$ wide, erect; spikes widely scattered, the lower nearly basal; perigynia little turgid.   137. *C. amphibola.*
    Leaves $2''$–$3''$ wide, spreading; lower spikes not nearly basal; perigynia turgid. 138. *C. grisea.*
  Leaves glaucous; larger spikes more than 12-flowered; lower scales shorter than perigynia.
    Perigynia $1\frac{1}{2}''$–$2''$ long, somewhat exceeding scales.   139. *C. glaucodea.*
    Perigynia $2\frac{1}{4}''$–$3''$ long, 2–3 times exceeding scales.   140. *C. flaccosperma.*

37. Gracillimae.

Plant glabrous; perigynia $1\frac{3}{4}''$ long or less.
  Perigynia rounded at apex, beakless.   141. *C. gracillima.*
  Perigynia sharp-pointed at apex, short-beaked.   142. *C. prasina.*
Sheaths and often foliage pubescent.
  Scales except lowest obtuse or acute; spikes all gynaecandrous.   143. *C. formosa.*
  Scales acuminate to cuspidate; lateral spikes pistillate.
    Bracts strongly sheathing; leaves $1\frac{1}{2}''$–$2''$ wide.

Upper scales acuminate or short-cuspidate; perigynia 1¾″-2″ long, less than 1″ wide; spikes linear-cylindric.   144. *C. oxylepis.*
Upper scales sharply cuspidate; perigynia 2″-2½″ long, more than 1″ wide; spikes linear-oblong or oblong-cylindric.   145. *C. Davisii.*
Lower bract only strongly sheathing; leaves 1″-2″ wide.
    Perigynia nearly 1″ wide, slightly inflated, strongly nerved; lateral spikes scarcely loosely flowered at base.   146. *C. aestivaliformis.*
    Perigynia about ½″ wide, obscurely nerved; lateral spikes alternately flowered at base.
                                                  147. *C. aestivalis.*

38. Debiles.

Perigynia smooth or puberulent.
    Scales tinged with reddish brown; perigynia firm, strongly several-nerved; upper sheaths puberulent.   148. *C. oblita.*
    Scales hyaline with green midrib or somewhat reddish-brown tinged; perigynia membranous, lightly nerved; upper sheaths glabrous.
        Perigynia sessile or subsessile; blades 1″-2½″ wide; scales obtuse to short-cuspidate.
            Perigynia 3″-5″ long, the beak subulate; scales white hyaline margined.   149. *C. debilis.*
            Perigynia 2¼″-3¼″ long, the beak less subulate; scales tawny tinged.   150. *C. flexuosa.*
        Perigynia strongly stipitate; blades 2½″-5″ wide; scales strongly cuspidate. 151. *C. arctata.*
Perigynia tuberculate-hispid.   152. *C. assiniboinensis.*

        39. Flexiles. Represented by 1 species.   153. *C. castanea.*
        40. Capillares. Represented by 1 species.   154. *C. capillaris.*

41. Longirostres.

Beak of perigynia much shorter than body; culms not fibrillose at base.   155. *C. cherokeensis.*
Beak of perigynia longer than body; culms strongly fibrillose at base.   156. *C. Sprengelii.*

42. Frigidae.

Perigynia rounded at base, its beak very short.   157. *C. atrofusca.*
Perigynia strongly narrowed at base, its beak conspicuous, serrulate.   158. *C. misandra.*

43. Virescentes.

Perigynia densely pubescent.
    Leaves exceeding culms; lowest bract setaceous, ¼″ wide; pistillate spikes oblong cylindric; perigynia obovoid.   159. *C. Swanii.*
    Culms exceeding leaves; lowest bract leaflet-like, ¼″-1½″ wide; pistillate spikes linear-cylindric; perigynia elliptic.   160. *C. virescens.*
Perigynia glabrous, at least at maturity.
    Perigynia much flattened, rounded at apex, lightly nerved.   161. *C. complanata.*
    Perigynia swollen, nearly orbicular in cross-section, pointed at apex, coarsely nerved.
        Perigynia 1″ long, brownish-green; scales not rough-cuspidate.   162. *C. caroliniana.*
        Perigynia longer, green; scales rough-cuspidate.   163. *C. Bushii.*

44. Pallescentes.

Perigynia beakless, lightly nerved.   164. *C. pallescens.*
Perigynia short-beaked, strongly nerved.   165. *C. abbreviata.*

        45. Trachychlaenae. Represented by 1 species.   166. *C. glauca.*
        46. Anomalae. Represented by 1 species.   167. *C. scabrata.*

47. Limosae.

Perigynia ovoid-elliptic, thick; pistillate scales broadly oval, obtuse or abruptly minutely pointed; culms obtusely triangular.   168. *C. rariflora.*
Perigynia broader, much flattened; pistillate scales ovate, tapering at apex to long cuspidate; culms sharply triangular
    Strongly stoloniferous; leaves involute, glaucous, 1½″ wide or less; scales little exceeding perigynia.   169. *C. limosa.*
    Tufted; leaves flat, not glaucous, wider; scales much exceeding perigynia.   170. *C. paupercula.*
        48. Scitae. Represented by 1 species.   171. *C. Barrattii.*

49. Atratae.

Scales not exceeding perigynia or but slightly so; sheaths not filamentose.
    Perigynia not flattened and two-edged.
        Perigynia 1¼″ long or less, nerveless, the beak minutely bidentate, the style not prominent.
                              172. *C. Halleri.*
        Perigynia longer, two-nerved; beak not bidentate, style prominent.   173. *C. stylosa.*
    Perigynia flattened and two-edged.
        Spikes all contiguous, sessile or short-peduncled, erect.   174. *C. Parryana.*
        Lower spike strongly stalked, often drooping and distant.   175. *C. atratiformis.*
Scales awned, noticeably exceeding perigynia; sheaths strongly filamentose.176. *C. Buxbaumii.*
        50. Shortianae. Represented by 1 species.   177. *C. Shortiana.*

51. Paludosae.

Perigynia not compressed-triangular, obovoid; staminate spike one.
    Perigynia strongly ribbed, squarrose, slightly glaucous; staminate scales tapering into short awn; basal sheaths not fibrillose.   178. *C. Joori.*
    Perigynia obscurely nerved, ascending, very glaucous; staminate scales abruptly awned; basal sheaths strongly fibrillose.   179. *C. glaucescens.*
Perigynia compressed-triangular, ovoid; staminate spikes two or three.   180. *C. acutiformis.*

52. Rigidae.

Culms aphyllopodic.
    Culms very strongly tufted, slender, very rough above; perigynia beak very short, not twisted.
        Perigynia elliptic, longer than or little exceeded by scales.   181. *C. stricta.*
        Perigynia obovate or orbicular, shorter than scales.   182. *C. Haydeni.*
    Culms tufted and short stoloniferous, stout at base, smooth above; perigynia beak prominent, twisted when dry.   183. *C. torta.*

Culms phyllopodic.
  Perigynia beak if present not bidentate.
    Culms single or in small clumps, strongly stoloniferous; scales very dark with slender mid-
      vein; strict and generally low species.
      Leaves with revolute margins; perigynia at most obscurely nerved. 184. *C. concolor.*
      Leaves with involute margins; perigynia conspicuously nerved.   185. *C. Goodenowii.*
    Culms densely cespitose, the stolons absent or not prominent; scales with broad light-colored
      center; tall slender species.
      Leaves ½″–1½″ wide; staminate spike one; rootstocks not sending out long stolons.
                                               186. *C. lenticularis.*
      Leaves broader; staminate spikes two or more; stolons long.   187. *C. aquatilis.*
  Perigynia beak markedly bidentate; perigynia strongly ribbed.   188. *C. nebraskensis.*

53. Cryptocarpae.

Pistillate spikes strictly erect, the upper sessile or nearly so; perigynia coriaceous.
  Leaves 1¼″ wide or less, the margins involute; lowest bract somewhat spathe-like; pistillate
    spikes few-flowered; culms 7′ or less high.   189. *C. subspathacea.*
  Leaves 1″–3″ wide, flat or with revolute margins; lowest bract not spathe-like; pistillate spikes
    many-flowered; culms usually taller.
    Scales little longer than perigynia, pale reddish; pistillate spikes usually 4′ long or less.
                                               190. *C. salina.*
    Scales much exceeding perigynia, dark purple, light brownish or chestnut; pistillate spikes
      1¼′–3′ long.   191. *C. recta.*
Pistillate spikes peduncled, drooping, widely spreading or suberect.
  Pistillate scales purplish brown; perigynia coriaceous.   192. *C. cryptocarpa.*
  Pistillate scales lighter colored; perigynia membranaceous.
    Awns of pistillate scales erect or appressed-ascending; perigynia not inflated; basal sheaths
      little or not at all fibrillose.   193. *C. maritima.*
    Awns of pistillate scales spreading or loosely-ascending; perigynia generally somewhat
      inflated; basal sheaths fibrillose.
      Sheaths rough hispid; lower pistillate scales tapering into awn. 194. *C. gynandra.*
      Sheaths smooth; lower pistillate scales abruptly contracted into awn. 195. *C. crinita.*

54. Hirtae.

Staminate scales not ciliate.
  Perigynia beak much shorter than body, the teeth ½″ long or less.
    Perigynia glabrous to sparsely pubescent.
      Leaves 3″–6″ wide, flat.
        Culms aphyllopodic, strongly purplish tinged and filamentose at base; mature peri-
          gynia strongly nerved.   196. *C. lacustris.*
        Culms phyllopodic, not purplish tinged or filamentose at base; mature perigynia im-
          pressed nerved.   197. *C. impressa.*
      Leaves 1″–2″ wide, becoming involute.   198. *C. Walteriana.*
    Perigynia densely or strongly pubescent.
      Perigynia nerves obscured by dense pubescence, the teeth of beak short.
        Perigynia beak with hyaline orifice at length somewhat bidentate; staminate spike
          usually one, sessile or short-stalked.   199. *C. vestita.*
        Perigynia beak with non-hyaline, strongly bidentate orifice; staminate spikes long-
          stalked.
          Leaves flat, more than 1″ wide.   200. *C. lanuginosa.*
          Leaves involute, 1″ wide or less.   201. *C. lasiocarpa.*
      Perigynia nerves prominent, the teeth of beak prominent, slender. 202. *C. Houghtonii.*
  Perigynia beak including teeth nearly as long as body, the teeth ¾″ long or more.
    Perigynia glabrous or pubescent, ovoid, the teeth less than 1″ long, erect or spreading; scales
      acute to aristate.   203. *C. trichocarpa.*
    Perigynia glabrous, lanceolate or ovoid-lanceolate, the teeth 1″–2″ long, widely spreading;
      scales long-aristate.   204. *C. atherodes.*
Staminate scales strongly ciliate.   205. *C. hirta.*

55. Flavae.

Bracts long-sheathing, erect; pistillate spikes scattered.   206. *C. fulvescens.*
Bracts except lower not long-sheathing, widely spreading; pistillate spikes, at least upper, approximate.
  Leaves involute; perigynia ascending, not yellowish.   207. *C. extensa.*
  Leaves not involute; perigynia squarrose, yellowish.
    Perigynia 1″–1½″ long, the beak scarcely half length of body.   208. *C. Oederi.*
    Perigynia 2″–3″ long, the beak about length of body.
      Scales hidden; perigynia 2″ long; beaks spreading or lower reflexed. 209. *C. lepidocarpa.*
      Scales conspicuous; perigynia 2½″–3″ long, the beaks spreading in all directions.
                                            210. *C. flava.*

    56. Collinsiae.   Represented by 1 species.   211. *C. Collinsii.*

57. Folliculatae.

Leaves 1″–2″ wide; scales acute or short acuminate.   212. *C. abacta.*
Leaves 2″–8″ wide; scales strongly awn-tipped.   213. *C. folliculata.*

58. Physocarpae.

1. Perigynia scarcely inflated; beak entire or emarginate; stigmas usually two; achenes lenticular.
  Leaves strongly involute, ½″–1½″ wide; rootstocks creeping, the culms arising one–several
    together; spikes sessile or short-peduncled; culms often somewhat filamentose at base.
    Pistillate spikes linear- or oblong-cylindric, 2″–2½″ wide; perigynia loosely ascending; leaves
      1″ wide or less.   214. *C. miliaris.*
    Pistillate spikes suborbicular to oblong, wider; perigynia appressed-ascending; leaves mostly
      wider.   215. *C. rhomalea.*
  Leaves flat, wider; rootstocks not creeping, but plants freely stoloniferous; culms not filamen-
    tose at base; spikes from nearly sessile to strongly peduncled.   216. *C. saxatilis.*

2. Perigynia from little to much inflated; beak bidentate; stigmas normally three; achenes triangular.
　　A. Pistillate scales (except rarely lowest) not rough-awned.
　　　a. Pistillate spikes oblong to cylindric, 15 to many-flowered.
　　Perigynia not reflexed; bracts not more than several times exceeding spikes.
　　　Achenes not excavated on one side or but little so.
　　　　Beak of perigynia smooth.
　　　　　Culms sharply triangular, rough above, scarcely spongy at base; perigynia ascending.
　　　　　　Perigynia 2″–2½″ long; culms often not filamentose at base.　217. *C. mainensis.*
　　　　　Perigynia longer; culms markedly filamentose at base.
　　　　　　Perigynia lanceolate, tapering into beak; spikes loosely flowered at base.
　　　　　　Perigynia ovoid or globose-ovoid; spikes not loosely flowered. 218. *C. Raeana.*
　　　　　　　Pistillate spikes less than 4½″ wide; perigynia globose-ovoid, abruptly contracted into beak.　　　　　　　　　　　　　　219. *C. monile.*
　　　　　　　Pistillate spikes 4½″–7½″ wide; perigynia ovoid, contracted into beak.
　　　　　　　　　　　　　　　　　　　　　　　　220. *C. vesicaria.*
　　　　Culms obtusely triangular, usually smooth above, often thick and spongy at base; peri-
　　　　　gynia often spreading, abruptly contracted into beak.
　　　　　　Leaves very narrow, involute.　　　　　　　221. *C. rotundata.*
　　　　　　Leaves broad, flat.
　　　　　　　Spikes 1¼′ long or less; perigynia strongly dark tinged, very membranous; culms
　　　　　　　　1° high or less.　　　　　　　　　　　222. *C. membranopacta.*
　　　　　　　Spikes much longer; perigynia little tinged, less membranous; culms much higher.
　　　　　　　　　　　　　　　　　　　　　　223. *C. rostrata.*
　　　　　　　　　　　　　　　　　　　　　　224. *C. bullata.*
　　Beak of perigynia rough.　　　　　　　　　　225. *C. Tuckermanii.*
　　　Achenes deeply excavated on one side.　　　　226. *C. retrorsa.*
　　Lower perigynia reflexed; bracts many times exceeding spikes.　227. *C. oligosperma.*
　　　b. Pistillate spikes globose or short-oblong, 5- to 15-flowered.
　　　　B. Pistillate scales rough-awned.
　　　　　Spikes cylindric, 7″–9″ thick; perigynia contracted into beak.　228. *C. lurida.*
　　　　　Spikes narrowly cylindric, 4″–6″ thick; perigynia abruptly contracted into beak.
　　　　　　　　　　　　　　　　　　　　　229. *C. Baileyi.*

59. Pseudo-Cypereae.

Perigynia suborbicular in cross-section, more or less inflated.
　Spikes linear-cylindric; staminate scales scarcely awned.　230. *C. Schweinitzii.*
　Spikes oblong or oblong-cylindric; staminate scales with short rough awns. 231. *C. hystricina.*
Perigynia obtusely triangular, scarcely inflated, closely enveloping achene.
　Perigynia teeth erect, ½″ long; body of beak ½″ long.　232. *C. Pseudo-cyperus.*
　Perigynia teeth recurved or spreading, ¾″–1″ long; body of beak ¾″–1″ long.
　　　　　　　　　　　　　　　　　　233. *C. comosa.*

60. Squarrosae.

Scales exceeding perigynia; terminal spike small, normally staminate.　234. *C. Frankii.*
Scales much shorter than perigynia; terminal spike gynaecandrous.
　Scales acuminate or awned; spikes oval.　　　235. *C. squarrosa.*
　Scales obtusish; spikes oblong-cylindric.　　236. *C. typhina.*

61. Lupulinae.

Pistillate spikes globose or subglobose; style straight.
　Scales usually strongly awned (varying to obtuse in few-flowered northern plants); pistillate
　　spikes 1–12-flowered; leaves 1¼″–3½″ wide, the sheath prolonged.　237. *C. intumescens.*
　Scales usually obtuse, varying to slightly cuspidate; pistillate spikes 6–30-flowered; leaves
　　2½″–4½″ wide, the sheath not prolonged.　　238. C. *Asa-Grayi.*
Pistillate spikes oblong or cylindric; style abruptly bent.
　Perigynia ascending, the beak less than twice length of body.
　　Culms arising singly from elongated rootstocks; leaves 1″–2½″ wide. 239. *C. louisianica.*
　　Culms cespitose; leaves 2″–5″ wide.
　　　Achenes longer than thick, the angles not prominently knobbed.　240. *C. lupulina.*
　　　Achenes not longer than thick, the angles prominently knobbed.　241. *C. lupuliformis.*
　Perigynia spreading at right angles, the beak 2–3 times length of body.　242. *C. gigantea.*

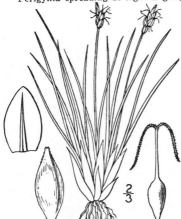

1. **Carex nàrdina** Fries. Nard Sedge. Fig. 868.

Carex nardina Fries, Mant. **2**: 55. 1839.

Culms very densely caespitose, filiform, smooth, erect, 2′–5′ tall, densely clothed at base with the old sheaths, many-leaved. Leaves filiform, erect, shorter than or longer than the culms; spike solitary, terminal, oblong, androgynous, bractless, 3″–7″ long, less than 2″ in diameter; perigynia oblong-elliptic, plano-convex, yellowish brown, faintly nerved, nearly erect, narrowed at both ends, 1½″–2″ long, ¾″ wide, slightly serrulate above, minutely and abruptly beaked, 2-toothed; scales broadly ovate, brown, with lighter center, thin, obtuse to slightly cuspidate, usually longer than the perigynia; racheola often present; stigmas 2 or 3.

Labrador and Hudson Bay to British Columbia and south to Washington and Colorado. Also in Europe and Asia. Summer.

2. Carex capitàta L.   Capitate Sedge.   Fig. 869.

Carex capitata L. Syst. Nat. Ed. 10, 1261. 1759.

Culms caespitose, slender, but stiff, strictly erect, 2′–18′ tall, slightly roughened above, few-leaved, the old sheaths not conspicuous, slightly fibrillose at base. Leaves filiform, involute, erect, shorter than the culm; spike solitary, terminal, short-ovoid, androgynous, bractless, 2″–4″ high, 1½″–2″ in diameter; perigynia membranous, broadly ovoid, ascending, whitish or tinged with brown, nerveless or nearly so, 1¼″ long, ¾″–1″ wide, abruptly contracted into a smooth, nearly entire, dark brown beak about ¼″ long, with hyaline orifice; scales broadly ovate, membranous, brown, with conspicuous white hyaline margins, obtuse, shorter than the perigynia; racheola often present; stigmas 2.

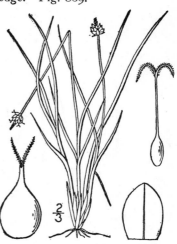

Greenland and Labrador to the Northwest Territory and Wyoming and on the higher summits of the White Mountains of New Hampshire. Also in Europe, Asia and South America. Summer.

3. Carex incúrva Lightf.   Curved Sedge. Fig. 870.

Carex incurva Lightf. Fl. Scot. 544. pl. 24. f. 1. 1777.

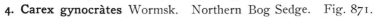

In small tufts, from elongated rootstocks, culms rather stiff, smooth, often curved, 1′–8′ long. Leaves less than 1″ wide, shorter than to exceeding the culm, usually curved; spikes 2–5, androgynous, sessile and closely aggregated into an ovoid or short oblong dense head, 2½″–8″ in diameter, appearing like a solitary spike; perigynia ovate, slightly swollen, compressed, scarcely margined, 1½″ long, 1″ wide, faintly several–many-nerved, contracted at the base and narrowed above into a short conic roughish beak; scales ovate, brownish, with silvery hyaline margins, acute or subacute, membranous, shorter than the perigynia; stigmas 2.

Greenland and Hudson Bay to British Columbia and Alaska. Also in Europe and Asia. Summer.

4. Carex gynocràtes Wormsk.   Northern Bog Sedge.   Fig. 871.

Carex gynocrates Wormsk.; Drejer, Rev. Crit. Car. 16. 1841.
Carex Redowskyana Fr. Schmidt Reisen im Amurl. 66. 1868.
Not C. A. Meyer, 1831.

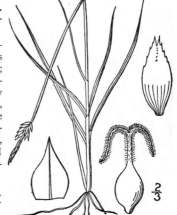

Culms very slender, stiff, erect, smooth, obtusely triangular, 3′–12′ tall, slenderly long-stoloniferous. Leaves almost bristle-form, erect, usually shorter than the culm; spike solitary, linear, terminal, erect, 2″–8″ long, androgynous, the pistillate part 2″–3″ thick, or sometimes wholly staminate or pistillate; perigynia ovoid-ellipsoid, biconvex, thin-edged, but margined above only, rounded and stipitate at base, dark brown at maturity, 1½″ long, ¾″ wide, spreading or reflexed when mature, strongly several-nerved, rough above, narrowed into a short, at length 2-toothed beak; scales ovate, light brown, spreading, acute to cuspidate, shorter than or equalling the perigynia; stigmas 2.

In bogs, Greenland to Alaska, south to Maine, New York, Pennsylvania, Michigan and in the Rocky Mountains to Colorado. Also in Europe and Asia. Summer.

5. Carex chordorrhìza Ehrh.   Creeping Sedge.   Fig. 872.

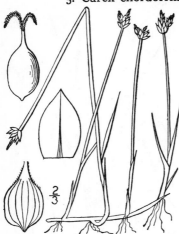

Carex chordorrhiza Ehrh. in L. f. Suppl. 414. 1781.

Culms elongated, the old ones prostrate, sending up from apical nodes (usually) fertile culms and from lower nodes sterile culms, the latter in succeeding seasons becoming prostrate and sending forth new culms from the nodes, the roots little developed. Fertile culms erect, smooth, 4′-18′ tall; leaves 1″-1½″ wide, shorter than the culm, somewhat involute in drying, straight, the lower ones reduced to short sheaths; spikes 2-4, bractless, androgynous, aggregated into a terminal ovoid or oblong head 3″-6″ long; perigynia broadly ovoid, ¼″ long, 1″ wide, flat on the inner side, convex on the outer, very thick, strongly many-nerved, slightly margined, abruptly tipped by a very short entire beak; scales ovate or ovate-lanceolate, brown, acute or acuminate, equalling the perigynia or a little longer; stigmas 2.

In sphagnum bogs and shallow water, Anticosti to Hudson Bay and the Northwest Territory, south to Maine, New York, northern Pennsylvania, Illinois and Iowa. Also in Europe and Asia. Summer.

6. Carex stenophýlla Wahl.   Involute-leaved Sedge.   Fig. 873.

Carex stenophylla Wahl. Kongl. Vet. Acad. Handl. (II.) 24 : 142. 1803.
?C. Eleocharis Bailey, Mem. Torr. Club 1 : 6. 1889.

Culms in small tufts from long creeping rootstocks, pale green, smooth, stiff, erect, 3′-8′ high. Leaves involute, about ½″ wide, shorter than to exceeding the culm; spikes 5 or 6, androgynous, sessile and aggregated into an ovoid dense head 3½″-7½″ long, 5″ wide or less, appearing like a solitary spike; perigynia ovate or ovate-oval, about 1½″ long and ¾″ wide, slightly margined above, faintly several-nerved, plano-convex, gradually narrowed into a short serrulate beak, the orifice oblique or in age slightly bidentate; scales ovate, brownish, membranous, obtuse to cuspidate, about equalling the perigynia; stigmas 2.

In dry soil, Manitoba to British Columbia, south to Iowa, Kansas and Colorado. Also in Europe and Asia. June-Aug.

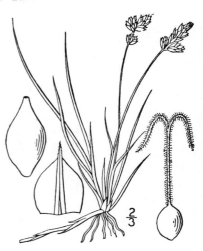

7. Carex campòrum Mackenzie.   Clustered Field Sedge.   Fig. 874.

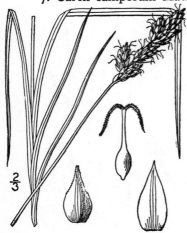

C. marcida Boott; Hook. Fl. Bor. Am. 2 : 212. pl. 213. 1840. Not J. F. Gmel. 1791.
Carex camporum Mackenzie, Bull. Torr. Club 37 : 244. 1910.

Light green, culms slender, sharply 3-angled, smoothish, or rough at least above, 1°-2° tall, from long creeping rootstocks. Leaves ¾″-2″ wide, flat or nearly so, much shorter than the culm; bracts short, subulate from a broader base, or wanting; spikes several, androgynous or gynaecandrous, clustered in a terminal linear-oblong to ovoid-oblong head 6″-18″ long, 3″-5″ wide, the lower ones rarely compound; perigynia ovate, dark brown at maturity, 1½″-2″ long, faintly nerved, sharp-margined, tapering into a flat serrate beak shorter than the body; scales ovate or ovate-lanceolate, brownish, hyaline-margined, acute or cuspidate, usually exceeding the perigynia; stigmas 2.

In dry soil, Michigan to British Columbia, south to Nebraska, Kansas, New Mexico and Nevada. Rarely adventive eastward. June-Sept.

8. Carex Douglásii Boott. Douglas' Sedge. Fig. 875.

Carex Douglasii Boott; Hook. Fl. Bor. Am. **2**: 213. *pl. 214.*
1840.

Light green, rootstocks extensively creeping, culms
normally dioecious, slender, erect, smooth or nearly so,
4'–12' tall. Leaves ¾"–1¼" wide, somewhat involute in
drying, shorter or longer than the culm, tapering to a
long tip; spikes linear or oblong, elliptic, 2½"–8"
long, several or numerous in a dense terminal oblong
or ovoid head 1'–2' long; perigynia ovate-lanceolate,
about 2" long, less than 1" wide, faintly several-nerved,
on both sides, rounded at base, the rough, at length
bidentate, tapering beak about one-half as long as the
body; scales pale greenish brown, or straw-colored,
lanceolate, scarious, smooth-awned, much longer than
the perigynia and completely concealing them;
stigmas 2.

In dry soil, Manitoba to Nebraska and New Mexico,
west to British Columbia and California. June–Aug.

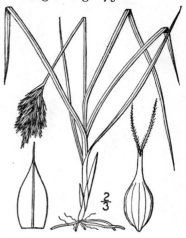

9. Carex Sartwéllii Dewey. Sartwell's Sedge. Fig. 876.

Carex Sartwellii Dewey, Am. Journ. Sci. **43**: 90.   1842.

Culms slender, stiff, erect, rough above, 3-angled,
1°–3° tall, from elongated dark rootstocks. Leaves
1"–2" wide, mostly shorter than the culm, long-
attenuate at the tip; bracts setaceous, usually very
small, or 1 or 2 of the lower sometimes elongated;
spikes numerous, ovoid or oblong, usually staminate
or androgynous, 2"–4" long, usually densely aggregated in a narrow but heavy head 1'–2' long and
5" wide, or the lower somewhat separated; perigynia
elliptic-lanceolate or ovate-lanceolate, 1¼"–2" long
and ¾"–1" wide, thin-margined, ascending, nerved on
both faces, tapering into a short 2-toothed beak;
scales ovate, obtuse or subacute, pale brown, scarious-
margined, about equalling the perigynia; stigmas 2.

In swamps, Ontario to British Columbia, south to central New York, Illinois, Arkansas and Utah. May–July.

10. Carex arenària L. Sand Sedge. Sand-star.
Fig. 877.

Carex arenaria L. Sp. Pl. 973. 1753.

Rootstock extensively creeping, culms erect, slender,
slightly scabrous above, 4'–15' high. Leaves ¾"–1¼"
wide, very long-pointed, shorter than the culm; lower
bract subulate, sometimes 1½' long; spikes oblong, 3"–6"
long, aggregated into a terminal ovoid head 1'–2' long,
the terminal commonly staminate, the middle ones staminate at the top, the lower usually wholly pistillate; perigynia lanceolate, 2"–2½" long, wing-margined above,
strongly several-nerved on both sides, the flat strongly
2-toothed serrulate beak nearly as long as the body and
decurrent on its summit; scales lanceolate, light brown,
long-acuminate or awned, about equalling the perigynia;
stigmas 2.

On sea beaches near Norfolk, Virginia. Adventive or
naturalized from Europe. Stare. Sea-sedge. Sea-bent. June–July.

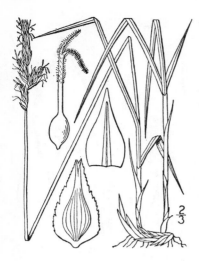

11. Carex siccàta Dewey.   Dry-spiked Sedge. Hillside Sedge.   Fig. 878.

Carex siccata Dewey, Am. Jour. Sci. **10** : 278.   1826.

Rootstocks long-creeping, and stout; culms slender, single or in small clumps, erect, rough above, 1°–2° tall.   Leaves erect, 1″–1½″ wide, usually shorter than the culm, the lower short; bracts short or the lowest bristle-form and elongated; head slender; spikes 3–8, oblong or subglobose, 2½″–4″ long, brownish or brown, clustered or more or less separated, usually gynaecandrous or staminate; perigynia ovate-lanceolate, much flattened but firm, 2½″–3″ long and 1″ wide, wing-margined, several-nerved on both sides, the inner face, concave by the incurved margins, the tapering rough beak nearly or fully as long as the body; scales ovate-lanceolate, membranous, acute or acuminate, nearly equalling the perigynia; stigmas 2.

In dry fields and on hills, Maine to Alaska, south to Rhode Island, New Jersey, Michigan, Arizona and California.   May–July.

12. Carex retrofléxa Muhl.   Reflexed Sedge. Fig. 879.

Carex retroflexa Muhl.; Willd. Sp. Pl. **4** : 235.   1805.
C. rosea var. *retroflexa* Torr. Ann. Lyc. N. Y. **3** : 389. 1836.

Culms very slender, erect, rather stiff, 8′–18′ tall, smooth or roughish above.   Leaves ½″–1¼″ in width, mostly shorter than the culm; lower bract bristle-form, sometimes 2′ long, usually shorter; spikes 4–8, normally androgynous, subglobose, 4–10-flowered, the upper all close together, the lower 1 to 3 separated; perigynia broadly ovoid with slightly raised margin, radiating or reflexed at maturity about 1½″ long and somewhat more than ¾″ wide, smooth, green-brown, compressed, but corky-thickened, biconvex, and finely nerved toward the base, tapering upwardly into a smooth 2-toothed beak more than one-third the length of the body; scales ovate, hyaline, acuminate, soon falling, about half as long as the perigynia; stigmas 2.

In woods and thickets, Massachusetts to Ontario, Michigan, Florida and Texas.   May–July.

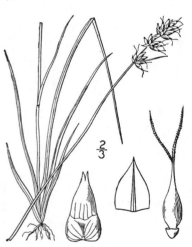

13. Carex texénsis (Torr.) Bailey.   Texas Sedge.   Fig. 880.

Carex rosea var. *texensis* Torr.; Ann. Lyc. N. Y. **3** : 389, hyponym.   1836.

Carex texensis Bailey, Mem. Torr. Club **5** : 97.   1894.

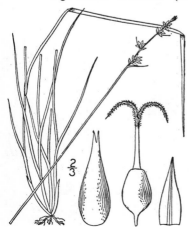

Similar to the preceding species, culms very slender, erect, smooth, 6′–18′ tall.   Leaves spreading or ascending, soft, about ½″ wide, shorter than the culm; lower bract commonly filiform, sometimes elongated; spikes 4–7, 4–8-flowered, all close together in a narrow head ½″–1½′ long, or the lower ones separated; perigynia narrowly lance-ovate or lanceolate, plano-convex, with slightly raised margin, corky-thickened and finely nerved towards base, green at maturity, radiating or widely spreading, about 1½″ long, ½″ wide, the smooth, tapering beak about one-half as long as the body; scales lanceolate or ovate, hyaline, acuminate, less than one-half as long as the perigynia; stigmas 2.

Southern Illinois and Missouri to South Carolina, Alabama and Texas.   April–May.

14. Carex ròsea Schk.   Stellate Sedge.   Fig. 881.

Carex rosea Schk.; Willd. Sp. Pl. 4 : 237.   1805.
Carex rosea var. *radiata* Dewey, Am. Journ. Sci. 10 : 276.
  1826.
Carex rosea var. *minor* Boott, Ill. Car. 2 : 81.   1860.

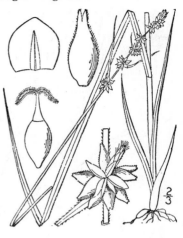

Rather bright green, culms very slender or filiform, erect or reclining, rough above, 1°–2½° long. Leaves flat, soft, spreading, ½″–1½″ wide, shorter than the culm; lower bract filiform or bristle-like, ½′–4′ long; spikes 2–8, androgynous, subglobose, 1½″–4″ in diameter, 2–15-flowered, the 2 to 4 upper close together, the others distant; perigynia narrowly to broadly ovoid-lanceolate, flat, bright green, stellately diverging or sometimes ascending, somewhat spongy at base and with a slightly raised margin, nerveless or nearly so, shining, 1″–2″ long, rather more than ⅓″ wide, tapering or contracted into a stout, rough, 2-toothed beak about one-fourth the length of the body; scales ovate-oblong to ovate-orbicular, obtuse or acutish, persistent, white-hyaline, half as long as the perigynia; stigmas 2.

In woods and thickets, Newfoundland to Manitoba, south to Georgia, Nebraska and Arkansas. Ascends to 2500 ft. in Virginia.   May–July.

15. Carex muricàta L.   Lesser Prickly Sedge.   Fig. 882.

Carex muricata L. Sp. Pl. 974 (in part).   1753.
Carex contigua Hoppe; Sturm, Deutschl. Fl. Heft 61.
  1835.

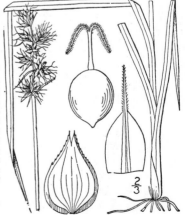

Bright green, culms slender, erect, roughish above, 1°–2½° tall, not wing-angled. Leaves 1″–1½″ wide, shorter than the culm, not conspicuously septate-nodulose; sheaths tight, not transversely rugulose; bracts short; spikes 5–10, 4–10-flowered, all clustered into an oblong head 7½″–20″ long, or the lower 1 or 2 little distant; perigynia ovate or ovate-lanceolate, dull green, 2″–3″ long, 1″ wide, smooth, shining, nerveless, ascending when young, spreading when mature, tapering into a rough-edged 2-toothed beak as long as the body; scales ovate or ovate-oblong, green or brownish, usually reddish-purple tinged, acute, somewhat shorter than the perigynia; stigmas 2.

In meadows and fields, southern Maine to Ohio and Virginia. Locally naturalized from Europe. Called also Greater prickly sedge.   June–Aug.

Carex echinàta Murr. a closely related European species, but with an elongated interrupted head, has been found in Kent County, New Brunswick, as a waif.

16. Carex Muhlenbérgii Schk.   Muhlenberg's Sedge.   Fig. 883.

Carex Muhlenbergii Schk.; Willd. Sp. Pl. 4 : 231.   1805.
Carex Muhlenbergii var. *enervis* Boott, Ill. 124.   1862.

Light green, culms slender but stiff and erect, sharply 3-angled, rough above, 1°–3° tall. Leaves 1″–2½″ wide, usually shorter than the culm, somewhat involute in drying; bracts bristle-form, not conspicuously enlarged at base, usually short; spikes 4–10, androgynous, ovoid or subglobose, distinct, the lower separate, but close together in an oblong head 11″–18″ long; perigynia spreading, broadly ovate-oval, 1½″ long, 1″ wide, from strongly nerved on both faces to nearly or quite nerveless, contracted into a 2-toothed beak nearly half length of body; scales hyaline with a green midvein, ovate-lanceolate, rough-cuspidate or short-awned, narrower than and about length of perigynia; stigmas 2.

In dry fields and on hills, Maine to Ontario and Minnesota, south to Florida and Texas.   May–July.

17. Carex austrìna (Small) Mackenzie.　Southern Sedge.　Fig. 884.

C. *Muhlenbergii* var. *australis* Olney; Bailey, Proc. Am. Acad. **22**: 141. 1886. Not *C. australis* T. Kirk, 1899.
C. *Muhlenbergii austrinus* Small, Fl. SE. U. S. 218. 1903.
C. *austrina* Mackenzie, Bull. Torr. Club **34**: 151. 1907.

Culms erect, slender, sharply triangular, rough above, 1°–2½° tall. Leaves 1¼″–2¼″ wide, usually noticeably shorter than culm; sheaths tight, thickened at mouth, not conspicuously septate-nodulose; bracts ½′–2′ long, dilated and much nerved at base, long-cuspidate and conspicuous; spikes numerous, androgynous, ovoid or subglobose, aggregated, the lower distinct but not separate, forming a head 7½″–15″ long, 4″–7½″ thick; perigynia ascending, 2″ long, 1½″ wide, the body suborbicular, nerved on outer and nearly nerveless on inner face, contracted into a 2-toothed beak half length of body; scales hyaline, strongly several-nerved, ovate, strongly awned, as wide as and longer than perigynia; stigmas 2.

In dry sunny places, Missouri and Kansas to Arkansas and Texas. April–July.

18. Carex mesochòrea Mackenzie.　Midland Sedge.　Fig. 885.

C. *mediterranea* Mackenzie, Bull. Torr. Club **33**: 439. 1906. Not C. B. Clarke, 1896.
C. *mesochorea* Mackenzie, Bull. Torr. Club **37**: 246. 1910.

Culms slender, erect, rough above, usually about 10′ tall. Leaves 1¼″–2″ wide, usually about half length of culm; bracts of lower spike bristle-form, short; spikes few, androgynous, only the lower distinguishable, subglobose, densely clustered in a terminal ovoid head 5″–8½″ long, 5″ thick; perigynia ascending or spreading, 1¾″ long, the body ovate, slightly more than 1″ wide, nerved on outer and nerveless on inner face, contracted into a 2-toothed beak half length of body; scales ovate, cuspidate, from slightly shorter to slightly longer and a little narrower than perigynia; stigmas 2.

In dry places, District of Columbia to Tennessee and Kansas. May–July.

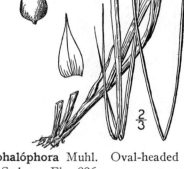

19. Carex cephalóphora Muhl.　Oval-headed Sedge.　Fig. 886.

Carex cephalophora Muhl.; Willd. Sp. Pl. **4**: 220. 1805.

Pale green, culms slender, erect, rough above, 8′–2° tall. Leaves 1″–2″ wide, sometimes overtopping the culm, usually shorter; bracts of the lower spikes bristle-form, usually short; spikes few, androgynous, only the lower distinguishable, subglobose, densely clustered in a terminal ovoid head 4″–7½″ long; perigynia narrowly to broadly elliptic-ovate, broadest just below the middle of body, 1¼″ long, ¾″ wide, nerveless or nearly so, narrowed into a 2-toothed beak shorter than the body; scales ovate, thin, green or slightly yellowish in age, acuminate to cuspidate, the body strongly exceeded by the perigynia; stigmas 2.

In dry fields and on hills, Maine and Ontario to Manitoba, south to Florida and Texas. Ascends to 2500 ft. in Virginia. May–July.

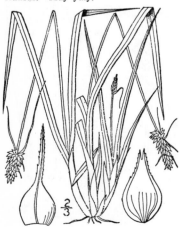

20. **Carex Leavenwórthii** Dewey. Leavenworth's Sedge. Fig. 887.

Carex Leavenworthii Dewey, Am. Jour. Sci. (II.) **2**: 246. 1846.

C. cephalophora var. *angustifolia* Boott, Ill. 123. 1862.

Similar to the preceding species but smaller, culms very slender or almost filiform, erect or spreading, roughish above, 6'–20' tall. Leaves much narrower, $\frac{1}{2}''$–$1\frac{1}{2}''$ wide, mostly shorter than the culm; bracts of the lower spikes short and bristle-form or wanting; spikes 4–7, androgynous, only the lower distinguishable, densely crowded in a short oblong head 4''–7$\frac{1}{2}''$ long, similar to that of *C. cephalophora* but usually smaller; perigynia orbicular-ovate, broadest near base, 1''–1$\frac{1}{4}''$ long and $\frac{5}{8}''$–$\frac{7}{8}''$ wide, narrowed into a 2-toothed beak, one-fourth length of body; scales ovate, acute or short-cuspidate, much shorter and narrower than the perigynia; stigmas 2.

In meadows, Ontario to District of Columbia, Iowa, Louisiana, Arkansas and Texas. May–June.

21. **Carex grávida** Bailey. Heavy Sedge. Fig. 888.

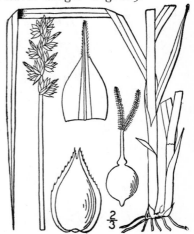

Carex gravida Bailey, Mem. Torr. Club **1**: 5. 1889.

Light green, culms slender, 1$\frac{1}{2}$°–3° tall, sharply 3-angled, erect, rough above. Leaves flat, 1$\frac{1}{2}''$–7'' wide, equalling or shorter than the culm; sheaths loose, conspicuously septate-nodulose, as are lower part of blades, the whitish part membranous, little if at all transversely rugulose; ligule prominent; bracts inconspicuous; spikes androgynous, several, in an oblong or ovoid-oblong dense head $\frac{3}{4}''$–1$\frac{1}{2}'$ long, pale, subglobose; perigynia flattened, spreading, dull green or light brownish-tinged, broadly ovate or suborbicular, 1$\frac{1}{2}''$–2'' long, 1''–1$\frac{1}{2}''$ wide, rounded at the base, narrowed into a 2-toothed beak, scarcely one-third as long as the body, several-nerved on the outer face or nerveless; scales ovate-lanceolate, dull green or brownish-tinged, acute to short-awned, about as long as the perigynia; achenes with suborbicular face, 1'' wide; stigmas 2.

Ohio to North Dakota, south to Kentucky, Missouri and Indian Territory. May–July.

22. **Carex cephaloìdea** Dewey. Thin-leaved Sedge. Fig. 889.

Carex muricata var. *cephaloidea* Dewey, Am. Jour. Sci. 11: 308. 1826.
Carex cephaloidea Dewey, Rep. Pl. Mass. 262. 1849.

Dark green, with green and white mottled sheaths; culms slender or stoutish, erect but not stiff, very rough above, 2°–3° tall. Leaves flat, 2''–4'' wide, thin and lax, somewhat shorter than the culm, the lower part septate-nodulose as are the loose membranous occasionally transversely rugulose sheaths; bracts usually not developed; spikes 4–8, androgynous, subglobose, aggregated in an oblong cluster 9''–20'' long; perigynia ovate or ovate-lanceolate, deep green, nearly 2'' long, 1'' wide, ascending, sharp-edged, nerveless or faintly few-nerved, tapering into a rough 2-toothed beak about half as long as the body; scale ovate, membranous, short-cuspidate or awned, about one-half as long as body of the perigynium; stigmas 2.

Woods and thickets, New Brunswick to Wisconsin and Pennsylvania. Local. May–July.

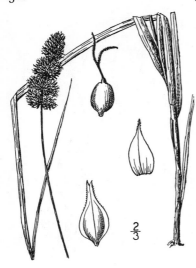

23. Carex aggregàta Mackenzie.   Glomerate Sedge.   Fig. 890.

Carex agglomerata Mackenzie, Bull. Torr. Club **33** : 442. 1906.   Not C. B. Clarke, 1903.

Carex aggregata Mackenzie, Bull. Torr. Club **37** : 246. 1910.

Culms slender, erect, triangular, rough beneath head only, 2° or less tall.   Leaves 1½″–2″ wide, shorter than the culm; bracts bristle-form, elongated but shorter than the head; spikes numerous, androgynous, ovoid or subglobose, densely aggregated, even the lowest but slightly separate, the head 12″–18″ long, 5″ thick; perigynia ascending or spreading, a little more than 1½″ long, the body ovate, 1″ wide, nerved on outer, nerveless on inner face, tapering into a 2-toothed beak about length of body; scales ovate, hyaline with green midrib, acuminate to cuspidate, narrower than and about length of body of perigynia; stigmas 2.

In dry woods, District of Columbia to Missouri.   May-June.

24. Carex sparganioìdes Muhl.   Bur-reed Sedge.   Fig. 891.

C. sparganioides Muhl.; Willd. Sp. Pl. **4** : 237.   1805.

Rather dark green with white and green mottled sheaths, culms stout or slender, rough above, sharply 3-angled, ½°–3° tall.   Leaves broad and flat, 2¼″–5″ wide, usually shorter than the culm, their lower part septate-nodulose as are the loose membranous transversely rugulose sheaths; spikes 6–12, deep green, oblong or subglobose, 2½″–4″ in diameter, 15–50-flowered, the upper aggregated, the lower 2–4 commonly separated, the lower bracts developed; perigynia flat, ovate, 1½″ long, 1″ wide, spreading or radiating, narrowly wing-margined to the rounded base, faintly few-nerved on the outer face, the rough 2-toothed beak one-half the length of the body; scales ovate, hyaline, acute or cuspidate, equalling body of perigynia; stigmas 2.

In woods and thickets, New Hampshire to Ontario and Michigan, south to Virginia, Kentucky and Kansas. Ascends to 2100 ft. in Virginia.   June–Aug.

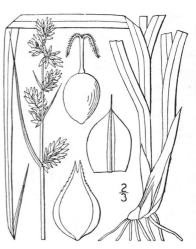

25. Carex alopecoìdea Tuckerm.   Foxtail Sedge. Fig. 892.

Carex cephalophora var. *maxima* Dewey, Am. Journ. Sci. **43** : 92.   1842.
Carex alopecoidea Tuckerm. Enum. Meth. 18.   1843.

Light green, culms stout but soft, sharply 3-angled, flattened, narrowly winged, erect or reclining, 1½°–3° long, roughish above.   Leaves flat, 1½″–4″ wide, shorter than or exceeding the culm, the sheath band strongly reddish-dotted but not transversely rugulose; bracts almost filiform, commonly short; spikes androgynous, several or numerous (10 or fewer) in a compact or somewhat interrupted head 1′–2′ long, 4″–5″ thick; perigynia ovate or ovate-lanceolate, rounded at base, short-stipitate, 1½″–2″ long, green or at maturity, yellowish brown, faintly few-nerved on the outer side, the tapering rough 2-toothed beak nearly as long as the body; scales ovate, brownish-tinged, acuminate or short-awned, about as long as the perigynia; stigmas 2.

In meadows, Maine to Pennsylvania and Michigan. Local.   June–July.

26. Carex conjúncta Boott.　Soft Fox Sedge.
Fig. 893.

Carex vulpina Carey, in A. Gray, Man. 541.　1848.　Not L. 1753.

Carex conjuncta Boott, Ill. **3**: 122.　1862.

Light green, culms roughish above, sharply 3-angled but flattened, somewhat winged, soft, erect, 1½°–3° tall. Leaves shorter than or sometimes equalling the culm, soft, flat, rough-margined, 2½″–5″ wide; bracts small and bristle-like or wanting; spikes androgynous, 10 or fewer, in a terminal elongated head 1′–3′ long, approximate, or the lower separated; perigynia ovate-lanceolate, rounded and slightly spongy at base, green even in age, 1½″–2″ long, thickened at the base, strongly several-nerved on outer face, tapering into a roughish 2-toothed beak shorter than the body; scales ovate to ovate-triangular, cuspidate or short-awned, about as long as the perigynia; stigmas 2.

In moist meadows and thickets, New York to District of Columbia, west to Minnesota and eastern Kansas. June–Aug.

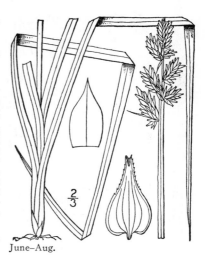

27. Carex vulpinoìdea Michx.　Fox Sedge.　Fig. 894.

Carex vulpinoidea Michx. Fl. Bor. Am. **2**: 169.　1803.

Culms slender, stiff, sharply 3-angled, very rough above, 1°–3° tall. Leaves 1″–2½″ wide, elongated, many exceeding the culm; sheaths tight, transversely rugulose; bracts bristle-like, sometimes 2′–3′ long; spikes ovoid-oblong, androgynous, densely flowered, 2″–4″ long, very numerous in a compact or somewhat interrupted narrow head, 1½′–5′ (usually 2′–3′) long, the lower ones distinguishable, sometimes compound, the upper confluent; perigynia narrowly to very broadly ovate, 1″–1¼″ long, rather more than ½″ wide, greenish yellow, flat, plano-convex, several-nerved on the outer face, nerveless or 1–3-nerved on the inner, ascending or spreading at maturity, tipped with a lanceolate 2-toothed beak about as long as the body; scales lanceolate, usually strongly awned, about as long as the upper and longer than lower perigynia, but narrower; stigmas 2.

In swamps and wet meadows, New Brunswick to Manitoba, south to Florida, Louisiana, Nebraska and Texas. Ascends to 2500 ft. in Virginia. June–Aug.

28. Carex annéctens Bicknell.　Yellow-fruited Sedge.　Fig. 895.

C. xanthocarpa Bicknell, Bull. Torr. Club **23**: 22.　1896. Not Regl. 1807.

Carex xanthocarpa var. *annectens* Bicknell, Bull. Torr. Club **23**: 22.　1896.

Carex annectens Bicknell, Bull. Torr. Club **35**: 492.　1908.

Culms stoutish, rough above, 1°–5° tall, exceeding the leaves. Leaves 1″–3½″ wide; head oblong or ovoid, usually dense, ¾′–2½′ long; sheaths tight, transversely rugulose; spikes androgynous, numerous, ovoid, many-flowered, short; bracts much less conspicuous than in the last; perigynia bright yellow, plano-convex, ovate to suborbicular, 1½″ long, with a narrowed or truncate base, and abruptly narrowed into a short minutely 2-toothed beak, nerveless, or obscurely few-nerved on the outer face; scales acuminate, short-awned.

In fields, Maine to New York, Iowa, Maryland and Missouri. June–Aug.

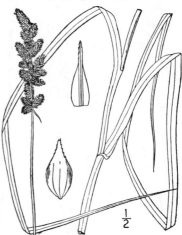

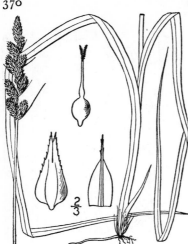

29. Carex setàcea Dewey. Bristly-spiked Sedge. Fig. 896.

Carex setacea Dewey, Am. Journ. Sci. **9**: 61. 1825.

Carex scabrior Sartw.; Boott, Ill. **3**: 125. 1862.

Culms 1½°–4° tall and slender, erect, rough above. Leaves 1°–2° long, 1″–3″ wide, shorter than the culm; sheaths red-dotted and rugulose; head narrowly oblong, 1¼′–2½′ long, 3″–5″ thick, sometimes branched at the base; bracts bristle-like, longer than the spikes or shorter; spikes androgynous, ovoid or ovoid-oblong, 2½″–4″ long, usually close together; perigynia dull at maturity, 1½″ long, lanceolate or ovate-lanceolate, tapering from a more or less truncate base to a narrow rough 2-toothed beak, few-nerved on outer face; scales acuminate, short-awned.

Vermont to Ontario, south to Maryland and Kentucky. June–Aug.

30. Carex diándra Schrank. Lesser Panicled Sedge. Fig. 897.

Carex diandra Schrank, in Acta Acad. Mogunt. 49. 1782.
Carex teretiuscula Gooden. Trans. Linn. Soc. **2**: 163. *pl.* *19.* 1794.

Loosely caespitose from short rootstocks, rather light green, culms slender, erect, very rough above, 1°–3° tall. Leaves ½″–1½″ wide, shorter than or sometimes equalling the culm, the lower sheaths reddish-brown dotted; bracts very small or scale-like; spikes several or numerous, staminate above, in a narrowly oblong compact or somewhat interrupted terminal cluster 1′–2′ long, 5″ thick or less; perigynia broadly ovoid, smooth, dark brown, very plump, hard, shining, strongly rounded and nerved on the outer side, slightly rounded and faintly nerved at base on the inner, 1″–1¼″ long, not margined, the body slightly more than ½″ long, suborbicular, truncate or rounded at the base, short-stalked, tapering into a flat conic beak nearly its own length; scales thin, ovate, brownish, acute or short-awned, about equalling the perigynia; stigmas 2.

In swamps and wet meadows, Nova Scotia to Alaska, south to Rhode Island, Pennsylvania, Nebraska and British Columbia. Also in Europe and Asia. May–July.

31. Carex pràirea Dewey. Prairie Sedge. Fig. 898.

Carex prairea Dewey, in Wood's Classbook, 578. 1855.
Carex teretiuscula var. *ramosa* Boott, Ill. Car. 145. 1867.
C. teretiuscula prairea Britton, Brit. & Br. Ill. Fl. **1**: 344. 1896.

Loosely caespitose from short rootstocks, the culms sharply triangular, slender, erect, rough above, 1½°–4° tall. Leaves ½″–1½″ wide, shorter than culm, lower sheaths reddish-brown dotted; bracts small or scale-like; spikes many, androgynous, clusters widely separate, lower usually compound, forming a flexuous nodding head 1½′–3′ long, often more than 5″ wide; perigynia ovoid, smooth, light brown, plump, hard, rounded and obscurely nerved on the outer side, flattish on inner, 1¼″–1½″ long, not margined, round-truncate at base, slightly stipitate, tapering into a flat beak shorter than body; scales thin, ovate, light brown with broad hyaline margins, acuminate or short-awned, usually exceeding perigynia; stigmas 2.

In wet meadows, Quebec to British Columbia, south to Connecticut, New Jersey, Kentucky and Utah. May–July.

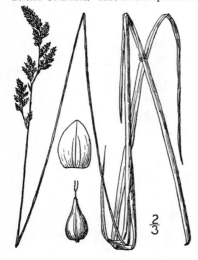
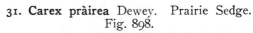

32. Carex decompósita Muhl.   Large-panicled Sedge.   Fig. 899.

Carex decomposita Muhl. Gram. 264. 1817.

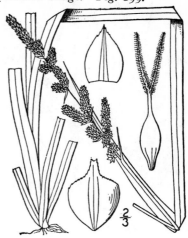

Dark green, culms smooth and very obtusely angled or terete below, roughened above, rather stout, erect, 1½°–3° tall. Leaves 2″–4″ wide, rough, rather stiff, longer or shorter than the culm, equitant at the base; spikes brownish, staminate above, small and very numerous in a terminal decompound cluster 2′–6′ long, the lower branches ascending and 1′–2′ long; bracts subulate or wanting; perigynia short-obovoid, 1″–1¼″ long, somewhat shining, dark brown, thick and hard, strongly rounded and strongly nerved on outer surface, slightly rounded and faintly nerved on inner surface, very narrowly margined, tapering at base, very abruptly tipped with a very short slightly 2-toothed beak; scales ovate, scarious-margined, nearly equalling the perigynia; stigmas 2.

In swamps, New York to Ohio and Michigan, south to Florida and Louisiana. May–Aug.

33. Carex stipàta Muhl.   Awl-fruited Sedge.   Fig. 900.

Carex stipata Muhl.; Willd. Sp. Pl. **4**: 233. 1805.

Culms erect or nearly so, sharply 3-angled before drying, slightly winged and strongly serrulate above, 1°–3½° tall. Leaves flat, 2″–4″ wide, usually shorter than the culm, the sheaths strongly transversely rugulose; bracts short, bristle-form or wanting; spikes numerous, androgynous, yellowish brown, crowded into a terminal oblong head 1′–4′ long, the lowest sometimes branched; perigynia lanceolate, strongly nerved, rounded and spongy at base, short-stipitate, 2″–2½″ long, about 1″ wide at the base, gradually tapering into a rough flattened 2-toothed beak 1–2 times as long as the body, giving the clusters a peculiarly bristly aspect; scales ovate or lanceolate, thin, hyaline, acuminate, much shorter than the perigynia; stigmas 2.

In swamps and wet meadows, Newfoundland to British Columbia, Florida, Tennessee, Missouri, New Mexico and California. Ascends to 4200 ft. in Virginia. May–July.

C. laevivaginàta (Küken.) Mackenzie, ranging from Maryland to North Carolina, differs in sheaths not transversely rugulose and thickened at the mouth.

34. Carex crús-córvi Shuttlw.   Raven's-foot Sedge.   Fig. 901.

Carex Crus-corvi Shuttlw.; Kunze, Riedg. Suppl. 128. *pl. 32.* 1844.
C. sicaeformis Boott, Journ. Bost. Nat. Hist. Soc. **5**: 113. 1845.
Carex Halei Dewey, Am. Journ. Sci. (II.) **2**: 248. 1846.

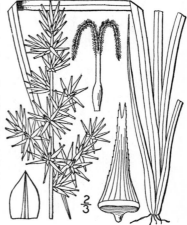

Pale green, culms in clumps, stout, 3-angled, very rough above, erect, 2°–4° tall. Leaves flat, 2½″–6″ wide, rough-margined, usually shorter than the culm, the sheaths conspicuously reddish dotted; spikes yellowish brown, staminate above, very numerous in a large compound branching terminal cluster 4′–12′ long, 1′–3′ thick; perigynia elongated-lanceolate, stipitate, strongly nerved, 1⅓″–4½″ long, strongly spongy and with a short hard disk-like base and a subulate rough 2-toothed beak 3 to 4 times as long as the body; scales ovate or lanceolate, thin, much shorter than perigynia; stigmas 2.

In swamps, Indiana to southern Minnesota and Nebraska, Florida, Louisiana and Texas. May–July.

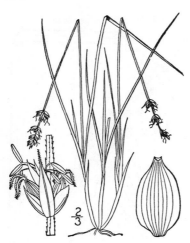

35. Carex dispérma Dewey. Soft-leaved Sedge. Fig. 902.

Carex tenella Schk. Riedgr. *23. f. 104.* 1801. Not Thuill. 1799.
Carex disperma Dewey, Am. Journ. Sci. **8**: 266. 1824.

Light green, rootstocks elongated, very slender, culms almost filiform, rough, commonly reclining, 6′–2° long. Leaves soft, ½″–¾″ wide, spreading, usually shorter than the culm; spikes very small, androgynous, only 1–5-flowered, distant or the upper close together, the bracts absent or bristle-form; perigynia ovoid-ellipsoid, very thick, hard, finely many-nerved, about 1″ long and ¾′ thick, tipped with a very minute, smooth, entire beak; scales ovate, hyaline, acute to cuspidate, shorter than or the lower equalling the perigynia; achene closely filling perigynium; stigmas 2.

In bogs, Newfoundland to British Columbia, New Jersey, Pennsylvania, Indiana, Michigan, Colorado and California. Also in Europe and Asia. June–Aug.

36. Carex trispérma Dewey. Three-fruited Sedge. Fig. 903.

Carex trisperma Dewey, Am. Journ. Sci. **9**: 63. 1825.
Carex trisperma var. *Billingsii* Knight, Rhodora **8**: 185. 1906.

Bright green, culms filiform, weak, often divaricate-spreading above lowest bract, usually reclining or spreading, very slightly roughened, 1°–2½° long, the rootstocks slender, often elongated. Leaves flaccid, flat, 1″ wide or less; spikes 1 to 3, only 1–5-flowered, gynaecandrous, widely separated, the lowest much exceeded by a bristle-form bract ½′–5′ long; perigynia oblong, ascending, green, 1¼″–2″ long, nearly 1″ wide, very finely many-nerved, narrowed at both ends and tipped with a very short, nearly entire beak, the margins smooth or nearly so; scales ovate or ovate-lanceolate, hyaline with a green midvein, acute, somewhat shorter than the perigynia; stigmas 2.

In swamps and wet woods, Newfoundland to Saskatchewan, south to Maryland, Ohio, Michigan and Nebraska. Ascends to 2500 ft. in Vermont. June–Aug.

37. Carex tenuiflòra Wahl. Sparse-flowered Sedge. Fig. 904.

Carex tenuiflora Wahl. Kongl. Vet. Acad. Handl. (II.) **24**: 147. 1803.

Light green, culms very slender or filiform, erect or reclining, rough above, 8″–2° long, loosely caespitose and stoloniferous. Leaves ¾″–1″ wide, flat, usually much shorter than the culm; spikes 2–4, gynaecandrous, subglobose, few-flowered, about 2½″ in diameter, usually bractless, densely aggregated into an ovoid or suborbicular head; perigynia pale, oblong-obovoid, densely puncticulate, coriaceous, obscurely nerved, narrowed at both ends, 1¼″–1¾″ long, a little more than ¾″ wide, almost beakless, spreading, smooth or nearly so; scales white with green midrib, acute or obtusish, about equalling the perigynia; achene nearly filling perigynium; stigmas 2.

In bogs, New Brunswick and Hudson Bay to Manitoba, south to Maine, Massachusetts, central New York and Minnesota. Local. Also in Europe and Asia. Summer.

38. Carex ursìna Dewey. Bear Sedge. Fig. 905.

Carex ursina Dewey, Am. Journ. Sci. **27** : 240. 1835.

C. glareosa var. *ursina* Bailey, Carex Cat. 3. 1884.

Culms low, tufted, erect or slightly curving, 2½′ or
less tall, smooth, from slender rootstocks. Leaves 1″
wide or less, involute toward apex, about equalling
the culm; bracts usually absent; spike solitary (or
rarely with a very small second one at base), brown,
obovoid or suborbicular, 2″–3½″ long, 2″ wide, gynaecandrous, but slightly clavate at base; perigynia 7–15,
appressed, ovate, not margined or serrulate, 1″ long,
a little more than ½″ wide, rounded and stipitate at base,
light colored, weakly nerved, abruptly tipped with a
very minute beak; scales ovate, strongly brownish-
tinged, obtuse, slightly shorter than perigynia; achene
filling perigynium; stigmas 2.

Circumboreal. Summer.

39. Carex Lachenàlii Schk. Arctic Hare's-foot Sedge. Fig. 906.

Carex Lachenalii Schk. Riedgr. 51, *pl. Y, f. 79.* 1801.
Carex lagopina Wahl. Kongl. Vet. Acad. Handl. (II.) **24:**
145. 1803.

Culms stiff, erect, smooth, except immediately beneath head, 3′–16′ tall, from slender rootstocks. Leaves
flat, not involute, ½″–1½″ wide, shorter than the culm,
bracts very short or wanting; spikes 2–6, gynaecandrous,
oblong, dark brown, narrowed at the base, 2½″–5″ long,
1½″–2½″ thick, densely many-flowered, clustered at
the summit or the lower somewhat separated; perigynia
appressed-ascending, elliptic or obovate, 1″–1¾″ long,
firm, lightly several-nerved, narrowed at the base, rather
abruptly tipped by the beak; scales ovate, brown,
hyaline-margined, obtuse to acutish, shorter than the
perigynia; achene filling perigynium; stigmas 2.

Circumboreal, extending south to Quebec and Labrador
and in the western mountains to Colorado and California.
Summer.

Carex hélvola Blytt, supposed to be a hybrid between
this species and *Carex canescens* L., is reported from
Greenland and Labrador.

40. Carex amphígena (Fernald) Mackenzie.
Northern Clustered Sedge. Fig. 907.

C. glareosa var. *amphigena* Fernald, Rhodora **8** : 47. 1906.
C. glareosa Wahl. Flora Danica **14** : *pl. 2430,* and of most
authors.
C. amphigena Mackenzie, Bull. Torr. Club **37** : 246. 1910.

Resembling *Carex glareosa* and *Carex Lachenalii.*
Culms weak and slender, 2′–18′ tall, smooth, except immediately beneath head, from slender rootstocks.
Leaves narrow, ¼″–¾″ wide, involute; spikes 2–8,
gynaecandrous, oblong or subglobose, 2½″–6″ long,
1½″–2″ wide, brown, subtended by very small scale-
like bracts; perigynia 5–10, broadly elliptic, 1¾″ long, ¾″
wide, strongly several-nerved, pale or brownish at
maturity; scales ovate, obtusish, brown with hyaline
margins; achene filling perigynium; stigmas 2.

Circumboreal, extending south along the coast to Labrador and Quebec. Summer. Erroneously illustrated in our
first edition as *Carex glareosa* Wahl.

41. Carex glareòsa Wahl. Weak Clustered Sedge. Fig. 908.

Carex glareosa Wahl. Kongl. Vet. Acad. Handl. (II.) **24**: 146. 1803.

Closely resembles the preceding species, but has weak spreading or reclining culms 2′–12′ long. Leaves narrower, ¼″–¾″ wide, involute; spikes 2 or 3, gynaecandrous, oblong or subglobose, few or several-flowered, 2½″–6″ long, about 1½″ in diameter, brown, subtended by very small scale-like bracts; perigynia lanceolate, 1¾″ long, ½″ wide, tapering at apex into the beak, strongly several-nerved; scales ovate, obtusish, brown, with hyaline margins, exceeded by the perigynia; achene filling perigynium; stigmas 2.

Brackish soil along St. Lawrence, Quebec.   **Very local.** Also in northern Europe.   Summer.

42. Carex Heleonástes Ehrh.   Hudson Bay Sedge.   Fig. 909.

Carex Heleonastes Ehrh.; L. f. Suppl. 414.   1781.

Culms slender, stiff, erect, sharply angled, very rough, 6′–18′ high, from slender, somewhat elongated rootstocks. Leaves rigid, erect, usually becoming involute, 1″ or less wide, shorter than the culm; bracts very short or none; spikes 2–5, subglobose, gynaecandrous, not long-clavate at base, brown, 2″–4½″ long, 2″–3″ wide, clustered at the summit; perigynia 5–10, appressed-ascending, broadly ovate or ovate-elliptic, blunt-edged, faintly several-nerved, 1½″ long, more than ½″ wide, tapering at apex into the short sharp beak; scales ovate, brown with broad hyaline margins, nearly as long as the perigynia; achene filling perigynium; stigmas 2.

Hudson Bay to the Canadian Rocky Mountains. Local. Also in Europe and Asia. Summer.

43. Carex norvégica Willd.   Norway Sedge.   Fig. 910.

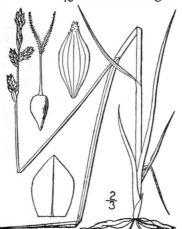

Carex norvegica Willd.; Schk. Riedgr. 50.   1801.

Culms slender but stiff and erect, smooth, 6′–16′ tall, from much elongated rootstocks, stoloniferous. Leaves 1¼″ wide or less, shorter than the culm, glaucous; bracts very short or wanting; spikes 3–6, gynaecandrous, brown, oblong or subglobose, the upper close together, the lower separate, densely many-flowered, 3″–6″ long, 1½″–3″ in diameter, the uppermost conspicuously clavate at base; perigynia ascending, 1¼″–1½″ long, 1″ wide, thick, coriaceous, broadly obovoid, blunt-edged, abruptly narrowed to a stipitate base, brownish, finely many-nerved, abruptly tipped with a very short smoothish beak; scales broadly ovate, reddish brown, obtuse, rather shorter than the perigynia; achene filling perigynium; stigmas 2.

Near salt meadows along coast, Maine and northward. Also in Europe and Asia. Summer.

Carex pseùdohélvola Kihlm., supposed to be a hybrid between this species and *Carex canescens* L., is reported from New Brunswick.

44. Carex canéscens L.   Silvery or Hoary Sedge.   Fig. 911.

Carex canescens L. Sp. Pl. 974. 1753.
Carex canescens var. *subloliacea* Laestad. Nov. Act. Soc.
    Sci. Ups. **11** : 282. 1839.
C. canescens var. *disjuncta* Fernald, Proc. Am. Acad. **37** :
    488. 1902.

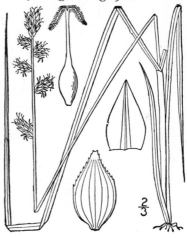

Pale green and somewhat glaucous, culms slender, erect, roughish above, 10'–2½° tall. Leaves flat, 1"–2" wide, usually shorter than the culm; bracts very short or none, or the lowest occasionally bristle-form and longer than its spike; spikes 4–9, gynaecandrous, short-oblong or subglobose, sessile, densely 10–many-flowered, 1½"–6" long, 1½"–2½" in diameter, scattered or the upper close together; perigynia oval or ovate-oval, silvery green or nearly white, faintly few-nerved, appressed-ascending, blunt-edged, from nearly 1" long to nearly 1½" long, ½"–⅛" wide, rough or sometimes smoothish above, tipped with a minute entire or emarginate beak; scales hyaline, ovate, acute or obtuse, slightly shorter than the perigynia; stigmas 2.

In swamps and bogs, Virginia and Ohio, north and north-westward to arctic circle, southward in western mountains. Also in Europe and Asia. Whitish sedge. May–July.

45. Carex brunnéscens (Pers.) Poir.   Brownish Sedge.   Fig. 912.

Carex curta var. *brunnescens* Pers. Syn. **2** : 539. 1807.
C. brunnescens Poir. in Lam. Encycl. Suppl. **3** : 286. 1813.
Carex canescens var. *vulgaris* Bailey, Bot. Gaz. **13** : 86. 1888.
Carex brunnescens gracilior Britton; Brit. & Br. Ill. Fl. **1** :
    351. 1896.

Rather dark green, not glaucous, culms slender, stiff, erect, roughish above, 8'–18' tall. Leaves ½"–1¼" wide, shorter than the culm; lower bract usually present, bristle-form; spikes 4–8, gynaecandrous, subglobose or short-oblong, 4–10-flowered, 2"–6½" long, somewhat scattered, or approximate; perigynia loosely spreading, brown-tinged, usually smaller than those of the preceding species, tipped with a manifest, minutely bidentate, roughish beak about one-fourth as long as the body; scales ovate, membranous, brownish, somewhat shorter than the perigynia; stigmas 2.

In wet or even dry places, mostly at high altitudes, Labrador to British Columbia, New York and New England, on the southern Alleghanies, and the Rocky Mountains. Also in Europe. Ascends to 6600 ft. in North Carolina. Summer.

46. Carex àrcta Boott.   Northern Clustered Sedge.   Fig. 913.

Carex canescens var. *polystachya* Boott; Richards. Arct.
    Exp. **2** : 344. 1852. Not *C. polystachya* Sw. 1803.
Carex arcta Boott, Ill. 155. *pl. 497.* 1867.

Rather light green but scarcely glaucous, culms caespitose, slender, usually strictly erect, 6'–2½° tall, rough above, often overtopped by the leaves which are flat and 1"–2" wide. Lower bract bristle-form and longer than its spike, or short, or wanting; spikes 5–15, oblong, or ovoid, many-flowered, gynaecandrous, 2"–5" long, 2"–3" in diameter, all aggregated into an oblong or ovoid head 7"–15" long; perigynia pale, ovate, broadest near base, many-nerved, ascending or somewhat spreading, 1"–1½" long, white-puncticulate, tapering into a serrulate bidentate beak about one-half as long as the body; scales membranous, usually pale brown, obtusish to short-cuspidate, shorter than the perigynia; stigmas 2.

In swamps and wet woods, Maine and New Brunswick to British Columbia, south to Massachusetts, New York, Minnesota and California. June–July.

47. Carex Deweyàna Schwein.   Dewey's Sedge.   Fig. 914.

Carex Deweyana Schwein. Ann. Lyc. N. Y. **1** : 65.   1824.

Pale green, culms densely caespitose, slender, spreading, slightly angled above, 6'–3° long. Leaves 1"–2½" wide, flat, soft, shorter than the culm; bracts bristle-form, the lower commonly elongated; spikes 2–7, gynaecandrous, ovate-oblong or subglobose, 3–15-flowered, about 2½" in diameter, sessile, distinctly separated or the upper ones contiguous; perigynia lanceolate or ovate-lanceolate, nerveless or nearly so, corky at base, 2"–2½" long, sharply margined above, ¾" wide, the inner face flat, the tapering rough strongly 2-toothed beak about one-half as long as the body; scales nearly white, hyaline with a green midvein, cuspidate or acuminate, equalling the perigynia, or shorter; achenes 1" long; stigmas 2.

In dry woods, Nova Scotia to British Columbia and Vancouver, south to Pennsylvania, Iowa, New Mexico and Arizona. May–July.

48. Carex bromoìdes Schk.   Brome-like Sedge.   Fig. 915.

C. bromoides Schk.; Willd. Sp. Pl. **4** : 258.   1805.

Bright green, culms densely caespitose, slender, erect, very rough above, 1°–2° high. Leaves 1" wide or less, flat, soft, equalling or shorter than the culm; bracts subulate or bristle-form, the lowest commonly elongated, sometimes overtopping the spikes; spikes 3–7, narrowly oblong-cylindric, 3½"–9" long, about 1½" thick, erect or ascending, mostly close together, loosely 6–15-flowered, the staminate flowers basal, or terminal, or both; perigynia narrowly lanceolate, little-margined above, firm, pale, noticeably or strongly several-nerved, 2"–2¾" long, ½" wide, corky at base, the inner face flat, the tapering rough 2-toothed beak at least one-half as long as the body; scales oblong-lanceolate, green, obtusish to acuminate, shorter than the perigynia, brownish-tinged; achenes ⅞" long or less; stigmas 2.

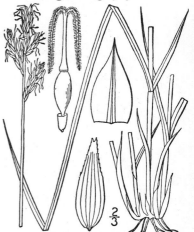

In bogs and swamps, Nova Scotia to Ontario and Michigan, south to Florida and Louisiana. June–Aug.

49. Carex exìlis Dewey.   Coast Sedge.   Fig. 916.

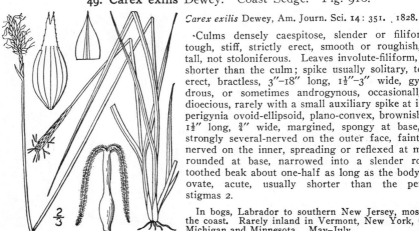

Carex exilis Dewey, Am. Journ. Sci. **14** : 351.   1828.

·Culms densely caespitose, slender or filiform, but tough, stiff, strictly erect, smooth or roughish, 10'–2° tall, not stoloniferous. Leaves involute-filiform, usually shorter than the culm; spike usually solitary, terminal, erect, bractless, 3"–18" long, 1½"–3" wide, gynaecandrous, or sometimes androgynous, occasionally quite dioecious, rarely with a small auxiliary spike at its base; perigynia ovoid-ellipsoid, plano-convex, brownish, about 1½" long, ¾" wide, margined, spongy at base, rather strongly several-nerved on the outer face, faintly few-nerved on the inner, spreading or reflexed at maturity, rounded at base, narrowed into a slender rough 2-toothed beak about one-half as long as the body; scales ovate, acute, usually shorter than the perigynia; stigmas 2.

In bogs, Labrador to southern New Jersey, mostly near the coast. Rarely inland in Vermont, New York, Ontario, Michigan and Minnesota. May–July.

50. Carex elachycàrpa Fernald.   Aroostook Sedge.   Fig. 917.

C. elachycarpa Fernald, Proc. Am. Acad. **37**: 492. 1902.

Kobresia elachycarpa Fernald, Rhodora **5**: 251. 1903.

Densely caespitose, culms slender, stiff, 6′–18′ high, roughened on the angles above. Leaves ¾″–1¼″ wide, flat or slightly involute, shorter than culm; head ½′–1′ long, narrow, of 2–6 approximate spikes, the terminal linear, staminate or gynaecandrous, the lateral suborbicular, 2″–3″ long, gynaecandrous or pistillate, with 7–15 perigynia; bracts not developed; young perigynia lanceolate, plano-convex, 1″ long, few-nerved, rounded at base, tapering into a rough, minutely bidentate beak, ⅓ length of body, the walls fragile, early ruptured by maturing achene elongating; scales ovate, acute, brown, concealing perigynia; achene yellow, nearly 1″ long, ¼″ wide, tipped by persistent style; stigmas 2.

Gravelly beaches of Aroostook River, Fort Fairfield, Maine.   July.   A critical species.

51. Carex intèrior Bailey.   Inland Sedge.   Fig. 918.

Carex interior Bailey, Bull. Torr. Club **20**: 426. 1893.

Similar to *C. Leersii,* culms caespitose, very slender, wiry, rather stiff, erect, 1°–2° tall, slightly roughened above. Leaves ½″–1″ wide, flat, shorter than the culm; bracts very short or lowest occasionally developed; spikes 2–4, the lateral usually pistillate with 1–10 widely spreading perigynia, nearly globular, somewhat separated, 2″ in diameter, the terminal one longer and gynaecandrous or staminate; perigynia brownish at maturity, plano-convex, ovate, broadest near base, 1¼″ long, about ¼″ wide, faintly few-nerved on the outer face, nearly nerveless on the inner, thickened, spongy and rounded at base, contracted into a rough 2-toothed beak one-fourth to one-third as long as the body, its teeth very short, erect, the suture on inner side inconspicuous; scales ovate, usually very obtuse, much shorter than the perigynia.

Wet soil, eastern Quebec to Hudson Bay, British Columbia, Florida and Arizona. May–July.

Carex stérilis Willd. (*C. scirpoides* Schk.) differing by rough-edged perigynia tapering into a very rough beak, and not much exceeding the obtusish scales, the plants often partly or wholly dioecious, occurs from New York and New Jersey to Ontario and Indiana.

52. Carex Hówei Mackenzie.   Howe's Sedge.   Fig. 919.

C. interior capillacea Bailey, Bull. Torr. Club **20**: 426. 1893.

C. scirpoides capillacea Fernald, Rhodora **10**: 47. 1908.

C. delicatula Bicknell, Bull. Torr. Club **35**: 495. 1908. Not C. B. Clarke, 1908.

C. Howei Mackenzie, Bull. Torr. Club **37**: 245. 1910.

Culms caespitose, capillary and slender, spreading, 6′–2° long, roughened above Leaves about ¼″ wide, usually involute, exceeding culms; bracts short; spikes 2–4, the lateral usually pistillate with 1–10 widely spreading perigynia, nearly globular, separated, 2″ in diameter, the terminal longer and gynaecandrous or staminate; perigynia green or brownish at maturity, plano-convex, ovate, broadest at base, 1¼″ long, about ¼″ wide, strongly nerved on outer face, less on inner, spongy and rounded at base, tapering into a rough 2-toothed beak about one-third as long as the body, its teeth very short, erect, the suture on inner side inconspicuous; scales ovate, obtuse to acutish, shorter than perigynia.

Wet soil, Massachusetts and New Hampshire to New York, New Jersey and Pennsylvania. June–July.

53. **Carex Leérsii** Willd.   Little Prickly Sedge.   Fig. 920.

C. Leersii Willd. Prodr. Fl. Berol. 28. 1787.

C. stellulata Good. Trans. Linn. Soc. **2**: 144. 1794.

C. echinata Murr.; Bailey, Proc. Am. Acad. **22**: 142. 1889.

C. cephalantha Bicknell, Bull. Torr. Club **35**: 493. 1908.

Culms slender to stoutish, stiff or in shade weak, erect or rarely spreading, 4′–3° tall, rough, at least above. Leaves ¼″–2″ wide, shorter than the culm; bracts very short or sometimes bristle-form; spikes 2–8, subglobose or short-oblong, closely contiguous to widely separated, about 2½″ thick, 3–40-flowered; staminate flowers basal; perigynia from lanceolate to broadly ovate, plano-convex, ascending when young, 1¼″–2″ long, ½″–1″ wide, spreading or reflexed when old, several-nerved on both faces, the nerves usually not conspicuous on inner face, thickened at base, tapering into a sharp-edged 2-toothed rough beak more than one-half as long as the body, the teeth and suture on inner side conspicuous; scales ovate, hyaline, acutish to acuminate, shorter than the perigynia; stigmas 2.

In moist soil throughout the continent north of Mexico; often locally absent. Also in Europe and Asia. Presenting many forms. May–July.

54. **Carex incompérta** Bicknell.   Prickly Bog Sedge.   Fig. 921.

C. sterilis Willd. Sp. Pl. **4**: 208 (in small part). 1805.

C. sterilis Willd.; Schk. Reidgr. f. 146 (in part). 1806.

C. incomperta Bicknell, Bull. Torr. Club **35**: 494. 1909.

Strongly resembling *C. atlantica,* but more slender, the culms acutely triangular and roughened above, 10′–24′ tall. Leaves usually less than 1″ wide, not stiff, flat or in drying involute, usually exceeding the spikes, the lower less conspicuously shortened; spikes 3–4, spreading, subglobose, 2¼″–3″ in diameter, 6–20-flowered, the staminate flowers numerous at base of terminal one; perigynia brownish at maturity, with suborbicular body, plano-convex, 1½″ long, about 1″ wide, sharp-margined, rounded at base, rather lightly nerved on both faces, spreading or reflexed at maturity, abruptly tipped with a stout, rough, 2-toothed beak, not half as long as the body; scales acute to short-acuminate; stigmas 2.

In boggy places, Massachusetts to Michigan, Pennsylvania and Florida. May–July.

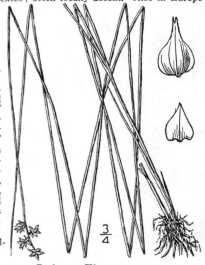

55. **Carex atlántica** Bailey.   Eastern Sedge.   Fig. 922.

Carex atlantica Bailey, Bull. Torr. Club **20**: 425. 1893.

C. sterilis Fernald, Proc. Am. Acad. **37**: 484. 1902.

Similar to *C. Leersii* but stouter, culms obtusely triangular below, more sharply triangular and roughish above, 1°–2½° tall. Leaves 1″–2″ wide, stiff, flat or in drying somewhat involute, the upper sometimes overtopping the spikes, the lower very short and acute; spikes 3–7, spreading, subglobose or short-cylindric, nearly 3″ in diameter, 15–50-flowered, the staminate flowers numerous at the base of the terminal one, or this rarely entirely staminate; perigynia green, with suborbicular body, plano-convex, sharp-margined, 1½″–1¾″ long, 1″–1½″ wide, rounded at the base, strongly nerved on both faces, spreading or reflexed at maturity, abruptly tipped with a stout, rough 2-toothed beak less than half as long as the body, the margins sometimes incurved; scales acutish to acute, shorter than the perigynia.

In swamps, near the coast, Newfoundland to Florida and Texas. Also very rarely inland in Quebec, Maine, New York and Pennsylvania, according to Fernald. June–July.

56. Carex rosaeoìdes E. C. Howe.   Weak Stellate Sedge.   Fig. 923.

C. *rosaeoides* E. C. Howe; Gord. & Howe, Fl. Renssalaer
Co. 33. 1894.

C. *seorsa* E. C. Howe; Gord. & Howe, loc. cit. 39. 1894.

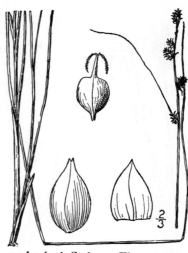

Culms caespitose, slender, weak, flattened, often
spreading or reclining, 7′–20′ tall, roughened on
angles. Leaves 1″–2″ wide, shorter than the culm;
bracts very short, or lowest occasionally developed;
spikes 3–7, the lateral usually pistillate, with 5–20
spreading perigynia, subglobose or short-oblong, more
or less separate, 2″–3½″ long, 2″–3″ wide, the terminal
gynaecandrous, or sometimes entirely staminate,
much longer and long-clavate at base; perigynia
green, plano-convex, ovoid-oval, broadest near middle,
1¼″ long, ¾″ wide, several-nerved on both faces,
spongy and round-tapering at base, abruptly narrowed
into the smooth beak ¼–⅓ length of body, its teeth
short, erect; scales ovate, hyaline, shorter than perigynia; achene in upper part of perigynium.

In swampy woodlands, Massachusetts to New York,
south to Stone mountain, Georgia. May–June.

57. Carex sychnocéphala Carey.   Dense Long-beaked Sedge.   Fig. 924.

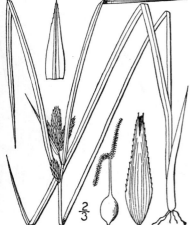

C. *sychnocephala* Carey, Am. Journ. Sci. (II.) 4: 24.
1847.

Culms erect, obtusely triangular, stoutish, smooth,
3′–18′ high. Leaves ¾″–2″ wide, usually shorter than
the culm; lower bracts similar to the leaves, much
elongated, 3′–12′ long, 1″–2½″ wide, nearly erect;
spikes 4–15, greenish or straw-colored, oblong,
densely many-flowered, the larger 4″–6″ long, 2½″–
3½″ wide, aggregated and confluent into an oblong or
ovoid head 1¼′ or less long; perigynia subulate,
substipitate, the margin at base nearly obsolete, 2½″–
3″ long, scarcely ½″ wide at the base, distended over
achene, tapering into a subulate rough 2-toothed beak
2–3 times as long as the few-nerved body; scales
linear-lanceolate, long-acuminate, hyaline, much
shorter and rather narrower than the perigynia.

In meadows and thickets, Ontario and central New
York to Iowa and British Columbia. July–Aug.

58. Carex oronénsis Fernald.   Orono Sedge.   Fig. 925.

C. *oronensis* Fernald, Proc. Am. Acad. 37: 471. 1902.

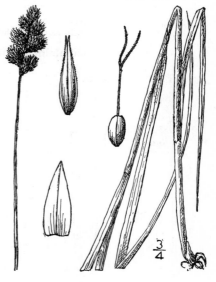

Culms erect, triangular, roughened above, slender, 20′–40′ high, in loose clumps. Leaves 1¼″–2″
wide, shorter than culm; lower one or two bracts
usually developed, but inconspicuous; spikes 3–9,
dark brownish, blunt, densely many-flowered,
obovoid-oblong, 2½″–4½″ long, 2″–3″ wide, loosely
aggregated in an oblong or linear-oblong head
10″–15″ long and 2½″–6″ thick; perigynia erect-
ascending, subulate, the margin at base nearly
obsolete, rounded at base, 2″–2¼″ long, ¼″–¾″ wide
at base, distended over achene, tapering into a
narrow rough 2-toothed beak shorter than the
lightly nerved body; scales dark brown with
lighter midrib and hyaline margins, about as
wide and long as the perigynia.

Dry open places, Orono and Bangor, Maine. June–
July.

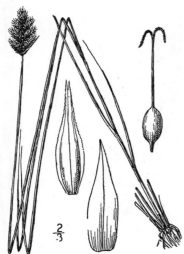

59. Carex Crawfórdii Fernald. Crawford's Sedge. Fig. 926.

C. scoparia var. *minor* Boott, Ill. Car. **3**: 116. *pl. 369.* 1862.
C. Crawfordii Fernald, Proc. Am. Acad. **37**: 469. 1902.
C. Crawfordii var. *rigens* Fernald, Proc. Am. Acad. **37**: 470. 1902.

Culms erect, acutely triangular and somewhat roughened above, slender but stiff, 5′–24′ high, in dense clumps. Leaves ½″–1½″ wide, shorter than the culm; lower one or two bracts usually developed, but inconspicuous, shorter than head; spikes 3–12, light brownish, blunt, densely many-flowered, oblong, 2½″–5½″ long, 1½″–2½″ wide, rather closely aggregated into an ovoid to linear-oblong head usually 6″–13″ long and 2″–4″ wide; perigynia erect-ascending, subulate, the margin at base nearly obsolete, rounded at base, 2″ long, about ½″ wide at the base, distended over achene, tapering into a narrow rough 2-toothed beak, shorter than the obscurely nerved body; scales lanceolate, acute or acuminate, light brown, dull, about as wide as but a little shorter than the perigynia; stigmas 2.

In open places, Newfoundland to British Columbia, Connecticut, Michigan and Washington. June–Sept.

60. Carex scopària Schk. Pointed Broom Sedge. Fig. 927.

Carex scoparia Schk.; Willd. Sp. Pl. **4**: 230. 1805.
C. scoparia var. *moniliformis* Tuckerm. Enum. Method. 8, 17. 1843.
C. scoparia var. *condensa* Fernald, Proc. Am. Acad. **37**: 468. 1902.

Culms slender, erect, roughish above, ½°–2½° tall. Leaves less than 1½″ wide, those of sterile shoots not very numerous, erect or ascending; lower bract bristle-form or wanting; spikes 3–10, oblong, narrowed at both ends, brownish or straw-colored, 3″–8″ long, 2″–3″ in diameter, densely many-flowered, varying from closely aggregated to scattered; staminate flowers basal; perigynia lanceolate, very thin, ascending or erect, 2″–3¼″ long, rather less than 1″ wide, the tips appressed, narrowly wing-margined, several-nerved on both faces, tapering into the serrulate 2-toothed beak; scales thin, brown, acute or acuminate, shorter than the perigynia; achenes ½″ long; stigmas 2.

In moist soil, Newfoundland to Washington, Florida and Colorado. Ascends to 6200 ft. in North Carolina. July–Sept.

61. Carex tribuloìdes Wahl. Blunt Broom Sedge. Fig. 928.

Carex tribuloides Wahl. Kongl. Vet. Acad. Handl. (II.) **24**: 145. 1803.
Carex lagopodioides Schk.; Willd. Sp. Pl. **4**: 230. 1805.
C. tribuloides var. *turbata* Bailey, Mem. Torr. Club **1**: 55. 1889.

Bright green, culms usually stout, erect, roughish above, 1′–3½° tall. Leaves flat, 1½″–4″ wide, shorter than or the uppermost overtopping the culm, those of sterile culms very numerous, widely spreading, the sheaths loose; lower bract bristle-form, sometimes elongated; spikes 6–20, generally obovoid or top-shaped, but varying to suborbicular, blunt, densely clustered or sometimes separated, 3½″–6″ long, 3″–4″ thick; staminate flowers basal; perigynia lanceolate, thin, sometimes distended over achene, greenish brown, flat, ascending or erect, the tips not spreading or recurved, 1¾″–2½″ long, about ¾″ wide, several-nerved on each face, with a sharply 2-toothed, rough wing-margined beak; scales lanceolate, straw-colored, acute, about half as long as the perigynia; achenes short-oblong, ¾″ long; stigmas 2.

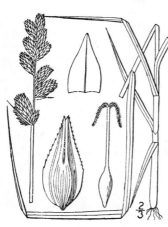

In meadows, New Brunswick to Saskatchewan, Florida and Arizona. Ascends to 2500 ft. in Virginia. July–Sept.

62. **Carex cristatélla** Britton. Crested Sedge. Fig. 929.

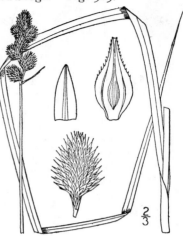

Carex cristata Schwein. Ann. Lyc. N. Y. **1** : 66. 1824.
Not Clairv. 1811.
Carex tribuloides var. cristata Bailey, Proc. Am. Acad.
22 : 148. 1886.
Carex cristatella Britton ; Brit. & Br. Ill. Fl. **1** : 357. 1896.

Culms rather stout, 1°–3° tall, stiff, erect, roughish
above, longer than the leaves. Leaves 1½″–3½″ wide,
those of sterile shoots numerous, spreading, the
sheaths loose ; lower bracts bristle-form, ¼″–1½′ long ;
heads 6–15, globose or subglobose, 2″–4″ in diameter,
all densely aggregated into an oblong head 1′ long or
more or the lower slightly separated ; staminate flowers basal ; perigynia rather broadly lanceolate, distended over achene, spreading or ascending, squarrose when mature, green or greenish brown, 1½″–2″
long, ¾″ wide, narrowly wing-margined, several-
nerved on both faces, tapering into a serrulate
2-toothed beak ; scales lanceolate, straw-colored,
much shorter than the perigynia ; achenes ¾″ long.

In meadows and thickets, eastern Massachusetts to
British Columbia, south to Virginia and Missouri. July–
Sept.

63. **Carex projécta** Mackenzie. Necklace Sedge. Fig. 930.

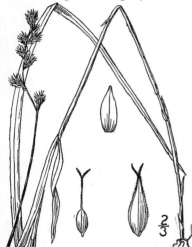

C. tribuloides var. reducta Bailey, Proc. Am. Acad. **22** :
148. 1886.
C. tribuloides moniliformis Britton ; Brit. & Br. Ill. Fl.
1 : 356. 1896.
C. projecta Mackenzie. Bull. Torr. Club **35** : 264. 1908.

Culms erect, triangular and roughened above,
slender and weak, 1½°–3° high, in large clumps.
Sterile culms leafy ; leaves with long loose sheaths,
blades 1½″–3½″ wide, shorter than culm ; lower
bracts inconspicuous ; spikes 8–15, straw-colored,
with 15–30 perigynia, suborbicular, blunt, clavate at
base, 2½″–4″ long, nearly as wide, alternately and
usually loosely arranged and forming a slender
flexuous head 1′–2′ long ; perigynia ascending-
spreading with divergent beaks, lanceolate, wing-
margined to the round-tapering base, 1½″–2½″ long,
¾″ wide at base, distended over achene, tapering into
a rough 2-toothed beak, shorter than the nerved
body ; scales ovate-lanceolate, obtuse to acutish,
straw-colored, narrower and shorter than the perigynia ; achene ¾″ long.

Damp soil, Nova Scotia to North Dakota, south to
District of Columbia and Illinois. May–July.

64. **Carex muskinguménsis** Schwein. Muskingum Sedge. Fig. 931.

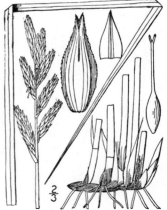

Carex muskingumensis Schwein. Ann. Lyc. N. Y. **1** : 66. 1824.
Carex arida Schwein. & Torr. Ann. Lyc. N. Y. **1** : 312. 1825.

Culm stout, stiff, erect, rough above, 2°–3° tall. Leaves
flat, long-pointed, 1½″–2½″ wide, subcordate at base, shorter
than the fertile culms, those of sterile culms very numerous, crowded near the summit, somewhat distichous ;
bracts very short and scale-like ; spikes 5–12, oblong-
cylindric, densely many-flowered, 7½″–13″ long, 2½″–3½″
in diameter, erect, approximate, pale brown, narrowed and
staminate at the base ; perigynia narrowly lanceolate,
closely appressed, 3½″–5″ long and 1¼″ wide, strongly
several-nerved, very flat, narrowed to both ends, scarious-
margined, rough-ciliate, the beak strongly bidentate ; scales
ovate-lanceolate, obtusish or acute, about one-half as long
as the perigynia ; achene linear-oblong, 1¼″ long ; stigmas 2.

In moist woods and thickets, Ohio to Manitoba, Missouri and
eastern Kansas. June–Aug.

65. Carex Bébbii Olney. Bebb's Sedge.
Fig. 932.

C. tribuloides var. *Bebbii* Bailey, Mem. Torr. Club 1 : 55.
1889.
C. Bebbii Olney ; Bailey, Bot. Gaz. 10 : 379. 1885.

Culms erect, acutely triangular and roughened above,
rather slender, 8′-2½° high, in dense clumps. Leaves
1″-2¼″ wide, shorter than the culm; lower one or two
bracts usually developed but inconspicuous; spikes
usually 5-10, brownish-tinged, blunt, densely many-
flowered, subglobose to broadly ovoid, 2″-4½″ long,
1½″-3″ wide, aggregated into an oblong or linear-
oblong head 7″-14″ long, 4″-6″ thick; perigynia ascending, narrowly ovate, wing-margined to the
rounded base, 1½″-2″ long, ¾″-1″ wide at base, distended over achene, tapering into a rough 2-toothed
beak, less than half length of the obscurely nerved
body; scales oblong-ovate, acute or short-acuminate,
brownish, nearly as wide as but shorter than perigynia;
stigmas 2.

In low grounds, Newfoundland to British Columbia and
northward, southward to New Jersey, Illinois and Colorado. June–August.

66. Carex stramínea Willd. Straw Sedge. Dog-
grass. Fig. 933.

Carex straminea Willd.; Schk. Riedgr. 49. *f. 34.* 1801.
Carex tenera Dewey, Am. Journ. Sci. 8 : 97. 1824.

Culms very slender, roughish above, 1°-2½° long, the
top often nodding. Leaves 1″ wide or less, long-pointed,
shorter than the culm; bracts short or the lower bristle-
form and exceeding its spike; spikes 3-8, subglobose or
slightly obovoid, 2″-2½″ thick, light brown or greenish,
separated on the commonly zigzag rachis, or contiguous,
usually clavate at base; staminate flowers basal; perigynia
widely spreading to ascending, narrowly to broadly ovate,
green, 1½″-2″ long, ¾″-1¼″ wide, strongly several-nerved
on the outer face, fewer-nerved or nerveless on the inner,
wing-margined, much distended over achene, tapering into
the rough 2-toothed beak; scales lanceolate, acute, somewhat shorter and narrower than the perigynia; stigmas 2.

In woods, New Brunswick to British Columbia, Kentucky,
Arkansas and California. June–July.

67. Carex normàlis Mackenzie. Larger Straw Sedge. Fig. 934.

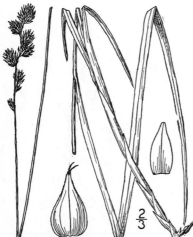

C. mirabilis Dewey, Am. Journ. Sci. 30 : 63, *pl. Bb. f. 92.*
1836. Not Host, 1809.
C. straminea var. *mirabilis* Tuckerm. Enum. Meth. 18.
1843.
C. mirabilis var. *perlonga* Fernald, Proc. Am. Acad. 37 :
473. 1902.
C. normalis Mackenzie, Bull. Torr. Club 37 : 244. 1910.

Culms erect or sometimes weak and spreading,
triangular, roughened above, 2°-3½° high, in dense
clumps. Leaves 1¼″-3″ (averaging 2″) wide, much
shorter than the culm; lower one or two bracts somewhat developed; spikes 4-12, green or brownish-
tinged, blunt, with 10-30 perigynia, subglobose, 3″-
4½″ long, 2½″-4″ wide, usually clavate at base, separate or aggregated into a head 1′-2′ long and 5″ wide
or occasionally in a moniliform head; perigynia
spreading, thickish, ovate, wing-margined, rounded at
base, 1½″-2″ long, ¾″-1″ wide near base, distended
over achene, conspicuously nerved on outer, fewer-
nerved or nerveless on inner face, tapering into a
rough 2-toothed beak about half the length of body;
scales ovate, nearly width of but shorter than perigynia.

Woodlands, Quebec to North Carolina, Kansas and Manitoba and in the western mountains.

68. Carex maclovìàna D'Urv. Falkland Island Sedge. Fig. 935.

C. macloviana D'Urv. Mém. Soc. Linn. Paris 4 : 599. 1826.

Strongly caespitose, the culms stout, stiff, 6″–15′ high, slightly roughened on the angles above. Leaves flat, 1½″–2″ wide, usually much shorter than the culm; head ¾′ long, short-oblong or ovoid, of 3–8 densely clustered ovoid-oblong or subglobose gynaecandrous spikes 2″–4″ long, 2″–3″ wide, each with 10–25 closely appressed perigynia; bracts small or not developed; perigynia ovate, brownish, much flattened and thin, but distended over achene, about 2″ long and 1″ wide, few-nerved on outer, nerveless on inner surface or nearly so, round-tapering at base, abruptly narrowed into a serrulate obscurely bidentate beak about one-third length of body; scales ovate, obtuse to acute, slightly shorter and narrower than perigynia, brown-ish-black with strongly developed white hyaline some-times incurved margins; stigmas 2.

Labrador and Greenland. Also in Lapland and in south-ern South America. July–August. Closely related to *Carex festiva* Dewey of the Rocky Mountain region.

69. Carex festucàceà Schkuhr. Fescue Sedge. Fig. 936.

Carex festucacea Schkuhr; Willd. Sp. Pl. 4: 242. 1805.
Carex straminea var. *brevior* Dewey, Am. Journ. Sci. 11: 158. 1826.
C. straminea var. *festucacea* Tuck. En. Meth. 18. 1843.

Culms slender or rather stout, smooth or roughened beneath head, stiff, strictly erect, 1°–4° tall. Leaves rather stiff, erect, 1″–2″ wide, shorter than the culm; sheaths with a conspicuous pale band and membranous auricle; spikes 3–10, green-brown or light-brown, oblong or nearly globular, clustered at the summit but not at all confluent, or the lower separate, 2″–4½″ in diameter, 3½″–7½″ long, rounded or clavate at base; bracts short or wanting; perigynia varying from orbicular to ovate, broadly wing-margined, 1¼″–1¾″ in diameter, 2″–2¾″ long, thickish, somewhat spreading or ascending, strongly nerved on outer face, faintly on inner face, the roughish beak about one-third the length of the body; scales lanceolate or ovate-lanceolate, acute or obtusish, rather shorter and narrower than the peri-gynia; stigmas 2.

In dry or moist soil, New Brunswick to British Columbia, south to Florida and Arkansas. May–July.

70. Carex Bicknéllii Britton. Bicknell's Sedge. Fig. 937.

Carex straminea var. *Crawei* Boott, Ill. 121. *pl. 388.* 1862. Not *C. Crawei* Dewey, 1846.
Carex Bicknellii Britton; Brit. & Br. Ill. Fl. 1 : 360. 1896.

Culms loosely tufted, 2°–4° high, erect or the top inclined, roughish above, much longer than the leaves. Leaves 4′–12′ long, 1¼″–2¼″ wide towards base; bracts usually very short; spikes 3–7, ovoid, subglobose, or somewhat obovoid, 4″–9″ long, 3″–6″ broad, approximate, or the lower separated, brown-ish or greenish or straw-colored, staminate at the base, the head stiff, erect; perigynia spreading-ascend-ing, very broadly ovate or suborbicular, thin, very prominently (about 16) nerved on both faces, 2¾″–3¾″ long, nearly 1½″–2″ wide, the membranous wing very broad, the rough 2-toothed beak one-fourth to one-half as long as the body; scales lance-ovate, obtuse or acute, straw-colored or brownish with hyaline margins, shorter and much narrower than perigynia; stigmas 2.

In dry soil, Maine to Manitoba, south to New Jersey, Arkansas and Nebraska. June–July.

71. **Carex hormathòdes** Fernald. Marsh Straw Sedge. Fig. 938.

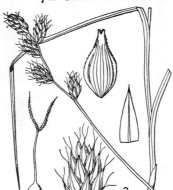

C. straminea var. *invisa* W. Boott, Coult. Bot. Gaz. **9**: 86. 1884.
C. tenera var. *Richii* Fernald, Proc. Am. Acad. **27**: 475. 1902.
C. hormathodes Fernald, Rhodora **8**: 165. 1906.

Culms very slender, erect or the summit nodding, slightly angled and often strongly roughened above, 1°–3° high. Leaves shorter than the culm, usually less than 1″ wide, tapering to a very long tip; bracts, when present, very narrow and bristle-form; spikes 3–9, ovoid, obtuse or short-pointed, densely many-flowered, separated or the upper contiguous, forming a slender moniliform head, greenish brown or brown at maturity, 3½″–8″ long, staminate and commonly much contracted at the base; perigynia narrowly to broadly ovate, ascending, or with somewhat spreading tips, 2″ to nearly 3″ long, 1″–1½″ wide, strongly about 10-nerved on both faces, wing-margined, the rough beak about half as long as the body; scales lanceolate, long-acuminate or aristate, nearly as long as the perigynia, but much narrower; stigmas 2.

In wet soil, chiefly near coast, Gulf of St. Lawrence to Virginia, locally inland to Ontario and recorded from Iowa; also on Pacific coast. May–June. Illustrated in our first edition as *C. ténera* Dewey.

72. **Carex suberécta** (Olney) Britton. Prairie Straw Sedge. Fig. 939.

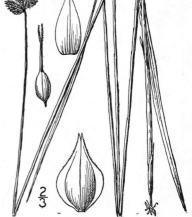

C. foenea var. *ferruginea* Gray, Man. Ed. 5, 580. 1867. Not *C. ferruginea* Scop.
C. tenera var. *suberecta* Olney; Bailey, Proc. Am. Acad. **22**: 149 (as synonym). 1889.
C. suberecta Britton, Man. Ed. 2, 1057. 1905.

Culms erect, slender, acutely triangular and strongly roughened, 2°–3° high. Leaves 1″–1½″ wide, shorter than the culm; lower one or two bracts usually developed, but shorter than the head and inconspicuous; spikes 2–5, silvery greenish or slightly brownish-tinged, short-pointed or rounded, densely many-flowered, ovoid, 3½″–7″ long, 2½″–4″ wide, approximate in a head 7″–15″ long, 4″–7″ thick; perigynia erect, strictly appressed, ovate, strongly margined, rounded at base, 2″–2½″ long, slightly more than 1″ to nearly 1½″ wide at base, distended over achene, tapering gradually into a rough 2-toothed beak ¼–⅓ the length of the nerveless or obscurely nerved body; scales ovate, short-acuminate to obtusish, silvery-green, or in age ferruginous, shorter and rather narrower than the perigynia; stigmas 2.

Moist places, Ontario and Ohio to Michigan, Illinois and Iowa. May–July.

73. **Carex alàta** Torr. Broad-winged Sedge. Fig. 940.

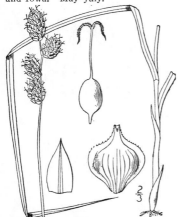

Carex alata Torr. Ann. Lyc. N. Y. **3**: 396. 1836.

Culms stiff, rather stout, strictly erect, slightly roughish above, 1°–3½° tall. Leaves grass-like, but somewhat rigid, 1″–2″ wide, shorter than the culm, sheaths green nearly to chartaceous auricle; spikes suborbicular to oblong-conic, whitish-green or in age brownish-green, very densely many-flowered, 5″–8″ long, 3″–5″ thick, the lateral rounded or little clavate at base, pointed or obtuse at the summit, all distinct but usually little separated, bractless, or short-bracted; perigynia orbicular or obovate-orbicular, very broadly winged, 2″–2½″ long, nearly 1½″ or more broad, firm, faintly few-nerved or almost nerveless on inner face, erect and appressed, or somewhat curved upward, the short, abrupt beak about one-third as long as the body; scales lanceolate, acuminate or aristate, scarcely shorter and much narrower than the perigynia; achene distinctly stipitate; stigmas 2.

In moist soil, New Hampshire to Florida, inland to Michigan. May–June.

74. Carex albolutéscens Schwein.   Greenish-white Sedge.   Fig. 941.

Carex albolutescens Schwein. Ann. Lyc. N. Y. 1 : 66.   1824.
Carex straminea var. *foenea* Torr. Ann. Lyc. N. Y. 3 : 395.
   1836.   Not *C. foenea* Willd. 1809.
Carex albolutescens var. *cumulata* Bailey, Bull. Torr. Club
   20 : 422.   1893.

Similar to the preceding species, but usually lower,
culms 1°–2½° tall, stout, strictly erect, slightly rough
above.   Leaves 1″–2″ wide, shorter than the culm;
bracts filiform or wanting; spikes 3–8, or sometimes
more numerous and somewhat compound, oblong, silvery green when young but becoming light brownish,
the lateral rounded or little clavate at base, 3″–6″ long,
mostly less than 3″ thick, clustered, but distinct, the
lowest sometimes separated, and very rarely stalked;
perigynia broadly ovate, firm, broadly winged, faintly
to strongly nerved on both faces, appressed, 1½″–2″
(rarely 2¼″) long, 1″–1½″ wide, the roughish beak about
one-third as long as the body; scales lanceolate, obtuse
or acutish, nearly as long as the perigynia, but much
narrower; achene nearly or quite sessile; stigmas 2.

In wet soil, along coast, New Brunswick to Venezuela;
also about the Great Lakes, on the Pacific coast, and at a few inland stations from Maine to New
York.   Bermuda.   May–July.

75. Carex silícea Olney.   Sea-beach Sedge.   Fig. 942.

C. straminea var. *moniliformis* Tuckerm. Enum. Meth. 17.   1843.
   Not *C. scoparia* var. *moniliformis* Tuckerm. 1843.
Carex foenea var. *sabulonum* A. Gray, Man. Ed. 5, 580.   1867.
   Not *C. sabulosa* Turcz. 1837.
Carex silicea Olney, Proc. Am. Acad. 7 : 393.   1868.

Culms slender, rather stiff, erect but the summit recurved or nodding, slightly roughish above, 1°–3° tall.
Leaves 1″–2″ wide, involute in drying, shorter than the
culm; bracts scale-like; spikes 3–8, or rarely more, ovoid-
conic or ovoid-oblong, silvery-green, nearly white or in age
becoming brownish, erect, conspicuously clavate and staminate at the base, 3″–10″ long, 2″–3½″ thick, all separated
or the uppermost close together, forming a flexuous moniliform head 1½′–3½′ long; perigynia with oval or obovate
body, firm, short-beaked, nerved on both faces, wing-margined, closely appressed, 2″–2½″ long, 1¼″–1½″ wide, longer
and much broader than the lanceolate acute scales;
stigmas 2.

In sands of the sea coast, Newfoundland to Virginia.   June–
Aug.

76. Carex leporìna L.   Hare's-foot Sedge.   Fig. 943.

Carex leporina L. Sp. Pl. 973.   1753.
C. ovalis Gooden. Trans. Linn. Soc. 2 : 148.   1794.

Culms slender, erect, stiff, roughish above, 6′–1½° tall,
caespitose.   Leaves 1″–1½″ wide, flat, shorter than the
culm, not bunched at base; bracts very short and scale-
like or wanting; spikes 3–7, ovoid or elliptic, blunt at
the summit, rounded and staminate at the base, 3½″–7″
long, 2½″–4″ thick, many-flowered, dark brown, shining,
clustered but distinct, in a terminal oblong head; perigynia appressed-ascending, ovate, 2″ long, nearly 1″
wide, rather narrowly wing-margined, several-nerved
on outer face, nerveless or lightly nerved on inner, the
rough tapering 2-toothed beak nearly as long as the
body; scales lanceolate, brown, with narrow hyaline
margin, acute, about as wide and as long as the perigynia; stigmas 2.

In dry places, Newfoundland to Massachusetts and New
York.   In ballast southward.   Europe and Asia.   June–Aug.

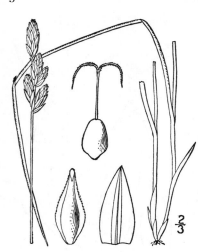

77. **Carex xerántica** Bailey.　White-scaled Sedge.　Fig. 944.

C. xerantica Bailey, Coult. Bot. Gaz. **17**: 151.　1892.
C. foenea var. *xerantica* Kükenthal, Pflanzenreich **38**: 205.　1909.

Culms caespitose, stiff, 1°–2° tall, strictly erect, smooth or little roughened above. Leaves 1″–1½″ wide, somewhat involute in drying, clustered toward the base, shorter than the culm; spikes 3–6, ellipsoid, densely many-flowered, close together or the lower slightly separated, 4″–7″ long, 2½″ in diameter, tapering at base, the staminate flowers basal; bracts scale-like; perigynia lanceolate-ovate, pale, 2″–2¾″ long, 1″–1¼″ wide, closely appressed, nerveless or nearly so on inner face, bright yellow at base, wing-margined, the rough tapering beak shorter than the body; scales with broad, silvery white margins and darker center, acute, equalling or a trifle longer than the perigynia and rather wider; stigmas 2.

Prairies, western Manitoba to Athabasca and Kansas.　May–July.

78. **Carex adústa** Boott.　Browned Sedge. Fig. 945.

Carex adusta Boott; Hook. Fl. Bor. Am. **2**: 215.　1840.
C. pinguis Bailey, Bull. Geog. Surv. Minn. **3**: 22.　1887.

Culms stout, stiff, erect, smooth, 1½°–2½° tall, caespitose. Leaves 1″–1½″ wide, long-pointed, shorter than the culm; bracts subulate, tapering from a broad nerved base, the lower 1 or 2 usually elongated; spikes 3–15, subglobose or short-oval, several-flowered, 3″–6″ long, 2″–3″ wide, densely clustered and apparently confluent, or slightly separated, brownish in age; staminate flowers basal; perigynia broadly ovate, firm, narrowly wing-margined, 2″–2½″ long, 1″–1½″ wide, narrowed into a 2-toothed rough beak, several-nerved on the outer face, nerveless on the inner, loosely ascending; scales ovate, acute or acuminate, about equalling the perigynia in length and width; achene 1″ broad; stigmas 2.

In dry soil, Newfoundland to southern Maine, Michigan, Minnesota and northwestward.　June–July.

79. **Carex praticola** Rydb.　Northern Meadow Sedge.　Fig. 946.

Carex pratensis Drejer, Rev. Crit. Car. 24.　1841. Not Host, 1797.
C. praticola Rydb. Mem. N. Y. Bot. Gard. **1**: 84.　1900.

Light green; culms slender, erect when young, the summit later nodding, slightly roughened above, caespitose, 10′–2° tall. Leaves ½″–1″ wide, shorter than the culm; lower bract bristle-form, usually short; head flexuous and moniliform; spikes 2–6, oblong, usually clavate at base, separated or the upper contiguous, silvery-brown and shining, 3″–8″ long, about 2½″ in diameter, several-flowered, the staminate flowers basal; perigynia lanceolate, closely appressed, thin, pale, nerveless or nearly so on the inner face, few-nerved on the outer, 2¼″–3¼″ long, nearly 1″ wide, wing-margined, tapering into a beak nearly as long as the body; scales brownish-tinged, with very broad white-hyaline margins, obtuse to acute, about as long and as wide as the perigynia.

Northern Maine to western Ontario, Michigan and Oregon, north to Greenland and Alaska, south in the Rocky Mountains to Colorado.　Summer.

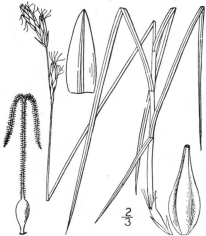

80. Carex aènea Fernald.   Fernald's Hay Sedge.
Fig. 947.

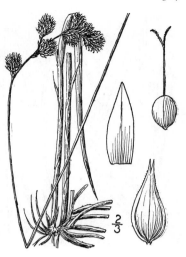

Carex foenea var. *sparsiflora* Howe, Rep. N. Y. Mus. Nat. Hist. **48** : 44. 1895. Not *C. sparsiflora* Fries.
Carex aenea Fernald, Proc. Am. Acad. **22** : 480. 1902.

Culms slender, nodding, $1\frac{1}{2}°-3°$ high, smooth except immediately below head. Leaves $1\frac{1}{4}''-2''$ wide, shorter than the culm; lower one or two bracts present but not conspicuous; spikes 3–12 in a moniliform or loose head $1\frac{1}{2}''-3'$ long, all separate or upper aggregated, oblong, $3\frac{1}{2}''-12''$ long, $2\frac{1}{2}''-3\frac{1}{2}''$ thick, rounded at apex, clavate at base, densely many-flowered; perigynia appressed-ascending, or loosely ascending in age, ovate, narrowly wing-margined, rounded at base, $2''-2\frac{1}{2}''$ long, $1''-1\frac{1}{4}''$ wide, tapering into a rough 2-toothed beak less than half the length of the nerveless or obscurely nerved body; scales ovate, acute or short-acuminate, white-hyaline with darker center, as wide and as long as perigynia; stigmas 2.

In dry places, Labrador to Connecticut, west to Michigan and British Columbia. May–July.

81. Carex foènea Willd.   Hay Sedge.   Fig. 948.

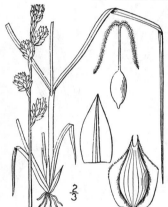

Carex foenea Willd. Enum. 957. 1809.
Carex argyrantha Tuckerm.; Wood, Class-book, 753. 1860.
Carex foenea var. *perplexa* Bailey, Mem. Torr. Club **1** : 27. 1889.

Rather light green, culm little roughened above, erect or the summit nodding, $1°-3\frac{1}{2}°$ tall. Leaves flat, soft, $1''-2''$ wide, shorter than the culm; bracts very short or wanting; inflorescence usually moniliform or flexuous, not stiff; spikes 4–15, subglobose or short-oblong, narrowed at the base, $3''-9''$ long, $2\frac{1}{2}''-3''$ in diameter, silvery green, all separated or the upper contiguous; staminate flowers basal; perigynia ovate, thin, $1\frac{1}{2}''-2\frac{1}{4}''$ long and about $1''$ wide, wing-margined, strongly several-nerved on both faces, tapering into a short rough 2-toothed beak half the length of body or less; scales silvery hyaline with darker center, ovate, acute to acuminate, about equalling the perigynia in length and concealing them; stigmas 2.

In dry woods, often on rocks, Newfoundland to British Columbia, south to Virginia and Iowa. May–July.

82. Carex Willdenòvii Schk.   Willdenow's Sedge.   Fig. 949.

Carex Willdenovii Schk.; Willd. Sp. Pl. **4** : 211. 1805.

Glabrous and pale green, culms from very short to 10' high, little serrulate. Leaves much elongated, nearly erect, rather stiff, $1''-1\frac{1}{2}''$ wide, $4'-15'$ long, very much overtopping the spikes, lowest reduced to bladeless sheaths; spikes 1–5, androgynous, or sometimes completely staminate, $1'$ long or less, the uppermost on filiform stalks $3'-7'$ long, the lower often appearing nearly basal, the stalks much shorter; pistillate flowers 3–9; body of the perigynium oblong, smooth, $1''-1\frac{1}{2}''$ long, rather less than $1''$ thick, narrowed into a flattened 2-edged rough beak of about its own length; scales acute, acuminate or awned, finely several-nerved, the lower 1 or 2 commonly bract-like and often foliaceous and overtopping the spike; stigmas 3.

In dry woods and thickets, Massachusetts to Ohio, Michigan and Manitoba, south to Florida, Kentucky and Texas. April–July.

83. Carex Jàmesii Schwein. James' Sedge. Fig. 950.

Carex Jamesii Schwein. Ann. Lyc. N. Y. 1: 67. 1824.
Carex Steudelii Kunth, Enum. 2: 480. 1837.

Similar to the preceding species, but the leaves rather narrower, soft, spreading or ascending, very much surpassing the spikes, the lowest mere clasping sheaths. Spikes androgynous, one or more of them filiform-stalked, the terminal staminate portion very slender, the pistillate flowers usually 2 or 3 and slightly separated; body of the perigynium subglobose, 1" in diameter, contracted at the base, abruptly tipped by a subulate rough beak of more than its own length; lower scales bract-like, foliaceous, commonly much overtopping the spike, the upper shorter and sometimes not exceeding the perigynia; stigmas 3.

In dry woods and thickets, southern Ontario and New York to Michigan and Iowa, south to West Virginia, Missouri and Kansas. April–May.

84. Carex durifòlia Bailey. Back's Sedge. Fig. 951.

Carex Backii Boott; Hook. Fl. Bor. Am. 2: 210. *pl. 209.* 1840. Not *C. Backana* Dewey, 1836.

Carex durifolia Bailey, Bull. Torr. Club 20: 428. 1893.

Similar to the preceding species, glabrous, culms from very short to 10' high. Leaves ascending or spreading, 6'–12' long, 1¼"–3" wide, very much overtopping the spikes; spikes 1–3, nearly basal, androgynous, 1 or 2 of them long-stalked, the staminate flowers few, inconspicuous, the pistillate 2–6, subtended by leafy bract-like elongated scales which nearly enclose the inflorescence; perigynia oval, smooth, gradually tapering into a stout two-edged beak nearly or quite as long as the body, which is about 1½" long and 1" thick; stigmas 3.

In woods and thickets, eastern Quebec to Assiniboia, south to Massachusetts, New York, Ohio and Nebraska. May–June.

85. Carex leptàlea Wahl. Bristle-stalked Sedge. Fig. 952.

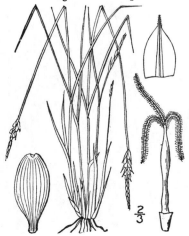

Carex leptalea Wahl. Kongl. Vet. Acad. Handl. (II.) 24: 139. 1803.
Carex polytrichoides Muhl.; Willd. Sp. Pl. 4: 213. 1805.
Carex Harperi Fernald, Rhodora 8: 181. 1906.

Light green and glabrous, culms filiform, smooth, erect or spreading, 6'–24' long. Leaves very narrow, mostly shorter than the culm; spike solitary, terminal, androgynous, narrowly linear, 2"–8" long, 1"–1½" thick; perigynia few, narrowly oblong, light green, finely many-nerved, narrowed at the base, obtuse and beakless at the summit, 1½"–2½" long, ½"–¾" thick; scales membranous, the lowest cuspidate, sometimes attenuated into a subulate awn nearly as long as the spike, the upper short-acuminate to very obtuse, much shorter than perigynia; stigmas 3.

In bogs and swamps, Newfoundland to Alaska, Florida, Louisiana, Texas, Colorado and Oregon. Ascends to 4300 ft. in North Carolina. June–Aug.

86. Carex paucifròra Lightf.   Few-flowered Sedge.   Fig. 953.

Carex pauciflora Lightf. Fl. Scot. 543. *pl. 6.* 1777.

Glabrous, culms from slender long running root-stocks, erect or assurgent, very slender, 3′–2° high, with two or three developed leaves. Leaves very narrow, usually shorter than the culm, the lowest reduced to sheaths; spike solitary, androgynous, the staminate and pistillate flowers each 1–6; perigynium green, narrow, scarcely inflated, 3″–4″ long, about ½″ in diameter, obscurely several-nerved, tapering from below the middle into a very slender beak with oblique orifice, strongly reflexed and readily detachable when mature, 2–3 times longer than the deciduous lanceolate or ovate scale; achene linear-oblong; stigmas 3.

In bogs, Newfoundland to Alaska, south to Connecticut, Pennsylvania, Michigan and Washington. June–Aug.

87. Carex microglòchin Wahl.   False Uncinia.   Fig. 954.

Carex microglochin Wahl. Kongl. Acad. Handl. (II.) 24 : 140. 1803.
Uncinia microglochin Spreng. Syst. 3 : 830. 1826.

Culms slender, from slender elongated rootstocks, weak, 4′–12′ high, with four to eight developed leaves. Leaves very narrow, shorter than the culm; spike solitary, 3½″–8″ long, androgynous, usually pistillate for more than one-half its length; scales oblong-lanceolate, 1-nerved, deciduous; perigynia 3–10, very narrowly lanceolate, 2″–3″ long, less than ½″ thick, strongly reflexed in fruit, obscurely nerved, tapering into the long smooth beak, the orifice oblique; achene linear-oblong, obtusely 3-angled, much shorter than the perigynium; racheola bristle-like, long-exserted beyond the orifice of the perigynium.

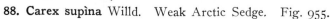

Greenland to James Bay and British Columbia; Colorado?. Also in the arctic and mountainous parts of Europe and Asia. Summer.

88. Carex supìna Willd.   Weak Arctic Sedge.   Fig. 955.

Carex supina Willd.; Wahl. Kongl. Vet. Acad. Handl. (II.) 24 : 158. 1803.

Glabrous, densely tufted, long-stoloniferous, culms slender, erect, sharply 3-angled, 3′–10′ tall. Leaves ½″–¾″ wide, flat, shorter than the culm, roughened toward apex; lower bract short, subulate; staminate spike solitary, sessile or very nearly so, 3″–7″ long; pistillate spikes 1–3, sessile, approximate, subglobose or short-oblong, usually 4–10-flowered, 2″–4″ long, 2″ wide, the upper one sometimes consisting of only 1–3 flowers; perigynia oval-obovoid, smooth, hard, shining, nerveless, 1¼″–1¾″ long, less than ½″ thick, obscurely 3-angled, tipped with a very short, obliquely cut beak; scales ovate, brownish with hyaline margins, obtuse to short-cuspidate, about length of perigynia; stigmas 3.

Northern Minnesota (according to Bailey) and Manitoba to arctic America and Greenland. Also in northern Europe and Asia. Summer.

89. Carex rupéstris All.   Rock Sedge.   Fig. 956.

Carex rupestris All. Fl. Ped. **2**: 264. *pl. 92. f. 1.*   1785.

Carex Drummondiana Dewey, Am. Journ. Sci. **29**: 251. 1836.

Culms slender, obtusely 3-angled, erect, 1′–6′ tall. Leaves ½″–1″ wide, involute in drying, often curved, shorter than or exceeding the culm; bract wanting; spike solitary, androgynous, slender, 6″–12″ long, the pistillate part loosely few-flowered; perigynia erect, smooth, oblong-obovoid, triangular, long-stipitate, firm, faintly nerved, 1½″–2″ long, abruptly very short-beaked, the beak truncate; scales purple-brown, ovate, obtuse or subacute, wider and longer than the perigynia; stigmas 3.

Quebec and Greenland to British Columbia, south in the Rocky Mountains to Colorado. Also in northern Europe and Asia.   Summer.

90. Carex filifòlia Nutt.   Thread-leaved Sedge.   Fig. 957.

Carex filifolia Nutt. Gen. **2**: 204.   1818.

Densely tufted, pale green and glabrous, culms slender but wiry, smooth, erect, 3′–14′ tall, equalling or longer than the leaves. Leaves filiform, rather stiff, scarcely ¼″ wide, their sheaths persistent and ultimately fibrillose; spike solitary, erect, bractless, androgynous, 3″–15″ long, the pistillate part about 2″ in diameter; perigynia 5–10, ovoid-oval, obtusely triangular, nearly nerveless, closely enveloping achene, puberulent at least above middle, 1½″ long, rather more than ½″ thick, tipped by a short cylindric hyaline entire beak; scales very broad and enveloping perigynia, concave with wide white scarious margins, obtuse or cuspidate, about as long as the perigynia; stigmas 3.

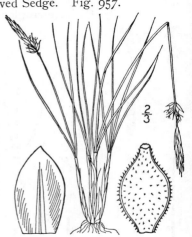

In dry soil, Manitoba to British Columbia, south to Kansas, Colorado and California.   May–July.

91. Carex scirpoìdea Michx.   Scirpus-like Sedge.   Fig. 958.

Carex scirpoidea Michx. Fl. Bor. Am. **2**: 171.   1803.

Dioecious, foliage glabrous, rather bright green, culms from elongated rootstocks, aphyllopodic, erect, slender but stiff, 6′–18′ tall. Leaves ½″–1″ wide, nearly erect, usually much shorter than the culm; spike solitary or rarely with an additional and very small one near its base, linear-cylindric, densely many-flowered, 8″–15″ long, 1½″–2½″ in diameter, subtended by a short or sometimes subulate bract; perigynia numerous, oval, two-nerved, obscurely triangular, densely pubescent, 1½″ long, ½″ thick, narrowed at the base, tipped with a very short entire or at length slightly bidentate beak; scales oblong-obovate, dark purple with a narrow green midvein, ciliate, obtusish, nearly as long as the perigynia.

In rocky soil, Greenland to Alaska, south to the higher mountains of New England, Lake Huron and British Columbia. Also in northern Europe.   Summer.

92. Carex caryophýllea Latourrette.   Vernal Sedge.   Fig. 959.

Carex praecox Jacq. Fl. Austr. **5** : 23. *pl. 446.* 1778. Not
 Schreb. 1771.
Carex caryophyllea Latourrette, Chlor. Lugdun, 27. 1785.
Carex verna Chaix, in Vill. Hist. Pl. Dauph. **2** : 204. 1787.

Dark green, stoloniferous, culms phyllopodic, erect
or reclining, smooth, 3'–12' long. Leaves ½"–1½" wide,
clustered near base and shorter than the culm; lower
bract subulate, ¼"–1' long, very short sheathing; staminate spike sessile or very short-stalked, usually large
and conspicuous; pistillate spikes 1–3, close together
or slightly separate, oblong, 5–20-flowered, 3"–6" long,
2"–3" in diameter, sessile or the lower short-stalked;
perigynia obovoid, sharply 3-angled, short-pubescent,
brown, about 1½" long, abruptly acute, tapering at base;
scales ovate, brownish with a lighter center, cuspidate
or the lower rough-awned, about equalling the perigynia.

Maine to District of Columbia, locally naturalized from
Europe. Native also of Asia. Pink-grass, Iron-grass. May–
June.

93. Carex commùnis Bailey.   Fibrous-rooted Sedge.   Fig. 960.

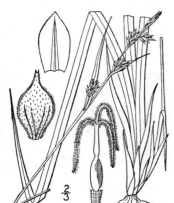

C. varia var. *pedicellata* Dewey, Am. Journ. Sci. **11** : 163. 1826.
Carex communis Bailey, Mem. Torr. Club **1** : 41. 1889.
Carex communis var. *Wheeleri* Bailey, Mem. Torr. Club **1** :
 41. 1889.
Carex pedicellata Britton, Mem. Torr. Club **5** : 87. 1894.
Carex pedicellata var. *Wheeleri* Britton, Mem. Torr. Club **5** :
 88. 1894.

Light green, not stoloniferous, fibrous-rooted, culms
usually well-developed, slender, aphyllopodic, roughish
above, erect or reclining, 6'–20' long. Leaves 1"–2" wide,
shorter than the culms; lower bract narrowly linear or
subulate, ¼"–2' long; staminate spike 2"–12" long, from
sessile to strongly peduncled; pistillate spikes 2–4, short-
oblong, 3–10-flowered, sessile and usually separated, or the
lowest short-stalked; perigynia obovoid, about 1" long
and a little more than ¼" in diameter, pale, short-pubescent,
slightly 1-ribbed on each side, tipped with a subulate
2-toothed beak one-fourth the length of the body; scales
green, ovate or narrower, acuminate to obtuse, nearly
equalling the perigynia; stigmas 3.

In dry soil, Nova Scotia to British Columbia, south to
Georgia, Ohio and Nebraska. Ascends to 5700 ft. in Virginia. May–July.

94. Carex pennsylvánica Lam.   Pennsylvania Sedge.   Fig. 961.

Carex pennsylvanica Lam. Encycl. **3** : 388. 1789.

Strongly stoloniferous, culms slender but strict, erect,
smoothish to very rough, 3'–15' tall. Leaves ½"–1½" wide,
the basal shorter than or sometimes exceeding the culm,
the old sheaths persistent and fibrillose; lower bract subulate
or scale-like, rarely over ¼' long; staminate spike sessile or
very short-stalked, ½'–1' long; pistillate spikes 1–4, short-
oblong, 4–20-flowered, sessile, contiguous or the lower somewhat distant; perigynia broadly obovoid, about 1" long and
more than ½" in diameter, short-pubescent, to nearly glabrous, 1-ribbed on two sides, strongly narrowed at the base,
tipped with a more or less bidentate beak from one-fourth
the length of to as long as the body; scales ovate, purplish,
acute or cuspidate, equalling or a little longer than the perigynia; stigmas 3.

In dry soil, New Brunswick to North Dakota, North Carolina
and Tennessee. Very variable. Ascends to 5000 ft. in North
Carolina. May–June.

Carex helióphila Mackenzie, of prairies and plains from Illinois to Alberta and New Mexico,
differs by larger perigynia, 1" wide, circular (not triangular) in cross-section.

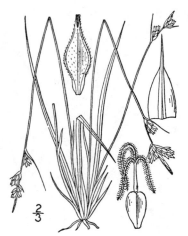

$\frac{2}{3}$

95. Carex vària Muhl. Emmons' Sedge. Fig. 962.

Carex varia Muhl.; Wahl. Kongl. Vet. Acad. Handl. (II.) 24: 159. 1803.
C. Emmonsii Dewey, Torr. Ann. Lyc. N. Y. 3: 411. 1836.
C. varia var. *colorata* Bailey, Mem. Torr. Club 1: 41. 1889.

Caespitose and little stoloniferous, culms filiform, erect to nearly prostrate, rough above, 3′–20′ long. Leaves elongated, $\frac{1}{2}''$–$1\frac{1}{4}''$ wide, from much shorter than to exceeding the culms; bracts short; staminate spike 2″–6″ long, sessile, sometimes not overtopping the upper pistillate one, but usually rather prominent; pistillate spikes 1–4, mostly close together and sessile, $1\frac{1}{2}''$–$3\frac{1}{2}''$ long, 4–12-flowered; perigynia oblong-ovoid, short-pubescent, about 1″ long, $\frac{1}{2}''$ thick, strongly narrowed at the base, tipped with a subulate minutely 2-toothed beak commonly one-half the length of the body; scales ovate, green or purplish-brown, acuminate or cuspidate, about as long as the perigynia; stigmas 3.

In dry soil, Nova Scotia to western Ontario and Manitoba, south to Georgia and Texas. May–July. Very variable.

96. Carex nòvae-ángliae Schwein. New England Sedge. Fig. 963.

C. novae-angliae Schwein. Ann. Lyc. N. Y. 1: 67. 1824.

Loosely caespitose and stoloniferous, culms filiform, erect or reclining, $2\frac{1}{2}''$–12″ long. Leaves about $\frac{1}{2}''$ wide, soft, elongated, usually exceeding the culms; staminate spike short-stalked, very narrow or almost filiform, 2″–8″ long, $\frac{1}{2}''$ wide or less; pistillate spikes 1–3, distant, subglobose, 2–10-flowered, sessile or the lower short-stalked; lower bract filiform, short or rarely overtopping the spikes; perigynia narrowly obovoid, about 1″ long and $\frac{1}{2}''$ thick, short-pubescent, tipped by a subulate 2-toothed beak about one-fourth the length of the body; scales ovate, greenish-brown, acute or cuspidate; stigmas 2 or 3.

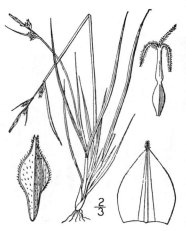

$\frac{2}{3}$

In wet shaded places, New Brunswick to Maine, Massachusetts and New York. Summer.

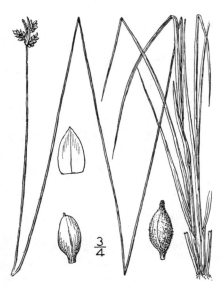

$\frac{3}{4}$

97. Carex álbicans Willd. White-tinged Sedge. Fig. 964.

C. albicans Willd.; Spreng. Syst. Veg. 3: 818. 1826.
C. Peckii E. C. Howe, Ann. Rep. N. Y. St. Museum 47: 177. 1895.

Loosely caespitose and short-stoloniferous, the culms smooth or nearly so, erect, 5′–20′ tall. Leaves $1\frac{1}{2}''$ wide or less, soft, much shorter than the culms, the lower sheaths but little fibrillose; lowest bract absent or short; staminate spike sessile and usually exceeded by upper pistillate spikes, $1\frac{1}{2}''$ long or less; pistillate spikes 2–4, subglobose, 2–8-flowered, closely contiguous or the lowest little distant; perigynia oblong-obovoid, $1\frac{1}{2}''$–2″ long, $\frac{1}{2}''$ in diameter, short gray-ish pubescent, 1-ribbed on two sides, strongly narrowed at base, abruptly tipped with a bidentate beak one-fourth the length of the body; scales broadly ovate, reddish-brown with broad white hyaline margins, short-acuminate to obtusish, all (except lower) but half the length of perigynia; stigmas 3.

Open woods and banks, Quebec to Alaska, south to Massachusetts, Pennsylvania, "Carolina" and Minnesota.

98. Carex defléxa Hornem.　Northern Sedge.
Fig. 965.

Carex deflexa Hornem. Plantel. Ed. 3, 1 : 938. 1821.
C. deflexa var. *Deanei* Bailey, Mem. Torr. Club 1 : 42. 1889.

Rootstocks slender, loosely branched and short-stoloniferous, culms filiform, erect or spreading, 1′–12′ long, shorter than or little exceeding the narrow bright green leaves. Bracts subulate or very narrowly linear, ½′–2′ long; staminate spike sessile, 1″–3″ long, inconspicuous; pistillate spikes 1–3, 1″–2½″ long, subglobose, 2–8-flowered, all sessile and closely contiguous or the lower somewhat separated, usually also 1 or 2 nearly basal filiform-stalked spikes from the lowest sheaths; perigynia oblong-obovoid, much narrowed at the base, short-pubescent, 1″ or less long, tipped with a flat, slightly 2-toothed beak about one-fourth the length of the body; scales ovate or ovate-lanceolate, acute or cuspidate; stigmas 3.

In open places, Greenland to Alaska, south to Massachusetts, Pennsylvania and Minnesota, mostly at high altitudes. Summer.

99. Carex Ròssii Boott.　Ross's Sedge.
Fig. 966.

C. Rossii Boott; Hook. Fl. Bor. Am. **2**: 222. 1840.
Carex deflexa var. *media* Bailey, Mem. Torr. Club
　1 : 43.　1889.　Not *C. media* R. Br. 1823.
C. deflexa Farwellii Britton; Brit. & Br. Ill. Fl. **1** :
　334.　1896.
C. Farwellii Mackenzie, Bull. Torr. Club **37**: 244.
　1910.

Rootstocks slender, loosely branched and stoloniferous; culms slender, erect, 8′–15′ long, shorter than or little exceeding the leaves. Leaves about 1′ wide; lowest bract conspicuous, often exceeding inflorescence; staminate spike sessile or nearly so, 3″–6″ long, conspicuous; pistillate spikes 2–3, 2½″–4″ long, short-oblong, 3–10-flowered, sessile or short-peduncled, approximate or the lower separate, filiform-stalked; basal spikes conspicuous; perigynia oblong-obovoid, much narrowed at base, short-pubescent, 2″ long, abruptly contracted into a bidentate beak from one-half length to nearly as long as the body; scales ovate, obtusish to short-cuspidate; stigmas 3.

Dry soil, Michigan to British Columbia, Oregon, and south in the Rocky Mountains.　June–July.

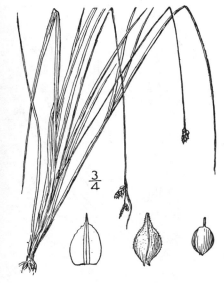

100. Carex nìgro-marginàta Schwein.　Black-edged Sedge.　Fig. 967.

C. nigro-marginata Schwein. Ann. Lyc. N. Y. 1 : 68.　1824.

Bright green, stoloniferous, culms filiform, wiry, aphyllopodic, erect or spreading, 2′–8′ long, very unequal in length. Leaves 1″–2″ wide, very much longer than the culms, rather stiff, often 12′ or more long; bracts very short and subulate or wanting; staminate spike sessile, inconspicuous, 2″–4″ long; pistillate spikes 1–3, few-flowered, sessile at the base of the staminate, about 3″ long; perigynia 1¾″ long, short-pubescent or nearly glabrous, the body oval, ½″ thick, stipitate, 1-ribbed on two sides, tipped with a cylindric-subulate 2-toothed beak one-third to one-half as long as the body; scales ovate, acute or cuspidate, green with purple margins or variegated, rather longer than the perigynia.

Dry soil, Connecticut to South Carolina. April–July.

Carex floridàna Schwein., distinguished by its light-colored scales, occurs from Virginia to Florida and Texas.

101. Carex umbellàta Schk.   Umbel-like Sedge.
Fig. 968.

Carex umbellata Schk.; Willd. Sp. Pl. **4**: 290. 1805.
Carex umbellata var. *vicina* Dewey, Am. Journ. Sci. **11**:
  317. *pl. D. f. 13.* 1826.

Rather light green, closely tufted and matted,
strongly fibrillose at base, stoloniferous, culms phyllopodic, filiform, 1'–6' long.   Leaves ½"–1½" wide, slender,
ascending, usually much exceeding the culm, sometimes 1° long; staminate spike solitary, terminal, ½' or
less long, commonly conspicuous; pistillate spikes 1–3,
all filiform-stalked from the basal sheaths or 1 or 2 of
them sessile or very nearly so at the base of the staminate, oblong, 6–20-flowered, 2"–6" long; perigynia usually less than 2' long; body oblong-orbicular, stipitate,
finely pubescent, pale, obtusely 3-angled, ¾"–1½" long,
tipped with a subulate 2-toothed beak of nearly its
length; scales ovate-lanceolate, acuminate or short-
awned, the lower partly hiding the perigynia; stigmas 3.

Dry soil, Nova Scotia to Michigan and Pennsylvania.
May–July.

Carex abdìta Bicknell, with short-beaked perigynia,
seems to be distinct. It ranges from Quebec to New York, Saskatchewan and Oklahoma.

102. Carex tónsa (Fernald) Bicknell.
Deep-green Sedge. Fig. 969.

C. umbellata var. *tonsa* Fernald, Proc. Am. Acad. **37**:
  507. 1902.
C. tonsa Bicknell, Bull. Torr. Club **35**: 492. 1908.

Deep green, closely tufted and matted, strongly
fibrillose at base, stoloniferous, culms phyllopodic, filiform, 1'–4' long. Leaves 1"–2" wide,
stiff, spreading in age, usually much exceeding
the culm but rarely more than 8' long; staminate
spike solitary, terminal, ½' or less long, commonly conspicuous; pistillate spikes 1–3, all filiform-stalked from the basal sheaths or 1 or 2
of them sessile or very nearly so at the base of
the staminate, oblong, 6–12-flowered, 2"–4" long;
perigynia 2" long or more, the body oblong-
orbicular, stipitate, glabrous, except beak of perigynia which is very sparsely hairy, pale, obtusely
3-angled, about 1" long, tipped with a subulate
2-toothed beak of its own length; scales ovate-
lanceolate, acuminate or short-awned, the lower
exceeding the perigynia; stigmas 3.

Dry soil, chiefly near the coast, Maine to New
York and New Jersey. May–June.

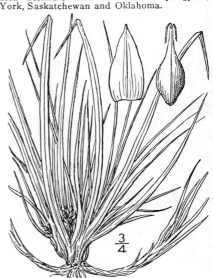

103. Carex hirtifòlia Mackenzie.   Pubescent
Sedge. Fig. 970.

Carex pubescens Muhl.; Willd. Sp. Pl. **4**: 281. 1805.
  Not Poir. 1789, nor Gilib. 1792.
C. hirtifolia Mackenzie, Bull. Torr. Club **37**: 244. 1910.

Pubescent all over, bright green, but reddened at
base, stoloniferous, culms aphyllopodic, weak, 1°–2°
long. Leaves flat, soft, elongated, usually shorter than
culm, 1½"–3½" wide; lower bracts 1'–3' long, occasionally overtopping the spikes, little if at all sheathing;
staminate spike sessile or nearly so, sometimes with
pistillate flowers at its base; pistillate spikes 2–4, oblong-cylindric, rather loosely flowered, erect, 3"–10"
long, 2"–2½" thick, the upper sessile, the lower separated and short-stalked; perigynia sharply 3-angled,
obovoid, narrowed to a stipe-like base, densely pubescent, and, including the subulate straight minutely
2-toothed beak, about 2" long; scales obovate, truncate, scarious-margined, rough-awned or cuspidate,
about as long as the perigynia.

In woods and thickets, Nova Scotia to North Dakota,
New Jersey, Kentucky and Kansas. May–Aug.

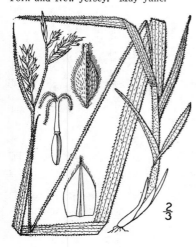

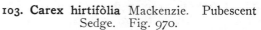

104. **Carex pícta** Steud. Boott's Sedge. Fig. 971.

Carex Boottiana Benth.; Boott, Bost. Journ. Nat. Hist.
5: 112. 1845. Not H. & A. 1841.
Carex picta Steud. Syn. Pl. Cyp. 184. 1855.

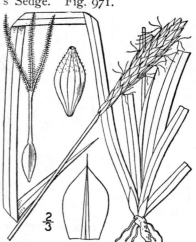

Dioecious, foliage glabrous, light green, culms phyllopodic, slender, smooth, erect or reclining, 4′–12′
long, usually much shorter than the leaves. Leaves
flat, 1″–3″ wide, the upper reduced to bladeless
sheaths, tinged with reddish purple; spike solitary or
rarely with a small accessory one near its base, erect,
densely many-flowered, the staminate generally 1′–2′
long, the pistillate cylindric but narrowed at the base,
1′–2½′ long, 2″–4″ thick; perigynia narrowly obovoid,
strongly many-nerved, puberulent at least toward the
obtuse summit, 2″–2½″ long, stipitate; scales reddish
purple, usually with green midvein and hyaline margins, shining, obovate, obtuse, acute or cuspidate,
longer and wider than the perigynia; stigmas 3.

In woods, central Indiana; Alabama and Louisiana.
Local. Summer.

105. **Carex pedunculàta** Muhl. Long-stalked Sedge. Fig. 972.

Carex pedunculata Muhl.; Willd. Sp. Pl. **4**: 222. 1805.

Densely matted, rather bright green, culms very
slender, roughish above, diffuse, 3′–12′ long, strongly
purple-tinged at base. Leaves flat, 1″–1½″ wide, the
basal commonly longer than the culms; upper sheaths
green, almost bladeless, the lower with short leaf-like
blades; terminal spike staminate, long-stalked, usually
with some pistillate flowers at its base; lateral spikes
2–4, pistillate or androgynous, 3″–6″ long, few-flowered,
filiform-stalked and spreading or drooping, scattered,
some of them appearing basal; perigynia obovoid,
sharply 3-angled above, puberulent or becoming glabrous, 2″ long, pale green, nerveless, narrowed below
into a stipe, tipped with a minute entire beak; scales
purplish, obovate, with green midrib, abruptly cuspidate
or the lower subulate-awned, nearly equalling or lower
exceeding perigynia; stigmas 3.

In dry woods, Anticosti to Saskatchewan, south to Virginia, Pennsylvania and Iowa. May–July.

106. **Carex concínna** R. Br. Low Northern Sedge. Fig. 973.

Carex concinna R. Br. Frank. Journ. 763. 1823.

Caespitose and stoloniferous, the culms slender,
nearly smooth, 2′–6′ tall. Leaves about 1″ wide, flat,
pale green, much shorter than the culm; bracts reduced
to green bladeless sheaths or occasionally with a short
blade; staminate spike solitary, sessile or nearly so,
1½′–3′ long; pistillate spikes 1–3, sessile and clustered
or the lower one somewhat distant and stalked, erect,
2″–4″ long, 1½″–2″ thick, compactly 5–10-flowered; perigynia oblong-ovoid, 3-angled, pubescent, very short-
beaked, obscurely nerved, about twice as long as the
broadly ovate obtuse dark scales; stigmas 3.

In rocky places, Quebec and New Brunswick to British
Columbia. Summer.

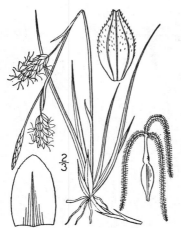

107. Carex Richardsònii R. Br. Richardson's Sedge. Fig. 974.

Carex Richardsonii R. Br. Frankl. Journ. 751. 1823.

Stoloniferous, the culms slender, rough, erect, 4′-12′ tall. Leaves flat, about 1″ wide, the basal shorter than or sometimes equalling the culm, those of the culm very short; bracts bladeless, sheathing, ¾′-1′ long, usually brown-purple with a white hyaline acute summit; staminate spike solitary, stalked, 6″-13″ long; pistillate spikes 1 or 2, erect, narrowly cylindric, short-stalked, 4″-10″ long, compactly many-flowered, close together, their stalks partly or wholly enclosed in the sheaths; perigynia obovoid, triangular, pubescent, about 1″ long, minutely beaked, obscurely nerved, mostly shorter than the ovate, subacute, purple, conspicuously white-margined scales; stigmas 3.

In dry soil, Ontario to Saskatchewan, Alberta and British Columbia, south to western New York, Illinois, Iowa and South Dakota. Summer.

108. Carex ebúrnea Boott. Bristle-leaved Sedge. Fig. 975.

C. alba var. *setifolia* Dewey, Am. Journ. Sci. **11** : 316. 1826.
C. eburnea Boott ; Hook. Fl. Bor. Am. **2** : 226. *pl. 225.* 1840.
Carex setifolia Britton ; Britton & Brown, Ill. Fl. **1** : 332. 1896.

Glabrous, pale green, culms filiform, smooth, weak, 4′-15′ long, from slender, elongated rootstocks. Leaves filiform, shorter than the culm, less than ¼″ wide; bracts reduced to bladeless sheaths 2″-5″ long; staminate spike solitary, sessile or very nearly so, 2″-4″ long; pistillate spikes 2-4, erect, slender-stalked, 2″-4″ long, rather less than 1″ thick, loosely few-flowered, the upper commonly overtopping the staminate, the lower one sometimes distant; perigynia oblong, pointed at both ends, 3-angled, 1″ long, ½″ or less thick, polished and nearly black when mature, very faintly few-nerved, tapering into a short entire beak; scales ovate, obtuse or the lower acute, thin, hyaline, shorter than the perigynia; stigmas 3.

In dry sandy or rocky soil, preferring limestone rocks, New Brunswick to Alberta, south to Virginia, Tennessee, Missouri and Nebraska. May-July.

109. Carex Hàssei Bailey. Hasse's Sedge. Fig. 976.

C. aurea var. *celsa* Bailey, Mem. Torr. Club **1** : 75. 1889.
Carex Hassei Bailey, Bot. Gaz. **21** : 5. 1896.
C. bicolor Robinson & Fernald, in A. Gray, Manual, Ed. 7, 232. 1909. Not All.

Similar to the following species, the culms slender, usually 6″-2° long, from slender elongated rootstocks. Leaves flat, 1″-2″ wide, generally shorter than culm; bracts similar to culm-leaves, the lower exceeding spikes, sheathing, not dark auricled; terminal spike short-stalked, gynaecandrous or frequently staminate; pistillate spikes 2-5, linear-oblong, the upper aggregated and sessile or short-stalked, the lower distant and long-stalked, loosely or somewhat compactly 6-20-flowered, 2″-10″ long, about 1¾″ thick; perigynia ellipsoid or narrowly obovoid, whitish pulverulent and not fleshy or translucent at maturity, less than 1″ in diameter, the nerves faint, tapering at base, beakless, the orifice entire; scales as in the next, but more often dark-tinged; stigmas 2.

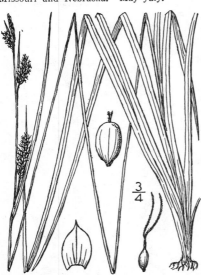

Labrador to north Maine, westward across the continent and south in the mountains of California. June-Aug.

110. **Carex aùrea** Nutt.　Golden-fruited Sedge.　Fig. 977.

Carex aurea Nutt. Gen. **2** : 205.　1818.

Glabrous, light green, culms very slender, erect or reclining, 2′–15′ long, from slender, elongated rootstocks. Leaves flat, 1″–2″ wide, the basal equalling or exceeding the culm; bracts similar to the culm-leaves, the lower commonly much overtopping the spikes, sheathing, not dark-auricled; terminal spike short-stalked, staminate or slightly gynaecandrous; pistillate spikes 2–4, oblong or linear-oblong, erect and clustered near the summit or the lower one distant, filiform-stalked, loosely or somewhat compactly 4–20-flowered, 2″–10″ long, about 1½″ thick; perigynia broadly obovoid or subglobose, white or nearly white when young, becoming fleshy and translucent, golden yellow or brown and about 1″ in diameter when mature, many-nerved, beakless, the orifice entire; scales very variable, ovate, membranous, blunt, acute, cuspidate or short-awned, shorter than or the lower exceeding the perigynia; stigmas mostly 2.

In wet meadows, springs and on wet rocks, Newfoundland to British Columbia, south to Massachusetts, Pennsylvania, Michigan, Utah and California.　Summer.

111. **Carex lívida** (Wahl.) Willd.　Livid Sedge.　Fig. 978.

Carex limosa var. *livida* Wahl. Kongl. Vet. Acad. Handl. (II.) **24** : 162.　1803.
Carex livida Willd. Sp. Pl. **4** : 285.　1805.

Glabrous, very glaucous, phyllopodic, long-stoloniferous, culms slender, strictly erect, smooth, 6′–1½° tall. Leaves 1″ wide or less, involute or folded, usually shorter than culm; bracts narrow, short-sheathing, usually short; staminate spike solitary, short-stalked; pistillate spikes 1–3, 5″–12″ long, about 2½″ thick, erect, approximate, sessile or short-peduncled, oblong, densely 5–15-flowered or looser at the base, the third, when present, distant or sometimes nearly basal, stalked; perigynia oblong, very pale, nearly 2″ long, less than 1″ thick, faintly nerved, straight, beakless, narrowed to an entire orifice; scales ovate, obtuse or the lower subacute, rather shorter than the perigynia, the margins colored; stigmas 3.

In bogs, Labrador and Hudson Bay to Alaska, south to Connecticut, the pine barrens of New Jersey, central New York, Michigan and California.　Also in Europe.　Summer.

112. **Carex panícea** L.　Grass-like Sedge.　Carnation-grass.　Fig. 979.

Carex panicea L. Sp. Pl. 977.　1753.

Glabrous, pale bluish green and glaucous, phyllopodic, long-stoloniferous, culms slender, smooth, erect, stiff, 6′–2° tall. Leaves flat, 1″–2″ wide, the lower usually shorter than the culm; bracts short, long-sheathing; staminate spike usually 1, its peduncle smooth; pistillate spikes 2 or 3, distant, stalked or the upper nearly sessile, erect, 1′ or less long, 2½″–3½″ thick, closely or at base loosely 8–25-flowered; perigynia oblong-obovoid, 1½″–2″ long and about 1″ in diameter, slightly swollen and obscurely 3-angled, yellow, purple or mottled, faintly few-nerved, tipped with a very short entire somewhat oblique beak; scales ovate, acute, purple or purple-margined, shorter than the perigynia; stigmas 3.

In fields and meadows, Nova Scotia to Connecticut. Locally naturalized from Europe.　Gilliflower-grass, Bluegrass, Pink-leaved sedge.　June–July.

113. Carex Mèadii Dewey. Mead's Sedge. Fig. 980.

Carex Meadii Dewey, Am. Journ. Sci. 43: 90. 1842.
Carex tetanica var. *Meadii* Bailey, Proc. Am. Acad. 22: 118. 1886.
Carex tetanica var. *Carteri* Porter, Proc. Acad. Phila. 1887: 76. 1887.
Carex tetanica var. *Canbyi* Porter, Proc. Acad. Phila. 1887: 76. 1887.

Similar to the following species, glabrous, culm stouter, strongly phyllopodic, not purplish-tinged, very rough above, 8′–18′ tall. Rootstocks elongated, deep-seated, slender; leaves flat, those of the fertile culm usually 6–10, 1½″–3½″ wide; bracts short, not overtopping the spikes; staminate spike long-stalked; pistillate spikes 1–3, sometimes staminate at the summit, rarely compound at the base, oblong-cylindric, densely 8–30-flowered, ½′–1½′ long, about 3″ in diameter, erect, stalked, or the upper one sessile, the lower often very long-stalked; perigynia obovoid, obtusely triangular, prominently many-nerved, green, 1½″ long, about 1″ in diameter, tipped with a minute slightly bent beak; scales ovate, green with reddish-brown or purple-brown margins, obtuse to cuspidate, and from much shorter than to exceeding perigynia; stigmas 3.

In swamps and wet meadows, New Jersey and Pennsylvania to Georgia, Michigan, Assiniboia, Nebraska and Arkansas. May–July.

114. Carex tetánica Schk. Wood's Sedge. Fig. 981.

C. tetanica Schk. Riedgr. Nachtr. 68. *figs. 100, 207.* 1806.
Carex tetanica var. *Woodii* Bailey, Mem. Torr. Club 1: 53. 1889.

Light green and glabrous, or sheaths slightly puberulent, culms slender, more or less strongly phyllopodic, not strongly purplish-tinged at base, erect, rough above, 6′–2° tall. Rootstocks elongated, deep-seated, white, slender; leaves flat, those of the fertile culm usually 3–5, 1″–1¾″ wide, shorter than the culm; bracts narrow, elongated, usually shorter than the spikes, sheathing; staminate spike long-stalked; pistillate spikes 1 to 3, erect, distant, narrowed at the base, compactly 6–20-flowered or attenuate at base, linear, 1′ long or less, the lower filiform-stalked and often drooping; perigynia oblong-obovoid, green, prominently many-nerved, about 1½″ long and 1″ thick, obtusely triangular, abruptly minutely beaked or beakless, the orifice entire; scales ovate, obtuse or mucronate, usually shorter than the perigynia; stigmas 3.

In meadows and wet woods, Massachusetts to Manitoba, south to District of Columbia and Missouri. June–July.

115. Carex coloràta Mackenzie. Purplish-tinged Sedge. Fig. 982.

Carex colorata Mackenzie, Bull. Torr. Club 37: 232. 1910.

Light green and glabrous, culms slender, aphyllopodic and strongly purplish-tinged at base, strongly stoloniferous, erect, roughened above, 1°–2° tall. Leaves flat, those of the fertile culm usually 2–4, 1¼″–2″ wide, shorter than the culm; bracts narrow, elongated, usually not exceeding the spikes, sheathing; staminate spike more or less peduncled; pistillate spikes 2–3, erect, distant, rather loosely 6–15-flowered, linear, 7″–18″ long, 2″ thick, on slender much-exserted peduncles; perigynia narrowly obovoid, obtusely triangular, lightly many-nerved, 1¾″–2″ long, ¾″ wide, tapering at both ends, the beak ¼″ long, slightly curving, the orifice oblique; scales ovate, obtuse to cuspidate, shorter than the perigynia; stigmas 3.

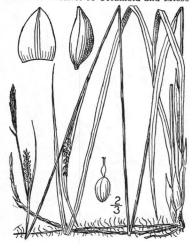

Woodlands, New York and Ontario to Michigan and Manitoba. May–June.

116. Carex vinàta Tausch.   Sheathed Sedge.   Fig. 983.

Carex vaginata Tausch, Flora 557. 1821.
Carex vaginata var. *altocaulis* Dewey, Am. Journ. Sci. (II.)
    41 : 227. 1866.
Carex saltuensis Bailey, Mem. Torr. Club 1 : 7. 1889.
Carex altocaulis Britton; Brit. & Br. Ill. Fl. 1 : 326. 1896.

Glabrous, light green, strongly stoloniferous, phyllo-podic, culms very slender, smooth, weak, diffuse, 6′–2½°
high. Leaves 1″–2½″ wide, much shorter than the culm,
the blades of the upper ones and of the long-sheathing
bracts usually very short; staminate spike long-stalked;
pistillate spikes 2 or 3, distant, slender-stalked or spread-ing, 4″–12″ long, loosely 3–20-flowered; perigynia
ovoid-oblong, 3-angled, scarcely inflated, narrowed at the
base, faintly nerved, about 2″′ long, nearly 1″ thick,
tipped with a beak about one-fourth the length of the
body, the orifice purplish-tinged, 2-toothed, oblique;
scales oval or ovate-lanceolate, purplish-tinged, acute
or the upper obtuse, usually shorter than the perigynia;
stigmas 3.

In boggy woods, Labrador to Alaska, northern New Eng-land, New York, Michigan, Minnesota and British Columbia, Europe and Asia.   June–Aug.

117. Carex polymórpha Muhl.   Variable Sedge. Fig. 984.

Carex polymorpha Muhl. Gram. 239. 1817.

Glabrous, rather dark green, from matted, elongated,
stout rootstocks, culms stiff, aphyllopodic and strongly
purplish-tinged at base, strictly erect, smooth or nearly
so, 1°–2° tall. Leaves flat, 1½″–2″ wide, nearly erect,
those of fertile culm short; bracts long-sheathing;
staminate spike 1 or rarely 2, long-stalked; pistillate
spikes commonly 1 or 2, erect, short-stalked, densely
12–25-flowered or sometimes looser at the base, 7″–1½′
long, 3″–4″ thick, often staminate at the summit; peri-gynia ovoid-oblong, obscurely 3-angled, 2½″ long, 1¼″ in
diameter, the beak one-half as long as the body, the
orifice oblique; scales red-brown, obtuse or the lower
acute, somewhat shorter than the perigynia; stigmas 3.

Wet meadows or borders of woods, southern Maine to
northern New Jersey, south to North Carolina. Local.
Ascends to 2000 ft. in Pennsylvania. June–Aug.

118. Carex plantagínea Lam.   Plantain-leaved Sedge.   Fig. 985.

Carex plantaginea Lam. Encycl. 3 : 392. 1789.

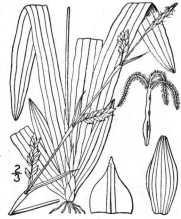

Glabrous, rather dark green, culms slender, erect or
reclining, 6′–2° long. Leaves of sterile culms 5″–13″
wide, persistent through the winter, those of fertile
culms with rudimentary blades, the sheaths strongly
reddened; bracts short; staminate spike long-stalked,
purple; pistillate spikes 3 or 4, erect, widely separated,
all stalked, 1′ or less long, loosely 4–8-flowered, the
stalks of the upper ones enclosed in the sheaths; peri-gynia oblong-elliptic, short-beaked, many-nerved, 1½″–
2¼″ long, nearly 1″ thick, longer than the ovate cuspi-date scales; stigmas 3.

In woods, New Brunswick and Ontario to Manitoba,
south to North Carolina and Illinois. Ascends to 2100 ft.
in Virginia. April–June.

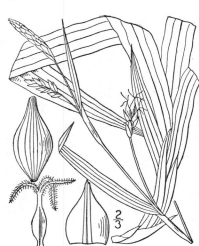

119. Carex Careyàna Torr.   Carey's Sedge.
Fig. 986.

Carex Careyana Torr.; Dewey, Am. Journ. Sci. **30**: 60. *f. 88.* 1836.

Glabrous, bright green, culms slender, erect or somewhat reclining, smooth or nearly so, 1°–2½° tall. Basal leaves flat, 3″–6″ wide, much shorter than the culm; bracts narrow, short, with very long purplish-tinged sheaths; staminate spike usually long-stalked, large; pistillate spikes 1–3, erect, loosely 3–8-flowered, less than 1′ long, the upper short-stalked and approximate, the lower long-stalked and remote; perigynia ovoid-elliptic, very sharply 3-angled, many-nerved, 2¾″ long or more and over 1″ thick, the short beak slightly oblique, entire; scales ovate with hyaline margins, cuspidate or awned, about half length of perigynia; stigmas 3.

In woods, New York to Michigan, south to District of Columbia, Virginia and extreme southern Missouri. May–June.

120. Carex platyphýlla Carey.   Broad-leaved Sedge.   Fig. 987.

Carex platyphylla Carey, Am. Journ. Sci. (II.) **4**: 23. 1847.

Glabrous, pale green and glaucous, culms slender, spreading or reclining, 4′–15′ long. Leaves of sterile culms flat and broad, 4½″–12″ wide, shorter than the culm, very smooth (except edges); those of fertile culms much reduced; bracts with long clasping sheaths, usually less than 2′ long; staminate spike stalked; pistillate spikes 2–4, distant, erect, all slender-stalked or the upper one nearly sessile, loosely 4–10-flowered, 5″–10″ long; perigynia ovoid-elliptic, 3-angled, many-nerved, slightly bent or nearly straight at the narrowed or short-beaked apex, 1¼″–1¾″ long, rather more than ½″ thick, somewhat longer than the broadly ovate, acute, cuspidate or short-awned scales; stigmas 3.

In woods and thickets, Quebec and Ontario to Michigan, south to Virginia and Illinois. Ascends to 2500 ft. in Virginia. May–June.

121. Carex abscóndita Mackenzie.   Thicket Sedge.   Fig. 988.

Carex ptychocarpa Steud. Syn. Pl. Cyp. 234. 1855. Not Link, 1799.
Carex abscondita Mackenzie, Bull. Torr. Club **37**: 244. 1910.

Glabrous, pale green and glaucous, culms erect, very slender, smooth, only 2′–8′ tall. Leaves flat, the larger 2″–4½″ wide, much longer than the culm; bract of second pistillate spike usually overtopping the culm; staminate spike small, 2″–4½″ long, sessile or nearly so; pistillate spikes 2 or 3, approximate, sessile or short-peduncled, or the lower one slender-stalked and nearly basal, all erect, loosely 5–10-flowered, 4″–8″ long; perigynia oblong-ovoid, pale, sharply 3-angled, finely many-nerved, about 1⅓″ long, rather more than ½″ thick, pointed at both ends; minutely straight-beaked or slightly bent, the orifice entire; scales ovate, thin, obtuse to short-cuspidate, about one-half as long as the perigynia; stigmas 3.

In moist woods and thickets, Massachusetts and New Jersey to Florida and Louisiana. Local. June–July.

122. Carex digitàlis Willd.   Slender Wood Sedge.   Fig. 989.

Carex digitalis Willd. Sp. Pl. **4** : 298.   1805.

Glabrous, bright green, not at all glaucous, culms weak, slender, smooth, usually reclining, 4′–18′ long. Leaves flat, 1″–2½″ wide, usually longer than the culm; bracts similar, the second exceeding the culm; staminate spike stalked; pistillate spikes 2–4, linear, loosely alternately 3–12-flowered, 5′–15″ long, the upper nearly sessile, the others filiform-stalked and widely spreading or drooping; perigynia ovoid-oblong, sharply triangular, many-nerved, brown when ripe, narrowed at both ends, scarcely 1½″ long, more than ⅓″ thick, the very short beak slightly oblique; scales lanceolate or ovate-lanceolate, scarious-margined, acute, acuminate or short-awned, shorter than or the lower about equalling the perigynia; stigmas 3.

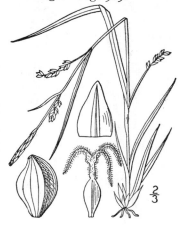

In woods and thickets, Maine and southern Ontario to Minnesota, south to Florida and Texas. Ascends to 3000 ft. in Virginia.   May–July.

123. Carex laxicúlmis Schwein.   Spreading Sedge.   Fig. 990.

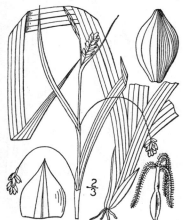

Carex laxiculmis Schwein. Ann. Lyc. N. Y. **1** : 70.   1824.
Carex retrocurva Dewey, Wood's Bot. 423.   1845.
Carex digitalis copulata Bailey, Mem. Torr. Club **1** : 47.   1889.

Glabrous, varying from strongly glaucous to deep green, culms filiform, smooth or very nearly so, ascending or diffuse, 6′–2° long. Sterile culm-leaves elongated, 2″–6″ wide, those of fertile culms shorter than the culms; bracts similar, usually short; staminate spike long-stalked; pistillate spikes 2–4, oblong, rather loosely 5–10-flowered, 3″–6″ long, about 2″ thick, drooping on long hair-like stalks or the upper short-stalked and erect; perigynia broadly ovoid-oblong, sharply 3-angled, many-nerved, 1½″–2″ long and rather more than ⅓″ thick, narrowed at both ends, but scarcely beaked, longer than the ovate, green, cuspidate or short-awned scales; stigmas 3.

In woods and thickets, Maine to southern Ontario, Michigan, Virginia and Missouri. Ascends to 5600 ft. in Virginia.   May–June.

124. Carex albursìna Sheldon.   White Bear Sedge.   Fig. 991.

Carex laxiflora var. *latifolia* Boott, Ill. 38.   1858.   Not *C. latifolia* Moench, 1794.
Carex albursina Sheldon, Bull. Torr. Club **20** : 284.   1893.

Glabrous, rather deep green, culms nearly smooth, strongly flattened and winged, erect or spreading, 8′–2° high. Basal leaves shorter than the culm, 3½″–1½′ wide; bracts similar to the narrower culm-leaves, the upper strongly overtopping the spikes, sheaths loose; staminate spike sessile or nearly so, the scales obtuse; pistillate spikes 2–4, distant and narrowly linear, stalked or the upper sessile and close together, ½′–1¼′ long, loosely flowered; perigynia obovoid, obtusely 3-angled, many-nerved, ¾″–2″ long, 1″ thick, tipped with a very short bent entire beak; scales broadly oblong, widely scarious-margined, very truncate or the lower rarely short-awned, shorter than the perigynia; stigmas 3.

In woods, Quebec to Minnesota, south to Virginia, Tennessee and southern Missouri. Ascends to 2300 ft. in Virginia.   June–Aug.   The specific name is in allusion to White Bear Lake, Minn.

125. Carex blánda Dewey. Woodland Sedge.
Fig. 992.

Carex blanda Dewey, Am. Journ. Sci. **10**: 45. 1826.
C. laxiflora var. *varians* Bailey, Mem. Torr. Club **1**: 32. 1889.

Glabrous, pale green, culms nearly smooth, flattened, usually coarsely cellular, loosely erect, 6′–2° high. Lower leaves usually shorter than culm, 1½″–7″ wide; the leaf-like bracts and upper leaves with loose sheaths, their margins much crisped; staminate spike usually short-stalked or sessile, the scales rarely reddish-brown-tinged; pistillate spikes 2–4, oblong or linear-oblong, 3″–13″ long, rather closely 8–25-flowered, the upper two contiguous to staminate spike and sessile or nearly so, the lower distant and stalked; perigynia obovoid, ascending, 1¼″–1½″ long, more than ½″ thick, narrowed at base, strongly nerved, contracted into a short stout outwardly bent entire beak; scales ovate with broad white scarious margins usually shorter than the perigynia, the lower strongly awned.

Maine and Ontario to Virginia, Arkansas and Kansas. Intergrades with the next. May–July.

126. Carex laxiflòra Lam. Loose-flowered Sedge. Fig. 993.

Carex laxiflora Lam. Encycl. **3**: 392. 1789.
Carex laxiflora var. *gracillima* Boott, Ill. Car. **1**: 37. 1858.

Glabrous, rather pale green, culms erect or reclining, slender, roughish above, 6′–2° long, scarcely if at all two-edged. Leaves 1½″–3½″ wide, soft, the lower mostly shorter than the culm, the bracts similar to the culm-leaves sometimes overtopping the spikes, their sheaths tight, the margins little crinkled; staminate spike usually stalked, its scales rarely reddish-brown tinged; pistillate spikes 2–4, distant, linear-cylindric, loosely several–many-flowered, ½′–1¼′ long, 1½″–2″ thick, more or less peduncled, spreading, or the upper erect; perigynia ascending, obovoid, more or less oblique, 1¼″–1½″ long, rather more than ½″ thick, narrowed at the base, strongly many-nerved, contracted into a short stout outwardly bent entire beak; scales ovate with broad white scarious margins, usually shorter than the perigynia, at least the lower strongly awned; stigmas 3.

In meadows and thickets, eastern Quebec and Ontario to Minnesota, south to Florida, Alabama and Texas. May–July.

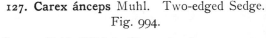

127. Carex ánceps Muhl. Two-edged Sedge.
Fig. 994.

C. anceps Muhl.; Willd. Sp. Pl. **4**: 278. 1805.
C. anceps var. *patulifolia* Dewey, Wood's Bot. 423. 1845.
C. laxiflora var. *patulifolia* Carey, in A. Gray, Man. Ed. 2, 524. 1856.
C. laxiflora var. *leptonervia* Fernald, Rhodora **8**: 184. 1906.

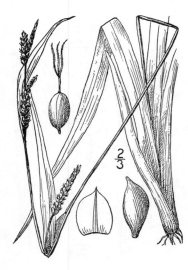

Glabrous, deep green, culms nearly smooth, often much flattened, loosely erect, 6′–2° high. Lower leaves shorter than culm, 2½″–10″ wide; bracts similar to narrower culm-leaves, the upper usually overtopping the spikes; staminate spike prominent, long- or short-stalked, the scales rarely reddish-brown tinged; pistillate spikes 2–4, loosely and alternately 5–15-flowered, distant and slenderly stalked or the upper approximate and short-stalked; perigynia appressed-ascending, elliptic-obovoid, obtusely 3-angled, strongly nerved or nearly nerveless, 1¾″ long, ¾″ wide, tipped with a short straight or slightly oblique beak; scales ovate, scarious, the lower abruptly short-awned; stigmas 3.

Woods, Newfoundland to Michigan, North Carolina and Tennessee. May–July.

128. Carex striátula Michx. Striate Sedge.
Fig. 995.

C. striatula Michx. Fl. Bor. Am. **2**: 173. 1803.
Carex laxiflora var. *divaricata* Bailey, Mem. Torr. Club **1**: 33. 1889.

Glabrous, pale green, culms loosely caespitose, nearly smooth, triangular, little flattened, erect, 1°–2° high. Leaves shorter than culm, 3½″–6″ wide; bracts short, usually exceeded by culms; staminate spike solitary, usually long-stalked, its scales commonly reddish-brown tinged; pistillate spikes usually two, widely separate, erect, peduncled, linear-oblong, 6″–15″ long, closely or at base loosely 6–20-flowered; perigynia obovoid-fusiform, divergent, obtusely triangular, strongly nerved, 2″–2½″ long, ¾″ wide, tipped with a straight or slightly oblique conspicuous beak; scales broadly ovate, short-cuspidate, hyaline-margined, slightly reddish-brown tinged, shorter than perigynia; stigmas 3.

New Jersey and Pennsylvania to Florida, Tennessee and Texas. April–June.

129. Carex stylofléxa Buckley. Bent Sedge.
Fig. 996.

Carex styloflexa Buckley, Am. Journ. Sci. **45**: 174. 1843.
Carex laxiflora var. *styloflexa* Boott, Ill. 37. 1858.

Glabrous, culms rather loosely caespitose, slender, triangular, often purplish at base, smooth, 1°–2½° tall. Leaves 1½″–3″ wide, flat, shorter than the culm; bracts short, usually exceeded by the spikes; staminate spike solitary, usually long-stalked but sometimes nearly sessile, its scales usually reddish-brown tinged; pistillate spikes 1–4, distant, loosely 4–12-flowered, 5″–10″ long, the lower drooping on elongated filiform stalks; perigynia elliptic-fusiform, triangular, many-nerved, 2″–2½″ long, ¾″ thick, somewhat divergent, tapering gradually to both ends and thus slender-beaked, the beak straight or little oblique; scales ovate or ovate-lanceolate, scarious-margined, reddish-brown-tinged, acute, cuspidate or short-awned, shorter than the perigynia; stigmas 3.

In woods and thickets, Connecticut to Florida and Texas. May–July.

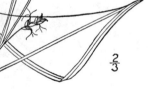
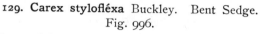

130. Carex Shríveri Britton. Shriver's Sedge. Fig. 997.

Carex Haleana Olney, Car. Bor. Am. 6. 1871. Not *C. Halei* Dewey, 1846.
C. *granularis Shriveri* Britton, in Britt. & Br. Ill. Fl. **1**: 322. 1896.
Carex Shriveri Britton, Manual 208. 1901.

Glabrous, light green and slightly glaucous, culms slender, erect, smooth or nearly so, 6′–2° tall. Leaves flat, 2½″–8″ wide, the basal shorter than the culm; bracts similar to the leaves, the lower rarely equalling the culm, strongly sheathing; staminate spike solitary, sessile or nearly so; pistillate spikes 2–5, distant, or the upper two contiguous, erect or somewhat spreading, linear-oblong, 3½″–14″ long, 2″–2½″ thick, densely 15–50-flowered, the lower at least exsert-peduncled; perigynia narrowly obovoid, little swollen, circular in cross-section, not strongly nerved, ascending, about 1″ long, ¾″ wide, contracted into a minute, usually entire, straight or rarely slightly bent beak, or essentially beakless; scales narrowly ovate, thin, acuminate or cuspidate, much shorter than the perigynia.

In moist meadows, Maine to North Dakota, Virginia and Indiana. May–July.

131. Carex granulàris Muhl.   Meadow Sedge.   Fig. 998.

Carex granularis Muhl.; Willd. Sp. Pl. 4: 179. 1805.
C. granularis recta Dewey; Wood's Class-book 763. 1860.

Glabrous, light green and slightly glaucous, culms slender, erect or ascending, smooth or nearly so, 6′–2½° tall. Leaves flat, roughish, 1½″–4½″ wide, the basal shorter than the culm; bracts similar to the culm-leaves, usually much exceeding the spikes, strongly sheathing; staminate spike solitary, sessile or short-stalked; pistillate spikes 2–5, distant or the upper two contiguous, erect or slightly spreading, narrowly oblong or cylindric, ½′–1¼′ long, 2½″ thick, densely 10–50-flowered, the lower at least exsert-peduncled; perigynia ovoid to obovoid, somewhat swollen and suborbicular in cross-section, strongly many-nerved, ascending, slightly more than 1″ long, ¾″–1″ wide, contracted into a short, usually entire, bent, or nearly straight beak; scales narrowly ovate, thin, acuminate or cuspidate, shorter than the perigynia; stigmas 3.

In moist meadows, New Brunswick to Manitoba, south to Florida and Louisiana. May–July.

132. Carex Cràwei Dewey.   Crawe's Sedge.   Fig. 999.

Carex Crawei Dewey, Am. Journ. Sci. (II.) 2: 246. 1846.
Carex heterostachya Torr. Am. Journ. Sci. (II.) 2: 248. 1846.

Glabrous, culms low, stiff, smooth or nearly so, erect, 3′–15′ tall, from long creeping rootstocks. Leaves rather stiff, 1″–2″ wide, erect or nearly so, shorter than the culm, the bracts similar, short, rarely overtopping the spikes; staminate spike usually 1, long-stalked; pistillate spikes 1–4, distant, oblong, erect, 5″–15″ long, 2″–3″ thick, densely 10–45-flowered, short-stalked or the upper sessile, the lowest often borne near the base of the culm; perigynia ovoid, ascending, obscurely many-nerved, usually minutely resinous dotted, 1½″–1¾″ long, suborbicular in cross-section, rounded at base, tapering into a very short entire, or emarginate beak; scales obovate, thin, acute or cuspidate, or the lowest blunt, shorter than the perigynia; stigmas 3.

In moist meadows and on banks, Cape Breton Island to Manitoba, south to northern Maine, Pennsylvania, Tennessee and Kansas. May–July.

133. Carex oligocàrpa Schk.   Few-fruited Sedge.   Fig. 1000.

Carex oligocarpa Schk.; Willd. Sp. Pl. 4: 279. 1805.

Glabrous, culms slender, spreading, roughish, 6′–20′ high. Leaves 1¼″–1¾″ wide, spreading, soft, the basal shorter than or exceeding the culm, the bracts similar, usually exceeding the spikes; sheaths smooth; staminate spike solitary, stalked or nearly sessile; pistillate spikes 2–4, erect or nearly so, distant, loosely 2–8-flowered, 4″–12″ long, about 2″ thick, erect, the lower filiform-stalked, the upper sessile; perigynia obovoid, firm, pale, finely many-striate, ascending, obtusely triangular, 1¾″–2″ long, abruptly narrowed into a short straight or oblique entire beak; scales ovate, cuspidate, or short-awned, longer or shorter than perigynia; stigmas 3.

In dry woods and thickets, Vermont and Ontario to Michigan, south to West Virginia, Kentucky and Oklahoma. May–July.

134. Carex Hitchcockiàna Dewey.   Hitch-cock's Sedge.   Fig. 1001.

Carex Hitchcockiana Dewey, Am. Journ. Sci. **10**: 274. 1826.

Culms slender, erect, somewhat rough, 8'–2½° tall. Leaves 1½''–3½'' wide, the upper and the similar bracts much overtopping the spikes; sheaths rough-pubescent; staminate spike stalked or nearly sessile; pistillate spikes 2–4, loosely 1–6-flowered, erect, rather distant, stalked or the upper sessile; perigynia obovoid, obtusely 3-angled, finely many-striate, ascending, 2¼''–2½'' long, 1'' thick, tipped with a short stout oblique entire beak; scales ovate, rough-awned, scarious-margined, longer or shorter than the perigynia; stigmas 3.

In woods and thickets, Vermont and Ontario to Michigan, south to West Virginia, Kentucky and western Missouri. May–July.

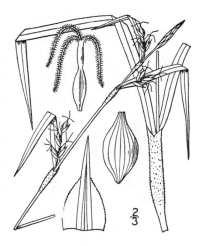

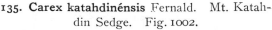

135. Carex katahdinénsis Fernald.   Mt. Katahdin Sedge.   Fig. 1002.

Carex katahdinensis Fernald, Rhodora **3**: 171. 1901.

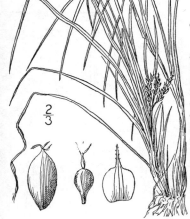

Glabrous, culms short, 2½' tall or less, roughened above. Leaves ¾''–1¼'' wide, much exceeding the culm; lower bracts similar to the culm-leaves, much exceeding spikes, their sheaths 1' long or less; staminate spike nearly sessile; pistillate spikes 3–4, closely approximate, erect, narrowly oblong, 3½''–7'' long, 2''–2½'' thick, rather closely 6–15-flowered, slightly exsert-peduncled; perigynia oval, suborbicular in cross-section, rounded to each end, finely many-striate, essentially beakless, 1¾'' long, slightly more than ½'' thick, the orifice entire; scales ovate, scarious-margined, more or less strongly cuspidate, as wide as but shorter than perigynia; stigmas 3.

Depot Pond, Mt. Katahdin, Maine; Lake St. John, Quebec. June–July. Possibly only a form of *Carex conoidea* Schk.

136. Carex conoìdea Schk.   Field Sedge.   Fig. 1003.

Carex conoidea Schk.; Willd. Sp. Pl. **4**: 280. 1805.

Glabrous, culms slender, rough, erect, 6'–30' tall. Leaves 1''–2'' wide, shorter than or but little exceeding the culm; lower bracts similar to the culm-leaves, sometimes but slightly overtopping the spike, their sheaths 1' long or less; staminate spike usually long-stalked; pistillate spikes 1–3, distant, erect, oblong or oblong-cylindric, 1½''–12'', usually 4''–8'', long, 2½'' thick, rather closely 8–25-flowered, the upper slightly exsert-peduncled, the lower strongly so; perigynia oval, suborbicular in cross-section, rounded to each end, finely many-striate, beakless, 1½'' long, slightly more than ½'' thick, the orifice entire; scales ovate, scarious-margined, acuminate tò rough-awned, the lower often longer than the perigynia, the upper shorter than or equalling them; stigmas 3.

In meadows, Nova Scotia to Ontario, south to Rhode Island, New Jersey, Ohio and Iowa, and in the mountains to North Carolina. May–June.

137. **Carex amphíbola** Steud. Narrow-leaved Sedge. Fig. 1004.

Carex amphibola Steud. Syn. Pl. Cyp. 234. 1855.
Carex grisea var. angustifolia Boott, Ill. 34. 1858.
Carex grisea var. (?) rigida Bailey, Mem. Torr. Club 1: 56. 1889.

Glabrous, culms very slender, slightly scabrous above, erect, or spreading, 1°-2° high. Leaves 1″-2″ wide, mostly erect and somewhat rigid, the basal shorter than the culm; bracts similar to the upper leaves, erect, not over 1″ wide, overtopping the spikes; staminate spike usually peduncled, sometimes sessile; pistillate spikes 2-4, erect, ½-1′ long, less than 2″ thick, loosely several-flowered, the upper sessile, the lower often from lower axils on long filiform stalks; perigynia oblong or obo-void, scarcely turgid, firm, pointed but beakless, 3-angled, many-striate, more or less 2-ranked, 2″ long, about 1″ thick, longer or lower shorter than the ovate, scarious-margined, awned spreading scales; stigmas 3.

In dry soil, New York to Iowa and Missouri, south to Florida and Texas. Intergrades with the next. April–June.

138. **Carex grísea** Wahl. Gray Sedge. Fig. 1005.

Carex grisea Wahl. Kongl. Vet. Acad. Handl. (II.) **24**: 154. 1803.

Glabrous, culms stoutish, erect or ascending, smooth or nearly so, 1°-2½° high. Leaves light green, some-times slightly glaucous, flat, usually soft and spread-ing, 2″-3½″ wide, the basal shorter than or exceed-ing the culm; bracts similar to the leaves, spread-ing, much overtopping the spikes; staminate spike solitary, sessile, or short-peduncled; pistillate spikes 3-5, dense, oblong, 5-15-flowered, 4″-12″ long, 2″-3½″ thick, the upper usually sessile and close together, the lower slender-stalked and distant, but not from lower axils; perigynia oblong, about 2½″ long, 1″ thick, nearly terete, turgid, subacute but beakless, finely many-striate, longer or the lower shorter than the ovate, scarious-margined, cuspidate or awned scales; stigmas 3.

In woods and thickets, Maine to Ontario and Minne-sota, south to North Carolina and Arkansas. May–July.

139. **Carex glaucòdea** Tuckerm. Glaucescent Sedge. Fig. 1006.

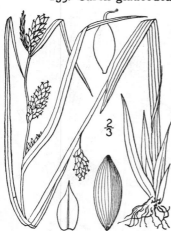

Carex grisea var. mutica Carey, in A. Gray, Man. 552. 1848. Not C. mutica R. Br. 1823.

Carex glaucodea Tuckerm.; Olney, Proc. Am. Acad. **7**: 395. 1868.

Similar in habit to *Carex grisea,* but pale and usu-ally very glaucous all over, culms smooth, erect or spreading, 4″-24′ high. Leaves 2″-5″ wide, the basal shorter or longer than the culm; bracts foliaceous, overtopping the spikes; staminate spike sessile or nearly so; pistillate spikes 3-5, 5″-15″ long, 2½″ thick, erect, densely 10-45-flowered, widely separate, the lower slender-stalked; perigynia oblong, many-striate, 1½″-2″ long, sub-acute, beakless, mostly nearly twice as long as the ovate scarious-margined, acute, cuspidate or short-awned scales; stigmas 3.

In open fields and meadows, Massachusetts to Ontario, Illinois, Virginia and Arkansas. Intergrades with the next. May–July.

140. Carex flaccospérma Dewey.   Thin-fruited Sedge.   Fig. 1007.

Carex laxiflora var. (?) *mutica* Torr. Ann. Lyc. N. Y. **3**:
414. 1836. Not *C. mutica* R. Br. 1823.
Carex flaccosperma Dewey, Am. Journ. Sci. (II.) **2**: 245.
1846.

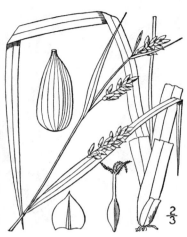

Similar to *Carex grisea* and *C. glaucodea;* slightly glaucous, rather deep green, culms erect, 6′–2° high. Leaves thin and flat, the basal ones 3″–9″ wide, usually shorter than the culm, the bracts leafy, much overtopping the spikes; staminate· spike sessile or nearly so; pistillate spikes 2–4, 6″–15″ long, 2″–3″ thick, densely 10–40-flowered, widely separate, oblong, erect, the lower slender-stalked; perigynia oblong, 3-angled, striate-nerved, subacute, 2¼″–3″ long; scales broadly ovate, green, with slightly scarious margins, acute, cuspidate or the upper obtuse, 2–3 times shorter than the perigynia; stigmas 3.

Southern Missouri to Texas, east to North Carolina and Florida. June–July.

141. Carex gracíllima Schwein.   Graceful Sedge.   Fig. 1008.

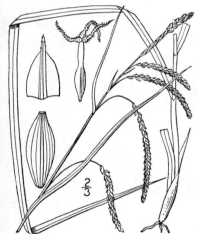

Carex gracillima Schwein. Ann. Lyc. N. Y. **1** : 66.   1824.
Carex gracillima var. *humilis* Bailey, Mem. Torr. Club
1 : 71. 1889.

Glabrous, culms slender, erect, smooth or nearly so, 1°–3° high. Leaves dark green, 1½″–4½″ wide, shorter than the culm, the basal ones wider than the upper; lower bract foliaceous, sometimes overtopping the spikes; spikes 3–5, narrowly cylindric, usually densely flowered except at the base, 1′–2½′ long, about 1½″ thick, or sometimes much reduced, filiform-stalked and drooping or ascending, the upper one partly or wholly staminate; perigynia ovoid-oblong, slightly swollen, few-nerved, glabrous, 1½″ long or less, rounded at apex; scales thin, ovate-oblong, very obtuse or the lower cuspidate, pale, scarious-margined, one-half as long as the perigynia or lower nearly as long; stigmas 3.

In moist woods and meadows, Newfoundland to Manitoba, North Carolina, Ohio and Michigan. May–July. A hybrid with *C. complanata* occurs at Philipstown, Putnam County, N. Y.

Carex Sullivántii Boott, is a hybrid with *C. hirtifolia,* found in Ohio, New York and Delaware.

142. Carex prásina Wahl.   Drooping Sedge.   Fig. 1009.

Carex prasina Wahl. Kongl. Vet. Acad. Handl. (II.) **24**:
161. 1803.
Carex miliacea Muhl.; Willd. Sp. Pl. **4**: 290. 1805.

Glabrous, rather light green, culms slender, slightly roughened above, sharply 3-angled, 1°–2½° high. Leaves shorter than or equalling the culm, flaccid, roughish, 1¼″–2½″ wide; lower bract similar, short-sheathing, commonly overtopping the spikes; staminate spike solitary, stalked, sometimes partly pistillate; pistillate spikes 2 to 4, narrowly linear-cylindric, drooping, the lower filiform-stalked, the upper short-stalked, 1′–2½′ long, 2″ in diameter, many-flowered, attenuate at base; perigynia light green, ovate-lanceolate, triangular, 1½″–2″ long, ¾″ wide, nerveless or nearly so, tapering into a smooth minutely 2-toothed or entire beak; scales ovate, acute, acuminate, or short-awned, hyaline, with green midrib, shorter than the perigynia; stigmas 3.

In meadows and moist thickets, Maine to Michigan, District of Columbia and Ohio, south in the Alleghanies to Georgia. Ascends to 4200 ft. in Virginia. May–July.

143. Carex formòsa Dewey. Handsome Sedge. Fig. 1010.

Carex formosa Dewey, Am. Journ. Sci. **8** : 98. 1824.

Culms slender, smooth, erect, 1°–2½° tall. Leaves flat, more or less pubescent, the basal 1½″–3½″ wide, shorter or longer than the culm; lower bract similar to the shorter culm-leaves; spikes 3–5, all gynaecandrous, oblong-cylindric, dense, ½′–1¼′ long, 2½″ in diameter, filiform-stalked, spreading or drooping, the lower distant; perigynia oblong-ovoid, glabrous, ascending, swollen, faintly few-nerved, 2″ long, 1″ thick, tipped with a very short, emarginate beak; scales lanceolate or ovate, green, with scarious margins, mostly acute or obtuse, and much shorter than the perigynia, or the lower short-awned and equalling them; stigmas 3.

In dry woods and thickets, Massachusetts and Vermont to southern Ontario, New York and Michigan. June–July.

144. Carex oxýlepis Torr. & Hook. Sharp-scaled Sedge. Fig. 1011.

Carex oxylepis Torr. & Hook. Ann. Lyc. N. Y. **3** : 409. 1836.

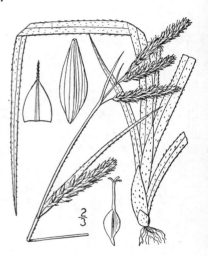

Culms slender, smooth, or nearly so, erect, 1°–2½° tall. Leaves flat, 1½″–3½″ wide, pubescent, especially on the sheaths, shorter than or exceeding the culm, the lower bract similar but narrower; spikes 4 or 5, linear-cylindric, ¾″–2′ long, 1½″–2″ in diameter, rather densely many-flowered, filiform-stalked and at maturity spreading or drooping, the terminal one normally gynaecandrous; perigynia oblong-oval, obscurely 3-angled, round-tapering at base, slightly swollen, 1¾″–2″ long, scarcely 1″ thick, punctate, several-nerved, rounded and abruptly minutely beaked, the orifice emarginate at apex; scales ovate-lanceolate with broad white scarious margins, acuminate or short-awned, about one-third shorter than the perigynia; stigmas 3.

Southern Missouri to Tennessee and South Carolina, south to Texas and Florida. April–May.

145. Carex Davísii Schwein. & Torr. Davis' Sedge. Fig. 1012.

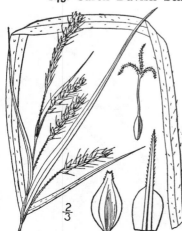

Carex Davisii Schwein. & Torr. Ann. Lyc. N. Y. **1** : 326. 1825.
Carex Torreyana Dewey, Am. Journ. Sci. **10** : 47. 1826.

Similar to *Carex oxylepis*, culms stouter, 1½°–3° tall. Leaves 1½″–3″ wide, flat, pubescent, especially on the sheaths, the basal ones shorter or longer than the culm; lower bract foliaceous, often overtopping the spikes; spikes 3–5, linear-oblong or oblong-cylindric, clustered near the summit or the lower one distant, dense, ¾′–1½′ long, 3″ or less in diameter, all filiform-stalked and at length spreading or drooping, the terminal gynaecandrous; perigynia oblong-ovoid, much swollen, glabrous, strongly several-nerved, 2″–2½″ long, rather more than 1″ thick, tipped with a very short, minutely 2-toothed beak; scales ovate-lanceolate, long-awned, spreading, shorter or longer than the perigynia; stigmas 3.

In moist thickets and meadows, Massachusetts to Minnesota, south to Georgia, Kentucky and Texas. May–July.

146. Carex aestivalifórmis Mackenzie.　False Summer Sedge.　Fig. 1013.

C. gracillima × *aestivalis?* Bailey, Bull. Torr. Club **20**: 419.　1893.
C. aestivalifórmis Mackenzie, Bull. Torr. Club **37**: 238. 1910.

Culms slender, erect, strongly roughened above, 1°– 2° high. Leaves flat, 1″–2″ wide, elongated, but usually shorter than the culms, rough, the lower sheaths (at least) short-pubescent, the blades sometimes slightly so; lower bract foliaceous, often overtopping the culm, strongly sheathing, the others smaller, little sheathing; spikes 3 to 5, narrowly linear-cylindric, 1′–1½′ long, 1½″ in diameter, many-flowered with perigynia rather close, filiform-stalked, and, at maturity, spreading or drooping, the terminal one gynaecandrous; perigynia oblong-ovoid, obscurely triangular, deep green, round-tapering at base, slightly swollen, somewhat less than 2″ long, nearly 1″ thick, strongly several-nerved, tapering and very short-beaked at apex, the orifice more or less hyaline and emarginate or shallowly bidentate; scales ovate, strongly hyaline-margined, somewhat shorter than the perigynia; stigmas 3.

In mountain meadows, New Jersey, New York and Pennsylvania. June–July.

147. Carex aestivàlis M. A. Curtis.　Summer Sedge.　Fig. 1014.

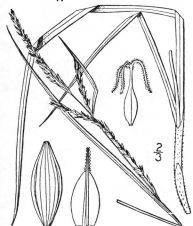

Carex aestivalis M. A. Curtis; A. Gray, Am. Journ. Sci. **42**: 28.　1842.

Culms very slender or filiform, erect or nearly so, roughish near the summit, 1°–2½° tall. Leaves flat, 1″–1½″ wide, elongated but usually shorter than the culm, the lower sheaths at least short-pubescent, the blades sometimes slightly so; lower bracts sheathing, similar to the leaves, but narrower, the others smaller, little sheathing; spikes 3–5, narrowly linear, erect or somewhat spreading, 1′–2′ long, about 1½″ thick, loosely many-flowered at the upper ones dense, the terminal one staminate at the base or also at the summit; perigynia narrowly elliptic, pointed at both ends, 3-sided, not inflated, glabrous, few-nerved, 1½″ long or less, ½″ thick, beakless, the orifice entire; scales ovate-oblong, obtuse, or the lower cuspidate or short-awned, green, thin, usually about one-half as long as the perigynia or more; stigmas 3.

In mountain woods, New Hampshire, Massachusetts and northern New York to Georgia.　Winter-grass. June–Aug.

148. Carex oblìta Steud.　Dark-green Sedge.　Fig. 1015.

Carex oblita Steud. Syn. Pl. Cyp. 231.　1855.
Carex glabra Boott, Ill. 93.　1860.
Carex venusta var. *minor* Boeckl. Linnaea **41**: 255.　1877.

Glabrous, culms slender, erect or lax, sharply 3-angled, smooth or very nearly so, 1°–3° high. Leaves 2″–3″ wide, shorter than the culm, slightly rough; lower bract similar to the culm-leaves but narrower, the sheaths puberulent; staminate spike solitary, filiform-stalked, sometimes partially pistillate; pistillate spikes 2–4, narrowly linear, 1′–2½′ long, about 2″ thick, loosely 4–18-flowered, slender-stalked, the upper mostly close together and spreading or ascending, the lower distant, drooping; perigynia dark green, coriaceous, 3-angled, glabrous, somewhat more than 2½″ long, less than 1″ thick, ascending, rather strongly about 10-nerved, tapering into a short hyaline-tipped, 2-toothed beak; scales obtuse or acute, strongly reddish-brown-tinged, one-third to one-half the length of the perigynia; stigmas 3.

In bogs, central New York, Pennsylvania and New Jersey to Alabama and Louisiana.　June–Aug.

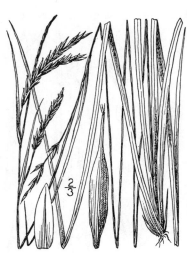

149. Carex débilis Michx.   White-edged Sedge. Fig. 1016.

Carex debilis Michx. Fl. Bor. Am. **2** : 172.   1803.
C. debilis var. *pubera* A. Gray, Man. Ed. 5, 593.   1867.
C. debilis var. *prolixa* Bailey, Proc. Am. Acad. **22** : 105.   1886.

Culms slender, slightly rough above, lax or erect, 1°–3° high. Leaves shorter than culm, light green, 1″–1½″ wide; lower bracts similar to the culm-leaves, the sheaths glabrous; staminate spike more or less stalked; pistillate spikes 2–4, narrowly linear, 1′–2′ long, 1½″–2″ wide, not approximate or but little so, weakly erect or drooping on slender peduncles, rather loosely 8–20-flowered; perigynia lanceolate, sessile, glabrous or puberulent, membranous, few-nerved, rather noticeably inflated, 3-angled, 3″–4½″ long, ¾″ wide, tapering into a subulate hyaline-tipped bidentate beak nearly 1″ long; scales ovate, obtuse, strongly white-hyaline-margined, one-third to one-half length of perigynia; stigmas 3.

Woods and copses, New Jersey to Tennessee, south to Florida and Texas.   Probably intergrades with the next. May–June.

150. Carex flexuòsa Muhl.   Slender-stalked Sedge.   Fig. 1017.

Carex tenuis Rudge, Trans. Linn. Soc. **7** : 97.   *pl. 9*.   1804. Not J. F. Gmel. 1791.
Carex flexuosa Muhl.; Willd. Sp. Pl. **4** : 297.   1805.
C. debilis var. *Rudgei* Bailey, Mem. Torr. Club **1** : 34.   1889.

Culms slender, rough above, erect or lax, 4′–3° high. Leaves usually shorter than the culm, light green, 1″–2″ wide; lower bracts similar to the culm-leaves, the sheaths glabrous; staminate spike short-stalked, sometimes partly pistillate; pistillate spikes 2–4, narrowly linear, ¾′–2½′ long, 1½″–2″ thick, loosely or alternately 12–25-flowered, filiform-stalked and spreading or drooping or sometimes erect; perigynia spindle-shaped, glabrous, membranous, few-nerved, from scarcely to noticeably inflated, 3-angled, 2¼″–3″ long, less than 1″ thick, tapering into a short, hyaline-tipped, 2-toothed beak; scales ovate or oblong, acute or short-cuspidate, scarious-margined, one-half as long as perigynia or longer, usually rusty-tinged; stigmas 3.

In woods, Newfoundland to Wisconsin, Virginia, the mountains of North Carolina and Kentucky.   Several slightly differing varieties have been described. Hybridizes with *C. Swànii*.   May–Aug.

151. Carex arctàta Boott.   Drooping Wood Sedge.   Fig. 1018.

Carex arctata Boott; Hook. Fl. Bor. Am. **2** : 227.   1840.
Carex arctata Faxoni Bailey, Bot. Gaz. **13** : 87.   1888.

Glabrous, culms slender, erect, 1°–2½° high, roughish above. Leaves flat, roughish-margined, much shorter than the culm, the basal ones 2½″–5″ wide; staminate spike solitary, short-stalked; pistillate spikes 2–5, linear, 1′–3′ long, 1½″–2″ thick, loosely 15–45-flowered, erect, ascending, or filiform-stalked, and at length drooping, the lower one usually remote; perigynia lanceolate, strongly stipitate, deep green, rather strongly few-nerved, narrowed at each end, 1¾″–2¼″ long, less than 1″ thick, 3-angled, tapering into a short, hyaline-tipped, 2-toothed beak; scales ovate, cuspidate or short-awned, about one-third shorter than the perigynia; stigmas 3.

In dry woods and thickets, Newfoundland and Quebec to Minnesota, Pennsylvania and Michigan.   May–June.

Carex Knieskèrnii Dewey, is probably a hybrid with *C. castanea*.   *C. arctàta* also hybridizes with *C. Swani*.

152. Carex assiniboinénsis W. Boott. Assiniboia Sedge. Fig. 1019.

C. assiniboinensis W. Boott, Bot. Gaz. **9** : 91. 1884.

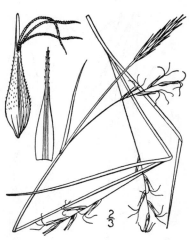

Glabrous and nearly smooth, culms slender, weak, aphyllopodic, 1°–2½° high, longer than the leaves, strongly reddened at base. Leaves and bracts ½″–1′ wide, the bracts sheathing; staminate spike long-stalked; pistillate spikes 2 or 3, widely separate, loosely and alternately 1–8-flowered, 4″–15″ long, drooping on filiform stalks or upper erect; perigynia very narrowly lanceolate, appressed, obtusely 3-angled, subulate-beaked above, 3″ long and 1″ thick above the base, densely short-tuberculate-hispid, the beak obliquely cut at orifice; scales lanceolate, scarious-margined, short-awned, nearly the length of the perigynia; stigmas 3.

In wet soil, northern Minnesota, North Dakota and Manitoba. May–July.

153. Carex castánea Wahl. Chestnut Sedge. Fig. 1020.

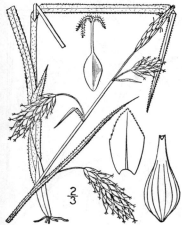

Carex castanea Wahl. Kongl. Vet. Acad. Handl. (II.) **24** : 155. 1803.
Carex flexilis Rudge, Trans. Linn. Soc. **7** : 98. *pl. 10.* 1804.

Culms slender, erect, rough above, 1°–3° tall, reddish-purple at base. Leaves 1¼″–3″ wide, pubescent, shorter than the culm; bracts linear-filiform, ½′–1½′ long; staminate spike short-stalked; pistillate spikes 1–4, approximate, oblong or oblong-cylindric, many-flowered, ½′–1′ long, about 3″ thick, drooping on filiform stalks, or upper spreading; perigynia glabrous, pale brown, ascending, oblong or oblong-lanceolate, slightly inflated, 3-angled, few-nerved, tapering gradually into a minutely 2-toothed beak one-half as long as the body; scales light chestnut, thin, ovate or ovate-lanceolate, acute or cuspidate, lacerate or entire, rather shorter than the perigynia; stigmas 3.

In dry thickets and on banks, Newfoundland to Minnesota, south to Connecticut, New York and the Great Lake region. June–July.

154. Carex capillàris L. Hair-like Sedge. Fig. 1021.

Carex capillaris L. Sp. Pl. 977. 1753.
Carex capillaris var. *elongata* Olney; Fernald, Proc. Am. Acad. **37** : 509. 1902.

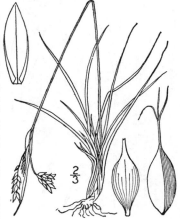

Glabrous, culms very slender or filiform, smooth, or a little roughish above, erect, 2′–20′ tall, often densely tufted. Leaves ½″–1″ wide, much shorter than the culm, flat or somewhat involute in drying; lower bract similar, the upper much narrower, all sheathing; spikes all filiform-stalked, the terminal staminate, usually exceeded by pistillate; pistillate spikes 1–3, narrowly oblong, 2″–6″ long, 1½″ thick, nodding, 2–12-flowered; perigynia oblong, 3-angled, light green, almost nerveless, about 1¼″ long, rather less than ½″ thick, the slender beak with oblique, hyaline orifice; scales ovate, scarious-margined, obtuse or acute, shorter than the perigynia; stigmas 3.

Greenland to Alaska, south to Maine, the White Mountains, northern New York, Michigan, and in the Rocky Mountains to Colorado and Utah. Also in Europe and Asia.

155. Carex cherokeénsis Schwein. Cherokee Sedge. Fig. 1022.

Carex recurva Muhl. Descr. Gram. 262. 1817. Not Huds. 1778.
Carex cherokeensis Schw. Ann. Lyc. N. Y. 1: 71. 1824.

Glabrous, light green, culms smooth, erect, 1°–2° high, darkened, but not fibrillose at base, phyllopodic, from stout rootstocks. Leaves thick, 1½″–3½″ wide, strongly striate, shorter than culm, the similar bracts shorter, long-sheathing; staminate spikes 1–3, peduncled; pistillate spikes 2–6, occasionally two from a sheath, oblong-cylindric with 10–50 closely arranged perigynia, 7½″–23″ long, 3″–4½″ wide, erect or somewhat drooping, widely separate, the lower long-peduncled, the upper short-peduncled; perigynia ovoid, appressed-ascending, smooth, slightly inflated, pale green, few-nerved, obtusely triangular, 2½″ long, 1¼″ wide, contracted into a short beak, scarious and obliquely cut at orifice, at length bidentate; scales lanceolate, acuminate, hyaline-margined, usually shorter than perigynia; stigmas 3.

In the low country from southeastern Missouri to Georgia, Florida and Texas. March–May.

156. Carex Sprengèlii Dewey. Long-beaked Sedge. Fig. 1023.

Carex longirostris Torr.; Schwein. Ann. Lyc. N. Y. 1: 71. 1824. Not Krock. 1814.
Carex Sprengelii Dewey; Spreng. Syst. 3: 827. 1826. .

Glabrous, light green, culms very slender, roughish above, erect, 10′–3° high, strongly fibrillose at base. Leaves flat, slightly scabrous, 1¼″–2″ wide, usually not exceeding the culm, the bracts similar, shorter, short-sheathing; staminate spikes 1–3, slender-stalked, rarely pistillate at the base; pistillate spikes 2–4, oblong-cylindric, 10–40-flowered, ½′–2′ long, 3″–5″ in diameter, pendulous or erect, all filiform-stalked, or the upper one nearly sessile; perigynia ascending or somewhat spreading, the body 1¼″–1½″ long, smooth, short-oblong, slightly inflated, pale, strongly 1-nerved on each side, contracted into a very slender beak once to twice its length, the beak obliquely cut, at length deeply bidentate; scales lanceolate, long-acuminate or acute, scarious-margined; stigmas 3.

On banks and in moist thickets, New Brunswick to Alberta, south to Massachusetts, New Jersey, Pennsylvania and Nebraska. May–July.

157. Carex atrofúsca Schk. Dark-brown Sedge. Fig. 1024.

C. atrofusca Schk. Riedgr. 1: 106. p. Y. *f. 82.* 1801.

C. ustulata Wahl. Vet.-Akad. Handl. 24: 156. 1803.

Glabrous, smooth, culms obtusely triangular, slender, erect, 4′–12′ high, short-stoloniferous. Leaves 1″–1½″ wide, mostly clustered at the base, much shorter than the culm, usually 1′–2½′ long; bracts sheathing, dark-tinged, the blades short; terminal spike slender-peduncled, staminate or gynaecandrous; pistillate spikes 2 or 3, approximate or a little separate, filiform-stalked, drooping, 4″–9″ long, 3½″–4½″ thick, 15–30-flowered; perigynia ovate-oval, flattened, triangular, blackish, 2½″ long, less than 1″ wide, rounded at base, appressed, abruptly very short-beaked, the beak minutely bidentate; scales oblong-ovate, obtuse or acute, black with lighter mid-rib, shorter than perigynia; stigmas 3.

Arctic America, Labrador and Greenland. Also in Europe and Asia. Summer.

158. Carex misándra R. Br.   Short-leaved Sedge. Fig. 1025.

Carex misandra R. Br. Suppl. Parry's Voy. 283. 1824.

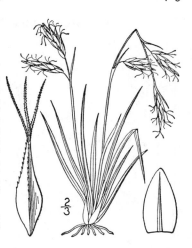

Glabrous and smooth, culms very slender, erect, 1′–15′ tall.   Leaves 1″–1½″ wide, clustered at the base, usually much shorter than the culm, seldom over 2½′ long; bracts narrowly linear, sheathing, with colored sheaths, not overtopping the spikes; terminal spike often partially pistillate at base or summit, slender-stalked; pistillate spikes 1 to 3, filiform-stalked, 3″–7″ long, about 2″ thick, rather few-flowered, drooping or weakly erect; perigynia narrowly lanceolate, tapering and beaked at the apex, narrowed at the base, 2¼″ long, ½″ wide, ascending, dark brown, serrulate above, the orifice oblique, at length bidentate; scales oval, acutish, purple-black with narrow white margins, somewhat shorter than the perigynia; stigmas 3, rarely 2.

Throughout arctic America, extending south in the Rocky Mountains to the higher summits of Colorado.   Also in arctic Europe and Asia. Summer.

159. Carex Swànii (Fernald) Mackenzie.   Swan's Sedge.   Fig. 1026.

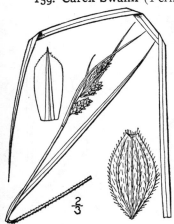

C. *viréscens* var. *Swanii* Fernald, Rhodora **8**: 183. 1906.
C. *Swanii* Mackenzie, Bull. Torr. Club **37**: 246. 1910.

Culms very slender, erect or somewhat reclining, rough above, 6′–20′ tall, little reddened at base, exceeded by leaves.   Leaves light green, 1¼″ wide, pubescent, especially on the sheaths, the uppermost usually 1½′–2′ below spikes; spikes 2–5, very short-stalked, erect or nearly so, oblong-cylindric, densely many-flowered, 3″–10″ long, 1½″–2½″ in diameter, the terminal gynaecandrous; lowest bract very slender, ¼″ wide, twice exceeding spikes; perigynia 3-sided, obovoid, about 1″ long, ascending, densely pubescent, few-nerved, green; beakless, the orifice entire; scales oblong-ovate, the lower cuspidate by the excurrent midvein, scarious-margined, slightly shorter than the perigynia; stigmas 3.

In dry woods and thickets, Nova Scotia to Michigan, North Carolina, Tennessee and Missouri.   Illustrated in our first edition as *C. viréscens.*   June–July.

160. Carex viréscens Muhl.   Ribbed Sedge.   Fig. 1027.

Carex virescens Muhl.; Willd. Sp. Pl. **4**: 251. 1805.
Carex virescens var. *costata* Dewey, Am. Journ. Sci. **9**: 260. 1825.
C. *costellata* Britton, Bull. Torr. Club **22**: 223. 1895.

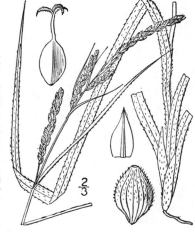

Similar to the preceeding species, but taller and more spreading; culms slender, 1½°–2½° long, strongly reddened at base, exceeding leaves.   Leaves 1½″–2″ wide, pubescent, especially on the sheaths, shorter than the culm, the upper one usually ¾′–1¼′ below spikes; spikes 2–5, linear-cylindric, many-flowered, rather loose, ½′–1½′ long, 1½″ in diameter, erect or slightly spreading, the terminal gynaecandrous, the lower one commonly filiform-stalked; lowest bract leaflet-like, ¼″–1″ wide, somewhat exceeding spikes; perigynia oblong-elliptic to rarely obovoid, densely pubescent, narrowed at each end, usually strongly several-ribbed, 1″ long, rather more than ½″ thick, beakless, the orifice entire; scales ovate, scarious-margined, acuminate or cuspidate, shorter than the perigynia; stigmas 3.

In woods, Maine and Ontario to Georgia and Kentucky. Ascends to 4000 ft. in North Carolina. June–Aug.

161. Carex complanàta Torr.   Hirsute Sedge.   Fig. 1028.

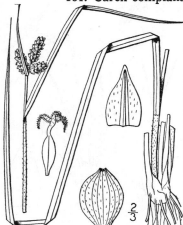

Carex triceps Michx. Fl. Bor. Am. **2**: 170. 1803. Not Schrank. 1789.
Carex hirsuta Willd. Sp. Pl. **4**: 252. 1805. Not Suter, 1802.
Carex complanata Torr. Ann. Lyc. N. Y. **3**: 408. 1836.
Carex triceps var. *hirsuta* Bailey, Mem. Torr. Club **1**: 35. 1889.

Light green, culms slender, rough above, erect or reclining, 6′–3° high. Leaves 1″–2″ wide, pubescent, at least on the sheaths, shorter than the culm, the lower bract similar, much exceeding the culm; spikes 2–5, oblong or oblong-cylindric, dense, erect, sessile or very nearly so, 3″–9″ long, 2½″–3½″ in diameter, usually clustered at the summit, the terminal one gynaecandrous; perigynia oval or obovoid, flattened, not inflated, green or greenish-brown, lightly nerved, imbricated, slightly pubescent when young, glabrous when mature, 1″ long, beakless and very rounded at apex, the orifice entire; scales ovate, scarious-margined, short-cuspidate to obtuse, rather shorter than the perigynia; achene elliptic-obovoid, its summit straight, but style often short-bent; stigmas 3.

In woods, fields and swamps, Maine to southern Ontario and Michigan, south to Florida and Texas. Ascends to 2500 ft. in Virginia. April–Aug.

162. Carex caroliniàna Schwein.   Carolina Sedge.   Fig. 1029.

Carex carolinana Schwein. Ann. Lyc. **1**: 67. 1824.
Carex Smithii Porter; Olney, Car. Bor. Am. 2, name only. 1871. Not Tausch. 1821.
C. triceps var. *Smithii* Bailey, Bot. Gaz. **13**: 88. 1888.

Culms very slender, erect, rough above, 1°–2½° tall. Leaves 1″–1½″ wide, rather dark green, glabrous, except on the sheaths, the upper and the similar but narrower bracts usually much overtopping the spikes; spikes 2–4, oblong, dense, sessile or nearly so, 4″–9″ long, 2″ in diameter, erect, clustered at the summit, the upper gynaecandrous; perigynia squarrose, obovoid, swollen, and orbicular in cross-section, 1″ long, about ½″ in diameter, not imbricated, coarsely nerved, brown, glabrous at least when mature, pointed at apex, the orifice entire or emarginate; scales brown, ovate, mucronate; achene pyriform, bent at the summit or tipped with a bent style; stigmas 3.

In meadows, New Jersey and Pennsylvania to North Carolina and Texas. May–July.

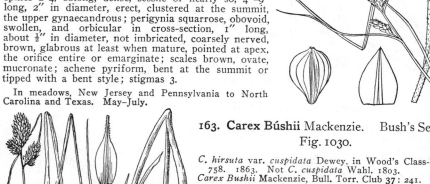

163. Carex Búshii Mackenzie.   Bush's Sedge. Fig. 1030.

C. hirsuta var. *cuspidata* Dewey, in Wood's Class-book, 758. 1863. Not *C. cuspidata* Wahl. 1803.
Carex Bushii Mackenzie, Bull. Torr. Club **37**: 241. 1910.

Culms slender, erect, roughish above, 1½°–2½° tall. Leaves 1″–1½″ wide, glabrous or pubescent beneath, the sheaths pubescent, the upper and the similar but reduced bracts much overtopping the spikes; spikes 2 or 3, oblong or oblong-cylindric, dense, sessile or nearly so, 3″–10″ long, 2½″–4″ in diameter, erect, clustered at the summit, the upper gynaecandrous; perigynia ascending, obovoid, orbicular in cross-section, 1¼″–1½″ long, swollen, not imbricated, about ¾″ in diameter, coarsely ribbed, green, glabrous when mature, pointed at apex, the orifice entire or emarginate; scales lanceolate, the middle and lower strongly rough-cuspidate, exceeding perigynia; achene obovoid, tipped with a bent style; stigmas 3.

In meadows, Rhode Island and New York to South Carolina and Oklahoma. May–July.

164. Carex palléscens L. Pale Sedge. Fig. 1031.

Carex pallescens L. Sp. Pl. 977. 1753.

Light green, culms slender, erect, sparsely hairy, $4'-2°$ tall. Leaves flat, $1''-1\frac{1}{2}''$ wide, short-pubescent, at least on the sheaths; lower bract similar to the culm-leaves, exceeding the spikes; staminate spike solitary, short-stalked; pistillate spikes 2–4, oblong, erect or somewhat spreading, short-stalked or the upper one nearly sessile, densely many-flowered, $2\frac{1}{2}''-10''$ long, $2''-3\frac{1}{2}''$ in diameter, mostly approximate; perigynia elliptic, slightly inflated, obscurely triangular, pale, $1\frac{1}{4}''-1\frac{1}{2}''$ long, $\frac{1}{2}''$ thick, obtuse, thin, faintly nerved, beakless, the orifice entire; scales ovate, membranous, acute to short-awned, a little longer or a little shorter than the perigynia; stigmas 3.

In fields and meadows, Newfoundland to New Jersey, Pennsylvania, Illinois and Wisconsin. Also in Europe and Asia. May–July.

165. Carex abbreviàta Prescott. Torrey's Sedge. Fig. 1032.

Carex Torreyi Tuckerm. Enum. Meth. 21. 1843. Not *C. Torreyana* Schwein. 1824.
Carex abbreviata Prescott; Boott, Trans. Linn. Soc. 20: 141. 1846.

Pale green, culms slender, rather stiff, erect, $10'-20'$ tall, finely pubescent. Leaves $\frac{3}{4}''-1\frac{1}{2}''$ wide, usually shorter than culm, rather densely short-pubescent; lower bract shorter or longer than spikes; staminate spike solitary, usually short-stalked; pistillate spikes 1–3, short-oblong, dense, $3''-8''$ long, about $3''$ thick, sessile or short-stalked, erect, clustered; perigynia obovoid, somewhat inflated and rather obscurely triangular, glabrous, $1\frac{1}{4}''-1\frac{1}{2}''$ long and slightly more than $\frac{1}{2}''$ thick, strongly many-nerved, depressed at apex and abruptly tipped by a short entire beak; scales ovate, acute to cuspidate, shorter than the perigynia; stigmas 3.

In dry soil, Minnesota to Saskatchewan, south in the Rocky Mountains to Colorado. Reported from New York and Pennsylvania. June–July.

166. Carex glàuca Murr. Heath-sedge. Fig. 1033.

Carex glauca Murr. Prodr. Stirp. Gotting. 76. 1770.

Carex flacca Schreb. Spic. Fl. Lips. App. 669. 1771.

Glabrous, pale green and glaucous, culms slender, erect, smooth or roughish above, $1°-2°$ tall, the rootstocks long and stout. Leaves usually shorter than the culm, $1\frac{1}{4}''-2\frac{1}{2}''$ wide, the sheaths scabrous; lower bract similar to the leaves, but narrower; staminate spikes mostly $1'-1\frac{1}{2}'$ long, stalked; pistillate spikes 2 or 3, ascending or at length drooping, slender-stalked, linear-cylindric, $1'-2'$ long, $3''-4''$ thick, densely many-flowered, commonly staminate at the summit; perigynia brown, ellipsoid, faintly few-nerved, or nerveless, minutely granulate or papillose, $1\frac{1}{2}''$ long, abruptly minutely beaked, the orifice entire; scales ovate or lanceolate, brown with a green midvein, obtusish, acute or acuminate, about as long as the perigynia; stigmas 3.

In meadows, Nova Scotia, Quebec and Ontario. Naturalized from Europe. Carnation-grass. Gilliflower-grass. Pink-grass. June–Aug.

167. Carex scabràta Schwein. Rough Sedge. Fig. 1034.

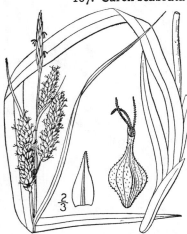

Carex scabrata Schwein. Ann. Lyc. N. Y. 1: 69. 1824.

Glabrous, strongly stoloniferous, culms sharply triangular, weakly erect, very rough above, leafy, 1°–3° high. Leaves very rough above, much elongated, 2½″–9″ wide, the bracts similar but narrower and usually exceeding the culm; staminate spike stalked; pistillate spikes 3–6, erect, the upper short-stalked, the lower sometimes spreading or drooping, narrowly cylindric, densely many-flowered, ½′–2½′ long, 2½″–4″ in diameter; perigynia greenish-brown, 1½″–2″ long, nearly 1″ wide, the body obovoid, slightly inflated, strongly nerved, papillose, abruptly contracted into a long beak with obliquely cut, at length slightly bidentate, hyaline orifice; scales lanceolate, acute or short-awned, strongly nerved, shorter than the perigynia; stigmas 3.

In moist woods and thickets, eastern Quebec to Ontario, Michigan, South Carolina and Tennessee. Ascends to 4200 ft. in Virginia. Hybridizes with *Carex crinìta*. May–Aug.

168. Carex rariflòra (Wahl.) J. E. Smith. Loose-flowered Alpine Sedge. Fig. 1035.

Carex limosa γ *rariflora* Wahl. Vet.-Akad. Handl. 24: 162. 1803.
Carex rariflora J. E. Smith, Engl. Bot. *pl. 2516.* 1813.

Glabrous, culms very slender, rather stiff, erect, obtusely triangular, 4′–14′ tall, smooth, from slender elongated rootstocks. Leaves ¾″–1¼″ wide, flat, green, shorter than the culm, the lower very short; bracts very short, purple at the base; staminate spike solitary, long-stalked, sometimes with a few pistillate flowers at the base; pistillate spikes 1 to 3, narrowly oblong, 3–18-flowered, 3″–8″ long, 2″–2½″ in diameter, nodding or ascending on filiform stalks; perigynia pale, ovoid-elliptic, thick, slightly inflated, tapering at base, 1¾″ long, nearly 1″ wide, slightly 2-edged, very obscurely nerved, rounded at apex and essentially beakless, the orifice entire; scales broadly oval, purple-brown with a greenish midvein, obtuse or short-mucronate, about equalling and half enveloping the perigynia; stigmas 3.

In wet places, Greenland and Labrador to Hudson Bay, locally south to Mt. Katahdin, Maine. Also in Europe and Asia. Summer.

169. Carex limòsa L. Mud Sedge. Fig. 1036.

Carex limosa L. Sp. Pl. 977. 1753.

Glabrous, glaucous, strongly long-stoloniferous, culms slender, rough above, sharply triangular, erect, 6′–2° tall. Leaves 1½″ wide, usually shorter than the culm, involute; bracts linear-filiform, the lower ½′–2½′ long, its auricles brownish; staminate spike solitary, long-stalked; pistillate spikes 1 to 3, filiform-stalked and drooping or the upper nearly erect, oblong, 5″–13″ long, 2½″–4″ thick, 8–30-flowered; perigynia glaucous-green, broadly ovate, strongly flattened and 2-edged, 1¼″ long, 1″ wide, few-nerved, tipped with a very minute entire beak, nearly as long as the ovate, usually dark-tinged, short-cuspidate or acute scales; stigmas 3.

In bogs, Labrador to British Columbia, south to Maine, New Jersey, Ohio, Iowa and Colorado. Also in Europe. Summer.

Carex macrochaèta C. A. Meyer (*Carex podocàrpa* Bailey, Proc. Am. Acad. 197, and of our 1st ed., p. 312, not R. Br.) is omitted as probably not found in our range; the real *Carex podocarpa* R. Br. is also omitted for the same reason.

170. Carex paupércula Michx.   Bog Sedge.   Fig. 1037.

Carex paupercula Michx. Fl. Bor.-Am. **2**: 172.   1803.
Carex irrigua Smith; Hoppe, Caric. 72.   1826.
C. paupercula var. *irrigua* and var. *pallens* Fernald,
   Rhodora 8: 76–77.   1906.

Glabrous, culms slender, tufted, sharply angled,
smooth or strongly roughened, erect, 4′–2½° tall.
Leaves flat, 1″–2″ wide, green, commonly shorter
than the culm, the lower bract similar and usually
overtopping the spikes; staminate spike usually solitary, frequently gynaecandrous; pistillate spikes 1–4,
oblong, filiform-stalked and drooping or somewhat
erect, 2″–10″ long, 2″–4″ in diameter; perigynia
glaucous-green, flattened, 2-edged, 1¼″–1½″ long, over
1″ wide, few-nerved, minutely granulate-papillose,
orbicular or broadly ovate, essentially beakless, the
orifice entire; scales lanceolate or ovate-lanceolate,
long-acuminate or awned, varying from dark green
and brownish-tinged to purplish-brown, 1½–2 times
as long as the perigynia; stigmas 3.

In bogs, Newfoundland and Labrador to British Columbia, Connecticut, Pennsylvania and Utah.   Also in
Europe and Asia.   Has been confused with *Carex magel-
lanica* Lam.   Ascends to 4600 ft. in the Adirondacks.
Summer.

171. Carex Barráttii Schw. & Torr.   Barratt's Sedge.   Fig. 1038.

Carex littoralis Schwein. Ann. Lyc. N. Y. **1**: 70.   1824.
   Not Krock. 1814.
Carex Barrattii Schwein. & Torr. Ann. Lyc. N. Y. **1**: 361.
   1825.

Glabrous, pale green, culms erect, slender, smooth,
1°–3° tall, aphyllopodic and strongly filamentose at
base.   Leaves 1½″–2″ wide, smooth, slightly glaucous,
long-attenuate, usually much shorter than the culm;
bracts not sheathing, the lower usually short and
narrow, its auricles prominent, usually dark-tinged;
staminate spikes 1–3, usually rather long-stalked;
pistillate spikes 2–4, drooping or the upper ascending, slender-stalked, linear-cylindric, ½′–2′ long, 3″ in
diameter, mostly androgynous; perigynia ovoid or
oval, dark at maturity, faintly few-nerved, 1¼″–1½″
long, slightly inflated and obscurely triangular, tipped
with a minute entire beak; scales brown-purple with
lighter margins, obtuse, usually shorter than the perigynia; stigmas 3.

In swamps, Connecticut to Pennsylvania and North
Carolina.   Mostly near the coast.   May–June.

172. Carex Hálleri Gunn.   Alpine Sedge.
Fig. 1039.

Carex Halleri Gunn. Fl. Norveg. n. 849.   1766–1772.
Carex alpina Sw.; Lilj. Sv. Flora, Ed. 2, 26.   1798.
Carex Vahlii Schk. Riedgr. 87.   1801.

Culms slender, phyllopodic, erect, leafy below the
middle, 6′–2° tall.   Leaves roughish, ½″–1½″ wide, usually shorter than the culm; spikes 2–4 (commonly
3), clustered at the summit, the terminal 1 or 2 gynaecandrous, oblong or globose, 2″–5″ long, closely 8–25-
flowered, sessile or the lower short-peduncled; perigynia oval, orbicular or obovoid, light green, 1¼″
long or less, tipped with a very short and minutely
2-toothed beak, nerveless or with a few very faint
nerves, slightly shorter than the ovate black or
purple-brown, light-margined, obtuse or acutish
scales; style short; stigmas 3.

In rocky places, Greenland to Alaska, eastern Quebec,
western Ontario, Lake Superior region and in the western mountains.   Also in Europe and Asia.   July–Aug.

173. Carex stylòsa C. A. Meyer.   Variegated Sedge.   Fig. 1040.

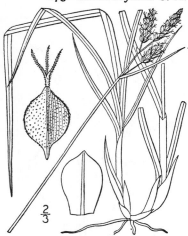

Carex stylosa C. A. Meyer, Mem. Acad. St. Petersb. Div. Sav. 1 : 222. *pl. 12.* 1831.

Culms phyllopodic, slender, erect, 6'–1½° tall, rough and leafless above. Leaves ¾"–1½" wide, usually shorter than the culm; staminate spike solitary, nearly sessile, sometimes partly pistillate; pistillate spikes 2 or 3, oblong-cylindric, erect, 3½"–9" long, 2"–3½" in diameter, the lowest slender-stalked and subtended by a linear-subulate bract; perigynia ovoid or ellipsoid, not flattened, slightly inflated, brown, minutely granulate, 1¼"–1¼" long, nerveless, except for two ribs, minutely apiculate, the orifice entire and closed by the protruding style, ½" long; scales black with light midvein, obtusish, shorter than the perigynia; stigmas 3.

Labrador and Greenland to Alaska.   Also in Europe and Asia.   Summer.

174. Carex Parryàna Dewey.   Parry's Sedge.   Fig. 1041.

C. Parryana Dewey, Am. Journ. Sci. **27** : 239. 1835.

Glabrous, culms phyllopodic, slender, smooth, erect, stiff, 5'–20' tall, leafless above. Leaves 1"–2" wide, much shorter than the culm, their margins revolute; spikes 1–4, dense, erect, contiguous, linear-cylindric, 3"–15" long, 1½" in diameter, the upper sessile and staminate below or throughout, the lowest short-stalked and subtended by an almost filiform bract; perigynia plano-convex, broadly obovate, two-edged, pale, 1¼" long, minutely papillose, faintly few-nerved, very minutely beaked, the orifice emarginate; scales ovate, obtuse to mucronulate, dark brown with lighter margins, nearly concealing perigynia; stigmas 3.

Hudson Bay to British Columbia, south to North Dakota and in the Rocky Mountains to Colorado.   Summer.

175. Carex atratifórmis Britton.   Black Sedge.   Fig. 1042.

Carex ovata Rudge, Trans. Linn. Soc. **7** : 96. *pl. 9.* 1804. Not Burm. 1768.
Carex atrata var. *ovata* Boott, Ill. 114. 1862.
Carex atratiformis Britton, Bull. Torr. Club **22** : 222. 1895.

Glabrous, culms phyllopodic, slender, erect, sharp-angled, roughish above, 8'–4° tall, usually leafy only below. Leaves smooth or roughish, 1"–2" wide, rarely over 6' long, much shorter than the culm; spikes 2–6, dense, oblong or oblong-cylindric, 4"–12" long, 2"–3" in diameter, the terminal gynaecandrous and short-peduncled, the others pistillate or gynaecandrous, slender-stalked, drooping when mature; lower bracts ½'–1½' long, very narrow, the upper ones subulate; perigynia flattened, ovate or nearly orbicular, puncticulate, ascending, about 1¼" long, tipped with a very short, 2-toothed beak; scales black or reddish-brown, oblong or obovate, obtuse or subacute, slightly narrower than the perigynia and about equalling them; stigmas 3.

Labrador and Newfoundland to the mountains of northern New England, west to Alberta.   June–Aug.

176. Carex Buxbaùmii Wahl. Brown Sedge. Fig. 1043.

Carex polygama Schkuhr, Reidgr. 1: 84. 1801. Not J. F. Gmel. 1791.
C. Buxbaumii Wahl. in Vet.-Akad. Handl. 24: 163. 1803.
Carex fusca Bailey, Mem. Torr. Club 1: 63. 1889. Not All. 1785.

Glabrous, culms aphyllopodic, strongly filamentose, stiff, erect, sharp-angled, rough above, 8′-3° tall. Leaves rough, erect, 1″-2″ wide, shorter than or exceeding the culm; spikes 2-7, oblong or cylindric, erect, all sessile and close together or the lowest sometimes distant and very short-stalked, 4″-20″ long, about 4″ in diameter when mature, the terminal staminate at base or rarely throughout; perigynia elliptic or somewhat obovate, flat, ascending, 1½″-2″ long, very light green, granular, faintly nerved, beakless, the apex minutely 2-toothed; scales ovate, awn-tipped, black or dark brown with a green midvein, longer than the perigynia; stigmas 3.

In bogs, Greenland to Alaska, south to Georgia, Kentucky, Missouri, Utah and California. Also in Europe and Asia. May–July.

177. Carex Shortiàna Dewey. Short's Sedge. Fig. 1044.

Carex Shortiana Dewey, Am. Journ. Sci. 30: 60. 1836.

Glabrous, culms rather slender, erect, rough above, 1°-3° tall, usually overtopped by the upper leaves. Leaves elongated, roughish, 2″-4″ wide; bracts short, narrow, the lowest little sheathing; spikes 3-7, gynaecandrous, linear-cylindric, densely many-flowered, ½′-1½′ long, 2″-2⅓″ in diameter, erect, the lower stalked, the uppermost staminate below for about one-half its length; perigynia spreading, orbicular or obovate, 1″ long, darkened at maturity, compressed, 2-edged, nerveless, slightly wrinkled, ridged at apex, abruptly and minutely beaked, the orifice entire or nearly so; scales ovate or oblong-lanceolate, scariousmargined, acute, acuminate, or obtusish and cuspidate, persistent, shorter or longer than perigynia; stigmas 3.

In moist meadows and thickets, Pennsylvania to Virginia and Tennessee, west to Iowa, eastern Kansas and Oklahoma. May–July.

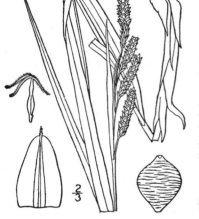

178. Carex Joòri Bailey. Cypress-swamp Sedge. Fig. 1045.

C. Joori Bailey, Proc. Am. Acad. 22: 12. 1886.

Glabrous, light green and glaucous, culms stout, phyllopodic, erect, much roughened on the angles above, 1½°-4° tall. Leaves 8-15 to a culm, flat or in drying somewhat involute, rough, 1½″-3″ wide, usually exceeded by the culm, tapering to a very long narrow tip, the basal sheaths not filamentose; lower bracts similar, shorter; staminate spike usually 1, long-stalked, the scales short-awned; pistillate spikes 3-5, cylindric, dense, 15-50-flowered, 7″-15″ long, 4″ thick, little separate, erect, sessile or the lower stalked; perigynia dark brown, slightly glaucous, squarrose, broadly ovoid, 3-angled, 2″ long, 1½″-1¾″ wide, strongly ribbed,. abruptly contracted into a sharp beak about one-fourth as long as the body, the orifice entire; scales oblong-ovate, scariousmargined, abruptly awned, from slightly exceeding to much shorter than perigynia; stigmas 3.

In swamps, Missouri to Florida and Texas. June–Aug.

Carex verrucòsa Muhl. (*C. macrokòlea* Steud.) admitted into our first edition, is a southern species not definitely known within our range.

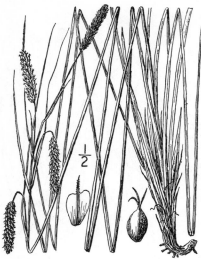

180. Carex acutifórmis Ehrh. Swamp or Marsh Sedge. Fig. 1047.

Carex acutiformis Ehrh. Beitr. **4**: 43. 1789.
C. paludosa Gooden. Trans. Linn. Soc. **2**: 202. 1794.

Culms stout, erect, sharp-angled, 2°–3° tall, smooth below, often rough above. Leaves 2½″–6″ wide, flat, glaucous-green, equalling or sometimes exceeding the culm; lower bracts similar to the leaves, the upper short and narrow; staminate spikes 1–4, stalked; pistillate spikes 3–5, narrowly linear-cylindric, 1½′–3′ long, 2″–2½″ thick, 40–100-flowered, the upper sessile or nearly so and erect, the others slender-stalked, spreading or drooping; perigynia ovoid, 1½″–1¾″ long, not inflated, strongly many-nerved, tapering into a very short and minutely 2-toothed beak; scales awn-tipped or acuminate, longer than the perigynia or the upper equalling them; stigmas 3.

In swamps and wet meadows, eastern Massachusetts, very locally naturalized from Europe. Lesser common sedge. Sniddle. June–Aug.

179. Carex glaucéscens Ell. Southern Glaucous Sedge. Fig. 1046.

Carex glaucescens Ell. Bot. S. C. and Ga. **2**: 553. 1824.

Glabrous, light green, glaucous, culms stout, phyllopodic, erect, somewhat roughened on the angles above, 1½°–4° tall. Leaves 5–10 to a culm, flat or involute towards base, rough, 1½″–2½″ wide, usually exceeded by the culm, long-tapering, the basal sheaths strongly filamentose; lower bracts similar, shorter; staminate spike one, stalked, the scales strongly cuspidate; pistillate spikes 3–4, cylindric, dense, many-flowered, 1′–2′ long, 3½″–5″ wide, slender-peduncled, at first erect, finely drooping; perigynia strongly glaucous, ascending, ovoid or obovoid, 3-angled, 1½″–1¾″ long, 1″ wide or more, obscurely nerved, tapering into a short beak with entire orifice; scales with obovate body, about length of perigynia, abruptly long-cuspidate, reddish-brown with green midrib, squarrose; stigmas 3.

In swamps, Virginia to Florida and Mississippi. July–Sept.

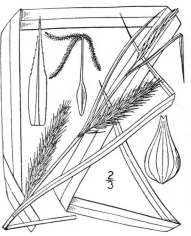

181. Carex strícta Lam. Tussock Sedge. Fig. 1048.

Carex stricta Lam. Encycl. **3**: 387. 1789.
C. stricta angustata Bailey, in A. Gray, Man. Ed. 6, 600. 1890.
C. xerocarpa S. H. Wright, Am. Journ. Sci. (II.) **42**: 334. 1866.

Glabrous, rather dark green, culms slender, aphyllopodic, stiff, erect, usually in dense clumps, sharply 3-angled and very rough above, 1°–4° tall; stolons little developed. Leaves long, rarely overtopping the culm, very rough on the margins, 1″–2″ wide, the lower sheaths becoming prominently filamentose; lower bract similar, sometimes equalling the culm; staminate spike solitary, or sometimes 2, stalked; pistillate spikes 2–5, very variable, linear-cylindric, or sometimes linear-oblong, often staminate at the top, very densely flowered, or loose at the base, ½′–4′ long, 1″–2″ thick, erect or somewhat spreading, all sessile or the lower stalked; perigynia ovate-elliptic, ascending, acute, faintly few-nerved or nerveless, 1¼″ long, minutely beaked, the orifice entire or nearly so; scales dark with green margins and midvein, oblong or lanceolate, obtuse to acuminate and from

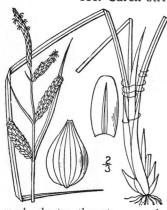

much shorter than to somewhat exceeding perigynia, appressed; stigmas 2.

In swamps, Newfoundland to Ontario, Nebraska, Georgia and Texas. Hybridizes with *C. lasiocarpa*. July–Sept.

182. Carex Hàydeni Dewey.   Hayden's Sedge. Fig. 1049.

Carex aperta Carey, in A. Gray, Man. 547. 1848. Not
Boott, 1840.
C. Haydeni Dewey, Am. Journ. Sci. (II.) **18**: 103. 1854.
C. stricta var. decora Bailey, Bot. Gaz. **13**: 85. 1888.

Gabrous, similar to small forms of C. stricta, culms
slender, rough above, seldom over 2° high. Leaves
1"–1½" wide, rough-margined, shorter than or sometimes a little overtopping the culm, the lower sheaths
slightly or not at all filamentose; lower bract foliaceous,
about equalling the culm; pistillate spikes linear-cylindric, 6"–15" long, about 2" in diameter, erect or somewhat spreading, all sessile or nearly so, sometimes with
a few staminate flowers at the summit; perigynia suborbicular or obovate, obtuse, about ¾" broad, faintly
2–4-nerved, minutely beaked, the orifice entire; scales
lanceolate, purplish, spreading, very acute, strongly exceeding the perigynia; stigmas 2.

Swamps, New Brunswick to Minnesota, south to New Jersey and Missouri.

183. Carex tórta Boott.   Twisted Sedge.   Fig. 1050.

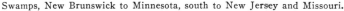

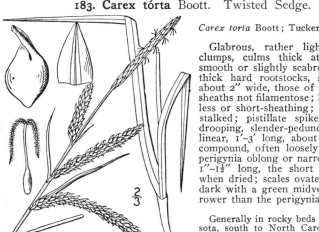

Carex torta Boott; Tuckerm. Enum. Meth. 11.   1843.

Glabrous, rather light green, in rather loose
clumps, culms thick at base, aphyllopodic, erect,
smooth or slightly scabrous above, 1½°–3° tall, from
thick hard rootstocks, short-stoloniferous. Leaves
about 2" wide, those of the fertile culm very short;
sheaths not filamentose; lower bract leaf-like, sheathless or short-sheathing; staminate spike usually one,
stalked; pistillate spikes 3–6, erect, spreading or
drooping, slender-peduncled or upper often sessile,
linear, 1'–3' long, about 2" in diameter, sometimes
compound, often loosely flowered toward the base;
perigynia oblong or narrowly ovate, green, nerveless,
1"–1½" long, the short beak more or less twisted
when dried; scales ovate-oblong, obtuse or subacute,
dark with a green midvein, shorter and mostly narrower than the perigynia; stigmas 2.

Generally in rocky beds of streams, Quebec to Minnesota, south to North Carolina and Missouri. Ascends
to 2600 ft. in Virginia.   June–July.

184. Carex cóncolor R. Br.   Bigelow's Sedge.   Fig. 1051.

Carex rigida Gooden. Trans. Linn. Soc. **2**: 193. pl. 22.
1794. Not Schrank, 1789.
Carex concolor R. Br. in Parry's Voy. App. 283. 1823.
Carex Bigelovii Torr.; Schwein. Ann. Lyc. N. Y. **1**: 67.
1824.
Carex hyperborea Drej. Rev. Crit. Car. 43. 1841.

Glabrous and smooth throughout or very nearly
so, culms phyllopodic, usually low and rigid, in
small clumps, sharp-angled, erect, 4'–18' tall, freely
short-stoloniferous, the rootstocks stout, scaly. Leaves
1½"–3½" wide, with revolute margins, not exceeding
the culm, the lower bracts similar, but shorter;
staminate spike stalked, sometimes pistillate at the
base; pistillate spikes 1–4, short-oblong to linear-
cylindric, usually loosely flowered at the base, dense
above, 3"–20" long, 1½"–2½" thick, the upper sessile,
the lower often slender-stalked; perigynia oval, 1¼"–
1½" long, ascending, faintly nerved or nerveless,
scarcely beaked, the orifice entire; scales purple-
brown with a narrow light midvein and often with
hyaline margins, obtuse or the lower acutish, equalling or a little exceeding the perigynia; stigmas 2,
rarely 3.

Greenland to Alaska, south to the higher mountains of northern New England and New York,
Colorado and California. Also in Europe and Asia. Very variable. Summer.

185. Carex Goodenòwii J. Gay.   Goodenough's Sedge.   Fig. 1052.

Carex caespitosa Gooden. Trans. Linn. Soc. **2**: 195. *pl. 21.* 1794. Not L. 1753.
C. Goodenowii J. Gay, Ann. Sci. Nat. (II.) **11**: 191. 1839.
Carex vulgaris E. Fries, Mant. **3**: 153. 1842.
Carex vulgaris var. *strictiformis* Bailey, Mem. Torr. Club **1**: 74. 1889.

Glabrous, culms stiff, erect, sharp-angled, smooth or sometimes rough above, 1°–3° tall, phyllopodic, strongly stoloniferous. Leaves elongated, glaucous, 1″–2″ wide, not exceeding the culm, the margins involute in drying; lower bracts usually foliaceous, sometimes equalling the culm; staminate spike stalked; pistillate spikes 2–4, all sessile or nearly so, erect, densely many-flowered, narrowly cylindric, 2″–2½″ in diameter, 5″–20″ long; perigynia flattened, broadly oval or ovate, finely nerved, green or dark-tinged, appressed, 1½″ long, minutely beaked, the orifice nearly entire; scales purple-brown to black with a slender green midvein, very obtuse (except lower), shorter than the perigynia; stigmas 2.

In wet grounds, Newfoundland to Massachusetts; Pennsylvania. Europe and Asia. Tufted or Common sedge. Torrets or Turrets. Stare or Star. June–Aug.

186. Carex lenticulàris Michx.   Lenticular Sedge.   Fig. 1053.

Carex lenticularis Michx. Fl. Bor. Am. **2**: 172. 1803.

Glabrous, pale green, culms very slender, erect, sharp-angled, slightly rough above, 1°–2° tall, phyllopodic, densely caespitose, not stoloniferous. Leaves elongated, rarely over 1″ wide, shorter than or rarely overtopping the culm, slightly rough-margined, their sheaths not fibrillose; lower bracts similar to the leaves, more or less sheathing, usually much overtopping the spikes; staminate spikes solitary or rarely 2, sessile or short-stalked, often pistillate above; pistillate spikes 2–5, clustered at the summit or the lower distant, sessile or the lower short-stalked, erect, linear-cylindric, 4″–2′ long, 1½″–2″ in diameter; perigynia ovate or elliptic, 1¼″ long, ¾″ wide, acute, minutely granulate, faintly few-nerved, appressed, tipped with a minute entire beak; scales dark-tinged between the broad green center and the hyaline margins, usually obtuse and much shorter than the perigynia; stigmas 2.

On shores, Labrador to Saskatchewan, south to Massachusetts, New York and Minnesota. Ascends to 4500 ft. in the White Mountains. June–Aug.

187. Carex aquátilis Wahl.   Water Sedge.   Fig. 1054.

Carex aquatilis Wahl. Kongl. Vet. Acad. Handl. (II.) **24**: 165. 1803.

Glabrous, somewhat glaucous and pale green, culms phyllopodic, caespitose, long-stoloniferous, rather stout, erect, sharp-angled above, smooth or somewhat rough above, 2°–5° tall. Leaves elongated, sometimes equalling the culm, 2″–4″ wide, their sheaths nodulose; bracts broad, similar to the leaves, the lower usually much overtopping the culm; staminate spikes 2–3, stalked; pistillate spikes 3–5, narrowly linear-cylindric, often staminate at the summit, erect or slightly spreading, 1′–3½′ long, 2″–3″ in diameter, many-flowered, sessile and dense, or the lower narrowed and loosely flowered at the base and short-stalked; perigynia elliptic or obovate, green, nerveless or faintly nerved, minutely beaked, the orifice entire; scales oblong, obtuse to acuminate, from much shorter than to exceeding the perigynia, but much narrower; stigmas 2.

In swamps and along streams, Newfoundland to Alaska, south to Maryland and Texas and in the western mountains. Also in Europe and Asia. Widely variable. June–Aug.

Carex Emóryi Dewey (*C. millegrana* Holm), with sharply trigonous culms, very rough involute-margined leaves and strongly nerved perigynia occurs from New Jersey and Maryland to North Dakota and New Mexico.

188. Carex nebraskénsis Dewey.   Nebraska Sedge.   Fig. 1055.

C. Jamesii Torr. Ann. Lyc. N. Y. **3**: 398. 1836. Not Schw. 1824.
Carex nebraskensis Dewey, Am. Journ. Sci. (II.) **18**: 102. 1854.

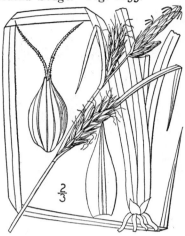

Glabrous, culms rather stout, erect, sharp-angled, smooth, or rough above, 1°–3° tall, phyllopodic, from stout, long-creeping rootstocks. Leaves pale green, 1½″–4″ wide, rough-margined, their sheaths more or less nodulose; lower bract foliaceous, shorter than to exceeding culm, the upper much shorter and narrower; staminate spikes commonly 2, stalked; pistillate spikes 2–5, dense, oblong-cylindric, erect, 9″–25″ long, 3″–4½″ in diameter, sessile or the lower short-stalked; perigynia ascending, elliptic or somewhat obovate, prominently several-ribbed when mature, short-beaked, the beak 2-toothed; scales ovate or lanceolate, obtusish to strongly acuminate, dark with a green midvein, the upper shorter than the perigynia; stigmas 2.

South Dakota and Nebraska to Oregon, California and New Mexico. May–Aug.

189. Carex subspathàcea Wormsk.   Hoppner's Sedge.   Fig. 1056.

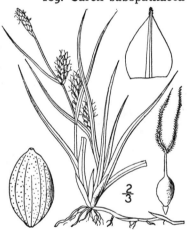

C. subspathacea Wormsk. Fl. Dan. **9** : 4. *pl. 1530.* 1816.

Carex Hoppneri Boott; Hook. Fl. Bor. Am. **2** : 219. *pl. 220.* 1840.

Glabrous, culm stiff, phyllopodic, smooth, 1′–7′ tall, from slender elongated rootstocks. Leaves rigid, smooth, ¾″–1¼″ wide, the margins strongly involute; lower bracts foliaceous, somewhat spathe-like; staminate spike solitary, stalked; pistillate spikes 1–3, erect, sessile or very short-stalked, 3″–6″ long, 1½″–2″ in diameter, closely 4–15-flowered; perigynia oval or ovoid, coriaceous, very short-beaked, pale green, faintly few-nerved, densely puncticulate, about 1½″ long; scales ovate or ovate-oblong, obtuse or acute, shorter than or equalling the perigynia; stigmas 2.

Greenland to Hudson Bay. Also in northern Europe and Asia. Intergrades with the next. July–Aug.

190. Carex salìna Wahl.   Salt-marsh Sedge.   Fig. 1057.

Carex salina Wahl. Kongl. Vet. Acad. Handl. (II.) **24** : 165. 1803.
C. cuspidata Wahl. loc. cit. 164. 1803. Not Host, 1801.

Glabrous, culm slender but stiff, phyllopodic, smooth, erect, 4′–18′ tall, from slender elongated rootstocks. Leaves very narrowly linear, 1″–2½″ wide, rarely overtopping the culm, the margins revolute; lower bract similar, not spathe-like; staminate spikes 1 or 2, stalked; pistillate spikes 2–4, slender-stalked or the upper sessile, erect, 15–40-flowered, 5″–15″ long, 2″–3″ thick; perigynia ovate-elliptic, coriaceous, pale, faintly few-nerved, densely puncticulate, ascending, 1½″ long, tapering into a very short entire beak; scales ovate, reddish-brown or chestnut, with a green midvein, acute or cuspidate, somewhat longer than the perigynia; stigmas 2.

In salt marshes, Greenland to Hudson Bay, south to Quebec. Also in arctic Europe. Intergrades with the next. July–Aug.

191. **Carex récta** Boott.   Cuspidate Sedge.   Fig. 1058.

C. recta Boott, in Hook. Fl. Bor.-Am. **2** : *220. pl. 222.* 1840.

Glabrous, culms phyllopodic, from long rootstocks, rather stout, smooth or rough above, erect, 1°–3° tall.   Leaves often equalling the culm, 1″–3″ wide, their margins revolute, smooth, their sheaths more or less nodulose; bracts similar, not spathe-like, usually overtopping the spikes; staminate spikes 1–3, stalked; pistillate spikes 2–4, approximate, narrowly cylindric, often staminate at the summit, 1′–2½′ long, erect, the upper often sessile, the lower stalked; perigynia elliptic, coriaceous, green, nerveless or 2–4-nerved, with a very short entire beak; scales dark-purple, brownish or chestnut with a green center, lanceolate, pale, acuminate or abruptly contracted into a serrate awn, much longer than the perigynia; stigmas 2.

In marshes, Labrador to the coast of Massachusetts. Also in Europe. Erroneously referred to *C. cuspidata* Wahl. in our first edition. July–Aug.

192. **Carex cryptocàrpa** C. A. Meyer.   Hidden-fruited Sedge.   Fig. 1059.

Carex cryptocarpa C. A. Meyer, Mem. Acad. St. Petersb. **1** : *226. pl. 14.* 1831.

Glabrous, stoloniferous, culm stout, erect, sharply 3-angled, rough above, 1½°–3° tall.   Leaves smooth, 1″–4″ wide, the basal shorter than or equalling the culm, the upper ones and the lower bract shorter; staminate spikes 2–4, stalked; pistillate spikes 2–5, all filiform-stalked and drooping, densely flowered, 1′–3′ long, 3″–4½″ in diameter; perigynia oblong or oval, puncticulate, green, several-nerved, 1½″ long, tipped with a very short entire beak; scales purple-brown, ascending, lanceolate, acute, acuminate or even cuspidate, from little to 2 or 3 times longer than the perigynia; stigmas 2.

Arctic America from Greenland to Alaska, south along the coast to Oregon and Washington. Europe and Asia. Summer.

193. **Carex marítima** Mueller.   Seaside Sedge.   Fig. 1060.

C. maritima Mueller, Fl. Dan. **4** : *fasc. 12, 5. pl. 703.* 1777.

Glabrous, light green, culms aphyllopodic, rather stout, erect, sharply 3-angled, smooth, or roughish above, little fibrillose at base, 1°–2½° tall, from stout stoloniferous rootstocks.   Leaves 1½″–5″ wide, roughish on the margins and midvein, rarely over-topping the culm, sterile culm-leaves longer; lower bracts exceeding culm; staminate spikes 1–3, slender-stalked; pistillate spikes 2–4, ovoid- to linear-oblong, often staminate at the summit, densely many-flowered, not flexuous, 1′–3′ long, 4″–8″ thick, droop-ing or widely spreading on filiform stalks; peri-gynia oval or nearly orbicular, ascending, pale, biconvex, not inflated, 1½″ long, few-nerved, tipped with a very short and nearly or quite entire beak; scales ascending, green, lanceolate-subulate, ciliate-scabrous, 2–8 times as long as the perigynia; stig-mas 2.

In salt meadows, Newfoundland and Labrador to Mas-sachusetts. Also in Europe. June–Aug.

194. Carex gynándra Schwein. Nodding Sedge. Fig. 1061.

Carex gynandra Schwein. Ann. Lyc. N. Y. **1** : 70. 1824.
Carex crinita var. *gynandra* Schwein. & Torr. Ann. Lyc. N. Y. **1** : 360. 1825.
Carex Porteri Olney, Car. Bor. Am. 12. 1871.
Carex crinita var. *simulans* Fernald, Rhodora **4** : 219. 1902.

Similar to the following species, culms stout, 2°–5° tall. Leaves 2″–6″ wide, their sheaths shortly rough-hispid ; pistillate spikes 1′–4′ long, narrowly cylindric, stalked, drooping ; perigynia obovoid to oval, ascending, compressed, less inflated, 1½″–2″ long, and about one-half as broad, round-tapering and minutely beaked, the orifice entire ; scales lanceolate, tapering into the long awn, the center strongly 3-ribbed and the body brownish-tinged, more ascending, 2–4 times as long as the perigynia ; stigmas 2.

In swamps, Newfoundland and Wisconsin, south to Georgia. Ascends to 5000 ft. in New Hampshire. June–Aug.

195. Carex crinìta Lam. Fringed Sedge. Sickle-grass. Fig. 1062.

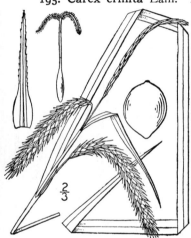

Carex crinita Lam. Encycl. **3** : 393. 1789.
Carex crinita var. *minor* Boott, Ill. Car. **1** : 18. 1858.

Glabrous, culms stout, 3-angled, aphyllopodic and filamentous at base, rough or very nearly smooth, erect or somewhat recurving, 2°–5° tall, from stout rootstocks. Leaves flat, rough-margined, 1½″–5″ wide, the upper sometimes overtopping the culm, the lowest very short and sheathing ; staminate spikes 1 or 2, stalked, often pistillate at the base or in the middle ; pistillate spikes 2–6, narrowly cylindric, densely very many-flowered, 1′–4½′ long, 3″–6″ in diameter, all stalked, drooping and commonly secund ; perigynia suborbicular or broadly obovoid, spreading, obtuse, 1″–1¾″ long, ½″–1″ wide, slightly inflated, the walls thin, nerveless, abruptly tipped by the very short entire beak ; scales abruptly long rough-cuspidate, the outer 1–3-ribbed and the body brownish-tinged, spreading, 2–6 times as long as the perigynia ; stigmas 2.

In swamps and wet woods, Newfoundland to Minnesota, south to Florida and Texas. June–Aug.

196. Carex lacústris Willd. Lake-bank Sedge. Fig. 1063.

Carex lacustris Willd. Sp. Pl. **4** : 306. 1805.

Carex riparia Muhl. Descr. Gram. 259. 1817. Not Curtis, 1783.

Culms generally stout and smooth, erect, 2°–3½° tall, strongly purplish-tinged and filamentose at base, the lower sheaths not blade-bearing. Leaves elongated, nodulose, usually more or less scabrous, somewhat glaucous, 2½″–6″ wide, usually exceeding the culm ; lower bract similar to the leaves, the upper reduced ; staminate spikes 1–5, linear ; pistillate spikes 2–5, cylindric, 1½′–4′ long, about 5″ in diameter, the upper erect, sessile or nearly so, the lower more or less stalked ; perigynia narrowly ovoid, 3″ long, 1¼″ wide, firm, strongly nerved, scarcely inflated, ascending, tapering gradually into a 2-toothed beak, the teeth erect or slightly divergent, ¼″ long ; scales lanceolate or oblanceolate, long-aristate or acute, the lower longer, the upper equalling or shorter than the perigynia, purplish-tinged ; stigmas 3.

In swamps, Newfoundland to James' Bay and Manitoba, south to Delaware, Iowa and Idaho. Great common-sedge. May–Aug.

197. Carex impréssa (S. H. Wright) Mackenzie. Hart Wright's Sedge. Fig. 1064.

Carex riparia var. impressa S. H. Wright, Bull. Torr. Club 9 : 151. 1882.
Carex impressa Mackenzie, Bull. Torr. Club 37 : 236. 1910.

Culms stout and generally smooth, erect, 1½°–3° tall, neither purplish-tinged nor filamentose at base, the lower sheaths blade-bearing. Leaves elongated, nodulose, usually more or less scabrous, somewhat glaucous, 2″–4″ wide, flat or folded at base, usually exceeding the culm; lower bract similar to the leaves, the upper reduced; staminate spikes 2–4, linear, their scales straw-colored or light purplish-tinged; pistillate spikes 2–4, cylindric, ½′–3′ long, about 5″ in diameter, usually widely separate, erect, short-peduncled; perigynia narrowly ovoid, 3″ long, 1¼″ wide, firm, impressed-nerved when young, appearing nearly nerveless when mature, scarcely inflated, tapering into a 2-toothed beak, the teeth erect or slightly divergent, ¼″ long; scales ovate, the lower aristate and exceeding perigynia, the upper more acute and shorter, straw-colored or light purplish-tinged; stamens 3.

In swamps, Ohio to Kansas, Florida and Texas.

198. Carex Walteriàna Bailey. Walter's Sedge. Fig. 1065.

Carex striata Michx. Fl. Bor. Am. 2 : 174. 1803. Not Gilib. 1792.
Carex Walteriana Bailey, Bull. Torr. Club 20 : 429. 1893.
C. striata var. brevis Bailey, Mem. Torr. Club 1 : 34. 1889.

Long-stoloniferous, the culms slender, strict, erect, slightly rough above, 1°–2½° tall. Leaves narrow and elongated, smooth or roughish, 1″–2″ wide, nodulose, becoming involute; lowest bract similar, often very long, the upper smaller and often almost filiform; staminate spikes 1 or 2, long-stalked; pistillate spikes 1 or 2, when 2 the lower remote from the upper, sessile or very short-stalked, oblong-cylindric, erect, ½′–2′ long, 3″–4″ in diameter, many-flowered; perigynia ovoid, many-nerved, slightly inflated, ascending, glabrous, or partly or wholly short-pubescent, 2″–2½″ long, 1¼″ in diameter, tapering into a short 2-toothed beak, the teeth short, variable; scales ovate, short-aristate, acute or obtuse, membranous, one-half to two-thirds length of perigynia.

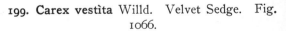

In pine-barren bogs, southeastern Massachusetts to Florida, near the coast. May–July.

199. Carex vestìta Willd. Velvet Sedge. Fig. 1066.

Carex vestita Willd. Sp. Pl. 4 : 263. 1805.

Carex vestita var. Kennedyi Fernald, Rhodora 2 : 170. 1900.

Strongly stoloniferous, culms strict, erect, 1½°–2½° tall, rough above, reddened and filamentose at base. Leaves distant, 1½″–2½″ wide, not overtopping the culm; bracts narrower, short, rough on the margins; staminate spike solitary, rarely 2, sessile or short-peduncled; pistillate spikes 1–5, oblong, 4″–14″ long, 3″–4″ in diameter, erect, commonly staminate at the summit, sessile or the lower very short-stalked; perigynia ovoid, ascending or the lower spreading, densely pubescent, less than 1″ in diameter, nerved, slightly shorter than or equalling the ovate-obtuse or acute scales; the beak white-hyaline at orifice, in age bidentate; stigmas 3.

In sandy woods, southern Maine to eastern New York and Pennsylvania, south to Georgia. May–July.

200. Carex lanuginòsa Michx. Woolly Sedge. Fig. 1067.

Carex lanuginosa Michx. Fl. Bor. Am. **2** : 175. 1803.
C. filiformis var. *latifolia* Boeckl. Linnaea **41** : 309. 1876.
Carex filiformis var. *lanuginosa* B.S.P. Prel. Cat. N. Y. 63. 1888.
Carex lanuginosa var. *kansana* Britton, in Britt. & Br. Ill. Fl. **1** : 305. 1896.

Culm usually rather stouter than that of *C. lasiocarpa* and less reddened and filamentose at base, sharp-angled and rough above. Leaves and lower bracts elongated, flat, not involute, 1″–2½″ wide, more or less strongly nodulose, sometimes overtopping the culm; staminate spikes 1–3, distant, sometimes pistillate at the base; pistillate spikes 1–3, usually distant, sessile or the lower slender-stalked, cylindric, 5″–25″ long, 2½″–3½″ in diameter; perigynia like those of *C. lasiocarpa,* but usually rather broader; scales acuminate or aristate.

In swamps and wet meadows, Nova Scotia to British Columbia, District of Columbia, southern Missouri, New Mexico and California. June–Aug.

201. Carex lasiocàrpa Ehrh. Slender Sedge. Fig. 1068.

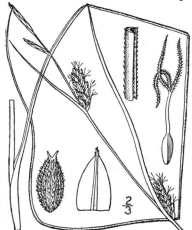

C. lasiocarpa Ehrh. in Hannov. Magaz. **9** : 132. 1784.
C. filiformis Good. Trans. Linn. Soc. **2** : 172. 1794. Not L.

Culms slender but stiff, smooth, obtusely angled, 2°–3° tall, strongly reddened and filamentose at base. Leaves very narrow and attenuate, prolonged, involute, 1″ wide or less, rough on the inrolled margins, not overtopping the culm; lower bract similar, often equalling the culm; upper bracts filiform; staminate spikes 1–3, commonly 2, distant; pistillate spikes 1–3, cylindric, 5″–25″ long, about 3″ in diameter, erect, sessile or the lower distant and short-peduncled; perigynia green, ascending, oval-ovoid, densely pubescent, obscurely nerved, about 1″ in diameter, tapering into a short 2-toothed beak; scales ovate, membranous, sometimes purplish, acute or short-awned, shorter than or equalling the perigynia; stigmas 3.

In wet meadows and swamps, Newfoundland to British Columbia, south to New Jersey, Pennsylvania, Iowa and Minnesota. Ascends to 2000 ft. in the Adirondacks. Also in Europe. June–Aug.

202. Carex Houghtònii Torr. Houghton's Sedge. Fig. 1069.

Carex Houghtonii Torr. Ann. Lyc. N. Y. **3** : 413. 1836.

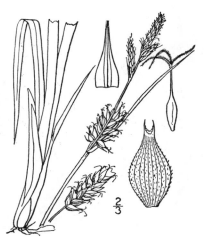

Long-stoloniferous, the culms rather stout, rough above, erect, 1°–2½° tall, exceeding the leaves. Leaves and lowest bract 1½″–3½″ wide, rough; upper bracts much shorter; staminate spikes 1–3, distant, sometimes pistillate at the base; pistillate spikes 2 or 3, oblong-cylindric, ½′–1½′ long, 3½″–6″ in diameter, erect, rather loosely 15–35-flowered, the upper sessile, the lower stalked; perigynia ovoid, 1½″ in diameter, light green, ascending, densely pubescent, prominently many-ribbed, narrowed into a short conspicuously 2-toothed beak; scales lanceolate, short-awned, hyaline-margined, somewhat shorter than the perigynia; stigmas 3.

In sandy or rocky soil, Nova Scotia to Saskatchewan, Athabasca, south to Maine, Ontario, Michigan and Minnesota. June–Sept.

203. **Carex trichocàrpa** Muhl.   Hairy-fruited Sedge.   Fig. 1070.

Carex trichocarpa Muhl.; Willd. Sp. Pl. **4**: 302. 1805.
Carex trichocarpa var. *turbinata* Dewey, Am. Journ. Sci. **11**:
159. 1826.
Carex laeviconica Dewey, Am. Journ. Sci. **24**: 47. 1857.
Carex trichocarpa var. *imberbis* A. Gray, Man. Ed. 5, 597.
1867.
Carex trichocarpa var. *Deweyi* Bailey, Coult. Bot. Gaz. **10**:
293. 1885.

Culm usually stout and tall, 2°–4° high, smooth below,
very rough above. Leaves elongated, glabrous, rough-
margined, 1½″–3″ wide, the upper ones and the similar
bracts commonly overtopping the culm; staminate spikes
2–6, long-stalked; pistillate spikes 2–4, cylindric, densely
flowered except at the base, 1′–4′ long, 5″–8″ in diameter,
the upper sessile or nearly so and erect, the lower slender-
stalked; perigynia ovoid, pubescent or glabrous, prominently many-ribbed, 4″–5″ long, 1½″–2″ in diameter, tapering gradually into the stout conspicuously 2-toothed beak,
the teeth erect or somewhat spreading, 1″ long; scales
hyaline, acute to aristate, about one-half as long as the
perigynia; stigmas 3.

In marshes and wet meadows, Quebec and Vermont to Oregon, south to Georgia, Missouri and Kansas. June–Aug.

204. **Carex atheròdes** Spreng.   Awned Sedge.   Fig. 1071.

Carex aristata R. Br. Frank. Journ. 751. 1823. Not
Houck. 1792.
Carex atherodes Spreng. Syst. Veg. **3**: 828. 1826.
Carex trichocarpa var. *aristata* Bailey, Coult. Bot. Gaz.
10: 294. 1885.

Culms stout, erect, smooth, or roughish above,
sharp-angled, 2°–5° tall. Leaves elongated, 2½″–6″
wide, more or less scabrous, often pubescent beneath
and on the sheaths; bracts similar, the lower often
overtopping the culm; staminate spikes as in the preceding species; pistillate spikes 3–5, remote, cylindric,
sessile or the lower short-stalked, loosely flowered
at the base, dense above, 1½′–4′ long, sometimes 8″
in diameter; perigynia ascending, lanceolate or ovoid-
lanceolate, glabrous, conspicuously many-ribbed, 4″–
6″ long, gradually tapering into the conspicuously
2-toothed beak, the teeth widely spreading, 1″–2″
long; scales oblong-lanceolate, strongly rough-awned,
thin-margined, one-half to two-thirds as long as the
perigynia; stigmas 3.

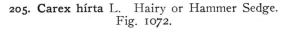

In bogs, Ontario to British Columbia, south to New
York, Missouri, Kansas, Utah and Oregon. Also in
Europe and Asia. June–Aug.

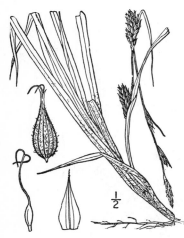

205. **Carex hírta** L.   Hairy or Hammer Sedge.
Fig. 1072.

Carex hirta L. Sp. Pl. 975. 1753.

Rootstocks extensively creeping, culms rather slender, erect, nearly smooth, 6′–2° tall. Leaves flat, pubescent, especially on the sheaths, rough, 1″–2½″ wide, the
basal ones much elongated, often exceeding the culm,
the upper and the similar bracts shorter; staminate
spikes 2 or 3, stalked, their scales ciliate; pistillate
spikes 2 or 3, widely separate, erect, oblong-cylindric,
7″–18″ long, 3″–4″ in diameter, rather loosely 10–many-
flowered; perigynia oblong-ovoid, green, ascending,
densely pubescent, 1¼″ in diameter, 2½″–4″ long, few-
ribbed, tapering into a stout prominently 2-toothed
beak, the teeth often as long as the beak; scales membranous, lanceolate, aristate, 3-nerved, somewhat
shorter than the perigynia; stigmas 3.

In fields and waste places, Massachusetts to New York,
New Jersey and Pennsylvania. Locally naturalized or ad-
ventive from Europe. Carnation- or Goose-grass. June–
Sept.

206. Carex fulvéscens Mackenzie.   Tawny Sedge.   Fig. 1073.

C. fulvescens Mackenzie, Bull. Torr. Club **37** : 239. 1910.

Glabrous, yellow-green, culms slender, erect, 6'–20' tall, smooth or slightly roughened on angles. Leaves ¾"–1¼" wide, flat, shorter than the culm, the lower bract shorter than the culm, ascending, long-sheathing; staminate spike solitary, strongly peduncled; pistillate spikes 1–3, oblong, erect, widely separate, the lower strongly exsert-peduncled, the upper short exsert-peduncled, densely 15–40-flowered, 4"–10" long, 3½"–5" thick; perigynia narrowly ovoid, yellowish-green, appressed-ascending, 2½" long, 1" wide, finely several-nerved, contracted into a rough bidentate beak half as long as body; scales ovate, acute or obtuse, brown with conspicuous white scarious margins, somewhat shorter than perigynia; stigmas 3.

In wet places, Anticosti, Miquelon and probably Newfoundland; also collected near Boston, Mass. Related to the European *Carex fulva* Good, and probably mistaken for it, but apparently distinct. July–Sept.

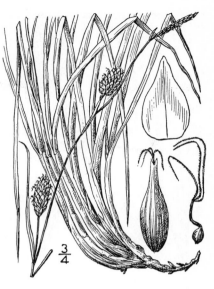

207. Carex exténsa Gooden.   Long-bracted Sedge.   Fig. 1074.

Carex extensa Gooden. Trans. Linn. Soc. **2** : 175. 1794.

Glabrous, bright green, culms stiff, erect, 10'–2° tall. Leaves about 1" wide, strongly involute, erect, the lower bract similar, much exceeding the spikes, sheathing, the upper shorter, sometimes spreading; staminate spike sessile or nearly so, rarely pistillate at the base; pistillate spikes 1–3, erect, sessile and close together or the lowest short-stalked and distant, oblong, densely 15–50-flowered, 3½"–10" long, 3"–4" thick; perigynia ovoid or ovoid-oblong, brown, 1¾" long, narrowed at the base, strongly several-ribbed and with thick walls, contracted into a short stout 2-toothed beak; scales ovate, acute, brown with a greenish midvein, shorter than the perigynia; stigmas 3.

Borders of salt meadows, Coney Island, N. Y., and near Norfolk, Va.   Naturalized from Europe.   June–Aug.

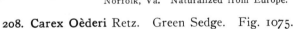

208. Carex Oèderi Retz.   Green Sedge.   Fig. 1075.

Carex Oederi Retz, Fl. Scand. Prodr. 179. 1779.
Carex viridula Michx. Fl. Bor. Am. **2** : 170. 1803.
C. flava var. *viridula* Bailey, Mem. Torr. Club **1** : 31. 1889.
Carex flava var. *cyperoides* Marss. Fl. Neuroys. 537. 1869.
Carex Oederi var. *pumila* (Coss. & Germ.) Fernald, Rhodora **8** : 201. 1906.

Glabrous, bright green, culms slender, smooth, erect, 3'–15' tall, often exceeded by the erect narrow basal leaves. Leaves 1½" or less wide, the bracts similar, usually strictly erect and much overtopping the spikes; staminate spike sessile or short-peduncled; pistillate spikes 2–10, all close together and sessile or scattered and short-stalked, oblong-cylindric to globose-oblong, 2"–6" long, 2"–3½" in diameter; perigynia ovoid-oval, 1"–1½" long, strongly few-nerved, narrowed at the base, abruptly contracted into a 2-toothed beak scarcely one-half as long as the body; scales ovate, much shorter than the perigynia and about as wide; stigmas 3.

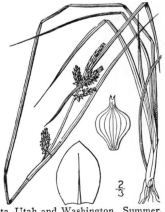

In bogs and on wet rocks, Newfoundland to Hudson Bay and the Northwest Territory, south to Maine, Pennsylvania, Minnesota, Utah and Washington.   Summer.

209. **Carex lepidocàrpa** Tausch.   Small Yellow Sedge.   Fig. 1076.

Carex lepidocarpa Tausch, Flora 129. 1834.
"*Carex flava* var. *rectirostra* Gaudin;" Fernald, Rho-
dora **8**: 201. 1906.
C. *flava* var. *graminis* Bailey, Mem. Torr. Club **1**: 30.
1889.
C. *flava* var. *elatior* Schlecht. Fl. Berol. **1**: 477. 1823.

Glabrous, yellow-green, culms slender, stiff, erect,
smooth or nearly so, 6′–18′ tall. Leaves 1½″ wide
or less, flat, usually shorter than the culm, the lower
bract elongated, spreading or ascending, sheathing;
staminate spike solitary, sessile or peduncled; pistillate spikes 1–4, oblong to subglobose, erect, sessile
and continuous or widely separate, the lower stalked,
densely 15–35-flowered, 5″–8″ long, about 3½″ thick
(extremes 3″–4½″); perigynia ovoid, yellow, spreading or the lower retrorse when mature, 2″ long, few-
nerved, abruptly contracted into a subulate 2-toothed
beak which is shorter than the body; scales ovate to
lanceolate, acute or acuminate, ⅓–½ length of perigynia, usually inconspicuous at maturity; stigmas 3.

Wet meadows, Newfoundland to Michigan, Rhode
Island and New Jersey. Also in Europe. June–Sept.

210. **Carex flàva** L.   Yellow Sedge.   Fig. 1077.

Carex flava L. Sp. Pl. 975. 1753.

Glabrous, yellow-green, culms slender but stiff and
erect, smooth or nearly so, 6′–2° tall. Leaves 1″–2½″
wide, flat, the lower shorter than or sometimes exceeding the culm, the lower bract elongated, spreading
or ascending, sheathing; staminate spike solitary, stalked
or sessile; pistillate spikes 1–4, globose-oblong, erect,
varying from sessile and close together to strongly
separate and the lower strongly stalked, densely 15–35-
flowered, 3″–9″ long, about 5″ thick (extremes 4½″–6″);
perigynia narrowly ovoid, yellow, and spreading in all
directions when mature, 2½″–3″ long, strongly several-
nerved, the subulate 2-toothed deflexed beak as long as
the body or longer; scales lanceolate or ovate, acute to
short-cuspidate, ⅓–½ length of perigynia, conspicuous
at maturity; stigmas 3.

In swamps and wet meadows, Newfoundland to British
Columbia, Rhode Island, New Jersey, Pennsylvania, Ohio
and Montana. Europe. Marsh hedgehog-grass. June–Sept.

211. **Carex Collínsii** Nutt.   Collins' Sedge.   Fig. 1078.

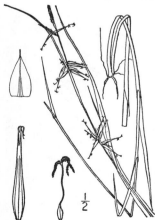

Carex subulata Michx. Fl. Bor. Am. **2**: 173. 1803. Not Gmel.
1791, nor Schum. 1801.
Carex Collinsii Nutt. Gen. **2**: 205. 1818.
Carex Michauxii Dewey, Am. Journ. Sci. **10**: 273. 1826. Not
Schwein. 1824.

Caespitose, glabrous, culms slender, weak, erect or reclining, 6′–2° long. Leaves narrow, soft, the broadest
about 2½″ wide; staminate spike terminal, short- or long-
stalked; pistillate spikes 2–4, distant, 2–8-flowered, short-
stalked, or the stalk of the lowest sometimes 1½′ long;
bracts similar to the upper leaves, elongated, strongly
sheathing; perigynia light green, scarcely inflated, subulate, 4″–7″ long, tapering from below the middle into an
almost filiform beak, faintly many-nerved, horizontal or
reflexed when mature, easily detached from rachis, about
3 times as long as the hyaline lanceolate-acuminate persistent scale, its teeth very strongly reflexed at maturity;
achenes linear-oblong; stigmas 3.

In bogs, Rhode Island to eastern Pennsylvania, south to
South Carolina and Georgia. Ascends to 2000 ft. in Pennsylvania. Attributed to Canada by Michaux. June–Aug.

212. Carex abácta Bailey. Yellowish Sedge. Fig. 1079.

Carex rostrata Michx. Fl. Bor. Am. **2** : 173. 1803. Not Stokes, 1787.
Carex Michauxiana Boeckl. Linnaea **40** : 336. 1877. Not *C. Michauxii* Schwein. 1824.
Carex abacta Bailey, Bull. Torr. Club **20** : 427. 1893.

Glabrous, whole plant yellowish, culm erect or slightly assurgent at the base, rather stiff, slender, 1°–2° high. Leaves narrow, the broadest about 2″ wide, the uppermost often exceeding the culm; staminate spike terminal, sessile or very nearly so; pistillate spikes 1 to 3, several–many-flowered, the upper sessile or very nearly so and closely approximate, the third, when present, remote and borne on a long stalk; bracts similar to the leaves, usually erect and overtopping the culm; perigynia lanceolate, 4″–7″ long, less than 1″ thick at the base, narrow, scarcely inflated, erect or spreading, tapering into a subulate 2-toothed beak, rather strongly many-nerved, about twice as long as the lanceolate or ovate, acute or short-acuminate scale; achenes oblong-obovoid, 1½″ long; stigmas 3.

In bogs and wet meadows, Labrador and Newfoundland to New Hampshire, New York and Pennsylvania, west to Michigan. Ascends to 5000 ft. in New Hampshire. Also in eastern Asia. June–Sept.

213. Carex folliculàta L. Long Sedge. Fig. 1080.

Carex folliculata L. Sp. Pl. 978. 1753.

C. xanthophysa Wahl. Vet. Akad. Handl. **24** : 152. 1803.

Glabrous, light green or yellowish, culm stout or slender, erect or reclining, 1½°–3½° long. Leaves broad and elongated, sometimes overtopping the culm, 2″–8″ wide; staminate spike stalked or nearly sessile; pistillate spikes 2–5, usually distant, all except the uppermost slender-stalked, several–many-flowered, the lower often nodding on a long stalk; bracts commonly overtopping the spikes; perigynia lanceolate, slightly inflated, ascending or spreading, rather prominently many-veined, 6″–8″ long, 1½″ in diameter near the base, tapering from below the middle into a slender 2-toothed beak, one-third to one-half longer than the awned, broadly scarious-margined, persistent scale; achenes oblong-obovoid, 1¾″ long; long; stigmas 3.

In swamps and wet woods, Newfoundland to Michigan, south to North Carolina. May–Sept.

214. Carex miliàris Michx. Northeastern Sedge. Fig. 1081.

Carex miliaris Michx. Fl. Bor. Am. **2** : 174. 1803.
Carex saxatilis var. *miliaris* Bailey, Bot. Gaz. **9** : 120. 1884.

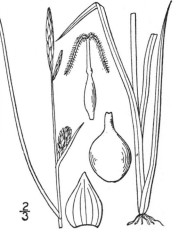

Glabrous, culm very slender, erect, smooth below inflorescence, little filamentose at base, 1°–2° tall, the rootstocks creeping. Leaves about 1″ wide, elongated, nodulose, the upper about equalling the culm; bracts similar to the leaves, often overtopping the culm; staminate spikes 1 or 2, stalked, narrowly linear; pistillate spikes 1–3, slender, oblong-cylindric, many-flowered, 4″–1′ long, about 2½″–3½″ thick; the upper sessile, the lowest more or less stalked; perigynia not inflated, ovoid, faintly few-nerved or nerveless, 1″–1½″ long, brown-tipped, tapering into a short, emarginate beak, slightly longer than the ovate or ovate-lanceolate, wholly or partly brown, obtusish to cuspidate scale; stigmas 2.

Borders of lakes and streams, Labrador and Hudson Bay, south to central Maine. Probably intergrades with the next. Summer.

215. Carex rhomàlea (Fernald) Mackenzie. Moosehead Lake Sedge. Fig. 1082.

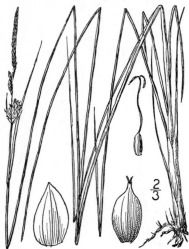

C. miliaris var. major Bailey, Mem. Torr. Club 1: 36. 1889.
C. saxatilis var. rhomalea Fernald, Rhodora 3: 50. 1901.
C. rhomalea Mackenzie, Bull. Torr. Club 37: 246. 1910.

Culms slender, sharply triangular and roughened above, reddened and but little if at all filamentose at base, 6′–2° tall; rootstocks creeping. Leaves ½″–1¼″ wide, strongly involute, usually shorter than the culm, obscurely nodulose, roughened towards apex; lowest bract narrower, shorter than or exceeding culm, erect or spreading; staminate spikes 1–3, short-stalked; pistillate spikes 1–3, ascending, suborbicular to oblong, 4″–9″ long, 2½″–3″ wide, sessile or the lower short-stalked; perigynia 1½″ long, oblong-ovoid, yellowish-green or dark-tinged, few-nerved, scarcely inflated, ascending, contracted into a short emarginate beak; scales ovate, obtuse, acute or acuminate, light-brown to strongly blackish-tinged, shorter than perigynia; stigmas usually 2.

On lake and river shores, central Maine to Newfoundland. Summer. Has been confused with *C. saxatilis* L. and with *C. rotundàta* Wahl.

216. Carex saxátilis L. Russet Sedge. Fig. 1083.

Carex saxatilis L. Sp. Pl. 976. 1753.
Carex pulla Gooden. Trans. Linn. Soc. 3: 78. 1797.

Glabrous, culms not filamentose at base, strongly stoloniferous, erect, slender, 3′–2° tall. Leaves flat, 1″–2½″ broad, obscurely nodulose, the upper not over-topping the culm; bracts short; staminate spike usually solitary, short-stalked; pistillate spikes 1–4, all stalked or the upper nearly or quite sessile, suborbicular to oblong-cylindric, 4″–12″ long, 3″–4½″ wide, densely 15–50-flowered; perigynia dark purple-brown or rarely straw-colored, ascending, ovoid, 1½″–2″ long, scarcely inflated, nerveless or very faintly few-nerved, tipped with a short emarginate beak; scales ovate, subacute, dark brown, shorter than or as long as the perigynia; stigmas usually 2, rarely 3.

Greenland and Labrador to Alaska. Also in arctic Europe and Asia. Summer.

Carex ambústa Boott, admitted into our first edition, is a far northwestern species not definitely known within our range.

217. Carex mainénsis Porter. Maine Sedge. Fig. 1084.

Carex pulla A. Gray, Man. Ed. 5, 602. 1867. Not Gooden. 1797.

Carex miliaris var. aurea Bailey, Mem. Torr. Club 1: 37. 1889. Not C. aurea Nutt. 1818.

Carex mainensis Porter; Britton, Manual 193. 1901.

Culms slender, smooth below inflorescence, little filamentose at base, 1½°–3° tall, the plant stoloniferous. Leaves ¾″–1½″ wide, flat, shorter than the culm, somewhat nodulose and scabrous on the margins, the lower bracts narrower, about equalling the culm; staminate spikes 1–4, slender-stalked; pistillate spikes 1–3, erect, cylindric, 6″–15″ long, 3″–4″ wide, sessile or the lower short-stalked; perigynia about 2″ long, oblong-ovoid, yellowish-green, few-nerved, slightly inflated, contracted into a rather conspicuous 2-toothed beak; scales lanceolate, acute or acuminate, yellowish or brown-margined, slightly shorter than the perigynia; stigmas 2 or 3.

On lake and river shores, central Maine to Labrador. Possibly a hybrid between *C. miliaris* and *C. vesicaria*. Illustrated as *Carex Raeana* in first edition. Summer.

218. Carex Reèàna Boott. Rae's Sedge.
Fig. 1085.

C. Raeana Boott; Richards, Arct. Exp. **2**: 344. 1857.
C. vesicaria var. *Raeana* Fernald, Rhodora **3**: 50. 1901.
C. miliaris var. *Raeana* Küken. Pflanzenreich **4**: 20: 719. 1909.

Culms slender, 1°–3° high, very rough below inflorescence, reddened and filamentose at base, the plant stoloniferous. Leaves 1″ wide, flat, shorter than or exceeding culm, little if at all nodulose; lower bracts narrow, exceeding culm; staminate spikes usually 2, slender-stalked; pistillate spikes 1–3, narrowly cylindric, 7″–24″ long, 2½″ wide, short-peduncled, loosely flowered at base; perigynia somewhat inflated, oblong-ovoid, yellowish-green, strongly several-nerved, 2½″–3″ long, abruptly contracted into a slender bidentate beak; scales lanceolate, sharply acuminate, narrower and somewhat shorter than perigynia; stigmas 3.

On lake and river shores, Maine to Quebec, west to Athabasca. Local. Intergrades with the next. Summer.

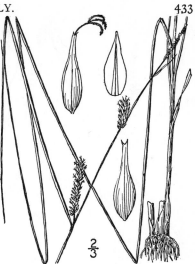

219. **Carex monìle** Tuckerm. Necklace Sedge.   Fig. 1086.

Carex monile Tuckerm. Enum. Meth. 20. 1843.
C. monile var. *monstrosa* Bailey, Mem. Torr. Club **1**: 40. 1889.
C. vesicaria vars. *monile* and *jejuna* Fernald, Rhodora **3**: 53. 1901.

Glabrous, culm slender, erect, 1½°–3° tall, not spongy at base, generally acutely angled and very rough above. Leaves elongated, 1½″–3″ wide, sometimes exceeding the culm, little nodulose; bracts similar, often overtopping the culm; staminate spikes 1–4, usually 2 or 3, slender-stalked, commonly subtended by short bracts; pistillate spikes 1–3, erect-spreading, cylindric, 1′–3′ long, about 3½″ in diameter, many-flowered with perigynia in several rows, rather loose at maturity, the upper sessile, the lower more or less slender-stalked and remote; perigynia yellowish-green, ascending, globose-ovoid, inflated, 2½″–4″ long, rather strongly 8–10-nerved, abruptly contracted into a slender 2-toothed beak; scales lanceolate, acuminate or awned, shorter than perigynia; stigmas 3.

In marshes and wet meadows, Newfoundland to Saskatchewan, south to New Jersey and Missouri. Also in Japan. Intergrades with the next. June–Aug.

220. **Carex vesicària** L. Inflated Sedge.
Fig. 1087.

Carex vesicaria L. Sp. Pl. 979. 1753.
Carex Vaseyi Dewey, Am. Journ. Sci. (II.) **29**: 347. 1860.

Glabrous, culm slender, erect, 1°–3° tall, not spongy at base, generally acutely angled and very rough above. Leaves elongated, 1½″–3″ wide, about equalling the culm, little nodulose; bracts similar, about equalling the culm; staminate spikes 2–4, slender-stalked, commonly subtended by short bracts; pistillate spikes 1–3, usually 2, erect, sessile or short-peduncled, oblong-cylindric, 1′–2½′ long, 4½″–7½″ thick, closely many-flowered, the perigynia in several rows; perigynia yellowish-green, ascending, ovoid, inflated, 3″–4″ long, rather strongly 8–10-nerved, contracted into a slender 2-toothed beak; scales ovate, acute or acuminate, narrower and shorter than perigynia; stigmas 3.

Quebec to British Columbia, south to Pennsylvania and Ohio. Also in Europe, Asia and north Africa. June–Aug.

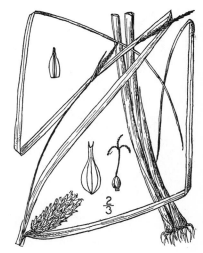

28

221. **Carex rotundàta** Wahl. Round-fruited Sedge. Fig. 1088.

C. rotundata Wahl. Vet.-Acad. Handl. **24**: 153. 1803.

Culms slender but stiff, obtusely triangular, smooth below inflorescence, not filamentose at base, 6′–1½° tall, caespitose but stoloniferous. Leaves 1½″ wide or less, strongly involute, usually exceeding the culm, not strongly nodulose, somewhat roughened towards apex, the lowest bract short, strongly exceeded by culm, usually widely divergent; staminate spikes 1–2, short-stalked; pistillate spikes 1–2, ascending, oblong, 5″–9″ long, 4″–5″ wide, sessile or very short-stalked; perigynia squarrose-spreading, dark-tinged, obsoletely nerved, broadly ovoid, inflated, 2″–3″ long, abruptly contracted into a short minutely 2-toothed beak; scales ovate, acute, narrower and shorter than perigynia, dark-tinged; stigmas 3.

Arctic America and Greenland. Also in northern Europe and Asia. July–Aug.

222. **Carex membranopácta** Bailey. Fragile Sedge. Fig. 1089.

Carex compacta R. Br. in Ross' Voy. App. cxliii. 1819. Not Krock. 1814.
Carex membranacea Hook. Parry's 2d Voy. App. 406. 1825. Not Hoppe.
C. membranopacta Bailey, Bull. Torr. Club **20**: 428. 1893.

Similar to the last, short-stoloniferous, the culms caespitose, 6′–18′ high. Leaves flat, not exceeding the culm, 1½″–2¼″ wide, the margin revolute; bracts short, the lower commonly longer than its spike, the upper subulate; staminate spikes 1–3, short-stalked; pistillate spikes 1–3 (commonly 2), sessile or the lower short-peduncled, approximate, narrowly oblong, densely 25–75-flowered, 6″–15″ long, 4″ in diameter; perigynia spreading, dark-tinged, broadly oval to obovoid, fragile, much inflated, about 2″ long and 1″ wide, tipped with a short minutely bidentate beak, little-nerved, rather longer than the ovate-oblong, brownish, hyaline-margined scales; stigmas usually 3.

Greenland and north Labrador to Alaska. Summer.

223. **Carex rostràta** Stokes. Beaked Sedge. Fig. 1090.

C. utriculata Boott; Hook. Fl. Bor. Am. **2**: 221. 1840.
C. rostrata Stokes; With. Arrang. Brit. Pl. **2**: 1059. 1787.

Glabrous, culms stout, erect, 1°–4° tall, thick at base, generally obtusely angled and smooth above. Leaves elongated, strongly nodulose, the upper mostly exceeding the culm, 1″–6″ wide, the midvein prominent; bracts overtopping the culm; staminate spikes 2–4, linear, stalked, the lower occasionally pistillate at the top and usually subtended by a very slender bract; pistillate spikes mostly 2 to 4, nearly erect, cylindric, densely many-flowered, the perigynia in many rows, or sometimes looser near the base, 2′–6′ long, 3″–10″ thick, the lower short-stalked, the upper sessile, sometimes staminate at the summit; perigynia spreading when old, ovoid, light green, somewhat inflated, few-nerved, 2″–4″ long, abruptly contracted into a short 2-toothed beak; scales lanceolate, the lower awned and slightly longer than the perigynia, the upper acute; stigmas 3.

Marshes, Labrador to British Columbia, Delaware, Ohio and California. Also in Europe and Asia. June–Sept.

224. Carex bullàta Schk.   Button Sedge.   Fig. 1091.

Carex bullata Schk.; Willd. Sp. Pl. 4: 309.   1805.

Carex Olneyi Boott, Ill. Car. 1: 15.   1858.

Carex Greenii Boeck. Flora 41: 649.   1858.

C. bullata var. *Greenii* Fernald, Rhodora 8: 202.   1906.

Glabrous, culms slender, erect, 1°–3½° high, roughened above on the sharp angles.   Leaves and bracts narrow and elongated, rather stiff, commonly overtopping the culm, rarely more than 2″ wide, rough-margined, sparingly nodulose; staminate spikes mostly 2, long-stalked; pistillate spikes 1–3, usually 2, light green, varying from almost sessile to long-stalked and spreading, many-flowered, oblong to cylindric, ¾′–2′ long, 4½″–9″ in diameter; perigynia much inflated, strongly nerved, dull or shining, ovoid, 2½″–4″ long, ascending-spreading, contracted into a subulate rough 2-toothed beak, longer than the lanceolate acuminate to obtusish scale; stigmas 3.

In swamps, Maine to Georgia.   June–Aug.

225. Carex Tuckermàni Dewey.   Tuckerman's Sedge.   Fig. 1092.

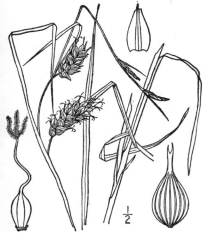

C. Tuckermani Dewey, Am. Journ. Sci. 49: 48.   1845.

Glabrous, culm very slender, roughened above on the angles, erect, 1½°–3½° tall.   Leaves and bracts much elongated, commonly much overtopping the culm, 1½″–2½″ wide, more or less nodulose; staminate spikes 2 or 3; pistillate spikes stout, cylindric, 1′–2′ long, 6″–9″ in diameter, the upper sessile or nearly so, the lower stalked and usually spreading; perigynia very much inflated and bladder-like, shining, broadly ovoid, prominently few-nerved, ascending, abruptly contracted into a smooth subulate 2-toothed beak; scales lanceolate, acute to short-cuspidate, less than half as long as the perigynia; achenes prominently excavated in middle; stigmas 3.

In bogs and meadows, New Brunswick to Minnesota, south to New Jersey, Indiana and Iowa.   June–Aug.

226. Carex retrórsa Schwein.   Retrorse Sedge.   Fig. 1093.

C. retrorsa Schwein. Ann. Lyc. N. Y. 1: 71.   1824.

Carex Hartii Dewey, Am. Journ. Sci. (II.) 41: 226.   1866.

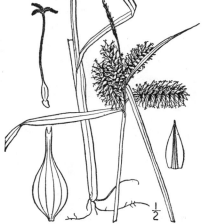

Glabrous, culm stout, erect, smooth or slightly rough above, 1°–3½° tall.   Leaves much elongated, thin, rough-margined, nodulose, 2½″–5″ wide, the upper commonly exceeding the culm, the bracts similar, usually very strongly overtopping the culm; staminate spikes 1–3, short-stalked; pistillate spikes 3–8, ascending or spreading, all close together at the summit and sessile or very nearly so, or the lowest distant and stalked, cylindric, densely many-flowered, 1′–3′ long, 7″–10″ in diameter; perigynia ovoid, membranous, strongly few-nerved, yellowish green, 3½″–5″ long, the lowest at least reflexed at maturity, tapering into a subulate 2-toothed beak; scales lanceolate, acute or acuminate, smooth, one-third to one-half as long as the perigynia; stigmas 3.

In swamps and wet meadows, Newfoundland to British Columbia, south to Pennsylvania, Iowa and Oregon.   Very variable.   Aug.–Sept.

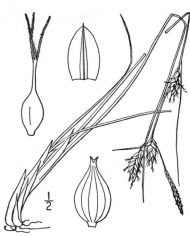

½

227. Carex oligospérma Michx. Few-seeded Sedge. Fig. 1094.

Carex oligosperma Michx. Fl. Bor. Am. **2**: 174. 1803.

Glabrous, culms very slender, erect, rather stiff, 1½°–3° tall, long-stoloniferous. Leaves about 1½″ wide, not exceeding the culm, involute when old; bracts similar, overtopping the culm; staminate spike stalked, narrowly linear; pistillate spikes 1 or 2, subglobose or short-oblong, 5″–10″ long, sessile or the lower short-stalked, 5–15-flowered, or the upper sometimes reduced to 1 or 2 perigynia and with a staminate summit; perigynia ovoid, erect, inflated, strongly few-nerved, yellowish green, shining, 2½″ long, about 1½″ in diameter, contracted into a minutely 2-toothed beak; scales acute or slightly mucronate, much shorter than the perigynia; stigmas 3.

In bogs, Labrador and Newfoundland to the Northwest Territory, south to Massachusetts, Pennsylvania and Michigan. June–Sept. Ascends to 4000 ft. in the Adirondacks.

228. Carex lùrida Wahl. Sallow Sedge. Fig. 1095.

C. lurida Wahl. Kongl. Acad. Handl. (II.) **24**: 153. 1803.
Carex tentaculata Muhl.; Willd. Sp. Pl. **4**: 266. 1805.
C. tentaculata var. *parvula* Paine, Cat. Pl. Oneida 105. 1865.
Carex lurida var. *flaccida* Bailey, Mem. Torr. Club **1**: 73. 1889.
Carex lurida var. *parvula* (Paine) Bailey, Bull. Torr. Club **20**: 418. 1893.
Carex lurida var. *exundans* Bailey; Britt. & Brown, Ill. Fl. **1**: 299. 1896.

Glabrous, culms erect, smooth or slightly scabrous above, 6′–3° tall. Leaves elongated, rough, 2″–3½″ wide, the upper and the similar bracts usually much overtopping the culm; staminate spike usually solitary, elongated, from nearly sessile to long-stalked; pistillate spikes 1–4, globose to oblong-cylindric, densely many-flowered, ½′–2½′ long, 7″–10″ in diameter, the upper sessile and erect, the lower peduncled and erect, spreading or drooping; perigynia inflated, shining, ovoid, strongly about 10-ribbed, contracted into a long subulate beak, ascending or the lower spreading, 3″–4½″ long, thin, yellowish green, longer than the rough-awned scale; stigmas 3.

In swamps and wet meadows, Nova Scotia to Minnesota, Nebraska, Florida and Texas. Very variable. June–Oct.

229. Carex Bàileyi Britton. Bailey's Sedge. Fig. 1096.

Carex tentaculata var. *gracilis* Boott, Ill. 94. 1860. **Not** C. gracilis R. Br. 1810.
Carex Baileyi Britton, Bull. Torr. Club **22**: 220. 1895.

Glabrous, culms erect, slender, minutely scabrous above, 1°–2° tall. Leaves roughish, elongated, 1″–2″ wide, the upper and the similar bracts exceeding the culm; staminate spike solitary, more or less strongly peduncled; pistillate spikes 1–3, narrowly cylindric, densely many-flowered, all erect or ascending, 9″–2′ long, about 4″–6″ in diameter, the upper sessile, the lower more or less stalked; perigynia inflated, ovoid, 2½″–3″ long, ascending, abruptly contracted into a subulate 2-toothed beak about as long as body, prominently about 10-ribbed, the lower about equalling, the upper longer than the linear-subulate rough-awned scale; stigmas 3.

Bogs, Maine and Vermont to Virginia and Tennessee. June–Aug.

½

230. Carex Schweinítzii Dewey. Schweinitz's Sedge. Fig. 1097.

Carex Schweinitzii Dewey; Schwein. Ann. Lyc. N. Y.
1 : 71. 1824.

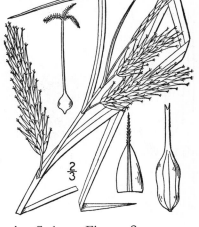

Glabrous, light green, culm erect, roughish above,
1°–2½° tall, from long-creeping rootstocks. Leaves
elongated, 2½″–5″ wide, the lower ones and the
similar bracts commonly overtopping the culm, those
of the culm mostly shorter; staminate spike solitary
or sometimes 2, slender-peduncled, the scales scarcely
awned; pistillate spikes 2–5, ascending, linear-cylindric, not very densely flowered, 1½′–3½′ long, about
4″–7″ thick, the upper usually sessile, the lower
stalked; perigynia light green, thin, somewhat inflated, ovoid-conic or oblong, contracted into the subulate, 2-toothed beak, 2½″–3½″ long and 1″ in diameter, ascending, rather prominently several-nerved,
equalling or the upper longer than the broad-based,
somewhat rough-awned or scabrous scale; stigmas 3.

In swamps and bogs, Vermont to Ontario, south to
Connecticut, New Jersey and Missouri. June–Aug.

231. Carex hystricìna Muhl. Porcupine Sedge. Fig. 1098.

Carex hystricina Muhl.; Willd. Sp. Pl. **4** : 282. 1805.

C. Cooleyi Dew. Am. Journ. Sci. 48 : 144. 1845.

C. hystricina Dudleyi Bailey, Mem. Torr. Club **1** : 54. 1889.

Caespitose, glabrous, light green, culms slender, erect,
rough above, 1°–3° tall, strongly reddened and occasionally
scabrous on lower sheaths. Leaves elongated, scabrous,
1½″–4″ wide, the upper and the similar bracts overtopping
the culm; staminate spike slender-stalked, the scales rough-
awned; pistillate spikes 1–4, oblong or oblong-cylindric,
densely many-flowered, ½′–2¼′ long, 5″–7″ in diameter, the
upper sessile or nearly so, the lower slender-stalked and
spreading or drooping; perigynia greenish, ascending, somewhat inflated, ovoid-conic, 2½″–3½″ long, strongly 15–20-
nerved, contracted into the subulate 2-toothed beak, equalling or the upper longer than the narrow rough-awned
scales; achene obovoid; stigmas 3.

In swamps and low meadows, Newfoundland to Alberta,
south to Georgia, New Mexico and Arizona. June–Aug.

232. Carex Pseùdo-Cypèrus L. Cyperus-like Sedge. Fig. 1099.

Carex Pseudo-Cyperus L. Sp. Pl. 978. 1753.

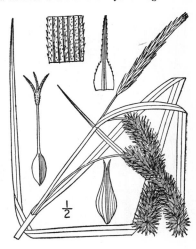

Glabrous, culms rather stout, rough on the sharp
angles, at least above, 2°–3½° high. Leaves elongated,
rough on the margins, nodulose, 2″–5″ wide, the
upper and the similar bracts overtopping the culm;
staminate spike short-stalked, the scales rough-awned;
pistillate spikes 2–5, linear-cylindric, densely many-
flowered, all slender-stalked and spreading or
drooping, 1′–2½′ long, 4″–6″ in diameter; perigynia
rigid, short-stipitate, scarcely inflated, lanceolate,
prominently and closely many-ribbed, somewhat flattened and triangular, at length reflexed, tapering into
a short 2-toothed beak, the short teeth slightly
spreading; scales linear with a broad base, rough-
awned, about equalling the perigynia; stigmas 3.

In bogs, Nova Scotia to Saskatchewan, south to Connecticut, New York and Michigan. Also in Europe and
Asia. June–Aug.

233. Carex comòsa Boott. Bristly Sedge. Fig. 1100.

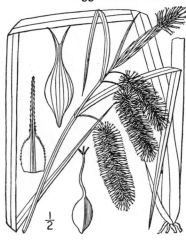

C. *furcata* Ell. Bot. S. C. and Ga. **2** : 552. 1824. Not Lapeyr. 1813.
Carex comosa Boott, Trans. Linn. Soc. **20**: 117. 1846.
Carex Pseudo-Cyperus var. *comosa* Boott, Ill. Car. **4**: 141. 1867.
Carex Pseudo-Cyperus var. *americana* Hochst.; Bailey, Mem. Torr. Club **1** : 54. 1889.

Similar to the preceding species, culms commonly stouter, sometimes 5° high, the leaves 3″-7″ wide. Staminate spike short-stalked, the scales rough-awned; pistillate spikes 2–6, usually 3–5, stalked or the uppermost nearly sessile, all spreading or drooping, stouter and bristly, 6″-7″ in diameter; perigynia lanceolate, rigid, scarcely inflated, somewhat flattened and triangular, strongly reflexed when mature, short-stipitate, tapering into a slender, prominently 2-toothed beak, the teeth subulate and recurved-spreading; scales mostly shorter than the perigynia, very rough-awned; stigmas 3.

In swamps and along the borders of ponds, Nova Scotia to Washington, south to Florida, Louisiana and California. May–Oct.

234. Carex Fránkii Kunth. Frank's Sedge. Fig. 1101.

Carex stenolepis Torr. Ann. Lyc. N. Y. **3** : 420. 1836. Not Less. 1831.
Carex Frankii Kunth, Enum. **2** : 498. 1837.

Glabrous, much tufted, culms stout, smooth, erect, very leafy, 1°–2½° tall. Leaves elongated, roughish, 1½″–4″ wide, the upper ones and especially the similar bracts overtopping the culm; staminate spike stalked or nearly sessile, occasionally pistillate at the summit, often small and inconspicuous; pistillate spikes 3–6, exceedingly dense, cylindric, erect, ½′–1½′ long, about 4″ in diameter, the upper nearly or quite sessile, the lower slender-stalked; perigynia green, slightly inflated, 2″ long, about 1″ in diameter, few-nerved, obconic, with a depressed summit from which arises the subulate 2-toothed beak; scales linear-subulate, very rough, longer than the perigynia; stigmas 3.

In swamps and wet meadows, eastern Pennsylvania to eastern Virginia and Georgia, west to Illinois, Missouri, Louisiana and Texas. June–Sept.

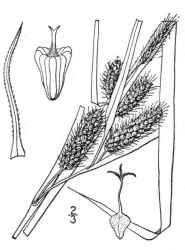

235. Carex squarròsa L. Squarrose Sedge. Fig. 1102.

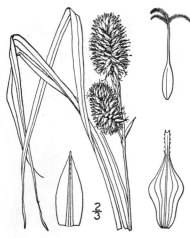

Carex squarrosa L. Sp. Pl. 973. 1753.

Glabrous, culms slender, erect, rough above on the angles, 2°–3° tall. Leaves elongated, 1½″–3″ wide, rough-margined, the upper somewhat overtopping the culm; bracts similar; spikes 1–3, generally 1, erect, stalked, oval, exceedingly dense, the pistillate portion 7″–15″ long, 6″–11″ in diameter, the upper one club-shaped, staminate at the base or sometimes for one-half its length or more; perigynia yellowish green, becoming tawny, squarrose or the lowest reflexed, somewhat inflated but firm, obovoid, about 1½″ in diameter, few-nerved, truncately contracted into the subulate minutely 2-toothed beak, twice as long as the scarious, lanceolate acuminate or awn-tipped scales; achene linear-oblong, 1½″ long, tapering into the stout, strongly flexuous style; stigmas 3.

In swamps and bogs, Ontario to Connecticut, Michigan, Nebraska, Georgia, Louisiana and Arkansas. June–Sept.

236. Carex týphina Michx. Cat-tail Sedge. Fig. 1103.

Carex typhina Michx. Fl. Bor. Am. **2**: 169. 1803.
Carex typhinoides Schwein. Ann. Lyc. **1**: 66. 1824.
Carex squarrosa var. typhinoides Dewey, Am. Journ. Sci. **11**: 316. 1826.

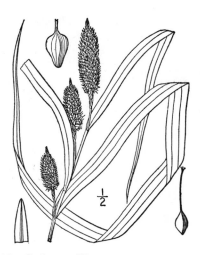

Similar to the preceding species, but the leaves generally much broader, 2″–5″ wide, the similar bracts much overtopping the culm; spikes 1–6, generally three, oblong-cylindric, very dense, the pistillate portion 1′–1¾′ long, 4″–8″ in diameter, often staminate at both ends, the terminal one commonly tapering to a conic summit; basal staminate flowers rather less numerous than in *C. squarrosa;* perigynia dull straw-color, obovoid, ascending or the lowest spreading or reflexed, inflated, truncately contracted into the slender 2-toothed beak, which is often upwardly bent; scales oblong-lanceolate, obtusish; achene obovoid, 1¼″ long, sharply 3-angled with concave sides, tipped with the slender style.

In swamps, Quebec to Virginia, Louisiana, Iowa and Missouri. July–Aug.

237. Carex intuméscens Rudge. Bladder Sedge. Fig. 1104.

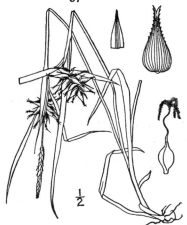

Carex intumescens Rudge, Trans. Linn. Soc. **7**: 97. *pl. 9. f. 3.* 1804.
Carex intumescens Fernaldi Bailey, Bull. Torr. Club. **20**: 418. 1893.

Glabrous, culms slender, commonly tufted, erect, 1½°–3° high. Leaves elongated, dark green, shorter than or sometimes equalling the culm, roughish, 1½″–3½″ wide, the sheath a little prolonged; bracts similar, overtopping the culm; staminate spike narrow, mostly long-stalked; pistillate spikes 2 (1–3), sessile or short-stalked, globular or nearly so; perigynia 1–12, spreading or the upper erect, 5″–10″ long, much inflated, about 2½′ in diameter above the rounded base, many-nerved, contracted into a subulate 2-toothed beak, the teeth somewhat spreading at maturity; scales narrowly lanceolate, aristate, or obtuse in few-flowered northern plants, about one-half as long as the perigynia; stigmas 3.

In swamps, bogs and wet woods, Newfoundland to Manitoba, south to Florida and Louisiana. May–Oct.

238. Carex Asa-Gràyi Bailey. Gray's Sedge. Fig. 1105.

Carex intumescens var. globularis A. Gray, Ann. Lyc. N. Y. **3**: 236. 1834. Not *C. globularis* L. 1753.
Carex Grayi Carey, Am. Journ. Sci. (II.) **4**: 22. 1847. Not *C. Grayana* Dewey, 1834.
Carex Grayi hispidula A. Gray; Bailey, Mem. Torr. Club **1**: 54. 1889.
Carex Asa-Grayi Bailey, Bull. Torr. Club **20**: 427. 1893.

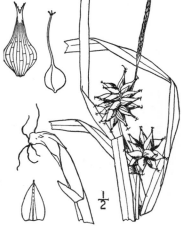

Glabrous, culms stout, erect, 2°–3° tall. Leaves elongated, dark green, 2½″–4½″ wide, the upper commonly overtopping the culm, the sheath not prolonged; bracts similar to the upper leaves, usually much overtopping the culm, short-sheathing; staminate spike mostly long-stalked; pistillate spikes 1 or 2, globose, dense, about 1′ in diameter; perigynia 6–30, 6″–9″ long, ovoid, glabrous or hispidulous, much inflated, many-ribbed, round-truncate at base, about 3½″ in diameter above the base, contracted into a sharp 2-toothed beak; scales ovate, obtuse to slightly cuspidate, scarious, about one-third as long as the perigynia; stigmas 3.

In swamps and wet meadows, Vermont to Michigan, south to Georgia and Missouri. June–Sept.

239. Carex louisiánica Bailey. Louisiana Sedge. Fig. 1106.

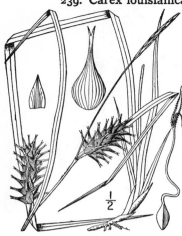

Carex Halei Carey; Chapm. Fl. S. States 543. 1860. Not Dewey, 1846.
C. louisianica Bailey, Bull. Torr. Club **20**: 428. 1893.
C. Eggertii Bailey, Bot. Gaz. **21**: 6. 1896.

Culms slender, erect, smooth or very nearly so, 1°–2° tall, arising singly from elongated rootstocks. Leaves 1″–2½″ wide, roughish, the upper overtopping the spikes; bracts similar to the upper leaves, rough, strongly sheathing; staminate spike solitary, long-stalked; pistillate spikes 1–4, oblong, about 1′ long, 8″–12″ thick, erect, the lower slender-stalked, the upper nearly sessile; perigynia ovoid, much inflated, smooth, strongly several-ribbed, shining when mature, 5″–6″ long, about 2½″ in diameter at the rounded base, contracted into the long 2-toothed beak, the small teeth slightly spreading; scales oblong-lanceolate, acute or acuminate, about one-half as long as the perigynia; stigmas 3.

Swamps, near Washington, south to Florida and Texas, north in Mississippi Valley to Missouri. June–Sept.

240. Carex lupulìna Muhl. Hop Sedge. Fig. 1107.

Carex lupulina Muhl.; Willd. Sp. Pl. **4**: 266. 1805.
Carex lupulina var. *pedunculata* Dewey, in Wood, Bot. & Flor. 376. 1870.
Carex lupulina Bella-villa Bailey, Mem. Torr. Club **1**: 12. 1889.
C. Bella-villa Dewey, Am. Journ. Sci. (II.) **41**: 229. 1866.

Glabrous, culms stout, erect, 1°–4° tall. Leaves elongated, nodulose, 2½″–8″ wide, the upper ones and the similar bracts much overtopping the culm; staminate spike solitary, rarely several, nearly sessile or slender-peduncled, rather stout; pistillate spikes 2–5, densely many-flowered, sessile to long-stalked, oblong, 1′–3′ long, often 1′ or more in diameter; perigynia ascending, much inflated, many-nerved, 5″–10″ long, 2″–3½″ in diameter just above the base, contracted into a subulate 2-toothed beak; achene longer than thick, thickened but not knobbed on angles; scales lanceolate, acuminate or aristate, generally much shorter than perigynia; stigmas 3.

In swamps and ditches, New Brunswick to Hudson Bay, western Ontario, Iowa, Florida and Texas. June–Sept.

241. Carex lupulifórmis Sartwell. Hop-like Sedge. Fig. 1108.

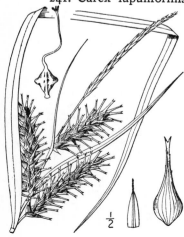

Carex lupulina var. *polystachya* Schw. & Torr. Ann. Lyc. **1**: 337. 1825. Not *C. polystachya* Sw.
Carex lupuliformis Sartw.; Dewey, Am. Journ. Sci. (II.) **9**: 29. 1850.

Glabrous, culm stout, erect, 1½°–3° tall. Leaves and bracts similar to those of the preceding species, much elongated; staminate spike solitary, stalked or nearly sessile, sometimes 4′ long; pistillate spikes 3–6, stalked or the upper nearly sessile, densely many-flowered, oblong-cylindric, 1½′–3½′ long, 10″–15″ in diameter, often staminate at the top; perigynia yellowish brown at maturity, at first appressed, later ascending, sessile, much inflated, several-nerved, 5″–10″ long, 2½″–3½″ in diameter above the base, contracted into a subulate 2-toothed beak; achene often thicker than long, its angles strongly knobbed; scales lanceolate, awned, generally shorter than the perigynia.

Swamps, Vermont to Minnesota, south to Delaware and Louisiana. June–Sept.

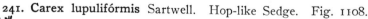

242. Carex gigantèa Rudge. Large Sedge.
Fig. 1109.

C. gigantea Rudge, Trans. Linn. Soc. **7** : 99. *pl. 10. f. 2.* 1804.
Carex grandis Bailey, Mem. Torr. Club **1** : 13. 1889.

Glabrous, culms slender, erect, 2°–3° high, long-stoloniferous. Leaves rather dark green, elongated, 3½″–8″ wide, the uppermost sometimes surpassing the culm; lower bracts similar to the leaves, much overtopping the culm, strongly sheathing; staminate spikes 1–3, sessile or peduncled; pistillate spikes 2–5, all stalked or the upper sessile, cylindric, 1′–3′ long, about 1′ thick, sometimes staminate at the summit; perigynia much swollen at the base, and 2″–3″ in diameter, 6″–9″ long, many-nerved, spreading at right angles at maturity, 3–4 times as long as the scarious lanceolate acuminate or aristate scale, abruptly contracted into a subulate 2-toothed beak 2–3 times as long as body; stigmas 3.

In swamps, Delaware to Kentucky and Missouri, south to Florida, Louisiana and Texas. June–Aug.

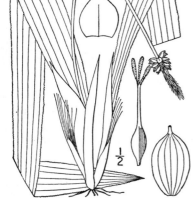

19. CYMOPHÝLLUS Mackenzie.

A perennial sedge with short rootstocks and flattened culms. Culms with four to six overlapping striate bladeless sheaths, and after flowering developing one large blade-bearing leaf without sheath, ligule or midrib, and with undulate margins appearing minutely serrulate. Spike one, bractless, androgynous, the flowers monoecious, solitary in the axils of the scales. Perianth none. Staminate flower of three stamens, the filaments filiform. Pistillate flowers of a single pistil, style and three stigmas enveloped by a bladder-like perigynium. Achene triangular. Racheola often developed. [Greek, referring to the undulate-margined leaves.]

A monotypic genus of the southeastern United States.

1. Cymophyllus Fràseri (Andr.) Mackenzie.
Fraser's Sedge. Fig. 1110.

Carex Fraseri Andr. Bot. Rep. *pl. 639.* 1811.
Carex Fraseriana Sims, Bot. Mag. *pl. 1391.* 1811.

Glabrous, culms smooth, slender, reclining, 6′–18′ long. Developed leaves 8′–16′ long, 1′–2′ wide, without midvein, sheath or ligule, perfectly flat, firm, spreading, finely many-nerved, subacute at the apex, their margins usually finely crumpled in drying, one to a culm, developed only after flowering; lower culm-leaves reduced to clasping basal sheaths; spike solitary, bractless, androgynous, ½′–1′ long, the pistillate portion dense, nearly ½′ in diameter in fruit; perigynia elliptic-ovoid, milk-white at maturity, diverging, thin and somewhat swollen, faintly nerved, 2½″–3″ long and rather more than 1″ in diameter, tipped with a short nearly truncate beak; scales ovate, obtusish, much shorter than the perigynia; stigmas 3.

In rich woods, southwestern Virginia, West Virginia, eastern Tennessee and North Carolina. May–July.

Family 12. ARÀCEAE Neck. Act. Acad. Theod. Palat. **2** : 462. 1770.
ARUM FAMILY.

Herbs mostly with basal long-petioled simple or compound leaves, and spathaceous inflorescence, the spathe enclosing or subtending a spadix. Rootstock tuberous or a corm, in our species mostly with an acrid or pungent sap. Spadix very densely flowered, the staminate flowers above, the pistillate below, or the plants wholly dioecious, or with perfect flowers in some species. Perianth wanting, or of 4–6 scale-like segments. Stamens 4–10 in our species; filaments very short; anthers 2-celled, commonly with a thick truncate connective, the sacs opening by dorsal pores or slits. Ovary 1–several-celled; ovules 1–several in each cell; style short or wanting; stigma terminal, mostly minute and sessile. Fruit a berry or utricle. Seeds various. Endosperm copious, sparse or none.

About 105 genera and 900 species, mostly of tropical regions, a few in the temperate zones.
Flowers without a perianth.
　Flowers monoecious or dioecious, borne at the base of the spadix.　　　1. *Arisaema.*
　Flowers monoecious, covering the whole spadix.　　　2. *Peltandra.*

Flowers perfect.　　　　　　　　　　　　　　　　　　3. *Calla.*
Flowers with a perianth.
　　Spadix enclosed in a shell-like fleshy spathe.　　　　4. *Spathyema.*
　　Spadix naked, terminating the scape.　　　　　　　　5. *Orontium.*
　　Spadix naked, borne at the base of a leaf-like spathe.　　6. *Acorus.*

1. ARISAÈMA Mart. Flora 14: 459. 1831.

Perennial herbs with acrid corms, simple scapes and 1 to 3 slender-petioled divided leaves unfolding with the flowers. Spadix included or exserted, bearing the flowers near its base. Spathe convolute, open or contracted at the throat. Flowers dioecious or monoecious, without any perianth, the staminate of 4 almost sessile 2–4-celled anthers which open by confluent slits at the apex, the pistillate with an ovoid or globose 1-celled ovary containing 1 or many orthotropous ovules; style short or none, stigma peltate-capitate. Fruit a cluster of globose red berries, conspicuous when ripe. Seeds with copious endosperm and an axial embryo. [Greek, referring to the red-blotched leaves of some species.]

About 50 species, mostly natives of temperate and subtropical Asia. Besides the following, 2 other species occur in the Southern States. Type species: *Arum nepenthoides* Wall.

Spathe hooded, open at the throat, enclosing the spadix.
　Leaves pale beneath; spadix club-shaped.　　　　　　　　1. *A. triphyllum.*
　Leaves green on both sides; spadix cylindric.
　　Spathe smooth, deep brown to black.　　　　　　　　　2. *A. pusillum.*
　　Spathe fluted, green or striped.　　　　　　　　　　　3. *A. Stewardsonii.*
Spathe convolute; summit of the spadix exserted.　　　　　4. *A. Dracontium.*

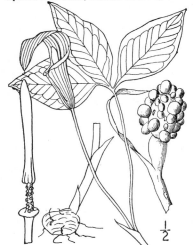

$\frac{1}{2}$

1. Arisaema triphýllum (L.) Torr. Jack-in-the-pulpit. Indian Turnip. Fig. 1111.

Arum triphyllum L. Sp. Pl. 965. 1753.
Arisaema atrorubens Blume, Rumphia 1: 97. 1835.
Arisaema triphyllum Torr. Fl. N. Y. 2: 239. 1843.

Leaves 1 or 2, nearly erect, 10′–3° high, usually exceeding the scape, 3-foliolate, the segments ovate, entire, or sometimes lobed, acute rounded or narrowed at the base, 3′–7′ long, 1½–3½′ wide, sessile or very short-stalked; flowers commonly dioecious, yellow, borne on the basal part of the spadix; spadix 2′–3′ long, its naked summit blunt, colored; spathe green, and purple-striped, curving in a broad flap over the top of the spadix, acuminate; filaments very short and thick; ovaries crowded; ovules 5 or 6; berries smooth, shining, about 5″ in diameter, forming a dense ovoid head 1′–3′ long.

In moist woods and thickets, Nova Scotia to Florida, Ontario, Minnesota, Kansas and Louisiana. Ascends to 5000 ft. in North Carolina. April–June. Fruit ripe June–July. The acrid bulb made edible by boiling. Three-leaved Indian turnip; Marsh, Pepper or Wild turnip. Bog-onion. Brown-dragon. Wake-robin. Starchwort.

2. Arisaema pusíllum (Peck) Nash. Peck's Jack-in-the-pulpit. Fig. 1112.

Arisaema triphyllum pusillum Peck, Rep. N. Y. State Mus. 51: 297. 1898.

Arisaema pusillum Nash; Britton, Man. 229. 1901.

Leaves 2 or sometimes 1, erect, mostly 8′–15′ high, 3-foliolate, the segments elliptic, ovate or ovate-lanceolate, entire, mostly dull, acuminate, or sometimes merely acute at the apex, narrowed at the base, 2′–5′ long, ¾–1¾′ wide, sessile or nearly so; spadix 1′–2′ long, the upper portion cylindric; spathe deep brown to black, the lower portion even, the apex short-acuminate; berries shining, 2″–2½″ in diameter, forming an ovoid head less than 1′ in diameter.

In bogs, New York to Georgia and Kentucky. May–July.

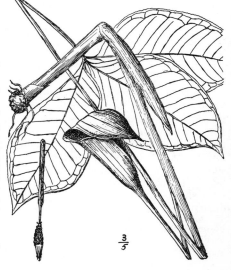

$\frac{3}{5}$

3. Arisaema Stewardsònii Britton. Stewardson Brown's Indian Turnip. Fig. 1113.

Arisaema Stewardsonii Britton, Man. Ed. 2, 1057. 1905.

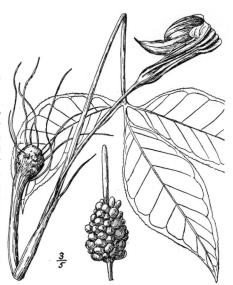

Leaves 2, or 1 in the case of small plants, 1°–2½° high, 3-foliolate, the segments lanceolate to ovate-lanceolate, erose-crenulate, shining, acuminate at the apex, narrowed at the base, 1½′–8′ long, 1′–3′ wide, sessile or nearly so; spadix 1½′–2′ long, the upper portion cylindric; spathe green or striped, the lower portion fluted, the apex rather long-acuminate; berries shining, 4″–5″ in diameter, forming an ovoid head over 1′ in diameter.

In wet woods, often among sphagnum, mountains of Pennsylvania and New Jersey. Reported from New England. June–Aug.

$\frac{3}{5}$

4. Arisaema Dracóntium (L.) Schott. Green Dragon. Dragon-root. Fig. 1114.

Arum Dracontium L. Sp. Pl. 964. 1753.
Arisaema Dracontium Schott, Melet. 1: 17. 1832.

Corms clustered. Leaves usually solitary, 8′–4° long, pedately divided into 5–17 segments, much longer than the scape; segments obovate or oblong, 3′–10′ long, 9″–4′ wide, abruptly acute at the apex, narrowed to a sessile or nearly sessile base, entire or the lateral ones somewhat lobed; scape sheathed by membranous scales at the base; spathe greenish or whitish, narrowly convolute, acuminate, 1′–2′ long, enwrapping the spadix, the upper part of which tapers into a slender appendage exserted 1′–7′ beyond its apex; inflorescence of the staminate plant nearly as long as the tubular part of the spathe; in the monoecious plant the pistillate flowers are borne on the lower part of the spadix; ovary turbinate, with 6–8 bottle-shaped ovules; stigmas depressed; berries reddish-orange in large ovoid heads.

$\frac{1}{2}$

Mostly in wet woods and along streams, but sometimes in dry soil, Maine to Ontario and Minnesota, south to Florida, Kansas and Texas. May–June.

2. PELTÁNDRA Raf. Journ. Phys. 89: 103. 1819.

Bog herbs, with entire sagittate acute or acuminate leaves, the long petioles sheathing the shorter scape at the base. Spathe elongated, convolute, or expanded above. Flowers monoecious, covering the whole spadix. Perianth none. Staminate flowers uppermost, consisting at first of irregularly 4-sided oblong flat-topped shields, from the edges of which appear 6–10 imbedded anthers opening by apical pores, the shields ultimately shrivelling and leaving the linear-oblong anthers nearly free. Ovaries ovoid, surrounded at base by 4 or 5 white fleshy scale-like staminodia, 1-celled; ovules solitary or few, amphitropous; style erect, short, thick, tipped with a small stigma. Fruit a green or red berry, 1–3-seeded, when ripe forming large globose heads at the extremity of the finally recurved scape, and enclosed in the persistent leathery base of the spathe. Seeds surrounded by a tenacious jelly; endosperm none. [Greek, referring to the shield-shaped staminate disks.]

The genus consists of two species, the following one being the type; the other inhabits marshes and springs from North Carolina to Florida.

1. Peltandra virgínica (L.) Kunth. Green Arrow-arum. Fig. 1115.

Arum virginicum L. Sp. Pl. 966. 1753.
Peltandra undulata Raf. Journ. Phys. **89**: 103. 1819.
Peltandra virginica Kunth, Enum. **3**: 43. 1841.

Leaves bright green, somewhat hastate-sagittate, 4'–30' long, 3'–8' wide, acute or acuminate at the apex, firm, strongly veined. Root a tuft of thick fibers; scape nearly as long as the leaves, recurving and immersing the fruiting spadix at maturity; spathe green, 4'–8' long, long-conic, closely investing the spadix throughout, the strongly involute margins undulate; spadix shorter than the spathe, the pistillate flowers covering about one-fourth of its length, the rest occupied by staminate flowers; ovaries globose-ovoid; style nearly ½" long; stigmas a little thicker than the style; berries green when ripe.

In swamps or shallow water, Maine to Ontario, Michigan, Florida, Louisiana and Missouri. Poison-arum. Virginia wake-robin. May–June.

Peltandra sagittaefólia (Michx.) Morong, admitted into our first edition, differs from *P. virginica* in a dilated spathe with a whitish summit; it is not definitely known north of North Carolina.

3. CÁLLA L. Sp. Pl. 968. 1753.

A bog herb with slender acrid rootstocks, broadly ovate or nearly orbicular cordate leaves, and a large white persistent spathe. Spathe ovate-lanceolate or elliptic, acuminate, open. Spadix cylindric, much shorter than the spathe, densely covered with flowers. Flowers perfect or the very uppermost staminate; perianth none. Stamens about 6; filaments linear, longer than the anthers; anther-sacs divaricate, opening by slits. Ovary ovoid, 1-celled; style very short; stigma small, flat, circular. Ovules 6–9, anatropous. Berries obconic, depressed. Seeds hard, smooth, oblong, striate toward the micropyle and pitted at the other end. Endosperm copious. [An ancient name, taken from Pliny.]

A monotypic genus of the cooler portions of the north temperate zone.

1. Calla palústris L. Water Arum. Wild Calla. Fig. 1116.

Calla palustris L. Sp. Pl. 968. 1753.

Petioles 4'–8' long, spreading or ascending. Blades thick, entire, 1½'–4' wide, cuspidate or abruptly acute at the apex, deeply cordate at the base; scapes as long as the petioles, sheathed at the base; rootstocks covered with sheathing scales and with fibrous roots at the nodes; spathe 1'–2½' long and about 1' wide, with an abruptly acuminate involute apex; spadix about 1' long; berries red, distinct, few-seeded, forming a large head when mature.

In bogs, Nova Scotia to Hudson Bay, Minnesota, New Jersey, Pennsylvania, Wisconsin and Iowa. Reported from Virginia. Also in Europe and Asia. Female or water-dragon. Water-lily. Swamp-robin. May–June. Fruit ripe July–Aug.

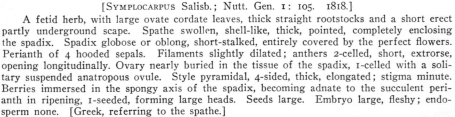

4. SPATHYÈMA Raf. Med. Rep. (II.) 5: 352. 1808.

[SYMPLOCARPUS Salisb.; Nutt. Gen. **1**: 105. 1818.]

A fetid herb, with large ovate cordate leaves, thick straight rootstocks and a short erect partly underground scape. Spathe swollen, shell-like, thick, pointed, completely enclosing the spadix. Spadix globose or oblong, short-stalked, entirely covered by the perfect flowers. Perianth of 4 hooded sepals. Filaments slightly dilated; anthers 2-celled, short, extrorse, opening longitudinally. Ovary nearly buried in the tissue of the spadix, 1-celled with a solitary suspended anatropous ovule. Style pyramidal, 4-sided, thick, elongated; stigma minute. Berries immersed in the spongy axis of the spadix, becoming adnate to the succulent perianth in ripening, 1-seeded, forming large heads. Seeds large. Embryo large, fleshy; endosperm none. [Greek, referring to the spathe.]

A monotypic genus of eastern North America and northeastern Asia.

1. Spathyema foètida (L.) Raf. Skunk Cabbage. Fig. 1117.

Dracontium foetidum L. Sp. Pl. 967. 1753.
Spathyema foetida Raf. Med. Rep. (II.) **5**: 352. 1808.
Symplocarpus foetidus Nutt. Gen. 1: 106. 1818.

Leaves numerous, in large crowns, 1°–3° long, often 1° wide, strongly nerved, abruptly acute at the apex, thin, entire, their petioles deeply channeled. Rootstock thick, descending, terminating in whorls of fleshy fibers; spathe preceding the leaves, erect, 3′–6′ high, 1′–3′ in diameter at the base, convolute, firm; purple-brown to greenish yellow, often mottled, its short scape usually subterranean, spadix about 1′ in diameter in flower, greatly enlarging and sometimes 6′ in diameter in fruit; mature seeds 4″–6″ long.

In swamps and wet soil, Nova Scotia to Ontario and Minnesota, North Carolina and Iowa. Feb.–April. Skunkweed. Meadow, Swamp- or Clumpfoot-cabbage. Pole-cat weed. Polk-weed. Collard. Fruit ripe Aug.–Sept.

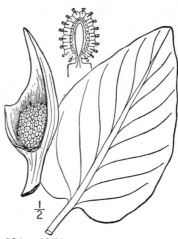

5. ORÓNTIUM L. Sp. Pl. 324. 1753.

Aquatic herbs, with thick rootstocks buried in the mud, oblong-elliptic nerved leaves without a distinct midvein, and slender terete scapes terminated by a cylindric spadix. Spathe enclosing the spadix when very young, soon parting and remaining as a sheathing bract at its base, or falling away. Flowers perfect, bright yellow, covering the whole spadix. Sepals 4–6, scale-like, imbricated upon the ovary (lower flowers commonly with 6, upper with 4). Stamens as many as the sepals; filaments linear, wider than the anthers, abruptly narrowed above; anthers small, with two diverging sacs opening by oblique slits. Ovary partly imbedded in the axis of the spadix, depressed, obtusely angled, 1-celled; ovule solitary, half-anatropous; stigma sessile. Fruit a green utricle. Endosperm none; embryo long-stalked. [Ancient name of some water plant, said to be from the Syrian river Orontes.]

A monotypic genus of eastern North America.

1. Orontium aquáticum L. Golden-club. Floating Arum. Fig. 1118.

Orontium aquaticum L. Sp. Pl. 324. 1753.

Leaves ascending or floating, depending on the depth of water, deep dull green above, pale beneath, the blade 5′–12′ long, 2′–5′ wide, entire, acute or cuspidate at the apex, narrowed at the base into a petiole 4′–20′ long. Scape 6′–24′ long, flattened near the spadix; spadix 1′–2′ long, 3″–4″ in diameter, frequently attenuate at the summit, much thickened in fruit; spathe bract-like, 2′–4′ long, 2-keeled on the back; usually falling away early; utricle depressed, roughened on top with 9 or 10 tubercles.

In swamps and ponds, Massachusetts to central Pennsylvania, south to Florida and Louisiana, mostly near the coast. Ascends to 2000 ft. on the Pocono plateau of Pennsylvania. Water-dock. Tawkin. April–May.

6. ÁCORUS L. Sp. Pl. 324. 1753.

Erect herbs, with very long horiontal branched rootstocks, sword-shaped leaves, and 3-angled scapes keeled on the back and channeled in front, and a seemingly lateral cylindric spadix, the scape appearing as if extending long beyond it, but this upper part is in reality a spathe. Flowers perfect, densely covering the whole spadix. Perianth of 6 membranous concave sepals. Stamens 6; filaments flattened, much longer than the anthers; anthers reniform or sagittate, 2-celled, the cells confluent at maturity. Ovary oblong, 2–4-celled with 2–8 orthotropous ovules in each cell; stigma sessile, depressed-capitate. Fruit a 2–3-celled gelatinous berry, few-seeded. Endosperm copious. [Name ancient.]

Two known species, the following widely distributed in the north temperate zone, the other Japanese. The following is the generic type.

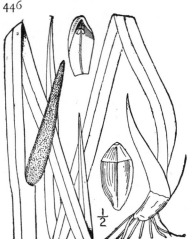

1. Acorus Cálamus L.   Sweet Flag.   Calamus or Flag-root.   Fig. 1119.

Acorus Calamus L. Sp. Pl. 324.   1753.

Leaves linear, erect, 2°–6° tall and 1' wide or less, sharp-pointed and sharp-edged, with a rigid mid-vein running their whole length, 2-ranked, closely sheathing each other and the scape below.   Spathe a leaf-like extension of the scape projecting 8'–30' beyond the spadix; spadix spike-like, 2'–3½' long, about ½' in diameter, compactly covered with minute greenish-yellow flowers.

In swamps and along streams, Nova Scotia to Ontario and Minnesota, south to Louisiana and Kansas.   Also in Europe and Asia.   In our territory fruit is rarely, if ever, formed.   The hard ovary is usually found to be imperfect, with 2 or 3 abortive cells and ovules.   The plant is propagated by its large rootstocks, which furnish the drug Calamus.   Interior of stalk sweet.   Myrtle-flag, -sedge or -grass.   Sweet-myrtle.   Sedge-grass, -cane, -root or -rush.   Sea-sedge.   Beewort.   May–July.

Family 13.   LEMNÀCEAE Dumort. Fl. Belg. 147.   1827.

DUCKWEED FAMILY.

Minute perennial floating aquatic plants, without leaves or with only very rudimentary ones.   The plant body consists of a disc-shaped, elongated or irregular thallus, which is loosely cellular, densely chlorophyllous and sometimes bears one or more rootlets.   The vegetative growth is by lateral branching, the branches being but slightly connected by slender stalks and soon separating.   In the autumn these disconnected branches fall to the bottom of the ditch or pond, but rise and again increase in size in the spring.   The inflorescence consists of one or more naked monoecious flowers borne on the edge or upper surface of the plant.   Each flower commonly consists of but a single stamen or a single flask-shaped pistil.   The anther has two to four pollen-sacs, containing spherical minutely barbellate grains.   The pistil is narrowed to the funnel-shaped scar-like stigmatic apex, and produces 1–6 erect or inverted ovules.   The fruit is a 1–6-seeded utricle.

Comprises the smallest of the flowering plants and contains 4 genera and about 26 species.

Thallus with one root or several.
   Roots several.   1. *Spirodela.*
   Root solitary.   2. *Lemna.*
Thallus rootless.
   Thallus thick, globose to subcylindric.   3. *Wolffia.*
   Thallus thin, ligulate.   4. *Wolffiella.*

1.   SPIRODÈLA Schleid. Linnaea 13: 391.   1839.

Thallus disc-shaped, 7–12-nerved.   The thinly-capped rootlets as well as the nerves are provided with a single bundle of vascular tissue. Spathe sac-like.   Anthers 2-celled.   The ovary produces two anatropous ovules.   Fruit rounded, wing-margined. [Greek, in allusion to the cluster of rootlets.] Two American species and *S. oligorhiza*, a native of southern Asia, Australia and the Fiji Islands.   The following is the type of the genus.

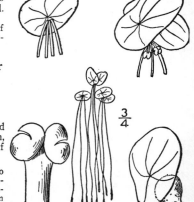

1. Spirodela polyrhìza (L.) Schleid.   Greater Duckweed.   Fig. 1120.

Lemna polyrhiza L. Sp. Pl. 970.   1753.

Spirodela polyrhiza Schleid. Linnaea 13: 392.   1839.

Thallus round-obovate, 2"–5" long, thick, flat and dark green above, slightly convex and purple beneath, palmately 5–11-nerved.   Each thallus bears a cluster of from 5–11 elongated rootlets.   Rootcap pointed.

In rivers, ponds, pools and shallow lakes, Nova Scotia to British Columbia, south to South Carolina, Texas, northern Mexico and Nevada.   Also in Jamaica.   Widely distributed in the Old World and in tropical America.   Seldom collected in flower.

2. LÉMNA L. Sp. Pl. 970. 1753.

Thallus disc-shaped, usually provided with a central nerve and with or without two or four lateral nerves. Each thallus produces a single rootlet, which is devoid of vascular tissue and is commonly provided with a thin blunt or pointed rootcap. The ovary contains from one to six orthotropous, amphitropous or anatropous ovules. Fruit ovoid, more or less ribbed. Endosperm in one or three layers. [Greek, in allusion to the growth of these small plants in swamps.]

About 8 species, in temperate and tropical regions. Besides the following, *Lemna angolensis* is a native of Lower Guinea. Type species: *Lemna trisulca* L.

Thalli long-stipitate.	1. *L. trisulca.*
Thalli short-stipitate or sessile.	
Spathe open.	
Thalli 1-nerved or nerveless.	
Thalli thin, without papules; rootcap strongly curved, tapering.	2. *L. cyclostasa.*
Thalli thick, with a row of papules along the nerves; rootcap little curved, cylindric.	
	3. *L. minima.*
Thalli 3-nerved; rootcap cylindric.	4. *L. perpusilla.*
Spathe sac-like.	
Thalli green or purplish beneath; fruit not winged.	5. *L. minor.*
Thalli pale beneath, usually strongly gibbous; fruit winged.	6. *L. gibba.*

1. Lemna trisúlca L. Ivy-leaved Duckweed. Star Duckweed. Fig. 1121.

Lemna trisulca L. Sp. Pl. 970. 1753.

Thallus lanceolate, submerged and devoid of stomata in the primary aquatic form, ovate to oblong-lanceolate, 6″–9″ long, floating and provided with stomata in the later flowering stage. The later and more common form is narrowed at the base to a slender stipe, thin, denticulate, with or without rootlets, and slightly 3-nerved. Several individuals often remain connected, so as to form a chain-like series. Rootcap pointed; fruit symmetrical; seed deeply 12–15-ribbed.

In ditches, springs, ponds, shallow lakes and sloughs, Nova Scotia to British Columbia, south to New Jersey, New Mexico and California. Also in Europe, Asia, Africa and Australia. July–Aug.

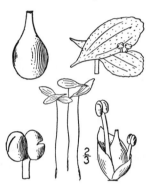

2. Lemna cyclóstasa (Ell.) Chev. Valdivia Duckweed. Fig. 1122.

L. minor cyclostasa Ell. Bot. S. C. and Ga. **2**: 518. 1824.
Lemna cyclostasa Chev. Fl. Paris **2**: 256. 1827.
Lemna valdiviana Philippi, Linnaea **33**: 239. 1864.

Thallus oblong-elliptical, 1″–1½″ long, thin, subfalcate and shortly stalked at the base, provided with numerous stomata, except on the borders, nerveless; rootcap short-tapering, curved; spathe reniform; fruit ovoid-oblong, unsymmetrical; seed prominently 12–29-ribbed.

In pools and rivers, Massachusetts to Florida, west to Illinois, Wyoming and California. Also in Jamaica and in South America. June–July.

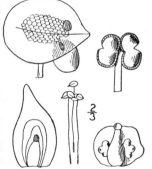

3. Lemna mínima Philippi. Least Duckweed. Fig. 1123.

Lemna minima Philippi, Linnaea **33**: 239. 1844.

Thallus oblong to elliptic, 1″–2″ long, obscurely 1-nerved, or nerveless, with a row of papules along the nerve, the lower surface flat, or slightly convex, the apex rounded; rootcap usually short, a little curved, rarely perfectly straight, cylindric, blunt; spathe open; pistil short-clavate; stigma concave; ovule solitary, obliquely orthotropous; seed oblong, pointed, about 16-ribbed, with many transverse striations.

Georgia and Florida to Kansas, Wyoming and California.

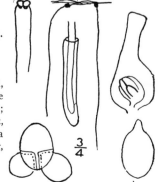

4. Lemna perpusílla Torr. Minute Duckweed. Fig. 1124.

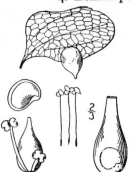

Lemna perpusilla Torr. Fl. N. Y. **2** : 245. 1843.

L. paucicostata Hegelm. in A. Gray, Man. Ed. 5, 681. 1868.

L. perpusilla trinervis Austin, in A. Gray, Man. Ed. 5, 479. 1867.

Thallus small, $1''$–$1\frac{1}{2}''$ long, oblong to obovate, often purplish-tinged beneath, unsymmetrical and abruptly narrowed to a very short stalk, provided throughout with numerous stomata, more or less 3-nerved; rootcap pointed; fruit ovoid, unsymmetrical; seed 12–16-ribbed, and transversely striated.

In ponds, rivers, springs and lakes, New York and New Jersey to Minnesota, Nebraska and Missouri. June–July.

5. Lemna mìnor L. Lesser Duckweed. Fig. 1125.

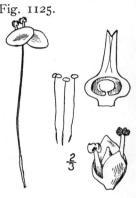

Lemna minor L. Sp. Pl. 970. 1753.

Thallus obovate or subcircular, $1''$–$3''$ long, thickish, rarely reddish or purplish-tinged, short-stalked when young, provided throughout with stomata, obscurely 3-nerved, very rarely 4–5-nerved. Rootcap obtuse or subtruncate; fruit symmetrical, subturbinate, wingless; seed with a prominent protruding hilum, deeply and unequally 12–15-ribbed.

In ponds, lakes and stagnant waters, throughout continental North America except the extreme north, and in the Bahamas. Also in Europe, Asia, Africa and Australia. Duck's-meat. Toad-spit. Water lentils. Mardling. Summer.

6. Lemna gíbba L. Gibbous Duckweed. Fig. 1126.

Lemna gibba L. Sp. Pl. 970. 1753.

Thallus slightly unsymmetrical, obovate or short-obovate, $1\frac{1}{2}''$–$3''$ long, thickish or more or less strongly gibbous beneath, short-stalked when young, soon separating, provided with stomata which are sparse beneath, obscurely 3–5-nerved; rootcap mostly short-pointed, rarely long-pointed or obtuse; fruit symmetrical, winged; seed thick, deeply and unequally ribbed.

In ponds and rivers, Nebraska, Texas, Arizona and California. Also in Mexico, Europe, Asia, Africa and Australia. June–July.

3. WÓLFFIA Horkel; Schleid. Linnaea, **13** : 389. 1839.

Thallus small, globose, ovoid-oblong, subcylindric or irregular, rootless, nerveless and leafless. The vegetative growth is from a cleft near one end of the plant, the branch being mostly sessile and soon detached. The ovary contains one orthotropous ovule. Fruit spherical or short-ovate, smooth. Endosperm in a single layer. [Name in honor of Nath. Matth. v. Wolff, 1724–1784, Polish physician and naturalist.]

A genus of few species, mostly in tropical and subtropical regions. Type species : *Lemna hyalina* Delile.

Thallus globose to ellipsoid, not punctate.   1. *W. columbiana*.
Thallus flattened above, at least on the margin, brown-punctate.
    Apex of thallus acute.   2. *W. punctata*.
    A large conic papule on the upper surface.   3. *W. papulifera*.

1. Wolffia columbiàna Karst.   Columbia Wolffia. Fig. 1127.

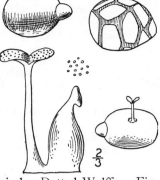

Wolffia columbiana Karst. Bot. Unters. **1**: 103. 1865–67.

Thallus spherical or subellipsoid, $\frac{1}{3}''-\frac{2}{3}''$ long, with a limited number of stomata (1 to 6), loosely cellular and clear green throughout, not dotted nor gibbous.

Floating as minute alga-like grains just beneath the surface of the water in stagnant ponds, pools and shallow lakes, Ontario to Connecticut and New Jersey, west to Minnesota and Missouri, south to Louisiana. Also in Mexico and South America. June–July.

2. Wolffia punctàta Griseb.   Dotted Wolffia. Fig. 1128.

W. punctata Griseb. Fl. Br. W. I. 512. 1864.

Thallus oblong, smaller than the last, $\frac{1}{4}''-\frac{1}{3}''$ long, flattish, densely cellular, with numerous stomata and dark green above, gibbous, more loosely cellular, with fewer stomata and paler beneath; brown-dotted throughout with minute pigment cells.

Floating on the surface of stagnant waters, Ontario to Michigan, south to Tennessee. Jamaica. Has been confused with *W. brasiliana* Wedd. June–July.

3. Wolffia papulífera Thompson.   Pointed Duckweed. Fig. 1129.

Wolffia papulifera Thompson, Ann. Rep. Mo. Bot. Gard. **9**: 20. *pl. 4D.* 1897.

Thallus slightly unsymmetrical, obliquely broadly ovate, about $\frac{1}{2}''$ broad, the apex rounded, the upper surface flat at the margin, gradually ascending into a prominent conic papule on the median line, brown-punctate, the under surface strongly gibbous, less punctate; stomata numerous on the upper surface; flower and fruit unknown.

Floating, with the entire upper surface exposed to the air, Kennett and Columbia, Mo.

4. WOLFFIÉLLA Hegelm. Engler's Bot. Jahrb. **21**: 303. 1895.

Thallus thin, unsymmetrical, rootless, curved in the form of a segment of a band, punctate on both surfaces with numerous brown pigment-cells. Pouch opening as a cleft in the basal margin of the thallus, a stipe attached to its margins. Flowers and fruit unknown. [Diminutive of *Wolffia*.]

About 7 species, mostly of tropical regions. Besides the following, 2 others are known from western North America. Type species: *Wolffiella oblonga* (Phil.) Hegelm.

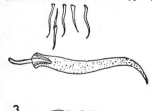

1. Wolffiella floridàna (J. D. Smith) Thompson. Florida Wolffiella. Fig. 1130.

Wolffia gladiata var. *floridana* J. D. Smith, Bull. Torr. Club **7**: 64. 1880.
Wolffia floridana J. D. Smith; Hegelm. Engler's Bot. Jahrb. **21**: 305. 1895.
Wolffiella floridana Thompson, Ann. Rep. Mo. Bot. Gard. **9**: 17. 1897.

Thalli solitary, or commonly coherent for several generations forming densely interwoven masses, strap-shaped, scythe-shaped, or doubly curved, tapering from the rounded oblique base to a long-attenuate apex, $2\frac{1}{2}''-5''$ long, 14–21 times as long as wide; basal portion of the long stipe persistent, the pouch elongated-triangular, or the upper angle rounded.

Georgia and Florida to Missouri, Arkansas and Texas.

Family 14. **MAYACÀCEAE** Walp. Ann. **3**: 662. 1853.

MAYACA FAMILY.

Slender branching aquatic moss-like herbs, with linear sessile 1-nerved entire soft leaves, notched at the apex. Flowers solitary, peduncled, white, perfect, and regular, the peduncles bracted at the base. Perianth persistent, consisting of 3 lanceolate green herbaceous sepals and 3 obovate white spreading petals. Stamens 3, hypogynous, alternate with the petals; filaments filiform; anthers oblong, somewhat 4-sided, 2-celled. Ovary superior, sessile, 1-celled with 3 parietal placentae; ovules several or numerous, orthotropous; style filiform; stigmas terminal, entire or with 3 short lobes. Capsule 1-celled, 3-valved. Seeds ovoid or globose, the testa reticulated; embryo at the apex of the mealy endosperm.

The family consists of the following genus.

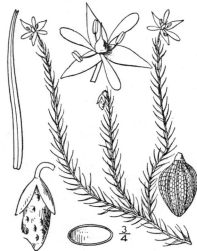

1. **MAYÀCA** Aubl. Pl. Guian. **1**: 42. 1775.

About 7 species known, natives of warm and tropical America. Only 2 species in the United States. Type species: *Mayaca fluviatilis* Aubl.

[Aboriginal name of these plants in Guiana.]

1. **Mayaca Aublèti** Michx. Mayaca.
Fig. 1131.

Mayaca Aubleti Michx. Fl. Bor. Am. **1**: 26. 1803.
Mayaca Michauxii Schott & Endl. Melet. **1**: 24. 1832.

Stems tufted, 3′–15′ long, usually little branched. Leaves densely clothing the stem and widely spreading, linear-lanceolate, translucent, 2″–3″ long, about ¼″ wide; peduncles 2″–6″ long, very slender, recurved in fruit; flowers 3″–4″ broad, axillary, but borne near the ends of branches, lateral, rarely more than one on each branch; capsule oblong-oval, about as long as the sepals, tipped until dehiscence by the subulate style.

In fresh water pools and streams, southeastern Virginia to Florida and Texas. May–July.

Family 15. **XYRIDÀCEAE** Lindl. Nat. Syst. Ed. 2, 388. 1836.

YELLOW-EYED GRASS FAMILY.

Perennial or annual tufted herbs with basal narrow equitant commonly 2-ranked leaves, and erect simple leafless scapes. Flowers perfect, mostly yellow, nearly or quite regular, solitary and sessile in the axils of coriaceous imbricated bracts (scales), forming terminal ovoid globose or cylindric heads. Sepals 3, the two lateral ones small, keeled, persistent, the other one larger, membranous (wanting in the South American genus *Abolboda*). Corolla inferior, with a narrow tube and 3 spreading lobes. Stamens 3, inserted on the corolla, usually alternating with as many plumose or bearded staminodia. Ovary sessile, 1-celled or incompletely 3-celled; ovules numerous or few, on 3 parietal placentae, orthotropous; style terminal (unappendaged in *Xyris*, in *Abolboda* appendaged at the base), 3-branched above; stigmas apical. Fruit an oblong 3-valved capsule. Seed-coat longitudinally striate. Embryo apical. Endosperm mealy or somewhat fleshy.

Two genera, *Xyris* L. and *Abolboda* H. & B., comprising some 60 species, mostly of tropical distribution in both the Old World and the New.

1. **XỲRIS** L. Sp. Pl. 42. 1753.

Characters of the family as given above. [Greek name for some plant with 2-edged leaves.]

Besides the following species there are some 9 others in the southern United States. Type species: *Xyris indica* L.

Lateral sepals wingless, the keel fringed with short hairs.   1. *X. flexuosa.*
Lateral sepals winged, the keel erose, fimbriate or lacerate.
   Plants not bulbous-thickened at the base; leaves flat or but slightly twisted.
      Lateral sepals about as long as the bracts, included, their keels erose to lacerate.
        Heads ovoid; bracts relatively few and in few series.
          Leaves linear or tapering from the base to the apex.

Lateral sepals nearly entire; heads narrowly ovoid, acute during anthesis.
                                                       2. *X. montana.*
    Lateral sepals toothed; heads broadly ovoid, obtuse.   3. *X. caroliniana.*
      Leaves broadest at the middle.   4. *X. communis.*
    Heads oblong or nearly cylindric; bracts numerous, in many series.   5. *X. elata.*
    Lateral sepals longer than the bracts, exserted, their keels fimbriate or erose-lacerate.
      Keels of the lateral sepals erose-lacerate; heads not plumose.   6. *X. Congdoni.*
      Keels of the lateral sepals long-fimbriate; heads somewhat plumose.   7. *X. fimbriata.*
    Plants conspicuously bulbous-thickened at the base; leaves spirally twisted.   8. *X. arenicola.*

1. Xyris flexuòsa Muhl. Slender Yellow-eyed Grass. Fig. 1132.

Xyris flexuosa Muhl. Cat. 5. 1813.

Xyris torta J. E. Smith, in Rees' Cyclop. 1818.

Scapes slender, straight or sometimes slightly twisted, 4'–18' tall, 2-edged above, bulbous-thickened at the base. Leaves narrowly linear, flat or becoming twisted when old, 1'–6' long, ½"–1½" wide; head globose, or short-oblong, obtuse, 3"–4" high; bracts broadly oval or slightly obovate, entire or somewhat lacerate at the apex; lateral sepals linear, about as long as the bracts, curved, finely fringed with short hairs on the wingless keel; expanded flowers 3"–4" broad.

In swamps and bogs, Maine to Minnesota, south to Georgia, Missouri and Texas. Yellow flowering-rush. July–Sept.

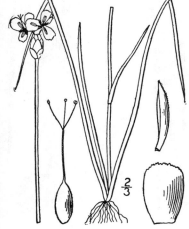

2. Xyris montàna H. Ries. Northern Yellow-eyed Grass. Fig. 1133.

Xyris flexuosa var. *pusilla* A. Gray, Man. Ed. 5, 548. 1867. Not *X. pusilla* R. Br. 1810.
Xyris montana H. Ries, Bull. Torr. Club **19**: 38. 1892.

Scapes very slender, straight or slightly twisted, 2-edged above, 2'–12' tall, not bulbous-thickened at the base. Leaves narrowly linear, 1'–6' long, ½"–1" wide, not at all twisted or but very slightly so; head ovoid, acute during anthesis, or narrowly subacute, 1½"–3" long; bracts oval or obovate, rounded and finely lacerate at the apex; lateral sepals linear, irregularly erose on the winged keel near the apex, about as long as the bracts.

In bogs, Nova Scotia to Ontario and Michigan, south to the Pocono Mountains of Pennsylvania. July–Aug.

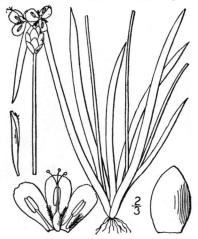

3. Xyris caroliniàna Walt. Carolina Yellow-eyed Grass. Fig. 1134.

Xyris caroliniana Walt. Fl. Car. 69. 1788.

Xyris Jupacai Michx. Fl. Bor. Am. **1**: 23. 1803.

Scapes mostly slender, straight or somewhat twisted, 2-edged above, 1°–2° tall, not thickened at the base. Leaves linear, flat, 4'–15' long, 1"–5" wide; head broadly ovoid, blunt, 4"–8" long; scales oval or slightly obovate, entire or somewhat lacerate; lateral sepals linear, about as long as the bracts, the narrowly winged keel serrate only above the middle.

In swamps and bogs, Maine and Massachusetts to Pennsylvania, Florida and Louisiana, mostly near the coast. Young states of this plant may be mistaken for *X. montana.* June–Aug.

4. **Xyris commùnis** Kunth. Southern Yellow-eyed Grass. Fig. 1135.

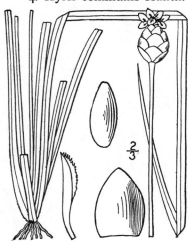

2/3

?Xyris macrocephala Vahl, Enum. **2** : 204. 1806.

Xyris communis Kunth, Enum. **4** : 12. 1843.

Xyris difformis Chapm. Fl. S. States 500. 1860.

Scapes slender, slightly twisted, 2-edged above, 1-edged below, not thickened at the base, 6'–18' tall. Leaves nearly linear or linear-lanceolate, flat, 3'–10' long, 1"–6" wide; head ovoid, or subglobose, blunt or subacute, about ½' long; scales ovate or oval, mostly entire; lateral sepals lanceolate, the winged keel fimbriate from the apex to below the middle; corolla-lobes obovate, 2"–3" long.

In bogs, Maryland to Florida and Louisiana. Widely distributed in tropical America. June–Aug.

5. **Xyris elàta** Chapm. Tall Yellow-eyed Grass. Fig. 1136.

Xyris elata Chapm. Fl. S. States 501. 1860.

Scapes rather stout, 1°–4½° tall, solitary or several together, 2-edged above. Leaves decidedly equitant, linear or nearly so, 8'–24' long; head oblong or nearly cylindric, 8"–12" long; scales suborbicular, numerous; lateral sepals 2"–2½" long, the narrowly winged keel toothed above the middle; corolla-lobes about 2" long.

In sandy swamps near the coast, southeastern Virginia to Florida and Louisiana. July–Oct.

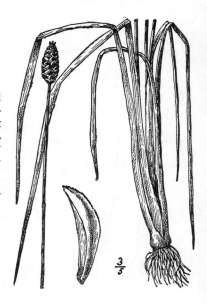

3/5

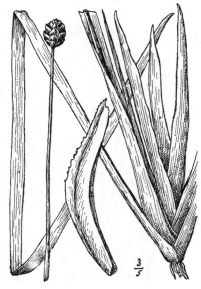

3/5

6. **Xyris Cóngdoni** Small. Congdon's Yellow-eyed Grass. Fig. 1137.

Xyris Congdoni Small; Britton, Man. Ed. 2, 1057. 1905.

Scapes relatively stout, smooth, flattened above, 1⅓°–2½° tall. Leaves over one-half as long as the scape, decidedly equitant, linear from a broad base which is commonly over 5" wide; head oval or nearly so, 6"–8" long; scales erose at the apex, less concave than those of *X. flexuosa;* lateral sepals 3"–3½" long, the broad keel erose-lacerate above the middle.

On boggy shores, near the coast, often in water, eastern Massachusetts to New Jersey. July–Sept.

7. Xyris fimbriàta Ell.   Fringed Yellow-eyed
Grass.   Fig. 1138.

Xyris fimbriata Ell. Bot. S. C. & Ga. **1**: 52.   1816.

Scapes rather stout, roughish, straight or somewhat twisted, strongly 2-edged above, 2°–4° high.
Leaves flat, one-half as long as the scapes or more,
3″–6″ wide; head oblong-cylindric, ½′–1′ long or
sometimes globose-ovoid and about ½′ in diameter;
scales obovate, their margins entire or the apex
lacerate; lateral sepals longer than the bracts, exserted, long-fringed on the winged keel above the
middle.

In wet pine barrens, southern New Jersey to Florida
and Mississippi, mostly near the coast.   July–Sept.

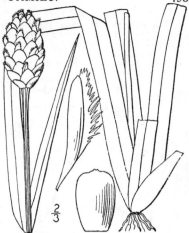

$\frac{2}{3}$

8. Xyris arenícola Small.   Twisted Yellow-
eyed Grass.   Fig. 1139.

Xyris torta Kunth, Enum. **4**: 14. 1843.   Not J. E. Smith.
Xyris arenicola Small, Fl. SE. U. S. 234.   1903.

Scapes stout, much spirally twisted, 1-edged below, or 2-edged at the summit, smooth or very nearly
so, the base conspicuously bulbous-thickened and with
the sheathing leaves sometimes 1′ in diameter.   Leaves
narrowly linear from a broad shining nearly black
base, rigid, rather shorter than the scapes, spirally
twisted (very markedly so when old); head oblong
or oblong-cylindric, acute or subacute, ½′–1′ long;
bracts oblong-obovate, minutely lacerate-serrulate at
the apex or entire; lateral sepals linear, exserted,
the winged keel fringed with rather short processes
above the middle.

In dry pine barrens, southern New Jersey to Florida,
west to Texas, mostly near the coast.   May–Aug.

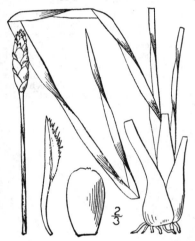

$\frac{2}{3}$

Family 16.   **ERIOCAULÀCEAE** Lindl. Veg. Kingd. 122.   1847.
PIPEWORT FAMILY.

Bog or aquatic herbs, perennial or perhaps sometimes annual, with fibrous
knotted or spongy roots, tufted grass-like basal leaves, and monoecious (androgynous) occasionally dioecious small flowers, in terminal solitary heads, on long
slender scapes.   Head of flowers involucrate by bracts, each flower borne in the
axil of a scarious scale.   Perianth of 2 series of segments, rarely of one series.
Stamens in staminate flowers as many or twice as many as the sepals.   Ovary
2–3-celled.   Ovules 2 or 3.   Fruit a 2–3-celled, 2–3-seeded capsule, loculicidally
dehiscent.   Seeds pendulous, orthotropous; endosperm farinaceous.

Nine genera and about 560 species, widely distributed in warm and tropical regions, a few ex-
tending into the temperate zones.   The family is most abundantly represented in South America.
Perianth of 2 series of segments; sepals and petals 2 or 3; stamens distinct; anthers 2-celled.
   Stamens twice as many as the inner perianth-segments (petals).   1. *Eriocaulon.*
   Stamens as many as the inner perianth-segments.   2. *Syngonanthus.*
Perianth simple, of 3 sepals; stamens 3, monadelphous below; anthers 1-celled.   3. *Lachnocaulon.*

1.   **ERIOCAÙLON** L. Sp. Pl. 87.   1753.

Acaulescent or very short-stemmed herbs, the scapes erect, or when immersed delicate,
angular, with a long sheathing bract at the base.   Leaves mostly short, spreading, acuminate,
parellel-nerved.   Head of flowers woolly, white, lead-colored or nearly black.   Staminate
flowers: Outer perianth-segments 2 or 3, distinct or sometimes connate, the inner united below in a tube, alternate with the outer ones, each with a minute spot or gland near its middle or apex; stamens mostly 4–6, one opposite each perianth-segment, the filaments of those
opposite the inner segments the longer; pistil small, rudimentary or none.   Pistillate flowers:
Outer perianth-segments as in the staminate flowers, the inner indistinct, narrow; stamens
wanting; ovary sessile or stalked; style columnar or filiform, stigmas 2 or 3, filiform.   Fruit
a thin-walled capsule.   Seeds oval, covered with minute processes.   [Greek, in allusion to
the wool at the base of the scape in some species.]

About 200 species, widely distributed, mostly in warm regions. Besides the following, 3 others occur in the southern United States, and 3 in Mexico. Type species: *Eriocaulon decangulare* L.

Leaf-blades as long as the sheaths, coarsely 3–9-fenestrate-nerved; scapes 7-angled; heads 1½″–3″ in diameter.

Mature head spheroidal, with a spreading or reflexed involucre; marginal flowers spreading or reflexed.   1. *E. septangulare.*

Mature head and involucre campanulate; marginal flowers permanently erect.   2. *E. Parkeri.*

Leaf-blades shorter than the sheaths, finely 6–20-fenestrate-nerved; scapes 10–12-angled, heads 3″–6″ in diameter.   3. *E. compressum.*

Leaf-blades much longer than the sheaths, finely 10–50-nerved; scapes 10–14-angled; heads 4″–8″ in diameter   4. *E. decangulare.*

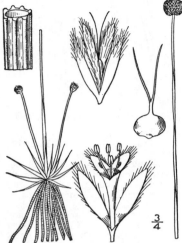

1. Eriocaulon septangulàre With. Seven-angled Pipewort. Fig. 1140.

E. septangulare With. Bot. Arr. Brit. Pl. 784. 1776.
Nasmythia articulata Huds. Fl. Angl. Ed. 2, 415. 1778.
E. articulatum Morong, Bull. Torr. Club **18**: 353. 1891.

Stem a mere crown. Leaf-blades pellucid, 3–8-fenestrate-nerved, ¼′–3′ long, usually as long as the sheaths; scapes weak, twisted, about 7-angled, 1′–8′ tall, or when submersed sometimes 4°–10° long; involucral bracts glabrous, or the innermost bearded at the apex, oblong, usually shorter than the flowers; marginal flowers usually staminate; scales of the receptacle spatulate or obovate, abruptly pointed, brown above, white-woolly; staminate flowers about 1¼″ high; pistillate flowers scarcely more than half as large; perianth-segments white-bearded.

In still water or on shores, Newfoundland to Ontario, Minnesota, Florida and Texas. Europe. July–Oct.

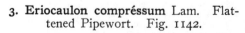

2. Eriocaulon Pàrkeri B. L. Robinson. Parker's Pipewort. Fig. 1141.

Eriocaulon Parkeri B. L. Robinson, Rhodora **5**: 175. 1903.

Stem very short. Leaf-blades pellucid, 7–9-fenestrate-nerved, 1¼′–2½′ long; scapes rather rigid, mostly 7-angled, smooth, 2′–4′ tall; involucral bracts glabrous, oval, about as long as the flowers; marginal flowers pistillate; scales of the receptacle glabrous or nearly so; staminate and pistillate flowers about 1″ high; perianth-segments of both kinds of flowers glabrous, or obscurely short-pubescent at the apex.

On the shores of the Delaware River, near Camden and Bordentown, New Jersey; Washington, D. C. July–Oct.

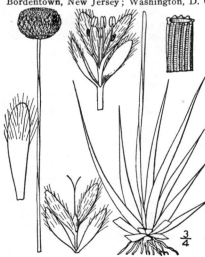

3. Eriocaulon compréssum Lam. Flattened Pipewort. Fig. 1142.

Eriocaulon compressum Lam. Encycl. **3**: 276. **1789.**
E. gnaphalodes Michx. Fl. Bor. Am. **2**: 165. 1803.

Leaf-blades 6–20-fenestrate-nerved, usually shorter than the sheaths and tapering to a long sharp point, rigid, or when submersed thin and pellucid. Stem a mere crown; scapes 6′–3° tall, smooth, flattened when dry, 10–12-angled; involucral bracts rounded, obtuse, scarious, shining, smooth, imbricated in 3 or 4 series; heads 3″–6″ in diameter, frequently dioecious; receptacle glabrous; flowers 1½″–2″ high.

In still shallow water, southern New Jersey to Florida and Texas. At flowering time the styles and stigmas are much exserted. May–Oct.

4. Eriocaulon decangulàre L.   Ten-angled Pipewort.   Fig. 1143.

Eriocaulon decangulare L. Sp. Pl. 87. 1753.

Stems short and thick, 1'–2' long. Leaf-blades finely many-nerved, tapering to a blunt point, 6'–20' long, 2"–8" wide, usually much longer than the sheaths; scapes stout, rigid, glabrous, 10–14-angled, 1°–3° tall; heads 4"–8" in diameter; involucral bracts ovate, often eroded, denticulate at the apex and pubescent below, imbricated in 4 or 5 series; receptacle pubescent with many-celled hairs; flowers 2" high, densely woolly at the base; scales longer than the flowers, acute, white-bearded; as are the spatulate perianth-segments.

In swamps, southern New Jersey and Pennsylvania to Florida and Texas. June–Oct.

2. SYNGONÁNTHUS Ruhland; Urban, Symb. Ant. 1: 487. 1900.

Perennial or rarely annual herbs, our species with much the habit of *Eriocaulon*. Stems very short. Leaves awl-shaped, tufted. Scapes slender, several-angled, erect, twisted in growth, sheathed at the base by a long acute bract. Flowers androgynous, in globular or hemispheric heads, each in the axil of a scale or the scales sometimes obsolete. Involucral bracts imbricated in 3 or 4 series. Perianth of 2 series, each of 2 or 3 segments in the staminate flowers, the outer segments distinct, the inner connate; stamens 2 or 3, inserted on the inner perianth and opposite its lobes. Pistillate flowers with the outer segments distinct, the inner often connate above the 2-celled, 2–3-ovuled ovary; style cleft into 2 or 3 entire or 2-cleft stigmas. Fruit a 2–3-celled, 2–3-seeded capsule, loculicidally dehiscent. [Greek, referring to the united petals of the pistillate flowers.]

About 80 species, mostly natives of tropical America. Only the following is known in the United States. Type species: *Eriocaulon umbellàtum* Lam.

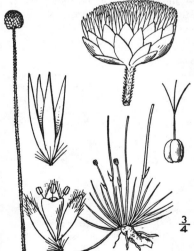

1. Syngonanthus flavídulus (Michx.) Ruhland.   Yellow Pipewort.   Fig. 1144.

Eriocaulon flavidulum Michx. Fl. Bor. Am. **2**: 166. 1803.
Paepalanthus flavidulus Kunth, Enum. **3**: 532. 1841.
Dupatya flavidula Kuntze, Rev. Gen. Pl. 745. 1891.
S. flavidulus Ruhland, in Engler, Pflanzenreich 4³⁰: 256. 1903.

Leaves 3–5-nerved, 1'–2' long, awl-shaped, woolly at the base, glabrous or sparingly pubescent above. Scapes numerous, 5-angled, pubescent, 4'–12' high; sheaths longer than the leaves, slightly inflated above, pubescent; involucral bracts straw-colored, glabrous, obtuse, oval, shining, somewhat pubescent at the base; receptacles glabrous or slightly pubescent; scales very thin, scarious-white, linear, slightly pubescent, about as long as the flowers; flowers about 1¼" high; perianth 6-parted; outer perianth of the staminate flowers stalked, woolly, the inner a campanulate tube with 3 stamens; pistillate flowers with both sets of perianth-segments distinct, the inner much narrower than the outer; style 3-parted.

Moist pine barrens, Virginia to Florida. March–July.

3. LACHNOCÀULON Kunth, Enum. **3**: 497. 1841.

Tufted herbs with the habit of *Eriocaulon*, the leaves linear. Scapes several-angled, sheathed at the base by an entire bract about as long as the leaves; heads globose. Receptacle pilose. Flowers androgynous. Perianth of 3 segments. Staminate flowers with 3 stamens united below into a thickened tube which is coalescent with a body, variously regarded as a corolla or as a rudimentary pistil, bearing at its apex 3 fimbriate or entire lobes alternate with the filaments; anthers 1-celled, minute. Pistillate flowers with a sessile 3-celled, 3-ovuled ovary surrounded by copious woolly hairs at the base; styles united below, spreading above into 3 divisions which are 2-parted, there being thus 6 stigmas. [Greek, referring to the woolly scapes of some species.]

Eight known species of the southern United States.   Type species: *Eriocaulon anceps* Walt.

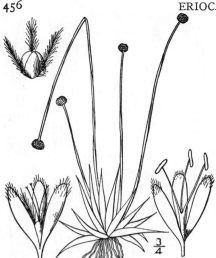

1. Lachnocaulon ánceps (Walt.) Morong. Hairy Pipewort. Fig. 1145.

Eriocaulon anceps Walt. Fl. Car. 83. 1788.
L. Michauxii Kunth, Enum. 3: 497. 1841.
L. anceps Morong, Bull. Torr. Club 18: 360. 1891.

Leaves glabrous or sparingly pubescent, 1'–3' long, tapering to an obtuse callous tip. Scapes slender, 2'–20' tall, 2–4-angled, clothed with long soft appressed upwardly pointed hairs; sheaths equalling or shorter than the leaves; heads globose, 1''–3'' in diameter; involucral bracts ovate or oblong, shorter than the flowers, usually brown; flowers about 1'' high; scales brown, spatulate, surrounded at the base by the yellowish silky hairs of the receptacle and white-bearded at the apex; perianth of the staminate flowers short-stalked, pubescent at the base, woolly and fimbriate at the summit; segments of the pistillate perianth white, glabrous, obtuse; ovary densely villous around the base; seeds strongly ribbed.

In moist pine barrens, Virginia to Florida. The white pistillate flowers mingled with the brown staminate ones impart a mixed gray and dark appearance to the heads. March–June.

Family 17. BROMELIÀCEAE J. St. Hil. Expos. Fam. 1: 122. 1805.
PINE-APPLE FAMILY.

Epiphytic herbs (some tropical species terrestrial), mostly scurfy, with elongated entire or spinulose-serrate leaves. Flowers spiked, panicled, or solitary, regular and perfect, usually conspicuously bracted. Perianth of 3 thin distinct or somewhat united sepals, and 3 clawed distinct or united petals. Stamens 6, usually inserted on the base of the corolla. Ovary inferior or superior, 3-celled; ovules numerous in each cell, anatropous; style short or elongated; stigmas 3. Capsule 3-valved in our species. Seeds numerous, the testa membranous. Embryo small, situated at the base of the copious endosperm.

About 35 genera and 900 species, all natives of tropical and subtropical America.

1. DENDROPÒGON Raf. Neogen. 3. 1825.

Epiphytic pendulous much-branched plants, with very narrow entire leaves and yellow or greenish flowers. Sepals distinct and separate or very nearly so. Petals distinct. Stamens long, the three inner ones inserted on the bases of the petals; filaments filiform; anthers linear. Ovary superior; style columnar; stigmas short. Capsule septicidally 3-valved. Seeds erect, narrow, supported on a long funiculus which splits up into fine threads. [Greek, meaning tree-beard.]

About 3 species; the following is the type of the genus.

1. Dendropogon usneoìdes (L.) Raf. Long Moss. Florida Moss. Fig. 1146.

Renealmia usneoides L. Sp. Pl. 287. 1753.
Tillandsia usneoides L. Sp. Pl. Ed. 2, 411. 1762.
Dendropogon usneoides Raf. Fl. Tel. 4: 25. 1838.

Stems very slender, thread-like, flexuous, hanging clustered in festoons from the branches of trees, 3°–20° long, gray and, like the filiform leaves, densely silvery-scurfy all over. Leaves scattered, 1'–3' long, scarcely ¼'' thick, their bases somewhat dilated; flowers sessile and solitary or rarely 2 together in the axils of the leaves; sepals about 3'' long, pale green; petals yellow, the blade about 2'' long; stamens about as long as the calyx; capsule linear, 9''–15'' long, at length splitting into 3 linear valves.

Eastern Virginia to Florida, west to Texas and Mexico. Very widely distributed in tropical America. Vegetable hair. Hanging, Spanish or Black moss. Long or Tree-beard.

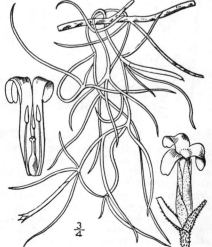

Family 18. **COMMELINÀCEAE** Reichenb. Consp. 57. 1828.

SPIDERWORT FAMILY.

Perennial or annual leafy herbs with regular or irregular perfect and often showy flowers in cymes, commonly subtended by spathe-like or leafy bracts. Perianth of 2 series; a calyx of mostly 3 persistent sepals, and a corolla of mostly 3 membranous and deciduous or fugacious petals. Stamens mostly 6, hypogynous, rarely fewer, all similar and perfect or 2 or 3 of them different from the others and sterile; filaments filiform or somewhat flattened; anthers 2-celled, mostly longitudinally dehiscent. Ovary superior, sessile or very nearly so, 2–3-celled; ovules 1 or several in each cell, anatropous or half anatropous; style simple; stigmas terminal, entire or obscurely 2–3-lobed. Seeds solitary or several in each cell of the capsule. Capsule 2–3-celled, loculicidally 2–3-valved. Embryo small. Endosperm copious.

About 25 genera and 350 species, mostly natives of tropical regions, a few in the temperate zones.

Perfect stamens 3, rarely 2; petals unequal; bracts spathe-like.	1. **Commelina.**
Perfect stamens 6, rarely 5; petals all alike; bracts leafy or minute.	
Cymes or cymules subtended by small minute bracts.	2. **Cuthbertia.**
Cymes or cymules subtended by leaf-like bracts.	3. **Tradescantia.**

1. **COMMELÌNA** L. Sp. Pl. 40. 1753.

Erect ascending or procumbent, somewhat succulent, branching herbs, with short-petioled or sessile leaves, and irregular mostly blue flowers in sessile cymes subtended by spathe-like bracts. Sepals somewhat unequal, the larger ones sometimes slightly united. Petals blue, unequal, 2 of them larger than the third. Perfect stamens 3, rarely 2, one of them incurved and its anther commonly larger. Sterile stamens usually 3, smaller, their anthers various. Filaments all glabrous. Capsule 2–3-celled. Seeds 1 or 2 in each cavity, the testa firm, roughened, smooth or reticulated. [Dedicated to Kaspar Commelin, 1667–1731, Dutch botanist.]

About 95 species of wide distribution in warm and temperate regions. Besides the following, some 3 others occur in the southern United States. Type species: *Commelina communis* L.

Spathes not united at the base.	
Spathes acuminate; capsules 3-celled, 5-seeded.	1. *C. nudiflora.*
Spathes acute; capsules 2-celled, 4-seeded.	2. *C. communis.*
Spathes with united bases.	
All three cavities of the ovary with 2 ovules.	
Sheaths, at least the lower ones, glabrous or nearly so; pubescence, if present, not hirsute.	3. *C. crispa.*
Sheaths more or less hirsute.	
Capsules 2-valved, dorsal cavity indehiscent.	4. *C. virginica.*
Capsules 3-valved, all cavities dehiscent.	5. *C. erecta.*
Ventral cavities of the ovary with 2 ovules, dorsal cavity with 1 ovule.	6. *C. hirtella.*

1. **Commelina nudiflòra** L.   Creeping Day-flower.   Fig. 1147.

Commelina nudiflora L. Sp. Pl. 41. 1753.
Commelina caroliniana Walt. Fl. Car. 68. 1788.
Commelina agraria Kunth. Enum. 4: 38. 1843.

Glabrous or very nearly so throughout, stems procumbent or creeping, rooting at the nodes, 1°–2½° long. Leaves lanceolate or ovate-lanceolate, 1'–3' long, 4''–8'' wide, acute or acuminate at the apex, their sheaths sometimes ciliate; spathe acute or acuminate, 8''–12'' long, peduncled, the 2 bracts not united by their margins; flowers few in each spathe, 3''–6'' broad; ventral cells of the ovary 2-ovuled, the dorsal 1-ovuled; capsule commonly 5-seeded (2 seeds in each of the ventral cells, 1 in the dorsal); seeds oblong, reticulated, about 1'' long.

Along streams and in waste places, New Jersey to Indiana, Missouri, Florida, Texas and through tropical America to Paraguay. Widely distributed in Asia and Africa. July–Oct.

2. Commelina commùnis L. Asiatic Day-flower. Fig. 1148.

Commelina communis L. Sp. Pl. 40. 1753.
Commelina Willdenovii Kunth, Enum. 4: 37. 1843.

Glabrous or nearly so, stems ascending or decumbent, rather slender, sometimes rooting at the nodes, 1°–3° long. Leaves lanceolate or oblong-lanceolate, 3'–5' long, 1'–1½' wide, acuminate at the apex, narrowed or rounded at the base, smooth; sheath white-membranous with green veins, sometimes ciliate, 8"–12" long; spathes few, peduncled, their 2 bracts acute or acuminate, nearly 1' long, glabrous or sometimes pubescent, distinct; flowers deep blue, ½' or more broad; ventral cells of the ovary 2-ovuled, the dorsal 1-ovuled; capsule 2-celled, each cell 2-seeded; seeds compressed, dark brown, roughened.

Southern New York to Pennsylvania, Georgia and Kentucky, and reported from farther north. Adventive or naturalized from Asia. July–Oct.

3. Commelina crispa Wooton. Curly-leaved Day-flower. Fig. 1149.

Commelina crispa Wooton, Bull. Torr. Club 25: 451. 1898.

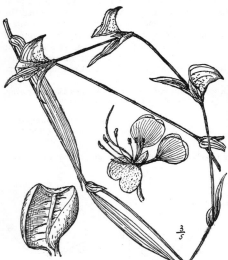

Finely villous or glabrous below; stems commonly branched at the base, the branches ½°–3° long. Leaves lanceolate or linear-lanceolate, 1¼'–3¼' long, acuminate, crisped, slightly contracted at the base; sheaths pale, 6"–7½" long, ciliate; spathes ¾'–1' long, acute or acuminate, pubescent, united at the base; sepals orbicular-elliptic; petals various, the 2 upper light blue, with slender claws and broadly reniform blades, 5"–7" broad, the third white, lanceolate, 1"–1½" long; capsule 2"–2½" long, 3-celled, 2 cavities dehiscent, third cavity indehiscent; seeds smooth.

In sandy and rocky soil, Missouri and Nebraska to Texas and New Mexico. July–Sept.

4. Commelina virgínica L. Virginia Day-flower. Fig. 1150.

Commelina virginica L. Sp. Pl. Ed. 2, 61. 1762.

Similar to *Commelina communis,* somewhat pubescent or glabrous, the stems diffusely branched, rather stouter, 1½°–3° high. Leaves lanceolate or linear-lanceolate, 3'–5' long, 5"–12" wide, acuminate at the apex; sheaths inflated, often pubescent, the orifice sometimes fringed; spathes several, usually peduncled, the 2 bracts acute or acuminate, 8"–12" long, distinct; flowers 1' broad or less, showy; capsule 3-celled, each cell 1-seeded, the dorsal one indehiscent and roughened.

In moist soil, southern New York to Illinois and Michigan, south to Florida, Kansas, Texas and through tropical America to Paraguay. June–Sept.

5. Commelina erécta L. Slender Day-flower. Fig. 1151.

Commelina erecta L. Sp. Pl. 41. 1753.

Somewhat pubescent or glabrous, the stems commonly tufted, erect or ascending, 1°–2° tall, the roots somewhat thickened. Leaves linear-lanceolate, 3′–6′ long, 4″–1′ wide, acuminate at the apex, narrowed at the base; sheaths ½′–1′ long, often pubescent; spathes peduncled or sessile, the 2 bracts more or less pubescent, acute or acuminate, distinct, 10″–20″ long; flowers ½′ or more broad; ovary 3-celled, each cell 1-ovuled; capsule papery, all its cells dehiscent, each 1-seeded; seeds ash-colored, nearly or quite smooth, puberulent.

In moist soil, southern Pennsylvania to Florida, Texas and in tropical America. Aug.–Oct.

6. Commelina hirtélla Vahl. Bearded Day-flower. Fig. 1152.

Commelina longifolia Michx. Fl. Bor. Am. **1**: 23. 1803. Not Lam. 1791.
Commelina hirtella Vahl. Enum. **2**: 166. 1806.

C. erecta A. Gray, Man. Ed. 2, 486. 1856. Not L. 1753.
Stems stout, erect or ascending, 2°–4° high. Leaves lanceolate, acuminate, roughish, 4′–7′ long, 1′–1½′ wide, their sheaths ½′–1′ long, fringed with long brownish hairs and sometimes pubescent; spathes sessile or short-peduncled, often clustered at the summits of the stem and branches, the 2 bracts acute, united by their margins, rather strongly cross-veined; ventral cells of the ovary 2-ovuled, the other 1-ovuled; capsule 5-seeded; seeds ellipsoid, brown, somewhat more than 1″ long, smooth, minutely puberulent.

In moist soil, southern New Jersey to Missouri, south to Florida and Texas. Aug.–Oct.

2. CUTHBÉRTIA Small, Fl. SE. U. S. 237. 1903.

Perennial herbs, with mostly tufted stems. Leaves alternate: blades very narrow and elongated. Cymes umbel-like, solitary at the ends of long peduncles, and subtended by very small bracts wholly unlike the leaves. Sepals 3. Petals 3, reddish, pink or rose-purple, distinct. Stamens 6: filaments pubescent. Capsule loculicidally 3-valved. Seed with a rough testa. [In honor of A. Cuthbert, of Augusta, Georgia, a diligent student of the southern flora.]

Two species, natives of the southeastern United States. Type species: *Cuthbertia rosea* (Vent.) Small.

1. Cuthbertia graminea Small. Grasslike Spiderwort. Fig. 1153.

Tradescantia rosea Chapm. Fl. S. U. S. 498. 1860. Not Vent. 1800.
C. graminea Small, Fl. SE. U. S. 237. 1903.

Stems erect, 2′–8′ tall, densely tufted, sometimes 100 or more together, commonly simple, slightly zigzag. Leaves quite numerous; blades narrowly linear or linear-filiform, 4′–6′ long, or shorter above, mostly ½″–1½′ broad, acute, mostly erect; sheaths ciliate: peduncles sometimes overtopping the leaves: cymes 3–15-flowered, simple: pedicels slightly thickened at the apex, becoming about 5″ long: sepals oblong-ovate or ovate, 2″–2½″ long, glabrous: corollas pink or rose-colored, 5″–10″ broad: capsules subglobose, about 1½″ in diameter.

On sand hills or in sandy woods, Maryland and Missouri to Florida and Texas. April–Aug.

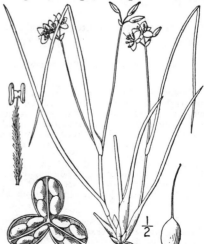

3. TRADESCÁNTIA L. Sp. Pl. 288. 1753.

Perennial, somewhat mucilaginous herbs, with mostly narrow and elongated leaves, and showy regular flowers in terminal or terminal and axillary umbels subtended by leaf-like or scarious bracts. Sepals 3, herbaceous. Petals 3, obovate, ovate or orbicular, sessile. Stamens 6, all alike and fertile, or those opposite the petals shorter; filaments bearded or glabrous. Ovary 3-celled, the cells 2-ovuled. Capsule 3-celled, loculicidally 3-valved, 3–12-seeded. [In honor of John Tradescant, gardener of Charles I, died 1638.]

About 35 species, natives of tropical and temperate America. Besides the following, some 6 others occur in the southern United States. Type species: *Tradescantia virginiana* L.

Leaves linear or linear-lanceolate, 12–50 times longer than broad.
  Stems 1′ to rarely 4′ long; bracts longer than the leaves.   1. *T. brevicaulis.*
  Stems elongated, 4′–3° long; bracts mostly shorter than the leaves.
    Foliage bright green; pedicels, like the sepals, pilose or villous.
      Pedicels and sepals pilose with gland-tipped hairs.
        Stems 4′–1° tall; bracts lanceolate, the bases sac-like, whitish, broader than the leaves.
                  2. *T. bracteata.*
        Stems 1°–2½° tall; bracts linear, the bases not sac-like, green, narrower than the
        leaves.   3. *T. occidentalis.*
      Pedicels and sepals villous with non-glandular hairs.   4. *T. virginiana.*
    Foliage glaucous; pedicels glabrous; sepals with a tuft of hairs at the apex.
                  5. *T. reflexa.*
Leaves lanceolate, 2–10 times longer than broad.
  Slender; stems strict; cymes solitary and terminal or on corymbed branches; species Alleghanian.
                  6. *T. montana.*
  Stout; stems zigzag; cymes terminal and sessile in the upper axils; species campestrian.
                  7. *T. pilosa.*

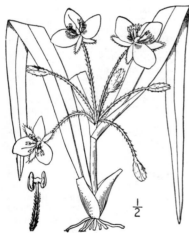

1. Tradescantia brevicaùlis Raf. Short-stemmed Spiderwort. Fig. 1154.

Tradescantia brevicaulis Raf. Atl. Journ. 150. 1832.
Tradescantia virginica var. *villosa* S. Wats. in A. Gray, Man. Ed. 6, 539. 1890.

Villous with long spreading hairs, stems only 1′–4′ high, the plant often appearing nearly acaulescent; leaves mostly basal, 6′–12′ long, 4″–8″ wide, grass-like, linear-lanceolate, acuminate, ciliate, at least at the base, glabrous or villous toward the apex; bracts similar to the leaves, but usually more elongated; umbel 4–12-flowered, sessile in the bracts, the pedicels slender, 1′–2′ long, villous; sepals oblong; corolla about 1′ broad, blue or rose-purple.

In dry soil, Indiana to Kentucky and Missouri. April–May.

2. Tradescantia bracteàta Small. Long-bracted Spiderwort. Fig. 1155.

Tradescantia bracteata Small; Britt. & Br. Ill. Fl. 3: 510. 1898.

Perennial, deep green, glabrous to the inflorescence, or nearly so. Stems erect, 4′–12′ tall, simple or sparingly branched; leaves linear to linear-lanceolate, 4′–8′ long, long-acuminate; sheaths paler than the leaf-blades, conspicuously ribbed, glabrous, or the upper ones sometimes ciliate; involucres of 2 bracts, these broader than the leaves, more or less strongly saccate at the base, ciliate and often sparingly villous on the back; umbel-like cymes few-flowered; pedicels glandular-pubescent, ½′–1′ long; sepals ovate-lanceolate to elliptic-lanceolate, glandular, more or less involute; corolla blue or reddish, about 1′ broad.

In sandy soil, Minnesota to South Dakota, south to Missouri, Kansas and Texas. Spring and summer. Ascends to 7500 ft. in the Black Hills.

3. Tradescantia occidentàlis (Britton) Smyth. Western Spiderwort. Fig. 1156.

T. virginiana occidentalis Britton; Britt. & Brown, Ill. Fl. 377. 1896.

T. occidentalis Smyth, Trans. Kans. Acad. Sci. 16: 163. 1899.

Mostly glabrous to the inflorescence, bright green. Stems solitary, erect, 1°–2½° tall, simple; leaves linear, 8′–12′ long, involutely folded, curved; sheaths 5″–20″ long, conspicuously ribbed, rarely with a few cilia; bracts of the involucre 2, linear, slightly unequal; pedicels rather slender, 5″–10″ long; sepals oblong or elliptic, apparently lanceolate by the involute edges, 4″–5″ long, glandular-pilose; petals blue or reddish, almost 5″ long, orbicular-ovate; capsules obovoid or oblong, 2½″– 3″ long, puberulent at the apex; seeds 1½″ long, pitted and ridged.

On sand-hills, Iowa to Texas, South Dakota and Colorado. June–Aug.

4. Tradescantia virginiàna L. Spiderwort. Trinity. Fig. 1157.

Tradescantia virginiana L. Sp. Pl. 288. 1753.

Glabrous or slightly pubescent, succulent, glaucous or green, stems stout, 8′–3° tall. Leaves more or less channeled, or in some forms nearly flat, linear or linear-lanceolate, long-acuminate, often more than 1° long, 4″–1′ wide; bracts foliaceous, commonly rather wider and shorter than the leaves; umbels solitary and terminal or rarely 2–4, loosely several–many-flowered; pedicels glabrous or pubescent, slender; flowers blue or purplish, rarely white, 1′–2′ broad, very showy; sepals oblong or oblong-lanceolate, acute or obtuse, glabrous or pubescent, 4″–10″ long, much longer than the capsule.

In rich soil, mostly in woods and thickets, southern New York to Ohio and South Dakota, south to Virginia, Kentucky and Arkansas. Escaped from cultivation farther north. Ascends to 4000 ft. in Virginia. Spider-lily. May–Aug.

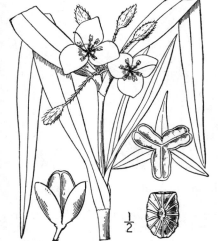

5. Tradescantia refléxa Raf. Reflexed Spiderwort. Fig. 1158.

Tradescantia reflexa Raf. Atl. Journ. 150. 1832.

Perennial, glabrous, glaucous. Stems erect, 1°–3° tall, nearly straight, commonly much branched; leaves linear, 8′–20′ long, straight, or somewhat curved, long-attenuate; sheaths large, 5″–15″ long; involucres of 2 unequal finally reflexed leaf-like bracts; umbel-like cymes usually dense at maturity; pedicels slender, 10″–13″ long, recurved; sepals oblong or elliptic, apparently lanceolate by the involute edges, 4″–5″ long, hooded, mostly with a tuft of hairs at the apex; corolla blue or red, 10″–15″ broad, the petals suborbicular; capsule ovoid to oblong, 2½″–3″ long, glabrous.

In sandy or clayey soil, Ohio to Minnesota, Florida and Texas. Spring and summer.

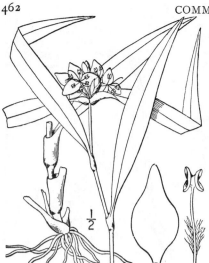

$\frac{1}{2}$

6. Tradescantia montàna Shuttlw. Mountain Spiderwort. Fig. 1159.

Tradescantia montana Shuttlw.; Britt. & Br. Ill. Fl. 1: 377. 1896.

Green and glabrous or somewhat pubescent, stems slender, simple or sparingly branched, 1°–2° tall. Leaves lanceolate or linear-lanceolate, 4'–10' long, 2''–6'' wide, mostly distant, their sheaths enlarged; bracts similar to the leaves but shorter; umbels mostly solitary and terminal, sessile in the bracts, rather densely flowered; pedicels and calyx glabrous or pubescent; flowers less than 1' broad; sepals lanceolate, acute, about 3'' long.

In woods and thickets, mountains of southwestern Virginia to Kentucky and Georgia. June–Aug.

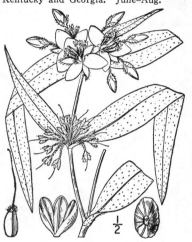

$\frac{1}{2}$

7. Tradescantia pilòsa Lehm. Zigzag Spiderwort. Fig. 1160.

T. pilosa J. G. C. Lehm. Sem. Hort. Hamb. 16. 1827.
Tradescantia flexuosa Raf. Atl. Journ. 150. 1832.

More or less puberulent or short-pilose, stem stout, 1°–3° high, commonly flexuous, often branched. Leaves broadly lanceolate, acuminate at the apex, mostly narrowed at the base, 6'–15' long, ½'–2' wide, dark green above, paler beneath; bracts usually narrower and shorter than the leaves; umbels 3–8, terminal and axillary or on short axillary branches, densely many-flowered; pedicels and calyx pubescent and more or less glandular, rarely nearly glabrous; corolla lilac-blue, 9''–15'' broad.

Southern Pennsylvania to Illinois and Missouri, south to Florida. June–Aug.

Family 19. **PONTEDERIÀCEAE** Dumort. Anal. Fam. 59. 1829.

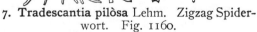

Pickerel-weed Family.

Perennial aquatic or bog plants, the leaves petioled, with thick blades, or long and grass-like. Flowers perfect, more or less irregular, solitary or spiked, subtended by leaf-like spathes. Perianth free from the ovary, corolla-like, 6-parted. Stamens 3 or 6, inserted on the tube or the base of the perianth; filaments filiform, dilated at the base or thickened at the middle; anthers 2-celled, linear-oblong or rarely ovate. Ovary 3-celled with axile placentae, or 1-celled with 3 parietal placentae; style filiform or columnar; stigma terminal, entire or minutely toothed; ovules anatropous, numerous, sometimes only 1 of them perfecting. Fruit a many-seeded capsule, or a 1-celled, 1-seeded utricle. Endosperm of the seed copious; embryo central, cylindric.

About 5 genera and 25 species, inhabiting fresh water in the warm and temperate regions of America, Asia and Africa.

Flowers 2-lipped, stamens 6; fruit a 1-seeded utricle.   1. *Pontederia.*
Flowers regular; stamens 3; fruit a many-seeded capsule.   2. *Heteranthera.*

1. **PONTEDÈRIA** L. Sp. Pl. 288. 1753.

Leaves thick with many parallel veins, the petioles long, sheathing, arising from a horizontal rootstock. Stem erect, 1-leaved, with several sheathing bract-like leaves at the base. Flowers blue, ephemeral, numerous, spiked, the spike (or spadix) peduncled and subtended by a thin bract-like spathe. Perianth 2-lipped, the upper lip of 3 ovate lobes, the middle lobe longest, the lower lip of 3 linear-oblong spreading lobes. Stamens 6, borne at unequal distances upon the perianth-tube, 3 of them opposite the lower lip, the others opposite the upper lip; anthers oblong, subversatile, introrse. Ovary 3-celled, 2 of the cells abortive and empty. Fruit a 1-seeded utricle, enclosed in the thickened tuberculate-ribbed base of the perianth. [In honor of Giulio Pontedera, 1688–1757, professor of botany in Padua.]

Seven or eight species, natives of America. The following is the type of the genus.

1. **Pontederia cordàta** L. Pickerel-weed. Fig. 1161.

Pontederia cordata L. Sp. Pl. 288. 1753.
Pontederia lancifolia Muhl. Cat. 34. 1813.
Pontederia cordata var. *angustifolia* Torr. Fl. N. U. S.
  1 : 343. 1824.
Pontederia cordata lancifolia Morong, Mem. Torr. Club
  5 : 105. 1894.

Stem rather stout, 1°–4° tall. Leaves ovate to
lanceolate, cordate-sagittate, truncate or narrowed at
the base, 2′–10′ long, ½′–6′ wide, the apex and basal
lobes obtuse; basal lobes often with long narrow
stipule-like appendages on the sheathing petiole;
spadix and inflorescence glandular-pubescent; perianth about 4″ long, it and the filaments, anthers and
style bright blue, its tube curved, slightly longer than
the lobes, middle lobe of the upper lip with 2 yellow
spots at the base within; ovary oblong, tapering into
the slender style; stigmas minutely 3–6-toothed.

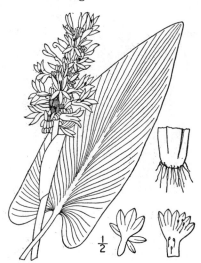

Borders of ponds and streams, Nova Scotia to Minnesota, south to Florida and Texas. After flowering the
lobes and upper part of the perianth-tube wither above,
while the persistent base hardens around the fruit. The
flowers are trimorphous. Includes several races, differing in width of leaves. June–Oct.

2. **HETERANTHÈRA** R. & P. Prodr. Fl. Per. 9. 1794.

[Schollera Schreb. Gen. 785. 1789. Not Roth. 1788.]

Herbs with creeping, ascending or floating stems, the leaves petioled, with cordate, ovate,
oval or reniform blades, or grass-like. Spathes 1-flowered or several-flowered. Flowers
small, white, blue or yellow. Lobes of the perianth nearly or quite equal, linear. Stamens
3, equal or unequal, inserted on the throat of the perianth. Ovary fusiform, entirely or incompletely 3-celled by the intrusion of the placentae; ovules numerous; stigma 3-lobed.
Fruit an ovoid many-seeded capsule, enclosed in the withered perianth-tube. Seeds ovoid,
many-ribbed. [Greek, referring to the unequal anthers of some species.]

About 10 species, 2 in tropical Africa, the others American; only the following in the United
States. Type species : *Heteranthera reniformis* R. & P.

Leaves reniform, oval or ovate.
  One of the filaments longer than the other two ; flowers few or several.
    Flowers white, little exserted from the sheath ; leaves reniform.   1. *H. reniformis.*
    Flowers blue, much exserted ; leaves ovate to subreniform.   2. *H. peduncularis.*
  Filaments broad, equal ; spathe 1-flowered ; leaves ovate to oval.   3. *H. limosa.*
Leaves linear, grass-like, floating.   4. *H. dubia.*

1. **Heteranthera renifórmis** R. & P. Mud Plantain. Fig. 1162.

Heteranthera reniformis R. & P. Fl. Per. 1 : 43. 1798.

Leptanthus reniformis Michx. Fl. Bor. Am. 1 : 25. 1803.

Stems creeping in the mud, rooting at the nodes. Leaves
long-petioled, the blades reniform, wider than long, 8″–20″
wide, rounded at the apex; petioles sheathing, 1′–4′ long;
spathe few-flowered; inflorescence little exserted; tube of
the perianth straight or slightly curved, slender, about 4″
long, its lobes shorter; flowers white or pale blue; anthers
basifixed, the 2 upper oval, the other on a longer filament
and linear; fruit oblong.

In mud or shallow water, Connecticut to New Jersey, Georgia,
Illinois and Louisiana, and in South and Central America and
the West Indies. July–Sept.

2. Heteranthera peduncularis Benth. Blue Mud Plantain. Fig. 1163.

Heteranthera peduncularis Benth. Pl. Hartw. 25. 1840.

Stem stout, creeping on the mud or floating, rooting at the nodes, much branched, often 1° long or longer. Leaves long-petioled, the blades ovate to subreniform, 2½' long or less, acute, obtuse or cuspidate at the apex, cordate at the base; spathe several-flowered, the inflorescence much exserted; flowers blue; perianth-tube nearly 5" long, straight or nearly so, the lobes much shorter than the tube; filaments slender, one of them longer than the other two; anthers basifixed, sagittate; fruit narrowly oblong, about 5" long.

Ponds, Nebraska(?), Missouri and Kansas to Mexico. July–Sept.

3. Heteranthera limòsa (Sw.) Willd. Smaller Mud Plantain. Fig. 1164.

Pontederia limosa Sw. Prodr. 57. 1788.
Heteranthera limosa Willd. Neue Schrift. Ges. Nat. Fr. Berlin 3 : 439. 1801.
Leptanthus ovalis Michx. Fl. Bor. Am. 1 : 25. 1803.

Stems commonly much branched from the base, 6'–15' long. Leaves numerous, oval or ovate, obtuse at the apex, rounded or slightly cordate at the base, 1' long or less; petioles 2'–5' long; spathes 1-flowered, often on peduncles 1' long or more; flowers white or blue, usually larger than those of the preceding species; filaments broad, equal or nearly so; anthers linear, often sagittate.

In mud or shallow water, Virginia to Kentucky, Missouri, Nebraska, Florida, Louisiana and throughout tropical America. July–Sept.

4. Heteranthera dùbia (Jacq.) MacM. Water Star-grass. Fig. 1165.

Commelina dubia Jacq. Obs. Bot. 3 : 9. *pl. 59.* 1768.
Leptanthus gramineus Michx. Fl. Bor. Am. 1 : 25. 1803.
Heteranthera graminea Vahl. Enum. 2 : 45. 1806.
Schollera graminea A. Gray, Man. 511. 1848.
Heteranthera dubia MacM. Met. Minn. 138. 1892.

Aquatic, stem slender, forked, often rooting at the nodes, 2°–3° long. Leaves linear, flat, elongated, acutish, finely parallel-nerved, their sheaths thin, furnished at the top with small acute stipule-like appendages; flowers light yellow, the perianth-segments narrow; stamens nearly equal; tube of the perianth very slender, 1'–1½' long; spathe 1–2-flowered; filaments dilated below; anthers linear, 2" long, sagittate; style shorter than the stamens; stigma several-lobed; capsule 1-celled with 3 parietal placentae, many-seeded.

In still water, Quebec to Oregon, south to Florida and Mexico. Also in Cuba. Occasionally occurs in a small form on muddy shores. July–Oct.

Family 20. JUNCÀCEAE Vent. Tabl. 2: 150. 1799.*

RUSH FAMILY.

Perennial or sometimes annual, grass-like, usually tufted herbs, commonly growing in moist places. Inflorescence usually compound or decompound, paniculate, corymbose, or umbelloid, rarely reduced to a single flower, bearing its flowers singly, or loosely clustered, or aggregated into spikes or heads. Flowers small, regular, with or without bractlets (prophylla). Perianth 6-parted, the parts glumaceous. Stamens 3 or 6, rarely 4 or 5, the anthers adnate, introrse, 2-celled, dehiscing by a slit. Pistil superior, tricarpous, 1-celled or 3-celled, with 3–many ascending anatropous ovules, and 3 filiform stigmas. Fruit a loculicidal capsule. Seeds 3–many, small, cylindric to subglobose, with loose or close seed-coat, with or without caruncular or tail-like appendages.

Eight genera and about 300 species, widely distributed.

Leaf-sheaths open; capsule 1- or 3-celled, many-seeded; placentae parietal or axial.   1. *Juncus.*
Leaf-sheaths closed; capsule 1-celled, 3-seeded, its placenta basal.   2. *Juncoides.*

1. JÚNCUS L. Sp. Pl. 325 (1753).

Usually perennial plants, principally of swamp habitat, with glabrous herbage, stems leaf-bearing or scapose, leaf-sheaths with free margins, and leaf blades terete, gladiate, grass-like, or channeled. Inflorescence paniculate or corymbose, often unilateral, sometimes congested, bearing its flowers either singly and with 2 bractlets (prophylla), or in heads and without bractlets, but each in the axil of a bract; bractlets almost always entire; stamens 6 to 3; ovary 1-celled or by the intrusion of the placentae 3-celled, the placentae correspondingly parietal or axial; seeds several–many, usually distinctly reticulated or ribbed, often tailed.

About 215 species, most abundant in the north temperate zone. Type species: *Juncus acùtus* L. The plants bloom in summer. [Latin, from *jungo,* to bind, in allusion to the use of these plants for withes.]

A. Lowest leaf of the inflorescence terete, not conspicuously channeled, erect, appearing like a continuation of the stem, the inflorescence therefore appearing lateral; stem leaves none.

1. Flowers bracteolate, inserted singly on the branches of the inflorescence. GENUINI.
   Perianth-parts green, or in age straw-colored.
     Perianth-parts equalling or exceeding the capsule, all acute.
       Stamens 3; leaf of the inflorescence much shorter than the stem.
         Capsule without a distinct apical papilla.   1. *J. effusus.*
         Capsule with a distinct apical papilla.   2. *J. conglomeratus.*
       Stamens 6; leaf of inflorescence about equalling the stem, or longer.   3. *J. filiformis.*
     Perianth-parts reaching only the middle of the capsule, inner obtuse.   4. *J. gymnocarpus.*
   Perianth-parts with a chestnut-brown stripe down either side of the midrib.   5. *J. balticus.*
2. Flowers not bracteolate, inserted in heads on the branches of the inflorescence. THALASSICI.
   Perianth-parts pale brown; seed tailless.   6. *J. Roemerianus.*
   Perianth-parts green, or in age straw-colored; seed tailed.   7. *J. maritimus.*

B. Lowest leaf of the inflorescence not appearing like a continuation of the stem, or if so, conspicuously channeled along the upper side, the inflorescence usually appearing terminal.

1. LEAF-BLADE TRANSVERSELY FLATTENED (INSERTED WITH ITS FLAT SURFACE FACING THE STEM), OR TERETE AND CHANNELED, NOT PROVIDED WITH SEPTA.

* Flowers bracteolate, inserted singly on the branches of the inflorescence, sometimes clustered or congested, but never in true heads. POIOPHYLLI.
   Annual; inflorescence, exclusive of its leaves, more than one-third the height of the plant.
                                               8. *J. bufonius.*
   Perennial; inflorescence, excluding leaves, not one-third the height of the plant.
     Leaf-blade flat, but sometimes involute in drying.
       Inflorescence 1–3-flowered; leaves with fimbriate auricles.   9. *J. trifidus.*
       Inflorescence, except in depauperate specimens, several–many-flowered; leaves with entire auricles.
         Cauline leaves 1 or 2, rarely wanting; perianth-parts obtuse.   10. *J. Gerardi.*
         Cauline leaves none; perianth-parts acute or acuminate.
           Auricles at top of leaf-sheath cartilaginous, yellow when dry.   11. *J. Dudleyi.*
           Auricles at top of leaf-sheath membranous, whitish or brownish.
             Inflorescence exceeded by its lowest leaf; flowers not conspicuously secund.
               Auricles of the upper leaves usually ½″–1½″ long, thin, membranous; perianth parts widely spreading.   12. *J. tenuis.*
               Auricles less than ½″ long, thin only at the margin; perianth parts appressed to the capsule.   13. *J. interior.*
             Inflorescence not exceeded by its lowest leaf; flowers conspicuously secund.
                                         14. *J. secundus.*

---

* Text contributed by Mr. FREDERICK V. COVILLE.

Leaf-blade terete, channeled along the upper side.
  Lowest leaf of inflorescence not four lengths of the panicle ; capsule oblong to obovoid.
    Seed tailed.
      Capsule as long as the perianth or longer.   15. *J. Vaseyi.*
      Capsule much shorter than the perianth.   16. *J. oronensis.*
    Seed not tailed.
      Perianth 1¼″–1½″ long, plainly exceeded by the capsule.   17. *J. Greenei.*
      Perianth 1¾″–2″ long, not exceeded by the capsule.   18. *J. dichotomus.*
  Lowest leaf of the inflorescence rarely less than four times as long as the panicle ;
    capsule globose-ovoid.   19. *J. setaceus.*
** Flowers not bracteolate, in true heads on branches of the inflorescence. Graminifolii.
  Stem erect ; capsule oblong or obovoid, obtuse at the apex.
    Stamens 3, with red-brown anthers ; capsule not mucronate.
      Inner perianth-parts obtuse or mucronate.
        Heads few, commonly 5–10-flowered.   20. *J. marginatus.*
        Heads numerous, commonly 2–5-flowered.   21. *J. aristulatus.*
      Inner perianth parts setiform-acuminate.   22. *J. setosus.*
    Stamens 6, with yellow anthers ; capsule mucronate.   23. *J. longistylis.*
  Stem creeping, floating, or ascending ; capsule subulate.   24. *J. repens.*

2. Leaf-blade not transversely flattened, commonly terete, hollow, provided
with septa.

* Leaf-blade usually channeled along the upper side ; septa usually imperfect, not externally evident ;
    inflorescence of 1–4 heads ; plants of arctic or alpine range. Alpini.
  Body of the seed ½″ in length or more.
    Leaf-sheath not auriculate.   25. *J. castaneus.*
    Leaf-sheath auriculate.   26. *J. stygius.*
  Body of the seed less than ½″ in length.
    Lowest leaf of inflorescence foliose, erect ; capsule deeply retuse at apex.
                                                                      27. *J. biglumis.*
    Lowest leaf of inflorescence membranous, spreading ; capsule obtuse and mucronate at the
      apex.   28. *J. triglumis.*
** Leaf-blade not channeled along the upper side (except in *J. bulbosus*), the septa perfect (except
    in *J. polycephalus*), and usually externally evident ; inflorescence, except in depauperate
    specimens, of several to many heads ; plants not of arctic-alpine range. Septati.

† Stamens 6, one opposite each perianth-part.

Heads reduced to one, or rarely two flowers.
  Plant erect ; flowers several–many, paniculate.   29. *J. pelocarpus.*
  Plant creeping or floating ; flowers 1 or 2, peduncled or sessile. 30. *J. subtilis.*
Heads 2–many-flowered.
  Epidermis not roughened.
    Plants with two kinds of leaves, one normal, the other basal, submersed, and capillary.
      Plant low, less than 10′ high.   31. *J. bulbosus.*
      Plant tall, more than 10′ high.   32. *J. militaris.*
    Plants without submersed capillary leaves.
      Capsule oblong, either abruptly acuminate or bluntly acute.
        Branches widely spreading ; capsule sharply acute, tapering into a conspicuous point.
                                                                      33. *J. articulatus.*
        Branches usually strict ; capsule broadly acute, or obtuse, with a short point.
                                                                      34. *J. alpinus.*
      Capsule subulate.
        Leaf-blades erect ; inner perianth-parts longer than the outer. 35. *J. nodosus.*
        Leaf-blades abruptly divergent from the stem ; outer perianth-parts longer than the
          inner.   36. *J. Torreyi.*
  Epidermis of the leaves roughened with minute tubercles.   37. *J. caesariensis.*

†† Stamens 3, none opposite the inner perianth-parts.

Capsule less than three-fourths as long as the perianth.   38. *J. brachycarpus.*
Capsule more than three-fourths as long as the perianth.
  Capsule tapering evenly into a prominent subulate beak.
    Leaf-blade vertically flattened and with incomplete septa, only rarely compressed and with
      complete septa.   39. *J. polycephalus.*
    Leaf-blade terete or nearly so, the septa complete.   40. *J. scirpoides.*
  Capsule obtuse to acute at the apex, sometimes mucronate, but not prolonged into a beak.
    Seed ⅓″–1″ long.
      Perianth about 1″ long, the fruiting head not more than 2″ high. 41. *J. brachycephalus.*
      Perianth 1½″–2″ long, or if shorter, the fruiting head 2″ high or more.
        Inflorescence broad ; capsule about as long as the perianth. 42. *J. canadensis.*
        Inflorescence narrow ; capsule much longer than the perianth. 43. *J. brevicaudatus.*
    Seed ⅛″–¼″ long.
      Perianth and mature capsule 1″–2″ long.
        Perianth 1¼″–2″ long, or if less the whole plant not 20′ high.
          Perianth equaling or longer than the capsule ; heads several–many-flowered.
                                                                      44. *J. acuminatus.*
          Perianth shorter than the capsule ; heads 2–7-flowered.   45. *J. debilis.*
        Perianth 1″–1¼″ long, the whole plant more than 20′ high.   46. *J. nodatus.*
      Perianth and mature capsules 2″–3″ long.   47. *J. diffusissimus.*

1. Juncus effùsus L.   Common Rush.   Bog Rush.   Soft Rush.   Fig. 1166.

Juncus effusus L. Sp. Pl. 326. 1753.

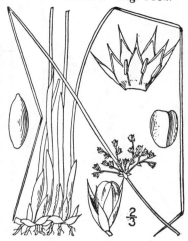

Plant 1½°–4° high, densely tufted, erect. Rootstock stout, branching, proliferous; stem soft, merely striate beneath the inflorescence; basal leaf-blades reduced to filiform rudiments; inflorescence many-flowered, 1′–4′ high, in one form congested into a still smaller compact cluster; lowest bract of the inflorescence 2′–10′ long, much shorter than the stem; perianth 1″–1½″ long, its parts green, lanceolate, acuminate; stamens 3, the anthers shorter than the filaments; capsule obovoid, 3-celled, muticous, regularly dehiscent; seed ⅛″–¼″ in length, obliquely oblong, reticulate in about 16 longitudinal rows, the reticulations smooth and two or three times broader than long.

In swamps and moist places, nearly throughout North America, except the arid and high northern portions. Ascends to 3000 ft. in Virginia. Also in Europe and Asia. Called also Water, Round, Hard, Candle and Pin-rush.

2. Juncus conglomeràtus L.   Glomerate or Staff Rush.   Fig. 1167.

Juncus conglomeratus L. Sp. Pl. 326. 1753.
Juncus Leersii Mars. Fl. Neu-Vorpom. 451. 1869.

Plant 1°–2½° high, densely tufted, erect. Rootstock stout, with proliferous branches; stem distinctly ribbed just beneath the inflorescence; leaf-blades wanting or reduced to minute filiform rudiments; inflorescence congested, seldom more than 10″ high; lowest bract of the inflorescence 2′–6′ long, much shorter than the stem; perianth 1¼″–2″ long, its parts green, lanceolate, acuminate; stamens 3; about two-thirds as long as the perianth; anthers shorter than the filaments; capsule nearly as long as the perianth, obovoid, obtuse or retuse at apex, tipped with the base of the style; seed ⅛″–¼″ in length, obliquely oblong, acute or abruptly apiculate at both ends, reticulate in about 16 longitudinal rows, the reticulations smooth and two or three times broader than long.

In sphagnum bogs, resembling in appearance specimens of *J. effusus* with congested inflorescence. Newfoundland, Nova Scotia, northern Europe and Asia. Pith-rush.

3. Juncus filifórmis L.   Thread Rush.   Fig 1168.

Juncus filiformis L. Sp. Pl. 326. 1753.

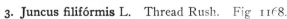

Perennial, stems 4′–25′ tall, erect, about ½″ in diameter, arising from a creeping rootstock; basal leaves reduced to bladeless sheaths; involucral leaf usually longer than the stem; inflorescence rarely with more than 20 flowers or more than 1′ high, commonly with less than 8 flowers and less than 10″ high; perianth 1¼″–1¾″ long, its parts nearly equal, green with hyaline margins, narrowly lanceolate, acute, or the inner obtuse; stamens 6, about half as long as the perianth; anthers shorter than the filaments; style very short; capsule obovoid, green, barely pointed, about three-fourths as long as the perianth, 3-celled; seed obliquely oblong, about ¼″ long, pointed at either end, with an irregularly wrinkled coat, seldom developing reticulations.

Newfoundland and Labrador to British Columbia, Pennsylvania, Michigan, and in the Rocky Mountains to Utah and Colorado. Also in Europe and Asia.

4. Juncus gymnocàrpus Coville. Pennsylvania Rush. Fig. 1169.

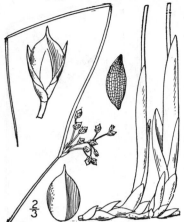

Juncus Smithii Engelm. Trans. St. Louis Acad. 2 : 444. 1866. Not Kunth, 1841.
J. gymnocarpus Coville, Mem. Torr. Club 5 : 106. 1894.

Stems erect, 1°–2½° high, about 1″ thick, arising at intervals from a creeping proliferous rootstock about 1½″ in diameter; basal leaves reduced to bladeless clasping sheaths; panicle commonly 7″–15″ high, spreading, its subtending leaf usually 4′–10′ long; perianth 1″ in length or a little less, its parts with a green midrib, equal, lanceolate, the outer acute, the inner obtuse; stamens 6, nearly as long as the perianth, the anthers shorter than the filaments; capsule almost twice as long as the perianth, broadly ovoid, conspicuously mucronate, brown and shining, barely dehiscent, 3-celled; seed obliquely obovoid or oblong, somewhat misshapen by compression in the capsule, about ⅓″ long, none with perfect markings seen.

In swamps, mountains of Schuylkill and Lebanon counties, Pennsylvania, and in Florida.

5. Juncus bàlticus Willd. Baltic Rush. Fig. 1170.

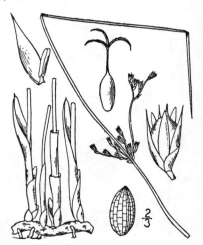

Juncus balticus Willd. Berlin Mag. 3 : 298. 1809.

Stems erect, 8′–36′ high, ½″–1¼″ thick, arising at intervals from a stout creeping rootstock 1″–1½″ thick; basal leaves reduced to bladeless sheaths; panicle commonly 1′–2½′ high; perianth 1½″–2¼″ long, its parts lanceolate, acute, or the inner sometimes obtuse, nearly equal, brown with a green midrib and hyaline margins; style ½″–1″ long; stigmas a little shorter; stamens 6, about two-thirds the length of the perianth; anthers about ¾″ in length, much longer than the filaments; capsule about as long as the perianth, pale to dark brown, narrowly ovoid, conspicuously mucronate, 3-celled; seeds usually with a loose coat, nearly ½″ long, oblong to narrowly obovoid, oblique, about 40-striate.

On shores, Newfoundland and Labrador to Alaska, Pennsylvania, Missouri and Nebraska; far south in the western mountains. Also in Europe and Asia. Consists of many races.

6. Juncus Roemeriànus Scheele. Roemer's Rush. Fig. 1171.

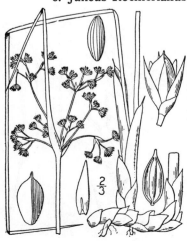

Juncus Roemerianus Scheele, Linnaea 22 : 348. 1849.

Stems 20′–4° tall, erect, arising singly from a tough scaly horizontal rootstock 2½″–5″ thick; inner sheaths bearing erect blades of about the same length as the stem; inflorescence 2¼′–6′ high, diffusely spreading, its leaf 4′–10′ long; heads 2–6-flowered; perianth pale brown, 1″–1¾″ long, the parts linear-oblong, the outer acuminate, the inner shorter and bluntly acute; flowers imperfectly dioecious; stamens 6, on fertile plants reduced to sterile staminodia; capsule brown, about as long as the perianth, narrowly obovoid, obtuse or truncate, mucronate, 3-celled; placenta very thick and spongy, about one-third as broad as the valve; seed dark brown, ¼″–⅜″ long, obovoid, abruptly apiculate, indistinctly reticulate or distinctly 20–26-ribbed and the intervening spaces imperfectly cross-lined.

In brackish marshes, New Jersey(?), Virginia to Florida and Texas.

7. Juncus marítimus Lam.   Sea Rush.   Fig. 1172.

Juncus martimus Lam. Encycl. 3: 264. 1789.

Stems 20′–40′ high, 1″–2″ thick, erect from a stout horizontal rootstock. Outer basal leaves reduced to bladeless sheaths, the innermost foliose, with a long terete stout blade about equalling the stem; leaf of the inflorescence erect, sometimes 1° long, sometimes barely exceeding the panicle; panicle 3′–8′ high, its branches stiff, erect; heads 2–6-flowered; perianth 1½″–1¾″ long, its parts green, lanceolate, with hyaline margins, the outer acuminate, the inner a little shorter; flowers perfect; stamens 6, two-thirds as long as the perianth; filaments about as long as the anthers; capsule 1½″–1¾″ long, narrowly ovoid, acute, mucronate, brown above, 3-celled, with thin placentae; seed brown, about ½″ long, the body narrowly and obliquely oblong, about ⅜″ in length, 20–30-ribbed, indistinctly reticulate, tailed at either end.

Coney Island, New York, the station now, perhaps, destroyed. Common on the coasts of the eastern hemisphere.

8. Juncus bufònius L.   Toad Rush.   Fig. 1173.

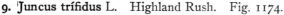

Juncus bufonius L. Sp. Pl. 328. 1753.

Plant branching from the base, annual, erect, seldom exceeding 8′ in height, the stems in large plants with 1 or 2 leaves below the inflorescence; leaf-blade flat, ¼″–½″ wide, in low plants often much narrower and filiform-involute; inflorescence about one-half as high as the plant, with blade-bearing leaves at the lower nodes; flowers inserted singly on its branches, in one form fasciculate; perianth-parts 2″–3½″ long, lanceolate, acuminate, equal; stamens usually 6, sometimes 3, seldom half as long as the perianth; anthers shorter than the filaments; capsule about two-thirds as long as the perianth; narrowly oblong, obtuse, mucronate, 3-celled; seed broadly oblong, with straight tips, ⅛″–¼″ long, minutely reticulate in 30–40 longitudinal rows, the areolae broader than long.

A cosmopolitan species, occurring throughout North America, except the extreme north, frequenting dried-up pools, borders of streams and roadsides in clayey soil. Frog, Toad or Coe-grass. Salt-weed.

9. Juncus trífidus L.   Highland Rush.   Fig. 1174.

Juncus trifidus L. Sp. Pl. 326. 1753.

Densely tufted, 4′–12′ high; stems closely set on a stout rootstock, erect, about ¼″ thick; basal leaves reduced to almost bladeless sheaths, the uppermost with a rudimentary blade and fimbriate auricles; stem-leaf 1, inserted near the inflorescence, with a narrower slender, flat or involute blade; inflorescence a cluster of 1–3 flowers, the lowest subtending bract similar to the stem-leaf, the succeeding one much smaller or wanting; perianth dark brown, 1¼″–1¾″ long; stamens 6; anthers about as long as the filaments; capsule equalling the perianth, coriaceous, 3-celled, obovoid with a conspicuously mucronate-aristate top; seeds few, narrowly obovoid, acute at the base, irregularly angled, minutely striate both longitudinally and transversely.

Greenland and Labrador, south on the higher mountains of New England and New York to Sam's Point, N. Y., and in North Carolina. Also in Europe and Asia.

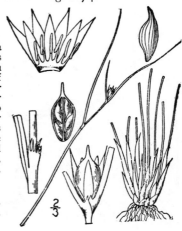

10. Juncus Geràrdi Lois. Black-grass. Fig. 1175.

Juncus Gerardi Lois. Journ. de Bot. **2**: 284. 1809.

Tufted, 8′–28′ high, with creeping rootstocks. Basal leaves with rather loosely clasping auriculate sheaths, the long blades flat, or when dry involute; 1 or 2 cauline leaves usually present, similar to the basal; inflorescence sometimes exceeded by its lowest bract; panicle erect, strict or slightly spreading; perianth 1″–1¼″ long, its parts oblong, obtuse, with green midrib and broad dark brown margins, straw-colored in age; stamens 6, barely exceeded by the perianth; anthers much longer than the filaments; capsule one-fourth to one-half longer than the perianth, obovoid, mucronate, dark brown, shining, 3-celled; seed dark brown, obovate, acute at base, obtuse and often depressed at the summit, marked by 12–16 conspicuous ribs, the intervening spaces cross-lined.

On salt meadows, Gulf of St. Lawrence to Florida; rare inland to western New York and the vicinity of the Great Lakes. Occurs also on the northwest coast, and in Europe.

Juncus compréssus Jacq., a similar European species, but glaucous and with filaments nearly as long as the anthers, has been found in Quebec.

11. Juncus Dúdleyi Wiegand. Dudley's Rush. Fig. 1176.

J. Dudleyi Wiegand, Bull. Torr. Club **27**: 524. 1900.

Plants 1°–4° high, pale green. Leaves basal; blades about half the length of the scapes or less, very narrowly linear but flat, frequently somewhat involute; scapes tufted, often relatively stout but wiry, striate-grooved: inflorescence 1′–2′ high, or rarely slightly larger, usually rather congested, considerably exceeded by its bract, few-flowered; perianth green or pale straw-colored, 2″–2½″ long, its parts firm, nearly equal, lanceolate-subulate, acute, more or less spreading, scarious-margined; stamens about half as long as the perianth; anthers slightly shorter than the filaments; capsules ovoid-oval, ¾–⅞ the length of the perianth, somewhat apiculate: seed oblong, ⅕″–¼″ long, apiculate at each end.

In damp soil and open places, Quebec and Maine to Ontario, Saskatchewan, British Columbia, Washington. Virginia, Tennessee and Mexico.

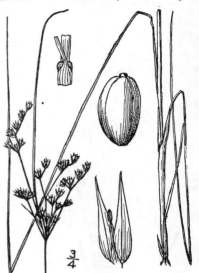

12. Juncus ténuis Willd. Slender Rush. Yard Rush. Fig. 1177.

Juncus tenuis Willd. Sp. Pl. **2**: 214. 1799.
Juncus monostichus Bartlett, Rhodora **7**: 50. 1905.
(?)*J. dichotomus platyphyllus* Wiegand, Bull. Torr. Club **30**: 448. 1903.

Tufted, 2′–30′ high; basal leaves with blades ¼″–¾″ wide, sometimes involute in drying, about half the height of the stem; the sheaths usually with broad scarious margins; inflorescence usually much exceeded by its lowest leaf, 4′ high or less, the flowers rarely secund; perianth 1¼″–2½″ long, its parts lanceolate, acuminate, exceeding the capsule, widely divergent, touching the capsule for about half their length; stamens 6, about half as long as the perianth; anthers shorter than the filaments; capsule oblong to obovoid, rounded at the top, imperfectly 3-celled; seed ⅕″–¼″ long, narrowly oblong to obovoid, with oblique ends, reticulated in about 16 rows, the areolae two or three times broader than long.

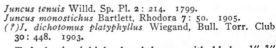

In dry or moist soil, especially on paths, almost throughout North America, now migrating to all parts of the world. Wire-grass. Poverty-grass.

13. Juncus intèrior Wiegand. Inland Rush. Fig. 1178.

Juncus interior Wiegand, Bull. Torr. Club **27**: 516. 1900.

Plants 1½°–3° high, light green. Leaves basal, several; blades about one-third the length of the plant, about ¼" wide, sometimes involute; scapes grooved: inflorescence 1'–4' long, many-flowered, the branches ascending; flowers scattered, rather distant; perianth straw-colored, 1½"–2" long, its parts nearly equal, lanceolate-subulate, slenderly acute or acuminate, appressed or erect, the petals margined to the apex: stamens 6, half as long as the perianth; anthers much shorter than the filaments; capsule oblong or rarely ovoid-oblong, about as long as the perianth, obtuse or barely apiculate: seeds oblong, ⅛"–¼" long.

In dry woods or on prairies, Illinois to North Dakota, Wyoming, Tennessee, Arkansas, Texas and Arizona.

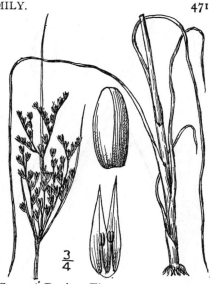

14. Juncus secúndus Beauv. Secund Rush. Fig. 1179.

Junus secundus Beauv.; Poir. Encycl. Sup. **3**: 160. 1813.
Juncus tenuis var. *secundus* Engelm. Trans. St. Louis Acad. **2**: 450. 1866.

Tufted, 6'–16' high; leaves usually less than one-third the height of the plant; inflorescence longer than its lowest leaf or only slightly exceeded by it, 10"–4' high, the flowers secund on the usually somewhat incurved branches; perianth-parts 1¼"–1¾" long, equalling or barely exceeding the capsule and appressed to it for about two-thirds their length, often reddish above; stamens 6, about one-half as long as the perianth; capsule narrowly ovoid, 3-sided above the middle with straight sides and a truncate apex, completely 3-celled, the placentae meeting in the axis; seed ¼"–⅓" long, narrowly oblong to ovoid, obliquely tipped, with 12–16 longitudinal rows of areolae two or three times broader than long.

In dry soil, Maine and Vermont to Pennsylvania, North Carolina, Illinois and Missouri.

15. Juncus Vàseyi Engelm. Vasey's Rush. Fig. 1180.

Juncus Vaseyi Engelm. Trans. St. Louis Acad. **2**: 448. 1866.

Stems erect, tufted, 1°–2½° high, ¼" in diameter or less; basal leaves with minutely auriculate sheaths, the uppermost bearing a terete channeled blade, half to three-fourths as long as the stem; stem-leaves none; inflorescence 1¾' in height or less, 4–40-flowered, the lowest bract usually not exceeding the inflorescence; perianth 1½"–2" long, the parts subulate-lanceolate, with hyaline margins, the inner slightly shorter; stamens 6, about two-thirds as long as the perianth; anthers shorter than the filaments; style almost wanting; stigmas short; capsule slightly exceeding the perianth, narrowly oblong, obtuse or truncate, with a short tip, 3-celled; seed long-tailed, with a linear-oblong oblique body about ¼" long, 20–24-ribbed, the intervening spaces with faint transverse markings.

New Brunswick to Maine, Ontario, Michigan, Illinois and Iowa.

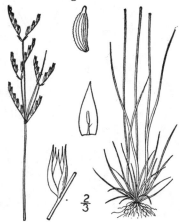

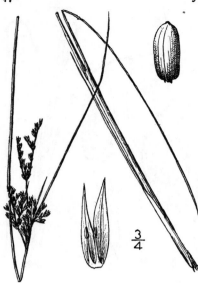

16. Juncus oronénsis Fernald. Maine Rush.
Fig. 1181.

Juncus oronensis Fernald, Rhodora **6**: 36. 1904.

Stems tufted, erect, 2½° high or less; basal leaves about half as long as the stem, the blades nearly terete, the sheaths slightly auricled; inflorescence 1′-3½′ long, the flowers somewhat secund on its erect or narrowly ascending branches; perianth 2″-2½″ long, its parts lanceolate-subulate, the outer slightly shorter than the inner; capsule narrowly oblong, trigonous, truncate or slightly emarginate, much shorter than the perianth; seed about ½″ long, the tail one-fourth as long as the body.

In thickets, known only from Maine.

17. Juncus Greènei Oakes & Tuckerm. Greene's Rush. Fig. 1182.

Juncus Greenei Oakes & Tuckerm. Am. J. Sci. **45**: 37. 1843.

Stems erect, densely tufted, 8′-2⅓° high. Basal leaves with slender terete channeled blades one-half or rarely two-thirds the length of the stem; stem-leaves none, or a single one below the inflorescence; panicle 10″-20″ high, rather compact, somewhat umbelloid, much exceeded by its lowest bract; perianth 1¼″-1½″ long, its parts stiff, lanceolate, sharply acute, with brownish red stripes and apex, the inner shorter; stamens 6, half to two-thirds as long as the perianth; anthers about as long as the filaments; style and stigmas very short; capsule one-fourth to one-half longer than the perianth, ovate-lanceolate in outline, truncate at the summit, 3-celled; seed obliquely oblong, ⅕″-¼″ long, slightly reticulated in about 20–24 rows, the areolae nearly square.

New Brunswick to New Jersey, near the coast; Michigan, Wisconsin, Iowa, Minnesota and Ontario.

18. Juncus dichótomus Ell. Forked Rush.
Fig. 1183.

Juncus dichotomus Ell. Bot. S. C. & Ga. **1**: 406. 1817.

Closely tufted, 1°-3° high; leaves all basal except those of the inflorescence; sheaths usually reddish, the blades terete, channeled along the upper side, about one-half the height of the stem; inflorescence paniculate, subsecund, 1½′-3¾′ high, usually exceeded by its lowest bract; perianth about 2″ long, its parts subulate-lanceolate, green when young, straw-colored when old; stamens 6, about one-half as long as the perianth, the anthers shorter than the filaments; capsule slightly shorter than the perianth, oblong, obtuse, mucronate, 1-celled, the placentae intruded half way to the center; seed oblong, dark brown, obliquely apiculate, less than ¼″ long, reticulate in about 14 longitudinal rows, the smooth areolae about as long as broad.

In dry soil, Connecticut to Florida and Texas, near the coast. Introduced in Jamaica.

19. Juncus setàceus Rostk. Awl-leaved Rush. Fig. 1184.

Juncus setaceus Rostk. Monog. Junc. 13. *pl. 1. f. 2.* 1801.

Densely tufted from stout branching rootstsocks. Stems terete, spreading and recurved above, $1\frac{1}{2}°-3°$ long; leaves all basal except those of the inflorescence, the uppermost sheath usually bearing a long terete blade similar to the stem, but channeled; the other sheaths with filiform blades less than $\frac{1}{2}'$ in length; involucral leaf appearing like a continuation of the stem, $4'-1°$ long; inflorescence appearing lateral, $2'$ long or less; perianth $1''-2\frac{1}{2}''$ long, its parts lanceolate, acuminate, rigid, widely divergent in fruit; stamens 6; anthers usually longer than the filaments; capsule globose, shining, mucronate, 1-celled, with intruded placentae, barely dehiscent; seed subglobose, $\frac{1}{4}''-\frac{1}{3}''$ long, reticulate in about 12 longitudinal rows, the areolae large.

In marshes, Delaware to Florida and Texas, near the coast, extending north in the Mississippi Valley to Missouri.

20. Juncus marginàtus Rostk. Grass-leaved Rush. Fig. 1185.

Juncus marginatus Rostk. Monog. Junc. 38. *pl. 2. f. 3.* 1801.
Juncus marginatus var. *paucicapitatus* Engelm. Trans. St. Louis Acad. **2**: 455. 1866.

Stems erect, tufted, $6'-30'$ high from branching rootstocks, somewhat bulbous at the base, compressed, 2-4-leaved. Leaf-sheaths auriculate; blades $\frac{1}{2}''-1\frac{1}{2}''$ broad, 2-4 conspicuous veins in addition to the midrib; inflorescence $4'$ high or less, the panicle composed of 2-20 turbinate to subspherical 5-10-flowered heads; perianth $1\frac{1}{4}''-1\frac{3}{4}''$ long, the outer parts ovate, acute, the inner slightly longer, obovate, obtuse, with hyaline margins; stamens 3, nearly as long as the perianth; anthers ovate, reddish brown when dry, much shorter than the filaments; capsule equalling the perianth, obovoid, truncate or retuse, almost 3-celled, the placentae deeply intruded; seed oblong, $\frac{1}{5}''-\frac{1}{4}''$ long, pointed at either end, 12-16-ribbed.

Grassy places, Maine to Ontario, Florida and Nebraska.

21. Juncus aristulàtus Michx. Large Grass-leaved Rush. Fig. 1186.

J. aristulatus Michx. Fl. Bor. Am. **1**: 192. 1803.
J. marginatus biflorus Wood, Class-book, Ed. 2, 725. 1861.
J. marginatus aristulatus Coville, Proc. Biol. Soc. Wash. **8**: 123. 1893.

Plants solitary or sparingly tufted, $10'-3°$ high or sometimes lower. Stems markedly bulbous-thickened at the base; leaves similar to those of *J. marginatus,* but sometimes $2\frac{1}{2}''$ broad; panicle $6'$ high or less, composed of numerous, usually 20-100, relatively small, 2-5-flowered heads; perianth about $1\frac{1}{4}''$ long; sepals acute or acuminate; petals oblong or obovate, obtuse, longer than the sepals; stamens as long as the perianth or longer; anthers much shorter than the filaments; capsules obovoid, about $1''$ long, truncate or depressed at the apex.

In moist soil or meadows, Massachusetts to Michigan, Florida, Texas and Mexico.

22. Juncus setòsus (Coville) Small. Awn-petaled Rush. Fig. 1187.

J. marginatus setosus Coville, Proc. Biol. Soc.
Wash. 8 : 124. 1893.
Juncus setosus Small, Fl. SE. U. S. 258. 1903.

Plants rather loosely tufted, 1°–2½° high,
bright green. Stems not much thickened at
the base; leaves with auriculate sheaths, sometimes quite numerous; blades 1″–2½″ wide,
similar to those of *J. marginatus;* panicle 1′–4′
high, composed of 20–100 heads, or smaller in
depauperate forms; perianth 1½″–2″ long, the
outer parts lanceolate-acuminate, the inner
slightly larger than the outer, lanceolate to
ovate-lanceolate, setiform-acuminate; stamens
3, much shorter than the perianth; anthers and
filaments about equal in length, the former
reddish-brown; capsules oblong, about as long
as the perianth, blunt.

In woods and wet places, Nebraska to Louisiana,
Arizona and Mexico.

23. Juncus longístylis Torr. Long-styled Rush. Fig. 1188.

Juncus longistylis Torr. Bot. Mex. Bound. 223. 1859.

Stems erect, loosely tufted, 8′–30′ high, rather stiff,
slender, compressed, 1–3-leaved. Leaf-blades ¾″–1½″ wide,
acute, striate, the midrib well defined; inflorescence 2′
high or less, usually of 2–10 irregular 3–8-flowered heads,
or reduced to a single larger one; perianth 2½″–3″ long,
the parts equal, brown, lanceolate, acuminate, with hyaline
margins; stamens 6, half to two-thirds as long as the perianth, the yellow linear anthers longer than the filaments;
style about ½″ long; stigmas 1″–1½″ long; capsule oblong,
brown, angled above, obtuse or depressed at the summit,
mucronate, 3-celled; seed oblong, white-tipped, about ¼″
long, 14–20-ribbed.

Newfoundland; western Ontario to Nebraska, British Columbia and New Mexico.

24. Juncus rèpens Michx. Creeping Rush. Fig. 1189.

Juncus repens Michx. Fl. Bor. Am. 1 : 191. 1803.

Perennial by prostrate rooting branches; stems tufted,
compressed, ascending, floating or prostrate, 2′–20′ long.
Leaves with compressed sheaths 10″ in length, auriculate, the blades 1′–3½′ long, ½″–1″ broad, filiform-acumi-
nate; inflorescence of 1–8 heads, one or more heads
often occurring also at the lower nodes; heads 5–10-
flowered; flowers 3″–5″ long, the outermost slightly
recurved; perianth-parts subulate-lanceolate, the outer
keeled, about one-third shorter than the inner; stamens
3, half to one-third the length of the perianth; filaments
longer than the yellow anthers; capsule subulate, beakless, about as long as the outer perianth-parts, 3-celled,
the valves membranous, breaking away from the axis in
dehiscence; seed oblong, acute at either end, ⅜″–½″ long,
finely reticulate in 25–40 longitudinal rows.

In swamps and streams, Delaware to Florida, Cuba and
Texas, and in Lower California.

25. Juncus castàneus Smith.   Chestnut Rush.   Clustered Alpine Rush.   Fig. 1190.

Juncus castaneus Smith, Fl. Brit. **1**: 383. 1800.

Stems erect, 4′–20′ high, terete, leafless, or with a single leaf, arising singly from a slender rootstock. Basal leaves 3–5, the outer sheaths short, loose, the inner clasping, sometimes 4′ long, not auriculate, their blades tapering from an involute-tubular base to a slender channeled acutish apex; inflorescence strict, usually exceeded by its lowest bract, the other bracts membranous and mostly equalling the flowers; heads 1–3, 3–12-flowered; pedicels ½″–1¼″ long; perianth brown or black, 2″–3½″ long, its parts lanceolate, acute; stamens nearly as long as the perianth; anthers about ½″ long; capsule brown, paler toward the base, 1½–2 times as long as the perianth, narrowly oblong, tapering to an acute summit, imperfectly 3-celled; seed 1¼″–2″ long, contracted into long slender tails, the body about ½″ long.

Newfoundland and Quebec to Alaska, south along the mountains to Colorado. Europe and Asia.

26. Juncus stýgius L.   Moor Rush.   Fig. 1191.

Juncus stygius L. Syst. Nat. Ed. 10, **2**: 987. 1759.
J. stygius var. *americanus* Buch. in Engler, Bot. Jahrb. **12**: 393. 1890.

Rootstock none; stems 3′–1° high, single, or few together, erect, 1–3-leaved below, leafless above; leaf-sheaths 5″–10″ long, clasping, nerved, auriculate; blades erect or nearly so, 10″–4′ long, slightly compressed, channeled on the upper side, tapering to a blunt point; inflorescence of 1–4 heads; heads 1–4-flowered; lowest bract usually exceeding the flowers; perianth 1½″–2½″ long, pale, its parts lanceolate, 3-nerved, equal, with membranous margins, obtuse or acute; stamens half as long as the perianth or more; anthers oblong, shorter than the filaments; capsule 3″–4″ long, pale brown, spindle-shaped, acute, mucronate, 3-celled below, few-seeded; seed spindle-shaped, 1¼″–1½″ long, with a loose coat, the body about ½″ long, narrowed into thick tails.

Newfoundland to Maine, northern New York, Michigan and Minnesota; also in British Columbia and Europe.

27. Juncus biglùmis L.   Two-flowered Rush.   Fig. 1192.

Juncus biglumis L. Sp. Pl. 328. 1753.

Stems 1′–8′ high, loosely tufted on a branched rootstock, erect, nearly terete. Leaves 1–5, all basal, the outermost sheath usually 4″ long or less, the innermost sometimes much longer, inconspicuously or not at all auriculate, the blades nearly terete; inflorescence a capitate cluster of 1–4 flowers, its lowest bract erect, foliose, green with brown membranous margins below; perianth 1½″–1¾″ long, dark brown, its parts membranous, oblong, obtuse, nearly equal; stamens equalling the perianth; anthers linear-oblong; capsule longer than the perianth, cylindric-oblong, 3-sided, retuse at the summit, with 3 keeled shoulders, purplish black, or with purple-margined valves, imperfectly 3-celled; seed ½″–¾″ long, fusiform, the body narrowed into short stout tails.

Baffin Bay to Alaska and British Columbia. Also in Europe and Asia.

28. Juncus triglùmis L.　Three-flowered Rush.　Fig. 1193.

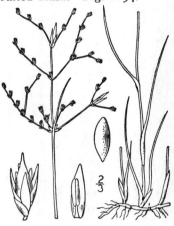

Juncus triglumis L. Sp. Pl. 328. 1753.

Stems 3′–7′ high, loosely tufted on a branched root-stock, erect, terete. Leaves 1–5, all basal, with sheaths clasping and conspicuously auriculate, the blades sub-terete, blunt, ¼″ in diameter, usually less than half the height of the plant; inflorèscence a capitate clus-ter of 1–5 (usually 3) flowers, the lowest 2 or 3 bracts nearly equal, divergent, about as long as the flowers, usually brown, obtuse and membranous; perianth 1½″–2″ long, its parts oblong-lanceolate, obtuse; sta-mens nearly as long as the perianth; anthers linear, short; capsule about equalling the perianth, oblong, obtuse, mucronate, 3-angled, imperfectly 3-celled; seed about 1″ long, its body oblong, abruptly contracted into long slender tails.

Labrador and Newfoundland to Alaska, south in the Rocky Mountains to Colorado. Also in Europe and Asia.

29. Juncus pelocàrpus E. Meyer.　Brown-fruited Rush.　Fig. 1194.

Juncus pelocarpus E. Meyer, Syn. Luz. 30. 1823.

Rootstock slender; stems 3′–20′ high, 1–5-leaved; basal leaves 2–4, with loose auriculate sheaths, mostly with slender terete blades seldom exceeding 5′ in length; stem leaves 1–5, similar to the basal; inflores-cence 4′ in height or less; secondary panicles rarely produced from the axils of the upper leaves; panicle loose, with distant heads of 1 or sometimes 2 flowers; perianth ¾″–1½″ long, the parts linear-oblong, green to reddish-green, obtuse or the inner sometimes acute, the outer usually the shorter, all of them frequently modi-fied into rudimentary leaves; stamens 6, about two-thirds as long as the perianth; anthers exceeding the filaments; style commonly ½″ and stigmas 1″ long; capsule subulate-linear, its slender beak exceeding the perianth, 1-celled; seed oblong to obovoid, ⅛″–¼″ long, reticulate in about 24 rows, the areolae smooth.

Newfoundland to New Jersey, Ontario and Minnesota.

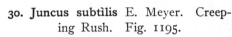

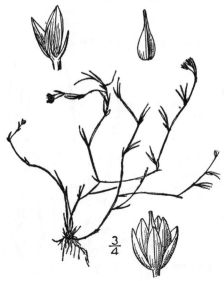

30. Juncus subtìlis E. Meyer.　Creep-ing Rush.　Fig. 1195.

Juncus subtilis E. Meyer, Syn. Luz. 31. 1823.

Juncus pelocarpus subtilis Engelm. Trans. St. Louis Acad. 2 : 456. 1866.

Tufted; stems filiform, creeping on mud, or floating, simple or branched, sometimes 5 dm. long, but usually much shorter, the leaves capillary, often fascicled at the nodes. Flow-ers only 1 or 2, axillary or terminal, short-peduncled or sessile; perianth about 1″ long, its parts linear-oblong, reddish, obtuse or acutish, the outer shorter than the inner; stamens 6, shorter than the perianth; anthers about as long as the filaments; capsule tri-gonous, slender-beaked, a little longer than the perianth.

Newfoundland to Quebec and Maine.

31. Juncus bulbòsus L.   Bulbous Rush.   Fig. 1196.

Juncus bulbosus L. Sp. Pl. 327. 1753.

Tufted, 2′–8′ high; stems erect, or procumbent and rooting at the joints, usually bulbous.   Leaves of two kinds, the basal mostly submersed, filiform, the cauline stouter, all with auriculate sheaths 10″ long or less, the septa of the blades inconspicuous; panicle of 1–10 heads; heads top-shaped to hemispheric, 4–15-flowered, some of the flowers often transformed into tufts of small leaves; perianth 1⅓″–1½″ long, its parts nearly equal, linear-lanceolate, obtuse, brown, or with a green midrib; stamens 3, shorter than the perianth; anthers a little shorter than the filaments; capsule narrowly oblong, obtuse, mucronate, slightly exceeding the perianth, brown above, 1-celled; seed narrowly oblong, about ¼″ long, acute at base, obtuse and apiculate above, 25–30-ribbed.

Labrador, Newfoundland and Nova Scotia.   Europe.

32. Juncus militàris Bigel.   Bayonet Rush.   Fig. 1197.

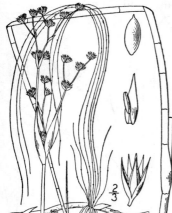

Juncus militaris Bigel. Fl. Bost. Ed. 2, 139.   1824.

Stems 20′–4° high, erect, stout, 1½″–3″ thick below, arising from a stout rootstock.   Leaves of two kinds, the submersed borne in dense fascicles on the rootstock and developing filiform, nodose blades sometimes 20′ long; basal leaves reduced to loose bladeless sheaths, sometimes 10′ long; stem leaves 1 or 2, the lower with a long stout terete blade 1″–2″ thick at the base, the upper, when present, reduced to a bladeless sheath; inflorescence 3′–6′ high, its bracts with obsolete blades; heads top-shaped to semiglobose, 6–12-flowered; perianth 1½″–1¾″ long, its parts narrowly linear-subulate, the inner longer than the outer; stamens 6, nearly as long; anthers slightly exceeding the filaments; capsule ovoid, acuminate, beaked, 1-celled, few-seeded, about equalling perianth; seed obovoid, about ¼″ long, reticulated in about 24 rows.

Shallow margins of lakes, ponds or streams, Nova Scotia to northern New York and Maryland.

33. Juncus articulàtus L.   Jointed Rush.   Spart. Fig. 1198.

Juncus articulatus L. Sp. Pl. 327.   1753.

Rootstock branching; stems erect or ascending, 8′–2° high, tufted, somewhat compressed, 2–4-leaved; basal blade-bearing leaves only 1 or 2, usually dying early; stem leaves with rather loose sheaths and conspicuously septate blades; inflorescence rarely exceeding 4′ in height, its branches spreading; heads hemispheric to top-shaped, 6–12-flowered; perianth 1″–1½″ long, the parts nearly equal, lanceolate, acuminate, reddish brown with a green midrib or green throughout; stamens 6, one-half to three-fourths as long as the perianth; anthers shorter than the filaments; capsule longer than the perianth, brown, 3-angled, sharply acute, tapering into a conspicuous tip, 1-celled; seed oblong-obovoid, about ¼″ long, reticulate in about 16–20 rows, the areolae finely cross-lined.

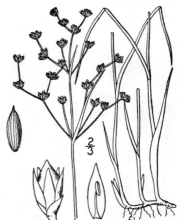

Labrador to Massachusetts, New York, Michigan and British Columbia.   Also in Europe and Asia.   On ballast ground about Philadelphia and Camden a form occurs with obtuse perianth-parts and broadly acute capsules, apparently introduced.

34. Juncus alpinus Vill. Richardson's Rush. Fig. 1199.

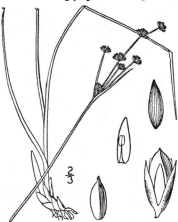

J. alpinus Vill. Hist. Pl. Dauph. **2** : 233. 1787.
J. Richardsonianus Schult. in R. & S. Syst. **7** : 201. 1829.
Juncus alpinus var. *insignis* Fries; Engelm. Trans. St. Louis
     Acad. **2** : 458. 1866.

Stems erect, 6′–20′ high in loose tufts, from creeping
rootstocks, 1–2-leaved; stem leaf or leaves usually borne
below the middle; panicle 2½′–8′ high, sparse, its
branches strict or slightly spreading; heads 3–12-flow-
ered; perianth 1″–1¼″ long, the inner parts shorter
than the outer, obtuse, usually purplish toward the
apex, the three outer paler, obtuse, mucronate or acute;
stamens 6, half to two-thirds as long as the perianth;
anthers much shorter than the filaments; capsule ovoid-
oblong, slightly exceeding the perianth, straw-color or
brown, broadly acute or obtuse, with a short tip; seed
about ¼″ in length, narrowly obovoid to oblong, apiculate, acute or acuminate at the base, lightly reticulate
in about 20 rows, the areolae finely cross-lined.

Greenland to British Columbia, south to Pennsylvania, Nebraska and Washington.

35. Juncus nodòsus L. Knotted Rush. Fig. 1200.

Juncus nodosus L. Sp. Pl. Ed. 2, 466. 1762.

Stems 6′–2° high, erect, arising singly from tuber-
like thickenings of a slender, nearly scaleless rootstock;
stem leaves 2–4, and like the basal ones with long erect
blades, the upper overtopping the inflorescence; panicle
shorter than its lowest bract, seldom exceeding 2¼′,
bearing 1–30 heads; heads spherical, several–many-
flowered, 3½″–6″ in diameter; perianth 1⅞″–1¾″ long, its
parts lanceolate-subulate, usually reddish brown above,
the inner longer than the outer; stamens 6, about one-
half as long as the perianth; anthers equalling the filaments; capsule lanceolate-subulate, 3-sided, 1-celled,
exceeding the perianth; seed oblong, acute below,
apiculate above, rarely more than ⅓″ long, reticulate in
20–30 rows, the areolae finely cross-lined.

Nova Scotia to Virginia, Nebraska and British Columbia
Also in Nevada.

36. Juncus Tórreyi Coville. Torrey's Rush. Fig. 1201.

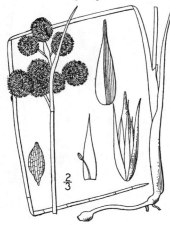

Juncus Torreyi Coville, Bull. Torr. Club **22** : 303. 1895.
J. nodosus var. *megacephalus* Torr. Fl. N. Y. **2** : 326. 1843.
Juncus megacephalus Wood, Bot. Ed. 2, 724. 1861. Not
     J. megacephalus M. A. Curtis, 1835.

Stems 8′–40′ high; rootstock slender, with tuberiform
thickenings at intervals, each supporting a single stem;
stem stout, 1–4-leaved; blade stout, terete, 5″–1′ thick,
abruptly divergent from the stem; inflorescence congested, consisting of 1–20 heads, exceeded by its lowest
bract; heads 5″–8″ in diameter; perianth 2″–2½″ long,
its parts subulate, the outer longer than the inner;
stamens 6, about half as long as the perianth; capsule
subulate, 3-sided, 1-celled, its beak ½″–¾″ long, exceeding
the perianth and holding the valves together throughout
dehiscence; seed ⅓″–¼″ in length, oblong, acute at both
ends, reticulate in about 20 longitudinal rows, the
areolae finely cross-lined.

Wet soil, Massachusetts to Ontario, Saskatchewan, Washington, Alabama, Texas and Arizona.

37. Juncus caesariénsis Coville. New Jersey Rush. Fig. 1202.

J. caesariensis Coville, Mem. Torr. Club **5**: 106. 1894.
Juncus asper Engelm. Trans. St. Louis Acad. **2**: 478. 1868.
Not Sauzé, 1864.

Stems 20′–40′ high, stout, erect, 1¼″ in thickness, slightly roughened; basal leaves few, the uppermost, like the cauline, with inconspicuously articulate sheaths and long erect terete roughened blades; inflorescence 1′–4′ high, with spreading branches, its lowest bract with a small blade sometimes 1½′ long; heads 2–5-flowered; perianth 2″–2½″ long, the parts lanceolate-acuminate, stiff, green, striate, the inner longer than the outer; stamens 6, about half as long as the perianth; filaments about equalling the anthers; style and stigmas long; capsule lanceolate-oblong, 3-sided, mucronate-acuminate, incompletely 3-celled; seed tailed at both ends, altogether about 1″ long, the body about ⅔″ long, closely striate, almost devoid of transverse lines.

Sandy swamps of southern New Jersey.

38. Juncus brachycàrpus Engelm. Short-fruited Rush. Fig. 1203.

Juncus brachycarpus Engelm. Trans. St. Louis Acad. **2**: 467. 1868.

Rootstocks bearing 1–6 stems; stems erect, 8′–36′ high, terete, 1–4-leaved; blades terete, 1″ thick or less, seldom exceeding 6′ in length, the upper much shorter; inflorescence sometimes 4′ high and with 20 spherical heads, or smaller and even reduced to a single head; perianth 1½″–2″ long, its parts subulate, the inner about three-fourths as long as the outer; stamens 3, about half as long as the perianth; capsule one-half to two-thirds as long as the perianth, oblong, acute, mucronate, 1-celled, dehiscent through the tip; seed oblong, acute at both ends, about ⅓″ long, reticulate in about 18 longitudinal rows, the areolae smooth and nearly square.

Southern Ontario, through the Mississippi Valley to Oklahoma, Texas and Mississippi; also from Massachusetts to Georgia.

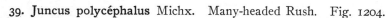

39. Juncus polycéphalus Michx. Many-headed Rush. Fig. 1204.

Juncus polycephalus Michx. Fl. Bor. Am. **1**: 192. 1803.
Juncus Engelmanni Buch. Krit. Verz. Junc. 67. 1880.

Stem stout, about 3° high, compressed, 2–4-leaved. Leaves 20′ in length or less, the upper shorter; blades vertically flattened, 1½″–4″ broad, the septa incomplete, or the blades rarely narrower, merely compressed, and with complete septa; inflorescence 3½′–12′ high, its leaves with nearly obsolete blades; heads globose, 3½″–5″ in diameter; perianth 1½″–2″ long, its parts subulate; stamens 3, one-half to three-fourths as long as the perianth; anthers shorter than the filaments; capsule subulate, 1-celled, exceeding the perianth, the valves remaining united by the slender beak, their margins finally involute; seed narrowly oblong, about ¼″ long, acute at each end, with nearly straight tips, reticulate in about 12 rows, the areolae smooth.

In swamps, Virginia to Missouri, Florida and Texas.

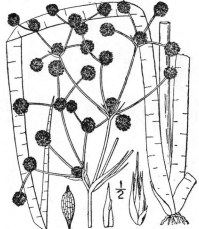

Juncus válidus Coville, which has beakless capsules and leaves with complete septa, is recorded from Missouri; Arkansas, Oklahoma, Mississippi and Texas.

40. Juncus scirpoìdes Lam.　Scirpus-like Rush.
Fig. 1205.

Juncus scirpoides Lam. Encycl. Meth. Bot. **3**: 267.　1789.

Stems 8′–3° high, erect, terete, in clusters from short, horizontal rootstocks.　Stem leaves 1–3; blades terete, 1″ thick or less, usually less than 4′ long, the septa perfect; basal leaves similar, but with longer blades; inflorescence strict or slightly spreading, sometimes 6′ in length; heads 2–30, either simple, globose, 3″–4″ in diameter in flower, and 4″–5½″ in fruit, or lobed, and of slightly greater diameter; perianth 1¼″–1¾″ long, its parts subulate, the inner somewhat shorter; stamens equalling the inner perianth-parts; capsule subulate, 1-celled, its long beak exceeding the perianth; seed oblong, abruptly apiculate at either end, ⅓″–¼″ long, reticulate in 14–20 longitudinal rows, the areolae smooth.

New York to Florida, Missouri and Texas.

Juncus megacéphalus M. A. Curtis, doubtfully admitted into our first edition from Virginia, is not definitely known to grow north of North Carolina.

41. Juncus brachycéphalus (Engelm.) Buch.　Small-headed Rush.　Fig. 1206.

Juncus brachycephalus Buch. in Engler, Bot. Jahrb. **12**: 268.　1890.
Juncus canadensis var. *brachycephalus* Engelm. Trans. St. Louis Acad. **2**: 474.　1868.

Stems 1°–2½° high, tufted from a branching rootstock, erect or occasionally reclining and rooting at the nodes, 2–4-leaved; leaves all with well developed blades, the lower commonly 4′–8′ long; inflorescence commonly 2½′–6′ high, with spreading branches, its lowest bract foliose; heads top-shaped, 2–5-flowered; perianth 1″–1¼″ long, its parts green, or reddish brown above, with hyaline margins, lanceolate, obtuse or sometimes acute, the outer shorter than the inner; stamens 3; anthers much shorter than the filaments; capsule reddish brown, about one-half longer than the perianth, ovoid-oblong, acute to obtuse, tipped, 3-sided, 1-celled; seed ⅓″–½″ long, with narrowly oblong body, short-tailed at either end, 20–30 ribbed somewhat cross-barred the intervening spaces finely cross-lined.

Maine to Pennsylvania, Missouri and Wisconsin.

42. Juncus canadénsis J. Gay.　Canada Rush.　Fig. 1207.

J. canadensis J. Gay; Laharpe, Monog. Junc. 134.　1825.
Juncus canadensis longicaudatus Engelm. Trans. St. Louis Acad. **2**: 474.　1868.
Juncus canadensis subcaudatus Engelm. loc. cit.　1868.

Stems 1°–4° high, erect, 2–4-leaved, few in a tuft, from a branched rootstock.　Basal leaves usually decayed at flowering-time; stem leaves with large loose auriculate sheaths commonly 2′–4′ long, and a stout erect blade usually 4′–10′ long; panicle 3′–10′ in height, the branches moderately spreading; heads top-shaped to hemispheric or subspheric, 5–40-flowered; perianth 1½″–2″ long, the parts narrowly lanceolate, acute, the inner longer than the outer; stamens 3, one-half to two-thirds as long as the perianth, anthers much shorter than the filaments; capsule lanceolate, acute, mucronate, 3-sided, 1-celled, reddish-brown, exceeding the perianth by ¼″ or less; seed ½″ to nearly 1″ long, tailed at either end, the body with a smooth shining coat, about 40-striate.

Newfoundland to Minnesota, Georgia and Louisiana.

43. Juncus brevicaudàtus (Engelm.) Fernald. Narrow-panicled Rush. Fig. 1208.

J. canadensis brevicaudatus Engelm. Trans. St. Louis Acad. **2**: 436. 1866.

J. canadensis coarctatus Engelm. loc. cit. 474. 1868.

J. brevicaudatus Fernald, Rhodora **6**: 35. 1904.

Plant 6′–2½° high, slender, tufted, the rootstocks short. Leaf-blades less than 1″ thick; inflorescence 1′–6′ long, with few or several 2–7-flowered heads on ascending branches; perianth a little more than 1″ long, its parts subulate-lanceolate, acute, or the inner obtusish, somewhat shorter than the outer or nearly equal; capsule dark brown, narrowly oblong, acute, longer than the perianth; tails of the seed about half as long as the body.

Wet ground, Newfoundland to Minnesota, New York, West Virginia and Michigan.

44. Juncus acuminàtus Michx. Sharp-fruited Rush. Fig. 1209.

Juncus acuminatus Michx. Fl. Bor. Am. **1**: 192. 1803.

Plant 10′–3° high; rootstock short and inconspicuous. Stems few or several in a tuft, erect, 1–3-leaved; blades of the lower leaves 4′–8′ long, ¼″–1″ thick, the upper shorter; inflorescence 2′–6′ high, and with 5–50 heads, rarely larger, or reduced even to a single head, its branches usually spreading; heads top-shaped, hemispheric or subspheric, 3–20-flowered; perianth 1¼″–1¾″ long, its parts lanceolate-subulate, nearly equal; stamens 3, about one-half as long as the perianth; anthers shorter than the filaments; capsule ovate-lanceolate, broadly acute, mucronate, 1-celled, equalling the perianth, light brown at maturity, the valves separating through the apex; seed oblong, about ¼″ in length, tipped at either end, reticulate in 16–20 longitudinal rows, the areolae transversely many-lined.

Maine to southern Ontario and Minnesota, south to Georgia and Mexico. Also on the northwest coast. Heads often proliferous. Knotty-leaved rush.

45. Juncus débilis A. Gray. Weak Rush. Fig. 1210.

Juncus debilis A. Gray, Man. 506. 1848.
J. acuminatus debilis Engelm. Trans. St. Louis Acad. **2**: 463. 1868.

Plants 8′–16′ high or sometimes with longer weak procumbent stems, often densely tufted. Leaves with blades usually less than half the length of the plant, more or less flattened in age or in drying; panicle with ascending or diffuse branches; heads mostly 2–6-flowered; perianth about 1″ long, its parts narrowly lanceolate the inner broader at the base and less concave; stamens much over half as long as the perianth; capsules linear-lanceolate in outline, about one-third exceeding the perianth.

In wet places and sandy shores, Rhode Island to Missouri, Florida, Mississippi and Arkansas.

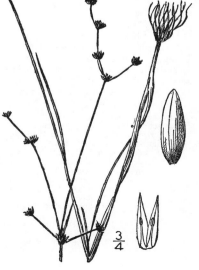

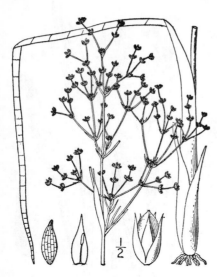

46. Juncus nodàtus Coville. Stout Rush.
Fig. 1211.

Juncus acuminatus var. *robustus* Engelm. Trans. St.
Louis Acad. 2: 463. 1868.
Juncus robustus Coville in Britton & Brown. Ill. Fl.
1: 395. 1896. Not S. Wats.

Plant about 3° high. Stems stout, commonly
1½″-2″ thick below, 1-2-leaved; blades erect,
terete, conspicuously many-septate, 8′-2° long,
1″-1¼″ thick, usually reaching or exceeding the
inflorescence; inflorescence 4′-10′ high, with in-
numerable (commonly 300-500) heads, the blade
of its lowest leaf sometimes half as long as the
inflorescence; heads 2-10-flowered; perianth 1″-
1¼″ long, its parts nearly equal, lanceolate-subulate; stamens 3, one-half to two-thirds as long
as the perianth; capsule equalling or one-third
exceeding the perianth, straw-colored at maturity,
narrowly to broadly oblong, obtuse with a short
tip, 3-sided when dry, 1-celled, the valves separate
and involute after dehiscence; seed nearly as in
J. acuminatus.

Southern Illinois to southeastern Kansas, Oklahoma, Louisiana and Texas.

47. Juncus diffusíssimus Buckley. Diffuse
Rush. Fig. 1212.

Juncus diffusissimus Buckley, Proc. Acad. Phila. 1862:
9. 1862.

Plant 1°-2° high. Stems few in a tuft, from a
short-branched inconspicuous rootstock, erect, slender, terete or slightly compressed, 2-4-leaved; blades
4′-8′ long, ½″-¾″ thick; inflorescence diffusely
branched, widely spreading, 4′-8′ high and broad,
its lowest bract with a blade either obsolete or sometimes nearly as long as the panicle; heads 3-12-flow-
ered; perianth 1¼″-1¾″ long, its parts subulate,
equal; stamens half to two-thirds as long as the
perianth; anthers shorter than the filaments; capsule narrowly linear-lanceolate in outline, 2″-2¼″
long, acute to obtuse at the apex, with a short tip,
3-sided, light brown, 1-celled; seed oblong to obovoid, ⅓″-¼″ long, acute at the base, abruptly tipped,
reticulate in about 16 rows, finely cross-lined.

Indiana to Kansas, Georgia and Texas.

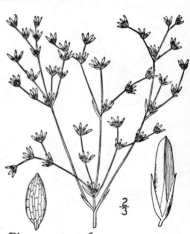

2. JUNCOÌDES Adans. Fam. Pl. 2: 47. 1763.
[Luzula DC. Fl. Fr. 3: 158. 1805.]

Perennial plants, with herbage either glabrous or sparingly webbed, stems leaf-bearing,
leaf-sheaths with united margins, and leaf-blades grass-like. Inflorescence umbelloid, paniculate, or corymbose, often congested; flowers always bracteolate, the bractlets usually
lacerate or denticulate; stamens 6 in our species; ovary 1-celled, its 3 ovules with basal insertion; seeds 3, indistinctly reticulate, sometimes carunculate at base or apex, but not distinctly tailed. [Greek, meaning like *Juncus.*]

About 65 species, widely distributed, mostly flowering in spring. Type species: *Juncus pilosus* L.

1. Juncoides carolìnae (S. Wats.) Kuntze. Hairy Wood-rush. Fig. 1213.

Luzula carolinae S. Wats. Proc. Am. Acad. **14**: 302. 1879.
Juncoides carolinae Kuntze, Rev. Gen. Pl. 724. 1891.
Luzula saltuensis Fernald, Rhodora **5**: 195. 1903.

Tufted, often somewhat stoloniferous. Stems erect, 2–4-leaved, $\frac{1}{2}°$–1° high; leaf-blades $1\frac{1}{2}''$–4'' wide, flat, slightly webbed, especially when young, acuminate into a blunt almost gland-like point; stem leaves with similar but successively shorter blades; inflorescence an umbelloid flower-cluster, with a bract 5''–12'' high, the filiform pedicels equal or nearly so, 1-flowered or sometimes 2-flowered; perianth $1\frac{1}{4}''$–$1\frac{1}{2}''$ long, its parts triangular-ovate, acuminate, brown with hyaline margins, about twice as long as the toothed bractlets; capsule about one-fourth exceeding the perianth, its valves ovate, acuminate; seed about $\frac{1}{4}''$ long, its body about 1'' in length, provided at the summit with a conspicuous hooked caruncle.

Newfoundland to Alaska, south to Georgia, Alabama, Michigan and Oregon. Formerly confused with the European *J. pilòsum* (L.) Kuntze.

2. Juncoides nemoròsum (Poll.) Kuntze. Forest Wood-rush. Fig. 1214.

Juncus nemorosus Poll. Hist. Pl. Pal. **1**: 352. 1776 .
Juncoides nemorosum Kuntze, Rev. Gen. Pl. 724. 1891.

Loosely tufted or somewhat stoloniferous. Stems 1°–$2\frac{1}{2}$° high, 1–6-leaved below the inflorescence; leaf-blades $1\frac{1}{2}''$–3'' wide, ciliate, flat, tapering to a slender sharp tip; inflorescence diffusely paniculate or corymbose, the few lower bracts foliose, and the lowermost branch often inserted 4' below the next or more; flowers in clusters of 3–8, the bractlets ovate, entire or sparingly denticulate above, about one-third as long as the perianth; perianth about $1\frac{1}{4}''$ in length, its parts from reddish brown with pale margins to dirty white, ovate-lanceolate, acute, the outer about one-fifth shorter than the inner; capsule ovoid, acuminate, barely equalling the perianth; seed obliquely ovoid, about $\frac{3}{8}''$ long.

A European species, naturalized at Riverdale, N. Y., and at Niagara Falls, Ontario.

3. Juncoides parviflòrum (Ehrh.) Coville. Small-flowered Wood-rush. Fig. 1215.

Juncus parviflorus Ehrh. Beitr. **6**: 139. 1791.
Luzula parviflora Desv. Journ. de Bot. **1**: 144. 1808.
J. parviflorum Coville, Contr. Nat. Herb. **4**: 209. 1893.

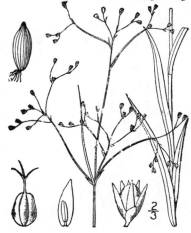

Stems single or few in a tuft, stoloniferous, erect, 10'–30' high, 2–5-leaved; leaves glabrous, their blades $1\frac{1}{2}''$–5'' wide, tapering to a sharp or blunt apex; inflorescence a nodding decompound panicle, commonly $1\frac{1}{2}'$–4' high, its lowest bract foliose, seldom more than one-fourth the length of the panicle; flowers borne singly, or sometimes 2 or 3 together, on the branches of the inflorescence, on slender pedicels; bractlets ovate, entire or rarely somewhat lacerate, perianth $\frac{3}{4}''$–$1\frac{1}{4}''$ in length, its parts ovate, acuminate, slightly exceeded by the green to brown ovoid capsule; seed narrowly oblong, $\frac{1}{2}''$–$\frac{3}{4}''$ in length, attached to its placenta by slender implexed fibers.

Labrador to Alaska, Massachusetts, New York and Minnesota; in the mountains to Arizona and California. Also in Europe and Asia.

4. Juncoides spicàtum (L.) Kuntze. Spiked Wood-rush. Fig. 1216.

Juncus spicatus L. Sp. Pl. 330. 1753.
Juncoides spicatum Kuntze, Rev. Gen. Pl. 725. 1891.
Luzula spicata DC. Fl. Fr. 3 : 161. 1805.

Closely tufted, without rootstocks. Stems erect, 4′-16′ high, distantly 1-3-leaved, tapering to a filiform summit; leaf-blades ½″-1½″ broad, often involute, especially above, tapering to a sharp apex, sparingly webby, especially at the base; inflorescence a nodding, spike-like, often interrupted panicle, commonly ½′-1′ in length, usually exceeded by its lowest involute-foliose bract; bractlets ovate-lanceolate, acuminate, equalling the perianth, sparingly lacerate; perianth brown, with hyaline margins, 1″-1½″ long, its parts lanceolate, aristate-acuminate; capsule broadly ovoid, bluntly acute, about two-thirds as long as the perianth; seed narrowly and obliquely obovoid, about 1½″ long.

Labrador to Quebec and Alaska; mountains of New England and New York; south in the western mountains to Colorado and California. Also in Europe and Asia.

5. Juncoides àrcticum (Blytt) Coville. Arctic Wood-rush. Fig. 1217.

Luzula arctica Blytt, Norg. Fl. 1 : 299. 1861.
Luzula campestris var. *nivalis* Laest. Kongl. Vet. Akad.
  Handl. 334. 1822.
Juncoides nivale Coville, Mem. Torr. Club 5 : 108. 1894.

Stems tufted, 2′-4′ high, erect, 1- or 2-leaved. Leaves with sheaths glabrous at the mouth, their blades 1″-2″ broad, seldom exceeding 1½′ in length, very minutely roughened on the back, at least toward the apex, flat and tapering to a usually blunt and callous tip; inflorescence an erect oblong to ovate, spike-like cluster, ½′ in height or less, exceeding its lowest semifoliaceous bract; bractlet and perianth dark purple, the former ovate and sparingly lacerate at the hyaline apex; perianth-parts ¾″-1″ in length, narrowly oblong, more or less broadly acute at the paler apex, sometimes denticulate above; capsule subspheric, obtuse or broadly acute, exceeding the perianth; seed narrowly oblong, about ½″ long.

Baffin Bay to Alaska. Also in arctic and alpine Europe and Asia.

6. Juncoides hyperbòreum (R. Br.) Sheldon. Northern Wood-rush. Fig. 1218.

Luzula hyperborea R. Br. Suppl. App. Parry's Voy. 183.
  1821.
Luzula confusa Lindeberg, Nya Bot. Not. 9. 1855.
Juncoides hyperboreum Sheldon, Bull. Geol. Surv. Minn. 9 :
  63. 1894.

Stems tufted, commonly 4′-8′ high, erect, 1-2-leaved above the base. Leaves with sheaths sparingly ciliate at the mouth, the blades erect, ½″-1½″ wide at the base, commonly 2½′-7′ long, usually involute in age, not roughened on the back, tapering into a very sharp point; inflorescence erect, exceeding its lowest foliose bract, consisting of a single oblong cluster ½′ in length or less, or its one or two lower divisions on peduncles ½′-1½′ long; bracts and bractlets membranous, fimbriate; perianth-parts brown, paler above, about 1½″ long, ovate-lanceolate, acuminate, denticulate, or slightly lacerate at the apex; capsule about three-fourths as long as the perianth, ovoid, obtuse; seed rather narrowly oblong, about ⅜″ long.

Arctic America, Labrador and the higher mountains of New England. Europe and Asia.

7. Juncoides campéstre (L.) Kuntze.   Common Wood-rush.   Fig. 1219.

Juncus campestris L. Sp. Pl. 329. 1753.
Luzula campestris DC. Fl. Fr. 3 : 161. 1805.
Juncoides campestre Kuntze, Rev. Gen. Pl. 722. 1891.

Stems densely tufted, erect, 4'–20' high, 2–4-leaved.
Leaf-blades flat, 1''–3½'' broad, tapering at the apex to
a blunt almost gland-like point, sparingly webbed when
young; inflorescence umbelloid; lower bracts foliose,
the lowest often exceeding the inflorescence, its several
branches straight, unequal, each bearing an oblong to
short-cylindric dense spike; floral bracts ovate, acuminate; bractlets similar but smaller, fimbriate at the apex;
perianth 1''–1½'' long, brown, its parts lanceolate-ovate,
acuminate; capsule obovoid or broadly oblong; seed with
an oblong body about ½'' in length, supported on a narrower white loosely cellular, strophiole-like base about
one-half as long.

In woodlands, almost throughout the United States and
British America.   Also in Europe and Asia.   Sweeps. Chim-
ney-sweeps.   Black-caps.   Good-Friday.   Black-head- or
Cuckoo-grass.   One of our earliest flowering plants, consisting of several slightly differing races.

Juncoides bulbòsum (Wood) Small, usually distinguishable from this plant by bearing bulblets
at the base of the stems, ranges from the District of Columbia to Georgia, Arkansas and Texas,
and may be specifically distinct.

Family 21.   MELANTHÀCEAE R. Br. Prodr. 1 : 272.   1810.

Bunch-flower Family.

Leafy-stemmed herbs (some exotic genera scapose), with rootstocks or rarely
with bulbs, the leaves broad or grass-like, parallel-veined, the veins often connected
by transverse veinlets.   Flowers perfect, polygamous, or dioecious, regular, racemose, panicled or solitary.   Perianth of 6 separate or nearly separate, usually
persistent segments.   Stamens 6, borne on the bases of the perianth-segments.
Anthers small, 2-celled, oblong or ovate, or confluently 1-celled and cordate or
reniform, mostly versatile and extrorsely dehiscent (introrse in *Tofieldia, Triantha*
and *Abama*).   Ovary 3-celled, superior or rarely partly inferior; ovules few or
numerous in each cavity, anatropous or amphitropous.   Styles 3, distinct or nearly
so.   Fruit a capsule with septicidal dehiscence (loculicidal in *Abama*).   Seeds
commonly tailed or appendaged.   Embryo small, in usually copious endosperm.

About 40 genera and 145 species, widely distributed.
Anthers oblong or ovate, 2-celled.
  Anthers introrsely dehiscent.
    Capsule septicidal ; flowers involucrate by 3 bractlets.
      Inflorescence centripetal ; seeds unappendaged.   1. *Tofieldia.*
      Inflorescence centrifugal ; seeds appendaged.   2. *Triantha.*
    Capsule loculicidal ; flowers not involucrate.   3. *Abama.*
  Anthers extrorsely dehiscent.
    Flowers perfect.
      Leaves basal, oblanceolate ; seeds numerous.   4. *Xerophyllum.*
      Stem very leafy ; leaves linear ; seeds few.   5. *Helonias.*
    Flowers dioecious ; stem leafy.   6. *Chamaelirium.*
Anthers cordate or reniform, confluently 1-celled.
  Plants glabrous.
    Perianth-segments not gland-bearing.
      Flowers perfect ; perianth-segments obtuse.   7. *Chrosperma.*
      Flowers polygamous ; perianth-segments acuminate.   8. *Stenanthium.*
    Perianth-segments bearing 1 or 2 glands, or a spot.
      Plant with a thick horizontal rootstock ; perianth-segments with 2 glands.
                                          9. *Zygadenus.*
      Plants bulbous, or with short erect rootstocks ; perianth-segments with 1 gland.
        Flowers perfect ; bulbs membranous-coated.
          Ovary partly inferior ; gland obcordate.   10. *Anticlea.*
         Ovary wholly superior ; gland obovate or half-orbicular. 11. *Toxicoscordion.*
        Flowers polygamous ; rootstocks fibrous-coated.   12. *Oceanoros.*
Stem and inflorescence pubescent.
  Perianth-segments distinctly clawed, glandless.   13. *Melanthium.*
  Perianth-segments not clawed or very short-clawed, 2-glandular. 14. *Veratrum.*

1. TOFIÈLDIA Huds. Fl. Angl. Ed. 2, 157. 1778.

Perennial herbs, with short erect or horizontal rootstocks, fibrous roots, slender erect stems leafless above or nearly so, linear somewhat 2-ranked and equitant leaves clustered at the base, and small perfect white or green flowers in a terminal raceme. Pedicels bracted at the base, solitary or clustered. Flowers usually involucrate by 3 scarious somewhat united bractlets below the calyx. Perianth-segments oblong or obovate, subequal, persistent, glandless. Stamens 6; filaments filiform; anthers ovate, sometimes cordate, introrse. Ovary sessile, 3-lobed at the summit; styles 3, short, recurved. Capsule 3-lobed, 3-beaked, septicidally dehiscent to the base, many-seeded. Seeds unappendaged. [Dedicated to Tofield, an English correspondent of Hudson.]

About 12 species, natives of the north temperate zone, 1 or 2 in the Andes of South America. Besides the following another occurs in the southeastern States and two in northwestern America. Type species: *Tofieldia palustris* Huds.

$\frac{2}{3}$

1. Tofieldia palústris Huds. Scottish Asphodel. False Asphodel. Fig. 1220.

Tofieldia palustris Huds. Fl. Angl. Ed. 2, 157. 1778.

Glabrous, stem slender, scape-like, leafless or bearing a few leaves near the base, 2'-10' tall. Leaves tufted, ½'-4' long, ½"-2" wide; raceme oblong or subglobose in flower, dense, elongating to an inch or less in fruit, the lower flowers first expanding; pedicels usually solitary, minutely involucrate, ½"-1" long in fruit; flowers greenish white, 1" broad; perianth-segments obovate, obtuse, much shorter than the oblong-globose minutely beaked capsule; seeds oblong, unappendaged.

Greenland and Labrador to Alaska, south to Quebec, the shores of Lake Superior, and the Canadian Rocky Mountains. Also in Europe and Asia. Lamb-lily. Summer.

2. TRIÁNTHA Nutt.; Baker, Journ. Linn. Soc. 17: 490. 1879.

Perennial herbs, with pubescent foliage, the leaves mainly basal, their blades narrow, flat. Flowers perfect, mostly clustered in 3's, in erect narrow centrifugal panicles; perianth-segments white or greenish, nearly equal, glandless, persistent; stamens 6, mainly hypogynous, the filaments slender, the anthers marginally dehiscent; ovules numerous in each cavity of the ovary; capsule 3-celled, dehiscent nearly or quite to the base; seeds with tail-like appendages at each end. [Greek, referring to the aggregation of the flowers in 3's.]

Three known species, the following and one in Japan. Type species: *Triantha glutinosa* (Michx.) Baker.

Capsule oblong, 3"-3½" long; perianth-segments thin. 1. *T. glutinosa.*
Capsule ovoid, 1½"-2" long; fruiting perianth-segments rigid. 2. *T. racemosa.*

1. Triantha glutinòsa (Michx.) Baker. Glutinous Triantha. Fig. 1221.

Narthecium glutinosum Michx. Fl. Bor. Am. 1: 210. 1803.
Tofieldia glutinosa Pers. Syn. 1: 399. 1805.
Triantha glutinosa Baker, Journ. Linn. Soc. 17: 490. 1879.

Stem viscid-pubescent with black glands, 6'-20' tall, bearing 2-4 leaves near the base. Basal leaves tufted, 2'-7' long, 1"-3" wide; raceme oblong and ½-1½' long in flower, longer in fruit, the upper flowers first expanding; pedicels commonly clustered in 3's (1's-4's), ascending, viscid-pubescent, becoming 2"-6" long in fruit; involucral bracts minute, united nearly or quite to their apices, borne just beneath the flowers; flowers 3"-4" broad; perianth-segments oblong, mostly obtuse, membranous; capsule oblong, about 3" high, 1½" in diameter, thin-walled, twice as long as the perianth, the beaks ½" long or less; seeds slender-tailed at each end.

In bogs, Newfoundland to Minnesota, Maine, Ohio, Michigan, and in the southern Alleghanies. May-June. False asphodel.

$\frac{3}{4}$

2. Triantha racemòsa (Walt.) Small. Viscid Tofieldia. Fig. 1222.

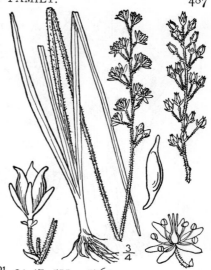

Melanthium racemosum Walt. Fl. Car. 126. 1788.
Narthecium pubens Michx. Fl. Bor. Am. **1** : 209. 1803.
Tofieldia racemosa B.S.P. Prel. Cat. N. Y. 55. 1888.
Triantha racemosa Small, Fl. SE. U. S. 249. 1903.

Similar to the preceding species but rather stouter and taller, stem 1°–3° high, the glutinous pubescence rougher. Leaves very narrowly linear, 6′–18′ long, 1½″–3″ wide; raceme 1′–4′ long in flower, often loose, somewhat longer in fruit, the uppermost flowers first expanding; pedicels mostly clustered in 3's, ascending, 2″–3″ long in fruit; involucral bractlets about ½″ long, united to above the middle, borne just beneath the flower; perianth-segments narrowly obovate, obtuse, rigid; capsule ovoid, 1½″ long, little longer than the calyx, its beaks ½″ long; seeds short-tailed at each end.

In swamps, southern New Jersey to Florida and Alabama. This and the preceding species are also known as False asphodel. June–Sept.

3. ABÀMA Adans. Fam. Pl. **2** : 47, 511. 1763.

[Narthecium Juss. Gen. 47. 1789.]

Perennial herbs, with creeping or horizontal rootstocks, fibrous roots, erect simple stems and linear grass-like basal leaves, those of the stem short and distant. Flowers small, greenish-yellow, perfect, borne in a terminal raceme. Pedicels bracted at base and usually bearing a small bractlet. Perianth-segments persistent, linear or linear-lanceolate, obscurely 3–5-nerved, glandless. Stamens 6; filaments subulate, woolly; anthers linear-oblong, erect, introrse. Ovary sessile; style very short or none; stigma slightly 3-lobed. Capsule oblong, loculicidally dehiscent, many-seeded, the linear seeds tailed at each end. [Greek, signifying without step, the plants reputed to cause lameness in cattle.]

Four known species, natives of the northern hemisphere. Besides the following, another occurs in northwestern America. Type species: *Anthericum ossifragum* L.

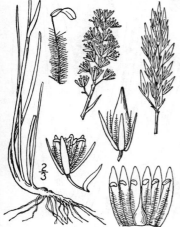

1. Abama americàna (Ker) Morong. American Bog-asphodel. Fig. 1223.

Narthecium americanum Ker, Bot. Mag. *pl. 1505.* 1812.
Narthecium ossifragum var. *americanum* A. Gray, Man. Ed. 5, 536. 1867.
Abama americana Morong, Mem. Torr. Club **5** : 109. 1894.

Glabrous, stems wiry, stiff, erect, 10′–18′ tall. Basal leaves 3′–8′ long, 1″ wide or less, finely 7–9-nerved; lower stem leaves ½′–2′ long, the upper much smaller; raceme 1′–2′ long, dense; perianth-segments narrowly linear, 2″–3″ long, slightly exceeding the stamens; filaments white-woolly; pedicels ascending, 3″–4″ long in fruit; capsule about 5″ long, 1″ in diameter at the middle, erect, nearly twice as long as the perianth-segments, tapering to a subulate beak; seeds, including the appendages, 3″–4″ long.

In pine barren swamps, southern New Jersey and Delaware. June–Sept. Yellow- or Moor-grass. Rosa-solis.

XEROPHÝLLUM Michx. Fl. Bor. Am. 1 : 210. 1803.

Tall perennial herbs, with thick short woody rootstocks, simple erect leafy stems, the leaves narrowly linear, rough-margined, the upper ones shorter than the lower. Flowers very numerous, medium-sized, white, in a large dense terminal raceme, the lower ones first expanding. Perianth withering-persistent, its segments oblong or ovate, 5–7-nerved, spreading, glandless. Stamens 6, rather shorter than the perianth-segments; filaments subulate, glabrous; anthers oblong. Ovary sessile, 3-grooved; styles 3, filiform, reflexed or recurved, stigmatic along the inner side; ovules only 2–4 in each cell. Capsule ovoid, 3-grooved, loculicidally and sometimes also septicidally dehiscent. Seeds 5, oblong, not at all appendaged, or only minutely so. [Greek, signifying a dry leaf.]

Three species, the following, which is the type of the genus, the others of western America.

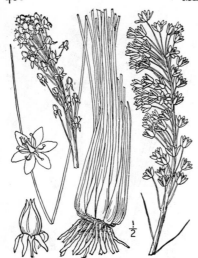

1. Xerophyllum asphodeloìdes (L.) Nutt.
Turkey-beard. Fig. 1224.

Helonias asphodeloides L. Sp. Pl. Ed. 2, 485. 1762.
X. setifolium Michx. Fl. Bor. Am. 1: 211. 1803.
Xerophyllum asphodeloides Nutt. Gen. 1: 235. 1818.

Stem stout, becoming stiff, 2½°–5° tall, densely
leafy below and at the base, sparsely leafy above.
Leaves very narrowly linear, slightly dilated at the
base, the lower 6′–18′ long, 1″ wide or less, except
at the broader base, the upper successively shorter
and narrower; flowering raceme 3′–6′ long, 2′–3′ in
diameter, its summit conic; flowering pedicels spreading, filiform, 9″–1″ long, in fruit erect; perianth-
segments ovate-oblong, obtuse, about 3′ long; styles
rather longer than the ovary; capsule ellipsoid, 2″
long, 1″–1½″ in diameter; seeds mostly 2 in each cell.

In dry pine barrens, southern New Jersey to eastern
Tennessee and Florida. May–July. Ascends to 5000
ft. in North Carolina.

5. HELÒNIAS L. Sp. Pl. 342. 1753.

A perennial glabrous bog herb, with a stout rootstock, thick fibrous roots, basal oblanceolate persistent leaves and rather large perfect purple flowers, racemed at the summit
of an erect hollow bracted scape. Perianth-segments spreading, spatulate, persistent.
Stamens 6, hypogynous, longer than the perianth-segments; filaments filiform; anthers ovate.
Ovary ovoid, 3-grooved, 3-celled, slightly 3-lobed, many-ovuled; styles 3, stigmatic along
the inner side, deciduous. Capsule obovoid, deeply 3-lobed, the lobes divergent, ventrally
dehiscent above. Seeds numerous, linear, white-appendaged at each end. [Name from the
Greek, in allusion to its growth in swamps.]

A monotypic genus of eastern North America.

1. Helonias bullàta L. Swamp Pink.
Fig. 1225.

Helonias bullata L. Sp. Pl. 342. 1753.

Leaves several or numerous, dark green, thin,
clustered at the base of the scape, 6′–15′ long,
½′–2′ wide, pointed or blunt, finely parallel-nerved.
Scape stout, bracted below, the bracts lanceolate,
acute or acuminate, membranous; raceme dense,
1′–3′ long in flower, becoming 4′–7′ long in fruit;
perianth-segments about 3″ long, equalling or
rather longer than the stout pedicels; capsules
about 3″ long, the valves papery; seeds 1½″–2″
long.

In bogs northern New Jersey, southern New York
and eastern Pennsylvania to North Carolina. Local.
The scape sometimes bears a few leaves at its base.
April–May. Stud-flower. Not definitely known to
grow wild in Pennsylvania at the present time.

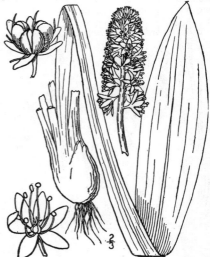

6. CHAMAELÍRIUM Willd. Mag. Nat. Fr. Berl. 2: 18. 1808.

An erect glabrous slightly fleshy herb, with a bitter tuberous rootstock. Basal leaves
spatulate, those of the stem lanceolate. Flowers small, white, dioecious, in a long narrow
bractless spike-like raceme. Perianth of 6 linear-spatulate 1-nerved segments. Staminate
flowers with 6 stamens, the filaments filiform, the anthers subglobose, 2-celled; pistillate
flowers with a 3-celled oblong ovary, 3 short styles, stigmatic along the inner side, and usually with 6 staminodia. Capsules erect, slightly 3-lobed, loculicidally 3-valved. Seeds 6–12
in each cavity, broadly winged at both ends, narrowly winged on the sides. [Greek, signifying a low lily.]

A monotypic genus of eastern North America.

1. Chamaelirium lùteum (L.) A. Gray. Blazing-star. Fig. 1226.

Veratrum luteum L. Sp. Pl. 1044. 1753.
Chamaelirium carolinianum Willd. Mag. Nat. Fr. Berl. **2**: 19. 1808.
Chamaelirium luteum A. Gray, Man. 503. 1848.
C. obovale Small, Torreya **1**: 108. 1901.

Staminate plant $1\frac{1}{2}°-2\frac{1}{2}°$ tall, the pistillate often taller, sometimes 4° high. Basal leaves $2'-8'$ long, $\frac{1}{2}'-1\frac{1}{2}'$ wide, mostly obtuse, tapering into a long petiole; stem leaves lanceolate, the or upper linear, acute or acuminate, sessile or the lower short-petioled; staminate raceme nodding or finally erect, $3'-9'$ long, pedicels spreading, $1''-2''$ long; pistillate raceme erect; flowers nearly $3''$ broad; capsule oblong or somewhat obovoid, $4''-7''$ long, $2''-3''$ in diameter.

In moist meadows and thickets, Massachusetts to southern Ontario and Michigan, south to Florida and Arkansas. Called also Devil's bit, unicorn-root or -horn, drooping starwort. False unicorn-plant. May–July.

7. CHROSPÉRMA Raf. Neog. 3. 1825.

[AMIANTHIUM A. Gray, Ann. Lyc. N. Y. **4**: 121. 1837.]

An erect glabrous herb, with an ovoid-oblong coated bulb, and numerous long blunt basal leaves, a few short ones on the stem. Flowers perfect, white, in a dense terminal raceme, the lower ones first expanding. Perianth of 6 distinct glandless persistent obtuse segments. Stamens inserted on the bases of the sepals; anthers small, reniform. Ovary ovoid, 3-lobed, 3-celled. Capsule 3-celled, dehiscent above the middle, the cavities 1–2-seeded, its 3 divergent lobes tipped with the subulate styles. Seeds ovoid, reddish brown. [Greek, referring to the colored seeds.]

A monotypic genus of eastern North America based on *Melanthium laètum* Soland.

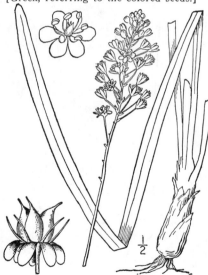

1. Chrosperma muscaetóxicum (Walt.) Kuntze. Fly-poison. Fig. 1227.

Melanthium muscaetoxicum Walt. Fl. Car. 125. 1788.
Melanthium laetum Soland. in Ait. Hort. Kew. **1**: 488. 1789.
Melanthium muscaetoxicum A. Gray, Ann. Lyc. N. Y. **4**: 122. 1837.
Chrosperma muscaetoxicum Kuntze, Rev. Gen. Pl. 708. 1891.

Bulb $1\frac{1}{2}'-2'$ long, nearly $1'$ in diameter. Stem $1\frac{1}{2}°-4°$ tall. Basal leaves $2''-15''$ wide, shorter than the stem, the upper few and distant, bract-like; raceme at first ovoid-conic, becoming cylindric, $2'-5'$ long; pedicels ascending, $4''-10''$ long; bractlets ovate, $1''-2''$ long; sepals ovate-oblong, obtuse, $2''-3''$ long; filaments filiform, about equalling the sepals; capsule $2''-3''$ in diameter above the middle, scarcely as long; seeds about $1\frac{1}{2}''$ long.

In dry sandy woods, Long Island and eastern Pennsylvania to Florida, Tennessee, Missouri and Arkansas. Ascends to 4000 ft. in Virginia and to 2100 ft. in Pennsylvania. Hellebore. Crow-poison. May–July.

8. STENÁNTHIUM Kunth, Enum. **4**: 189. 1842.

Erect glabrous bulbous herbs, with leafy stems and small white or greenish, polygamous flowers in an ample terminal penicle. Leaves narrowly linear, keeled. Perianth-segments narrowly lanceolate, acuminate, glandless, spreading, persistent, adnate to the base of the ovary. Stamens shorter than the perianth-segments, inserted on their bases; anthers small, cordate or reniform. Ovary ovoid. Capsule ovoid-oblong, 3-lobed, finally dehiscent to the base, the lobes with short slightly divergent beaks. Seeds about 4 in each cavity, oblong, angled, somewhat flattened. [Greek, in allusion to the narrow perianth-segments.]

The genus comprises only the two following species of which the first is the type.

Leaves $2''-3''$ wide; capsule reflexed.   1. *S. gramineum.*
Leaves $3''-10''$ wide; capsule erect.   2. *S. robustum.*

1. Stenanthium gramíneum (Ker) Morong. Grass-leaved Stenanthium. Fig. 1228.

Helonias graminea Ker, Bot. Mag. *pl. 1599.* 1813.

Veratrum angustifolium Pursh, Fl. Am. Sept. 242. 1814.

Stenanthium angustifolium Kunth, Enum. 4: 190. 1843.

Stenanthium gramineum Morong, Mem. Torr. Club 5: 110. 1894.

Stem slender, 3°–4° tall. Leaves grass-like, some of them often 1° long or more, 2″–3″ wide, the upper, reduced to small linear lanceolate bracts subtending the branches of the panicle; panicle open, simple or somewhat compound, 1°–2° long, its branches nearly filiform, often flexuous, spreading or drooping; bracts ½″–1″ long, equalling or longer than the pedicels; flowers 4″–6″ broad; perianth-segments linear-lanceolate; capsule ovoid-oblong, with a top-shaped base, 3″–4″ long, reflexed.

In dry soil, Virginia to Kentucky, Missouri, Florida and Alabama. Ascends to 6000 ft. in North Carolina. Fruit apparently scarce. Aug.–Sept.

2. Stenanthium robústum S. Wats. Stout Stenanthium. Fig. 1229.

Stenanthium robustum S. Wats. Proc. Am. Acad. 14: 278. 1879.

Stems stout, 3°–5° tall, usually very leafy. Leaves often 1° long or more, the lower 4″–10″ wide, the upper reduced to bracts; panicle denser than that of the preceding species, commonly longer, usually compound, its branches spreading or ascending; flowers greenish or white, 6″–8″ broad; capsule ovoid-oblong, 4″–6″ long, erect, longer than its pedicel, the very short beaks recurved-spreading.

In moist soil, Pennsylvania and Ohio to South Carolina, Tennessee and Missouri. July–Sept. Apparently distinct from the preceding species, though closely related.

9. ZYGADÈNUS Michx. Fl. Bor. Am. 1: 213. 1803.

A glabrous erect perennial herb with a thick rootstock and a leafy stem. Leaves narrowly linear. Flowers perfect, white, in a terminal panicle. Perianth withering-persistent, its segments lanceolate, separate, bearing 2 glands just above the narrowed base. Stamens free from the perianth segments and about equalling them in length; anthers cordate or reniform. Capsule 3-lobed, 3-celled, the cavities not diverging, dehiscent to the base. Seeds numerous in each cavity, oblong or linear, angled. [Greek, referring to the two glands.]

A monotypic genus of southeastern North America.

1. Zygadenus glabérrimus Michx. Large-flowered Zygadenus. Fig. 1230.

Z. glaberrimus Michx. Fl. Bor. Am. 1: 214. 1803.

Rather dark green, slightly glaucous, stem stout, 2°-4° tall, from a thick rootstock. Leaves 3″-6″ wide, long-acuminate, channelled, often 1° long or more, the upper gradually smaller, appressed, passing into the short ovate bracts of the panicle; panicle 6′-12′ long, its branches rather stout, stiff, ascending; pedicels stout, longer than the bractlets; flowers white, mostly perfect, 1′-1½′ broad, perianth-segments lanceolate or oblong-lanceolate, narrowed into a short claw, bearing 2 orbicular glands; styles subulate; capsule narrowly ovoid, shorter than the perianth.

In swamps, Virginia to Florida, near the coast. July–Sept.

10. ANTICLÈA Kunth, Enum. 4: 191. 1843.

Glabrous perennial herbs, with membranous-coated bulbs, leafy stems, and rather large greenish or yellowish-white flowers in terminal racemes. Leaves linear. Flowers perfect. Perianth withering-persistent, adnate to the lower part of the ovary, its segments bearing a single obcordate gland. Stamens distinct from the perianth-segments. Capsule 3-celled, the cavities dehiscent to the base. Seeds numerous. [Named for the mother of Ulysses.]

About 6 species, natives of North America and northern Asia. Type species: *A. sibírica* (L.) Kunth.

1. Anticlea élegans (Pursh) Rydb. Glaucous Anticlea. Fig. 1231.

Zygadenus elegans Pursh, Fl. Am. Sept. 241. 1814.
Melanthium glaucum Nutt. Gen. 1: 232. 1818.
Zygadenus glaucus Nutt. Journ. Acad. Phila. 7: 56. 1834.
Z. chloranthus Richards. Frank. Journ. 736. 1821.
A. chlorantha Rydb. Bull. Torr. Club 30: 273. 1903.

Plant glaucous; bulb ovoid, about 1′ long, its coats membranous. Stem slender, 6′-3° tall; leaves 2″-7″ wide, keeled, the lower 4′-12′ long, the upper much shorter; bracts lanceolate, rather large, green or purplish; inflorescence a simple raceme or a large panicle, sometimes 1° long, open, its branches slender, ascending; flowers greenish or yellowish, 8″-10″ broad; perianth-segments oval or obovate, obtuse, bearing a large obcordate gland just above the short claw; capsule oblong, nearly 1′ long, exceeding the perianth.

In moist places, New Brunswick to Alaska, Vermont, New York, Missouri, and in the Rocky Mountains to New Mexico. June–Aug.

11. TOXICOSCÓRDION Rydb. Bull. Torr. Club 30: 272. 1903.

Glabrous, poisonous perennial herbs, with membranous-coated bulbs, narrowly linear conduplicate leaves and small perfect or polygamous flowers in racemes or panicles. Perianth wholly inferior, free from the ovary, its segments distinct, short-clawed, each bearing an obovate or semi-orbicular gland at or above the base. Stamens mostly adnate to the bases of the perianth-segments; anthers subreniform, confluently 1-celled. Ovary 3-celled. Capsule 3-beaked, 3-celled, containing numerous seeds.

About 7 species, natives of North America. Type species: *Zygadenus intermedius* Rydb. [Greek, poison-onion.]

Leaves 3″-8″ wide; flowers mostly perfect.   1. *T. Nuttallii.*
Leaves 2″-3″ wide; flowers polygamous.   2. *T. gramineum.*

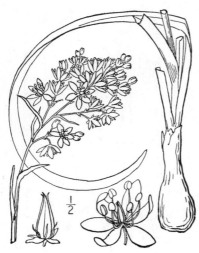

1. Toxicoscordion Nuttàllii (A. Gray) Rydb.
Nuttall's Camass.   Fig. 1232.

Amianthium Nuttallii A. Gray, Ann. Lyc. N. Y. **4**: 123.
   1837.
Zygadenus Nuttallii S. Wats. Proc. Am. Acad. **14**: 279.
   1879.
T. Nuttallii Rydb. Bull. Torr. Club **30**: 272.   1903.

Light green, scarcely glaucous, stem 1°–2° high.
Bulb large, coated; leaves 3″–8″ wide, shorter than
the stem, strongly conduplicate, the upper very short;
inflorescence racemose or paniculate bracts membranous, scarious, shorter than the slender pedicels;
flowers mostly perfect, about 6″ broad; perianth-
segments oval or ovate, obtuse, free from the ovary,
thin, short-clawed, bearing a roundish spot-like
gland; capsule 4″–6″ long.

On prairies, Kansas, Colorado and Arkansas. May–
June.

2. Toxicoscordion gramíneum Rydb.   Death-
camass.   Fig. 1233.

Zygadenus venenosus Rydb. Contr. Dept. Agric. **3**: 525.
   1896.   Not S. Wats.
Zygadenus gramineus Rydb. Bull. Torr. Club **27**: 535.
   1900.
T. gramineum Rydb. Bull. Torr. Club **30**: 272.   1903.

Pale green, stem slender, 6′–2° tall, from a small
coated bulb. Leaves conduplicate, roughish, 2″–3″
wide, shorter than the stem, the upper small and
distant; inflorescence a simple or somewhat
branched raceme, 2′–4′ long in flower, elongating in
fruit, the slender pedicels longer than the scarious
lanceolate bracts; flowers yellow or yellowish,
polygamous, about 4″ wide; perianth-segments
ovate or elliptic, obtuse or acutish, short-clawed, free
from the ovary, bearing a roundish gland with an
irregular margin; fruiting pedicels erect; capsule
longer than the perianth.

South Dakota to Saskatchewan, Idaho, Nebraska and Utah. Hog's-potato. Death-camass.
Roots poisonous. May–June.

12.   OCEANÒRUS Small, Fl. SE. U. S. 252.   1903.

A glabrous perennial herb, with erect, fibrous-coated bulb-like rootstocks. Leaves narrow, borne on the lower part of the stem, the outer ones reduced to sheathing scales. Flowers polygamous, in terminal panicles, the lower ones of each branch of the inflorescence
fertile. Perianth-segments each bearing a thick yellow gland at the base. Stamens somewhat shorter than the perianth. Ovary 3-celled, each cavity with several ovules. Capsule
conic, erect, tipped by the short styles. [Greek, the plant inhabiting both coastal and mountainous regions.]

A monotypic genus of the southeastern United States.

1. Oceanorus leimanthoìdes (A. Gray)
Small.   Pine-barren Oceanorus.
Fig. 1234.

Amianthium leimanthoides A. Gray, Ann. Lyc.
  N. Y. 4: 125. 1837.
Zygadenus leimanthoides S. Wats. Proc. Am.
  Acad. 14: 280. 1879.
Oceanorus leimanthoides Small, Fl. SE. U. S. 252.
  1903.

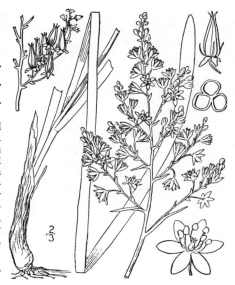

Stem slender, 1°–4° high, its base sheathed
by short leaves which soon become fibrous.
Leaves 2″–4″ wide, green on both sides, often
1° long, blunt, or the upper acuminate and
much shorter; panicle 4′–12′ long, its branches
densely many-flowered, spreading or ascending; bractlets much shorter than the slender
pedicels; lower flowers perfect, white or greenish, about 4″ broad; perianth-segments oblong,
obtuse, sessile, not clawed, adnate to the very
base of the ovary; capsule ovoid, 4″ high,
much longer than the perianth.

In swamps or wet soil, Long Island to Georgia,
North Carolina and Tennessee. July–Aug.

13. MELÁNTHIUM L. Sp. Pl. 339. 1753.

Tall leafy herbs, perennial by thick rootstocks, the stem, at least its upper part, and the
inflorescence, pubescent. Leaves narrow, oblanceolate or linear, sheathing or the upper sheathless. Flowers greenish, white or cream-colored, darker in withering, monoecious or polygamous, slender pedicelled in a large terminal panicle. Perianth of 6 spreading separate
persistent clawed 2-glandular segments, mostly free from the ovary. Stamens shorter than
the segments and adnate to them; anthers cordate or reniform, their sacs confluent. Ovary
ovoid; styles 3, subulate, spreading. Capsule 3-lobed, 3-celled, the cavities several-seeded,
tipped by the style. Seeds very flat and broadly winged, several in each cavity. [Greek,
signifying black flower.]

The genus comprises the following species, and one in the southern states, the first being the type.

Blade of the perianth-segments oblong, entire; leaves linear.   1. *M. virginicum.*
Blade of the perianth-segments nearly orbicular, undulate; leaves oblanceolate.   2. *M. latifolium.*

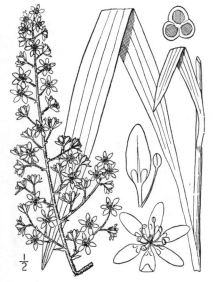

1. Melanthium virgínicum L.   Bunch-
flower.   Fig. 1235.

Melanthium virginicum L. Sp. Pl. 339. 1753.

Stems rather stout, 2½°–5° high. Leaves linear,
acuminate, often 1° long, 4″–12″ wide, the lower
sheathing, the upper smaller, sessile, the uppermost very small; panicle 6′–18′ long, usually much
longer than the ovate-oblong bracts; flowers 6″–
10″ broad, greenish yellow, turning brown;
perianth-segments obtuse, the blade oblong, flat,
entire, sometimes obcordate, at least twice as long
as the claw, bearing 2 dark glands at its base; capsule 5″–7″ long, the persistent styles erect, 1″–1½″
long; seeds 8–10 in each cavity, 2″–3″ long.

In meadows, wet woods and marshes, Rhode Island
to New York and Minnesota, south to Florida and
Texas. Black or bunch-flower. Quafodil. June–Aug.

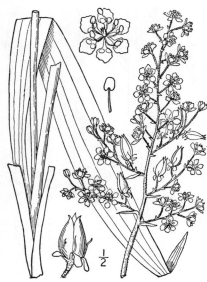

2. Melanthium latifòlium Desr. Crisped Bunch-flower. Fig. 1236.

?Melanthium hybridum Walt. Fl. Car. 125. 1788.

M. *latifolium* Desr. in Lam. Encycl. **4**: 25. 1797.

Melanthium latifolium longipedicellatum A. Brown, Bull. Torr. Club **23**: 152. 1896.

Stem stout or slender, 2°–4° tall. Leaves oblanceolate, acute, 6″–2′ wide, the lower clasping, the upper sessile and much smaller; panicle usually 1° long or more, its branches ascending or spreading; flowers 6″–8″ broad, greenish white, turning darker; blade of the perianth-segments orbicular or ovate, undulate and crisped, longer than the claw or about equalling it, bearing 2 glands at the base; capsule 6″–8″ long, its cavities 4–8-seeded; seeds rather larger than those of the preceding species; flowers fragrant.

In dry woods and on hills, Connecticut to Pennsylvania and South Carolina. Ascends to 2000 ft. in North Carolina. Pedicels 3″–8″ long. July–Aug.

14. VERÀTRUM L. Sp. Pl. 1044. 1753.

Tall perennial herbs, with thick short poisonous rootstocks, the leaves mostly broad, clasping, strongly veined and plaited, the stem and inflorescence pubescent. Flowers greenish or yellowish or purple, rather large, polygamous or monoecious, on short stout pedicels in large terminal panicles. Perianth-segments 6, glandless or nearly so, not clawed, sometimes adnate to the base of the ovary. Stamens opposite the perianth-segments and free from them, short, mostly curved. Anthers cordate, their sacs confluent. Ovary ovoid; styles 3, persistent. Capsule 3-lobed, 3-celled, the cavities several-seeded. Seeds very flat, broadly winged. [Ancient name of the Hellebore.]

About 12 species, natives of the north temperate zone. Besides the following another occurs in southern United States and 3 in western North America. Type species: *Veratrum album* L.

Flowers yellowish green; perianth-segments pubescent, ciliate.   1. *V. viride.*
Flowers purple or greenish; perianth-segments glabrous or nearly so.
   Flowers purple; pedicels short.   2. *V. Woodii.*
   Flowers greenish; pedicels filiform.   3. *V. parviflorum.*

1. Veràtrum víride Ait. American White Hellebore. Indian Poke. Poke-root. Fig. 1237.

Veratrum viride Ait. Hort. Kew. **3**: 422. 1789.

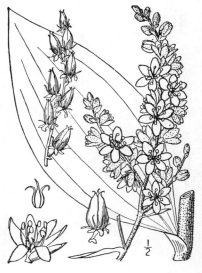

Rootstock erect, 2′–3′ long, 1′–2′ thick, with numerous fibrous-fleshy roots. Stem stout, 2°–8° tall, very leafy; leaves acute, the lower broadly oval or elliptic, 6′–12′ long, 3′–6′ wide, short-petioled or sessile, sheathing, the upper successively narrower, those of the inflorescence small; panicle 8′–2° long, densely many-flowered, its lower branches spreading or somewhat drooping; pedicels 1″–3″ long, mostly shorter than the bracts; flowers yellowish green, 8″–12″ broad; perianth-segments oblong or oblanceolate, ciliate-serrulate, twice as long as the stamens; ovary glabrous; capsule 10″–12″ long, 4″–6″ thick, many-seeded; seed 4″–5″ long.

In swamps and wet woods, New Brunswick and Quebec to Ontario, south to Georgia, Tennessee and Minnesota. Ascends to 4000 ft. in the Adirondacks. Big, Swamp or False hellebore. Duck-retten. Earth-gall. Devil's-bite. Bear-corn. Poor Annie. Itch-weed. Tickle-weed. May–July.

2. Veratrum Woòdii Robbins. Wood's False Hellebore. Fig. 1238.

Veratrum Woodii Robbins in Wood, Classbook, Ed. 41, 557. 1855.

Rootstock short, erect. Stem slender, 2°–5° tall; leaves mostly basal, oblong or oblanceolate, often 1° long, 2'–4' wide, narrowed into sheathing petioles about as long as the blade; upper leaves small and linear-lanceolate; panicle open, 1°–2° long, its branches ascending; pedicels shorter than the perianth, about as long as the bracts; flowers 6''–8'' broad, purple; perianth-segments oblanceolate, obtuse, nearly or quite glabrous, entire, little longer than the stamens; ovary pubescent when young, becoming glabrous; capsule 6''–8'' long, few-seeded.

In dry woods and on hills, southern Indiana to Missouri. Indian poke-weed. June–July.

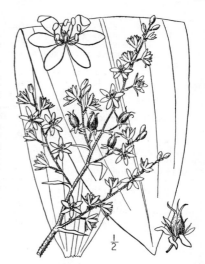

3. Veratrum parviflòrum S. Wats. Small-flowered Veratrum. Fig. 1239.

V. parviflorum Michx. Fl. Bor. Am. **2**: 250. 1803.
Melanthium parviflorum S. Wats. Proc. Am. Acad. **14**: 276. 1879.

Stem slender, 2°–5° tall. Lower leaves broadly oval or oblanceolate, acute, 4'–8' long, 1½'–4' wide, with narrow sheathing bases, the upper narrowly linear-lanceolate, acuminate; panicle 1°–2° long, loose and open, its very slender branches divergent or ascending; pedicels filiform, much longer than the bracts, somewhat longer than the perianth-segments; flowers 4''–6'' broad, greenish; perianth-segments oblanceolate, glandless, short-clawed or sessile; capsule 5''–6'' long, the cavities 4–6-seeded; seeds 3''–4'' long.

Dry woods, mountains of Virginia to South Carolina. June–Aug.

Family 22. LILIÀCEAE Adans. Fam. Pl. 42. 1763.

LILY FAMILY.

Scapose or leafy-stemmed herbs from bulbs or corms, or rarely with rootstocks or a woody caudex (*Yucca*), the leaves various. Flowers solitary or clustered, regular, mostly perfect. Perianth parted into 6 distinct or nearly distinct segments, or these more or less united into a tube, inferior, or partly superior (*Aletris*). Stamens 6, hypogynous or borne on the perianth or at the bases of its segments; anthers 2-celled, mostly introrse, sometimes extrorse. Ovary 3-celled; ovules few or numerous in each cavity, anatropous or amphitropous; styles united; stigma 3-lobed or capitate. Fruit a loculicidal capsule (septicidal in *Calochortus*), or in *Yucca* sometimes fleshy and indehiscent. Embryo in copious endosperm.

About 125 genera and 1300 species, widely distributed.

* Plants bulbous, or with rootstocks, or fibrous-fleshy roots.

† Ovary superior, not adnate to the perianth.

Roots fibrous-fleshy; scape tall; flowers orange or yellow.	1. *Hemerocallis.*
Low fleshy herb with a short rootstock; flowers white.	2. *Leucocrinum.*
Plants with bulbs or corms.	
Flowers umbelled.	
Perianth 6-parted.	
Odor characteristically onion-like; ovules 1 or 2 in each cavity.	3. *Allium.*
Odor not onion-like; ovules several in each cavity.	4. *Nothoscordum.*
Perianth funnelform, the tube about as long as the lobes.	5. *Androstephium.*
Flowers solitary, racemed, corymbed or panicled.	
Anthers not introrse.	
Perianth-segments all alike or nearly so; capsule loculicidal.	
Anthers versatile; tall herbs.	6. *Lilium.*
Anthers not versatile; low herbs.	

Stem leafy.
    Flowers nodding; perianth-segments with a nectary at the base.   7. *Fritillaria.*
    Flowers erect; perianth-segments without a nectary.   8. *Tulipa.*
    Leaves only 2, appearing basal; flowers bractless.   9. *Erythronium.*
  Outer segments narrower than the inner; capsule septicidal.   10. *Calochortus.*
Anthers introrse.
  Perianth of 6 separate segments.
    Filaments filiform.   11. *Quamasia.*
    Filaments flattened.   12. *Ornithogalum.*
  Perianth globose, oblong or urn-shaped.   13. *Muscari.*
    †† Ovary half inferior; roots fibrous; flowers racemed.   14. *Aletris.*
** Stem a woody caudex; leaves rigid, mostly bearing marginal fibers.   15. *Yucca.*

1. HEMEROCÁLLIS L. Sp. Pl. 324. 1754.

    Tall glabrous herbs, with fibrous roots, basal linear leaves and large erect or spreading mostly orange or yellow flowers clustered at the ends of leafless scapes. Perianth funnelform, its lobes oblong or spatulate, much longer than the cylindric tube. Stamens 6, inserted at the summit of the perianth-tube, shorter than the lobes, declined; filaments filiform; anthers linear-oblong, the sacs introrsely dehiscent. Ovary oblong, 3-celled; ovules numerous in each cavity; style slender, declined, tipped with a small capitate stigma. Capsule oblong or ovoid, thick-walled, 3-angled, wrinkled, loculicidally 3-valved. [Greek, beautiful for a day.]

    About 5 species, natives of Europe and Asia. Type species: *H. Lilio-Asphódelus* L.

1. Hemerocallis fúlva L. Day Lily. Fig. 1240.

Hemerocallis fulva L. Sp. Pl. Ed. 2, 462. 1762.

    Scapes 3°–6° high, stout, mostly longer than the leaves. Leaves 4″–6″ wide, channeled, tapering to an acute tip; scape bearing several short bracts above; flowers 6–15, short-pedicelled, tawny orange, panicled, 4′–5′ long, opening for a day; tube of the perianth 1′–1½′ long, the lobes oblong, somewhat spreading, netted-veined; the three outer nearly flat, acutish; the 3 inner undulate and blunt.

    In meadows and by streams, New Brunswick to Virginia and Tennessee. Europe and Asia. Escaped from cultivation. Eve's-thread. Lemon-lily. June–Aug.

    Hemerocallis flàva L., the Yellow day-lily, with yellow flowers, their lobes parallel-veined, is occasionally found near old gardens, and on roadsides.

    Niobe coerùlea (Andr.) Nash, and **N. japónica** (Thunb.) Nash, Plantain-lilies, with drooping flowers and broad leaves, common in gardens, are occasionally established on roadsides.

2. LEUCÓCRINUM Nutt.; A. Gray, Ann. Lyc. N. Y. 4: 110. 1837.

    A low acaulescent rather fleshy herb, from a short rootstock, the roots thick, fibrous. Outer leaves membranous, acute, short; inner leaves linear, elongated, the innermost reduced to bracts. Flowers large, white, umbellate from the subterranean axils. Pedicels filiform. Perianth with a very narrow tube and a salverform limb, persistent, the 6 linear-oblong lobes spreading, nerved, shorter than the tube. Stamens borne near the top of the perianth-tube, shorter than the lobes; filaments filiform; anthers linear, their sacs introrsely dehiscent. Ovary ovoid, 3-celled; style filiform, stigma small. Capsule oval or obovoid, 3-angled, sessile, loculicidal. Seeds several in each cavity, angled. [Greek, white lily.]

    A monotypic genus of northwestern North America.

1. Leucocrinum montànum Nutt. Sand-lily.
Fig. 1241.

L. montanum Nutt.; A. Gray, Ann. Lyc. N. Y. 4: 110. 1837.

    Root-fibers very thick, numerous. Inner leaves 2′–10′ long, 1″–3″ wide; flowers 3–8; pedicels ½′–2′ long; perianth-limb about ½′ broad, the lobes acute; perianth-tube 1′–2′ long, less than 1″ in diameter; filaments 3″–4″ long; anthers coiled at least when dry; capsule 3″–4″ long, erect, leathery; seeds black.

    In sandy soil, South Dakota and Nebraska to Montana, Oregon, Colorado and California. April–June.

3. ÁLLIUM L. Sp. Pl. 294. 1753.

Bulbous herbs, odorous (alliaceous); bulbs solitary, or clustered on short rootstocks. Leaves narrowly linear, rarely lanceolate or oblong, sheathing, basal, or sometimes also on the stem. Stem (usually a scape) simple, erect. Flowers white, purple, pink or green, in a terminal simple umbel, subtended by 2 or 3 membranous separate or united bracts. Pedicels slender, not jointed. Perianth persistent; segments 6, separate, or united by their bases. Stamens on the bases of the perianth-segments; filaments sometimes toothed; anther-sacs introrsely dehiscent. Ovary nearly sessile, 3-celled; style filiform, jointed, deciduous; stigmas small; ovules 1-6 in each cavity. Capsule loculicidal. [Latin for garlic.]

About 300 species of wide distribution. Besides the following, some 50 others occur in the western United States. Type species: *Allium sativum* L.

Leaves oblong-lanceolate, absent at flowering time. [VALIDALLIUM Small.]   1. *A. tricoccum.*
Leaves linear, present at flowering time.
  Bulb-coats membranous, not fibrous-reticulated.
    Umbel capitate; pedicels shorter than the flowers.   2. *A. sibiricum.*
    Umbel loose; pedicels much longer than the flowers.
      Flowering umbel nodding.
        Perianth campanulate, white or pink, its outer segments acute.   3. *A. cernuum.*
        Perianth urn-shaped, purple, outer segments obtuse or notched.   4. *A. alleghaniense.*
      Flowering umbel erect.
        Leaves flat or channeled, all nearly basal.   5. *A. stellatum.*
        Leaves terete, hollow, some on stem; flowers often replaced by bulblets.
          Filaments with a tooth on each side.   6. *A. carinatum.*
          Filaments simple, not toothed.   7. *A. vineale.*
  Bulb-coats fibrous-reticulated.
    Capsule not crested.
      Flowers mostly replaced by bulblets; scape 8'-2° tall.   8. *A. canadense.*
      Flowers rarely replaced by bulblets.
        Scape 1°-2° tall; pedicels 8"-12" long; perianth-segments thin.   9. *A. mutabile.*
        Scape 4'-8' tall; pedicels 4"-6" long; perianth-segments rigid in fruit.   10. *A. Nuttallii.*
    Capsule-valves with 2 short crests.   11. *A. reticulatum.*

1. Allium tricóccum Ait. Wild Leek.
Fig. 1242.

Allium tricoccum Ait. Hort. Kew. 1: 428. 1789.

Bulbs ovoid, clustered, 1'-2' high, seated on a short rootstock, their coats fibrous-reticulated. Leaves oblong-lanceolate or elliptic, appearing early in the spring, but withering before flowering time, 6'-12' long, 1'-2' wide, narrowed at both ends, tapering into a long petiole; scape 4'-15' tall; bracts of the umbel usually 2, at first enclosing the flowers, acuminate, deciduous; umbel erect; pedicels becoming rigid, 6"-10" long; flowers white; perianth-segments oblong, obtuse, 2"-3" long; filaments lanceolate-subulate; ovule 1 in each cavity; capsule deeply 3-lobed, about 3" broad, 1½"-2" high, its valves not crested; seeds globose, black, smooth.

In rich woods, New Brunswick to Minnesota, North Carolina, Tennessee and Iowa. Often grows in large beds. Three-seeded leek. Ramps. June-July.

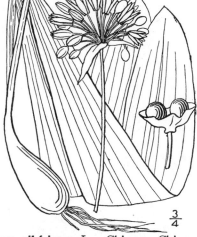

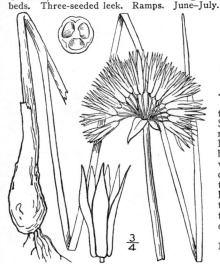

2. Allium sibíricum L. Chives. Chive- or Rush-garlic. Fig. 1243.

Allium sibiricum L. Mant. 2: 562. 1771.

Bulbs narrowly ovoid, clustered, 1' high or less, their membranous coats not fibrous-reticulated. Scape rather stout, 8'-2° high, bearing below the middle 1 or 2 elongated linear terete hollow leaves about ½" in diameter, or the leaves all basal; bracts of the umbel 2, broadly ovate, veiny; umbel many-flowered, capitate, the pedicels 1"-3" long; flowers rosé-color, longer than the pedicels; perianth-segments 4"-6" long, lanceolate, acuminate; stamens much shorter than the perianth; filaments subulate, half-terete; ovules 2 in each cavity of ovary; capsule obtusely 3-lobed, half as long as the perianth.

In moist soil, Newfoundland to Alaska, Maine, New York, Michigan, Wyoming and Washington.

32

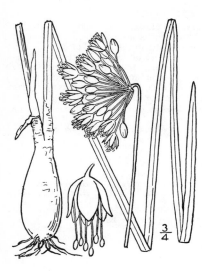

Europe and Asia. Civet. Shore-onion. June–July.
Closely related to *A. Schoenóprasum* L., to which it
was referred in our first edition.

3. Allium cérnuum Roth. Nodding Wild Onion. Fig. 1244.

A. cernuum Roth; Roem. Arch. 1: Part 3, 40. 1789.

Bulbs usually clustered on a short rootstock, narrowly ovoid, with a long neck, 1′–2½′ high, the
coats not fibrous-reticulated. Scape slender, slightly
ridged, 1°–2° high; leaves linear, channeled or nearly
flat, 1″–2″ wide, mostly shorter than the scape,
bluntish, umbel many-flowered, nodding in flower,
subtended by 2 short-deciduous bracts; pedicels filiform, 8″–15″ long; flowers campanulate, white or
rose; perianth-segments ovate-oblong, acute or obtusish, 2″–3″ long; stamens longer than the perianth;
filaments nearly filiform; ovules 2 in each cavity of
the ovary; capsule 3-lobed, rather shorter than the
perianth, each valve bearing 2 short processes near
the summit.

On banks and hillsides, New York to Minnesota and
British Columbia, West Virginia, Kentucky, South
Dakota, and in the Rocky Mountains to New Mexico.
July–Aug.

4. Allium alleghaniénse Small. Alleghany Onion. Fig. 1245.

Allium alleghaniense Small, Bull. N. Y. Bot. Gard.
1: 279. 1899.

Bulbs ovoid. Leaves few, with narrowly
linear blades 6′–12 long; scapes 1°–2° tall,
2-edged at least at maturity; umbel nodding,
12–40-flowered; pedicels 7″–13″ long, filiform,
becoming thicker in age; perianth purple,
mostly urn-shaped, 2″–2½″ high, the segments
oval, the outer obtuse or notched, shorter than
the inner; capsule 2″–2½″ high, with 2 large
processes on each valve.

Rocky soil and on cliffs, Virginia to Tennessee
and Georgia. July–Aug.

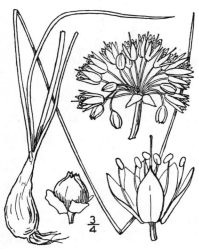

5. Allium stellàtum Ker. Prairie Wild Onion. Fig. 1246.

Allium stellatum Ker, Bot. Mag. *pl. 1576.* 1813.

Bulbs solitary or several together, narrowly ovoid,
1′–2′ long, their coats membranous. Scape slender,
8′–18′ tall, somewhat ridged above; leaves linear,
¾″–1¼″ wide, nearly flat; umbel several–many-
flowered, erect, subtended by 2 lanceolate or ovate
acuminate bracts; pedicels filiform, 6″–10″ long;
flowers rose-color; perianth-segments ovate-oblong,
acute, 2″–3″ long, equalling or rather shorter than
the stamens; filaments filiform, slightly widened at
the base; capsule shorter than the perianth, 3-lobed,
about 6-seeded, each valve bearing 2 erect processes
or crests below the apex.

On rocky banks, Illinois and Minnesota to Missouri,
Nebraska and Kansas. July–Aug.

6. Allium carinàtum L. Keeled Garlic.
Fig. 1247.

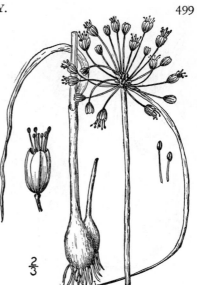

Allium carinatum L. Sp. Pl. 297. 1753.

Similar to *Allium vineale*. Bulb ovoid, its coats membranous. Stem terete, leafy up to about the middle, 8'–20' tall; leaves linear, channeled below, flat toward the apex, prominently 3–5-nerved; bracts of the umbel 2, narrowly linear, one much longer than the other; umbel erect, bearing either bulbs or capsules; pedicels filiform, 10″–20″ long; flowers about 3″ long, violet to rose; filaments not toothed.

Roadsides, New Jersey and southeastern Pennsylvania. Adventive from Europe.

$\frac{2}{3}$

7. Allium vineàle L. Wild Garlic. Field Garlic. Crow Garlic. Fig. 1248.

Allium vineale L. Sp. Pl. 299. 1753.

Bulb ovoid, 1' high or less, its coats membranous. Stem 1°–3° tall, bearing 2–4 narrowly linear terete hollow somewhat channeled leaves below the middle at flowering time, the early basal leaves similar, numerous, 4'–10' long; bracts of the umbel 2, lanceolate, acuminate, deciduous; umbel few–many-flowered, erect, the flowers often wholly or in part replaced by small ovoid bulblets which are tipped with a long capillary appendage; pedicels 3″–12″ long, filiform, the lower spreading or drooping; flowers green or purple, about 2″ long; perianth-segments ovate-lanceolate, stamens included or slightly exserted; filaments flattened, broad, the 3 interior ones bearing a tooth on each side just below the anther; capsule 3-lobed, shorter than the perianth.

In fields and meadows, Rhode Island to Virginia, Tennessee and Missouri. Naturalized from Europe. A troublesome weed in the Middle States, infesting pastures, and tainting the flavor of spring butter. June–July.

8. Allium canadénse L. Meadow Garlic.
Fig. 1249.

Allium canadense L. Sp. Pl. 1195. 1753.

Bulb ovoid, solitary, usually less than 1' high, the outer coats fibrous-reticulated. Scape terete, 8'–2° tall; leaves basal or nearly so, narrowly linear, flat or flattish above, slightly convex beneath, 1″–1½″ wide, usually shorter than the scape; bracts of the umbel 2 or 3, white, broadly ovate, acuminate; flowers usually or often replaced by ovoid bulblets; pedicels, when present, about ½' long; flowers pink or white, the perianth-segments oblong-lanceolate, acute, about as long as the stamens; filaments widened at the base, none of them toothed; capsule valves not crested.

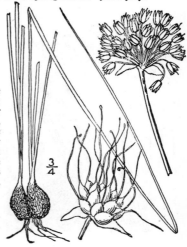

$\frac{3}{4}$

In moist meadows and thickets, New Brunswick to Minnesota, south to Florida, Louisiana, Texas and Colorado. Ascends to 2500 ft. in Virginia. Wild garlic. May–June.

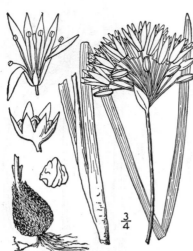

3/4

9. Allium mutábile Michx. Wild Onion.
Fig. 1250.

Allium mutabile Michx. Fl. Bor. Am. **1**: 195. 1803.

Bulbs ovoid, solitary or several together, 1' high or less, their coats prominently fibrous-reticulated. Scape terete, 1°–2° tall; leaves basal, channeled, 1″–2″ wide, shorter than the scape; bracts of the umbel 2 or 3, long-acuminate; umbel erect, many-flowered, rarely bulblet-bearing; pedicels filiform, 8″–12″ long; flowers pink, rose or white, 2½″–4″ long; perianth-segments lanceolate or ovate-lanceolate, acute, acuminate, or obtusish, thin, longer than the stamens; filaments somewhat widened below; capsule rather shorter than the perianth, its valves not crested.

In moist soil, North Carolina to Missouri, Nebraska, south to Florida and Texas. April–June.

10. Allium Nuttàllii S. Wats. Nuttall's Wild Onion. Fig. 1251.

Allium Nuttallii S. Wats. Proc. Am. Acad. **14**: 227. 1879.

Allium Helleri Small, Fl. SE. U. S. 264. 1903.

Bulbs usually solitary, ovoid, ½′–1′ high, their coats fibrous-reticulated. Culm slender, terete or nearly so, 4′–8′ tall; leaves basal, ½″–1″ wide, shorter than the scape or sometimes equalling it; bracts of the umbel 3 or 2, ovate or ovate-lanceolate, acute or acuminate; umbel several-flowered; no bulblets seen; pedicels slender, 4″–6″ long; flowers rose or white, about 3″ long; perianth-segments ovate, acute or acuminate, firm, becoming rigid in fruit; stamens shorter than the perianth; capsule shorter than the perianth, its valves not crested.

On prairies, South Dakota to Colorado, Texas and Arizona. April–June.

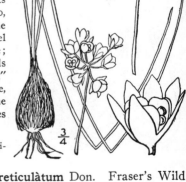

3/4

11. Allium reticulàtum Don. Fraser's Wild Onion. Fig. 1252.

Allium reticulatum Nutt. Fraser's Cat. Name only. 1813.

Allium reticulatum Don, Mem. Wern. Soc. **6**: 36. 1826–31.

Similar to the preceding species, the bulb rather larger, its coats prominently fibrous-reticulated. Scape 3′–10′ tall, slender; leaves usually less than 1″ wide; bracts of the several-flowered umbel mostly 2, acuminate; pedicels slender, 3″–6″ long; flowers white or pink, 2½″–3″ long; perianth-segments longer than the stamens, thin; capsule shorter than the perianth, each of its valves bearing 2 short crests just below the summit.

Saskatchewan to South Dakota, Montana, New Mexico and Arizona. May–July.

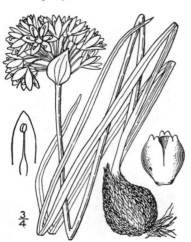

3/4

4. NOTHOSCÓRDUM Kunth, Enum. 4: 457. 1843.

Scapose herbs, similar to the onions, but without alliaceous odor, with membranous-coated bulbs, narrowly linear basal leaves and small yellow or yellowish-green flowers in an erect terminal simple 2-bracted umbel. Perianth 6-parted to the base, withering-persistent, its segments 1-nerved. Stamens 6, inserted on the bases of the perianth-segments; filaments filiform or subulate; anther-sacs introrsely dehiscent. Ovary sessile, 3-celled; ovules several in each cavity; style filiform, jointed near the base, but commonly persistent; stigma small, capitate. Capsule 3-lobed, loculicidal. Seeds angled or flattish, black. [Greek, signifying false garlic.]

About 10 species, the following in the United States, West Indies and Mexico, 8 or 9 in tropical and South America, 1 Chinese. Type species: *Nothoscordum pulchellum* Kunth.

1. Nothoscordum biválve (L.) Britton.
Yellow False Garlic. Fig. 1253.

Ornithogalum bivalve L. Sp. Pl. 306. 1753.
Allium ornithogaloides Walt. Fl. Car. 121. 1788.
Allium striatum Jacq. Coll. Suppl. 51. 1796.
Nothoscordum striatum Kunth, Enum. 4: 459. 1843.

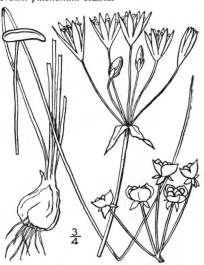

Bulb globose, less than 1′ in diameter, its coats membranous. Leaves ½″–2½″ wide, flat, blunt or acutish, shorter than the scape or equalling it; bracts of the umbel lanceolate, acuminate, membranous, persistent; umbel 6–12-flowered; pedicels filiform, usually unequal, becoming rather rigid and 1′–2′ long in fruit; flowers 5″–6″ long; perianth-segments thin, oblong-lanceolate, acute, longer than the stamens; capsule obovoid or somewhat depressed, obtusely 3-lobed, 2″–3″ high, the style as long or slightly longer.

In sandy soil, Virginia to Ohio, Tennessee, Nebraska, Florida, Texas and Mexico. Ascends to 1500 ft. in Georgia. Also in Bermuda and in Jamaica. March–July.

5. ANDROSTÈPHIUM Torr. Bot. Mex. Bound. Surv. 218. 1859.

Scapose herbs from a small membranous-coated corm. Leaves basal, narrowly linear. Flowers rather large, blue, in a terminal erect several-bracted umbel. Perianth funnelform, withering-persistent, the tube about as long as the 6 oblong lobes. Stamens 6, inserted on the throat of the perianth; filaments dilated, united to the middle or above into an erect crown-like tube with toothed lobes alternating with the linear-oblong anthers. Ovary sessile, 3-celled; ovules several in each cavity; style filiform; stigmas 3-grooved. Capsule membranous, 3-angled, loculicidal. Seeds few, large, oval, black. [Greek, referring to the crown.]

Two species, natives of the southwestern United States; the following is the type of the genus.

1. Androstephium coerùleum (Scheele) Greene.
Androstephium. Fig. 1254.

Milla coerulea Scheele, Linnaea 25: 260. 1852.
Androstephium violaceum Torr. Bot. Mex. Bound. Surv. 219. 1859.
Androstephium coeruleum Greene, Pittonia 2: 57. 1890.

Corm subglobose, less than 1′ in diameter. Scape 2′–8′ tall, simple; leaves 1″–2″ wide, half terete, equalling the scape, or sometimes longer, bracts of the umbel 2–4, scarious, lanceolate, acuminate, persistent, shorter than the pedicels; umbel 2–7-flowered; pedicels rather stout, ¾′–1½′ long; perianth 10″–14″ long, the lobes about as long as the tube; filament-tube about 5″ long, its lobes exceeding the anthers; style about as long as the filament-tube; capsule 4″–6″ high; seeds nearly 3″ long, very thin, narrowly winged.

Prairies, Kansas to Texas. March–April.

6. LÍLIUM L. Sp. Pl. 302. 1753.

Tall bulbous herbs, with simple leafy stems, and large erect or drooping showy flowers. Perianth funnelform or campanulate, deciduous, of 6 separate spreading or recurved segments, each with a nectar-bearing groove at its base within. Stamens 6, mostly shorter than the perianth, hypogynous, slightly attached to the segments; filaments filiform or subulate; anthers linear, versatile, their sacs longitudinally dehiscent. Ovary 3-celled; ovules numerous; style long, somewhat club-shaped above; stigma 3-lobed. Capsule oblong or obovoid, loculicidally dehiscent. Seeds numerous, flat, horizontal, packed in 2 rows in each cavity. [Latin, from the Greek name of the Lily, said to be from the Celtic *li*, white.]

About 45 species, natives of the north temperate zone. Besides the following, some 8 others occur in western North America. Type species: *Lilium cándidum* L.

Flower or flowers erect; perianth-segments narrowed into long claws.
　Perianth-segments merely acute.
　　Leaves lanceolate, nearly all verticillate. .... 1. *L. philadelphicum.*
　　Leaves narrowly linear, nearly all alternate. .... 2. *L. umbellatum.*
　Perianth-segments long-acuminate; leaves all alternate, appressed. .... 3. *L. Catesbaei.*
Flowers drooping or spreading; perianth-segments not clawed.
　Leaves or most of them verticillate, their axils not bulbiferous; native species.
　　Leaves finely roughened on the veins beneath.
　　　Perianth-segments recurved or spreading. .... 4. *L. canadense.*
　　　Perianth-segments not recurved; mountain species. .... 5. *L. Grayi.*
　　Leaves perfectly smooth; perianth-segments recurved.
　　　Leaves lanceolate; stem 3°–10° tall; flowers 1–40. .... 6. *L. superbum.*
　　　Leaves oblanceolate; stem 2°–3° tall; flowers 1–3. .... 7. *L. carolinianum.*
　Leaves alternate, crowded; upper axils bulb-bearing; escaped from gardens. .... 8. *L. tigrinum.*

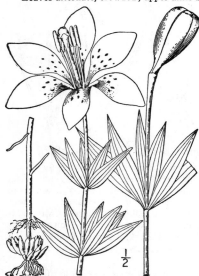

1. Lilium philadélphicum L. Red Lily. Wood Lily. Philadelphia Lily. Fig. 1255.

Lilium philadelphicum L. Sp. Pl. Ed. 2, 435. 1762.

Bulb 1′ in diameter or less, composed of narrow jointed fleshy scales. Stem 1°–3° tall, with a few distant scales below, leafy above; leaves lanceolate, acute at both ends or the lower sometimes obtuse, 1′–4′ long, 3″–7″ wide, all verticillate in 3's–8's, or a few of them alternate, thin, the margins finely roughened; flowers 1–5, erect, 2½′–4′ high; perianth reddish orange, its segments spatulate, somewhat spreading, acute or obtusish, the blade ½′–1′ wide, rather gradually narrowed into the claw, purple spotted below; capsule obovoid-oval, 1¼′–2′ high; seeds 3″–4″ long, narrowly winged.

In dry woods and thickets, Maine to Ontario, south to North Carolina and West Virginia. Ascends to 4000 ft. in Virginia. Glade-, flame- or huckleberry-lily. Wild orange- or tiger-lily. June–July.

2. Lilium umbellàtum Pursh. Western Red Lily. Fig. 1256.

L. andinum Nutt. Fras. Cat. Without description. 1813.
Lilium umbellatum Pursh, Fl. Am. Sept. 299. 1814.
L. lanceolatum Fitzpatrick, Iowa Nat. 2: 30. 1907.

Bulb similar to that of the preceding species, the stem usually more slender, 1°–2° tall. Leaves linear, blunt or the upper acute, ascending, or sometimes appressed, 1′–3′ long, 1″–2½″ wide, all alternate or the uppermost verticillate, their margins finely roughened; flowers 1–3, erect, 2′–3′ high; perianth segments red, orange or yellow, narrowed into the claw, acute, spotted below, the claw shorter than the blade; capsule oblong, 3′–4′ long, about 8″ thick; seeds like those of *L. Philadelphicum,* of which species it may be a narrow-leaved race.

In dry soil, Ontario to Ohio, Minnesota, British Columbia, Missouri, Arkansas and Colorado. Ascends to 4000 ft. in the Black Hills. June–July.

3. Lilium Catesbaèi Walt.   Southern Red Lily.
Fig. 1257.

Lilium Catesbaei Walt. Fl. Car. 123. 1788.

Bulb ½′–1′ high, composed of narrow leaf-bearing scales, their leaves narrowly linear, 2′–4′ long, often falling away before the plant flowers. Stem slender, 1°–2° high; stem leaves all alternate, narrowly linear or linear-lanceolate, acute or acuminate, erect or appressed, 1′–3′ long, 1″–3″ wide; flower (always?) solitary, erect; perianth-segments scarlet with a yellow purple-spotted base and a slender claw, spreading or somewhat recurved, 3′–5′ long, ½′–1′ wide, long-acuminate, wavy-margined; capsule 1′ high or less; seeds 2″–3″ long.

In moist pine barrens, North Carolina to Florida and Alabama. Reported from Kentucky, southern Illinois and Missouri. July–Aug.

4. Lilium canadénse L.   Wild Yellow Lily.
Canada or Nodding Lily. Fig. 1258.

Lilium canadense L. Sp. Pl. 303. 1753.

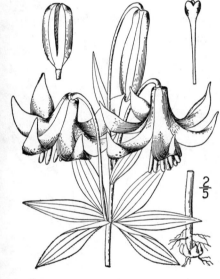

Bulbs subglobose, 1′–2′ in diameter, borne on a stout rootstock, composed of numerous thick white scales. Stem 2°–5° tall, slender or stout; leaves lanceolate or oblong-lanceolate, verticillate in 4′s–10′s or some of them alternate, acuminate, 2′–6′ long, 3″–15″ wide, finely roughened on the margins and on the veins beneath; flowers 1–16, nodding on long peduncles; peduncles sometimes bearing a small leaf-like bract; perianth-segments 2′–3′ long, yellow or red, usually thickly spotted below, recurved or spreading, not clawed; capsule oblong, erect, 1½′–2′ long.

In swamps, meadows and fields, Nova Scotia to Ontario, Minnesota, Georgia, Alabama, Missouri and Nebraska. Ascends to 6000 ft. in North Carolina. Red-flowered races with slightly spreading perianth-segments resemble the following species, and races with strongly recurved segments, *L. superbum.* Field- or meadow-lily. June–July.

5. Lilium Gràyi S. Wats.   Asa Gray's Lily.
Fig. 1259.

Lilium Grayi S. Wats. Proc. Am. Acad. 14: 256. 1879.

Rootstock bearing small subglobose bulbs with thick ovate scales. Stem slender, 2°–3° high; leaves oblong-lanceolate, acute or acuminate at the apex, narrowed at the base, 2′–4′ long, ½′–1′ wide, verticillate in 3′s–8′s or the lowest commonly smaller and scattered, all finely roughened on the veins beneath; flowers 1–3, long-peduncled, spreading or slightly drooping, 2′–3′ long, red or tinged with yellow at the base; perianth-segments oblong-spatulate, not clawed, acute, spotted; capsule fig-shaped, about 1½′ high.

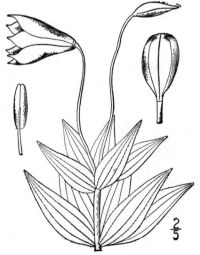

Peaks of Otter, Virginia, and on the higher mountain summits in North Carolina. July–Aug.

6. Lilium supérbum L. Turk's-cap Lily.
Fig. 1260.

Lilium superbum L. Sp. Pl. Ed. 2, 434. 1762.

Bulbs globose, 1'–2' in diameter, borne on short rootstocks, their scales white, thick, ovate. Stem stout or slender, 3°–8° high; leaves lanceolate or linear-lanceolate, smooth on both sides, acuminate at both ends, 2'–6' long, ¼'–1½' wide, verticillate in 3's–8's or the upper alternate, the veinlets not prominently anastomosing; flowers orange, orange-yellow or rarely red, 3–40, or rarely solitary, nodding, long-peduncled, forming, when numerous, a large panicle; perianth-segments 2½'– 4' long, lanceolate, acuminate, purple-spotted, at length usually strongly recurving from below the middle; capsule obovoid, 1½'–2' high.

In meadows and marshes, New Brunswick to Ontario, Minnesota, North Carolina, Tennessee and Missouri. Ascends to 5000 ft. in Virginia. Turk's-head-, nodding- or wild tiger-lily. July–Aug.

7. Lilium caroliniànum Michx. Carolina Lily.
Fig. 1261.

Lilium carolineanum Michx. Fl. Bor. Am. **1**: 197. 1803.
Lilium superbum var. *carolineanum* Chapm. Fl. S. States, 484. 1860.

Bulbs borne on short rootstocks, globose, 1'–2' in diameter, composed of numerous fleshy scales. Stem 2°–3° high, slender; leaves oblanceolate or obovate, smooth, verticillate or the upper and lower alternate, acute, obtuse or short-acuminate at the apex, narrowed at the base, the veinlets prominently anastomosing; flowers 1–3, orange-red, 3'–4' long, long-peduncled, nodding; perianth-segments lanceolate, acuminate, purple-spotted below, strongly arched backward with their tips sometimes connivent.

In dry woods, Virginia to Florida and Louisiana. Ascends to 3500 ft. in Virginia. Aug.

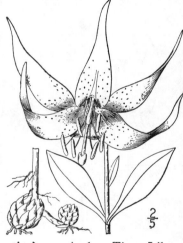

8. Lilium tigrìnum Andr. Tiger Lily.
Fig. 1262.

Lilium tigrinum Andr. Bot. Rep. **9**: errata. 1809.

Bulb solitary, globose, about 1½' in diameter, composed of numerous oblong-lanceolate, appressed scales. Stem stout, purple or nearly black, white-pubescent above, 2°–5° tall, leafy nearly to the base; leaves lanceolate, all alternate, glabrous or slightly pubescent, 4'–6' long, 5''–10'' wide, the upper bearing blackish bulblets, of 3 or 4 scales, in their axils, which sometimes emit roots while attached to the plant; flowers 5–25, orange-red, nodding, 3'–4½' long; perianth-segments lanceolate, papillose, recurved, purple-spotted.

Escaped from gardens, Maine, Massachusetts and eastern Pennsylvania. Native of China and Japan. Summer.

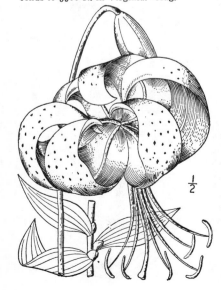

7. FRITILLÀRIA [Tourn.] L. Sp. Pl. 303. 1753.

Bulbous herbs with simple leafy stems, and rather large nodding solitary or racemed leafy-bracted flowers. Perianth mostly campanulate, deciduous, of 6 separate and nearly equal oblong or ovate segments, each with a nectar-pit or spot at the base. Stamens 6, hypogynous; filaments filiform or somewhat flattened; anthers linear or oblong. Ovary nearly or quite sessile, 3-celled; ovules numerous in each cavity; style slender or filiform, 3-lobed or 3-cleft, the lobes stigmatic along the inner side. Capsule obovoid or globose, 6-angled, loculicidally dehiscent. Seeds numerous, flat, obovate or suborbicular, margined or winged. [Latin, from *fritillus*, a dice-box or chess-board, in allusion to the form or to the checkered markings of the perianth in some species.]

About 50 species, natives of the north temperate zone. Besides the following, about 12 others occur in western North America. Type species: *Fritillaria pyrenàica* L.

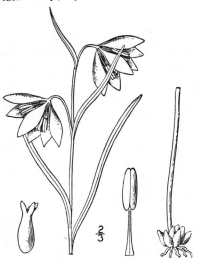

1. Fritillaria atropurpùrea Nutt. Purple Fritillaria. Fig. 1263.

F. atropurpurea Nutt. Journ. Acad. Phila. 7: 54. 1834.

Bulb ½' in diameter or less. Stem 6'–15' high, slender, leafless below; leaves linear, alternate, sessile, 1½'–3½' long, 1½''–2'' wide or less; flowers 1–6, purple or purplish green and mottled; perianth-segments narrowly oblong, obtusish, 6''–10'' long; peduncles ½'–1' long; stamens one-half to two-thirds as long as the perianth; style 3-cleft to about the middle, the lobes linear; capsule erect, acutely angled, 5''–6'' high.

North Dakota to Nebraska, Montana and California. June–July.

8. TÙLIPA (Tourn.) L. Sp. Pl. 305. 1753.

Bulbous herbs with erect leaf-bearing stems and large solitary (rarely 2) erect flowers. Perianth campanulate, the segments distinct, erect or erect-spreading, deciduous, usually with a spot at the base, but without a nectar-gland; stamens 6, hypogynous, shorter than the perianth; anthers erect, basifixed; ovary nearly or quite sessile, 3-celled; ovules numerous; capsule oblong or globose; seeds numerous, flat. [Ancient name.]

Fifty species or more, natives of Europe and Asia. Type species: *Tulipa sylvestris* L.

1. Tulipa sylvéstris L. Wild Tulip. Fig. 1264.

Tulipa sylvestris L. Sp. Pl. 305. 1753.

Bulb ovoid, 1' long or less, covered with dark brown scales. Stem about 2° high, bearing 1–3 linear-lanceolate grayish-green acuminate leaves; flower yellow, 3'–4' broad; inner perianth-segments somewhat broader than the outer, acute; filaments pubescent at the base; capsule narrowly oblong, 3-angled; seeds obovoid, brown.

Meadows, Bucks County, Pennsylvania. Adventive from Europe. April–June.

9. ERYTHRÒNIUM L. Sp. Pl. 305. 1753.

Low herbs, from deep membranous-coated corms, sometimes propagated by offshoots, the stem simple, bearing a pair of broad or narrow unequal leaves, usually below the middle, the leaves thus appearing basal. Flowers large, nodding, bractless, solitary, or several in some western species. Many plants are flowerless and 1-leaved, these leaves often wider and longer petioled than those of the stem. Perianth-segments separate, lanceolate, oblong or oblanceolate, deciduous, with nectariferous groove, and sometimes 2 short processes at the base. Stamens 6, hypogynous, shorter than the perianth; anthers linear oblong, not versatile. Ovary sessile, 3-celled; ovules numerous or several in each cavity; style filiform 'or thickened above, 3-lobed or 3-cleft. Capsule obovoid or oblong, somewhat 3-angled, loculicidal. Seeds compressed, or somewhat angled and swollen. [Greek, in allusion to the red flowers of some species.]

About 12 species, all but one North American. The species are erroneously called *Dog's-tooth Violet*. Type species: *Erythronium Dens-canis* L.

Stem with no offshoot; flowers 10″–2′ long.
  Offshoots produced at the base of the corm; perianth-segments recurved.
    Flowers yellow; stigmas very short.   1. E. americanum.
    Flowers white, blue or purple; stigmas 1″–1½″ long, recurved.   2. E. albidum.
  No offshoots, propagating by basal corms; perianth-segments not recurved. 3. E. mesachoreum.
Stem with fleshy offshoot below the leaves; flowers rose, about ½′ long.   4. E. propullans.

braska and Arkansas. Ascends to 5500 ft. in Virginia. Yellow- or Trout-lily. Trout-flower. Yellow-bells. Yellow snowdrop. Rattlesnake- or Dog's-tooth violet. Lamb's- or Deer's-tongue. Scrofula-root. Snake-root. March–May.

2. Erythronium álbidum Nutt. White Adder's-tongue. Fig. 1266.

Erythronium albidum Nutt. Gen. 1: 223. 1818.

Similar to the preceding species, the plant propagating by offshoots from the base of the corm, the leaves mottled, or green all over, sometimes rather narrower. Flower white, blue or purple; perianth-segments oblong, recurved, none of them auricled at the base; style somewhat thickened upward; stigmas linear, finally recurving, 1″–1½″ long; capsule obovoid or oblong, 5″–9″ high.

In moist woods and thickets, Ontario to Minnesota, south to Georgia, Tennessee and Texas. Not common eastward. Spring-lily. Deer's-tongue. White Dog's-tooth violet. March–May.

1. Erythronium americànum Ker. Yellow Adder's-tongue. Fig. 1265.

Erythronium americanum Ker, Bot. Mag. *pl. 1113.* 1 Je. 1808.
Erythronium angustatum Raf. Med. Rep. (II.) 5: 354. 20 Jl. 1808.
Erythronium bracteatum Bigel.; Beck, Bot. N. & Mid. States 365. 1833.

Corm ovoid, 6″–10″ high, producing offshoots from its base. Stem ½°–1° long; leaves oblong or oblong-lanceolate, 3′–8′ long, ½′–2′ wide, acute or short-acuminate at the apex, flat, usually mottled with brown, but sometimes green all over, narrowed into clasping petioles; peduncle about as long as the leaves, rarely bearing a bract; flower yellow, or rarely purplish tinged; perianth-segments oblong, 10″–2′ long, 3″–4″ wide, recurved, dotted within, the 3 inner auricled at the base; style club-shaped, with 3 very short stigmatic ridges; capsule obovoid, contracted into a short stipe, 6″–10″ high; seeds curved, rounded on the back, about 1½″ long, pointed at both ends.

In moist woods and thickets, Nova Scotia to Ontario and Minnesota, south to Florida, Ne-

3. Erythronium mesachòreum Knerr.
Midland Adder's-tongue. Fig. 1267.

Erythronium mesachoreum Knerr, Midland Col-
lege Monthly **2** : 5. 1891.

Corm ovoid, 10" high or less, not developing
offshoots, the new corms formed at or within
the base of the old one. Leaves narrowly
oblong or linear-oblong, not mottled, 4'-10'
long, ¼'-1' wide, somewhat folded; flower
lavender tinted, 1'-2' long; perianth-segments
not recurved, sometimes a little spreading;
style slender; stigmas recurved; capsule obovoid, larger than that of *E. albidum,* ½'-1½'
high.

On prairies, Iowa to Missouri, Nebraska and
Kansas. Blooms before *E. albidum* when the two
grow in proximity. The flowering plants are said
to appear before the 1-leaved flowerless ones.

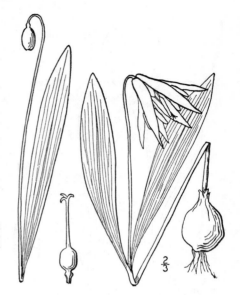

4. Erythronium propúllans A. Gray.
Minnesota Adder's-tongue. Fig. 1268.

Erythronium propullans A. Gray, Am. Nat. 298.
pl. 74. 1871.

Corm ovoid, 10" high or less, not developing offshoots. Stem ascending, 6'-8' long,
bearing a fleshy curved offshoot 1'-2' long
from a slit near the base of the petiole-sheath;
leaves oblong, acute, 2'-4' long, slightly mottled
or green; flower rose or pink, about ½' long,
borne on a filiform peduncle shorter than the
leaves, perianth-segments with a yellow base,
apparently not recurved, none of them auricled;
stigmas mere ridges.

In rich woods, Minnesota. May.

10. CALOCHÓRTUS Pursh, Fl. Am. Sept. 240. 1814.

Branched or simple herbs, with coated corms, narrowly linear leaves and large showy
peduncled flowers, erect in the following species. Perianth-segments separate, spreading or
connivent, yellow, blue, purple, white or variegated; the 3 outer sepal-like, narrow; the 3
inner petaloid, gland-bearing, and barbed or spotted within, sometimes with a nectar-pit
near the base. Stamens 6, hypogynous; filaments short, subulate; anthers erect, linear or
oblong. Ovary 3-celled; ovules numerous; style very short or none; stigmas 3, recurved.
Capsule oblong or linear, 3-angled, mostly septicidal, the valves sometimes 2-cleft. Seeds
flat. [Greek, signifying beautiful herb.]

About 35 species, of western North America and Mexico. Type species : *C. elegans* Pursh.

Anthers obtuse; gland of inner perianth-segments orbicular or oval.   1. *C. Nuttallii.*
Anthers acute; gland transverse, curved or reniform.   2. *C. Gunnisoni.*

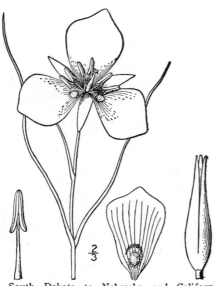

$\frac{2}{3}$

1. Calochortus Nuttàllii T. & G. Nuttall's Mariposa Lily. Fig. 1269.

Fritillaria alba Nutt. Gen. 1: 222. 1818?
Calochortus Nuttallii T. & G. Pac. R. R. Rep. 2: 124. 1855.

Corm ovoid-oblong, 6″–10″ high. Stem slender, few-leaved, branched or sometimes simple, 3′–15′ tall; leaves 1′–3′ long, 1″–2½″ wide, the lowest commonly bearing a bulb in its axil; peduncles 2′–6′ long; outer perianth-segments lanceolate or ovate-lanceolate, green with lighter margins, acute or acuminate, shorter than the inner, sometimes with a dark or hairy spot within; inner perianth-segments broadly obovate-cuneate, 1′–1½′ long, 10″–12″ wide, white, lilac or yellowish, with a yellow base and a purple or purplish spot, the gland orbicular or oval and more or less pubescent; filaments 3″–4″ long, about equalling the oblong obtuse sagittate anthers; capsule about 1½′ long, 3″–4″ thick, acuminate, the valves obliquely cross-lined.

South Dakota to Nebraska and California. June–July.

2. Calochortus Gunnisòni S. Wats. Gunnison's Mariposa Lily. Butterfly-lily. Fig. 1270.

Calochortus Gunnisoni S. Wats. Bot. King's Exp. 348. 1871.

Stem slender, often simple, 6′–15′ high. Leaves usually less than 1′ wide, involute, at least when dry, none of the axils bulb-bearing in any specimen seen; peduncles 1′–4′ long; outer perianth-segments lanceolate or oblong-lanceolate, scarious-margined, acuminate; inner perianth-segments similar to those of the preceding species, lilac, yellowish below the middle, purple-lined and banded, the gland transverse, oblong, curved or reniform, pubescent; anthers acute; capsule narrowly oblong, narrowed at both ends, about 1¼′ long.

South Dakota and Nebraska to Arizona and New Mexico. June–July.

$\frac{2}{3}$

11. QUAMÀSIA Raf. Am. Month. Mag. 2: 265. 1818.

[CAMÁSSIA Lindl. Bot. Reg. *pl. 1846.* 1832.]

Scapose herbs, with membranous-coated edible bulbs, linear basal leaves, and rather large, blue, purple or white bracted flowers in a terminal raceme. Perianth of 6 separate equal spreading persistent 3–7-nerved segments. Pedicels jointed at the base of the flower. Stamens inserted at the bases of the perianth-segments; filaments filiform; anthers oblong or linear-oblong, versatile, introrse. Ovary 3-celled, sessile; ovules numerous in each cavity; style filiform, its base persistent; stigma 3-lobed. Capsule oval, 3-angled, loculicidal. Seeds black, shining. [From quamash, the Indian name.]

About 4 species, natives of North America. Type species: *Quamasia esculenta* Raf.

1. Quamasia hyacínthina (Raf.) Britton. Wild Hyacinth. Fig. 1271.

Scilla esculenta Ker, Bot. Mag. *pl. 1754.* 1813.
Lemotrys hyacinthina Raf. Fl. Tell. **3**: 51. 1836.
Scilla Fraseri A. Gray, Man. Ed. *2*, 469. 1856.
Camassia Fraseri Torr. Pac. R. R. Rep. **4**: 147. 1857.
Quamasia esculenta Coville, Proc. Biol. Soc. Wash. **11**: 64. 1897. Not Raf.

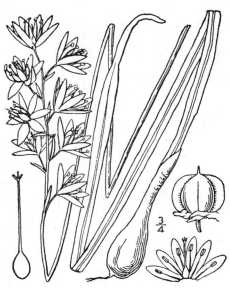

Bulb ovoid, $1'-1\frac{1}{2}'$ long, its outer coat usually nearly black. Scape slender, $1°-2°$ tall, sometimes bearing 1 or 2 short linear scarious leaves; basal leaves narrowly linear, acuminate, shorter than the scape, $1\frac{1}{2}''-4''$ wide; raceme open, $3'-8'$ long in flower, longer in fruit; flowers several or many; pedicels filiform, $6''-10''$ long, about as long as the bracts and the perianth-segments; bracts long-acuminate; perianth-segments narrowly oblong, 3-5-nerved, blue or nearly white, longer than the stamens; capsule about $4''$ high, $5''-6''$ thick, the valves transversely veined.

In meadows and along streams, Pennsylvania to Minnesota, Georgia and Texas. Ascends to 2100 ft. in Virginia. Eastern camass. April–May.

12. ORNITHÓGALUM L. Sp. Pl. 306. 1753.

Scapose herbs, with coated bulbs, narrow basal fleshy leaves, and large white or yellow flowers in a terminal bracted corymb or raceme. Perianth-segments equal or nearly so, separate, white, or sometimes green without, persistent, faintly several-nerved. Stamens hypogynous; filaments flattened, often broad; anthers versatile, introrse. Ovary 3-celled, sessile; ovules several or numerous in each cavity; style short or columnar, 3-sided; stigma capitate, 3-lobed or 3-ridged. Capsule subglobose, 3-sided or 3-lobed, loculicidal. Seeds black. [Greek, signifying bird's milk, said to be in allusion to the egg-white color of the flowers in some species.]

About 75 species, natives of Europe, Asia and Africa. Type species: *O. arabicum* L.

Flowers corymbose, erect; pedicels long, slender. 1. *O. umbellatum.*
Flowers racemose, drooping; pedicels very short, stout. 2. *O. nutans.*

1. Ornithogalum umbellàtum L. Star-of-Bethlehem. Summer Snow-flake. Star-flower. Fig. 1272.

Ornithogalum umbellatum L. Sp. Pl. 307. 1753.

Tufted, bulbs ovoid, $\frac{1}{2}'-1\frac{1}{2}'$ long, the coats membranous. Scape slender, $4'-12'$ high; leaves narrowly linear, $1''-2\frac{1}{2}''$ wide, dark green with a light midvein, blunt, equalling or longer than the scapes; flowers corymbose, opening in sunshine; bracts membranous, linear-lanceolate, mostly shorter than the pedicels; pedicels erect or ascending, the lower $1'-3'$ long; perianth-segments oblong-lanceolate, acute, white above, green with white margins beneath, $6''-10''$ long, about twice as long as the stamens; filaments somewhat flattened, not toothed.

In fields and meadows, New Hampshire to Pennsylvania and Virginia. Locally very abundant. Naturalized from Europe. Nap-at-noon. Sleepy Dick. Ten o'clock-, Eleven o'clock-lady. May–June.

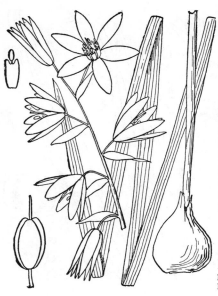

2. Ornithogalum nùtans L. Drooping Star-of-Bethlehem. Fig. 1273.

Ornithogalum nutans L. Sp. Pl. 308. 1753.

Bulb ovoid, 1'–2' long. Scape stout, 1°–2° high; leaves usually equalling the scape or longer, blunt, 2"–4" wide; flowers several or numerous, racemose, nodding; raceme 3'–8' long, loose; pedicels stout, 2"–6" long; bracts lanceolate, long-acuminate, much longer than the pedicels, often as long as the flowers; perianth-segments thin, oblong-lanceolate, about 1' long and 4" wide, nearly twice as long as the stamens; filaments broad, flat, 2-toothed at the apex.

Escaped from gardens in eastern and southern Pennsylvania, and· in the District of Columbia. Native of Europe. April–May. The bulbs of this and other species have for centuries past been a portion of the food of Italy, the Levant and other parts of the Old World.

13. MUSCÀRI Mill. Gard. Dict. Abr. ed. 4. 1754.

Low bulbous scapose herbs, with basal linear fleshy leaves, and nodding bracted racemose flowers, deep blue (rarely white) in the following species. Bulbs membranous-coated. Perianth globose, urn-shaped, or oblong, with 6 teeth or short lobes, tardily deciduous. Stamens 6, inserted on the perianth-tube, included; anthers ovate, versatile, introrse. Ovary 3-celled, sessile; ovules 2 in each cavity; style short; stigmas 3-lobed. Capsule 3-sided or 3-winged, usually 6-seeded, loculicidal. Seeds black, angled. [From the musk-like odor of the flowers of some species.]

About 40 species, natives of Europe, Asia and Africa. Type species: *Hyacinthus botryoides* L.

Perianth globose, 1"–1½" in diameter; leaves erect.   1. *M. botryoides.*
Perianth oblong, urn-shaped, 2"–3" long; leaves recurved.   2. *M. racemosum.*

1. Muscari botryoìdes (L.) Mill. Grape-Hyacinth. Fig. 1274.

Hyacinthus botryoides L. Sp. Pl. 318. 1753.
Muscari botryoides Mill. Gard. Dict. Ed. 8, no. 1. 1768.

Bulb 1' high or less. Scape 4'–10' high; leaves about as long as the scape, erect or nearly so, 1"–4" wide, channeled, blunt or acutish; raceme oblong-cylindric, 1'–1½' long, dense, or becoming longer and looser in fruit; pedicels shorter than the faintly odorous flowers; bracts very short; perianth globose, 1"–1½" in diameter, 6-toothed, the teeth white, recurved; valves of the capsule obovate.

In meadows and thickets and along roadsides, escaped from gardens, New Hampshire to Ohio and Virginia. Naturalized or adventive from southern Europe. Native also of Asia. Grape-flower. Baby's-breath. Blue-bells. Blue-bottle. April–June.

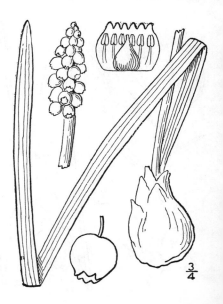

$\frac{3}{4}$

2. Muscari racemòsum (L.) Mill.   Starch Grape-Hyacinth.   Fig. 1275.

Hyacinthus racemosus L. Sp. Pl. 318: 1753.

Muscari racemosum Mill. Gard. Dict. Ed. 8, no. 2. 1768.

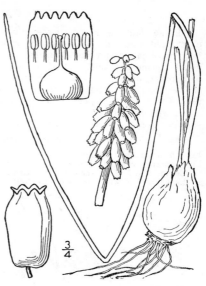

Similar to the preceding species. Leaves 1″–2″ wide, recurved or spreading, channeled above; raceme oblong or ovoid, many-flowered, dense, 1′–2½′ long; pedicels shorter than the starchy-scented flowers or sometimes equalling them, slender, much longer than the bracts; perianth oblong, urn-shaped, constricted at the throat, 2″–3″ long, with 6 deltoid recurved white teeth; capsule-valves suborbicular, retuse.

Escaped from gardens, Connecticut and southern New York to Pennsylvania and Virginia. Native of southern Europe. Grape-flower. Pearls-of-Spain. Starch-hyacinth. April–May.

14. ÁLETRIS L. Sp. Pl. 319.   1753.

Scapose perennial bitter fibrous-rooted herbs, with basal spreading lanceolate leaves, and small white or yellow bracted perfect flowers in a terminal spike-like raceme. Perianth oblong or campanulate, roughened without, 6-lobed, its lower part adnate to the ovary. Stamens 6, inserted on the perianth at the bases of the lobes, included; filaments short; anthers introrse. Ovary 3-celled; ovules numerous, anatropous; style subulate, or short, 3-cleft above; stigmas minutely 2-lobed. Capsule ovoid, enclosed by the persistent perianth, 3-celled, many-seeded, loculicidal. Seeds oblong, ribbed. Embryo small. Endosperm fleshy. [Greek, signifying to grind corn, apparently in allusion to the rough, mealy flowers.]

About 8 species, natives of eastern North America and eastern Asia. Type species: *A. farinosa* L.

A genus of uncertain affinity, which has been placed by authors in Haemodoraceae and in Amaryllidaceae.

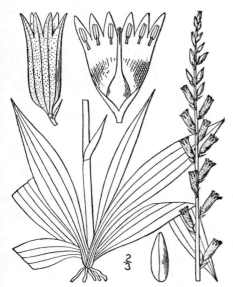

1. Aletris farinòsa L.   Star-grass.   Ague or Colic-root.   Fig. 1276.

Aletris farinosa L. Sp. Pl. 319.   1753.

Roots numerous, tough, scape 1½°–3° tall, slender, terete, striate, bearing several or numerous small distant bract-like leaves. Basal leaves several, lanceolate or linear-lanceolate, acuminate at the apex, narrowed to the base, spreading, pale yellowish green, 2′–6′ long, 3″–10″ wide; raceme 4′–12′ long in flower, or longer in fruit, dense, erect, pedicels 1″ long or less; bracts subulate, longer than the pedicels, sometimes 2 to each flower; perianth tubular-oblong, white, or the short lobes yellowish, 3″–4″ long, about 1½″ thick; style subulate; capsule ovoid, about 2″ long, loculicidal above, each of its 3 valves tipped with a subulate portion of the style.

In dry, mostly sandy soil, Maine to Ontario and Minnesota, south to Florida and Arkansas. Ascends to 3500 ft. in Virginia. Ague-grass, Blazing-star, Bitter-grass, Bitter-plant, Crow-corn. Mealy-starwort. Aloe-, Star- or Husk-root. Unicorn-root or -horn. May–July.

Aletris àurea Walt., admitted into our first edition, is not certainly known to grow north of South Carolina. It has been mistaken in New Jersey for yellowish-flowered races of *A. farinosa*.

15. YÚCCA L. Sp. Pl. 319. 1753.

Large plants, with a short sometimes subterranean caudex, or tall woody and leafy stem, or bracted scape, the leaves linear or lanceolate, usually rigid and sharp-pointed, bearing long marginal thread-like fibers in our species. Flowers large, bracted, nodding in a terminal raceme or panicle. Perianth campanulate, or nearly globular, white in our species, of 6 ovate, or ovate-lanceolate separate or slightly united segments. Stamens hypogynous, shorter than the perianth; filaments thickened above, often papillose; anthers small, versatile. Ovary sessile, 3-celled; or imperfectly 6-celled; ovules numerous; style columnar, short, with 3 stigmatic lobes. Fruit a loculicidal or septicidal capsule, or fleshy, or spongy and indehiscent. Seeds numerous, flattened, horizontal. [The Haytien name.]

About 16 species, natives of North and Central America. Type species: *Yucca aloifolia* L.

Fruit fleshy, indehiscent, drooping.   1. *Y. baccata.*
Fruit an erect capsule.
    Leaves 2″–6″ wide; scape short, bearing a long raceme.   2. *Y. glauca.*
    Leaves 10″–2′ wide; scape 2°–10° high, bearing a large panicle.   3. *Y. filamentosa.*

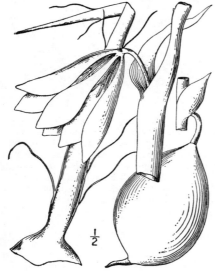

1. Yucca baccàta Torr. Spanish Bayonet or Dagger. Fig. 1277.

Yucca baccata Torr. Bot. Mex. Bound. Surv. **221.** 1859.

Caudex very short, or sometimes 2°–8° tall, covered with the reflexed dead leaves. Leaves 1½°–3° long, 1′–2′ wide with a much wider base, acuminate, with a stout brown point, concave, the marginal fibers 2′–5′ long; panicle peduncled; pedicels stout, 8″–20″ long; flowers 4′–5′ broad; perianth-segments 2½′–3½′ long, 8″–12″ wide; style slender, as long as the ovary, or shorter; fruit oval, dark purple, fleshy, indehiscent, edible, drooping, 2′–3′ long, 1½′–2′ in diameter, with a 6-grooved beak of one-half its length or less; seeds 3″–8″ long, 1″–1½″ thick.

Western Kansas (?), southern Colorado to Texas, California and Mexico. Hosh-kawn. April–June. Fruit ripe Sept.–Oct.

2. Yucca glaùca Nutt. Bear-grass. Soapweed. Fig. 1278.

Yucca glauca Nutt. Fraser's Cat. 1813.
Yucca angustifolia Pursh, Fl. Am. Sept. 227. 1814.

Caudex very short, the leaves all basal, narrowly linear, smooth, very stiff, sharp-pointed, 1°–3° long, 3″–6″ wide, with a broader base, concave, at least when dry, the marginal fibers filiform, usually numerous; scape short; flowers greenish-white, 1½′– 3′ broad, racemose or in a little-branched panicle 1°–6° long; perianth-segments ovate, 1′–1½′ long; style short, green; stigmas shorter than the ovary; pedicels stout, erect and 1′–1½′ long in fruit; capsule oblong, 2′–3′ long, about 1′ thick, 6-sided; seeds very flat, about ½′ broad.

In dry soil, Iowa and South Dakota to Montana, south to Missouri, Texas and Arizona. Adam's-needle. Palmillo. May–June.

Yucca arkansàna Trelease, with grass-like flexible leaves, growing from Arkansas to Texas, may occur in southern Missouri.

3. Yucca filamentòsa L. Adam's Needle. Silk- or Bear-grass. Fig. 1279.

Yucca filamentosa L. Sp. Pl. 319. 1753.

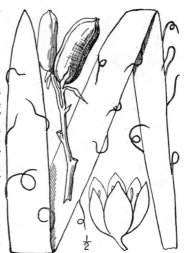

Caudex very short, or sometimes 1° high. Leaves lanceolate, narrowed above the broad base, acuminate and sharp-pointed, flat, roughish, 1°–2½° long, 9″–2′ wide; scape 2°–10° high; panicle large, its branches divergent or ascending, the lower often 1° long or more; flowers numerous; perianth-segments 1½′–2¼′ long, ovate; stigmas slender, but shorter than the ovary; pedicels rarely more than ½′ long; capsule oblong, 1½′–2′ long, about 10″ thick.

In sandy soil, Maryland to Florida, Tennessee and Louisiana. Much cultivated for ornament. Escaped from gardens in southern Pennsylvania. Bear's-thread, thread-and-needle. Eve's-darning-needle. May–July.

Family 23. **CONVALLARIÀCEAE** Link. Handb. 1: 184. 1829.

Lily-of-the-Valley Family.

Scapose or leafy-stemmed herbs, with simple or branched rootstocks, never with bulbs or corms. Flowers solitary, racemose, panicled or umbelled, regular and perfect. Leaves broad, parallel-veined and sometimes with cross-veinlets, alternate, verticillate or basal, or in *Asparagus* and its allies reduced to scales bearing filiform or flattened branchlets in their axils. Perianth inferior, 4–6-parted with separate segments, or oblong, cylindric or urn-shaped and 6-lobed or 6-toothed. Stamens 6, rarely 4, hypogynous or borne on the perianth; anthers introrsely, extrorsely or laterally dehiscent. Ovary 2–3-celled, superior; ovules anatropous or amphitropous; style slender or short; stigma mostly 3-lobed. Fruit a fleshy berry, rarely a capsule. Seeds few or numerous. Embryo small.

About 23 genera and 215 species, widely distributed.

Leaves reduced to scales; leaf-like bractlets filiform.	1. *Asparagus.*
Leaves broad; stems simple or somewhat branched.	
Leaves basal; flowers umbelled or solitary.	2. *Clintonia.*
Leaves alternate (solitary in flowerless plants of no. 4).	
Perianth-segments separate.	
Flowers racemed, umbelled, panicled or solitary, terminal.	
Flowers racemed or panicled.	
Perianth-segments 6.	3. *Vagnera.*
Perianth-segments 4.	4. *Unifolium.*
Flowers umbelled or solitary.	
Fruit a berry.	5. *Disporum.*
Fruit a capsule.	6. *Uvularia.*
Flowers solitary or two together, axillary.	7. *Streptopus.*
Perianth cylindric or oblong, 6-toothed.	8. *Polygonatum.*
Leaves nearly basal; flowers racemed; perianth 6-toothed.	9. *Convallaria.*

1. **ASPÁRAGUS** L. Sp. Pl. 313. 1753.

Stem at first simple, fleshy, scaly, at length much branched; the branchlets filiform and mostly clustered in the axils of the scales in the following species, flattened and linear, lanceolate or ovate in some others. Flowers small, solitary, umbelled or racemed. Perianth-segments alike, separate or slightly united at the base. Stamens inserted at the bases of the perianth-segments; filaments mostly filiform; anthers ovate or oblong, introrse. Ovary sessile, 3-celled; ovules 2 in each cavity; style slender, short; stigmas 3, short, recurved. Berry globose. Seeds few, rounded. [Ancient Greek name.]

About 100 species, natives of the Old World, the following being the generic type.

1. **Asparagus officinàlis** L. Asparagus. Fig. 1280.

Asparagus officinalis L. Sp. Pl. 313. 1753.

Rootstock much branched. Young stems succulent, edible, stout, later branching, and becoming 3°–7° tall, the filiform branchlets 3″–9″ long, less than ¼″ thick, mostly clustered in the axils of minute scales. Flowers mostly solitary at the nodes, green, drooping on filiform jointed peduncles; perianth campanulate, about 3″ long, the segments linear, obtuse; stamens shorter than the perianth; berry red, about 4″ in diameter.

Escaped from cultivation and naturalized, especially along salt marshes, New Brunswick to Virginia, and locally in waste places in the interior. Native of Europe. Sperage. Sparrow-grass. May–June, or flowering also in the autumn.

2. **CLINTÒNIA** Raf. Journ. Pys. **89**: 102. 1819.

Somewhat pubescent scapose herbs, with slender rootstocks, erect simple scapes, and few broad petioled sheathing basal leaves, the bractless flowers umbelled at the summit of the scape in our species. Perianth-segments distinct, equal or nearly so, erect-spreading. Stamens 6, inserted at the bases of the perianth-segments; filaments filiform; anthers oblong, laterally dehiscent. Ovary 2–3-celled; ovules 2–several in each cavity; style stout or slender; stigma obscurely 2–3-lobed. Berry globose or oval. [Name in honor of De Witt Clinton, 1769–1828, American naturalist, Governor of the State of New York.]

Six species, the following of eastern North America, 2 of western North America, 2 Asiatic. Type species: *Clintonia borealis* (Ait.) Raf.

Flowers greenish-yellow, drooping, 8″–10″ long; berry blue.   1. *C. borealis.*
Flowers white, not drooping, 4″–5″ long; berry black.   2. *C. umbellulata.*

1. **Clintonia boreàlis** (Ait.) Raf. Yellow Clintonia. Fig. 1281.

Dracaena borealis Ait. Hort. Kew 1: 454. 1789.
Clintonia borealis Raf. Atl. Journ. 120. 1832.

Scape 6′–15′ high, pubescent above or nearly glabrous. Leaves 2–5, usually 3, oval, oblong or obovate, thin, shorter than the scape, 1½′–3½′ wide, ciliate, short-acuminate or cuspidate; umbel 3–6-flowered; flowers drooping, greenish yellow, 8″–10″ long; pedicels 3″–15″ long, slender, pubescent, erect or ascending in fruit; perianth-segments obtuse or acutish; stamens about as long as the perianth; ovary 2-celled; ovules numerous, in 2 rows in each cavity, style slender, somewhat thickened above, about equalling the stamens; berry oval, blue, several-seeded, about 4″ in diameter.

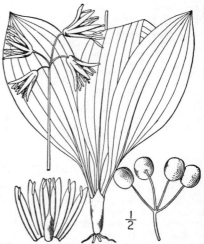

In moist woods and thickets, Newfoundland to Ontario and Manitoba, south to North Carolina and Wisconsin. Ascends to 4500 ft. in Virginia. A flower is occasionally borne on the scape below the umbel, and rarely a small leaf. Clinton's-lily. Heal-all. Wild lily-of-the-valley. Bear- or cow-tongue. Northern lily. Dogberry. Wild corn. May–June.

2. Clintonia umbellulàta (Michx.) Torr. White Clintonia. Fig. 1282.

Dracaena umbellulata Michx. Fl. Bor. Am. 1: 202. 1803.

Clintonia ciliata Raf. Journ. Phys. 89: 102. 1819.

C. umbellata Torr. Fl. N. Y. 2: 301. 1843.

Scape more or less pubescent, 8′–18′ high, sometimes bearing a small leaf. Leaves 2–5, oblong, oblanceolate or obovate, shorter than the scape or equalling it, acute or cuspidate, ciliate on the margins and sometimes also on the midvein beneath, 1½′–4′ wide; umbel several–many-flowered; pedicels ascending or erect, slender, pubescent, at first short, becoming ½′–1½′ long in fruit; flowers white, odorous, often purplish dotted, 4″–5″ long; perianth-segments obtusish; ovary 2-celled; ovules 2 in each cavity; style slender; berry globose, black, about 3″ in diameter, few-seeded.

In woods, New York and New Jersey to Georgia and Tennessee. Ascends to 4000 ft. in Virginia. Dog-plum. May–June.

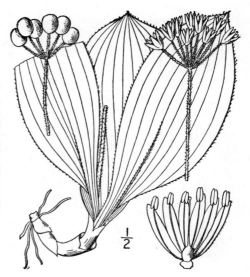

3. VÀGNERA Adans. Fam. Pl. 2: 496. 1763.

[SMILACINA Desf. Ann. Mus. Paris 9: 51. 1807.]

Rootstocks slender, or short and thick. Stem simple, scaly below, leafy above, the leaves alternate, short-petioled or sessile ovate, lanceolate or oblong. Inflorescence a terminal raceme or panicle. Flowers white or greenish white, small. Perianth of 6 separate spreading equal segments. Stamens 6, inserted at the bases of the perianth-segments; filaments filiform or slightly flattened; anthers ovate, introrse. Ovary 3-celled, sessile, subglobose; ovules 2 in each cavity; style short or slender, columnar; stigma 3-grooved or 3-lobed. Berry globular. Seeds usually 1 or 2, subglobose. [Named in honor of Wagner.]

About 25 species, natives of North America, Central America and Asia. Besides the following, one or two others occur in the western United States. Type species: *Convallaria stellata* L.

Flowers numerous, panicled.   1. *V. racemosa.*
Flowers few–several, racemose.
    Plant 8′–20′ high; leaves numerous.   2. *V. stellata.*
    Plant 2′–15′ high; leaves 2–4.   3. *V. trifolia.*

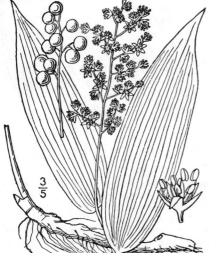

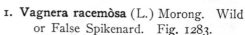

1. Vagnera racemòsa (L.) Morong. Wild or False Spikenard. Fig. 1283.

Convallaria racemosa L. Sp. Pl. 315. 1753.

Smilacina racemosa Desf. Ann. Mus. Paris 9: 51. 1807.

V. racemosa Morong, Mem. Torr. Club 5: 114. 1894.

Rootstock rather thick, fleshy. Stem somewhat angled, slender or stout, erect or ascending, leafy, finely pubescent above, or nearly glabrous, sometimes zigzag, 1°–3° high. Leaves oblong-lanceolate or oval, sessile or the lower short-petioled, 3′–6′ long, 1′–3′ wide, acuminate, finely pubescent beneath and sometimes also above, their margins minutely ciliate; panicle densely many-flowered, 1′–4′ long, peduncled; pedicels shorter than the flowers, or equalling them; flowers about 2″ broad; perianth-segments oblong, equalling the ovary; berry red, aromatic, speckled with purple, 2″–3″ in diameter.

In moist woods and thickets, Nova Scotia to British Columbia, south to Georgia, Missouri and Arizona. Ascends to 2500 ft. in Virginia. Job's-tears. Golden-seal. Small or Zigzag Solomon's-seal. May–July.

Vagnera amplexicaùlis (Nutt.) Greene, of western North America, distinguished from this by its clasping leaves and longer style, may occur in western Nebraska.

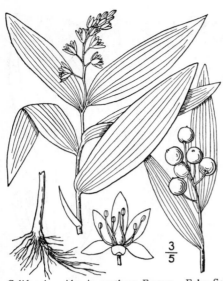

2. **Vagnera stellàta** (L.) Morong. Star-flowered Solomon's Seal. Fig. 1284.

Convallaria stellata L. Sp. Pl. 316. 1753.
Smilacina stellata Desf. Ann. Mus. Paris **9**: 52. 1807.
Vagnera stellata Morong, Mem. Torr. Club **5**: 114. 1894.

Rootstock stout, fleshy. Stem rather stout, erect, glabrous, 8′–20′ tall, straight or somewhat zigzag, leafy. Leaves oblong-lanceolate or lanceolate, sessile and somewhat clasping, minutely pubescent beneath, 2′–5′ long, ½′–1½′ wide, acute, acuminate, or blunt at the apex, flat or somewhat concave; raceme sessile or short-peduncled, 1′–2′ long, several-flowered; pedicels 1″–4″ long, usually shorter than the flowers; perianth-segments oblong, obtuse, longer than the stamens; style about as long as the ovary; berry green with 6 black stripes or black, 3″–5″ in diameter.

In moist soil, Newfoundland to British Columbia, south to Virginia, Kentucky, Kansas and California. Also in northern Europe. False Solomon's-seal. May–June.

3. **Vagnera trifòlia** (L.) Morong. Three-leaved Solomon's Seal. Fig. 1285.

Convallaria trifolia L. Sp. Pl. 316. 1753.
Smilacina trifolia Desf. Ann. Mus. Paris **9**: 52. 1807.
Vagnera trifolia Morong, Mem. Torr. Club **5**: 114. 1894.

Glabrous, rootstock slender. Stem slender, erect, 2′–15′ high, 2–4-leaved (usually 3-leaved); leaves oval, oblong or oblong-lanceolate, sessile, sheathing, 2′–5′ long, ½′–2′ wide, acute or acuminate at the apex, narrowed at the base; raceme few-flowered, peduncled, 1′–2′ long; perianth-segments oblong or oblong-lanceolate, obtuse, finally somewhat reflexed, longer than the stamens; style about as long as the ovary; berry dark red, 2½″–3″ in diameter.

In bogs and wet woods, Newfoundland to British Columbia, south to Connecticut, New Jersey, Pennsylvania and Michigan. Also in northern Asia. May–June.

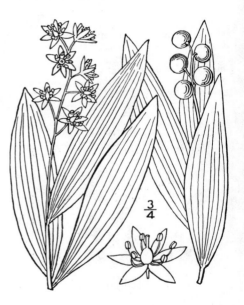

4. **UNIFÒLIUM** Haller; Zinn, Cat. Plant. Hort. Goett. 104. 1757.

[MAIANTHEMUM Wigg. Prim. Fl. Hols. 14. 1780.]

Low herbs, with slender rootstocks, erect simple few-leaved stems, petioled or sessile leaves and small white flowers in a terminal raceme, the pedicels commonly 2–3 together. Perianth of 4 separate spreading segments. Stamens 4, inserted at the bases of the segments; filaments filiform; anthers introrse. Ovary sessile, globose, 2-celled; ovules 2 in each cavity; style about as long as the ovary, 2-lobed or 2-cleft. Berry globular, 1–2-seeded. [Many plants bear only a solitary long-petioled leaf, arising from the rootstock, whence the Latin name.]

Two known species, the following of eastern North America, the other of Europe, Asia and northwest America. Type species: *Convalaria bifolia* L.

1. Unifolium canadénse (Desf.) Greene. False or Wild Lily-of-the-valley. Two-leaved Solomon's Seal. Fig. 1286.

Maianthemum canadense Desf. Ann. Mus. Paris
9: 54. 1807.
Smilacina bifolia var. *canadensis* A. Gray Man.
Ed. 2, 467. 1856.
Unifolium canadense Greene, Bull. Torr. Club **15**:
287. 1888.

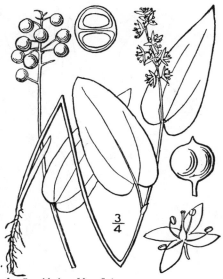

Glabrous or pubescent. Stem slender, erect,
often zigzag, 1–3-leaved (usually 2-leaved),
2′–7′ high; leaves ovate or ovate-lanceolate,
1′–3′ long, acute, acuminate, or blunt and cuspidate at the apex, cordate at the base with a
narrow or closed sinus, sessile, short-petioled,
or the lowest sometimes with a petiole ½′
long; solitary leaves of the stemless plants on
petioles 1′–4′ long; raceme rather dense, many-
flowered, 1′–2′ long; pedicels mostly longer
than the flowers; perianth-segments oblong,
obtuse, becoming reflexed, about 1″ long,
rather longer than the stamens; berry pale red,
speckled, about 2″ in diameter.

In moist woods and thickets, Newfoundland to
the Northwest Territory, south to North Carolina,
Tennessee, Iowa and South Dakota. Ascends to
5000 ft. in Virginia. Cowslip. Bead-ruby. One-leaf. One-blade. May–July.

5. DÍSPORUM Salisb. Trans. Hort. Soc. 1: 331. 1812.

[Prosartes Don, Ann. Nat. Hist. **4**: 341. 1840.]

More or less pubescent herbs with slender rootstocks, branching stems, scaly below,
leafy above, and alternate somewhat inequilateral sessile or clasping leaves, the flowers terminal, drooping, whitish or greenish yellow, solitary or few in simple umbels. Perianth
of 6 narrow equal separate deciduous segments. Stamens 6, hypogynous; filaments filiform
or somewhat flattened, longer than the anthers; anthers oblong, or linear, extrorse. Ovary
3-celled; ovules 2 or sometimes several in each cavity; style slender; stigma 3-cleft or entire. Berry ovoid or oval, obtuse. [Greek, referring to the 2 ovules in each cavity of the
ovary, in most species.]

About 15 species, natives of North America and Asia. Besides the following, some 5 others
occur in western North America. Type species: *Disporum pullum* Salisb.

Stamens shorter than the perianth; fruit smooth, 2–6-seeded.	1. *D. lanuginosum.*
Stamens as long as the perianth; fruit roughened, 4–18-seeded.	2. *D. trachycarpum.*

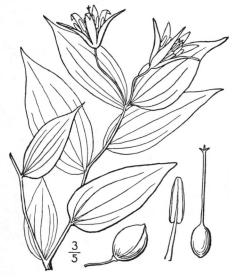

1. Disporum lanuginòsum (Michx.) Nichols. Hairy Disporum. Fig. 1287.

Streptopus lanuginosus Michx. Fl. Bor. Am. **1**:
201. 1803.

Prosartes lanuginosa Don, Trans. Linn. Soc. **18**:
532. 1841.

Disporum lanuginosum Nichols. Dict. Gard. **1**:
485. 1884.

Finely and rather densely pubescent, 1½°–
2½° high. Leaves ovate-lanceolate, or oblong-
lanceolate, 2′–4½′ long, 1′–2′ wide, long-acuminate at the apex, rounded at the base, 7–15-
nerved; flowers solitary or 2–3 together,
greenish, 6″–9″ long; pedicels filiform, about
1′ long; perianth narrowly campanulate, its
segments linear-lanceolate, acuminate, somewhat spreading, glabrous, one-third to one-
half longer than the stamens; ovary oblong;
style slender, longer than the stamens or
equalling them, 3-cleft; berry oval, red, pulpy,
2–6-seeded, 5″–7″ long.

In woods, Ontario to western New York,
Georgia and Tennessee. Ascends to 4000 ft. in
Virginia. May–June.

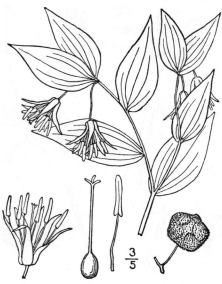

2. Disporum trachycàrpum (S. Wats.)
B. & H.   Rough-fruited Disporum.
Fig. 1288.

Prosartes trachycarpa S. Wats. Bot. King's Exp.
344. 1871.
Disporum trachycarpum B. & H. Gen. Pl. 3: 832.
1883.

Puberulent, at least when young, 1°–2°
high. Leaves ovate, oval or oblong-lanceolate, 1½'–3½' long, 1'–2½' wide, acute or short-
acuminate at the apex, rounded or subcordate
at the base, 5–11-nerved; flowers solitary or
2–3 together, yellowish-white, 4''–7'' long;
pedicels ½'–1' long; perianth narrowly campanulate, its segments narrowly oblong or
oblanceolate, acute, little spreading, about
equalling the stamens, ovary depressed-
globose; style slender, about equalling the
stamens, 3-lobed; berry roughened, depressed-
globose or somewhat obovoid, 4''–5'' in diameter, apparently leathery rather than pulpy,
4–18-seeded.

Manitoba to Alberta, British Columbia, South
Dakota, Nebraska, Washington and Arizona.
May–Aug.

6. UVULÀRIA L. Sp. Pl. 304. 1753.

Erect forked herbs, perennial by rootstocks. Stem leafy above, scale-bearing below, the
leaves alternate, sessile or perfoliate. Flowers large, solitary at the ends of the branches or
rarely 2 together, peduncled, drooping. Perianth bell-shaped or narrower; segments distinct, deciduous, sometimes bearing a nectary at the base. Stamens 6, free, or adnate to the
very bases of the perianth-segments; filaments filiform; anthers linear, the sacs longitudinally dehiscent. Ovary 3-lobed, 3-celled, short-stalked or sessile; styles united to about
the middle, stigmatic along the inner side above; ovules several in each cell. Capsule ovoid
or obovoid, 3-angled or 3-winged, loculicidally dehiscent. Seeds globose, 1–3 in each cavity.
[Name Latin, from *ùvula*, a palate, in allusion to the hanging flowers.]

Five or six species, natives of eastern North America. Type species: *Uvularia perfoliata* L.

Capsule obtusely 3-angled, truncate or rounded; leaves perfoliate.
  Glabrous, glaucous; perianth-segments papillose within.   1. *U. perfoliata.*
  Leaves pubescent beneath; perianth-segments smooth or nearly so.   2. *U. grandiflora.*
Capsule acutely 3-angled or 3-winged, acute at each end; leaves sessile. (OAKESIELLA Small.)
  Leaves thin, slightly rough-margined, narrowed at both ends.   3. *U. sessilifolia.*
  Leaves firm, manifestly rough-margined, sometimes subcordate.   4. *U. puberula.*

1. Uvularia perfoliàta L.   Perfoliate Bellwort.
Wild Oat. Fig. 1289.

Uvularia perfoliata L. Sp. Pl. 304. 1753.

Glabrous and glaucous or pale green. Stems 6'–20'
high, slender, forked above the middle, usually with
1–3 leaves below the fork; leaves oval, oblong or
ovate-lanceolate, acute at the apex, rounded or sometimes narrowed at the base, smooth-margined, 2'–5'
long when mature, small at flowering time; flowers
10''–16'' long, pale yellow; peduncle becoming ½'–1'
long in fruit; perianth-segments granular-papillose
within, sometimes but slightly so; stamens shorter than
the styles or equalling them, the connective sharp-
tipped; capsule obovoid, truncate, thicker than long,
4''–5'' long, obtusely 3-angled, with concave sides and
grooved angles, its lobes dehiscent above.

In moist woods and thickets, Quebec and Ontario to
Florida and Mississippi. Ascends to 3500 ft. in Virginia.
Flowers fragrant. Mealy bellwort. Straw-bell. Mohawk-
weed. May–June.

2. Uvularia grandiflòra J. E. Smith. Large-flowered Bellwort. Fig. 1290.

Uvularia grandiflora J. E. Smith, Ex. Bot. 1: 99. *pl. 51.* 1804–5.

Stems rather stouter than that of the preceding species, naked or with 1 or 2 leaves below the fork. Leaves perfoliate, oblong, oval or ovate, pubescent beneath, at least when young, glabrous above, becoming 2′–5′ long; flowers lemon-yellow, 1′–1½′ long; perianth-segments smooth on both sides or very slightly granular within; stamens exceeding the styles, the connective blunt; capsule obtusely 3-angled, truncate, 4″–5″ long, the lobes dehiscent above.

In rich woods, Quebec to Ontario, Minnesota, Georgia, Tennessee and Kansas. April–June.

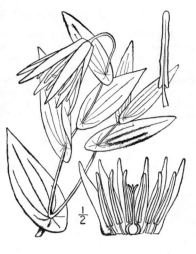

3. Uvularia sessilifòlia L. Sessile-leaved Bellwort. Fig. 1291.

Uvularia sessilifolia L. Sp. Pl. 305. 1753.
Oakesia sessilifolia S. Wats. Proc. Am. Acad. 14: 269. 1879.

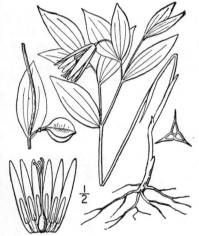

Glabrous, stem slender, naked or bearing 1 or 2 leaves below the fork. Leaves oblong or oblong-lanceolate, 1½′–3′ long when mature, thin, sessile, acute at each end, roughish-margined, pale or glaucous beneath; flowers greenish yellow, 8″–15″ long; perianth-segments smooth; styles exceeding stamens; anthers blunt; peduncle ½′–1′ long in fruit; capsule sharply 3-angled, narrowed at both ends, short-stipitate, about 1′ long, 6″–8″ thick.

In moist woods and thickets, New Brunswick and Ontario to Minnesota, south to Georgia and Arkansas. Wild oat. Straw-lilies. May–June.

4. Uvularia pubérula Michx. Mountain Bellwort. Fig. 1292.

Uvularia puberula Michx. Fl. Bor. Am. 1: 199. 1803.
Oakesia puberula S. Wats. Proc. Am. Acad. 14: 269. 1879.

Stem rather stout, sparingly rough-pubescent with short hairs, at least on the forks. Leaves oblong, oval or ovate; rough-margined, firm and 1½′–3′ long when mature, sessile, acute at the apex, obtuse, subcordate or sometimes narrowed at the base, shining, green on both sides, the midvein sometimes pubescent; flowers light yellow, about 1′ long; styles about equalling the stamens; capsule sharply 3-angled, acute at both ends, sessile or very nearly so on the short peduncle, 10″–12″ long.

In mountain woods, Virginia and West Virginia to South Carolina. Ascends to 5000 ft. in Virginia. May–June.

Uvularia nítida (Britton) Mackenzie, of the pine-barrens of New Jersey, differs in having the styles exceeding the stamens and a smaller capsule; it may be specifically distinct.

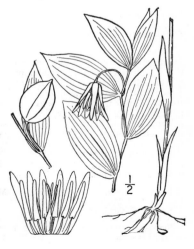

7. STRÉPTOPUS Michx. Fl. Bor. Am. 1 : 200. 1803.

Branching herbs, with stout or slender rootstocks, thin sessile or clasping alternate many-nerved leaves, the flowers solitary or 2 together, extra-axillary, slender-peduncled, greenish or purplish, small, nodding. Peduncles bent or twisted at about the middle. Perianth somewhat campanulate, its 6 separate segments recurved or spreading, deciduous, the outer flat, the inner keeled. Stamens 6, hypogynous; filaments short, flattened; anthers sagittate, extrorse. Ovary 3-celled; ovules numerous in 2 rows in each cavity; style slender, 3-cleft, 3-lobed or entire. Berry globose or oval, red, many-seeded. [Greek, twisted-stalk, in reference to the bent or twisted peduncles.]

About 5 species, natives of the north temperate zone. Besides the following, another occurs on the Pacific Coast. Type species: *Streptopus roseus* Michx.

Leaves glaucous beneath, clasping; flowers greenish-white.   1. *S. amplexifolius*.
Leaves green on both sides, sessile; flowers purple or rose.   2. *S. roseus*.

3/5

1. Streptopus amplexifòlius (L.) DC. Clasping-leaved Twisted-stalk. Liver-berry. Fig. 1293.

Uvularia amplexifolia L. Sp. Pl. 304. 1753.
Streptopus amplexifolius DC. Fl. France 3 : 174. 1805.

Rootstock short, stout, horizontal, covered with thick fibrous roots. Plant 1½°–3° high; stem glabrous, usually branching below the middle, leaves 2′–5′ long, 1′–2′ wide, acuminate at the apex, cordate-clasping at the base, glabrous, glaucous beneath; peduncles 1′–2′ long, 1–2-flowered; flowers greenish white, 4″–6″ long; perianth-segments narrowly lanceolate, acuminate; anthers subulate-pointed; stigma simple, obtuse or truncate; berry oval, 5″–8″ long.

In moist woods, Greenland to Alaska, south to North Carolina, Ohio, Michigan and New Mexico. Ascends to 4000 ft. in the Adirondacks. May–July.

2. Streptopus ròseus Michx. Sessile-leaved Twisted-stalk. Fig. 1294.

Streptopus roseus Michx. Fl. Bor. Am. 1 : 201. 1803.
Streptopus longipes Fernald, Rhodora 8 : 71. 1906.

Plant 1°–2½° high, from a short stout rootstock covered with fibrous roots, sometimes stoloniferous. Branches sparingly pubescent; leaves 2′–4½′ long, acuminate at the apex, sessile, rounded, or slightly clasping at the base, green on both sides, or somewhat paler beneath, their margins finely ciliate; peduncles ½′–1′ long, usually pubescent, 1-flowered, rarely 2-flowered; flowers purple or rose, 4″–6″ long; perianth-segments lanceolate, acuminate; anthers 2-horned; style 3-cleft, the spreading branches stigmatic along the inner side; berry 5″–6″ in diameter.

In moist woods, Newfoundland to Manitoba, Georgia and Michigan. Ascends to 5600 ft. in Virginia. Liver-berry. May–July.

Streptopus oreópolus Fernald is apparently a hybrid between this and the preceding species.

3/5

8. POLYGONÀTUM [Tourn.] Mill. Gard. Dict. Abr. Ed. 4. 1754.

[SALOMÒNIA Heist.; Fabr. Enum. Pl. Hort. Helmst. 20. 1759.]

Glabrous or pubescent herbs, with thick, horizontal jointed and scarred rootstocks, simple arching or erect stems, scaly below, leafy above, the leaves ovate or lanceolate, sessile and alternate in our species (opposite or verticillate in some exotic ones). Flowers greenish or pinkish, axillary, drooping, peduncled, solitary or 2–10 in an umbel, the pedicels jointed at the base of the flower. Perianth tubular or oblong-cylindric or somewhat expanded above the base, 6-lobed, the short lobes not spreading. Stamens 6, included; filaments adnate to the perianth for half their length or more; anthers sagittate, introrse. Ovary 3-celled; ovules 2–6 in each cavity; style slender; stigmas small, capitate or slightly 3-lobed. Berry globular, pulpy, dark blue or nearly black, with a bloom, in our species. [Genus dedicated to Salomon.]

About 20 species, natives of the north temperate zone. Type species: *Convallaria polygonatum* L.

Leaves pubescent beneath; filaments filiform, roughened.   1. *P. biflorum.*
Plant glabrous throughout; filaments smooth, somewhat flattened.   2. *P. commutatum.*

1. Polygonatum biflòrum (Walt.) Ell. Hairy Solomon's Seal. Fig. 1295.

Convallaria biflora Walt. Fl. Car. 122. 1788.
Polygonatum biflorum Ell. Bot. S. C. & Ga. 1: 393. 1817.
Salomonia biflora Farwell, Rep. Com. Parks Detroit 11: 53. 1900.

Stem slender, glabrous, often zigzag above, 8'–3° high. Leaves lanceolate, oval to ovate, 2'–4' long, ½'–2' wide, acute or acuminate at the apex, narrowed or sometimes obtuse at the base, pubescent especially on the veins and pale beneath, glabrous above, the upper commonly narrower than the lower; peduncles 1–4-flowered (often 2-flowered), glabrous; perianth 4"–6" long, about 1½" thick, filaments filiform, adnate to the perianth for about three-fourths its length, papillose-roughened; berry 3"–4" in diameter.

In woods and thickets, New Brunswick to Ontario and Michigan, south to Florida, West Virginia and Tennessee. Recorded from Kansas and Texas. Sealwort. Dwarf Solomon's-seal. Conquer-john. April–July.

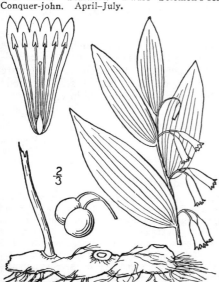

2. Polygonatum commutàtum (R. & S.) Dietr. Smooth Solomon's Seal. Fig. 1296.

Convallaria commutata R. & S. Syst. 7: 1671. 1830.
Polygonatum commutatum Dietr.; Otto & Dietr. Gartenz. 3: 223. 1835.
Polygonatum giganteum Dietr.; Otto & Dietr. Gartenz. 3: 222. 1835.
Salomonia commutata Britton, Man. 273. 1901.

Glabrous throughout, stem stout or slender, 1°–8° high. Leaves lanceolate, oval or ovate, 1½'–6' long, 3'–4' wide, rather darker green above than beneath, acute, acuminate or blunt at the apex, narrowed, rounded or somewhat clasping at the base, the upper often narrower than the lower; peduncles 1–8-flowered, glabrous; perianth 6"–10" long, 1½"–2" thick; filaments somewhat flattened, smooth, adnate to the perianth for half its length or more; berry 4"–6" in diameter.

In moist woods and along streams, rarely in dry soil, Rhode Island to New Hampshire, Ontario and Manitoba, south to Georgia, Louisiana, Utah, New Mexico and Arizona. Sealwort. Giant Solomon's-seal. May–July.

9. CONVALLÀRIA L. Sp. Pl. 314. 1753.

A low glabrous herb, with horizontal rootstocks, very numerous fibrous roots, and 2 or sometimes 3 erect broad leaves, narrowed into sheathing petioles, the lower part of the stem bearing several sheathing scales. Flowers white, racemed, fragrant, nodding. Raceme 1-sided. Perianth globose-campanulate, 6-lobed, deciduous, the short lobes recurved. Stamens 6, included; filaments short, adnate to the lower part of the perianth; anthers oblong, introrse. Ovary 3-celled; ovules several in each cavity; style slender, 3-grooved; stigma small, capitate, slightly 3-lobed. Berry globose, pulpy. [Latin from *Convallis*, valley, and the Greek for lily.]

A monotypic genus of Europe, Asia and the higher Alleghanies.

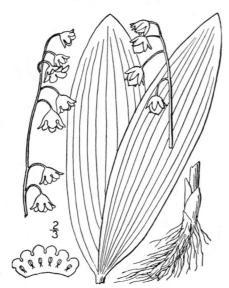

1. Convallaria majàlis L. Lily-of-the-valley. Fig. 1297.

Convallaria majalis L. Sp. Pl. 314. 1753.

Convallaria majuscula Greene, Rep. Nov. Spec. 5: 46. 1907.

Stem 4′–9′ high. Leaves oblong, or oval, appearing nearly basal, acute at both ends, 5′–12′ long, 1′–2½′ wide; basal scales large, 1′–4′ long, one of them subtending an erect angled scape shorter than the leaves; raceme 1′–3½′ long, loosely several-flowered; pedicels filiform, recurved, 3″–6″ long, exceeding or sometimes shorter than the lanceolate bracts; perianth 3″–4″ long, its lobes 1″ long or less; filaments shorter than the anthers; berry about 3″ in diameter.

On the higher mountains of Virginia, North Carolina and South Carolina. Common in cultivation. Consists of several slightly differing races. May blossoms. Wood-lily. Conval-lily. May-lily. May–June.

Family 24. TRILLIÀCEAE Lindl. Nat. Syst. ed. 2, 347. 1836.

WAKE-ROBIN FAMILY.

Somewhat fleshy herbs, perennial by rootstocks. Leaves cauline, whorled, or sometimes solitary long-petioled ones are borne on the rootstock. Flowers terminal, solitary or umbelled, sessile or pedicelled, perfect. Perianth of 3 separate sepals and 3 separate petals. Stamens 6; anthers 2-celled. Ovary sessile, 3-celled; styles 3, stigmatic along the inner side; ovules several or numerous in each cavity. Fruit a globose or 3-lobed berry.

Three genera and about 25 species, natives of the north temperate zone.

Leaves in 2 (rarely 3) whorls; flowers umbelled.   1. *Medeola*.
Leaves in 1 whorl; flowers solitary.   2. *Trillium*.

1. MEDÈOLA L. Sp. Pl. 339. 1753.

A slender erect unbranched herb, loosely provided with deciduous wool. Rootstock thick, white, tuber-like, with somewhat the odor and taste of cucumbers, the slender fibrous roots numerous. Leaves of flowering plants in 2 whorls (rarely 3 whorls); lower whorl of 4–10 oblong-lanceolate or obovate leaves; upper whorl of 3–5 ovate or oval leaves, subtending, like an involucre, the sessile umbel of small greenish yellow declined flowers. Perianth of 6 separate equal oblong recurved segments. Stamens 6, hypogynous; filaments slender, smooth, longer than the oblong extrorse anthers, the sacs laterally dehiscent. Ovary 3-celled; ovules several in each cavity; styles 3, recurved, stigmatic along the inner side. Berry globose, pulpy. [Name from *Medea*, a sorceress, referring to the supposed healing properties.]

A monotypic genus of eastern North America.

1. Medeola virginiàna L. Indian Cucumber-root. Fig. 1298.

Medeola virginiana L. Sp. Pl. 339. 1753.

Rootstock 1'–3' long. Stem 1°–2½° tall, bearing the lower whorl of leaves above the middle, or in flowerless plants at the summit; leaves of the lower whorl sessile, 2½'–5' long, 1'–2' wide, acuminate at the apex, narrowed at the base, 3–5-nerved and reticulate-veined; leaves of the upper whorl 1'–2' long, ½'–1' wide, short-petioled or sessile; umbel 2–9-flowered; pedicels filiform, 1' long or less, declined in flower, erect or ascending in fruit; perianth-segments 3''–5'' long, obtuse; berry dark purple, 4''–7'' in diameter.

In moist woods and thickets, Nova Scotia to Ontario, Minnesota, Florida and Tennessee. May–June.

2. TRÍLLIUM L. Sp. Pl. 339. 1753.

Glabrous erect unbranched herbs, with short scarred rootstocks and 3 leaves whorled at the summit of the stem, subtending the sessile or peduncled solitary bractless flower. Solitary long-petioled leaves are sometimes borne on the rootstock. Perianth of 2 distinct series of segments, the outer 3 (sepals) green, persistent, the inner 3 (petals) white, pink, purple or sometimes greenish, deciduous or withering. Stamens 6, hypogynous; filaments short; anthers linear, mostly introrse. Ovary sessile, 3–6-angled or lobed, 3-celled; ovules several or numerous in each cavity; styles 3, stigmatic along the inner side. Berry many-seeded. Seeds horizontal. [Latin, in allusion to the 3-parted flowers and the 3 leaves.]

About 24 species of North America and Asia. Besides the following, some 7 others occur in southern and western North America. Known as *Three-leaved Nightshade* and *Birthroot*. Phyllody, *i. e.*, the reversion of petals or sepals to leaves, is occasional, and the floral parts are sometimes in 4's. Type species: *Trillium cernuum* L.

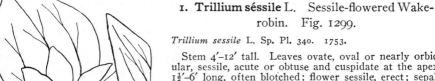

Flower sessile.
  Leaves sessile; sepals not reflexed.
    Flowers purple; petals lanceolate. 1. *T. sessile.*
    Flowers green; petals linear. 2. *T. viride.*
  Leaves petioled; sepals reflexed. 3. *T. recurvatum.*
Flower peduncled.
  Leaves oval or ovate, obtuse or obtusish, 1'–2' long. 4. *T. nivale.*
  Leaves broadly ovate or rhombic, acuminate, 2'–7' long.
    * Leaves sessile, or narrowed at the base and short-petioled.
      Petals obovate or oblanceolate, 1½'–2½' long. 5. *T. grandiflorum.*
      Petals ovate or lanceolate, ½'–1½' long.
        Peduncle 1¼'–4' long, erect or declined; petals spreading.
          Petals brown-purple, rarely white; filaments two-thirds as long as the anthers, or longer. 6. *T. erectum.*
          Petals white; filaments not more than half as long as anthers. 7. *T. declinatum.*
        Peduncle 1¼' long or less, recurved; petals recurved. 8. *T. cernuum.*
    ** Leaves distinctly petioled, obtuse or rounded at the base. 9. *T. undulatum.*

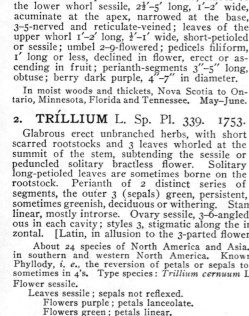

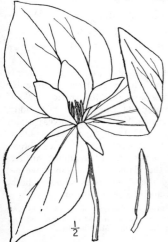

1. Trillium séssile L. Sessile-flowered Wake-robin. Fig. 1299.

Trillium sessile L. Sp. Pl. 340. 1753.

Stem 4'–12' tall. Leaves ovate, oval or nearly orbicular, sessile, acute or obtuse and cuspidate at the apex, 1½'–6' long, often blotched; flower sessile, erect; sepals lanceolate, acute or obtuse, spreading, ½'–2' long, petals lanceolate, acute or obtuse, somewhat longer than the sepals, erect-spreading, purple or green; anthers 3''–7'' long, longer than filament, the connective prolonged beyond the sacs; berry globose, 6-angled, about ½' in diameter.

In moist woods and thickets, Pennsylvania to Ohio and Minnesota, south to Florida, Mississippi and Arkansas. Flowers pleasantly odorous. Three-leaved nightshade. April–May.

2. Trillium víride Beck.   Green Wake-robin.
Fig. 1300.

Trillium viride Beck, Am. Journ. Sci. **11**: 178.   1826.

Perennial by a short corm-like rootstock, light green. Stems solitary, or several together, 4'–15' tall, rough-pubescent near the top, or glabrous in age; leaves oblong to ovate, 2'–4' long, obtuse or acutish, 3–5-nerved, usually blotched, more or less pubescent on the nerves beneath; flowers sessile; sepals linear or linear-lanceolate, 1'–2' long, bright green, acute or obtuse; petals clawed, the blades linear or nearly so, surpassing the sepals, light green or purplish green, the claws sometimes brown or purple; stamens about ⅓ as long as the petals; filaments flattened, ⅓–½ shorter than the anthers.

In woods and glades, Kansas to Missouri, Tennessee, Mississippi and Arkansas. Spring.

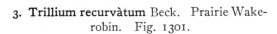

Trillium viridéscens Nutt., a species distinguished from *T. víride* by its purplish petals and acuminate leaves, occurs in Arkansas and is reported from Kansas.

3. Trillium recurvàtum Beck.   Prairie Wake-robin.   Fig. 1301.

Trillium recurvatum Beck, Am. Journ. Sci. **11**: 178. 1826.

Stem 6'–18' tall. Leaves ovate, oval or oblong, 1½'–4' long, acute at the apex, narrowed into petioles 3''–9'' long, sometimes blotched; flower sessile, erect; sepals lanceolate, acuminate, 6''–15'' long, reflexed between the petioles; petals spatulate or oblong, nearly erect, clawed, acute or acuminate, equalling the sepals or somewhat longer; anthers 4''–7'' long, much longer than the filaments, the connective prolonged beyond the sacs; berry ovoid, 6-winged above, about 9'' long.

In woods and thickets, Ohio to Minnesota, Mississippi and Arkansas. April–June.

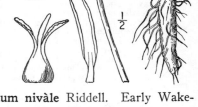

4. Trillium nivàle Riddell.   Early Wake-robin.   Fig. 1302.

Trillium nivale Riddell, Syn. Fl. W. States, 93. 1835.

Stem 2'–6' high. Leaves ovate, oval or nearly orbicular, 1'–2' long, obtuse at the apex, rounded or narrowed at the base, petioled; petioles 2''–6'' long; flowers peduncled; peduncle ½'–1' long, erect, bent, or recurved beneath the leaves; sepals narrowly oblong or oblong-lanceolate, obtuse, ½'–1' long; petals white, oblong or oval, obtuse, longer than the sepals, erect-spreading; anthers about as long as the filaments, the connective not prolonged beyond the sacs; styles slender; berry globose, 3-lobed, about ½' in diameter.

In woods and thickets, Pennsylvania to Ohio and Minnesota, south to Kentucky and Nebraska. Showy or Dwarf white wake-robin. March–May.

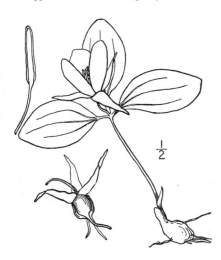

5. Trillium grandiflòrum (Michx.) Salisb.
Large-flowered Wake-robin. Fig. 1303.

Trillium rhomboideum var. *grandiflorum* Michx. Fl. Bor.
Am. 1: 216. 1803.
Trillium grandiflorum Salisb. Par. Lond. 1: *pl. 1.* 1805.

Stems usually stout, 8′–18′ high. Leaves broadly
rhombic-ovate or rhombic-oval, 2½′–6′ long, acuminate
at the apex, narrowed to the sessile or nearly sessile
base; peduncle erect or somewhat inclined, 1½′–3′ long;
sepals lanceolate or oblong-lanceolate, acuminate but
sometimes bluntish, 1′–2′ long, spreading; petals erect-
spreading, oblanceolate; obovate or rarely ovate-oblong,
obtuse or cuspidate, strongly veined, white or pink, thin,
longer than the sepals; anthers about ½′ long, longer
than the filaments; styles slender, 3″–4″ long, ascending or erect; berry globose, black, slightly 6-lobed, 8″–
12″ in diameter.

In woods, Quebec to Ontario and Minnesota, south to
North Carolina and Missouri. Reported from farther south.
Ascends to 5000 ft. in Virginia. White lilies. Bath-flower.
Trinity-lily. May–June.

A monstrous form, with two long-petioled leaves, was collected by Dr. Pitcher in Michigan.

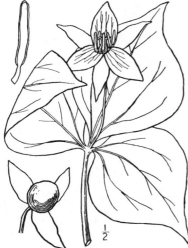

6. Trillium eréctum L. Ill-scented Wake-robin. Birth-root. Fig. 1304.

Trillium erectum L. Sp. Pl. 340. 1753.

Stem stout, 8′–16′ high. Leaves very broadly
rhombic, 3′–7′ long, often as wide or wider, sessile,
acuminate at the apex, narrowed at the base;
peduncle 1¼′–4′ long, erect, or nearly so; sepals
lanceolate, acuminate, spreading, ¾′–1½′ long; petals
lanceolate or ovate, acute or acutish, spreading,
equalling the sepals or a little longer, dark purple,
pink, greenish or white; anthers longer than the
filaments, sometimes twice as long, exceeding the
stigmas; ovary purple; styles short, spreading or
recurved; berry ovoid, somewhat 6-lobed, reddish,
8″–12″ long.

In woods, Nova Scotia to Ontario, south to North
Carolina and Tennessee. Flowers unpleasantly scented.
Indian balm. Red or purple trillium, or wake-robin.
Bath-, Beth- or Squaw-flower. Beth- or bumble-bee-
root. Lamb's-quarters. Nosebleed. Daffy-down-dilly.
Red benjamin. Shamrock. True-love. Orange-blossom.
April–June.

7. Trillium declinàtum (A. Gray) Glea-son. Drooping Wake-robin. Fig. 1305.

T. erectum var. *declinatum* A. Gray, Man. Ed. 5,
523. 1878.

T. declinatum Gleason, Bull. Torr. Club 33: 389.
1906.

Stem stout, 1¼° high or less. Leaves broadly
rhombic, 3′–5′ long and about as wide as long,
short-acuminate at the apex, narrowed at the
base; peduncle horizontal or declined, 1½′–2½′
long; petals oblong-ovate, white, about as long
as the sepals; filaments 1″–2½″ long, not more
than half as long as the anthers; ovary white
or pinkish; stigmas nearly sessile.

Woods, Ohio to Michigan, Minnesota and Missouri. April–June.

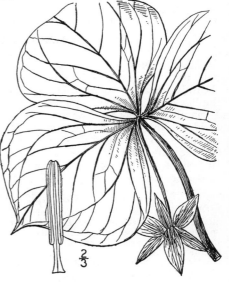

½

8. Trillium cérnuum L. Nodding Wake-robin. Fig. 1306.

Trillium cérnuum L. Sp. Pl. 339. 1753.

Stem usually slender, 8′–20′ high. Leaves similar to those of the preceding species, broadly rhombic, acuminate at the apex, narrowed at the base, sessile, or with the petioles 1″–2″ long; peduncles ½′–1¼′ long, recurved beneath the leaves, the flower drooping; sepals lanceolate or ovate-lanceolate, acuminate, 6″–12″ long; petals white or pink, ovate-lanceolate or oblong-lanceolate, rolled backward, wavy-margined, equalling the sepals, or a little longer; anthers about as long as the subulate filaments; ovary whitish; styles rather stout, recurved; berry ovoid, red-purple, pendulous, 8″–10″ long.

In rich woods, Newfoundland to Ontario and Manitoba, south to Georgia and Missouri. Ground-lily. Cough-root. Rattlesnake-root. Jewsharp-plant. White benjamin. Snake-bite. April–June.

9. Trillium undulàtum Willd. Painted Wake-robin. Fig. 1307.

Trillium undulatum Willd. Neue Schrift. Gesell. Nat. Fr. Berlin 3: 422. 1801.
Trillium erythrocarpum Michx. Fl. Bor. Am. 1: 216. 1803.

Stem usually slender, 8′–2° high. Leaves ovate, 3′–8′ long, 2′–5′ wide, petioled, long-acuminate at the apex, obtuse or rounded at the base; petioles 2″–10″ long; flowers peduncled, erect or somewhat inclined, 1′–2½′ long; sepals lanceolate, acuminate, 9″–15″ long, spreading; petals ovate or ovate-lanceolate, acuminate or acute, white with purple veins or stripes, thin, longer than the sepals, widely spreading, wavy-margined; anthers about equalling the filaments; styles slender; berry ovoid, obtuse, bluntly 3-angled, bright red, shining, 6″–10″ in diameter.

In woods, Nova Scotia to Ontario and Wisconsin, south to Georgia and Missouri. Ascends to 5600 ft. in Virginia. Wild pepper. Sarah. Benjamin. May–June.

½

Family 25. SMILÀCEAE Vent. Tabl. 2: 146. 1799.*
SMILAX FAMILY.

Mostly vines, with woody or herbaceous, often prickly stems. Leaves alternate, netted-veined, usually punctate or lineolate, several-nerved, petioled. Petiole sheathing, bearing a pair of slender tendril-like appendages (stipules?), persistent, the blade falling away. Flowers small, mostly green, dioecious, in axillary umbels. Perianth-segments 6. Stamens mostly 6, distinct; filaments ligulate; anthers basi-fixed, 2-celled, introrse. Ovary 3-celled, the cavities opposite the inner perianth-segments; ovules 1 or 2 in each cavity, orthotropous, suspended; style very short or none; stigmas 1–3. Fruit a globose berry containing 1–6 seeds. Seeds brownish; endosperm horny, copious; embryo small, oblong, remote from the hilum.

Genera 3; species about 200, in warm and temperate regions; only the following in North America.

1. SMÌLAX L. Sp. Pl. 1028. 1753.

Rootstocks usually very large and tuberous, stems usually twining, and climbing by means of the spirally coiling appendages of the petiole. Lower leaves reduced to scales; upper leaves entire or lobed. Flowers regular. Perianth-segments distinct, deciduous. Pedicels borne on a globose or conic receptacle, inserted in small pits, generally among minute bractlets. Filaments inserted on the bases of the perianth-segments. Staminate flowers without an ovary. Pistillate flowers usually smaller than the staminate, with an

* Text contributed to the first edition by the late Rev. THOMAS MORONG.

ovary and usually with 1–6 abortive stamens.   Berry black, red or purple (rarely white), with 3 strengthening bands of tissue running through the outer part of the pulp, connected at the base and apex.   Embryo lying under a tubercle at the upper end of the seed.   [Ancient Greek name, perhaps not originally applied to these plants.]

About 225 species of wide distribution, most abundant in tropical America and Asia.   Besides the following, about 12 others occur in the southern United States and 1 in California and Oregon. Type species: *Smilax áspera* L.

Stem annual, herbaceous, unarmed.   [NEMEXIA Raf.]
  Petioles tendril-bearing; stems climbing.
    Leaves usually ovate, thin.   1. *S. herbacea.*
    Leaves usually hastate, coriaceous.   2. *S. tamnifolia.*
  Petioles without tendrils or nearly so; stems erect.   3. *S. ecirrhata.*
Stem perennial, woody, usually armed with prickles
  Berries black or bluish-black.
    Fruit ripening the first year.
      Leaves glaucous.   4. *S. glauca.*
      Leaves green on both sides.
        Leaves rounded or lanceolate, 5-nerved.   5. *S. rotundifolia.*
        Leaves ovate, 7-nerved.   6. *S. hispida.*
        Leaves round-ovate, often narrowed at the middle, 7–9-nerved. 7. *S. Pseudo-China.*
        Leaves deltoid or deltoid-hastate, 5–7-nerved, often with 1 or 2 additional nerves on
          each side.   8. *S. Bona-nox.*
    Fruit ripening the second year; leaves elliptic or lanceolate, evergreen. 9. *S. laurifolia.*
  Berries red.
    Leaves ovate or ovate-lanceolate, base rounded; berries bright red. 10. *S. Walteri.*
    Leaves lanceolate, acute at the base; berries dull red. 11. *S. lanceolata.*

1. Smilax herbàcea L.   Carrion-flower. Jacob's Ladder.   Fig. 1308.

Smilax herbacea L. Sp. Pl. 1030.   1753.
S. pulverulenta Michx. Fl. Bor. Am. **2**: 238.   1803.
Coprosmanthus herbaceus Kunth, Enum. **5** : 264.   1850.

Tubers short, thick, scarred, numerous.   Stem herbaceous, glabrous, terete or obtusely angled, unarmed, commonly much branched.   Petioles 4″–3½′ long; tendrils numerous; leaves ovate, rounded or lanceolate, acute, acuminate or cuspidate at the apex, obtuse or cordate at the base, thin, frequently downy beneath, 7–9-nerved, 1½′–5′ long, 1′–3½′ wide, the margins entire or denticulate; peduncles 4′–9′ long, usually 6–10 times as long as the petioles, flattened; umbels 15–80-flowered; pedicels 3″–8″ long; flowers carrion-scented when open; stamens sometimes 5 or 7; filaments 2–3 times as long as the anthers; berries bluish black, 2–4-seeded, 3″–4″ in diameter.

In woods and thickets, New Brunswick to Ontario, Manitoba, Florida, Louisiana, Nebraska and Oklahoma.   April–June.

½

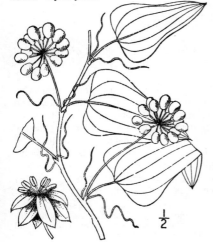

½

2. Smilax tamnifòlia Michx.   Halberd-leaved Smilax.   Fig. 1309.

S. tamnifolia Michx. Fl. Bor. Am. **2**: 238.   1803.
Coprosmanthus tamnifolius Kunth, Enum. **5**: 267. 1850.

Glabrous, herbaceous; stem and branches terete or obtusely angled, unarmed.   Petioles ½′–1½′ long, the sheath tendril-bearing, very short or none; leaves coriaceous, mostly ovate-hastate, with broad obtuse lobes at the base, slightly narrowed at about the middle, acute, obtuse or acuminate at the apex, truncate or subcordate at the base, entire, green on both sides, 1¼′–3′ long, ½′–2′ wide, 5–7-nerved; peduncles 1–3 from the same axil, 1′–4′ long, usually much longer than the leaves, often flattening in drying; umbels 10–30-flowered; pedicels 2″–3″ long; segments of the staminate flowers slightly pubescent; filaments 1–2 times as long as the anthers; berries black, 2″–3″ in diameter, 1–3-seeded.

In dry soil, Long Island to New Jersey, Pennsylvania, South Carolina and Tennessee.   May–July.

3. Smilax ecirrhàta (Engelm.) S. Wats. Upright Smilax. Fig. 1310.

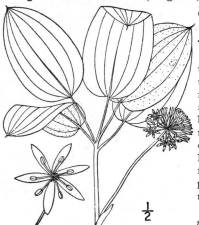

Coprosmanthus herbaceus var. *ecirrhata* Engelm.; Kunth, Enum. **5**: 266. 1850.

S. ecirrhatus S. Wats. in A. Gray, Man. Ed. 6, 520. 1890.

Stem herbaceous, glabrous, simple, erect, 6′–2° tall. Tendrils none, or sometimes present on the uppermost petioles; leaves often whorled at the summit of the stem, ovate, acute, obtuse, cuspidate or acuminate at the apex, rounded or cordate at the base, thin, 5–9-nerved, 2½′–5½′ long, 1¼′–4′ wide, sometimes larger, more or less pubescent beneath and erose-denticulate on the margins; petioles 10″–18″ long; peduncles long, often pubescent; staminate flowers commonly not more than 25 in the umbels; pedicels 2″–5″ long; anthers shorter than the filaments or equalling them.

In dry soil, Maryland to Minnesota, Florida and Missouri. May–June.

4. Smilax glaùca Walt. Glaucous-leaved Greenbrier. Fig. 1311.

Smilax glauca Walt. Fl. Car. 245. 1788.
S. spinulosum J. E. Smith; Torr. Fl. N. Y. **2**: 303. 1843.

Rootstock deep, knotted and tuberous. Stem terete; branches and twigs angled, armed with rather stout numerous or scattered prickles, or sometimes unarmed; petioles 3″–6″ long, tendril-bearing; leaves ovate, acute or cuspidate at the apex, sometimes cordate at the base, entire, glaucous beneath and sometimes also above, mostly 5-nerved, 1½′–6′ long, ½′–5′ wide; peduncles flattened 6″–16″ long; umbels 6–12-flowered; pedicels 2″–4″ long; berries bluish black, ripening the first year, about 3″ in diameter, 2–3-seeded.

In dry sandy soil, eastern Massachusetts to Florida, Ohio, Kansas and Texas. Sarsaparilla. May–June.

5. Smilax rotundifòlia L. Greenbrier. Catbrier. Horsebrier. Fig. 1312.

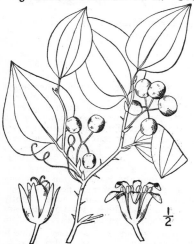

Smilax rotundifolia L. Sp. Pl. 1030. 1753.
Smilax caduca L. Sp. Pl. 1030. 1753.
Smilax quadrangularis Willd. Sp. Pl. **4**: 775. 1806.

Rootstocks long, sparingly tuberous. Stem woody, terete, the branches and young shoots often 4-angled, glabrous; prickles scattered, stout, straight or a little curved, sometimes none; petioles 3″–6″ long; leaves thick and shining when mature, thin when young, ovate, nearly orbicular, or lanceolate, acute or acuminate at the apex, obtuse or cordate at the base, entire or the margins erose-denticulate, 5-nerved, 2′–6′ long, 10″–6′ wide; peduncles flattened 3″–1′ long; umbels 6–25-flowered; pedicels 1″–4″ long; perianth-segments pubescent at the tip; filaments 2–3 times as long as the anthers; berries black, 1–3-seeded, about 3″ in diameter, maturing the first year.

In woods and thickets, Nova Scotia to Minnesota, south to Florida and Texas. Biscuit-leaves. Wait-a-bit. Nigger-head. Bamboo-brier. Devil's-hop-vine. Bread-and-butter. Hungry-vine. April–June.

6. Smilax híspida Muhl.   Hispid Greenbrier. Bristly Sarsaparilla.   Fig. 1313.

Smilax hispida Muhl.; Torr. Fl. N. Y. **2** : 302.   1843.

Glabrous, stem terete below, and commonly thickly hispid with numerous slender straight prickles, the branches more or less angled; petioles 4″–9″ long, tendril-bearing, rarely denticulate; leaves thin, green on both sides, ovate, abruptly acute and cuspidate at the apex, obtuse or subcordate at the base, 7-nerved, or the older ones sometimes with an additional pair of faint nerves, 2′–5′ long, 1′–5′ wide, the margins usually denticulate; peduncles flattened, 9″–2′ long; umbels 10–26-flowered; pedicels slender, 2″–3″ long; filaments a little longer than the anthers; berries bluish black, about 3″ in diameter, maturing the first year.

In thickets, Connecticut to Ontario, Minnesota, Nebraska, North Carolina and Texas. May–July.

7. Smilax Pseùdo-Chìna L.   Long-stalked Greenbrier.   Fig. 1314.

Smilax Pseudo-China L. Sp. Pl. 1031.   1753.

Glabrous throughout, rootstock often bearing large tubers, stem terete, the branches angled. Lower part of the stem beset with straight needle-shaped prickles, the upper part and the branches mostly unarmed; petioles 3″–12″ long; leaves firm, or becoming quite leathery when old, green on both sides or occasionally glaucous beneath, ovate, often narrowed at about the middle or lobed at base; acute or cuspidate at the apex, 7–9-nerved, 2½′–5′ long, 1½′–3½′ wide, often denticulate on the margins; peduncles flattened, 1′–3′ long; umbels 12–40-flowered; pedicels 3″–4″ long; stamens 6–10; anthers as long as the filaments or longer; berries black, 8–16 in the umbels, 2″–3″ in diameter, 1–3-seeded, maturing the first year.

In dry or sandy thickets, Maryland to Illinois, Nebraska, Kansas, Florida and Texas. American or False china-root. Sarsaparilla. Bull-brier. March–Aug. Recorded from New Jersey.

8. Smilax Bòna-nóx L.   Bristly Greenbrier.   Fig. 1315.

Smilax Bona-nox L. Sp. Pl. 1030.   1753.
Smilax hastata Willd. Sp. Pl. **4** : 782.   1806.
Smilax tamnoides A. Gray, Man. 485.   1848.   Not L.

Rootstocks bearing large tubers, stem terete or slightly angled, the branches often 4-angled. Prickles scattered or numerous, stout or needle-like, often wanting on the branches; petioles 3″–6″ long, often prickly; leaves thick, ovate or commonly deltoid-hastate, sometimes narrowed at the middle, glabrous, green and usually shining on both sides, often spiny on the margins and on the veins beneath, acute or abruptly cuspidate at the apex, obtuse, truncate or cordate at the base, 5–9-nerved, 1½′–4½′ long, 8″–3′ wide; peduncles slender, flattened, 7″–15″ long; umbels 15–45-flowered; pedicels 2″–4″ long; stigmas 1–3; berries 8–20 in the umbels, 2″–3″ in diameter, mostly 1-seeded, ripening the first year.

In thickets, New Jersey(?); Virginia to Kentucky, Kansas, Florida, Texas and Mexico. Fiddle-shaped greenbrier. April–July. Erroneously recorded from Massachusetts.

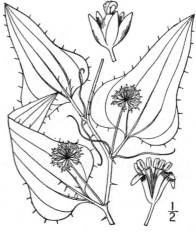

9. Smilax laurifòlia L. Laurel-leaved Greenbrier. Fig. 1316.

Smilax laurifolia L. Sp. Pl. 1030. 1753.

Rootstocks bearing tubers sometimes 6′ thick, stem stout, high-climbing, terete, striate, armed with strong straight prickles, the branches angled, mostly unarmed. Petioles stout, 3″–8″ long; leaves leathery, evergreen, elliptic or oblong-lanceolate, acute or abruptly cuspidate at the apex, narrowed at the base, entire, 3-nerved, or sometimes with an additional pair of nerves near the margins, 2′–4½′ long, ½′–2′ wide; peduncles stout, angled, 2″–10″ long; umbels 6–30-flowered; pedicels 2″–3″ long; anthers usually about one-third shorter than the filaments; stigma 1, sometimes 2; berries black, ovoid, 2″–3″ thick, not ripening until the second year.

In moist woods and thickets, southern New Jersey to Florida and Texas, north in the Mississippi Valley to Arkansas. Bahamas. Bamboo-vine. March–Sept.

10. Smilax Wàlteri Pursh. Walter's Greenbrier. Red-berry Bamboo. Fig. 1317.

Smilax Walteri Pursh, Fl. Am. Sept. 249. 1814.

Glabrous, stem angled, prickly below, the branches commonly unarmed. Petioles 2″–6″ long, stout, angled; leaves ovate or ovate-lanceolate, rarely lobed at the base, cordate or subcordate, obtuse or abruptly acute at the apex, entire, 5–7-nerved, 2′–5′ long, 10″–3¼′ wide; peduncles 2″–5″ long, flattened, thickening in age; umbels 6–15-flowered; pedicels very slender, 2″–3″ long; berries globose, coral-red (rarely white), 3″–4″ in diameter, 2–3-seeded, ripening the first year.

In wet soil, pine barrens of New Jersey to Florida, Tennessee and Louisiana. Sarsaparilla. April–June.

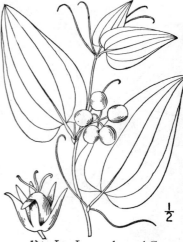

11. Smilax lanceolàta L. Lance-leaved Greenbrier. Fig. 1318.

Smilax lanceolata L. Sp. Pl. 1031. 1753.

Glabrous, stem terete, usually prickly, the branches slender, long, slightly angled, mostly unarmed. Petioles 1″–2″ long; leaves rather thin, lanceolate, acute or acuminate at the apex, narrowed at the base, entire, shining above, 5–7-nerved, 2′–3½′ long, 6″–20″ wide; peduncles thick, angled, 3″–8″ long; umbels 8–40-flowered; pedicels 2″–7″ long; filaments longer than the anthers; berries dark red, globose, 2″–3″ in diameter, usually 2-seeded, ripening the first year.

In thickets, Virginia to Arkansas, Florida and Texas. March–Aug.

Family 26. HAEMODORÀCEAE R. Br. Prodr. Fl. Nov. Holl. 1: 299. 1810.
BLOODWORT FAMILY.

Perennial herbs with erect stems, narrowly linear leaves, and regular or somewhat irregular small perfect flowers in terminal cymose panicles. Perianth 6-parted or 6-lobed, adnate to the ovary, persistent. Stamens 3, opposite the 3 inner perianth-segments. Ovary wholly or partly interior, 3-celled or rarely 1-celled; ovules usually few in each cavity, half-anatropous; style mostly slender; stigma small, entire or 3-grooved. Fruit a loculicidally 3-valved capsule. Seeds few or rarely numerous; embryo small in fleshy endosperm.

About 9 genera and 35 species, mostly natives of South Africa and Australia, a few in tropical America; only the following genus in the north temperate zone.

1. GYROTHÈCA Salisb. Trans. Hort. Soc. 1: 327. 1812.

[LACHNANTHES Ell. Bot. S. C. & Ga. 1: 47. 1816.]

A rather stout herb, with a short rootstock, red fibrous roots and equitant leaves, the basal ones longer than those of the stem. Flowers numerous, yellowish, small, in a dense terminal woolly cymose panicle. Perianth 6-parted to the summit of the ovary, the outer segments smaller than the inner. Filaments filiform, longer than the perianth; anthers linear-oblong, versatile. Ovary 3-celled; ovules few in each cavity, borne on fleshy placentae; style very slender, declined. Capsule enclosed by the withering-persistent perianth, nearly globular, 3-valved. Seeds about 6 in each cavity, flattened, nearly orbicular, peltate. [Greek, referring to the round fruit.]

A monotypic genus of southeastern North America and the West Indies.

1. Gyrotheca tinctòria (Walt.) Salisb.
Red root. Paint root. Fig. 1319.

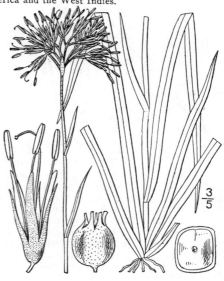

Gyrotheca tinctoria Salisb. Trans. Hort. Soc. 1: 327. 1812.

Lachnanthes tinctoria Ell. Bot. S. C. & Ga. 1: 47. 1816.

Gyrotheca capitata Morong. Bull. Torr. Club 20: 472. 1893. Not *Anonymo capitata* Walt.

Stem 1½°–2½° tall, glabrous below, pubescent or woolly above. Leaves 2″–5″ wide, acuminate, the basal ones shorter than the stem, the upper reduced to bracts; panicle 2′–5′ broad when expanded, dense and almost capitate when young, white-woolly; flowers 4″–5″ broad, bracteolate, the perianth yellow and glabrous within; style about as long as the stamens; pedicels stout, about as long as the capsule, rather shorter than the bractlets; capsule about 3″ in diameter.

In swamps, eastern Massachusetts to New Jersey and Florida, mostly in pine barrens near the coast. Also in Cuba. Carolina or Indian redroot. Spirit-weed. July–Sept.

Family 27. AMARYLLIDÀCEAE Lindl. Nat. Syst. Ed. 2: 328. 1836.

AMARYLLIS FAMILY.

Perennial herbs (some tropical species woody or even arboreous), with bulbs or rootstocks, scapose or sometimes leafy stems and usually narrow and entire leaves. Flowers perfect, regular or nearly so. Perianth 6-parted or 6-lobed, the segments or lobes distinct, or united below into a tube which is adnate to the surface of the ovary (adnate only to the lower part of the ovary in *Lophiola*). Stamens 6 in our genera, inserted on the bases of the perianth-segments or in the throat of the perianth opposite the lobes. Anthers versatile or basifixed, 2-celled, the sacs usually longitudinally dehiscent. Ovary wholly or partly inferior, usually 3-celled. Style filiform, entire, lobed, or divided into 3 stigmas at the summit. Ovules usually numerous, rarely only 1 or 2 in each cavity of the ovary, anatropous. Fruit capsular, rarely fleshy. Seeds mostly black, the embryo small, enclosed in fleshy endosperm.

About 70 genera and 800 species, principally natives of tropical and warm regions, some in the temperate zones.

Bulbous herbs with flowers on scapes.
  Flower solitary (in our species); perianth with a crown.   1. *Narcissus.*
  Flower solitary; perianth without a crown.
    Anthers versatile; tube of the perianth not greatly elongated.   2. *Atamasco.*
    Anthers erect; tube of the perianth several times the length of its lobes.   3. *Cooperia.*
  Flowers clustered; perianth with a membranous crown connecting the lower parts of the filaments.   4. *Hymenocallis.*
Bulbless herbs, with rootstocks or corms.
  Perianth adnate to the whole surface of the ovary; leaves mostly basal.
    Tall, fleshy-leaved; anthers versatile.   5. *Manfreda.*
    Low, linear-leaved; anthers not versatile.   6. *Hypoxis.*
  Perianth adnate only to the lower part of the ovary; stem leafy; flowers woolly.   7. *Lophiola.*

1. NARCÍSSUS L. Sp. Pl. 289. 1753.

Bulbous herbs, the flowers solitary or several on leafless scapes, the leaves linear, basal. Flowers subtended by a deciduous spathe; perianth 6-parted, bearing a cup-like funnelform or cylindric crown in the throat. Stamens inserted on the tube of the perianth; ovary 3-celled; ovules numerous in each cavity; capsule thin-walled.

About 20 species, natives of the Old World. Type species: *Narcissus poeticus* L.

1. Narcissus Pseùdo-Narcíssus L. Daffodil. Fig. 1320.

Narcissus Pseudo-Narcissus L. Sp. Pl. 289.   1753.

Scape about 1° high, 2-edged. Leaves narrowly linear, about as long as the scape; flower bright yellow, 2′–3′ broad; crown crenate, rather longer than the perianth.

Escaped from cultivation, Pennsylvania and New Jersey. Native of Europe. Called also Daffy, Daffodilly, Daffodowndilly. April–May. Flowers often double.

Narcissus poéticus L., Poets' Narcissus, with white flowers, the crown shorter than the perianth; has also, locally, escaped from cultivation; it is a native of Europe.

2. ATAMÓSCO Adans. Fam. Pl. 2: 57, 524. 1763.

[ZEPHYRÁNTHES Herb. App. Bot. Reg. 36. 1821.]

Glabrous herbs with coated bulbs, narrow leaves, and erect 1-flowered scapes, the flower large, erect, pink, white or purple. Perianth funnelform, naked in the throat, with 6 membranous equal erect-spreading lobes united below into a tube, subtended by an entire or 2-cleft bract. Stamens inserted on the throat of the perianth, equal or nearly so; anthers versatile. Ovary 3-celled; style long, filiform, 3-cleft at the summit; ovules numerous, in 2 rows in each cavity of the ovary. Capsule thin-walled, subglobose or depressed, 3-lobed loculicidally 3-valved. Seeds mostly flattened, black or nearly so. [Greek, signifying wind-flower.]

About 30 species, natives of America. Besides the following, 5 others occur in the southern United States. Type species: *Amaryllis Atamasco* L.

1. Atamosco Atamásco (L.) Greene. Atamasco Lily. Stagger-grass. Fig. 1321.

Amaryllis Atamasco L. Sp. Pl. 292.   1753.
Zephyranthes Atamasco Herb. App. Bot. Reg. 36.   1821.
Atamosco Atamasco Greene, Pittonia 3: 187.   1897.

Bulb ovoid, about 1′ long. Leaves fleshy, somewhat concave, shining, 6′–15′ long, about 1½″–3″ wide, blunt, usually shorter than the scape; scape terete, erect, 2″–3″ in diameter; bract membranous, 2-cleft into acuminate lobes, longer than the ovary; flowers 2′–3½′ high, white with a purplish tinge or sometimes light purple; perianth-segments oblong-lanceolate, acute, shorter than the tube; stamens shorter than the perianth; style longer than the stamens; capsule depressed, about ½′ high.

In moist places, southern Pennsylvania to eastern Virginia, Florida and Alabama. Perianth rarely 8-lobed. Swamp-, Fairy- or Easter-lily. March–June.

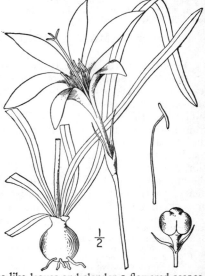

3. COOPÈRIA Herb. Bot. Reg. pl. *1835.* 1836.

Low herbs with coated bulbs, very narrow grass-like leaves and slender 1-flowered scapes, the flower large, long, erect, subtended by a membranous spathe-like bract. Perianth salverform with 6 oval or ovate spreading lobes united into a tube several times their length, the tube cylindric or slightly dilated at the summit. Stamens inserted on the throat of the perianth; filaments short; anthers linear, erect. Ovary 3-celled; style filiform; stigma slightly 3-lobed; ovules numerous, in 2 rows in each cavity of the ovary. Capsule depressed, globose or obovoid, 3-lobed, loculicidally 3-valved. Seeds numerous, horizontal, black. [In honor of Daniel Cooper, 1817?–1842, Curator, Botanical Society of London.]

Two known species, natives of the southwestern United States and Mexico, the following being the type.

1. Cooperia Drummóndii Herb. Drummond's Cooperia. Fig. 1322.

C. Drummondii Herb. Bot. Reg. *pl. 1835.* 1836.

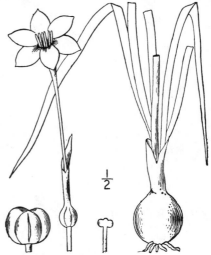

Bulb globose, about 1′ in diameter. Leaves 6′–12′ long, 2″–3″ wide, erect; scape slender, hollow, about as long as the leaves; spathe-like bract 1′–2′ long, 2-cleft above into acuminate lobes 4″–6″ long; flower 3′–5′ high, white or pinkish; tube of the perianth very slender, about 1½″ in diameter, slightly expanded just below the limb; segments oblong, obtuse and cuspidate or acutish, nearly 1′ long, 3″–4″ wide, ovary sessile; capsule somewhat obovoid, about ½′ in diameter, deeply lobed.

On prairies, Kansas to Louisiana, Texas, Mexico and New Mexico. Prairie-lily. April–July.

4. HYMENOCÁLLIS Salisb. Trans. Hort. Soc. 1: 338. 1812.

Mostly tall bulbous herbs with usually lanceolate or linear-oblong leaves, and large white sessile or short-pedicelled umbelled flowers on erect solid scapes, each flower subtended by 2 long membranous bracts. Perianth of 6 spreading or recurved narrow equal elongated lobes, united below into a long cylindric tube. Stamens inserted in the top of the perianth-tube, the lower parts of the long filaments connected by a membranous cup-like crown; anthers linear, versatile. Ovary 3-celled; ovules only 1 or 2 in each cavity; style filiform, long-exserted; stigma small, entire or nearly so. Capsule ovoid or globose, rather fleshy. Seeds usually only 1 or 2, large, green, fleshy. [Greek, beautiful membrane, referring to the crown.]

About 30 species, all American. Besides the following, 10 others occur in the southern States. Type species: *Hymenocallis littoralis* Salisb.

1. Hymenocallis occidentàlis (Le Conte) Kunth. Hymenocallis. Fig. 1323.

Pancratium carolinianum L. Sp. Pl. 291. 1753?.
Pancratium occidentale LeConte, Ann. Lyc. N. Y. 3: 146. 1830.
Hymenocallis occidentalis Kunth, Enum. 5: 856. 1850.

Bulb large. Leaves linear-oblong, narrowed at each end, fleshy, glaucous, 1°–2° long, 9″–2′ wide; scape stout, equalling or longer than the leaves; bracts linear-lanceolate, 1½′–2½′ long; umbel several-flowered; perianth-tube 1½″–2″ in diameter, 3′–5′ long, the linear lobes nearly as long; crown funnelform, narrowed below, 1′–1¼′ long, its margins entire, erose or 2-toothed between the filaments; free part of the filaments about 1′ long, white; anthers about ½′ long and ½″ wide, yellow; style extending for 2′–3′ beyond the crown, green; fruit 6″–9″ in diameter.

In moist soil, Georgia to Alabama, southern Illinois, Missouri and Arkansas. July–Sept.

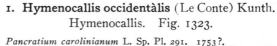

5. MÁNFREDA Salisb. Gen. Pl. Fragm. 78. 1866.

Fleshy herbs with bulbiferous rootstocks and bracted scapes, the leaves basal, and large bracted flowers in terminal spikes or racemes. Perianth tubular or funnelform, withering-persistent, of 6 erect or spreading equal or nearly equal lobes, united below into a tube. Stamens inserted on the perianth, exserted; filaments flattened or filiform; anthers versatile. Ovary 3-celled, style slender, exserted, 3-lobed; ovules numerous, in 2 rows in each cavity of the ovary. Capsule ovoid, subglobose or oblong, 3-lobed, 3-celled, thick-walled, many-seeded. Seeds compressed. [Named for Manfred, an ancient Italian writer.]

About 25 species, all American. Besides the following, some 3 others occur in the southern states. Type species: *Agave virginica* L.

1. Manfreda virgínica (L.) Salisb. False Aloe. Fig. 1324.

Agave virginica L. Sp. Pl. 323. 1753.
M. virginica Salisb.; Rose, Contr. U. S. Nat. Herb. **8**: 19. 1903.

Glabrous throughout, rootstock a short crown with numerous fibrous roots. Scape 2°–6° tall, rather slender, sometimes nearly ½′ in diameter at the base, its bracts distant, long-acuminate, the lower 3′–6′ long; leaves narrowly oblong, ½°–2° long, ½′–2½′ wide, acuminate, their margins entire or denticulate; spike 1°–2° long, loose; flowers greenish yellow, odorous, solitary in the axils of short bracts, sessile or the lowest distinctly pedicelled; perianth nearly tubular, slightly expanded above, 8″–12″ long, the tube about twice as long as the erect lobes; filaments at length about as long as the perianth; capsule 5″–8″ in diameter, slightly longer than thick, abruptly contracted into a short stalk.

In dry soil, Maryland to Indiana and Missouri, south to Florida and Texas. Rattlesnake's-master.

6. HYPÓXIS L. Syst. Ed. 10, **2**: 986. 1759.

Low, mostly villous herbs with a corm or short rootstock, grass-like leaves and slender few-flowered scapes, the flowers rather small. Perianth 6-parted, its segments equal or nearly so, separate to the summit of the ovary, spreading, withering-persistent, the 3 outer ones greenish on the lower side in our species. Stamens inserted on the bases of the perianth-segments; filaments short; anthers erect, sagittate or entire. Ovary 3-celled; style short; stigmas 3, erect; ovules numerous, in 2 rows in each cavity. Capsule subglobose or oblong, thin-walled, not dehiscent by valves. Seeds globular, black, laterally short-beaked by their stalks. [Greek, originally given to some plant with sour leaves.]

About 50 species, widely distributed. Besides the following, 3 others occur in the southern states. Type species: *Hypoxis erectum* L.

1. Hypoxis hirsùta (L.) Coville. Yellow Star-grass. Fig. 1325.

Ornithogalum hirsutum L. Sp. Pl. 306. 1753.
Hypoxis erectum L. Syst. Ed. 10, **2**: 986. 1759.
Hypoxis hirsuta Coville, Mem. Torr. Club **5**: 118. 1894.

Corm ovoid, oblong or globose, ¼′–½′ in diameter, with numerous fibrous roots. Leaves basal, narrowly linear, 1″–2½″ wide, more or less villous, mostly longer than the scapes; scapes slender, erect, villous above, usually glabrous below, 2′–6′ high; flowers 1–6, umbellate; bracts subulate, shorter than the pedicels; perianth-segments narrowly oblong, spreading, mostly obtuse, bright yellow within, greenish and villous without, 3″–5″ long; stamens somewhat unequal; style rather shorter than the stamens, 3-angled, the stigmas decurrent on the angles; capsule about 1½″ in diameter; seeds angled, black.

In dry soil, Maine to Ontario, Assiniboia, Kansas, Florida and Texas. Ascends to 3000 ft. in Virginia. Star-of-Bethlehem. May–Oct.

7. LOPHÌOLA Ker, Bot. Mag. *pl. 1596.* 1814.

An erect perennial herb with slender rootstocks, fibrous roots, erect sparingly leafy stems, the leaves narrowly linear and mostly basal, and numerous small yellowish flowers in a terminal woolly cymose panicle. Perianth campanulate, persistent, of 6 nearly equal woolly erect-spreading segments, slightly united at the base, and adnate to the lower part of the ovary. Stamens inserted on the bases of the perianth-segments; filaments filiform, short; anthers basifixed. Ovary 3-celled; style subulate, at length 3-cleft; ovules numerous, in 2 rows in each cavity. Capsule ovoid, tipped with the style, finally loculicidally 3-valved at the summit. Seeds oblong, ribbed. [Greek, referring to the tufts of wool on the perianth.]

A monotypic genus of southeastern North America, by some authors placed in the Haemodoraceae.

1. Lophiola americàna (Pursh) Coville. Lophiola. Fig. 1326.

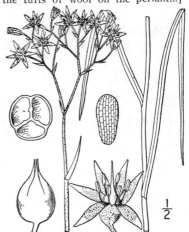

Conostylis americana Pursh, Fl. Am. Sept. 224. 1814.
Lophiola aurea Ker, Bot. Mag. *pl. 1596.* 1814.
Lophiola americana Coville, Mem. Torr. Club 5: 118. 1894.

Stem stiff, erect, terete, glabrous below, white-woolly above, 1°–2° tall. Leaves equitant, glabrous, much shorter than the stem, the upper ones reduced to bracts; panicle densely white-woolly, composed of numerous few–several-flowered cymes; pedicels short, rather stout, erect or ascending; perianth-segments linear-lanceolate, about 2″ long, woolly outside, longer than the stamens and with a tuft of wool at the base within; capsule about as long as the persistent style, shorter than the perianth.

Pine barren bogs, New Jersey to Florida. June–Aug.

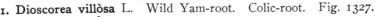

Family 28. DIOSCOREÀCEAE Lindl. Nat. Syst. Ed. 2: 359. 1836.
YAM FAMILY.

Herbaceous or slightly woody twining vines with fleshy or woody rootstocks, slender stems, petioled, mostly cordate, several-nerved and reticulate-veined leaves, alternate or the lower opposite or verticillate, and small inconspicuous dioecious or monoecious (in some exotic genera perfect) regular flowers in spikes, racemes or panicles. Perianth 6-parted, that of the pistillate flowers persistent. Staminate flowers with 6 or 3 stamens, sometimes with a rudimentary ovary. Pistillate flowers with an inferior 3-celled ovary, 3 styles and 3 terminal stigmas, sometimes also with 3 or 6 staminodia; ovules 2 (rarely 1) in each cavity of the ovary, pendulous, anatropous or amphitropous. Fruit a 3-valved, 3-angled capsule in the following genus. Endosperm of the seed fleshy or cartilaginous, enclosing the small embryo.

About 9 genera and 175 species, mostly natives of America, a few in the Old World.

1. DIOSCORÈA [Plum.] L. Sp. Pl. 1032. 1753.

Characters of the family as defined above. [Name in honor of the Greek naturalist Dioscorides.]

About 160 species, numerous in tropical regions, a few in the temperate zones. The large fleshy rootstocks of several tropical species furnish the yams of commerce. Type species: *D. sativa* L.

1. Dioscorea villòsa L. Wild Yam-root. Colic-root. Fig. 1327.

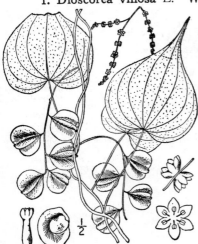

Dioscorea villosa L. Sp. Pl. 1033. 1753.

Rootstock slender or stout, simple or branched, horizontal, woody, ¼–¾′ thick. Stem 6°–15° long, twining or rarely suberect; leaves ovate, entire, slender-petioled, alternate or the lower opposite or in 4's, acuminate at the apex, cordate at the base, 2′–6′ long, 1′–4′ wide, 9–13-nerved, thin, green and glabrous or nearly so above, pale and pubescent or sometimes glabrous beneath; petioles often longer than the blades; flowers greenish yellow, nearly sessile, the staminate 1″–1½″ broad in drooping panicles 3′–6′ long, the pistillate about 3″ long in drooping spicate racemes; capsules membranous, yellowish green, 7″–12″ long, strongly 3-winged, with 2 or sometimes only 1 thin-winged seed in each cavity.

In moist thickets, Rhode Island to Ontario, Minnesota, Kansas, Florida and Texas. Rheumatism-root. June–July. Fruit ripe Sept., persistent on the vines into the winter. Consists of several races, by some regarded as distinct species, differing in the amount of pubescence and in the rootstock. The plant identified as *D. paniculata* Michx., corresponds closely with the Linnaean type.

Family 29. **IRIDÀCEAE** Lindl. Nat. Syst. Ed. 2: 382. 1836.

IRIS FAMILY.

Perennial herbs with narrow equitant 2-ranked leaves and perfect regular or irregular mostly clustered flowers subtended by bracts. Perianth of 6 segments or 6-lobed, its tube adnate to the ovary, the segments or lobes in two series, convolute in the bud, withering-persistent. Stamens 3, inserted on the perianth opposite its outer series of segments or lobes; filaments filiform, distinct or united; anthers 2-celled, extrorse. Ovary inferior, mostly 3-celled; ovules mostly numerous in each cell, anatropous; style 3-cleft, its branches sometimes divided. Capsule 3-celled, loculicidally dehiscent, 3-angled or 3-lobed (sometimes 6-lobed), many-seeded. Seeds numerous, in 1 or 2 rows in each cavity of the capsule. Endosperm of the seed fleshy or horny; embryo straight, small.

About 57 genera and 1000 species, of wide distribution, in temperate and tropical regions of both hemispheres.

Style-branches opposite the anthers, very broad, petal-like.
Style-branches alternate with the anthers, slender or filiform.
  Style-branches 2-cleft; plants bulbous.
  Style-branches undivided; plants not bulbous.
    Filaments all distinct; seeds fleshy.
    Filaments united; seeds dry.

1. *Iris.*
2. *Nemastylis.*

3. *Gemmingia.*
4. *Sisyrinchium.*

1. **ÌRIS** [Tourn.] L. Sp. Pl. 38. 1753.

Herbs with creeping or horizontal, often woody and sometimes tuber-bearing rootstocks, erect stems, erect or ascending equitant leaves, and large regular terminal sometimes panicled flowers. Perianth of 6 clawed segments united below into a tube, the 3 outer dilated, spreading or reflexed, the 3 inner narrower, smaller, usually erect, or in some species about as large as the outer. Stamens inserted at the base of the outer perianth-segments; anthers linear or oblong. Ovary 3-celled; divisions of the style petal-like, arching over the stamens, bearing the stigmas immediately under their mostly 2-lobed tips; style-base adnate to the perianth-tube. Capsule oblong or oval, 3–6-angled or lobed, mostly coriaceous. Seeds numerous, vertically compressed, in 1 or 2 rows in each cell. [Greek, rainbow, referring to the variegated flowers.]

About 100 species, mostly in the north temperate zone. Besides the following, some 8 others occur in the southern and western parts of North America. The names *Flower-de-luce* and *Fleur-de-lis* are applied to the species, many of which are in cultivation, and highly esteemed for their beautiful flowers. Type species: *Iris germanica* L.

Stems tall, usually several-flowered, leafy; outer perianth-segments distinctly larger than the inner, native species.
  Flowers blue, variegated with yellow, white or green (rarely all white).
    None of the perianth-segments crested; native species.
      Leaves ½'–1' wide.
        Leaves somewhat glaucous.
        Leaves bright green, not glaucous.
          Outer perianth-segments 3'–4' long; flowers sessile.
          Outer perianth-segments 2½'–3' long; flowers pedicelled.
            Capsule 3-angled; seeds in 1 row in each cavity.
            Capsule 6-angled; seeds in 2 rows in each cavity.
      Leaves much narrower, 2"–5" wide.
        Capsule obtusely angled, 3–6-lobed.
          Capsule 3-lobed; northern.
          Capsule 6-lobed; western.
        Capsule sharply 3-angled.
    Outer perianth-segments strongly crested; introduced and widely cultivated species.
  Flowers reddish or red-brown; native species.
  Flowers bright yellow; introduced species.
Stems low, seldom over 6' tall, 1–3-flowered; outer and inner perianth-segments nearly equal.
  Outer perianth-segments crested; leaves lanceolate.
    Perianth-tube very slender, exceeding the bracts.
    Perianth-tube expanded above, not exceeding the bracts.
  Outer perianth-segments crestless, claws slightly pubescent; leaves narrowly linear.

1. *I. versicolor.*

2. *I. hexagona.*

3. *I. georgiana.*
4. *I. foliosa.*

5. *I. Hookeri.*
6. *I. missouriensis.*
7. *I. prismatica.*
8. *I. germanica.*
9. *I. fulva.*
10. *I. Pseudacorus.*

11. *I. cristata.*
12. *I. lacustris.*

13. *I. verna.*

1. Iris versícolor L.   Larger Blue-flag.   Poison- or Water-flag.   Fig. 1328.

Iris versicolor L. Sp. Pl. 39.   1753.

Iris virginica L. Sp. Pl. 39.   1753.

Rootstock horizontal, thick, fleshy, covered with the fibrous roots. Stems terete or nearly so, straight or flexuous, 2°–3° tall, often branched above, leafy; leaves erect, shorter than the stem, somewhat glaucous, 6″–12″ wide; bracts commonly longer than the pedicels, the lower one sometimes foliaceous; flowers several, violet-blue, variegated with yellow, green and white; perianth-segments glabrous, crestless, the outer ones spatulate, 2′–3′ long, longer and wider than the inner; perianth-tube dilated upward, shorter than the ovary; capsule oblong, obscurely 3-lobed, about 1½′ long and 8″ in diameter; seeds 2″–3″ broad, in 2 rows in each cell, the raphe indistinct.

In marshes, thickets and wet meadows, Newfoundland to Manitoba, south to Florida and Arkansas. Snake-, Liver- or Flag-lily.   May–July.

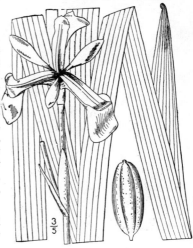

2. Iris hexagòna Walt.   Southern Blue-flag. Fig. 1329.

Iris hexagona Walt. Fl. Car. 66.   1788.

Rootstock stout, thick. Stems terete, usually simple, straight or flexuous, leafy, 1°–3° tall; leaves ½′–1½′ wide, green, not glaucous, the lower often 2°–3° long; flowers solitary in the upper axils, sessile, similar to those of *I. versicolor*, but larger, the broader outer crestless perianth segments often 4′ long and over 1′ wide, much wider than the erect inner ones; perianth-tube rather longer than the ovary, a little dilated upward; capsule oblong-cylindric, 6-angled, about 2′ long; seeds in 2 rows in each cavity.

In swamps, South Carolina to Florida, Kentucky(?) and Texas. Not certainly known within our area.   April–May.

3. Iris georgiàna Britton.   Carolina Blue-flag. Fig. 1330.

Iris caroliniana S. Wats. in A. Gray, Man. Ed. 6, 514.   1890. Not *I. carolina* Radius, 1822.

Rootstock stout, fleshy. Stem rather stout, simple or branched 2°–3° tall, equalled or exceeded by the bright green leaves which are 8″–1¼′ wide; flowers solitary or 2 or 3 together, lilac, variegated with yellow, purple and brown, pedicelled; pedicels somewhat shorter than the bracts; outer perianth-segments broadly spatulate, 2½′–3′ long, with narrow claws, the inner narrower and nearly erect; perianth-tube about ½′ long above the ovary; capsule oblong, obtusely 3-angled, 1½′–2′ long; seeds in 1 row in each cavity, 4″–5″ broad.

In swamps, southern Virginia and eastern North Carolina to Georgia and Louisiana.   May–June.

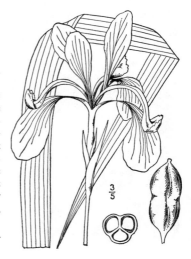

4. Iris foliòsa Mackenzie & Bush. Leafy Blue-flag. Fig. 1331.

Iris foliosa Mackenzie & Bush, Trans. Acad. St. Louis **12**: 81. 1902.

Rootstock stout. Stem terete, zigzag, $1°-1\frac{1}{2}°$ tall; leaves conspicuously overtopping the inflorescence, green, $\frac{3}{4}-1\frac{1}{4}'$ wide, the larger ones about $2°$ long; flowers axillary, on pedicels $10''-14''$ long; perianth-segments bluish, spreading, $2\frac{1}{4}'$ long, about $1'$ wide, not crested; capsule strongly 6-angled, oblong-cylindric, about $1\frac{1}{2}'$ long, short-beaked; seeds in 2 rows in each cavity.

In swamps, Kentucky to Illinois, Missouri and Kansas. June–July.

5. Iris Hóokeri Penny. Hooker's Blue-flag. Fig. 1332.

Iris Hookeri Penny; Steud. Nomencl. Ed. 2, Part 1, 822. 1840.
I. setosa canadensis Foster, Rhodora **5**: 158. 1903.

Rootstock rather slender. Stems slender, simple or branched, terete, $10'-2°$ tall. Leaves mostly basal, narrowly linear, bright green, shorter than or equalling the stem, $2''-5''$ wide; flowers solitary or 2 together, violet-blue and white, pedicelled, the pedicels shorter than the bracts; perianth-segments glabrous, crestless, the inner ones involute, oblanceolate, much shorter and smaller than the outer; capsule short-oblong, blunt, $1'-1\frac{1}{2}'$ long, $7''-10''$ in diameter, thin-walled, transversely veined, obtusely 3-lobed; seeds in 2 rows in each cavity, about $1\frac{1}{2}''$ broad, the raphe prominent.

On river shores, Newfoundland and Labrador to Quebec and Maine. Closely related to the Asiatic *Iris setòsa* Pall. Summer.

6. Iris missouriénsis Nutt. Western Blue-flag. Fig. 1333.

I. missouriensis Nutt. Journ. Acad. Phila. **7**: 58. 1834.

Rootstock stout. Stem rather slender, usually simple, terete, $6'-2°$ tall, 1–2-flowered; leaves mostly basal, green, sometimes purplish below, shorter than or about equalling the stem, $2''-4''$ wide; flowers pale blue and variegated, pedicelled; pedicels slender, $\frac{1}{2}'-2'$ long, usually shorter than the scarious bracts; perianth-segments glabrous, crestless, the outer ones $2'-2\frac{1}{2}'$ long, the inner somewhat shorter, the tube $3''-4''$ long above the ovary; capsule oblong, $1'-1\frac{1}{2}'$ long, about $\frac{1}{2}'$ in diameter, obtusely 6-angled, faintly veined.

In wet soil, South Dakota to Montana, Idaho and Nevada, south to Colorado and Arizona. May–July.

7. Iris prismática Pursh.   Narrow Blue-flag.   Poison Flag-root.   Fig. 1334.

Iris virginica Muhl. Cat. 4. 1813. Not L. 1753.
Iris prismatica Pursh, Fl. Am. Sept. 30. 1814.
Iris gracilis Bigel. Fl. Bost. 12. 1814.
?Iris carolina Radius, Schrift. Naturf. Ges. Leipzig 1: 158. 1822.

Rootstock rather slender, tuberous-thickened. Stems slender, often flexuous, 1°–3° tall, usually simple, bearing 2 or 3 leaves; leaves almost grass-like, 1½″–2½″ wide, mostly shorter than the stem; flowers solitary or 2 together, blue veined with yellow, slender-pedicelled; pedicels commonly longer than the bracts; outer perianth-segments 1½′–2′ long, glabrous and crestless, the inner smaller and nar-rower, the tube 2″–3″ long above the ovary; capsule narrowly oblong, acute at each end, sharply 3-angled, 1′–1½′ long, 3″–4″ thick; seeds about 1″ broad, thick, borne in 1 row in each cavity.

In wet grounds, Nova Scotia to Pennsylvania and Georgia, mainly near the coast. May–June.

8. Iris germánica L.   Fleur-de-lis.   Fig. 1335.

Iris germanica L. Sp. Pl. 38. 1753.

Rootstock thick. Stems stout, usually branched and several-flowered, 2°–3° tall, bearing several leaves. Leaves glaucous, 8″–2′ wide, the basal ones mostly shorter than the stem; bracts scarious; flowers nearly sessile in the bracts, large and very showy, deep violet-blue veined with yellow and brown or sometimes white; outer perianth-segments broadly obovate, 3′–4′ long, their claws strongly crested; inner perianth-segments narrower, arching.

Escaped from gardens to roadsides in Massachusetts and Virginia. Native of Europe. May–June.

Iris Duerínckii Buckley, Am. Journ. Sci. **45**: 176, de-scribed from specimens collected at St. Louis, Mo., but doubtless cultivated, appears to be *I. aphýlla* L., a native of central Europe.

9. Iris fúlva Ker.   Red-brown Flag.   Fig. 1336.

Iris fulva Ker, Bot. Mag. *pl. 1496.* 1812.

Iris cuprea Pursh, Fl. Am. Sept. 30. 1814.

Rootstock stout, fleshy. Stems rather slender, 2°–3° tall, simple or branched, several-flowered and bearing 2–4 leaves; leaves pale green and some-what glaucous, shorter than or equalling the stem, 3″–8″ wide; pedicels ½″–1′ long, shorter than the bracts; flowers reddish brown, variegated with blue and green; perianth-segments glabrous, crestless, the outer ones 1½′–2′ long, the inner smaller, spread-ing; style-branches 2″–3″ wide.

In swamps, southern Illinois to Georgia and Louisiana, west to Missouri, Arkansas and Texas. May–June.

10. Iris Pseudácorus L. Yellow or Sword-flag, Corn-flag. Fig. 1337.

Iris Pseudacorus L. Sp. Pl. 38. 1753.

Rootstock thick. Stems $1\frac{1}{2}°-3°$ high, usually several-flowered; leaves pale green and glaucous, stiff, $4''-8''$ wide, the lower equalling or longer than the stem; flowers bright yellow, short-pedicelled; perianth-segments glabrous and crestless, the outer broadly obovate, $2'-2\frac{1}{2}'$ long, the inner oblong, nearly erect, scarcely longer than the claws of the outer ones; capsule oblong, $2'-3'$ long.

In marshes, Massachusetts to southern New York and New Jersey. Naturalized or adventive from Europe. False Sweet-flag. Yellow Water-flag or -skegs. Jacob's-sword. Daggers. Flagons. May–July.

Iris orientàlis Mill., native of Asia, with pale yellow flowers and short stem-leaves, is reported as escaped from cultivation into marshes on the coast of Connecticut.

11. Iris cristàta Ait. Crested Dwarf Iris. Fig. 1338.

Iris cristata Ait. Hort. Kew. 1 : 70. 1789.

Rootstock slender, branched, creeping, tuberous-thickened. Stems only $1'-3'$ high, 1–2-flowered; leaves lanceolate, bright green, $4'-9'$ long, $3''-9''$ wide, much exceeding the stems; scape flattened, flowers blue, pedicelled; perianth-segments obovate, $1'-1\frac{1}{2}'$ long, the outer crested, little longer than the naked inner ones, the tube very slender, $1\frac{1}{2}'-2\frac{1}{2}'$ long above the ovary, longer than the bracts; capsule oval, sharply triangular, narrowed at each end, $6''-9''$ high, $4''-5''$ thick.

On hillsides and along streams, Maryland to southern Ohio and Indiana, south to Georgia, Tennessee and Missouri. April–May.

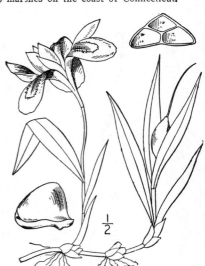

12. Iris lacústris Nutt. Dwarf Lake-iris. Fig. 1339.

Iris lacustris Nutt. Gen. 1 : 23. 1818.

Similar to the preceding species in size and foliage, or the leaves rather narrower, sometimes wavy-margined. Flowers blue; perianth-tube only $\frac{1}{2}'-1'$ long, shorter than the bracts and the sometimes yellowish perianth-segments somewhat expanded upward; capsule ovoid, about $8''$ high, borne on a pedicel of about its own length.

Shores of Lakes Huron, Michigan and Superior. Local. May. Perhaps a race of the preceding species.

13. Iris vérna L. Dwarf or Spring Iris.
Fig. 1340.

Iris verna L. Sp. Pl. 39. 1753.

Rootstock slender. Stems 1′-3′ high, usually 1-flow-ered. Leaves narrowly linear, 3′-8′ high, 2″-5″ wide; flowers violet-blue or rarely white, pedicelled; perianth-segments crestless, the outer about 1½′ long, obovate, narrowed into slightly pubescent slender yellow claws, the inner somewhat smaller, glabrous; capsule obtusely triangular, short.

On shaded hillsides and in woods, southern Pennsylvania to Virginia, Kentucky, Alabama and Georgia. Rootstock described as "pungently spicy." Slender Blue-flag. April–May.

2. NEMÁSTYLIS Nutt. Trans. Am. Phil. Soc. (II.) 5: 157. 1833–37.

[EUSTYLIS Engelm. & Gray, Bost. Journ. Nat. Hist. 5: 235. 1845.]

Bulbous herbs with erect slender terete usually branched stems and elongated linear folded leaves. Flowers rather large, in our species blue or purple, solitary or several together, fugacious, subtended by 2 herbaceous bracts. Perianth of 6 spreading nearly equal obovate segments, distinct nearly or quite to the summit of the ovary. Filaments more or less united; anthers short; style short, its branches alternate with the anthers, each slen-derly 2-parted; stigmas small, terminal. Capsule oblong, ovoid or obovoid, loculicidally dehiscent at the summit. [Greek, referring to the thread-like style-branches.]

About 10 species, natives of America. Besides the following, some 3 others occur in the south-ern United States. Type species: *Nemastylis coelestìna* Nutt.

1. Nemastylis acùta (Bart.) Herb. Northern Nemastylis. Fig. 1341.

Ixia acuta Bart. Fl. N. A. 2: 89. *pl. 66.* 1822.
Nemastylis gemmiflora Nutt. Trans. Am. Phil. Soc. (II.) 5: 157. 1833–37.
Nemastylis acuta Herb. Bot. Mag. *pl. 3779.* 1839–40.

Bulb dark colored, ovoid, scaly, 1′ or less long. Stem 1°-2° tall, bearing 3 or 4 leaves, 3′-10′ long, 1½″-2½″ wide; bracts lanceolate, each pair subtend-ing 1 or 2 flowers; flowers light blue or purple, 1′-2′ broad; pedicels slender, rather shorter than the bracts; perianth-segments oblong-obovate, obtuse; style-branches exserted between the free parts of the filaments, their filiform divisions 2″-3″ long; cap-sule obovoid, 5″-6″ high, 3″-4″ in diameter.

On prairies, Tennessee to Missouri, Kansas, Arkansas, Louisiana and Texas. April–June.

Nemastylis coelestìna Nutt., ranging from Georgia to Arkansas and Texas, may occur in south-ern Missouri; it differs from *N. acuta* in having more broadly obovate perianth-lobes.

3. GEMMÍNGIA Fabr. Enum. Pl. Hort. Helm. 1759.

[BELAMCANDA Adans. Fam. Pl. 2: 60. 1763.]

[PARDANTHUS Ker, in Konig & Sims, Ann. Bot. 1: 246. 1805.]

An erect perennial herb, with short stout rootstocks and *Iris*-like leaves. Flowers in terminal bracted clusters, rather large, orange and purple-mottled. Perianth of 6 oblong spreading nearly equal withering-persistent segments, distinct very nearly to the summit of the ovary. Stamens inserted on the bases of the segments; filaments distinct; anthers linear-oblong. Style very slender, enlarged above, the 3 slender undivided branches alternate with the anthers. Capsule fig-shaped, obovoid, thin-walled, loculicidally 3-valved, the valves recurving, finally falling away, exposing the mass of black fleshy seeds, borne on a central axis.

A monotypic genus of eastern Asia, based on *Ixia chinensis* L.

1. Gemmingia chinénsis (L.) Kuntze. Blackberry Lily. Fig. 1342.

Ixia chinensis L. Sp. Pl. 36. 1753.
Belamcanda chinensis DC. in Red. Lil. **3**: *pl. 121.* 1807.
Pardánthus chinensis Ker, in Konig & Sims, Ann. Bot. **1**: 246. 1805.
G. chinensis Kuntze, Rev. Gen. Pl. 701. 1891.

Stem rather stout, $1\frac{1}{2}°-4°$ tall, leafy. Leaves pale green, nearly erect, equitant, folded, 8′–10′ long, 8″–12″ wide, the two sides united above the middle; bracts lanceolate, much shorter than the leaves, the upper ones scarious; flowers several or numerous, $1\frac{1}{2}′-2′$ broad; perianth-segments obtuse at the apex, narrowed at the base, persistent and coiled together on the ovary after flowering, mottled with crimson and purple on the upper side; capsule about 1′ high and rather more than $\frac{1}{2}′$ in diameter, truncate or rounded at the summit; mass of globose seeds erect, resembling a blackberry, whence the common name.

On hills and along roadsides, Connecticut to Georgia, Indiana and Kansas. Naturalized from Asia. Leopard-flower. Dwarf tiger-lily. June–July. Fruit ripe July–Sept.

4. SISYRÍNCHIUM L. Sp. Pl. 954. 1753.

Perennial tufted slender herbs, with short rootstocks, simple or branched 2-edged or 2-winged stems, linear grass-like leaves, and rather small mostly blue terminal flowers umbellate from a pair of erect green bracts. Perianth-tube short or none, the 6 spreading segments oblong or obovate, equal, mostly aristulate. Filaments united to above the middle in our species. Ovary 3-celled, each cavity several-ovuled. Style-branches filiform, undivided, alternate with the anthers. Capsule globose, oval or obovoid, loculicidally 3-valved. Seeds subglobose or ovoid, smooth or pitted, dry.

About 150 species, all American. Besides the following, many others occur in the southern and western states, some in Mexico and a few in the West Indies. Type species: *Sisyrinchium Bermudiana* L., which has larger flowers and fruit than any of ours, and is found only in Bermuda.

Filaments free above; anthers about 2″ long; spathes 2, sessile.	1. *S. hastile.*
Filaments united to the top; anthers about 1″ long.	
Spathes 2 together, sessile; stem simple.	2. *S. albidum.*
Spathes solitary at the end of the stem or branches.	
Stems simple (occasionally branched in *S. angustifolium*).	
Capsules 2″–3″ high; pedicels ascending.	3. *S. angustifolium.*
Capsules less than 2″ high; leaves very narrow.	
Margins of the outer bract united-clasping below.	4. *S. mucronatum.*
Margins of the outer bract separate to the base.	5. *S. campestre.*
Stems branched above, the several spathes long-stalked.	
Basal leaf-sheaths persistent as tufts of fibers.	6. *S. arenicola.*
Basal leaf-sheaths not persistent.	
Stems broadly winged; pedicels spreading.	7. *S. graminoides.*
Stems narrowly winged.	
Peduncles strictly erect.	8. *S. strictum.*
Peduncles diverging or ascending.	9. *S. atlanticum.*

1. Sisyrinchium hastìle Bicknell. Spear-like Blue-eyed Grass. Fig. 1343.

S. hastile Bicknell, Bull. Torr. Club **26**: 297. 1899.

Very slender and stiffly erect, dull green, about 1° high. Stems ½″ wide or less, compressed-subterete and bluntly two-edged, not at all winged, closely striate, minutely granulose-roughened; leaves similar to the stems, usually shorter, thick-edged, obtusely slender-pointed, the conduplicate broadened base smooth and membranous; spathes usually two in a close pair at the top of the stem, each usually 4-bracted, the bracts lanceolate-attenuate, the inner ones about 1′ long; primary bract much prolonged; interior scales ample, sometimes slightly exserted; flowers few, pedicels not longer than the inner bracts; young capsule longer than thick. Fruit and color of flower unknown.

Sandy shores of Belle Isle in the Detroit River, Mich. Much resembles *S. Pringlei* Rob. & Greenm. from Jalisco, Mexico. Early June.

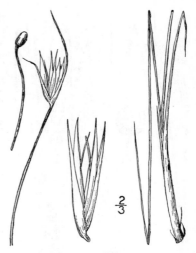

2. Sisyrinchium álbidum Raf. White Blue-eyed Grass. Fig. 1344.

S. albidum Raf. Atl. Journ. 17. 1832.
S. versicolor Bicknell, Bull. Torr. Club **26**: 606. 1899.

Green and glaucescent, the spathes often purplish, 2° high or less. Leaves half the height of the stems or longer, ½″–2″ wide, very acute, mostly smooth-edged; stems usually broadly winged, the edges serrulate to smooth; spathes sessile in a close terminal pair, the prominent outer bract 1′–3′ long, its edges free to the base; inner bracts acuminate, often ciliolate on the keel, much shorter than the outer; flowers white to pale blue; 4″–6″ long; capsules pale, depressed-subglobose, 1″–1½″ high, on erect-spreading pedicels usually shorter than the inner bracts; seeds globose, ½″ or less in diameter, umbilicate, distinctly pitted.

Grassy places, Ontario to Wisconsin, Ohio, North Carolina, Arkansas and Mississippi. Also in Connecticut and in southern New York, apparently introduced. April–June.

Sisyrinchium flaviflòrum Bicknell, is a little known, yellow flowered species from Missouri, perhaps *S. campestre* with yellow flowers, as suggested by Mr. B. F. Bush, the collector.

3. Sisyrinchium angustifòlium Mill. Pointed Blue-eyed Grass. Fig. 1345.

S. angustifolium Mill. Gard. Dict. Ed. 8. 1768
S. montanum Greene, Pittonia 4: 33. 1899.

Stiff and erect, pale and glaucous, 4′–2° high. Leaves half the height of the stem or longer, ½″–2″ wide, acute, the edges minutely serrulate; stem simple or rarely branched, winged, the edges minutely serrulate; spathes green or slightly purplish, the outer bract rarely less than twice the length of the inner one, 2½′ long or less, obscurely hyaline-margined, united-clasping at base; inner bract attenuate; flowers deep violet-blue, 5″–6″ long; capsules broadly oval to globose, 2″–3″ high, dull brown to whitish, often purplish-tinged, on erect pedicels usually shorter than the inner bract.

Fields and hillsides, Newfoundland to Saskatchewan, British Columbia, Virginia, Nebraska, Colorado and Utah. Blue-eyed Mary. Star-eyed grass. Grass-flower. Pigroot. Blue-grass. May–July.

Sisyrinchium septentrionàle Bicknell, a diminutive species with large capsules, enters our northwestern limits in North Dakota.

4. Sisyrinchium mucronàtum Michx. Michaux's Blue-eyed Grass. Fig. 1346.

S. *mucronatum* Michx. Fl. Bor. Am. **2** : 33. 1803.
S. *intermedium* Bicknell, Bull. Torr. Club **26** : 498. 1899.

More caespitose than *S. angustifolium* and decidedly more slender and delicate, with smaller spathes and capsules, sometimes scarcely glaucescent and the spathes often bright red-purple. Stem and leaves from capillaceous to ¾″ wide, rarely wider, the stems narrowly winged or merely margined; bracts thin, glabrous, hyaline-margined, the outer one slenderly prolonged, united-clasping at base; inner bracts scarious, obtuse to attenuate, gradually emerging from the outer one, flowers deep violet-blue, rarely white, 3″–7″ long; capsules pale and thin-walled, subglobose to obovate-oblong, 1″–2″ high, on slender subspreading exserted pedicels.

Fields and meadows, Massachusetts to Michigan, Pennsylvania and Virginia. May–June.

5. Sisyrinchium campéstre Bicknell. Prairie Blue-eyed Grass. Fig. 1347.

S. *campestre* Bicknell, Bull. Torr. Club **26** : 341. 1899.
S. *campestre kansanum* Bicknell, loc. cit. 344. 1899.

Similar to *S. mucronatum* Michx., but often stiffer and more glaucous, with always smooth-edged stems and leaves, usually broader, more gibbous spathes, and pale blue or frequently white flowers. Stem ½″–1″ wide, the leaves rather broader; spathes green to dull pink-purple, the bracts commonly scabrous-puberulent all over, but sometimes glabrous, the outer one usually less slenderly prolonged than in *S. mucronatum,* and not united-clasping at base, or but slightly so, 1′–2′ long, rarely more than twice longer than the inner bract, which emerges more abruptly from the base of the spathe than in *S. mucronatum,* and has more broadly hyaline edges; capsules pale, trigonous-subglobose, 1″–2″ high.

Prairies, Wisconsin. to Missouri, Louisiana, North Dakota, Kansas and New Mexico. May–June.

6. Sisyrinchium arenícola Bicknell. Sand Blue-eyed Grass. Fig. 1348.

S. *arenicola* Bicknell, Bull. Torr. Club **26** : 496. 1899.

Deep green or glaucescent, often purplish-tinged, the base of the tufts coarsely fibrous. Leaves stiff, attenuate, ½″–1½″ wide, closely striate, usually serrulate; stem erect, often curved, 8′–20′ high, stiff, the firm wings striate, mostly serrulate; node commonly only one, its erect leaf conspicuous, attenuate-acute, the compressed broadened base strongly striate; peduncles 2–4, curved; spathes erect, the acute bracts firm and strongly striate, slightly unequal; flowers sometimes as many as 12, deep violet-blue, 4″–5″ long; capsules dark and thick-walled on erect slightly exserted pedicels, broadly oval or obovoid, 2″–3″ high.

Sandy soil, Massachusetts to New Jersey and North Carolina, mostly near the coast. June–July.

Sisyrinchium Farwéllii Bicknell, known only from southeastern Michigan, differs in not turning black in drying and has flexuous rather than straight pedicels and slightly smaller capsules.

7. **Sisyrinchium graminoìdes** Bicknell. Stout Blue-eyed Grass. Fig. 1349.

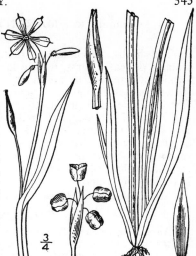

Sisyrinchium anceps S. Wats. in A. Gray, Man. Ed. 6, 515. 1890. Not Cav.
S. graminoides Bicknell, Bull. Torr. Club **23** : 133. 1896.
S. gramineum Curtis, Bot. Mag. *pl. 464.* 1789. Not Lam.

Rather light green, somewhat glaucous; stem broadly 2-winged, stout, erect, or reclining, 8′–18′ tall, usually terminating in two unequal branches subtended by a conspicuous grassy leaf. Basal leaves equalling or shorter than the stem, 1″–3″ wide; often lax and grass-like; edges of stems and leaves usually perceptibly rough-serrulate; bracts 1′ long or less, green, nearly or quite equal but the outer one occasionally prolonged; umbels 2–4-flowered; pedicels filiform, 8″–12″ long, exceeding the bracts, finally often spreading or recurved; flowers 6″–9″ broad; petals sparsely pubescent on outer surface; capsule subglobose, 2½″–3″ in diameter when mature; seeds black, globose, about ½″–⅔″ in diameter, pitted.

In grassy places, in moist or dry soil, sometimes in woods, Nova Scotia to Florida, Minnesota, Arkansas and Texas. Santo Domingo. Plant dark in drying. Pepper-grass. April–June.

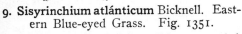

8. **Sisyrinchium strìctum** Bicknell. Strict Blue-eyed Grass. Fig. 1356.

S. strictum Bicknell, Bull. Torr. Club **26** : 299. 1899.

About 1° high in erect tufts, not fibrose at base, pale light green and glaucous, not changing color when dry. Stems and leaves ½″–1″ wide, mostly serrulate, the leaves thin, tapering-acute, over half the height of the stems; node only one, its erect leaf closely united-clasping below, subequal with the two short approximate peduncles; spathes erect, narrow, 8″–10″ long, the bracts subequal, sharp-pointed; flowers rather small, deep violet-blue; capsules pale and thin-walled, somewhat obovoid, about 2″ high, on suberect, slightly exserted margined pedicels.

In sandy soil, Montcalm Co., Mich. June.

9. **Sisyrinchium atlánticum** Bicknell. Eastern Blue-eyed Grass. Fig. 1351.

S. atlanticum Bicknell, Bull. Torr. Club **23** : 134. 1896.
S. apiculatum Bicknell, Bull. Torr. Club **26** : 300. 1899.

Glaucous-green, tufted, not drying black, the stem slender, rather narrowly 2-winged, very smooth-edged, sometimes 2° long, and reclining, terminating in two or three mostly subequal branches, often also with one or two lateral ones; peduncles slender and wiry, often recurved and forming a distinct angle with the floral bracts. Leaves rarely over 1″ wide, the basal ones usually much shorter than the stem; bracts nearly or quite equal, narrow, mostly somewhat scarious, often purplish; flowers 3″–6″ long; capsules thick-walled, on generally erect pedicels 7″–10″ long, oval, 1″–2″ long and ¾″–1½″ in diameter, sometimes apiculate; seeds oval, subglobose, ¼″–½″ in diameter, dark, faintly pitted or nearly smooth.

In moist fields, meadows and brackish marshes, often in sandy soil, Maine to Florida and Mississippi. Also in Michigan. Pepper-grass. May–June.

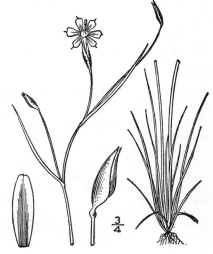

Family 30. MARANTÀCEAE Lindl. Nat. Syst. 1830.

ARROWROOT FAMILY.

Tall herbs, perennial by rootstocks or tubers, or sometimes annual, with scapose or leafy stems, mostly large entire long-petioled sheathing leaves, often swollen at the base of the blade, the veins pinnate, parallel. Flowers perfect or sometimes polygamous, irregular, in panicles, racemes or spikes. Perianth superior, its segments distinct to the summit of the ovary or united into a tube, normally in 2 series of 3, the outer (sepals) usually different from the inner (petals). Perfect stamen 1; anthers 1–2-celled. Staminodia mostly 5, often petal-like, separate or united by their bases, very irregular. Ovary 1–3-celled, inferior; ovule 1 in each cavity, anatropous; style slender, curved, terminal; stigma simple. Fruit capsular or berry-like, 1–3-celled. Seed solitary in each cavity. Embryo central, in copious endosperm.

About 12 genera and 160 species, mostly in the tropics, a few in warm-temperate regions.

1. THÀLIA L. Sp. Pl. 1193. 1753.

Annual (or perennial?) herbs, with large long-petioled basal leaves, erect simple scapes and terminal panicled spikes of bracted usually purple flowers. Sepals 3, membranous, separate, equal. Petals 3, separate or somewhat coherent at the base. Staminodia slightly united below, one of them (labellum) broad, crested. Anther 1-celled. Ovary 1-celled or with 2 additional small empty cavities. Base of the style adnate to the base of the stamen-tube. Stigma 2-lipped, dorsally appendaged. Capsule globose or ovoid. Seed erect. Embryo strongly curved. [In honor of Johann Thalius, German naturalist of the sixteenth century.]

About 7 species, all American. Besides the following, another occurs in the southern States. Type species: *Thalia geniculata* L.

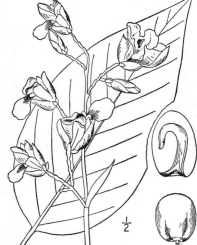

1. Thalia dealbàta Roscoe. Powdery Thalia.

Fig. 1352.

Thalia dealbata Roscoe, Trans. Linn. Soc. 8: 340. 1807.

Plant finely white-powdery nearly all over. Scapes rather stout, terete, 3°–6° tall; petioles 1°–2½° long, terete; leaves ovate-lanceolate, acute or acuminate at the apex, rounded, narrowed or subcordate at the base, ½°–1° long, 3′–5′ wide; panicle 8′–18′ long, its spikes numerous, usually erect or ascending; bracts of the panicle narrow, deciduous, not longer than the spikes; bractlets ovate, unequal, coriaceous, about ½′ long; flowers purple, longer than the bractlets; capsule ovoid, about 4″ in diameter.

In ponds and swamps, South Carolina to Louisiana, Missouri and Texas.

Family 31. BURMANNIÀCEAE Blume, Enum. Pl. Jav. 1: 27. 1830.*

BURMANNIA FAMILY.

Low annual herbs, with filiform stems and fibrous roots. Leaves basal or reduced to cauline scales or bracts. Flowers regular, perfect, the perianth with 6 small thick lobes, its tube adnate to the ovary. Stamens 3 or 6, included, inserted on the tube of the perianth; anthers 2-celled, the sacs transversely dehiscent. Style slender; stigmas 3, dilated; ovary inferior, with 3 central or parietal placentae. Ovules numerous. Capsule many-seeded. Seeds minute, oblong; endosperm none.

Ten genera and about 70 species, widely distributed in tropical regions. The family is represented in North America by the following genus and by *Apteria* of the Gulf States.

* Text contributed to the first edition by the late Rev. THOMAS MORONG.

1. BURMÁNNIA L. Sp. Pl. 287. 1753.

Erect herbs, with simple stems and small alternate scale-like leaves. Tube of the perianth strongly 3-angled or 3-winged, the 3 outer lobes longer than the inner. Stamens 3, opposite the inner perianth-lobes. Filaments very short; connective of the anthers prolonged beyond the sacs into a 2-cleft crest. Ovary 3-celled, with 3 thick 2-lobed central placentae; stigmas globose or 2-lobed. Capsule crowned by the perianth, opening by irregular lateral ruptures. [In honor of Johann Burmann, Dutch botanist of the eighteenth century.]

About 20 species, natives of warm regions. Besides the following, another occurs in the southeastern States. Type species: *Burmannia dísticha* L.

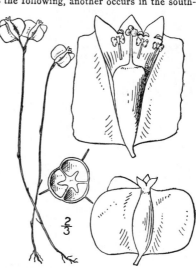

1. Burmannia biflòra L. Northern Burmannia. Fig. 1353.

Burmannia biflora L. Sp. Pl. 287. 1753.

Tripterella coerulea Nutt. Gen. 1: 22. 1818.

Stems very slender, 2′–6′ high, from a few fibrous roots, simple or forked above. Flowers 1 or several, often 2, terminal. Angles of the perianth-tube conspicuously winged, the outer lobes ovate, acute, the inner linear and incurved; seeds very numerous, oblong-linear, sparingly striate, escaping through irregular fissures in the sides of the capsule.

In swamps and bogs, Virginia to Florida and Louisiana. Sept.–Nov.

Family 32. ORCHIDÀCEAE Lindl. Nat. Syst. Ed. 2: 336. 1836.

ORCHID FAMILY.

Perennial herbs, with corms, bulbs or tuberous roots, sheathing entire leaves, sometimes reduced to scales, the flowers perfect, irregular, bracted, solitary, spiked or racemed. Perianth superior, of 6 segments, the 3 outer (sepals) similar or nearly so, 2 of the inner ones (petals) lateral, alike; the third inner one (lip) dissimilar, often markedly so, usually larger, often spurred, sometimes inferior by torsion of the ovary or pedicel. Stamens variously united with the style into an unsymmetrical column; anther 1, or in a few genera 2, 2-celled; pollen in 2–8 pear-shaped usually stalked masses (pollinia), united by elastic threads, the masses waxy or powdery, attached at the base to a viscid disk (gland). Style often terminating in a beak (rostellum) at the base of the anther or between its sacs. Stigma a viscid surface, facing the lip beneath the rostellum, or in a cavity between the anther-sacs (clinandrium). Ovary inferior, usually long and twisted, 3-angled, 1-celled; ovules numerous, anatropous, on 3 parietal placentae. Capsule 3-valved. Seeds very numerous, minute, mostly spindle-shaped, the loose coat hyaline, reticulated; endosperm none; embryo fleshy.

About 430 genera and over 5000 species, of wide distribution, most abundant in the tropics, many of those of warm regions epiphytes.

Anthers 2; lip a large inflated sac. (CYPRIPEDIEAE.)
　Lip not fissured; stems leafy.　　　　　　　　　　　　　　　　1. *Cypripedium.*
　Lip fissured in front; leaves 2, basal.　　　　　　　　　　　　2. *Fissipes.*
Anther solitary.
　Pollinia with a caudicle, which is attached at the base to a viscid disk or gland. (ORCHIDEAE.)
　　Gland enclosed in a pouch.
　　　Sepals free; lip 3-lobed.　　　　　　　　　　　　　　　　3. *Orchis.*
　　　Sepals united above into a hood; lip entire.　　　　　　　4. *Galeorchis.*
　　Gland not enclosed in a pouch.
　　　Lip not fringed nor cut-toothed.
　　　　Stem leafy; anther sacs mostly parallel.
　　　　　Valves of the anthers dilated at the base enclosing the gland below. 5. *Perularia.*
　　　　　Valves not dilated at the base.
　　　　　　Gland surrounded by a thin membrane.　　　　　　6. *Coeloglossum.*
　　　　　　Gland naked.
　　　　　　　Beak of the stigma with 2 or 3 appendages.　　　7. *Gymnadeniopsis.*
　　　　　　　Beak of the stigma not appendaged.

Lateral sepals free ; anther-sacs opening in front. 8. *Limnorchis.*
Bases of the lateral sepals adnate to the claw of the lip ; anther-sacs
opening laterally. 9. *Piperia.*
Stem scapiform ; leaves 1–2, basal ; anther-sacs divergent.
Basal leaves 2 ; ovary straight. 10. *Lysias.*
Basal leaf 1 ; ovary arcuate. 11. *Lysiella.*
Lip fringed or parted and cut-toothed. 12. *Blephariglottis.*
Pollinia not produced into a caudicle (except apparently in No. 25).
Pollinia granulose or powdery. (Neottieae.)
Flowers comparatively large, solitary or few ; anthers incumbent on a long column.
Leaves not grass-like ; lip free.
Flowers terminal ; lip crested.
Leaves alternate. 13. *Pogonia.*
Stem-leaves whorled. 14. *Isotria.*
Flowers axillary ; lip not crested. 15. *Triphora.*
Leaves grass-like.
Flower solitary ; lip adherent to the base of the column. 16. *Arethusa.*
Flowers racemose ; lip free. 17. *Limodorum.*
Flowers numerous, in spikes or racemes ; anthers erect, jointed to a short column.
Anther operculate ; leaves broad, alternate. 18. *Serapias.*
Anther not operculate.
Leaves green, borne on the stem.
Leaves alternate ; spike mostly twisted. 19. *Ibidium.*
Leaves 2, opposite ; spike not twisted. 20. *Ophrys.*
Leaves white-reticulated, basal. 21. *Peramium.*
Pollinia smooth and waxy. (Epidendreae.)
Plants with corms or solid bulbs, rarely with coralloid roots ; leaves basal or cauline.
Leaves unfolding before or with the flowers.
Leaf cauline ; lip ovate, or auricled at the base. 22. *Malaxis.*
Leaf or leaves basal.
Leaves 2 ; lip flat ; flowers racemed. 23. *Liparis.*
Leaf 1 ; lip saccate ; flower solitary. 24. *Cytherea.*
Leaf 1, basal, unfolding after the flowering time.
Flowers long-spurred ; lip 3-lobed. 25. *Tipularia.*
Flowers not spurred ; lip 3-ridged. 26. *Aplectrum.*
Plants with coralloid roots, bulbless ; the leaves reduced to scales.
Pollinia 4, in 2 pairs ; flowers gibbous or spurred. 27. *Corallorrhiza.*
Pollinia 8, united ; flowers not gibbous nor spurred. 28. *Hexalectris.*

1. CYPRIPÈDIUM L. Sp. Pl. 951. 1753.

Glandular-pubescent herbs, with leafy stems and tufted roots of thick fibres. Leaves large, broad, many-nerved. Flowers solitary or several, drooping, large, showy. Sepals spreading, separate, or 2 of them united under the lip. Lip a large inflated sac. Column declined, bearing a sessile or stalked anther on each side and a dilated petaloid sterile stamen above, which covers the summit of the style. Pollinia granular, without a caudicle or glands. Stigma terminal, broad, obscurely 3-lobed. [Name Greek, Venus' sock or buskin.]

About 20 species, natives of the north temperate zone. Besides the following, some 4 others occur in western North America. Type species: *Cypripedium Calceolus* L.

Sepals separate ; stem leafy, 1-flowered. 1. *C. arietinum.*
Lateral sepals more or less united.
Sepals and petals not longer than the lip.
Plant 6′–10′ high ; lip about ½′ long. 2. *C. passerinum.*
Plant 1°–2½° high ; lip 1′–2′ long. 3. *C. reginae.*
Sepals and petals equalling or longer than the lip.
Sterile stamen lanceolate ; lip white. 4. *C. candidum.*
Sterile stamen triangular ; lip yellow. 5. *C. parviflorum.*

1. Cypripedium arietìnum R. Br. Ram's-head Ladies'-slipper. Fig. 1354.

Cypripedium arietinum R. Br. in Ait. Hort. Kew. Ed. 2, 5: 222. 1813.

Stem 8′–12′ high, 1-flowered. Leaves 3 or 4, elliptic or lanceolate, 2′–4′ long, ½′–3′ wide; sepals separate, lanceolate, 8″–10″ long, longer than the lip; petals linear, greenish brown, about as long as the sepals; lip 7″–8″ long, red and white, veiny, prolonged at the apex into a long blunt spur, somewhat distorted at the upper end which resembles a ram's head, whence the specific name.

In cold and damp woods, Quebec to Manitoba, Massachusetts, New York and Minnesota. Ram's-head. American valerian. May–Aug.

½

2. Cypripedium passerìnum Richards. Northern Ladies'-slipper. Fig. 1355.

Cypripedium passerinum Richards, App. Frank. Journ. 34.
1823.

Stem villous-pubescent, leafy, 6'-10' high, bearing 1
or 2 flowers at the top. Leaves oblong to elliptic-lanceolate, acute; sepals and petals shorter than the lip, the
upper sepal broad, nearly orbicular, yellowish; lip nearly
spherical, magenta, deeper magenta within toward the
base.

Woods and along streams, Ontario to Alberta and the
Yukon Territory. Summer.

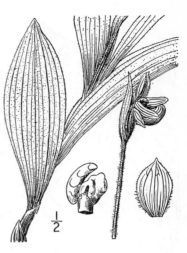

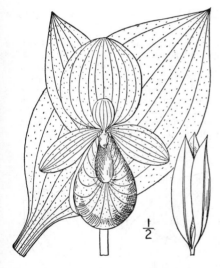

3. Cypripedium regìnae Walt. Showy Ladies'-slipper. Fig. 1356.

?C. hirsutum Mill. Gard. Dict. Ed. 8, no. 3. 1768.
Cypripedium reginae Walt. Fl. Car. 222. 1788.
Cypripedium album Ait. Hort. Kew. 3: 303. 1789.
Cypripedium spectabile Salisb. Trans. Linn. Soc. 1:
78. 1791.

Stem stout, villous-hirsute, 1°-2½° high, leafy
to the top. Leaves elliptic, acute, 3'-7' long, 1'-4'
wide, flowers 1-3; sepals round-ovate, white, not
longer than the lip, the lateral ones united for
their whole length; petals somewhat narrower
than the sepals, white; lip much inflated, 1'-2'
long, white, variegated with crimson and white
stripes; stamens cordate-ovate.

In swamps and woods, Newfoundland to Ontario,
Minnesota and Georgia. Nerve-root. Ducks. Whippoor-will's-shoe. June–Sept.

4. Cypripedium cándidum Willd. Small White Ladies'-slipper. Fig. 1357.

Cypripedium candidum Willd. Sp. Pl. 4: 142. 1805.

Stem 6'-12' high, leafy. Leaves 3 or 4, elliptic or
lanceolate, acute or acuminate, 3'-5' long, 8"-16"
wide, with several obtuse sheathing scales below
them; bracts 1'-2' long, lanceolate; flower solitary;
sepals lanceolate, equalling or longer than the lip,
greenish, purple spotted; petals somewhat longer and
narrower than the sepals, wavy-twisted, greenish;
lip white, striped with purple inside, about 10" long;
sterile stamen lanceolate.

In bogs and meadows, New York and New Jersey to
Kentucky, Minnesota, Missouri and Nebraska. Ducks.
May–July.

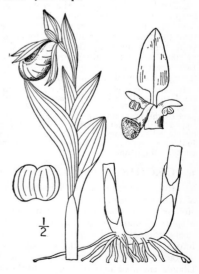

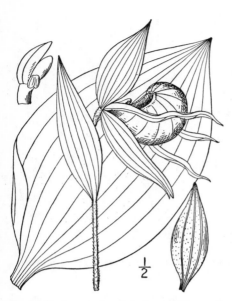

$\frac{1}{2}$

5. Cypripedium parviflòrum Salisb.
Yellow or Downy Ladies'-slipper.
Fig. 1358.

C. *parviflorum* Salisb. Trans. Linn. Soc. **1**: 77.
1791.
C. *pubescens* Willd. Sp. Pl. **4**: 143. 1805.

Stems leafy, 1°–2½° high. Leaves oval or elliptic, 2′–6′ long, 1½′–3′ wide, acute or acuminate; sepals ovate-lanceolate, usually longer than the lip, yellowish or greenish, striped with purple; petals narrower, usually twisted; lip much inflated, 8″–2′ long, pale yellow with purple lines; its interior with a tuft of white jointed hairs at the top; sterile stamen triangular; stigma thick, somewhat triangular, incurved.

In woods and thickets, Nova Scotia to Ontario and Minnesota, Alabama and Nebraska. Ascends to 4000 ft. in Virginia. Consists of several races, differing mainly in the size of the flowers. Whippoor-will's shoe. Yellows. Slipper-root. Indian shoe. Yellow moccasin-flower. Noah's-ark. Ducks. American valerian. May–July.

2. FÍSSIPES Small, Fl. SE. U. S. 311. 1903.

Acaulescent herbs, with fleshy-fibrous roots and glandular-pubescent foliage. Leaves 2, basal; blades ample, plaited, spreading. Scape simple. Flower usually solitary. Perianth irregular. Sepals greenish, narrowed upward. Lateral petals about as long as the sepals, linear, greenish. Lip a large drooping inflated sac with a closed fissure down its whole length in front. Column declined, glandular-pubescent, bearing a sessile anther on each side, and a rhomboidal glandular-pubescent sterile stamen above. Stamens spreading, the free tips at right angles to the column. Pollen granular, without glands or tails. Stigma broadest at the apex. Capsule ascending. [Latin, in allusion to the cleft lip.]

A monotypic genus of eastern North America.

1. Fissipes acaùlis (Ait.) Small. Moccasin Flower. Stemless Ladies'-slipper. Fig. 1359.

Cypripedium acaule Ait. Hort. Kew. **3**: 303. 1789.
Fissipes acaulis Small, Fl. SE. U. S. 311. 1903.

Scape 6′–15′ high, rather stout. Leaves 2, basal, elliptic, 6′–8′ long, 2′–3′ wide, thick; occasionally a smaller leaf is borne on the scape; sepals greenish purple, spreading, 1½′–2′ long, lanceolate, the 2 lateral ones united; petals narrower and somewhat longer than the sepals; lip often over 2′ long, somewhat obovoid, folded inwardly above, pink with darker veins or sometimes white, the upper part of its interior surface crested with long white hairs; sterile stamen triangular, acuminate, keeled inside.

In sandy or rocky woods, Newfoundland to Manitoba, south to North Carolina, Tennessee and Minnesota. Ascends to 4500 ft. in Virginia. The hairs on the lower part of the bract and on the base of the ovary are often tipped with scarlet glands. Flower fragrant. Pink or purple ladies'-slipper. Nerve-root. Noah's-ark. Camel's-foot. Squirrel's-shoes. Two-lips. Indian-moccasin. Old-goose. May–June.

$\frac{1}{2}$

3. ÓRCHIS [Tourn.] L. Sp. Pl. 939. 1753.

Roots tuberous, or of numerous fleshy fibres; stems in our species scape-like, 1-leaved at the base. Flowers in short terminal spikes. Sepals separate, subequal, spreading. Petals similar to the sepals. Lip connate with the base of the column, 3-lobed, produced below into a spur. Column short, scarcely extending beyond the base of the lip. Anther 2-celled, the sacs contiguous and slightly divergent; pollinia granulose, 1 large mass in each sac, produced into a slender caudicle, the end of which is attached to a small gland. Stigma a hollowed surface between the anther-sacs, the rostellum a knob-like projection under the anther. Glands enclosed in a pouch. Capsule oblong, erect, without a beak. [Name ancient.]

About 80 species, natives of the north temperate zone. Only the following is known in North America. Type species: *Orchis militâris* L.

1. **Orchis rotundifòlia** Pursh. Small Round-leaved Orchis. Fig. 1360.

Orchis spectabilis Pursh, Fl. Am. Sept. 588. 1814. Not L.

Platanthera rotundifolia Lindl. Gen. & Sp. Orch. 292. 1835.

Stem slender, 8′–10′ high, 1-leaved near the base. Leaf varying from nearly orbicular to oval, $1\frac{1}{2}$′–3′ long, 1′–2′ wide, with 1 or 2 sheathing scales below it; spike 2–6-flowered; flowers 6″–8″ long, subtended by small bracts; lateral sepals spreading, sometimes longer than the petals; sepals and petals oval, rose-color; lip white, purple spotted, longer than the petals, 3-lobed, the middle lobe larger, dilated, 2-lobed or notched at the apex; spur slender, shorter than or about equalling the lip.

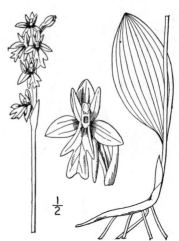

In damp woods, Greenland to the Rocky Mountains, Maine, New York and Wisconsin. June–July.

4. **GALEÓRCHIS** Rydb. in Britton, Man. 292. 1901.

Rootstock short, with numerous fleshy roots. Stem scape-like, with 2 large leaves at the base. Flowers in a short loose spike, subtended by large bracts. Sepals united above, forming a hood. Petals connivent, somewhat adnate to the sepals. Lip wavy, produced into a spur. Column short, scarcely extending beyond the base of the lip. Anther 2-celled, its sacs divergent; pollinia granulose, 1 large mass in each sac, with a slender caudicle. Glands enclosed in a pouch. [Greek, referring to the hood-like united sepals.].

A monotypic genus of North America.

1. **Galeorchis spectábilis** (L.) Rydb. Showy Orchis. Fig. 1361.

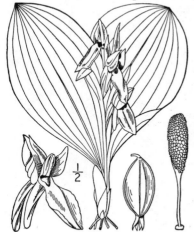

Orchis spectabilis L. Sp. Pl. 943. 1753.
Galeorchis spectabilis Rydb. in Britton, Man. 292. 1901.

Stems 4′–12′ high, thick, fleshy, 5-angled. Leaves 2, near the base of the stem, with 1 or 2 scales below them, obovate, sometimes 8′ long and 4′ wide, but usually smaller, clammy to the touch; spike 3–6-flowered; flowers about 1′ long, violet-purple mixed with lighter purple and white; bracts foliaceous, sheathing the ovaries; sepals united in an arching galea; petals connivent under the sepals, more or less attached to them; lip whitish, divergent, entire, about as long as the petals; spur obtuse, about 8″ long; column violet on the back; capsule about 1′ long, strongly angled.

In rich woods, New Brunswick to Ontario, Dakota, Georgia, Kentucky, Missouri and Nebraska. Ascends to 4000 ft. in Virginia. Purple, gay- or spring-orchis. April–June.

5. **PERULÀRIA** Lindl. Bot. Reg. **20**: under *pl. 1701.* 1835.

Leafy-stemmed plants, from a cluster of thick fibrous roots. Flowers small, greenish, in a long open spike with long bracts. Sepals and petals broad, spreading. Lip lanceolate, with a tooth on each side and a tubercle at the middle of the base or nearly orbicular. Spur slender, straight, longer than the lip, but shorter than the ovary. Valves of the anthers horizontal, opening upward, dilated at the base so as to form an oblong cavity, enclosing the orbicular incurved gland. Pollinia granulose, produced at the base into a caudicle. [Latin, a little wallet.]

About 4 species, of the north temperate zone. Besides the following, another occurs in the southeastern States.

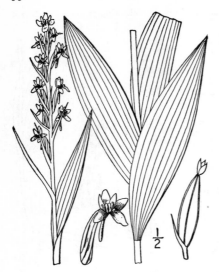

1. Perularia flàva (L.) Farwell. Tuber-cled Orchis. Small Pale-green Orchis. Fig. 1362.

Orchis flava L. Sp. Pl. 942. 1753.
Orchis virescens Willd. Sp. Pl. 4 : 37. 1805.
Habenaria virescens Spreng. Syst. 3 : 688. 1826.
H. flava A. Gray, Am. Journ. Sci. 38 : 308. 1840.
P. flava Farwell, Ann. Rep. Parks Detroit 11 : 54. 1900.

Stem rather stout, 1°–2° high, leafy. Leaves lanceolate or elliptic, acute or obtuse, 4′–12′ long, 8″–3′ wide; spike 2′–6′ long; bracts acuminate, longer than the ovaries; petals greenish; sepals and petals ovate or roundish, about 3″ long; sepals greenish yellow, lip a little longer than the petals, entire or crenulate, mostly with an obtuse tooth on each side and a tubercle at the middle of the base; anther-sacs parallel, the sides forming a rounded cavity, in which lie the orbicular in-curved glands; capsule about 4″ long.

In moist soil, Nova Scotia and Ontario to Minne-sota, Florida and Louisiana and Missouri. Yellow or greenish orchis. Green rein-orchis. Races differ in the shape of the lip. June–July.

6. COELOGLÓSSUM Hartm. Handb. Scand. Fl. 323. 1820.

Leafy plants, with biennial 2-cleft tubers. Flowers greenish in a long leafy-bracted spike. Sepals free, somewhat arcuate, bent together and forming a hood. Petals narrow. Lip oblong, obtuse, 2–3-toothed at the apex. Spur much shorter than the lip, blunt, sac-like. Column short. Pollinia with long caudicles. Glands small, scarcely wider than the caudicle, surrounded by a thin membrane. [Latin, heaven-tongue.]

A boreal genus of 2 or 3 species, only the following in North America. Type species: *Coelo-glossum víride* (L.) Hartm.

1. Coeloglossum bracteàtum (Willd.) Parl. Long-bracted Orchis. Fig. 1363.

Orchis bracteata Willd. Sp. Pl. 4 : 34. 1805.
H. bracteata R. Br. in Ait. Hort. Kew. Ed. 2, 5 : 192. 1813.
Habenaria viridis var. *bracteata* Reichenb. Ic. Fl. Germ. 13 : 130. *f. 435.* 1851.
C. bracteatum Parl. Fl. Ital. 3 : 409. 1858.

Stem slender or stout, leafy, 6′–2° high. Leaves lanceolate, ovate or oval, or the lowest sometimes obovate, obtuse or acute, 2′–5′ long, the upper much smaller; bracts longer than the ovaries, the lower ones 2 or 3 times as long; spike 3″–5′ long, loosely flowered; flowers green or greenish; sepals ovate-lanceolate, spreading, dilated or somewhat gibbous at the base, about 3″ long; petals very narrow, sometimes thread-like; lip 3″–4″ long, oblong-spatulate, 2–3-toothed or lobed at the apex, more than twice as long as the white sac-like spur; anther-sacs divergent at the base.

In woods and meadows, Nova Scotia to Alaska, North Carolina and Nebraska. Also in Europe. Vegetable satyr. Bracted green orchis. May–Sept.

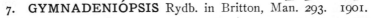

7. GYMNADENIÓPSIS Rydb. in Britton, Man. 293. 1901.

Leafy plants, with fleshy fibrous or somewhat tuberous roots, and a short spike of small flowers. Sepals free and spreading. Lip entire or 3-toothed at the apex, much exceeded by the long filiform or clavate spur. Beak of the stigma with 2 or 3 oblong or clavate appen-dages. Anther-sacs parallel and approximate, their glands naked and contiguous. Pollinia granular, with short caudicles.

A North American genus, formerly included in *Habenaria*. It is closely related to the Euro-pean genus *Gymnadenia*, from which it differs in the appendages of the stigma; hence the name. Type species: *Gymnadeniopsis nívea* (Nutt.) Rydb.

Lip entire ; stigma with 2 appendages ; stem several-leaved.
    Ovary not twisted ; spur longer than the ovary ; flowers white.   1. *G. nivea.*
    Ovary twisted ; spur shorter than the ovary ; flowers orange.   2. *G. integra.*
Lip 3-toothed ; stigma with 3 appendages ; leaves 1 or 2.   3. *G. clavellata.*

1. Gymnadeniopsis nívea (Nutt.) Rydb. Southern Small White Orchis. Fig. 1364.

Orchis nivea Nutt. Gen. **2** : 188. 1818.
Habenaria nivea Spreng. Syst. **3** : 689. 1826.
G. nivea Rydb. in Britton, Man. 293. 1901.

Stem slender, angled, 12′–15′ high. Leaves linear-lanceolate, acuminate, 4′–8′ long, the upper much shorter and passing into the bracts of the spike; spike 2′–4′ long, loosely many-flowered; flowers small, white; lateral sepals broadly oblong, dilated or slightly eared at the base, spreading, about 3″ long; petals and upper sepal smaller; spur capillary, as long as the ovary or longer; stigma appendaged by 2 small horns affixed to the back of the anther; ovary straight.

In pine barren bogs, New Jersey to Florida and Alabama. Aug.

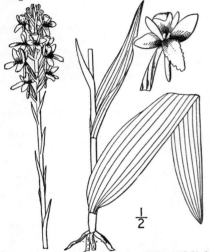

2. Gymnadeniopsis íntegra (Nutt.) Rydb. Small Southern Yellow Orchis. Fig. 1365.

Orchis integra Nutt. Gen. **2** : 188. 1818.
Habenaria integra Spreng. Syst. **3** : 689. 1826.
G. integra Rydb. in Britton, Man. 293. 1901.

Stem 1°–2° high, angled, with 1–3 linear-lanceolate leaves below, and numerous bract-like ones above. Lower leaves 2′–8′ long, acute; spike 1′–3′ long, densely flowered; flowers orange-yellow; upper sepals and petals connivent; lateral sepals longer, oval or obovate, spreading; lip oblong, mostly crenulate or erose, sometimes entire; spur straight, longer than the lip, shorter than the ovary; stigma with 2 lateral fleshy appendages and a narrow beak.

In wet pine barrens, New Jersey to Florida and Louisiana. The upper surface of the leaves is often reticulated with hexagonal cells. July.

3. Gymnadeniopsis clavellàta (Michx.) Rydb. Small Green Wood Orchis. Fig. 1366.

Orchis clavellata Michx. Fl. Bor. Am. **2** : 155. 1803.
Orchis tridentata Willd. Sp. Pl. **4** : 41. 1805.
Habenaria tridentata Hook. Exot. Fl. **2** : *pl. 81.* 1825.
Habenaria clavellata Spreng. Syst. **3** : 689. 1826.
G. clavellata Rydb. in Britton, Man. 293. 1901.

Stem 8′–18′ high, angled, 1-leaved near the base, often with several small bract-like leaves above, or one of these larger. Basal leaf oblanceolate, 2′–6′ long; bracts shorter than the ovaries; spikes ½′–2′ long, loosely flowered; flowers small, greenish or whitish; sepals and petals ovate, lip dilated and 3-toothed at the apex, the teeth often small and inconspicuous; spur longer than the ovary, incurved, clavate; stigma with 3 club-shaped appendages; anther-sacs nearly parallel; capsule ovoid, 3″–4″ long, nearly erect.

In wet or moist woods, Newfoundland to Minnesota, south to Florida and Louisiana. Three-toothed or rein-orchis. July–Aug.

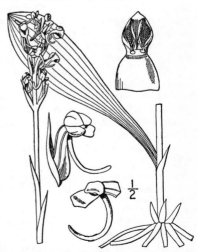

Gymnadenia conópsea (L.) R. Br., otherwise known only from the Old World, has been collected at Litchfield, Connecticut. The flower has a broad 3-lobed lip and a slender spur much longer than the ovary.

8. LIMNORCHIS Rydb. Mem. N. Y. Bot. Gard. 1: 104. 1900.

Leafy plants with thick fleshy roots and small greenish or whitish flowers in a long spike. Sepals and petals free and spreading. Lip entire. Beak of the stigma without appendages. Anther-sacs nearly parallel, wholly adnate. Glands naked. Pollinia granular. [Greek, marsh-orchis.]

A North American genus of about 15 species, differing from *Lysias* in the general habit and the almost parallel anther-sacs. Type species: *Limnorchis hyperborea* (L.) Rydb.

Flowers greenish; base of the lip little dilated.   1. *L. hyperborea.*
Flowers white; base of the lip much dilated.   2. *L. dilatata.*

1. Limnorchis hyperbòrea (L.) Rydb. Tall Leafy Green Orchis. Fig. 1367.

Orchis hyperborea L. Mant. 121. 1767.
Habenaria hyperborea R. Br. in Ait. Hort. Kew. Ed. 2,
    5: 193. 1813.
Orchis huronensis Nutt. Gen. 2: 189. 1818.
Limnorchis huronensis Rydb. in Britton, Man. 294. 1901.

Stem rather stout, 8′-3° high. Leaves lanceolate, mostly acute, 2′-12′ long, 6″-18″ wide; spike narrow, 3′-8′ long; flowers small, greenish or greenish yellow; sepals and petals ovate, obtuse, 2″-3″ long; upper sepal slightly crenulate at the apex; lip lanceolate, entire, obtuse, about 3″ long; spur about equalling the lip, shorter than the ovary, blunt, slightly incurved, sometimes clavate; anther-sacs parallel, diverging at the base; glands small; ovary more or less twisted.

In bogs and wet woods, Greenland to Alaska, New Jersey, Colorado and Oregon, Iceland. Northern green orchis. May–Aug.

L. mèdia Rydb. is probably a hybrid of this species and the next. L. major (Lange) Rydb. is, apparently, restricted to Greenland.

2. Limnorchis dilatàta (Pursh) Rydb. Tall White Bog Orchis. Fig. 1368.

Orchis dilatata Pursh, Fl. Am. Sept. 588. 1814.
Habenaria dilatata Hook. Exot. Fl. 2: *pl. 95.* 1825.
L. fragrans Rydb. in Britton, Man. 294. 1901.
L. dilatata Rydb. in Britton, Man. 294. 1901.

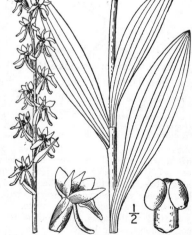

Stem slender, leafy, 1°-2° high. Leaves lanceolate, 3′-12′ long, 4″-10″ wide, obtuse or acute; spike 2′-10′ long; bracts acute, the lower longer than the ovary, the upper shorter; flowers small, white, sometimes fragrant; sepals ovate to lanceolate, nearly 3″ long; petals acute, lanceolate; lip entire, dilated or obtusely 3-lobed at the base, obtuse at the apex, about as long as the blunt incurved spur; anther-sacs nearly parallel; glands close together, strap-shaped, nearly as long as the pollinia and caudicle; stigma with a trowel-shaped beak between the bases of the anther-sacs.

In bogs and wet woods, Nova Scotia to Alaska, south to Maine, New York and Oregon. Northern white orchis. June–Sept. Consists of several races, differing in the shape of the sepals and petals and width of leaves.

Limnorchis graminifòlia Rydb. of the northwest, with much narrower leaves, is recorded from Quebec.

9. PIPÈRIA Rydb. Bull. Torr. Club 28: 269. 1901.

Herbs resembling *Limnorchis* in habit, but with short rounded tubers. Leaves mainly near the base of the stem, early withering. Spike strict. Flowers white, greenish, purplish or yellowish. Sepals 1-nerved, the lateral ones adnate to the claw of the lip. Petals 1-nerved. Lip with a median ridge, truncate or hastate at the base. Anther-sacs opening laterally. [Dedicated to Professor C. V. Piper, of Washington.]

Three species or more, natives of northern North America. Type species: *Piperia elegans* (Lindl.) Rydb.

1. Piperia unalaschénsis (Spreng.) Rydb. Alaska Piperia. Fig. 1369.

Spiranthes unalaschensis Spreng. Syst. 3: 708. 1826.

Habenaria unalaschensis S. Wats. Proc. Am. Acad. 12: 277. 1877.

P. unalaschensis Rydb. Bull. Torr. Club 28: 270. 1901.

Stem strict, 1°-2° tall; lower leaves oblanceolate, 4′-6′ long, obtuse or acutish; upper leaves lanceolate to linear-lanceolate, alternate; spike 4′-12′ long; bracts lanceolate or ovate-lanceolate, shorter than the flowers; flowers greenish; lateral sepals oblong-lanceolate; petals lanceolate, nearly as long as the lateral sepals; lip oblong-lanceolate, somewhat hastately dilated at the base.

In moist woods, Alaska to California, Colorado, Quebec and Ontario. June–Sept.

10. LÝSIAS Salisb. Trans. Hort. Soc. 1: 288. 1812.

Plants with tubers or fleshy roots; stem scapose. Leaves 2, basal. Flowers greenish or white; sepals free, large and spreading; petals small and narrow; lip entire, linear or nearly so; spur long and slender, generally longer than the elongated, straight ovary. Beak of the stigma without appendages. Anther-sacs widely diverging, their narrow beak-like bases projecting forward; stalk of the pollen-mass laterally affixed to the back of the orbicular gland, whose face is turned inward. Pod cylindric-clavate, distinctly stipitate. [Named for Lysias, an Attic orator.]

A circumboreal genus of about half a dozen species. Type species: *Lysias bifolia* (L.) Salisb.

Scape bracted.   1. *L. orbiculata.*
Scape naked.   2. *L. Hookeriana.*

1. Lysias orbiculàta (Pursh) Rydb. Large Round-leaved Orchis. Fig. 1370.

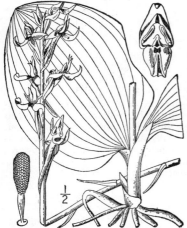

Orchis orbiculata Pursh, Fl. Am. Sept. 588. 1814.
Habenaria macrophylla Goldie, Edinb. Phil. Journ. 6: 331. 1822.
Habenaria orbiculata Torr. Comp. 318. 1826.
L. orbiculata Rydb. in Britton, Man. 294. 1901.

Scape stout, bracted, 1°-2° high, occasionally bearing a small leaf. Basal leaves 2, orbicular, spreading flat on the ground, shining above, silvery beneath, 4′-7′ in diameter; raceme loosely many-flowered; pedicels nearly ½′ long, the fruiting ones erect; flowers greenish white; upper sepal short, rounded; lateral sepals spreading, falcate-ovate, obtuse 4″-5″ long; petals smaller; lip oblong-linear, entire, obtuse, white, about 6″ long; spur longer than the ovary, often 1½′ long; anther-sacs prominent, converging above; glands small, orbicular, nearly ¼′ apart, their faces turned toward the axis.

In rich woods, Newfoundland to British Columbia, North Carolina and Minnesota. Ascends to 4500 ft. in Virginia. Large two-leaved orchis. Heal-all. July–Aug. Races differ in the size of leaves and in length of the spur.

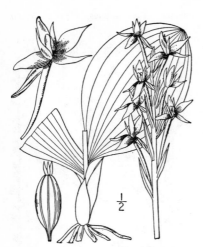

2. Lysias Hookeriàna (A. Gray) Rydb.
Hooker's Orchis. Fig. 1371.

Habenaria Hookeriana A. Gray, Ann. Lyc. N. Y. **3** : 229.
1836.
Habenaria orbiculata Goldie, Edinb. Phil. Journ. **6** : 331.
1822. Not *Orchis orbiculata* Pursh, 1814.
Habenaria Hookeri var. *oblongifolia* Paine, Cat. Pl.
Oneida, 83. 1865.
L. Hookeriana Rydb. in Britton, Man. 295. 1891.

Scape 8′–15′ high, not bracted. Leaves 2, basal,
fleshy, shining, spreading or ascending, oval, orbicular or obovate, 3′–5½′ long; raceme rather loosely
many-flowered, 4′–8′ long; bracts acute, about as long
as the yellowish green flowers; lateral sepals greenish, lanceolate, acute, spreading, about 4″ long; petals
narrowly linear or awl-shaped; lip linear-lanceolate,
acute, 4″–5″ long; anther-sacs widely diverging
below; glands small, their faces turned inward; spur
slender, acute, 8″ long or more, as long as the ovary
or considerably longer.

In woods, Nova Scotia to Minnesota, New Jersey,
Pennsylvania and Iowa. Small two-leaved orchis.
Solomon's-seal. June–Sept.

11. LYSIÉLLA Rydb. in Britton, Man. 295. 1901.

A small plant with a short rootstock and thick root-fibers. Stem scapose, naked, with a
single obovate leaf at the base; flowers greenish yellow. Upper sepal round-ovate, erect,
surrounding the broad column; lateral sepals reflexed, spreading; petals lanceolate, smaller;
lip entire, linear-lanceolate, deflexed; spur slightly curved, shorter than the ovary. Beak of
stigma not appendaged. Anther-sacs widely diverging, wholly adnate, arcuate; glands small,
their faces turned inward. Pod obovoid. [Name diminutive of *Lysias*.]

A monotypic genus of North America and northern Norway.

1. Lysiella obtusàta (Pursh) Richards.
Small Northern Bog Orchis. Fig. 1372.

Orchis obtusata Pursh, Fl. Am. Sept. 588. 1814.
H. obtusata Richards, App. Frank. Journ. 750. 1823.
Lysiella obtusata Rydb. in Britton, Man. 295. 1901.

Scape slender, naked, 4′–10′ high, 4-angled. Leaf
solitary, basal obovate, 2′–5′ long, 5″–12″ wide; spike
1′–2½′ long, loose flowers greenish yellow, about 3″
long; upper sepal erect, round-ovate, green with
whitish margins; lateral sepals spreading, oblong,
obtuse; petals shorter, dilated or obtusely 2-lobed at
the base, connate with the base of the column; lip
entire, lanceolate, obtuse, deflexed, about 3″ long;
spur about as long as the lip, slender, nearly straight,
blunt; anther-sacs widely divergent below, glands
small, rather thick.

In bogs, New Brunswick to British Columbia, Maine,
New York and Colorado. Arctic Norway. Dwarf or
one-leaved orchis. July–Sept.

12. BLEPHARIGLÓTTIS Raf. Fl. Tell. 2 : 38. 1836.

Plants with tall and leafy stems and fleshy or tuberous roots. Flowers several or
numerous in an open spike with foliaceous bracts; corolla white, yellow or purple; sepals
broad and spreading or reflexed; lip variously fringed or 3-parted and cut-toothed; spur
longer than the lip. Anther-sacs widely separated and usually diverging, their narrow beak-
like bases supported by the arms of the stigma, strongly projecting forward and upward;
gland naked; pollinia granular. [Greek, fringed throat.]

About 10 species of North America. Type species: *Blephariglottis albiflora* Raf.

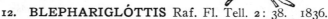

Lip not 3-parted, pectinately fringed.
  Spur half as long as the ovary; flowers yellow.   1. *B. cristata.*
  Spur longer than the ovary.
    Flowers bright yellow.   2. *B. ciliaris.*
    Flowers white.   3. *B. blephariglottis.*
Lip 3-parted.
  Segments of the lip deeply fringed.
    Segments narrow; fringe of a few threads.   4. *B. lacera.*
    Segments broadly fan-shaped; fringe copious.
      Segments fringed to the middle or deeper; flowers white.   5. *B. leucophaea.*

Segments not fringed beyond the middle; flowers lilac, rarely white.
     Raceme 4–5 cm. thick; lip 1–2 cm. broad.   6. *B. grandiflora.*
     Raceme 1–3 cm. thick; lip 8–12 mm. broad.   7. *B. psycodes.*
Segments of the lip cut-toothed; flowers violet-purple.   8. *B. peramoena.*

1. Blephariglottis cristàta (Michx.) Raf.
Crested Yellow Orchis. Fig. 1373.

Orchis cristata Michx. Fl. Bor. Am. **2**: 156. 1803.
Habenaria cristata R. Br. in Ait. Hort. Kew. Ed. 2, **5**:
    194. 1813.
Blephariglottis cristata Raf. Fl. Tell. **2**: 39. 1836.

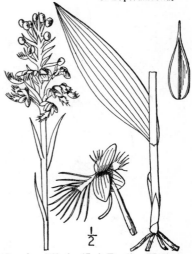

Stem slender, angled, 8′–2° high. Leaves narrowly
lanceolate, 2′–8′ long, 3″–8″ wide, the upper much
smaller, similar to the bracts; bracts as long as the
flowers; spike 2′–4′ long, dense; flowers orange;
sepals roundish-ovate, about 1½″ long, the lateral
ones spreading; petals narrower, pectinate-fringed;
lip slightly longer than the sepals, not 3-parted, but
deeply fringed to the middle or beyond; spur 2″–3″
long, about half as long as the ovary; anther-sacs
divergent at the base, widely separated.

In bogs, New Jersey to Florida, Arkansas and Louisiana. July–Aug. A hybrid between this species and *B.
blephariglottis,* from Delaware, is known as *Habenaria
Cánbyi* Ames.

2. Blephariglottis ciliàris (L.) Rydb. Yellow-
fringed Orchis. Fig. 1374.

Orchis ciliaris L. Sp. Pl. 939. 1753.
Habenaria ciliaris R. Br. in Ait. Hort. Kew. Ed. 2, **5**:
    194. 1813.
B. ciliaris Rydb. in Britton, Man. 296. 1901.

Stem slender, 1°–2½° high. Leaves lanceolate,
acute, 4′–8′ long, 6″–18″ wide, the upper smaller;
spike closely many-flowered, 3′–6′ long, sometimes
nearly 3′ thick; flowers orange or yellow, large,
showy; sepals orbicular or broadly ovate, oblique at
the base, 2″–4″ long; the lateral ones mostly reflexed; petals much smaller, oblong or cuneate, usually
toothed; lip oblong, 5″–7″ long, copiously fringed
more than half-way to the middle; spur 1–1½′ long,
very slender; anther-sacs large, divergent at the base,
bearing a small white tubercle on the outer side.

In meadows, Vermont and Ontario to Michigan, Missouri, Florida and Texas. July–Aug.

3. Blephariglottis blephariglóttis (Willd.)
Rydb. White-fringed Orchis. Fig. 1375.

Orchis blephariglottis Willd. Sp. Pl. **4**: 9. 1805.
Habenaria blephariglottis Torr. Comp. 317. 1826.
Platanthera holopetala Lindl. Gen. & Sp. Orch. 291.
    1835.
B. albiflora Raf. Fl. Tell. **2**: 38. 1836.
Habenaria blephariglottis holopetala A. Gray, Man. ed.
    5, 502. 1867.
H. ciliaris var. *alba* Morong, Bull. Torr. Club **20**: 38. 1893.
B. blephariglottis Rydb. in Britton, Man. 296. 1901.

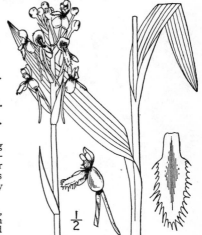

Stems and leaves similar to those of the preceding
species. Spikes densely or rather loosely many-
flowered; flowers pure white, usually a little smaller
than those of *B. ciliaris;* lip narrower, oblong; petals
toothed or somewhat fringed at the apex, rarely
entire; fringe of the lip copious or sparse.

In bogs and swamps, Newfoundland to Minnesota,
Florida and Mississippi. Blooms a few days earlier than
ciliaris where the two grow together. Feather-leaved
orchis. July–Aug.

4. Blephariglottis lácera (Michx.) Farwell. Ragged or Green-fringed Orchis. Fig. 1376.

Orchis lacera Michx. Fl. Bor. Am. **2**: 156. 1803.
Habenaria lacera R. Br. Prodr. Fl. Nov. Holl. **1**: 312. 1810.
B. lacera Farwell, Ann. Rep. Mich. Acad. Sci. **2**: 42. 1901.

Stem rather slender, 1°–2° high. Leaves firm, lanceolate, 5′–8′ long, 10″–18″ wide, the upper gradually smaller; spike 2′–6′ long, loose; flowers greenish yellow; sepals ovate, obtuse, about 3″ long, the upper one a little broader than the others, petals linear, entire, obtuse, about as long as the sepals; lip 3-parted, the segments narrow, deeply fringed, the fringe of a few threads about ½′ long; spur 7″–8″ long, curved, shorter than the ovary, clavate at the apex; anther-sacs divergent at the base, their bases beaked and projecting upward; glands oblong-linear, hyaline, as long as the caudicle.

In swamps and wet woods, Newfoundland to Minnesota, south to Georgia and Arkansas. June–July. Hybridizes with *B. psycodes*.

5. Blephariglottis leucophaèa (Nutt.) Farwell. Prairie White-fringed Orchis. Fig. 1377.

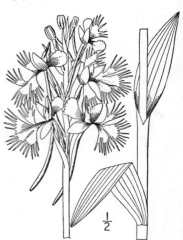

Orchis leucophaea Nutt. Trans. Am. Phil. Soc. (II.) **5**: 161. 1833–37.
Habenaria leucophaea A. Gray, Man. Ed. 5, 502. 1867.
B. leucophaea Farwell, Ann. Rep. Mich. Acad. Sci. **2**: 42. 1901.

Stem stout, angled, 1½°–2½° high. Leaves lanceolate, 4′–8′ long; spike 3′–5′ long, very thick, loosely flowered; flowers large, white, fragrant, sometimes tinged with green; sepals broadly ovate; petals obovate, minutely cut toothed, about 3″ long; lip 3-parted, 6″–7″ long, the segments broadly wedge-shaped and copiously fringed; spur 1′–1½′ long, longer than the ovary; anther-sacs widely diverging at the base; caudicles long and slender; glands transversely oval; ovary often recurved.

On moist prairies, Nova Scotia to Minnesota, Kentucky, Louisiana and Nebraska. Western greenish-fringed orchis. July.

6. Blephariglottis grandiflòra (Bigel.) Rydb. Large or Early Purple-fringed Orchis. Meadow-pink. Fig. 1378.

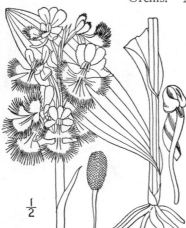

Orchis fimbriata Willd. Sp. Pl. **4**: 39. 1805. Not Dryand. 1789.
Habenaria grandiflora Torr. Comp. 319. 1826.
Orchis grandiflora Bigel. Fl. Bost. Ed. 2, 321. 1824.
Habenaria fimbriata A. Gray, Man. Ed. 5, 503. 1867.
B. grandiflora Rydb. in Britton, Man. 296. 1901.

Stem 1°–5° high. Leaves oval or lanceolate, 4′–10′ long, 10″–3′ wide, obtuse, or the upper smaller and acute; raceme 3′–15′ long, sometimes 2½′ thick, densely flowered; flowers lilac or purplish, sometimes white or nearly so, fragrant; upper sepal and petals erect, connivent; petals oblong or oblanceolate, more or less toothed, ¼′ long; lip 3-parted, ½′–1′ broad, about ½′ long, the segments broadly fan-shaped, copiously fringed to about the middle, anther-sacs divergent at the base; glands orbicular, turned inward; spur filiform, clavate, 1′–1½′ long.

In rich woods and meadows, Newfoundland to Ontario, south to North Carolina. Perhaps a large-flowered race of the following species. Tattered fringed orchis. June–Aug.

7. Blephariglottis psycòdes (L.) Rydb.   Smaller Purple-fringed Orchis.
Fig. 1379.

Orchis psycodes L. Sp. Pl. 943. 1753.
Orchis fimbriata Ait. Hort. Kew. 3 : 297. 1789.
Habenaria psycodes Spreng. Syst. 3 : 693. 1826.
Blephariglottis psycodes Rydb. in Britton, Man. 296. 1901.

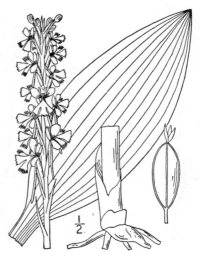

Stem rather slender, 1°–3° high. Leaves oval, elliptic or lanceolate, 2′–10′ long, 8″–3′ wide, the upper smaller; raceme 2′–6′ long, 1′–1½′ thick, loosely or densely several–many-flowered; flowers lilac, rarely white, fragrant; lower sepals ovate, obtuse, about 4″ long, the upper one a little narrower; petals oblong or oblanceolate, toothed on the upper margin; lip 3-parted, 4″–6″ broad, the segments fan-shaped and copiously fringed, the fringe of the middle segment shorter than that of the lateral ones; spur somewhat clavate at the apex, about 8″ long, longer than the ovary.

In meadows, swamps and wet woods, Newfoundland to Minnesota, North Carolina and Tennessee. Pink-fringed or flaming-orchis. Soldier's-plume. July–Aug.

8. Blephariglottis peramoèna (A. Gray) Rydb.   Fringeless Purple Orchis.
Fig. 1380.

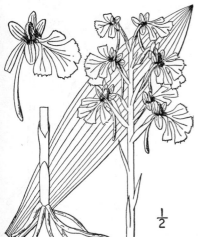

Orchis fissa Pursh, Fl. Am. Sept. 589. 1814. Not Willd. 1805.

H. peramoena A. Gray, Am. Journ. Sci. **38** : 310. 1840.

B. peramoena Rydb. in Britton, Man. 297. 1901.

Stem 1°–2½° high. Leaves elliptic or lanceolate, 4′–8′ long, ½′–1½′ wide, the upper gradually smaller; spike 2′–7′ long, 1′–2½′ thick, many-flowered; flowers showy, violet-purple; lateral sepals round-ovate, 3″–4″ long, the upper one smaller; petals smaller, round-obovate, clawed, entire, or slightly erose; lip 7″–10″ long, 3-parted, the segments fan-shaped, cut-toothed, not fringed, the middle one 2-lobed; spur about as long as the ovary, curved, clavate; anther-sacs widely divergent, little separated; glands orbicular, oblique.

In moist meadows, New Jersey to Illinois, Missouri, Virginia, Alabama and Tennessee. Great purple orchis. July–Aug.

13. POGÒNIA Juss. Gen. Pl. 65. 1789.

Mostly low herbs, the flowers terminal, solitary, the leaves alternate. Sepals and petals separate, erect or ascending. Lip erect from the base of the column, spurless. Column elongated, club-shaped at the summit. Anther terminal, stalked, attached to the back of the column, its sacs parallel; pollinia 2, 1 in each sac, powdery-granular, without a caudicle. Stigma a flattened disk below the anther. Capsule oblong or ovoid, erect. [Greek, bearded, from the bearded lip of the type species.]

A few species of the north temperate zone ; only the following known in North America. Type species : *Arethusa ophioglossoides* L.

Sepals and petals nearly equal and alike ; lip bearded.   1. *P. ophioglossoides.*
Sepals longer and narrower than the petals ; lip not bearded.   2. *P. divaricata.*

1. Pogonia ophioglossoìdes (L.) Ker. Rose Pogonia. Snake-mouth. Fig. 1381.

Arethusa ophioglossoides L. Sp. Pl. 951. 1753.

Pogonia ophioglossoides Ker, in Lindl. Bot. Reg. *pl. 148.* 1816.

Stem 8'–15' high, 1–3-leaved, not rarely with a long-petioled basal leaf. Stem leaf or leaves ½'–3' long, lanceolate or ovate, erect, bluntly acute; flowers fragrant, pale rose-color, slightly nodding, large, solitary or occasionally in pairs, subtended by a foliaceous bract; sepals and petals about equal, elliptic or oval, 6''–10'' long; lip spatulate, free or somewhat appressed to the column below, crested and fringed; column much shorter than the petals, thick, club-shaped.

In meadows and swamps, Newfoundland to Ontario, Minnesota, Florida, Kansas and Texas. Also in Japan. Roots fibrous. Propagates by runners. Adder's-mouth pogonia. June–July.

2. Pogonia divaricàta (L.) R. Br. Spreading Pogonia. Fig. 1382.

Arethusa divaricata L. Sp. Pl. 951. 1753.

Pogonia divaricata R. Br. in Ait. Hort. Kew. Ed. 2, **5**: 203. 1813.

Stem 1°–2° high, bearing a leaf near the middle, and a foliaceous bract near the flower. Leaf lanceolate, or narrowly elliptic, obtuse, clasping, 2'–4' long; flower terminal, solitary, about 1' long; sepals linear, longer and narrower than the petals, diverging, dark colored; petals flesh-color, lanceolate, narrowed at the apex, lip as long as the petals, 3-lobed, crenulate or wavy-margined, greenish, veined with purple, crested, but not bearded, the upper lobe long.

In swamps, southern New Jersey to Florida and Alabama. Ascends to 4000 ft. in North Carolina. July.

14. ISÒTRIA Raf. Med. Rep. II. 5: 357. 1808.

Low herbs with a rootstock and fibrous roots. Flowers terminal. Leaves 5 in whorl near the top of the plant. Sepals and petals separate, ascending, the former generally longer than the latter. Lip erect from the base of the column, crested, spurless, sessile. Anthers and pollinia as in *Pogonia*. Capsule oblong, erect. [Greek, in equal threes.]

Only the following species are known, the first being the type:

Lip crested along a narrow line down the face; peduncle as long as the capsule or longer.
                                                                    1. *I. verticillata.*
Lip crested over the whole face and lobes; peduncle shorter than the capsule.   2. *I. affinis.*

1. Isotria verticillàta (Willd.) Raf. Whorled Pogonia. Fig. 1383.

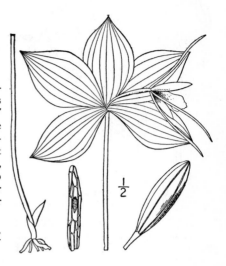

Arethusa verticillata Willd. Sp. Pl. **4**: 81. 1805.
Pogonia verticillata Nutt. Gen. **2**: 192. 1818.
I. verticillata Raf. Med. Rep. II. **5**: 357. 1808.

Stem 10′–12′ high, from long fleshy roots, bearing a whorl of 5 leaves at the summit. Leaves obovate, abruptly pointed at the apex, sessile, 1′–3′ long; flower solitary, erect or declined; peduncle 6″–8″ long, in fruit usually equalling or exceeding the capsule; sepals linear, 1½′–2′ long, about 1″ wide, spreading, dark purple; petals linear, erect, obtuse, greenish yellow, about 10″ long; lip 3-lobed, crested along a narrow band, the upper part expanded, undulate; capsule erect, 1′ or more long.

In moist woods, Ontario to Massachusetts, Indiana, Michigan and Florida. Ascends to 4500 ft. in Virginia. Whorled snake-mouth. May–June.

2. Isotria affìnis (Austin) Rydb. Smaller Whorled Pogonia. Fig. 1384.

Pogonia affinis Austin; A. Gray, Man. Ed. 5, 507. 1867.

I. affinis Rydb. in Britton, Man. 297. 1901.

Smaller than the preceding species, stem 8′–10′ high. Leaves in a whorl of 5 at the summit, 1′–2′ long; flowers 2 or solitary, greenish yellow; peduncle 2″–4″ long, much shorter than the ovary and capsule; sepals equalling the petals, or but little longer, somewhat narrowed at the base; lip crested over nearly the whole face and lobes; capsule erect, 1′ long or less.

In moist woods, Vermont and Massachusetts to southern New York and Pennsylvania. Rare and local. Our figure is taken from Mr. Austin's original sketches. June.

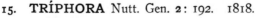

15. TRÍPHORA Nutt. Gen. **2**: 192. 1818.

Low herbs, with fleshy tubers and few axillary flowers. Sepals and petals separate, nearly equal. Lip erect, slightly clawed, somewhat 3-lobed, crestless and spurless. Column club-shaped above. Anther terminal, stalked, attached to the back of the column, its sacs parallel; pollinia 1 in each anther-sac, powdery-granular, without caudicles. Stigma a flattened disk below the anther. Capsule oval, drooping. [Greek, bearing 3 flowers.]

About 10 species, natives of America, only the following, the generic type, found in the United States.

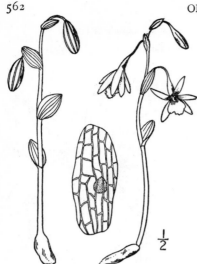

$\frac{1}{2}$

1. Triphora trianthóphora (Sw.) Rydb.
Nodding Pogonia. Fig. 1385.

Arethusa trianthophora Sw. Kongl. Vet. Acad. Handl. (II.) **21**: 230. 1800.
Triphora pendula Nutt. Gen. **2**: 193. 1818.
Pogonia pendula Lindl. Bot. Reg. *pl. 908.* 1825.
Pogonia trianthophora B.S.P. Prel. Cat. N. Y. 52. 1888.
Triphora trianthophora Rydb. in Britton, Man. 298. 1901.

Stems glabrous, 3′–8′ high, from a tuberous root, often clustered. Leaves 2–8, alternate, ovate, 3″–9″ long, clasping; flowers 1–7, on axillary peduncles, pale purple, at first nearly erect, soon drooping; perianth 6″–8″ long; sepals and petals about equal, connivent, elliptic, obtuse; lip clawed, somewhat 3-lobed, roughish or crisped above, not crested, about as long as the petals; capsule oval, drooping, about 6″ long.

In rich woods, Canada(?), Maine to Rhode Island, Florida, Wisconsin, Missouri and Kansas. Ascends to 3500 ft. in North Carolina. Local. Three-birds. Aug.–Sept.

16. ARETHÙSA L. Sp. Pl. 950. 1753.

Low herbs, with small bulbs and mostly solitary flowers on bracted scapes, the solitary leaf linear, hidden at first in the upper bract, protruding after flowering. Sepals and petals about equal, connivent and hooded above, coherent below. Lip dilated and recurved-spreading at the apex, crested on the face with straight somewhat fleshy hairs, slightly gibbous at the base. Column adherent to the lip below, linear, narrowly winged and dilated at the summit. Anther operculate, of 2 approximated sacs incumbent upon the column; pollinia 4, 2 in each sac, powdery-granular. Capsule erect, ellipsoid, strongly angled. [Dedicated to the nymph Arethusa.]

Two known species, the following, the generic type, occurring in North America, the other in Japan.

1. Arethusa bulbòsa L. Arethusa. Dragon's-mouth. Wild-pink. Fig. 1386.

Arethusa bulbosa L. Sp. Pl. 950. 1753.

Scape glabrous, 5′–10′ high, bearing 1–3 loose sheathing bracts. Leaf linear, many-nerved, becoming 4′–6′ long; flower solitary (rarely 2), arising from between a pair of small unequal scales, rose-purple, 1′–2′ high; sepals and petals linear to elliptic, obtuse, arched over the column; lip usually drooping beneath the sepals and petals, the apex broad, rounded, often fringed or toothed, variegated with purplish blotches, bearded, crested down the face in three white hairy ridges; capsule about 1′ long, ellipsoid, strongly 6-ribbed, rarely maturing.

In bogs, Newfoundland to Ontario and Minnesota, South Carolina and Indiana. May–June.

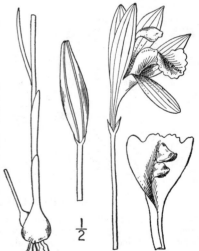

$\frac{1}{2}$

17. LIMODÒRUM L. Sp. Pl. 950. 1753.
[Calopògon R. Br. in Ait. Hort. Kew. Ed. 2, **5**: 204. 1813.]

Scapose herbs, with round solid bulbs which arise from the bulb of the previous year, a leaf appearing the first season, succeeded in the following year by the scape. Flowers several in a loose terminal spike or raceme. Sepals and petals nearly alike, separate, spreading. Column elongated, 2-winged above. Anther terminal, operculate, sessile; pollinia solitary, 1 in each sac, loosely granular. Lip spreading, raised on a narrow stalk, dilated at the apex, bearded on the upper side with long club-shaped hairs. [Greek, a meadow-gift.]

Five species, natives of the eastern United States, Cuba and the Bahamas. Type species: *Limodorum tuberosum* L.

1. Limodorum tuberòsum L. Grass-pink.
Calopogon. Fig. 1387.

Limodorum tuberosum L. Sp. Pl. 950. 1753.
Cymbidium pulchellum Willd. Sp. Pl. 4: 105. 1805.
Calopogon pulchellum R. Br. in Ait. Hort. Kew. Ed. 2,
    5: 204. 1813.

Scape slender, naked, 1°–1½° high. Leaf linear-
lanceolate, 8′–12′ long, 3″–10″ wide, sheathing, with
several scales below it; spike 4′–15′ long, 3–15-flow-
ered; flowers about 1′ long, purplish pink, subtended
by small acute bracts; sepals obliquely ovate-lanceolate, acute, about 10″ long; petals similar; column
incurved; anther-sacs parallel, attached by a slender
thread to the back of the column; lip as long as the
column, broadly triangular at the apex, crested along
the face with yellow, orange and rose-colored hairs;
capsule oblong, nearly erect.

In bogs and meadows, Newfoundland to Ontario and
Minnesota, south to Florida and Missouri. Bearded-
pink. Swamp-pink. June–July.

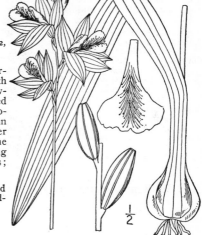

18. SERÀPIAS L. Sp. Pl. 949. 1753.

[EPIPÁCTIS (Hall.) Zinn, Cat. Pl. Hort. Goett. 85. 1757.]

Tall stout herbs with fibrous roots and simple leafy stems. Leaves ovate or lanceolate,
plicate, clasping. Flowers leafy-bracted, in terminal racemes. Sepals and petals all separate.
Spur none. Lip free, sessile, broad, concave below, constricted near the middle, the upper
portion dilated and petal-like. Column short, erect. Anther operculate, borne on the margin
of the clinandrium, erect, ovate or semiglobose, its sacs contiguous. Pollinia 2-parted,
granulose, becoming attached to the glandular beak of the stigma. Capsule oblong, beakless.
[Named for Serapis, an Egyptian deity.]

About 10 species, widely distributed. Besides the following typical species, another occurs in
the western United States.

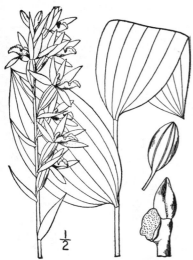

1. Serapias Helléborine L. Helleborine.
Bastard Hellebore. Fig. 1388.

Serapias Helleborine L. Sp. Pl. 949. 1753.
Serapias viridiflora Hoffm. Deutsch. Fl. 2: 182. 1804.
Epipactis latifolia var. *viridiflora* Irm. Linnaea 16: 451.
    1842.
Epipactis viridiflora Reichb. Fl. Exc. 134. 1830.

Stem 1°–2° high, glabrous below, pubescent above.
Leaves ovate or lanceolate, obtuse or acute, 1½′–3′
long, 9″–1½′ wide; flowers greenish yellow to purple;
pedicels 2″–3″ long; sepals 4″–5″ long, lanceolate;
petals narrower; lip expanded into a slightly undulate
apex, tapering to a point; bracts lanceolate, longer
than the flowers.

Quebec and Ontario to Massachusetts and Pennsylvania. Local; probably introduced. Widely distributed
in Europe. July–Aug.

19. IBÍDIUM Salisb. Trans. Hort. Soc.
London 1: 291. 1812.

[GYRÓSTACHYS Pers. Syn. 2: 511, as subgenus. 1807.]
[SPIRÁNTHES L. C. Richard, Mem. Mus. Paris 4:
    42. 1818.]

Erect herbs, with fleshy-fibrous or tuberous roots and slender stems or scapes, leaf-
bearing below or at the base. Flowers small, spurless, spiked, 1–3-rowed, the spikes more or
less twisted. Sepals free, or more or less coherent at the top, or united with petals into a
galea. Lip sessile or clawed, concave, erect, embracing the column and often adherent to
it, spreading and crisped, or rarely lobed or toothed at the apex, bearing minute callosities
at the base. Column arched below, obliquely attached to the top of the ovary. Anther without a lid, borne on the back of the column, erect. Stigma ovate, prolonged into an acuminate
beak, at length bifid, covering the anther and stigmatic only underneath. Pollina 2, 1 in each
sac, powdery. Capsule ovoid or oblong, erect. [The anther has a fancied resemblance to the
head of an Ibis.]

About 55 species, widely distributed in tropical and temperate regions. Besides the following, about 5 others occur in the southern States, and one in California. The flowers are often fragrant. Type species: *Ophrys spiràlis* J. E. Smith.

*Flowers 3–several-ranked; rachis not conspicuously twisted.
Sepals and petals coherent and connivent into a hood.   1. *I. strictum.*
Lateral sepals free and separate.
   Lip of quadrate type; callosities imperfect, mostly in edge of lip-base; vernal-flowering.
           2. *I. plantagineum.*
   Lip of ovate type; callosities prominent, mostly curved; autumnal-flowering.
     Petals linear, not dilated at the base; spike stout, over 7″ thick.   3. *I. cernuum.*
     Petals lanceolate, dilated at the base; spike slender, less than 7″ thick.   4. *I. ovale.*
**Flowers merely alternate, appearing secund from the spiral twisting of the rachis.
Stem leafy below; leaves narrow, elongate, persistent.
   Lip pubescent without, of an ovate type, the base dilated.   5. *I. vernale.*
   Lip glabrous without, of an oblong type, the base not dilated.   6. *I. praecox.*
Stem merely scaly; leaves basal, broad, short, fugacious.
   Root solitary; lip white, of an ovate type, erose-crisped from apex to the middle.   7. *I. Beckii.*
   Roots clustered; lip green, except the crisped margin, of an oblong type, the apex wavy-crisped.
                                          8. *I. gracile.*

1. Ibidium stríctum (Rydb.) House. Hooded Ladies'-tresses. Fig. 1389.

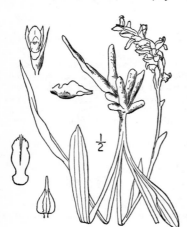

Gyrostachys stricta Rydb. Mem. N. Y. Bot. Gard. **1**: 107. 1900.
I. strictum House, Bull. Torr. Club **32**: 381. 1905.

Stem 6′–15′ high, glabrous, leafy below, bracted above, the inflorescence rarely puberulent. Lower leaves 3′–8′ long, linear or linear-oblanceolate; spike 2′–4′ long, 4″–7″ thick; flowers in 3 rows, white or greenish, ringent, 4″–5″ long, spreading horizontally, very fragrant; sepals and petals broad at the base, all more or less connivent into a hood; lip oblong or ovate-oblong, broad at the base, contracted below the dilated crisped apex, thin, transparent, veined; callosities mere thickenings of the basal margins of the lip, or none.

In bogs, Newfoundland to Alaska, south to Maine, Pennsylvania, Minnesota and California. July–Aug. Confused in our first edition with the Alaskan plant described as *Spiranthes Romanzoffiàna* Cham. which has narrower, long-acuminate sepals and petals and an ovate, pointed end to the lip.

2. Ibidium plantagíneum (Raf.) House. Wide-leaved Ladies'-tresses. Fig. 1390.

Neottia plantaginea Raf. Am. Month. Mag. **2**: 206. 1818.

N. lucida H. H. Eaton, Trans. Journ. Med. **5**: 107. 1832.

Spiranthes plantaginea Torr. Fl. N. Y. **2**: 284. 1843.

Spiranthes lucida Ames, Orch. **2**: 258. 1908.

I. plantagineum House, Bull. Torr. Club **32**: 381. 1905.

Stem 4′–10′ high, glabrous or pubescent, bracted above, bearing 4 or 5 lanceolate or oblanceolate leaves below. Leaves 1′–5′ long; spike 1′–2′ long, 4″–5″ thick, dense; floral bracts mostly much shorter than the flowers; flowers spreading, about 3″ long; petals and sepals white, lateral sepals free, narrowly lanceolate, the upper somewhat united with the petals; lip pale yellow on the face, oblong, not contracted in the middle, the wavy apex rounded, crisped or fringed, the base short-clawed; callosities none, or mere thickenings of the lip margins.

Moist banks and woods, Nova Scotia to Minnesota, south to Virginia and Wisconsin. June–Aug.

3. Ibidium cérnuum (L.) House.   Nodding or drooping Ladies'-tresses.
Fig. 1391.

Ophrys cernua L. Sp. Pl. 946. 1753.
Spiranthes cernua L. C. Rich. Orch. Ann. 37. 1817.
Gyrostachys cernua Kuntze, Rev. Gen. Pl. 664. 1891.
Spiranthes odorata Lindl. Gen. & Sp. Orch. 467. 1840.
Gyrostachys ochroleuca Rydb. in Britton, Man. 300. 1901.
I. incurvum Jennings, Ann. Car. Mus. 3 : 483. 1906.
I. cernuum House, Bull. Torr. Club 32 : 381. 1905.

Stem 6'–25' high (rarely taller), usually pubescent
above, mostly bearing 2–6 acuminate bracts. Leaves
nearly basal, linear-oblanceolate or linear, 3'–14'
long, the blade narrow, the petiole 2'–10' long; spike
4'–5' long, 6"–7" thick; flowers white or yellowish,
fragrant, nodding or spreading, about 5" long, in
3 rows; lateral sepals free, the upper arching and
connivent with the petals; lip oblong, or sometimes
ovate, the broad apex rounded, crenulate or crisped;
callosities nipple-shaped, straight, hairy.

In wet meadows and swamps, Newfoundland to Ontario, South Dakota, Florida and New Mexico. Wild
tube-rose. Screw-auger. Aug.–Oct.

4. Ibidium ovàle (Lindl.) House.   Small-flow-
ered Ladies'-tresses.   Fig. 1392.

Spiranthes ovalis Lindl. Gen. & Sp. Orch. 466. 1840.

S. cernua parviflora Chapm. Fl. S. U. S. Ed. 3, 488. 1897.

Gyrostachys parviflora Small, Fl. SE. U. S. 318. 1903.

Spiranthes parviflora Ames, Orch. 137. 1909.

Ibidium ovale House, Muhlenbergia 1 : 128. 1906.

Stem 4'–15' tall, leafy below, minutely pubescent
above. Leaves broadly linear to linear-oblong, 1½'–6½'
long, or the upper smaller; spike slender, compact,
tapering upward, the bracts shorter than the flowers;
flowers small, about 2" long, white, nodding; lateral
sepals free, lanceolate; lip about 2" long, ovate, narrow
at the apex or acute; callosities slender, curved.

In woods and swamps, Ohio to Missouri, Georgia and
Louisiana. Sept.–Oct.

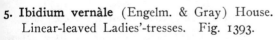

5. Ibidium vernàle (Engelm. & Gray) House.
Linear-leaved Ladies'-tresses.   Fig. 1393.

Spiranthes vernalis Engelm. & Gray, Bost. Journ. Nat. Hist.
5 : 236. 1845.

S. neglecta Ames, Rhodora 6 : 30. *pl. 51.* 1904.

Gyrostachys linearis Rydb. in Britton, Man. 300. 1901.

Ibidium vernale House, Bull. Torr. Club 32 : 381. 1905.

Stem slender, 6'–22' high, usually copiously glandular-
pubescent above, leafy. Leaves linear, or somewhat
tapering to both ends, mostly 2½'–6½' long, persistent;
spike strongly spiral, 2½'–6' long, mostly 4"–6" thick;
bracts much longer than the ovaries; flowers yellowish;
lip 3"–3½" long, ovate, much shorter than the median
sepal; callosities slender, often hooked at the tip.

In dry or wet soil, Massachusetts to Florida and New
Mexico, north in the Mississippi Valley to Illinois and Kansas. Aug.–Sept. A hybrid with *I. gracile* has been described.

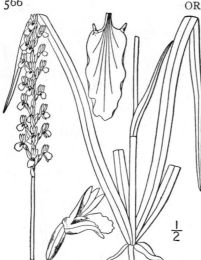

7. Ibidium Béckii (Lindl.) House, Muhlenbergia 1: 128. 1906. Little Ladies'-tresses. Fig. 1395.

Spiranthes Beckii Lindl. Gen. & Sp. Orch. 472. 1840
S. simplex A. Gray, Man. Ed. 5, 506. 1867. Not Griseb.
Gyrostachys simplex Kuntze, Rev. Gen. Pl. 664. 1891.
Ibidium Beckii House, Muhlenbergia 1: 128. 1905.

Stems very slender, 5'–9' high, from a solitary spindle-shaped tuberous root, with small deciduous bracts above. Leaves basal, ovate or oblong, short, abruptly narrowed into a petiole, mostly disappearing at or before the flowering time; spike slender, about 1' long and 3" thick, glabrous, little twisted; flowers white, 1"–1½" long; lip thin, striped, ovate to orbicular-ovate, erose-crisped from below the middle to the apex, short-clawed; callosities nipple-shaped, slender, usually curved or hooked at the tip.

In dry sandy soil, Massachusetts to New Jersey, Florida, Kentucky, Arkansas and Texas. Aug–Sept.

6. Ibidium praècox (Walt.) House. Grass-leaved Ladies'-tresses. Fig. 1394.

Limodorum praecox Walt. Fl. Car. 221. 1788.
Spiranthes graminea var. *Walteri* A. Gray, Man. Ed. 5, 505. 1867.
Spiranthes praecox S. Wats. in A. Gray, Man. Ed. 6, 503. 1890.
G. praecox Kuntze, Rev. Gen. Pl. 663. 1891.
Ibidium praecox House, Muhlenbergia 1: 129. 1906.

Stem slender, 10'–30' high, sparingly glandular-pubescent above, leafy. Leaves linear, 4'–12' long, with narrow grass-like blades and long sheathing petioles, mostly persistent through the flowering season, the upper smaller; spike usually much twisted, 2'–8' long, 4"–6" thick; bracts about as long as the ovaries; flowers white, 3"–4" long, spreading; lip 3½"–4" long, oblong, contracted above, the dilated apex obtuse, crenulate; callosities stout, usually straight.

In grassy places, southern New York to Florida and Texas.   July–Aug.

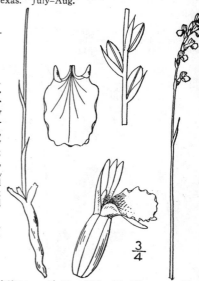

8. Ibidium grácile (Bigel.) House. Slender Ladies'-tresses. Fig. 1396.

Neottia gracilis Bigel. Fl. Bost. Ed. 2, 322. 1824.
Spiranthes gracilis Beck, Bot. 343. 1833.
Gyrostachys gracilis Kuntze, Rev. Gen. Pl. 664. 1891.
I. gracile House, Bull. Torr. Club 32: 381. 1905.

Stem slender, 8'–2° high, from a cluster of spindle-shaped tuberous roots, glabrous, or rarely pubescent above, bearing small deciduous bracts. Leaves basal, obovate, or ovate-lanceolate, petioled, the blades ½'–2' long, 4"–10" wide, mostly perishing before the flowering season; spike 1'–3' long, 4"–6" thick, loose, usually much twisted; flowers white, fragrant, 2"–2½" long; sepals a little longer than the lip, the lateral ones free; lip about 2" long, oblong, dilated and crenulate or wavy-crisped at the apex, usually thick and green in the middle, white and hyaline on the margins, slightly clawed at the base; callosities small, nipple-shaped, stout, straight.

In dry fields and open woods, Nova Scotia to Manitoba, Florida, Louisiana and Texas. Ascends to 2500 ft. in North Carolina.   Twisted-stalk.   Corkscrew-plant.   Aug.–Oct.

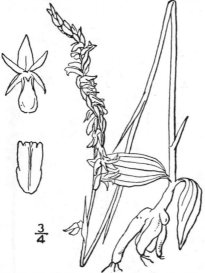

20. ÓPHRYS [Tourn.] L. Sp. Pl. 945. 1753.

[LÍSTERA R. Br. in Ait. Hort. Kew. Ed. 2, **5**: 201. 1813.]

Small herbs, with fibrous or sometimes rather fleshy-fibrous roots, bearing a pair of opposite green leaves near the middle, and 1 or 2 small scales at the base of the stem. Flowers in terminal racemes, spurless. Sepals and petals nearly alike, spreading or reflexed, free. Anther without a lid, erect, jointed to the column. Pollinia 2, powdery, united to a minute gland. Capsule ovoid or obovoid. [Greek, the eyebrow.]

About 12 species, natives of the north temperate and arctic zones. Besides the folowing, another occurs in northwestern North America. Type species: *Ophrys ovàta* L.

Lip broadly wedge-shaped, retuse or 2-lobed at the apex.
   Leaves oval; pedicels and ovaries glandular.   1. *O. convallarioides.*
   Leaves reniform; pedicels and ovaries glabrous.   2. *O. Smallii.*
Lip oblong or linear.
   Lip broad, 2-cleft ¼–⅓ its length; base auricled.   3. *O. auriculata.*
   Lip 2-cleft about ½ its length.
      Lip twice as long as the petals, with lateral teeth.   4. *O. cordata.*
      Lip 4–8 times as long as the petals, with auricles at the base.   5. *O. australis.*

1. Ophrys convallarioìdes (Sw.) W. F. Wight.   Broad-lipped Twayblade. Fig. 1397.

Epipactis convallarioides Sw. Kongl. Vet. Acad. Handl.
  (II.) **21**: 232. 1800.
Listera convallarioides Torr. Comp. 320. 1826.
O. convallarioides W. F. Wight, Bull. Torr. Club **32**: 380.
  1905.

Stem 4'–10' high, glandular-pubescent above the leaves. Leaves smooth, round-oval or ovate, obtuse or cuspidate at the apex, sometimes slightly cordate or reniform at the base, 3–9-nerved. Raceme 1½'–3' long, loosely 3–12-flowered; flowers greenish yellow, pedicels filiform, bracted, 3''–4'' long; petals and sepals linear-lanceolate, much shorter than the lip; lip broadly wedge-shaped, with 2 obtuse lobes at the dilated apex, generally with a tooth on each side at base; column elongated, but shorter than the lip, a little incurved, with 2 short projecting wings above the anther; capsule obovoid, about 3'' long.

In woods, Newfoundland to Alaska and California, south to Vermont and Michigan. June–Aug.

2. Ophrys Smàllii (Wiegand) House.   Kidney-leaf Twayblade.   Fig. 1398.

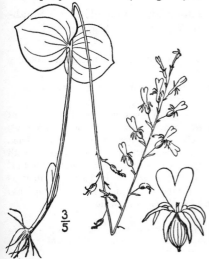

Listera reniformis Small, Bull. Torr. Club **24**: 334.
  1897. Not D. Don.
L. Smallii Wiegand, Bull. Torr. Club **26**: 169. 1890.
O. Smallii House, Bull. Torr. Club **32**: 379. 1905.

Perennial, deep green. Stems erect, 4'–12' tall, slender, glabrous below, densely glandular-pubescent above; leaves reniform, or ovate-reniform, 5''–14'' broad, apiculate or short-acuminate, pubescent beneath, cordate or subcordate, sessile; racemes ¾'–4' long; bracts lanceolate to ovate-lanceolate, 1½''–2½'' long, acute; pedicels 2''–4'' long; sepals oblong or linear-oblong, about 1½'' long, reflexed; corolla greenish, the lip wedge-shaped, 3''–3½'' long, with 2 prominent teeth near the base, sharply cleft, the segments obtuse; capsules oval, 2''–2½'' long.

In damp thickets in the mountains, Pennsylvania to Virginia and North Carolina. Also in eastern Asia. Formerly confused with the preceding species. Spring and summer.

3. Ophrys auriculàta (Wiegand) House. Auricled Twayblade. Fig. 1399.

Listera auriculata Wiegand, Bull. Torr. Club **26**: 166. 1899.

O. auriculata House, Bull. Torr. Club **32**: 379. 1905.

Stem slender, 4′–7′ high, glabrous below, glandular above the leaves. Leaves large, 1½′–2′ long, oval or elliptic-ovate, borne above the middle of the stem; raceme many-flowered; rachis pubescent; pedicels and ovaries glabrous; sepals lance-ovate; petals oblong-linear, longer than the ovary, spreading, obtuse; lip slightly ciliate, oblong, broadest at the auricled base, cleft ¼–⅛ its length; column rather stout, a little over 1″ long.

Cedar swamps and wet banks, Quebec, Maine, New Hampshire and Vermont. July.

4. Ophrys cordàta L. Heart-leaved Twayblade. Double-leaf. Fig. 1400.

Ophrys cordata L. Sp. Pl. 946. 1753.

Listera cordata R. Br. in Ait. Hort. Kew. Ed. 2, **5**: 201. 1813.

Stem very slender, glabrous or nearly so, 3′–10′ high. Leaves sessile, cordate, ovate, mucronate, ½′–1′ long; racemes rather loose, ½′–2′ long, 4–20-flowered; flowers purplish, minute; pedicels bracted, about 1″ long; sepals and petals oblong-linear, scarcely 1″ long; lip narrow, often with a subulate tooth on each side at the base, twice as long as the petals, 2-cleft, the segments setaceous and ciliolate; column very small, the clinandrium just appearing above the anther; capsule ovoid, 2″ long.

In moist woods, Labrador to Alaska, New Jersey, Michigan, Colorado and Oregon. Also in Europe and Asia. Twi-foil. June–Aug.

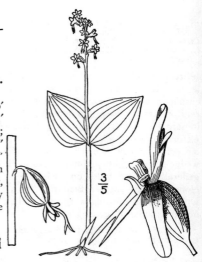

5. Ophrys austràlis (Lindl.) House. Southern Twayblade. Fig. 1401.

Listera australis Lindl. Gen. & Sp. Orch. 456. 1840.

O. australis House, Bull. Torr. Club **32**: 379. 1905.

Stem slender, 4′–10′ high, more or less pubescent above. Leaves ovate, acutish, mucronate, glabrous, shining, 8″–10″ long, 3–7-nerved; raceme 2′–3′ long, loosely 8–15-flowered; flowers yellowish green with purplish stripes; sepals and petals minute; lip ¼′–½′ long, 2-parted, split nearly to the base, 4–8 times as long as the petals, its segments linear-setaceous; column very small; capsule ovoid.

In bogs, Ontario, New York and New Jersey to Florida, Alabama and Louisiana. A third leaf is rarely borne below the flowers.

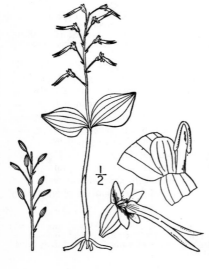

21. PERÀMIUM Salisb. Trans. Hort. Soc. 1: 301. 1812.

[GOODYERA R. Br. in Ait. Hort. Kew. Ed. 2, **5**: 197. 1813.]

Herbs with bracted erect scapes, the leaves basal, tufted, often blotched with white, the roots thick fleshy fibers. Flowers in bracted spikes. Lateral sepals free, the upper one united with the petals into a galea. Lip sessile, entire, roundish ovate, concave or saccate; without callosities, its apex reflexed. Anther without a lid, erect or incumbent, attached to the column by a short stalk; pollinia one in each sac, attached to a small disk which coheres with the top of the stigma, composed of angular grains. [Greek, referring to the pouch-like lip.]

About 25 species, widely distributed in temperate and tropical regions. Type species: *Peramium repens* (L.) Salisb.

Spike 1-sided or loosely spiral; lip elongated.
  Margin of the lip recurved or flaring.
    Anther blunt; beak of stigma shorter than the body.   1. *P. ophioides.*
    Anther acuminate; beak of stigma as long as body or longer.   2. *P. tesselatum.*
  Margin of the lip involute.   3. *P. decipiens.*
Spike dense, cylindric; lip short-tipped.

1. Peramium ophiòides (Fernald) Rydb.
Lesser Rattlesnake Plantain. Fig. 1402.

Goodyera repens var. *ophioides* Fernald, Rhodora **1**: 6. 1899.
P. ophioides Rydb. in Britton, Man. 302. 1901.
Epipactis repens ophioides A. A. Eaton, Proc. Biol. Soc. Wash. **21**: 65. 1908.

Scape 5'–10' high, glandular-pubescent, bearing several small scales. Leaves ovate, the blade 6″–15″ long, 4″–8″ wide, somewhat reticulated or blotched with white, tapering into a sheathing petiole; spike short, 1-sided; flowers greenish white; perianth 1¾″–2″ long; galea concave, ovate, with a short spreading or slightly recurved tip; lip saccate, with a narrow recurved or spreading apex; column very short; anther 2-celled; pollinia not prolonged into a caudicle.

In woods, Newfoundland to Yukon, South Carolina, Michigan and Colorado. Ascends to 5000 ft. in Virginia. Creeping-root plant. White-plantain. Squirrel-ear. *P. repens* (L.) Salisb. is restricted to Europe and Asia. July–Aug.

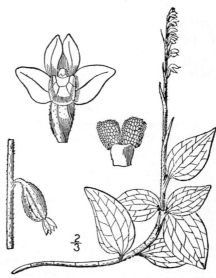

2. Peramium tesselàtum (Lodd.) Heller. Loddiges' Rattlesnake Plantain. Fig. 1403.

Goodyera tesselata Lodd. Bot. Cab. **10**: *pl. 952.* 1824.

Epipactis tesselata A. A. Eaton, Proc. Biol. Soc. Wash. **21**: 66. 1908.

P. tesselatum Heller, Cat. N. A. Pl. Ed. 2, 4. 1900.

Scape 6'–12' high, glandular-pubescent, scaly. Leaves ovate, oblong-ovate or ovate-lanceolate, the blade 10″–20″ long, 4″–11″ wide, bright green, more or less conspicuously marked with white, usually abruptly narrowed into the sheathing petiole; spike 1½'–5' long, loosely spiral; flowers whitish, larger than those of *P. ophioides*; perianth 2″–2½″ long; galea broad, the tip recurved; lip slightly saccate at the base, the long tip somewhat recurved; column short.

Mostly in coniferous woods, Newfoundland to Ontario, Pennsylvania and Michigan. July–Sept.

3. Peramium decípiens (Hook.) Piper. Menzies' Rattlesnake Plantain.
Fig. 1404.

Spiranthes decipiens Hook. Fl. Bor. Am. 2: 203. 1839.
Goodyera Menziesii Lindl. Gen. & Sp. Orch. 492. 1840.
P. Menziesii Morong, Mem. Torr. Club 5: 124. 1894.
P. decípiens Piper, Contr. U. S. Nat. Herb. 11: 208. 1906.
Epipactis decípiens Ames, Orchidaceae 2: 261. 1908.

Scape stout, 8′–15′ high, glandular-pubescent. Leaves ovate-lanceolate, 1½–2½′ long, 8″–15″ wide, the blade acute at both ends, often without white blotches or reticulations; spike not 1-sided; flowers greenish white; perianth 4″–4½″ long; galea concave, ovate-lanceolate, the tip long, usually recurved, lip swollen at the base, with a long narrow recurved or spreading apex; anther ovate, pointed, on the base of the column, which is prolonged above the stigma into a gland-bearing awl-shaped beak.

In woods, Quebec to British Columbia, New Hampshire, Michigan, Arizona and California. Aug.

4. Peramium pubéscens (Willd.) MacM. Downy Rattlesnake Plantain. Networt. Fig. 1405.

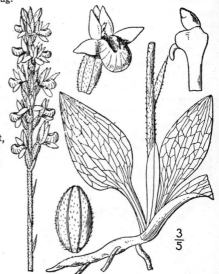

Neottia pubescens Willd. Sp. Pl. 4: 76. 1805.
Goodyera pubescens R. Br. in Ait. Hort. Kew. Ed. 2, 5: 198. 1813.
Peramium pubescens MacM. Met. Minn. 172. 1892.
Epipactis pubescens A. A. Eaton, Proc. Biol. Soc. Wash. 21: 65. 1908.

Scape 6′–20′ high, densely glandular-pubescent, bearing 5–10 lanceolate scales. Leaves 1′–2′ long, 8″–1′ wide, strongly white-reticulated, oval or ovate; spike not 1-sided; flowers greenish white; perianth 2½″–3″ long; lateral sepals ovate; galea ovate, its short tip usually not recurved; lip strongly saccate with a short broad obtuse recurved or spreading tip.

In dry woods, Maine to Ontario and Minnesota, south to Florida and Tennessee. Ascends to 4000 ft. in North Carolina. Adder's-violet. Net-leaf or spotted-plantain. Rattlesnake-leaf. Rattlesnake- or scrofula-weed. Ratsbane. July–Aug.

22. MALÁXIS Soland. Sw. Prodr. 119. 1788.
[Achroánthes Raf. Med. Rep. (II.) 5: 352. 1808.]
[Micróstylis Nutt. Gen. 2: 196. 1818.]

Low herbs, from a solid bulb, most species 1-leaved, and with 1–several scales at the base of the stem. Flowers small, white or green, in a terminal raceme. Sepals spreading, separate, the lateral ones equal at the base. Petals filiform or linear, spreading. Lip cordate or eared at the base, embracing the column. Anther erect between the auricles, 2-celled; pollinia 4, smooth and waxy, 2 in each sac, the pairs cohering at the summit, without caudicles or glands. Capsule oval, sometimes nearly globose, beakless. [Greek, perhaps in allusion to the soft tissues.]

About 45 species, widely distributed. Besides the following, about 4 others occur in the southern and western parts of North America. Type species: Malaxis spicàta Sw.

Leaf sheathing the base of the stem.   1. A. monophylla.
Leaf clasping the stem near the middle.   2. A. unifolia.

1. Malaxis monophýlla (L.) Sw. White Adder's-mouth. Fig. 1406.

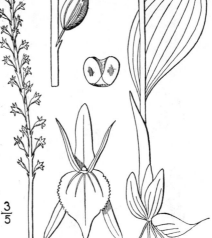

Ophrys monophyllos L. Sp. Pl. 947. 1753.

Malaxis monophyllos Sw. Vet. Akad. Nya Handl. **21**: 234. 1800.

Microstylis monophylla Lindl. Bot. Reg. *pl. 1290.* 1829.

Achroanthes monophylla Greene, Pittonia **2**: 183. 1891.

Stem slender, 4'–6' high, smooth, glabrous, striate. Leaf sheathing the stem at its base, the blade 1'–2' long, ½'–1½' wide; raceme 1'–3' long, narrow, 3"–5" thick; flowers whitish, about 1" long; pedicels nearly erect, bracted, 1"–2" long; sepals acute; lip triangular or ovate, acuminate, the lateral lobes obtuse; capsule oval, about 3" long.　　$\frac{3}{5}$

In woods, Quebec to Manitoba, Pennsylvania and Nebraska. July.

2. Malaxis unifòlia Michx. Green Adder's-mouth. Fig. 1407.

M. unifolia Michx. Fl. Bor. Am. **2**: 157. 1803.
Achroanthes unifolia Raf. Med. Rep. (II.) **5**: 352. 1808.
Microstylis ophioglossoides Nutt. Gen. **2**: 196. 1818.

Stem glabrous, striate, 4'–10' high. Leaf clasping the stem near the middle, oval or nearly orbicular, 1'–2½' long, 10"–1½' wide; raceme 1'–3' long, sometimes 1' thick; flowers greenish, about 1" long, the pedicels very slender, spreading, 3"–5" long; sepals oblong; lip broad, 3-toothed at the apex; capsule oval or subglobose.

In woods and thickets, Newfoundland to Ontario and Manitoba, south to Florida, Alabama and Missouri. Ascends to 4000 ft. in North Carolina. July.

Malaxis paludòsa (L.) Sw. (*Sturmia paludosa* Reichb.), a small species, with several basal leaves and very small flowers, otherwise known only from Europe and Asia, has been found in Otter Tail County, Minnesota.

$\frac{2}{3}$

23. LÍPARIS L. C. Richard, Mem. Mus. Paris **4**: 43, 60. 1817.

[LEPTÓRCHIS Thouars, Nouv. Bull. Soc. Philom. **1**: 317. Hyponym. 1809.]

Low herbs, with solid bulbs, the base of the stem sheathed by several scales and 2 broad shining leaves. Flowers in terminal racemes. Sepals and petals nearly equal, linear, spreading, petals usually very narrow. Column elongated, incurved, thickened and margined above. Pollinia 2 in each sac of the anther, smooth and waxy, the pairs slightly united, without stalk, threads or glands. Lip nearly flat, often bearing 2 tubercles above the base. [Greek, fat, referring to the texture of the leaves.]

About 100 species, widely distributed in temperate and tropical regions; only the following known to occur in North America. Type species: *Liparis Loeselii* (Willd.) L. C. Rich.

The name *Leptorchis*, used for this genus in our first edition, appears, after an examination of its first publication, to be intended only as a French designation, not Latin.

Raceme many-flowered; lip as long as the petals.　　　1. *L. liliifolia.*
Raceme few-flowered; lip shorter than the petals.　　　2. *L. Loeselii.*

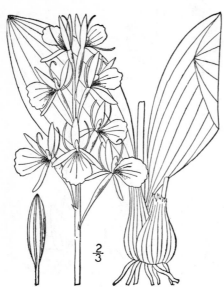

$\frac{2}{3}$

1. Liparis liliifòlia (L.) L. C. Rich. Large Twayblade. Fig. 1408.

Ophrys liliifolia L. Sp. Pl. 946. 1753.
Liparis liliifolia L. C. Rich.; Lindl. Bot. Reg. *pl. 882.* 1825.
Leptorchis liliifolia Kuntze, Rev. Gen. Pl. 671. 1891.

Scape 4'–10' high, 5–10-striate. Leaves ovate or oval, 2'–5' long, 1'–2½' wide, obtuse, keeled below, the sheaths large and loose. Raceme sometimes 6' long; flowers numerous, showy; sepals and petals somewhat reflexed; petals very narrow or thread-like; lip erect, large, 5"–6" long, about as long as the petals, wedge-obovate; column 1½" long, incurved, dilated at the summit; pedicels slender, ascending or spreading, 4"–8" long; capsule somewhat club-shaped, abous 6" long, the pedicels thickened in fruit.

In moist woods and thickets, Maine to Minnesota, Georgia and Missouri. Ascends to 3000 ft. in Virginia. May–July.

2. Liparis Loesèlii (L.) L. C. Rich. Fen Orchis. Loesel's Twayblade. Fig. 1409.

Ophrys Loeselii L. Sp. Pl. 947. 1753.
Liparis Loeselii L. C. Rich. Mem. Mus. Paris 4: 60. 1817.
Leptorchis Loeselii MacM. Met. Minn. 173. 1892.

Scape 2'–8' high, strongly 5–7-ribbed. Leaves elliptic or elliptic-lanceolate, 2'–6' long, ½'–2' wide, obtuse; raceme few-flowered; flowers greenish, smaller than those of the preceding species, 2"–3" long; sepals narrowly lanceolate, spreading; petals linear, somewhat reflexed; lip obovate, pointed, rather shorter than the petals and sepals, its tip incurved; column half as long as the lip or less; capsule about 5" long, wing-angled, on a thickened pedicel.

In wet thickets and on springy banks, Nova Scotia to Saskatchewan, south to Alabama and Missouri. Also in Europe. May–July.

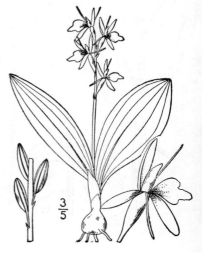

$\frac{3}{5}$

24. CYTHERÈA Salisb. Trans. Hort. Soc. 1: 301. 1812.

[CALÝPSO Salisb. Par. Lond. *pl. 89.* 1807. Not Thouars. 1805.]

Bog herb, with a solid bulb and coralloid roots, the low 1-flowered scape sheathed by 2 or 3 loose scales and a solitary petioled leaf at the base. Flower large, showy, terminal, bracted. Sepals and petals similar, nearly equal. Lip large, saccate or swollen, 2-parted below. Column dilated, petal-like, bearing the lid-like anther just below the summit. Pollinia 2, waxy, each 2-parted, without caudicles, sessile on a thick gland, the stigma at the base. [Surname of Venus.]

A monotypic species of the cooler portions of the north temperate zone.

1. Cytherèa bulbòsa (L.) House. Calypso.
Fig. 1410.

Cypripedium bulbosum L. Sp. Pl. 951. 1753.
Calypso borealis Salisb. Par. Lond. *pl. 89.* 1807.
Calypso bulbosa Oakes, Cat. Vermont Pl. 28. 1842.
Cytherea bulbosa House, Bull. Torr. Club **32**: 382. 1905.

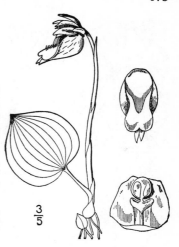

Bulb 5″ in diameter or less. Scape 3′–6′ high; leaf round-ovate, 1′–1½′ long, nearly as wide, obtusely pointed at the apex, rounded or subcordate at the base, the petiole 1′–2′ long; flowers variegated, purple, pink and yellow, the peduncle jointed; petals and sepals linear, erect or spreading, 5″–7″ long, with 3 longitudinal purple lines; lip large, saccate, 2-divided below, spreading or drooping, with a patch of yellow woolly hairs; column erect, broadly ovate, shorter than the petals; capsule about ½′ long, many-nerved.

Labrador to Alaska, south to Maine, Michigan, California, and in the Rocky Mountains to Arizona. Also in Europe. Flower somewhat resembling that of a small *Cypripedium.* May–June.

$\frac{3}{5}$

25. TIPULÀRIA Nutt. Gen. **2**: 195. 1818.

Slender scapose herbs, with solid bulbs, several generations connected by offsets, the flowers in a long loose terminal raceme. Leaf solitary, basal, unfolding long after the flowering season (in autumn), usually after the scape has perished. Scape with several thin sheathing scales at the base. Flowers green, nodding, bractless. Sepals and petals similar, spreading. Lip 3-lobed, produced backwardly into a very long spur. Column erect, wingless or very narrowly winged. Anther terminal, operculate, 2-celled. Pollinia 4, ovoid, waxy, 2 in each anther-sac, separate, affixed to a short stipe, which is glandular at the base. [Latin, similar to *Tipula,* a genus of insects, in allusion to the form of the flower.]

Two known species, the following of eastern North America being the generic type, the other Himalayan.

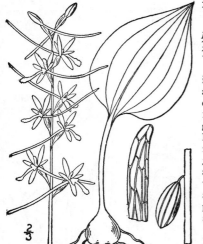

1. Tipularia unifòlia (Muhl.) B.S.P. Cane-fly Orchis. Fig. 1411.

Limodorum unifolium Muhl. Cat. 81. 1813.
Tipularia discolor Nutt. Gen. **2**: 195. 1818.
Tipularia unifolia B.S.P. Prel. Cat. N. Y. 51. 1888.

Scape glabrous, 15′–20′ high, from a hard, often irregular solid bulb or corm. Leaf arising in autumn from a fresh lateral corm, ovate, 2′–3′ long, dark green, frequently surviving through the winter, 1′–2′ wide. Raceme 5′–10′ long, very loose; flowers green, tinged with purple; pedicels filiform, bractless, 4″–6″ long; sepals and petals 3″–4″ long, narrow; lip shorter than the petals or equalling them, 3-lobed, the middle lobe narrow, prolonged, dilated at the apex, the lateral lobes short, triangular; spur very slender, straight or curved, often twice as long as the flower; column narrow, erect, shorter than the petals, the beak minutely pubescent; capsule ellipsoid, 6-ribbed, about 6″ long.

In woods, Massachusetts to Pennsylvania, Florida, Kentucky, Arkansas and Louisiana. Reported from Vermont and Michigan. Tallow-root. July–Aug.

$\frac{2}{3}$

26. APLÉCTRUM Nutt. Gen. **2**: 197. 1818.

Scapose herbs, from a corm, produced from the one of the previous season by an offset, sometimes with coralloid fibres, the scape clothed with several sheathing scales. Leaf solitary, basal; developed in autumn or late summer, broad, petioled. Flowers in terminal racemes, the pedicels subtended by small bracts. Petals and sepals similar, narrow. Lip clawed, somewhat 3-ridged; spur none. Column free, the anther borne a little below its summit. Pollinia 4, lens-shaped, oblique. [Greek, meaning without a spur.]

A monotypic North American genus.

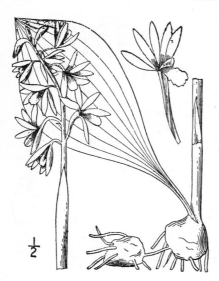

½

1. Aplectrum hyemàle (Muhl.) Torr.
Adam-and-Eve. Putty-root. Fig. 1412.

Arethusa spicata Walt. Fl. Car. 222. 1788.
Cymbidium hyemale Muhl.; Willd. Sp. Pl. **4**: 107. 1805.
Aplectrum hyemale Torr. Compend. 322. 1826.
Aplectrum spicatum B.S.P. Prel. Cat. N. Y. 51. 1888.
Not *Arethusa spicata* Walt.
A. Shortii Rydb. in Britton, Man. 305. 1901.

Scape glabrous, 1°–2° high, bearing about 3 scales. Leaf arising from the corm, at the side of the scape, elliptic or ovate, 4′–6′ long, ½′–3′ wide, usually lasting over winter; raceme 2′–4′ long, loosely several-flowered; flowers dull yellowish brown mixed with purple, about 1′ long, short-pedicelled; sepals and petals linear-lanceolate, about ½′ long; lip shorter than the petals, obtuse, somewhat 3-lobed and undulate; column slightly curved, shorter than the lip; capsule oblong-ovoid, angled, about 10″ long.

In woods and swamps, Ontario to Saskatchewan and Oregon, south to Georgia, Missouri and California. Several old corms usually remain attached to the latest one. May–June.

27. CORALLORRHÌZA (Haller) Chatelain, Spec. Inaug. 8. 1760.

Scapose, yellowish or purplish herbs, saprophytes or root-parasites, with large masses of coralloid branching rootstocks, the leaves all reduced to sheathing scales. Flowers in terminal racemes. Sepals nearly equal, the lateral ones united at the base with the foot of the column, forming a short spur or gibbous protuberance, the other one free, the spur adnate to the summit of the ovary. Petals about as long as the sepals, 1–3-nerved. Lip 1–3-ridged. Column nearly free, slightly incurved, somewhat 2-winged. Anther terminal, operculate. Pollinia 4, in 2 pairs, oblique, free, soft-waxy. [Greek, from the coral-like roots.]

About 15 species, widely distributed in the north temperate zone. Besides the following, some 4 others occur in southern and western North America. Type species: *Corallorrhiza trifida* Chatelain.

Lip 3-lobed.
  Lateral lobes of lip very small; spur a small protuberance.   1. *C. Corallorrhiza.*
  Lateral lobes of lip large; spur prominent.   2. *C. maculata.*
Lip not lobed, entire, notched or undulate.
  Perianth 3″–6″ long; spur evident, sometimes small.
    Lip long-clawed, notched.   3. *C. Wisteriana.*
    Lip short-clawed or sessile, not notched.
      Perianth about 2″ long; lip spotted.   4. *C. odontorrhiza.*
      Perianth 5″–6″ long; lip not spotted.   5. *C. ochroleuca.*
    Perianth 8″–9″ long; no spur.   6. *C. striata.*

1. Corallorrhiza Corallorrhìza (L.) Karst.
Early Coral-root. Fig. 1413.

Ophrys Corallorrhiza L. Sp. Pl. 945. 1753.
C. trifida Chatelain, Spec. Inaug. 8. 1760.
C. Neottia Scop. Fl. Carn. Ed. 2, **2**: 207. 1772.
C. innata R. Br. in Ait. Hort. Kew. Ed. 2, **5**: 209. 1813.
C. Corallorrhiza Karst. Deutsch. Fl. 448. 1880–83.

Scape glabrous, 4′–12′ high, clothed with 2–5 closely sheathing scales. Raceme 1′–3′ long, ‹–12-flowered; flowers mainly dull purple, on very short minutely bracted pedicels; sepals and petals narrow, about 3″ long; lip shorter than the petals, oblong, whitish, 2-toothed or 2-lobed above the base; spur a sac or small protuberance adnate to the summit of the ovary; capsule 4″–6″ long, oblong or somewhat obovoid.

In woods, Newfoundland to Alaska, south to New Jersey, in the mountains to Georgia, and to Ohio, Nebraska and Colorado. Ascends to 3000 ft. in Vermont. Also in Europe. May–June.

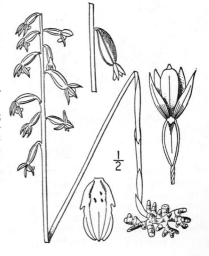

½

2. Corallorrhiza maculàta Raf.   Large Coral-root.   Fig. 1414.

C. maculata Raf. Am. Month. Mag. **2**: 119.   1817.
Corallorrhiza multiflora Nutt. Journ. Acad. Phila. **3**: 138. *pl. 7.*   1823.

Scape 8'–20' high, purplish, clothed with several appressed scales.   Raceme 2'–8' long, 10–30-flowered; flowers mainly brownish purple, short-pedicelled; sepals and petals somewhat connivent at the base, linear-lanceolate, about 3" long; lip white, spotted and lined with crimson, oval or ovate in outline, deeply 3-lobed, crenulate, bearing two narrow lamellae, the middle lobe broader than the lateral ones, its apex curved; spur manifest, yellowish; capsule ovoid or oblong, 5"–8" long, drooping.

In woods, Nova Scotia to British Columbia, south to Florida, Missouri, New Mexico and California. A race with yellow scapes and flowers occurs occasionally. Dragon's-claws. July–Sept.

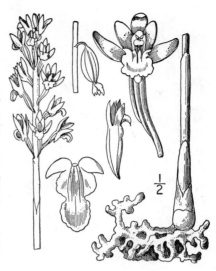

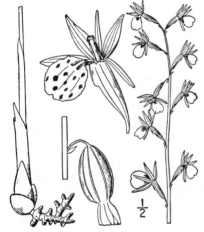

3. Corallorrhiza Wisteriàna Conrad.   Wister's Coral-root.   Fig. 1415.

C. Wisteriana Conrad, Journ. Acad. Phila. **6**: 145.   1829.

Stem slender, but sometimes stouter than that of the preceding species, 8'–16' high, bearing several sheathing scales.   Raceme 2'–5' long, loose, 6–16-flowered; flowers slender-pedicelled, ascending or erect; sepals and petals 2½"–3½" long; lip broadly oval or obovate, 3"–4" long, abruptly clawed, white with crimson spots, crenulate, notched at the apex; lamellae, 2 short prominent ridges; spur, a somewhat conspicuous protuberance adnate to the top of the ovary; column strongly 2-winged toward the base; capsule elliptic-oblong or oblong-obovoid, about 5" long, drooping when ripe.

In woods, New England; Pennsylvania to Ohio, Florida and Texas. Feb.–May.

4. Corallorrhiza odontorhìza (Willd.) Nutt.   Small or Late Coral-root. Fig. 1416.

Cymbidium odontorrhizon Willd. Sp. Pl. **4** : 110.   1805.

Corallorrhiza odontorrhiza Nutt. Gen. **2** : 197.   1818.

Scape slender, purplish, 6'–15' high.   Raceme 2'–4' long, 6–20-flowered; flowers mainly purplish; sepals and petals lanceolate, 2" long or less, marked with purple lines; lip about as long as the petals, broadly oval or obovate, entire or denticulate, narrowed at the base, not notched, whitish, spotted with purple; spur, a small sac adnate to the top of the ovary; wings of the column very narrow.

In woods, Maine to Ontario, Michigan, Florida and Missouri. Ascends to 3000 ft. in North Carolina.   Turkey-claw. Dragon's-claw. Crawley-root. July–Sept.

5. Corallorrhiza ochroleùca Rydb. Yellow Coral-root. Fig. 1417.

Corallorrhiza ochroleuca Rydb. Bull. Torr. Club **31**: 402. 1904.

Scape stout, yellow, 7'–16' high. Raceme 3'–5' long, 10–15-flowered; flowers yellow; sepals and petals oblong to oblong-lanceolate, 5''–6'' long, not striped; lip yellow, shorter than the petals, ovate, entire, rounded at the base, obtuse; spur small, adnate to the ovary; wings of the column narrow.

In cañons, western Nebraska and Colorado. June–July.

6. Corallorrhiza striàta Lindl. Striped Coral-root. Fig. 1418.

C. striata Lindl. Gen. & Sp. Orch. 534. 1840.

Corallorrhiza Macraei A. Gray, Man. Ed. 2, 453. 1856.

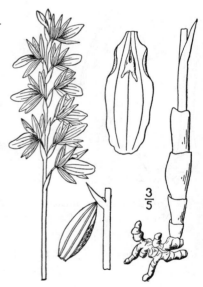

Scape stout, purplish, 8'–20' high. Raceme 2'–6' long, 10–25-flowered. Flowers dark purple; sepals and petals narrowly elliptic, striped with deeper purple lines, 6''–7'' long, spreading; lip oval or obovate, striate-veined, entire or a little undulate, somewhat narrowed at the base, about as long as the petals, bearing two short lamellae; spur none, but the perianth has a gibbous saccate base; capsule ellipsoid, reflexed, 8''–10'' long.

In woods, Ontario and northern New York to Michigan, Oregon and California. July.

28. HEXALÉCTRIS Raf. Neog. 4. 1825.

Scapose herbs, from thick scaly rootstocks and fleshy coralloid roots, the leaves reduced to purplish scales, sheathing the scape. Flowers bracted in a loose terminal raceme. Perianth not gibbous or spurred at the base, the petals and sepals similar, nerved, spreading. Lip obovate, with several crested ridges down the middle, somewhat 3-lobed, the middle lobe a little concave. Column free, thick, slightly incurved. Pollinia 8, united in a cluster. Capsule ellipsoid, the fruiting pedicels thick. [Greek, signifying six crests.]

Two known species, the following typical one and another in Mexico.

1. Hexalectris spicàta (Nutt.) Barnhart.
Crested Coral-root. Fig. 1419.

Arethusa spicata Walt. Fl. Car. 222. 1788.
Bletia aphylla Nutt. Gen. 2: 194. 1818.
Hexalectris squamosus Raf. Fl. Tell. 4: 48. 1836.
H. aphyllus Raf.; A. Gray, Man. Ed. 6, 501. 1890.
H. spicata Barnhart, Torreya 4: 121. 1904.

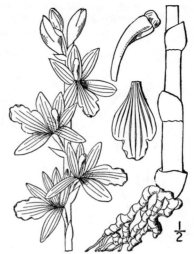

Scape stout, 8′–20′ high, its upper scales lanceolate, the lower sheathing and truncate or acute. Raceme 4′–7′ long, 8–12-flowered; flowers large, brownish purple, 1′ high or more; pedicels short, stout; sepals and petals narrowly elliptic, obtuse or acutish, spreading, striped with purple veins, 6″–9″ long, longer than the broad lip; middle lobe of the lip rounded or crenulate, the lateral ones shorter, rounded; column slightly spreading at the summit, shorter than the lip; capsule ellipsoid, nearly 1′ long, the fruiting pedicels 4″–5″ long.

In rich woods, North Carolina to Kentucky, Missouri, Florida, Mississippi and Texas. Aug.

Sub-class 1. *DICOTYLÉDONES.*

Embryo of the seed with two cotyledons (in a few genera one only, as in *Cyclamen, Pinguicula* and some species of *Ranunculaceae* and *Capnoides*), the first leaves of the germinating plantlet opposite. Stem exogenous, of pith, wood and bark (endogenous in structure in Nymphaeaceae), the wood in one or more layers surrounding the pith, traversed by medullary rays and covered by the bark. Leaves usually pinnately or palmately veined, the veinlets forming a network. Parts of the flower rarely in 3's or 6's.

Dicotyledonous plants are first definitely known in Cretaceous time. They constitute between two-thirds and three-fourths of the living angiospermous flora.

Series 1. *Choripetalae.*

Petals separate and distinct from each other, or wanting.

The series is also known as Archichlamideae, and comprises most of the families formerly grouped under Apetalae (without petals) and Polypetalae (with separate petals). Exceptions to the typical feature of separate petals are found in the Leguminosae, in which the two lower petals are more or less united; in the Fumariaceae, where the two inner petals or all four of them are sometimes coherent; the Polygalaceae, in which the three petals are united with each other, and with the stamens; *Oxalis* in Geraniaceae; and Ilicaceae, whose petals are sometimes joined at the base.

Family 1. SAURURÀCEAE Lindl. Nat. Syst. Ed. 2, 184. 1836.
LIZARD'S-TAIL FAMILY.

Perennial herbs with broad entire alternate petioled leaves, and small perfect incomplete bracteolate flowers, in peduncled spikes. Perianth none. Stamens 6–8, or sometimes fewer, hypogynous; anthers 2-celled, the sacs longitudinally dehiscent. Ovary 3–4-carpelled, the carpels distinct or united, 1–2-ovuled; ovules orthotropous. Fruit capsular or berry-like, composed of 3–4 mostly indehiscent carpels. Seeds globose or ovoid, the testa membranaceous. Endosperm copious, mealy. Embryo minute, cordate, borne in a small sac near the end of the endosperm.

Three genera and 4 species, natives of North America and Asia. The family differs from the Piperaceae in having more than one carpel to the ovary. It is represented in North America by the following and by *Anemópsis,* occurring in California and Arizona.

1. SAURÙRUS L. Sp. Pl. 341. 1753.

Marsh herbs, with slender rootstocks, jointed stems and cordate leaves, their petioles sheathing the stem at the nodes, and small white flowers, in 1 or 2 dense elongated spikes opposite the leaves. Bractlets adnate to the flowers or to their minute pedicels. Stamens 6–8. Filaments filiform, distinct. Carpels united at the base. Styles as many as the carpels, recurved, stigmatic along the inner side. Fruit rugose, depressed-globose, separating into 3 or 4 one-seeded carpels. [Name Greek, lizard's tail, alluding to the long slender spike.]

Two species, the following typical one of eastern North America, the other of eastern Asia.

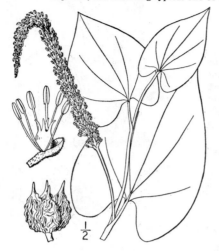

1. Saururus cérnuus L.   Lizard's-tail.

Fig. 1420.

Saururus cernuus L. Sp. Pl. 341.   1753.

Somewhat pubescent when young, becoming glabrous; stem rather slender, erect, sparingly branched, 2°–5° high. Leaves ovate, thin, palmately 5–9-ribbed and with a pair of strong ribs above, which run nearly to the apex, dark green, entire, deeply cordate at the base, acuminate, 3′–6′ long, 2′–3½′ wide; petioles stout, shorter than the blades, striate; spikes few, very dense, longer than their peduncles, 4′–6′ long, the apex drooping in flower; flowers fragrant; stamens white, spreading, about 2″ long; fruit slightly fleshy, 1½″ in diameter, strongly wrinkled when dry.

In swamps and shallow water, Rhode Island to Florida, west to southern Ontario, Minnesota and Texas. Swamp-lily. Breast-weed. June–Aug.

Family 2.   JUGLANDÀCEAE Lindl. Nat. Syst. Ed. 2, 180.   1836.

WALNUT FAMILY.

Trees with alternate pinnately compound leaves, and monoecious bracteolate flowers, the staminate in long drooping aments; the pistillate solitary or several together. Staminate flowers consisting of 3–numerous stamens with or without an irregularly lobed perianth adnate to the bractlet, very rarely with a rudimentary ovary. Anthers erect, 2-celled, the sacs longitudinally dehiscent; filaments short. Pistillate flowers bracted and usually 2-bracteolate, with a 3–5-lobed (normally 4-lobed) calyx or with both calyx and petals, and an inferior 1-celled or incompletely 2–4-celled ovary. Ovule solitary, erect, orthotropous; styles 2, stigmatic on the inner surface. Fruit in our genera a drupe with indehiscent or dehiscent, fibrous or woody exocarp (husk; ripened calyx; also regarded as an involucre), large, 2–4-lobed. Endosperm none. Cotyledons corrugated, very oily. Radicle enclosing the bony endocarp or nut which is incompletely 2–4-celled. Seed minute, superior.

Six genera and about 35 species, mostly of the warmer parts of the north temperate zone, extending in America south along the Andes to Bolivia. The young leaves in the bud are stipulate in at least two species of *Hicoria*. The family is not closely related to the other ament-bearing ones; its affinity is with the Anacardiaceae.

Husk indehiscent; nut rugose.   1. *Juglans.*
Husk at length splitting into segments; nut smooth or angled.   2. *Hicoria.*

1.   JÙGLANS L. Sp. Pl. 997.   1753.

Trees with spreading branches, superposed buds, fragrant bark, and odd-pinnate leaves, with nearly or quite sessile leaflets, the terminal one sometimes early perishing. Staminate flowers in drooping cylindric aments, borne on the twigs of the previous year; perianth 2–6-lobed; stamens 8–40 in 2 or more series. Pistillate flowers solitary or several together on a terminal peduncle at the end of shoots of the season, the calyx 4-lobed, with 4 small petals adnate to the ovary at the sinuses; styles fimbriate, very short. Drupe large, globose or ovoid, the exocarp somewhat fleshy, fibrous, indehiscent, the endocarp bony, rugose or sculptured, 2–4-celled at the base, indehiscent, or in decay separating into 2 valves. [Name a contraction of the Latin *Jovis glans,* the nut of Jupiter.]

About 8 species, natives of the north temperate zone, one in the West Indies, 1 or 2 in the Andes of South America. Besides the following 3 others occur in the southwestern United States. Type species: *Juglans regia* L.

Fruit globose, obtuse, not viscid; petioles puberulent.   1. *J. nigra.*
Fruit oblong, pointed, viscid; petioles pubescent.   2. *J. cinerea.*

1. Juglans nìgra L.　Black Walnut.　Fig. 1421.

Juglans nigra L. Sp. Pl. 997. 1753.

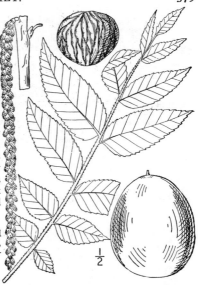

A large forest tree with rough brown bark, maximum height about 150°, trunk diameter 8°, the twigs of the season and petioles puberulent, the older twigs glabrous or very nearly so. Leaflets 13–23, ovate-lanceolate, more or less inequilateral, acuminate at the apex, rounded or subcordate at the base, serrate with low teeth, glabrous or very nearly so above, pubescent beneath, 3′–5′ long, 1′–2′ wide; staminate aments solitary in the axils of leaf-scars of the preceding season, 3′–5′ long; drupes usually solitary or 2 together, globose or a little longer than thick, 1½′–3′ in diameter, glabrous but papillose, not viscid; nut corrugated, slightly compressed, 4-celled at the base and apex.

In rich woods, Massachusetts to southern Ontario and Minnesota, south to Florida and Texas. Wood strong, hard, rich brown; weight per cubic foot 38 lbs. April–May. Fruit ripe Oct.–Nov.

2. Juglans cinèrea L.　Butternut.　White or Lemon Walnut.　Oil-nut.　Fig. 1422.

Juglans cinerea L. Sp. Pl. Ed. 2, 1415. 1763.

A forest tree, resembling the Black Walnut, but smaller, rarely over 100° high and 3° in trunk diameter, the bark gray, smoother, the twigs, petioles and leaflets viscid-pubescent, at least when young. Leaflets 11–19, oblong-lanceolate, acuminate at the apex, scarcely inequilateral, obtuse, rounded or truncate at the base, serrate with low teeth; drupes racemed, oblong, densely viscid-pubescent, 2′–3′ long and about one-half as thick, pointed; nut 4-ribbed, deeply sculptured, and with sharp longitudinal ridges, firmly adherent to the husk, 2-celled at the base.

In rich or rocky woods. New Brunswick and Ontario to North Dakota, south to Delaware, in the Alleghanies to Georgia, and to Mississippi, Arkansas and Kansas. Ascends to 2500 ft. in Virginia. Wood soft, rather weak, light brown; weight per cubic foot 25 lbs. April–May. Fruit ripe Oct.–Nov.

2. HICÒRIA Raf. Med. Rep. (II.) 5: 352. 1808.

[CARYA Nutt. Gen. 2: 221. 1818.]

Trees, with close or shaggy bark, odd-pinnate leaves and serrate or serrulate leaflets. Staminate flowers in slender drooping aments, borne mostly in 3's on a common peduncle at the base of the shoots of the season, or clustered and sessile or nearly so in the axils of leaf-scars at the summit of twigs of the preceding year; calyx adnate to the bract, 2–3-lobed or 2–3-cleft; stamens 3–10; filaments short. Pistillate flowers 2–6, together on a terminal peduncle; bract fugacious or none; calyx 4-toothed; petals none; style 2 or 4, papillose or fimbriate, short. Fruit subglobose, oblong or obovoid, the husk separating more or less completely into 4 valves; nut bony, smooth or angled, incompletely 2–4-celled; seed sweet and delicious or very bitter and astringent. [From the aboriginal name Hicori.]

About 15 species of eastern North America, one in Mexico. Type species: *Juglans alba* L.

Bud-scales few, valvate; lateral leaflets lanceolate or oblong-lanceolate, falcate.
　Nut not compressed or angled; seed sweet.　　　　　　　　　　　　1. *H. Pecan.*
　Nut somewhat compressed or angled; seed intensely bitter.
　　　Leaflets 5–9; nut smooth.　　　　　　　　　　　　　　　　2. *H. cordiformis.*
　　　Leaflets 9–13; nut angled.　　　　　　　　　　　　　　　　3. *H. aquatica.*
Bud-scales imbricate; lateral leaflets not falcate.
　Husk of the fruit freely splitting to the base; bract of the staminate calyx at least twice as
　　long as the lobes.

Bark shaggy, separating in long plates; foliage glabrous or puberulent.
Leaflets 3–5 (rarely 7); nut rounded at the base, 6″–10″ long.
Leaflets oval to oblong-lanceolate, puberulent.   **4. H. ovata.**
Leaflets lanceolate, glabrous or glaucous beneath.   **5. H. carolinae-septentrionalis.**
Leaflets 7–9; nut usually pointed at both ends, 1′–1½′ long.   **6. H. lacinosa.**
Bark close, rough; foliage very pubescent and fragrant.
Rachis of the leaves and staminate catkins densely hirsute.   **7. H. alba.**
Rachis and staminate aments scurfy, at least when young.   **8. H. pallida.**
Husk of fruit not freely splitting to the base; bract of the staminate calyx about as long as the lobes.
Fruit nearly globular; nut thin-shelled; bark shaggy, at least when old.
Fruit little flattened; bract of staminate calyx short.   **9. H. microcarpa.**
Fruit much flattened; bract of staminate calyx long.   **10. H. borealis.**
Fruit obovoid; nut thick-shelled; bark close.
Foliage glabrous, or little pubescent; anther-sacs acute.   **11. H. glabra.**
Foliage pubescent or scurfy; anther-sacs obtuse.   **12. H. villosa.**

1. Hicoria Pècan (Marsh) Britton. Pecan. Illinois Nut. Soft-shell Hickory. Fig. 1423.

Juglans Pecan Marsh. Arb. Am. 69. 1785.
Carya olivaeformis Nutt. Gen. 2 : 221. 1818.
Hicoria Pecan Britton, Bull. Torr. Club 15 : 282. 1888.

A large slender tree, with somewhat roughened bark, maximum height of 170° and trunk diameter 6°. Young twigs and leaves pubescent; mature foliage nearly glabrous; bud-scales few, small, valvate; leaflets 11–15, falcate, oblong-lanceolate or ovate-lanceolate, short-stalked, inequilateral, acuminate, 4′–7′ long; staminate aments sessile or nearly so in the axils of leaf-scars near the end of twigs of the preceding season or sometimes on the young shoots, 5′–6′ long; bract of the staminate calyx linear, much longer than the broadly oblong lateral lobes; fruit oblong-cylindric, 1½′–2½′ long; husk thin, 4-valved; nut smooth, oblong, thin-shelled, pointed, 2-celled at base, dissepiments thin, very astringent; seed delicious.

In moist soil, especially along streams, Indiana to Iowa and Kansas, south to Alabama and Texas. Wood hard, brittle, light brown; weight 45 lbs. April–May. Fruit ripe Sept.–Oct.

2. Hicoria cordifórmis (Wang.) Britton. Bitter-nut. Swamp Hickory. Fig. 1424.

Juglans alba minima Marsh. Arb. Am. 68. 1785.
Juglans cordiformis Wang. Nordam. Holz. 25, *pl. 10, f. 25.* 1787.
Carya amara Nutt. Gen. 2 : 222. 1818.
Hicoria minima Britton, Bull. Torr. Club 15 : 284. 1888.
Hicoria cordiformis Britton, N. A. Trees 228. 1908.

A slender tree, sometimes 100° high, with trunk 3° in diameter, the bark close and rough. Bud-scales 6–8, small, valvate, caducous, young foliage puberulent, becoming nearly glabrous; leaflets 7–9, sessile, long-acuminate, lanceolate or oblong-lanceolate, 3′–6′ long, ¾′–1½′ wide, the lateral ones falcate; staminate aments slightly pubescent, peduncled in 3's at the bases of shoots of the season or sometimes on twigs of the previous year; lobes of the staminate calyx about equal, the bract narrower; fruit subglobose, narrowly 6-ridged 1′–1½′ in diameter; husk thin, tardily 4-valved; nut little compressed, not angled, short-pointed, 9″–12″ long, thin-shelled; seed very bitter.

In moist woods and swamps, Quebec to southern Ontario, Minnesota, Florida and Texas. Ascends to 3500 ft. in Virginia. Wood hard and strong, dark brown; weight per cubic foot 47 lbs. Bitter pignut. Bitter or pig-hickory. May–June. Fruit ripe Sept.–Oct.

3. Hicoria aquática (Michx. f.) Britton. Water or Swamp Hickory. Fig. 1425.

Juglans aquatica Michx. f. Hist. Arb. Am. **1**: 182. *pl. 5.* 1810.

Carya aquatica Nutt. Gen. **2**: 222. 1818.

Hicoria aquatica Britton, Bull. Torr. Club **15**: 284. 1888.

A swamp tree, attaining a maximum height of about 100° and a trunk diameter of 3°, the bark close, the young foliage pubescent, becoming nearly glabrous when mature. Leaflets 9–13, lanceolate, or the terminal one oblong, long-acuminate at the apex, narrowed at the base, 3′–5′ long, ½′–1′ wide, the lateral strongly falcate; staminate aments and calyx as in the preceding species; fruit oblong, ridged, 1′–1½′ long, pointed; husk thin, tardily splitting; nut oblong, thin-shelled, angular; seed bitter.

In wet woods and swamps, Virginia to Florida, west to Illinois, Missouri, Arkansas and Texas. Wood soft, strong, dense, dark brown; weight per cubic foot 46 lbs. Bitter pecan. Water-bitternut. March–April. Fruit ripe Sept.–Oct.

4. Hicoria ovàta (Mill.) Britton. Shag-bark. Shell-bark Hickory. Fig. 1426.

Juglans ovata Mill. Gard. Dict. Ed. 8, No. 6. 1768.

Carya alba Nutt. Gen. **2**: 221. 1818. Not *Juglans alba* L.

Hicoria ovata Britton, Bull. Torr. Club **15**: 283. 1888.

A large tree, sometimes 120° high, with a trunk diameter of 4°; bark shaggy in narrow plates; young twigs and leaves puberulent, becoming glabrous. Leaflets 5, or sometimes 7, oblong, oblong-lanceolate or the upper obovate, acuminate at the apex, narrowed to the sessile base, 4′–6′ long, those of young plants much larger, bud-scales 8–10, imbricated, the inner becoming very large and tardily deciduous; staminate aments in 3's, on slender peduncles at the bases of shoots of the season; bract of the staminate calyx linear, longer than the lateral lobes; fruit subglobose, 1¼′–2½′ long; husk thick, soon splitting into 4 valves; nut white, somewhat compressed, 4-celled at the base, 2-celled (rarely 3-celled) above, pointed, angled, thin-shelled; seed sweet.

In rich soil, Quebec to southern Ontario and Minnesota, south to Florida, Kansas and Texas. Wood strong and tough, light brown; weight per cubic foot 52 lbs. Walnut. Sweet or white walnut. King-nut. Upland or white hickory. Red-heart hickory. May. Fruit ripe Sept.–Nov.

5. Hicoria carolìnae-septentrionàlis Ashe. Southern Shag-bark. Fig. 1427.

Hicoria carolinae-septentrionalis Ashe, Notes on Hickories. 1896.

A small tree attaining a maximum height of about 80°, and diameter of 2½°, with gray bark hanging in long loose strips. Bud-scales 8–10, imbricated, the inner greatly enlarging in leafing, and tardily deciduous; terminal bud ovate-lanceolate, truncate, the scales spreading, barely ¾′ long; lateral buds oblong; twigs very slender, ⅛′ thick, glaucous, smooth, purplish-brown; staminate aments in threes, glabrous on short peduncles, at base of shoots of the season; stamens glabrous; ovary glabrous; young foliage blackening in drying, glabrous, ciliate, with few resinous globules; leaflets 3–5, the 2 upper ¾′–1¼′ wide, 4′–6′ long, lanceolate; lower pair often smaller; fruit subglobose, ¾′–1¼′ long; husk soon falling into 4 pieces; nut white or brownish, much compressed, angled, cordate or subcordate at top, thin-shelled; seed large and sweet.

Sandy or rocky woods, rarely entering "bottoms," Delaware to Georgia and Tennessee.

6. Hicoria laciniòsa (Michx. f.) Sarg.   Big Shag-bark.   King-nut.   Fig. 1428.

Carya sulcata Nutt. Gen. **2** : 221. 1818. Not *Juglans sulcata* Willd. 1796.
Juglans laciniosa Michx. f. Hist. Arb. Am. **1** : 199. *pl. 8.* 1810.
Hicoria sulcata Britton, Bull. Torr. Club **15** : 283. 1888.
Hicoria laciniosa Sarg. Mem. Torr. Club **5** : 354. 1894.

A large tree, reaching about the size of the preceding species, the bark separating in long narrow plates, the young foliage densely puberulent, the mature leaves somewhat so beneath. Leaflets 7–9 (rarely 5), acute or acuminate, oblong-lanceolate or the upper obovate, sometimes 8′ long by 5′ wide; staminate aments peduncled in 3′s at the base of shoots of the season; bract of the staminate calyx linear, twice as long as the lateral lobes; fruit oblong, 2′–3′ long; husk thick, soon splitting to the base; nut oblong, somewhat compressed, thick-shelled, pointed at both ends, yellowish-white; seed sweet.

In rich soil, New York and Pennsylvania to Indiana, Iowa, Tennessee, Kansas and Oklahoma. Wood strong, tough, darker than the preceding; weight 50 lbs. per cubic foot. May. Western or thick shell-bark. Fruit ripe Sept.–Oct.

7. Hicoria álba (L.) Britton.   White-heart Hickory.   Mocker-nut.   Fig. 1429.

Juglans alba L. Sp. Pl. 997. 1753.
Juglans tomentosa Lam. Encycl. **4** : 504. 1797.
Carya tomentosa Nutt. Gen. **2** : 221. 1818.
Hicoria alba Britton, Bull. Torr. Club **15** : 283. 1888.

A large tree, maximum height 100°, and trunk diameter 3½°, the foliage and twigs persistently tomentose-pubescent, fragrant when crushed, the bark rough and close; bud-scales very large, imbricated; leaflets 7–9, oblong-lanceolate or the upper oblanceolate or obovate, sessile, long-acuminate, narrowed or rounded and somewhat inequilateral at the base; staminate aments peduncled in 3′s, tomentose; bract of the staminate calyx linear, much longer than the lateral lobes; fruit globose or oblong-globose, 1½′–3½′ long; husk thick, freely splitting to the base; nut grayish-white, angled, pointed at the summit, little compressed, thick-shelled, 4-celled at the base; seed sweet.

In rich soil, eastern Massachusetts to southern Ontario, Illinois and Nebraska, south to Florida and Texas. Ascends to 3500 ft. in Virginia. Wood very hard and tough, dark brown; weight per cubic foot 51 lbs. Fragrant or hard-bark hickory. King-nut. Bull-nut. White-bark, black or red hickory. May–June. Fruit ripe Oct.–Nov.

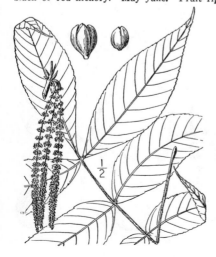

8. Hicoria pállida Ashe.   Pale Hickory. Fig. 1430.

Hicoria pallida Ashe, Notes on Hickories. 1896.

A forest tree, reaching a maximum height of nearly 100°, with a trunk up to 3° in diameter, the bark rough and furrowed. Bud-scales 5–9, imbricated; leaves scurfy, at least when young; leaflets 7–9, lanceolate to ovate-lanceolate, 2′–6′ long, acuminate; staminate catkins scurfy, 3¾′–8′ long; bract of the staminate flower longer than the lateral lobes; fruit subglobose to obovoid, the rather thin husk splitting tardily into 4 valves; nut flattened, nearly white, rather thin-shelled; seed sweet.

Dry soil, Virginia and Tennessee to Florida and Alabama. April–May.

9. Hicoria microcàrpa (Nutt.) Britton. Small-fruited Hickory. Fig. 1431.

Juglans alba odorata Marsh. Arb. Am. 68. 1785.
Carya microcarpa Nutt. Gen. 2: 221. 1818.
H. microcarpa Britton, Bull. Torr. Club 15: 283. 1888.
H. glabra var. odorata Sarg. Silva 7: 167. pl. 354. 1895.

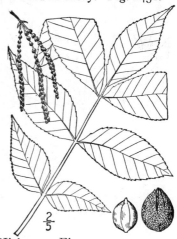

A forest tree, reaching a maximum height of about
90° and a trunk diameter of 3½°, the bark close, when
older separating in narrow plates, the foliage glabrous
throughout. Bud-scales 6–8, imbricated, the inner ones
somewhat enlarging; leaflets 5–7, oblong, or ovate-
lanceolate, acuminate at the apex, narrowed or sometimes rounded at the base, 3½′–5′ long; staminate aments
glabrous, peduncled in 3's at the base of shoots of the
season; bract of the staminate calyx equalling or somewhat longer than the lateral lobes; fruit globose or
globose-oblong, less than 1′ long, the husk thin, tardily
and incompletely splitting to the base; nut subglobose,
nearly white, slightly compressed, not angled, thin-
shelled, pointed; seed sweet.

In rich woods, Massachusetts to Michigan, Virginia,
Georgia, Illinois and Missouri. Wood hard, strong, tough,
light brown. Small or little pignut. Little shag-bark.
May–June. Fruit ripe Sept.–Oct.

10. Hicoria boreàlis Ashe. Northern Hickory. Fig. 1432.

Hicoria borealis Ashe, Notes on Hickories. 1896.

A small tree, with rough furrowed bark when
young, becoming shaggy in long narrow strips with
age. Bud-scales 8–10, imbricated, the inner bright-
colored and sericeous, enlarging in leafing and
tardily deciduous; terminal bud ovate-lanceolate, ½′
long; twigs very slender, ⅛′ thick, glabrous, bright
brownish red; staminate aments in 3's at base of
shoots of season; bract of staminate calyx much prolonged; young foliage blackening in drying, pubescent when young, becoming smooth, ciliate, with few
resinous globules on lower surface; leaflets 5, occasionally 3, lanceolate, the upper ¾′–1¼′ wide, 3½′–6′
long; lower pair often smaller; fruit ovoid, much
flattened, ¾′ or more long; husk very thin, rugose,
coriaceous, usually not splitting; nut white, somewhat angled; shell thin and elastic; seed large, sweet
and edible.

A small tree of dry uplands, growing with oaks and
Hicoria microcarpa of which it is, perhaps, a northern
race. Southern and eastern Michigan, east to Belle Isle,
Detroit river.

11. Hicoria glàbra (Mill.) Britton. Pig-nut Hickory. Fig. 1433.

Juglans glabra Mill. Gard. Dict. Ed. 8, No. 5. 1768.
Carya porcina Nutt. Gen. 2: 222. 1818.
Hicoria glabra Britton, Bull. Torr. Club 15: 284. 1888.
H. glabra hirsuta Ashe, Notes on Hickories. 1896.

A tree, sometimes 120° high and with a trunk diameter of 5°, bark close, rough; foliage glabrous, or somewhat pubescent. Bud-scales 8–10, imbricated, the inner
ones enlarging; leaflets 3–7, rarely 9, oblong, oblong-
lanceolate or the upper obovate, sessile, acuminate at
the apex, mostly narrowed at the base, 3′–6′ long, in
young plants much larger; staminate aments glabrous,
peduncled in 3's; lobes of the staminate calyx about
equal in length, the bract narrower; fruit obovoid or
obovoid-oblong, 1¼′–2′ long; husk thin, the valves very
tardily dehiscent; nut brown, angled, pointed, very
thick-shelled; seed astringent and bitter, not edible.

In dry or moist woods, Maine to southern Ontario and
Minnesota, south to Florida and Texas. Wood hard,
strong, tough, rather dark brown; weight per cubic foot
51 lbs. Brown, red, white or black hickory. Broom-
hickory. May–June. Fruit ripe Oct.–Nov.

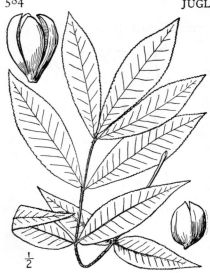

½

12. Hicoria villòsa (Sarg.) Ashe. Scurfy Hickory. Woolly Pignut. Fig. 1434.

H. glabra var. *villosa* Sarg. Sylva 7: 167. 1895.
H. villosa Ashe, Bull. Torr. Club 24: 11, 530. 1897.

A small or medium sized tree reaching a maximum height of about 80°, and a diameter of 2°, with deeply furrowed dark gray bark. Buds of 6–8 imbricated scales, the outer usually thickly dotted with resinous globules, the inner somewhat enlarging in leafing; terminal bud ovate, ¼′ long, lateral buds mostly short-stiped; staminate aments pubescent, and gland and scurf covered, peduncled in 3's at base of shoots of the season; twigs slender, ⅛′ thick or less, usually glabrous, bright purple-brown; petiole pubescent; leaflets 5–9, at first thickly covered beneath with silvery peltate glands, mixed with resinous globules, generally pubescent; fruit about 1′ long, obovoid or subglobose, the husk dotted with resinous globules, ⅛′ thick and partly splitting; nut thick-shelled, angled; seed small, but sweet.

Sandy or rocky soils, Delaware to Georgia and Missouri. Wood hard, dark brown; weight per cubic foot 50 lbs. Perhaps a race of the preceding.

Family 3. **MYRICÀCEAE** Dumort. Anal. Fam. 95. 1829.

BAYBERRY FAMILY.

Shrubs or trees with alternate, mostly coriaceous and aromatic simple leaves and small monoecious or dioecious flowers, in linear, oblong or globular bracted aments. Flowers solitary in the axils of the bracts. Perianth none. Staminate flower with 2–16 (usually 4–8) stamens inserted on the receptacle; filaments short, distinct or somewhat united; anthers ovate, 2-celled, the sacs longitudinally dehiscent. Pistillate flowers with a solitary 1-celled ovary, subtended by 2–8 bractlets; ovule solitary, orthotropous; style very short; stigmas 2, linear. Fruit a small oblong or globose drupe or nut, the exocarp often waxy. Seed erect. Endosperm none. Cotyledons plano-convex. Radicle very short.

Two genera and 35 species of wide geographic distribution.
Ovary subtended by 2–4 bractlets; leaves serrate or entire, exstipulate.　　　　1. *Myrica.*
Ovary subtended by 8 linear persistent bractlets; leaves pinnatifid, stipulate.　　2. *Comptonia.*

1. **MYRÌCA** L. Sp. Pl. 1024. 1753.

Shrubs or small trees with entire, dentate or lobed, mostly resinous-dotted leaves, our species usually dioecious. Staminate aments oblong or narrowly cylindric, expanding before or with the leaves. Stamens 4–8. Pistillate aments ovoid or subglobose; ovary subtended by 2–4, mostly short, deciduous or persistent bractlets. Drupe globose or ovoid, its exocarp waxy. [Ancient Greek name of the Tamarisk.] Type species, *Myrica Gale* L.

Besides the following species, another occurs in the Southern States and 2 on the Pacific coast.
Bractlets of pistillate aments persistent, clasping the drupes; low bog shrub.　　1. *M. Gale.*
Bractlets of pistillate aments deciduous, the ripe drupes separated.
　　Leaves mostly acute, narrow; drupe less than 1″ in diameter.　　　　2. *M. cerifera.*
　　Leaves mostly obtuse, broader; drupe 1″–1½″ in diameter.　　　　3. *M. carolinensis.*

1. **Myrica Gàle** L. Sweet Gale. Fig. 1435.

Myrica Gale L. Sp. Pl. 1024. 1753.

A shrub, usually strictly dioecious, the twigs dark brown. Leaves oblanceolate, obtuse and dentate at the apex, narrowed to a cuneate entire base, short-petioled, dark green and glabrous above, pale and puberulent or glabrous beneath, 1′–2½′ long, 5″–10″ wide, unfolding after the aments; staminate aments linear-oblong, 6″–10″ long, crowded; pistillate aments ovoid-oblong, obtuse, about 4″ long and 2″ in diameter in fruit, their bracts imbricated; drupe resinous-waxy, not longer than the 2 ovate persistent bractlets, which clasp it and are adnate to its base.

½

In swamps and along ponds and streams, Newfoundland to Alaska, southern New York, Virginia, Michigan and Washington. Also in Europe and Asia. Ascends to 3000 ft. in the Adirondacks. Fern or scotch-gale. Sweet willow. Bay-bush. Meadow-fern. Golden osier. Moor-, bog-, Dutch- or Burton-myrtle. April–May.

2. Myrica cerífera L.　Wax-myrtle.　Fig. 1436.

Myrica cerifera L. Sp. Pl. 1024.　1753.

A slender tree, or a shrub, maximum height about 40°, trunk diameter 1½°, the bark gray, nearly smooth. Leaves narrow, oblong or oblanceolate, mostly acute at the apex, entire or sparingly dentate, narrowed or somewhat cuneate at the base, fragrant when crushed, short-petioled, dark green above, paler and sometimes pubescent beneath; golden-resinous, 1'–3' long, 3"–9" wide, unfolding with or before the aments; staminate aments cylindric; pistillate aments short, oblong; ripe drupes separated, globose, bluish-white, waxy, less than 1" in diameter, tipped with the minute base of the style, long-persistent, the bracts and bractlets deciduous.

In sandy swamps or wet woods, southern New Jersey to Florida and Texas, north to Arkansas. Also in the West Indies. March–April. Leaves mostly persistent through the winter. Wood light, brown; weight per cubic foot 35 lbs. Waxberry. Tallow-bayberry. Candleberry. Tallow-shrub. Sweet oak. Candleberry-myrtle.

3. Myrica carolinénsis (Mill.).　Small Waxberry.　Bayberry.　Fig. 1437.

Myrica carolinensis Mill. Gard. Dict. Ed. 8, no. 3. 1768.

A shrub, 2°–8° high, with smooth gray bark, the twigs glabrous or often pubescent. Leaves oblanceolate or obovate, glabrous above, often pubescent beneath, resinous, 2'–4' long, 6"–18" wide, serrate with a few low teeth above the middle, or entire, obtuse or sometimes acute at the apex, narrowed at the base, short-petioled; staminate aments cylindric or oblong, 3"–9" long; pistillate aments short, oblong; ripe drupes separated, globose, bluish white, very waxy, 1"–1½" in diameter, long-persistent, the bracts and bractlets deciduous.

In dry or moist sandy soil, Nova Scotia to Florida and Louisiana and on the shores of Lake Erie. Occurs also in bogs in northern New Jersey and Pennsylvania. April–May. The fruit was much used as a source of wax by the early settlers of the eastern United States, and is still utilized along the coast of New England.

2.　COMPTÒNIA Banks; Gaertn. Fr. & Sem. 2: 58. *pl. 90.*　1791.

A low, monoecious or dioecious branching shrub with terete brown branches and narrow, deeply pinnatifid, stipulate leaves, the young foliage pubescent. Aments expanding with the leaves, the staminate ones and their flowers as in *Myrica.* Fertile aments globose-ovoid, on monoecious plants appearing below the staminate, several-flowered. Ovary subtended by 8 linear-subulate persistent bractlets, which form an involucre to the ovoid-oblong bony nut. [Name in honor of Rev. Henry Compton, 1632–1713, bishop of Oxford.]

A monotypic genus of eastern North America.

1. Comptonia peregrìna (L.) Coulter. Sweet Fern. Fern-gale. Fig. 1438.

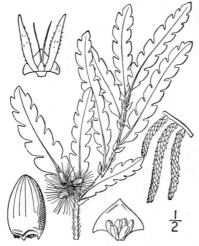

Liquidambar peregrina L. Sp. Pl. 999. 1753.

Myrica asplenifolia L. Sp. Pl. 1024. 1753.

Liquidambar asplenifólia L. Sp. Pl. Ed. 2, 1418. 1763.

C. asplenifolia Gaertn. Fr. & Sem. 2 : 58. 1791.

C. peregrina Coulter, Mem. Torr. Club 5 : 127. 1894.

A shrub, 1°–3° tall, the branches erect or spreading. Leaves linear-oblong or linear-lanceolate in outline, short-petioled, obtuse or subacute at the apex, deeply pinnatifid into numerous oblique rounded entire or sparingly dentate lobes, 3'–6' long, ¼'–½' wide, fragrant when crushed, the sinuses very narrow; stipules semi-cordate, mostly deciduous; staminate aments clustered at the ends of the branches, 1' or less long, their bracts reniform, acute; pistillate aments bur-like in fruit, the subulate bractlets longer than the light brown, shining, striate, obtuse nut.

In dry soil, especially on hill-sides, Nova Scotia to Saskatchewan, south to North Carolina, Indiana and Michigan. Ascends to 2000 ft. in Virginia. Meadow- or shrubby-fern. Sweet-bush or -ferry. Fern- or spleenwort-bush. Canada sweet-gale. April–May.

Family 4. LEITNERIÀCEAE Drude, Phanerog. 407. 1879.

CORK-WOOD FAMILY.

Dioecious shrubs or small trees, with large entire petioled alternate exstipulate (or sometimes stipulate?) leaves, and flowers of both sexes in aments formed at the end of the season, which expand before the leaves. Staminate flowers with no perianth; stamens 3–12, inserted on the receptacle; filaments short, distinct; anthers oblong, erect, 2-celled, the sacs longitudinally dehiscent. Pistillate flowers with a solitary 1-celled ovary, subtended by 3 or 4 minute glandular-lacerate bractlets; style terminal, simple, grooved and flattened, slender, recurved and stigmatic above, caducous; ovule solitary, laterally affixed to the ovary wall, amphitropous. Fruit an oblong drupe with thin exocarp and hard endocarp. Testa thin. Endosperm thin, fleshy. Cotyledons flat, cordate at the base; radicle short, superior.

A family related morphologically to the Myricaceae, but its anatomical characteristics point to affinity with Liquidambar and Platanus. It comprises only the following monotypic genus.

1. LEITNÈRIA Chapm. Fl. S. States, 427. 1860.

Characters of the family. [In honor of Dr. E. F. Leitner, a German naturalist, killed in Florida during the Seminole war.]

1. Leitneria floridàna Chapm. Leitneria. Cork-wood. Fig. 1439.

Leitneria floridana Chapm. Fl. S. States, 428. 1860.

A shrub or small tree, attaining a maximum height of about 20° and a trunk diameter of 5', the bark gray and rather smooth, the young twigs, leaves and aments densely pubescent. Leaves oblong or elliptic-lanceolate, acute, obtuse or cuspidate at the apex, narrowed at the base, bright green, firm, 3'–6' long, 1'–3' wide, when mature, glabrous or nearly so above, finely pubescent, at least on the veins, and rugose-reticulated beneath; petioles 9"–15" long; staminate aments many-flowered, ascending, 1'–2' long, their bracts triangular-ovate, acute, tomentose; pistillate aments shorter, few-flowered; drupe slightly compressed, about 10" long, 3"–4" thick, rugose-reticulated.

In swamps, southern Missouri to Texas, and in Florida. Propagates by suckers. Wood lighter than cork and probably the lightest wood known, weighting only about 12½ lbs. per cubic foot. March.

Family 5. **SALICÀCEAE** Lindl. Nat. Syst. Ed. 2, 186. 1836.

WILLOW FAMILY.

Dioecious trees or shrubs with light wood, bitter bark, brittle twigs, alternate stipulate leaves, the stipules often minute and caducous. Flowers of both sexes in aments, solitary in the axil of each bract. Aments expanding before or with the leaves. Staminate aments often pendulous; staminate flowers consisting of from one to numerous stamens inserted on the receptacle, subtended by a gland-like or cup-shaped disk; filaments distinct or more or less united; anthers 2-celled, the sacs longitudinally dehiscent. Pistillate aments pendulous, erect or spreading, sometimes raceme-like; pistillate flowers of a sessile or short-stipitate 1-celled ovary subtended by a minute disk; placentae 2–4, parietal; ovules usually numerous, anatropous; style short, slender, or almost wanting; stigmas 2, simple or 2–4-cleft. Fruit an ovoid, oblong or conic 2–4-valved capsule. Seeds small or minute, provided with a dense coma of long, mostly white, silky hairs. Endosperm none.

The family includes only the 2 following genera, consisting of 200 or more species, mostly natives of the north temperate and arctic zones.

Bracts fimbriate or incised; stamens numerous; stigmas elongated.   1. *Populus.*
Bracts entire; stamens 2–10; stigmas short.   2. *Salix.*

1. **PÓPULUS** L. Sp. Pl. 1034. 1753.

Trees with scaly resinous buds, terete or angled twigs and broad or narrow, usually long-petioled leaves, the stipules minute, fugacious. Bracts of the aments fimbriate or incised. Disk cup-shaped, oblique, lobed or entire. Staminate aments dense, pendulous. Staminate flowers with from 4–60 stamens, their filaments distinct. Pistillate aments sometimes like through the elongation of the pedicels, pendulous, erect or spreading. Ovary sessile; style short, stigmas 2–4, entire or 4-lobed. [Name ancient, used for these trees by Pliny.]

About 30 species, natives of the northern hemisphere. Besides the following, some 8 others occur in the western parts of North America. Type species: *Populus alba* L.

* **Petioles terete or channeled, scarcely or not at all flattened laterally.** (POPLARS.)
Leaves persistently and densely white-tomentose beneath.   1. *P. alba.*
Leaves glabrous or very nearly so when mature, crenulate.
   Foliage glabrous or nearly so; capsule very short-pedicelled.
     Leaves broadly ovate, rounded or cordate at the base.
       Petioles glabrous; leaves rounded or truncate at the base.   2. *P. balsamifera.*
       Petioles ciliate; leaves mostly cordate.   3. *P. candicans.*
     Leaves lanceolate or ovate-lanceolate, mostly narrowed at the base.
       Leaves acute, short-petioled.   4. *P. angustifolia.*
       Leaves acuminate, long-petioled.   5. *P. acuminata.*
   Foliage densely tomentose when young; capsules slender-pedicelled.   6. *P. heterophylla.*
** **Petioles strongly flattened laterally.** (ASPENS.)
Leaves coarsely undulate-dentate.   7. *P. grandidentata.*
Leaves crenulate-denticulate.
   Leaves ovate or suborbicular, short-pointed.   8. *P. tremuloides.*
   Leaves broadly deltoid, abruptly acuminate.
     Leaves obtuse at the base; capsules nearly sessile.   9. *P. nigra.*
     Leaves truncate at the base; capsules slender-pedicelled.
       Pedicels as long as the capsules or longer.   10. *P. deltoides.*
       Pedicels shorter than the capsules.   11. *P. Sargentii.*

1. Populus álba L. Abele. White or Silver-leaf Poplar. Aspen. Fig. 1440.

Populus alba L. Sp. Pl. 1034. 1753.

A large tree, with smooth light gray bark, attaining a maximum height of about 120° and a trunk diameter of 6°. Young foliage densely white-tomentose, the leaves becoming glabrate and dark green above, persistently tomentose beneath, broadly ovate or nearly orbicular in outline, apex acute, base truncate or subcordate, 3–5-lobed or irregularly dentate, 2½′–4′ long; petioles nearly terete, shorter than the blade; staminate aments 1′–2′ long.

In yards and along roadsides, springing up from suckers of older trees. New Brunswick to Ontario and Virginia. Native of Europe and Asia. Wood soft, nearly white; weight 38 lbs. per cubic foot. Abel or rattler-tree. White or great aspen. Dutch beech. White-bark. March–May.

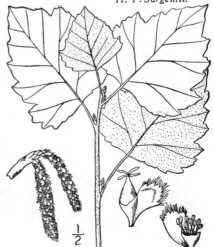

2. Populus balsamífera L.　Tacamahac.　Balsam or Carolina Poplar.　Fig. 1441.

Populus balsamifera L. Sp. Pl. 1034. 1753.

A large tree, with nearly smooth gray bark, reaching a maximum height of about 80° and a trunk diameter of 7°, the branches stout, spreading, the large buds very resinous, the foliage glabrous. Leaves broadly ovate, dark green and shining above, pale beneath, acute or acuminate at the apex, rounded or subcordate at the base, crenulate, 3′–5′ long, petioles terete; aments and bracts somewhat pubescent; stamens 18–30; lobes of the stigmas broad; capsules ovoid, short-pedicelled.

In moist or dry soil, especially along streams and lakes, Newfoundland to Hudson Bay and Alaska, south to Connecticut, New York, Michigan, South Dakota and Oregon. Wood soft, weak, brown, compact; weight per cubic foot 23 lbs. Ontario or rough-bark poplar. April.

3. Populus cándicans Ait.　Balm of Giliad.　Fig. 1442.

P. candicans Ait. Hort. Kew. **3**: 406. 1789.
P. balsamifera var. *candicans* A. Gray, Man. Ed. *2*, 419. 1856.

A large tree, sometimes nearly 100° high, with a trunk up to 6½° in diameter, the old bark gray, ridged, the young twigs slightly pubescent, the buds resinous, pointed. Leaves broadly ovate, 2¼′–6′ long, cordate to narrowed at the base, acute at the apex, dark green above, pale beneath, crenulate, pubescent when young, and somewhat so on the veins beneath when mature, the petioles terete, pubescent or ciliate; aments 6′ long or less, their bracts lacerate; capsules narrowly ovoid, acute, short-pedicelled, 3″–4″ long.

Roadsides and along streams, Newfoundland to Virginia, Michigan, South Dakota and Alaska; in the east mostly or wholly escaped from cultivation. Wood soft, weak, brown; weight per cubic foot about 24 lbs. April–May.

4. Populus angustifòlia James.　Narrow-leaved Cottonwood.　Fig. 1443.

Populus angustifolia James, Long's Exp. **1**: 497. 1823.
Populus balsamifera var. *angustifolia* S. Wats. Bot. King's Exp. 327. 1871.

A slender tree, maximum height about 65°, trunk diameter 2°; crown narrowly pyramidal, branches ascending, foliage glabrous. Twigs terete, gray; leaves lanceolate, ovate-lanceolate or ovate, spreading, drying brownish, gradually acuminate or acute at the apex or some of them obtuse, narrowed, rounded or rarely subcordate at the base, 2′–4½′ long, ½′–1¼′ wide, finely crenulate from base to apex; petioles plano-convex, not flattened laterally; ¼′–½′ long; lateral veins 8–15 on each side of the blade; staminate aments oblong-cylindric, 1′–2½′ long; lobes of the stigmas broad; capsules ovoid, short-pedicelled.

In moist soil, especially along streams, Assiniboia to South Dakota, Nebraska, New Mexico, and Chihuahua. Wood soft, weak, brown, compact; weight per cubic foot 24 lbs. Black or willow-cottonwood. April–May.

5. Populus acumináta Rydberg. Black Cottonwood. Fig. 1444.

P. acuminata Rydberg, Bull. Torr. Club **20**: 50. 1893.

Populus coloradensis Dode, Mem. Soc. Hist. Nat. Autun **18**: [reprint 58]. 1905.

A slender tree, with terete twigs, reaching approximately the dimensions of the preceding species, the crown broadly pyramidal with spreading branches, the foliage glabrous. Leaves rhomboid-lanceolate, spreading or drooping, drying green, abruptly or gradually long-acuminate at the apex, cuneate, obtuse or rounded at the base, 2′–6′ long, 1′–2½′ wide, crenulate or the base entire; petioles slender, 1′–2½′ long; staminate aments about 1½′ long; pistillate aments slender, drooping, 3′–5′ long; capsules ovoid, obtuse, distinctly pedicelled.

Borders of lakes and streams, North Dakota to Assiniboia, western Nebraska, New Mexico and Nevada. April–May.

6. Populus heterophýlla L. Swamp or Downy Poplar. Fig. 1445.

Populus heterophylla L. Sp. Pl. 1034. 1753.

An irregularly branching tree, sometimes 80° high and with a trunk 3° in diameter, the bark rough. Young foliage densely tomentose. Leaves long-petioled, broadly ovate, obtuse or subacute at the apex, rounded, truncate or subcordate at the base, crenulate-denticulate, 5′–6′ long, or those of young plants much larger, glabrous or somewhat floccose beneath when mature; petioles terete; bracts glabrous or nearly so; staminate aments stout, 3′–4′ long, 9″–12″ in diameter, drooping; stamens numerous; pistillate aments raceme-like, peduncled, erect or spreading, loosely flowered; capsules ovoid, acute, 2-valved, 4″–6″ long, shorter than or equalling their pedicels.

In swamps, Connecticut to Georgia, west to Louisiana, north in the Mississippi Valley to Ohio, Indiana, Missouri and Arkansas. Wood soft, weak, compact, brown, weight per cubic foot 26 lbs. River- or swamp-cottonwood. Balm-of-gilead. April–May.

7. Populus grandidentáta Michx. Large-toothed Aspen. Fig. 1446.

Populus grandidentata Michx. Fl. Bor. Am. **2**: 243. 1803.

A forest tree with smooth, greenish-gray bark, maximum height about 75°, and trunk diameter 2½°. Leaves ovate-orbicular, those of very young plants densely white-tomentose beneath, sometimes 1° long, with irregularly denticulate margins, those of older trees tomentose when young, glabrous when mature, short-acuminate, coarsely undulate-dentate, obtuse or truncate at the base, 2½′–4′ long; petioles slender, flattened laterally; bracts silky, irregularly 4–7-cleft; staminate aments 2′–4′ long, about 5″ in diameter, drooping; pistillate aments somewhat pubescent, dense, 3′–5′ long in fruit, also drooping; stigma-lobes narrow; capsule conic, acute, 2-valved, about 3″ long, rather less than 1″ in diameter, papillose.

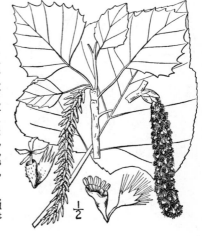

In rich woods, Nova Scotia to Ontario and Minnesota, south to Delaware, North Carolina and Tennessee. Wood soft, weak, light brown, compact; weight per cubic foot 29 lbs. White poplar. April.

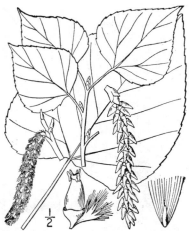

8. Populus tremuloìdes Michx. American Aspen. Quiver-leaf. Fig. 1447.

Populus tremuloides Michx. Fl. Bor. Am. **2**: 243. 1803.

A slender tree, with smooth, light green bark, reaching a maximum height of about 100° and a trunk diameter of 3°, the young foliage glabrous, excepting the ciliate margins of the leaves. Petioles very slender, flattened laterally, causing the leaves to quiver in the slightest breeze; leaves broadly ovate or orbicular, short-acuminate at the apex, finely crenulate all around, truncate, rounded or subcordate at base, 1′–2½′ broad, or those of very young plants much larger; bracts silky, deeply 3–5-cleft into linear lobes; aments drooping, the staminate 1½′–2½′ long, 3″–4″ in diameter, the pistillate longer, dense; stigma-lobes linear; capsule like that of the preceding species, but somewhat smaller.

In dry or moist soil, Newfoundland to Hudson Bay and Alaska, New Jersey, Pennsylvania, Kentucky, Nebraska, in the Rocky Mountains to Mexico and to Lower California. Ascends to 3000 ft. in the Adirondacks. American, trembling or white poplar. Quaking or mountain asp. Wood soft, weak, light brown; weight per cubic foot 25 lbs. March–May.

9. Populus nìgra L. Black Poplar. Willow Poplar. Fig. 1448.

Populus nigra L. Sp. Pl. 1034. 1753.

A large tree, sometimes 100° tall and the trunk 4° in diameter, usually much smaller. Twigs terete; young foliage somewhat pubescent, the mature leaves firm, nearly or quite glabrous; petioles slender, flattened laterally; leaves broadly deltoid, abruptly acuminate at the apex, broadly cuneate or obtuse at the base, crenate, 2′–4′ long; staminate aments 1′–2′ long; stamens about 20; pistillate aments 2′–5′ long in fruit, spreading; capsules oblong, very obtuse, borne on pedicels of much less than their own length.

Valleys of the Hudson and Delaware Rivers, naturalized from Europe. Cat-foot poplar. Devil's-fingers. Old English poplar. April–May.

The Lombardy poplar (**Populus itálica** Moench (*Populus dilatàta* Ait.), commonly planted for ornament, occasionally spreads by sending up shoots from its subterranean parts. Poplar-pine.

10. Populus deltoìdes Marsh. Cottonwood. Necklace Poplar. Fig. 1449.

Populus deltoides Marsh, Arb. Am. 106. 1785.
Populus carolinensis Moench, Verz. Pl. 81. 1785.
Populus monilifera Ait. Hort. Kew. **3**: 406. 1789.
Populus angulata Ait. Hort. Kew. **3**: 407. 1789.

A large tree, the greatest of the poplars, attaining a maximum height of 150° and a trunk diameter of 7½°, the bark grayish-green somewhat rough when old. Foliage glabrous; leaves broadly deltoid-ovate, abruptly acuminate at the apex, crenulate, truncate at the base, 4′–7′ long; petiole flattened laterally, stout, about as long as the blade; bracts glabrous, deeply fimbriate; staminate aments drooping, 3′–5′ long, 5″–6″ in diameter; pistillate aments loosely flowered, becoming 6′–10′ long in fruit; capsules ovoid, acute, 4″–5″ long, 2–4-valved, shorter than or equalling their pedicels.

In moist soil, especially along streams and lakes, Quebec to Manitoba, south to Connecticut, Florida and Tennessee. Wood soft, weak, dark brown; weight per cubic foot 24 lbs. April–May. Carolina poplar. Water- or

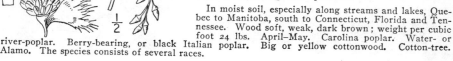

river-poplar. Berry-bearing, or black Italian poplar. Big or yellow cottonwood. Cotton-tree. Alamo. The species consists of several races.

11. Populus Sargentii Dode. Western Cottonwood. Fig. 1450.

P. deltoides occidentalis Rydb. Mem. N.
Y. Bot. Gard. **1**: 115. 1900.
P. Sargentii Dode, Mem. Soc. Hist. Nat.
Autun **18**: [reprint 40]. 1905.
Populus occidentalis Britton; Tillotson,
Rep. Board. Agric. Nebr. 1906–7: 218.

Similar to *P. deltoides,* the bark
thick and gray, the twigs smooth,
greenish to light yellow. Leaves
glabrous on both sides when mature,
broadly triangular-ovate, often wider
than long, subcordate to truncate at
the base, rather long-acuminate at the
apex, the margins coarsely and
bluntly toothed; petioles flattened,
about as long as the blades; staminate aments 2′–3½′ long, not very
dense; ripe pistillate aments 5′ long
or more; capsules ovoid, 5″–7″ long,
longer than their pedicels.

River bottoms, Saskatchewan to North
Dakota, Nebraska, Kansas and New
Mexico. Wood soft and brownish;
weight per cubic foot about 22 lbs.
March–April.

$\frac{2}{3}$

2. SÀLIX [Tourn.] L. Sp. Pl. 1015. 1753.

Trees or shrubs, with single-scaled buds, the scales with an adherent membrane within,
mostly narrow and short-petioled leaves and persistent or early deciduous broad or minute
stipules. Bracts of the aments entire. Disk gland-like, small or minute. Staminate aments
dense, erect, spreading or drooping. Staminate flowers with 1–10, mostly 2, stamens, their
filaments distinct or sometimes united. Pistillate aments usually erect or spreading. Ovary
sessile or short-stipitate. Style short or filiform. Stigmas 2, entire or 2-cleft. Capsule mostly
2-valved. [Name ancient.]

About 200 species, of wide geographic distribution throughout the north temperate and arctic
zones, a few in the southern hemisphere. Besides the following, some 60 others occur in the northern
and western parts of North America. Many hybrids are known. Type species: *Salix alba* L.

A. Trees, or large shrubs mostly more than 3° high (no. 24 smaller).

1. Capsule glabrous, or in nos. 7 and 8, silky-pubescent, not tomentose.

Stamens 3–7 (sometimes 2 in no. 6); filaments hairy at the base; bracts caducous, light yellow.
  Pedicels slender, 3–5 times as long as the gland.
    Petioles and stipules without glands.
      Leaves green or pale beneath.
        Leaves narrowly lanceolate; petioles very short.   1. *S. nigra.*
        Leaves broadly lanceolate; petioles slender.   2. *S. amygdaloides.*
      Leaves whitish beneath.   3. *S. Wardi.*
    Petioles and stipules with prominent glands.
      Summer-fruiting; leaves green beneath.   4. *S. lucida.*
      Autumn-fruiting; leaves pale beneath.   5. *S. serissima.*
  Pedicels about twice as long as the gland.   6. *S. fragilis.*
Stamens 2.
  Filaments hairy at the base; bracts caducous, yellow.
    Pedicels in fruit 1–3 mm. long; native shrubs or small trees.
      Leaves entire, finely and almost permanently silky.   7. *S. exigua.*
      Leaves denticulate, coarsely silky when young, glabrate in age. 8. *S. interior.*
    Pedicels in fruit less than 1 mm. long; large introduced trees.
      Branches not drooping; leaves lanceolate.   9. *S. alba.*
      Branches drooping; leaves linear-lanceolate.   10. *S. babylonica.*
  Filaments glabrous; bracts persistent.
    Mature leaves glabrous.
      Length of leaf-blade less than three times its breadth.
        Mature leaves thin, dull.   11. *S. pyrifolia.*
        Mature leaves thick, firm, dark green and shining above. 12. *S. glaucophylla.*
      Length of leaf-blade three times its breadth or more.   13. *S. cordata.*
    Mature leaves densely silky-pubescent; capsule subsessile.   14. *S. adenophylla.*

2. Capsule tomentose.

Filaments united; capsule sessile; style none.   15. *S. purpurea.*
Filaments distinct.

Capsule subsessile ; style long.   16. *S. viminalis.*
Capsule distinctly pedicelled.
    Style filiform, longer than the stigmas.
       Leaves white-tomentose beneath..   17. *S. candida.*
       Leaves silvery-velvety beneath.   18. *S. pellita.*
    Style short, or none.
       Leaves finely and sharply serrulate, expanding with the aments.
          Capsule short-pedicelled, blunt.   19. *S. sericea.*
          Capsule long-pedicelled, pointed.   20. *S. petiolaris.*
       Leaves crenate, crenulate or subentire.
          Leaves slender-petioled, expanding with the aments.   21. *S. Bebbiana.*
          Leaves short-petioled, expanding after the aments.
             Mature leaves glabrous, or somewhat loosely hairy beneath.   22. *S. discolor.*
             Leaves persistently tomentose beneath.
                Leaves distinctly petioled ; aments ¾'–1½' long ; middle-sized shrub.
                                                     23. *S. humilis.*
                Leaves short-petioled ; aments ½' long ; low shrub.   24. *S. tristis.*
   B. Low or depressed, mainly arctic, subarctic and alpine shrubs, mostly less than 3°
      high (no. 25 sometimes higher; no. 29 extending south to New Jersey and Iowa).
Aments sessile on the branches of the previous year.
    Capsule tomentose.   25. *S. phylicifolia.*
    Capsule glabrous.   26. *S. obtusata.*
Aments on short lateral leafy branches.
    Capsule glabrous, or sometimes loosely hairy, not tomentose.
       Leaves toothed.
          Leaves obovate to oblong.   27. *S. Uva-Ursi.*
          Leaves nearly orbicular ; branches spreading.   28. *S. herbacea.*
       Leaves entire.
          Style very short ; bracts not herbaceous.   29. *S. pedicellaris.*
          Style long ; bracts large, herbaceous.   30. *S. chlorolepis.*
    Capsule tomentose or villous (sometimes glabrous in no. 33).
       Style very short or none.
          Leaves glabrous beneath when mature.   31. *S. reticulata.*
          Leaves permanently silky-hairy.   32. *S. vestita.*
       Style manifest.
          Capsule distinctly pedicelled.
             Mature leaves glabrous or nearly so.   33. *S. Barkleyi.*
             Leaves persistently silvery silky beneath.   34. *S. argyrocarpa.*
          Capsule subsessile ; leaves entire.
             Leaves rounded and obtuse at the apex.
                Bracts obovate to oblong, dark brown or blackish.   35. *S. arctica.*
                Bracts oblong, yellow.   36. *S. Waghornei.*
             Leaves, or some of them, pointed or acute at apex.
                Bracts yellow, yellowish or brownish.
                    Capsules 2"–2½" long ; leaves green in drying.   37. *S. desertorum.*
                    Capsules 3"–4" long ; leaves blackening in drying.   38. *S. glauca.*
                Bracts dark brown or blackish.   39. *S. anglorum.*

1. **Salix nìgra** Marsh.   Black or Swamp Willow.   Fig. 1451.

Salix nigra Marsh. Arb. Am. 139. 1785.

Salix falcata Pursh, Fl. Am. Sept. 2 : 614. 1814.

S. nigra falcata Torr. Fl. N. Y. 2 : 209. 1843.

A tree, with rough flaky dark brown bark, attaining a maximum height of about 120° and a trunk diameter of 3°. Leaves narrowly lanceolate, acute or acuminate at the apex, narrowed at the base, often falcate, short-petioled, serrulate, somewhat pubescent when young, glabrous and green above, somewhat paler, and sometimes pubescent on the veins beneath when mature, 2½'–5' long, 2"–9" wide; stipules various, persistent or deciduous; aments expanding with the leaves, on short lateral branches, the staminate 1'–2' long, the pistillate 1½'–3' long and spreading in fruit; stamens 3–7, distinct, their filaments pubescent below; bracts deciduous; stigmas nearly sessile; capsule ovoid, acute, glabrous, about twice as long as its pedicel.

Along streams and lakes, New Brunswick to western Ontario, North Dakota, Florida and Texas. Hybridizes with *S. alba.* Wood soft, weak, light brown ; weight per cubic foot 28 lbs. Scythe-leaved or pussy-willow. April–May.

2. Salix amygdaloìdes Anders. Peach-leaved Willow. Fig. 1452.

Salix amygdaloides Anders, Ofv. Handl. Vet. Akad. **1858**: 114. 1858.

A small tree, similar to the preceding species, sometimes 70° high and the trunk 2° in diameter, the brown bark scaly. Leaves lanceolate or ovate-lanceolate, pubescent when young, glabrous when old, dark green above, paler and slightly glaucous beneath, long-acuminate at the apex, narrowed at the base, 3½–5′ long, about 1′ wide, sharply serrulate, slender-petioled; petioles 3″–7″ long, glandless; stipules commonly fugacious; aments appearing with the leaves, terminal on short lateral branches, the staminate 1′–2′ long, the pistillate loose, spreading and 2½′–4′ long in fruit; stamens more than 2; filaments distinct, pubescent at the base; bracts deciduous; stigmas nearly sessile; capsule narrowly ovoid, acute, glabrous, at length about as long as its filiform pedicel.

On lake and river shores, Quebec to British Columbia, New York, Ohio, Missouri and New Mexico. Wood soft, weak, light brown; weight 28 lbs. Black or almond-leaved willow. April–May.

3. Salix Wàrdi Bebb. Ward's Willow. Fig. 1453.

Salix nigra var. *Wardi* Bebb; Ward, Bull. U. S. Nat. Mus. **22**: 114. 1881.
Salix Wardi Bebb, Gard. & For. **8**: 363. 1895.

A tree, sometimes 30° high, the trunk reaching 8′ in diameter, the branches spreading or drooping, the bark dark reddish brown. Leaves lanceolate or oblong-lanceolate, long-acuminate or acute at the apex, rounded, subcordate, or narrowed at the base, 2½′–7′ long, ½′–1½′ wide, bright green above, silvery white and usually somewhat pubescent beneath; stipules often large, sometimes persistent; aments expanding with the leaves, terminal, the staminate 2′–4′ long, the pistillate as long or shorter; stamens 3–6, separate; filaments pilose at the base; bracts villous without, deciduous; capsule conic, glabrous, about twice as long as its pedicel.

Along streams and lakes, Maryland to Kansas, Florida and Arkansas. Wood dark brown. March–May. Has been confused with *S. lóngipes* Shuttlw.

4. Salix lùcida Muhl. Shining Willow. Glossy Willow. Fig. 1454.

Salix lucida Muhl. Neue Schrift. Ges. Nat. Fr. Berlin, **4**: 239. *pl. 6. f. 7.* 1803.

A tall shrub, or sometimes a tree 20° high, the bark smooth or slightly scaly, the twigs yellowish-brown, shining. Leaves lanceolate, or ovate-lanceolate, mostly long-acuminate, narrowed or rounded at the base, sharply glandular-serrulate all around, green and glossy on both sides or bearing a few scattered hairs when very young, 3′–5′ long, 1′–1½′ wide when mature or sometimes persistently pubescent; stipules small, semi-cordate or oblong, very glandular, commonly persistent; petioles stout, 3″–6″ long, glandular at the base of the blade; aments hairy-stalked on short, lateral leafy branches, the staminate stout, 1′–2′ long, the pistillate 2′–3′ long in fruit; bracts deciduous; stamens about 5; filaments pubescent below; stigmas stalked or nearly sessile; capsule narrowly ovoid, acute, glabrous, much longer than its pedicel.

In swamps and along streams and lakes, Newfoundland to Athabasca, New Jersey, Kentucky and Nebraska. April–May.

$\frac{1}{2}$

5. Salix seríssima (Bailey) Fernald. Autumn Willow. Fig. 1455.

S. lucida serissima Bailey; Arthur, Bull. Geol. Surv. Minn. **3**: 19. 1887.

S. serissima Fernald, Rhodora **6**: 6. 1903.

A shrub, up to 12° high, the bark brown or yellowish-brown, shining. Leaves oblong-lanceolate, lanceolate or elliptic-lanceolate, mostly acute or short-acuminate at the apex, narrowed at the base, glabrous, glandular-serrulate, dark green and shining above, pale beneath, 1½′-4′ long, firm in texture; petioles slender, 6″ long or less, with 2 to 6 glands at the base of the blade; aments terminating leafy branches, the stalk and rachis pilose; staminate aments about ½′ long, the pistillate 1′-2′ long in fruit, persistent until autumn; scales deciduous; filaments hairy below; stigmas distinctly stalked; capsule conic-subulate, glabrous, 3″-6″ long, short-pedicelled.

In bogs and swamps, Quebec to New Jersey, Alberta and Wisconsin. May–July.

6. Salix frágilis L. Crack Willow. Brittle or Snap Willow. Fig. 1456.

Salix fragilis L. Sp. Pl. 1017. 1753.

A tall, slender tree, with roughish gray bark, attaining a maximum height of about 80° and a trunk diameter of 7°, twigs reddish green, very brittle at the base. Leaves lanceolate, long-acuminate, narrowed at the base, sharply serrulate, glabrous on both sides, rather dark green above, paler beneath, 3′-6′ long, ½′-1′ wide; glandular at the base of the blade; petioles 3″-8″ long, glandular above; stipules semicordate, fugacious; staminate aments 1′-2′ long; stamens 2, or sometimes 3-4; filaments pubescent below, distinct; pistillate aments 3′-5′ long in fruit, rather loose; stigmas nearly sessile; capsule long-conic, glabrous, 2½″-3″ long, short-pedicelled.

Escaped from cultivation, Newfoundland to New Jersey and Kentucky. Native of Europe. Hybridizes with *Salix alba*. The twigs break away and grow into new plants. Stag's-head, red-wood or varnished willow. April–May.

$\frac{2}{3}$

Salix pentándra L., the bay-leaved willow of Europe with shining ovate-oblong, short-acuminate leaves, and smooth long-conic capsules rounded or impressed at the base, is much planted for ornament, and occasionally escapes from cultivation.

7. Salix exígua Nutt. Slender Willow. Fig. 1457.

Salix exigua Nutt. Sylva 1: 75. 1842.

S. fluviatilis exigua Sargent, Silva 9: 124. 1896.

S. luteosericea Rydb. in Britton, Man. 316. 1901.

A shrub or small tree up to 20° high, the twigs often permanently pubescent. Leaves linear to linear-oblanceolate, small, usually not more than 3′ long and 3″ wide, short-petioled, entire, acute at each end, often permanently silky-hairy; stipules early deciduous; aments borne on lateral leafy branches, 2′ long or less; bracts obtuse, pubescent, deciduous; stamens 2, the filaments hairy below; capsule silky when young, becoming about 2½″ long and glabrous when mature, very short-stalked; style very short, shorter than the stigmas.

River and lake shores, Nebraska to Wyoming, British Columbia, Texas and California.

$\frac{2}{3}$

8. Salix interior Rowlee.   Sandbar Willow.   River-bank Willow.   Fig. 1458.

Salix longifolia Muhl. Neue Schrift. Ges. Nat. Fr. Berlin
  4: 238. *pl. 6. f. 6.* 1803. Not Lam. 1778.
S. interior Rowlee, Bull. Torr. Club **27**: 253. 1900.
S. linearifolia Rydb. in Britton, Man. 316. 1901.

A much-branched shrub, 2°–12° high, forming thickets, or sometimes a slender tree, 20°–30° tall, and with
a trunk 1° in diameter, the young foliage silky-
pubescent, the mature leaves glabrous, or nearly so,
those of seedlings pinnately dentate or lobed. Leaves
linear-lanceolate or linear-oblong, 2½'–4' long, 1½"–
5" wide, acuminate, remotely denticulate with somewhat spreading teeth, short-petioled, bright green;
petioles not glandular; stipules minute or none;
aments on short, leafy branches, linear-cylindric, the
staminate dense, 1'–1½' long, the pistillate looser, about
2' long in fruit; bracts deciduous; stamens 2; filaments
pubescent, distinct; stigmas broad, sessile; capsule
ovoid-conic, glabrous or silky, about 2" long.

Along streams and lakes, Quebec to Athabasca, Virginia,
Kentucky and Texas. Has been confused with *Salix fluviá-
tilis* Nutt. Wood soft, reddish-brown; weight per cubic
foot 31 lbs. Long- or Narrow-leaf willow. Red or white
willow. Osier- or shrub-willow. April–May.

Salix Wheèleri (Rowlee) Rydb., of lake and river shores from New Brunswick to Illinois, differs
in having the leaves permanently silky.

9. Salix álba L.   White or Common Willow. Huntingdon or European Willow. Fig. 1459.

Salix alba L. Sp. Pl. 1021. 1753.
Salix vitellina L. Sp. Pl. Ed. 2, 1442. 1763.

A large tree, sometimes 90° tall and a trunk diameter of 8°; bark gray, rough; twigs brittle at the
base. Leaves lanceolate or oblong-lanceolate, acute
or acuminate, narrowed at the base, serrulate, silky-
pubescent on both sides when young, less so and pale
or glaucous beneath when mature, 2'–4½' long, 4"–
8" wide; stipules ovate-lanceolate, deciduous; petioles 2"–4" long, glandless or sparingly glandular;
aments on short lateral leafy branches; stamens 2;
filaments distinct, pubescent at the base; pistillate
aments linear-cylindric, 1½'–2½' long; stigmas nearly
sessile; capsule ovoid, acute, glabrous, short-pedicelled or sessile.

In moist soil, especially along streams, Nova Scotia
to Ontario, North Carolina and Iowa, escaped from cultivation. Native of Europe. Composed of several races,
with twigs green to yellow. Duck-willow. April–May.

10. Salix babylónica L.   Weeping Willow. Drooping or Ring Willow.   Fig. 1460.

Salix babylonica L. Sp. Pl. 1017. 1753.

A large tree, with rough gray bark, sometimes attaining a height of 70° and a trunk diameter of 6°, the
twigs slender, green, elongated, drooping. Leaves
narrowly lanceolate, long-acuminate at the apex, serrulate all around, narrowed at the base, sparsely pubescent when young, glabrous when mature, green above,
paler beneath, 4'–7' long, 3"–6" wide, sometimes curl-
ing into rings; petioles 3"–6" long, glandular above;
aments appearing on short lateral leafy branches;
bracts ovate-lanceolate, obtuse, deciduous; stamens 2;
style almost none; capsule ovoid-conic sessile, glabrous.

Widely cultivated and sometimes spreading by the distribution of its twigs. Connecticut to Michigan and Virginia.
Garb-willow. Native of Asia. April–May.

11. Salix pyrifòlia Anders. Balsam Willow. Fig. 1461.

Salix cordata var. balsamifera Hook. Fl. Bor. Am. **2**:
149. 1839.
S. pyrifolia Anders. Vet. Acad. Handl. **6**[1]: 162. 1867.
S. balsamifera Barratt; Hook. loc. cit. As synonym.
1839. Bebb. Bot. Gaz. **4**: 190. 1879.

A shrub, 4°–10° high, the twigs glabrous, shining, the youngest foliage pubescent. Mature leaves
elliptic, ovate-oval or obovate, thin, glabrous, acute
or some of them obtuse at the apex, rounded or
subcordate at the base, dark green above, glaucous
and prominently reticulate-veined beneath, 2′–3′ long,
1′–1½′ wide, slightly crenulate-serrulate the minute
teeth glandular; stipules minute or none; petioles
slender, 3″–6″ long; aments expanding with the
leaves, leafy at the base, cylindric, the staminate
dense, about 1′ long, the pistillate rather loose, 2′–3′
long in fruit; bracts villous, persistent; stamens 2;
filaments glabrous; style almost none; capsules very
narrow, acute, glabrous, 2″–2½″ long, slender-
pedicelled.

In swamps, Newfoundland to British Columbia, south
to Maine, New York, Michigan and Minnesota. May.

12. Salix glaucophýlla Bebb. Broad-leaved Willow. Fig. 1462.

S. glaucophylla Bebb, in A. Gray, Man. Ed. 6, 485. 1889.

A shrub, 4°–10° high; foliage glabrous or when
young sparingly pubescent. Mature leaves ovate,
obovate or oblong-lanceolate, firm, dark green and
shining above, white-glaucous beneath, short-acuminate, the base rounded or acute, serrulate with gland-
tipped teeth, 2′–4′ long, ¾′–2′ wide; stipules large,
persistent; petioles stout, 3″–6″ long; aments expanding before the leaves, leafy at the base, the
staminate 1′–2′ long, the pistillate 1½′–3′ long in fruit;
bracts densely white-villous, persistent; stamens 2;
filaments glabrous; style filiform; capsule beaked
from an ovoid base, acute, glabrous, 3″–5″ long,
slender-pedicelled.

On sand dunes, Quebec to Alberta, Maine, northern
Ohio and Illinois and Wisconsin. April.

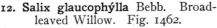

13. Salix cordàta Muhl. Heart-leaved Wil-low. Missouri or Diamond Willow. Fig. 1463.

Salix cordata Muhl. Neue Schrift. Ges. Nat. Fr. Berlin
4: 236. pl. 6. f. 3. 1803.
Salix angustata Pursh, Fl. Am. Sept. 613. 1814.
S. cordata angustata Anders. Vet. Acad. Handl. **6**[1]: 159.
1867.
S. missouriensis Bebb, Gard. & For. **8**: 373. 1895.
S. acutidens Rydb. in Britton, Man. 315. 1901.

A shrub, 5°–12° high, or a tree up to 50° tall, the
twigs puberulent or glabrous; young leaves pubescent. Mature leaves oblong-lanceolate to linear-
lanceolate, green on both sides or paler beneath,
acuminate at the apex, narrowed, obtuse or subcordate at the base, serrulate with glandular teeth,
green in drying; stipules oblique, serrulate, usually
large and persistent; petioles 4″–9″ long; aments
leafy at the base, expanding before the leaves, the
staminate about 1′ long, the pistillate 1½′–2½′ in fruit;
bracts silky, persistent; stamens 2; filaments glabrous; style short; capsules narrowly ovoid, acute,
glabrous, 2″–4″ long; short-pedicelled.

In wet soil, New Brunswick to British Columbia, Virginia, Missouri, Colorado and California.
Hybridizes with S. sericea and other species. April–May.

Salix Mackenzìàna Barrett, a small tree, with young leaves glabrous or merely puberulent,
cuneate, finely serrate, and pedicels 2–4 times as long as the bracts, occurs from Manitoba westward.

Salix lútea Nutt., with light yellow twigs, is apparently otherwise inseparable from S. cordata.
It ranges from western Nebraska to Assiniboia.

14. Salix adenophýlla Hook. Furry Willow. Fig. 1464.

Salix adenophylla Hook. Fl. Bor. Am. **2**: 146. 1839.
Salix syrticola Fernald, Rhodora **9**: 225. 1907.

A straggling shrub, 3°–8° high, the twigs, petioles, stipules and leaves densely silky-tomentose, the silky hairs falling away from the leaves when old. Leaves ovate, acute or short-acuminate, or the lower obtuse at the apex, cordate or rounded at the base, finely serrulate with gland-tipped teeth, 1′–4′ long, 8″–2′ wide; petioles stout, 1½″–3″ long, dilated at the base; stipules ovate-cordate, obtuse, serrulate, persistent; aments dense, expanding with the leaves, the staminate about 1′ long, the pistillate 1½′–4′ long in fruit; bracts villous, persistent; stamens 2; filaments glabrous; style filiform, longer than the stigmas; capsule nearly sessile or ovoid-conic, acute, 1½″–2½″ long.

On lake and river shores. Labrador to James Bay, Ontario, Pennsylvania and Michigan. Hybridizes with *S. cordata.* April–May.

15. Salix purpùrea L. Purple Willow. Fig. 1465.

Salix purpurea L. Sp. Pl. 1017. 1753.

A slender shrub or small tree, with purplish flexible twigs, maximum height about 12°; branches often trailing; bark smooth and very bitter. Leaves oblanceolate or spatulate, acute, serrulate, narrowed at the base, short-petioled, glabrous, green above, paler and somewhat glaucous beneath, 1½′–3′ long, 2½″–4″ wide, some of them commonly subopposite; stipules minute; petioles 1″–2″ long, not glandular; aments appearing before the leaves, dense, leafy at the base, the staminate about 1′ long, the pistillate 1′–2′ long, sessile or nearly so; stamens 2; filaments and sometimes also the anthers united, pubescent; bracts purple, persistent; stigmas very nearly sessile; capsules ovoid-conic, obtuse, tomentose, 2½″ long.

Sparingly escaped from cultivation in the Atlantic States, Ontario and Ohio. Native of Europe. Also called bitter-, rose or whipcord-willow. April–May.

16. Salix viminàlis L. Osier or Basket Willow. White or Velvet-Osier. Fig. 1466.

Salix viminalis L. Sp. Pl. 1021. 1753.

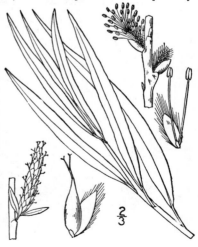

A small slender tree or shrub, with terete green twigs. Leaves elongated-lanceolate or linear-lanceolate, long-acuminate at the apex, sparingly repand-crenulate or entire, revolute-margined, short-petioled, dark green and glabrous above, persistently silvery-silky beneath, 3′–6′ long, 2″–8″ wide; stipules narrow, deciduous; aments expanding before the leaves, dense, the pistillate 2′–3′ long and nearly ½′ in diameter in fruit; stamens 2; filaments glabrous; style longer than the stigmas; capsule narrowly ovoid-conic, acute, silky-pubescent, about 3″ long, very short-pedicelled.

Cultivated for wicker-ware and occasionally escaped into wet places, Newfoundland to Pennsylvania. Native of Europe and Asia. Common osier. Twigwithy. Ausier. Wilgers. April–May.

17. Salix cándida Fluegge. Hoary or Sage Willow. Fig. 1467.

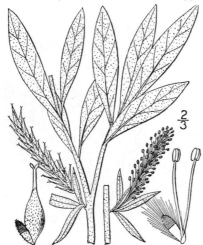

Salix candida Fluegge; Willd. Sp. Pl. 4: 708. 1806.

An erect shrub, 2°–5° tall, the older twigs red or purple and terete, the younger densely white-tomentose. Leaves mostly persistently white-tomentose beneath, green and loosely tomentose or becoming glabrate above when mature, oblong or oblong-lanceolate, thick, sparingly repand-denticulate or entire, acute at both ends or the lower obtuse at the apex, 2′–4′ long, 3″–8″ wide, their margins slightly revolute; petioles 1½″–2″ long; stipules lanceolate-subulate, about equalling the petioles, deciduous; aments expanding before the leaves, dense, cylindric, the staminate about 1′ long, the pistillate 1′–2′ long in fruit; bracts villous, persistent; stamens 2; filaments glabrous; style filiform, red, three times as long as the stigmas; capsule ovoid-conic, acute, densely tomentose, 2½″–3″ long, very short-pedicelled.

In bogs, Newfoundland to Athabasca, Wyoming, New Jersey, Pennsylvania, Iowa and South Dakota. Hybridizes with S. petiolàris and S. cordata. May.

18. Salix pellìta Anders. Satiny Willow. Fig. 1468.

S. pellita Anders. Vet. Acad. Handl. 6: 139. 1867.

A shrub, 5°–8° high, or sometimes a small tree, the twigs reddish to olive-brown. Leaves oblong, lanceolate or oblanceolate, 1½′–4′ long, entire or obscurely crenulate, acute at both ends, or bluntish at the apex, bright green and glabrous above, pale and satiny-pubescent or nearly glabrous beneath, short-petioled; aments on short leafy branches, the pistillate ones becoming 2′ long in fruit; stamens 2, glabrous; bracts villous; capsule densely white-pubescent, 2″–2½″ long, short-conic with a rounded base, very short-pedicelled; style longer than the stigmas.

Along rivers and swamps, Quebec to Manitoba, Maine and Vermont.

19. Salix serícea Marsh. Silky Willow. Fig. 1469.

Salix sericea Marsh. Arb. Am. 140. 1785.
Salix coactilis Fernald, Rhodora 8: 22. 1906.

A shrub, 5°–12° tall, with slender purplish puberulent twigs, the young leaves densely silky-pubescent. Mature leaves glabrous or pubescent, lanceolate, acuminate, narrowed or obtuse at the base, serrulate all around with gland-tipped teeth, dark green, paler and somewhat glaucous beneath, turning brown or black in drying, 2½′–4′ long, 5″–10″ wide; stipules narrow, deciduous; petioles 2″–7″ long, sometimes glandular; aments expanding before the leaves, sessile, the pistillate with a few leaves at the base, dense, the staminate about 1′ long, the pistillate 1′–1½′ long in fruit; bracts villous, persistent; stamens 2; filaments glabrous; style short; capsule ovoid-oblong, obtuse, pubescent, short-pedicelled, about 1½″ long.

In swamps and along streams, New Brunswick to Michigan, North Carolina and Ohio. May.

Salix subserícea (Anders.) Schneider, of eastern Massachusetts, has leaf and capsule characters intermediate between this species and the following one, and has been regarded both as a hybrid and as a distinct species.

20. Salix petiolàris J. E. Smith. Slender Willow. Fig. 1470.

S. petiolaris J. E. Smith, Trans. Linn. Soc. **6**: 122. 1802.
S. gracilis Anders. Proc. Am. Acad. **4**: 67. 1858.

A shrub, similar to the preceding species, but the young leaves only slightly silky, the branches slender, upright or ascending. Mature leaves lanceolate, acuminate at both ends, serrulate with blunt cartilaginous teeth, remaining green in drying, 4″–8″ wide; petioles 2″–5″ long; stipules deciduous; aments expanding before the leaves, the pistillate short-peduncled, usually rather loose, about 1′ long in fruit; bracts villous, oblong to obovate; stamens 2; filaments glabrous; stigmas nearly sessile; capsule tapering from an ovoid or oblong base, pubescent, 2½″–4″ long, usually about twice as long as the filiform pedicel.

Swamps, New Brunswick to Manitoba, Tennessee, Michigan and North Dakota. Dark long-leaved willow. May.

21. Salix Bebbiàna Sarg. Beaked, Livid or Bebb's Willow. Fig. 1471.

Salix rostrata Richards. Frank. Journ. App. 753. 1823. Not Thuill. 1799.
Salix Bebbiana Sarg. Gard. & For. **8**: 463. 1895.

A shrub, 6°–18° tall, or sometimes a tree 25° high, the twigs pubescent or puberulent, terete. Leaves elliptic, oblong or oblong-lanceolate, acute, acuminate or some of them blunt at the apex, rounded or narrowed at the base, sparingly serrate or entire, dull green and puberulent above, pale, reticulate-veined and tomentose beneath or nearly glabrous on both sides when very old; petioles 2″–6″ long; stipules semicordate, acute, deciduous; aments sessile, expanding with or before the leaves, dense, the staminate 1′–1½′ long, the pistillate 2′ long in fruit; bracts villous; stamens 2; filaments distinct, glabrous; stigmas nearly sessile; capsule very narrowly long-conic, densely pubescent, twice as long as the filiform pedicel.

Dry soil and along streams, Newfoundland to Alaska, New Jersey, Nebraska and Utah. April–May.

Salix perrostràta Rydb., inhabiting hillsides and stream-banks from Nebraska and South Dakota to New Mexico and Yukon Territory, differs in having leaves thinner, glabrous when mature.

22. Salix díscolor Muhl. Pussy, Glaucous or Silver Willow. Fig. 1472.

Salix discolor Muhl. Neue Schrift. Ges. Nat. Fr. Berlin **4**: 234. *pl. 6. f. 1.* 1803.
Salix eriocephala Michx. Fl. Bor. Am. **2**: 225. 1803.
Salix prinoides Pursh, Fl. Am. Sept. 613. 1814.
Salix laurentiana Fernald, Rhodora **9**: 221. 1907.

A shrub or low tree, maximum height 25°, trunk diameter 1°; twigs glabrous or pubescent; young leaves sometimes pubescent. Mature leaves usually glabrous, bright green above, glaucous and nearly white beneath, oblong, oblong-lanceolate or oblanceolate, acute at both ends, irregularly serrate or nearly entire, slender-petioled, 3′–5′ long, 8″–18″ wide; petioles 3″–12″ long; stipules obliquely lanceolate or semicordate, commonly deciduous; aments unfolding much before the leaves, dense, the pistillate 1½′–3′ long in fruit; bracts persistent, brown-purple, villous; stamens 2; filaments glabrous; stigmas nearly sessile; capsule narrowly conic, tomentose, 2½″–3″ long, much longer than its pedicel.

In swamps or on moist hillsides, Nova Scotia to Saskatchewan, Delaware and Missouri. Consists of several races differing in pubescence and in leaf-forms. Wood soft, weak, yellow-brown; weight per cubic foot 27 lbs. Bog- or swamp-willow. March–May.

Salix squamàta Rydb., with fruiting aments persistent until September on leafy branches, the yellowish scales longer than the pedicels, is probably a state of the preceding species.

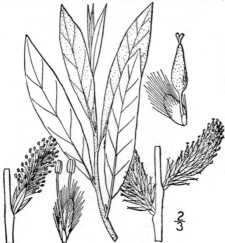

23. Salix hùmilis Marsh. Prairie Willow. Fig. 1473.

Salix humilis Marsh. Arb. Am. 140. 1785.

A shrub, 2°–8° tall, the twigs tomentose or pubescent, terete. Leaves mostly oblanceolate, petioled, rather firm in texture, 2'–6' long, 4"–8" wide, acute at both ends or the lower broader and obtuse at the apex, sparingly denticulate, the margins slightly revolute, the upper surface dark green, dull, puberulent or glabrous, the lower densely and persistently gray-tomentose; petioles 2"–5" long; stipules obliquely lanceolate or ovate, acute, commonly persistent; aments unfolding much before the leaves, sessile, ovoid-oblong, dense, the pistillate about 1' long in fruit; stamens 2; filaments glabrous; stigmas nearly sessile; capsule narrowly conic, densely pubescent, much longer than its pedicel.

In dry soil, Newfoundland to Ontario, North Carolina, Tennessee and Kansas. Hybridizes with *S. discolor.* Low or bush-willow. April–May.

24. Salix trístis Ait. Dwarf Gray Willow. Sage Willow. Fig. 1474.

Salix tristis Ait. Hort. Kew. 3: 393. 1789.
S. humilis tristis Griggs, Proc. Ohio Acad. 4: 301. 1905.

A tufted, slender shrub, 1°–2° tall, the twigs terete, puberulent, the roots long and thick. Leaves oblanceolate or linear-oblong, acute or obtusish, somewhat undulate, green and puberulent or glabrous above, persistently white-tomentose beneath, crowded, 1'–2' long, their margins revolute; petioles 1"–3" long; stipules minute, deciduous; aments expanding much before the leaves, dense, very small, few-flowered, sessile, the pistillate globose-ovoid and about ½' long in fruit; bracts persistent; stamens 2; filaments glabrous; stigmas sessile or nearly so; capsule ovoid with a long, slender beak, tomentulose, about 3" long, much longer than its filiform pedicel.

In dry soil, Nova Scotia (?), Maine to Minnesota, Florida, Tennessee and Nebraska. March–April.

25. Salix phylicifòlia L. Tea-leaved Willow. Fig. 1475.

Salix phylicifolia L. Sp. Pl. 1016. 1753.

A shrub 1°–10° high, much branched, the twigs glabrous, dark purple-green, sometimes glaucous. Leaves oblong, lanceolate or elliptic, acute or obtuse at the apex, minutely repand-crenulate or entire, narrowed at the base, bright green and shining above, pale and glaucous beneath, 1'–3' long, ½'–1' wide; petioles 3"–8" long; stipules minute, fugacious, or wanting; aments sessile, dense, oblong-cylindric, the staminate 1' or less long, the pistillate 1'–2' long in fruit; bracts villous, persistent; stamens 2; filaments glabrous; style rather longer than the stigmas; capsule conic, acute, pubescent or tomentose, 2½" long, much longer than its pedicel.

Swamps, Labrador to Alaska and the mountains of Maine, New Hampshire and Vermont. Europe. Summer.

Salix chlorophýlla Anders., does not appear to be specifically distinct.

26. Salix obtusàta Fernald. Blunt-leaved Willow. Fig. 1476.

Salix obtusata Fernald, Rhodora **9**: *223.* 1907.

A shrub 1½°–3° high, with smooth brown shining twigs. Leaves oblong to oblong-orbicular, thin, rounded at the apex, obtuse or subcordate at the base, ¾'–2' long, 1¼' wide or less, closely dentate, somewhat pubescent beneath when young, glabrous when mature, the petioles 3''–6'' long, slender, the small cordate stipules persistent; aments sessile, borne on twigs of the previous season, the pistillate 10'' long or less, 3''–4'' thick; bracts obovate, obtuse, villous; capsule conic, glabrous, about 1½'' long; gland very short; style not longer than the stigmas.

Gravelly shores, Quebec. Summer.

27. Salix Uva-úrsi Pursh. Bearberry Willow. Fig. 1477.

Salix Uva-ursi Pursh, Fl. Am. Sept. 610. 1814.
Salix Cutleri Tuckerm. Am. Journ. Sci. **45**: 36. 1843.

A depressed or prostrate glabrous shrub, the terete brown branches 6'–12' long, diffuse from a deep central root. Leaves obovate or elliptic, obtuse or acute at the apex, narrowed at the base, crenulate-denticulate, 4''–10'' long, 2''–5'' wide, prominently veined, deep green and shining above, pale beneath; petioles 1''–2'' long; aments on short leafy branches, dense, about ½' long in flower, the pistillate 1'–2' long in fruit; bracts persistent, obovate, obtuse, densely silky; stamens usually solitary, rarely 2; filaments glabrous; style short; capsule ovoid-conic, acute, glabrous, very short-pedicelled.

Labrador and Hudson Bay to Alaska, south to the summits of the mountains of New York and New England. May–June.

Salix myrtillifòlia Anders., of high boreal regions, differs in having nearly erect branches and larger leaves.

28. Salix herbàcea L. Dwarf Willow. Herb-like Willow. Fig. 1478.

Salix herbacea L. Sp. Pl. 1018. 1753.

A depressed matted shrub, with very slender angled twigs 1'–6' long, the youngest foliage somewhat pubescent. Mature leaves glabrous, suborbicular, rounded or retuse at the apex, cordate or rounded at the base, thin, crenulate-denticulate all around, finely reticulate-veined, bright green and shining on both sides, 5''–10'' in diameter; petioles very slender, 2''–4'' long; aments terminating 2-leaved branchlets, 4–10-flowered, 2''–4'' long; bracts obovate, obtuse, persistent, glabrous or nearly so; stamens 2; filaments glabrous; style rather longer than the 2-cleft stigmas; capsule narrowly conic, glabrous, nearly sessile.

Labrador and Quebec, through arctic America, and on the White Mountains of New Hampshire and Mt. Katahdin, Maine. Also in Europe and Asia. Summer.

29. Salix pedicellàris Pursh.   Bog Willow.   Fig. 1479.

Salix pedicellaris Pursh, Fl. Am. Sept. 611. 1814.
S. myrtilloides pedicellaris Anders. Vet. Acad. Handl. 6¹:
96. 1867.

An erect slender glabrous shrub, 1°–3° high, the
twigs light brown, terete. Leaves linear-oblong,
elliptic-oblanceolate or obovate, obtuse or acute at
the apex, entire, mostly narrowed at the base, 1′–3′
long, 3″–8″ wide, short-petioled, bright green above,
pale or glaucous beneath, their margins slightly
revolute; aments expanding with the leaves, leafy at
the base, rather dense, 1′ or less long, or the pistillate longer in fruit; bracts persistent, obtuse, slightly
villous; stamens 2; filaments glabrous; style shorter
than or equalling the stigmas; gland short; capsule
oblong-conic, obtuse, glabrous, 2½″ long, 2–3 times
as long as the filiform pedicel which slightly exceeds
the scale.

In bogs, New Brunswick and Quebec to British Columbia, New Jersey, Pennsylvania, Ohio and Iowa. April–
May. Has been confused with the similar *S. myrtil-
loìdes* L., of Europe.

Salix fuscéscens Anders., occurring on Mt. Albert,
Quebec, in Alaska, and in western Siberia, differs in a
shorter pedicel and longer-pointed gland, the capsule glabrous or loosely pubescent.

30. Salix chlorólepis Fernald.   Green-scaled
Willow.   Fig. 1480.

S. chlorolepis Fernald, Rhodora 7· 186. 1905.

A shrub, 3° high or less, the branches smooth,
nearly erect. Leaves oblong to oblong-obovate, entire, mostly obtuse at the apex, narrowed or rounded
at the base, 5″–12″ long, 3″–5″ wide, slightly pubescent when young, glabrous when mature, pale beneath, the petioles 2″–7″ long, the stipules deciduous; aments on short leafy branches, 7″ long or less,
2″–3″ thick; bracts oblong to obovate, glabrous,
green, subtruncate, about 1½″ long; filaments glabrous; capsule conic, glabrous, very short-pedicelled,
2″ long; style slender, twice as long as the stigmas.

On rocks and in alpine meadows, Mount Albert,
Quebec.

Salix calcícola Fernald & Wiegand, another recently
described boreal species with glabrous capsules and long
styles, has large terminal and lateral catkins and nearly
orbicular leaves; pale beneath.

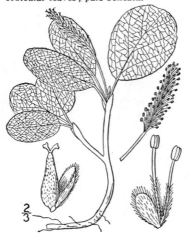

31. Salix reticulàta L.   Net-veined Willow.
Fig. 1481.

Salix reticulata L. Sp. Pl. 1018. 1753.
Salix orbicularis Anders. in DC. Prodr. 16²: 300. 1868.

A procumbent shrub, 3′–10′ high, often sending
out roots from the twigs, the young shoots 4-sided,
purple-green. Leaves elliptic or obovate, thick, obtuse, narrowed, rounded or subcordate at the base,
slender-petioled, glabrous or somewhat silky-pubescent when young, dark green above, not shining,
glaucous and strongly reticulate-veined beneath, 1′–
2′ long; petioles 4″–12″ long, channeled, not glandular; leaves obscurely crenulate or entire; stipules
oblong, obtuse; aments terminal, long-stalked, dense;
bracts obtuse; stamens 2; filaments distinct, pubescent at the base; stigmas sessile; capsule ovoid-
conic, sessile, tomentose, about 3″ long.

Labrador and Quebec to Alaska, south in the Rocky
Mountains to Colorado. Also in northern Europe and
Asia. Wrinkled-leaf willow. June.

32. Salix vestìta Pursh. Hairy Willow. Fig. 1482.

Salix vestita Pursh, Fl. Am. Sept. 610. 1814.
Salix Fernaldii Blankenship, Mont. Agric. Coll. Stud.
Bot. 1: 46. 1905.

A low shrub, similar to the preceding species, the twigs 4-sided, green. Leaves obovate, thick, mostly retuse or emarginate at the apex, slightly crenulate, narrowed or rounded at the base, dark green and glabrous above, persistently tomentose-silky beneath, short-petioled, 1'–3' long; petioles 2''–4'' long, channeled, not glandular; aments small, terminal, unfolding after the leaves, stalked; stamens 2; filaments distinct; capsules narrowly ovoid-conic, sessile, densely silky-tomentose, about 3'' long.

Labrador and Quebec to British Columbia and Montana. June.

33. Salix Bàrclayi Anders. Barclay's Willow. Fig. 1483.

S. Barclayi Anders. Ofv. Handl. Vet. Akad. 1858: 125. 1858.
Salix Barclayi latiuscula Anders. in DC. Prodr. 16: Part 2, 255. 1868.
Salix latiuscula Anders. Vet. Acad. Handl. 6¹: 165. 1867.

A low shrub, with dark brown glabrous twigs, the young shoots pubescent. Leaves obovate, oval, or oval-lanceolate, short-pointed at the apex, very minutely serrulate, floccose-pubescent when young, when mature glabrous, bright green above, pale beneath, 1'–2' long, ½'–1' wide; petioles 1''–2'' long; stipules ovate, acute, deciduous; aments unfolding with the leaves, borne at the ends of short branches, dense, spreading or erect, the staminate 1' long, the pistillate 2'–3' long in fruit; bracts persistent, slightly villous; stamens 2; filaments distinct; capsule narrowly conic, tomentose or glabrous, acute, 3'' long; style longer than the stigmas.

Newfoundland and Quebec. Northwestern arctic America. Summer.

34. Salix argyrocàrpa Anders. Silver Willow. Fig. 1484.

S. argyrocarpa Anders. Mon. Sal. 107. *f.* 60. 1867.

An erect or diffuse shrub, 6'–2° high, the twigs dark green, nearly terete, shining. Leaves oblong or oblanceolate, acute at each end or the lower obtuse, short-petioled, entire or crenulate, bright green and glabrous above, persistently silvery-silky beneath, 1'–2' long, 3''–6'' wide, the margins slightly revolute; aments unfolding with the leaves, leafy at the base, dense, 1' or less long; bracts persistent, villous; stamens 2, distinct, their filaments glabrous; style slender, longer than the stigmas; capsule oblong-conic, densely silvery, acute, 1''–1½'' long, about twice as long as its pedicel.

Labrador and Quebec to the mountains of Maine and New Hampshire. Hybridizes with *S. phylicifolia.* June–July.

35. Salix àrctica Pall. Arctic Willow. Ground Willow. Fig. 1485.

Salix arctica Pall. Fl. Ross. 1 : Part 2, 86. 1788.

A low branching shrub, rarely 6' high, the twigs terete or nearly so. Leaves glabrous, elliptic or broadly obovate, entire, obtuse and usually rounded at the apex, narrowed or rounded at the base, long-petioled, pale, glaucous and reticulate-veined beneath, 1'–2' long, ½'–1½' wide, often darkening in drying; petioles slender, ½'–1½' long; aments borne at the ends of short leafy branches, very dense, the pistillate 1'–2' long in fruit; bracts dark purple-brown, oblong to obovate, obtuse, white-villous, persistent; stamens 2; filaments glabrous; style filiform, longer than the stigmas; capsule conic, villous, very short-pedicelled.

Quebec; Arctic America and Asia. Summer.

Salix callicarpaèa Trautv., of Labrador and recorded from Quebec, has broadly obovate leaves which do not darken in drying.

36. Salix Waghórnei Rydberg. Waghorne's Willow. Fig. 1486.

S. cordifolia Hook. Fl. Bor. Am. 2 : 152. 1840. Not Pursh, 1814.
S. Waghornei Rydberg, Bull. N. Y. Bot. Gard. 1 : 271. 1899.

A low shrub, 1° high or less, with smooth shining twigs, sparingly pubescent when young. Leaves obovate or elliptic-obovate, rather thin, entire, about 1' long, ½'–¾' wide, sparingly long-hairy when young, soon glabrous, remaining green in drying, the petioles 1''–4'' long, the stipules deciduous; aments on lateral leafy branches, the pistillate about 1' long in fruit, the staminate shorter; bracts yellow, oblong, obtuse, silky-hairy; capsule short-stalked, white-tomentose, conic, 2''–3'' long; style about as long as the stigmas.

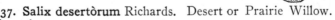

Labrador. Summer.

37. Salix desertòrum Richards. Desert or Prairie Willow. Fig. 1487.

Salix desertorum Richards. Frank. Jour. App. 371. 1823.

A shrub, 6'–12' high, with purplish-green glabrate twigs. Leaves oblong or oblanceolate, acute at the apex and cuneate at the base or the lower obtuse at both ends, entire or very nearly so, very short-petioled, tomentose beneath or glabrate when old, ½'–2' long, 2''–4'' wide; stipules fugacious; aments expanding with the leaves, dense, ½' or less long, leafy at the base; stamens 2; filaments glabrous; style about as long as the deeply 2-cleft stigmas; capsule ovoid-conic, acute, densely tomentose, about 2'' long, very short-pedicelled.

Anticosti and Quebec to western arctic America. Summer.

Salix brachycárpa Nutt. (*S. stricta* (Anders.) Rydb.), differs in leaves tomentose on both sides and hairy twigs. It occurs in Quebec, and ranges otherwise from Saskatchewan to Colorado.

38. Salix glaùca L. Northern Willow.
Fig. 1488.

Salix glauca L. Sp. Pl. 1019. 1753.
S. atra Rydb. Bull. N. Y. Bot. Gard. 1 : 272. 1899.
S. labradorica Rydb. Bull. N. Y. Bot. Gard. 1 : 274. 1899.

A low arctic shrub, with terete brown twigs, the young shoots and leaves densely tomentose, becoming glabrate when old. Leaves elliptic or elliptic-lanceolate, entire, obtuse or acute at the apex, narrowed at the base, 1′–3′ long, ¼′–1′ wide; petioles 1″–5″ long; stipules deciduous; aments borne on short leafy branches, the staminate dense, about 1′ long, the pistillate 2′–3′ long in fruit, rather loose; stamens 2; filaments distinct; bracts persistent, yellowish or brownish, densely white-villous; capsule ovoid-conic, densely white-tomentose, sessile or very short-pedicelled, 3″–4″ long; style about as long as the stigmas.

Greenland and Labrador to Alaska. Also in arctic and alpine Europe and Asia. The American races differ slightly from those of the Old World. Summer.

39. Salix anglòrum Cham. Brown's Willow.
Fig. 1489.

Salix arctica R. Br. Ross' Voy. cxliv. 1819. Not Pall.
Salix anglorum Cham. Linnaea 6 : 541. 1831.
S. Brownii Lundst. Nov. Act. Soc. Sci. Ups. 16 : 6. 1877.

A low, much branched shrub, the twigs 4-angled, slender. Leaves oblong or lanceolate, glabrous or sometimes ciliolate, mostly acute at the apex, entire, narrowed at the base, short-petioled, 1′–3′ long, 3″–12″ wide, blackening in drying, the lower surface pale or glaucous, the margins not revolute; stipules narrow, deciduous; aments borne on short leafy branches, large, the pistillate 1′–2½′ long in fruit; bracts villous, persistent, obovate, obtuse, dark brown; stamens 2; filaments glabrous; style filiform, much longer than the stigmas; capsule ovoid-conic, tomentose, short-pedicelled, acute, 2″–4″ long.

Labrador to Alaska, and in the Rocky Mountains to Colorado. Summer.

Salix Macoùni Rydberg (S. vacciniförmis Rydberg), of Labrador, Hudson Bay and Quebec, differs in having smaller aments and leaves remaining green in drying.

Salix groenlándica (Anders.) Lundst., of high arctic regions has smaller leaves, darkening in drying.

Family 6. BETULÀCEAE Agardh, Aphor. 208. 1825.
BIRCH FAMILY.

Monoecious or very rarely dioecious trees or shrubs, with alternate petioled simple leaves, and small flowers in linear-cylindric oblong or subglobose aments. Stipules mostly fugacious. Staminate aments pendulous. Staminate flowers 1–3 together in the axil of each bract, consisting of a membranous 2–4-parted calyx, or none, and 2–10 stamens inserted on the receptacle, their filaments distinct, their anthers 2-celled, the anther-sacs sometimes distinct and borne on the forks of the 2-cleft filaments. Pistillate aments erect, spreading or drooping, spike-like or capitate. Pistillate flowers with or without a calyx adnate to the solitary 2-celled ovary; style 2-cleft or 2-divided; ovules 1 or 2 in each cavity of the ovary, anatropous, pendulous. Fruit a small compressed or ovoid-globose, mostly 1-celled and 1-seeded nut or samara. Endosperm none. Cotyledons fleshy.

Six genera and about 75 species, mostly natives of the northern hemisphere.

Staminate flowers solitary in the axil of each bract, destitute of a calyx; pistillate flowers with a calyx.
  Staminate flowers with no blactlets; pistillate aments spike-like; nut small, subtended by or enclosed in a large bractlet.
    Fruiting bractlet flat, 3-cleft and incised.   1. Carpinus.
    Fruiting bractlet bladder-like, closed, membranous.   2. Ostrya.
  Staminate flowers with 2-bractlets; pistillate flowers 2–4, capitate; nut large, enclosed by a leafy involucre.   3. Corylus.
Staminate flowers 3–6 together in the axil of each bract, with a calyx; pistillate flowers without a calyx.
  Stamens 2; filaments 2-cleft; fruiting bracts 3-lobed or entire, deciduous.   4. Betula.
  Stamens 4; anther-sacs adnate; fruiting bracts woody, erose or 5-toothed, persistent.   5. Alnus.

1. CARPÌNUS (Tourn.) L. Sp. Pl. 998. 1753.

Trees or shrubs, with smooth gray bark, furrowed and ridged stems and straight-veined leaves, the primary veins terminating in the larger teeth. Aments expanding before the leaves. Staminate aments linear-cylindric, sessile at the ends of short lateral branches of the preceding season, their flowers solitary in the axil of each bract, consisting of 3–12 stamens; filaments short, 2-cleft, each fork bearing an anther-sac. Pistillate flowers in small terminal aments, 2 to each bract, consisting of a 2-celled ovary adnate to a calyx and subtended by a flat persistent bractlet, which becomes much enlarged, foliaceous and lobed or incised in fruit, the bracts deciduous; style slender or almost none; stigmas 2, subulate. Nut small, ovoid, nerved, acute, borne at the base of the large bractlet. [The ancient name.]

About 12 species, only the following American. Type species: *Carpinus Betulus* L.

1. Carpinus caroliniàna Walt. American Hornbeam. Blue Beech. Water-beech. Fig. 1490.

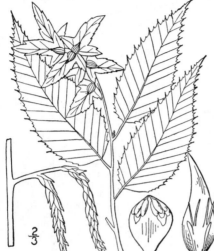

Carpinus caroliniana Walt. Fl. Car. 236. 1788.

A small tree, with slender terete gray twigs; maximum height about 40°, trunk diameter of 2½°. Leaves ovate-oblong, acute or acuminate at the apex, sharply and doubly serrate all around, rounded or subcordate at the base, somewhat inequilateral, 2½′–4′ long, 1′–1½′ wide, green on both sides, glabrous above, slightly pubescent on the veins beneath, petioles very slender, 4″–7″ long; staminate aments 1′–1½′ long, their bracts triangular-ovate, acuminate, puberulent; anther-sacs villous at the summit; bractlet of the pistillate flowers 3-lobed at the base, firm-membranous, strongly veined and about 1′ long when mature, its middle lobe lanceolate, acute, 2–4 times as long as the lateral ones, incised-dentate on one side, often nearly entire on the outer; nut 2″ long.

In moist woods and along streams, Nova Scotia to Ontario and Minnesota, Kansas, Florida and Texas. Wood very hard and strong, durable, light brown; weight per cubic foot 45 lbs. April–May, the fruit ripe Aug.–Sept. Water-beech. Iron-wood.

2. ÓSTRYA (Micheli) Scop. Fl. Carn. 414. 1760.

Trees similar to the Hornbeams, the primary veins of the leaves simple or forked, the aments expanding with or before the leaves. Staminate aments sessile at the ends of branchlets of the preceding season, their flowers as in *Carpinus,* solitary in the axil of each bract; filaments 2-cleft. Pistillate aments small, terminal, erect, the flowers 2 to each bract, subtended by a tubular, persistent bractlet which enlarges into a membranous, nerved, bladder-like sac in fruit. Style slender; stigmas 2, subulate. Nut ovoid-oblong, compressed, smooth, sessile. Mature pistillate ament hop-like. [The ancient name.]

Six species, the following, 2 in the southwest, 1 in Europe and Asia, 1 in Mexico, 1 Japanese. Type species: *Ostrya Ostrya* (L.) MacM.

1. Ostrya virginiàna (Mill.) Willd. Hop-hornbeam. Iron-wood. Fig. 1491.

Carpinus virginiana Mill Gard. Dict. Ed. 8. 1768.
Ostrya virginica Willd. Sp. Pl. 4: 469. 1805.

A tree, with a maximum height of about 50°, trunk diameter of 2°, twigs of the season pubescent. Leaves ovate or oblong-ovate, the apex acuminate, the base rounded or inequilateral, sharply and doubly serrate, sparingly pubescent and green above, pubescent or tomentose beneath, 2½′–4′ long, 1′–1½′ wide; petioles rarely more than 2″ long; staminate aments 1½′–3′ long, their bracts triangular-ovate, acuminate; anther-sacs villous at the summit; bractlet of each fertile flower forming a sac 6″–8″ long and 4″–5″ in diameter in fruit, acute, cuspidate, pubescent, villous near and at the base with bristly hairs, parellel-veined and finely reticulated; nut 2½″ long, compressed, shining; ripe fertile aments erect or spreading, 1½′–2½′ long, resembling hops.

In dry woods, Cape Breton Island to Manitoba, Nebraska, Florida and Texas. Wood similar to that of the Hornbeam, but heavier; weight per cubic foot 51 lbs. Bark scaly. April–May. Fruit ripe July–Aug. Hard-hack. Indian-cedar or black-hazel. Lever- or deer-wood.

3. CÓRYLUS (Tourn.) L. Sp. Pl. 998. 1753.

Shrubs or small trees, with broad thin serrulate or incised leaves. Staminate aments sessile at the ends of twigs of the previous season, expanding much before the leaves, the flowers solitary in the axil of each bract, of about 4 stamens and 2 bractlets; filaments 2-cleft or 2-divided, each fork bearing an anther-sac, which is villous at the summit. Calyx none. Pistillate flowers from scaly buds, clustered at the ends of short branches of the season, each in the axil of a bract, consisting of an incompletely 2-celled ovary adnate to a calyx, a short style and 2 slender stigmas; bractlets 2, enlarged in fruit, forming a leaf-like involucre to the nut, remaining nearly distinct or united into a tubular beak. Nut ovoid or oblong, sometimes compressed, large, bony. [Name Greek, from the helmet-like involucre.]

Species 7, in the northern hemisphere. Besides the following, another occurs in California. Type species: *Corylus Avellàna* L.

Involucre of 2 broad laciniate bractlets; leaves serrulate.　　　　　　　　1. *C. americana.*
Involucral bractlets united, prolonged into a tubular bristly beak.　　　　　2. *C. rostrata.*

1. Corylus americàna Walt. Hazel-nut.
Filbert. Fig. 1492.

Corylus americana Walt. Fl. Car. 236. 1788.

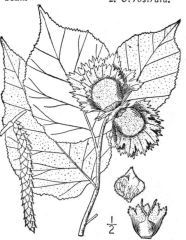

A shrub, 3°–8° tall, the young shoots russet-brown, densely hispid-pubescent with pinkish hairs, the twigs becoming glabrous. Leaves ovate or broadly oval, acute or acuminate at the apex, serrulate all around, cordate or obtuse at the base, glabrous or nearly so above, finely tomentose beneath, 3'–6' long, 2'–4½' wide; petioles 2''–4'' long; staminate aments mostly solitary, 3'–4' long; involucre of the nut compressed, composed of the 2 nearly distinct finely pubescent leaf-like bractlets, which are lacinate on their margins, commonly broader than high and exceeding the nut; nut compressed, light brown, striate, ½' high.

In thickets, Maine and Ontario to Saskatchewan, Florida and Kansas. March–April. Nuts ripe July–Aug.

2. Corylus rostràta Ait. Beaked Hazel-nut. Fig. 1493.

Corylus rostrata Ait. Hort. Kew. 3: 364. 1789.

A shrub, similar to the preceding species, but the foliage usually less pubescent. Leaves ovate or narrowly oval, acuminate at the apex, cordate or obtuse at the base, incised-serrate and serrulate, glabrous, or with some scattered appressed hairs above, sparingly pubescent at least on the veins beneath, 2½'–4' long, 1'–2½' wide; petioles 2''–4'' long; involucral bractlets bristly hairy, united to the summit and prolonged into a tubular beak about twice the length of the nut, laciniate at the summit; nut ovoid, scarcely compressed, striate, 5''–7'' high.

In thickets, Nova Scotia to British Columbia, south to Georgia, Tennessee, Kansas and Oregon. April–May. Fruit ripe Aug.–Sept.

4. BÉTULA (Tourn.) L. Sp. Pl. 982. 1753.

Aromatic trees or shrubs, with dentate or serrate leaves, scaly buds and flowers of both kinds in aments expanding before or with the leaves, the pistillate erect or spreading. Staminate flowers about 3 together in the axil of each bract, consisting of a membranous, usually 4-toothed perianth, 2 stamens, and subtended by 2 bractlets; filaments short, deeply 2-cleft, each fork bearing an anther-sac. Pistillate flowers 2 or 3 (rarely 1) in the axil of each bract, the bracts 3-lobed, or sometimes entire, deciduous with the fruits; perianth none; ovary sessile, 2-celled; styles 2, stigmatic at the apex, mostly persistent. Nut small, compressed, membranous-winged on each side (a samara), shorter than the bracts. [The ancient name.]

About 35 species of the north temperate and arctic zones. Type species: *Betula alba* L.

Trees (except mountain and boreal races of no. 3).
  Fruiting aments peduncled.
    Fruiting aments not tomentose.
      Leaves long-acuminate.
        Leaves bright green, irregularly toothed; bark not readily peeling, chalky white.
                                                      1. *B. populifolia.*
          Leaves dull green; bark readily peeling, white to bronze.   2. *B. coerulea.*
          Leaves merely acute; bark white to bronze, readily peeling.   3. *B. papyrifera.*
      Fruiting aments tomentose.   4. *B. nigra.*
    Fruiting aments sessile.
      Fruiting bracts 2″-2½″ long; leaves mostly cordate.
        Fruiting bracts glabrous; bark dark brown, close.   5. *B. lenta.*
        Fruiting bracts ciliate; bark gray to yellow brown, close or peeling.   6. *B. alleghanensis.*
      Fruiting bracts 3½″-5″ long, ciliate; bark yellow-gray, freely peeling.   7. *B. lutea.*
Shrubs (nos. 8 and 10 sometimes forming small trees).
  Twigs glandular-warty.
    Samara-wings broader than the nut.   8. *B. fontinalis.*
    Samara-wings narrower than the nut.   9. *B. glandulosa.*
  Twigs not glandular-warty.
    Samara-wings broader than the nut.   10. *B. Sandbergi.*
    Samara-wings narrower than the nut.
      Young foliage densely pubescent; fruiting bracts all 3-lobed.   11. *B. pumila.*
      Leaves glabrous; at least the upper fruiting bracts entire.   12. *B. nana.*

1. Betula populifòlia Marsh. American White Birch. Fig. 1494.

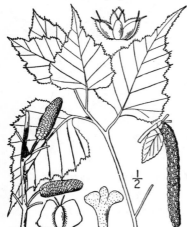

Betula populifolia Marsh. Arb. Am. 19. 1785.
B. alba var. *populifolia* Spach, Ann. Sci. Nat. (II.) **15**: 187. 1841.

A slender tree with very white smooth bark, tardily separating in thin sheets; maximum height 45°; trunk diameter 1½°; the twigs russet, warty. Leaves deltoid, pubescent on the veins when young, nearly glabrous when old, minutely glandular, bright green above, light green beneath, long-acuminate, sharply irregularly dentate and commonly somewhat lobed, obtuse or truncate at the base, 1½′-2½′ long, 1′-2′ wide, slender-petioled; petioles channeled; staminate aments 2′-3′ long; pistillate aments cylindric, in fruit 9″-18″ long, 3″-5″ in diameter, slender-peduncled; fruiting bracts puberulent, 1″-2″ long, lateral lobes divergent, larger than the middle one; nut narrower than its wings.

In moist or dry soil, Nova Scotia to southern Ontario, Pennsylvania and Delaware. Wood soft, weak, light brown; weight per cubic foot 36 lbs. Leaves tremulous like those of the aspens. Gray-, poverty- or old-field-birch. Broom- or pin-birch. May.

2. Betula coerùlea Blanchard. Blue Birch. Fig. 1495.

Betula coerulea Blanchard, Betula 1. 1904.
B. coerulea-grandis Blanchard, loc. cit. 1904.

A tree, attaining a maximum height of about 65° and a trunk diameter of 2°, the bark white, readily peeling off in thin layers, the young twigs somewhat pubescent, becoming glabrous. Leaves ovate, 2′-4½′ long, serrate, long-acuminate at the apex, broadly cuneate at the base, when mature glabrous and dull bluish-green above, slightly pubescent on the veins beneath, the petioles slender, about 1′ long; staminate aments 1½′-3′ long; pistillate aments cylindric, about 1′ long, on stalks about one-half as long; fruiting bracts with divergent lateral lobes; nut much narrower than its wings.

Hillsides, Quebec to Manitoba, Maine and Vermont. Resembles the European *Betula pendula* Roth.

3. Betula papyrifera Marsh. Paper or Canoe Birch. Fig. 1496.

Betula papyrifera Marsh. Arb. Am. 19. 1785.
Betula papyracea Ait. Hort. Kew. **3**: 337. 1789.
Betula papyrifera minor Tuckerm. Am. Journ. Sci. **45**: 31. 1843.

A large forest tree with maximum height of about 80° and trunk diameter of 3°, or on mountains reduced to a low shrub; bark, except of the young wood, peeling in thin layers. Leaves ovate, acute or acuminate, dentate and denticulate, subcordate, truncate or obtuse at the base, dark green and glabrous above, glandular and pubescent on the veins beneath, slender-petioled, 1½′–4½′ long, 1′–3′ wide; petioles ½′–1½′ long; staminate aments 2′–4′ long; pistillate aments cylindric, slender-peduncled, 1′–2′ long, ¼′–½′ in diameter in fruit; fruiting bracts 2″–3″ long, puberulent or ciliate; nut narrower than its wings.

Newfoundland to Alaska, Pennsylvania, Michigan, Nebraska and Washington. Similar to the Old World *B. alba* L. Wood hard, strong, reddish-brown; weight per cubic foot 37 lbs. The chalky-white outer bark interesting to tourists. Silver-, bolean- or white-birch. Spool-wood. April–May.

Betula cordifòlia Regel, differs in having distinctly cordate leaves, but scarcely otherwise.

4. Betula nìgra L. River Birch. Red Birch. Fig. 1497.

Betula nigra L. Sp. Pl. 982. 1753.

A slender tree, sometimes 90° high and the trunk 2½° in diameter; bark reddish or greenish-brown, peeling in very thin layers; twigs reddish. Young shoots, petioles and lower surfaces of the leaves tomentose; leaves rhombic-ovate, apex acute or obtuse, irregularly serrate or somewhat lobed, base cuneate, when mature dark green and glabrous above, pale and glabrous or somewhat tomentose beneath, 1½′–3′ long; petioles 3″–8″ long; staminate aments mostly clustered in 2's or 3's, 2½′–3½′ long; pistillate aments oblong-cylindric, spreading, peduncled, 1′–2′ long, 5″–6″ in diameter in fruit; fruiting bracts tomentose, about equally 3-lobed, 3″–5″ long; nut broadly obovate, wider than its wings, pubescent at the base.

Along streams and lakes, Massachusetts to Illinois, Minnesota, Kansas, Florida and Texas. Wood hard, strong, brown; weight per cubic foot 36 lbs. The bark of the branches peels off in almost membranous layers. Water- or black-birch. April–May.

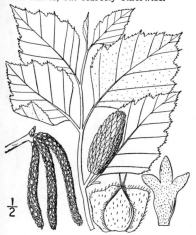

5. Betula lénta L. Cherry, Black or Sweet Birch. Fig. 1498.

Betula lenta L. Sp. Pl. 983. 1753.

A large forest tree, sometimes 80° high, with dark brown close smooth bark, becoming furrowed, not separating in layers; foliage aromatic; twigs smooth, warty, young leaves silky. Mature leaves ovate or ovate-oblong, acute or acuminate, the base cordate or rounded, sharply serrulate, bright green, and shining above, dull green and pubescent on the veins beneath, 2½′–4′ long, 1′–2′ wide; petioles 3″–6″ long; staminate aments clustered, 2½′–4′ long; pistillate aments sessile, dense, oblong, about 1′ long and ½′ in diameter in fruit, nearly erect; bracts glabrous or minutely puberulent, not ciliate, appressed, about 2″ long, nearly equally 3-lobed, the lateral lobes somewhat divergent; nut oblong, broader than its wings.

New England to western Ontario, Florida and Tennessee. Wood hard, strong, dark brown; weight per cubic foot 47 lbs. The aromatic oil of the branches and foliage (same as oil of wintergreen) is distilled in quantities and is an important article of commerce. Tree much resembles the cherry. A hybrid with *B. pumila* is *B. Jackii* Schneider. Spice-, river- or mahogany-birch. Mountain-mahogany. April–May.

6. Betula alleghanénsis Britton. Southern Yellow Birch. Fig. 1499.

B. alleghanensis Britton, Bull. Torr. Club **31** : 166.   1904.

Similar to *Betula lenta* and to *Betula lutea,* the bark either close and fissured, or peeling off in thin layers on young trunks and branches. Leaves ovate or ovate-oblong, usually acuminate at the apex, cordate or rounded at the base, 5′ long or less, sharply and rather coarsely toothed, hairy when young, glabrous when old, except on the veins beneath ; staminate aments clustered ; ripe pistillate aments oblong-cylindric, short-stalked or sessile, about 1′ long, their bracts nearly or quite as wide as long, more or less pubescent, 3-lobed to about the middle, the margins ciliate ; nut narrowly oblong to obovate, broader than its wings.

Woodlands, Massachusetts to Quebec, Michigan and Georgia.   May.

7. Betula lùtea Michx. f.   Yellow Birch.   Gray Birch.   Fig. 1500.

Betula lutea Michx. f. Arb. Am. **2** : 152. *pl. 5.*   1812.

A large forest tree, reaching a maximum height of about 100° and a trunk diameter of 4°, the bark yellowish or gray, separating in thin layers or close, the twigs gray-brown. Leaves ovate or oblong-ovate, mostly acuminate at the apex, rounded, obtuse or rarely subcordate at the base, sharply serrulate all around, dark green and dull above, pubescent on the veins beneath, 1½′–4′ long, petioles 4″–9″ long ; staminate aments usually 2–4 together ; pistillate aments sessile, oblong or oblong-cylindric, 1½′ or less long, 7″–9″ thick in fruit, rather loose ; bracts nearly equally 3-lobed to somewhat above the middle, ciliolate, the lateral lobes ascending ; nut broadly oblong, wider than its wings.

Newfoundland to Manitoba, Massachusetts, Pennsylvania and Wisconsin.   Wood hard, strong, light brown ; weight per cubic foot 41 lbs.   Swamp- or silver-birch.   April–May.

8. Betula fontinàlis Sargent.   Western Red Birch.   Cherry Birch.   Fig. 1501.

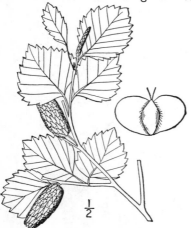

B. fontinalis Sargent, Bot. Gaz. **31** : 239.   1901.

A shrub or rarely a tree attaining a height of about 40° and a trunk diameter of 1½°, the bark smooth, dark bronze, the twigs gray-brown, warty. Leaves broadly ovate or nearly orbicular, acute or obtuse at the apex, sharply serrate, rounded or obtuse at the base, short-petioled, glabrous on both sides or sparingly pubescent on the veins beneath, 1′–2′ long ; petioles slender, 2″–6″ long ; pistillate aments penduncled, cylindric, spreading or pendent, 1′–1¼′ long, about 5″ in diameter in fruit ; fruiting bracts ciliolate, about 3″ long, their lateral lobes ascending ; nut much narrower than its wings.

South Dakota to western Nebraska, British Columbia, California and New Mexico, and on Mt. Albert, Quebec. Has been confused with *Betula occidentalis* Hook., and referred to the Asiatic *B. microphylla* Bunge.   Black, gray, sweet or water-birch.   April–May.

9. Betula glandulòsa Michx. Glandular or Scrub Birch. Fig. 1502.

Betula glandulosa Michx. Fl. Bor. Am. **2** : 180. 1803.

A shrub, 1°–6° high, the twigs brown, glandular-warty, not pubescent. Leaves orbicular, reniform, oval or obovate, glabrous, rounded at the apex, rounded, narrowed or cuneate at the base, crenate-dentate, bright green above, pale green and glandular-dotted beneath, short-petioled, ¾′–1′ long; petioles 1″–3″ long; staminate aments commonly solitary, about ½′ long; pistillate aments cylindric, erect, peduncled, 5″–12″ long and about 2″ in diameter in fruit; fruiting bracts 2″–3″ long, the lateral lobes rather shorter than the middle one; nut oblong to nearly orbicular, its wings mostly narrow.

Newfoundland to Alaska, the higher mountains of Maine and northern New York, Michigan, Minnesota, in the Rocky Mountains to Colorado, and to California. Also in Asia. Dwarf birch. June–July.

10. Betula Sandbérgi Britton. Sandberg's Birch. Fig. 1503.

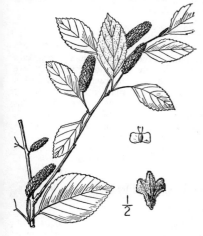

Betula Sandbergi Britton, Bull. Torr. Club **31** : 166. 1904.

A shrub or small tree, the close bark dark reddish-brown, the young twigs finely pubescent. Leaves oval to rhombic-ovate, acute at both ends, serrate, 1′–2½′ long, firm in texture, dull green above, paler and sparingly pubescent beneath, the slender petioles about ½′ long; staminate aments 2′–2½′ long; fruiting pistillate aments about 1′ long, cylindric, slender-stalked; bracts about 2″ long and broad, pubescent, the middle lobe obtuse or acute, longer than the obtuse lateral ones; nut narrower than its wings.

Swamps, Minnesota to Saskatchewan and Montana. Spring.

11. Betula pùmila L. Low Birch. Fig. 1504.

Betula pumila L. Mant. 124. 1767.

A bog shrub, 2°–15° tall, the twigs brown, becoming glabrous, the young foliage densely brownish-tomentose. Leaves obovate, broadly oval or orbicular, rounded at both ends or some of them cuneate-narrowed at the base, rather coarsely dentate, when mature glabrous and dull green above, pale, persistently tomentose or becoming glabrous beneath and prominently reticulate-veined, ½′–1½′ long; petioles 1½″–3″ long; fruiting pistillate aments oblong-cylindric, erect, peduncled, 1′ long or less, about 3″ in diameter; bracts puberulent or ciliolate, the lateral lobes spreading at right angles, shorter than the middle one; nut oblong, mostly rather broader than its wings.

In bogs, Newfoundland to western Ontario and the Northwest Territory, south to New Jersey, Ohio and Minnesota. Also in Europe and Asia. Dwarf birch. Tag-alder. May–June.

    Betula glandulífera (Regel) Butler, differs in having leaves smooth or nearly so on both sides. It grows from Michigan and western Ontario to British Columbia.

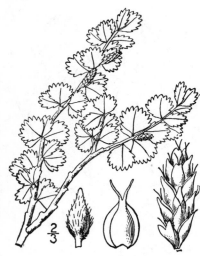

12. Betula nàna L.　Dwarf or Alpine Birch.
Fig. 1505.

Betula nana L. Sp. Pl. 983. 1753.

B. Michauxii Spach, Ann. Sci. Nat. (II.) **15**: 195. 1841.

A low diffuse shrub, similar to the preceding species, but the twigs glandless, puberulent or glabrous. Leaves orbicular, obovate, or reniform and wider than long, bright green, firm, glabrous, on both sides when mature, deeply and incisely crenulate, rounded at the apex, rounded, obtuse or cuneate at the base, 3″–10″ long; petioles rarely more than 1″ long; staminate aments ½′–1′ long, solitary or clustered; pistillate aments oblong, sessile or short-peduncled, erect or somewhat spreading, 3″–5″ long; fruiting bracts glabrous, the lower usually 3-lobed, the upper ovate or lanceolate, mostly entire; nut oblong, wingless or narrowly winged.

Greenland and Labrador to Hudson Bay.　Also in northern Europe and Asia.　May–June.

5.　ÁLNUS (Tourn.) Hill, Brit. Herb. 510.　1756.

Shrubs or trees, with dentate or serrulate leaves, few-scaled or naked buds, and flowers of both kinds in aments, expanding before, with or after the leaves, making in most species their first appearance during the preceding season, the staminate pendulous, the pistillate erect, clustered.　Staminate flowers 3 or sometimes 6 in the axil of each bract, consisting of a mostly 4-parted perianth, 4 stamens and subtended by 1 or 2 bractlets; filaments short, simple; anther-sacs adnate.　Pistillate flowers 2–3 in the axil of each bract, without a perianth, but subtended by 2–4 minute bractlets; ovary sessile, 2-celled; styles 2; bracts woody, persistent, 5-toothed or erose.　Nut small, compressed, wingless or winged.　[Ancient Latin name derived from the Celtic, in allusion to the growth of these trees along streams.]

About 14 species, natives of the northern hemisphere and the Andes of South America.　Besides the following, some 4 others occur in western North America.　Type species: *Alnus vulgaris* Hill.

Nut bordered by a membranous wing on each side.　　　　　　　　　1. *A. Alnobetula.*
Nut acute-margined, wingless.
　　Leaves obovate, broadly oval or suborbicular, dull; aments expanding long before the leaves.
　　　　Leaves finely pale-tomentose or glaucous beneath.　　　　2. *A. incana.*
　　　　Leaves green and glabrous or pubescent beneath, obovate to suborbicular.
　　　　　Leaves finely serrulate; native.　　　　　　　　　　　3. *A. rugosa.*
　　　　　Leaves dentate-serrate; twigs glutinous; introduced tree.　4. *A. Alnus.*
　　Leaves mostly oblong, bright green and shining above; aments expanding in late summer or
　　　　autumn.　　　　　　　　　　　　　　　　　　　　　　5. *A. maritima.*

1. Alnus Alnobétula (Ehrh.) K. Koch.　Green or Mountain Alder.　Fig 1506.

Betula Alnobetula Ehrh. Beitr. **2**: 72. 1788.
Alnus viridis DC. Fl. Fr. **3**: 304. 1805.
A. crispa Pursh, Fl. Am. Sept. 623. 1814.
A. Mitchelliana M. A. Curtis; A. Gray, Am. Journ. Sci.
　　42: 42. 1842.
Alnus Alnobetula K. Koch, Dendr. **2**¹: 625. 1872.
Alnus mollis Fernald, Rhodora **6**: 162. 1904.

A shrub, 2°–10° high, the young leaves glutinous and more or less pubescent, the twigs glabrous or pubescent.　Leaves oval or ovate, obtuse or acute, sharply and more or less irregularly serrulate or incised-serrulate, when mature dark green and glabrous above, light green and glabrous or pubescent beneath, 2′–5′ long, 1½′–3′ wide; petioles 4″–12″ long; aments expanding with the leaves, the staminate slender, naked, 1½′–2½′ long, the pistillate oblong or ovoid-oblong, slender-peduncled, becoming 4″–10″ long and 4″–5″ in diameter in fruit, their bracts irregularly 5-toothed; nut oblong, the thin wings about as broad as the body.

Newfoundland to Manitoba, Massachusetts, New York, Michigan, and in the higher Alleghanies to North Carolina.　Also in Europe and Asia.　June.　Consists of many races, differing mainly in the amount of pubescence.

2. Alnus incàna (L.) Willd. Spreckled or Hoary Alder. Fig. 1507.

Betula Alnus var. *incana* L. Sp. Pl. Ed. 2, 1394. 1763.
Alnus incana Willd. Sp. Pl. 4: 335. 1805.

A shrub, or rarely a small tree, 8°–40° high, the twigs glabrous, the young shoots pubescent. Leaves oval or ovate, acute or sometimes obtuse at the apex, finely serrulate or dentate, with the teeth serrulate, obtuse or some of them acute at the base, dark green above, pale or glaucous and pubescent, at least on the veins beneath, 2′–5′ long, 1½′–4′ wide, the veins prominent on the lower surface; stipules oblong-lanceolate, deciduous; petioles 4″–12″ long; aments unfolding much before the leaves, the staminate 1½′–3′ long, the pistillate ovoid, about ½′ long and 3″–5″ in diameter in fruit, their bracts 5-toothed; nut orbicular, coriaceous-margined.

In wet soil, Newfoundland to Saskatchewan, New York, Pennsylvania and Nebraska. Also in Europe and Asia. Wood soft, light brown; weight per cubic foot 28 lbs. Black or tag-alder. April–May.

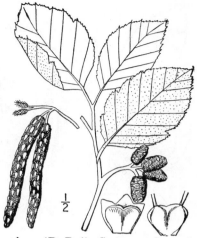

3. Alnus rugòsa (DuRoi) Spreng. Smooth or Hazel Alder. Fig. 1508.

Betula Alnus rugosa DuRoi, Harbk. 1: 112. 1771.
Alnus serrulata Willd. Sp. Pl. 4: 336. 1805.
Alnus rugosa Spreng. Syst. 3: 848. 1826.

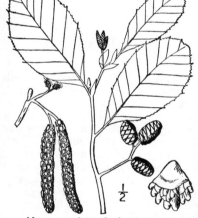

A shrub 5°–20° tall, or sometimes a small tree, attaining a maximum height of 40° and a trunk diameter of 6′, the bark smooth, the younger shoots somewhat pubescent. Leaves green on both sides, obovate or oval, mostly obtuse and rounded at the apex, narrowed or rounded at the base, sharply and minutely serrulate, when mature glabrous above, usually pubescent beneath, at least on the veins, 3′–5′ long, stipules oval, deciduous; petioles 4″–12″ long; aments unfolding much before the leaves (or in the South after the leaves), the staminate 2′–4′ long, the pistillate ovoid, 6″–9″ long in fruit; nut ovate, narrowly coriaceous-margined.

In wet soil, or on hillsides, Maine to Ohio and Minnesota, Florida and Texas. Wood soft, light brown; weight per cubic foot 29 lbs. Common, tag, american or green alder. March–April.

Alnus noveboracénsis Britton, has leaves acute at both ends and densely pubescent on the veins beneath. It may be a race of this species. It is known certainly only from a Staten Island tree, now destroyed.

4. Alnus Álnus (L.) Britton. European Alder. Eller. Fig. 1509.

Betula Alnus and var. *glutinosa* L. Sp. Pl. 983. 1753.
Alnus vulgaris Hill, Brit. Herb. 510. 1756.
Alnus glutinosa Gaertn. Fr. & Sem. 2: 54. 1791.

A tree, reaching a maximum height of about 75° and a trunk diameter of 2½°, the bark smooth, the branches nearly horizontal, the foliage glutinous. Leaves broadly oval, orbicular or obovate, thick, dark green, dull, often obtuse at both ends, coarsely dentate and the teeth denticulate, glabrous above, pubescent on the veins beneath, 2′–5′ long; petioles ½′–1′ long; aments appearing from naked buds, expanding much before the leaves, the staminate 3′–4′ long, the pistillate ovoid-oblong, 6″–9″ long in fruit; nut wingless, coriaceous-margined.

In wet places, Newfoundland to New Jersey and Illinois, escaped from cultivation. Native of Europe. Wood soft, brown; weight per cubic foot 35 lbs. Irish mahogany. Hollard. Ooler. Black alder. April.

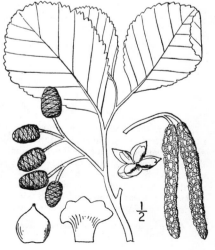

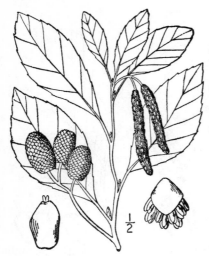

5. Alnus marítima (Marsh.) Muhl. Seaside Alder. Fig. 1510.

Betula Alnus maritima Marsh. Arb. Am. 20. 1785.
Alnus maritima Muhl.; Nutt. Sylva **1**: 34. *t. 102.* 1865.

A small tree, sometimes 30° tall and the trunk 6′ in diameter, glabrous or very nearly so throughout. Leaves oblong, ovate-oblong or obovate, firm, acute at both ends, bright green and shining above, pale green and dull beneath, sharply serrulate, 2′–4′ long, 1′–2′ wide; petioles 3″–10″ long; aments unfolding long after the leaves, their buds developing during the season, the staminate 1′–2½′ long, the pistillate oblong, 9″–12″ long, 5″–7″ in diameter in fruit; nut oblong-obovate, wingless, coriaceous-margined.

In wet soil, southern Delaware and eastern Maryland; also in Oklahoma. Closely related to *A. japónica* of northeastern Asia. Wood soft, light brown; weight per cubic foot 31 lbs. Aug.–Sept.

Family 7. FAGÀCEAE Drude, Phan.´409. 1879.

BEECH FAMILY.

Trees or shrubs. Leaves alternate, petioled, simple, dentate, serrate, lobed, cleft or entire, pinnately veined, the stipules, if any, deciduous. Flowers small, monoecious, the staminate in pendulous erect or spreading aments, or capitate, the pistillate solitary or several together, subtended by an involucre of partly or wholly united bracts, which becomes a bur or cup. Petals none. Staminate flowers with a 4–7-lobed perianth and 4–20 stamens; filaments slender, distinct, simple; anther-sacs adnate, longitudinally dehiscent. Pistillate flowers with a 4–8-lobed urn-shaped or oblong perianth, adnate to the 3–7-celled ovary; ovules 1–2 in each cavity, only 1 in each ovary ripening, pendulous, anatropous; styles as many as the cavities of the ovary, linear, terminally or longitudinally stigmatic. Fruit a 1-seeded nut, with a coriaceous or somewhat bony exocarp. Testa thin. Endosperm none; cotyledons large, fleshy, often rugose; radicle short.

About 5 genera and 375 species, of very wide geographic distribution.

Staminate flowers capitate; nut sharply triangular.   1. *Fagus.*
Staminate flowers in slender aments; nut rounded or plano-convex.
    Pistillate flowers 2–5 in each involucre; involucre becoming globose and very prickly in fruit, enclosing the nuts.   2. *Castanea.*
    Pistillate flower 1 in each involucre; involucre of numerous scales forming a cup in fruit and subtending the acorn.   3. *Quercus.*

1. FÀGUS (Tourn.) L. Sp. Pl. 997. 1753.

Trees, with smooth light gray bark, and serrate straight-veined leaves. Flowers appearing with the leaves, the staminate in slender-peduncled pendulous globose heads, the pistillate about 2 together in short-peduncled subulate-bracted involucres in the upper axils. Staminate flowers yellowish-green, subtended by deciduous bracts, consisting of a campanulate 4–8-lobed calyx, and 8–16 stamens with filiform filaments. Pistillate flowers with a 6-lobed perianth adnate to a 3-celled ovary; ovules 2 in each cavity, usually 1 only of each ovary maturing; styles 3, filiform. Nut coriaceous, sharply 3-angled, enclosed in the 4-valved bur. [Name from the Greek, to eat, referring to the esculent nuts.]

About 4 species of the northern hemisphere. Only the following is native in North America. Type species: *Fagus sylvatica* L.

1. Fagus grandifòlia Ehrh. American Beech. Fig. 1511.

Fagus americana latifolia Muench. Hausv. **5**: 162. 1770.
Fagus grandifolia Ehrh. Beytr. Naturk. **3**: 22. 1788.
Fagus ferruginea Ait. Hort. Kew. **3**: 362. 1789.
Fagus americana Sweet, Hort. Brit. 370. 1826.
F. grandifolia caroliniana Fernald & Rehder, Rhodora **9**: 114. 1907.

A large forest tree, with maximum height of about 120°, and a trunk diameter of 4½°, the lower branches spreading. Leaves ovate, ovate-oblong or oblong-obovate, firm, acuminate at the apex, obtuse, subcordate or narrowed at the base, 2′–4½′ long, 1′–3′ wide, densely silky when young, glabrous or somewhat pubescent when mature, green on both sides, not shining, rather coarsely serrate; petioles 2″–6″ long; heads of staminate flowers 6″–9″ in diameter, hanging on peduncles 1′–3′ long; bur 6″–10″ high, densely tomentose, its soft, long or short prickles recurved or spreading; nut brown; seed sweet.

In rich soil, Nova Scotia to Ontario, Minnesota, Missouri, Florida and Texas. Wood hard, strong, tough, close-grained; color light or dark red; weight 43 lbs. per cubic foot. April–May. Nuts ripe Sept.–Oct. Leaves of seedlings and young shoots are sometimes pinnatifid. Red or white beech.

2. CASTÀNEA (Tourn.) Hill, Brit. Herbal 509. 1756.

Trees or shrubs, with serrate straight-veined leaves, their teeth sharply acuminate. Flowers appearing after the leaves, the staminate in erect or spreading, narrowly cylindric, interrupted axillary yellowish aments, several in the axil of each bract, the bracts fugacious, the pistillate borne in prickly involucres at the bases of the staminate aments or in separate axils. Staminate flowers 2-bracteolate, consisting of a mostly 6-lobed campanulate perianth and numerous stamens, sometimes also with an abortive ovary; filaments filiform, long-exserted. Pistillate flowers 2–5 (commonly 3) in each involucre, consisting of an urn-shaped 6-lobed perianth adnate to the mostly 6-celled ovary, and usually with 4–12 abortive stamens; ovules 2 in each cavity, 1 ovule only of each ovary usually maturing; styles as many as the cavities of the ovary, slender, exserted; stigmas minute. Pistillate involucre enlarging and becoming a globose mostly 4-valved very prickly bur in fruit, enclosing 1–several nuts. Nut rounded or plano-convex, 1-seeded, the shell coriaceous. Seed large, sweet. Style mostly persistent. [Name Greek, from a city in Thessaly.]

Four or five species, natives of the northern hemisphere. Besides the following, another occurs in the southeastern United States. Type species: *Castanea vulgàris* Hill.

Leaves green on both sides; nuts usually 2–5 in each involucre; large tree.   1. *C. dentata.*
Leaves densely white-tomentose beneath; nut usually solitary; shrub or small tree.   2. *C. pumila.*

1. Castanea dentàta (Marsh.) Borkh. American Chestnut. Fig. 1512.

Fagus Castanea dentata Marsh. Arb. Am. 46. 1785.
Castanea dentata Borkh. Handb. Forstb. **1**: 741. 1800.
C. vesca var. *americana* Michx. Fl. Bor. Am. **2**: 193. 1803.

A large forest tree, with gray bark rough in longitudinal plates, reaching a maximum height of about 100° and a trunk diameter of 14°; lower branches spreading. Leaves oblong-lanceolate, glabrous, firm, acuminate at the apex, narrowed or rounded at the base, coarsely serrate, with very sharp-pointed ascending teeth, rather dark green above, lighter beneath, 5′–12′ long, 1½′–3′ wide; petioles stout, ½′–1′ long; staminate aments erect, numerous, borne solitary in the upper axils, 6′–12′ long, 4″–5″ in diameter; burs 1½′–4′ in diameter, solitary or 2–4 together, enclosing 1–5 nuts; nuts puberulent, dark brown, plano-convex or angled on the face, or when solitary ovoid.

In rich soil, Maine and Ontario to Michigan, Georgia and Arkansas. Wood coarse-grained, durable, brown; weight per cubic foot 28 lbs. Involucre sometimes suppressed and the nuts naked. June–July. Nuts ripe Sept.–Oct. Sardian nut. Prickly bur.

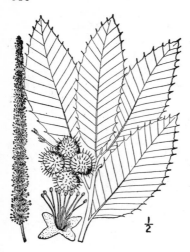

2. Castanea pùmila (L.) Mill. Chinquapin.
Fig. 1513.

Fagus pumila L. Sp. Pl. 998. 1753.
Castanea pumila Mill. Gard. Dict. Ed. 8, no. 2. 1768.

A shrub or small tree, sometimes 45° high and with a trunk 3° in diameter, the young shoots puberulent. Leaves oblong, acute at both ends, sharply serrate with ascending or divergent teeth, dark green and glabrous above, densely white-tomentulose beneath, 3′–6′ long, 1′–2½′ wide; staminate aments erect or somewhat spreading, 3′–5′ long, 3″–4″ in diameter; burs 1½′ in diameter or less, commonly spicate, enclosing a solitary ovoid brown nut (rarely 2); seed very sweet.

In dry soil, New Jersey and Pennsylvania to Missouri, Florida and Texas. Wood strong, coarse-grained, dark brown; weight per cubic foot 37 lbs. June. Nuts ripe Sept.

3. QUÉRCUS (Tourn.) L. Sp. Pl. 994. 1753.

Trees or shrubs, with pinnatifid lobed dentate crenate or entire leaves, deciduous or in some species persistent. Flowers very small, green or yellowish, appearing with or before the leaves, the staminate numerous in slender mostly drooping aments, the pistillate solitary in many-bracted involucres borne on the twigs of the preceding season or on the young shoots. Staminate flowers subtended by caducous bracts, consisting of a mostly 6-lobed campanulate perianth and 3–12 stamens with filiform filaments, sometimes also with an abortive pilose ovary. Pistillate flowers involucrate, with an urn-shaped or oblong calyx, adnate to a mostly 3-celled ovary; ovules 2 in each cavity of the ovary, rarely more than 1 in each ovary maturing; styles as many as the ovary-cavities, short, erect or recurved. Fruit consisting of the imbricated and more or less united bracts of the involucre (cup), subtending or nearly enclosing the ovoid, oblong or subglobose 1-seeded coriaceous nut (acorn). [The ancient Latin name, probably of Celtic derivation, signifying "beautiful tree."]

About 220 species, natives of the northern hemisphere. Besides the following, some 40 others occur in the western and southern sections of North America. Our species hybridize freely. Type species: *Quercus Ròbur* L.

* Leaves or their lobes bristle-tipped, deciduous; acorns maturing in autumn of second year.

† Leaves pinnatifid or pinnately lobed.

Leaves green on both sides.
    Cup of the acorn saucer-shaped, much broader than high.
        Cup 8″–12″ broad; acorn ovoid; leaves dull.   1. *Q. rubra.*
        Cup 4″–8″ broad; leaves shining.
            Acorn subglobose or short-ovoid; northern.   2. *Q. palustris.*
            Acorn ovoid; southern.   3. *Q. Schneckii.*
    Cup of the acorn turbinate or hemispheric.
        Inner bark gray to reddish; leaves deeply lobed.
            Leaves dull, not shining, pale beneath.   4. *Q. borealis.*
            Leaves shining above.
                Cup of acorn 5″–7″ wide.   5. *Q. ellipsoidalis.*
                Cup of acorn 8″–12″ wide.   6. *Q. coccinea.*
        Inner bark orange; leaves pubescent in the axils of the veins beneath.   7. *Q. velutina.*
Leaves white or gray-tomentulose beneath.
    Large tree; leaf-lobes lanceolate or linear-lanceolate, long.
        Leaves rounded or obtuse at the base, 3–5-lobed.   8. *Q. triloba.*
        Leaves cuneate, obtuse or truncate at the base, 5–11-lobed. 9. *Q. pagodaefolia.*
    Shrub or low tree; leaf-lobes triangular-ovate, short.   10. *Q. nana.*

†† Leaves 3–5-lobed above the middle or entire, obovate or spatulate in outline.

Leaves obovate-cuneate, brown-floccose beneath.   11. *Q. marilandica.*
Leaves spatulate to obovate, glabrous both sides.   12. *Q. nigra.*

††† Leaves entire, oblong, lanceolate or linear-oblong (sometimes lobed in no. 14).

Leaves linear-oblong, green and glabrous on both sides.   13. *Q. Phellos.*
Leaves oblong, glabrous, dark green and shining above.   14. *Q. laurifolia.*
Leaves oblong or lanceolate, brown-tomentulose beneath.   15. *Q. imbricaria.*

** *Leaves or their lobes not bristle-tipped, deciduous; acorns maturing in autumn of first year.*
† **Leaves pinnatifid or pinnately lobed.**

Mature leaves pale, or glaucous and glabrate beneath; cup shallow.   16. *Q. alba.*
Mature leaves tomentulose beneath; cup one-third to fully as long as the acorn.
  Upper scales of the cup not awned.
    Leaves yellowish-brown-tomentulose beneath; acorn ovoid.   17. *Q. stellata.*
    Leaves white-tomentulose beneath; acorn depressed-globose.   18. *Q. lyrata.*
  Upper scales awned, forming a fringe around the acorn.   19. *Q. macrocarpa.*

†† **Leaves crenate or dentate, not lobed.**

Fruit peduncled.
  Peduncle much longer than petioles; leaves white-tomentulose beneath.   20. *Q. bicolor.*
  Peduncle equalling or shorter than the petioles; leaves gray-tomentulose beneath.
    Teeth of the leaves acute or mucronulate.   21. *Q. Michauxii.*
    Teeth of the leaves rounded.   22. *Q. Prinus.*
Fruit sessile or nearly so.
  Tall tree; leaves oblong, obovate or lanceolate.   23. *Q. Muhlenbergii.*
  Shrub or low tree; leaves oval or obovate.   24. *Q. prinoides.*
*** *Leaves entire (rarely with a few bristle-tipped lobes), evergreen.*   25. *Q. virginiana.*

1. Quercus rùbra L.   Red Oak.   Fig. 1514.

Quercus rubra L. Sp. Pl. 996. 1753.

A large forest tree, with a maximum height of
about 140°, and a trunk diameter of 7°, the bark
dark gray, slightly roughened. Leaves oval or somewhat obovate in outline, deciduous, when mature
glabrous, or pubescent in the axils of the veins beneath, 4′–8′ long, 3′–6′ wide, dull green above, paler
beneath, sinuses rounded, lobes triangular-lanceolate,
tapering from a broad base to an acuminate apex,
1–4-toothed or entire, teeth and apices tipped with
filiform bristles; petioles 1½′–3′ long; styles slender,
spreading; fruit maturing the second autumn, sessile
or nearly so; cup saucer-shaped, its base flat or
slightly convex, 8″–12″ broad, bracts ovate or ovate-
lanceolate, obtuse or the upper acute, appressed;
acorn ovoid, about 1′ long, 2–4 times as long as
the cup.

Nova Scotia to Ontario and Minnesota, south to
Florida and Kansas. Wood hard, strong, coarse-grained;
color light reddish-brown; weight per cubic foot 41 lbs.
May–June. Acorns ripe Oct.–Nov. Champion, Black
or Spanish oak.

Quercus rùbra runcinàta A. DC. with leaf-lobes
nearly entire and acorn ½′ high, from near St. Louis, is perhaps a hybrid with *Q. triloba* Michx.

Quercus Catésbaei Michx., a tree of the southeastern states, differs from all related species by
having the inner scales of the deep cup of the acorn inflexed. It has been observed northward to
southern Virginia.

2. Quercus palústris DuRoi.   Swamp Oak.   Pin Oak.   Fig. 1515.

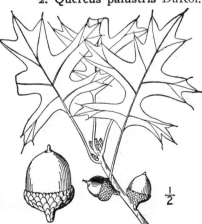

Q. palustris DuRoi, Harbk. 2: 268. *pl. 5. f. 4.* 1772.

A forest tree, maximum height about 120° and
trunk diameter 5°, the lower branches deflexed;
bark brown, rough when old. Leaves broadly
oblong or obovate in outline, deeply pinnatifid,
sometimes almost to the midrib, bright green
glabrous and shining above, duller, glabrous or
with tufts of hairs in the axils of the veins beneath, 3′–5′ long, the lobes oblong, lanceolate
or triangular-lanceolate, divergent, 1–4-toothed
or entire, teeth and apices tipped with filiform
bristles; styles slender; fruit maturing in the second autumn; cup saucer-shaped, 4″–6″ broad,
base flat, bracts triangular-ovate, acute or obtuse,
appressed; acorn subglobose or ovoid, 4″–7″
high, often striate, 2–3 times as long as the cup.

In moist ground, Massachusetts to Michigan, Virginia and Arkansas. Wood hard, very strong, coarse-
grained; color light brown; weight per cubic foot 43
lbs. May–June. Acorns ripe Sept.–Oct. Swamp
Spanish oak.

$\frac{1}{2}$

3. Quercus Schnéckii Britton.   Schneck's Oak. Fig. 1516.

Quercus Schneckii Britton, Manual 333. 1901.

A forest tree, attaining a maximum height of about 200° and a trunk diameter of 8°, usually smaller. Bark reddish-brown, with broad ridges broken into plates; leaves mostly obovate in outline, bright green and shining above, paler and with tufts of wool in the axils beneath, 2'–6' long, truncate or broadly wedge-shaped at the base, deeply pinnatifid into 5–9 oblong or triangular lobes, which are entire or coarsely few toothed, the lobes and teeth bristle-tipped; styles short; fruit maturing in the autumn of the second season; cup deeply saucer-shaped, 6"–8" broad, its scales obtusish or acute, appressed; acorn ovoid, $\frac{1}{2}$'–1' long, 2–3 times as high as the cup.

North Carolina to Indiana, Iowa, Missouri, Florida and Texas. Wood hard, light red-brown; weight per cubic foot 57 lbs. April–May. Acorns ripe Sept.–Oct. Has been confused with *Quercus texana* Buckley.

4. Quercus boreàlis Michx. f. Gray Oak. Fig. 1517.

Q. borealis Michx. f. N. A. Sylv. 1: 98. *pl.* 26. 1817.

A forest tree, reaching at least 90° in height, with a trunk diameter up to 3°, the bark rough in plates and ridges. Leaves ovate to obovate in outline, 7–13-lobed, 4'–7' long, deep green and dull above, paler green and with tufts of hairs in the axils of the leaves beneath, the acute lobes bristle-tipped, the slender leaf-stalk 2' long or less; pistillate flowers with long spreading styles; fruit maturing in the autumn of the second season; cup depressed-hemispheric, $\frac{1}{2}$'–$\frac{3}{4}$' wide, embracing one-third to one-half of the acorn, its scales obtuse.

Quebec and Ontario to New York and Pennsylvania.

$\frac{1}{2}$

5. Quercus ellipsoidàlis E. J. Hill. Hill's Oak. Fig. 1518.

Quercus ellipsoidalis E. J. Hill, Bot. Gaz. **27**: 204. 1899.

A tree, becoming about 65° high, with a trunk diameter up to nearly 4°, the rather thin, shallowly fissured bark grayish-brown without, yellow within, the twigs appressed-pubescent when young. Leaves broadly oval or obovate in outline, deeply pinnatifid with rounded sinuses, bright green and shining above, paler, and with tufts of hairs in the axils of the veins beneath, 2$\frac{1}{2}$'–6' long, the lobes and teeth bristle-tipped; fruit ripening in the autumn of the second season; cup turbinate, 5"–7" wide, embracing one-third to one-half of the ellipsoid to globose-ovoid acorn, its scales ovate, blunt.

In clayey soils, Illinois to Minnesota. May. Acorns ripe Oct. Also called yellow oak and black oak. Perhaps a hybrid between *Q. velùtina* and *Q. coccinea*.

$\frac{1}{2}$

6. Quercus coccínea Wang. Scarlet Oak. Fig. 1519.

Q. coccinea Wang. Amer. 44. *pl. 4. f. 9.* 1787.

A forest tree, attaining a maximum height of about 160°, the trunk diameter sometimes 5°; inner bark pale reddish or gray; foliage turning scarlet in autumn. Leaves deeply pinnatifid, glabrous, bright green and shining above, paler and sometimes pubescent in the axils of the veins beneath, 4′–8′ long, rather thin, the lobes oblong or lanceolate, divergent or ascending, few-toothed, the teeth and apices bristle-tipped; fruit maturing in the autumn of the second season; styles slender, recurved-spreading; cup hemispheric or top-shaped, its bracts triangular-lanceolate, appressed or the upper slightly squarrose, mostly glabrous; acorn ovoid or ovoid-globose, 6″–10″ high, about twice as long as the cup.

In dry soil, Maine and Ontario to Minnesota, North Carolina and Missouri. Wood hard, strong, light brown or red; weight 46 lbs. per cubic foot. May–June. Acorns ripe Sept.–Oct. Black, red or Spanish oak.

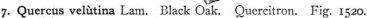

7. Quercus velùtina Lam. Black Oak. Quercitron. Fig. 1520.

Quercus velutina Lam. Encycl. 1: 721. 1783.
Q. tinctoria Bartram, Travels, 37. Name only. 1791.
Q. coccinea var. *tinctoria* A. Gray, Man. Ed. 5, 454. 1867.

A large forest tree, similar to *Q. coccinea*, maximum height about 150°, trunk diameter 5°; outer bark very dark brown, rough in low ridges, the inner bright orange. Leaves pinnatifid or lobed to beyond the middle, firm, brown-pubescent or sometimes stellate-pubescent when young, when mature glabrous and dark dull green above, pale green and usually pubescent on the veins beneath, the broad oblong or triangular-lanceolate lobes and their teeth bristle-tipped; fruit maturing in the autumn of the second season; cup hemispheric or top-shaped, ¾′–1′ broad, commonly narrowed into a short stalk, its bracts mostly pubescent, the upper somewhat squarrose; acorn ovoid, ½′–1′ high, longer than the cup.

Maine to Ontario, Minnesota, Florida and Texas. Wood reddish-brown; weight per cubic foot 44 lbs. May–June. Consists of several races differing in leaf-lobing, amount of pubescence and size of acorns. Dyer's or spotted oak. Yellow-bark oak.

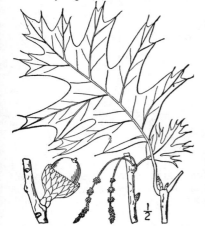

8. Quercus tríloba Michx. Spanish or Water Oak. Fig. 1521.

Quercus nigra digitata Marsh. Arb. Am. 121. 1785.
Quercus triloba Michx. Hist. Chen. Am. *pl. 26.* 1801.
Q. falcata Michx. Hist. Chen. Am. 16. *pl. 28.* 1801.
Quercus digitata Sudw. Gard. & For. 5: 99. 1892.

A tree, with maximum height of about 95°, and trunk diameter of 5°. Leaves dark green and glabrous above, gray-tomentulose beneath, deeply pinnatifid into 3–7 linear or lanceolate, often falcate, acuminate, entire or dentate lobes; teeth and apices bristle-tipped; terminal lobes commonly elongated; styles slender; fruit maturing during the second autumn; cup saucer-shaped with a turbinate base, 5″–7″ broad, its bracts ovate, obtuse, appressed; acorn subglobose or depressed, about twice as high as the cup.

In dry soil, Long Island(?), New Jersey to Florida, Indiana, Missouri and Texas. Wood hard, strong, reddish-brown; weight per cubic foot 43 lbs. May–June. Acorns ripe Sept.–Oct. Red or turkey-oak.

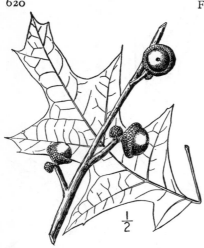

9. Quercus pagodaefòlia (Ell.) Ashe.
Elliott's Oak. Fig. 1522.

Q. falcata pagodaefolia Ell. Bot. S. C. & Ga. **2**: 605. 1824.
Q. pagodaefolia Ashe, Bot. Gaz. **24**: 375. 1897.

A tree, attaining a maximum height of about 100°, with a trunk diameter up to nearly 5°, the thick, close, scaly bark grayish-brown, the young twigs velvety-pubescent. Leaves ovate to oblong in outline, 6'–12' long, dark green and shining above, pale and persistently tomentose beneath, 5–11-lobed, the lobes and teeth bristle-tipped, the petiole 2' long or less; styles long, spreading; fruit maturing the second autumn, very short-stalked; cup shallowly top-shaped, its scales oblong, pubescent; acorn subglobose, about 5" long and twice as long as the cup.

Borders of swamps and streams, Massachusetts to Florida, Illinois, Missouri and Arkansas.

10. Quercus ilicifòlia Wang.   Bear or Scrub Oak.   Fig. 1523.

Quercus rubra nana Marsh. Arb. Am. 123. 1785.
Quercus ilicifolia Wang. Amer. 79. *pl. 6. f. 17.* 1787.
Quercus nana Sarg. Gard. & For. **8**: 93. 1895.

A shrub or rarely a small tree, often forming dense thickets, maximum height about 25°, and trunk diameter 6'; bark gray, nearly smooth. Leaves mostly obovate, 2'–5' long, short-petioled, dark green and glabrous above, grayish-white tomentulose beneath, 3–7-lobed, the lobes triangular-ovate, acute, bristle-tipped; styles recurved; fruit maturing the second autumn; cup saucer-shaped, 4"–6" broad, with a turbinate or rounded base; its bracts lanceolate, appressed; acorn globose-ovoid, more or less longer than the cup.

In sandy or rocky soil, Maine to Ohio, North Carolina and Kentucky. Wood hard, strong, light brown. May. Acorns ripe Oct.–Nov. Holly, bitter or barren oak. Bitter-bush or black scrub-oak. Dwarf black-oak.

A hybrid of this, presumably with *Q. coccinea,* was found by Dr. Robbins at Uxbridge, Mass.

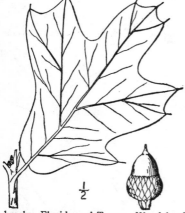

11. Quercus marilándica Muench.   Black-Jack or Barren Oak.   Fig. 1524.

Quercus nigra β L. Sp. Pl. 995. 1753.
Quercus marilandica Muench. Hausv. **5**: 253. 1770.

A tree, sometimes 60° high, usually lower; maximum trunk diameter 2°; bark nearly black, very rough in ridges. Leaves obovate in outline, stellate-pubescent above and brown-tomentose beneath when young, 3–5-lobed toward the broad usually nearly truncate apex, cuneate below, the lobes short, entire or sparingly toothed, bristle-tipped; mature leaves dark green, glabrous above, paler and more or less floccose beneath, 3'–7' long, 2'–5' wide; fruit maturing the second autumn; styles recurved; cup deep, 5"–8" broad, its bracts oblong-lanceolate, appressed, pubescent; acorn ovoid, 2–3 times as high as the cup.

In dry soil, Long Island, N. Y., to Minnesota, Nebraska, Florida and Texas. Wood hard, strong, dark brown; weight per cubic foot 46 lbs. May–June. Acorns ripe Uct.–Nov. Iron or jack-oak.

Quercus Rúdkini Britton, Bull. Torr. Club **9**: 14, a hybrid of this with *Q. Phellos,* occurs from Staten Island, N. Y., to North Carolina.

Quercus Bríttoni W. T. Davis, Scien. Am. **67**: 145, is a hybrid with *O. ilicifolia.* Staten Island, N. Y.

12. Quercus nìgra L.   Water or Black-Jack Oak.   Fig. 1525.

Quercus nigra L. Sp. Pl. 995. 1753.
Quercus nigra var. *aquatica* Lam. Encycl. 1: 721. 1783.
Quercus aquatica Walt. Fl. Car. 234. 1788.

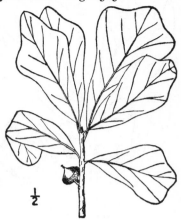

A tree, with maximum height of about 80° and trunk diameter of 4°; bark gray, rough in ridges. Leaves spatulate or obovate, 1–3-lobed at the apex, or some of them entire and rounded, coriaceous, short-petioled, rather bright green and shining on both sides, finely reticulate-veined, glabrous when mature except tufts of hairs in the axils of the veins beneath, 1½′–3′ long, the lobes low, usually obtuse and bristle-tipped; styles recurved; fruit maturing the second autumn; cup saucer-shaped with a rounded base, 5″–7″ broad, its bracts appressed; acorn globose-ovoid, 2–3 times as high as the cup.

Along streams and swamps or sometimes on the upland, Delaware to Kentucky, Missouri, Florida and Texas. Wood hard, strong, close-grained, light brown; weight per cubic foot 45 lbs. April–May. Acorns ripe Sept.–Oct. Leaves of seedlings and young shoots incised or pinnatifid, very bristly. Duck-, spotted-, barren-, punk- or possum-oak.

13. Quercus Phéllos L.   Willow Oak.
Peach or Sand-Jack Oak.   Fig. 1526.

Quercus Phellos L. Sp. Pl. 994. 1753.

A tree, with slightly roughened reddish-brown bark, attaining a maximum height of about 80° and a trunk diameter of 3°. Leaves narrowly oblong or oblong-lanceolate, entire, acute at both ends, very short-petioled, bristle-tipped, glabrous or very slightly pubescent in the axils of the veins beneath when mature, 2′–4′ long, 4″–12″ wide; styles slender, recurved-spreading; fruit maturing in the autumn of the second season; cup saucer-shaped, nearly flat on the base, 4″–6″ broad; acorn subglobose, 4″–6″ high.

In moist woods, Long Island, N. Y., to Florida, Kentucky, Missouri and Texas. Wood strong, rather soft and close-grained, reddish-brown; weight per cubic foot 46 lbs. April–May. Acorns ripe Sept.–Oct.

Quercus heterophýlla Michx. f. Hist. Am. 2: 87, *pl. 16*, the Bartram oak, a hybrid of *Q. Phellos* with *Q. rubra*, intermediate in leaf and fruit characters between the two, occurs from Staten Island to North Carolina. *Q. Phellos* hybridizes also with *Q. ilicifolia*.

14. Quercus laurifòlia Michx.   Laurel or Swamp Oak.   Fig. 1527.

Quercus laurifolia Michx. Hist. Chen. Am. no. 10. *pl. 17*. 1801.

Trunk sometimes 100° tall, reaching 4° in diameter at the base; bark nearly black, with flat ridges. Leaves oblong or oblong-obovate, often somewhat falcate, tardily deciduous, shining above, paler beneath, glabrous when mature, 1½′–6′ long, 5″–2′ wide, entire, or those of young shoots undulate-lobed, the apex bristle-tipped; styles rather short, recurving; fruit maturing in the autumn of the second season; abortive ovules in the summit of the acorn; cup saucer-shaped, 4″–6″ wide, its base somewhat rounded, its scales ovate, rounded, appressed; acorn ovoid or nearly hemispheric, about 3 times as long as the cup.

Along streams and swamps, southeastern Virginia to Florida and Louisiana, mostly near the coast. Closely related to the willow oak. Wood dark reddish-brown, strong; weight per cubic foot 48 lbs. Water-oak.

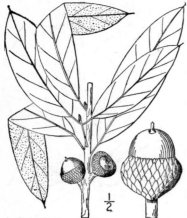

15. Quercus imbricària Michx. Shingle Oak. Fig. 1528.

Q. imbricaria Michx. Hist. Chen. Am. 9. *pl. 15, 16.* 1801.

A forest tree, with maximum height about 100°, and trunk diameter of 3½°. Leaves oblong or lanceolate, entire, coriaceous, acute at both ends, short-petioled, bristle-tipped, dark green above, persistently brown-tomentulose beneath, 3′-7′ long, 9′-2′ wide; styles recurved; fruit maturing the second autumn; cup hemispheric or turbinate, 5″-7″ broad, its bracts appressed; acorn subglobose, 5″-7″ high.

Central Pennsylvania to Michigan, Nebraska, Georgia, Tennessee and Arkansas. Reported from eastern Massachusetts. Wood hard, coarse-grained, light reddish-brown; weight per cubic foot 47 lbs. April–May. Lea-, Jack- or Laurel-oak.

Quercus Leàna Nutt. Sylva 1: 134, *pl. 5b,* is a hybrid of this and *Q. velutina,* with intermediate characters. Ohio to Missouri and District of Columbia.

Quercus tridentàta Engelm. *Q. nigra* var. *tridentata* A. DC. Prodr. 16: Part 2, 64, is a hybrid with *Q. marilandica.* Illinois and Pennsylvania. A hybrid with *Q. palustris* has been found near St. Louis, Mo., and in Iowa.

16. Quercus álba L. White Oak. Fig. 1529.

Quercus alba L. Sp. Pl. 996. 1753.

A forest tree, with light gray bark scaling off in thin plates; maximum height about 150°, trunk diameter up to 8°. Leaves obovate in outline, green above, pale and more or less glaucous beneath, pubescent when young, nearly glabrous when old, thin, pinnatifid into 3–9 oblong obtuse ascending toothed or entire lobes, 4′-7′ long, 2′-4½′ wide; petioles about ½′ long; styles short, erect; fruit maturing the first season, peduncled; cup depressed-hemispheric, 7″-10″ broad, its bracts thick, obtuse, woolly or at length glabrate, closely appressed; acorn ovoid-oblong, 1′ high or less, 3-4 times as high as the cup.

Maine to Ontario, Minnesota, Florida and Texas. Wood hard, strong, tough, close-grained; color brown; weight per cubic foot 46 lbs. May–June. Acorns ripe Sept.–Oct.

Hybrids with *Q. macrocarpa* have been observed in Illinois; with *Q. stellata,* from Illinois to Virginia and South Carolina, and with *Q. Prinus,* near Washington, D. C.. and New York. Stave-oak.

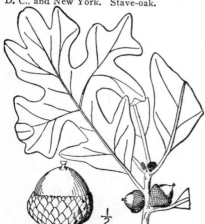

17. Quercus stellàta Wang. Post or Iron Oak. Fig. 1530.

Quercus alba minor Marsh. Arb. Am. 120. 1785.
Quercus stellata Wang. Amer. 78. *pl. 6. f. 15.* 1787.
Q. obtusiloba Michx. Hist. Chen. Am. 1. *pl. 1.* 1801.
Quercus minor Sargent, Gard. & For. 2: 471. 1889.

A tree, with rough gray bark, or sometimes a shrub; maximum height about 100° and trunk diameter 4°. Leaves broadly obovate in outline, deeply lyrate-pinnatifid into 3-7 broad rounded often deeply undulate or toothed lobes, when mature firm, glabrous, dark green and shining above, brown-tomentulose beneath, 5′-8′ long, 4′-6′ wide or smaller; petioles stout, ½′-1′ long; fruit maturing the first season, nearly or quite sessile; styles short; cup hemispheric, 6″-8″ broad, base narrowed, its bracts lanceolate, subacute, slightly squarrose; acorn ovoid, 6″-10″ high, 2-3 times as long as the cup.

In dry soil, Massachusetts to New York, Iowa, Florida and Texas. Wood hard, close-grained, very durable, brown; weight per cubic foot 52 lbs. May–June. Acorns ripe Sept.–Oct. Brash, white, rough or turkey-oak. Box or rough white-oak.

Quercus Margarétta Ashe, ranging from Virginia to Florida and Alabama, has similar but smaller acorns and leaves with rounded lobes; it is probably a race of this species, or a hybrid.

18. Quercus lyràta Walt. Overcup or Swamp Post Oak. Fig. 1531.

Quercus lyrata Walt. Fl. Car. 235. 1753.

A large tree, maximum height about 100° and trunk diameter 3½°; bark gray or reddish, in thin plates. Leaves obovate in outline, mostly narrowed at the base, 6'–8' long, lyrate-pinnatifid or lobed to beyond the middle, thin, when mature bright green, glabrous and shining above, densely white-tomentulose or becoming glabrate beneath, the lobes lanceolate or oblong, rounded or subacute, entire or toothed, the upper pair the larger and usually divergent; petioles 3"–9" long; fruit maturing the first season, peduncled; styles short; cup depressed-globose, 1'–1½' in diameter, ½'–1' high, its bracts broad, thin, cuspidate; acorn depressed-globose, ½'–1½' high, nearly or quite immersed in the cup.

In swamps or along streams, New Jersey to Indiana and Missouri, Florida and Texas. Wood hard, strong, tough, close-grained, very durable, dark brown; weight per cubic foot 52 lbs. April–May. Water white-oak.

19. Quercus macrocàrpa Michx. Mossy-cup, Blue or Bur Oak. Fig. 1532.

Q. macrocarpa Michx. Hist. Chen. Am. 2. *pl. 23.* 1801.
Q. olivaeformis Michx. f. Hist. Arb. Am. 2: *pl. 2.* 1812.

A large tree, with gray flaky bark; maximum height about 160°, and trunk diameter 8°. Leaves obovate or oblong-obovate in outline, rather thin, irregularly lobed, pinnatifid, or some coarsely crenate; when mature bright green and shining above, grayish-white-tomentulose beneath, 4'–8' long, the lobes toothed or entire, rounded, ascending or somewhat divergent; petioles ½'–1' long; fruit short-peduncled or sessile, maturing the first season; styles short; cup hemispheric or subglobose, 8"–2' in diameter, its bracts floccose, thick, hard, ovate or lanceolate, the lower acute, the upper subulate-tipped, the tips forming a fringe around the acorn; acorn 8"–1½' long, ovoid, 1–2 times as high as the cup.

In rich soil, Nova Scotia to Manitoba, Wyoming, Massachusetts, Georgia, Kansas and Texas. Wood hard, strong, tough, close-grained; color dark brown; weight 46 lbs. Mossy-cup white-oak. Scrub-oak. May–June.

20. Quercus bìcolor Willd. Swamp White Oak. Fig. 1533.

Q. Prinus platanoides Lam. Encycl. 1: 720. 1783.

Quercus bicolor Willd. Neue Schrift. Ges. Nat. Fr. Berlin, 3: 396. 1801.

Q. platanoides Sudw. Rep. Secy. Agric. 1892: 327. 1893.

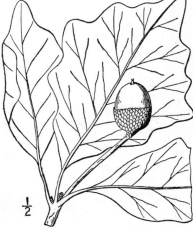

A large tree, with flaky gray bark; maximum height about 110° and trunk diameter 9°. Leaves obovate, or oblong-obovate, coarsely toothed or sometimes lobed nearly to the middle, narrowed or rounded at the base, firm, when mature 4'–7' long, 3½'–4½' wide, dark green, dull and glabrous above, densely white-tomentulose beneath; petioles stout, 3"–9" long; fruit maturing the first year; peduncles 2–5 times as long as the petioles; cup hemispheric, its bracts pubescent, lanceolate, appressed, the lower obtuse, the upper acute or acuminate; acorn oblong-ovoid, about 1' high; cup about 6" high; seed rather sweet.

In moist or swampy soil, Quebec to Minnesota, Georgia and Arkansas. Wood hard, strong, tough, close-grained, light brown; weight 48 lbs. per cubic foot. Swamp-oak. May–June. Acorns ripe Sept.–Oct.

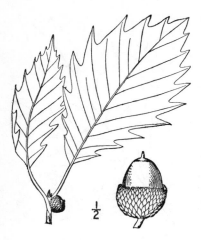

21. Quercus Michaùxii Nutt. Cow Oak. Basket Oak. Fig. 1534.

Quercus Michauxii Nutt. Gen. **2** : 215. 1818.

A large tree, with gray flaky bark; maximum height about 100° and trunk diameter 7°. Leaves obovate or broadly oblong, apex acute or acuminate, base narrowed, rounded or subcordate, when mature bright green, shining above, pale and gray tomentulose beneath, sharply toothed, 4′-7′ long, 2½′-4½′ wide, the teeth acute or mucronulate; petioles slender, ½′-1½′ long; fruit maturing the first season, short-peduncled or sessile; styles very short; cup depressed-hemispheric, 1′-1½′ broad, its bracts thick, ovate or lanceolate, appressed; acorns ovoid, 1′-1½′ high, about 3 times as high as the cup.

In moist soil, Delaware to Indiana, Missouri, Florida and Texas. Wood hard, strong, tough, dense, durable; color light brown; weight 50 lbs. per cubic foot. April–May. Acorns ripe Sept.–Oct., sweet and edible. Swamp chestnut-oak.

22. Quercus Prìnus L. Rock Chestnut Oak. Fig. 1535.

Quercus Prinus L. Sp. Pl. 996. 1753.

A large forest tree; maximum height about 100°, and trunk diameter 5°; lower branches spreading; bark brown, ridged, slightly flaky. Leaves coarsely crenate, oblong, oblong-lanceolate or obovate, when mature dark green, glabrous and feebly shining above, finely gray-tomentulose beneath, 5′-8′ long, 1½′-4′ wide; petioles slender, ½′-1½′ long; fruit maturing the first season; peduncles equalling or shorter than the petioles; styles very short; cup hemispheric, ½′-1½′ broad, its bracts tomentose, triangular-ovate, acute or cuspidate, appressed; acorn ovoid, 1′-1½ high, 2–3 times as high as the cup; seed edible, but not very sweet.

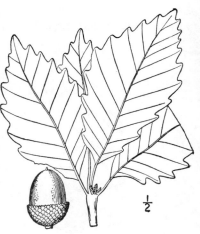

In dry soil, Maine to southern Ontario, Alabama and Tennessee. Wood hard, strong, close-grained, durable; color dark brown; weight per cubic foot 47 lbs. May–June. Acorns ripe Oct.–Nov. Swamp or white chestnut-oak. Rock-, tan-bark- or mountain-oak.

23. Quercus Muhlenbérgii Engelm. Chestnut or Yellow Oak. Fig. 1536.

Quercus Prinus acuminata Michx. Hist. Chen. Am. no. 5. *pl. 8.* 1801.

Quercus Muhlenbergii Engelm. Trans. St. Louis Acad. **3** : 391. 1877.

Quercus acuminata Sarg. Gard. & For. **8** : 93. 1895.

A tree with close or flaky bark, much resembling the chestnut; maximum height about 160°, and trunk diameter 3½°. Leaves oblong, lanceolate or obovate, apex acuminate or acute, base narrowed or rounded, coarsely toothed, when mature dark green and shining above, pale gray-tomentulose and prominently veined beneath, 4′-6′ long, 1′-2½′ wide; petioles slender, ½′-1′ long; fruit sessile or very short-peduncled, maturing the first season; cup hemispheric, 5″-8″ broad, its bracts floccose, ovate, acute or cuspidate, appressed; acorn ovoid, 6″-10″ high, about twice as high as the cup.

In dry soil, preferring limestone ridges, Vermont and Ontario to Minnesota, Nebraska, Alabama and Texas. Wood hard, strong, dense, close-grained, durable, dark brown; weight per cubic foot 54 lbs. May–June. Acorns ripe Oct.–Nov., edible. Pin-, shrub-, scrub-, chinkapin- or yellow chestnut-oak.

Quercus Alexánderi Britton, at first supposed to be separable from *Q. Muhlenbergii* by its obovate leaves and flaky bark; does not now appear to be specifically distinct from that species.

24. Quercus prinoìdes Willd.   Scrub or Dwarf Chestnut Oak.   Fig. 1537.

Quercus prinoides Willd. Neue Schrift. Ges. Nat. Fr. Berlin **3**: 397. 1801.
Q. prinoides rufescens Rehder, Rhodora **9**: 61. 1907.

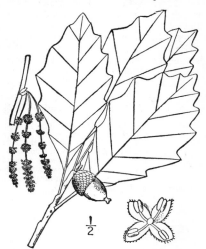

A shrub, $2°-15°$ tall, sometimes tree-like, the bark gray, the twigs glabrous or pubescent. Leaves obovate, coarsely toothed, when mature bright green and somewhat shining above, gray-tomentulose beneath, $2\frac{1}{2}'-5'$ long, $2'-3'$ wide, mostly acute or short-acuminate at the apex, narrowed at the base, the teeth short, triangular, subacute or obtuse; petioles slender, $3''-9''$ long; fruit sessile, maturing the first season; cup hemispheric, thin, about $\frac{1}{2}'$ broad and one-half as high, its bracts floccose, triangular-ovate or oblong-lanceolate, appressed; acorn ovoid, obtuse, 2–3 times as long as the cup; seed sweet.

In dry sandy or rocky soil, Maine to Minnesota, south to Alabama and Texas. April–May. Acorns ripe Sept.–Oct. Chinkapin or running white-oak.

25. Quercus virginiàna Mill.   Live Oak.   Fig. 1538.

Quercus virginiana Mill. Gard. Dict. Ed. 8, no. 16. 1768.
Quercus virens Ait. Hort. Kew. **3**: 356. 1789.

A tree, with rough brown bark, attaining a maximum height of about $60°$ and trunk diameter of $7°$, but often shrubby, the young shoots puberulent. Leaves evergreen, coriaceous, oblong, elliptic or oblanceolate, apex obtuse, base narrowed or rounded, entire or with a few bristle-tipped teeth, bright green and glabrous above, pale green and puberulent or becoming glabrous beneath, $1'-3'$ long; petioles stout, $1''-3''$ long; fruit peduncled, maturing the first season; peduncle $\frac{1}{4}'-1'$ long; cup turbinate, $5''-8''$ broad, its bracts closely appressed; ovate or lanceolate; acorn ovoid-oblong, about twice as high as the cup; seed not edible; cotyledons united.

In dry soil, Virginia to Florida, Texas and Mexico, mostly near the coast. Also in Cuba. Wood very hard; tough, close-grained and dense; color yellow-brown; weight per cubic foot 59 lbs. March–April. Acorns ripe Sept.–Oct.

Family 8. ULMÀCEAE Mirbel, Elém. **2**: 905. 1815.

ELM FAMILY.

Trees or shrubs, with alternate simple serrate petioled pinnately veined stipulate leaves, the stipules usually fugacious. Flowers small, monoecious, dioecious, perfect or polygamous, lateral or axillary, clustered, or the pistillate solitary. Perianth 3–9-parted or of 3–9 distinct sepals. Petals none. Stamens in our species as many as the perianth-lobes or sepals and opposite them; filaments straight; anthers ovate or oval, erect in the bud, longitudinally dehiscent. Ovary 1-celled (rarely 2-celled), mostly superior; ovule solitary, pendulous, anatropous or amphitropous; styles or stigmas 2. Fruit a samara, drupe or nut. Endosperm of the seed little or none. Embryo straight or curved; cotyledons mostly flat.

About 13 genera and 140 species, widely distributed in temperate and tropical regions.

Flowers borne in clusters on twigs of the preceding season; fruit a samara, or nut-like.
   Flowers mostly expanding before the leaves; calyx 4–9-cleft; fruit a samara.   1. *Ulmus.*
   Flowers expanding with the leaves; calyx 4–5-cleft; fruit nut-like.   2. *Planera.*
Flowers borne on twigs of the season, the pistillate mostly solitary; fruit a drupe.   3. *Celtis.*

1. ÚLMUS (Tourn.) L. Sp. Pl. 225. 1753.

Trees, with 2-ranked straight-veined inequilateral serrate leaves, with thin caducous stipules. Flowers perfect or polygamous, fascicled or racemose, greenish, mostly axillary on the twigs of the preceding season. Calyx campanulate, 4–9-lobed, persistent, its lobes imbricated. Filaments erect, slender, exserted. Ovary sessile or stalked, compressed, 1–2-celled. Styles 2, divergent, stigmatic along the inner margin, ovule 1 in each cavity of the ovary, suspended, anatropous. Fruit a 1-seeded flat orbicular or oval samara, its membranous wings continuous all around except at the apex, commonly as broad as or broader than the body. Embryo straight. [The ancient Latin name of the elm; Celtic, *elm*.]

About 16 species, natives of the northern hemisphere. Besides the following two others occur in the southern United States and one in Mexico. Type species: *Ulmus campestris* L.

Flowers appearing in the spring long before the leaves.
    Leaves smooth or slightly rough above; samara densely ciliate.
        None of the branches corky-winged; samara-faces glabrous.   1. *U. americana.*
        Some or all of the branches corky-winged; samara-faces pubescent.
            Leaves 2′–5′ long; flowers racemose; northern.   2. *U. Thomasi.*
            Leaves 1′–3′ long; flowers fascicled; southern.   3. *U. alata.*
    Leaves very rough above; samara not ciliate; twigs not corky-winged.   4. *U. fulva.*
Flowers appearing in the autumn.   5. *U. serotina.*

1. Ulmus americàna L. American, White or Water Elm. Fig. 1539.

Ulmus americana L. Sp. Pl. 226. 1753.

A large tree, with gray flaky bark, and glabrous or sparingly pubescent twigs and buds; maximum height about 120°, and trunk diameter 11°; the branches not corky-winged, terete. Leaves oval or obovate, apex abruptly acuminate, base obtuse or obtusish, and very inequilateral, sharply and usually doubly serrate, smooth or rough above, pubescent or becoming glabrous beneath, 2′–5′ long, 1½′–3′ wide; flowers fascicled; pedicels filiform, drooping, jointed; calyx 7–9-lobed, oblique, its lobes oblong, rounded; samara ovate-oval, reticulate-veined, 5″–6″ long, its faces glabrous, its margins densely ciliate; styles strongly incurved.

In moist soil, especially along streams, Newfoundland to Manitoba, Florida and Texas. Wood hard, strong, close-grained, compact, dark brown; weight per cubic foot 40.5 lbs. March–April. Samaras ripe in May. The species consists of many slightly differing races. Swamp- or rock-elm.

2. Ulmus Thómasi Sarg. Cork or Rock Elm. Fig. 1540.

Ulmus racemosa Thomas, Am. Journ. Sci. **19**: 170. 1831. Not Borckh.
Ulmus Thomasi Sargent, Silva **14**: 102. 1902.

A large tree, attaining a maximum height of about 100° and a trunk diameter of 4°, the young twigs puberulent; bud-scales ciliate and somewhat pubescent; branches, or some of them, winged by narrow plates of cork. Leaves similar to the preceding, but less sharply serrate, smooth above; flowers racemose; pedicels filiform, drooping, jointed; calyx-lobes oblong, rounded; samara oval, 6″–8″ long, its faces pubescent or puberulent, its margins densely ciliate; persistent styles strongly incurved and overlapping.

In rich soil, Quebec to Ontario, Minnesota, Kentucky, Missouri and Nebraska. Wood hard, strong, tough, compact; color light reddish-brown; weight per cubic foot 45 lbs. April. Called also Cliff-, hickory- or swamp-elm. Racemed or corky white-elm. Wahoo.

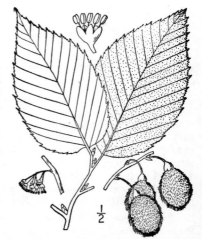

Ulmus campéstris L., from Europe, distinguished by its nearly sessile flowers, nearly or quite glabrous, not ciliate samaras, and wingless branches, rarely escapes from cultivation.

3. Ulmus alàta Michx. Winged Elm.
Wahoo. Fig. 1541.

Ulmus alata Michx. Fl. Bor. Am. 1: 173. 1803.

A small tree, sometimes 50° high and with a trunk diameter of 2½°; the branches, or most of them, with corky wing-like ridges. Twigs and buds glabrous or nearly so; leaves oblong, oblong-lanceolate or oblong-ovate, acute, doubly serrate, base obtuse, inequilateral and sometimes subcordate, roughish above, pubescent beneath, at least on the veins, 1′–3′ long, ½–1¼′ wide, the veins ascending, some of them commonly forked; flowers fascicled; pedicels filiform; calyx-lobes obovate, rounded; samara oblong, 4″–5″ long, pubescent on the faces, the margins densely ciliate; styles very slender.

In dry or moist soil, southern Virginia to Florida, west to southern Illinois, Missouri and Texas. Wood hard, weak, compact; color brown; weight per cubic foot 47 lbs. March. Water-, cork- or witch-elm.

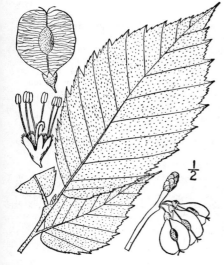

4. Ulmus fúlva Michx. Slippery, Red or
Moose Elm. Fig. 1542.

Ulmus pubescens Walt. Fl. Car. 111. 1788.?
Ulmus fulva Michx. Fl. Bor. Am. 1: 172. 1803.

A tree, with rough gray fragrant bark, maximum height about 70°, and trunk diameter 2½°; twigs rough-pubescent; branches not corky-winged; bud-scales densely brown-tomentose. Leaves ovate, oval or obovate, very rough with short papillae above, pubescent beneath, sharply doubly serrate, acuminate at the apex, obtuse, inequilateral and commonly cordate at the base, 4′–8′ long, 2′–2½′ wide; flowers fascicled; pedicels 2″–3″ long, spreading, jointed near the base; calyx-lobes lanceolate, subacute; samara oval-orbicular, 6″–9″ long, pubescent over the seed, otherwise glabrous, the margins not ciliate, retuse.

In woods, on hills and along streams, Quebec to North Dakota, Florida and Texas. Wood hard, strong, compact, durable; color dark reddish-brown; weight per cubic foot 43 lbs. Foliage and mucilaginous inner bark very fragrant in drying. March–April. Indian or sweet elm. Rock-elm.

5. Ulmus serótina Sargent. Red Elm.
Fig. 1543.

Ulmus serotina Sargent, Bot. Gaz. 27: 92. 1899.

A tree, reaching a maximum height of about 50° with a trunk diameter up to 3°, the bark thin, shallowly fissured, light brown, the young twigs smooth or nearly so. Leaves ovate to obovate, acuminate, 2′–4′ long, pubescent on the veins beneath, rather coarsely serrate, firm in texture but thin; flowers in short racemes in the axils of leaves of the season; calyx-lobes spatulate; samara oblong to elliptic, stalked, 5″–6″ long, fringed with long white hairs.

Banks and bluffs, Kentucky to Georgia and Alabama. Autumn.

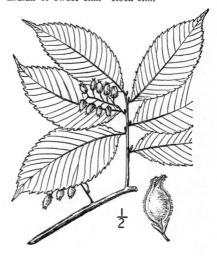

2. PLÁNERA J. F. Gmel. Syst. **2**: Part 1, 150. 1791.

A tree, similar to the elm, the flowers monoecious or polygamous, unfolding with the leaves. Staminate flowers fascicled on twigs of the preceding season, the pistillate or perfect ones in the axils of the leaves of the year. Calyx 4–5-cleft, campanulate, persistent, the lobes imbricated. Filaments filiform, straight, exserted. Ovary stalked, ovoid, slightly compressed, 1-celled. Styles 2, spreading, stigmatic along the inner side. Fruit nut-like; coriaceous, obliquely ovoid, compressed, ridged on the back, covered with short fleshy processes. Embryo straight. [Name in honor of Johann Jakob Planer, 1743–1789, Professor of Botany in Erfurt.]

A monotypic genus of southeastern North America.

1. Planera aquática (Walt.) J. F. Gmel. Planer-tree. Water Elm. Fig. 1544.

Anonymos aquatica Walt. Fl. Car. 230. 1788.
Planera aquatica J. F. Gmel. Syst. **2**: Part 1, 150. 1791.

A small tree, sometimes 40° high, and with a trunk 2° in diameter, the foliage nearly glabrous. Leaves ovate or oblong-lanceolate, acute at the apex, obtuse or cordate and usually somewhat inequilateral at the base, serrate, 1′–2′ long; petioles 1½″–2″ long; stipules lanceolate, about as long as the petioles, deciduous; staminate flowers fascicled and somewhat racemose from scaly buds borne at the axils of leaves of the preceding season; perfect or pistillate flowers on short branches; fruit 2″–3″ long, about equalling its stalk, its soft processes ½″ long.

In swamps, Missouri to southern Indiana, Kentucky and North Carolina, south to Texas and Florida. Wood soft, weak, compact, light brown; weight per cubic foot 33 lbs. April–May. Sycamore. (N. C.)

3. CÉLTIS (Tourn.) L. Sp. Pl. 1043. 1753.

Trees or shrubs, with serrate or entire pinnately veined or in some species 3–5-nerved leaves, and polygamous or monoecious (rarely dioecious?) flowers, borne in the axils of leaves of the season, the staminate clustered, the fertile solitary or 2–3 together. Calyx 4–6-parted or of distinct sepals. Filaments erect, exserted. Ovary sessile. Stigmas 2, recurved or divergent, tomentose or plumose. Fruit an ovoid to globose drupe, the exocarp pulpy, the endocarp bony Seed-coat membranous. Embryo curved. [Name ancient, used by Pliny for an African Lotus-tree.]

About 60 species, of temperate and tropical regions. Besides the following, some three others occur in southern and western North America. Type species: *Celtis australis* L.

1. Celtis occidentàlis L. Hackberry. Sugar-berry. Fig. 1545.

Celtis occidentalis L. Sp. Pl. 1044. 1753.
Celtis pumila Pursh, Fl. Am. Sept. 200. 1814.

A tree or shrub, attaining a maximum height of about 90° and a trunk diameter of 3°, the bark dark, rough, often corky. Leaves ovate or ovate-lanceolate, sharply serrate, mostly thin, acute or acuminate at the apex, inequilateral, 1½′–4′ long, 1′–2½′ wide, smooth and glabrous above, pubescent, at least on the veins, beneath; staminate flowers numerous; pistillate flowers usually solitary, slender-peduncled; calyx-segments linear-oblong, deciduous; drupes globose to globose-oblong, purple, or nearly black when mature, or orange, 4″–5″ in diameter, sometimes edible, on stalks usually twice their length or longer.

In dry, often rocky, soil, Quebec to Manitoba, North Carolina, Missouri and Oklahoma. Wood soft, weak, coarse-grained; color light yellow; weight per cubic foot 40 lbs. April–May. Fruit ripe Sept. Nettle-tree.   False or bastard-elm.   Beaver-wood. Juniper-tree.   One-berry.   Rim- or hoop-ash.

Celtis canìna Raf., differing by relatively longer, narrower and usually longer tipped leaves, and growing in rich soil, within the range of *C. occidentalis,* may be a race of that species.

2. Celtis crassifòlia Lam. Rough-leaved Hackberry. Fig. 1546.

Celtis crassifolia Lam. Encycl. 4: 138. 1797.

A large tree, sometimes 125° high and with a trunk diameter of 3° or more, with rough, usually corky-thickened bark, the young twigs pubescent. Leaves ovate to ovate-lanceolate, acute or short-acuminate, rather coarsely toothed, 3′–6′ long, scabrous above, rough-pubescent, especially on the veins, beneath, the petioles rather short; drupes short-oblong or nearly globular, about 5″ in diameter, on stalks 8″–12″ long.

In rich soil, especially in river valleys, Massachusetts(?), New Jersey to Indiana, South Dakota, South Carolina and Colorado. April–May. Fruit ripe Aug.–Sept. Probably not specifically distinct from the preceding species.

3. Celtis mississippiénsis Bosc. Southern Hackberry. Fig. 1547.

Celtis mississippiensis Bosc, Dict. Agric. 10: 41. 1810.

A tree, reaching a maximum height of about 100°, the trunk up to 3° in diameter, the bark light gray, rough and warty. Leaves ovate or lanceolate, firm, shining, entire or with a few low sharp teeth, 3-nerved and prominently pinnately veined, glabrous on both sides, long-acuminate at the apex, inequilateral and obtuse or sometimes cordate at the base, 1′–3′ long, ½′–1½′ wide; peduncles mostly shorter than those of the preceding species; drupes globose, purple-black, 2½″–3″ in diameter.

In dry soil, Virginia to southern Illinois and Missouri, south to Florida and Texas. Bermuda. April. Fruit ripe July–Aug.

Celtis Smàllii Beadle, of the southern United States, differs by sharply serrate leaves and ranges north to western Kentucky.

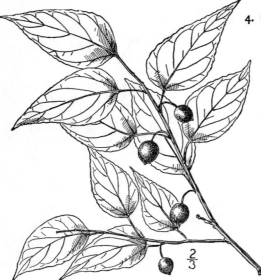

4. Celtis georgiàna Small. Georgia Hackberry. Fig. 1548.

C. georgiana Small, Bull. Torr. Club **24**: 439. 1897.

A shrub, or small tree, up to 30° high, the young twigs slender, pubescent becoming purple-brown and glabrous. Leaves ovate, small, rarely over 2′ long, firm when mature, acute or bluntish, serrate, or sometimes nearly or quite entire, the upper surface bright green, scabrous, the under side pubescent, at least on the veins; drupe globose, 3″–4″ in diameter, red-purple to yellowish, borne on short peduncles 2″–4″ long.

Rocky or gravelly soil, New Jersey to Missouri, Florida and Alabama. April–May.

5. Celtis reticulàta Torr. Thick-leaved Hackberry. Fig. 1549.

C. reticulata Torr. Ann. Lyc. N. Y. **1**: 247. 1824.

A small tree, up to 45° high, the bark rough with corky warts and ridges sometimes 1′ high or more, the young twigs pubescent or nearly glabrous, green, becoming brown. Leaves thick, ovate, 3′ long or less, strongly reticulate-veined, acute, scabrous or nearly smooth above, pubescent beneath, entire or serrate, the stout petioles 2″–5″ long; drupe globular, red, 4″–6″ in diameter, on peduncles usually longer than the petioles.

Along rivers in rocky or gravelly soil, Kansas to Texas, Colorado, Nevada and southern California. Reported to extend into Lower California. May.

Family 9. MORÀCEAE Lindl. Veg. Kingd. 266. 1847.

MULBERRY FAMILY.

Trees or shrubs with milky sap, alternate petioled stipulate leaves, the stipules fugacious, and small monoecious or dioecious axillary clustered flowers, or the pistillate flowers solitary in some exotic genera. Calyx mostly 4–5-parted, becoming fleshy in fruit, inferior. Petals none. Staminate flowers, spicate or capitate, the stamens as many as the calyx-segments. Filaments inflexed in the bud. Pistillate flowers capitate or spicate. Ovary superior, 1-celled in our genera. Ovule solitary, pendulous, anatropous. Styles 1 or 2. Fruit mostly aggregate. Embryo curved or spiral.

About 55 genera and 1,000 species, natives of temperate and tropical regions. The largest genus is *Ficus,* the Fig, of which there are over 600 known species.

Staminate and pistillate flowers spiked ; leaves dentate or lobed.	1. *Morus.*
Staminate flowers racemose or spiked ; pistillate capitate.	
Pistillate perianth deeply 4-cleft ; leaves entire.	2. *Toxylon.*
Pistillate perianth 3–4-toothed ; leaves various.	3. *Papyrius.*

1. MÒRUS (Tourn.) L. Sp. Pl. 986. 1753.

Trees or shrubs, with milky sap, alternate dentate and often lobed, 3-nerved leaves, fugacious stipules, and small monoecious or dioecious flowers, in axillary ament-like spikes, the pistillate spikes ripening into a succulent aggregate fruit. Staminate flowers with a 4-parted perianth, its ovate segments somewhat imbricated, and 4 stamens, the filaments inflexed in the bud, straightening and exserted in anthesis. Pistillate flowers with a 4-parted persistent perianth, which becomes fleshy in fruit, a sessile ovary, and 2 linear spreading stigmas. Fruiting perianth enclosing the ripened ovary (achene) the exocarp succulent, the endocarp crustaceous. Endosperm scanty; embryo curved. [The ancient name of the mulberry; Celtic *mor*.]

About 10 species, natives of the northern hemisphere. Besides the following, two others occur in the southern United States. Type species: *Morus nigra* L.

Leaves rough above, pubescent beneath ; fruit purple; spikes 1'–1½' long.   1. *M. rubra*.
Leaves smooth and glabrous, or very nearly so, on both sides; fruit nearly white; spikes 5"–7" long.
                                                                                  2. *M. alba*.

1. Morus rùbra L. Red Mulberry.
Fig. 1550.

Morus rubra L. Sp. Pl. 986. 1753.

A tree, attaining a maximum height of about 65° and a trunk diameter of 7°, the bark brown and rough. Leaves ovate or nearly orbicular in outline, scabrous above, persistently pubescent beneath, or when young almost tomentose, acuminate at the apex, rounded, truncate or cordate at the base, serrate-dentate or 3–7-lobed, 3'–5' long; petioles slender, 7"–18" long; staminate spikes drooping, 1½'–3' long; pistillate spikes spreading or pendulous in fruit, 1'–1½' long, 4"–5" in diameter when mature, slender-peduncled, dark purple-red, delicious.

In rich soil, Vermont and Ontario to Michigan and South Dakota, south to Florida and Texas. Wood soft, weak, compact, durable; color light yellow; weight per cubic foot 37 lbs. April–May. Fruit ripe in June–July.

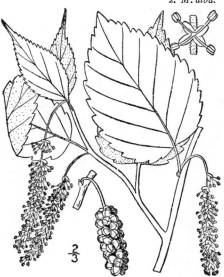

2. Morus álba L. White Mulberry.
Fig. 1551.

Morus alba L. Sp. Pl. 986. 1753.

A small tree, sometimes 40° high and with a trunk 3° in diameter, the bark light gray, rough, the branches spreading. Leaves ovate, thin, smooth, glabrous and somewhat shining on both sides, acute or abruptly acuminate at the apex, rounded, truncate or cordate at the base, varying from serrate to variously lobed, 2'–6' long; petioles slender, shorter than the blades; staminate spikes slender, drooping, about 1' long; pistillate spikes oblong or subglobose, drooping, 5"–7" long, 3" in diameter and white or pinkish. when mature, not as succulent as those of the preceding species.

Sparingly escaped from cultivation, Maine and Ontario to Florida. Introduced from the Old World for feeding silkworms. May. Fruit ripe July–Aug.

Morus nìgra L., the black-mulberry of Europe, with smooth leaves and black fruit, has escaped from cultivation in the southeastern states and has been recorded from as far north as New York.

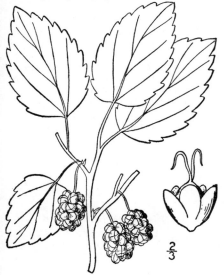

2. TÓXYLON Raf. Am. Month. Mag. 2: 118. 1817.
[Maclura Nutt. Gen. 2: 233. 1818.]

A tree, with milky sap, thick entire dark green alternate petioled pinnately veined leaves, stout axillary spines, caducous stipules and dioecious axillary flowers, the staminate racemose, the pistillate capitate. Staminate flowers with a 4-parted calyx, its segments valvate, and 4 stamens, the filaments inflexed in the bud, straightening and somewhat exserted in anthesis. Pistillate flowers with a 4-cleft calyx enclosing the sessile ovary, and a filiform simple long-exserted style, the calyces becoming fleshy and enlarged in fruit, densely aggregated into a large globular head. Endosperm none; embryo curved. [Name Greek, signifying bow-wood.]

A monotypic genus of the south-central United States; its name originally printed, by typographical error, *Ioxylon*.

1. Toxylon pomíferum Raf. Osage or Wild Orange. Fig. 1552.

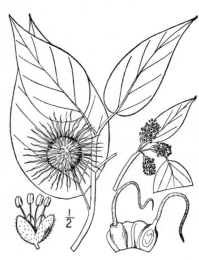

T. pomiferum Raf. Am. Month. Mag. **2** : 118. 1817.
Maclura aurantiaca Nutt. Gen. **2** : 234. 1818.
Maclura pomifera Schneider, Handb. Laubh. **1** : 806. 1906.

A tree with ridged brown bark, and spreading branches; maximum height about 60°, and trunk diameter 2½°; foliage puberulent when young, glabrous when mature. Leaves ovate, ovate-lanceolate or ovate-oblong, glossy, entire, 3′–6′ long, apex acuminate, base obtuse, truncate or subcordate; petioles ½′–2′ long; axillary spines straight, sometimes 3′ long; staminate racemes ½′–1′ long, usually numerous; flowers about 1″ broad; head of pistillate flowers peduncled, pendulous, about 1′ in diameter, ripening into a hard yellowish tubercled syncarp 2′–6′ in diameter.

In rich soil, Missouri and Kansas to Texas. Wood hard, very strong, dense, durable; color bright orange; weight per cubic foot 48 lbs. Much planted for hedges and occasionally spontaneous in the East. May–June. Fruit ripe Oct.–Nov. Bow- or yellow-wood. Hedgeplant. Osage. Osage-apple.

3. PAPÝRIUS Lam. Encycl. **3** : 382. 1797. Tabl. Encycl. *pl. 762.* 1798.

[Broussonétia L'Her; Vent. Tabl. **3** : 547. 1799. Not Ortega, 1798.]

Trees, with milky sap, the leaves alternate, petioled, entire, serrate, or 3–5-lobed, 3-nerved at the base. Flowers dioecious, the staminate in cylindric ament-like spikes, the pistillate capitate. Staminate flowers with a deeply 4-cleft perianth, 4 stamens, and a minute rudimentary ovary. Pistillate flowers with an ovoid or tubular 3–4-toothed perianth, a stalked ovary and a 2-cleft style. Head of fruit globular, the drupes red, exserted beyond the persistent perianth. [Name in allusion to the use of the bark in paper-making.]

About 4 species, natives of eastern Asia, the following being the type.

1. Papyrius papyrífera (L.) Kuntze. Paper Mulberry. Fig. 1553.

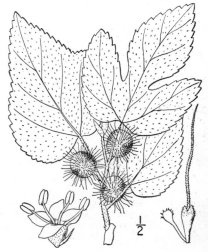

Morus papyrifera L. Sp. Pl. 986. 1753.
Broussonetia papyrifera Vent. Tabl. **3** : 548. 1799.
P. papyrifera Kuntze, Rev. Gen. Pl. 629. 1891.

A small tree, sometimes 40° high, the young shoots hirsute-tomentose. Leaves mostly ovate, thin, long-petioled, serrate nearly all around, often deeply 3-lobed, sometimes with a lobe on one side only, as in *Sassafras,* rarely 5-lobed, rough above, tomentose beneath, 3′–8′ long, the sinuses rounded; petioles ½′–3′ long, hirsute-tomentose, at least when young; spikes of staminate flowers 2′–3′ long; peduncled; heads of pistillate flowers ½′–1′ in diameter, stout-peduncled.

Escaped from cultivation, southern New York to Georgia and Missouri. May–June. Otaheite Mulberry. Cut-paper.

Ficus Càrica L., the Fig, a shrub with deeply lobed leaves and hollow pear-shaped receptacles lined with minute imperfect flowers, is occasionally spontaneous or persistent after cultivation from Virginia and West Virginia to Florida and Texas.

Family 10. **CANNABINÀCEAE** Lindl. Veg. Kingd. 265. 1847.

HEMP FAMILY.

Annual or perennial herbs, the stems erect or twining. Leaves opposite or sometimes alternate, toothed, lobed, or divided, petioled, the stipules persistent. Flowers dioecious; staminate flowers in panicled racemes, usually 5-parted; pistillate flowers in bracted spikes, the perianth entire, the ovary 1-celled, the styles or stigmas 2, the ovule pendulous. Fruit an achene, with crustaceous pericarp. Endosperm fleshy; embryo curved or coiled.

Twining vines; pistillate flowers in ament-like clusters.   1. *Humulus.*
Erect, tall herbs; pistillate flowers spicate.   2. *Cannabis.*

1. **HÙMULUS** L. Sp. Pl. 1028. 1753.

Twining herbaceous perennial rough vines, with broad opposite thin petioled palmately veined serrate 3–7-lobed or undivided leaves, lanceolate membranous persistent stipules, and dioecious axillary flowers, the staminate panicled, the pistillate in ament-like drooping clustered spikes. Staminate flowers with a 5-parted calyx, the segments distinct and imbricated, and 5 short erect stamens. Pistillate flowers in 2's in the axil of each bract of the ament, consisting of a membranous entire perianth, clasping the ovary, and 2 filiform caducous stigmas. Fruiting aments cone-like, the persistent bracts subtending the compressed ovate achenes. Embryo spirally coiled. [Name said to be the diminutive of the Latin *humus,* earth.]

Two species, widely distributed through the north temperate zone. Type species: *Humulus Lupulus* L.

1. Humulus Lùpulus L. Hop. Fig. 1554.

Humulus Lupulus L. Sp. Pl. 1028. 1753.

A dextrorsely twining or prostrate vine, often 25° long, very rough with stiff reflexed hairs. Leaves orbicular or ovate in outline, slender-petioled, deeply 3–7-cleft or some of the upper ones ovate, acute and merely serrate; petioles ½–3′ long; stipules reflexed, ovate or lanceolate, acuminate, 4″–12″ long; panicles of staminate flowers 2′–5′ long; ripe pistillate clusters (hops) 1′–2½′ long; fruiting bracts broadly ovate, concave, thin, glabrous or nearly so, obtuse, much longer than the achenes; fruiting calyx and achene strongly resinous-aromatic.

In thickets and on river banks, Nova Scotia to Manitoba, south to southern New York, Pennsylvania, Georgia, Kansas, Arizona and New Mexico. Extensively escaped from cultivation. Native also of Europe and Asia. July–Aug. Fruit ripe Sept.–Oct.

Humulus japónicus Sieb. & Zucc., the Japanese hop, with deltoid acuminate, not resinous, bracts, has escaped from cultivation from Connecticut to New Jersey.

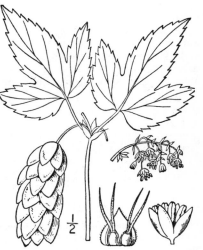

2. **CÀNNABIS** (Tourn.) L. Sp. Pl. 1027. 1753.

A stout erect rough and puberulent herb, with alternate and opposite petioled digitately 5–11-divided thin leaves, persistent subulate stipules, and greenish dioecious axillary flowers, the staminate panicled, the pistillate spicate. Staminate flowers with a 5-parted calyx, the sepals distinct and imbricated, and 5 short stamens. Pistillate flowers solitary in the axils of foliaceous bracts, consisting of a thin entire calyx clasping the sessile ovary, and 2 filiform caducous stigmas. Fruit a compressed achene. Embryo curved. [The classic name of hemp.]

A monotypic genus of central Asia.

1. **Cannabis satìva** L. Hemp. Red-root. Fig. 1555.

Cannabis sativa L. Sp. Pl. 1027. 1753.

An annual branching herb, 3°–10° tall, the inner fibrous bark very tough, the branches nearly erect. Leaves divided to the base, the segments lanceolate or linear-lanceolate, acuminate at both ends, sharply and coarsely serrate, 3′–6′ long, ¼′–1′ wide; staminate panicles narrow, loose, peduncled, 3′–5′ long; pedicels filiform, bracteolate, 1″–3″ long; pistillate spikes erect, leafy-bracted, 1′ long or less in fruit; achene crustaceous, ovoid-oblong, about 2″ high.

In waste places, New Brunswick to Ontario and Minnesota, south to North Carolina, Tennessee and Kansas. Widely distributed in all temperate regions through cultivation, and occasionally a troublesome weed. Native of Europe and Asia. July–Sept. Gallow-grass. Neckweed.

Family 11. **URTICÀCEAE** Reichenb. Consp. 83. 1828.

NETTLE FAMILY.

Herbs (some tropical species shrubs or trees), with watery sap, alternate or opposite mostly stipulate simple leaves, and small greenish dioecious, monoecious or polygamous flowers, variously clustered. Calyx 2–5-cleft, or of distinct sepals. Petals none. Stamens in the staminate flowers as many as the lobes or segments of the calyx (sepals) and opposite them, the filaments inflexed and anthers reversed in the bud, straightening at anthesis. Ovary superior, 1-celled; style simple; stigma capitate and penicillate, or filiform; ovule solitary, erect or ascending, orthotropous, or in some genera partly amphitropous. Fruit an achene. Endosperm oily, usually not copious; embryo straight.

About 40 genera and 550 species of wide geographic distribution.
Herbs with stinging hairs.

Leaves opposite; both kinds of flowers 4-parted; achene straight.   1. *Urtica.*
Leaves alternate; staminate flowers 5-parted; achene oblique.   2. *Urticastrum.*
Herbs without stinging hairs.
    Flower-clusters panicled or spiked, not involucrate; leaves mostly opposite.
        Pistillate calyx 3-parted or of 3 sepals.   3. *Pilea.*
        Pistillate calyx 2-4-toothed or entire.   4. *Boehmeria.*
    Flower-clusters involucrate by leafy bracts; leaves alternate.   5. *Parietaria.*

1. **URTÌCA** (Tourn.) L. Sp. Pl. 983. 1753.

Annual or perennial simple or branching herbs, with stinging hairs, opposite 3–7-nerved petioled dentate or incised leaves, and distinct or connate stipules. Flowers very small and numerous, axillary, cymose-paniculate, or glomerate, dioecious, monoecious or androgynous. Staminate flowers with a deeply 4-parted calyx and 4 stamens. Pistillate calyx 4-parted, the segments unequal, the exterior ones usually smaller than the inner; ovary straight; stigma sessile or nearly so; ovule erect, orthotropous. Achene compressed, ovate or oblong, enclosed by the persistent membranous or slightly fleshy calyx. Seed-coat thin; endosperm little; cotyledons broad. [The ancient Latin name.]

About 30 species of wide distribution. Type species: *Urtica dioica* L.

Perennials, 2°–7° tall; flower-clusters large, compound.
    Leaves ovate, cordate at base, pubescent.   1. *U. dioica.*
    Leaves lanceolate to ovate-lanceolate, rounded to cordate at base, nearly glabrous.
                                                                                   2. *U. gracilis.*
Annuals, 6′–3° tall; flower-clusters small, mostly glomerate.
    Leaves oval to ovate, laciniate-dentate; plant leafy to the top.   3. *U. urens.*
    Leaves ovate or lanceolate, or the lower orbicular, crenate; upper leaves small.
                                                                                   4. *U. chamaedryoides.*

1. Urtica dioìca L. Stinging or Great Nettle. Fig. 1556.

Urtica dioica L. Sp. Pl. 984. 1753.

Perennial, densely beset with stinging hairs, stem rather stout, 2°–4° tall, puberulent above. Leaves thin, ovate, long-petioled, acute or acuminate at the apex, cordate at the base, sharply or incisely serrate with triangular or lanceolate acute teeth, pubescent beneath, 3–5-nerved, 3′–5′ long, 1′–3′ wide; petioles slender, much shorter than the blades; stipules lanceolate; flower-clusters large, compound, cymose-paniculate; flowers dioecious or androgynous.

In waste places, Newfoundland to Ontario, Minnesota, South Carolina, Missouri and Colorado. Naturalized from Europe. Native also of Asia. Plant lower, stouter and much more stinging than the following species. July–Sept.

2. Urtica grácilis Ait. Slender or Tall Wild Nettle. Fig. 1557.

Urtica gracilis Ait. Hort. Kew. 3: 341. 1789.

Perennial, sparingly armed with stinging hairs, stem usually slender, erect, simple or with few erect branches, 2°–7° tall. Leaves lanceolate or ovate-lanceolate, slender-petioled, long-acuminate at the apex, narrowed or sometimes rounded at the base (rarely subcordate), sharply serrate, 3–5-nerved, sparingly pubescent, 3′–6′ long, ½′–1½′ wide; petioles shorter than the blades, usually bristly; stipules lanceolate; flower-clusters compound, smaller than those of the preceding species, but commonly longer than the petioles; flowers dioecious or androgynous.

In dry soil, Newfoundland to British Columbia, south to North Carolina, Louisiana and Colorado. Races with ovate-lanceolate, coarsely serrate, cordate leaves have been referred to *Urtica Lyallii* S. Wats. June–Oct.

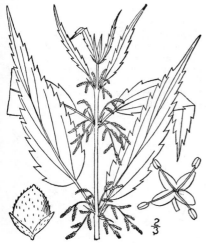

3. Urtica ùrens L. Small or Dwarf Nettle. Fig. 1558.

Urtica urens L. Sp. Pl. 984. 1753.

Annual, stem rather stout, 6′–18′ high, ascending or erect, it and its slender branches stinging-bristly. Leaves thin, glabrous or very nearly so, elliptic, oval or ovate in outline, deeply incised or sometimes doubly serrate, with acute, ascending or spreading teeth, 3–5-nerved, obtuse at both ends, or acutish, 1′–3′ long, slender-petioled; petioles often as long as the blades; stipules short; flower-clusters oblong, rather dense, mostly shorter than the petioles; flowers androgynous.

In waste places, Newfoundland to Manitoba, northern New York, New Jersey and Florida. Also on the Pacific Coast and in Bermuda. Naturalized from Europe. May–Sept. Burning or stinging nettle.

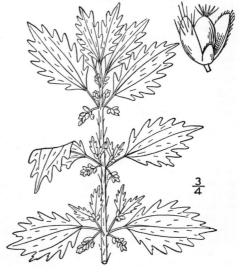

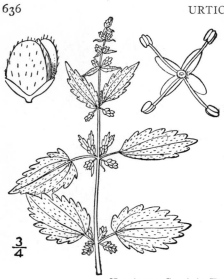

4. Urtica chamaedryoìdes Pursh.
Weak Nettle. Fig. 1559.

Urtica chamaedryoides Pursh, Fl. Am. Sept. 113. 1814.

Annual, sparingly stinging-bristly but otherwise nearly or quite glabrous, stem very slender, weak, ascending, simple or branched, 6'–3° long. Leaves slender-petioled, thin, crenate-dentate, the lower broadly ovate or orbicular, obtuse at the apex and usually cordate at the base, $\frac{1}{2}'–1\frac{1}{2}'$ wide, the upper ovate or lanceolate, acute or acuminate at the apex, rounded or narrowed at the base, the uppermost very small; stipules lanceolate-subulate; flower-clusters small, glomerate, shorter than the petioles; flowers androgynous.

In thickets, Kentucky to Arkansas, south to Georgia and Texas. April–Aug.

2. URTICÁSTRUM Fabr. Enum.
204.   1759.

[LAPÒRTEA Gaud. in Freyc. Voy. Bot. 498. 1826.]

Perennial herbs, armed with stinging hairs, the leaves broad, alternate, serrate, petioled, the flowers monoecious or dioecious, sessile in loose axillary compound cymes. Staminate flowers in our species with 5 imbricated sepals, 5 stamens and a rudimentary ovary. Pistillate flowers with 4 unequal sepals, the outer 1 or 2 minute, an oblique or nearly straight compressed ovary and a subulate slender persistent style; ovule erect. Achene very oblique, flat, reflexed. Seed-coat membranous. Endosperm scanty or wanting. [Latin, star nettle.]

About 25 species, mostly of tropical distribution, only the following, the generic type, North American.

1. Urticastrum divaricàtum (L.) Kuntze.
Wood or Canada Nettle. Fig. 1560.

Urtica divaricata L. Sp. Pl. 985. 1753.
Urtica canadensis L. Sp. Pl. 985. 1753.
Laportea canadensis Gaud. in Freyc. Voy. Bot. 498. 1826.
U. divaricatum Kuntze, Rev. Gen. Pl. 635. 1891.

Stem rather stout, erect or ascending, $1\frac{1}{2}°–4°$ tall. Leaves thin, ovate, long-petioled, acuminate or acute at the apex, sharply serrate, 3-nerved and pinnately veined, glabrous or with some stinging hairs, 3'–7' long, 2'–5' wide; petioles very slender, $1\frac{1}{2}'–5'$ long; stipule solitary, small, lanceolate, 2-cleft, commonly deciduous; flower-clusters large and loose, often longer than the petioles, the lower staminate, the upper pistillate, divergent, 2'–6' broad in fruit; ultimate branches of the fruiting clusters flat, cuneate, emarginate; achene twice as long as the calyx, glabrous, $1\frac{1}{2}''$ long.

In rich woods, Nova Scotia to Ontario and North Dakota, south to Florida and Kansas. Ascends to 3000 ft. in the Adirondacks. July–Aug. Albany-hemp.

3. PÌLEA Lindl. Coll. *pl. 4.* 1821.

[ADÍCEA Raf. Ann. Nat. 179. Hyponym. 1815.]

Annual or perennial, glabrous or pubescent stingless herbs, with opposite petioled mostly 3-nerved leaves, connate stipules, and small numerous monoecious or dioecious flowers in axillary cymose or glomerate clusters. Staminate flowers mostly 4-parted (sometimes 2- or 3-parted) and with a rudimentary ovary. Pistillate flowers 3-parted, the segments in most species unequal, each subtending a staminodium in the form of a concave scale; ovary straight; stigma sessile, penicillate. Achene compressed, ovate or suborbicular. Seed-coat thin. Endosperm scanty or none. [Name referring to the larger sepal of the type species which is something like a cap.]

About 200 species, chiefly in the tropics, most abundant in tropical America. Besides the following, another occurs in the southern United States. Type species: *Pilea muscòsa* Lindl.

1. Pilea pùmila (L.) A. Gray. Clearweed. Richweed. Coolweed. Fig. 1561.

Urtica pumila L. Sp. Pl. 984. 1753.

Adicea pumila Raf.; Torr. Fl. N. Y. **2**: 223. As synonym. 1843.

Pilea pumila A. Gray, Man. 437. 1848.

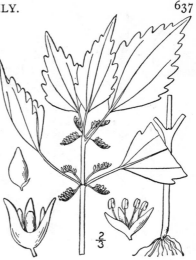

Annual, stems pellucid, erect, usually branched, glabrous, succulent, 6′–2° high. Leaves membranous, ovate, slender-petioled, acuminate or acute at the apex, rounded or narrowed at the base, 3-nerved, coarsely dentate, 1′–5′ long, sparingly pubescent with scattered hairs; petioles often as long as the blades and much longer than the pistillate flower-clusters; sepals of the pistillate flowers lanceolate, nearly equal; achene ovate, acute, ½″ long.

In swampy, shaded situations, often on old logs, New Brunswick to western Ontario and Minnesota, Florida, Louisiana, Nebraska and Kansas. Ascends to 3000 ft. in Virginia. Also in Japan. July–Sept. Stingless-nettle.

4. BOEHMÈRIA Jacq. Stirp. Am. 246. *pl. 157.* 1763.

Perennial stingless herbs (some tropical species shrubs or even trees), with opposite or alternate petioled 3-nerved leaves, distinct or connate stipules, and small monoecious or dioecious flowers, glomerate in axillary spikes or heads, the fertile clusters sometimes leafy at the summit. Staminate flowers mostly 4-parted or the calyx of 4 distinct sepals, usually with a rudimentary ovary. Pistillate calyx tubular or urn-shaped, 2–4-toothed or entire, enclosing the sessile or stalked ovary; stigma subulate, papillose or pubescent along one side. Achene enclosed by the withering-persistent pistillate calyx. [In honor of Georg Rudolph Boehmer, 1723–1803, Professor in Wittenberg.]

About 50 species, mostly natives of tropical regions, the following of eastern North America. Type species: *Boehmeria ramiflora* Jacq.

1. Boehmeria cylíndrica (L.) Sw. False Nettle. Fig. 1562.

Urtica cylindrica L. Sp. Pl. 984. 1753.
Urtica capitata L. Sp. Pl. 985. 1753.
Boehmeria cylindrica Sw. Prodr. 34. 1788.
Boehmeria cylindrica scabra Porter, Bull. Torr. Club **16**: 21. 1889.
Boehmeria scabra Small, Fl. SE. U. S. 358. 1903.

A perennial rough pubescent or nearly smooth and glabrous erect branching herb, 1°–3° tall. Stem stiff; leaves ovate, ovate-oblong or ovate-lanceolate, thin, petioled, opposite, or some alternate, coarsely dentate, 1′–3′ long, ½–1½′ wide; petioles shorter than the blades; stipules lanceolate-subulate, distinct; flowers dioecious or androgynous; staminate spikes usually interrupted, the pistillate mostly continuous, ¾′–1½′ long, often with small leaves at the top; achene ovate-oval, acute, rather less than 1″ long.

In moist soil, Quebec and Ontario to Minnesota, Florida, Kansas and Texas. Bermuda and other West Indies. Consists of numerous races. July–Sept.

5. PARIETÀRIA L. Sp. Pl. 1052. 1753.

Annual or perennial stingless diffuse or erect herbs, with alternate entire 3-nerved petioled leaves, no stipules, and axillary glomerate polygamous flowers, involucrate by leafy bracts. Calyx of the staminate flowers 4-parted or of 4 (rarely 3) distinct sepals. Fertile flowers with a tubular or campanulate 4-lobed calyx investing the ovary, a short or slender style, and a penicillate stigma. Achene ovoid, enclosed by the withering-persistent pistillate calyx. [Ancient Latin, referring to the growth of some species on walls.]

About 7 species, widely distributed; besides the following another occurs in the southern United States and one in western North America. Type species: *Parietaria officinalis* L.

1. Parietaria pennsylvánica Muhl.   Pennsylvania Pellitory.   Fig. 1563.

P. pennsylvanica Muhl.; Willd. Sp. Pl. **4**: 955.   1806.

Annual, pubescent, stem weak, simple or sparingly branched, ascending or reclining, very slender, 4′–15′ long.   Leaves lanceolate or oblong-lanceolate, membranous, dotted, acuminate at the apex, narrowed at the base, 3-nerved and with 1–3 pairs of weaker veins above, slender-petioled, 1′–3′ long, ¼′–½′ wide; petioles ¼′–1′ long, almost filiform; flowers glomerate in all except the lowest axils, the clusters shorter than the petioles; bracts of the involucre linear, 2–3 times as long as the flowers; style almost none; achene about ½″ long.

On dry rocks and banks, Maine to Ontario, British Columbia, Nevada, Tennessee and Mexico.   June–Aug. Hammerwort.

Family 12.   **LORANTHÀCEAE** D. Don, Prodr. Fl. Nepal. 142.   1825.

MISTLETOE FAMILY.

Parasitic green shrubs or herbs, containing chlorophyll, growing on woody plants and absorbing food from their sap through specialized roots called haustoria (a few tropical species terrestrial).   Leaves in the following genera opposite, in *Razoumofskya* reduced to opposite scales.   Flowers regular, terminal or axillary, clustered or solitary, dioecious or monoecious, and perianth simple, or in some exotic genera perfect, and with perianth of both calyx and corolla. Calyx-tube adnate to the ovary, its limb entire, toothed or lobed.   Stamens 2–6; anthers 2-celled or confluently 1-celled.   Ovary solitary, erect; style simple or none; stigma terminal, undivided, obtuse.   Fruit a berry.   Seed solitary, its testa indistinguishable from the endosperm, which is usually copious and fleshy.

About 21 genera and 500 species, widely distributed; most abundant in tropical regions.

Leaves scale-like, united at the base; anthers 1-celled; berry peduncled.   1. *Razoumofskya.*
Leaves thick, flat; anthers 2-celled; berry sessile.   2. *Phoradendron.*

1.   RAZOUMÓFSKYA Hoffm. Hort. Mosq.   1808.

[ARCEUTHOBIUM Bieb. Fl. Taur. **3**: 629.   1819.]

Small or minute fleshy glabrous plants, parasitic on the branches of coniferous trees, their branches 4-angled, and leaves reduced to opposite connate scales.   Flowers dioecious, not bracted, solitary or several together in the axils of the scales.   Staminate flowers with a 2–5-parted calyx and usually an equal number of stamens, the 1-celled anthers sessile on the segments.   Pistillate flowers with the ovary adnate to the tube of the calyx, the calyx-limb 2-parted.   Disk present in both kinds of flowers.   Berry fleshy, ovoid, more or less flattened, borne on a short somewhat recurved peduncle.   Embryo enclosed in the copious endosperm. [In honor of Alexis Razoumofski, Russian botanist.]

About 10 species.   Besides the following, 7 or 8 others occur in western North America and Mexico, 2 in Europe and Asia.   Type species: *Razoumofskya caucásica* Hoffm.

1. Razoumofskya pusílla (Peck) Kuntze. Small Mistletoe.   Fig. 1564.

Arceuthobium pusillum Peck, Rep. N. Y. State Mus. **25**: 69. 1873.
Arceuthobium minutum Engelm. Bull. Torr. Club **2**: 43. Without description.   1871.
Razoumofskya pusilla Kuntze, Rev. Gen. Pl. 587.   1891.

Plant inconspicuous, stems 2″–10″ long, nearly terete when fresh, somewhat 4-angled when dry, simple or sparingly branched, greenish-brown, slender.   Scales suborbicular, appressed, obtuse, about ½″ wide, connate at the base; flowers strictly dioecious (the staminate and pistillate plants sometimes on different trees), solitary in most of the axils, longer than the scales; berry ovoid-oblong, acute, about 1″ long, nodding on a slightly exserted peduncle; seeds enclosed in a viscid mucus.

On twigs of spruces and tamarack, Newfoundland to Connecticut, New York, Pennsylvania and Michigan.   June.

2. PHORADÉNDRON Nutt. Journ. Acad. Phila. (II.) 1: 185. 1847–50.

Shrubs, parasitic on trees, with opposite coriaceous flat entire or undulate faintly nerved leaves, terete or angled, usually jointed and brittle twigs, and dioecious axillary spicate bracted small flowers, solitary or several in the axil of each bract. Staminate flowers with a 3-lobed (rarely 2-4-lobed) globose or ovoid calyx, bearing a sessile transversely 2-celled anther at the base of each lobe. Pistillate flowers with a similar calyx adnate to the ovoid inferior ovary. Style none or very short, stigma obtuse or capitate. Fruit a sessile ovoid or globose fleshy berry. Endosperm copious. [Greek, tree-thief, from its parasitic habit.]

About 100 species, all American. Besides the following, 5 or 6 others occur in the western states and 1 in Florida. Type species: *Phoradendron californicum* Nutt.

1. Phoradendron flavéscens (Pursh) Nutt. American Mistletoe. Fig. 1565.

Viscum flavescens Pursh, Fl. Am. Sept. 114. 1814.

P. flavescens Nutt.; A. Gray, Man. Ed. 2, 383. 1856.

A branching glabrous or slightly pubescent shrub, the twigs rather stout, terete, brittle at the base. Leaves oblong or obovate, rounded at the apex, narrowed into short petioles, 3–5-nerved, entire, 1′–2′ long, 5″–10″ wide, dark green, coriaceous; petioles 1″–4″ long; spikes solitary, or 2 or 3 together in the axils, linear, shorter than the leaves; berry globose, white, about 2″ in diameter.

Parasitic on deciduous leaved trees, notably on the tupelo and red maple, central New Jersey to Ohio, Indiana and Missouri, south to Florida, Texas and New Mexico. May–July.

Family 13. SANTALÀCEAE R. Br. Prodr. Fl. Nov. Holl. 1: 350. 1810

SANDALWOOD FAMILY.

Herbs or shrubs (some exotic genera trees), with alternate or opposite entire exstipulate leaves. Flowers clustered or solitary, axillary or terminal, perfect, monoecious or dioecious, mostly greenish. Calyx adnate to the base of the ovary, or to the disk, 3–6-lobed, the lobes valvate. Petals none. Stamens as many as the calyx-lobes and inserted near their bases, or opposite them upon the lobed or annular disk; filaments slender or short. Ovary 1-celled; ovules 2–4, pendulous from the summit of the central placenta; style cylindric, conic or sometimes none; stigma capitate. Fruit a drupe or nut. Seed 1, ovoid or globose. Testa none; endosperm copious, fleshy; embryo small, apical.

About 26 genera and 250 species, mostly of tropical distribution, a few in the temperate zones.

Perennial herbs; flowers perfect, cymose or solitary.   1. *Comandra.*
Shrubs; flowers imperfect, mostly dioecious.
     Leaves alternate; flowers racemose.   2. *Pyrularia.*
     Leaves opposite; flowers umbellate or solitary.   3. *Nestronia.*

1. COMÁNDRA Nutt. Gen. 1: 157. 1818.

Glabrous erect perennial herbs, some (or all?) parasitic on roots of other plants. Leaves alternate, oblong, oval, lanceolate or linear, entire, pinnately veined. Flowers perfect, terminal or axillary, rarely solitary, cymose, bractless. Calyx campanulate, the base of its tube adnate to the ovary, its limb 5-lobed (rarely 4-lobed). Stamens 5, or rarely 4, inserted at the base of the calyx-lobes and between the lobes of the disk, attached to the middle of the lobes by tufts of hairs. Anthers ovate, 2-celled. Fruit drupaceous, globose or ovoid, crowned by the persistent calyx. [Greek, referring to the hairy attachments of the anthers.]

Four known species, the following North American, one European. Type species: *Comandra umbellata* (L.) Nutt.

Cymes mostly corymbose-clustered at the summit of the stem; leaves acute, sessile; style slender.
     Leaves oblong, green; fruit globose-urn-shaped.   1. *C. umbellata.*
     Leaves lanceolate or linear, glaucous; fruit ovoid.   2. *C. pallida.*
Peduncles few, axillary; leaves oval, obtuse, short-petioled; style short.   3. *C. livida.*

1. Comandra umbellàta (L.) Nutt. Bastard Toad-flax. Fig. 1566.

Thesium umbellatum L. Sp. Pl. 208. 1753.
Comandra umbellata Nutt. Gen. 1 : 157. 1818.
C. Richardsiana Fernald, Rhodora 7 : 48. 1905.

Stem slender, very leafy, usually branched, 6′–18′ tall. Leaves oblong or oblong-lanceolate, green, acute or subacute at both ends, sessile, ascending, ½′–1¼′ long, the lower smaller; cymes several-flowered, corymbose at the summit of the plant or also axillary, their branches divergent or ascending; peduncles filiform, ¼′–1′ long; pedicels very short; calyx greenish-white or purplish, about 2″ high; style slender; drupe globose, 2½″–3″ in diameter, crowned by the upper part of the calyx-tube and its 5 oblong lobes.

In dry fields and thickets, Cape Breton Island to Ontario and Assiniboia, south to Georgia, Kansas and Arkansas. April–July.

2. Comandra pállida A. DC. Pale Comandra. Fig. 1567.

Comandra pallida A. DC. Prodr. 14 : 636. 1857.

Similar to the preceding species but paler and glaucous, usually much branched, the leaves narrower, linear or linear-lanceolate, acute or the lowest and those of the stem oblong-elliptic; cymes few-several-flowered, corymbose-clustered at the summit; peduncles usually short; pedicels about 1″ long; calyx purplish, about 2″ high; fruit ovoid-oblong, 3″–4″ high and 2″–2½″ in diameter, crowned by the short upper part of the calyx-tube and its 5 oblong lobes.

In dry soil, Manitoba to British Columbia, south to Minnesota, Kansas, Texas, New Mexico and California. April–July.

3. Comandra lívida Richards. Northern Comandra. Fig. 1568.

Comandra livida Richards. App. Frank. Journ. 734. 1823.

Stem slender, usually quite simple, 4′–12′ high. Leaves oval, thin, obtuse or rounded at the apex, narrowed at the base, short-petioled, ½′–1′ long, ¼′–½′ wide; petioles 1″–2″ long; cymes axillary, few (often only 1 to each plant), 1–5-flowered; peduncle shorter than its subtending leaf, filiform; flowers sessile; style very short; drupe globose-oblong, about 3″ in diameter, red, edible, crowned by the ovate calyx-lobes.

In moist soil, Newfoundland to Hudson Bay, Alaska, New Hampshire, Vermont, Ontario, Michigan, Idaho and Washington. June–July.

2. PYRULÀRIA Michx. Fl. Bor. Am. 2 : 231. 1803.

A branching oil-bearing shrub (the Asiatic species trees), with thin alternate pinnately-veined entire short-petioled deciduous leaves, and dioecious or polygamous small greenish racemose flowers. Staminate flowers with a campanulate 3–5-cleft calyx, the lobes valvate, recurved or spreading, pubescent at the base within; disk of 3–5 distinct glands or scales; stamens 4 or 5, inserted between the glands and opposite the calyx-lobes; filaments short; anthers ovate. Pistillate and perfect flowers with a top-shaped calyx adnate to the obovoid ovary; style short, stout; stigma capitate, depressed. Fruit a pear-shaped or oval drupe, the endocarp thin and endosperm of the seed very oily. [Name from *Pyrus,* the pear, from the similar shape of the fruit.]

Three species, the following typical one, and two Asiatic.

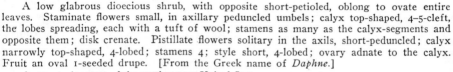

1. Pyrularia pùbera Michx. Oil-nut. Buffalo-nut. Elk-nut. Fig. 1569.

Pyrularia pubera Michx. Fl. Bor. Am. **2**: 233. 1803.
Hamiltonia oleifera Muhl.; Willd. Sp. Pl. **4**: 1120. 1805.
Pyrularia oleifera A. Gray, Man. Ed. 2, 382. 1858.

A straggling or erect much branched shrub, 3°–15° tall, with terete twigs, the young foliage pubescent. Leaves oblong, oblong-lanceolate or somewhat obovate, nearly glabrous when mature, acute or acuminate at both ends, 3'–5' long, 8''–1½' wide; petioles 2''–4'' long; racemes terminating short branches, the staminate many-flowered, 1'–2' long, the pistillate few-flowered and shorter; pedicels slender, 1½''–2'' long; staminate flowers about 2'' broad; calyx 3–5-cleft; drupe about 1' long, crowned by the ovate acute calyx-lobes.

In rich woods, southern Pennsylvania to Georgia, mostly in the mountains. May. Fruit ripe Aug.–Sept.

3. NESTRÒNIA Raf. New Flora 3: 12. 1836.

[Darbya A. Gray, Am. Journ. Sci. (II) **1**: 388. 1846.]

A low glabrous dioecious shrub, with opposite short-petioled, oblong to ovate entire leaves. Staminate flowers small, in axillary peduncled umbels; calyx top-shaped, 4–5-cleft, the lobes spreading, each with a tuft of wool; stamens as many as the calyx-segments and opposite them; disk crenate. Pistillate flowers solitary in the axils, short-peduncled; calyx narrowly top-shaped, 4-lobed; stamens 4; style short, 4-lobed; ovary adnate to the calyx. Fruit an oval 1-seeded drupe. [From the Greek name of *Daphne*.]

A monotypic genus of the southeastern United States.

1. Nestronia umbéllula Raf. Nestronia. Fig. 1570.

Nestonia umbellula Raf. New Flora **3**: 13. 1836.
Darbya umbellulata A. Gray, Am. Journ. Sci. (II.) **1**: 388. 1846.

Shrub 1°–3° high, branching. Leaves thin, 1'–2' long, acute or obtuse at the apex, narrowed or rounded at the base, pinnately veined, bright green above, petioles 1''–3'' long; peduncles of the staminate 3–9-flowered umbels filiform, nearly or quite one-half as long as the leaves, the pedicels about 2'' long, equalling the green calyx; stamens shorter than the oblong-ovate calyx-segments; pistillate calyx glaucescent, about 3'' long, the lobes much shorter than the tube; drupe globose, about ⅓' in diameter.

In woods and along streams, parasitic on tree-roots, Virginia to Georgia and Alabama. April–May.

Family 14. ARISTOLOCHIÀCEAE Blume, Enum. Pl. Jav. 1: 81. 1830.

Birthwort Family.

Herbs or shrubs, acaulescent, or with erect or twining and leafy stems. Leaves alternate or basal, petioled, mostly cordate or reniform, exstipulate. Flowers axillary or terminal, solitary or clustered, perfect, mostly large, regular or irregular. Calyx-tube (hypanthium) mostly adnate to the ovary, its limb 3-lobed, 6-lobed or irregular. Petals none. Stamens 5–many, inserted on the pistil, the anthers 2-celled, extrorse, their sacs longitudinally dehiscent. Ovary wholly or partly inferior, mostly 6-celled; ovules numerous in each cavity, anatropous, horizontal or pendulous. Fruit a many-seeded mostly 6-celled capsule. Seeds ovoid or oblong, angled or compressed, the testa crustaceous, smooth or wrinkled, usually with a fleshy or dilated raphe; endosperm copious, fleshy.

Six genera and about 200 species, of wide distribution. The family is not closely related to those that precede and follow it in the arrangement followed in this work; its affinity is uncertain.

Acaulescent herbs; perianth regular, 3-lobed, persistent; filaments distinct.
  Calyx-tube wholly adnate to the ovary; styles united.   1. *Asarum.*
  Calyx-tube nearly free from the ovary; styles distinct.   2. *Hexastylis.*
Leafy erect herbs or twining vines; perianth irregular, deciduous; anthers sessile, adnate to the
  stigma.   3. *Aristolochia.*

1. ÁSARUM (Tourn.) L. Sp. Pl. 442. 1753.

Acaulescent perennial often clustered herbs, with slender aromatic branched rootstocks, thick fibrous-fleshy roots, long-petioled cordate, mostly ovate or orbicular entire leaves, and solitary large peduncled purple-brown or mottled flowers, borne very near or upon the ground. Calyx campanulate or hemispheric, adnate to the ovary at least below, regularly 3-lobed, the lobes valvate. Stamens 12, inserted on the ovary; filaments short, stout; connective of the anther-sacs more or less continued beyond them as a tip. Ovary partly or wholly inferior, 6-celled, the parietal placentae intruded; ovules numerous, horizontal or pendulous. Capsule coriaceous, crowned by the withering-persistent calyx and stamens, subglobose or hemispheric, at length bursting irregularly or longitudinally dehiscent. Seeds compressed. [The ancient name, its meaning obscure.]

About 20 species, natives of the north temperate zone. Besides the following, 3 others occur in western North America. Type species: *Asarum europaeum* L. The plants are known as *Asarabacca.*
Calyx-segments lanceolate-acuminate, longer than the tube, not reflexed.
  Calyx-segments slightly longer than the tube, the tubular portion 2″–4″ long; species mainly
    Alleghanian.   1. *A. canadense.*
  Calyx-segments much longer than the tube, the tubular portion 5″–10″ long; species campes-
    trian.   2. *A. acuminatum.*
Calyx-segments triangular, merely acute, about as long as the tube, reflexed.   3. *A. reflexum.*

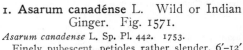

1. Asarum canadénse L. Wild or Indian Ginger. Fig. 1571.

Asarum canadense L. Sp. Pl. 442. 1753.

Finely pubescent, petioles rather slender, 6′–12′ long. Leaves commonly 2 to each plant, reniform, thin, short-pointed at the apex, 4′–7′ broad, dark green, not mottled, the basal sinus deep and open; flower slender-peduncled from between the bases of the petioles, 1′ broad or more when expanded, brownish purple; calyx ovoid, its tube completely adnate to the ovary, its lobes inflexed in the bud, ovate-lanceolate, acute or long-acuminate, spreading, a little longer than the tube; filaments longer than the anthers; stigmas radiating; capsule 6″–8″ in diameter.

In rich woods, New Brunswick to Manitoba, south to North Carolina, Missouri and Kansas. Ascends to 3000 ft. in Virginia. Called also Canada snakeroot. April–May. Rootstocks with the flavor of ginger. False coltsfoot. Colic-root. Heart-, Vermont or southern snakeroot. Asarabacca.

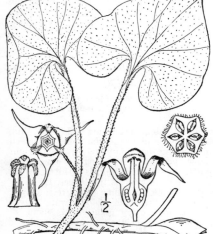

2. Asarum acuminàtum (Ashe) Bicknell. Long-tipped Wild Ginger. Fig. 1572.

A. canadense var. *acuminatum* Ashe, Contr. 1 : 2. 1897.
Asarum acuminatum Bicknell; Britton & Brown, Ill. Fl. 3: 513. 1898.

Similar to *A. canadense* but more pubescent, at least when young. Leaves thin and membranous, reniform-cordate and acutely short-pointed or broadly reniform and blunt, at first densely cinereous-tomentose on the lower surface, less so when old, the larger veins often densely divaricate-pubescent, giving the leaves beneath a coarsely white-reticulated appearance; calyx-lobes much longer than in *A. canadense* and duller brownish-purple, caudate-acuminate, or flagellate, the slender terminations recurved-spreading, often flexuous, 5″–10″ long.

Rich woods, Minnesota and Wisconsin to Iowa, Indiana, Kentucky and Ohio. May–June.

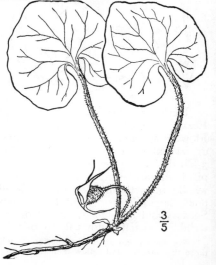

3. **Asarum refléxum** Bicknell.   Short-
lobed Wild Ginger.   Fig. 1573.

Asarum reflexum Bicknell, Bull. Torr. Club **24**: 533.
  pl. 317. 1897.
Asarum reflexum ambiguum Bicknell, Bull. Torr. Club
  24: 535. 1897.

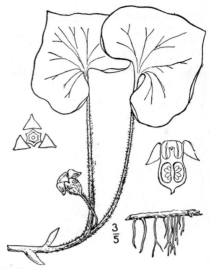

Similar to *A. canadense,* more loosely pubescent, rootstocks more elongated, slender.   Leaves
reniform, broader than long, the basal sinus shallow or deep, obtusely pointed, the upper surface
commonly nearly glabrous, the petioles often
nearly glabrous in age; flowers smaller than
those of *A. canadense,* the calyx-tube white
within; lobes of the calyx-limb early reflexed,
purplish-brown, 4″–8″ long, about as long as the
tube, triangular, with a straight obtuse tip, 1″–4″
long.

In rich woods, along streams or river valleys, often
forming large patches, Connecticut and southeastern
New York to Iowa, Michigan, North Carolina, Missouri and Kansas.   April–May.

2.   HEXÁSTYLIS Raf. Neog. 3: 1825.

Perennial evergreen acaulescent herbs, the foliage glabrous or nearly so.   Rootstocks
dichotomously branched, the roots fleshy.   Leaves leathery, often mottled, petioled.   Flower
solitary, its peduncle subtended by a membranous bract.   Calyx glabrous without, the tube
sometimes inflated, the 3 segments short, valvate.   Corolla none.   Stamens 12; filaments
shorter than the anthers, or wanting.   Ovary mainly free from the calyx-tube; styles distinct, prolonged beyond the extrorse stigmas into cleft appendages.   Capsule free from the
calyx.   Seeds flattened.   [Greek, referring to the six styles.]

About 6 species, natives of eastern North America.   Type species: *Asarum arifolium* Michx.

Leaves ovate or suborbicular.
  Calyx 1′–2′ long, much longer than thick.   1. *H. Shuttleworthii.*
  Calyx less than 1′ long, little longer than thick.
      Fruiting calyx campanulate, its lobes about half as long as the tube.   2. *H. virginica.*
      Fruiting calyx urn-shaped, its lobes about one third as long as tube.   3. *H. Memmingeri.*
Leaves hastate.   4. *H. arifolia.*

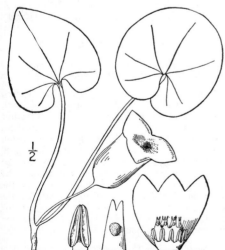

1. **Hexastylis Shuttlewórthii** (Britten
and Baker) Small.   Large-flowered
Hexastylis.   Fig. 1574.

Hexastylis Shuttleworthii Small; Britton, Man.
  348. 1891.
Asarum grandiflorum Small, Mem. Torr. Club **4**:
  150. 1893. Not Kl.
Asarum macranthum Small, Mem. Torr. Club **5**:
  136. 1894. Not Hooker.
Asarum Shuttleworthii Britten & Baker, Journ.
  Bot. **36**: 98. 1898.

Glabrous; rootstocks branched.   Leaves 1
or 2 to each plant or branch, broadly ovate or
suborbicular, dark green and usually mottled
above, paler beneath, 2′–4′ long, 1½′–3′ wide,
obtuse or subacute at the apex, the basal sinus
mostly narrow; petioles 3′–8′ long, ascending;
calyx tubular-campanulate, 8″–20″ long, not
or scarcely contracted at the throat, the lobes
obtuse, mottled with violet on the inner side,
one-third to one-half as long as the tube; peduncle 8″–20″ long; filaments shorter than the
anthers; anthers equally 4-ribbed, not pointed; styles 6, each 2-cleft.

In rich mountain woods. Virginia and North Carolina.   May–July.

2. Hexastylis virgínica (L.) Small. Virginia Hexastylis. Fig. 1575.

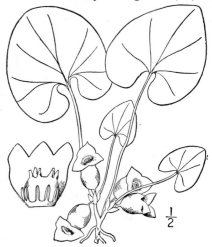

Asarum virginicum L. Sp. Pl. 442. 1753.
Asarum heterophyllum Ashe, Contr. Herb. 1 : 3. 1897.
H. virginica Small; Britton, Man. 348. 1901.

Rootstocks slender, scaly, clustered, simple or branched. Leaves 1-3 to each plant or branch, coriaceous, glabrous, orbicular or broadly ovate, rounded at the apex, 1½'-3' wide, usually mottled, the basal sinus open or nearly closed; petioles pubescent along one side or glabrous, 3'-7' long, ascending; flower short-peduncled, purple, 6''-8'' long; calyx campanulate to turbinate, narrowed at the throat, its tube adnate to the lower part of the ovary, free above, the lobes ovate or nearly semicircular, about one-half as long as the tube; peduncle ¼'-½' long; filaments much shorter than the anthers; anthers not pointed; styles 6, each 2-lobed, the stigmas sessile below the lobes; capsule hemispheric, about 4'' high.

In rick woods, Virginia and West Virginia to Georgia and South Carolina. May–June. Southern wild ginger. Black snakeweed. Heart-leaf.

3. Hexastylis Mémmingeri (Ashe) Small. Memminger's Hexastylis. Fig. 1576.

Asarum Memmingeri Ashe, Contr. Herb. 1 : 3. 1897.

Hexastylis Memmingeri Small; Britton, Man. 348. 1901.

Slender. Leaf-blades suborbicular or ovate, 1½'-3' long, mostly obtuse or retuse, sometimes mottled, with a slightly open sinus; petioles about as long as the blades or much longer; calyx 5''-7'' long, urn-shaped, the tube more or less constricted at the throat; the segments rarely over 1½'' long, obtuse; peduncle as long as the calyx or shorter; prolongations of the styles slender, usually deeply cleft; capsule conspicuously distending the calyx; seed sharply triangular.

In sandy woods, Virginia and West Virginia, south to Georgia. May and June.

4. Hexastylis arifòlia (Michx.) Small. Halberd or Heart-leaved Hexastylis.

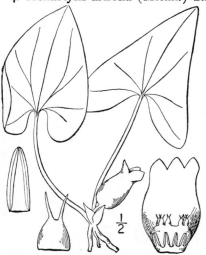

A. arifolium Michx. Fl. Bor. Am. 1 : 279. 1803.
H. arifolia Small; Britton, Man. 348. 1901.

Pubescent, at least on the veins of the leaves, rootstocks slender, usually branched and with 1 or 2 leaves to each branch. Leaves rather thick, usually mottled, 2'-5' long, some of them hastate, some suborbicular, the basal sinus often broad; petioles more or less pubescent, 3'-8' long; flower stout-peduncled, about 1' long; calyx urn-shaped, much contracted at the throat, the lobes rounded, about one-fifth as long as the tube, which is adnate to the lower half of the ovary; anthers nearly sessile, short-pointed; styles 6, 2-cleft, with a sessile stigma below the cleft; capsule subglobose, about 8'' in diameter.

In woods, Virginia to Tennessee, Florida and Alabama. Ascends to 3000 ft. in Virginia. April–June.

Hexastylis Rùthii (Ashe) Small, differing in the calyx not constricted in the throat, ranges from southwestern Virginia and Tennessee to Alabama.

3. **ARISTOLÒCHIA** (Tourn.) L. Sp. Pl. 960. 1753.

Perennial herbs or twining vines. Leaves alternate, mostly petioled and entire (some exotic species 3–7-lobed), cordate, palmately 3–many-nerved. Flowers irregular, solitary or clustered. Calyx adnate to the ovary, at least to its base, the tube narrow, usually inflated around the style and contracted at the throat, the limb spreading or reflexed, entire, 3–6-lobed or appendaged. Stamens mostly 6; anthers sessile, adnate to the short style or stigma, 2-celled, the sacs longitudinally dehiscent. Ovary partly or wholly inferior, mostly 6-celled with 6 parietal placentae. Style 3–6-lobed. Capsule naked, septicidally 6-valved. Seeds very numerous, horizontal, compressed, their sides flat or concave. [Named for its supposed medicinal properties.]

About 200 species, widely distributed in tropical and temperate regions. Besides the following, some 6 others occur in the southern and western United States.   Type species : *Aristolochia rotunda* L.

Erect herbs.
    Calyx-tube bent ; flowers solitary, on basal scaly branches.   1. *A. Serpentaria.*
    Calyx-tube straight ; flowers axillary, clustered.   2. *A. Clematitis.*
Long twining vines ; flowers axillary ; calyx-tube bent.
    Leaves minutely pubescent ; calyx-limb flat, spreading.   3. *A. macrophylla.*
    Leaves tomentose ; calyx-limb rugose, reflexed.   4. *A. tomentosa.*

1. **Aristolochia Serpentària** L.   Virginia Snakeroot.   Serpentary.   Fig. 1578.

Aristolochia Serpentaria L. Sp. Pl. 961. 1753.

A perennial pubescent nearly erect herb, 10′–3° tall, with short rootstocks and fibrous aromatic roots. Leaves ovate, ovate-lanceolate or oblong-lanceolate, thin, green on both sides, acuminate at the apex, cordate or hastate at the base, 1½′–5′ long, ½′–2′ wide ; petioles ¼′–1′ long ; lowest leaves reduced to scales ; flowers solitary and terminal, on slender basal scaly branches ; tube of the calyx curved like the letter S, enlarged at the ovary and at its throat, the limb short, spreading, slightly 3-lobed ; anthers contiguous in pairs ; stigma 3-lobed ; capsule subglobose, ridged, about ½′ in diameter. Flowers sometimes cleistogamous.

In dry woods, Connecticut and New York to Michigan, Florida, Louisiana and Missouri. Ascends to 2500 ft. in Virginia. June–July. Fruit ripe Sept. Sangrel snakeweed. Sangree-root. Black snakeroot. Pellican-flower.

Aristolochia hastàta Nutt. (*A. Nashii* Kearney, of the southern states), with narrowly lanceolate or linear, sagittate or hastate leaves, is reported to range as far north as Virginia.

2. **Aristolochia Clematìtis** L.   Birthwort. Fig. 1579.

Aristolochia Clematitis L. Sp. Pl. 962. 1753.

Herbaceous, perennial ; stem erect, glabrous, zigzag, striate, 1°–2° tall. Leaves dark green, reniform, subacute or obtuse at the apex, glabrous or their margins minutely spinulose-ciliate, strongly reticulate-veined, 2′–5′ wide ; petioles shorter than the blades ; flowers fascicled in the axils, 1′–1¾′ long ; tube of the calyx yellowish green, straight, enlarged around the ovary, the 6 lobes appendaged ; anthers equidistant.

Roadsides and thickets, New York to Maryland. Escaped from cultivation. Native of southern Europe. Summer. Upright birthwort.

3. Aristolochia macrophýlla Lam. Dutchman's Pipe. Fig. 1580.

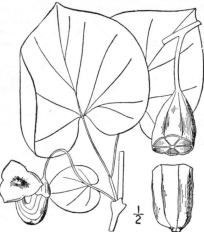

4. Aristolochia tomentòsa Sims. Woolly Pipe-vine. Fig. 1581.

A. tomentosa Sims, Bot. Mag. *pl. 1369.* 1811.

A twining vine, similar to the preceding, but the twigs, petioles, leaves and peduncles persistently tomentose. Leaves suborbicular or broadly ovate, obtuse or rounded at the apex, 3'–6' broad when mature; petioles rather stout, 1'–3' long; peduncles axillary, mostly solitary, slender, bractless; calyx densely tomentose, the tube sharply curved, yellowish-green, about 1½' long, its throat nearly closed, the limb becoming reflexed, wrinkled, dark purple, 3-lobed; anthers contiguous in pairs beneath the 3 spreading lobes of the stigma; capsule oblong-cylindric.

In woods, Missouri and southern Illinois to North Carolina, Kansas, Arkansas, Alabama and Florida. May–June.

Aristolochia macrophylla Lam. Encycl. 1 : 255. 1783.
Aristolochia Sipho L'Her. Stirp. Nov. 13. 1784.

A twining vine, the stem sometimes 1' in diameter and 30° long, the branches very slender, terete, green, glabrous. Leaves thin, broadly reniform or suborbicular, densely pubescent beneath when young, glabrous or nearly so and 6'–15' broad when mature; petioles slender, 1'–4' long; peduncles solitary or 2 or 3 together in the axils, about as long as the petioles, each with a suborbicular clasping bract at about the middle ½'–1' in diameter; calyx-tube strongly curved, 1' or more long, inflated above the ovary, contracted at the throat, yellowish-green, veiny, the limb flat, spreading, purple-brown, somewhat 3-lobed; anthers contiguous in pairs under the 3 lobes of the stigma; capsule oblong-cylindric, strongly parallel-nerved, 2'–3' long, 8"–10" in diameter.

In rich woods, southern Pennsylvania to Minnesota, Georgia, Tennessee and Kansas. Ascends to 4500 ft. in Virginia. May–June. Fruit ripe Sept. Pipe-vine. Wild ginger. Big sarsaparilla.

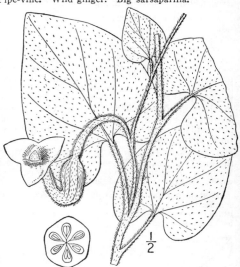

Family 15. POLYGONÀCEAE Lindl. Nat. Syst. Ed. 2, 211. 1836.*
BUCKWHEAT FAMILY.

Herbs, vines, shrubs or trees, with alternate or sometimes opposite or whorled simple mostly entire leaves, jointed stems, and usually sheathing united stipules (ocreae). Flowers small, perfect, dioecious, monoecious or polygamous, spicate, racemose, corymbose, umbellate or panicled. Petals none. Calyx inferior, free from the ovary, 2–6-cleft or 2–6-parted, the segments or sepals more or less imbricated, sometimes petaloid, sometimes developing wings in fruit. Stamens 2–9, inserted near the base of the calyx, or in staminate flowers crowded toward the centre; filaments filiform or subulate, or often dilated at the base, distinct or united into a ring; anthers 2-celled, the sacs longitudinally dehiscent. Pistil solitary; ovary superior, 1-celled; ovule solitary, orthotropous, erect or pendulous; style 2- or 3-cleft or 2- or 3-parted (rarely 4-parted), sometimes very short; stigmas capitate or tufted, rarely 2-cleft or toothed; fruit a lenticular, 3-angled or rarely 4-angled achene, usually invested by the persistent calyx; endosperm mealy or horny; cotyledons accumbent or incumbent, flat; embryo straight or curved.

About 40 genera and 800 species, of wide geographic distribution.

* Text contributed by DR. JOHN KUNKEL SMALL.

Flowers or flower-clusters subtended by involucres.
Ocreae present; calyx 2–4-parted; stamens 3 or fewer.   1. *Macounastrum.*
Ocreae none; calyx 6-cleft or 6-parted; stamens 9; achene 3-angled.   2. *Eriogonum.*
Flowers or flower-clusters not involucrate; stamens 4–8.
  Stigmas tufted; sepals of the inner row usually accrescent.
    Calyx 6-parted; style 3-parted; achene 3-angled.   3. *Rumex.*
    Calyx 4-parted; style 2-parted; achene lenticular.   4. *Oxyria.*
  Stigmas capitate, 2-cleft or toothed; sepals of the outer row often accrescent or reflexed.
    Floral tube obscurely or only slightly developed; stigmas not 2-cleft; ovule erect.
      Internodes of the stem and branches not adnate; plants not heath-like; sepals of the outer
        row often winged .
        Leaf-blades jointed at the base; ocreae 2-lobed, becoming lacerate; filaments, at least
          the inner, dilated.   5. *Polygonum.*
        Leaf-blades not jointed at the base; ocreae various, not 2-lobed; filaments slender.
          Ocreae cylindric, truncate.
            Sepals 4; calyx curved; stamens 4.   6. *Tovara.*
            Sepals 5, if fewer, the stamens more than 5; calyx not curved.   7. *Persicaria.*
          Ocreae oblique, more or less open on the side facing the leaf.
            Plants with fleshy rootstocks, scaly caudices and simple flower-stems; leaves
              mostly basal; inflorescence simple; styles elongated.   8. *Bistorta.*
            Plants with fibrous roots or slender rootstocks, without scaly caudices, with
              branching stems; leaves not basal; inflorescence branched.
              Sepals neither keeled nor winged.
                Racemes collected into terminal corymbs; embryo dividing the endosperm
                  by an S-shaped curve; plants smooth.   9. *Fagopyrum.*
                Racemes not in terminal corymbs; embryo slender, at one side of the
                  endosperm; plants prickle-armed.   10. *Tracaulon.*
              Sepals, at least the outer ones, keeled or winged.
                Stigmas rounded capitate; styles short, erect, or none.   11. *Tiniaria.*
                Stigmas dilated, toothed; styles divaricate.   12. *Pleuropterus.*
      Internodes of the stem and branches partially adnate; plants heath-like; sepals of the inner
        row often winged.   13. *Polygonella.*
    Floral tube well developed, enclosing the achene, winged; stigmas 2-cleft; ovule pendulous.
      14. *Brunnichia.*

1. MACOUNÁSTRUM Small in Britt. & Brown, Ill. Fl. 1: 541. 1895.

[Koenigia L. Mant. 35. 1767. Not *Konig* Adans. 1763.]

Low glabrous annual herbs, with fibrous roots, erect or spreading simple or forked stems, alternate or opposite entire leaves, funnelform membranous ocreae, and minute perfect terminal clustered flowers, subtended by a several-leaved involucre. Calyx 2–4-parted (usually 3-parted), greenish-white, the segments valvate, equal; pedicels short, subtended by transparent bracts; stamens 2 or 4, alternate with and often protruding between the calyx-segments; filaments short, stout; anthers ovoid. Style 2- or 3-parted; stigmas capitate; achene ovoid, 3-angled or lenticular, exceeding the persistent calyx; embryo eccentric, accumbent.

Two or three species, the following typical one circumboreal, the others of the higher Himalayas.

1. Macounastrum islándicum (L.) Small.
    Macounastrum.   Fig. 1582.

Koenigia islandica L. Mant. 35. 1767.
M. islandicum Small in Britt. & Brown, Ill. Fl. 1: 542. 1896.

Stems very slender, 1′–4′ long, sometimes tufted. Leaves obovate, oblong or almost orbicular, 1″–5″ long, fleshy, obtuse at the apex, sessile or short-petioled; ocreae about ½″ long; involucre consisting of 3–6 obovate or orbicular leaves more or less united at their bases; flowers fascicled in the involucres, short-pedicelled; calyx ½″ long, the segments ovate-lanceolate, rather obtuse; stamens very short; style-branches short; achene less than 1″ long, brown, often slightly curved, striate, its faces convex.

Greenland and Labrador to Hudson Bay and Alaska. Also in arctic Europe and Asia. Summer.

2. ERIÓGONUM Michx. Fl. Bor. Am. 1: 246. 1803.

Annual or perennial acaulescent or leafy-stemmed herbs, some species very woody at the base, with simple or branched, often tufted stems, and entire alternate opposite or whorled leaves. Flowers small, fascicled, cymose, umbellate or capitate, subtended by 5–8-toothed or cleft campanulate top-shaped or almost cylindric involucres. Calyx 6-cleft or 6-parted, usually colored, the segments equal or the outer ones larger. Stamens 9, included or exserted; filaments filiform, often villous. Style 3-parted; stigmas capitate. Achene pyramidal, 3-angled, more or less swollen near the base, invested by the calyx-segments, or winged. Embryo axial or somewhat eccentric. [Greek, referring to the woolly and jointed stems.]

Over 200 species, natives of America, mostly of the western United States. Type species: *Eriogonum tomentosum* Michx.

Achenes 3-winged.   1. *E. alatum.*
Achenes merely angled, never winged.
  Calyx contracted into a stipe-like base.
    Caulescent; stems topped by compound cymes.
      Stem leaves alternate.   2. *E. longifolium.*
      Stem leaves opposite or whorled.
        Calyx yellow; basal leaves rounded at the base; Alleghanian.   3. *E. Alleni.*
        Calyx white or pink; basal leaves narrowed at the base; campestrian. 4. *E. Jamesii.*
    Scapose; stems topped by simple umbels.   5. *E. flavum.*
  Calyx jointed to the pedicel without a stipe-like base.
    Scapes or peduncles topped by capitate clusters.
      Calyx glabrous; achenes over 2 mm. long.   6. *E. pauciflorum.*
      Calyx villous; achenes less than 2 mm. long.   7. *E. multiceps.*
    Scapes or peduncles topped by more or less compound cymes.
      Involucres erect or at least never drooping.
        Ovaries and achenes completely clothed with wool; leaf-blades silky above with silvery hairs.   8. *E. lachnogynum.*
        Ovaries and achenes glabrous or villous at the top; leaf-blades more or less floccose.
          Annual; herbaceous throughout and usually simple at the base. 9. *E. annuum.*
          Perennial; shrubby and much branched at the base.
            Calyx yellow.   10. *E. campanulatum.*
            Calyx white, pink or reddish.
              Leaf-blades copiously pubescent on both sides.   11. *E. corymbosum.*
              Leaf-blades densely pubescent beneath, inconspicuously so above.
                Inflorescence 2–3 times compound; branches spreading.
                    12. *E. microthecum.*
                Inflorescence 5–7 times compound; branches erect or nearly so.
                    13. *E. effusum.*
      Involucres on drooping or deflexed peduncles.   14. *E. cernuum.*

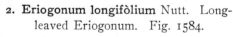

1. Eriogonum alàtum Torr. Winged Eriogonum. Fig. 1583.

E. alatum Torr. Sitgreaves' Rep. 168. *pl. 8.* 1853.

Perennial by a long thick root, stem rather stout, erect, strigose, paniculately branched, somewhat angled, 1°–3° tall. Leaves mostly basal, spatulate, oblanceolate or narrowly obovate, 1′–3′ long, those of the stem alternate, nearly linear, short-petioled, all obtuse or subacute at the apex, glabrous or pubescent and with midrib prominent beneath, ciliate; panicle open; bracts lanceolate or subulate; involucres cymose at the ends of the branches, campanulate, 5-toothed, 1″–1½″ long, the segments obtuse and somewhat reflexed; calyx yellowish, 1″ long, campanulate; stamens slightly exserted; achene long-pointed, 2½″–3″ long, reticulated, closely invested by 3 wings.

On plains, western Nebraska to Texas, west to Colorado and New Mexico. June–Sept.

2. Eriogonum longifòlium Nutt. Long-leaved Eriogonum. Fig. 1584.

Eriogonum longifolium Nutt. Trans. Am. Phil. **Soc.** (II.) **5**: 164. 1833–37.
E. Texanum Scheele, Linnaea **22**: 150. 1849.

Perennial, strigose throughout, stem stout, erect, paniculately or corymbosely branched, leafy, finely grooved, 2°–4° tall. Leaves narrowly oblong or linear-oblong, obtuse at the apex, more or less tomentose beneath, the upper sessile, the lower narrowed into petioles with dilated and sheathing bases; bracts lanceolate or subulate; involucres turbinate-campanulate, 1½″–2″ long; peduncles 1′ long or less; calyx oblong-campanulate, 2″–3″ high, 6-parted to near the base, very villous; stamens and style-branches exserted; achene 2½″ long, much enlarged at the base, villous, loosely invested by the calyx-segments, not winged.

Southern Missouri to Texas. June–Nov.

3. Eriogonum Álleni S. Wats. Allen's Eriogonum. Fig. 1585.

E. Alleni S. Wats. in A. Gray, Man. Ed. 6, 734. 1890.

Perennial, flocose-tomentose throughout; stem rather stout, erect, sparingly branched above, 1°–1½° tall. Leaves oblong, or ovate-oblong, 1′–3′ long, the basal long-petioled, obtuse at both ends, those of the stem in whorls of 3–5 at the somewhat swollen nodes, short-petioled, narrowed at the base, the upper small and bract-like; inflorescence compoundly cymose; involucres top-shaped, 5-toothed, 2½″–3″ long, the teeth obtusish; bracts leaf-like, spatulate; calyx yellow, 1½″ long, broadly campanulate, its segments obovate or orbicular; stamens and style-branches exserted.

Stony mountain-sides, Virginia and West Virginia. July–Aug.

4. Eriogonum Jàmesii Benth. James' Eriogonum. Fig. 1586.

Eriogonum Jamesii Benth. in DC. Prodr. 14: 7. 1856.
Eriogonum sericeum Torr.; T. & G. Proc. Am. Acad. 8: 155. 1870. Not Pursh, 1814.

Perennial, base woody, scaly, somewhat branched, stem usually spreading, branched, to-mentose, slender, 6′–18′ long, the branches erect, 3 or 4 times forked, light brown or reddish. Leaves mostly basal, spatulate or oblong, 1′–4′ long, long-petioled, the upper smaller, sessile in whorls of 3 or 4 at the somewhat swollen nodes, all obtuse or subacute, dark green and sparingly tomentose above, densely gray-tomentose beneath, their margins sometimes slightly revolute and crisped; inflorescence compoundly cymose; involucres turbinate-campanulate, 2″ long, 5-toothed, the teeth rounded; bracts foliaceous, narrowly oblong or oblong-spatulate; calyx white or pink, campanulate, villous, 2½″–3″ long, its segments oblong or obovate; stamens and style-branches exserted; achene 2½″–3″ long, smooth, villous near the apex, its angles margined.

On plains, Kansas and Colorado to Texas, Arizona and Chihuahua. July–Oct.

5. Eriogonum flàvum Nutt. Yellow Eriogonum. Fig. 1587.

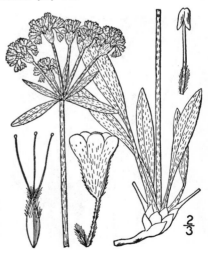

Eriogonum flavum Nutt. Fras. Cat. 1813.
E. sericeum Pursh, Fl. Am. Sept. 277. 1814.

Perennial, scapose, white-tomentose throughout, root short, scaly, spindle-shaped, stem very short and thick, simple and solitary or tufted and creeping, woody. Scapes 2′–12′ tall, erect; leaves crowded on the short stem, linear-oblong or ob-long-spatulate, 1′–3′ long, mostly obtuse at the apex, flat, narrowed into petioles; petioles dilated at the base and imbricated; inflorescence regularly umbellate; involucres top-shaped, 2″–2½″ long, nearly entire, rather densely clustered; peduncles ¼′–1½′ long; bracts spatulate, foliaceous; calyx yellow, 3″ high, top-shaped, very villous, the segments obovate; stamens and style-branches ex-serted; achene constricted at the middle, 2″ long, villous at the summit, the angles undulate, the faces swollen.

Nebraska and Kansas to Alberta and Arizona. June–Sept.

6. **Eriogonum pauciflòrum** Pursh. Few-flowered Eriogonum. Fig. 1588.

E. pauciflorum Pursh, Fl. Am. Sept. 735. 1814.
Eriogonum parviflorum Nutt. Gen. 1: 261. 1818.

Perennial, root long and slender, stems very short, simple or sparingly branched, loosely tufted, covered by the scarious dilated bases of the petioles. Scapes erect, slender, simple, slightly tomentose, 2′–6′ high; leaves linear or linear-spatulate, 1′–3′ long, rather obtuse, but apparently acute from the strongly revolute margins, glabrous or sparingly pubescent above, white-tomentose or cottony beneath, narrowed into slender petioles; inflorescence capitate; involucres 4–10, 1½″ long, turbinate-campanulate, 5-toothed, the teeth obtuse, more or less reflexed; calyx white, campanulate, 1½″ long, glabrous, the segments ovate; achene 1¼″ long, its faces swollen at about the middle, inconspicuously striate-reticulated.

On dry plains Nebraska and Colorado. July–Sept.

7. **Eriogonum múlticeps** Nees. Branched Eriogonum. Fig. 1589.

E. multiceps Nees, Max. Reise N. A. 2: 446. 1841.
E. gnaphaloides Benth. Kew. Journ. Bot. 5: 263. 1853.

Perennial by a slender root, scapose, densely white-tomentose throughout; stems short, tufted, much branched, sometimes several inches long. Scapes simple, 1′–5′ high; leaves spatulate, ½′–2′ long, numerous, obtuse at the apex, narrowed below into petioles; inflorescence capitate; involucres 3–12, sessile, 1½″ long, 5–6-toothed, the teeth acute; bracts foliaceous, spatulate; calyx white or rose-color, 1½″–2½″ long, campanulate, somewhat villous, 6-cleft to about the middle, the segments cuneate, obtuse or emarginate; stamens and style-branches exserted; achene ½″ long.

On dry plains, Nebraska and Colorado. June–Aug.

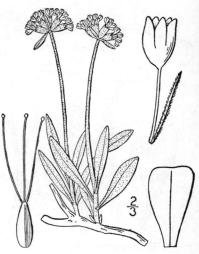

8. **Eriogonum lachnógynum** Torr. Long-rooted Eriogonum. Fig. 1590.

Eriogonum lachnogynum Torr.; Benth. in DC. Prodr. 14: 8. 1856.

Perennial, scapose, root long, fusiform, stems stout and short, tufted, much branched, covered with the dilated petiole-bases. Scape erect, slender, 4′–12′ tall, white-tomentose, sparingly branched above; leaves numerous, crowded, lanceolate or narrowly oblong, ½′–1′ long, acute at the apex, narrowed at the base, silky above, white-tomentose beneath, long-petioled, their margins somewhat revolute; inflorescence irregularly umbellate or paniculate; involucres broadly campanulate or nearly hemispheric, 1½″–2″ high, sessile or peduncled, 5-toothed, teeth obtuse; bracts small, lanceolate; calyx campanulate, 1½″ long, villous; stamens and style-branches exserted.

Western Kansas to Colorado and Arizona. May–Nov.

9. Eriogonum ánnuum Nutt. Annual Eriogonum. Fig. 1591.

Eriogonum annuum Nutt. Trans. Am. Phil. Soc. (II.) **5**: 164. 1833–37.
E. Lindheimerianum Scheele, Linnaea **22**: 149. 1849.

Annual, white floccose-tomentose throughout, simple or branched, leafy below, naked above, 1°–3° tall. Leaves oblong, oblong-lanceolate or oblanceolate, acute or obtuse at the apex, narrowed or acuminate at the base, petioled, the margins somewhat revolute or crisped; inflorescence cymose; involucres top-shaped, 1″–1½″ long, secund, erect, 5-toothed, the teeth obtuse; bracts triangular, not foliaceous; calyx white or whitish, ½″–1″ long, campanulate, 6-cleft to beyond the middle, the lobes obovate; achene pointed, less than 1″ long, its angles smooth, its base almost globular.

On plains, Nebraska to Texas, west to New Mexico, extending into Mexico. July–Sept.

10. Eriogonum campanulàtum Nutt. Narrow-leaved Eriogonum. Fig. 1592.

Eriogonum campanulatum Nutt. Journ. Acad. Phila. (II.) **1**: 163. 1848.
Eriogonum brevicaule Nutt. Journ. Acad. Phila. (II.) **1**: 163. 1848.
Eriogonum micranthum Nutt. Journ. Acad. Phila. (II.) **1**: 164. 1848.

Perennial, scapose, stem short, thick and woody, more or less tomentose; scapes erect or nearly so, glabrous, 4′–12′ tall; leaves crowded, narrowly oblanceolate, spatulate or nearly linear, 1′–3′ long, obtuse at the apex, narrowed into long petioles, white-tomentose on both sides, the margins sometimes revolute; inflorescence compoundly cymose; involucres oblong-turbinate, 1″ long, 5-toothed, teeth obtuse; bracts triangular, not foliaceous; calyx yellow, ovoid-campanulate, about 1″ long, 6-cleft, the lobes oblong or fiddle-shaped, emarginate; stamens and style-branches exserted; achene 1½″ long, enlarged at base.

Nebraska and Kansas to Oregon, Utah and New Mexico. July–Sept.

11. Eriogonum corymbòsum Benth. Crisp-leaved Eriogonum. Fig. 1593.

E. corymbosum Benth. in DC. Prodr. **14**: 17. 1856.

Perennial, woody, densely floccose-tomentose throughout; stem erect, branched, leafy below, naked above, 6′–12′ tall. Leaves oblong, obtuse at the apex, narrowed at the base, petioled, ¾′–1½′ long, their margins more or less crisped; inflorescence compoundly cymose; involucres short-campanulate, 5-toothed, about 1½″ long, the teeth subacute; bracts triangular or triangular-lanceolate, not foliaceous; calyx broadly campanulate, 1″–1½″ long, constricted near the middle, 6-cleft, the segments fiddle-shaped, emarginate, the 3 inner ones shorter than the outer; style-branches exserted; achene 1″ long, enlarged at the base, rough on the angles.

Western Nebraska and Kansas to New Mexico, Utah and Arizona. Aug.–Sept.

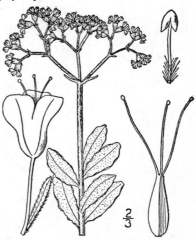

12. Eriogonum microthècum Nutt. Slender Eriogonum. Fig. 1594.

Eriogonum microthecum Nutt. Journ. Acad. Phila. (II.) 1 : 172. 1848.

Perennial, woody, especially below, more or less floccose-tomentose throughout; stem erect or ascending, branched, especially from the base, leafy below, naked above, 6'–12' high. Leaves oblong or oblanceolate, obtuse at the apex, narrowed into short petioles, ½'–2' long, the upper bract-like; inflorescence compoundly cymose; involucres top-shaped, 1½'' long, 5-toothed, the teeth obtusish; bracts triangular; calyx white, pink or reddish, 1½'' long, campanulate, at length constricted near the middle; stamens and style-branches included; achene pointed, 1'' long, rough on the angles.

Western Nebraska to Washington, south to New Mexico and California. July–Oct.

13. Eriogonum effùsum Nutt. Effuse Eriogonum. Fig. 1595.

Eriogonum effusum Nutt. Journ. Acad. Phila. (II.) 1 : 164. 1848.
Eriogonum microthecum effusum T. & G. Proc. Am. Acad. 8 : 172. 1870.

Perennial, shrubby, 6'–16' high, white floccose-tomentose; stem stout, diffusely branched. Leaves linear or narrowly oblong, ½'–1½' long, revolute; peduncles ¾'–4' long, topped by 5–7 times compound stiff corymbose cymes; bracts scale-like; involucres narrowly campanulate, 1''–1½'' high, the tube slightly angled, the teeth rather obtuse; calyx 1'' high, white, glabrous, the outer segments cuneate-obovate, rounded at the apex, the inner notched; filaments villous, especially below the middle; achenes 3-angled.

On dry plains or prairies, South Dakota and Nebraska to Wyoming, south to New Mexico. Summer.

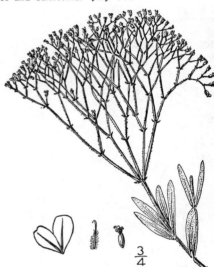

14. Eriogonum cérnuum Nutt. Nodding Eriogonum. Fig. 1596.

E. cernuum Nutt. Journ. Acad. Phila. (II.) 1 : 162. 1848.

Annual, low, stem very short; scape erect, usually much branched, 6'–12' high. Leaves confined to the short stem, orbicular or oblong-orbicular, less than 1' long, obtuse or slightly apiculate, flat, floccose-tomentose, especially beneath, petioled; inflorescence paniculate; involucres campanulate, slightly more than ½'' long, solitary on slender deflexed peduncles 1' long or less, 5-cleft to near the middle, the lobes obtuse; bracts triangular or lanceolate, not foliaceous; calyx whitish, campanulate, ½'' long, 6-parted, slightly constricted near the summit, the segments fiddle-shaped; stamens and style-branches included; achene ½'' long, nearly globular at the base, rough on the angles.

Nebraska and Kansas to New Mexico and Utah. July–Sept.

Eriogonum Górdoni Benth., with the habit of *E. cernuum;* but with glabrous leaves, erect slender peduncles and oblong calyx-segments, extends from Colorado into South Dakota.

3. RÙMEX L. Sp. Pl. 333. 1753.

Perennial or annual, leafy-stemmed herbs, some species slightly woody, the leaves in some mainly basal. Stem grooved, mostly branched, erect, spreading or creeping. Leaves entire or undulate, flat or crisped, the ocreae usually cylindric, brittle and fugacious, the inflorescence consisting of simple or compound, often panicled racemes. Flowers green, perfect, dioecious, or polygamo-monoecious, whorled, on jointed pedicels. Corolla none. Calyx 6-parted, the 3 outer sepals unchanged in fruit, the 3 inner ones mostly developed into wings, one or all three of which usually bears a callosity (tubercle) ; wings entire, dentate, or fringed with bristle-like teeth. Stamens 6, included or exserted; filaments very short, glabrous. Style 3-parted; stigmas peltate, tufted; achene 3-angled, the angles more or less margined. Embryo curved or nearly straight, borne in one of the faces of the 3-angled seed. [The ancient Latin name.]

About 140 species, of wide geographic distribution. Besides the following, some 15 others occur in the southern and western parts of North America. Type species: *Rumex Patientia* L.

* Leaves hastate; flowers dioecious; foliage acid; low species.

Inner sepals not developing wings in fruit ; achene granular.	1. *R. Acetosella.*
Fruiting inner sepals developing wings ; achene smooth.	
Basal leaves numerous ; wings orbicular-cordate.	2. *R. hastatulus.*
Basal leaves few ; wings broadly oblong-cordate.	3. *R. Acetosa.*

** Leaves not hastate; flowers perfect or polygamo-dioecious; foliage scarcely or not at all acid; tall species.

Leaves flat, bright or light green, or glaucescent.	
Wings ½–1½' broad, reddish ; no tubercles.	4. *R. venosus.*
Wings small, not red, bearing tubercles.	
Tubercles usually 3.	
Pedicels little longer than the wings.	
Tubercles broad, nearly or quite as long and often nearly as wide as the wings.	
	5. *R. pallidus.*
Tubercles narrow, much shorter and narrower than the wings.	6. *R. mexicanus.*
Pedicels several times longer than the wings.	7. *R. verticillatus.*
Tubercle usually 1 ; pedicels equalling the wings.	8. *R. altissimus.*
Leaves wavy-margined or crisped, dark green, not glaucescent.	
Wings entire, more or less undulate.	
Lower leaves narrowed or acuminate at the base.	
Tubercle 1.	9. *R. Patientia.*
Tubercles 3.	10. *R. Britannica.*
Lower leaves cordate or rounded at the base.	
Tubercles wanting.	11. *R. occidentalis.*
Tubercles mostly 3.	
Inflorescence not leafy ; pedicels long.	12. *R. crispus.*
Inflorescence leafy ; pedicels short.	13. *R. conglomeratus.*
Tubercle 1 ; inflorescence not leafy ; pedicels short.	14. *R. sanguineus.*
Wings toothed or fringed.	
Lower leaves cordate.	
Wings ovate or oblong-ovate ; tubercles mostly 2.	15. *R. pulcher.*
Wings hastate or ovate-hastate ; tubercle 1.	16. *R. obtusifolius.*
Lower leaves mostly narrowed at base ; wings with 4 spreading bristle-like teeth.	
	17. *R. persicarioides.*

1. Rumex Acetosélla L.   Field, Wood, Red or Sheep Sorrel.   Fig. 1597.

Rumex Acetosella L. Sp. Pl. 338. 1753.

Annual or perennial, glabrous, dioecious, stem slender, erect or nearly so, simple or branched, the rootstock woody, horizontal or creeping. Leaves narrowly hastate, 1'–4' long, obtuse or acute at the apex, usually widest above the middle, petioled, the basal auricles entire or 1–2-toothed, or the uppermost leaves nearly linear and not auricled, all papillose; ocreae silvery, 2-parted, soon lacerate; flowers in erect panicled racemes; racemes interrupted; calyx green, ½'' long, pedicelled; stamens exserted; achene less than 1'' long, very granular, exceeding the persistent calyx, its angles not margined.

In dry fields and on hillsides throughout North America except the extreme north. In large part naturalized from Europe. Sometimes a troublesome weed. Foliage very acid. Native also of Asia. Ascends to 6000 ft. in North Carolina. May–Sept. Sour-dock. Sour-grass or -leek. Mountain- or gentleman's-sorrel. Horse-, cow- or toad-sorrel. Red weed.

¾

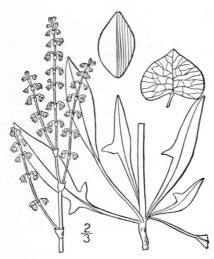

2. Rumex hastátulus Muhl. Engelmann's Sorrel. Fig. 1598.

Rumex hastatulus Muhl. Cat. Ed. 2, 37. 1818.
R. Engelmanni Meisn. in DC. Prodr. 14: 64. 1856.

Perennial from a woody base, glabrous, dioecious; stem rather strict, simple or branched, erect, 5′–20′ tall. Leaves hastate, oblong or oblanceolate, 1′–5′ long, the basal numerous, more or less auricled at the base, subacute, petioled, those of the stem linear, all papillose; ocreae silvery, 2-parted, at length lacerate; racemes ascending, at length interrupted; calyx green, slender-pedicelled, winged in fruit; pedicels equalling or longer than the wings; wings orbicular, mostly broader than high, cordate, 1¼″–1¾″ long; stamens slightly exserted; achene reddish, smooth, shining, less than 1″ long, invested by the calyx-wings, its angles margined.

On the sea-coast, Massachusetts to Florida and on the plains from Illinois to Kansas and Texas, a geographic distribution nearly the same as that of *Chenopodium leptophyllum*. March–Aug.

3. Rumex Acetòsa L. Green Sorrel. Sour or Sharp Dock. Fig. 1599.

Rumex Acetosa L. Sp. Pl. 337. 1753.

Perennial, glabrous, dioecious; stem erect, simple, grooved, 1°–3° tall. Leaves oblong-hastate or ovate-sagittate, 1′–5′ long, acute at the apex, crisped or erose on the margins, the basal few, long-petioled, the upper subsessile, the acute auricles entire or 1-toothed and more or less reflexed; ocreae lacerate; racemes nearly erect, crowded, at length interrupted; calyx green, 1″ long, pedicelled, winged in fruit; pedicels equalling or shorter than the wings, jointed near the middle; wings broadly ovate or orbicular, cordate, 2″–2½″ long; achene rather more than 1″ long, pointed, smooth, shining, blackish, invested by the calyx-wings.

Labrador to Alaska. Naturalized from Europe in Vermont, New York and Pennsylvania. Native also of Asia. Summer. Sour-grass. Green sauce. Meadow, English or cock-sorrel. Red shank.

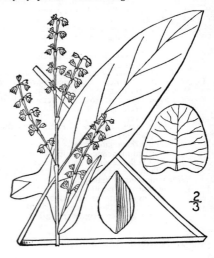

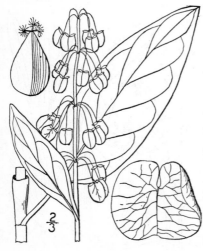

4. Rumex venòsus Pursh. Veined or Winged Dock. Fig. 1600.

Rumex venosus Pursh, Fl. Am. Sept. 733. 1814.

Perennial by a woody rootstock, glabrous; stem rather stout, erect, somewhat flexuous, 6′–15′ tall, grooved, branched. Leaves ovate, ovate-lanceolate or oblong, 1′–5′ long, acute at both ends or acuminate at the base, petioled, rather coriaceous; ocreae funnelform, thin, brittle; racemes mostly erect, soon interrupted; calyx red, pedicelled, very conspicuously winged in fruit; pedicels at maturity rather stout, slightly shorter than the wings, jointed at about the middle; wings large, ½′–1½′ broad, suborbicular with a deep sinus at the base, veiny, reddish; style-branches divergent in fruit; achene 3″ long, smooth, shining, its faces concave, its angles margined.

Saskatchewan to Oregon and Washington, south to Missouri and Nevada. May–Aug.

5. Rumex pállidus Bigel. Large-tubercled Dock. Fig. 1601.

Rumex pallidus Bigel. Fl. Bost. Ed. 3, 153. 1840.

Perennial, glabrous, somewhat glaucous; stem depressed or ascending, often zigzag, 1°–2½° long. Leaves oblong to narrowly lanceolate or almost linear, acute or acuminate at both ends, petioled, rather fleshy; racemes, at least the lower ones, spreading or reflexed, very dense; calyx pale; pedicels shorter than the wings, jointed below the middle; wings deltoid or ovate-deltoid, 1½″–2″ long, undulate, each bearing a large broadly ovoid tubercle; achene 1″–1½″ long, red, its angles narrowly margined.

On beaches, rocks and in salt marshes along the coast, Nova Scotia, New Brunswick and New England. June–Sept.

6. Rumex mexicànus Meisn. White, Pale or Willow-leaved Dock. Fig. 1602.

R. salicifolius Hook. Fl. Bor. Am. 2: 129. 1840. Not Weinm. 1821.
R. mexicanus Meisn. in DC. Prodr. 14: 45. 1856.

Perennial, glabrous, pale green; stem erect or ascending, simple or branched, grooved, flexuous, 1°–3° high. Leaves lanceolate, linear-lanceolate or the lower oblong, acute or acuminate at both ends, or rarely obtuse at the apex, petioled; racemes erect or ascending, dense, in fruit interrupted below; flowers in dense clusters; calyx pale green, 1″ long, pedicelled, winged in fruit; pedicels slightly longer than the wings, jointed near the base; wings triangular-ovate, 1½″ long, undulate or subdentate, each bearing a narrowly ovoid or oblong tubercle; achene 1″ long, dark red, smooth, shining, its faces concave, its angles slightly margined.

In moist, rich or rocky soil, Labrador and Newfoundland to Maine, British Columbia, Texas and Mexico. Occasionally introduced eastward. May–Sept.

7. Rumex verticillàtus L. Swamp Dock. Fig. 1603.

Rumex verticillatus L. Sp. Pl. 334. 1753.

Perennial, glabrous, rather bright green; stem stout, grooved, simple or nearly so, erect, ascending or decumbent, 2°–5° long, more or less flexuous when old. Leaves narrowly oblong, oblong-lanceolate or lanceolate, 2′–12′ long, narrowed at both ends or obtusish at the apex, slightly papillose, long-petioled; racemes interrupted below, spreading in fruit; flowers in rather dense whorls; calyx green, 1″ long, winged in fruit; pedicels stout, thickened above, jointed near the base, 1–5 times as long as the wings; wings broadly deltoid, 2″ long, more or less decurrent on the pedicel, each bearing a narrowly ovoid tubercle; achene 1¼″ long, reddish, pointed, smooth, shining, its faces concave.

In swamps, Quebec and Ontario to Iowa, south to Florida and Texas. May–July.

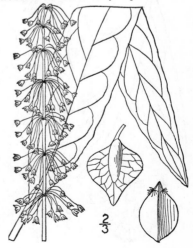

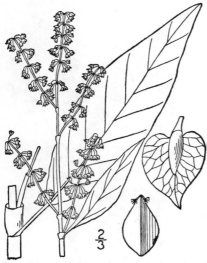

$\frac{2}{3}$

8. Rumex altíssimus Wood. Tall or Peach-leaved Dock. Fig. 1604.

Rumex altissimus Wood, Class-book, 477. 1853.
Rumex Brittannica Meisn. in DC. Prodr. 14: 47. 1856. Not L. 1753.

Perennial, glabrous, rather pale green; stem stout, erect, simple or sparingly branched above, grooved, 2°–4° tall. Leaves lanceolate, oblong-lanceolate or ovate-lanceolate (sometimes oblanceolate), 2′–10′ long, acute at both ends, papillose; panicle rather open; racemes slightly interrupted in fruit; flowers densely whorled; calyx light green, 1″ long, winged in fruit; pedicels slender, jointed near the base, as long as the wings; wings triangular-cordate, 2″–2½″ long, usually one of them only bearing an ovoid tubercle; achene 1½″ long, dark red, smooth, shining, its faces concave.

Along streams and in swamps, Connecticut to Iowa, Nebraska, Maryland and Texas. Pale dock. April–June.

9. Rumex Patiéntia L. Patience Dock. Garden Patience. Fig. 1605.

Rumex Patientia L. Sp. Pl. 333. 1753.

Perennial, glabrous, stem erect, simple or sparingly branched, grooved, 2°–5° tall. Lower leaves ovate-lanceolate, long-petioled, 4′–16′ long, the upper oblong-lanceolate or oblong-elliptic, acute or obtusish, the uppermost lanceolate; fruiting panicle dense; racemes erect, somewhat interrupted in fruit; flowers densely whorled; calyx green; pedicels slender, 2–4 times as long as the calyx-wings, jointed below the middle; wings orbicular-cordate, 2″–3″ long, one of them bearing a prominent but small ovoid callosity; achene 1½″ long, light brown, smooth, shining, its faces concave, its angles obscurely margined.

In waste places, Newfoundland and Ontario to Wisconsin, Connecticut, Pennsylvania and Kansas. Also in the Far West. Naturalized from Europe. Passions. Monk's-rhubarb. May–June.

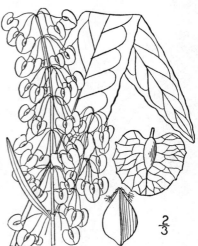

$\frac{2}{3}$

10. Rumex Británnica L. Great Water-Dock. Fig. 1606.

Rumex Britannica L. Sp. Pl. 334. 1753.
Rumex Hydrolapathum var.? *americanum* A. Gray, Man. Ed. 2, 377. 1856.
R. orbiculatus A. Gray, Man. Ed. 5, 420. 1867.

Perennial, glabrous, dark green, stem stout, erect, more or less branched, grooved, 3°–6° tall. Leaves lanceolate or oblong-lanceolate, the lower 1°–2° long, long-petioled, the upper 2′–6′ long, short-petioled; fruiting panicle dense; racemes nearly erect, more or less interrupted; flowers densely whorled; calyx light green; pedicels slender, conspicuously jointed above the base, ½–2 times as long as the calyx-wings; wings broadly cordate, 3″ long, irregularly denticulate, each bearing a callosity; achene ovoid-oblong, or oblong, 2″ long, pointed at both ends, brown, smooth, shining, its faces concave, its angles slightly margined.

In swamps and wet soil, Newfoundland and New Brunswick to Ontario, Minnesota, New Jersey, Pennsylvania, Illinois and Kansas. Horse-sorrel. July–Aug.

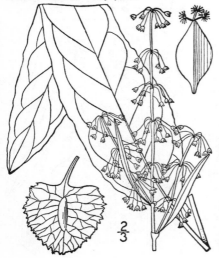

$\frac{2}{3}$

11. Rumex occidentàlis S. Wats. Western Dock. Fig. 1607.

R. occidentalis S. Wats. Proc. Am. Acad. **12** : 253. 1876.

Perennial, glabrous, stem stout, strict, erect or nearly so, strongly grooved, simple or sparingly branched, 2°–3° high. Leaves lanceolate or ovate-lanceolate, bluish-green, somewhat crisped and wavy-margined, papillose, the lower 8′–12′ long, obtuse or subacute at the apex, more or less cordate at the base, long-petioled, the upper smaller and usually lanceolate; panicle rather dense, leafless or nearly so, erect; racemes usually not interrupted; flowers loosely whorled; calyx pale green, 1″ long; pedicels obscurely jointed below the middle, 2–3 times longer than the calyx-wings; wings triangular-ovate, 2½″–4″ long, somewhat dentate or undulate, bearing no tubercles; achene oblong, 2″–2½″ long, short-pointed, chestnut-brown, smooth, shining.

In wet places, Labrador to Alaska, Ontario, Maine and in the Rocky Mountains to Texas and to California. May–Aug.

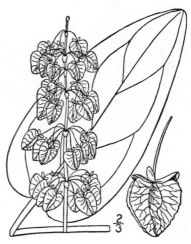

12. Rumex críspus L. Curled or Narrow Dock. Fig. 1608.

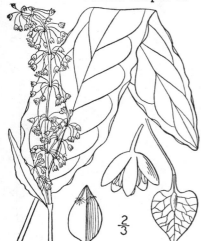

Rumex crispus L. Sp. Pl. 335. 1753.
?*Rumex elongatus* Guss. Pl. Rar. Neap. 150. 1826.

Perennial, glabrous, dark green; stem rather slender, erect, simple or branched above, grooved, 1°–3½° tall. Leaves crisped and wavy-margined, the lower oblong or oblong-lanceolate, 6′–12′ long, long-petioled, the upper narrowly oblong or lanceolate, 3′–6′ long, short-petioled, all cordate or obtuse at the base, more or less papillose; panicle rather open; racemes simple or compound, by the elongation of the pedicels apparently continuous in fruit; flowers rather loosely whorled; calyx dark green; fruiting pedicels 1½–2 times as long as the calyx-wings, jointed near the base; wings cordate, 1½″–2″ long, truncate or notched at base, erose-dentate, or nearly entire, each bearing a tubercle; achene 1″ long, dark brown, shining.

In fields and waste places nearly throughout the United States and southern British America. Often a troublesome weed. Sour or yellow dock. Also in the West Indies and Mexico. Naturalized from Europe. Native also of Asia. Hybridizes with *R. obtusifolius* L. June–Aug.

13. Rumex conglomeràtus Murr. Clustered or Smaller Green Dock. Fig. 1609.

R. conglomeratus Murr. Prodr. Fl. Goett. 52. 1770.

Perennial, glabrous, pale green; stem slender, erect, simple or branched, grooved, 1°–3° tall. Leaves ovate, oblong or lanceolate, 1′–5′ long, some of them slightly fiddle-shaped, acute at the apex, obtuse at the base, crenulate and slightly crisped on the margins, petioled; panicle loose and open in fruit; racemes leafy, slender, ascending, much interrupted; flowers loosely whorled; calyx small, green; pedicels shorter than or equalling the calyx-wings, jointed near the base; wings ovate, fiddle-shaped, 1½″ long, toothed near the base, each bearing a large oblong callosity; achene less than 1″ long, pointed, red, smooth, its faces convex.

In waste places, Virginia to South Carolina. Also in California and Washington. Naturalized from Europe. May–July.

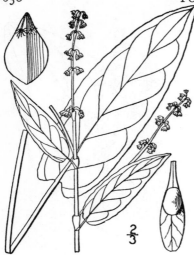

14. Rumex sanguíneus L. Bloody or Red-veined Dock. Bloodwort. Fig. 1610.

Rumex sanguineus L. Sp. Pl. 334. 1753.

Perennial, glabrous, stem slender, erect, grooved, simple or branched, 1°–3° high. Leaves oblong, oblong-lanceolate or lanceolate, 1′–6′ long, the lower long-petioled, cordate at the base, acute or obtuse at the apex, usually red-veined, the upper short-petioled; panicle loose; racemes slender, spreading, not leafy, interrupted; flowers loosely whorled; calyx very small; pedicels slender, 1–1½ times as long as the calyx-wings, jointed at the base; wings oblong, 1½″ long, one of them bearing a spherical-oblong callosity; achene less than 1″ long, sharp-pointed, dark red, smooth, shining, its faces convex.

In waste places and ballast, Massachusetts to southern New York, Virginia and Louisiana. Naturalized or adventive from Europe. Olcott-root. May–Aug.

15. Rumex púlcher L. Fiddle Dock. Fig. 1611.

Rumex pulcher L. Sp. Pl. 336. 1753.

Perennial, dark green; stem slender, erect or procumbent, grooved, diffusely branched, 1°–3° long, the branches spreading. Leaves oblong, or some of the lower fiddle-shaped, 1′–6′ long, long-petioled, obtuse at the apex, cordate at the base; upper oblong or oblong-lanceolate, 1′–3′ long, short-petioled, usually narrowed at both ends; petioles more or less pubescent; panicle loose; racemes long, divergent, sometimes reflexed, much interrupted, rather leafy; flowers few in the whorls; calyx very small, green; pedicels equalling the calyx-wings, jointed at or below the middle; wings ovate or oblong-ovate, 2″ long, truncate at the base, one larger than the others or all three of different sizes, fringed with spine-like teeth, usually two, sometimes one or all three bearing tubercles; achene 1″ long, pointed, reddish, smooth, shining.

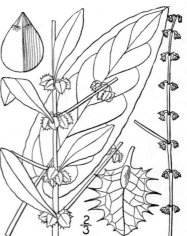

In waste places, Virginia to Florida and Louisiana. Also on the Pacific Coast and in ballast about the northern seaports. Naturalized from Europe. June–Sept.

16. Rumex obtusifòlius L. Broad-leaved or Bitter Dock. Fig. 1612.

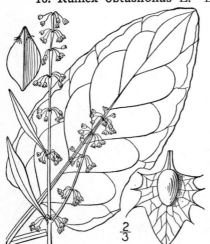

Rumex obtusifolius L. Sp. Pl. 335. 1753.

Perennial, glabrous, dark green; stem stout, erect, simple or sparingly branched, grooved, more or less scurfy above, 2°–4° tall. Lower leaves oblong-lanceolate, 6′–14′ long, long-petioled, all cordate or rounded at the base, obtuse or acute at the apex, the upper lanceolate or oblong-lanceolate, 2′–6′ long, short-petioled, the margins somewhat undulate or crisped; panicle rather open; racemes nearly erect, continuous or interrupted below; flowers loosely whorled; pedicels slender, somewhat longer than the calyx-wings, jointed below the middle; wings hastate, 2″–2½″ long, fringed with a few spreading spiny teeth, one of them bearing an oblong tubercle; achene 1″ long, pointed, dark red, smooth, shining, its angles slightly margined.

In waste places, Nova Scotia and New Brunswick to British Columbia, Oregon, Florida and Texas. Also in the West Indies. Naturalized from Europe. Native also of Asia. Blunt-leaved or butter-dock. Celery-seed. June–Aug.

17. Rumex persicarioìdes L.　Golden Dock.
Fig. 1613.

Rumex persicarioides L. Sp. Pl. 335. 1753.

Annual, pubescent, pale green; stem rather stout, erect and simple, or diffusely branched, 1°–3° high, or sometimes spreading or creeping, very leafy. Leaves lanceolate, or oblong, 1′–12′ long, narrowed at the base, or sometimes cordate, or sagittate, acute at the apex, the margins undulate and more or less crisped; panicle simple or compound; racemes erect, leafy-bracted, mostly interrupted; flowers densely whorled; pedicels slender, 1–1½ times as long as the calyx-wings, jointed at the base; calyx very small; wings oblong, 1″ long, with 1–3 bristles on each margin, each bearing an ovoid or oblong callosity; achene less than 1″ long, pointed, reddish, smooth, shining, its faces convex, its angles slightly margined.

On sandy shores, New Brunswick to Virginia, extending across the continent to British Columbia, south in the interior to Kansas and New Mexico and on the Pacific Coast to California. Has been confounded with *R. maritimus* L. of the Old World. July–Oct.

4. OXÝRIA Hill, Veg. Syst. 10: 24. 1765.

Low fleshy glabrous perennial herbs, with erect stems. Leaves mostly basal, long-petioled, reniform or orbicular, cordate, palmately nerved, with cylindric ocreae. Flowers perfect, small, green, in terminal panicled racemes. Calyx unequally 4-parted, the outer segments smaller than the inner; stamens 6, included; filaments subulate. Ovary 1-celled; ovule solitary; style short, 2-parted, its branches divergent; stigmas fimbriate, persistent on the large wings of the fruit. Achene-body ovate, lenticular, broadly winged. Embryo straight, borne in the centre of the endosperm. [Greek, sour, from the acid leaves.]

Two known species, the following typical; one in the Himalayas.

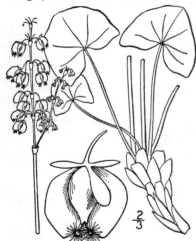

1. Oxyria dígyna (L.) Hill.　Mountain Sorrel.　Fig. 1614.

Rumex digynus L. Sp. Pl. 337. 1753.
O. digyna Hill, Hort. Kew. 158. 1768.
Oxyria reniformis Hook. Fl. Scot. 111. 1821.

Rootstock large, chaffy; stems scape-like, simple or sparingly branched, leafless or nearly so, 2′–12′ tall. Leaves reniform or orbicular-reniform, ½′–1½′ wide, undulate, sometimes emarginate at the apex, the basal long-petioled; ocreae oblique, loose, those on the stem bearing flowers; racemes many-flowered; flowers slender-pedicelled; segments oblong, the inner erect, the outer reflexed in fruit; achene-body pointed, smooth, surrounded by a broad membranous wing.

Greenland and Labrador to Alaska, south to the White Mountains of New Hampshire and in the Rocky Mountains to Colorado and to California. Also in northern Europe and Asia. Sour dock. July–Sept.

5. POLÝGONUM [Tourn.] L. Sp. Pl. 359. 1753.

Annual or perennial, often somewhat shrubby herbs, with terete, but usually striate erect or prostrate stems. Leaves alternate, leathery or somewhat fleshy, sometimes plicate, articulated to the ocreae. Ocreae lobed when young, at length lacerate, hyaline, not fringed. Inflorescence axillary, consisting of clusters bearing normally several flowers at each node throughout the plant or confined to the branches and branchlets. Sepals 5 or 6, mostly green with white or pink or yellow margins, 2 wholly interior, 2 wholly exterior and 1 with one edge exterior and one edge interior. Stamens varying from 3 to 8, often 5 or 6, included; filaments, at least the inner ones, dilated; Styles 3, usually distinct, sometimes very short. Achenes 3-angled, included or slightly exserted, brown or black, granular or smooth and shining. Endosperm horny. Cotyledons incumbent. [Greek, many-knees, from the swollen joints of some species.]

About 100 species, of wide geographic distribution. Besides the following, some 10 others occur in the western parts of North America. Type species: *Polygonum aviculare* L.

* Stem and branches terete and usually striate.

†Achenes much exserted from the calyx.

Plants prostrate; achene broad.
 Sepals decidedly petaloid, very broad, much overlapping and lax at maturity; achene acutish or blunt.   1. *P. maritimum.*
 Sepals slightly petaloid, rather narrow, scarcely overlapping and appressed; achene acuminate.
    2. *P. Fowleri.*

Plants erect; achene narrow.
 Flowers hidden in the ocreae; sepals about ½″ long; achene slender.   3. *P. leptocarpum.*
 Flowers exserted; sepals 1″–1½″ long; achene stout.   4. *P. exsertum.*

††Achenes included in the calyx, or exposed at the tip.

Sepals with white or pink margins.
 Pedicels not exserted from the ocreae; sepals less than 1¼″ long at maturity.
  Achenes with striate faces.
   Mature sepals over ¾″ long; achenes acute.   5. *P. aviculare.*
   Mature sepals less than ¾″ long; achenes acuminate.   6. *P. neglectum.*
  Achenes with granular or nearly smooth faces.
   Plants prostrate; leaves broad; mature sepals about ¾″ long.   7. *P. buxiforme.*
   Plants erect or nearly so; leaves narrow; mature sepals about 1″ long.
     8. *P. prolificum.*
 Pedicels exserted; sepals over 1¼″ long at maturity.   9. *P. atlanticum.*
Sepals with yellowish or greenish margins.
 Leaves broad; achene mostly dull.   10. *P. erectum.*
 Leaves narrow; achene mostly shining.
  Achenes much longer than wide; mature calyx tapering at the base.   11. *P. ramosissimum.*
  Achenes about as wide as long; mature calyx auriculate-cordate at the base.
    12. *P. triangulum.*

** Stem and branches angled.

Leaves plicate; fruiting pedicels short, erect.   13. *P. tenue.*
Leaves flat, margins revolute; fruiting pedicels long, deflexed.   14. *P. Douglasii.*

1. Polygonum marítimum L. Seaside Knotweed. Fig. 1615.

Polygonum maritimum L. Sp. Pl. 361. 1753.
Polygonum glaucum Nutt. Gen. 1: 254. 1818.

Perennial, glaucous, often nearly white, glabrous, root usually deep, woody, stem prostrate or ascending, branched, 8′–20′ long, deeply striate. Leaves oblong, elliptic or sometimes ovate, mostly equalling or longer than the internodes, 3″–12″ long, fleshy, veined beneath, somewhat rugose above, the margins often revolute; ocreae large, silvery, at length lacerate, becoming brown at the base; flowers 1–3 together in the axils, becoming slender-pedicelled; sepals white or pinkish, the margins decidedly pink; achene 3-angled, ovoid, 1½″–2″ long, acute or blunt, smooth, shining, longer than the calyx.

In sands of the seashore, Massachusetts to Florida. Also on the coast of Europe. Coast knot-grass. July–Sept.

2. Polygonum Fówleri Robinson. Fowler's Knotweed. Fig. 1616.

Polygonum Fowleri Robinson, Rhodora 4: 67. 1902.

Perennial, glabrous, pale green or slightly glaucous, stem 3′–24′ long, prostrate, usually much branched, striate. Leaves ovate-lanceolate, oblong or obovate, 3″–15″ long, short-petioled, obtuse or abruptly pointed at the apex, veined beneath, inconspicuoulsy so above, shorter than the internodes or equalling them; ocreae becoming lacerate, silvery, brown and glaucous at the base when old; flowers 2–4 together in the axils; sepals greenish, or the margins white or pinkish; achene ovoid, 3-angled, 2″–2½″ long, slightly granular but shining, acuminate, exceeding the calyx.

In waste places, New Brunswick, Anticosti and Quebec to Maine. Also from Alaska to Washington. May–Sept. *P. Rayi* Babington, with which this was confused in our first edition, is not definitely known to occur on this continent.

3. Polygonum leptocàrpum Robinson. Narrow-pointed Knotweed. Fig. 1617.

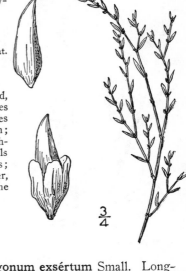

Polygonum leptocarpum Robinson, Proc. Bost. Soc. Nat.
  Hist. **31**: 263. 1904.

Annual, glabrous, stem very slender, copiously branched,
about 1° tall, nearly terete, the branches angled. Leaves
various, the lower ones early deciduous the upper ones
linear-lanceolate to linear, 3″–4″ long, acute, pale green;
ocreae becoming very finely lacerate, brown or reddish-
brown at the base; flowers sessile or nearly so; sepals
4 or 5, green and with whitish or reddish margins;
stamens 4 or 5, included; style 3-parted; achene slender,
3-angled, lanceolate, less than 1½″ long, curved at the
apex.

Kansas.   Sept.–Oct.

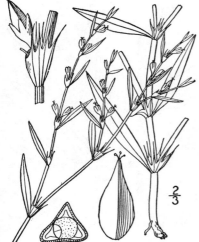

4. Polygonum exsértum Small. Long-fruited Knotweed. Fig. 1618.

P. exsertum Small, Bull. Torr. Club **21**: 172. 1894.

Annual, glabrous, sometimes slightly glaucous,
stem slender, brownish, erect or nearly so, conspicuously striate, branched, 1½°–3° tall. Leaves lanceolate, rarely oblanceolate, ½′–1½′ long, acute or
cuspidate at the apex, acuminate at the base, nearly
sessile; ocreae soon lacerate, silvery, becoming brownish; sepals 6, greenish, with white margins; stamens
5 or 6, included; achene stout, 3-angled, 2½″–3″ long,
more or less constricted above the middle, chestnut-
brown, smooth, shining, 2–3 times as long as the
calyx, at length twisted.

In brackish marshes, New Brunswick to New Jersey;
and on prairies and sand bars, Saskatchewan to Illinois
and Nebraska. Aug.–Oct. This is considered by some a
" semiviviparous " state of *P. ramosíssimum*.

5. Polygonum aviculàre L. Knot-grass. Door-weed. Pink-weed. Fig. 1619.

Polygonum aviculare L. Sp. Pl. 362. 1753.
P. monspeliense Pers. Syn. **1**: 439. 1805.

Annual, pale green or bluish green, stem low
or ascending, simple or branched, 4′–2° long.
Leaves broad, oblong to obovate-oblong, 13″–20″
long, short-petioled, narrowed at the base, usually
acute at the apex or sometimes obtuse; ocreae
silvery, at length lacerate; clusters 1–5-flowered;
flowers relatively large, pedicelled; sepals 5, green,
except the white, pink or purplish borders, 1″–1½″
long; stamens 5–8; achene 3-angled, ovoid, 1½″–2″
long, acute, striate-reticulated.

A weed in cultivated and waste grounds, common
almost throughout North America, Asia and Europe.
Bird-, beggar-, stone- or wire-weed. Door-, wire-,
way- or crab-grass. Cow-, goose-, swine- or bird-
grass. Bird's knot-grass. Bird's- or sparrow-tongue.
Ninety-knot. June–Oct.

5. Polygonum negléctum Besser. Narrow-leaved Knotweed. Fig. 1620.

P. neglectum Besser, Enum. Pl. Vol. 45. 1822.
P. aviculare angustissimum Meisn., in DC. Prodr. 14: 98. 1856.

Annual, dull green, stem usually widely much-branched at the base, mostly ½°–2° long. Leaves narrow, elliptic to elliptic-lanceolate or linear, 3″–10″ long, acute or somewhat acuminate at both ends, sessile or nearly so, those of the branchlets often conspicuously small and inclined to be obtuse; ocreae silvery-brown; flowers often 2 together, sessile or short-pedicelled; sepals green, except the pale pink to purplish-red margins, about ½″ long; achene 3-angled, ovoid, about 1″ long, acuminate, striate-reticulated.

A weed in yards, cultivated and waste grounds, nearly throughout North America, and common in nearly all parts of the north temperate zone. June–Nov.

7. Polygonum buxifórme Small. Shore Knotweed. Fig. 1621.

P. buxiforme Small, Bull. Torr. Club 33: 56. 1906.

Annual, stout, bright green or slightly glaucous, stem 1°–4° long, prostrate, diffusely branched from a woody base, striate. Leaves broad, oblong, oblong-lanceolate or oblanceolate or oval, 2″–9″ long, obtuse or subacute at the apex, thick, short-petioled; ocreae at length lacerate; flowers 2–6 in each cluster; sepals green, except the whitish margins, or carmine; stamens 8; achene broadly ovoid, 3-angled, 1″–1½″ long, more or less contracted at the apex, dark-brown, the faces often granular.

On shores and in waste places, New Brunswick to Minnesota and British Columbia, south to Virginia, Illinois and New Mexico. Aug.–Sept. Mistaken in first edition for *P. littoràle* Link, a European species with reddish-brown narrower and shining achenes, which has recently been reported from Mt. Desert, Maine.

Polygonum provinciàle C. Koch, a more slender plant than *P. buxiforme,* with narrower, revolute acute leaves, smaller flowers and narrower achenes, occurs locally in our range. It is native of Europe.

8. Polygonum prolíficum (Small) Robinson. Proliferous Knotweed. Fig. 1622.

P. ramosissimum prolificum Small, Bull. Torr. Club 21: 171. 1894.
P. prolificum Robinson, Rhodora 4: 68. 1902.

Annual, glabrous, stem erect or nearly so, 1½° tall or less, usually much branched. Leaves narrow, spatulate, linear-spatulate or nearly oblong, rather slender-petioled, 5″–15″ long, obtuse or acutish; ocreae inconspicuous, early much-lacerate; flowers exceeded by the ocreae; calyx sessile or nearly so, the sepals pinkish, slightly enlarged in fruit; stamens mostly 5; achene rhombic-ovoid, about 1¼″ long, 3-angled, slightly roughened, abruptly contracted at the apex.

In brackish marshes and on the sea-shore, Quebec and Maine to Virginia; and in the interior from South Dakota to Missouri and Colorado. July–Oct.

9. Polygonum atlánticum (Robinson) Bicknell. Atlantic Coast Knotweed. Fig. 1623.

Polygonum ramosissimum atlanticum Robinson, Rhodora **4**: 72. 1902.
P. atlanticum Bicknell, Bull. Torr. Club **36**: 450. 1909.

Annual, glabrous, bright-green, stem erect, 1°–3° tall, terete, but striate. Leaves narrowly elliptic or linear-elliptic, or broadest above or below the middle, slender-petioled, 10″–25″ long, acute or slightly acuminate at both ends; ocreae brown, early much-lacerate; calyx long-pedicelled, the sepals 5 or rarely 6, with pink margins; stamens mostly 5; achene ovoid or rhombic-ovoid, about 1½″ long, 3-angled, smooth and shining.

In salt marshes and low brackish grounds, Maine to Rhode Island. Aug.–Sept.

Polygonum Bellàrdi All., admitted into our first edition, appears in our range only as a waif from the Old World.

$\frac{3}{4}$

$\frac{2}{3}$

10. Polygonum eréctum L. Erect Knotweed. Fig. 1624.

Polygonum erectum L. Sp. Pl. 363. 1753.

Annual, glabrous, yellowish-green, stem erect or ascending, 8′–2° high, terete, nearly simple or much branched. Leaves oval, oblong or obovate, subsessile or short-petioled, 3″–18″ long, obtuse or subacute at the apex, conspicuously jointed to the ocreae; ocreae oblique, soon lacerate, silvery when young; flowers 1–2 together in the axils; sepals greenish-yellow or yellowish, enlarged in fruit; stamens 6 (sometimes 5); achene ovoid-pyramidal, 3-angled, 1″–1½″ long, dull, invested by the calyx, or the apex slightly protruding.

In moist or dry soil, Ontario to the Northwest Territory, Tennessee and Arkansas. July–Sept.

11. Polygonum ramosíssimum Michx. Bushy Knotweed. Fig. 1625.

P. ramosissimum Michx. Fl. Bor. Am. **1**: 237. 1803.

Annual, yellowish or yellowish-green, glabrous, stem erect or ascending, usually very much branched, slender, striate, usually rigid, 4′–4° tall. Leaves lanceolate or linear-oblong, short-petioled, 3″–20″ long, acute or acuminate at both ends; ocreae few-nerved, becoming deeply lacerate; flowers several together in the axillary clusters, short-pedicelled; sepals 5 or 6, yellow or yellowish, 1″–1½″ long; stamens 6 or fewer; achene 3-angled, acute, sometimes slightly protruding beyond the calyx, nearly 1½″–2″ long, black.

In saline soil, Minnesota to the Northwest Territory, New Mexico and California. July–Sept.

Polygonum campòrum Meisn., admitted into our first edition, is here omitted as not certainly known within our area.

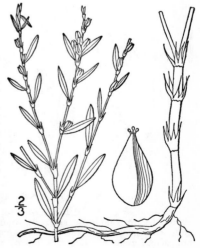

$\frac{2}{3}$

12. Polygonum triángulum Bicknell. Missouri Knotweed. Fig. 1626.

Annual, bright green or somewhat yellowish-tinged, similar to *P. ramosissimum* in habit, but smaller, stem rather sparingly branched, usually 2°-4° tall. Leaves mostly oblong or elliptic or slightly broadened upward, 4″-14″ long, acute, rather persistent; ocreae becoming very much lacerate; flowers sparingly clustered, short-pedicelled; sepals mostly 5, yellowish-margined, ½″-¾″ long; stamens mostly 5; achene 3-angled, acute, about 1″ long, broadly ovoid.

Common near Atherton, Missouri. Aug.–Sept.

13. Polygonum ténue Michx. Slender Knotweed. Fig. 1627.

Polygonum tenue Michx. Fl. Bor. Am. 1 : 238. 1803.

Annual, glabrous, somewhat rough about the nodes, stem very slender or filiform, erect, simple or branched, somewhat 4-angled, 4′-12′ tall. Leaves linear or linear-lanceolate, sessile, acuminate at the apex, 2″-12″ long, articulated to the ocreae, 1-ribbed with a lateral impression on each side of the rib, the margins minutely scabrous or serrulate; ocreae funnelform, soon lacerate; flowers several in the axillary clusters, green, subsessile; sepals whitish; stamens 8; fruit erect; achene 3-angled, black, 1″-1½″ long, reticulated on the angles, the centre of its faces smooth.

Dry soil, Ontario to Minnesota, Nebraska, Georgia and Arkansas. July–Sept.

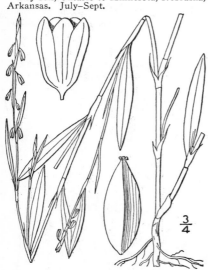

14. Polygonum Douglásii Greene. Douglas' Knotweed. Fig. 1628.

Polygonum Douglasii Greene, Bull. Cal. Acad. (II.) 1 : 125. 1885.

Annual, similar to the preceding species, glabrous, somewhat rough at the nodes, sometimes slightly glaucous, stem erect, 8′-18′ tall, simple or usually much branched, almost terete. Leaves oblong or narrowly lanceolate, ½′-2′ long, subsessile, rather thin, flat or revolute, with no lateral impressions parallel to the midrib; ocreae oblique, short, soon lacerate; clusters axillary, several-flowered; the flowers and fruit deflexed; sepals green with white or rose-colored margins; stamens 8; achene 3-angled, 1¼″-2″ long, oblong or ovoid-oblong, black, smooth and shining.

Northwest Territory and British Columbia to New Mexico, Nebraska and Oklahoma, east through Ontario and New York to Vermont. June–Sept.

6. TOVÀRA Adans. Fam. Pl. 2 : 276. 1763.

Annual or perennial herbaceous plants, becoming somewhat woody below. Stem mostly erect, virgate, simple or virgately branched. Leaves alternate, membranous, acute at both ends, continuous with the ocreae. Ocreae cylindric, fringed with bristles. Flower-clusters not dense, remote. Racemes linear, very long and wand-like, conspicuously interrupted.

Calyx more or less colored, somewhat curved. Sepals 4, the 2 lateral ones overlapping the others. Stamens 4, and alternating with the sepals, or 5, the fifth one opposite the lower sepal; filaments barely flattened, erect or slightly spreading. Styles 2, conspicuously exserted, recurved or curled. Achenes lenticular, strongly biconvex, brown or cream-colored, smooth and shining. Endosperm horny. Cotyledons accumbent. [Derivation uncertain.]

Two known species, the following typical, the other Japanese.

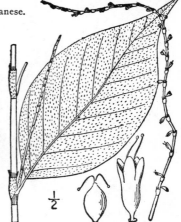

1. **Tovara virginiàna** (L.) Raf.   Virginia Knotweed. Fig. 1629.

Polygonum virginianum L. Sp. Pl. 360. 1753.
Tovara virginiana Raf. Fl. Tell. 3: 12. 1836.

Plants nearly glabrous or strigose-pubescent, stem erect or arching, simple or branched above, 1°–4° tall. Leaves ovate or elliptic-ovate or ovate-lanceolate, short-petioled, acuminate at the apex, 2′–6′ long, sparingly ciliate; ocreae cylindric, strigose, fringed with short bristles; racemes spicate, erect, terminal and axillary, naked, greatly elongated and interrupted, sometimes 12′ long; calyx curved, greenish or rose-color, 4-cleft; stamens 5; style long, exserted, 2-parted to the base, its branches at length curled; achene 2″ long, ovate-oblong, lenticular, strongly biconvex, dark brown or cream-colored, smooth, shining.

In woods, Nova Scotia to Minnesota, south to Florida and Texas. Ascends to 4000 ft. in North Carolina. July–Nov.

7. **PERSICARIA** (Tourn.) Mill. Gard. Dict. Abr. Ed. 4. 1754.

Annual or perennial, often pubescent or glandular caulescent herbs, various in habit, never twining nor climbing. Leaves alternate, entire, continuous with the ocreae, often glandular-punctate. Ocreae cylindric, mostly membranous, truncate, naked, ciliate or fringed with bristles. Racemes spike-like, varying from linear to ovoid, dense and erect, or few-flowered, lax and drooping. Ocreolae funnelform, naked, ciliate or fringed. Pedicels rather stout, articulated at the base of the calyx. Calyx more or less colored, varying from white and green to red, often glandular-punctate, investing the achene. Sepals mostly 5, 2 wholly exterior, 2 wholly interior and 1 with one margin interior and the other exterior, none of them winged or keeled. Stamens varying from 4 to 8, included or exserted; filaments not dilated, erect or nearly so. Styles mostly 2, sometimes 3, usually partially united, included or exserted; stigmas capitate. Achene mostly lenticular, sometimes 3-angled and lenticular on the same plant, usually black, smooth or granular. Endosperm horny. Cotyledons accumbent. [From *Persica,* from the resemblance of the leaves to those of the peach.]

About 125 species, widely distributed. Type species: *Polygonum Persicaria* L.

Racemes solitary or paired.
  Raceme short and stout; leaves obtuse or merely acute.   1. *P. amphibia.*
  Raceme long and slender; leaves acuminate.   2. *P. Muhlenbergii.*
Racemes several or numerous.
  Ocreae not fringed with bristles.
    Racemes drooping.   3. *P. lapathifolia.*
    Racemes erect.
      Style and stamens included or slightly exserted; achene-faces concave.
        Achene biconvex, broadly oblong; plants perennial.   4. *P. portoricensis.*
        Achene concave, orbicular; plants annual.   5. *P. pennsylvanica.*
      Style or stamens conspicuously exserted; achene-faces swollen. 6. *P. longistyla.*
  Ocreae bristle-fringed.
    Ocreae without spreading borders; leaves relatively narrow, short-petioled or sessile; stigmas minute.
      Stem, branches and peduncles rough-glandular.   7. *P. Careyi.*
      Stem, branches and peduncles not rough-glandular.
        Sepals not glandular-punctate.
          Racemes short, stout and compact.
            Ocreae inconspicuously fringed; achene narrowly ovoid. 8. *P. Persicaria.*
            Ocreae conspicuously fringed; achene broadly ovoid. 9. *P. persicarioides.*
          Racemes slender, elongated and lax or interrupted.
            Ocreae strigose, fine-bristly; leaves glabrous or somewhat strigose.
              Calyx greenish-white; ocreolae copiously long-bristly. 10. *P. opelousana.*
              Calyx white, pink or purplish-pink; ocreolae sparingly fine-bristly.
                          11. *P. hydropiperoides.*
            Ocreae hirsute or appressed-hirsute, coarse-bristly; leaves conspicuously appressed-hirsute. 12. *P. setacea.*
        Sepals glandular-punctate.
          Achene granular and dull; racemes drooping.   13. *P. Hydropiper.*
          Achene smooth and shining; racemes erect.   14. *P. punctata.*
    Ocreae with spreading border; leaves broad, long-petioled; stigmas large.
                                15. *P. orientalis.*

$\frac{3}{5}$

1. Persicaria amphíbia (L.) S. F. Gray.
Water Persicaria. Willow-weed.
Fig. 1630.

Polygonum amphibium L. Sp. Pl. 361. 1753.
Persicaria amphibia S. F. Gray, Nat. Am. Brit. Pl. **2**: 268. 1821.
Polygonum Hartwrightii A. Gray, Proc. Am. Acad. **8**: 294. 1870.

Aquatic, perennial, glabrous or pubescent; stem floating or submersed, simple or sparingly branched, 4°–20° long. Leaves oblong, elliptic or elliptic-lanceolate or narrowly lanceolate, 1½′–4′ long, petioled, obtuse or subacute at the apex, slightly inequilateral, rounded or narrowed at the base, sometimes ciliate; ocreae cylindric, those of the branches often longer than the internodes, their limbs sometimes spreading, usually glabrous; raceme terminal, usually solitary, ½′–1′ long, dense, erect, oblong or ovoid; calyx rose-color, 5-parted; stamens 5, exserted; style 2-cleft, exserted; achene orbicular-oblong, 1½″ long, biconvex, black, smooth and shining, or granular.

In ponds and lakes, Quebec to Alaska, New Jersey, Kentucky, Colorado and California. Europe. Ascends to 2000 ft. in the Adirondacks. Ground-willow. Willow-grass. Red shanks. Heartsease. July–Aug.

Several species, reducible to this and the following, have been described by Dr. E. L. Greene (Leaflets **1**: 26–45).

2. Persicaria Muhlenbérgii (S. Wats.)
Small. Swamp Persicaria. Fig. 1631.

Polygonum amphibium var. *emersum* Michx. Fl. Bor. Am. **1**: 240. 1803.
P. Muhlenbergii S. Wats. Proc. Am. Ac. **14**: 295. 1879.
P. emersum Britton, Trans. N. Y. Acad. Sci. **8**: 73. 1889.
Persicaria Muhlenbergii Small; Rydb. Fl. Colo. 111. 1906.

Perennial by long creeping or horizontal rootstocks, glabrous or strigose-pubescent; stem erect or assurgent, commonly simple, channeled, enlarged at the nodes, 1°–3° high. Leaves ovate-lanceolate or oblong-lanceolate, or the upper sometimes narrowly lanceolate, 2½′–8′ long, acute or usually acuminate at the apex, rounded or cordate at the base, petioled, the lateral nerves prominent, sometimes forking; ocreae cylindric, becoming loose, not ciliate; racemes 1 or 2, erect, 1′–3′ long, linear-oblong, dense; calyx dark rose-color, 5-parted; stamens 5, exserted; style 2-cleft, exserted; achene broadly obovate or orbicular, 1½″ long, very convex, lenticular, black and slightly granular, but shining.

In swamps and moist soil, Ontario to British Columbia, Virginia, Louisiana and Mexico. July–Sept.

$\frac{1}{2}$

$\frac{1}{2}$

3. Persicaria lapathifòlia (L.) S. F. Gray.
Dock-leaved or Pale Persicaria. Fig. 1632.

Polygonum lapathifolium L. Sp. Pl. 360. 1753.
P. incarnatum Ell. Bot. S. C. & Ga. **1**: 456. 1817.
P. lapathifolia S. F. Gray, Nat. Arr. Brit. Pl. **2**: 270. 1821.
P. lapathifolium incanum Koch, Syn. Fl. Germ. 711. 1837.
P. lapathifolium nodosum Small, Mem. Torr. Club **5**: 140. 1894.

Annual, stem simple or much branched, erect or ascending, swollen at the nodes, 1°–3° high, the peduncles and pedicels often glandular. Leaves lanceolate or oblong-lanceolate, 2′–10′ long, usually broader than those of the preceding, attenuate to the apex, acuminate at the base, short-petioled, ciliate, glabrous or pale-pubescent, inconspicuously punctate; ocreae cylindric, ribbed or striate; racemes panicled, 1′–4′ long, drooping, narrow, rather dense; calyx pink, greenish or white, 5-parted; stamens 6; achene broadly oblong or ovoid, lenticular, 1″ long, brownish or black, slightly reticulated but shining, its faces concave.

In waste places, throughout temperate North America. Naturalized from Europe. Sometimes a troublesome weed. Native also of Asia. Willow-weed. June–Sept.

Persicaria tomentòsa (Schrank) Bicknell (*Polygonum tomentosum* Schrank) has been separated from *P. lapathifolia* by the pubescent leaves and slightly larger flowers and fruits.

4. Persicaria portoricénsis (Bertero) Small. Dense-flowered Persicaria. Fig. 1633.

Polygonum densiflorum Meisn. in Mart. Fr. Bras. **5**: Part 1, 13. 1855. Not Blume, 1825–26.
Polygonum portoricense Bertero; Meisn. in DC. Prodr. 14: 121. 1856.
Persicaria portoricensis Small, Fl. SE. U. S. 377. 1903.

Perennial, more or less scurfy; stem erect, decumbent or floating, 3°–5° long or longer, branched, enlarged at the nodes, often dark brown. Leaves lanceolate or linear-lanceolate, 1½′–12′ long, acuminate at both ends, obscurely punctate, short-petioled, the nerves prominent beneath; ocreae cylindric, sometimes bristly when young, naked when mature, sometimes hispid; racemes spicate, paniculate, often in pairs, 1′–5′ long, dense, erect; calyx white or whitish, 5-parted to near the base; stamens 6–8, included; style 2–3-cleft, somewhat exserted; achene lenticular and strongly biconvex or 3-angled, 1″–1½″ long, broadly oblong, orbicular or even broader than high, black, smooth and shining, or minutely granular.

In wet soil, southern New Jersey and Missouri to Florida, Texas, the West Indies and South America. May–Nov.

5. Persicaria pennsylvánica (L.) Small. Pennsylvania Persicaria. Fig. 1634.

Polygonum pennsylvanicum L. Sp. Pl. 362. 1753.
Persicaria pennsylvanica Small, Fl. SE. U. S. 377. 1903.

Annual, glabrous below; stem erect, simple or branched, 1°–3° tall, the upper parts, the peduncles and pedicels glandular. Leaves lanceolate, acuminate at the apex, petioled, ciliate, 2′–11′ long, the upper sometimes glandular beneath; ocreae cylindric, thin, naked, glabrous; racemes panicled, erect, thick, oblong or cylindric, dense, 1′–2′ long; calyx dark pink or rose-color, 5-parted; stamens 8 or fewer, style 2-cleft to about the middle; achene orbicular or mostly broader than high, 1½″ long, short-pointed, lenticular, smooth, shining.

In moist soil, Nova Scotia to Ontario, Minnesota, Florida and Texas. Ascends to 2000 ft. in Virginia. July–Sept.

Persicaria omíssa (Greene) Small, differing by an ovoid achene and copiously glandular peduncles, recently found in western Kansas, was first discovered in Colorado.

6. Persicaria longístyla Small. Long-styled Persicaria. Fig. 1635.

Polygonum longistylum Small, Bull. Torr. Club **21**: 169. 1894.
Persicaria longistyla Small, Fl. SE. U. S. 377. 1903.

Annual or perennial, glabrous except the glandular upper branches and peduncles; stem erect, rather slender, 1°–3° tall, becoming somewhat woody below. Leaves lanceolate or ovate-lanceolate, 1′–6′ long, acuminate at both ends, petioled, their margins undulate, slightly crisped, more or less ciliolate; ocreae cylindric, entire, brittle, soon falling away; racemes panicled, sometimes geminate, 1′–4′ long, rather dense, erect; calyx lilac, spicuously exserted; stigmas black; achene broadly 5-parted to below the middle, the lobes petaloid; stamens 6–8, included; style 2-parted, slender, conovoid, lenticular, slightly gibbous on both sides, long-pointed, black, granular, but somewhat shining, 1¼″ long.

In moist soil, southern Illinois, Missouri and Kansas to Louisiana and New Mexico. Aug.–Oct.

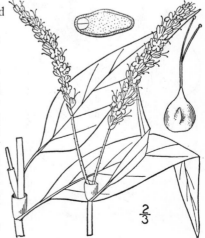

7. Persicaria Càreyi (Olney) Greene. Carey's Persicaria. Fig. 1636.

Polygonum Careyi Olney, Proc. Providence Franklin Soc. 1: 29. 1847.
Persicaria Careyi Greene, Leaflets 1: 24. 1904.

Annual, rough-glandular throughout, stem erect, 1°–3° tall, simple or sparingly branched above. Leaves oblong-lanceolate or linear-lanceolate, the uppermost nearly linear, 2'–11' long, short-petioled the midrib, ciliate, sparingly punctate; ocreae cylin- or nearly sessile, acuminate at both ends, hispid on dric, sparsely hispid, fringed with long brïstles; racemes several, narrow, terminal, loosely-flowered, drooping, 1'–2½' long; calyx purplish; stamens 5 or sometimes 8; style 2-parted to below the middle; achene lenticular, broadly ovoid or obovoid, 1¼"–1½" long, short-pointed, thick, smooth and shining.

In marshes, Maine and Ontario to Rhode Island, New Jersey, Delaware and Pennsylvania. Ascends to 2000 ft. in Pennsylvania. Also in Michigan. July–Sept.

8. Persicaria Persicària (L.) Small. Lady's Thumb. Heartweed. Fig. 1637.

Polygonum Persicaria L. Sp. Pl. 361. 1753.
Persicaria Persicaria Small, Fl. SE. U. S. 378. 1903.

Annual, glabrous or puberulent; stem erect or ascending, simple or much branched, ½°–2° high. Leaves lanceolate or linear-lanceolate, 1'–6' long, short-petioled or nearly sessile, acuminate at both ends, conspicuously punctate, usually with a dark triangular or lunar blotch near the centre, their margins entire or slightly eroded, often ciliate; ocreae cylindric, nearly glabrous, fringed with short bristles; racemes solitary or panicled, ½'–2' long, ovoid or oblong, dense, erect; calyx pink or dark purple; stamens mostly 6; style 2–3-parted to below the middle; achene broadly ovate and lenticular, often gibbous or 3-angled, 1"–1¼" long, smooth and shining.

In waste places, throughout North America, except the extreme north. Naturalized from Europe. Often an abundant weed. Common persicary. Spotted-knotweed. Red- or pink-weed. Heartsease. Peach-wort. Willow-weed. Red-shanks. Lover's-pride. Black-heart. June–Oct.

9. Persicaria persicarioìdes (H.B.K.) Small. Southwestern Persicaria. Fig. 1638.

Polygonum persicarioides H.B.K. Nov. Gen. 2: 179. 1817.
Persicaria persicarioides Small, Fl. SE. U. S. 378. 1903.

Perennial, glabrous or minutely pubescent; stem erect, decumbent or creeping, simple or branched, 1°–3° long. Leaves lanceolate or linear-lanceolate, acuminate at both ends, punctate, short-petioled or subsessile, 1½'–10' long; ocreae cylindric, glabrous or sparingly strigillose, fringed with short bristles; spicate racemes more or less panicled, erect, 1'–3' long, narrowly oblong or linear, loosely-flowered; calyx rose-color tinged with green, 5-parted to below the middle; stamens 8 or fewer, included; style 2–3-parted to near the base; achene lenticular and biconvex, or 3-angled, more or less gibbous, 1" long, ovoid or broadly oblong, short-pointed, black, minutely granular, but shining.

Nebraska to Mexico; widely distributed in tropical America. June–Sept.

10. Persicaria opelousàna (Riddell)
Small. Opelousas Persicaria.
Fig. 1639.

Polygonum opelousanum Riddell; Small, Bull. Torr.
  Club **19** : 354. 1892.
Persicaria opelousana Small, Fl. SE. U. S. 378. 1903.

Perennial, glabrous or nearly so throughout,
stem slender, erect or ascending, sparingly or
considerably branched, 1°–3° tall, becoming woody
below. Leaves linear or linear-lanceolate, 1½′–5′
long, sessile, ciliate; ocreae cylindric, strigose,
fringed with slender bristles; spicate racemes
panicled, erect, often geminate, ½′–2′ long, not
densely flowered; calyx greenish-white, pedicelled,
5-parted to below the middle; stamens 8 or fewer,
included; style deeply 3-parted; achene 3-angled
or rarely 4-angled, broadly ovoid or obovoid, ¾″–
1″ long, black, smooth and shining.

In wet soil, Massachusetts and Missouri to Louisiana, Texas and Mexico. July–Sept.

11. Persicaria hydropiperoìdes (Michx.)
Small. Mild Water Pepper. Fig. 1640.

P. hydropiperoides Michx. Fl. Bor. Am. **1** : 239. 1803.
Polygonum mite Pers. Syn. **1** : 440. 1805.
Persicaria hydropiperoides Small, Fl. SE. U. S. 378. 1903.
Polygonum hydropiperoides Macouni Small, Mem. Dept.
  Bot. Col. Coll. **1** : 81. 1895.

Perennial, glabrous or strigillose, stem erect, decumbent or prostrate, simple or branched above, slender,
1°–3° long. Leaves narrowly lanceolate or oblong-
lanceolate, varying to linear-lanceolate, 2′–6′ long,
short-petioled, ciliate, pubescent with appressed hairs
on the midrib beneath; ocreae cylindric, loose, strigose,
fringed with slender bristles; racemes panicled, terminal, erect, narrow, more or less interrupted, 1½′–3′
long; calyx white, pink or purplish-pink; stamens 8;
style 3-parted to below the middle; achene 3-angled,
ovoid or oblong, 1″–1¼″ long, smooth, shining.

In swamps and wet soil, New Brunswick to Minnesota
and California, south to Florida and Mexico. June–Sept.

12. Persicaria setàcea (Baldw.) Small.
Bristly Persicaria. Fig. 1641.

Polygonum setaceum Baldw.; Ell. Bot. S. C. & Ga.
  1 : 455. 1817.
Persicaria setacea Small, Fl. SE. U. S. 379. 1903.

Perennial, appressed-hirsute, stem 2°–4° high,
erect, simple or sparingly branched; leaves lanceolate or oblong-lanceolate, 2′–9′ long, mostly
short-petioled, acuminate at both ends, ciliate,
conspicuously pubescent on both sides, inconspicuously punctate; ocreae cylindric, long, hirsute or appressed-hirsute, fringed with very long,
stout bristles; racemes few, terminal, 1′–2½′ long,
erect, linear-oblong, sometimes geminate, rather
loosely flowered; calyx white or pink; stamens 8;
style 3-parted to below the middle; achene 3-
angled, oblong or obovoid, short, thick-pointed,
1″–1½″ long, minutely reticulated and rather dull,
or smooth and shining.

In swamps, Massachusetts and Missouri, Louisiana
and Florida. June–Sept.

13. Persicaria Hydrópiper (L.) Opiz. Smart-weed. Water Pepper. Fig. 1642.

Polygonum Hydropiper L. Sp. Pl. 361. 1753.
Persicaria Hydropiper Opiz, Seznam 72. 1852.

Annual, glabrous, stem erect, simple or branched, red or reddish, sometimes green, 8′–24′ tall. Leaves lanceolate or oblong-lanceolate, 1′–4′ long, short-petioled, acute or acuminate at the apex, undulate or slightly crisped, punctate, ciliate, very acrid; ocreae cylindric, fringed with short bristles, sometimes slightly pubescent, usually swollen at the base by the development of several flowers within; racemes panicled, 1′–3′ long, narrow, drooping, interrupted; calyx green, 3–5-parted (usually 4-parted) conspicuously punctate; stamens 4 or sometimes 6; style short, 2–3-parted; achene lenticular or 3-angled, broadly oblong or ovoid, slightly gibbous, 1¼″–1½″ long, granular, dull.

⅔

In moist waste places, almost throughout North America. Naturalized from Europe in our area, perhaps indigenous in the far Northwest. Biting-persicaria or -knotweed. Bite-tongue. Snake- or sickle-weed. Pepper-plant. Red-shanks. Red-knees. July–Sept.

14. Persicaria punctàta (Ell.) Small. Dotted or Water Smart-weed. Fig. 1643.

P. punctatum Ell. Bot. S. C. & Ga. **1**: 455. 1817.
P. acre H.B.K. Nov. Gen. **2**: 179. 1817. Not Lam.
Persicaria punctata Small, Fl. SE. U. S. 379. 1903.
Polygonum punctatum leptostachyum (Meisn.) Small, Bull. Torr. Club **19**: 356. 1892.
Polygonum acre var. *leptostachyum* Meisn. in DC. Prodr. **14**: 108. 1856.

Annual or perennial, glabrous or very nearly so, stem erect or ascending, rarely prostrate, simple or branched, 1°–3° long. Leaves linear-lanceolate to oblong-lanceolate, 1′–8′ long, acuminate at both ends, petioled, ciliate, conspicuously punctate, acrid, the midrib often with a few scattered hairs; ocreae cylindric; falling away at maturity; fringed with long bristles; racemes terminal, narrow, erect or slightly drooping; loosely flowered, ½′–3′ long; calyx greenish; stamens 8; style 2–3-parted to the base; achene oblong, short, thick, lenticular or 3-angled, 1″–1¼″ long, smooth, shining.

⅔

In swamps and wet places, throughout North America except the extreme north. Also in the West Indies. Turkey-troop. Water-pepper. June–Oct.

Persicaria robustior (Small) Bicknell (*Polygonum punctatum robustius* Small) seems to be only a stout form of the above species with somewhat larger flowers and fruits.

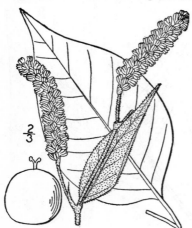

15. Persicaria orientàlis (L.) Spach. Prince's Feather. Fig. 1644.

Polygonum orientale L. Sp. Pl. 362. 1753.
Persicaria orientalis Spach, Hist. Nat. Veg. **10**: 535. 1841.

Annual, more or less hispid, stem erect, 1°–8° tall, branched. Leaves ovate or broadly oblong, 3′–12′ long, petioled, acuminate at the apex, ciliate; petioles slightly winged; ocreae cylindric, loose, with or without a spreading border, ciliate; racemes panicled, oblong-cylindric, 1′–4′ long, dense, drooping; flowers large for the genus; calyx dark rose-color or crimson; stamens 7, exserted; style 2-cleft to above the middle, included; stigmas large; achene orbicular or broader than long, lenticular, flat, nearly 1½″ in diameter, finely reticulated and rather dull.

⅔

In waste places, escaped from gardens throughout eastern North America. Native of India. Ragged-sailor. Gentleman's-cane. Aug.–Sept.

8. BISTÓRTA Adans. Fam. Pl. 2: 277. 1763.

Perennial herbs with corm-like scaly rootstocks. Stems erect, simple. Leaves alternate, mostly basal, narrow or sometimes rather broad, continuous with the ocreae. Ocreae elongated, oblique at the top, persistent. Flower-clusters contiguous, dense. Raceme oblong or cylindric, solitary, sometimes with bulblets replacing the lower flower-clusters. Calyx white or pale. Sepals 5, slightly enlarged but otherwise unchanged at maturity. Stamens 5–8, exserted; filaments very slender. Styles 3, slender, exserted, nearly straight; stigmas capitate. Achenes 3-angled, loosely included in the calyx. [Latin, double-twisted, referring to the shape of the rootstock.]

About 8 species, natives of boreal regions. Besides the following, two or more others occur in western North America. Type species: *Polygonum Bistorta* L.

1. Bistorta vivípara (L.) S. F. Gray.   Alpine Bistort.   Serpent-grass.   Fig. 1645.

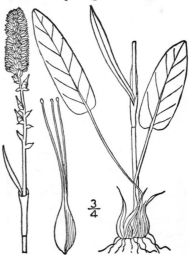

Polygonum viviparum L. Sp. Pl. 360. 1753.

Bistorta vivipara S. F. Gray, Nat. Arr. Brit. Pl. 2: 268. 1821.

Perennial, mostly glabrous and somewhat glaucous; stems solitary or clustered, erect, simple, slender, 2′–10′ tall. Basal leaves oblong or lanceolate, 1′–8′ long, rather acute at the apex, cordate or subcordate at the base, long-petioled; stem leaves narrowly lanceolate or linear, 1′–3′ long, the lower petioled, the upper sessile, their margins often revolute: ocreae long, clasping below, open above; raceme solitary, terminal, narrow, rather dense, bearing a number of dark colored bulblets about its base; calyx 5-parted, pale rose-color or white; stamens 8, exserted; style 3-parted, its branches exserted; achene oblong, 3-angled.

Greenland and Labrador to Alaska, south to the high summits of the mountains of New England, in the Rocky Mountains to Colorado and to Washington. Also in arctic and alpine Europe and Asia. June–Aug.

$\frac{3}{4}$

9. FAGOPÝRUM Gaertn. Fr. & Sem. 2: 182. 1791.

Annual or perennial rather fleshy, usually glabrous leafy herbs, with erect, simple or branched, striate or grooved stems. Leaves alternate, petioled, hastate or deltoid, with oblique, cylindric or funnelform ocreae. Flowers small, white or green, in terminal or axillary usually paniculate racemes, perfect, borne solitary or several together from each ocreola, slender-pedicelled. Calyx about equally 5-parted, persistent and unchanged in fruit, the segments petaloid, shorter than the achene. Stamens 8, included; filaments filiform, glabrous; anthers oblong. Ovary 1-celled, 1-ovuled; style 3-parted; stigmas capitate. Achene 3-angled. Embryo central, curved, dividing the mealy endosperm into two parts; cotyledons broad. [Greek, beech-wheat, from the similarity of the grain.]

About 6 species, natives of Europe and Asia. Type species: *Fagopyrum tataricum* Gaertn.

Racemes panicled or corymbose; angles of the achene not crested.   1. *F. Fagopyrum.*
Racemes mostly simple; angles of the achene crested, undulate.   2. *F. tataricum.*

1. Fagopyrum Fagopyrum (L.) Karst. Buckwheat. Fig. 1646.

Polygonum Fagopyrum L. Sp. Pl. 364. 1753.
Fagopyrum esculentum Moench, Meth. 290. 1794.
F. Fagopyrum Karst. Deutsch. Fl. 522. 1880–83.

Annual, glabrous except at the nodes, stem strongly grooved when old, 1°–3° high. Leaves hastate, 1'–3' long, abruptly narrowed above the middle, acuminate, the nerves on the lower surface slightly scurfy; ocreae brittle and fugacious; racemes mostly panicled, sometimes corymbose, many-flowered, erect or inclined to droop; pedicels as long as the calyx; segments white or whitish; stamens included; style-branches deflexed in fruit; achene acute, 2½" long, about twice as long as the calyx, its faces pinnately-striate when mature, the angles acute, entire.

In waste places, and persistent in fields after cultivation. Reported from almost all parts of the northern United States and southern British America. Also in the West Indies. Native of eastern Europe or western Asia. Brank. Corn-heath. Beech-wheat. Crap. Saracen's-corn or -wheat. June–Sept.

2. Fagopyrum tatáricum (L.) Gaertn. Tartary Buckwheat. Fig. 1647.

Polygonum tataricum L. Sp. Pl. 364. 1753.

Fagopyrum tataricum Gaertn. Fr. & Sem. 2: 182. pl. 189. f. 6. 1791.

Annual, similar to the preceding species, but the leaves deltoid-hastate or oblong-hastate, often broader than long, 1'–4' wide, acute or short-acuminate at the apex; racemes terminal and axillary, mostly solitary, simple and few-flowered, long-peduncled; flowers whitish, short-pedicelled; achene subacute, 2½" long, its angles crested with 3 prominent lobes above the middle, its faces pinnately sulcate from a conspicuous groove.

In waste places, eastern Canada and New England. Rough buckwheat. Adventive from Asia. Summer.

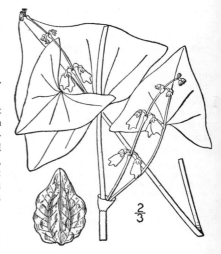

10. TRACÀULON Raf. Fl. Tell. 3: 13. 1836.

Annual or sometimes perennial prickle-armed herbs, with reclining-climbing 4-angled stems. Leaves alternate, truncate, hastate or cordate, membranous, the petiole, midrib and principal nerves armed with small recurved prickles. Ocreae oblique, finely nerved, variously roughened about the base. Flowers in terminal and axillary spike-like racemes, these usually somewhat interrupted, or in capitate clusters. Sepals somewhat colored, 4–5, neither keeled nor winged, enveloping the achene. Stamens varying from 5 to 8, included; filaments not dilated. Styles 2 or 3, partially united, included. Achenes lenticular or 3-angled, variously colored, strongly biconvex or three-angled, smooth and shining. Endosperm horny. Cotyledons accumbent. [Greek, rough-stem.]

About 18 species, natives mostly of North America and Asia. Type species: Tracaulon arifolium Raf.

Leaves sagittate; achene 3-angled.   1. T. sagittatum.
Leaves halberd-shaped; achene lenticular.   2. T. arifolium.

1. Tracaulon sagittàtum (L.) Small. Arrow-leaved Tear-thumb. Fig. 1648.

Polygonum sagittatum L. Sp. Pl. 363. 1753.
Tracaulon sagittatum Small, Fl. SE. U. S. 381. 1903.

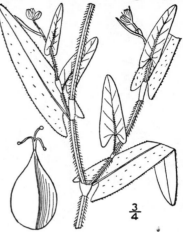

Annual, light green, stem slender, weak, decumbent, or climbing over other plants by the abundant sharp recurved prickles which arm its 4 prominent angles. Leaves lanceolate-sagittate or oblong-sagittate, ½'–3' long, obtuse or acute at the apex, slightly rough on the margins, the lower petioled, the upper subsessile; petioles and lower surface of the midribs prickly; ocreae oblique, not ciliate, fringed at the base by a few bristle-like prickles; flowers in rather dense terminal heads or racemes; calyx greenish or rose-colored; stamens usually 8; style 3-parted to below the middle; achene 3-angled, oblong-pyramidal, thick-pointed, 1½" long, dark red, smooth, shining.

In wet soil, Newfoundland and Nova Scotia to the Northwest Territory, south to Florida and Kansas. Ascends to 3000 ft. in Virginia. July–Sept.

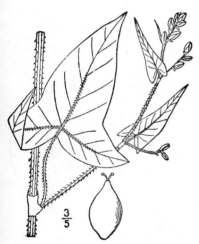

2. Tracaulon arifòlium (L.) Raf. Halberd-leaved Tear-thumb. Fig. 1649.

Polygonum arifolium L. Sp. Pl. 364. 1753.
Tracaulon arifolium Raf. Fl. Tell. 3: 13. 1836.

Perennial, stem ridged, reclining, 2°–6° long, the ridges armed with recurved prickles. Peduncles and pedicels glandular or pubescent; leaves broadly hastate, long-petioled, 1'–10' long, pubescent or glabrous beneath, the apex and basal lobes acuminate; petioles and stronger nerves prickly; ocreae oblique, fringed at the summit with short bristles and at the base with slender prickles; flowers in terminal and axillary heads or racemes; calyx rose-color or greenish, 4-parted; stamens 6; style 2-parted; achene lenticular, broadly obovate, 2" long, strongly biconvex, dark brown, smooth, shining.

In moist or wet soil, New Brunswick and Ontario to Minnesota, south to Georgia. Sickle-grass. Scratch-grass. July–Sept.

11. TINIÀRIA Webb & Moq.; Webb & Berth. Hist. Nat. Canar. 3: 221. 1836–40.

Annual or perennial often scurfy unarmed twining vines. Leaves alternate, membranous or leathery, cordate or hastate. Ocreae oblique, naked or fringed at the top or the base. Racemes loosely flowered, axillary or terminal, often paniculate, leafy-bracted or naked. Sepals 5, green, white or yellowish, 2 exterior, 2 interior and 1 with one edge interior and one edge exterior, this sepal and the two outer keeled or strongly and conspicuously winged. Pedicels slender, reflexed and articulated. Stamens 8, included; filaments short, converging. Styles 3, short or almost wanting, distinct or rarely united. Achenes 3-angled, dark brown or black, included, smooth and shining or granular and dull. Endosperm horny. Cotyledons accumbent. [Latin, worm, referring to the habit of the plants.]

About 8 species, natives of North America and Asia. Type species: *Tiniaria Convolvulus* (L.) Webb. & Moq.

Outer segments of the calyx unchanged, or keeled in fruit.
  Achene granular and dull; ocreae not bristly.   1. *T. Convolvulus.*
  Achene smooth and shining; ocreae bristly.   2. *T. cilinodis.*
Outer segments of the calyx conspicuously winged in fruit.
  Calyx-wings not incised.
    Fruiting calyx 5"–6" long, the wings crisped.   3. *T. scandens.*
    Fruiting calyx 3"–4" long, the wings rather flat.   4. *T. dumetorum.*
  Calyx-wings incised.   5. *T. cristata.*

1. Tiniaria Convólvulus (L.) Webb & Moq.
Black or Corn Bindweed. Fig. 1650.

Polygonum Colvolvulus L. Sp. Pl. 364. 1753.
T. Convolvulus Webb. & Moq. loc. cit. 1836–40.

Annual, glabrous, scurfy, stem twining or trailing, 6′–3° long, mostly branched, the internodes elongated. Leaves ovate-sagittate or the uppermost lanceolate-sagittate, long-petioled, acuminate at the apex, slightly ciliate, ½′–3′ long; ocreae oblique, short, rough on the margin; axillary clusters or racemes loosely flowered; flowers greenish, pendulous on slender pedicels; calyx 5-parted, closely investing the achene, the outer lobes slightly or not at all keeled; stamens 8; style short, nearly entire; stigmas 3; achene 3-angled, obovoid-pyramidal, 1½″ long, thick-pointed, black, granular, rather dull.

In waste and cultivated grounds, nearly throughout North America except the extreme north. Also in the West Indies. Naturalized from Europe. Native of Asia. Sometimes a troublesome weed. Calyx rarely 6-parted. Bearbind. Ivy- or climbing-bindweed. Cornbind. Devil's-tether. Knot- or blackbird-bindweed. July–Sept.

2. Tiniaria cilinòdis (Michx.) Small.
Fringed Black Bindweed. Fig. 1651.

P. cilinode Michx. Fl. Bor. Am. 1: 241. 1803.
P. cilinode erectum Peck, N. Y. State Mus. Rep. 46: 129. 1893.
T. cilinodis Small, Fl. SE. U. S. 382. 1903.

Perennial, sparingly pubescent, stem red or reddish, twining or prostrate, or nearly erect, 1°–10° long. Leaves broadly ovate or somewhat hastate, acuminate at the apex, cordate at the base, rather long-petioled, undulate, finely ciliate, 1′–4′ long, or the upper smaller; ocreae small, armed with reflexed bristles near the base; racemes mostly panicled, axillary and terminal, interrupted; calyx whitish; style short, 3-parted to the base; achene 3-angled, oblong-pyramidal or ovoid, nearly 1½″ long, very smooth and shining.

In rocky places, Nova Scotia to Ontario, Minnesota and Pennsylvania, south in the Alleghanies to North Carolina. Ascends to 2000 ft. in tĥe Catskills. June–Sept.

3. Tiniaria scándens (L.) Small. Climbing
False Buckwheat. Fig. 1652.

Polygonum scandens L. Sp. Pl. 364. 1753.
Tiniaria scandens Small, Fl. SE. U. S. 382. 1903.

Perennial, glabrous, stem climbing, 2°–20° long, rather stout, striate, branched, rough on the ridges. Leaves ovate, acuminate, cordate at the base, 1′–6′ long or the upper smaller, the larger long-petioled, finely punctate, the margins scabrous; ocreae oblique, smooth and glabrous; racemes usually numerous and panicled, interrupted, leafy, 2′–8′ long; flowers yellowish-green, long-pedicelled; calyx 5-parted, the three outer segments very strongly winged and decurrent on the pedicels, especially in fruit; stamens 8; style almost none; stigmas 3; fruiting calyx 5″–6″ long, the wings crisped, not incised; achene 2″–2½″ long, 3-angled, rather blunt at both ends, smooth, shining.

In woods and thickets, Nova Scotia to Ontario and British Columbia, south to Florida, Nebraska and Texas. Aug.–Sept.

4. Tiniaria dumetòrum (L.) Opiz. Copse or Hedge Buckwheat. Fig. 1653.

Polygonum dumetorum L. Sp. Pl. Ed. 2, 522.　1762.
Tiniaria dumetorum Opiz, Seznam 98.　1852.

Perennial, glabrous, similar to the preceding species, stem extensively twining, 2°–12° long, striate,
much branched. Leaves ovate or somewhat hastate, and sometimes inequilateral, acuminate at the
apex, cordate at the base, 1′–2½′ long, long-petioled, or the upper smaller and nearly sessile; ocreae
oblique, smooth; racemes mostly axillary, numerous, much interrupted, leafy-bracted, 2′–5′ long;
flowers yellowish-green, pendulous; calyx 5-parted,
the three outer segments winged or keeled and much
enlarged in fruit; stamens 8; style short, 3-parted;
fruiting calyx 3″–4″ long, the wings nearly flat, not
incised; achene oblong, 3-angled, 2″ long, inclined to
be pointed at both ends, black, smooth, shining.

About thickets and in woods, locally throughout the
northeastern United States. Naturalized from Europe.
False buckwheat. July–Sept.

5. Tiniaria cristàta (Engelm. & Gray) Small. Crested False Buckwheat. Fig. 1654.

Polygonum cristatum Engelm. & Gray, Bost. Journ.
Nat. Hist. 5: 259.　1847.
Tiniaria cristata Small, Fl. SE. U. S. 382.　1903.

Perennial, scurfy, stem slender, twining, 2°–10°
long, more or less branched. Leaves triangular
or ovate, 1′–5′ long, acuminate at the apex, undulate, truncate or cordate at the base, rather long-
petioled; ocreae cylindric funnelform; flowers in
axillary simple or compound often naked racemes 1′–5′ long; pedicels about 2½″ long, jointed
near the middle; calyx greenish-white, 2″–2½″
long, 5-parted to near the base, the 3 outer segments keeled and at maturity winged; stamens
8, included; style none; stigmas 3; fruiting calyx
3″–4″ long, its wings incised; achene 3-angled,
oblong, black, smooth, shining, about 1½″ long.

Sandy woods and rocky banks, Massachusetts to
Florida, Oklahoma and Texas. Aug.–Oct. This may
be a form of the preceding species.

12. PLEURÓPTERUS Turcz. Bull. Soc. Nat. Moscou 21¹: 587.　1848.

Perennial, often large, many-stemmed herbs with wide-spreading rootstocks, the very
stout stems hollow. Leaves alternate, broad, commonly large, usually truncate or cordate
at the base, petioled. Ocreae oblique, fugacious as in Rumex. Flowers numerous, in short,
axillary, sessile or nearly sessile panicles with short spreading branches. Sepals mostly 5,
pale, usually white, enveloping the achene, the 3 outer ones winged at maturity, the wings
firm. Stamens 6–8, included; filaments slender. Styles 3, short, distinct. Achenes 3-angled,
much shorter than the sepal-wings. [Greek, referring to the calyx-wings.]

About 4 species, natives of eastern Asia. Type species: Pleuropterus cordàtus Turcz. = Polygonum multiflòrum Thunb.

1. Pleuropterus Zuccarínii Small. Japanese Knotweed. Fig. 1655.

Polygonum cuspidatum Sieb. & Zucc. Fl. Jap. Fam. Nat.
2 : 84. 1846. Not Willd. 1825.
Polygonum Zuccarinii Small, Mem. Dept. Bot. Col. Coll.
1 : 158. *pl. 66.* 1895.

Perennial, glabrous, more or less scurfy, stem stout, erect, woody below, terete or slightly angled, much branched, 4°–8° tall. Leaves oblong-ovate or ovate-lanceolate, petioled, 2′–6′ long, acuminate-cuspidate at the apex, truncate or subcordate at the base, reticulate-veined on both surfaces, their margins undulate; ocreae oblique, smooth, fugacious; racemes mostly terminal, panicled, 2′–4′ long, or axillary, many-flowered, more or less pubescent; flowers greenish-white, long-pedicelled; outer segments of the 5-parted calyx very broadly winged in fruit; stamens 8; style 3-parted; achene 3-angled, narrowly oblong or oblong-pyramidal, 1¼″–1½″ long, black, smooth, shining.

Escaped from cultivation locally throughout our range. Native of Japan. July–Oct.

13. POLYGONÉLLA Michx. Fl. Bor. Am. 2 : 240. 1803.

Annual or perennial glabrous herbs, sometimes slightly woody, with erect branched usually conspicuously jointed stems, alternate narrow leaves articulated to the ocreae, and small white or greenish flowers in slender panicled racemes. Calyx unequally 5-parted, persistent, its segments petalloid, loosely investing the achene or its base in fruit, the three inner calyx-segments often winged. Stamens 8, included; filaments filiform, or much dilated or auricled at the base. Style 3-parted, short or almost wanting; stigmas capitate; ovary 1-celled; ovule solitary. Achene 3-angled, smooth. Embryo slender, nearly straight, in one of the angles of the seed. [Diminutive of Polygonum.]

About 8 species, natives of eastern North America. Type species : *Polygonella parvifolia* Michx.

Annual; inner sepals not winged in fruit; pedicels reflexed.   1. *P. articulata.*
Perennial; inner sepals winged in fruit; pedicels divergent.   2. *P. americana.*

1. Polygonella articulàta (L.) Meisn. Coast Jointweed. Fig. 1656.

Polygonum articulatum L. Sp. Pl. 363. 1753.

Polygonella articulata Meisn. Gen. 2 : 228. 1836–43.

Annual, glaucous, stem slender, wiry, erect or sometimes diffusely spreading, simple or branched, striate or slightly angled, 4′–10′ long. Leaves linear or linear-subulate (apparently filiform from the revolute margins), sessile, 4″–20″ long, jointed to the summits of the ocreae, cylindric, slightly expanded at the summit; racemes numerous, erect, many-flowered, 1′–1½′ long; ocreolae crowded or imbricated; pedicels reflexed; calyx-segments white with a dark midrib, loosely investing the achene, not winged in fruit; achene narrowly ovoid-pyramidal, pointed, 1″ long, brown, smooth, shining.

In sands of the seashore and sandy soil along the coast, Maine to Florida, and on the shores of the Great Lakes. Sand-grass. July–Oct.

2. Polygonella americàna (F. & M.) Small.   Southern Jointweed.   Fig. 1657.

Gonopyrum americanum F. & M. Mem. Acad. St. Petersb.
(VI.) **4**: 144. 1840.
Polygonella ericoides Engelm. & Gray, Bost. Journ. Nat.
Hist. **5**: 230. 1845.
P. americana Small, Mem. Torr. Club **5**: 141. 1894.

Perennial by a long slender root, slightly glaucous, stem erect or ascending, wiry, somewhat flexuous, 1½°–4° high, simple or slightly branched, covered with a ridged more or less scaly bark.   Leaves linear or linear-spatulate, ¼′–1′ long, often fascicled on short branches, sessile, rather fleshy, obtuse and revolute at the apex; ocreae scarious-margined, split on one side; racemes 1′–3′ long, dense, divergent; calyx white or pink, its three inner segments developing orbicular cordate wings, the two outer reflexed in fruit; pedicels divergent, jointed below the middle; achene elliptic-oblong, 1¼″ long, chestnut-brown, pointed at both ends, smooth, shining.

In dry soil, Missouri to Texas, east to Georgia and Alabama.   Aug.–Oct.

14. BRUNNÍCHIA Banks; Gaertn. Fr. & Sem. **1**: 213. *pl. 45. f. 2.*   1788.

Perennial vines with elongated, grooved much branched stems climbing by tendrils, and alternate entire broad petioled leaves, the ocreae obscure or wanting, and small perfect flowers in panicled terminal and axillary racemes, the flowers fascicled in the axils of lanceolate-subulate bracts.   Pedicels slender, jointed near the base.   Calyx 5-parted, the segments spreading when fresh, converging when dry, the flower-tube much enlarged, coriaceous and winged on one side in fruit, closely investing the achene.   Stamens 7–10, mostly 8; filaments filiform, much dilated at the base.   Style 3-parted, the stigmas 2-cleft; ovule solitary, pendulous.   Achene 3-angled.   Seed irregularly 6-grooved, the embryo in one of its angles.   [Name in honor of M. T. Brunnich, Norwegian naturalist.]

Two known species, the following typical one of southeastern North America, the other of tropical Africa.

1. Brunnichia cirrhòsa Banks.   Brunnichia.   Fig. 1658.

Brunnichia cirrhosa Banks; Gaertn. Fr. & Sem. **1**: 213.
pl. 45. f. 2. 1788.

Rajania ovata Walt. Fl. Car. 247. 1788.

Stem 6°–20° long, somewhat woody, rather tough, slender, grooved.   Tendrils filiform; leaves ovate or ovate-lanceolate, acute or acuminate at the apex, truncate or subcordate at the base, 1′–6′ long, petioled, slightly pubescent beneath; ocreae obsolete or represented by a ring of short bristles; racemes 2′–6′ long; flowers in fascicles of from 2–5; calyx greenish, 5-parted; stamens exserted; achene oblong-ovoid, 3″ long, brown, smooth, closely invested by the persistent and coriaceous flower-tube which becomes 1′ or more in length.

On banks of streams, southern Illinois to Arkansas and Texas, east to South Carolina and Florida.   May–June.   Fruit mature in August.

INDEX OF LATIN GENERA IN VOLUME I.

[Classes, Families and Tribes in SMALL CAPITALS ; genera in Roman ; synonyms in *Italics*.]